ADVANCED ENGINEERING MATHEMATICS

Reena Garg
Chandrika Prasad

KHANNA BOOK PUBLISHING CO. (P) LTD.
Publisher of Engineering and Computer Books
4C/4344, Ansari Road, Darya Ganj, New Delhi-110002
Phone : 011-23244447-48 **Mobile:** +91-9910909320
E-mail : contact@khannabooks.com
Website : www.khannabooks.com

Dear Readers,

To prevent the piracy, this book is secured with HIGH SECURITY HOLOGRAM on the front title cover. In case you don't find the hologram on the front cover title, please write us to at contact@khannabooks.com or whatsapp us at +91-99109 09320 and avail special gift voucher for yourself.

Specimen of Hologram on front Cover title:

Moreover, there is a SPECIAL DISCOUNT COUPON for you with EVERY HOLOGRAM.

How to avail this SPECIAL DISCOUNT:

Step 1: Scratch the hologram
Step 2: Under the scratch area, your "coupon code" is available
Step 3: Logon to www.khannabooks.com
Step 4: Use your "coupon code" in the shopping cart and get your copy at a special discount
Step 5: Enjoy your reading!

Thanks,
Khanna Publishing Team

ISBN: 978-93-86173-52-2

Advanced Engineering Mathematics
by Reena Garg, Chandrika Prasad

First Edition: 2018
Reprint Edition: 2025, 2026

To view complete list of books, please scan the QR Code:

Published by:
Khanna Book Publishing Co. (P) Ltd.
CIN: U22110DL1998PTC095547

Visit us at: **www.khannabooks.com**
Write us at: **contact@khannabooks.com**

Printed in India.

PREFACE

This book provides a comprehensive, thorough and up to date treatment of mathematics in engineering and sciences. This is intended to introduce students of engineering, physics, mathematics, computer sciences and other related fields to those areas of applied mathematics that are most relevant for solving practical problems. Practice is the key word in the learning process of mathematics . The aim of this book is to provide a vast knowledge of mathematics and its diverse practical use in daily lives. The course contents in this book are the sole pre-requisites. The experience of the author of more than a decade in teaching at under graduate, post graduate level and in the research areas of mathematics in University makes this book useful.

In this book all the topics and related concepts have been given in a lucid and simple way filling every gap between students and mathematics. A lot of worked examples are given so as to help the readers understand better.

Last but not the least, I earnestly hope that the present book will serve the students better. I extend my heartiest gratitude to the publisher for publishing the book in the interest of mathematics students.

Authors

This book provides a comprehensive, thorough and up-to-date treatment of mathematics in engineering and sciences. This is intended to introduce students of engineering, physics, mathematics, computer sciences and other related fields to those areas of applied mathematics that are most relevant for solving the practical problems. Practice is the key word in the learning process of mathematics. The aim of this book is to provide a vast knowledge of mathematics and its diverse practical use in daily lives. The course contents in this book are the sole prerequisites. The experience of the author of more than a decade in teaching at under graduate/post graduate level and in the research areas of mathematics in University makes this book useful.

In this book all the topics and related concepts have been given in a lucid and simple way [illegible] every [illegible] between students and mathematics. A lot of worked examples are given so as to help the readers understand both [illegible]

Last but not the least, I earnestly hope that the present book will serve the students [illegible] I extend my heartiest gratitude to the publisher for publishing the book in the interest of mathematics students.

Authors

ACKNOWLEDGEMENT

I express my sincere gratitude to friends, faculty members and well wishers for their contribution, guidance and suggestions to write this book. We are especially indebted to my most respected Prof. Dinesh Aggarwal, Vice Chancellor, J.C. Bose University of Science & Technology, YMCA Faridabad. We are indebted to Prof. Renu Chugh (Professor, Maharshi Dayanand University, Rohtak); Prof. Bhudev Sharma (former Professor, Atlanta University, USA); Prof. Babu Ram (former Professor, Maharshi Dayanand University, Rohtak); Dr. Madhu Jain (Associate Professor, I.I.T. Roorkee); Dr. S.C. Malik (Professor, Maharshi Dayanand University, Rohtak); Dr Deepankar Sharma (Director DJ College of Engineering, Modinagar), Prof. Kuldeep Bansal (Professor, Guru Jambheshwar University of Science & Technology, Hissar), Prof. R. S. Goel (Ex-Principal Aggarwal College, Ballabgarh), Prof. Satish Bansal (DAV College Faridabad), Dr Vandana Mohla (Principal K L Mehta Dayanand College for women, Faridabad) & Professor Alka Choubey (Jaypee University, Noida)

I am extremely thankful and grateful to Mr. Mukul Seth, MD, Khanna Publishing House, New Delhi, for [illegible] continued support. We are thankful to Editorial team of Khanna Publishing House for bringing out thse book in the present form.

Last but not the least, I want to thank my family members and two loving children, Vedant Garg and Vaishnavi Garg, without their constant support I would not be able to achieve this milestone of my life.

I am thankful to Almighty God for giving the opportunity to serve the society.

Authors

CONTENTS

Preface iii

Acknowledgement v

1. Matrices and Its Applications **1**

1.1 Matrices 1
1.2 Some Special Matrices 1
1.3 Associated (Related) Matrices 3
1.4 Vectors 6
1.5 Linear Dependence and Independence of Vectors 6
1.6 Linear Transformations 8
1.7 Orthogonal Transformation 8
1.8 Elementary Transformation (or Operations) on a Matrix 11
1.9 Gauss-Jordan Method of Finding the Inverse 12
1.10 Normal Form of a Matrix 12
1.11 Rank of Matrix 15
1.12 Solution of Linear System of Equations 21
1.13 Consistency of a Linear System of Equations 21
1.14 Characteristic Equation 27
1.15 Eigen Vectors 27
1.16 Properties of Eigen Values 28
1.17 Cayley-Hamilton Theorem 33
1.18 Reduction of a Matrix to Diagonal Form 40
1.19 Reduction of Quadratic Form to Canonical Form 42

2. Application to Geometry **46**

2.1 Tangent to a Curve 46
2.2 Normal 47
2.3 Polar Coordinates 49
2.4 Angle Between Radius Vector and Tangent 50
2.5 Differential Coefficient of Length of Arc 53

2.6	Radius of Curvature	54
2.7	Formula for Polar Equations	55
2.8	Curve Tracing	57
2.9	Nature of Curve Near Origin	57
2.10	Asymptotes	58
2.11	Rules for Finding Asymptotes	59
2.12	Procedure for Tracing Cartesian Equations	61
2.13	Polar Equations	62
2.14	Some Well-Known Curves	64
3.	**Coordinate Geometry of Three Dimensions**	**67**
3.1	Rectangular Coordinates	67
3.2	Cylindrical and Spherical Coordinates	67
3.3	Vectors	68
3.4	Addition of Vectors	69
3.5	Multiplication By a Scalar	69
3.6	Resolution of a Vector	70
3.7	Coordinates in Space	70
3.8	Scalar Product	72
3.9	Vector Product	73
3.10	Application to Geometry	76
3.11	Direction Cosines	77
3.12	Angle Between Two Lines	78
3.13	Distance of a Point From a Line	79
3.14	Equation of a Plane	80
3.15	Normal Form	81
3.16	Distance of a Point From a Plane	82
3.17	Examples on Planes	82
3.18	Scalar Triple Product	83
3.19	Vector Triple Product	84
3.20	Equation of a Line	87
3.21	Intersection of Two Planes	89
3.22	A Line and a Plane	91
3.23	Plane Through the Intersection of Two Planes	92
3.24	Intersection of Three Planes	94
3.25	Shortest Distance	95
3.26	Equation of a Sphere	98
3.27	Tangent Plane	99
3.28	Section of a Sphere	100
3.29	Length of the Tangent	101
3.30	Equation of a Cylinder	104
3.31	Oblique Cylinder	105
3.32	Equation of a Cone	106

3.33 Right Circular Cone 106
3.34 Cone Through a Given Curve 107
3.35 Quadratic Surfaces 108

4. Expansion of Functions (Taylor's and Maclaurin's Series) 112

4.1 Finite and Infinite Series 112
4.2 Taylor's Infinite Series 112
4.3 Working Method for Expansion as Taylor's Infinite Series 113
4.4 Failure of Taylor's Series 113
4.5 Maclaurin's Infinite Series 120
4.6 Working Method of Expansion as Maclaurin's Series 121
4.7 Failure of Maclaurin's Series 121
4.8 Some Useful Expansions 125
4.9 Successive Differentiation 128
4.10 Leibnitz's Rule 128
4.11 Taylor's Series for a Function of Two Variables 134
4.12 Branch of a Curve 142
4.13 Asymptote 142
4.14 Asymptotes Parallel to Coordinate Axis 142
4.15 Asymptotes Parallel to Co-ordinate Axes for $f(x, y) = 0$ *i.e.,* Algebraic Curve 142
4.16 Oblique Asymptotes 144
4.17 To Find the Oblique Asymptotes of the General Algebraic Curve 145
4.18 Working Rule for Finding Oblique Asymptotes of an Algebraic Curve of – nth Degree 147
4.19 Intersection of a Curve and Its Asymptotes 153
4.20 Asymptotes in Polar Coordinate 156
4.21 Working Method for Finding Polar Asymptotes 158
4.22 Mathematical Definition of a Curvature 162
4.23 Curvature of a Circle 162
4.24 Radius of Curvature of Cartesian Curve 163
4.25 Radius of Curvature of Polar Curves 176
4.26 Radius of Curvature of Pedal Equations 178
4.27 Radius of Curvature At the Origin 183
4.28 Centre of Curvature, Circle of Curvature and Evolute 188
4.29 Curve and Its Forms 192
4.30 Procedure of Curve Tracing for Cartesian Curves 192
4.31 Procedure of Curve Tracing for Polar Curves 201
4.32 Procedure of Curve Tracing for Parametric Curves 206

5. Partial Differentiation 213

5.1 Dependent and Independent Variables 213
5.2 Partial Derivatives 213
5.3 Which Variable is to be Treated as Constant 222
5.4 Homogeneous Function 225
5.5 Euler's Theorem on Homogeneous Function of Two Variables 225

5.6 Total Derivatives 238
5.7 Composite Functions 238
5.8 Differentiation of Composite Functions 238
5.9 Differentiation of Implicit Functions 239
5.10 Concept of a Jacobian 253
5.11 Properties of Jacobians 253
5.12 Concept of Maxima and Minima 261
5.13 Stationary and Extreme Points 261
5.14 Necessary Condition for Existence of a Maximum or Minima of $f(x, y)$ at $x = a, y = b$ 261
5.15 Sufficient Condition for Maxima or Minima 262
5.16 Working Rules for Maxima and Minima 262
5.17 Lagrange's Method of Undetermined Multipliers 275
5.18 Differentiation Under the Integral Sign 288
5.19 Leibnitz's Rule for Differentiation Under the Integral Sign 288

6. Reduction Formulae: Definite Integrals 302

6.1 Reduction Formulae 302
6.2 Integration of $\text{Sin}^m x \text{Cos}^n x$ 303
6.3 Reduction Formula for $\int \text{Sin}^m x \text{Cos}^n x \, dx$ 305
6.4 The Integral $\int_0^{\pi/2} \text{Sin}^m x \text{Cos}^n x \, dx$ 305
6.5 Integration of $1/(a + b \cos x)$ 308
6.6 Functions of $\sin x$ and $\cos x$ 309
6.7 Other Transcendental Functions 310
6.8 Reduction Formulae 312
6.9 Definite Integrals 314
6.10 The Integral as a Sum 317
6.11 Fundamental Theorem of Integral Calculus 317
6.12 The Integral as an Area 318
6.13 Integrals With Infinite Limits 319
6.14 Improper Integrals 320
6.15 Differentiation Under Sign of Integration 320
6.16 Solid of Revolution and Surface of Revolution 323
6.17 (A) Volume of Solid of Revolution (for Certain Curves) 323
(B) Volume of Solid of Revolution (for Parametric Curves) 323
(C) Volume of the Solid of Revolution (for the Polar Curves) 324
(D) Surface Area of Solid of Revolution 324
6.18 Multiple Integrals and Their Applications 332
6.19 Evaluation of Double Integrals 332
6.20 Double Integration in Polar Co-ordinates 339
6.21 Change of Order of Integration 343
6.22 Volume as a Double Integral 352
6.23 Triple Integral 357

6.24 Change of Variables 361
6.25 Volume as Triple Integral 362
6.26 Gamma Function 367
6.27 Reduction Formula for Γ(n) 368
6.28 Beta Function 369
6.29 Some Important Deductions 369

7. Vector Calculus 383

7.1 Scalar and Vector Quantities 383
7.2 Scalar and Vector Product, Angular Velocities 383
7.3 Differentiation of Vectors 384
7.4 Formulae of Differentiation 384
7.5 Scalar and Vector Point Function 384
7.6 Vector Differential Operator Del *i.e.* ∇ 384
7.7 Geometrical Interpretation of Gradient 385
7.8 Directional Derivative of 'ϕ' in the Direction of $\overrightarrow{PQ}$ 385
7.9 Del Applied to Vector Point Functions 390
7.10 Physical Interpretation of Divergence 391
7.11 Physical Interpretation of Curl 392
7.12 Repeated Operation by ∇ 392
7.13 Properties of Divergence and Curl 393
7.14 Some Basic Concepts 400
7.15 A Line Integral 401
7.16 Surface Integral 406
7.17 Volume Integral 407
7.18 Green's Theorem 415
7.19 Stoke's Theorem 420
7.20 Gauss's Divergence Theorem 429

8. Infinite Series 440

8.1 Sequence 440
8.2 Limit 440
8.3 Real Sequence 440
8.4 Range of Sequence 440
8.5 Constant Sequence 440
8.6 Bounded and Unbounded Sequence 441
8.7 Convergent, Divergent and Oscillatory Sequence 441
8.8 Monotonic Sequence 442
8.9 Some Standard Limits 444
8.10 Series 444
8.11 Convergent, Divergent and Oscillatory Series 445
8.12 Properties of Infinite Series 445
8.13 Partial Sums 445

8.14 Geometric Series 448
8.15 Positive Term Series 449
8.16 Necessary Condition for Convergent Series 449
8.17 Cauchy's Fundamental Test for Divergence 449
8.18 Hyper Harmonic Series Test, P - Series Test for $\Sigma\frac{1}{n^p}$ 450
8.19 Comparison Test 452
8.20 D'alembert's Ratio Test 459
8.21 Raabe's Test (Higher Ratio Test) 465
8.22 Logarithmic Test 472
8.23 Gauss's Test 479
8.24 Cauchy's Root Test 482
8.25 Integral Test 487
8.26 Alternating Series 491
8.27 Alternating Convergent Series 496

9. Fourier Series 504

9.1 Fourier Series 504
9.2 Dirichlet's Conditions for Fourier Series 505
9.3 Even and Odd Functions 516
9.4 Fourier Series for Discontinuous Functions 517
9.5 Change of Interval (Change of Scale) 526
9.6 Half Range Series 527

10. Differentiation 546

10.1 Functions 546
10.2 Graphical Representation 546
10.3 Limits 546
10.4 Infinite Limits 548
10.5 Continuity 548
10.6 Differential Coefficient 548
10.7 Differentiation of Simple Functions 549
10.8 Differentiation of a Sum, Product or Quotient 550
10.9 Differentiation of a Function of a Function 551
10.10 Inverse Function 552
10.11 Hyperbolic Functions 553
10.12 Logarithmic Differentiation 554
10.13 Implicit Functions 554
10.14 Parametric Equations 554
10.15 Differential Coefficient as Rate of Change 555
10.16 Successive Differentiation 557
10.17 Standard Results 558
10.18 Leibnitz Theorem 559
10.19 Exact Differential Equations 561

10.20 Equations Reducible to Exact Differential Equation 568
10.21 Linear Differential Equation 578
10.22 The Operator 578
10.23 Theorems 579
10.24 Auxiliary Equation (A.E.) 580
10.25 Rules for Finding the Complementary Function 580
10.26 The Inverse Operator $\frac{1}{f(D)}$ 586
10.27 Rules for Finding the Particular Integral 587
10.28 Working Procedure to Solve the Equation 590
10.29 Method of Variation of Parameters to Find Particular Integral 603
10.30 Cauchy's Homogeneous Linear Equation 608
10.31 Legender's Linear Equation 609
10.32 Simultaneous Linear Equation With Constant Coefficients 620

11. Partial Differential Equations 631

11.1 Partial Derivative 631
11.2 Formation of Partial Differential Equations 631
11.3 Equations Solvable By Direct Integration 637
11.4 Linear Partial Differential Equation of First Order 639
11.5 Lagrange's Linear Equation 639
11.6 Solution of Lagrange's Linear Equation 639
11.7 Working Rule to Solve Lagrange's Liner Equation $(Pp + q = R)$ 640
11.8 Non-Linear Partial Differential Equation of 1st Order 646
11.9 Charpit's Method 654
11.10 Application of Second Order Partial Differential Equations 659
11.11 Method of Separation of Variables 659
11.12 Wave Equation-Vibration of Stretched String 663
11.13 Solution of Wave Equation 664
11.14 One Dimensional Heat Flow 673
11.15 Solution of Heat Equation 673

12. Function of a Complex Variable 684

12.1 Complex Number 684
12.2 Limit of a Function 684
12.3 Continuity of a Function 684
12.4 Differentiability of a Function of a Complex Variable 685
12.5 Analytic Function 686
12.6 Necessary and Sufficient Conditions for $f(z)$ to be Analytic 686
12.7 Cauchy-Riemann Equations in Polar Co-ordinates 688
12.8 Harmonic Functions 689
12.9 Orthogonal System 689
12.10 Milne-Thomson's Method 693
12.11 Complex Integration 698

12.12 Simply and Multiply Connected Regions 702
12.13 Cauchy's Integral Theorem 703
12.14 Cauchy's Integral Formula 704
12.15 Singular Points, Residues 707
12.16 Residue Theorem 708
12.17 Calculation of Residues 709
12.18 Series of Complex Terms 713
12.19 Taylor's Series 714
12.20 Laurent's Series 715
12.21 Application of Residues to Evaluate Real Integrais 722
12.22 Integrals of Type 724

13. Laplace Transforms and Its Applications 729

13.1 Definition of Laplace Transform 729
13.2 Some Important Results 729
13.3 Properties of Laplace Transforms 730
13.4 Conditions for Existence of Laplace Transforms 736
13.5 Laplace Transform of $t^n f(t)$ 736
13.6 Laplace Transform of $\left[\frac{f(t)}{t}\right]$ 737
13.7 Second Shifting Theorem 738
13.8 Laplace Transform of Derivatives 742
13.9 Laplace Transform of Integrals 743
13.10 Inverse Laplace Transform 747
13.11 Important Formulae of Inverse Laplace Transform 747
13.12 Methods of Finding Inverse Laplace Transforms 753
13.13 Convolution Theorem 754
13.14 Unit Step Function (or Heaviside's Unit Function) 762
13.15 Laplace Transform of Unit Step Function 762
13.16 Second Shifting Property for Inverse Laplace 763
13.17 Unit Impulse Function (or Dirac-Delta Function) 766
13.18 Laplace Transform of Unit Impulse Function 767
13.19 Definition of Periodic Function 769
13.20 Laplace Transform of a Periodic Function $f(t)$ 769
13.21 Application of Laplace Transforms to Solve Ordinary Differential Equations with Constant and Simultaneous Differential Equations 774
13.22 Simultaneous Differential Equations 782

14. Fourier Transform and Its Application 791

14.1 Fourier Transform 791
14.2 Fourier Integral Transform of a Function 791
14.3 Fourier Integral Formula 792
14.4 Infinite Fourier Transform and Inverse Formula 796

14.5 Finite Transform and Inversion 806
14.6 Properties of Fourier Transform 809
14.7 Definition of Convolutions 814
14.8 Relation Between Fourier and Laplace Transforms 816
14.9 Fourier Transforms of the Derivatives 820
14.10 Unit Step Function 824
14.11 Unit Impulse or Dirac-Delta Functions 824
14.12 Fourier Transform of Dirac-Delta Function 824

15. Empirical Laws and Curve Fitting 826

15.1 Introduction 826
15.2 Graphical Method 826
15.3 Laws Reducible to Linear Laws 826
15.4 Method of Group Averages 829
15.5 Laws Containing Three Constants 832
15.6 Principle of Least Squares 838
15.7 Fitting a Straight Line 839
15.8 Fitting a Parabola or a Second Degree Curve 842
15.9 Fitting of Other Curves 846
15.10 Curve Fitting By a Sum of Exponents 852

16. Solution to Numerical, Algebraic and Transcendental Equation 854

16.1 Introduction 854
16.2 Bisection Method (Bolzano Method or Interval Halving Method) 855
16.3 Method of False Position (Regula-Falsi Method) 860
16.4 Newton-Raphson Method 865
16.5 Convergence of Newton-Raphson Method 866
16.6 Generalized Newton's Method for Multiple Roots 872
16.7 Secant Method 874
16.8 Fixed-Point Iteration Method 875
16.9 Muller's Method 881

17. Numerical Differentiation and Integration 886

17.1 Introduction 886
17.2 Formula for Derivatives 886
17.3 Maxima and Minima of a Tabulated Function 889
17.4 Cubic Spline Method 897
17.5 Numerical Integration 899
17.6 Quadrature Formula 899
17.7 Newton-Cotes' Formula 902
17.8 Errors in Quadrature Formula 915
17.9 Romberg's Method of Integration 918
17.10 Euler-Maclaurin Formula 921

17.11 Gaussian Integration 924
17.12 Numerical Double Integration 928

18. Numerical Solution of Ordinary Differential Equations 934

18.1 Introduction 934
18.2 Picard's Method 934
18.3 Taylor's Series Method 941
18.4 Euler's Method 947
18.5 Modified Euler's Method 947
18.6 Runge-Kutta (RK) Methods 955
18.7 Predictor-Corrector Methods 961
18.8 Milne's Method 962
18.9 Adams-Bashforth Method 969

19. Numerical Solution of Partial Differential Equations 975

19.1 Introduction 975
19.2 Finite Difference Approximation to Various Derivatives 975
19.3 Finite Difference Approximation to Partial Derivatives 976
19.4 Classification of Partial Differential Equations 978
19.5 Elliptic Equation 978
19.6 Solution of Laplace Equation by Liebmann's Iteration Process 980
19.7 Wave Equation 996
19.8 Solution of One-Dimensional Heat Equation 1003

MATRICES AND ITS APPLICATIONS

This chapter is devoted to the study of special matrices, associated matrices, vectors, linear dependence and independence of vectors, linear and orthogonal transformations. Gauss-Jordan method of finding the inverse, normal form of matrix, consistency of a linear system of equations, characteristics equation, eigenvectors, diagonal and canonical forms of matrices.

1.1 MATRICES

An array of *m n* numbers arranged in *m* rows and *n* columns and bounded by square bracket [ ] brackets ( ) or in the form || || is called *m* by *n* matrix and is represented as a matrix.

$$A = \begin{bmatrix} a_{11} & a_{12} & \dots & a_{1n} \\ a_{21} & a_{22} & \dots & a_{2n} \\ a_{31} & a_{32} & \dots & a_{3n} \\ \dots & \dots & \dots & \dots \\ a_{i1} & a_{i2} & \dots & a_{in} \\ \dots & \dots & \dots & \dots \\ a_{m1} & a_{m2} & \dots & a_{mn} \end{bmatrix} \text{ or } \begin{pmatrix} a_{11} & a_{12} & \dots & a_{1n} \\ a_{21} & a_{22} & \dots & a_{2n} \\ \dots & \dots & \dots & \dots \\ a_{i1} & a_{i2} & \dots & a_{in} \\ \dots & \dots & \dots & \dots \\ a_{m1} & a_{m2} & \dots & a_{mn} \end{pmatrix} \quad \dots(1)$$

(1) is known as $m \times n$ matrix in which there are *m* rows and *n* columns. Each number of $m \times n$ matrix is known as an element of the matrix. In general, we take a_{ij} as an element of (1) in which its position is in i^{th} row and j^{th} column, so that as $A = [a_{ij}]$.

1.2 SOME SPECIAL MATRICES

Row and Column Matrices

A single row matrix is known as row matrix, *i.e.,* [1 0 2 3 4] is a row matrix *i.e.,* 1×5 matrix in this case and $\begin{bmatrix} 1 \\ 2 \\ 3 \end{bmatrix}$ is a column matrix with $m = 3$ and $n = 1$, *i.e.,* 3×1 matrix.

These row and column matrices are also known as row vectors or column vectors respectively which we will use in the section known as vectors and characteristics vectors.

Square Matrix

In a $m \times n$ matrix when $m = n$, we say that the matrix is square matrix of order *n form.*

Examples:

$$A = \begin{bmatrix} 1 & 2 \\ 2 & 1 \end{bmatrix}, A_1 = \begin{bmatrix} a & h & g \\ h & b & f \\ g & f & c \end{bmatrix} \text{ etc. are square matrices.}$$

In A_1 the elements a, b and c are called leading or principal diagonal of the square matrix and the sum of the diagonal elements of a square matrix A is called the trace of A.

Singular and Non-Singular Matrices

A square matrix A is called a singular matrix if $|A| = 0$ *i.e.*, if determinant formed by elements of A is zero. If $|A| \neq 0$, then matrix is known as non-singular matrix.

Examples:

$$A = \begin{bmatrix} 3 & 0 & 0 \\ 0 & 5 & 0 \\ 0 & 0 & -7 \end{bmatrix}, B = \begin{bmatrix} 5 & 0 & 0 \\ 0 & 5 & 0 \\ 0 & 0 & 5 \end{bmatrix}, C = \begin{bmatrix} 1 & 0 & 0 \\ 0 & 2 & 0 \\ 0 & 0 & 1 \end{bmatrix}$$

The matrix B in the above examples is called scalar matrix as all the elements of leading diagonal are equal and matrices A and C are diagonal matrices.

Unit Matrix

A diagonal matrix in which element of the diagonal matrix are unity.

For example:

$$A_2 = \begin{bmatrix} 1 & 0 \\ 0 & 1 \end{bmatrix}, A_3 = \begin{bmatrix} 1 & 0 & 0 \\ 0 & 1 & 0 \\ 0 & 0 & 1 \end{bmatrix} \text{ and } A_n = \begin{bmatrix} 1 & 0 & 0 & \dots & 0 \\ 0 & 1 & 0 & \dots & 0 \\ 0 & 0 & 1 & 0 \dots & 0 \\ \dots & \dots & \dots & \dots & \dots \\ \dots & \dots & \dots & \dots & \dots \\ 0 & 0 & 0 & 0 \dots & 1 \end{bmatrix}$$

A_2 is unit matrix of order 2, A_3 is unit matrix of order 3 and A_n is unit matrix of order n.

Null Matrix

If each element of a matrix is zero then it is called a null matrix. Null matrix may be a square matrix or $m \times n$ matrix. For example

$$\begin{bmatrix} 0 & 0 & 0 & 0 & 0 \\ 0 & 0 & 0 & 0 & 0 \\ 0 & 0 & 0 & 0 & 0 \end{bmatrix} \text{ and } \begin{bmatrix} 0 & 0 & 0 \\ 0 & 0 & 0 \\ 0 & 0 & 0 \end{bmatrix} \text{ are null matrices.}$$

Symmetric Matrix

A square matrix $A = [a_{ij}]$ is said to be symmetric if $a_{ij} = a_{ji}$ for all i and j.

Examples $\begin{bmatrix} a & h & g \\ h & b & f \\ g & f & c \end{bmatrix}, \begin{bmatrix} 1 & 2 & 3 \\ 2 & 5 & 4 \\ 3 & 4 & 7 \end{bmatrix}$

Skew Symmetric Matrix

If a square matrix A has its elements such that $a_{ij} = -a_{ji}$ for i and j and the leading diagonal elements are zeroes, then matrix A is known as skew symmetric matrix. For example

$$\begin{bmatrix} 0 & 2 & -3 \\ -2 & 0 & 1 \\ 3 & -1 & 0 \end{bmatrix}, \begin{bmatrix} 0 & h & g \\ -h & 0 & -f \\ -g & f & 0 \end{bmatrix} \text{ are skew matrices.}$$

Triangular Matrix

A square matrix, whose elements, below the leading diagonal are all zeroes, is called an upper triangular matrix. For example

$$A = \begin{bmatrix} a & h & g \\ 0 & b & f \\ 0 & 0 & e \end{bmatrix}$$

A square matrix whose elements, above the leading diagonal, are all zeroes is called a lower triangular matrix. For example

$$A_1 = \begin{bmatrix} a & 0 & 0 \\ h & b & 0 \\ g & f & e \end{bmatrix}, A_2 = \begin{bmatrix} 1 & 0 & 0 & 0 \\ -2 & 2 & 0 & 0 \\ 3 & -4 & -1 & 0 \\ -4 & 5 & 6 & 7 \end{bmatrix}$$

Equality of Matrices

If A and B are two matrices, then they are said to be equal if and only if (*i*) matrices are of the same order $m \times n$ and (*ii*) each element a_{ij} of A is equal to corresponding element b_{ij} of B.

1.3 ASSOCIATED (RELATED) MATRICES

I. ***Transpose of a Matrix***

Any matrix which is obtained from a given matrix A, by interchanging rows and columns is called the transpose of A and this new matrix is represented by A'.

So, $$\text{matrix } A_1 = \begin{bmatrix} 1 & 2 & 3 \\ a & b & c \\ l & m & n \end{bmatrix} \Rightarrow A' = \begin{bmatrix} 1 & a & l \\ 2 & b & m \\ 3 & c & n \end{bmatrix}$$

and for $$A_1 = \begin{bmatrix} 1 & 2 \\ 3 & 4 \\ 5 & 6 \end{bmatrix} \Rightarrow A' = \begin{bmatrix} 1 & 3 & 5 \\ 2 & 4 & 6 \end{bmatrix}$$

Hence, the transpose of given $m \times n$ matrix is $n \times m$ matrix. So, the transpose of A is A' and therefore transpose of A' is $(A')' = A$.

If A is symmetric matrix then $A = A'$.

For example,

$$A = \begin{bmatrix} a & h & g \\ h & b & f \\ g & f & c \end{bmatrix} \Rightarrow A' = \begin{bmatrix} a & h & g \\ h & b & f \\ g & f & c \end{bmatrix} = A$$

Properties of Transpose of a Matrix

If A and B be two matrices then we represent by A' and B' the transpose of A and B respectively, then

(*a*) To show that $A = (A')'$ *i.e.*, the transpose of transpose of a matrix A is matrix A itself. We represent the matrix A by $[a_{ij}]_{m \times n}$. So order, of transpose of A (= A') is then $n \times m$ and order of transpose of A' *i.e.*, $(A')'$ is $m \times n$. This shows that order of matrices A and $(A')'$ are same and their corresponding elements are the same. This implies $A = (A')'$.

(*b*) To show $(A+B)' = A' + B'$ i.e., the transpose of the sum of two matrices is equal to the sum of their transpose. If $A = [a_{ij}]_{m\times n}$ and $B = [b_{ij}]_{m\times n}$ be given matrices so that $A' = [a_{ij}]_{n\times m}$ and $B' = [b_{ij}]_{n\times m}$ which are transpose of A and B respectively.

$\therefore$ Order of $A + B$ is $m \times n$ and so order of $(A+B)'$ is $n \times m$. Also order of $A' + B'$ is of order $n \times m$ which implies that order of $A' + B'$ is same as order of $(A+B)'$ and their corresponding elements are equal such that $(A+B)' = A' + B'$.

(*c*) To prove that $(AB)' = B'A'$ *i.e.*, the transpose of the product of two matrices is equal to the product of their transposes, taken in reverse order.

$A = [a_{ij}]_{m\times n}$ and $B = [b_{ij}]_{n\times p}$ be two matrices conformal for the product AB then order of AB is $m \times p$.

$\Rightarrow$ order of $(AB)'$ is $p \times m$. As A is of order $m \times n$ and so that A' is of order $n \times m$ and B is of order $n \times p$ so B' is of order $p \times n$.

$\Rightarrow$ $B'A'$ is order of $p \times m$.

Therefore $(AB)'$ and $(B'A')$ are of the same order.

So,

$(i, j)^{th}$ element of $(AB)' = (j, i)^{th}$ element of AB

= sum of the products of the elements of jth row of A with the corresponding elements of ith column of B.

= sum of the products of the elements i^{th} column of B with the corresponding elements of jth row of A.

= sum of the products of the elements i^{th} row of transpose matrix B' with the corresponding elements of jth column of transpose matrix $A' = (i, j)^{th}$ element of the matrix $B'A'$.

Now, we see that order of the matrices $(AB)'$ and $B'A'$ are the same and their corresponding elements are equal. This implies $(AB)' = B'A'$.

Theorem: Every square matrix can be expressed as the sum of symmetric matrix and a skew-symmetric matrix in one and only one way.

Solution. If A be any square, then we consider

$$B = \frac{1}{2}(A + A') \text{ and } C = \frac{1}{2}(A - A')$$

$$\Rightarrow \quad B + C = \frac{1}{2}(A + A') + \frac{1}{2}(A - A') = \frac{1}{2}(A + A' + A - A') = A$$

Similarly,

$$B' = \frac{1}{2}[(A + A')]' = \frac{1}{2}[A' + (A')'] = \frac{1}{2}[A' + A] = \frac{1}{2}(A + A') = B$$

$B' = B$. This implies B is symmetric matrix.

Now, we consider

$$C' = \frac{1}{2}[A - A']' = \frac{1}{2}[A' - (A')'] = \frac{1}{2}[A' - A]$$

$$= -\frac{1}{2}(A - A') = -C$$

i.e., $C' = -C$

Hence C is a skew symmetric matrix.

Therefore, every square matrix A can be expressed as $A = B + C$, where $B = \frac{1}{2}(A + A')$, which is symmetric matrix and $C = \frac{1}{2}(A - A')$ is skew-symmetric matrix ($\because C' = -C$). The same process can be used to show that the result is unique.

II. *Adjoint of a Square Matrix*

If A is $\begin{bmatrix} a_1 & b_1 & c_1 \\ a_2 & b_2 & c_2 \\ a_3 & b_3 & c_3 \end{bmatrix}$ then determinant of A is $|A|$

i.e.,
$$|A| = \Delta = \begin{vmatrix} a_1 & b_1 & c_1 \\ a_2 & b_2 & c_2 \\ a_3 & b_3 & c_3 \end{vmatrix}$$

We form a matrix with the cofactors of elements of A which is

$$\begin{bmatrix} A_1 & B_1 & C_1 \\ A_2 & B_2 & C_2 \\ A_3 & B_3 & C_3 \end{bmatrix} \qquad ...(1)$$

The transpose of this matrix (1), denoted by

$$\begin{bmatrix} A_1 & A_2 & A_3 \\ B_1 & B_2 & B_3 \\ C_1 & C_2 & C_3 \end{bmatrix} \qquad ...(2)$$

is called adjoint of A and is denoted as adj A, *i.e.*, the adjoint of a matrix A is the transpose matrix of co-factors of matrix A.

III. *Inverse of a Matrix*

If A be any given matrix and B be another matrix (if it exists), such that

$$AB = BA = 1 \qquad ...(3)$$

Then B is called the inverse matrix of matrix A and is denoted by $B = A^{-1}$

As AB and BA exist and are equal and therefore they must be square matrices of the same order.

$\therefore$ determinant of $AB = |AB| = |A|\,|B| = |1| = 1 \neq 0$

i.e. $|A| \neq 0$ and $|B| \neq 0$

This indicates that the matrix A and its inverse $B(A^1)$ must be non singular

$\therefore$ Using (3), we get $AA^{-1} = A^{-1}A = 1$.

To show $A^{-1} = \dfrac{1}{|A|}$ adj. A

We consider,

$$A(\text{adj. } A) = \begin{bmatrix} a_1 & b_1 & c_1 \\ a_2 & b_2 & c_2 \\ a_3 & b_3 & c_3 \end{bmatrix}\begin{bmatrix} A_1 & A_2 & A_3 \\ B_1 & B_2 & B_3 \\ C_1 & C_2 & C_3 \end{bmatrix}$$

$$= \begin{bmatrix} a_1A_1 + b_1B_1 + c_1C_1 & a_1A_2 + b_1B_2 + c_1C_2 & a_1A_3 + b_1B_3 + c_1C_3 \\ a_2A_1 + b_2B_1 + c_2C_1 & a_2A_2 + b_2B_2 + c_2C_2 & a_2A_3 + b_3B_3 + c_2C_3 \\ a_3A_1 + b_3B_1 + c_3C_1 & a_3A_2 + b_3B_2 + c_3C_2 & a_3A_3 + b_3B_3 + c_3C_3 \end{bmatrix}$$

$$= \begin{bmatrix} |A| & 0 & 0 \\ 0 & |A| & 0 \\ 0 & 0 & |A| \end{bmatrix} = |A| \begin{bmatrix} 1 & 0 & 0 \\ 0 & 1 & 0 \\ 0 & 0 & 1 \end{bmatrix}$$

$$\Rightarrow \frac{1}{|A|} A(\text{adj. } A) = 1 \quad \frac{1}{|A|} \text{ adj. } A = A^{-1} \text{ which proves (3)}$$

Observation I.

We may show that inverse of a matrix A is unique. If possible let B and C be inverse matrices of matrix A, then

$$AB = BA = I \quad ...(i)$$

and $$AC = CA = I \quad ...(ii)$$

We consider

$$CAB = (CA)B = IB = B$$

and $$CAB = C(AB) = CI = C$$

The above two results show that $CAB = B = C$

i.e., B and C (if inverse of A) are the same.

Observation II.

To show that $(AB)^{-1} = B^{-1} A^{-1}$

If A and B taken as two matrices compatible with the product AB, then $(AB)^{-1}$ is the reciprocal of their product AB.

Now we consider $(AB)\, B^{-1} A^{-1} = A(BB^{-1})\, A^{-1} = AI\, A^{-1} = AA^{-1} = I$ and again $(B^{-1} A^{-1})\,(AB) = B^{-1}\,(A^{-1} A)\, B = B^{-1} B = I$

$(B^{-1} A^{-1})\,(AB) = (AB)\,(B^{-1} A^{-1}) = I$

Hence $B^{-1} A^{-1}$ is reciprocal of AB, *i.e.*, $B^{-1} A^{-1} = (AB)^{-1}$

1.4 VECTORS

Any vector in two dimension is represented by $X = (x_1, x_2)$ and in three dimension as $X = (x_1, x_2, x_3)$. Similarly, we may represent vector in 4th dimensional space as $X = (x_1, x_2, x_3, x_4)$ and so on.

In n dimensional space vector X which is represented by such numbers (called ordered n-tuple numbers) *i.e.*, $X = (x_1, x_2, x_3, ..., x_n)$ is called an n-vector in which x_i is fixed for $i = 1, 2, 3, ..., n$.

Numbers $x_1, x_2, x_3, ..., x_n$ with in brackets are called components of vector $X = (x_1, x_2, x_3, ..., x_n)$.

A vector is written either in rows or columns such as

$$X = \begin{bmatrix} x1 \\ x2 \\ x3 \end{bmatrix} \text{ or } X = [x_1, x_2, x_3]$$

A null vector consists of all elements as zeroes and is represented by O, *i.e.*, $O = (0, 0, 0, ..., 0)$

If we take two vectors in ordered n-tuples as

$$X = (x_1, x_2, x_3, ..., x_n) \text{ and } Y = (y_1, y_2, y_3, ..., y_n)$$

Then above two vectors are defined as equal if each element of X is equal to corresponding element of Y.

i.e., $$x_i = y_i \text{ for all } i = 1, 2, ..., n, \text{ then } X = Y$$

and sum of the above two vectors is defined as follows :

$$X + Y = (x_1 + y_1, x_2 + y_2, x_3 + y_3, ..., x_n + y_n)$$

and scalar multiple of vector X is defined as

$$kX = (kx_1, kx_2, kx_3, ..., kx_n)$$

such that

$$mX + nY = (mx_1 + ny_1, mx_2 + ny_2, mx_3 + ny_3, ..., mx_n + ny_n)$$

1.5 LINEAR DEPENDENCE AND INDEPENDENCE OF VECTORS

Let $X_1, X_2, X_3, ..., X_r$ be n-tupled vectors r in numbers, if corresponding to above r vectors, we could find r scalars K_i ($i = 1, 2, 3, 4, ..., r$), in which all are not zeroes (a few may be zero) such that

$$K_1X_1 + K_2X_2 + K_3X_3 + ... + K_rX_r = 0 \quad ...(1)$$

then these vectors $X_i (i = 1, 2, ..., r)$ are said to be dependent.

If (1) is satisfied such that

$$K_1 = K_2 = \dots = K_r = 0, \text{ then}$$

The vector $X_i (i = 1, 2, \dots, r)$ called independent vectors.

Example **1.1.** Are the following vectors linearly dependent? If so, find the relation between them, $x_1 = (3, 2, 7)$, $x_2 = (2, 4, 1)$ and $x_3 = (1, -2, 6)$.

Solution. Let λ_1, λ_2 and λ_3 be the scalars such that

$$\lambda_1 x_1 + \lambda_2 x_2 + \lambda_3 x_3 = 0 \quad \dots(1)$$

$$\Rightarrow \quad \lambda_1(3, 2, 7) + \lambda_2(2, 4, 1) + \lambda_3(1, -2, 6) = 0$$

$$\Rightarrow \quad 3\lambda_1 + 2\lambda_2 + \lambda_3 = 0 \quad \dots(A)$$

$$2\lambda_1 + 4\lambda_2 - 2\lambda_3 = 0 \quad \dots(B)$$

$$7\lambda_1 + \lambda_2 + 6\lambda_3 = 0 \quad \dots(C)$$

eliminating λ_3 from (*A*) and (*B*) we get $\lambda_1 + \lambda_2 = 0$ and also putting $\lambda_1 + \lambda_2 = 0$ in (*C*), we get $\lambda_2 - \lambda_3 = 0$.

$$\Rightarrow \quad \lambda_2 = \lambda_3 = -\lambda_1$$

so, equation (1) becomes $\lambda_1(x_1 - x_2 - x_3) = 0$, but $\lambda_1 \neq 0$

so, $x_1 - x_2 - x_3 = 0 \Rightarrow x_1 = x_2 + x_3$, *i.e.*, vectors are linearly dependent.

Example **1.2.** Are the following vectors linearly dependent? If so, find the relation between them, $x_1 = (1, 1, 2, 3)$, $x_2 = (1, 2, 3, 4)$ and $x_3 = (2, 3, 4, 9)$.

Solution. Let λ_1, λ_2 and λ_3 be scalars

such that $$\lambda_1 x_1 + \lambda_2 x_2 + \lambda_3 x_3 = 0 \quad \dots(1)$$

$$\Rightarrow \lambda_1(1, 1, 2, 3) + \lambda_2(1, 2, 3, 4) + \lambda_3(2, 3, 4, 9)$$

$$\Rightarrow \quad \lambda_1 + \lambda_2 + 2\lambda_3 = 0 \quad \dots(A)$$

$$\lambda_1 + 2\lambda_2 + 3\lambda_3 = 0 \quad \dots(B)$$

$$\lambda_1 + 3\lambda_2 + 4\lambda_3 = 0 \quad \dots(C)$$

$$3\lambda_1 + 4\lambda_2 + 9\lambda_3 = 0 \quad \dots(D)$$

Now from (*B*) – (*A*), we get $\lambda_2 + \lambda_3 = 0$ ... (E)

And (*C*) – (*B*) provides $\lambda_1 + \lambda_3 = 0$ ... (F)

From (*E*) and (*F*), using in (*D*), we get

$$3\lambda_1 - 4\lambda_1 - 9\lambda_1 = 0 \Rightarrow \lambda_1 = 0$$

$\therefore$ From (*E*) and (*F*) $\lambda_2 = 0$ and $\lambda_3 = 0$

i.e., $\lambda_1 = 0, \lambda_2 = 0$ and $\lambda_3 = 0$

$\Rightarrow$ Vectors x_1, x_2 and x_3 represented by (1) are linearly independent vectors (and not dependent vectors).

EXERCISE 1.1

Are the following vectors linearly dependent? If so, find the relation between them.

1. $x_1 = (1, 2, 4)$, $x_2 = (2, -1, 3)$, $x_3 = (0, 1, 2)$ and $x_4 = (-3, 7, 2)$

Ans. Yes; $8x_1 - 12x_2 + 5x_3 - 5x_4 = 0$

2. $x_1 = (1, 3, 4, 2)$, $x_2 = (3, -5, 2, 2)$ and $x_3 = (2, -1, 3, 2)$

Ans. Yes; $x_1 + x_2 - 2x_3 = 0$

3. $x_1 = (2, -1, 3, 2)$, $x_2 = (1, 3, 4, 2)$ and $x_3 = (3, -5, 2, 2)$

Ans. Yes; $2x_1 - x_2 - x_3 = 0$

4. $x_1 = (2, 3, 1, -1)$, $x_2 = (2, 3, 1, -2)$ and $x_3 = (4, 6, 2, 1)$

Ans. Yes; $5x_1 - 3x_2 - x_3 = 0$

5. $x_1 = (1, 2, 1)$, $x_2 = (2, 1, 4)$ and $x_3 = (4, 5, 6)$

Ans. Yes; $x_3 = 2x_1 + x_2$

1.6 LINEAR TRANSFORMATIONS

If $P(x, y)$ be a point in the dimensions with x, y as axes and these axes are rotated by an angle θ. Keeping the origin O fixed such that the new co-ordinate of P is (x', y') with reference to the new axes ox' and oy' respectively, then

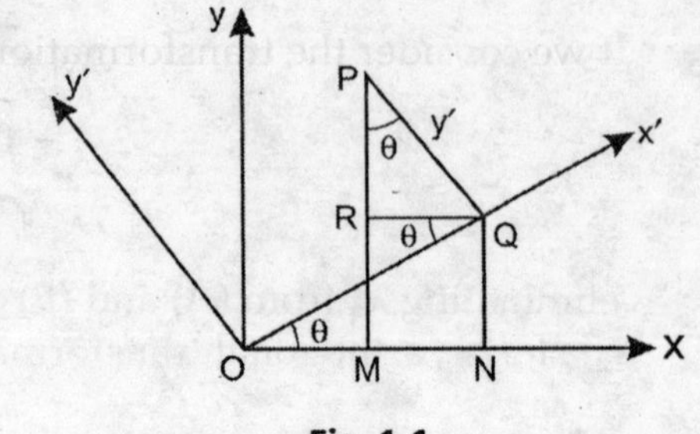

$$OM = x \text{ and } PM = y$$

$$OQ = x' \text{ and } PQ = y'$$

Fig. 1.1

so that

$$OM = ON - MN$$

$$x = x' \cos\theta - y' \sin\theta \qquad ...(1)$$

and

$$PM = PR + RM = PR + QN = y' \cos\theta + x' \sin\theta$$

$$y = x' \sin\theta + y' \cos\theta \qquad ...(2)$$

From (1) and (2), we get linear relation in the form

$$x' = a_1x + b_1y$$

and

$$y' = a_2x + b_2y$$

which can be represented in the form

$$\begin{bmatrix} x' \\ y' \end{bmatrix} = \begin{bmatrix} a_1 & b_1 \\ a_2 & b_2 \end{bmatrix} \begin{bmatrix} x \\ y \end{bmatrix} \Rightarrow X' = AX \qquad ...(3)$$

This transformation (3) is linear transformation in two dimensions.

Similarly, if we take the relation of the form.

$$x' = a_1x + b_1y + c_1z$$

$$y' = a_2x + b_2y + c_2z$$

$$z' = a_3x + b_3y + c_3z$$

which can be represented in the matrix form as

$$X' = AX \text{ where } X = \begin{bmatrix} x \\ y \\ z \end{bmatrix}, X' = \begin{bmatrix} x' \\ y' \\ z' \end{bmatrix} \text{ and } A = \begin{bmatrix} a_1 & b_1 & c_1 \\ a_2 & b_2 & c_2 \\ a_3 & b_3 & c_3 \end{bmatrix}$$

Then $X' = AX$ represents a linear transformation in three dimension.

1.7 ORTHOGONAL TRANSFORMATION

The linear transformation $Y = AX$ is called orthogonal if it transforms $y_1^2 + y_2^2 + ... + y_n^2$ into $x_1^2 + x_2^2 + ... + x_n^2$. The matrix A of this transformation is called orthogonal matrix.

i.e., if

$$X = \begin{bmatrix} x_1 \\ x_2 \\ ... \\ x_n \end{bmatrix} \text{ and } X' = [x_1\ x_2\ ...\ x_n]$$

so that

$$X' X = [x_1\, x_2 \dots x_n] \times \begin{bmatrix} x_1 \\ x_2 \\ \dots \\ x_n \end{bmatrix}$$

$$= x_1^2 + x_2^2 + \dots x_n^2$$

and similarly

$$Y'Y = y_1^2 + y_2^2 \dots + y_n^2$$

If we consider the transformation $Y = AX$ to be an orthogonal transformation, then

$$Y'Y = (AX)'(AX) = (X'A')(AX) = X'A'AX$$

$$= X'(A'A)X = X'IX = X'X \text{ is true only.}$$

if $\quad A'A = I \Rightarrow A' = A^{-1}$ since $A^{-1}A = I$

i.e., for an orthogonal transformation $A' = A^{-1}$. Hence for any square matrix A, A is said to orthogonal.

if $\quad AA' = A'A = I$ where $A' = A^{-1}$

Also since

$$|A'| = |A|,\ A'A = I \Rightarrow |A'|\,|A| = |I| = 1$$

$\Rightarrow \quad |A|\,|A| = 1 \Rightarrow |A|^2 = 1 \Rightarrow |A| = \pm 1$

Let $n \times n$ matrix A be $= \begin{bmatrix} a_{11} & a_{12} & \dots & a_{1n} \\ a_{21} & a_{22} & \dots & a_{2n} \\ a_{31} & a_{32} & \dots & a_{3n} \\ \dots & \dots & \dots & \dots \\ a_{n1} & a_{n2} & \dots & a_{nn} \end{bmatrix}$ and column matrix X be

$X = \begin{bmatrix} x_1 \\ x_2 \\ \dots \\ x_n \end{bmatrix}$ and column matrix $Y = \begin{bmatrix} y_1 \\ y_2 \\ \dots \\ y_n \end{bmatrix}$ such that

$$Y = AX$$

Then (1) defines, a linear transformation, which transforms any vector X into another vector Y, using the matrix A. Matrix A is said to be the linear operator of transformation from vector X to vector Y because the linear property of vectors is satisfied.

i.e., if X_1 and X_2 are two vector such that

$$Y_1 = AX_1 \text{ and } Y_2 = AX_2 \Rightarrow aY_1 + bY_2 = A(aX_1 + bX_2)$$

If $|A| \neq 0$, then the transformation matrix A is called non-singular and then the linear transformation is known as regular (or non-singular) and if $|A| = 0$, then the transformation of matrix A is called singular and the linear transformation is known as singular. If the transformation of matrix A is singular (non-singular) then the transformation is accordingly called singular (non-singular). Also for a $Y = AX$ (a non-singular transformation), $X = A^{-1}Y$ is the inverse transformation.

Further, if a vector $X = (x_1, x_2, x_3)$ is transformed by A to vector $Y = (y_1, y_2, y_3)$ such that $Y = AX$ and by another transformation B vector Y is transformed to vector $Z = (z_1, z_2, z_3)$ so that we have $Z = BY$ then $Z = BY = B(AX) = (BA)X$.

This transformation BA is called composite transformation.

Example **1.2.** Is the matrix $A = \begin{bmatrix} 2 & -3 & 1 \\ 4 & 3 & 1 \\ -3 & 1 & 9 \end{bmatrix}$ orthogonal?

Solution. Transpose of A is $A' = \begin{bmatrix} 2 & 4 & -3 \\ -3 & 3 & 1 \\ 1 & 1 & 9 \end{bmatrix}$

Consider

$$A'A = \begin{bmatrix} 2 & 4 & -3 \\ -3 & 3 & 1 \\ 1 & 1 & 9 \end{bmatrix} \times \begin{bmatrix} 2 & -3 & 1 \\ 4 & 3 & 1 \\ -3 & 1 & 9 \end{bmatrix}$$

$$= \begin{bmatrix} 4+16+9 & -6+12-3 & 2+4-27 \\ -6+12-3 & 9+9+1 & -3+3+9 \\ 2+4-27 & -3+3+9 & 1+1+81 \end{bmatrix}$$

$$= \begin{bmatrix} 29 & 3 & -21 \\ 3 & 19 & 9 \\ -21 & 9 & 83 \end{bmatrix} \neq 1$$

So, matrix A is not orthogonal..

Example **1.3.** Is the matrix $A = \begin{bmatrix} \cos\theta & 0 & \sin\theta \\ 0 & 1 & 0 \\ -\sin\theta & 0 & \cos\theta \end{bmatrix}$ orthogonal?

Solution. The transpose matrix $A' = \begin{bmatrix} \cos\theta & 0 & -\sin\theta \\ 0 & 1 & 0 \\ \sin\theta & 0 & \cos\theta \end{bmatrix}$

So,

$$A'A = \begin{bmatrix} \cos\theta & 0 & -\sin\theta \\ 0 & 1 & 0 \\ \sin\theta & 0 & \cos\theta \end{bmatrix} \times \begin{bmatrix} \cos\theta & 0 & \sin\theta \\ 0 & 1 & 0 \\ -\sin\theta & 0 & \cos\theta \end{bmatrix}$$

$$= \begin{bmatrix} \cos^2\theta + 0 + \sin^2\theta & 0+0+0 & \cos\theta\sin\theta - \cos\theta\sin\theta \\ 0+0+0 & 0+1+0 & 0+0+0 \\ \cos\theta\sin\theta - \cos\theta\sin\theta & 0+0+0 & \sin^2\theta + 0 + \cos^2\theta \end{bmatrix}$$

$$= \begin{bmatrix} 1 & 0 & 0 \\ 0 & 1 & 0 \\ 0 & 0 & 1 \end{bmatrix} = 1$$

Hence the given matrix is orthogonal.

Example **1.4.** Show the transformation $y_1 = 2x_1 + x_2 + x_3$, $y_2 = x_1 + x_2 + 2x_3$, $y_3 = x_1 - 2x_3$ is regular. Write down the inverse transformation.

Solution. If the transformation be taken as $Y = AX$ such that $X = A^{-1} Y$,

where $X = \begin{bmatrix} x_1 \\ x_2 \\ x_3 \end{bmatrix}$, $Y = \begin{bmatrix} y_1 \\ y_2 \\ y_3 \end{bmatrix}$ and $A = \begin{bmatrix} 2 & 1 & 1 \\ 1 & 1 & 2 \\ 1 & 0 & -2 \end{bmatrix}$

$\therefore$ The determinant of A is $|A| = \begin{vmatrix} 2 & 1 & 1 \\ 1 & 1 & 2 \\ 1 & 0 & -2 \end{vmatrix} = -1 \neq 0$. So that transformed matrix A is non-singular and the transformation is regular.

Now, if $$A = \begin{bmatrix} a_1 & b_1 & c_1 \\ a_2 & b_2 & c_2 \\ a_3 & b_3 & c_3 \end{bmatrix}$$ then to find A^{-1} we find

Co-factors of $|A|$ corresponding to $a_1, b_1, c_1, a_2, b_2, c_2$ etc., which are

$$A_1 = -2, B_1 = 4, C_1 = -1, A_2 = 2, B_2 = -5, C_2 = 1,$$
$$A_3 = 1, B_3 = -3 \text{ and } C_3 = 1.$$

Now, we know that

$$A^{-1} = \frac{1}{|A|} \text{adj.}A = \frac{1}{(-1)} \begin{bmatrix} A_1 & A_2 & A_3 \\ B_1 & B_2 & B_3 \\ C_1 & C_2 & C_3 \end{bmatrix} = \begin{bmatrix} 2 & -2 & -1 \\ -4 & 5 & 3 \\ 1 & -1 & -1 \end{bmatrix}$$

∴ Inverse transformation

∴ $$X = A^{-1} Y \Rightarrow \begin{bmatrix} x_1 \\ x_2 \\ x_3 \end{bmatrix} = \begin{bmatrix} 2 & -2 & -1 \\ -4 & 5 & 3 \\ 1 & -1 & -1 \end{bmatrix} \begin{bmatrix} y_1 \\ y_2 \\ y_3 \end{bmatrix}$$

from which we get

$$x_1 = 2y_1 - 2y_2 - y_3$$
$$x_2 = -4y_1 + 5y_2 + 3y_3$$

and $$x_3 = y_1 - y_2 - y_3$$

which is the inverse transformation.

1.8 ELEMENTARY TRANSFORMATION (OR OPERATIONS) ON A MATRIX

The following operations on a matrix are called elementary transformations (*i.e.*, *E* operations or *E*-transformations)

(*a*) The interchange of ith and j^{th} rows is represented by R_{ij}, and the interchange of ith and j^{th} columns is represented by C_{ij}.

(*b*) Multiplication of each element of a row or a column by a non-zero number *k*.
i.e., the multiplication of ith row by *m* is represented by mR_i and the multiplication of ith column by *n* is represented by nC_i and

(*c*) Addition of *m* times the elements of a row (or column) to the corresponding elements of another row (or column) multiplied by *n*, where $m \neq 0, n \neq 0$.

The addition of *m* times *i*th row to the *n* times *j*th row is represented by $mR_i + nR_j$.

If a matrix *B* is obtained from matrix *A* by such transformation, then the matrix *B* is called the equivalent matrix to matrix *A*.

Two such equivalent matrices *A* and *B* are denoted as $A \sim B$ and the symbol ~ is used for equivalence.

So, two equivalent matrices have the same order and same rank.

Elementary Matrices

An elementary matrix is that which is obtained from a unit matrix, by subjecting it to any of the elementary transformations.

If $$I_3 = \begin{bmatrix} 1 & 0 & 0 \\ 0 & 1 & 0 \\ 0 & 0 & 1 \end{bmatrix} \text{ then } R_{23} = \begin{bmatrix} 1 & 0 & 0 \\ 0 & 0 & 1 \\ 0 & 1 & 0 \end{bmatrix} = C_{23}$$

i.e., R_2 and R_3 have been interchanged. Similarly, C_2 and C_3 have been interchanged. Similarly,

$$kR_3 = \begin{bmatrix} 1 & 0 & 0 \\ 0 & 1 & 0 \\ 0 & 0 & k \end{bmatrix} \text{ and } R_2 \to 2R_2 + 3R_3 = \begin{bmatrix} 1 & 0 & 0 \\ 0 & 2 & 3 \\ 0 & 0 & 1 \end{bmatrix}$$

$$R_2 \to R_2 + R_3 = \begin{bmatrix} 1 & 0 & 0 \\ 0 & 1 & 1 \\ 0 & 0 & 1 \end{bmatrix}$$

Theorem: Elementary row (column) transformations of a matrix A can be obtained by pre-multiplying (post-multiplying) the matrix A by the corresponding elementary matrices.

If we take matrix

$$A = \begin{bmatrix} a_1 & b_1 & c_1 \\ a_2 & b_2 & c_2 \\ a_3 & b_3 & c_3 \end{bmatrix} \text{ and } I_3 = \begin{bmatrix} 1 & 0 & 0 \\ 0 & 1 & 0 \\ 0 & 0 & 1 \end{bmatrix}$$

Then

$$R_{23} = \begin{bmatrix} 1 & 0 & 0 \\ 0 & 0 & 1 \\ 0 & 1 & 0 \end{bmatrix}$$

Then

$$R_{23} \times A = \begin{bmatrix} 1 & 0 & 0 \\ 0 & 0 & 1 \\ 0 & 1 & 0 \end{bmatrix} \times \begin{bmatrix} a_1 & b_1 & c_1 \\ a_2 & b_2 & c_2 \\ a_3 & b_3 & c_3 \end{bmatrix} = \begin{bmatrix} a_1 & b_1 & c_1 \\ a_3 & b_3 & c_3 \\ a_2 & b_2 & c_2 \end{bmatrix}$$

So, we see that if A has been pre-multiplied by R_{23} then the result obtained is that in which second and 3rd rows of A have been interchanged.

Similarly, by post-multiplication, the elementary column transformation will be performed.

1.9 GAUSS-JORDAN METHOD OF FINDING THE INVERSE

The elementary row transformations which reduce a given square matrix A to the unit matrix, when applied to unit matrix I give, the inverse of A.

By successive row transformations, the matrix A is reduced to unit matrix I, which is obtained from pre-multiplication by elementary matrices $R_1, R_2, ..., R_i$ such that

$$R_i R_{i-1} \ldots R_2 R_1 A = I$$

Then by post-multiplication of both sides, we get

$$R_i R_{i-1} \ldots R_2 R_1 AA^{-1} = IA^{-1} = A^{-1}$$

$$\Rightarrow \quad R_i R_{i-1} \ldots R_2 R_1 (AA^{-1}) = A^{-1}$$

$$\Rightarrow \quad R_i R_{i-1} \ldots R_2 R_1 I = A^{-1}$$

which gives the required result.

1.10 NORMAL FORM OF A MATRIX

If A is an $m \times n$ matrix and by a series of elementary (*row or column or both*) operations, it can be put into one of the following forms (called normal forms):

$\left[\frac{Ir : O}{O : O}\right]$, $[I_r : O]$, $[I_r]$, where I_r is the unit matrix of order r.

***Example* 1.5.** Use Gauss-Jordan reduction method (*Elementary matrices method*) to compute the inverse of the matrix. $\begin{bmatrix} 3 & -3 & 4 \\ 2 & -3 & 4 \\ 0 & -1 & 1 \end{bmatrix}$

Solution. Given matrix $A = \begin{bmatrix} 3 & -3 & 4 \\ 2 & -3 & 4 \\ 0 & -1 & 1 \end{bmatrix}$

We will use elementary row transformation, which may reduce $A = IA$ to $I = PA$ such that P is the inverse matrix of A. Now,

$$\begin{bmatrix} 3 & -3 & 4 \\ 2 & -3 & 4 \\ 0 & -1 & 1 \end{bmatrix} = \begin{bmatrix} 1 & 0 & 0 \\ 0 & 1 & 0 \\ 0 & 0 & 1 \end{bmatrix} \times A$$

Operating $R_1 \to \frac{1}{3} R_1$, we get

$$\begin{bmatrix} 1 & -1 & 4/3 \\ 2 & -3 & 4 \\ 0 & -1 & 1 \end{bmatrix} = \begin{bmatrix} 1/3 & 0 & 0 \\ 0 & 1 & 0 \\ 0 & 0 & 1 \end{bmatrix} A$$

Next operating $R_2 \Rightarrow R_2 - 2R_1$, we get

$$\begin{bmatrix} 1 & -1 & 4/3 \\ 0 & -1 & 4/3 \\ 0 & -1 & 1 \end{bmatrix} = \begin{bmatrix} 1/3 & 0 & 0 \\ -2/3 & 1 & 0 \\ 0 & 0 & 1 \end{bmatrix} A$$

Next, multiplying R_2 by (-1), we get

$$\begin{bmatrix} 1 & -1 & 4/3 \\ 0 & 1 & -4/3 \\ 0 & -1 & 1 \end{bmatrix} = \begin{bmatrix} 1/3 & 0 & 0 \\ 2/3 & -1 & 0 \\ 0 & 0 & 1 \end{bmatrix} A$$

Again operating $R_1 \to R_1 + R_2$ and $R_3 \to R_3 + R_2$, we get

$$\begin{bmatrix} 1 & 0 & 0 \\ 0 & 1 & -4/3 \\ 0 & 0 & -1/3 \end{bmatrix} = \begin{bmatrix} 1 & -1 & 0 \\ 2/3 & -1 & 0 \\ 2/3 & -1 & 1 \end{bmatrix} A$$

Next operating $R_3 \to -3R_3$, we get

$$\begin{bmatrix} 1 & 0 & 0 \\ 0 & 1 & -4/3 \\ 0 & 0 & 1 \end{bmatrix} = \begin{bmatrix} 1 & -1 & 0 \\ -2/3 & -1 & 0 \\ -2 & 3 & -3 \end{bmatrix} A$$

Now, operating $R_2 \to -R_2 + \frac{4}{3} R_3$, we get

$$\begin{bmatrix} 1 & 0 & 0 \\ 0 & 1 & 0 \\ 0 & 0 & 1 \end{bmatrix} = \begin{bmatrix} 1 & -1 & 0 \\ -2 & 3 & -4 \\ -2 & 3 & -3 \end{bmatrix} A \Rightarrow I = PA$$

$\Rightarrow$ $P = A^{-1} = \begin{bmatrix} 1 & -1 & 0 \\ -2 & 3 & -4 \\ -2 & 3 & -3 \end{bmatrix}$ is the required result.

Example 1.6 Find the non-singular matrices P and Q such that PAQ is in the normal form for the matrix $A = \begin{bmatrix} 1 & -1 & 1 \\ 1 & 1 & 1 \\ 3 & 1 & 1 \end{bmatrix}$

Solution. Matrix A may be written as

$$A = \begin{bmatrix} 1 & -1 & 1 \\ 1 & 1 & 1 \\ 3 & 1 & 1 \end{bmatrix} = \begin{bmatrix} 1 & 0 & 0 \\ 0 & 1 & 0 \\ 0 & 0 & 1 \end{bmatrix} A \begin{bmatrix} 1 & 0 & 0 \\ 0 & 1 & 0 \\ 0 & 0 & 1 \end{bmatrix} \quad ...(1)$$

We take up row operations on pre-factor and column operations on post factor. Operating $C_2 + C_1$ and $C_3 + C_1$ in (1), we get

$$A \sim \begin{bmatrix} 1 & 0 & 0 \\ 1 & 2 & 2 \\ 3 & 4 & 4 \end{bmatrix} = \begin{bmatrix} 1 & 0 & 0 \\ 0 & 1 & 0 \\ 0 & 0 & 1 \end{bmatrix} A \begin{bmatrix} 1 & 1 & 1 \\ 0 & 1 & 0 \\ 0 & 0 & 1 \end{bmatrix}$$

Operating $R_2 - R_1$ and $R_3 - 3R_1$, we get

$$A \sim \begin{bmatrix} 1 & 0 & 0 \\ 0 & 2 & 2 \\ 0 & 4 & 4 \end{bmatrix} = \begin{bmatrix} 1 & 0 & 0 \\ -1 & 1 & 0 \\ -3 & 0 & 1 \end{bmatrix} A \begin{bmatrix} 1 & 1 & 1 \\ 0 & 0 & 0 \\ 0 & 0 & 1 \end{bmatrix}$$

Again operating $C_3 - C_2$ and $\frac{C_2}{2}$, we get

$$A \sim \begin{bmatrix} 1 & 0 & 0 \\ 0 & 1 & 0 \\ 0 & 2 & 0 \end{bmatrix} = \begin{bmatrix} 1 & 0 & 0 \\ -1 & 1 & 0 \\ -3 & 0 & 1 \end{bmatrix} A \begin{bmatrix} 1 & 1/2 & 0 \\ 0 & 1/2 & -1 \\ 0 & 0 & 1 \end{bmatrix}$$

Next, we operating $R_3 - 2R_2$ and get

$$A \sim \begin{bmatrix} 1 & 0 & 0 \\ 0 & 1 & 0 \\ 0 & 0 & 0 \end{bmatrix} = \begin{bmatrix} 1 & 0 & 0 \\ -1 & 1 & 0 \\ -1 & -2 & 1 \end{bmatrix} A \begin{bmatrix} 1 & 1/2 & 0 \\ 0 & 1/2 & -1 \\ 0 & 0 & 1 \end{bmatrix} = PAQ$$

$$\therefore \quad \begin{bmatrix} I_r & 0 \\ 0 & 0 \end{bmatrix} = PAQ \text{ where } P \begin{bmatrix} 1 & 0 & 0 \\ -1 & 1 & 0 \\ -1 & -2 & 0 \end{bmatrix} \text{ and } Q = \begin{bmatrix} 1 & 1/2 & 0 \\ 0 & 1/2 & -1 \\ 0 & 0 & 1 \end{bmatrix}$$

are the required result.

Example 1.7 Find the non-singular matrices P and Q such that PAQ is in normal form for the matrix $\begin{bmatrix} 1 & 2 & 3 & -2 \\ 2 & -2 & 1 & 3 \\ 3 & 0 & 4 & 1 \end{bmatrix}$

Solution. Let the matrix A equal to $\begin{bmatrix} 1 & 2 & 3 & -2 \\ 2 & -2 & 1 & 3 \\ 3 & 0 & 4 & 1 \end{bmatrix}$

We may write $A = IAI$

$$A = \begin{bmatrix} 1 & 2 & 3 & 2 \\ 2 & -2 & 1 & 3 \\ 3 & 0 & 4 & 1 \end{bmatrix} = \begin{bmatrix} 1 & 0 & 0 \\ 0 & 1 & 0 \\ 0 & 0 & 1 \end{bmatrix} \begin{bmatrix} 1 & 2 & 3 & -2 \\ 2 & -2 & 1 & 3 \\ 3 & 0 & 4 & 1 \end{bmatrix} \begin{bmatrix} 1 & 0 & 0 & 0 \\ 0 & 1 & 0 & 0 \\ 0 & 0 & 1 & 0 \\ 0 & 0 & 0 & 1 \end{bmatrix}$$

Operating $R_2 \to R_2 - 2R_1$ and $R_3 \to R_3 - 3R_1$ in pre-factor, we get

$$\begin{bmatrix} 1 & 2 & 3 & -2 \\ 0 & -6 & -5 & 7 \\ 0 & -6 & -5 & 7 \end{bmatrix} = \begin{bmatrix} 1 & 0 & 0 \\ -2 & 1 & 0 \\ -3 & 0 & 1 \end{bmatrix} A \begin{bmatrix} 1 & 0 & 0 & 0 \\ 0 & 1 & 0 & 0 \\ 0 & 0 & 1 & 0 \\ 0 & 0 & 0 & 1 \end{bmatrix}$$

Now, operating $C_2 \to C_2 - 2C_1$, $C_3 \to C_3 - 3C_1$ and $C_4 \to C_4 + 2C_1$ in post-factor, we get

$$\begin{bmatrix}1 & 0 & 0 & 0\\0 & -6 & -5 & 7\\0 & -6 & -5 & 7\end{bmatrix} = \begin{bmatrix}1 & 0 & 0\\-2 & 1 & 0\\-3 & 0 & 1\end{bmatrix} A \begin{bmatrix}1 & -2 & -3 & 2\\0 & 1 & 0 & 0\\0 & 0 & 1 & 0\\0 & 0 & 0 & 1\end{bmatrix}$$

Now dividing the columns (in post-factor) C_2 by -6, C_3 by -5 and C_4 by 7, we get

$$\begin{bmatrix}1 & 0 & 0 & 0\\0 & 1 & 1 & 1\\0 & 1 & 1 & 1\end{bmatrix} = \begin{bmatrix}1 & 0 & 0\\-2 & 1 & 0\\-3 & 0 & 1\end{bmatrix} A \begin{bmatrix}1 & 1/3 & 3/5 & 2/7\\0 & -1/6 & 0 & 0\\0 & 0 & -1/5 & 0\\0 & 0 & 0 & 1/7\end{bmatrix}$$

Now, operating $C_3 \to C_3 - C_2$ and $C_4 \to C_4 - C_2$ (in post-factor), we get

$$\begin{bmatrix}1 & 0 & 0 & 0\\0 & 1 & 0 & 0\\0 & 1 & 0 & 0\end{bmatrix} = \begin{bmatrix}1 & 0 & 0\\-2 & 1 & 0\\-3 & 0 & 1\end{bmatrix} A \begin{bmatrix}1 & 1/3 & 4/15 & -1/21\\0 & -1/6 & 1/6 & 1/6\\0 & 0 & -1/5 & 0\\0 & 0 & 0 & 1/7\end{bmatrix}$$

Finally taking $R_3 - R_2$ in pre factor, we get

$$\begin{bmatrix}1 & 0 & 0\\0 & 1 & 0\\0 & 0 & 0\end{bmatrix} = \begin{bmatrix}1 & 0 & 0\\-2 & 1 & 0\\-1 & -1 & 1\end{bmatrix} A \begin{bmatrix}1 & 1/3 & 4/15 & -1/21\\0 & -1/6 & 1/6 & 1/6\\0 & 0 & -1/5 & 0\\0 & 0 & 0 & 1/7\end{bmatrix}$$

which is in normal form

where $P = \begin{bmatrix}1 & 0 & 0\\-2 & 1 & 0\\-1 & -1 & 1\end{bmatrix}$ and $Q = \begin{bmatrix}1 & 1/3 & 4/15 & -1/21\\0 & -1/6 & 1/6 & 1/6\\0 & 0 & -1/5 & 0\\0 & 0 & 0 & 1/7\end{bmatrix}$

are required results.

1.11 RANK OF MATRIX

Definition : Let A be $m \times n$ matrix. So, A has sub-matrices of various orders. The determinant of any such sub-matrices are defined as minor of matrix. A of order r where $r < m$ and $r < n$ [or $r \le m$, $r \le n$]. If all minors of order $(r + 1)$ are zeroes and we have at least one non-zero minor of order r, then it is said that the rank of matrix A is r and rank of r is represented $\rho(A)$ = rank of $A = r$.

Thus from the above definition of the rank of a matrix A, we have the following observations :

(*a*) If $A = \begin{bmatrix}0 & 0 & 0\\0 & 0 & 0\\0 & 0 & 0\end{bmatrix}$ = null matrix then rank of $A = \rho(A) = 0$.

(*b*) If $A_1 = \begin{bmatrix}a & b & 0\\0 & 0 & 0\\0 & 0 & 0\end{bmatrix}$ or $A_2 = \begin{bmatrix}a & b & 0\\c & 0 & 0\\0 & 0 & 0\end{bmatrix}$ etc, then rank of matrix

(c) If A is $m \times n$ matrix then $A = \begin{vmatrix} 1 & 0 & 0 & 0 & 0 & \dots & 0 \\ 0 & 1 & 0 & 0 & 0 & \dots & 0 \\ 0 & 0 & 1 & 0 & 0 & \dots & 0 \\ \dots & \dots & \dots & \dots & \dots & \dots & \dots \\ 0 & 0 & 0 & 1 & \dots & \dots & 0 \\ 0 & 0 & 0 & 0 & \dots & \dots & 0 \\ 0 & 0 & 0 & 0 & 0 & \dots & 1 \end{vmatrix} = 1 \neq 0$

i.e., rank of $A = \rho(A) = n$.

(d) If A is $m \times n$ matrix, then $\rho(A) \leq$ min of m and n

(e) If all minors of order r are equal to zero then rank of $A = \rho(A) < r$

To Determine the Rank of Matrix A

If A is $m \times n$ matrix and by various elementary row (or column) transformations, A could be arranged in terms of an equivalent matrix in the following form (known as normal forms).

$$\begin{bmatrix} I_r & 0 \\ 0 & 0 \end{bmatrix} \text{i.e.} \begin{bmatrix} 1 & 0 & 0 & 0 & 0 & 0 \\ 0 & 1 & 0 & 0 & 0 & 0 \\ 0 & 0 & 1 & 0 & 0 & 0 \\ 0 & 0 & 0 & 1 & 0 & 0 \\ 0 & 0 & 0 & 0 & 1 & 0 \\ 0 & 0 & 0 & 0 & 0 & 0 \end{bmatrix} \text{Here } r = 5$$

As the rank of matrix does not change as a result of elementary transformations (or operations).

so $\delta(A) = \delta$ (in the above example $\delta = 5$) and $|I_5| = I \neq 0$

***Example* 1.8** Determine the rank of the matrix $A = \begin{bmatrix} 1 & 4 & 5 \\ 2 & 6 & 8 \\ 3 & 7 & 22 \end{bmatrix}$

Solution. Operating $R_2 \to R_2 - 2R_1$ and $R_3 \to R_3 - 3R_1$, we get

$$\sim \begin{bmatrix} 1 & 4 & 5 \\ 0 & -2 & 2 \\ 0 & -5 & 7 \end{bmatrix}$$

Again operating $R_2 \to -\frac{1}{2} R_2$, we get

$$\sim \begin{bmatrix} 1 & 4 & 5 \\ 0 & 1 & 1 \\ 0 & -5 & 7 \end{bmatrix}$$

Next, operating $R_3 \to R_3 + 5R_2$, we get

$$\sim \begin{bmatrix} 1 & 4 & 5 \\ 0 & 1 & 1 \\ 0 & 0 & 12 \end{bmatrix}$$

This implies that rank of $A = 3$.

***Example* 1.9.** Determine the rank of the matrix $A = \begin{bmatrix} 1 & 2 & 3 \\ 1 & 4 & 2 \\ 2 & 6 & 5 \end{bmatrix}$

Solution. Operating $R_2 \to R_2 - R_1$ and $R_3 \to R_3 - 2R_1$

$$\sim \begin{bmatrix} 1 & 2 & 3 \\ 0 & 2 & -1 \\ 0 & 2 & -1 \end{bmatrix}$$

Next, operating $R_3 \to R_3 - R_2$, we get

$$\sim \begin{bmatrix} 1 & 2 & 3 \\ 0 & 2 & -1 \\ 0 & 0 & 0 \end{bmatrix}$$ Here 3rd order minor is zero.

But 2nd order minor exist, *i.e.*, $\begin{vmatrix} 2 & 3 \\ 2 & -1 \end{vmatrix} = -2 - 6 = -8 \neq 0$

So, the rank of matrix $A = 2$

Example **1.10.** Find the rank of the matrix $\begin{bmatrix} 2 & 3 & -1 & -1 \\ 1 & -1 & -2 & -4 \\ 3 & 1 & 3 & -2 \\ 6 & 3 & 0 & -7 \end{bmatrix}$

Solution. Let the given matrix be A, then by operations R_{12} (interchange 1st and 2nd rows).

$$A \sim \begin{bmatrix} 1 & -1 & -2 & -4 \\ 2 & 3 & -1 & -1 \\ 3 & 1 & 3 & -2 \\ 6 & 3 & 0 & -7 \end{bmatrix}$$

Next operating $R_2 - 2R_1$, $R_3 - 3R_1$ and $R_4 - 6R_1$, we get

$$A \sim \begin{bmatrix} 1 & -1 & -2 & -4 \\ 0 & 5 & 3 & 7 \\ 0 & 4 & 9 & 10 \\ 0 & 9 & 12 & 17 \end{bmatrix}$$

Again using the operation $C_2 + C_1$, $C_3 + 2C_1$ and $C_4 + 4C_1$, we get

$$A \sim \begin{bmatrix} 1 & 0 & 0 & 0 \\ 0 & 5 & 3 & 7 \\ 0 & 4 & 9 & 10 \\ 0 & 9 & 12 & 17 \end{bmatrix}$$

Next, operating $R_4 - R_2 - R_3$, we get

$$A \sim \begin{bmatrix} 1 & 0 & 0 & 0 \\ 0 & 5 & 3 & 7 \\ 0 & 4 & 9 & 10 \\ 0 & 0 & 0 & 0 \end{bmatrix}$$

Again, operating $R_2 - R_3$, we get

$$A \sim \begin{bmatrix} 1 & 0 & 0 & 0 \\ 0 & 1 & -6 & -3 \\ 0 & 4 & 9 & 10 \\ 0 & 0 & 0 & 0 \end{bmatrix}$$

Now, operating $R_3 - 4R_2$, we get

$$A \sim \begin{bmatrix} 1 & 0 & 0 & 0 \\ 0 & 1 & -6 & -3 \\ 0 & 0 & 33 & 22 \\ 0 & 0 & 0 & 0 \end{bmatrix}$$

Next, operating $C_3 + 6C_2$ and $C_4 + 3C_2$, we get

$$A \sim \begin{bmatrix} 1 & 0 & 0 & 0 \\ 0 & 1 & 0 & 0 \\ 0 & 0 & 33 & 22 \\ 0 & 0 & 0 & 0 \end{bmatrix}$$

Operating $\frac{1}{33} C_3$, we get

$$A \sim \begin{bmatrix} 1 & 0 & 0 & 0 \\ 0 & 1 & 0 & 0 \\ 0 & 0 & 1 & 22 \\ 0 & 0 & 0 & 0 \end{bmatrix}$$

Again operating $C_4 - 22C_3$, we get

$$A \sim \begin{bmatrix} 1 & 0 & 0 & 0 \\ 0 & 1 & 0 & 0 \\ 0 & 0 & 1 & 0 \\ 0 & 0 & 0 & 0 \end{bmatrix} \sim \begin{bmatrix} I_3 & 0 \\ 0 & 0 \end{bmatrix}$$

Hence rank of $A = \rho(A) = 3$ is the required result.

***Example* 1.11.** Find the rank of the matrix $A = \begin{bmatrix} 3 & -4 & -1 & 2 \\ 1 & 7 & 3 & 1 \\ 5 & -2 & 5 & 4 \\ 9 & -3 & 7 & 7 \end{bmatrix}$

Solution. Operation R_{12}, we get

$$A \sim \begin{bmatrix} 1 & 7 & 3 & 1 \\ 3 & -4 & -1 & 2 \\ 5 & -2 & 5 & 4 \\ 9 & -3 & 7 & 7 \end{bmatrix}$$

Now, operating $R_2 \to R_2 - 3R_1$, $R_3 \to R_3 - 5R_1$, $R_4 \to R_4 - 9R_1$, we get

$$\sim \begin{bmatrix} 1 & 7 & 3 & 1 \\ 0 & -25 & -10 & -1 \\ 0 & -37 & -10 & -1 \\ 0 & -66 & -20 & -2 \end{bmatrix}$$

Next operating $R_3 \to R_3 - \frac{37}{25} R_2$, $R_4 \to R_4 - \frac{66}{25} R_2$

$$\sim \begin{bmatrix} 1 & 7 & 3 & 1 \\ 0 & -25 & -10 & -1 \\ 0 & 0 & -10 + \frac{74}{5} & -1 + \frac{37}{25} \\ 0 & 0 & -20 + \frac{132}{5} & -2 + \frac{66}{25} \end{bmatrix} = \begin{bmatrix} 1 & 7 & 3 & 1 \\ 0 & -25 & -10 & -1 \\ 0 & 0 & \frac{24}{5} & \frac{12}{25} \\ 0 & 0 & \frac{32}{5} & \frac{16}{25} \end{bmatrix}$$

Again operating $R_4 - \frac{4}{3} R_3$, we get

$$\sim \begin{bmatrix} 1 & 7 & 3 & 1 \\ 0 & -25 & -10 & -1 \\ 0 & 0 & \frac{24}{5} & \frac{12}{25} \\ 0 & 0 & 0 & 0 \end{bmatrix}$$

$\Rightarrow$ $\rho(A)$ = rank of $A = 3$ is the required result.

Example **1.12.** Find the rank of the matrix $A = \begin{bmatrix} -1 & 2 & 3 & -2 \\ 2 & -5 & 1 & 2 \\ 3 & -8 & 5 & 2 \\ 5 & -12 & -1 & 6 \end{bmatrix}$

Solution. Operating $R_2 \to R_2 + 2R_1$, $R_3 \to R_3 + 3R_1$ and $R_4 \to R_4 + 5R_1$, we get

$$A \sim \begin{bmatrix} 1 & 2 & 3 & -2 \\ 0 & -1 & 7 & -2 \\ 0 & -2 & 14 & -4 \\ 0 & -2 & 14 & -4 \end{bmatrix}$$

Again, operating $R_3 \to R_3 - 2R_2$, $R_4 \to R_4 - 2R_2$, we get

$$\sim \begin{bmatrix} -1 & 2 & 3 & -2 \\ 0 & -1 & 7 & -2 \\ 0 & 0 & 0 & 0 \\ 0 & 0 & 0 & 0 \end{bmatrix}$$

Hence only the second order co-factors exist

i.e., $\begin{bmatrix} -1 & 2 \\ 0 & -1 \end{bmatrix} = 1 \neq 0$ $\therefore$ rank of $A = 2$

Example **1.13.** Find the rank of matrix after reducing the matrix to normal form $\begin{bmatrix} 8 & 1 & 3 & 6 \\ 0 & 3 & 2 & 2 \\ -8 & -1 & -3 & 4 \end{bmatrix}$

Solution. If the given matrix be $A = \begin{bmatrix} 8 & 1 & 3 & 6 \\ 0 & 3 & 2 & 2 \\ -8 & -1 & -3 & 4 \end{bmatrix}$ then operating $C_1 \to \frac{1}{8} C_1$, we get

$$A \sim \begin{bmatrix} 1 & 1 & 3 & 6 \\ 0 & 3 & 2 & 2 \\ -1 & -1 & -3 & 4 \end{bmatrix}$$

Again operating $R_3 \to R_3 + R_1$, we get

$$\sim \begin{bmatrix} 1 & 1 & 3 & 6 \\ 0 & 3 & 2 & 2 \\ 0 & 0 & 0 & 10 \end{bmatrix}$$

Next operating $C_2 \to C_2 - C_1$, $C_3 \to C_3 - 3C_1$, $C_4 \to C_4 - 6C_1$, we get

$$\sim \begin{bmatrix} 1 & 0 & 0 & 0 \\ 0 & 3 & 2 & 2 \\ 0 & 0 & 0 & 10 \end{bmatrix}$$

Now, operating $C_2 \to \frac{1}{3} C_2$, we get

$$\sim \begin{bmatrix} 1 & 0 & 0 & 0 \\ 0 & 1 & 2 & 2 \\ 0 & 0 & 0 & 10 \end{bmatrix}$$

Again operating $C_3 \to C_3 - 2C_2$ and $C_4 \to C_4 - 2C_2$, we get

$$\sim \begin{bmatrix} 1 & 0 & 0 & 0 \\ 0 & 1 & 0 & 0 \\ 0 & 0 & 0 & 10 \end{bmatrix}$$

Operating $\frac{1}{10} C_4$, we get

$$\sim \begin{bmatrix} 1 & 0 & 0 & 0 \\ 0 & 1 & 0 & 0 \\ 0 & 0 & 0 & 1 \end{bmatrix}$$

Again operating C_{34}, we get

$$\sim \begin{bmatrix} 1 & 0 & 0 & 0 \\ 0 & 1 & 0 & 0 \\ 0 & 0 & 1 & 0 \end{bmatrix}$$

Hence rank of $A = 3$

which is the required result.

EXERCISE 1.2

1. Find non-singular matrices P and Q such that PAQ is in the normal form for the matrices.

(i) $A = \begin{bmatrix} 1 & 1 & 2 \\ 1 & 2 & 3 \\ 0 & -1 & -1 \end{bmatrix}$ (ii) $A = \begin{bmatrix} 1 & 2 & 3 \\ 3 & 2 & 1 \\ 1 & 3 & 2 \\ 2 & 1 & 3 \end{bmatrix}$ (iii) $A = \begin{bmatrix} 1 & -1 & -1 \\ 1 & 1 & 1 \\ 3 & 1 & 1 \end{bmatrix}$ (iv) $A = \begin{bmatrix} 1 & 2 & 3 & -2 \\ 2 & -2 & 1 & 3 \\ 3 & 0 & 4 & 1 \end{bmatrix}$

Ans. (i) $P = \begin{bmatrix} 1 & 0 & 0 \\ -1 & 1 & 0 \\ -1 & 1 & 1 \end{bmatrix}$ and $Q = \begin{bmatrix} 1 & -1 & -1 \\ 0 & 1 & -1 \\ 0 & 0 & 1 \end{bmatrix}$

(ii) $P = \begin{bmatrix} 1 & 0 & 0 & 0 \\ 3/4 & -1/4 & 0 & 0 \\ 7/12 & -1/12 & -4/12 & 0 \\ -3/2 & -1/2 & 1 & 1 \end{bmatrix}$ and $Q = \begin{bmatrix} 1 & -2 & -1 \\ 0 & 1 & -2 \\ 0 & 0 & 1 \end{bmatrix}$

(iii) $P = \begin{bmatrix} 1 & 0 & 0 \\ -1/2 & 1/2 & 0 \\ 1/2 & 1 & -1/2 \end{bmatrix}$ and $Q = \begin{bmatrix} 1 & 1 & -1 \\ 0 & 1 & 0 \\ 0 & 0 & 1 \end{bmatrix}$

(iv) $P = \begin{bmatrix} 1 & 0 & 0 \\ -2 & 1 & 0 \\ -1 & -1 & 1 \end{bmatrix}$ and $Q = \begin{bmatrix} 1 & 1/3 & -4/3 & -1/3 \\ 0 & -1/6 & -5/6 & 7/6 \\ 0 & 0 & 1 & 0 \\ 0 & 0 & 0 & 1 \end{bmatrix}$

2. Use Gauss-Jordan method to find the inverse of the following matrices.

(i) $A = \begin{bmatrix} 2 & 1 & -1 \\ 0 & 2 & 1 \\ 5 & 2 & -3 \end{bmatrix}$ (ii) $A = \begin{bmatrix} 1 & 1 & 3 \\ 1 & 3 & -3 \\ -2 & -4 & -4 \end{bmatrix}$ (iii) $A = \begin{bmatrix} 2 & 0 & -1 \\ 5 & 1 & 0 \\ 0 & 1 & 3 \end{bmatrix}$ (iv) $A = \begin{bmatrix} 8 & 4 & 3 \\ 2 & 1 & 1 \\ 1 & 2 & 1 \end{bmatrix}$

Ans. (i) $\begin{bmatrix} 8 & -1 & -3 \\ -5 & 1 & 2 \\ 10 & -1 & -4 \end{bmatrix}$ (ii) $\begin{bmatrix} 3 & 1 & 3/2 \\ -5/4 & -1/4 & -3/4 \\ -1/4 & -1/4 & -1/4 \end{bmatrix}$ (iii) $\begin{bmatrix} 3 & -1 & 1 \\ -15 & 6 & -5 \\ 5 & -2 & 2 \end{bmatrix}$ (iv) $\frac{1}{21}\begin{bmatrix} 1 & 10 & -7 \\ 1 & -11 & 14 \\ -3 & 12 & 0 \end{bmatrix}$

3. Determine the rank of the matrix $A = \begin{bmatrix} 6 & 1 & 3 & 8 \\ 4 & 2 & 6 & -1 \\ 10 & 3 & 9 & 7 \\ 16 & 4 & 12 & 15 \end{bmatrix}$ by reducing it to normal form.

Ans. $\rho(A) = 2$

4. By reducing the following matrices in normal form, find their ranks.

(i) $A = \begin{bmatrix} 1 & 2 & 0 & -1 \\ 3 & 4 & 1 & 2 \\ -2 & 3 & 2 & 5 \end{bmatrix}$ (ii) $A = \begin{bmatrix} 0 & 2 & 3 & 4 \\ 2 & 3 & 5 & 4 \\ 4 & 8 & 13 & 12 \end{bmatrix}$

Ans.

(i) $\begin{bmatrix} I_3 & 0 \\ 0 & 0 \end{bmatrix}$ and rank = 3. (ii) $\begin{bmatrix} I_2 & 0 \\ 0 & 0 \end{bmatrix}$, $\rho(A) = 2$

5. Find the ranks of the following matrices

(i) $A = \begin{bmatrix} 2 & -4 & 6 \\ -1 & 2 & -3 \\ 3 & -6 & 9 \end{bmatrix}$ (ii) $A = \begin{bmatrix} 3 & -1 & 2 \\ -6 & 2 & 4 \\ -3 & 1 & 2 \end{bmatrix}$ (iii) $A = \begin{bmatrix} 1 & 2 & 3 \\ 2 & 4 & 7 \\ 3 & 6 & 10 \end{bmatrix}$ (iv) $A = \begin{bmatrix} 5 & 6 & 7 & 8 \\ 6 & 7 & 8 & 9 \\ 11 & 12 & 13 & 14 \\ 16 & 17 & 18 & 19 \end{bmatrix}$

(v) $A = \begin{bmatrix} 2 & 4 & 3 & -2 \\ -3 & -2 & -1 & 4 \\ 6 & -1 & 7 & 2 \end{bmatrix}$ (vi) $A = \begin{bmatrix} 1 & 2 & 3 & 4 \\ 2 & 3 & 4 & 5 \\ 3 & 4 & 5 & 6 \\ 4 & 5 & 6 & 7 \end{bmatrix}$ (vii) $A = \begin{bmatrix} 0 & 1 & -3 & -1 \\ 1 & 0 & 1 & 1 \\ 3 & 1 & 0 & 2 \\ 1 & 1 & -2 & 0 \end{bmatrix}$ (viii) $A = \begin{bmatrix} 1 & 3 & 4 & 3 \\ 3 & 9 & 12 & 3 \\ 1 & 3 & 4 & 1 \end{bmatrix}$

Ans.

(i) $\rho(A) = 1$ (ii) $\rho(A) = 2$ (iii) $\rho(A) = 2$ (iv) $\rho(A) = 2$

(v) $\rho(A) = 3$ (vi) $\rho(A) = 2$ (vii) $\rho(A) = 2$ (viii) $\rho(A) = 2$

1.12 SOLUTION OF LINEAR SYSTEM OF EQUATIONS

Types of Linear Equations

I. ***Consistent system of linear equation***

A system of equations is said to be consistent if they have one or more solutions. For example

$$x + 2y = 5$$
$$3x - y = 1 \text{ has unique solution } x = 1, y = 2$$

but
$$x + 2y = 5$$
and
$$2x + 4y = 10$$

does not have a unique solution but infinite number of solution i.e., for $x = 0$, $y = \frac{5}{2}$; for $x = 1$, $y = 2$; for $x = -1$, $y = 3$ and so on.

II. ***Inconsistent system of linear equations***

If a system of equations has no solution, then it is known as inconsistent system of linear equations.

1.13 CONSISTENCY OF A LINEAR SYSTEM OF EQUATIONS

If
$$a_{11}x_1 + a_{12}x_2 + \dots + a_{1n}x_n = b_1$$
$$a_{21}x_1 + a_{22}x_2 + \dots + a_{2n}x_n = b_2 \qquad \dots(1)$$

....

....

$$a_{m1}x_1 + a_{m2}x_2 + ... + a_{mn}x_n = b_n$$

be given system of m linear equations then (1) may be written as

$$\begin{bmatrix} a_{11} & a_{12} & \dots & a_{1n} \\ a_{21} & a_{22} & \dots & a_{2n} \\ \dots & \dots & \dots & \dots \\ \dots & \dots & \dots & \dots \\ a_{m1} & a_{m2} & \dots & a_{mn} \end{bmatrix} \begin{bmatrix} x_1 \\ x_2 \\ \dots \\ \dots \\ x_n \end{bmatrix} = \begin{bmatrix} b_1 \\ b_2 \\ \dots \\ \dots \\ b_m \end{bmatrix}$$

$$\Rightarrow \qquad AX = B \text{ and } C = [A : B] = \begin{bmatrix} a_{11} & a_{12} & \dots & a_{1n} : b_1 \\ a_{21} & a_{22} & \dots & a_{2n} : b_2 \\ \dots & \dots & \dots & \dots \\ a_{m1} & a_{m2} & \dots & a_{mn} : b_m \end{bmatrix}$$

then $[A : B]$ or C is called augmented matrix. Sometimes we also write $A : B$ for $[A : B]$

Consistent Equations

The equations are consistent if rank of $[A]$ = rank of $[A : B]$ and

(*i*) have unique solutions when $\rho(A) = \rho(A : B) = r = n$ $\{\rho = \text{rank}\}$

(*ii*) have infinite number of solutions, when $\rho(A) = \rho(A : B) = r < n$

Inconsistent Equations

The equation are inconsistent if $\rho(A) \neq \rho(A : B)$ and have no solution.

***Example* 1.14.** Discuss the consistency of the following system of equation $2x + 3y + 4z = 11$, $x + 5y + 7z = 15$, $3x + 11y + 13z = 25$, if consistent, obtain the solution.

Solution. The augmented matrix $[A : B] = \begin{bmatrix} 2 & 3 & 4:11 \\ 1 & 5 & 7:15 \\ 3 & 11 & 13:25 \end{bmatrix}$

R_{12} operation is done so $\sim \begin{bmatrix} 1 & 5 & 7:15 \\ 2 & 3 & 4:11 \\ 3 & 11 & 13:25 \end{bmatrix}$

Next operating $R_2 - R_2 - 2R_1$ and $R_3 \to R_3 - 3R_1$, we get

$$\sim \begin{bmatrix} 1 & 5 & 7:15 \\ 0 & -7 & 10:-19 \\ 0 & -4 & -8:-20 \end{bmatrix}$$

Again, operating $R_2 \to -\frac{1}{7} R_2$ and $\to -\frac{1}{4} R_3$, we get

$$\sim \begin{bmatrix} 1 & 5 & 7:15 \\ 0 & 1 & \frac{10}{7}:\frac{19}{7} \\ 0 & 1 & 2:5 \end{bmatrix}$$

Next, operating $R_3 \to R_3 - R_2$, we get

$$\begin{bmatrix} 1 & 5 & 7:15 \\ 0 & 1 & \frac{10}{7}:\frac{19}{7} \\ 0 & 0 & \frac{4}{7}:\frac{16}{7} \end{bmatrix}$$

Operating $R_3 \to \frac{7}{4} R_3$, we get

$$\begin{bmatrix} 1 & 5 & 7:15 \\ 0 & 1 & \frac{10}{7}:\frac{19}{7} \\ 0 & 0 & 1:4 \end{bmatrix}$$

$$\therefore \qquad \rho(A) = 3 \text{ and } \rho(A:B) = 3$$

Hence the equation are consistent, we get $z = 4 \quad y + \frac{10}{7}z = \frac{19}{7}$

$$y = -3$$

$$x + 5y + 7z = 15$$

$$\left.\begin{aligned} x &= 2 \\ y &= -3 \\ z &= 4 \end{aligned}\right]$$

***Example* 1.15.** Show that if $\lambda \neq -5$, the system of equation $3x - y + 4z = 3$, $x + 2y - 3z = -2$ and $6x + 5y + \lambda z = -3$ have a unique solution. If $\lambda = -5$, show that the equations are consistent. Determine the solution, in each case.

Solution. The given equations are

$$3x - y + 4z = 3$$

$$x + 2y - 3z = -2 \qquad \text{...(1)}$$

and $$6x + 5y + \lambda z = 3$$

If $$A = \begin{bmatrix} 3 & -1 & 4 \\ 1 & 2 & -3 \\ 6 & 5 & \lambda \end{bmatrix}, X = \begin{bmatrix} x \\ y \\ z \end{bmatrix} \text{ and } B = \begin{bmatrix} 3 \\ -2 \\ -3 \end{bmatrix} \text{ such that } AX = B \text{ from (1)}$$

Then augmented matrix $[A:B] = \begin{bmatrix} -3 & -1 & 4:3 \\ 1 & 2 & -3:-2 \\ 6 & 5 & \lambda:-3 \end{bmatrix}$ ...(2)

Operating R_{12} (i.e., interchanging R_1 and R_2)

$$[A:B] = \begin{bmatrix} 1 & 2 & -3:-2 \\ 3 & -1 & 4:3 \\ 6 & 5 & \lambda:-3 \end{bmatrix}$$

Now operating $R_2 - 3R_1$ and *i.e.*, $R_3 - 6R_1$, we get

$$[A:B] \sim \begin{bmatrix} 1 & 2 & 3:-2 \\ 0 & -7 & 13:9 \\ 0 & -7 & \lambda+18:9 \end{bmatrix}$$

Next operating $R_3 - R_2$, we get

$$\sim \begin{bmatrix} 1 & 2 & -3:-2 \\ 0 & -9 & 13:9 \\ 0 & 0 & \lambda+5:0 \end{bmatrix}$$

(*i*) If $\lambda = -5$ then $\rho(A) = \rho(A:B) = 2$ which is less than 3, (the number of unknowns) and hence the equations will be inconsistent and will have infinite number of solutions.

Next, operating, $R_1 + \frac{2}{7} R_2$, we get

$$\sim \begin{bmatrix} 1 & 0 & \frac{5}{7}:\frac{4}{7} \\ 0 & -7 & 13:9 \\ 0 & 0 & \lambda+5:0 \end{bmatrix} \text{ from this matrix, if } \lambda \neq -5$$

(*ii*) If $\lambda \neq -5$; then $\rho(A) = \rho(A : B) = 3$

therefore, rank is 3 and the equation will be consistent and we get

$$x + \frac{5}{7} z = \frac{4}{7}; -7y + 13z = 9 \text{ and } (\lambda + 5)z = 0 \text{ i.e., } z = 0$$

$$\Rightarrow \quad -7y = 9 \Rightarrow y = -\frac{9}{7} \text{ and } x + 0 = \frac{4}{7} \text{ i.e., } x = \frac{4}{7}$$

i.e., unique solution is $x = \frac{4}{7}, y = -\frac{9}{7}, z = 0$, which is required result.

If $\lambda = -5$, then from (2), we have $x + 2y - 3z = -2, -7y + 13z = 9$ ...(3)

If we take $z = k$ then from (3).

$$y = \frac{13k-9}{7} \text{ and } x = 3k - 2\left(\frac{13k-9}{7}\right) - 2 = \frac{4-5k}{7}$$

***Example* 1.16.** Examine whether the following equations are consistent and solve them if they are consistent $2x + 6y + 11 = 0$, $6x + 20y - 6z + 3 = 0$ and $6y - 18z + 1 = 0$

Solution. The above equations may be written in the form

$$AX = B \text{ which is } \begin{bmatrix} 2 & 6 & 0 \\ 6 & 20 & -6 \\ 0 & 6 & -18 \end{bmatrix} \begin{bmatrix} x \\ y \\ z \end{bmatrix} = \begin{bmatrix} -11 \\ -3 \\ -1 \end{bmatrix} \quad ...(1)$$

Now the augmented matrix may be written as

$$[A : B] = \begin{bmatrix} 2 & 6 & 0 & : & -11 \\ 6 & 20 & -6 & : & -3 \\ 0 & 6 & -18 & : & -1 \end{bmatrix} \quad ...(2)$$

Operating $R_2 \to R_2 - 3R_1$, we get

$$[A : B] \sim \begin{bmatrix} 2 & 6 & 0 & : & -11 \\ 0 & 2 & -6 & : & 30 \\ 0 & 6 & -18 & : & -1 \end{bmatrix}$$

Now, operating $R_3 - R_3 - 3R_2$, we get

$$\Rightarrow \quad \sim \begin{bmatrix} 2 & 6 & 0 & : & -11 \\ 0 & 2 & -6 & : & 30 \\ 0 & 0 & 0 & : & -91 \end{bmatrix}$$

Hence rank of $A = \rho(A) = 2$ and $\rho(A : B) = 3$. So, $\rho(A) = 2 < 3$ (number of variables). This indicated that given equations are inconsistent and so it has no solution.

***Example* 1.17.** Solve the following system of equations by matrix method $x + y + z = 8$; $x - y + 2z = 6$ and $3x + 5y - 7z = 14$.

Solution. The above equation written in the form $AX = B$.

$$\text{where} \quad A = \begin{bmatrix} 1 & 1 & 1 \\ 1 & -1 & 2 \\ 3 & 5 & -7 \end{bmatrix}, X = \begin{bmatrix} x \\ y \\ z \end{bmatrix} \text{ and } B = \begin{bmatrix} 8 \\ 6 \\ 14 \end{bmatrix}$$

So, we may write augmented matrix as

$$[A : B] = \begin{bmatrix} 1 & 1 & 1 & : & 8 \\ 1 & -1 & 2 & : & 6 \\ 3 & 5 & -7 & : & 14 \end{bmatrix} \qquad \text{...(1)}$$

Operating $R_2 \to R_2 - R_1$ and $R_3 \to R_3 - 3R_1$, we have

$$[A : B] \sim \begin{bmatrix} 1 & 1 & 1 & : & 8 \\ 1 & -2 & 1 & : & -2 \\ 0 & 2 & -10 & : & -10 \end{bmatrix} \qquad \text{...(2)}$$

Again $R_3 \to R_3 + R_2$, we have

$$\sim \begin{bmatrix} 1 & 1 & 1 & : & 8 \\ 0 & -2 & 1 & : & -2 \\ 0 & 0 & -9 & : & -12 \end{bmatrix} \qquad \text{...(3)}$$

This implies that

$$\begin{aligned} x + y + z &= 8 \\ -2y + z &= -2 \\ \text{and} \quad -9z &= -12 \end{aligned} \qquad \text{...(4)}$$

$$\Rightarrow \quad z = \frac{4}{3} \text{ and } 2y = z + 2 = \frac{4}{3} + 2 = \frac{10}{3}$$

$$\therefore \quad y = \frac{5}{3}$$

Using 1st equation of (3), we get $x + y + z = 8$

$\Rightarrow x + \frac{5}{3} + \frac{4}{3} = 8 \Rightarrow x = 8 - 3 = 5$

From (2) we see that $\rho(A) = 3$ = number of variables so the system of equations are consistent and solutions are $x = 5$, $y = \frac{5}{3}$, $z = \frac{4}{3}$.

***Example* 1.18.** Determine for what values of λ and μ the following equations have (*i*) No solution (*ii*) a unique solution (*iii*) infinite number of solution: $x + y + z = 6$, $x + 2y + 3z = 10$ and $x + 2y + \lambda z = \mu$.

Solution. The above equations may be written in the form $AX = B$.

i.e.,

$$\begin{bmatrix} 1 & 1 & 1 \\ 1 & 2 & 3 \\ 1 & 2 & \lambda \end{bmatrix} \begin{bmatrix} x \\ y \\ z \end{bmatrix} = \begin{bmatrix} 6 \\ 10 \\ \mu \end{bmatrix}$$

The augmented matrix $[A : B] = \begin{bmatrix} 1 & 2 & 1 & : & 6 \\ 1 & 2 & 3 & : & 10 \\ 1 & 2 & \lambda & : & \mu \end{bmatrix}$

Operating $R_2 \to R_2 - R_1$ and $R_3 \to R_3 - R_1$, we get

$$\sim \begin{bmatrix} 1 & 1 & 1 & : & 6 \\ 0 & 1 & 2 & : & 14 \\ 0 & 1 & \lambda - 1 & : & \mu - 6 \end{bmatrix}$$

Again operating $R_3 - R_2$, we get

$$\sim \begin{bmatrix} 1 & 1 & 1 & : & 6 \\ 0 & 1 & 2 & : & 4 \\ 0 & 0 & \lambda-3 & : & \mu-10 \end{bmatrix}$$

$\Rightarrow$ We get $x + y + z = 6$, $y + 2z = 4$ and $(\lambda - 3)z = \mu - 10$.

(*i*) If $\lambda = 3$, $\mu \neq 10$ then $\rho(A) = 2$ $\rho(A : B) = 3$ $\rho(A) \neq \rho(A : B)$, the equations have no solution.

(*ii*) If $\lambda \neq 3$, $\mu - 10$ may have any value, then $\rho(A) = 3$ $\rho(A : B) = 3$ $\rho(A) = \rho(A : B) = 3$, the equations have a unique solution.

(*iii*) If $\lambda = 3$ and $\mu = 10$ then $\rho(A) = \rho(A : B) = 2 < 3$ the number of unknown, the equations have an infinite number of solutions.

EXERCISE 1.3

Test for consistency of the following equations and if consistent solve them :

1. $4x - 5y + z = 2$; $3x + y - 2z = 9$; $x + 4y + z = 5$.

Ans. Consistent and $x = 2$, $y = 1$, $z = -1$

2. $3x + 3y + 2z = 1$; $x + 2y = 4$; $19y + 3z = -2$; $2x - 3y - z = 5$.

Ans. Consistent and $x = 2$, $y = 1$, $z = -4$

3. $x_1 - x_2 + x_3 - x_4 + x_5 = 1$; $2x_1 - x_2 + 3x_3 + 4x_5 = 2$; $3x_1 - 2x_2 + 2x_3 + x_4 + x_5 = 1$; $x_1 + x_3 + 2x_4 + x_5 = 0$.

Ans. Consistent and $x_1 = -3k_1 + k_2 - 1$, $x_2 = -3k_1 - 1$, $x_3 = k_1 - 2k_2 + 1$, $x_4 = k_1$, $x_5 = k_2$.

4. $3w - 6x - y - z = 0$; $w - 2x + 5y - 3z = 0$; $2w - 4x + 3y - z = 3$.

Ans. Consistent and $x = k$; $y = 1$, $z = 2$, $w = 2k + 1$.

5. Find the value of λ for which the system of equations $x + y + 4z = 1$; $x + 2y - 2z = 1$ and $\lambda x + y + z = 1$ will have unique solution.

Ans. $\lambda \neq \frac{7}{10}$

6. For what values λ and μ, the following system of equations $2x + 3y + 5z = 9$; $7x + 3y - 2z = 8$; $2x + 3y + lz = m$ have (*i*) unique solution (*ii*) no solution.

Ans. (*i*) $\lambda \neq 5$; (*ii*) $\lambda \neq 5$, $\mu \neq 9$

7. Show that the equation $3x + 4y + 5z + a$; $4x + 5y + 6z = b$ and $5x + 6y + 7z = c$ don't have a solution unless $a + c = 2b$. Solve the equation when $a = b = c = -1$.

Ans. $x = k + 1$, $y = -(2k + 1)$, $z = k$

8. Find the value of k such that the system of equations $2x + 3y - 2z = 0$; $3x - y + 3z = 0$; $7x + ky - z = 0$ have non-trivial solution. Find the solution.

Ans. $k = 5$, $x = -\frac{7a}{11}$, $y = \frac{12a}{11}$, $z = a$.

9. Find the values of k, such that the system of equations $4x + 9y + z = 0$, $kx + 3y + kz = 0$ and $x + 4y + 2z + 0$ has non-trivial solution. Hence find the solution of the equation.

Ans. $k = 1$, $x = 2\lambda$, $y = -\lambda$, $z = \lambda$.

10. Find the values of k for which the following system of equations $4x + 2y - 5z = 0$; $x + ky + 2z = 0$; $2x + y - z = 0$ has a non-zero solution.

Ans. $k = \frac{1}{2}$

11. Determine the values of a and b for which the system

$$\begin{bmatrix} 3 & -2 & 1 \\ 5 & -8 & 9 \\ 2 & 1 & a \end{bmatrix}\begin{bmatrix} x \\ y \\ z \end{bmatrix} = \begin{bmatrix} b \\ 3 \\ -1 \end{bmatrix}$$ has (*i*) a unique solution (*ii*) has no solution (*iii*) has infinite solutions.

Ans. Students solve by themselves.

12. Find the values of λ for which the equation $(2-\lambda)x + 2y + 3 = 0$, $2x + (4-\lambda)y + 7 = 0$ and $2x + 5y + 6 - 1 = 0$ are consistent and find the values of x and y corresponding to each value of λ.

Ans. $\lambda = -1, 1, 12; x = -\frac{1}{11}, y = -\frac{15}{11}; x = -5\ y = 1; x = \frac{1}{2}, y = 1$

1.14 CHARACTERISTIC EQUATION

If A be a square matrix of order n, then we can form the matrix $A - \lambda I$, where I is the unit matrix of order n and λ is scalar. The determinant corresponding to this matrix equated to zero is called the characteristic equation, i.e., if $A - \lambda I$ be the matrix, then

$$|A - \lambda I| = \begin{vmatrix} a_{11}-\lambda & a_{12} & a_{13} & \dots & a_{1n} \\ a_{21} & a_{22}-\lambda & a_{23} & \dots & a_{2n} \\ a_{31} & a_{32} & a_{33}-\lambda & \dots & a_{3n} \\ \dots & \dots & \dots & \dots & \dots \\ \dots & \dots & \dots & \dots & \dots \\ a_{n1} & a_{n2} & a_{n3} & \dots & a_{nn} \end{vmatrix} = 0 \qquad \dots(1)$$

is the characteristic equation of A.

On expanding the determinant (1), the characteristic equation may be written as

$(-1)^n \lambda^n + a_1\lambda^{n-1} + a_2\lambda^{n-2} + \dots + a_{n-1}\lambda + a_n = 0$

which is nth degree equation in λ.

The roots of (1) are called eigen values or characteristics roots or latent roots of the matrix A.

1.15 EIGEN VECTORS

We take the matrix

$$A = \begin{bmatrix} a_{11} & a_{12} & a_{13} & \dots & a_{1n} \\ a_{21} & a_{22} & a_{23} & \dots & a_{2n} \\ a_{31} & a_{32} & a_{33} & \dots & a_{3n} \\ \dots & \dots & \dots & \dots & \dots \\ \dots & \dots & \dots & \dots & \dots \\ a_{n1} & a_{n2} & a_{n3} & \dots & a_{nn} \end{bmatrix} \qquad \dots(1)$$

and if $X = \begin{bmatrix} x_1 \\ x_2 \\ x_3 \\ \dots \\ x_n \end{bmatrix}$ where $x_1, x_2, \dots, x_n$ are vectors.

then the linear transformation $Y = AX$ ... (2), transforms the column vector X into the column vector Y. Generally, it is required to find such vectors which either transform into themselves or to a scalar multiple of themselves. If X be such a vector which is transformed into λX using the transformation (2) then $\lambda X = AX \Rightarrow AX - \lambda X = 0$

i.e., $[A - \lambda I]X = 0$ ...(3)

The matrix equation (3) represents n homogeneous linear equations.

$(a_{11} - \lambda)x_1 + a_{12}x_2 + a_{13}x_3 + \dots + a_{1n}x_n = 0$

$$a_{21}x_1 + (a_{22} - \lambda)x_2 + a_{23} + x_3 + \ldots + a_{2n}x_n = 0$$

$$a_{31}x_1 + a_{32}x_2 + (a_{33} - \lambda)x_3 + \ldots + a_{3n}x_n = 0 \qquad \ldots(4)$$

....

$$a_{n1}x_1 + a_{n2}x_2 + a_{n3}x_3 + \ldots + (a_{nn} - \lambda)x_n = 0$$

This equation (4) will have a non-trivial solution only if the co-efficient matrix is singular, *i.e.*, if the determinant $|A - \lambda I| = 0$.

This equation is also called characteristic equation of the transformation and is also the same as the characteristic equation (1) of matrix *A*. This characteristic equation has *n* roots which are eigen values of *A* corresponding to each root of (4), also the equations (3) and (4) have non-zero solution.

$$X = \begin{bmatrix} x1 \\ x2 \\ x2 \\ \ldots \\ xn \end{bmatrix}$$

which is known as an eigen vector. So, if *X* is a solution of (3) then *KX* is also a solution, where *K* is an arbitrary constant. So, we see that the eigen vector corresponding to an eigen value is not unique.

Example **1.19.** Find the eigen values and eigen vectors of the matrices $A = \begin{bmatrix} 1 & 2 \\ 2 & 4 \end{bmatrix}$

Solution. The characteristic equation of the given matrix is $|A - \lambda I| = 0$

$$\Rightarrow \qquad \begin{vmatrix} 1-\lambda & 2 \\ 2 & 4-\lambda \end{vmatrix} = 0$$

i.e., $(1 - \lambda)(4 - \lambda) - 4 = 0 \Rightarrow \lambda^2 - 5\lambda = 0 \Rightarrow \lambda(\lambda - 5) = 0$

i.e., $\lambda = 0, 5$ $\therefore$ eigen values of *A* are 0 and 5.

So, corresponding to $\lambda = 0$ eigen vectors are given by $\begin{vmatrix} 1-0 & 2 \\ 2 & 4-0 \end{vmatrix} \begin{bmatrix} x1 \\ x2 \end{bmatrix} = 0$

i.e., $x_1 + 2x_2 = 0$ and $2x_1 + 4x_2 = 0$

i.e., single equation $x_1 + 2x_2 = 0 \Rightarrow \frac{x1}{2} = \frac{x2}{-1}$ so for $\lambda = 0$ eigen vectors are (2, – 1) and for

$$\lambda = 5, \text{ we have } \begin{vmatrix} 1-5 & 2 \\ 2 & 4-5 \end{vmatrix} \begin{bmatrix} x1 \\ x2 \end{bmatrix} = 0$$

$$\Rightarrow \qquad -4x_1 + 2x_2 = 0 \text{ and } 2x_1 - x_2 = 0$$

i.e., eigen vectors are $\frac{x1}{1} = \frac{x2}{2}$, *i.e.*, (1, 2) are eigen vectors corresponding to $\lambda = 5$

1.16 PROPERTIES OF EIGEN VALUES

(i) The sum of the eigen values of a matrix is the sum of the elements of the principal diagonal. We will prove this property for a matrix of order 3 and the method can be extended for the matrices of any finite order.

Let
$$A = \begin{bmatrix} a11 & a12 & a13 \\ a21 & a22 & a23 \\ a31 & a32 & a33 \end{bmatrix} \qquad \ldots(1)$$

Then characteristic matrix $|A - \lambda I| = 0$

$$\Rightarrow \begin{vmatrix} a_{11}-1 & a_{12} & a_{13} \\ a_{21} & a_{22}-1 & a_{23} \\ a_{31} & a_{32} & a_{33}-1 \end{vmatrix} = 0$$

$$\Rightarrow -\lambda^3 + \lambda^2(a_{11} + a_{22} + a_{33}) - \lambda(\,) + \ldots = 0 \qquad \ldots(2)$$

If λ_2, λ_2 and λ_3 be eigen values of A then

$$|A - \lambda I| = -\lambda_3^3 + \lambda_3^2(\lambda_1 + \lambda_2 + \lambda_3) - \ldots + (-1)^3 \lambda_1 \lambda_2 \lambda_3 \qquad \ldots(3)$$

Equating the co-efficients of λ^2 from (2) and (3), we get

$\lambda_1 + \lambda_2 + \lambda_3 = a_{11} + a_{22} + a_{33}$ which is the required result.

(ii) The product of the eigen values of a matrix A is equal to its determinant. If we put $\lambda = 0$ in (3) then, we get $|A - 0| = -\lambda_1 \lambda_2 \lambda_3$ *i.e.*, $|A| = -\lambda_1 \lambda_2 \lambda_3$ which is the required result.

(iii) If λ is an eigen values of a matrix A, then $\frac{1}{\lambda}$ is the eigen value of inverse matrix A^{-1}. If X be the eigen vector corresponding to the eigen value λ then $AX = \lambda X$. ...(4)

Pre-multiplying (4) by A^{-1}, we get $A^{-1} AX = A^{-1} \lambda X$

i.e., $IX = \lambda A^{-1} X \Rightarrow X = \lambda(A^{-1} X) \Rightarrow A^{-1} X = \frac{1}{\lambda} X$

This is of the same form as that in (1) from which we get that $\frac{1}{\lambda}$ is an eigen value of the inverse matrix A^{-1}.

(iv) If λ is an eigen value of a matrix A, then $\frac{1}{\lambda}$ is an eigen value of A^{-1}. As A is an orthogonal matrix so A^{-1} will be same as the transpose of matrix A. i.e., $A' = A^{-1}$. So, $\frac{1}{\lambda}$ is an eigen value of A'. But the matrix A and A' have the same eigen values. [since we know that $|A - \lambda I| = |A' - \lambda I|$]. Hence $\frac{1}{\lambda}$ is also an eigen value of A.

(v) If $\lambda_1, \lambda_2, \ldots, \lambda_n$ are eigen values of a matrix A then A^m has the eigen values $\lambda_1^m, \lambda_2^m, \ldots, \lambda_n^m$ where m is a positive integer. If λ_i be the eigen value of A and X_i be the corresponding eigen vector, then

$$AX_i = \lambda_i X_i \qquad \ldots(1)$$

Consider $A^2 X_i = A(AX_i) = A(\lambda_i X_i) = \lambda_i(AX_i) = \lambda_i(\lambda_i X_i) = \lambda_i^2 X_i$ similarly, we proceed and find $A^3 X_i = \lambda_i^3 X_i$ and so on such that in general we get

$$A^m X_i = \lambda_i^m X_i \qquad \ldots(2)$$

which has the same form as (1). Hence λ_i^m is an eigen-value of A^m and the corresponding eigen vector is the same as that of X_i.

Example **1.20.** Find the characteristic roots and characteristic vectors of the matrix $A = \begin{vmatrix} 8 & -6 & 2 \\ -6 & 7 & -4 \\ 2 & -4 & 3 \end{vmatrix}$

Solution. The characteristic equation of matrix A is $|A - \lambda I| = 0$ i.e.,

i.e., $(8 - \lambda)[(7 - \lambda)(3 - \lambda) - 16] + 6[(-6)(3 - \lambda) + 8] + 2[24 - 2(7 - \lambda)] = 0$

i.e., $(8 - \lambda)[(2t + \lambda^2 - 10\lambda - 16] + 6[-10 + 6\lambda] + 2[24 - 14 + 2\lambda] = 0$

i.e., $\lambda^3 - 18\lambda^2 + 45\lambda = 0$ i.e., $\lambda = 0, 3, 15$

$\therefore$ Corresponding to $\lambda = 0$, eigen vectors are given by

$$\begin{bmatrix} 8 & -6 & 2 \\ -6 & 7 & -4 \\ 2 & -4 & 3 \end{bmatrix} \begin{bmatrix} x_1 \\ x_2 \\ x_3 \end{bmatrix} = 0$$

i.e., equations are

$$8x_1 - 6x_2 + 2x_3 = 0 \quad ...(1)$$
$$-6x_1 + 7x_2 - 4x_3 = 0 \quad ...(2)$$
$$2x_1 - 4x_2 + 3x_3 = 0 \quad ...(3)$$

From (2) and (3) we get

$$\frac{x_1}{21-16} = \frac{x_2}{-8+18} = \frac{x_3}{24-14} \text{ i.e., } \frac{x_1}{1} = \frac{x_2}{2} = \frac{x_3}{3}$$

i.e., eigen vector is (1, 2, 2)

Similarly from (1) and (2) we get the same vectors.

Now for $\lambda = 3$, eigen vectors are obtained from $\begin{bmatrix} 8-3 & -6 & 2 \\ -6 & 7-3 & -4 \\ 2 & -4 & 3-3 \end{bmatrix}\begin{bmatrix} x_1 \\ x_2 \\ x_3 \end{bmatrix} = 0$

i.e. $\begin{bmatrix} 5 & -6 & 2 \\ -6 & 4 & -4 \\ 2 & -4 & 0 \end{bmatrix}\begin{bmatrix} x_1 \\ x_2 \\ x_3 \end{bmatrix} = 0$

i.e.
$$5x_1 - 6x_2 + 2x_3 = 0 \quad ...(4)$$
$$-6x_1 + 4x_2 - 4x_3 = 0 \quad ...(5)$$
and
$$2x_1 - 4x_2 = 0 \quad ...(6)$$

From (4) and (5), we get

$$\frac{x_1}{24-8} = \frac{x_2}{-12+20} = \frac{x_3}{20-36}$$

i.e.,
$$\frac{x_1}{16} = \frac{x_2}{8} = \frac{x_3}{-16} \Rightarrow \frac{x_1}{2} = \frac{x_2}{1} = \frac{x_3}{-2}$$

i.e., eigen vectors are (2, 1, – 2) and for $\lambda = 15$, eigen vectors are given by

$$\begin{bmatrix} 8-15 & -6 & 2 \\ -6 & 7-15 & -4 \\ 2 & 4 & 3-15 \end{bmatrix}\begin{bmatrix} x_1 \\ x_2 \\ x_3 \end{bmatrix} = 0 \Rightarrow \begin{bmatrix} -7 & -6 & 2 \\ -6 & -8 & -4 \\ 2 & 4 & 12 \end{bmatrix}\begin{bmatrix} x_1 \\ x_2 \\ x_3 \end{bmatrix} = 0$$

i.e., equation are
$$-7x_1 - 6x_2 + 2x_3 = 0 \qquad 7x_1 + 6x_2 - 2x_3 = 0 \quad ...(7)$$
$$6x_1 + 8x_2 + 4x_3 = 0 \text{ or } 3x_1 + 4x_2 + 2x_3 = 0 \quad ...(8)$$
and
$$2x_1 - 4x_2 + 2x_3 = 0 \qquad x_1 - 2x_2 + x_3 = 0 \quad ...(9)$$

From (7) and (8), we get

$$\frac{x_1}{12+8} = \frac{x_2}{-6-14} = \frac{x_3}{28-18} \text{ i.e. } \frac{x_1}{20} = \frac{x_2}{-20} = \frac{x_3}{10}$$

i.e. eigen vectors are (2, – 2, 1) corresponding to $\lambda = 15$.

***Example* 1.21.** Find the eigen values and eigen vectors of the matrix $\begin{bmatrix} 6 & -2 & 2 \\ -2 & 3 & -1 \\ 2 & -1 & 3 \end{bmatrix}$

Solution. Let the given matrix be $A = \begin{bmatrix} 6 & -2 & 2 \\ -2 & 3 & -1 \\ 2 & -1 & 3 \end{bmatrix}$

So, the characteristic equation of A is $|A - \lambda I| = 0$

i.e.,
$$\begin{vmatrix} 6-\lambda & -2 & 2 \\ -2 & 3-\lambda & -1 \\ 2 & -1 & 3-\lambda \end{vmatrix} = 0 \quad \text{...(1)}$$

$$\Rightarrow (6-\lambda)[(3-\lambda)^2 - 1] + 2[-2(3-\lambda) + 2] + 2[2 - 2(3-\lambda)] = 0$$

$$\Rightarrow (6-\lambda)[9 - 6\lambda + \lambda^2 - 1] + 2[2\lambda - 4] + 2[2\lambda - 4] = 0$$

$$\Rightarrow -\lambda^3 + \lambda^2[6-6] - \lambda[36 - 8 + 8] + [48 - 8 - 8] = 0$$

$$\Rightarrow \lambda^3 - 12\lambda^2 + 36\lambda - 32 = 0$$

$$\Rightarrow \lambda^3 - 2\lambda^2 - 10\lambda^2 + 20\lambda + 16\lambda - 32 = 0$$

$$\Rightarrow (\lambda - 2)^2 (\lambda - 8) = 0 \quad i.e. \quad \lambda = 2, 2 \text{ and } 8$$

which are the characteristic roots of (1)

Now corresponding to the eigen values $\lambda = 2, 2, 8$ the given eigen vectors are obtained from $[A - \lambda]X = 0$.

i.e.,
$$\begin{bmatrix} 6-\lambda & -2 & 2 \\ -2 & 3-\lambda & -1 \\ 2 & -1 & 3-\lambda \end{bmatrix} \begin{bmatrix} x_1 \\ x_2 \\ x_3 \end{bmatrix} = \begin{bmatrix} 0 \\ 0 \\ 0 \end{bmatrix} \quad \text{...(2)}$$

(2) may be written as

$$(6-\lambda)x_1 - 2x_2 + 2x_3 = 0, \quad \text{...(A)}$$

$$-2x_1 + (3-\lambda)x_2 - x_3 = 0, \quad \text{...(B)}$$

and
$$2x_1 - x_2 + (3-\lambda)x_3 = 0 \quad \text{...(C)}$$

We, now consider different cases.

Case I. When $\lambda = 2$, then (A), (B) and (C) may be written as

$$\left.\begin{aligned} 4x_1 - 2x_2 + 2x_3 &= 0 \\ -2x_1 + x_2 - x_3 &= 0 \\ 2x_1 - x_2 + x_3 &= 0 \end{aligned}\right\} \Rightarrow 2x_1 - x_2 + x_3 = 0 \quad \begin{aligned} &\text{...}(A_1) \\ &\text{...}(B_1) \\ &\text{...}(C_1) \end{aligned}$$

If $x_3 = 0$, then from (A_1), (B_1) and (C_1), we get

$$-2x_1 + x_2 = 0 \; i.e. \; \frac{x_1}{1} = \frac{x_2}{2}$$

and so eigen vector for $\lambda = 2$, for $x_3 = 0$ is $X_1 = \begin{bmatrix} 1 \\ 2 \\ 0 \end{bmatrix}$

and when $x_2 = 0$, then from (A_1) and (B_1) for $\lambda = 2$,

$$2x_1 + x_3 = 0 \Rightarrow \frac{x_1}{1} = \frac{x_3}{-2}$$

$\therefore$ another eigen vector for $\lambda = 2$ is $X_2 = \begin{bmatrix} 1 \\ 0 \\ -2 \end{bmatrix}$

Case II: When $\lambda = 8$, equations (A), (B) and (C) becomes

$$-2x_1 - 2x_2 + 2x_3 = 0 \quad \text{...}(A_{11})$$

$$-2x_1 - 5x_2 - x_3 = 0 \quad \text{...}(B_{11})$$

$$2x_1 - x_2 + 5x_3 = 0 \quad \text{...}(C_{11})$$

eliminating (A_{11}) and (B_{11}), we get

$$x_1 + 2x_2 = 0 \quad \textit{i.e.,} \quad \frac{x_1}{2} = \frac{x_2}{-1} \quad \text{...(M)}$$

and by eliminating x_1 from (A_{11}) and (B_{11}), we get

$$x_2 + x_3 = 0 \quad \textit{i.e.,} \quad \frac{x_2}{-1} = \frac{x_3}{1} \quad \text{...(N)}$$

Using (M) and (N), we get $\frac{x_1}{2} = \frac{x_2}{-1} = \frac{x_3}{1}$

i.e., corresponding to $\lambda = 8$, eigen vector is $X_3 = \begin{bmatrix} 2 \\ -1 \\ 1 \end{bmatrix}$

Example **1.22.** Find the eigen values and eigen vectors of the matrix $A = \begin{bmatrix} -2 & 2 & -3 \\ 2 & 1 & -6 \\ -1 & -2 & 0 \end{bmatrix}$

Solution. The characteristic equation of the given matrix is

$$|A - \lambda I| = \begin{vmatrix} -2-\lambda & 2 & -3 \\ 2 & 1-\lambda & -6 \\ -1 & -2 & -\lambda \end{vmatrix} = 0$$

i.e., $\lambda^3 + \lambda^2 - 21\lambda - 45 = 0 \Rightarrow (\lambda+3)(\lambda+3)(\lambda-5) = 0$

i.e., eigen values are $\lambda = -3, -3, 5$

$\therefore$ If x, y and z be the eigen vectors, corresponding to the eigen values λ.

(I) then for $\lambda = 5$ we have $\begin{vmatrix} -2-\lambda & 2 & -3 \\ 2 & 1-\lambda & -6 \\ -1 & -2 & -\lambda \end{vmatrix} = 0$ ...(1)

$\Rightarrow$ $-7x + 2y - 3z = 0$ and $2x - 4y - 6z = 0$

i.e., $\frac{x}{-12-12} = \frac{y}{-6-42} = \frac{z}{28-4} \Rightarrow \frac{x}{1} = \frac{y}{2} = \frac{z}{-1}$

Hence eigen vector is $(1, 2, -1)$ for $\lambda = 5$.

(II) If $\lambda = -3$, then from (1), we get $\begin{bmatrix} 1 & 2 & -3 \\ 2 & 4 & -6 \\ -1 & 2 & 3 \end{bmatrix} \begin{bmatrix} x \\ y \\ z \end{bmatrix} = 0$ which gives only one independent equation $x + 2y - 3z = 0$

if we take

$$y = 0, \text{ we get} \quad \text{...(3)}$$

$$x - 3z = 0 \Rightarrow \frac{x}{3} = \frac{y}{0} = \frac{z}{1}$$

$\therefore$ for $\lambda = -3$, eigen vector is $(3, 0, 1)$ when $y = 0$.

when $\quad z = 0$, (3) gives $x + 2y = 0 \Rightarrow \frac{x}{2} = \frac{y}{-1} = \frac{z}{0}$

i.e., eigen vector in this case is (2, – 1, 0). The eigen vectors obtained are (1, 2, –1), (3, 0, 1) and (2, – 1, 0) which are the required result.

Example 1.23. Find the sum and the product of eigen values of $A = \begin{bmatrix} 2 & 3 & -2 \\ -2 & 1 & 1 \\ 1 & 0 & 2 \end{bmatrix}$

Solution. The characteristic equation of matrix A is $|A - \lambda I| = 0$

i.e. $\begin{bmatrix} 2-\lambda & 3 & -2 \\ -2 & 1-\lambda & 1 \\ 1 & 0 & 2-\lambda \end{bmatrix} = 0$

i.e. $(2 - \lambda)(1 - \lambda)(2 - \lambda) + 3[1 + 2(2 - \lambda) + (2)(0 - \overline{1 - 1}) = 0$

$\Rightarrow (2 - \lambda)(\lambda^2 - 3\lambda + 2 + 3) - 6\lambda + 15 + 2 - 2\lambda = 0$

$\Rightarrow - \lambda^3 + 5\lambda^2 - 11\lambda + 10 - 6\lambda + 15 + 2 = 2\lambda = 0$

$\Rightarrow \lambda^3 - 5\lambda^2 + 19\lambda + 19 = 0$

$\therefore$ sum of the eigen value $\lambda_1 + \lambda_2 + \lambda_3 = -(-5) = 5$

and the product of the eigen values of $\lambda_1 \lambda_2 \lambda_3 = -19$.

1.17 CAYLEY-HAMILTON THEOREM

Every square matrix satisfies characteristic equation i.e., if A be the given square matrix of order n then its characteristic equation is $|A - \lambda I| = 0$.

i.e., $(-1)^n \lambda^n + K_1 \lambda^{n-1} + K_2 \lambda^{n-2} + K_3 \lambda^{n-3} + ... + + K_{n-1} \lambda + K_n = 0$ ...(M_1)

then A will satisfy (M_1) *i.e.*, $(-1)^n A^n + K_1 A^{n-1} + K_2 A^{n-2} + K_3 A^{n-3} + ... + K_n I = 0$ will hold good.

We take adjoint of the matrix $A - \lambda I$ as P *i.e.*, P = adj. $(A - \lambda I)$. Also, each element of P is a co-factor of elements of $|A - \lambda I|$ so these co-factors are polynomials of degree $(n - 1)$ or less in λ.

So P can be split up into a number of matrices, which are co-efficients of the same power of λ and can be written as

$$P = P_1 \lambda^{n-1} + P_2 \lambda^{n-2} + P_3 \lambda^{n-3} + ... + P_{n-1} \lambda + P_n \qquad ...(M_2)$$

where $P_1, P_2, P_3, ..., P_n$ are all square matrix of order n, whose elements are functions of matrix A.

By matrix property it is known that if A is a square matrix then

$A \times$ adj. $A = |A| \times I$, where I is unit matrix of same order as that of A.

$\therefore [A - \lambda I] \times P = |A - \lambda I| \times I$

so using (M_1) and (M_2) we may write it as

$$[A - \lambda I] \times [P_1 \lambda^{n-1} + P_2 \lambda^{n-2} + P_3 \lambda^{n-1} + ... + P_{n-1} \lambda + P_n]$$
$$= [(-1)^n \lambda^n + K_1 \lambda^{n-1} + ... + K_{n-1} \lambda + K_n] I \qquad ...(M_3)$$

From (M_3), equating the co-efficients of powers of λ, we get

$$(-1) P_1 = (-1)^n I, \qquad ...(M_4)$$
$$AP_1 - P_2 = K_1 I, \qquad ...(M_5)$$
$$AP_2 - P_3 = K_2 I, \qquad ...(M_6)$$
$$AP_3 - P_4 = K_3 I, \qquad ...(M_7)$$

$$\cdots \qquad \cdots$$

$$AP_{n-1} - P_n = K_{n-1}\, I, \qquad \text{...}(M_{n+3})$$

and
$$AP_n = K_n\, I, \qquad \text{...}(M_{n+4})$$

Next, pre-multiplying the equations (M_4) by A^n, (M_5) by A_{n-1}, ... and (M_{n+3}) by P and (M_{n+4}) by I, we get

$$-A^n P_1 = (-1)^n A^n, \qquad \text{...}(R_1)$$

$$A^n P_1 - A^{n-1} P_2 = K_1 - A^{n-1} \qquad \text{...}(R_2)$$

$$A^{n-1} P_2 - A^{n-2} P_3 = K_2 - A^{n-2} \qquad \text{...}(R_3)$$

$$\cdots \qquad \cdots$$

$$A^2 P_{n-1} - AP_n = K_{n-1} A \qquad \text{...}(R_n)$$

$$AP_n = K_n I \qquad \text{...}(R_{n+1})$$

Now, adding them, we get

$$O = (-1)^n A^n + K_1 - A^{n-1} + K_2 - A^{n-2} + \ldots + K_{n-1} A + K_n I \qquad \text{...(X)}$$

As left hand side terms cancel.

To find the inverse of any matrix A, we multiply both sides of (X) by A^{-1} and get $(-1)n\, A^{n-1} + K_1\, A^{n-2} + K_2\, A^{n-3} + \ldots + K^{n-1} + K_n\, A^{-1} = 0$

$$\Rightarrow \qquad A^{-1} = -\frac{1}{K_n}\,[(-1)^n A^{n-1} + K_1 A^{n-2} + K_2 A^{n-3} + \ldots + K_{n-1}]$$

***Example* 1.24.** Using Cayley - Hamilton theorem, find the inverse of matrix $A = \begin{bmatrix} 5 & 3 \\ 3 & 2 \end{bmatrix}$

Solution. Let the characteristic equation be $|A - \lambda I| = 0$

i.e.,
$$\begin{bmatrix} 5-\lambda & 3 \\ 3 & 2-\lambda \end{bmatrix} = 0$$

$$\Rightarrow (5-\lambda)(2-\lambda) - 9 = 0 \Rightarrow 10 + \lambda^2 - 7\lambda - 9 = 0$$

$$\Rightarrow \qquad \lambda^2 - 7\lambda + 1 = 0$$

Now
$$A^2 = A \times A = \begin{bmatrix} 5 & 3 \\ 3 & 2 \end{bmatrix} \times \begin{bmatrix} 5 & 3 \\ 3 & 2 \end{bmatrix} = \begin{bmatrix} 34 & 21 \\ 21 & 13 \end{bmatrix}$$

Consider

$$\therefore \qquad A^2 - 7A + I = \begin{bmatrix} 34 & 21 \\ 21 & 13 \end{bmatrix} - 7\begin{bmatrix} 5 & 3 \\ 3 & 2 \end{bmatrix} + \begin{bmatrix} 1 & 0 \\ 0 & 1 \end{bmatrix}$$

$$= \begin{bmatrix} 34-35+1 & 21-21 \\ 21-21 & 13-14+1 \end{bmatrix} = \begin{bmatrix} 0 & 0 \\ 0 & 0 \end{bmatrix} \qquad \text{...(2)}$$

$\therefore$ Cayley Hamilton is satisfied. Now, multiplying both sides of (2) by A^{-1}, we get

$$A - 7I + A^{-1} = 0$$

$$\therefore \qquad A^{-1} = 7I - A = \begin{bmatrix} 7 & 0 \\ 0 & 7 \end{bmatrix} - \begin{bmatrix} 5 & 3 \\ 3 & 2 \end{bmatrix} = \begin{bmatrix} 7-5 & 3 \\ -3 & 7-2 \end{bmatrix}$$

$$\therefore \qquad A^{-1} = \begin{bmatrix} 2 & -3 \\ -3 & 5 \end{bmatrix}$$

***Example* 1.25.** Using Cayley-Hamilton theorem, find the inverse of the matrix $A = \begin{bmatrix} 1 & 0 & 3 \\ 2 & 1 & -1 \\ 1 & -1 & 1 \end{bmatrix}$

Solution. The characteristic equation of the matrix A is

$$|A-\lambda I| = 0 \text{ i.e. } \begin{bmatrix} 1-\lambda & 0 & 3 \\ 2 & 1-\lambda & -1 \\ 1 & -1 & 1-\lambda \end{bmatrix} = 0$$

This may be written as $(1-\lambda)[(1-\lambda)^2 - 1] + 3(-2+\lambda-1) = 0$

$$-\lambda^3 + 3\lambda^2 + \lambda - 9 = 0$$

i.e., $\lambda^3 - 3\lambda^2 - \lambda + 9 = 0$ ...(1)

As (1) satisfies Cayley-Hamilton theorem, (given by question) we have

$A^3 - 3A^2 - A + 9I = 0$ ...(2)

Multiplying (2) by A^{-1}, we get $A^2 - 3A - I + 9A^{-1} = 0$ ...(3)

So, we require

$$A^2 = \begin{bmatrix} 1 & 0 & 3 \\ 2 & 1 & -1 \\ 1 & -1 & 1 \end{bmatrix} \times \begin{bmatrix} 1 & 0 & 3 \\ 2 & 1 & -1 \\ 1 & -1 & 1 \end{bmatrix}$$

$$= \begin{bmatrix} 1+0+3 & 0+0-3 & 3+0+3 \\ 2+2-1 & 0+1+1 & 6-1+1 \\ 1-2+1 & 0-1-1 & 3+1+1 \end{bmatrix}$$

$$\therefore \quad A^2 = \begin{bmatrix} 4 & -3 & 6 \\ 3 & 2 & 4 \\ 0 & -2 & 5 \end{bmatrix}$$

$\therefore$ (2) may be written as $A^{-1} = \frac{1}{9}[I + 3A - A^2]$

i.e.
$$A^{-1} = \frac{1}{9}\left\{\begin{bmatrix} 1 & 0 & 0 \\ 0 & 1 & 0 \\ 0 & 0 & 1 \end{bmatrix} + \begin{bmatrix} 3 & 0 & 9 \\ 6 & 3 & -3 \\ 3 & -3 & 3 \end{bmatrix} - \begin{bmatrix} 4 & -3 & 6 \\ 3 & 2 & 4 \\ 0 & -2 & 5 \end{bmatrix}\right\}$$

$$= \frac{1}{9}\begin{bmatrix} 0 & 3 & 3 \\ 3 & 2 & -7 \\ 3 & -1 & -1 \end{bmatrix} = \begin{bmatrix} 0 & 1/3 & 1/3 \\ 1/3 & 2/9 & -7/9 \\ 1/3 & -1/9 & -1/9 \end{bmatrix}$$

is the required inverse matrix of A.

Example **1.26.** Verify Cayley-Hamilton theorem, for the matrix $A = \begin{bmatrix} 1 & 2 \\ 2 & -1 \end{bmatrix}$ and then find A^8.

Solution. The characteristic equation is $|A - \lambda I| = 0$

i.e.
$$\begin{vmatrix} 1-\lambda & 2 \\ 2 & 1-\lambda \end{vmatrix} = 0 \Rightarrow \lambda^2 - 1 - 4 = 0 \Rightarrow \lambda^2 - 5.$$

Now,
$$A^2 = \begin{bmatrix} 1 & 2 \\ 2 & -1 \end{bmatrix} \times \begin{bmatrix} 1 & 2 \\ 2 & -1 \end{bmatrix} = \begin{bmatrix} 5 & 0 \\ 0 & 5 \end{bmatrix} = 5I$$

i.e., $A^2 = 5I.$

Hence Cayley-Hamilton theorem is satisfied.

Next, $A^2 = 5I \Rightarrow A^4 = 25I \Rightarrow (A^4)^2 = (25I)^2$

$\Rightarrow \quad A^8 = 625I.$

Example **1.27.** Find the characteristic equation of the matrix $A = \begin{bmatrix} 2 & 1 & 1 \\ 0 & 1 & 0 \\ 1 & 1 & 2 \end{bmatrix}$ and hence find the matrix represented by $A^8 - 5A^7 - 7A^6 - 3A^5 + A^4 - 5A^3 + 8A^2 - 2A + I.$

Solution. The characteristic equation of the matrix A is $|A - \lambda I| = 0$

i.e.,
$$\begin{bmatrix} 2-\lambda & 1 & 1 \\ 0 & 1-\lambda & 0 \\ 1 & 1 & 2-\lambda \end{bmatrix} = 0$$

$\Rightarrow$
$$\lambda^2 - 5\lambda^2 + 7\lambda - 3 = 0 \qquad ...(1)$$

Using Cayley-Hamilton theorem, we know that matrix A satisfies the eqn. (1)

i.e.
$$A^3 - 5A^2 + 7A - 3I = 0 \qquad ...(2)$$

Now, we consider the matrix

$$A^8 - 5A^7 - 7A^6 - 3A^5 + A^4 - 5A^3 + 8A^2 - 2A + I \qquad ...(3)$$

and arrange it, in such a manner that (2) is used to reduce (3) in simple form, *i.e.* (3) may be arranged as

$$(A^8 - 5A^7 - 7A^6 - 3A^5) + (A^4 - 5A^3 + 7A^2 - 3A) + A^2 + A + I$$

$$= A^5 (A^3 - 5A^2 + 7A - 3I) + A(A^3 - 5A^2 + 7A - 3I) + A^2 + A + I$$

Using (2),

$$= A^2 + A + I = \begin{bmatrix} 5 & 4 & 4 \\ 0 & 1 & 0 \\ 4 & 4 & 5 \end{bmatrix} + \begin{bmatrix} 2 & 1 & 1 \\ 0 & 1 & 0 \\ 1 & 1 & 2 \end{bmatrix} + \begin{bmatrix} 1 & 0 & 0 \\ 0 & 1 & 0 \\ 0 & 0 & 1 \end{bmatrix}$$

$$= \begin{bmatrix} 8 & 5 & 5 \\ 0 & 3 & 0 \\ 5 & 5 & 8 \end{bmatrix}$$

which is the required result.

Since,
$$A^2 = \begin{bmatrix} 2 & 1 & 1 \\ 0 & 1 & 0 \\ 1 & 1 & 2 \end{bmatrix} \times \begin{bmatrix} 2 & 1 & 1 \\ 0 & 1 & 0 \\ 1 & 1 & 2 \end{bmatrix} = \begin{bmatrix} 4+0+1 & 2+1+1 & 2+0+2 \\ 0+0+0 & 0+1+0 & 0+0+0 \\ 2+0+2 & 1+1+2 & 1+0+4 \end{bmatrix}$$

$$= \begin{bmatrix} 5 & 4 & 4 \\ 0 & 1 & 0 \\ 4 & 4 & 5 \end{bmatrix}$$

***Example* 1.28.** Find the characteristic equation of the matrix $A = \begin{bmatrix} 1 & 3 & 7 \\ 4 & 2 & 3 \\ 1 & 2 & 1 \end{bmatrix}$. Show that the equation is satisfied by A and hence obtain the inverse of the given matrix.

Solution. The characteristic equation is $|A - \lambda I| = 0$

i.e.,
$$\begin{bmatrix} 1-\lambda & 3 & 7 \\ 4 & 2-\lambda & 3 \\ 1 & 2 & 1-\lambda \end{bmatrix} = 0$$

$\Rightarrow$
$$\lambda^2 - 4\lambda^2 - 20\lambda - 35 = 0 \qquad ...(1)$$

We have to show that A satisfies (1) *i.e.,* $A^3 - 4A^2 - 20A - 35I = 0$ ...(2)

Consider

$$A^2 = A \cdot A = \begin{bmatrix} 1 & 3 & 7 \\ 4 & 2 & 3 \\ 1 & 2 & 1 \end{bmatrix} \begin{bmatrix} 1 & 3 & 7 \\ 4 & 2 & 3 \\ 1 & 2 & 1 \end{bmatrix}$$

$$= \begin{bmatrix} 1+12+7 & 3+6+14 & 7+9+7 \\ 4+8+3 & 12+4+6 & 28+6+3 \\ 1+8+1 & 3+4+2 & 7+6+1 \end{bmatrix}$$

$$\Rightarrow \qquad A^2 = \begin{bmatrix} 20 & 23 & 25 \\ 15 & 22 & 37 \\ 10 & 9 & 14 \end{bmatrix}$$

$$A^3 = A^2\,A = \begin{bmatrix} 20 & 23 & 23 \\ 15 & 22 & 37 \\ 10 & 9 & 14 \end{bmatrix}\begin{bmatrix} 1 & 3 & 7 \\ 4 & 2 & 3 \\ 1 & 2 & 1 \end{bmatrix}$$

$$= \begin{bmatrix} 20+92+23 & 60+46+46 & 140+69+23 \\ 15+88+37 & 45+44+74 & 105+66+37 \\ 10+36+14 & 30+18+28 & 70+27+14 \end{bmatrix}$$

$$= \begin{bmatrix} 135 & 152 & 232 \\ 140 & 163 & 208 \\ 60 & 76 & 111 \end{bmatrix}$$

Now, we consider $A^3 - 4A^2 - 20A - 35I$, which is

$$= \begin{bmatrix} 135 & 152 & 232 \\ 140 & 163 & 208 \\ 60 & 76 & 111 \end{bmatrix} - 4\begin{bmatrix} 20 & 23 & 23 \\ 15 & 22 & 37 \\ 10 & 9 & 14 \end{bmatrix} - 20\begin{bmatrix} 1 & 3 & 7 \\ 4 & 2 & 3 \\ 1 & 2 & 1 \end{bmatrix} - \begin{bmatrix} 35 & 0 & 0 \\ 0 & 35 & 0 \\ 0 & 0 & 35 \end{bmatrix}$$

$$= \begin{bmatrix} 135-80-20-35 & 152-92-60 & 232-92-140 \\ 140-60-80 & 163-88-40-35 & 208-148-60 \\ 60-40-40 & 76-36-20 & -56-20-35 \end{bmatrix}$$

$$= \begin{bmatrix} 0 & 0 & 0 \\ 0 & 0 & 0 \\ 0 & 0 & 0 \end{bmatrix}$$

$\therefore$ Equation (2) is satisfied and $A^{-1} = \frac{1}{35}\,[A^2 - 4A - 20I]$

i.e.,

$$A^{-1} = \frac{1}{35}\left\{\begin{bmatrix} 20 & 23 & 23 \\ 15 & 22 & 37 \\ 10 & 9 & 14 \end{bmatrix} - 4\begin{bmatrix} 1 & 3 & 7 \\ 4 & 2 & 3 \\ 1 & 2 & 1 \end{bmatrix} - \begin{bmatrix} 20 & 0 & 0 \\ 0 & 20 & 0 \\ 0 & 0 & 20 \end{bmatrix}\right\}$$

$$= \frac{1}{35}\begin{bmatrix} 20-4-20 & 23-12 & 23-28 \\ 15-16 & 23-8-20 & 37-12 \\ 10-14 & 9-8 & 14-4-20 \end{bmatrix}$$

$$= \frac{1}{35}\begin{bmatrix} -1 & 11 & 5 \\ -1 & -6 & 25 \\ 6 & 1 & -10 \end{bmatrix}$$

i.e.,

$$A^{-1} = \frac{1}{35}\begin{bmatrix} -4 & 11 & -5 \\ -1 & -6 & 25 \\ 6 & 1 & -10 \end{bmatrix} \text{ is the required result.}$$

***Example* 1.29.** Find the characteristic equation of the matrix $A = \begin{bmatrix} 1 & 2 & 3 \\ 2 & -1 & 4 \\ 3 & 1 & -1 \end{bmatrix}$. Show that the equation is satisfied by A and hence find A^{-1}.

Solution. The matrix $A = \begin{bmatrix} 1 & 2 & 3 \\ 2 & -1 & 4 \\ 3 & 1 & -1 \end{bmatrix}$

So, its characteristic equation is $|A - \lambda I| = \begin{bmatrix} 1-\lambda & 2 & 3 \\ 2 & -2-\lambda & 4 \\ 3 & 1 & -1-\lambda \end{bmatrix} = 0$

Expanding, we get $\lambda^3 + \lambda^2 - 18\lambda - 40 = 0$ ...(1)

We have to show that A satisfies (1), *i.e.*, $A^3 + A^2 - 18A - 40I = 0$ ...(2)

So,

$$A^2 = \begin{bmatrix} 1 & 2 & 3 \\ 2 & -1 & 4 \\ 3 & 1 & -1 \end{bmatrix}\begin{bmatrix} 1 & 2 & 3 \\ 2 & -1 & 4 \\ 3 & 1 & -1 \end{bmatrix}$$

$$= \begin{bmatrix} 1+4+9 & 2-2+3 & 3+8-3 \\ 2-2+12 & 4+1+4 & 6-4-4 \\ 3+2-6 & 6-1-1 & 9+4+1 \end{bmatrix} = \begin{bmatrix} 14 & 3 & 8 \\ 12 & 9 & -2 \\ 2 & 4 & 14 \end{bmatrix}$$

Similarly

$$A^3 = A^2 A = \begin{bmatrix} 14 & 3 & 8 \\ 12 & 9 & -2 \\ 2 & 4 & 14 \end{bmatrix}\begin{bmatrix} 1 & 2 & 3 \\ 2 & -1 & 4 \\ 3 & 1 & -1 \end{bmatrix}$$

$$= \begin{bmatrix} 14+6+24 & 28-3+8 & 42+12-8 \\ 12+18-6 & 24-9-2 & 36+36+2 \\ 2+8+42 & 4-4+14 & 6+16-14 \end{bmatrix} = \begin{bmatrix} 44 & 33 & 46 \\ 24 & 13 & 74 \\ 52 & 14 & 8 \end{bmatrix}$$

Now we consider $A^3 + A^2 - 18A - 40I$

$$= \begin{bmatrix} 44 & 33 & 46 \\ 24 & 13 & 74 \\ 52 & 14 & 8 \end{bmatrix} + \begin{bmatrix} 14 & 3 & 8 \\ 12 & 9 & -2 \\ 2 & 4 & 14 \end{bmatrix} - 18\begin{bmatrix} 1 & 2 & 3 \\ 2 & -1 & 4 \\ 3 & 1 & -1 \end{bmatrix} - \begin{bmatrix} 40 & 0 & 0 \\ 0 & 40 & 0 \\ 0 & 0 & 40 \end{bmatrix}$$

$$= \begin{bmatrix} 44+14-18-40 & 33+3-36+0 & 46+8-54+0 \\ 24+12-36+0 & 13+9+18-40 & 74-2+72+0 \\ 52+2-54+0 & 14+4-18+0 & 8+14+18-40 \end{bmatrix}$$

$$= \begin{bmatrix} 0 & 0 & 0 \\ 0 & 0 & 0 \\ 0 & 0 & 0 \end{bmatrix}$$

Hence, we see that (2) is satisfied using (3)

i.e., $A^3 + A^2 - 18A - 40I = 0 \quad d \quad A^2 + A - 18I = 40\,A^{-1}$

i.e.,

$$A^{-1} = \frac{1}{40}\left\{\begin{bmatrix} 14 & 3 & 8 \\ 12 & 9 & -2 \\ 2 & 4 & 14 \end{bmatrix} + \begin{bmatrix} 1 & 2 & 3 \\ 2 & -1 & 4 \\ 3 & 1 & -1 \end{bmatrix} - \begin{bmatrix} 18 & 0 & 0 \\ 0 & 18 & 0 \\ 0 & 0 & 18 \end{bmatrix}\right\}$$

$$= \frac{1}{40}\begin{bmatrix} 14+1-18 & 3+2+0 & 8+3+0 \\ 12+2+0 & 9-1-18 & -2+4+0 \\ 2+3+0 & 4+1+0 & 14-1-18 \end{bmatrix}$$

$$= \frac{1}{40}\begin{bmatrix} -3 & 5 & 11 \\ 14 & -10 & 2 \\ 5 & 5 & -5 \end{bmatrix}$$ is the required solution.

EXERCISE 1.4

1. Find the sum and product of the eigen value of

(i) $\begin{bmatrix} 2 & 1 & 1 \\ 0 & 1 & 0 \\ 1 & 1 & 2 \end{bmatrix}$ (ii) $\begin{bmatrix} 1 & 1 & 1 \\ 1 & 2 & 2 \\ 1 & 2 & 3 \end{bmatrix}$ (iii) $\begin{bmatrix} 1 & 2 & -2 \\ 1 & 0 & 3 \\ -2 & -1 & -3 \end{bmatrix}$

(iv) $\begin{bmatrix} 6 & -2 & 2 \\ -2 & 3 & -1 \\ 2 & -1 & 3 \end{bmatrix}$ (v) $\begin{bmatrix} -2 & 2 & -3 \\ 2 & 1 & -6 \\ -1 & -2 & 0 \end{bmatrix}$

Ans.

(i) 5 ; 3 (ii) 6 ; 1 (iii) − 2, − 1 (iv) 12, 32 (v) − 1, 45

2. Find the eigen values and eigen vectors of the matrix.

(i) $\begin{bmatrix} 5 & 4 \\ 1 & 2 \end{bmatrix}$ (ii) $\begin{bmatrix} 2 & 0 & 1 \\ 0 & 2 & 0 \\ 1 & 0 & 2 \end{bmatrix}$ (iii) $\begin{bmatrix} 2 & 1 & 1 \\ 1 & 2 & 1 \\ 0 & 0 & 1 \end{bmatrix}$

Ans.

(i) 1, 6 ; (1, − 1) ; (4, 1) (ii) 1, 2, 3 ; (1, 0, 1) ; (1, 0, − 1) ; (0, 1, 0); (iii) 1, 1, 3 ; (1, − 2, 1) ; (1, 1, 0)

3. Find the product of the eigen values of $\begin{bmatrix} 7 & 2 & 2 \\ -6 & -1 & 2 \\ 6 & 2 & -1 \end{bmatrix}$

Ans. − 21

4. Using Cayley-Hamilton theorem, find the inverse of

(i) $\begin{bmatrix} 1 & 1 & 3 \\ 1 & 3 & -3 \\ 2 & -4 & -4 \end{bmatrix}$ (ii) $\begin{bmatrix} 7 & 2 & -2 \\ -6 & -1 & 2 \\ 6 & 2 & -1 \end{bmatrix}$ (iii) $\begin{bmatrix} 1 & 0 & 2 \\ 0 & 2 & 1 \\ 2 & 0 & 3 \end{bmatrix}$

(iv) $\begin{bmatrix} 2 & -1 & 3 \\ 1 & 0 & 2 \\ 4 & -2 & 1 \end{bmatrix}$ (v) $\begin{bmatrix} 5 & -1 & 5 \\ 0 & 2 & 0 \\ -5 & 3 & -15 \end{bmatrix}$

Ans.

(i) $\frac{1}{4}\begin{bmatrix} 12 & 4 & 6 \\ -5 & -1 & -3 \\ -1 & -1 & -1 \end{bmatrix}$ (ii) $\frac{1}{3}\begin{bmatrix} -3 & -2 & 2 \\ 6 & 5 & -2 \\ -6 & -2 & 5 \end{bmatrix}$ (iii) $\begin{bmatrix} -3 & 0 & 2 \\ 1 & 1/2 & 1/2 \\ 2 & 0 & -1 \end{bmatrix}$

(iv) $-\frac{1}{5}\begin{bmatrix} 4 & -5 & -2 \\ 7 & -10 & -1 \\ -2 & 0 & 1 \end{bmatrix}$ (v) $\frac{1}{10}\begin{bmatrix} 3 & 0 & 1 \\ 0 & 5 & 0 \\ -1 & 1 & -1 \end{bmatrix}$

5. Show that the following matrices satisfies Cayley-Hamilton theorem and hence find their inverse

(i) $A = \begin{bmatrix} 1 & 1 & 2 \\ 3 & 1 & 1 \\ 2 & 3 & 1 \end{bmatrix}$ (ii) $A = \begin{bmatrix} 1 & -2 & 2 \\ 1 & 2 & 3 \\ 0 & -1 & 2 \end{bmatrix}$

Ans.

(*i*) $\frac{1}{11}\begin{bmatrix} -2 & 5 & -1 \\ -1 & -3 & 5 \\ 7 & -1 & -2 \end{bmatrix}$ (*ii*) $\frac{1}{9}\begin{bmatrix} 7 & 2 & -10 \\ -2 & 2 & -1 \\ -1 & 1 & 4 \end{bmatrix}$

6. Using Cayley Hamilton theorem, find A^6 if $A = \begin{bmatrix} 2 & 1 \\ 5 & -2 \end{bmatrix}$

1.18 REDUCTION OF A MATRIX TO DIAGONAL FORM

Theorem : ***If a square matrix A of order n has n linearly independent eigen vectors, then a matrix P can be found such that $P^{-1}AP$ is a diagonal matrix and denote it by D.***

First, we consider a square matrix A of order 3, such that $\lambda_1, \lambda_2, \lambda_3$ are its eigen values and

$$X_1 = \begin{bmatrix} x_1 \\ y_1 \\ z_1 \end{bmatrix}, X_2 = \begin{bmatrix} x_2 \\ y_2 \\ z_2 \end{bmatrix} \text{ and } X_3 = \begin{bmatrix} x_3 \\ y_3 \\ z_3 \end{bmatrix}$$

be the eigen vectors corresponding to eigen values λ_1, λ_2 and λ_3 respectively.

Next, we write the square matrix P as

$$P = [X_1\ X_2\ X_3] = \begin{bmatrix} x_1 & x_2 & x_3 \\ y_1 & y_2 & y_3 \\ z_1 & z_2 & z_3 \end{bmatrix}$$

And
$$AP = A[X_1\ X_2\ X_3] = [AX_1, AX_2, AX_3] = [\lambda_1 X_1, \lambda_2 X_2, \lambda_3 X_3]$$

$$= \begin{bmatrix} \lambda_1 x_1 & \lambda_2 x_2 & \lambda_3 x_3 \\ \lambda_1 y_1 & \lambda_2 y_2 & \lambda_3 y_3 \\ \lambda_1 z_1 & \lambda_2 z_2 & \lambda_3 z_3 \end{bmatrix} = \begin{bmatrix} x_1 & x_2 & x_3 \\ y_1 & y_2 & y_3 \\ z_1 & z_2 & z_3 \end{bmatrix} \times \begin{bmatrix} \lambda_1 & 0 & 0 \\ 0 & \lambda_2 & 0 \\ 0 & 0 & \lambda_3 \end{bmatrix}$$

$$AP = PD \text{ where } D = \begin{bmatrix} \lambda_1 & 0 & 0 \\ 0 & \lambda_2 & 0 \\ 0 & 0 & \lambda_3 \end{bmatrix} = \text{diagonal matrix}$$

$$P^{-1}AP = P^{-1}PD = ID = D \text{ which proves the theorem}$$

Note. The matrix $P = \begin{bmatrix} x_1 & x_2 & x_3 \\ y_1 & y_2 & y_3 \\ z_1 & z_2 & z_3 \end{bmatrix}$ which diagonalizes the matrix A is called the modal matrix of A and the diagonal matrix thus obtained is known as spectral matrix of matrix A.

***Example* 1.30.** Diagonalize $A = \begin{bmatrix} -1 & 2 & -2 \\ 1 & 2 & 1 \\ -1 & -1 & 0 \end{bmatrix}$ and obtain the modal matrix.

Solution. The characteristic equation of the given matrix is $|A - \lambda I| = 0$

i.e.,
$$\begin{bmatrix} -1-\lambda & 2 & -2 \\ 1 & 2-\lambda & 1 \\ -1 & -1 & -\lambda \end{bmatrix} = 0 \qquad ...(1)$$

Solving (1), we get

$$\lambda^3 - \lambda^2 - 5\lambda + 5 = 0$$

$$\Rightarrow \quad \lambda^3 - \lambda^2 - 5(\lambda - 1) = 0 \Rightarrow (\lambda - 1)|\lambda^2 - 5| = 0$$

i.e., $\lambda = 1, \pm\sqrt{5}$

Now, corresponding to these eigen values, eigen vectors will be obtained from the equation

$$\begin{bmatrix} -1-\lambda & 2 & -2 \\ 1 & 2-\lambda & 1 \\ -1 & -1 & -\lambda \end{bmatrix}\begin{bmatrix} x_1 \\ x_2 \\ x_3 \end{bmatrix} = 0 \quad ...(2)$$

Case I. For $\lambda = 1$, equation are

$$-2x_1 + 2x_2 - 2x_3 = 0 \quad i.e. \quad x_1 - x_2 + x_3 = 0 \quad ...(A)$$

$$x_1 + x_2 + x_3 = 0 \quad i.e. \quad x_1 + x_2 + x_3 = 0 \quad ...(B)$$

$$-x_1 - x_2 - x_2 = 0 \quad i.e. \quad x_1 + x_2 + x_3 = 0 \quad ...(C)$$

For $x_2 = 0, x_1 + x_3 = 0$ *i.e.,* $\frac{x_1}{1} = \frac{x_2}{0} = \frac{x_3}{-1}$ *i.e.,* eigen vector for $\lambda = 1$, is

$$X_1 = \begin{bmatrix} 1 \\ 0 \\ -1 \end{bmatrix}$$

Case II. For $\lambda = \sqrt{5}$, equation are

$$(-1-\sqrt{5})x_1 + 2x_2 + 2x_3 = 0 \quad ...(A_1)$$

$$x_1 + (2-\sqrt{5})x_2 + x_3 = 0 \quad ...(B_1)$$

and $$-x_1 - x_2 - \sqrt{5}\,x_3 = 0 \quad ...(C_1)$$

Adding (B_1) and (C_1), we get $(1-\sqrt{5})x_2 + (1-\sqrt{5})x_3 = 0$

$$\Rightarrow \quad \frac{x_2}{1} = \frac{x_3}{-1} \quad ...(M)$$

Now, multiplying (B_1) by $\sqrt{5}$ and then adding with (C_1), we get

$$x_1(\sqrt{5}-1) + (2\sqrt{5}-5-1)x_2 = 0$$

$$\Rightarrow \quad \frac{x_1}{6-2\sqrt{5}} = \frac{x_2}{\sqrt{5}-1} \quad ...(N)$$

(M) may be written then as $\frac{x_2}{\sqrt{5}-1} = \frac{x_3}{-(\sqrt{5}-1)}$ ...(M_1)

$\therefore$ Using (N) and (M_1), we get

$$\frac{x_1}{(1-\sqrt{5})^2} = \frac{x_2}{(\sqrt{5}-1)} = \frac{x_3}{-(\sqrt{5}-1)}$$

$$\Rightarrow \quad \frac{x_1}{\sqrt{5}-1} = \frac{x_2}{1} = \frac{x_3}{-1}$$

$\Rightarrow$ Eigen vector corresponding to $\lambda = \sqrt{5}$ is $X_2 = \begin{bmatrix} \sqrt{5}-1 \\ 1 \\ -1 \end{bmatrix}$

Next, corresponding value $\lambda = -\sqrt{5}$, we can calculate in similar way and get eigen vector

$$X_3 = \begin{bmatrix} \sqrt{5}+1 \\ -1 \\ 1 \end{bmatrix} \text{ since } x_2 + x_3 = 0$$

$$\Rightarrow \qquad \frac{x_2}{1} = \frac{x_3}{-1} \text{ and } \frac{x_1}{\sqrt{5}+1} = \frac{x_2}{-1}$$

$\therefore$ Required modal matrix

$$P = [X_1\ X_2\ X_3] = \begin{bmatrix} 1 & \sqrt{5}-1 & \sqrt{5}+1 \\ 0 & 1 & -1 \\ -1 & -1 & 1 \end{bmatrix}$$

$$\Rightarrow \qquad D = \text{diagonal matrix} = P^{-1}AP = \begin{bmatrix} \lambda_1 & 0 & 0 \\ 0 & \lambda_2 & 0 \\ 0 & 0 & \lambda_3 \end{bmatrix} = \begin{bmatrix} 1 & 0 & 0 \\ 0 & \sqrt{5} & 0 \\ 0 & 0 & -\sqrt{5} \end{bmatrix}$$

which is the required result.

1.19 REDUCTION OF QUADRATIC FORM TO CANONICAL FORM

A second degree homogeneous expression, in any number of variables is called a quadratic form

e.g., If $A = \begin{bmatrix} a & h & g \\ h & b & f \\ g & f & c \end{bmatrix}, X = \begin{bmatrix} x \\ y \\ z \end{bmatrix}$ *i.e.* $X' = [x\ y\ z]$ then

$$X'AX = ax^2 + by^2 + cz^2 + 2hxy + 2fyz + 2gxz \qquad \text{...(M)}$$

is a quadratic form.

If λ_1, λ_2 and λ_3 be the eigen values of the matrix A and then

$$X_1 = \begin{bmatrix} x_1 \\ y_1 \\ z_1 \end{bmatrix}, X_2 = \begin{bmatrix} x_2 \\ y_2 \\ z_2 \end{bmatrix} \text{ and } X_3 = \begin{bmatrix} x_3 \\ y_3 \\ z_3 \end{bmatrix}$$

are corresponding eigen vectors in normalized form and

$$P^{-1}AP = \begin{bmatrix} \lambda_1 & 0 & 0 \\ 0 & \lambda_2 & 0 \\ 0 & 0 & \lambda_3 \end{bmatrix} \text{ and } P = \begin{bmatrix} x_1 & x_2 & x_3 \\ y_1 & y_2 & y_3 \\ z_1 & z_2 & z_3 \end{bmatrix}$$

and the quadratic form (M) is reduced to a sum of squares $\lambda_1 x^2 + \lambda_2 y^2 + \lambda_3 z^2$ and P is a matrix of transformation which is orthogonal matrix.

This method of reduction in canonical form is called orthogonal transformation.

Example 1.31. Reduce the quadratic form

$$3x^2 + 5y^2 + 3z^2 - 2yz + 2zx - 2xy \qquad \text{...(1)}$$

to the canonical form. Also specify the matrix of transformation.

Solution. Here $a = 3, b = 5, c = 3, h = -1, g = 1$ and $f = -1$

So, the matrix of the given quadratic form (1) may be written as

$$A = \begin{bmatrix} a & h & g \\ h & b & f \\ g & f & c \end{bmatrix} = \begin{bmatrix} 3 & -1 & 1 \\ -1 & 5 & -1 \\ 1 & -1 & 3 \end{bmatrix}$$

Characteristic equation of A is $|A - \lambda I| = 0$

i.e. $$\begin{vmatrix} 3-\lambda & -1 & 1 \\ -1 & 5-\lambda & -1 \\ 1 & -1 & 3-\lambda \end{vmatrix} = 0$$

$\Rightarrow (3-\lambda)[(5-\lambda)(3-\lambda) - 1] + [\lambda - 3 + 1] + [1 + \lambda - 5] = 0$

$\Rightarrow \lambda^3 - 11\lambda^2 + 36\lambda - 36 = 0 \Rightarrow (\lambda - 2)(\lambda - 3)(\lambda - 6) = 0$

$\therefore$ eigen values of this characteristic equation are $\lambda = 2, 3, 6$.

Characteristic vectors corresponding to these eigen values are obtained from the equation

$$\begin{bmatrix} 3-\lambda & -1 & 1 \\ -1 & 5-\lambda & -1 \\ 1 & -1 & 3-\lambda \end{bmatrix}\begin{bmatrix} x_1 \\ x_2 \\ x_3 \end{bmatrix} = 0$$

$\therefore$ Characteristic vectors can be obtained from the equations

$$(3-\lambda)x_1 - x_2 + x_3 = 0 \quad \text{...(A)}$$

$$-x_1 + (5-\lambda)x_2 - x_3 = 0 \quad \text{...(B)}$$

$$x_1 - x_2 + (3-\lambda)x_3 = 0 \quad \text{...(C)}$$

The characteristic vector for $\lambda = 2$ is $X_1 = \begin{bmatrix} x_1 \\ x_2 \\ x_3 \end{bmatrix}$ which is obtained from (A), (B) and (C), *i.e.*,

when $\lambda = 2$

$$x_1 - x_2 + x_3 = 0 \quad \text{...}(A_1)$$

$$x_1 - 3x_2 + x_3 = 0 \quad \text{...}(B_1)$$

$$x_1 - x_2 + x_3 = 0 \quad \text{...}(C_1)$$

Putting $x_2 = 0$ and solving, we get $x_1 + x_3 = 0$, *i.e.*, $\dfrac{x_1}{1} = \dfrac{x_2}{0} = \dfrac{x_3}{-1}$

$\therefore$ $X_1 = \begin{bmatrix} 1 \\ 0 \\ -1 \end{bmatrix}$ and the normalized form is $X_1 = \begin{bmatrix} 1/\sqrt{2} \\ 0 \\ -1/\sqrt{2} \end{bmatrix}$

Similarly characteristic vector corresponding to $\lambda = 3$, is obtained from

$$x_1 - x_3 = 0 \quad \text{...}(A_{11})$$

$$x_1 - 2x_2 + x_3 = 0 \quad \text{...}(B_{11})$$

and $$x_1 - x_2 = 0 \quad \text{...}(C_{11})$$

i.e. $x_1 = x_2 = x_3$

$\therefore$ $X_2 = \begin{bmatrix} 1 \\ 1 \\ 1 \end{bmatrix}$ and the normalized form is $X_2 = \begin{bmatrix} \frac{1}{\sqrt{3}} \\ \frac{1}{\sqrt{3}} \\ \frac{1}{\sqrt{3}} \end{bmatrix}$

$\therefore$ Modal matrix $$P_1 = \begin{bmatrix} 1 & 1 & 1 \\ 0 & 1 & -2 \\ -1 & 1 & 1 \end{bmatrix}$$

and the matrix of transformation is

$$P = \begin{bmatrix} \frac{1}{\sqrt{2}} & \frac{1}{\sqrt{3}} & \frac{1}{\sqrt{6}} \\ 0 & \frac{1}{\sqrt{3}} & -\frac{2}{\sqrt{6}} \\ -\frac{1}{\sqrt{2}} & \frac{1}{\sqrt{3}} & \frac{1}{\sqrt{6}} \end{bmatrix}$$

Also the canonical form of the given quadratic equation may be written as

$\lambda_1 x_1^2 + \lambda_2 x_2^2 + \lambda_3 x_3^2$

i.e., $2x_1^2 + 3x_2^2 + 6x_3^2$ where $\lambda_1 = 2, \lambda_2 = -3$ and $\lambda_3 = 6$.

OBJECTIVE QUESTIONS

1. If A is a nilpotent matrix, then what is the eigen value of A?

Ans. 0

2. If 2, 3, 4 be the eigen values of M, then what is the determinant of M?

Ans. 24

3. If A is an $n \times 1$ null matrix and B is an $1 \times n$ non-null matrix shows that rank $(AB) = 1$.

4. If λ is an eigen value of M, then what is the value of $|M-1I|$.

Ans. 0

5. Can a matrix having linearly independent eigen vectors be diagonal?

Ans. No.

6. What is the sum of eigen values of matrix $\begin{bmatrix} 3 & 4 \\ x & 1 \end{bmatrix}$ for real and negative values of x?

Ans. Greater than 0

7. What is the sum and product of eigen values of the matrix $\begin{bmatrix} 1 & 1 & 3 \\ 1 & 5 & 1 \\ 3 & 1 & 1 \end{bmatrix}$

Ans. 7, – 36

8. What are the eigen values of $\begin{bmatrix} 4 & -2 \\ -2 & 1 \end{bmatrix}$

Ans. 0 and 5

9. If A is a square matrix of order n and rank = $(n-1)$, then what is the rank of (adj. A)?

Ans. 1

10. If $A = \begin{bmatrix} 0 & 1 \\ 0 & -2 \end{bmatrix}$ then what is e^{At}?

Ans. $\begin{bmatrix} 0 & \frac{1}{2}(1-e^{-2t}) \\ 0 & e^{-2t} \end{bmatrix}$

11. The system of equations $4x + 6y = 8$, $7x + 8y = 9$ and $3x + 2y = 1$ has how many solutions?

Ans. Only one solution

12. If A be a 3×3 matrix with eigen values $1, -1, 0$. Then what is determinant of $I + A^{100}$?

Ans. 4

13. Show that are transformation

$y_1 = \cos\theta\, x_1 + \sin\theta\, x_2$, $y_2 = \sin\theta\, x_1 + \cos\theta\, x_2$, $y_3 = x_3$ is regular. Write the inverse transformation.

Ans. Inverse transformation does not exist

14. If λ is an eigen value of matrix A, then what is the eigen value of A^{-1}?

Ans. $\frac{1}{\lambda}$

15. Using Cayley-Hamilton Theorem find A^{10} if $A = \begin{bmatrix} 2 & 1 \\ 1 & -2 \end{bmatrix}$

Ans. $5^5 I$

16. Verify Cayley-Hamilton Theorem for $A = \begin{bmatrix} 3 & 2 \\ 0 & 5 \end{bmatrix}$

Application to Geometry

2.1 TANGENT TO A CURVE

Let P be a given point on a curve $y = f(x)$, and let Q be any other point on it. As Q tends to P, the straight line PQ tends, in general, to a definite straight line PT, which is called the *tangent* to the curve at P.

Let P be the point (x, y) and Q be $(x + h, y + k)$. Then

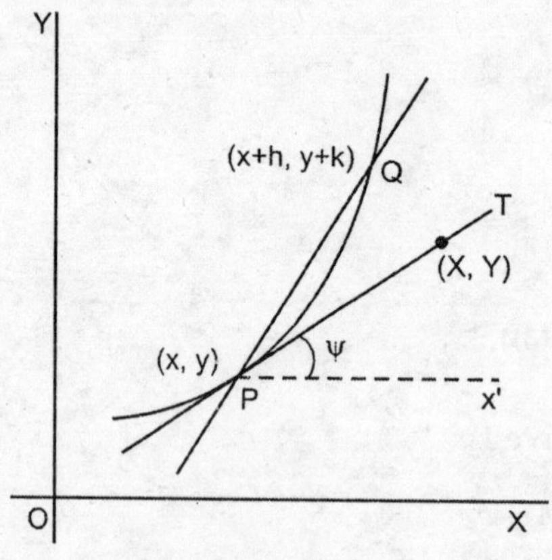

Fig. 2.1

$$\tan QPX' = \frac{(y+k)-y}{(x+h)-x} = \frac{f(x+h)-f(x)}{h}$$

As Q tends to P, PQ tends to the tangent PT. So $\angle QPX'$ tends to ψ, the angle PT makes with the x-axis. Hence

$$\tan \psi = \lim_{h \to 0} \frac{f(x+h)-f(x)}{h} = \frac{d}{dx} f(x).$$

If the current coordinates of a point on the tangent PT are (X, Y) the equation of the tangent is

$$Y - y = \tan \psi \, (X - x),$$

or

$$Y - y = \frac{dy}{dx} (X - x).$$

Example 1. Find the tangent at the point on the ellipse

$$x = a \cos t, \; y = b \sin t.$$

Here $$\frac{dy}{dx} = \frac{dy}{dt} \Big/ \frac{dx}{dt} = \frac{b\cos t}{-a\sin t}$$

Hence the tangent at t is

$$Y - b\sin t = -\frac{b\cos t}{a\sin t}(X - a\cos t),$$

i.e., $$aY\sin t + bX\cos t = ab(\cos^2 t + \sin^2 t) = ab.$$

Example 2. Find the angle of intersection of the curves

$$y = 4 - x^2 \quad \text{...(1)}$$

and $$y = x^2 \quad \text{...(2)}$$

Subtracting (2) from (1), we see that for the point of intersection

$$4 - 2x^2 = 0, \text{ or } x = \pm\sqrt{2}.$$

For the first curve

$$\frac{dy}{dx} = -2x = 2\sqrt{2} \text{ at } x = \sqrt{2};$$

and for the second curve.

$$\frac{dy}{dx} = -2x = 2\sqrt{2} \text{ at } x = \sqrt{2}.$$

If these make angles ψ_1 and ψ_2 with the x-axis, then the angle of intersection is $\psi_2 - \psi_1$, where

$$\tan(\psi_2 - \psi_1) = \frac{\tan\psi_2 - \tan\psi_1}{1 + \tan\psi_2\tan\psi_1}$$

$$= \frac{2\sqrt{2} - (-2\sqrt{2})}{1 + (2\sqrt{2})(-2\sqrt{2})} = \frac{4\sqrt{2}}{-7}.$$

So the angle between the two curves is $\tan^{-1}(4\sqrt{2}/7)$, *i.e.* 38.9°. [$\tan^{-1} 4\sqrt{2}/7$ will give the obtuse angle between the curves.]

The other intersection $x = -\sqrt{2}$, will give the same angle.

2.2 NORMAL

A normal to the curve $y = f(x)$ at (x, y) is the straight line through (x, y) perpendicular to the tangent there. Its equation will therefore be

$$Y - y = m(X - x),$$

where

$$m.\frac{dy}{dx} = -1.$$

Hence the equation of the normal at (x, y) is

$$\frac{dy}{dx}(Y - y) + X - x = 0$$

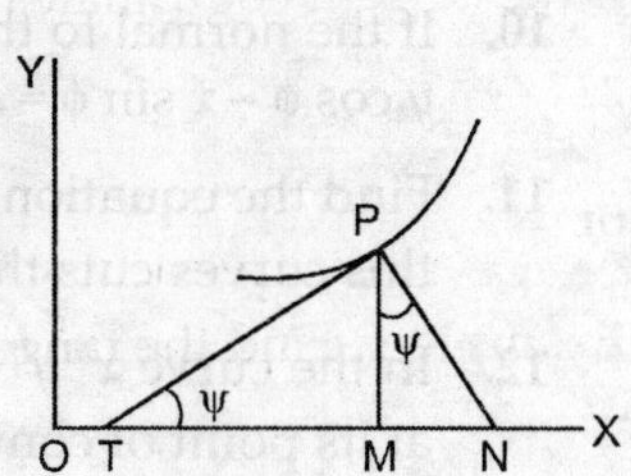

Fig. 2.2

Let the tangent and normal at a point P on the curve meet the x-axis in T and N respectively, and let PM be the ordinate. Then TM is called the *subtangent* and MN the *subnormal*. Also PT is called the *length of the tangent* and PN the *length of the normal*.

Let ψ be the angle PTM; then $\tan\psi = dy/dx$. Also $PM = y$. Therefore

Subtangent $= TM = y \cot \psi = y \Big/ \left(\dfrac{dy}{dx}\right)$;

subnormal $= MN = y \tan \psi = y \dfrac{dy}{dx}$.

Example: Find the equation of the normal at point t of the ellipse $x = a \cos t,\ y = b \sin t$.

Here $$\frac{dy}{dx} = \frac{dy}{dt} \Big/ \frac{dx}{dt} = \frac{b \cos t}{-a \sin t}$$

Hence the normal at t is

$$-\frac{b \cos t}{a \sin t}(Y - b \sin t) + (X - a \cos t) = 0,$$

or $a X \sin t - b Y \cos t = (a^2 - b^2) \sin t \cos t$

or $a X \sec t - b Y \operatorname{cosec} t = a^2 - b^2$.

Examples :

1. Find the equations of the tangent and normal at the point, (x, y) on each of the following curves:
   (*i*) $y^2 = 4ax$. (*ii*) $xy = a^2$.
   (*iii*) $x^m/a^m + y^m/b^m = 1$. (*iv*) $y = a \cosh (x/a)$.
2. Find the equation of the tangent and normal at the point t on each of the following curves:
   (*i*) $x = t^2 - a,\ y = t^3 - b$. (*ii*) $x = a(t + \sin t),\ y = a(1 - \cos t)$. [*Roorkee, Arch*, 1973]
   (*iii*) $x = a \sin^3 t,\ y = b \cos^3 t$.
3. Prove that $x/a + y/b = 1$ touches the curve $y = be^{-x/a}$ at the point where the curve crosses the axis of y.
4. Prove that the curve $(x/a)^n + (y/b)^n = 2$ touches the straight line $x/a + y/b = 2$ at the point (a, b) whatever be the value of n. [*Banaras* 1980]
5. Prove that all points of the curve $y^2 = 4a\left\{x + a \sin\left(\frac{x}{a}\right)\right\}$ at which the tangent is parallel to the axis of x lie on a parabola. [*Bhopal*, 1977]
6. Tangents are drawn from the origin to the curve $y = \sin x$. Prove the their points of contact lie on $x^2y^2 = x^2 - y^2$. [*Roorkee*, 71]
7. If $p = x \cos \alpha + y \sin \alpha$ touches the curve $\left(\frac{x}{a}\right)^{n/(n-1)} + \left(\frac{y}{b}\right)^{n/(n-1)} = 1$, prove that $p^n = (a \cos \alpha)^n + (b \sin \alpha)^n$ [Gorakhpur, 1972]
8. Find the angle of the intersection of the curves
   (*i*) $x^2 + y^2 + 4x = 16$ and $y^2 = 2x$; [*Roorkee, Arch*, 1971]
   (*ii*) $x^2 + 4ay$ and $2y^2 = ax$; [*Roorkee, Arch*, 1973]
   (*iii*) $xy = a^2$ and $x^2 + y^2 = 2a^2$.
9. Find the condition that the curves $ax^2 + by^2 = 1$ and $a_1x^2 + b_1y^2 = 1$ should intersect orthogonally. [*Bhopal*, 1977]
10. If the normal to the curve $x^{2/3} + y^{2/3} = a^{2/3}$ makes an angle ϕ with the axis of x, show that its equation is $y \cos \phi - x \sin \phi = a \cos 2\phi$.
11. Find the equation of the tangent at the point (am^2, am^3) on the curve $ay^2 = x^3$. The tangent at a point P on this curves cuts the curve again at Q, and is normal to it there, find the value of m at P. [*Roorkee*, 1967]
12. In the curve $x^m y^n = a^{m+n}$, prove that the portion of the tangent intercepted between the axes, is divided at its point of contact into segments which are in a constant ratio. [*Allahabad*, 70]
13. Find the equation of the tangent line to the curve $x^{2/3} + y^{2/3} = a^{2/3}$ at any point (x_1, y_1) and show that the length of the tangent line between the coordinate axes is constant. [*Roorkee, Arch*, 1970]

14. In the tractrix

$$x = a\left(\cos t + \log \tan \frac{1}{2} t\right),$$
$$y = a \sin t,$$

prove that the portion of the tangent intercepted between the curve and the axis of x is a constant length. [*Roorkee, Arch.*, 1960]

15. If the tangent to the curve $x^{1/2} + y^{1/2} = a^{1/2}$ at any point on it cuts the axes OX, OY at P, Q respectively, prove that $OP + OQ = a$. [*Roorkee*, 1978]

16. In the catenary $y = a \cosh (x/a)$, prove that the length of the portion of the normal intercepted between the curve and the axis of x is y^2/a. [*Bhopal, Arch.*, 1978]

17. Prove that in the ellipse $x^2/a^2 + y^2/b^2 = 1$, the length of the normal varies inversely as the perpendicular from the origin on the tangent. [*Allahabad*, 1969]

18. Show that for the curve $y = ae^{kx}$ the subtangent is constant, and the subnormal varies as the square of the ordinate. [*Gorakhpur*, 1972]

19. Show that the subtangent at any point of the curve $x^m y^n = a^{m+n}$ varies as the abscissa. [*Bhopal*, 1978]

2.3 POLAR COORDINATES

The position of a point P on a plane can also be indicated by stating (*i*) its distance r from a fixed point O, and (*ii*) the inclination θ of OP to a fixed straight line OA. r and θ are called the *polar coordinates* of P; r is called the *radius vector*, θ the vectorial angle, O the *pole*, and OA the *initial line*.

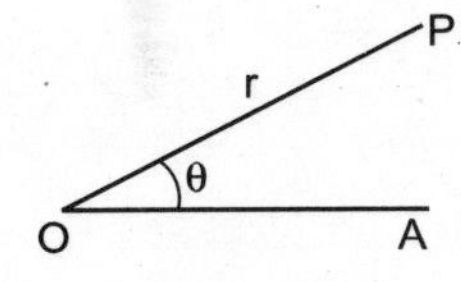

Fig. 2.3

r is considered to be positive when measured away from O along the line bounding the vectorial angle, and θ is considered to be positive when measured in the anti-clockwise direction.

If r and θ are given, there is only one point which will have the coordinates (r, θ). But if P be given, not only may the coordinates be stated as (r, θ), but also as $(r, \theta \pm 2\pi)$, $(r, \theta \pm 4\pi)$, ..., or as $(-r, \theta \pm \pi)$, $(-r, \theta \pm 3\pi)$, ... The student should remember this, otherwise he would commit mistakes.

When converting polar coordinates to Cartesians, or vice versa, it is customary to take the pole as the origin and the initial line as the x-axis. Then the formulae of conversion are

$$x = r \cos \theta,\ y = r \sin \theta.$$

An equation $r = f(\theta)$ will represent a curve in the polar coordinates. It is usual to regard θ as the independent variable.

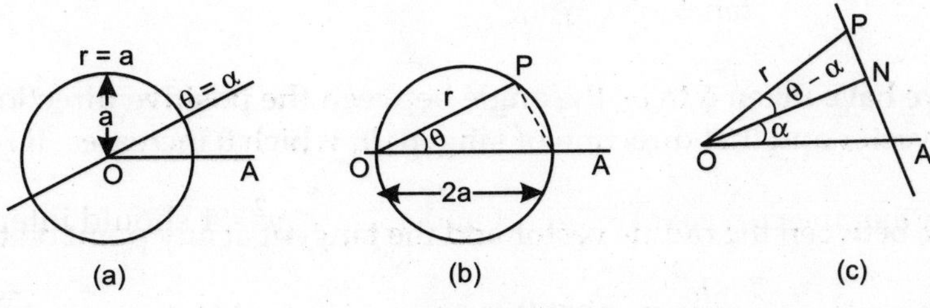

Fig. 2.4

In particular $r = a$ is a circle with O as centre and radius a, and $\theta = \alpha$ is a straight line through O making an angle a with OA (Fig. 2.4 a). It is easy to see from Fig. 2.4 b that $r = 2a \cos \theta$ is a circle of diameter $2a$. Similarly, $r \cos (\theta - \alpha) = p$ will be a straight line at distance ON $(= p)$ from origin, ON making an angle α with OA (Fig. 2.4 c).

2.4 ANGLE BETWEEN RADIUS VECTOR AND TANGENT

Let P be a given point on the curve $r = f(\theta)$ and let Q be any other point on the curve. Let the polar coordinates of P and Q be (r, θ) and $(r + k, \theta + h)$. Then the straight line PQ tends to the tangent line PT as $Q \to P$. Draw QM perpendicular to OP, and let ϕ be the angle TPM (between the tangent and radius vector).

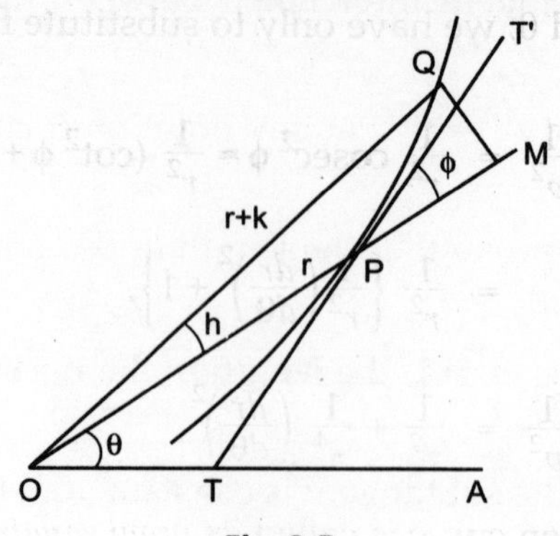

Fig. 2.5

Then

$$\tan\phi = \tan \lim_{h\to 0} \angle QPM = \lim_{h\to 0} \tan QPM.$$

$$= \lim_{h\to 0} \frac{QM}{PM} = \lim_{h\to 0} \frac{(r+k)\sin h}{(r+k)\cos h - r}$$

$$= \lim_{h\to 0} \frac{(r+k)\sin h}{k\cos h - r(1-\cos h)}$$

$$= \lim_{h\to 0} \frac{(r+k)\dfrac{\sin h}{h}}{\dfrac{k}{h}\cos h - r\dfrac{2\sin^2\frac{1}{2}h}{h}} = \frac{(r+0).1}{\dfrac{dr}{d\theta} - r.0}$$

Since $\qquad \lim_{h\to 0} \dfrac{\sin h}{h} = 1, \lim_{h\to 0} \dfrac{k}{h} = \dfrac{dr}{d\theta},$

and $\qquad \lim_{h\to 0} \dfrac{2\sin^2\frac{1}{2}h}{h} = \lim_{h\to 0} \dfrac{\sin\frac{1}{2}h}{\frac{1}{2}h} \sin\frac{1}{2}h = 1 \times 0 = 0.$

Therefore $\qquad \tan\phi = r\dfrac{d\theta}{dr}.$

In the above diagram we have taken ϕ to be the angle between the positive direction of the radius vector and that direction of the radius vector and that direction of tangent in which θ increases. If $r\, d\theta/dr$ is negative, it means that ϕ is obtuse.

Example. Find the angle between the radius vector and the tangent at any point on the cardioid $r = a(1 - \cos\theta)$.

Here $\qquad \dfrac{dr}{d\theta} = a\sin\theta,$

Therefore, $\qquad \tan\phi = r\dfrac{d\theta}{dr} = \dfrac{a(1-\cos\theta)}{a\sin\theta} = \dfrac{2\sin^2\frac{1}{2}\theta}{2\sin\frac{1}{2}\theta\cos\frac{1}{2}\theta}$

$$= \tan\frac{1}{2}\theta.$$

Hence $\qquad \phi = \dfrac{1}{2}\theta.$

3.2.2. Pedal Equation

If p be the length of the perpendicular OT from the pole O to the tangent at any point P on the curve, then, from the figure,

$$p = r \sin \phi.$$

Fig. 2.6

If we want the result in terms of r and θ, we have only to substitute for $\sin \phi$ from the equation $\tan \phi = r\, d\theta/dr$.

Thus
$$\frac{1}{p^2} = \frac{1}{r^2} \operatorname{cosec}^2 \phi = \frac{1}{r^2} (\cot^2 \phi + 1)$$

$$= \frac{1}{r^2} \left\{ \frac{1}{r^2} \left(\frac{dr}{d\theta} \right)^2 + 1 \right\},$$

or
$$\frac{1}{p^2} = \frac{1}{r^2} + \frac{1}{r^4} \left(\frac{dr}{d\theta} \right)^2 \qquad ...(1)$$

The relation between p and r for a given curve is called its *pedal equation*. The pedal equation for a curve $f(r, \theta) = 0$ can be obtained by eliminating θ between this equation and (1). For certain curve the pedal equation is very simple.

When the equation of the given curve is in Cartesian coordinates :

$$f(x, y) = 0 \qquad ...(2)$$

the tangent at (x, y) is

$$Y - y - \frac{dy}{dx}(X - x) = 0$$

Since p is the perpendicular on this from the origin ($X = 0$, $Y = 0$),

$$p = \frac{x\dfrac{dy}{dx} - y}{\sqrt{\left\{1 + \left(\dfrac{dy}{dx}\right)^2\right\}}} \qquad ...(3)$$

Also
$$r^2 = x^2 + y^2. \qquad ...(4)$$

Hence we get the pedal equation by eliminating x and y between the equations (2), (3) and (4).

Example. 1. Find the pedal equation of the cardioid $r = a(1 - \cos \theta)$. [*Allahabad*, 1970]

For the cardioid
$$\phi = \frac{1}{2}\theta$$

Therefore
$$p = r \sin \phi = a(1 - \cos \theta) \sin \frac{1}{2}\theta$$

$$= 2a \sin^3 \frac{1}{2}\theta.$$

Also,
$$r = a(1 - \cos \theta) = 2a \sin^2 \frac{1}{2}\theta$$

Therefore
$$r^3 = 2ap^2.$$

This is the required pedal equation.

Example 2. Find the pedal equation of the parabola $y^2 = 4a(x + a)$. [*Gorakhpur*, 1971]

Differentiating,
$$2y\frac{dy}{dx} = 4a, \text{ or } \frac{dy}{dx} = \frac{2a}{y}.$$

Hence the tangent at (x, y) is $Y - y = (2a/y)(X - x)$.

Therefore, $$p = \frac{x.2a/y - y}{\sqrt{(1 + 4a^2/y^2)}} = \frac{2ax - y^2}{\sqrt{(y^2 + 4a^2)}}$$

$$= \frac{2ax - 4a(x + a)}{\sqrt{(4ax + 4a^2 + 4a^2)}} = \frac{-2a(x + 2a)}{\sqrt{\{4a(x + 2a)\}}}$$

$$= -\sqrt{\{a(x + 2a)\}}.$$

Also $$r^2 = x^2 + y^2 = x^2 + 4a(x + a) = (x + 2a)^2.$$

Therefore, $$p^2 = ar,$$

which is the required pedal equation.

Examples :

1. Find the angle between the radius vector and the tangent at (r, θ), for the following curves:
   (i) $l/r = 1 + e \cos \theta$; (ii) $r^n = a^n \sec (n\theta + \alpha)$; (iii) $a\theta = (r^2 - a^2)^{1/2} - a \cos^{-1} (a/r)$.
2. Find the angle of intersection of the curves:
   (i) $r = \sin \theta + \cos \theta$, and $r = 2 \sin \theta$; [Roorkee, 1996]
   (ii) $r = 2 \sin \theta$ and $r = 2 \cos \theta$;
   (iii) $r^n = a^n \sin n\theta$ and $r^n = b^n \cos n\theta$. [Roorkee, 1973]
3. Show that the cardioids intersect at right angles. [Roorkee, 1979]

$$r = a(1 + \cos \theta) \text{ and } r = b(1 - \cos \theta).$$

4. Show that in the equiangular spiral $r = ae^{\theta \cot \alpha}$ the tangent is inclined at a constant angle to the radius vector.
5. In the parabola $r = 2a/(1 + \cos \theta)$, prove that
   (i) $\phi = \frac{1}{2} \pi - 2 \frac{1}{2} \theta$,
   (ii) the polar subtangent is $2a \operatorname{cosec} \theta$, and
   (iii) $p = a \sec \frac{1}{2} \theta$. [Roorkee, Arch., 1972]

[If the tangent and normal at a point P on the curve meet the line through O, perpendicular to OP, at T and N respectively, then OT is the *polar subtangent* and ON the *polar subnormal*.]

6. Prove that the length of the perpendicular from the pole on the tangent to the ellipse $l/r = 1 + e \cos \theta$ is given by

$$\frac{1}{p^2} = \frac{1}{l^2}\left(\frac{2l}{r} - 1 + e^2\right)$$ ...[Allahabad, 1970]

7. Find the pedal equation of the following curves
   (i) $r^2 = a^2 \cos 2\theta$; [Roorkee, 1996]
   (ii) $r^2 \cos 2\theta = a^2$;
   (iii) $r^m = a^m \cos m\theta$. [Roorkee, 1978]
8. Show that the pedal equation of the ellipse

$$\frac{x^2}{a^2} + \frac{y^2}{b^2} = 1 \text{ is } a^2 + b^2 - r^2 = \frac{a^2b^2}{p^2}.$$ [Roorkee, 1973]

2.5 DIFFERENTIAL COEFFICIENT OF LENGTH OF ARC

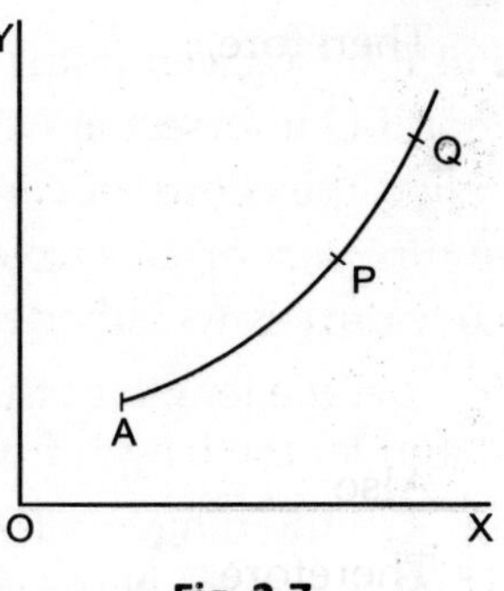

Fig. 2.7

Let the length of the arc AP of a curve, measured from a fixed point A on it be s. Then s is evidently some function of x.

To obtain ds/dx, take another point Q on the curve. Let the coordinates of P and Q be (x, y) and $(x + h, y + k)$. We shall assume as an axiom that

$$\lim_{h\to 0} \frac{\text{arc } PQ}{\text{chord } PQ} = 1$$

Then
$$\frac{ds}{dx} = \lim_{h\to 0} \frac{\text{arc } PQ}{h} = \lim_{h\to 0} \frac{\text{chord } PQ}{h}$$

$$= \lim_{h\to 0} \frac{\sqrt{(h^2+k^2)}}{h} = \lim_{h\to 0} \sqrt{\left(1+\frac{k^2}{h^2}\right)},$$

i.e.
$$\frac{ds}{dx} = \sqrt{\left\{1+\left(\frac{dy}{dx}\right)^2\right\}}.$$

When the equation of the curve is given in polar coordinates (as in the marginal figure),

$$PQ^2 = PM^2 + MQ^2.$$
$$= (r \sin h)^2 + (r + k - r\cos h)^2$$
$$= r^2 \sin^2 h + k^2 + 2rk(1-\cos h) + 2(1-\cos h)^2.$$

Fig. 2.8

Therefore,
$$\frac{ds}{d\theta} = \lim_{h\to 0} \frac{\text{arc } PQ}{h} = \lim_{h\to 0} \frac{\text{chord } PQ}{h}$$

$$= \lim_{h\to 0} \sqrt{\left\{r^2 \frac{\sin^2 h}{h^2} + \frac{k^2}{h^2} + 2r\frac{k}{h}\frac{2\sin^2\frac{1}{2}h}{h} + \frac{4r^2\sin^4\frac{1}{2}h}{h^2}\right\}}$$

$$= \sqrt{\left\{r^2 . 1 + \left(\frac{dr}{d\theta}\right)^2 + 2r . \left(\frac{dr}{d\theta}\right) . 0 + 0\right\}}.$$

Thus
$$\frac{ds}{d\theta} = \sqrt{\left\{r^2 + \left(\frac{dr}{d\theta}\right)^2\right\}}.$$

When the equation of the curve is given in the parametric form: $x = f(t)$, $y = g(t)$; then

$$\frac{ds}{dt} = \frac{ds}{dx}\cdot\frac{dx}{dt} = \sqrt{\left\{1+\left(\frac{dy}{dx}\right)^2\right\}}.\frac{dx}{dt}$$

Therefore
$$\frac{ds}{dt} = \sqrt{\left\{\left(\frac{dx}{dt}\right)^2 + \left(\frac{dy}{dt}\right)^2\right\}}$$

Example. Find ds/dt for the cycloid $x = a(1 - \cos t)$, $y = a(t + \sin t)$.

Here
$$\frac{dx}{dt} = a \sin t, \quad \frac{dy}{dt} = a(1 + \cos t).$$

Therefore
$$\frac{ds}{dt} = \sqrt{\{(a\sin t)^2 + a^2(1+\cos t)^2\}}$$

$$= a\sqrt{\{\sin^2 t + 1 + 2\cos t + \cos^2 t\}}$$

$$= a\sqrt{\{2+2\cos t\}} = a\sqrt{\left\{4\cos^2\frac{1}{2}t\right\}}$$

$$= 2a \cos\frac{1}{2}t.$$

2.6 Radius of Curvature

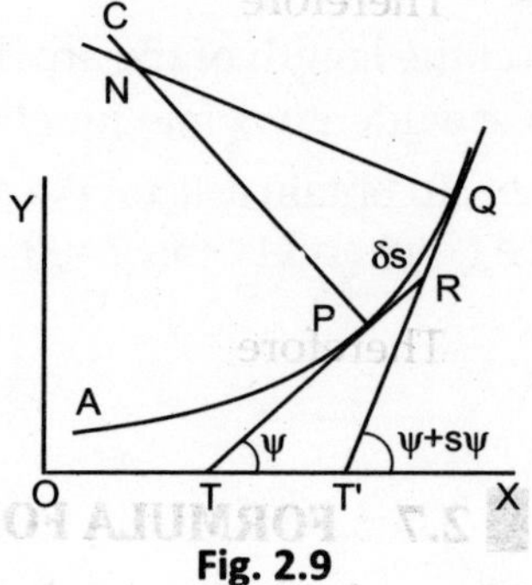

Fig. 2.9

Let P be a given point on a curve, and Q any other point on it. Let the normals at P and Q intersect at N. If N tends to a definite position C as Q tends to P, then C is called the centre of *curvature* of the curve at P. The distance CP is called the *radius of curvature* of the curve at P, and its reciprocal is called the *curvature*. The radius of curvature is usually denoted by the Greek letter ρ.

Let the length of the arc up of P, measured from a fixed point A on the curve, be s, and let the length from A up to Q be $s + \delta s$.

Let the tangents PT and QT to the curve make angle $\psi + \delta\psi$ with the axis of x. Then the angle between the normals PN and QN is equal to the angle $\delta\psi$ between the tangents PT and QT'. So from the triangle PQN,

$$\frac{PN}{PQ} = \frac{\sin \angle NQP}{\sin \delta\psi}.$$

Hence the radius of curvature at P

$$= \lim_{\delta s \to 0} PN = \lim_{\delta\psi \to 0} \frac{PQ \sin \angle NQP}{\sin \delta\psi}$$

$$= \lim_{\delta\psi \to 0} \frac{PQ}{\delta s} \cdot \frac{\delta s}{\delta\psi} \cdot \frac{\delta\psi}{\sin \delta\psi} \cdot \sin \angle NQP$$

$$= 1. \frac{ds}{d\psi}.1.1,$$

i.e.

$$\rho = \frac{ds}{d\psi}.$$

We thus get a formula for the radius of curvature of any curve for which a relation between s and ψ is given. Such a relation is called *intrinsic equation*.

When the Cartesian equation is given,

$$\frac{dy}{dx} = \tan \psi.$$

Therefore,

$$\frac{d^2y}{dx^2} = \sec^2 \psi . \frac{d\psi}{dx} = \sec^2 \psi . \frac{d\psi}{ds} \cdot \frac{ds}{dx}$$

$$= \left\{1 + \left(\frac{dy}{dx}\right)^2\right\} . \frac{1}{\rho} . \left\{1 + \left(\frac{dy}{dx}\right)^2\right\}^{1/2}.$$

Hence

$$\rho = \frac{\left\{1 + \left(\frac{dy}{dx}\right)^2\right\}^{3/2}}{\frac{d^2y}{dx^2}}.$$

***Example* 1.** Find ρ for the catenary whose intrinsic equation is

$$s = c \tan \psi.$$

We have

$$\rho = ds/d\psi = c \sec^2 \psi.$$

***Example* 2.** Find the radius of curvature at the point t on the curve $x = a \cos t$, $y = b \sin t$.

Here

$$\frac{dx}{dt} = -a \sin t, \frac{dy}{dt} = b \cos t.$$

Hence

$$\frac{dy}{dx} = -\frac{b}{a} \cot t.$$

Therefore $$\frac{d^2y}{dx^2} = \frac{d}{dx}\left(\frac{dy}{dx}\right) = \frac{d}{dt}\left(\frac{dy}{dx}\right) \cdot \frac{dt}{dx} = \frac{d}{dt}\left(-\frac{b}{a}\cot t\right) \cdot \frac{dt}{dx}$$

$$= +\frac{b}{a}\operatorname{cosec}^2 t \cdot \left(-\frac{1}{a}\right) \operatorname{cosec} t.$$

Therefore $$\rho = \frac{\{1+(b/a)^2\cot^2 t\}^{3/2}}{(b/a^2)\operatorname{cosec}^3 t} = \frac{(a^2\sin^2 t + b^2\cos^2 t)^{3/2}}{ab}$$

2.7 FORMULA FOR POLAR EQUATIONS

We see from the marginal figure that

$$\psi = \theta + \phi.$$

Therefore $$\frac{d\psi}{d\theta} = 1 + \frac{d\phi}{d\theta}. \quad \text{...(1)}$$

Fig. 2.10

But $$\phi = \tan^{-1}\left(r\frac{d\theta}{dr}\right)$$

$$= \tan^{-1}\frac{r}{r'}.$$

Therefore $$\frac{d\phi}{d\theta} = \frac{1}{1+(r/r')^2} \cdot \frac{r'.r' - rr''}{(r')^2} = \frac{(r')^2 - rr''}{r^2+(r')^2}. \quad \text{...(2)}$$

From (1) and (2), we have

$$\frac{d\psi}{d\theta} = \frac{r^2 + 2(r')^2 - rr''}{r^2+(r')^2}.$$

Also $$\frac{ds}{d\theta} = \sqrt{\{r^2+(r')^2\}}.$$

Hence $$\rho = \frac{ds}{d\psi} = \frac{\{r^2+(r')^2\}^{3/2}}{r^2+2(r')^2 - rr''}.$$

In this formula $r' = dr/d\theta$ and $r'' = d^2r/d\theta^2$.

Another formula (for pedal equations) is

$$\rho = r\frac{dr}{dp}.$$

***Example* 1.** Find the radius of curvature for the cardioid $r = a(1 - \cos\theta)$.

Here $$\frac{dr}{d\theta} = a\sin\theta \text{ and } \frac{d^2r}{d\theta^2} = a\cos\theta.$$

Therefore, $$\rho = \frac{\{a^2(1-\cos\theta)^2 + (a\sin\theta)^2\}^{3/2}}{a^2(1-\cos\theta)^2 + 2(a\sin\theta)^2 - a(1-\cos\theta)a\cos\theta}$$

$$= a\frac{\{2-2\cos\theta\}^{3/2}}{3-3\cos\theta} = \frac{2}{3}a\{2-2\cos\theta\}^{1/2}$$

$$= \frac{4}{3}a\sin\frac{1}{2}\theta.$$

***Example* 2.** Prove the formula $$\rho = r\frac{dr}{dp}.$$

Since $p = r\sin\phi$, therefore

$$\frac{dp}{dr} = \sin\phi + r\cos\phi\,\frac{d\phi}{dr}$$

$$= \frac{r + r\dfrac{dr}{d\theta}\cdot\dfrac{d\phi}{dr}}{\sqrt{\left\{r^2 + \left(\dfrac{dr}{d\theta}\right)^2\right\}}}$$

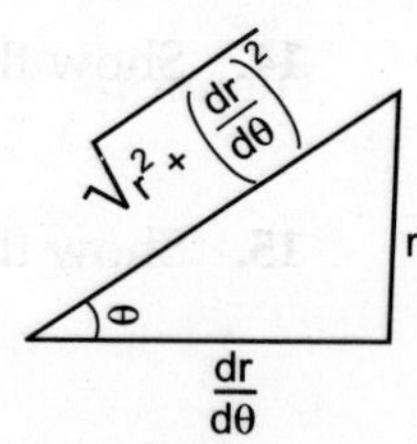

Fig. 2.11

since $\tan\phi = r/\dfrac{dr}{d\theta}$ (see fig.),

$$= \frac{r(1 + d\phi/d\theta)}{ds/d\theta} = \frac{r\,d\psi/d\theta}{ds/d\theta}, \text{ by (1),}$$

$$= r\frac{d\psi}{ds} = \frac{r}{\rho}.$$

Hence $\rho = r\,dr/dp.$

Examples :

1. Calculate ds/dx for the following curves:
   (i) $y^2 = 4ax$; (ii) $y = \log\cos x$; (iii) $y = a\cosh(x/a)$.
2. Calculate: ds/dt for the following curves:
   (i) $x = t^2, y = t - 1$; (ii) $x = a\sec t, y = b\tan t$; (iii) $x = a\cos t, y = b\sin t$.
3. Calculate $ds/d\theta$ for the following curves:
   (i) $r = a(1 - \cos\theta)$; (ii) $r = 2a/(1 + \cos\theta)$
4. Prove that for any curve $\dfrac{ds}{d\theta} = \dfrac{r^2}{p}$. [*Allahabad*, 1969]
5. Find the radius of curvature at the point (s, ψ) on the following curves:
   (i) $s = 4a\sin\psi$ (cycloid); (ii) $s = c\log\sec y$ (tractrix).
6. Find the radius of curvature at the point (x, y) on the following curves:
   (i) $y^2 = 4ax$; (ii) $xy = c^2$; (iii) $y = a\cosh(x/a)$.
7. Show that the curvature at the point $\left(\frac{3}{2}a, \frac{3}{2}a\right)$ on the Folium $x^3 + y^3 = 3axy$ is $-8\sqrt{2}/3a$. [*Roorkee*, 1978]
8. Prove that for the ellipse $x^2/a^2 + y^2/b^2 = 1$,
   $$\rho = a^2b^2/p^3,$$
   a being the perpendicular from the centre upon the tangent at (x, y). [Banaras, 1978]
9. Show that the radius of curvature at a point $(a\cos^3\theta, a\sin^3\theta)$ on the curve $x^{2/3} + y^{2/3} = a^{2/3}$ is $3a\sin\theta\cos\theta$.
10. Find the radius of curvature at any point of the curve
    $$x = a(t - \sin t),\ y = a(1 - \cos t)$$ [*Roorkee*, 1973]
11. If CP, CD be a pair of conjugate semi-diameters of an ellipse, prove that the radius of curvature at P is CD^3/ab, a and b being the lengths of the semi-axes of the ellipse. [*Roorkee*, 1971]
12. If ρ, ρ' be the radii of curvature at the extremities of two conjugate diameters of an ellipse, prove that $(\rho^{2/3} + \rho'^{2/3})(ab)^{2/3} = a^2 + b^2$. [*Allahabad*, 1969]
13. Find the radius of curvature at the point (r, θ) on the following curves:
    (i) $r = a(2\cos\theta - 1)$; (ii) $r(1 + \cos\theta) = a$; (iii) $r^n = a^n\sin n\theta$. [Gorakhpur, 1996]

14. Show that in the rectangular hyperbola $r^2 \cos 2\theta = a^2$,

$$\rho = r^3 / a^2.$$

15. Show that for the cardioid $r = a(1 + \cos\theta)$,

$$\rho = (4a/3) \cos \frac{1}{2}\theta.$$

16. Show that for the curve $r^n = a^n \cos n\theta$, the radius of curvature is $a^n r^{-n+1}/(n+1)$.

17. Show that at any point on the equiangular spiral $r = ae^{\theta \cot \alpha}$, $\rho = r \operatorname{cosec} \alpha$, and that it subtends a right angle at the pole. [Bhopal, 1977]

18. Find the radius of curvature at the point (p, r) on the following curves:

(i) $r^3 = 2ap^2$ (cardioid); (ii) $r^3 = a^2p$ (lemniscate); (iii) $pr = a^2$ (hyperbola).

2.8 CURVE TRACING

The object of curve tracing is to find the approximate shape of a curve without the labour of plotting a large number of points.

Some of the steps which shorten the labour are given below in § 3.7. They are applicable to Cartesian equations which can be solved for y. If an equation can be solved for x (and not for y), the same procedure is applicable with x and y interchanged. In some cases the equation cannot be solved for x or y, but a conversion to polar coordinates gives r in terms of θ. Such equations can then be traced by the method of § 2.13.

For tracing a curve properly it is often of importance to determine the nature of the curve near the origin and for large values of x (or y). We discuss this in the next two sections.

2.9 NATURE OF CURVE NEAR ORIGIN

Let the equation of a curve (arranged in ascending powers) be

$$a_0 + a_1x + a_2y + b_1x^2 + b_2xy + b_3y^2 + c_1x^3 + \ldots = 0. \qquad \ldots(1)$$

The curve passes through the origin if $a_0 = 0$.

Let $a_0 = 0$; then near the origin, since x and y are small, curve (1) and the straight line.

$$a_1x + a_2y = 0. \qquad \ldots(2)$$

are nearly coincident. So (2) must be the tangent at the origin at the curve (1).

If $a_0 = a_1 = a_2 = 0$; then the lowest degree terms in (1) are

$$b_1x^2 + b_2xy + b_3y^2 = 0. \qquad \ldots(3)$$

For small x and y the curves (1) and (3) nearly coincide. But (3) represents a pair of straight lines. These must therefore be tangential to the curve (1).

This shows that *the lowest degree in terms in* (1) *equated to zero give the equation of the tangent (or tangents) at the origin.*

If the lowest degree terms are of second degree, then the origin is a *double point* since two branches of the curve pass through it. If the two branches cut at an angle, the double point is called a *node.* If these branches touch, the point is called a *cusp.* The figures below show double cusps of the first and second kinds.

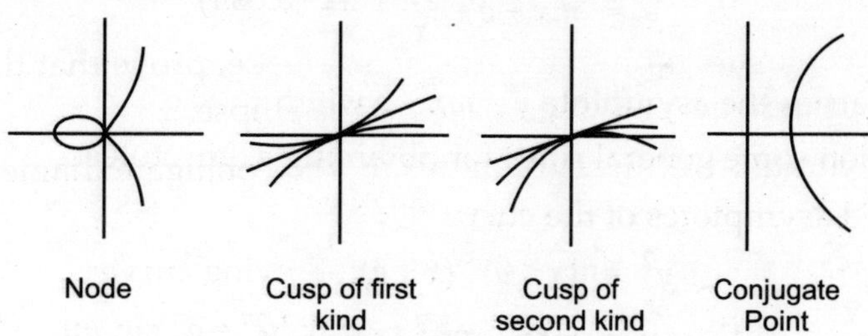

Fig. 2.12

If the curve lies only on one side of the normal, we get a single cusp. A third case arises when there are no real points on the curve in the neighbourhood of the double point. This is called a *conjugate point* (or an *isolated point*).

The nature of a double point can be determined by solving for y and considering its values for small x. The nature of the double point is also indicated by the equation of the tangents there. The tangents at a node are real and distinct and at a cusp coincident. The tangents at a conjugate point are imaginary, or sometimes real and coincident.

Example. Show that the curve $y^3 = (x - a)^2(2x - a)$ has a single cusp of the first kind at the point $(a, 0)$.

O O' (a, o)

Fig. 2.13

Transferring the origin to the point $(a, 0)$, the equation of the curve becomes

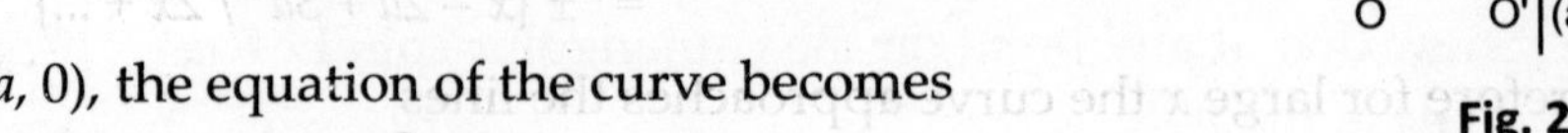

$$y^3 = (x + a - a)^2 (2x + 2a - a),$$

or

$$y^3 = x^2(2x + a) \quad ...(1)$$

The tangents at the new origin are $x^2 = 0$. For small positive x, equation (1) gives a positive y. For small negative x also, (1) gives a positive y. So the curve near $(a, 0)$ is as shown. This is a single cusp of the first kind.

2.10 ASYMPTOTES

Consider the curve given by the equation

$$y = \frac{x+1}{x-1}. \quad ...(1)$$

When $x = 5$, $y = 1.5$, when $x = 11$, $y = 1.2$, when $x = 21$, $y = 1.1$ In fact writing (1) as

$$y = 1 + \frac{2}{x-1},$$

we see that $y \to 1$ as $x \to \infty$. Thus the curve (1) approaches the straight line $y = 1$ when $x \to \infty$. This line is called an *asymptote* of the curve.

Again, when $x \to 1$ from the right, y tends to infinity. Evidently, the curve approaches the straight line $x = 1$, as $y \to \infty$. So $x = 1$ is another asymptote of curve (1). The graph of curve (1) and its asymptotes is shown in the figure below.

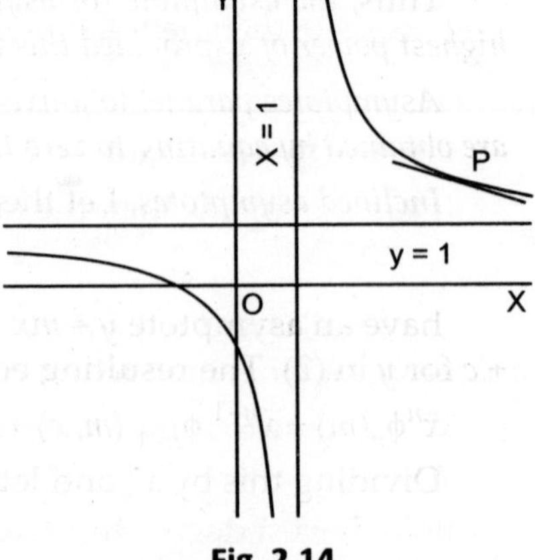

Fig. 2.14

Thus an asymptote is a straight line, the distance between which and a point on the curve tends to zero as the point moves farther and farther away from the origin. It is clear from the figure above that if we draw a tangent to the curve at a point P, this tangent will tend to the asymptote as P moves away from the origin.

There are no real asymptotes for closed curves like ellipse or circle. The asymptotes exist only for curves with infinite branches; and even such curves may not always have an asymptote. The asymptote may be parallel to the axes (as in the above example), or may be inclined. A curve of degree n will have at most n asymptotes.

Asymptotes parallel to the axes can often be found by inspection. An inclined asymptote can be easily found if the equation of the curve can be solved for y in the form

$$y = mx + c + \frac{A}{x} + \frac{B}{x^2} + ... \quad ...(2)$$

Evidently curve (2) approaches the asymptote $y = mx + c$ when $x \to \infty$.

We derive in the next section some general rules for obtaining asymptotes.

Example. Find the inclined asymptotes of the curve

$$y^2 = (x - a)^3/(x + a).$$

Since

$$y = \pm (x - a)^{3/2} (x + a)^{-1/2}$$

$$= \pm x(1 - a/x)^{3/2}(1 + a/x)^{-1/2}$$

$$= \pm x\left(1 - \frac{3}{2}\cdot\frac{a}{x} + \frac{\frac{3}{2}\cdot\frac{1}{2}}{1.2}\cdot\frac{a^2}{x^2} - \ldots\right)\left(1 - \frac{1}{2}\cdot\frac{a}{x} + \frac{\frac{1}{2}\cdot\frac{3}{2}}{1.2}\frac{a^2}{x^2} + \ldots\right)$$

$$= \pm x\left\{1 - 2\frac{a}{x} + \left(\frac{3}{8} + \frac{3}{8} + \frac{3}{4}\right)\frac{a^2}{x^2} + \ldots\right\}$$

$$= \pm\{x - 2a + 3a^2 / 2x + \ldots\}$$

therefore for large x the curve approaches the lines

$$y = \pm(x - 2a).$$

Hence the inclined asymptotes are

$$y - x + 2a = 0 \text{ and } y + x - 2a = 0.$$

2.11 RULES FOR FINDING ASYMPTOTES

Asymptotes parallel to x-axis. Let the equation of the curve arranged according to powers of x be

$$a_0x^n + (a_1y + b_0)x^{n-1} + (a_2y^2 + b_1y + c_0)x^{n-2} + \ldots = 0. \quad \ldots(1)$$

If the curve has an asymptote $y = c$ parallel to x-axis, then $y \to c$ as $x \to \infty$. Hence (1) gives, on dividing by x^n and letting $x \to \infty$,

$$a_0 = 0.$$

Taking $a_0 = 0$, dividing (1) by x^{n-1} and letting $x \to \infty$, we obtain

$$a_1y + b_0 = 0.$$

This is the asymptote of curve (1) parallel to the x-axis.

If $a_0 = a_1 = 0$, a similar argument shows that $b_0 = 0$ for the existence of an asymptote parallel to x-axis. The asymptotes in this case are

$$a_2y^2 + b_1y + c_0 = 0.$$

Thus, *the asymptote (or asymptotes) parallel to the axis of x can be obtained by equating to zero the coefficient of the highest power of x, provided this is not merely a constant.*

Asymptotes parallel to y-axis. Interchanging the axes of x and y, we see that the *asymptotes parallel to the axis of y are obtained by equating to zero the coefficient of the highest power of y, provided this is not merely a constant.*

Inclined asymptotes. Let the curve

$$f(x, y) = 0 \quad \ldots(2)$$

have an asymptote $y = mx + c$. Then y, as given by (2), will tend to $mx + c$ when $x \to \infty$. Substitute therefore $mx + c$ for y in (2). The resulting equation $f(m, mx + c) = 0$ will be of the form

$$x^n\phi_n(m) + x^{n-1}\,\phi_{n-1}(m, c) + x^{n-2}\,\phi_{n-2}(m, c) + \ldots = 0. \quad \ldots(3)$$

Dividing this by x^n and letting $x \to \infty$, we get

$$\phi_n(m) = 0 \quad \ldots(4)$$

This equation gives the values of m for the asymptotes.

If m has one of these values then (3) becomes

$$x^{n-1}\,\phi_{n-1}(m, c) + x^{n-2}\,\phi_{n-2}(m, c) + \ldots = 0.$$

Dividing this by x^{n-1} and letting $x \to \infty$, we obtain

$$\phi_{n-1}(m, c) = 0, \quad \ldots(5)$$

This gives the values of c for the asymptotes.

When equation (4) has repeated roots, it may happen that (5) is identically satisfied. In such cases dividing (3) by x^{n-2} and letting $x \to \infty$, we get

$$\phi_{n-2}(m, c) = 0$$

which will give the values of c. If this also is identically zero, $\phi_{n-3}(m, c) = 0$ will give c. Hence we have the following rule for determining inclined asymptotes.

Substitute $mx + c$ for y in the equation of the curve and equate to zero the coefficients of the two highest powers of x. Determine m and c from these. If $m_1, c_1; m_2, c_2;$ are the values of m and c thus obtained, the asymptotes are

$$y = m_1x + c_1; y = m_2x + c_2; \ldots$$

If any value of m derived from the first equation makes the second identically zero, then the corresponding values of c should be determined from the equation obtained by equating to zero the coefficient of the next highest power of x.

***Example* 1.** Find the asymptotes of the curve

$$x^4 + x^2y^2 - a^2(a^2 + y^2) = 0$$

which are parallel to the axes.

The highest power of x is x^4, whose coefficient is a constant. The coefficient of the highest power of y (viz. y^2) equated to zero gives $x^2 - a^2 = 0$. Hence the required asymptotes are

$$x = a, x = -a.$$

***Example* 2.** Find the asymptotes of $y^3 + x^2y + 2xy^2 - y + 1 = 0$.

Putting $y = mx + c$, we have

$(mx + c)^3 + x^2(mx + c) + 2x(mx + c)^2 - (mx + c) + 1 = 0,$

or $x^3(m^3 + m + 2m^2) + x^2(3m^2c + c + 4mc) + x(3mc^2 + 2c^2 - m) + \ldots = 0.$

Equating to zero the coefficients of x^3 and x^2, we get

$$m^3 + 2m^2 + m = 0, \qquad \ldots(1)$$

and

$$(3m^2 + 4m + 1)c = 0 \qquad \ldots(2)$$

The first equation gives $m(m + 1)^2 = 0$, i.e. $m = 0, -1, -1$.

The second equation then gives $c = 0$ when $m = 0$. Hence one asymptote is $y = 0$. (This could also be obtained by the rule for asymptotes parallel to axes.)

When $m = -1$, equation (2) becomes $0. c = 0$ and fails to give c. Equating to zero the coefficient of next highest power, we have

$$(3m + 2)c^2 - m = 0.$$

For $m = -1$ this gives $c = \pm 1$. Hence the other two asymptotes are

$$y = -x + 1 \text{ and } y = -x - 1.$$

Hence the 3 asymptotes of the given curve are

$$y = 0, y + x - 1 = 0, y + x + 1 = 0.$$

***Examples* :**

1. Write down the equations to the tangents at the origin for the following curves:
   (i) $x^4 + 3x^3y + 2xy - y^2 = 0;$ *(ii)* $x^3 + 3xy + 7x^2 = 0.$
2. Find the nature of the origin on the following curves:
   (i) $a^4y^2 = x^4(x^2 - a^2);$ *(ii)* $y^2(a^2 + x^2) = x^2(a^2 - x^2).$
3. Find the asymptotes of the following curves which are parallel to the axes:
   (i) $x^3 + 3xy^2 + y^2 + 2x + y = 0;$ *(ii)* $y^4 + x^2y^2 + 2xy^2 - 4x^2 - y + 1 = 0;$
   (iii) $x^2y^3 + x^3y^2 = x^3 + y^3.$
4. Obtain the inclined asymptotes of $y^3 = x^3 + ax^2$, by expanding y in descending powers of x.
5. Show that the parabola $y^2 = 4ax$ has no asymptote. *[Roorkee, 1973]*

   Find the asymptotes of the following curves:
6. $x^3 + y^3 = a^3.$ *[Roorkee, Arch, 1978]*
7. $x^3 + 2x^2y - xy^2 - 2y^3 + 3xy + 3y^2 + x + 1 = 0.$ *[Gorakhpur, 1971]*

8. $2x^3 + 3x^2y - 3xy^2 - 2y^3 + 3x^2 - 3y^2 + y = 3.$

9. $x^3 + x^2y - xy^2 - y^3 + x^2 - y^2 = 2.$

10. $x^3 + 3x^2y - 4y^3 - x + y + 3 = 0.$ [*A.M.I.E.*, 1977]

11. $y^2(x^2 - a^2) = x.$

12. $y^2(x - b) = x^3 + a^3$

13. $(x^2 - a^2)y^2 = x^2(x^2 - 4a^2).$

14. $x(x^2 - y^2) + x^2 + y^2 + x + y = 0.$ [*Aligarh*, 1980]

2.12 PROCEDURE FOR TRACING CARTESIAN EQUATIONS

The following procedure should be adopted for tracing a curve whose cartesian equation $f(x, y) = 0$ is given.

1. **Symmetry:** Notice if the curve is symmetrical, by applying the following rules:

   (*i*) If the powers of y which occur in the equation are all even, the curve is symmetrical about the axis of x.

   (*ii*) If the powers of x are all even, the curve is symmetrical about the axis of y. A curve might, of course, be symmetrical about both axes.

   (*iii*) If x and y can be interchanged without altering the equation, the curve is symmetrical about the line $y = x$.

   (*iv*) If on changing the signs of x and y both, the equation to the curve is not altered, the curve after being turned through two right angles will coincide with its old trace. (Symmetry in opposite quadrants.)

2. **Origin:** Notice if the curve passes through the origin. If it does, write down the equation of the tangent, or tangents, there. If the origin is a double point, find its nature.

3. **Solve for y:** Solve for y and consider how y will vary when x increases from 0 (or some other convenient value x_1) and then tends to infinity. Pay special attention to those values of x for which $y = 0$, or $\to \infty$.

   If on solving for y we get more than one value, consider each value separately. However, in case about the x-axis, only positive values of y need be considered. The curve for negative values of y can be drawn from symmetry.

4. **Consider all values of x:** Starting from $x = 0$ (or x_1), repeat the above procedure as x decreases and then tends to $-\infty$.

   Of course, if the curve is symmetrical about the y-axis (or in opposite quadrants), it can be drawn for negative values of x by symmetry; so such values of x need not be considered afresh.

5. **Imaginary values of y:** In the above procedure if y is found to be imaginary for a certain range of values of x, say for values of x between a and b, it would mean that the curve does not exist in the region bounded by the lines $x = a$, $x = b$.

6. **Asymptotes:** The procedure of paras 3 and 4 will indicate the asymptotes parallel to the axes, and the position of the curve with respect to them.

   Expansion of y in descending powers of x will generally give the oblique asymptotes. If necessary they may be found separately.

7. **Special Points:** Find the coordinates of a few points on the curve if it appears necessary.

   For example, if $y = 0$ at $x = a$ and again at $x = b$, it might be desirable to find the point where y is maximum or minimum (i.e. $dy/dx = 0$). Alternatively, we can find y at $x = \frac{1}{2}(a + b)$ say. A few representative points will help in locating the curve.

The student should realise that having solved for y, he can find points on the cuve as close to each other as he likes. His task is to save time by picking out the most important values of x and noticing how y varies for the intermediate values.

***Example* 1.** Trace the curve $y = \dfrac{x^2}{1-x^2}$. [*Roorkee*, 1978]

(*i*) This curve is symmetrical about the axis of *y*.

(*ii*) The curve $[y(1-x^2) = x^2]$ passes through the origin. The tangent there is $y = 0$ (i.e. *x*-axis).

(*iii*) The given equation is already solved for *y*. When $x = 0$, $y = 0$. When *x* increases *y* increases (because numerator increases and denominator decreases). When $x \to 1$ from left $y \to \infty$.

When $x \to 1$ from right, $y \to \infty$. Writing *y* as $y = -1 - \dfrac{1}{x^2-1}$, we see that

for $x > 1$, *y* is negative and numerically decreases when *x* increases. When

$x \to \infty, y \to -1$.

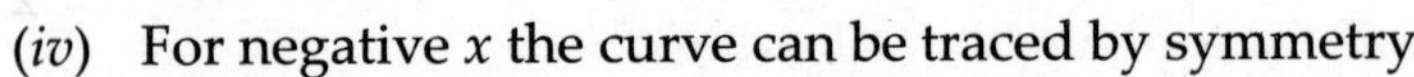

(*iv*) For negative *x* the curve can be traced by symmetry.

(*v*) The asymptotes are $x = \pm 1$ and $y = 1$.

(*vi*) To trace the curve more minutely, we find the following points on it.

x:	0.5	0.8	1.5	2.0
y:	0.33	1.78	– 1.8	– 1.33

Hence the curve is as shown in the figure.

Fig. 2.15

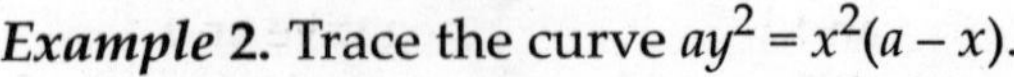

***Example* 2.** Trace the curve $ay^2 = x^2(a-x)$. ...(1) [*Gorakhpur*, 1978]

(*i*) The curve is symmetrical about the axis of *x*.

(*ii*) The curve passes through the origin. The tangents are given by $ay^2 = ax^2$, *i.e.* $y = \pm x$. Hence the origin is a node.

(*iii*) Solving for *y* and considering only the positive value,

$$y = x\frac{\sqrt{(a-x)}}{\sqrt{a}}$$

When $x = 0$, $y = 0$; *y* is zero again when $x = a$. When $x > a$, *y* is imaginary. ...(2)

Differentiating (1), we get

$$\frac{dy}{dx} = \frac{2ax - 3x^2}{2ay} \quad ...(3)$$

So $dy/dx = 0$ when $x = \dfrac{2}{3}a$. The maximum *y* is $2a/3\sqrt{3}$ (= 0.4*a* roughly).

Thus the curve starts from the origin, touching the line $y = x$, reaches a maximum at $x = \dfrac{2}{3}a$, and then comes down to $(a, 0)$. By (3), the tangent is vertical there. So the curve from $x = 0$ to $x = a$ is as shown by *OAB*. The curve does not exist for $x > a$, since *y* is imaginary there.

(*iv*) When *x* is negative, *y* as given by (2) is real and negative. When $x = -a$, $y = -a\sqrt{2}$. When *x* decreases, *y* decreases. When $x \to -\infty$, $y \to -\infty$. This part of the curve is shown by *OC*.

(*v*) For large *x*, y^2 varies as $-x^3$. So there is no linear asymptote. Taking symmetry into account the complete curve is as shown in the figure.

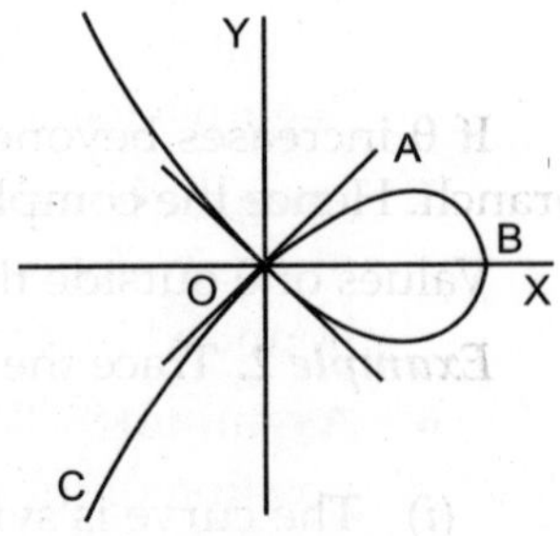

Fig. 2.16

2.13 POLAR EQUATIONS

1. **Solve for *r*.** Solve the equation for *r* and consider how *r* varies as θ increases from 0 (or some convenient value of θ_1) to $+\infty$, and also as θ diminishes from 0 (or θ_1) to $-\infty$. This is sufficient to trace the curve. If necessary, or convenient, *form a table of corresponding values of* θ *and r*.
2. **Limited Values Needed:** In most of the polar equations only periodic function (sin θ, cos θ, etc.) occur, and so values of θ from 0 to 2π (or sometimes some multiple or submultiple of 2π) need alone be considered. The remaining values of θ give no new branches of the curve.

3. **Symmetry:** If changing the sign of θ does not change the value of r, the curve is symmetrical about the initial line.

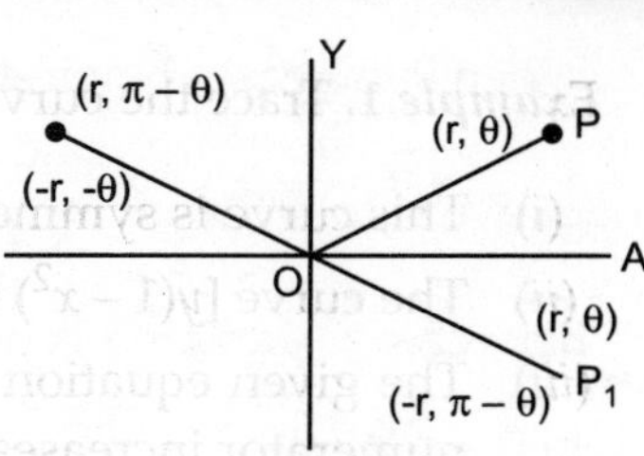

Fig. 2.17

Sometimes this test fails, then we can try the following alternative test: if changing θ to π – θ and r to – r does not alter the given equation, the curve is symmetrical about the initial line. The reasons are clear from the marginal diagram (on the next page), where the coordinates of a point P and its point of symmetry P_1 are written down.

Similarly, the rule for symmetry about the line OY ($\theta = \frac{1}{2}\pi$) is either of the following:

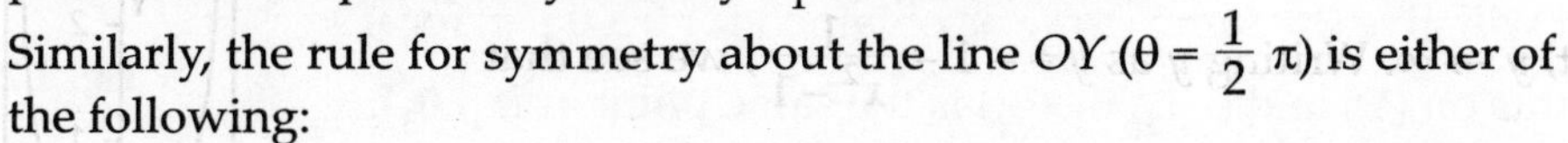

(*i*) If changing r to – r and θ to – θ does not alter the given equation; or

(*ii*) If changing θ to π – θ does not after the value of r, the curve is symmetrical about the line $\theta = \frac{1}{2}\pi$.

Example **1.** Trace the curve $r = \sin 3\theta$.

The following table gives corresponding values of θ and r/a.

$3\theta = 0$	Inter mediate values	$\frac{1}{2}\pi$	Inter-mediate values	π	Inter-mediate values	$\frac{3}{2}\pi$	Inter-mediate values	2π
$\theta = 0$	"	$\pi/6$	"	$2\pi/6$	"	$3\pi/6$	"	$4\pi/6$
$r/a = 0$	Positive and increasing	1	Positive and decreasing	0	negative and numerically increasing	–1	negative and numerically decreaseing	0

It is evident that the greatest positive value of r is a. Hence the curve between θ = 0 and θ = 120° is as shown in Fig. 2.18. As θ increases further, r/a goes through the same cycle of changes as before, being a maximum at $3\theta = 2\pi + \frac{1}{2}\pi$, i.e., θ = 150°.

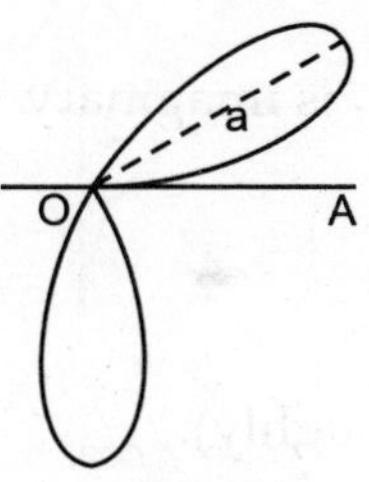

Fig. 2.18

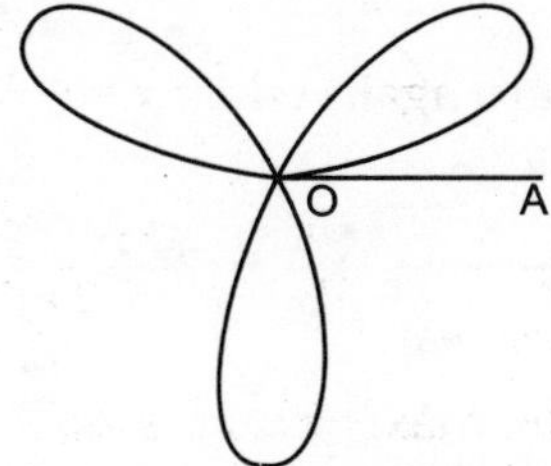

Fig. 2.19

If θ increases beyond π up to 2π, the same branches of the curve are repeated and we do not get any new branch. Hence the complete curve is as shown is Figure 2.19.

Values of θ outside the range (0, 2π) need not be considered, as the value of r/a is periodic.

Example **2.** Trace the curve

$$x^3 + y^3 - 3axy = 0.$$

(*i*) The curve is symmetrical about the line $y = x$.

(*ii*) The curve passes through the origin. The tangents there are $3axy = 0$, i.e. $x = 0$ and $y = 0$. So the origin is a node.

(*iii*) The given equation cannot be solved for y or x. Transforming it to polars, by putting $x = r\cos\theta$ and $y = r\sin\theta$, we get

$$r = \frac{3a\cos\theta\sin\theta}{\cos^3\theta + \sin^3\theta} \qquad ...(1)$$

When θ = 0, r = 0; r is again 0 when $\theta = \frac{1}{2}\pi$. Between these values r is positive. This gives a loop in the first quadrant. To trace it more accurately, notice that $r = 3a\sqrt{2}$ when θ = π/4.

When θ increases beyond $\frac{1}{2}\pi$ (but is less than $\frac{3}{4}\pi$), r increases numerically but has a negative sign. When $\theta \to \frac{3}{4}\pi$, $r \to -\infty$.

Between $\frac{3}{4}\pi$ and π, r is positive $r \to \infty$ when $\theta \to \frac{3}{4}\pi$; and r decreases as θ increases. When $\theta = \pi$, $r = 0$.

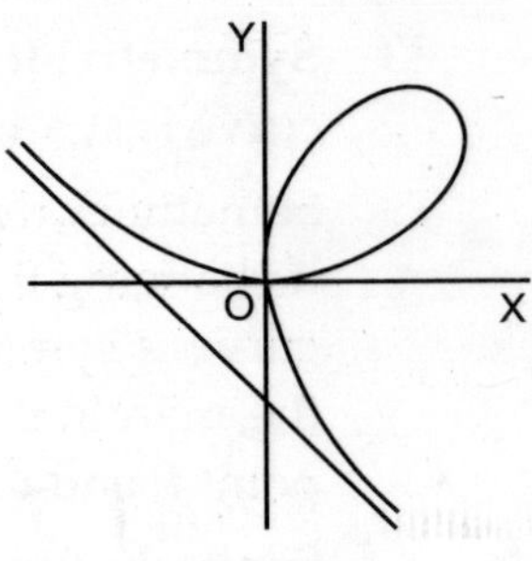

Fig. 2.20

Values of θ between π and 2π will give the same curve as obtained for $0 \le \theta \le \pi$; for if (r_1, θ_1) and $(r_2, \theta_1 + \pi)$ be two points on (1), and $(-r_1, \theta_1 + \pi)$ is the same point as (r_1, θ_1).

$$r_2 = \frac{3a\cos(\theta_1+\pi)\sin(\theta_1+\pi)}{\cos^3(\theta_1+\pi)+\sin^3(\theta_1+\pi)} = -\frac{3a\cos\theta_1\sin\theta_1}{\cos^3\theta_1+\sin^3\theta_1} = -r_1,$$

(*iv*) The Cartesian equation shows that the asymptote is $x + y + a = 0$.

Hence the curve is as shown in the figure.

[This curve is called "Folium of Descartes". Rene Descartes (1596-1650) was the great French mathematician and philosopher after whom the Cartesian coordinates are named.

2.14 SOME WELL-KNOWN CURVES

It is desirable that the student should be familiar with some of the well-known curves. Their equations and shapes are given below, but the student should first trace them independently and then compare his result with the figures given here. The student is supposed to be familiar with the ellipse, parabola, hyperbola and the graphs of the circular, logarithmic and exponential functions.

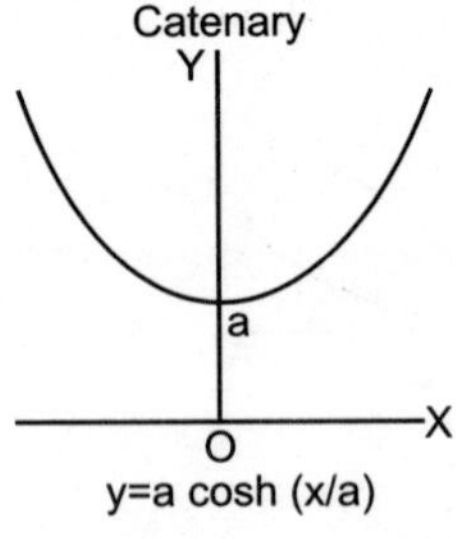

Fig. 2.21

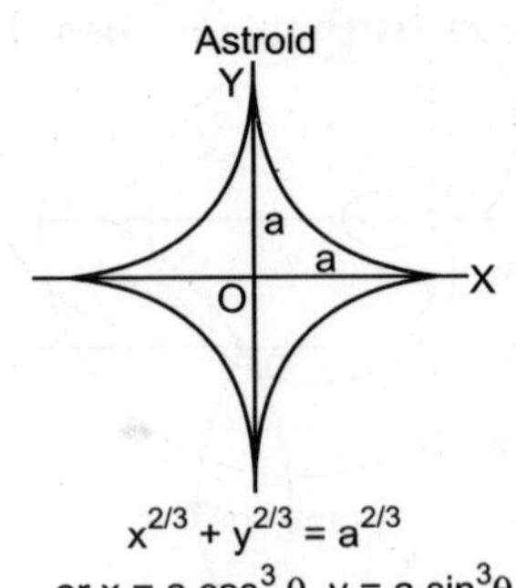

Fig. 2.22

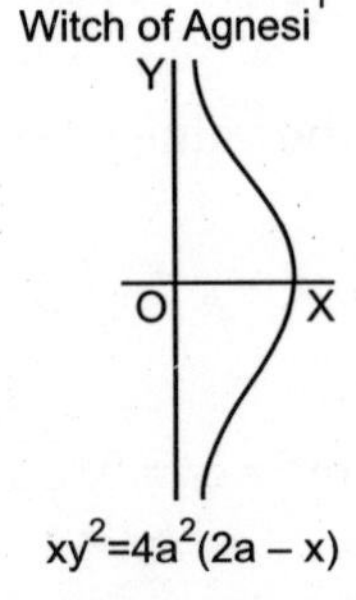

Fig. 2.23

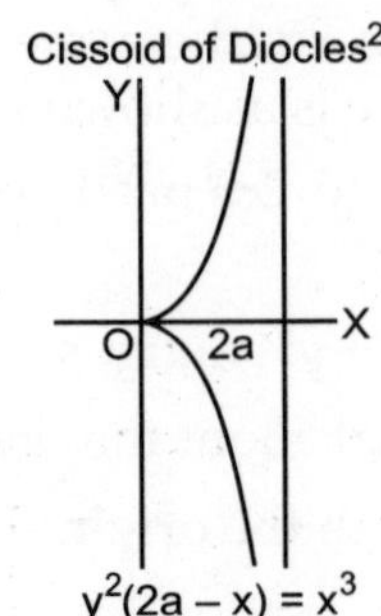

Fig. 2.24

---

1. Called after the Italian lady mathematician Maria Gaetna Agensi (1718-1799) who treated it in her book *Insitiuzioni Analitiche*.

2. Diocles was a Greek mathematician (flourished about 180 B.C.). This curve he used for finding two mean proportionals between two given straight lines.

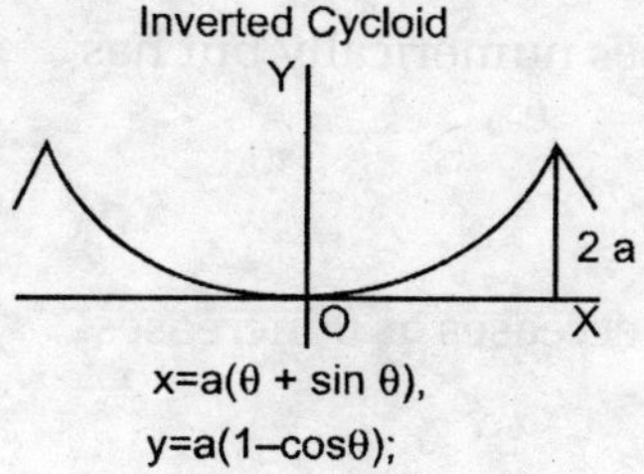

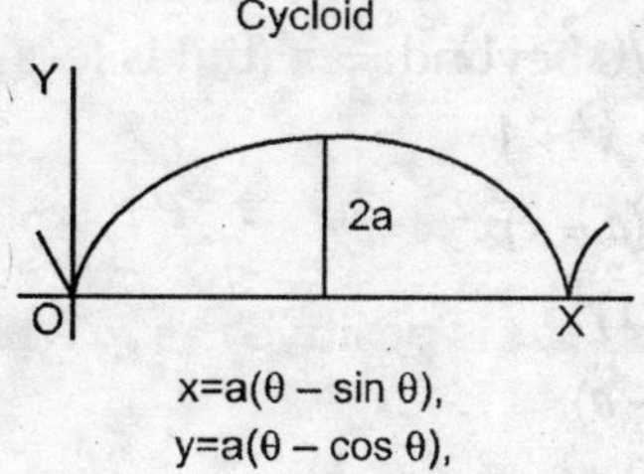

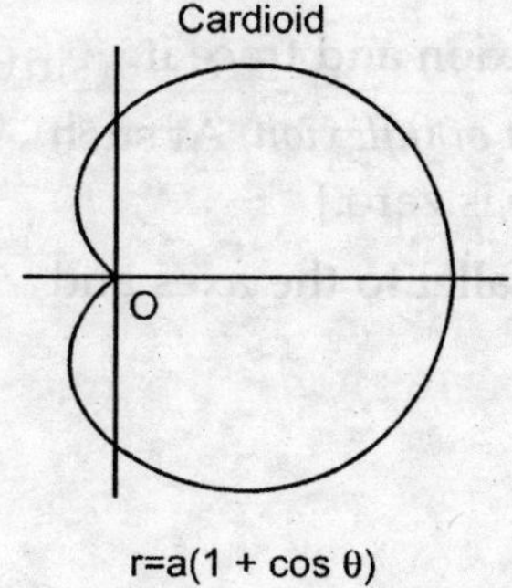

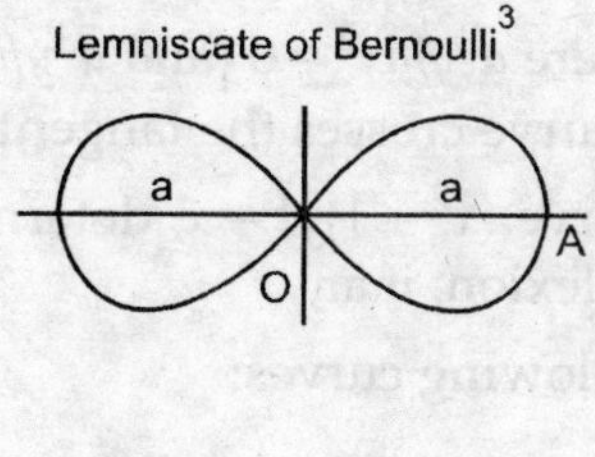

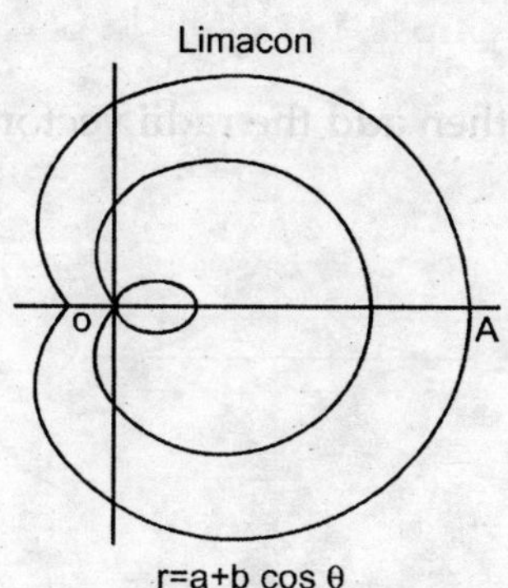

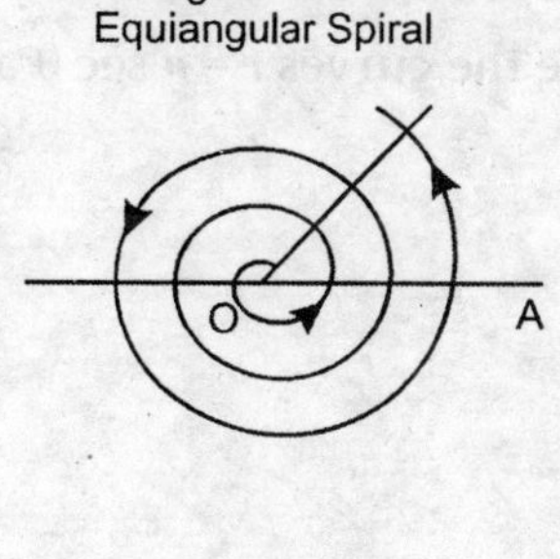

Fig. 2.25

The two curves shown in the figure correspond to $a < b$ and $a > b$ respectively. The Cardioid is only a particular case of this.

***Examples* :**

Trace the following curves :

1. $y^2 = x^3$.
2. $y = x(x^2 - 1)$.
3. $y = (x - 2)(x + 1)^2$.
4. $x = (y - 1)(y - 2)(y - 3)$. [*Gorakhpur*, 1971]
5. $y(x^2 + 4a^2) = 8a^3$.

---

[3] Called after Jakob Bernoulli (1654-1705), professor of mathematics at the University of Basel (Switzerland). He and his brother Johann Bernoulli were staunch friends of Leibnitz and they enriched the calculus immensely in its early days. The lemniscate was first considered by Jakob Bernoulli in the *Acta eruditorum* (1694). He also considered the logarithmic spiral, which so fascinated him that he willed the curve to be engraved upon his tombstone.

6. $y = (x^2 + 1)/(x^2 - 1)$.
7. $y^2(x^2 - 1) = x^2 + 1$. [*Roorkee*, 1970]
8. $y^2(a + x) = (a - x)x^2$. [*Banaras*, 1965]
9. $x^2 = y^2(x + 1)^3$. [*Gorakhpur*, 1972]
10. $ay^2 = x^2(x - a)$.
11. $y^2 = (x - a)^3 / (x + a)$.
12. $y^3 = a^2x - x^3$.
13. $a^2y^2 = x^3(2a - x)$.
14. Show that the curve $(a^2 + x^2)y = a^2x$ has three points of inflexion and trace it.
    [A point where $d^2y/dx^2 = 0$ (and $d^3y/dx^3 \neq 0$) is called a *point of inflexion*. At such a point the curve crosses the tangent line and the curvature is zero.]

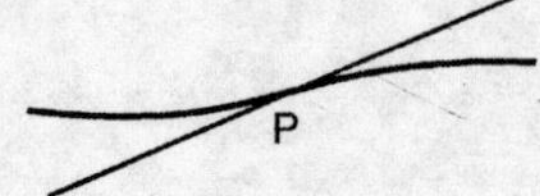

Fig. 2.26

15. Trace the curve $(x^2 - 1)y^2 = x$, determining its tangents parallel to the axes and points of inflexion, if any.

Trace the following curves:

16. $r = a\theta$.
17. $r = a \sin 2\theta$.
18. $r = a \cos 3\theta$.
19. $xy^2 + x^2y = a^3$.
20. $r = a(\sec\theta + \cos\theta)$. [*Roorkee*, 1979]
    [Hint. Trace the curves $r = a \sec\theta$ and $r = a \cos\theta$ first, and then add the radii vectors.]

Coordinate Geometry of Three Dimensions

3.1 RECTANGULAR COORDINATES

The coordinate geometry of three dimensions is an extension of the coordinate geometry of the plane. To indicate the position of a point in space, we evidently used need three coordinates instead of the two required in the coordinate geometry of the plane.

The most commonly used system of coordinates is the rectangular Cartesian coordinates. Let $X' OX$ and $Y' OY$ be any two perpendicular line intersecting at O and let $Z' OZ$ be a third line through O perpendicular to the plane containing of the first two. Then the three lines OX, OY, OZ from a set of mutually perpendicular lines (*i.e.*, OX and OZ, OY is perpendicular to OX and OZ, etc.) We can take these lines as our axes of coordinates. The three planes XOY, YOZ and ZOX are called the coordinate planes. O is the origin.

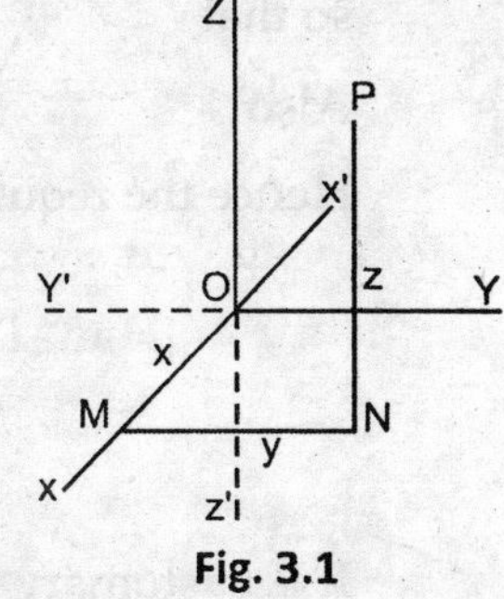

Fig. 3.1

Let P be any point in space. Draw from P the perpendicular PN to the plane XOY. From the foot N of the perpendicular draw the line NM parallel to OY cutting OX in M. Let x, y, z be the lengths of OM, MN, NP respectively. Then x, y, z are the rectangular Cartesian coordinates of P, and are generally written as (x, y, z). We also say that P is the point (x, y, z).

It should be noted that the coordinates x, y, z indicate respectively the displacements parallel to OX, OY and OZ, which must be given to a point at O to reach P. If the displacements are in the positive directions OX, OY, OZ the coordinates are positive. If any of the displacements is in the opposite direction (i.e., in the direction OX', OY' or OZ') the corresponding coordinate has a negative value.

Each coordinate plane divides the space into two parts such that one of the coordinates has opposite signs on the opposite sides of the plane. For example, XOY divides the space into two such parts: the z coordinate is positive above XOY and negative below it. The three coordinates planes together divide the space into eight parts called octants. The octant in which all the three coordinates are positive is called the first or the positive octant.

The system of axes shown in the figure above is a right-handled system. It is so called because a right-handed screw moving in the direction OX, will rotate from OY towards OZ. The other system of coordinates, namely the left-handed system, will be obtained if we interchange the axes of x and y. Throughout the book we shall use the right-handed system of coordinates.

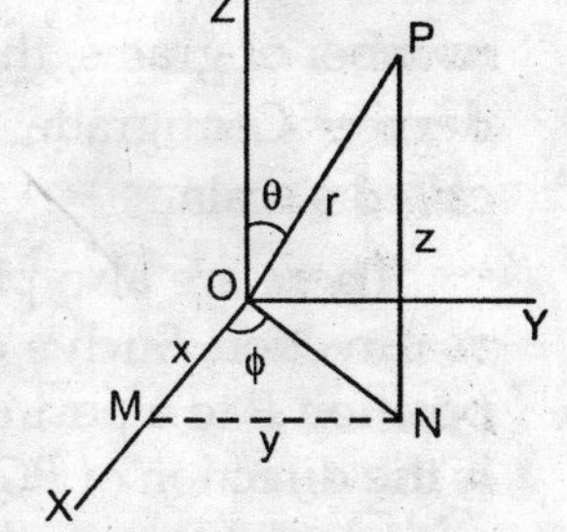

Fig. 3.2

3.2 CYLINDRICAL AND SPHERICAL COORDINATES

The other two systems of coordinates commonly used to locate a point in space are the cylindrical and the spherical coordinates. The cylindrical coordinates (r, ϕ, z) of a point P are the distance ON, the angle XON and the length PN of the perpendicular from P on XOY. Since the triangle ONM is right-angled at M, we have

$$x = \rho \cos \phi, \quad y = \rho \sin \phi. \qquad \text{...(1)}$$

This gives the relation between Cartesian and cylindrical coordinates. The third coordinate is the same in both the systems.

The spherical or polar coordinates (ρ, θ, ϕ) of a point P are the distance OP, the angle ZOP and the angle XON. In the triangle OPN, the angle P is θ and the angle N a right angle. Therefore

$$z = r\cos\theta, \quad \text{and} \quad \rho = r\sin\theta. \qquad ...(2)$$

This gives the relation between cylindrical and spherical coordinates. From (1) and (2) we see that the relation between Cartesian and spherical coordinates is

$$x = r\sin\theta\cos\phi,\ y = r\sin\theta\sin\phi,\ z = r\cos\theta. \qquad ...(3)$$

Example. Find r, θ, ϕ in terms of x, y, z.

Squaring the three relations in (3) and adding, we have

$$x^2 + y^2 + z^2 = r^2\sin^2\theta(\cos^2\phi + \sin^2\phi) + r^2\cos^2\theta$$

$$= r^2\sin^2\theta + r^2\cos^2\theta = r^2.$$

Hence $$r = \sqrt{(x^2 + y^2 + z^2)} \qquad ...(4)$$

Again $$\frac{x^2 + y^2}{z^2} = \frac{r^2\sin^2\theta(\cos^2\phi + \sin^2\phi)}{r^2\cos^2\theta} = \tan^2\theta,$$

so that $$\tan\theta = \sqrt{(x^2 + y^2)}/z. \qquad ...(5)$$

Also $$\tan\phi = y/x.$$

Hence the required relations are

$$r = \sqrt{(x^2 + y^2 + z^2)},$$

$$\theta = \tan^{-1}\{\sqrt{(x^2 + y^2)}/z\},$$

$$\phi = \tan^{-1}(y/x).$$

It is customary to take r positive and θ between 0 and π. Thus we take the positive root in both the equations (4) and (5). The signs of x and y then determine the quadrant in which ϕ lies, and the sign of z that of θ.

3.3 VECTORS

Some physical quantities are completely determined when their magnitudes are given (in terms of some conveniently chosen units). Thus, the mass of a body can be described by the number of grams, the surface area by the number of square centimetres, the temperature by degrees Centigrade, and so on. A quantity which is completely specified by its magnitude is called a scalar.

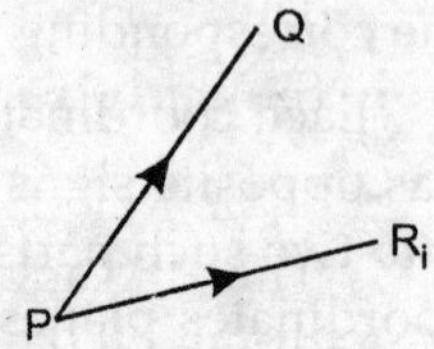

Q
R
P
Fig. 3.3

There are also physical quantities which need for their specification magnitude as well as direction. Such a quantity is called a vector. For example, if a particle is displaced from a position P to a position Q, then the displacement PQ is a vector. Its magnitude is the length PQ, and its direction is the direction of PQ. The displacement PQ is different from displacement PR, even when the lengths PQ and PR are equal, because their directions are different. The student should distinguish displacement from length. The former is a vector quantity while the latter is a scalar. Other familiar examples of vectors are force and velocity.

A vector is represented on a diagram by a line segment whose length is proportional to the magnitude of the vector and whose direction is same as that of the vector. An arrowhead is used to indicate the sense of the vector (i.e. in which of the two directions of the line the vector lies). For the complete specification of some vectors it is also necessary to specify the point at which the vector acts. Such vectors are called localised vectors. A force is a localised vector since its effect depends, in general, on the point where it acts. In many other cases the point where the vector acts is immaterial, and vectors having the same magnitude and direction and acting at different positions in the same line or parallel lines are equivalent. Such vectors are called free vectors. A couple, for example, is a free vector. Unless specified otherwise, we shall be dealing with free vectors.

To distinguish vectors from scalars, bold face type (e.g. **A, B, a, PQ**) is used for vectors. In writing, one usually puts a 'tilde' (~) below a vector quantity (e.g. $\underset{\sim}{A}$). Other notations used in writing vectors are

$$\vec{A}, \overline{A}, \underline{A}.$$

The magnitude (or length) of a vector A is generally denoted by A or $|A|$. The magnitude is always a positive number.

Two vectors A and B are said to be equal (*i.e.* A = B) when they have the same magnitude and the same direction.

If A is a vector, its negative — A is a vector having the same magnitude as A, but an exactly opposite direction.

A vector of unit magnitude is called a unit vector. Two different unit vectors will differ in direction only.

A vector of zero magnitude is called a null vector or zero vector. We denote it by 0.

3.4 ADDITION OF VECTORS

The sum of two **vectors** A and B is defined as follows:

Draw a line segment PQ representing the vector A in magnitude and direction, and from Q draw a line segment QR representing B in magnitude and direction, then the sum A + B is the vector represented in magnitude and direction by the line segment PR. Thus

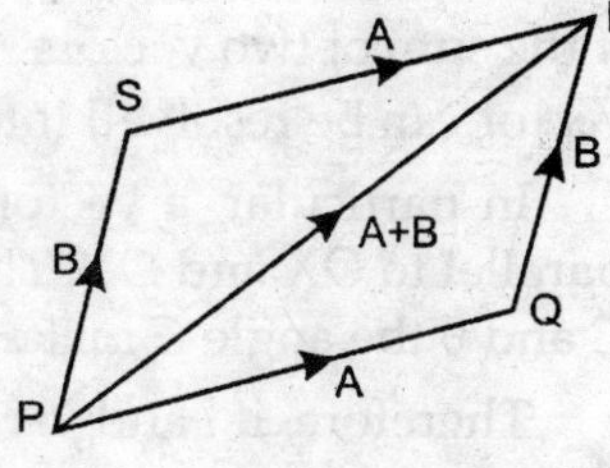

Fig. 3.4

$$\overrightarrow{PQ} + \overrightarrow{QR} = \overrightarrow{PR},$$

or,
$$A + B = C,$$

where C denotes the vector sum **PR**. This is the triangle law of addition of vectors.

The motivation for this rule of addition of vectors is provided by displacements. If a particle at P is displaced to Q and then from Q to R, its resultant displacement is PR. Thus the sum of the displacements PQ and QR is PR.

If we complete the parallelogram PQRS, then PS = QR = B. So A + B is represented by the diagonal of the parallelogram whose sides are A and B. We see from this parallelogram that

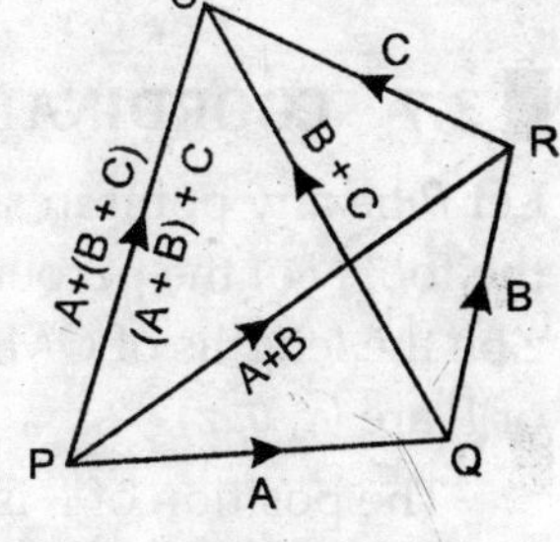

Fig. 3.5

$$A + B = PQ + QR = PR = PS + SR$$

$$= B + A.$$

Thus the addition of vectors is commutative.

Similarly, we can easily show from the above diagram that

$$(A + B) + C = A + (B + C).$$

Thus the addition of vectors is also associative. It should be noted that this is true even when PQRS is a skew quadrilateral (i.e. when A, B, C are non-coplanar vectors).

The difference of A – B of two vectors is defined as the sum

A + (– B),

where – B is the negative of B. It is evident from the definition of the null vector that

$$A - A = 0 \text{ and } A + 0 = A.$$

3.5 MULTIPLICATION BY A SCALAR

If A be a vector and m a real positive number, then mA is defined to be a vector having the same direction as A and m times its magnitude. If m be negative, equal to – n say, then mA is a vector in a direction opposite to A and having a magnitude n times that of A. In particular, (– 1)A = – A, a vector equal and opposite to A.

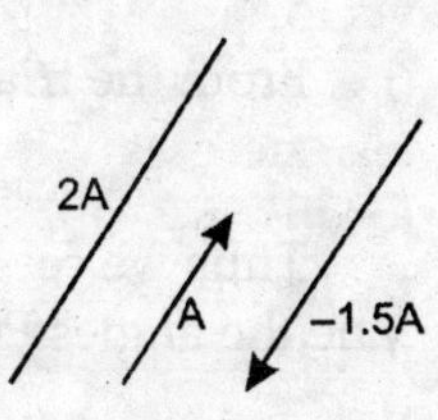

Fig. 3.6

From the above definition we see that if A_1 be a unit vector having the same direction as A, then

$$A = |A| A_1 = AA_1, \text{ and } A_1 = A/A.$$

It is easily seen by drawing a diagram that

$$m(A + B) = mA + mB.$$

If two vectors have the same direction, their addition reduces to ordinary algebraic addition; for example

$$A + A = 2A; 3A + 5A = 8A; 3A - 5A = -2A.$$

In general $\quad mA + nA = (m + n)A.$

Also $\quad m(nA) = (mn)A.$

Thus the definition of addition and multiplication by a number supplement each other, and obey the usual laws of algebraic addition and multiplication.

3.6 RESOLUTION OF A VECTOR

If the sum of two vectors A and B is a third vector C, we call A and B the components of C. It is evident that a given vector can be resolved into.

In particular, a vector C in the plane XOY can be resolved into two components OQ and OR, respectively parallel to OX and OY. The magnitude of these components are C cos θ and C sin θ, where C is the magnitude of C and θ the angle C makes with OX.

Therefore, if i and j are unit vectors in the directions OX and OY, we can write

$$C = (C \cos \theta)i + (C \sin \theta)j,$$

i.e. $\quad C = C_x i + C_y j,$

where $\quad C_x = C \cos \theta, C_y = C \sin \theta.$

3.7 COORDINATES IN SPACE

Let P be any point in space. Draw from P the perpendicular PN to the plane XOY. From the foot N of the perpendicular draw the line NM parallel to OY cutting OX in M. Let x, y, z be the lengths of OM, MN, NP respectively. Then the rectangular Cartesian coordinates of P are (x, y, z).

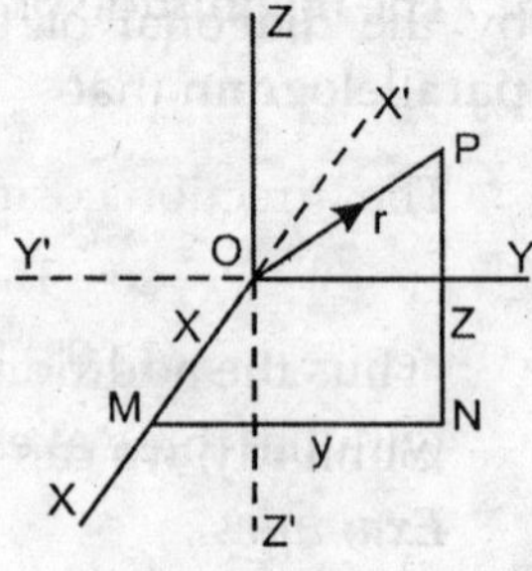

Fig. 3.7

The position of P is also indicated by the vector OP. If $OP = r$, the vector r is called the position vector of P. We see that

$$r = OP = OM + MN + NP$$
$$= xi + yj + zk,$$

where i, j, k are unit vectors parallel to the axes of x, y, z respectively.

Similarly, any vector A represented by the line segment PQ can be represented in terms of i, j, k by drawing planes through P and Q parallel to the coordinate planes. These places form a rectangular parallelopiped whose diagonal is PQ (see figure). If Ax, Ay, Az are the lengths of the three edges PM, MN, NQ, we see that

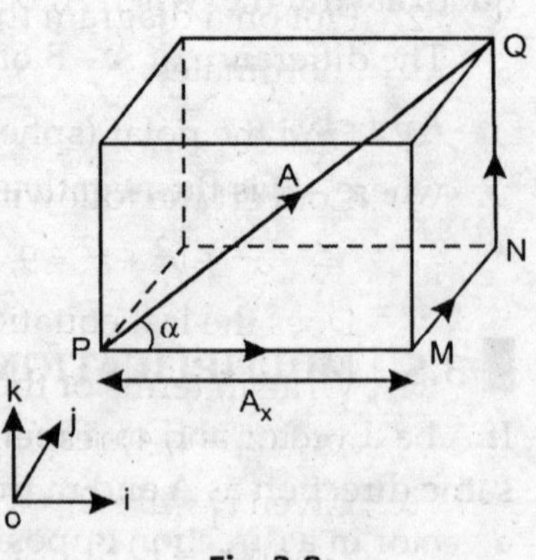

Fig. 3.8

$$PQ = PM + MN + NQ,$$

i.e. $\quad A = A_x + A_y + A_z$

From the triangle PQM, right angled at M,

$$PM = PQ \cos \alpha, \quad \text{i.e.} \quad A_x = A \cos \alpha.$$

Thus we see that A_x, A_y, A_z are the projections of A on the three axes. If α, β, γ be the angles PQ makes with the three axes, then

$$A_x = A \cos \alpha, A_y = A \cos \beta, A_z = A \cos \gamma.$$

The three angles α, β, γ any line (or vector) makes with the axes of coordinates, are called *direction angles*; and their cosines are called *direction cosines*.

Similarly,

$$r = xi + yj + zk$$
$$= r\cos\alpha\, i + r\cos\beta\, j + r\cos\gamma\, k,$$

since x, y, z are the projections of r on the axes.

The three vectors i, j, k in terms of which all the vectors are expressed, are called base vectors. In fact any three vectors a, b, c (not in the same plane) can be chosen as base vectors, and other vectors expressed in terms of them. But the choice of unit vectors in the directions of the three axes as base vectors is particularly convenient.

In terms of the base vectors, the sum of two vectors can be written as

$$A + B = (A_x i + A_y j + A_z k) + (B_x i + B_y j + B_z k)$$
$$= (A_x + B_x)i + (A_y + B_y)j + (A_z + B_z)k.$$

Similarly, the product of a vector by a scalar can be written as

$$mA = m(A_x i + A_y j + A_z k)$$
$$= mA_x i + mA_y j + mA_z k.$$

The magnitude $|A|$ of a vector $A = A_x i + A_y j + A_z k$ is equal to its length (PQ in Figure 3.8 on page 70) and is therefore given by

$$|A| = \sqrt{(A_x^2 + A_y^2 + A_z^2)}\,.$$

Example: The position vectors of points P and Q are $2i + 4j - k$ and $8i + j + k$. Find its magnitude and direction of the vector PQ.

Let O be the origin, then $OP + PQ = OQ$. Therefore

$$PQ = OQ - OP$$
$$= 8i + j + k - (2i + 4j - k)$$
$$= 6i - 3j + 2k.$$

The magnitude of PQ

$$= \sqrt{(6^2 + 3^2 + 2^2)} = \sqrt{(36 + 9 + 4)} = \sqrt{(49)} = 7.$$

The direction cosines of PQ are therefore 6/7, – 3/7, 2/7. Hence PQ makes the angles

cos–1 (6/7), cos–1 (– 3/7), cos–1 (2/7),

i.e. 31.0°, 115.4°, 73.4°.

with the three axes.

Examples :

1. Plot on a diagram the points (2, 3, 4) and (3, – 2, 1).
2. Plot on a diagram the point $\left(2, \frac{1}{4}\pi, 3\right)$ in cylindrical coordinates; and the point $\left(3, \frac{1}{3}\pi, \frac{1}{4}\pi\right)$ in spherical coordinates.
3. Find the polar (spherical) coordinates of the points $(3, \sqrt{3}, 2)$ and $(-2, 1, -2)$.
4. Convert the following equations to spherical coordinates :
   (*i*) $x^2 + y^2 + z^2 = 9$, (*ii*) $x^2 + y^2 = 3z^2$, (*iii*) $x^2 + y^2 = 4$.
   Does the last equation have a simple form in cylindrical coordinates?
5. Write in terms of the base vectors i, j, k the vectors OP, OQ and PQ, when P and Q are the points (3, 1, 0) and (2, – 1, 4) respectively. Also find the magnitude and the direction cosines of PQ.
6. Given $r_1 = 4i - 2j + k$ and $r_2 = 2i - 4j - 4k$, find the magnitudes and the direction cosines of the vectors $r_1 + r_2$ and $r_1 - r_2$.
7. Find a unit vector parallel to the sum of the vectors
   $$r_1 = 2i + 4j - 5k,\ r_2 = i + 2j + 3k.$$ **[Gorakhpur, 1973]**

8. ABCDEF is a regular hexagon. Forces AB, AC, AD, AE and AF act at A. Show that their resultant is 3AD.
9. If A and B are non-zero vectors not in the same direction, show that mA + nB is a vector lying in the plane determined by A and B. [Punjabi, 1967]
10. If mA + nB = 0, show that either vectors A and B are parallel, or $m = n = 0$. [Gorakhpur, 1996]
11. If mA + nB + pC = 0, show that either vectors A, B and C are coplanar, or $m = n = p = 0$.
12. The position vectors of A and B are a and b. Show that the position vector of the point P dividing AB in the ratio $m : n$ is $(na + mb)/(m + n)$.
13. Prove that the vectors can form the sides of a triangle. Also find the length of the median bisecting the vector c. [Gorakhpur, 1977]
$$a = 3i + j - 2k,\ b = -i + 3j + 4k,\ c = 4i - 2j - 6k$$
14. The position vectors of P, Q, R, S are
$$i + j + k,\ 2i + 5j,\ 3i + 2j - 3k,\ i - 6j - k$$
respectively. Prove that PQ and RS are parallel.
15. Show that the following sets of vectors are coplanar
(i) $a - 2b + 3c, -2a + 3b - 4c, -b + 2c$; [Alld, 1973]
(ii) $5a + 6b + 7c, 7a - 8b + 9c, 3a + 20b + 5c$;
where a, b, c are non-coplanar.

3.8 SCALAR PRODUCT

The scalar product of two vectors A and B is defined as the product of their magnitudes and the cosine of the angle between them. It is denoted by $A\,.\,B$. Thus

$$A.\,B = AB\cos\theta$$

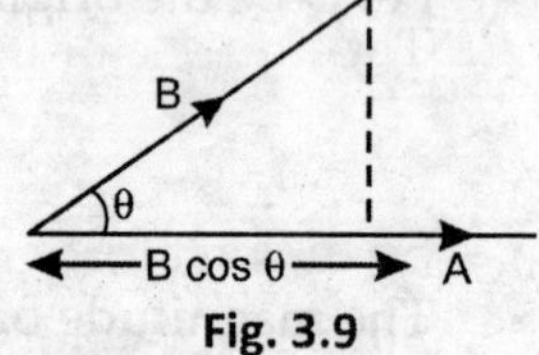

Fig. 3.9

where θ is the angle between the vectors A and B. The scalar product is also called the dot product. It is evident that the scalar product of two vectors is a scalar quantity.

We see that

$$A\,.\,B = A\,(B\cos\theta) \quad ...(1)$$

$$= (\text{mag. of } A) \times (\text{projection of } B \text{ on } A).$$

A physical example of the scalar product of $F.\,x$, the work done by a constant force F when its point of application undergoes a displacement x.

It follows from the definition of a scalar product that

$$A.\,B = B.\,A,$$

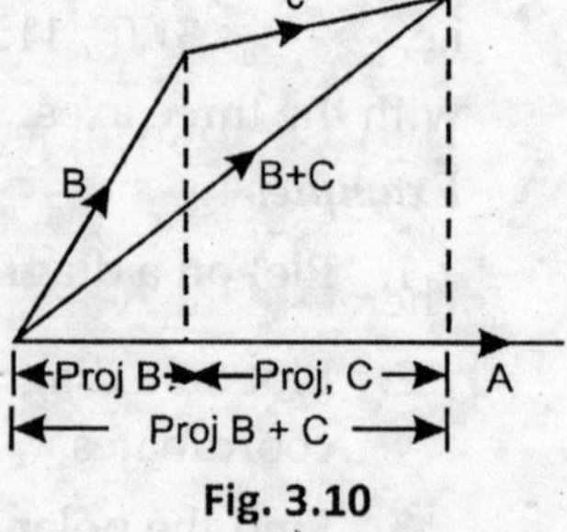

Fig. 3.10

i.e., the scalar product is commutative. We see from the adjoining figure that $A \times$ (projection of $B + C$) = $A \times$ projection of $B + A \times$ projection of C,

i.e., $$A\,.\,(B + C) = A.\,B + A\,.\,C,$$

i.e., the scalar product is also distributive. Hence algebraic operations involving scalar multiplication can be freely performed on vectors in the same way as for ordinary numbers.

The scalar product is zero when the angle between the two vectors is 90°. Hence the condition of perpendicularly of two vectors A and B is A . B = 0.

In particular, we see that

$$i\,.\,j = j\,.\,k = k\,.\,i = 0 \quad ...(2)$$

Also $$i\,.\,i = 1\,.\,1\cos 0 = 1.$$

Similarly $$j\,.\,j = k\,.\,k = 1. \quad ...(3)$$

The product $A\,.\,A$ is often written as A^2. Evidently $A^2 = |A|^2$.

If the vectors A and B are written as

$$A = A_x \mathrm{i} + A_y \mathrm{j} + A_z k$$

and
$$B = B_x \mathrm{i} + B_y \mathrm{j} + B_z k,$$

then
$$A \,.\, B = (A_x \mathrm{i} + A_y \mathrm{j} + A_z k) \,.\, (B_x \mathrm{i} + B_y \mathrm{j} + B_z k)$$
$$= A_x B_x + A_y B_y + A_z B_z,$$

on opening the brackets and using relations (2) and (3).

Example: 1.Given A = $2i + 2j - k$, $B = 6i - 3j + 2k$,

$$A \,.\, B = (2i + 2j - k) \,.\, (6i - 3j + 2k)$$
$$= 2 \,.\, 6 - 2 \,.\, 3 - 1 \,.\, 2 = 4,$$

since
$$i^2 = j^2 = k^2 = 1 \text{ and } i \,.\, j = i \,.\, k = \text{etc.} = 0,$$

By (1), the angle θ between A and B is given by

$$\cos\theta = A \,.\, \mathrm{B/AB} = 4/\sqrt{(2^2 + 2^2 + 1)}\sqrt{(6^2 + 3^2 + 2^2)}$$
$$= 4/3 \,.\, 7 = 4/21.$$

Hence
$$\theta = \cos{-1}\, \frac{4}{21} = 79.0°.$$

Example. 2.**Forces of magnitudes 5, 3, 1 units act in the directions 6i + 2j + 3k, 3i – 2j + 6k, 2i – 3j – 6k respectively on a particle which is displaced from the point (2, – 1, – 3) to (5, – 1, 1). Find the work done by the forces. [A.M.I.E., 1977]**

The unit vector in the direction $6i + 2j + 3k$ is

$$(6i + 2j + 3k)/\sqrt{(6^2 + 2^2 + 3^2)}, \text{ i.e. } (6i + 2j + 3k)/7.$$

So the first force is $5(6i + 2j + 3k)/7$.

Similarly the other two forces are

$$3(3i - 2j + 6k)/7 \text{ and } (2i - 3j + 6k)/7.$$

Hence the sum of the forces is

$$F = [5(6i + 2j + 3k) + 3(3i - 2j + 6k) + (2i - 3j + 6k)]/7.$$
$$= (41i + j + 27k)/7.$$

The displacement is

$$x = (5 - 2)i + \{-1 - (-1)\}j + \{1 - (-3)\}k = 3i + 4k$$

Therefore the work done by the forces is

$$F \,.\, x = \frac{1}{7}(41i + j + 27k) \,.\, (3i + 4k)$$
$$= \frac{1}{7}(41 \,.\, 3 + 1 \,.\, 0 + 27 \,.\, 4) = 231/7 = 33.$$

3.9 VECTOR PRODUCT

The vector product of two vectors A and B, denoted by A × B, is a vector normal to A and B of magnitude AB sin θ, where θ is the angle between A and B. Its sense is such that A, B and A × B form a right-handed system. Thus

$$A \times B = (AB \sin\theta)n, \qquad ...(1)$$

where n is a unit vector normal to A and B such that A, B, n form a right-handed system.

The vector product is also called the cross product.

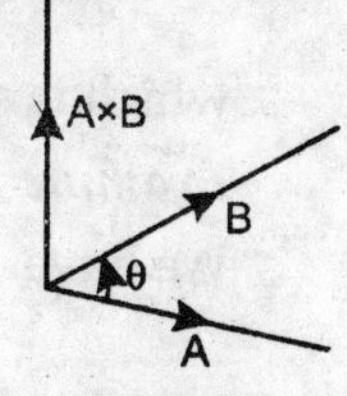

Fig. 3.11

A physical example of the vector product is $r \times F$ the moment about O of a force F acting at a point P whose radius vector is r. The moment measures the tendency of the force F to turn the body about O. It is a vector quantity, its direction being parallel to the axis about which the body tends to turn.

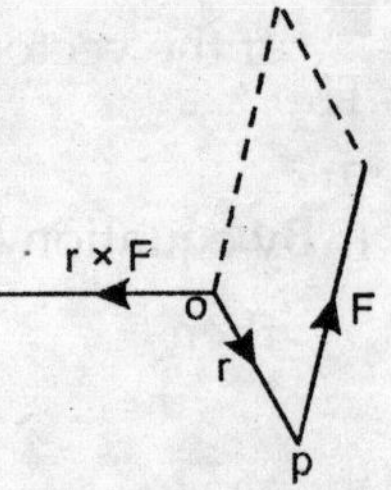

Fig. 3.12

The area of a plane figure can be represented by a vector whose magnitude is equal to the area and whose direction is perpendicular to the plane of the area. With this interpretation $A \times B$ will represent the area of the parallelogram whose two adjacent sides are A and B.

We see from the definition of vector product that $B \times A$ is a vector of magnitude AB sin θ normal to A and B and in a direction such that B, A, $B \times A$ form a right-handed system. Therefore $B \times A$ is equal in magnitude but opposite in direction to $A \times B$, that is

$$B \times A = -A \times B.$$

Thus the vector product is not commutative, a reversal in the order of factors changes the sign of the product. It can be proved, however that the vector product is distributive, viz.

$$A \times (B + C) = A \times B + A \times C.$$

Thus the vector product will obey ordinary laws of algebra, except that the order of the factors should be kept unchanged.

If A and B are two vectors having the same direction, then evidently $A \times B$ is a vector of zero magnitude. This is also true when A and B have opposite directions. A case of special importance is

$$A \times A = 0.$$

In particular, we see that

$$i \times i = j \times j = k \times k = 0.$$

We also see that $i \times j$ is a vector of unit magnitude perpendicular to i and j. Hence

$$i \times j = k$$

Also $\quad j \times i = -k.$

Similarly $\quad j \times k = i, k \times j = -i,$

and $\quad k \times i = j, i \times k = -j.$

Note that we have a positive sign when i, j, k are in the same cyclic order, and a negative sign otherwise.

When A and B are given in terms of base vectors, viz.

$$A = A_x i + A_y j + A_z k,$$

and
$$B = B_x i + B_y j + B_z k,$$

then
$$\begin{aligned} A \times B &= (A_x i + A_y j + A_z k) \times (B_x i + B_y j + B_z k) \\ &= A_x B_x i \times i + A_x B_y i \times j + A_x B_z i \times k + \text{two sets of similar terms} \\ &= 0 + A_x B_y k - A_x B_z j + \text{similar terms} \\ &= (A_y B_z - A_z B_y) i + (A_z B_x - A_x B_z) j + (A_x B_y - A_y B_x) k. \end{aligned}$$

This can also be written as

$$A \times B = \begin{vmatrix} \mathbf{i} & \mathbf{j} & \mathbf{k} \\ A_x & A_y & A_z \\ B_x & B_y & B_z \end{vmatrix}$$

which is easier to remember.

Example 1. Given A = $2i + 2j - k$, B = $6i - 3j + 2k$, find $A \times B$ and the unit vector perpendicular to both A and B.

We have

$$A \times B = \begin{vmatrix} \mathbf{i} & \mathbf{j} & \mathbf{k} \\ 2 & 2 & -1 \\ 6 & -3 & 2 \end{vmatrix}$$

$$= i(4-3) - j(4+6) + k(-6-12)$$
$$= i - 10j - 18k.$$

By equation (1), the unit vector normal to A and B

$$= A \times B/AB \sin q = A \times B / |A \times B|$$
$$= (i - 10j - 18k) / \sqrt{(1 + 10^2 + 18^2)}$$
$$= \frac{1}{5\sqrt{17}} i - \frac{2}{\sqrt{17}} j - \frac{18}{5\sqrt{17}} k.$$

Example 2. A force $F = 3i + 2j - 4k$ is applied at the point $(1, -1, 2)$. Find the moment of the force about the point $(2, -1, 3)$. [Gorakhpur, 1995]

Denote $(2, -1, 3)$ by A and $(1, -1, 2)$ by B; then the vector AB

$$= (1-2)i + \{-1-(-1)\}j + (2-3)k = -i - k.$$

The moment of force F about A is

$$AB \times F = (-i-k) \times (3i + 2j - 4k)$$
$$= -2i \times j + 4i \times k - 3k \times i - 2k \times j$$
$$= -2k - 4j - 3j + 2j = 2i - 7j - 2k.$$

Examples:

1. If $A = 4i - 2j + 4k$ and $B = 3i - 6j - 2k$, find $A.B$ and the angle between A and B.
2. Determine a so that the vectors $2i + aj + k$ and $4i - 2j - 2k$ are perpendicular. [Gorakhpur, 1973]
3. If $A = i + 2j + 3k$, $B = -i + 2j + k$ and $C = 3i + j$, find t such that $A + tB$ is perpendicular to C. [Allahabad, 1971]
4. Show that the vectors $A = 3i - 2j + k$, $B = i - 3j + 5k$, $C = 2i + j - 4k$ form a right-angled triangle. [Roorkee, 1972]
5. If $A = 2i + 2j - k$, $B = 6i - 3j + 2k$, find $A.B$ and $A \times B$. [Allahabad, 1974]
6. Find *a* unit vector perpendicular to the vectors A and B of the previous question. [Delhi, 1976]
7. If $a = 4i + 3j + k$, $b = 2i - j + 2k$, find *a* unit vector n perpendicular to vectors a and b such that a, b, n form a right-handed system. Also find the angle between the vectors a and b. [Roorkee, 1971]
8. Find the projection of $i - 2j + k$ on $4i - 4j + 7k$. [Gorakhpur, 1977]
9. Show that $(a - b) \times (a + b) = 2a \times b$; and give a geometrical interpretation.
10. If $A = 3i - 2j - 2k$ and $B = 2i + 3j + k$ evaluate $(A + B) \times (A - B)$ and $(A + B) . (A - B)$. [A.M.I.E., 78]
11. Show that the diagonals of a rhombus are at right angles.
12. Find the area of the parallelogram having the diagonals
$$A = 3i + j - 2k, B = i - 3j + 4k.$$ [Gorakhpur, 1973]
13. Show that the area of a triangle whose sides are represented by the vectors a, b, c is $\frac{1}{2}|a \times b|$. Hence find the area of the triangle ABC whose vertices are
$$A(1, -2, 3), B(3, 1, 2), C(2, 3, -1).$$
14. If $a + b + c = 0$, prove that and intercept is geometrically. [Delhi, 1972]
$$a \times b = b \times c = c \times a,$$

15. Does $A \times B = A \times C$ imply $B = C$? Given reasons for your answer. [Gorakhpur, 1996]

16. Show that if $A \neq 0$ and both the conditions.

(*a*) $A . B = A . C$ and (*b*) $A \times B = A \times C$

hold simultaneously then $B = C$, but if only one of these conditions holds then $B \neq C$ necessarily. [Roorkee, 1972]

17. Constant forces $P = 2i - 5j + 6k$ and $Q = -i + 2j - k$ act on a particle. Determine the work done when the particle is displaced from A to B, the position vectors of A and B being $4i - 3j - 2k$ and $6i + j - 3k$ respectively. [Roorkee, 1971]

18. Find the torque about the point $2i + j - k$ of *a* force represented by $4i + k$ acting through the point $i - j + 2k$. [Gorakhpur, 1994]

3.10 APPLICATION TO GEOMETRY

The use of vectors simplifies the derivation of a number of formulae in solid geometry. We shall therefore derive the formulae by vector methods whenever convenient. We shall now derive such formulae:

Distance between two point $P(x_1, y_1, z_1)$ and $Q(x_2, y_2, z_2)$. Since OP + PQ = OQ, therefore

$$\begin{aligned} PQ &= |PQ| = |OQ - OP| \\ &= |x_2 i + y_2 j + z_2 k - (x_1 i + y_1 j + z_1 k)| \\ &= |(x_2 - x_1)i + (y_2 - y_1)j + (z_2 - z_1)k| \\ &= \sqrt{\{(x_2 - x_1)^2 + (y_2 - y_1)^2 + (z_2 - z_1)^2\}}. \end{aligned}$$

Division of the line joining $P(x_1, y_1, z_1)$ and $Q(x_2, y_2, z_2)$ in the ratio $m_1 : m_2$. Let $R(x, y, z)$ be the point which divides PQ in ratio $m_1 : m_2$; then

$$PR = OR - OP = (x - x_1)i + (y - y_1)j + (z - z_1)k,$$

and

$$RQ = OQ - OR = (x_2 - x)i + (y_2 - y)j + (z_2 - z)k,$$

Now PR : RQ $= m_1 : m_2$, *i.e.*, $m_2 PR = m_1 RQ$,

therefore $m_2\{(x - x_1)i + (y - y_1)j + (z - z_1)k\} = m_2\{(x_2 - x)i + (y_2 - y)j + (z_2 - z)k\}$.

Equating separately the coefficients of *i*, *j*, *k* this gives

$$x = \frac{m_2 x_1 + m_1 x_2}{m_1 + m_2},\ y = \frac{m_2 y_1 + m_1 y_2}{m_1 + m_2},\ z = \frac{m_2 z_1 + m_1 z_2}{m_1 + m_2}$$

Example. Show that the points (1, 4, – 3), (2, 5, – 4) and (4, 7, – 6) are collincar.

Denote the points by A, B and C; then

$$AB = (2 - 1)i + (5 - 4)j + (-4 + 3)k = i + j - k,$$

and

$$BC = (4 - 2)i + (7 - 5)j + (-6 + 4)k = 2i + 2j - 2k.$$

Since these are parallel vectors, *AB* and *BC* are in the same straight line, i.e. *A*, *B*, *C* are collinear.

Examples :

1. Find the distance between the points (2, – 4, 1) and $\left(\frac{1}{2}, 2, 3\right)$.
2. Find the distance between the feet of perpendicular drawn from the point (2, 3, 4) on planes XOY and XOZ.
3. Show that the points (0, – 7, – 10), (4, – 9, – 6), (1, – 6, – 6) form a right-angled isosceles triangle. [Roorkee, 1976]
4. Prove that the triangle formed by the three points (1, – 2, – 3), (2, – 3, – 1) and (3, – 1, – 2) is an equilateral triangle. [Banaras, 1975]

5. Show that the point $\left(-\frac{1}{2}, 2, 0\right)$ is the circumcentre of the triangle formed by the points (1, 1, 0), (1, 2, 1), (– 2, 2, – 1). [Roorkee, Arch., 1973]
6. Find the coordinates of the point dividing the line joining (5, 8, – 3) and (1, 0, – 3) in the ratio 1 : 3.
7. Show that the points (5, 4, 2), (6, 2, – 1) and (8, – 2, – 7) are collinear.
8. *A* and *B* are the points (1, 2, 3) and (2, – 1, 3), and *P* is a variable point (*x*, *y*, *z*). Find the equation satisfied by the locus of *P*, if
   (*i*) PA = PB, 6 (*ii*) PA = 4.
9. Find the coordinates of the centroid of a triangle having the vertices (x_1, y_1, z_1), (x_2, y_2, z_2), (x_3, y_3, z_3).
10. *A*(3, 2, 0), *B*(5, 3, 2), *C*(– 9, 6, – 3) are three points forming a triangle. AD, bisector of angle BAC, meets BC in D. Find the coordinates of D. [Roorkee, 1974]
    [Hint : BD : DC = AB : AC].

3.11 DIRECTION COSINES

The cosines of the three angles a line makes with the positive directions of the axes of coordinates are called the direction cosines of the line.

Let *OP* be a line making angles α, β, γ respectively with *OX*, *OY* and *OZ*. Then the direction cosines of *OP* are

$$\cos\alpha, \cos\beta, \cos\gamma.$$

The angles α, β, γ are called the direction angles of *OP*.

Any line *RS*, parallel to *OP*, will make the same angles with the axes of coordinates and will have the same direction cosines. The line *PO*, in the direction opposite to *OP*, will make angles π – α, π – β, π – γ with the axes and will have the direction cosines

$$-\cos\alpha, -\cos\beta, -\cos\gamma.$$

Fig. 3.13

The three direction cosines are not independent but are connected by the relation

$$\cos^2\alpha + \cos^2\beta + \cos^2\gamma = 1. \quad ...(1)$$

To prove this, let *r* be the position vector of *P* (*x*, *y*, *z*) then $r = xi + yj + zk$. Therefore

$$r \,.\, i = (xi + yj + zk) \,.\, i = x,$$

i.e. $$r \,.\, 1.\cos a = x, \text{ or } \cos a = x/r. \quad ...(2)$$

Similarly $$\cos b = y/r \text{ and } \cos c = z/r. \quad ...(3)$$

Again $$r \,.\, r = (xi + yj + zk) \,.\, (xi + yj + zk),$$

i.e. $$r^2 = x^2 + y^2 + z^2.$$

Dividing by r^2 and substituting from (2) and (3), the result follows.

Directions Ratios: Any three numbers proportional to the direction cosines of a line are called direction ratios of the line. Let a, b, c be directions ratios of a line and cos α, cos β, cos γ its direction cosines then,

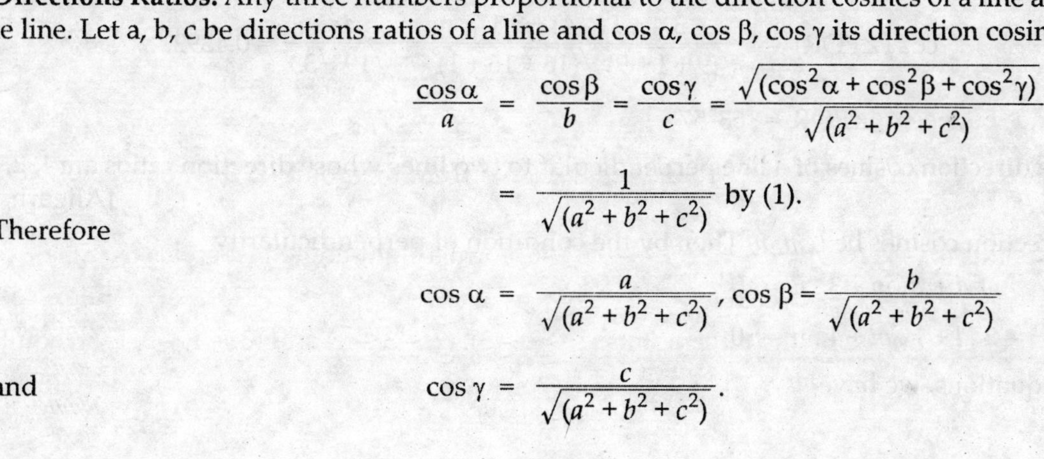

$$\frac{\cos\alpha}{a} = \frac{\cos\beta}{b} = \frac{\cos\gamma}{c} = \frac{\sqrt{(\cos^2\alpha + \cos^2\beta + \cos^2\gamma)}}{\sqrt{(a^2 + b^2 + c^2)}}$$

$$= \frac{1}{\sqrt{(a^2 + b^2 + c^2)}} \text{ by (1).}$$

Therefore

$$\cos\alpha = \frac{a}{\sqrt{(a^2 + b^2 + c^2)}}, \cos\beta = \frac{b}{\sqrt{(a^2 + b^2 + c^2)}}$$

and $$\cos\gamma = \frac{c}{\sqrt{(a^2 + b^2 + c^2)}}.$$

Direction Cosines of Line Joining : (x_1, y_1, z_1) and (x_2, y_2, z_2). Let P and Q be the given points and α, β, γ the angles PQ makes with the axes. Then, in the figure of heading 3.7 $\alpha = \angle QPM$. Therefore

$$\cos\alpha = PM/PQ = (x_2 - x_1)/r,$$

where $r = \sqrt{\{(x_2 - x_1)^2 + (y_2 - y_1)^2 + (z_2 - z_1)^2\}}$. Similarly

$$\cos\beta = (y_2 - y_1)/r, \cos\gamma = (z_2 - z_1)/r.$$

If follows that the line joining (x_1, y_1, z_1) and (x_2, y_2, z_3) has direction ratios

$$x_1 - x_2, y_2 - y_1, z_2 - z_1.$$

3.12 ANGLE BETWEEN TWO LINES

To find the angle between two lines OP and OQ, whose direction cosines are $\cos\alpha$, $\cos\beta$, $\cos\gamma$ and $\cos\alpha'$, $\cos\beta'$, $\cos\gamma'$, respectively.

Let r and r' be the position vector of $P(x, y, z)$ and $Q(x', y', z')$, and let θ be the angle between OP and OQ. Then

$$r = xi + yj + zk \text{ and } r = x'i + y'j + z'k. \quad \text{...(1)}$$

Therefore $\quad r . i = x$ and $r' . i = x'$,

i.e. $\quad r\cos\alpha = x$ and $r'\cos\alpha' = x'$.

Similarly

$$r\cos\beta = y, r\cos\gamma = z \text{ and } r'\cos\beta' = y', r'\cos\gamma' = z'.$$

From (1) : $\quad r . r' = xx' + yy' + zz'$,

i.e. $\quad rr'\cos\theta = xx' + yy' + zz'$

$$= rr'(\cos\alpha + \cos\alpha' + \cos\beta + \cos\beta' + \cos\gamma + \cos\gamma').$$

Hence $\quad \cos\theta = \cos\alpha + \cos\alpha' + \cos\beta + \cos\beta' + \cos\gamma + \cos\gamma'.$

COROLLARY 1. The two lines are perpendicular if

$$\cos\theta = \cos\alpha + \cos\alpha' + \cos\beta + \cos\beta' + \cos\gamma + \cos\gamma' = 0.$$

COROLLARY 2. The angle θ between two lines whose direction ratios a, b, c and a', b', c', is given by.

$$\cos\theta = \frac{aa' + bb' + cc'}{\sqrt{(a^2 + b^2 + c^2)}\sqrt{(a'^2 + b'^2 + c'^2)}}$$

The lines are perpendicular if

$$aa' + bb' + cc' = 0$$

The lines are parallel if $\alpha = \alpha'$, $\beta = \beta'$, $\gamma = \gamma'$, i.e.

$$\frac{a}{a'} = \frac{b}{b'} = \frac{c}{c'}$$

***Examples* 1.** The coordinates of P, Q and R are (1, 2, 3), (– 2, 1, 0) and (2, – 3, 1). Find the angle PQR. [Gorakhpur, 1975]

Direction ratios of QP are 1 – (– 2), 2 – 1, 3 – 0, i.e. 3, 1, 3, and those of QR are 4, – 4, 1. Therefore

$$\cos(\angle PQR) = \frac{3.4 + 1(-4) + 3.1}{\sqrt{(9+1+9)}\sqrt{(16+16+1)}} = \frac{11}{\sqrt{19}\sqrt{33}} = 0.4392,$$

and $\quad \angle PQR = 63°57'.$

***Example* 2.** Find the direction cosines of a line perpendicular to two lines whose direction ratios are 1, 2, 3 and – 2, 1, 4. [Aligarh, 1967]

Let the required direction cosines be l, m, n. Then by the condition of perpendicularity

$$1 . l + 2 . m + 3 . n = 0$$

and

$$-2 . l + 1 . m + 4 . n = 0$$

Solving these two equations, we have

$$\frac{l}{8-3} = \frac{-m}{4+6} = \frac{n}{1+4},$$

or $$\frac{l}{1} = \frac{m}{-2} = \frac{n}{1} = \frac{1}{\sqrt{(1+4+1)}}.$$

So the required direction cosines are

$$\frac{1}{\sqrt{6}}, -\frac{2}{\sqrt{6}}, \frac{1}{\sqrt{6}}.$$

3.13 DISTANCE OF A POINT FROM A LINE

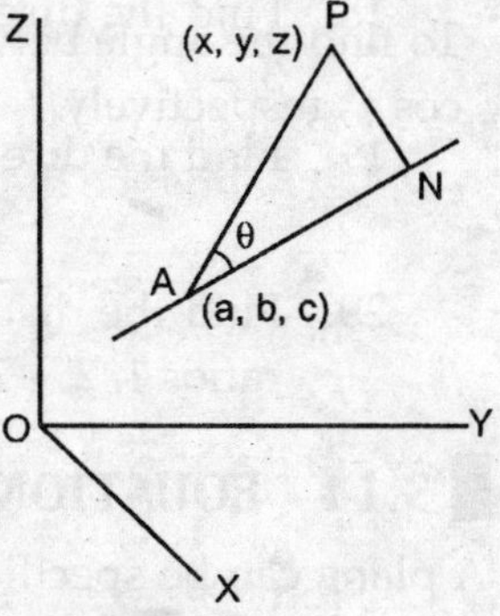

Fig. 3.14

Let P be a point (x, y, z) and AN a line through A (a, b, c) having the direction cosines $\cos \alpha, \cos \beta, \cos \gamma$.

If θ be the angle PAN and u be unit vector in direction AN, then

$$PN = AP \sin \theta = |AP \times u|.$$

But $$AP = (x-a)i + (y-b)j + (z-c)k,$$

and $$u = (\cos \alpha)i + (\cos \beta)j + (\cos \gamma)k;$$

so that $AP \times u = \{(y-b)\cos \gamma - (z-c)\cos b\}i + \{(z-c)\cos \alpha - (x-a)\cos \gamma\}j + \{(x-a)\cos \beta - (y-b)\cos \alpha\}k.$

Therefore $$PN = \sqrt{[\Sigma\{(y-b)\cos\gamma - (z-c)\cos\beta\}^2]}.$$

***Examples*:**

1. If a line makes an angle of 45° with x-axis and 60° with y-axis, what angle does it make with z-axis?
2. Find the direction cosines of a line making equal angles with the axes.
3. The projections of a line on the axes are 2, 3, 6. What is the length of the line?
   [The projection of a line segment AB on a line CD is AB cos θ, where θ is the angle between the two lines.]
4. If a line makes the angles α, β, γ with the axes, prove that
   $\sin^2 \alpha \sin^2 \beta + \sin^2 \gamma = 2.$ [Roorkee, Arch., '71]
5. Find the direction cosines and direction angles of a line directed upwards and having direction ratios 2, – 2, 1.
6. Find the direction cosines of the line joining the following points
   (*i*) (1, 3, 4), (– 2, 3, 7); (*ii*) (3, 5, 4), (– 6, 1, 2).
7. Find the angle between the lines whose direction ratios are (1, 2, 3) and (3, – 2, 1).
8. Show that the triangle ABC having the vertices (3, 5, – 4), (– 1, 1, 2) and (– 5, – 5, – 2) is isosceles. Find the angle at the vertex. [Allahabad, 1971]
9. Show that the line joining the points (1, 2, 3) and (4, 5, 7) is parallel to the line joining (– 2, 4, – 3) and (4, 10, 5).
10. P, Q, R, S are the points (2, 3, – 1), (3, 5, – 3), (1, 2, 3), (3, 5, 7). Prove that PQ is perpendicular to RS.
11. A, B, C, D are the points (6, – 6, 0), (– 1, – 7, 6), (3, – 4, 4), (2, – 9, 2). Find the projection to AB on CD. What conclusion do you draw about the two lines?
12. Find the angle between the diagonals of a cube. [Gorakhpur, 1965]
13. A line makes angles $\alpha, \beta, \gamma, \delta$ with the four diagonals of cube. Prove that
    $\cos^2 \alpha + \cos^2 \beta + \cos^2 \gamma + \cos^2 \delta = \frac{4}{3}.$ [Roorkee, 1975]
14. Lines are drawn from the origin with direction cosines proportional to (1, 2, 2), (2, 3, 6), (3, 4, 12). Show that the axis of the right circular cone through them has direction cosines

$\left(-\frac{1}{\sqrt{3}}, \frac{1}{\sqrt{3}}, \frac{1}{\sqrt{3}}\right)$ [Roorkee, 1974]

15. Show that a line with direction ratios α, β, 0 is perpendicular to the z-axis. Write down the direction ratios of any line perpendicular to the *x*-axis.
16. If l_1, m_1, n_1 and l_2, m_2, n_2 are the direction cosines of two mutually perpendicular straight lines, prove that the direction cosines of a straight line perpendicular to both of them are.

$$m_1n_2 - m_2n_1,\ n_1l_2 - n_2l_1,\ l_1m_2 - l_2m_1.$$ [A.M.I.E., 1977]

17. For what value of k will the line with direction ratios 2, k, – 3 be perpendicular to the line with direction ratios 5, – 8, 10.
18. Find the direction ratios of a line perpendicular to two lines whose direction ratios are – 1, 2, 3 and 2, 3, – 2.
19. Find the direction cosines of a line perpendicular to two lines having direction ratios 3, 4, 1 and 6, 2, – 1. [Banaras, 1967]
20. Find the distance of the point (4, 4, 4) from the line passing through (– 2, – 3, 3) and having direction ratios 1, 2, – 2.

3.14 EQUATION OF A PLANE

A plane can be specified by its orientation and a given point on it. The orientation is generally specified by the direction of the normal to the plane. Thus, there is only one plane normal to a given line and passing through a given point. The equation of the plane will be a relation satisfied by the coordinates (x, y, z) of any point on it.

I. To find the equation of a plane passing through a point P_1 (x_1, y_1, z_1) and perpendicular to a line AN having direction ratios *a, b, c*.

Let $P(x, y, z)$ be any point on the plane. Then AN is perpendicular to P_1P. Now the direction ratios of P1P are

$$x - x_1,\ y - y_1,\ z - z_1.$$

Fig. 3.15

Therefore,

$$a(x - x_1) + b(y - y_1) + c(z - z_1) = 0. \qquad ...(1)$$

This is the required equation.

II. To find the vector equation of a plane normal to a vector *n* and passing through a point P_1 having the position vector r_1.

Let *r* be the position vector of any point *P* on the plane. Then the vector P_1P lies in the plane and is therefore perpendicular to *n*.

Now $\qquad P_1P = r - r_1$. Hence

$$n \cdot (r - r_1) = 0 \qquad ...(2)$$

This is the vector equation of the given plane.

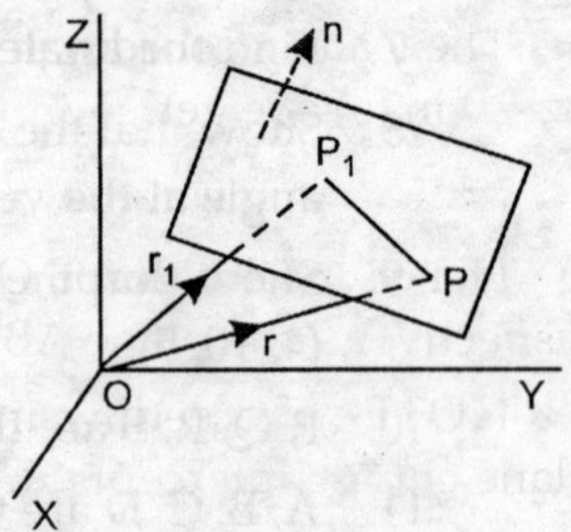

Fig. 3.16

III. Equations (1) and (2) are equivalent.

Put $\qquad r = xi + yj + zk,\ r_1 = x_1i + y_1j + z_1k,$

and $\qquad n = ai + bj + ck,$

then

$$n \cdot (r - r_1) = (ai + bj + ck) \cdot \{(x - x_1)i + (y - y_1)j + (z - z_1)k\}$$
$$= a(x - x_1) + b(y - y_1) + c(z - z_1).$$

Hence equations (1) and (2) are equivalent.

IV. The general equation of the first degree represents a plane.

The general equation of the first degree is

$$ax + by + cz + d = 0, \quad \text{...(3)}$$

which contains first degree terms in x, y, z and a constant term. Let x_1, y_1, z_1 be a set of values satisfying it, then

$$ax_1 + by_1 + cz_1 + d = 0. \quad \text{...(4)}$$

Subtracting (4) from (3), we get

$$a(x - x_1) + b(y - y_1) + c(z - z_1) = 0.$$

This shows that if P be a variable point whose coordinates (x, y, z) satisfy equation (3), the line joining P to a fixed point P_1 (x_1, y_1, z_1) is always normal to a fixed line having direction ratios a, b, c. Thus P lies on the plane through P_1 perpendicular to the fixed line. Hence the proposition.

We may note the following particular cases of the general equation (3).

(*i*) If $d = 0$, the plane passes through the origin.

(*ii*) If $a = 0$, the plane is parallel to the x-axis.

(*iii*) If $a = b = 0$, the plane is parallel to the xy-plane.

Other cases can be similarly interpreted. The last case is specially important. The planes parallel to XOY, YOZ, ZOX are respectively of the form z = constant, x = constant, y = constant. In particular the three coordinate planes have the equations

$$z = 0, x = 0, y = 0.$$

COROLLARY 1. The normal to the plane $ax + by + cz + d = 0$ has direction ratios a, b, c.

COROLLARY 2. A plane parallel to the plane $ax + by + cz + d = 0$ is $ax + by + cz + d' = 0$.

COROLLARY 3. If $ax + by + cz + d = 0$ and $a'x + b'y + c'z + d' = 0$ be two planes, the angle θ between them is equal to the angle between their normals. Hence

$$\cos\theta = \frac{aa' + bb' + cc'}{\sqrt{(a^2 + b^2 + c^2)}\sqrt{(a'^2 + b'^2 + c'^2)}}$$

Example. Find the intercepts the plane $2x + 3y + 4z - 12 = 0$ makes on the axes of coordinates, and hence object it on a diagram.

Transposing the constant term on the right, and dividing the whole equation by it, we can write the equation of the plane as

$$\frac{x}{6} + \frac{y}{4} + \frac{z}{3} = 1 \quad \text{...(5)}$$

The y and z coordinates of the point A where the plane cuts OX, are zero. Putting $y = z = 0$ in (5), we get

$$x = 6.$$

Hence the intercept $OA = 6$. Similarly the intercepts on OY and OZ are 4 and 3 respectively. The plane ABC cutting these intercepts is drawn in the diagram above.

NOTE : Proceeding in a similar manner with equation (3), we can show that the plane cutting intercepts A, B, C from the axes, has the equation

$$\frac{x}{A} + \frac{y}{B} + \frac{z}{C} = 1$$

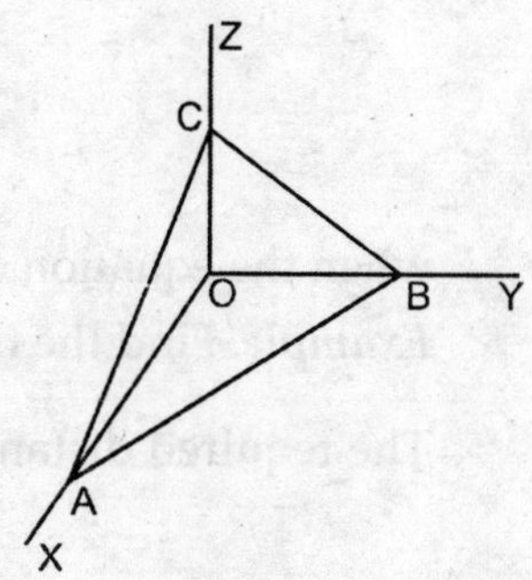

Fig. 3.17

3.15 NORMAL FORM

To find the equation of a plane in terms of p, the perpendicular from origin to the plane, and its direction cosines $\cos a$, $\cos b$, $\cos c$.

Let p be the vector ON perpendicular to the given plane, of length p and making angles α, β, γ with the axes (N being on the plane) ; then

$$p = (p\cos\alpha)i + (p\cos\beta)j + (p\cos\gamma)k.$$

Now the given plane passes through the point p and the vector normal to the plane is also p. So its equation is

$$p.(r-p) = 0, \quad \text{or} \quad p.r = p^2, \qquad ...(1)$$

or $\{(p\cos\alpha)i + (p\cos\beta)j + (p\cos\gamma)k\}.\{xi + yj + zk\} = p^2,$

or $$x\cos\alpha + y\cos\beta + z\cos g = p. \qquad ...(2)$$

Since $p/p = n$, the unit vector normal to the plane, we can write (1) as

$$n.r = p.$$

This is the vector form of equation (2).

If a given plane has the equation $ax + by + cz + d = 0$, a comparison with (2) gives

$$\frac{\cos\alpha}{a} = \frac{\cos\beta}{b} = \frac{\cos\gamma}{c} = \frac{-p}{d} = \frac{\sqrt{(\cos^2\alpha + \cos^2\beta + \cos^2\gamma)}}{\sqrt{(a^2+b^2+c^2)}}$$

$$= 1/\sqrt{(a^2+b^2+c^2)}. \qquad ...(3)$$

3.16 DISTANCE OF A POINT FROM A PLANE

To find the perpendicular distance of a point $P(x_1, y_1, z_1)$ from a given plane.

(i) Let the given plane be $(r - r_0).n = 0$ where r_0 is the position vector of a point A on the plane and n is the unit vector normal to the plane.

Draw PN_1 perpendicular to the plane; then the perpendicular distance of P from the plane

$$= N_1P = \text{projection of } AP \text{ on the line } N_1P$$

$$= AP.n = (r_1 - r_0).n, \qquad ...(4)$$

where r_1 is the position vector of P.

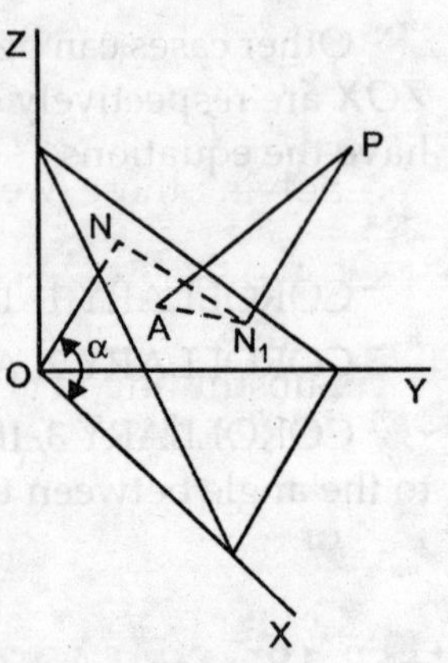

Fig. 3.18

(ii) Draw ON perpendicular to the plane. Let ON make angles α, β, γ with the axes. Then

$$n = (\cos\alpha)i + (\cos\beta)j + (\cos\gamma)k.$$

Also $$r_1 = x_1 i + y_1 j + z_1 k.$$

The position vector of N is pn, where $p = ON$. So taking N in place of A, (4) gives

$$N_1P = (r_1 - pn).n = r_1.n - pn^2 = r_1.n - p$$

$$= x_1\cos\alpha + y_1\cos\beta + z_1\cos\gamma - p$$

$$= \frac{ax_1 + by_1 + cz_1 + d}{\sqrt{(a^2+b^2+c^2)}}, \text{ by (3) of Heading 3.18}$$

when the equation of the plane is $ax + by + cz + d = 0$.

Example. Find the distance of the point (2, 3, 4) from the plane $3x - 6y + 2z + 12 = 0$. [Allahabad; 1969]

The required distance $$= \frac{3.2 - 6.3 + 2.4 + 12}{\sqrt{(3^2+6^2+2^2)}}$$

$$= (6 - 18 + 8 + 12)/\sqrt{49} = 8/7.$$

3.17 EXAMPLES ON PLANES

Though four constants appear in the general equation of the plane, they reduce to only three arbitrary contants after division by one of them. Hence a plane can be found to satisfy essentially three given conditions. Some examples illustrating this are given below.

Example. 1. Find the equation of the plane through the points (1, 1, 0), (1, 2, 1), (– 2, 2, – 1).

A plane through the first point is

$$a(x-1)+b(y-1)+cz = 0$$

It passes through the other two points if

$$a(1-1)+b(2-1)+c\,.\,1 = 0,$$

and $$a(-2-1)+b(2-1)+c(-1) = 0,$$

i.e. $$b+c = 0 \text{ and } -3a+b-c=0.$$

These give $a:b:c::2:3:-3$. Hence the required equation is

$$2(x-1)+3(y-1)-3z = 0, \text{ or } 2x+3y-3z=5.$$

Example. 2. Find the equation of the plane passing through the point (1, 2, – 1) and perpendicular to the planes $x+y-2z=5$ and $3x-y+4z=12$. [Punjab, 1980]

The equation of a plane through the given point is

$$a(x-1)+b(y-2)+c(z+1) = 0 \qquad ...(1)$$

It is perpendicular to the given planes, that

$$a+b-2c = 0,$$

$$3a-b+4c = 0.$$

Solving these, we get

$$\frac{a}{4-2} = \frac{-b}{4+6} = \frac{c}{-1-3} \text{ or } \frac{a}{1} = \frac{b}{-5} = \frac{c}{-2}.$$

Substituting in (1), the equation of the plane is found to be

$$(x-1)-5(y-2)-2(z+1) = 0,$$

or $$x-5y-2z+7 = 0.$$

3.18 SCALAR TRIPLE PRODUCT

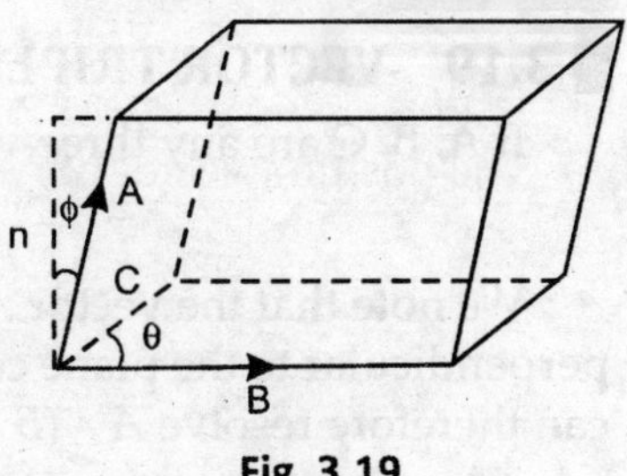

Fig. 3.19

When we have more than two vectors we can obtain their continued product by repeated multiplication. Since the multiplication can be either the multiplication of a vector by a number, the scalar multiplication of two vectors or their vectors multiplication, the same vectors will give a variety of continued products. For example, some of the products of vectors *A*, *B* and *C* are

$$(A\,.\,B)C, \quad A\,.\,(B\times C), \quad A\times(B\times C).$$

The first of these is a scalar multiple of the vector *C*. The second is the scalar product of *A* and *B* × C, and is called a scalar triple product. The third is the vector product of *A* and *B* × C, and is called a vector triple product. We proceed to obtain their values.

Let *A*, B and C be any three vectors. Construct a parallelopiped with its three coterminus edges the same magnitude and directions as *A*, *B* and C respectively. Let *q* be the angle between *B* and C and *f* the angle between A and n, the unit vector normal to *B* and C. Then the scalar triple product

$$A\,.\,(B\times C) = A\,.\,\{n(BC\sin\theta)\}$$

$$= (A\,.\,n)(BC\sin\theta)$$

$$= (A\cos\phi)(BC\sin\theta).$$

Now *BC* sin θ is the are of the parallelogram formed by the vectors *B* and C, i.e. the base of the parallelopiped (see figure). Also *A* cos ϕ is the height of the parallelopiped. Their product gives the volume of the parallelopiped. Hence the scalar triple product *A* . (*B* × C) represents the volume of the parallelopiped formed by the vectors *A*, *B* and C.

The geometrical representation also shows that

$$A\,.\,(B \times C) = (A \times B)\,.\,C \qquad ...(1)$$

For though we take in the latter product the area of the face formed by A and B multiplied by the height perpendicular to it, we get the same volume. We express this by saying that the positions of the dot and the cross in a scalar triple product are interchangeable.

By a similar reasoning we see that

$$A\,.\,(B \times C) = B\,.\,(C \times A) = C\,.\,(A \times B).$$

The last result also follows from (1) using the commutative property of dot product. Using this property several other results could also be written. It should be noted, however, that

$$A\,.\,(B \times C) = -A\,.\,(C \times B)$$

since the cross product is not commutative. In fact the sign of the scalar triple product changes as soon as the cyclic order of A, B, C in the product changes.

It also follows that the scalar triple product $A\,.\,(B \times C)$ is zero when any two of the vectors are parallel (or equal), or when the three vectors are coplanar.

Some authors use the symbol [ABC] to denote the scalar product $A\,.\,(B \times C)$.

Ex. If A, B, C are three vectors with components Ax, Ay, Az, etc., show that

$$A\,.\,(B \times C) = \begin{vmatrix} Ax & Ay & Az \\ Bx & By & Bz \\ Cx & Cy & cz \end{vmatrix}$$

We have $\quad A = Ax\,i + Ay\,j + Az\,k,$

and $\quad B \times C = (B_yC_z - B_zC_y)i + (B_zC_x - B_xC_z)j + (B_xC_y - B_yC_x)k$, by Heading 3.9

Therefore $A\,.\,(B \times C)$

$$= A_x(B_yC_z - B_zC_y) + A_y(B_zC_x - B_xC_z) + A_z(B_xC_y - B_yC_x)$$

$$= \begin{vmatrix} Ax & Ay & Az \\ Bx & By & Bz \\ Cx & Cy & Cz \end{vmatrix}$$

3.19 VECTOR TRIPLE PRODUCT

If A, B, C are any three vectors, then

$$A \times (B \times C) = (A\,.\,C)\,B - (A\,.\,B)C.$$

We note that the vector $A \times (B \times C)$ is perpendicular to the vector $B \times C$. But the latter is perpendicular to the plane containing B and C. Hence the former must lie in this plane. We can therefore resolve $A \times (B \times C)$ into components parallel to B and C, and write

$$A \times (B \times C) = mB + nC, \qquad ...(1)$$

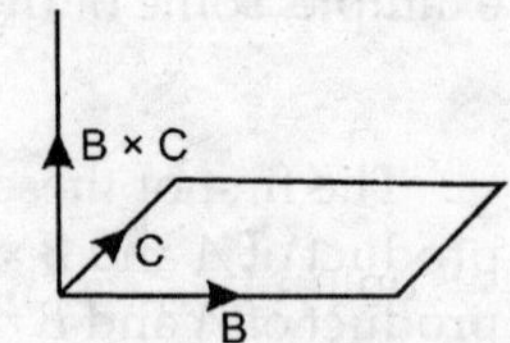

Fig. 3.20

where m and n are suitable numbers.

Multiply both sides scalarly by A, then

$$A\,.\,\{A \times (B \times C)\} = m(A\,.\,B) + n(A\,.\,C). \qquad ...(2)$$

The left-hand side is a scalar triple product of A, A and B × C. This is zero since two of the vectors are equal. Hence

$$m(A\,.\,B) + n(A\,.\,C) = 0,$$

or

$$\frac{m}{A.C} = \frac{n}{-(A.B)} = p, \text{ say.}$$

Putting those values of m and n in (1), we get

$$A \times (B \times C) = p(A\,.\,C)\,B - p(A\,.\,B)C. \qquad ...(3)$$

We notice that both the sides in (3) are equally balanced in A, B, C. Hence p must be some numerical constant independent of A, B, C. To determine it, take the special case when $A = B = i$ and $C = j$, then

$$i \times (i \times j) = p(i \cdot j)\, i - p\, (i \cdot i) j,$$

or

$$i \times k = 0 - pj,$$

or

$$-j = -pj,$$

which gives $p = 1$. Substituting in (3) we get the required result.

COR.
$$(A \times B) \times C = -C \times (A \times B)$$
$$= [(C \cdot B)A - (C \cdot A)B]$$
$$= (A \cdot C)B - (B \cdot C)A.$$

Example 1. Prove that

$A \times (B \times C) + B \times (C \times A) + C \times (A \times B) = 0.$ [Gorakhpur, 1976]

We have

$$A \times (B \times C) = (A \cdot C)B - (A \cdot B)C,$$
$$B \times (C \times A) = (B \cdot A)C - (B \cdot C)A,$$

and
$$C \times (A \times B) = (C \cdot B)A - (C \cdot A)B.$$

Adding these, the terms on right cancel in pairs and we get the result.

Example 2. Prove that

$$(A \times B) \cdot (C \times D) = \begin{vmatrix} A.C & B.C \\ A.D & C.D \end{vmatrix}.$$ [Gorakhpur, 1995]

$$\text{L.H.S.} = (A \times B) \cdot P, \text{ where } P = C \times D,$$
$$= A \cdot (B \times P), \text{ since dot and cross are interchangeable in scalar triple product,}$$
$$= A \cdot \{B \times (C \times D)\}$$
$$= A \cdot \{C(B \cdot D) - D(B \cdot C)\}$$
$$= (A \cdot C)(B \cdot D) - (A \cdot D)(B \cdot C) = \text{R.H.S}$$

Example. 3. Prove that

$$(A \times B) \cdot \{(B \times C) \times (C \times A)\} = \{A \cdot (B \times C)\}^2.$$ [A.M.I.E., 1977]

Putting
$$B \times C = P, \text{ we see that}$$
$$(B \times C) \times (C \times A) = P \times (C \times A) = (P \cdot A)C - (P \cdot C)A$$
$$= \{(B \times C) \cdot A\}\, C - \{(B \times C) \cdot C\}A = mC,$$

on putting
$$(B \times C) \cdot A = m, \text{ and noting that } (B \times C) \cdot C = 0$$

Hence
$$(A \times B) \cdot \{(B \times C) \times (C \times A)\}$$
$$= (A \times B) \cdot mC$$
$$= A \cdot (B \times C)m, \text{ as . and cross are interchangeable,}$$
$$= \{A \cdot (B \times C)\}^2.$$

Example. 4. Find the vector equation of a plane through three point with position vector a, b, and c. Also find the length of the perpendicular from the origin on this plane. [Roorkee, 1979]

Let the three points with vectors a, b and c be A, B and C; then

$$AB = b - a \text{ and } AC = c - a.$$

Since AB and AC lie in the plane, a normal to the plane is

$$n = AB \times AC = (b - a) \times (c - a)$$
$$= b \times c - b \times a - a \times c + a \times a$$
$$= b \times c + c \times a + a \times b. \quad ...(1)$$

Also the plane passes through A. Therefore its equation is

$$n \cdot (r - a) = 0, \text{ or } (b \times c + c \times a + a \times b) \cdot (r - a) = 0,$$

or $$(b \times c + c \times a + a \times b) \cdot r = a \cdot (b \times c), \quad ...(2)$$

since the other scalar triple products are zero.

If n_1 denotes the unit vector normal to the plane, then $n_1 = n/|n|$. So dividing (2) by $|n|$ we can write it as

$$n_1 \cdot r = a \cdot (b \times c)/|n|.$$

Then the perpendicular on the plane from origin

$$= \frac{a.(b \times c)}{|n|} = \frac{a.(b \times c)}{|b \times c + c \times a + a \times b|}.$$

Examples :

1. Find the equation of the plane through the point (2, – 3, 5) and normal to the line with direction ratios 4, 6, – 3.
2. Find the equation of the plane perpendicular to the line joining (– 3, 2, 1) to (7, 6, 3) and passing through its middle point.
3. Find the equation of the plane through the point (2, – 3, 1) and parallel to the plane $3x - 4y + 2z = 5$.
4. Find the equation of the plane perpendicular to the vector $2i + 3j - 6k$ and passing through a point whose position vector is $i - 3j + 2k$.
5. Find the equation of the plane parallel to the vectors $i + j - 2k$ and $3i - j + 4k$ and passing through the point (1, 2, – 1).
6. Find the equation of the plane passing through three points whose position vectors are $2i + 3j + 4k$, $3i + j + 3k$ and $-3i + 5j + k$.
7. Find the intercepts on the coordinate axes made by the plane $x + 2y - z = 9$. Also find the length and the direction cosines of the perpendicular from the origin on it.
8. A plane meets the coordinates axes at A, B, C such that the centroid of the triangle ABC is the point (a, b, c). Show that the equation of the plane is
$$x/a + y/b + z/c = 3.$$
[Gorakhpur, 1972]
9. Find the acute angle between the planes $2x - y + z = 6$, $x + y + 2z = 3$. [Roorkee, 1967]
10. Find the distance of the origin from the plane of Ex. 4.
11. Find the distance of the point whose position vector is 2i + j – 3k from the plane of Ex. 6.
12. Find the distances of the points (2, 3, – 5), (3, 4, 7) from the plane $x + 2y - 2z = 9$. Are the points on the same side of the plane? [Punjab, 1980]
13. Find the perpendicular distance of a point a from a plane parallel to b and c and passing through a point d. Find the distance when
$$a = (1, 1, 3), b = (2, 1, -1), c = (1, -1, -1), d = (1, 0, 2).$$
[Roorkee, 1978]
14. Find the equations of the planes which bisect the angles between the planes Identify the one which bisects the acute angle between these two planes. [Punjab, 1978]
$$3x - 2y + 6z + 8 = 0 \text{ and } 2x - y + 2z + 3 = 0.$$
15. Show that the four points (0, – 1, 0), (2, 1, – 1), (1, 1, 1) and (3, 3, 0) are coplanar. Find the equation of the plane through them. [Gorakhpur, 1975]

16. Find the equation of the plane through the point (1, 0, – 2) and perpendicular to the planes $2x + y - z = 2$ and $x - y - z = 3$. [Kurukshetra, 1969]
17. Find the equation of the plane through the points (2, 2, 1), (1, – 2, 3) and perpendicular to the plane $x - 2y + 3z + 4 = 0$.
18. Find the equation of the plane through the points (2, 2, 1), (1, – 2, 3) and parallel to x-axis. [Allahabad, 1963]
19. Find the equation of the plane through the points (2, 2, 1), (1, – 2, 3) and parallel to the line joining the points (2, 1, – 3), (– 1, 5, – 8).
20. If the plane $ax + by + cz + d = 0$ is parallel to the line with direction ratios l, m, n, show that $al + bm + cn = 0$.
21. Show that the equation of the plane passing through the three points (x_1, y_1, z_1), (x_2, y_2, z_2), (x_3, y_3, z_3) is

$$\begin{vmatrix} x & y & z & 1 \\ x_1 & y_1 & z_1 & 1 \\ x_2 & y_2 & z_2 & 1 \\ x_3 & y_3 & z_3 & 1 \end{vmatrix} = 0$$ [Allahabad, 1973]

22. *A* variable plane which remains at a constant distance $3p$ from the origin cuts the coordinate axes at *A*, *B*, *C*. Show that the locus of the centroid of the triangle ABC is

$$\frac{1}{x^2} + \frac{1}{y^2} + \frac{1}{z^2} = \frac{1}{p^2}.$$ [Roorkee, 1970]

23. A variable plane passes through a fixed point (a, b, c) and meets the coordinate axes in A, B, C. Show that the locus of the point common to the planes through A, B, C parallel to the coordinate planes is

$$\frac{a}{x} + \frac{b}{y} + \frac{c}{z} = 1.$$ [Allahabad, 1972]

24. Evaluate $A \cdot (B \times C)$ if $A = 2i - 3j$, $B = i + j - k$ and $C = 3i - k$.
25. Find a such that the vectors $2i - j + k$, $i + 2j - 3k$, and $3i + aj + 5k$ are coplanar. [Gorakhpur, 1995]
26. If $A = 3j - 2j + 2k$, $B = 6i + 4j - 2k$, $C = -3i - 2j - 4k$, then compute the values of $A \cdot (B \times C)$ and $A \times (B \times C)$.

Prove the following identities:

27. $(A + B) \cdot (B + C) \times (C + A) = 2A \cdot B \times C$. [Roorkee, 1972]
28. $i \times (a \times i) + j \times (a \times j) + k \times (a \times k) = 2a$. [Gorakhpur, 1994]
29. $A \times \{B \times (C \times D) = (B \cdot D)(A \times C) - (B \cdot C)(A \times D)$. [Allahabad, 1973]
30. $(A \times B) \times (C \times D) = (A \times B \cdot D)C - (A \times B \cdot C)D$. [Gorakhpur, 1996]

3.20 EQUATION OF A LINE

A line can be specified by its direction and a point on it. Thus there is only one line with given direction cosines and passing through a given point. The equation of the line will be a relation (or relations) satisfied by the coordinates (x, y, z) of any point on it.

I. *To find the equation of a line passing through a point P_1 (x_1, y_1, z_1) and having direction cosines* cos α, cos β, cos γ.

Let P (x, y, z) be any point on the line, and let P_1P be of length r. Then the direction cosines of P_1P are given by

$$\cos\alpha = \frac{x - x_1}{r}, \cos\beta = \frac{y - y_1}{r}, \cos\gamma = \frac{z - z_1}{r}. \quad ...(1)$$

These relations give

$$\frac{x - x_1}{\cos\alpha} = \frac{y - y_1}{\cos\beta} = \frac{z - z_1}{\cos\gamma}. \quad ...(2)$$

These two equations in x, y, z are the required equations of the line.

COROLLARY 1. *The equations of a line through* (x, y, z) *having direction ratios* l, m, n *are*

$$\frac{x-x_1}{l} = \frac{y-y_1}{m} = \frac{z-z_1}{n}. \quad ...(3)$$

This is known as the symmetric form of the equations of a line.

Putting each of the expressions in (3) equal to t, we get the parametric form of the equations of a line:

$$x = x_1 + lt, y = y_1 + mt, z = z_1 + nt.$$

COROLLARY 2. *The coordinates of a point on the line at a distance r from* (x_1, y_1, z_1) *are*

$$x = x_1 + r\cos\alpha, y = y_1 + r\cos\beta, z = z_1 + r\cos\gamma.$$

This follows from equations (1).

II. *To find the vector equation of a line passing through a point* P_1 *with position vector* $\mathbf{r}_1$ *and having the same direction as the vector* $\mathbf{m}$.

Let the position vector of any point P on the line be $\mathbf{r}$. Then the vector $\mathbf{P_1P}$ and $\mathbf{m}$ are parallel; therefore

$$\mathbf{r} - \mathbf{r}_1 = \mathbf{m}t, \text{ or } \mathbf{r} = \mathbf{r}_1 + \mathbf{m}t, \quad ...(4)$$

where t is a parameter. This is the vector form of the parametric equation of the line.

Forming the vector product of (4) with $\mathbf{m}$, we get the vector equation of the line in the form:

$$(\mathbf{r} - \mathbf{r}_1) \times \mathbf{m} = 0. \quad ...(5)$$

If $\mathbf{m} = l\mathbf{i} + m\mathbf{j} + n\mathbf{k}$, then equating the coefficients of $\mathbf{i}, \mathbf{j}, \mathbf{k}$ equation (4) gives the scalar parametric form:

Fig. 3.21

$$x = x_1 + lt, y = y_1 + mt, z = z_1 + nt,$$

which (5) gives the symmetric form

$$\frac{x-x_1}{l} = \frac{y-y_1}{m} = \frac{z-z_1}{m}.$$

***Example.* 1.** Find the equations of the line through (1, 4, – 2) and parallel to the planes $6x + 2y + 2z + 3 = 0$ and $x + 2y - 6z + 4 = 0$. [*Banaras*, 1969]

The line, being parallel to the planes, must be perpendicular to their normals. Therefore, if its direction ratios are l, m, n we have

$$6l + 2m + 2n = 0$$

and

$$l + 2m - 6n = 0$$

giving

$$\frac{l}{-16} = \frac{-m}{-38} = \frac{n}{10} \text{ or } \frac{l}{-8} = \frac{m}{19} = \frac{n}{5}.$$

Hence the required line is

$$\frac{x-1}{-8} = \frac{y-4}{19} = \frac{z+2}{5}.$$

***Example.* 2.** Find the equation of the line perpendicular to the two lines $r = r_1 + mt$, $r = r_2 + nt$, and through a point P with position vector r_0.

The direction of the required line is given by $m \times n$ which is perpendicular to both the given lines. Hence the equation of the line is

$$r = r_0 + m \times nt.$$

3.21 INTERSECTION OF TWO PLANES

We have seen about that a line is given by two equations. In fact, if we have two planes $ax + by + cz + d = 0$ and $a'x + b'y + c'z + d' = 0$, their line of intersection has the equations: taken simultaneously. Only the points on the line of intersection satisfy both the equations.

$$ax + by + cz + d = 0, a'x + b'y + c'z + d' = 0 \quad ...(1)$$

In order to write the equations (1) in the symmetric form, we shall need the direction ratios of the line, say l, m, n, and the coordinates of a point on it, say $(x_1, y_1, 0)$.

Since the line of intersection lies in the plane $ax + by + cz + d = 0$, it is perpendicular to the normal to the plane. Therefore

$$al + bm + cn = 0$$

Similarly $$a'l + b'm + c'n = 0.$$

Solving these we get $l : m : n$.

Again, since $(x_1, y_1, 0)$ lies on the two planes, we get

$$ax_1 + by_1 + d = 0$$

and $$a'x_1 + b'y_1 + d' = 0$$

These gives x_1 and y_1. It should be noted that instead of $(x_1, y_1, 0)$, we could have chosen the point as $(0, y_1, z_1)$ or $(x_1, 0, z_1)$; or in fact, any one of the three coordinates could have been chosen arbitrarily.

Example 1. Obtain the equations of the line of intersection of the plane $x + y - 2z = 8$ and $3x - y + 4z = 12$ in the symmetric form.

Let l, m, n be the direction ratios of the line of intersection and $(x_1, y_1, 0)$ a point on it; then

$$l + m - 2n = 0,$$

and $$3l - m + 4n = 0;$$

giving $$\frac{l}{4-2} = \frac{-m}{4+6} = \frac{n}{-1-3},$$

or $l : m : n :: 1 : -5 : -2$.

Also $$x_1 + y_1 = 8,$$

and $$3x_1 - y_1 = 12;$$

giving $$4x_1 = 20, \text{ i.e. } x_1 = 5 \text{ and } y_1 = 3.$$

Hence the line of intersection is

$$\frac{x-5}{1} = \frac{y-3}{-5} = \frac{z}{-2}.$$

Example 2. Show that the points (1, 2, 3), (– 5, 4, 3) lie on the planes $(x - 1)/3 = (y - 2)/(-1)$ and $z - 3 = 0$. Hence given an interpretation of the symmetric form of equations of the line joining the two points.

Direction ratios of the line joining the two points are 1 – (– 5), 2 – 4, 3 – 3, or 3, – 1, 0. Hence by the previous article the equation of the line are

$$\frac{x-1}{3} = \frac{y-2}{-1} = \frac{z-3}{0}. \quad ...(2)$$

Since division by zero is not defined, a question of interpreting this equation arises.

By substitution we see that the two points lie on the plane $(x - 1)/3 = (y - 2)/(-1)$ and also on the plane $z - 3 = 0$. Hence the line joining the two points is their line of intersection:

$$\frac{x-1}{3} = \frac{y-2}{-1}, z - 3 = 0. \quad ...(3)$$

Thus the equations (2) should always be interpreted as equations (3). are to be interpreted as the equations.

NOTE. Similarly, the equation

$$\frac{x-a}{1} = \frac{y-b}{0} = \frac{z-c}{0}$$

$$y - b = 0, z - c = 0$$

Example 3. Find the equation of the line of intersection of the planes $r \cdot n_1 = d_1$ and $r \cdot n_2 = d_2$.

The line of intersection lies in both the planes. So it is normal to n_1 and n_2. Hence it is parallel to $n_1 \times n_2$.

If r_1 be the position vector of a point on the line of intersection, then $r_1 \cdot n_1 = d_1$ and $r_1 \cdot n_2 = d_2$.

Therefore the equation of the line of intersection is

$$(r - r_1) \times (n_1 \times n_2) = 0,$$

or

$$\begin{aligned} r \times (n_1 \times n_2) &= r_1 \times (n_1 \times n_2) \\ &= (r_1 \cdot n_2)n_1 - (r_1 \cdot n_1)n_2 \\ &= d_2 n_1 - d_1 n_2. \end{aligned}$$

Examples:

1. Write down the equations of the line through the point (2, – 3, – 7) and having direction ratios 3, – 4, 5.
2. Write down the equations of the line through the point (1, 2, 3) which makes equal angles with the axes.
3. Find the equations of the line passing through the point (1, – 2, 0) and parallel to the line
$$\frac{x-1}{2} = \frac{y+2}{-5} = \frac{z-7}{6}.$$
4. Find the equations of the line passing through the point (– 1, 2, 4) and perpendicular to the plane $3x - y + 2z = 5$.
5. Find the equations of the line passing through the points (– 1, 2, 4) and (3, 1, – 2).
6. Prove that the points (3, 2, 4), (4, 5, 2) and (5, 8, 0) are collinear. Find the equations of the line through them. [*Ranchi*, 73]
7. Prove that the equations of the line passing through the points (x_1, y_1, z_1) and (x_2, y_2, z_2) are
$$\frac{x-x_1}{x_2-x_1} = \frac{y-y_1}{y_2-y_1} = \frac{z-z_1}{z_2-z_1}.$$
8. Find the equation of the line passing through the points whose position vectors are **a** and **b**. Find the line when **a** = 3**i** + 2**j** + 4**k** and **b** = 2**i** + **j** – **k** and show that the point (5, 4, 14) lies on it.
9. Find the coordinates of the point in which the line joining (– 9, 4, 5) and (11, 0, – 1) is met by the perpendicular from the origin. [*Roorkee*, 1965]
10. Find the equations of the line joining (2, 4, 1) to (– 3, 5, 1) in the symmetric form. Show that they may also be written as $x + 5y = 22$, $z = 1$.
11. Write the equations $x + 2y + 4z = 0 = 2x + 4y - z$ of a line in the symmetric form.
12. Obtain the equations of the line of intersection of the planes $4x + 4y - 5z = 12$ and $8x + 12y - 13z = 32$ in the symmetric form. [*Roorkee, Arch.*, 1973]
13. Show that the line $2x + 2y - z - 6 = 0 = 2x + 3y - z - 8$ is parallel to the plane $y = 0$. Find the point where it meets the plane $x = 0$. [*Allahabad*, 1973]
14. Find the equations of the line through the point (1, 4, – 2) and parallel to the planes $6x + 2y + 2z + 3 = 0$ and $x + 2y - 6z + 4 = 0$.
15. Show that the straight lines $3x + 2y + z - 5 = 0 = x + y - 2z - 3$ and $8x - 4y - 4z = 0 = 7x + 10y - 8z$ are right angles. [Delhi, 1968]

16. Show that the perpendicular distance of a point a from the line passing through b and parallel to the vector m is $\sqrt{[(a-b)^2 - \{(a-b).m\}^2/m^2]}$. [*Roorkee*, 1973]

17. Find the perpendicular distance of the point (1, 1, 1) from the line

$$\frac{x-1}{2} = \frac{y-2}{3} = \frac{z}{-1}.$$ [*Ranchi*, 1973]

Also find the coordinates of the foot of the perpendicular.

18. Find the coordinates of the foot of the perpendicular from (1, 0, 2) to the line through (– 1, 2, – 1) and having its direction ratios 3, – 2, 1. Also find the length of the perpendicular and its equations. [*Delhi*, 1975]

19. Find the image of the point (– 2, 1, 3) in the plane $x + y - 2z + 1 = 0$. [*Roorkee*, 1994]

20. Assuming the plane $4x - 3y + 7z = 0$ to be horizontal, find the equations of the greatest slope through the point (2, 1, 1) in the plane $2x + y - 5z = 0$. [*Punjab*, 1980]

[Hint. The line of greatest slope if perpendicular to the line of intersection of the given plane with the horizontal plane.

3.22 A LINE AND A PLANE

I. Let

$$\frac{x-x_1}{l} = \frac{y-y_1}{m} = \frac{z-z_1}{n} \qquad ...(1)$$

be the equations of a given line and

$$ax + by + cz + d = 0 \qquad ...(2)$$

the equation of a given plane. Keeping in mind that l, m, n give the direction ratios of the line and a, b, c those of the normal to the plane, we see that

(*i*) The line is *perpendicular* to the plane when

$$l/a = m/b = n/c, \qquad ...(3)$$

(*ii*) The line is *parallel* to the plane when

$$al + bm + cn = 0, \qquad ...(4)$$

(*iii*) The line *lies in the plane* when

$$al + bm + cn = 0 \text{ and } ax_1 + by_1 + cz_1 + d = 0. \qquad ...(5)$$

Barring the cases (*ii*) and (*iii*), the line will intersect the plane in one point. To find its coordinates, take any point

$$(x_1 + lt, y_1 + mt, z_1 + nt) \qquad ...(6)$$

on the line (1). If this point lies on the plane (2) also, we get

$$a(x_1 + lt) + b(y_1 + mt) + c(z_1 + nt) + d = 0.$$

This gives t. Substituting in (6) we get the point of intersection.

II. If the plane have the equation $(r - r_1) . n = 0$ and the line $r = r_2 + mt$, then their point of intersection is given by

$$(r_2 + mt/ - r_1) . n = 0,$$

or
$$t = \frac{(r_1 - r_2) . n}{m . n}.$$

i.e.
$$r = r_2 + \frac{(r_1 - r_2) . n}{m . n} m.$$

The line is parallel to the plane if $m \cdot n = 0$. It lies in the plane if

$$m \cdot n = 0 \text{ and } (r_1 - r_2) \cdot n = 0.$$

Example **1.** Does the line

$$\frac{x-1}{5} = \frac{y+2}{6} = \frac{z-3}{4}$$

intersect any of the planes (*i*) $2x + 3y - 7z + 29 = 0$, (*ii*) $2x + 3y - 7z + 25 = 0$, or (*iii*) $2x + 3y - 6z + 20 = 0$? If it does, find the point of intersection.

The line has direction ratios 5, 6, 4 and the normal to plane (*i*) has direction ratios 2, 3, – 7. These satisfy the relation

$$5.2 + 6.3 + 4(-7) = 0 \quad \text{...(7)}$$

Hence the line is parallel to the plane (i). It does not lie on the plane, since (1, – 2, 3) does not lie on the plane.

For the plane (ii) the condition (7) is satisfied. In addition

$$2.1 + 3(-2) - 7.3 + 25 = 0.$$

Hence the line lies on the plane (ii).

For the plane (*iii*) no such conditions are satisfied. Taking a point on the line as

$$x = 5t + 1, y = 6t - 2, z = 4t + 3, \quad \text{...(8)}$$

and substituting in the equation of the plane, we get

$$2(5t + 1) + 3(6t - 2) - 6(4t + 3) + 20 = 0,$$

or

$$4t - 2 = 0, \text{ or } t = \frac{1}{2}.$$

This is the value of t for the point of intersection. From (8), we get its coordinates : $\left(3\frac{1}{2}, 1, 5\right)$.

Example **2.** Find the equation of the plane passing through the line and the point (4, 3, 7).

$$\frac{x-1}{5} = \frac{y+2}{6} = \frac{z-3}{4}$$

A plane through the given point is

$$a(x - 4) + b(y - 3) + c(z - 7) = 0$$

If it passes through the given line, we get

$$5a + 6b + 4c = 0,$$

and

$$a(1 - 4) + b(-2 - 3) + c(3 - 7) = 0,$$

i.e.,

$$3a + 5b + 4c = 0.$$

These give $a : b : c :: 4 : -8 : 7$. Hence the required plane is

$$a(x - 4) - 8(y - 3) + 7(z - 7) = 0,$$

or

$$4x - 8y + 7z = 41.$$

3.23 PLANE THROUGH THE INTERSECTION OF TWO PLANES

Let

$$ax + by + cz + d = 0 \quad \text{...(1)}$$

and

$$a'x + b'y + c'z + d' = 0 \quad \text{...(2)}$$

be two given planes. Then a plane through their line of intersection is given by

$$ax + by + cz + d + \lambda(a'x + b'y + c'z + d') = 0. \quad \text{...(3)}$$

The reason is that the coordinates of a point on the line of intersection satisfy both (1) and (2). Therefore they satisfy equation (3) also, whatever λ may be.

This method is simpler to apply than that of Ex. 2 above when the equation of the line is not given in the symmetric form. The constant λ will be determined from any additional condition the plane may have to satisfy.

Example. Find the equation of the plane through the line of intersection of $2x + y - z = 3$ and $5x - 3y + 4z + 9 = 0$ and parallel to the line $\frac{x-1}{2} = \frac{y-3}{4} = \frac{z-5}{5}$. [*Punjabi*, 1968]

A plane through the line of intersection of the given planes is

$$2x + y - z - 3 + \lambda(5x - 3y + 4z + 9) = 0.$$

Direction ratios of the normal to this plane are

$$2 + 5\lambda,\ 1 - 3\lambda,\ -1 + 4\lambda.$$

Hence this plane is parallel to the given line if

$$2(2 + 5\lambda) + 4(1 - 3\lambda) + 5(4\lambda - 1) = 0$$

i.e. if $18\lambda + 3 = 0$, or $\lambda = -\frac{1}{6}$.

Therefore the required equation is

$$2x + y - z - 3 - \frac{1}{6}(5x - 3y + 4z + 9) = 0,$$

or $$6(2x + y - z - 3) - (5x - 3y + 4z + 9) = 0,$$

or $$7x + 9y - 10z - 27 = 0.$$

Examples:

1. Prove that the line $\frac{x-4}{2} = \frac{y-2}{3} = \frac{z-3}{6}$ lies in the plane $3x - 4y + z = 7$. [*Banaras*, 1967]

2. Show that the line $\frac{x-1}{3} = \frac{y+2}{-2} = \frac{z-1}{2}$ is parallel to the plane $2x + 2y - z = 6$, and find the distance between them. [*Roorkee*, 1968]

3. Find the point of intersection of the line

$$\frac{x-1}{2} = \frac{y-2}{-3} = \frac{z+3}{4}$$

with the plane $2x + 4y - z + 1 = 0$.

4. Find the point of intersection of the plane $x + y + z + 1 = 0$ with the straight line

$$x + 2y + 3z = 4,\ x - y + 2z = 2.$$

Also find the angle that the line makes with the plane. [*Delhi*, 1967]

5. Find the distance of the point $(1, -2, 3)$ from the plane $x - y + z = 5$ measured parallel to the line $x/2 = y/3 = z/(-6)$.

6. Find the equation of the line drawn through the point $(-4, 3, 1)$ parallel to the plane $x + 2y - z = 5$. so as to intersect the line

$$\frac{x+1}{-3} = \frac{y-3}{2} = \frac{z-2}{-2}.$$ [*A.M.I.E.*, 1976]

7. Find the perpendicular distance of the point $(1, 1, 1)$ from the line

$$\frac{x-2}{3} = \frac{y+3}{2} = \frac{z}{-1}.$$ [*Roorkee*, 1976]

Find also the coordinates of the foot of the perpendicular and the equations of the perpendicular. [*Punjab*, 1961]

8. Find the equation of the plane through the points (2, – 1, 0) and (3, – 4,), parallel to the line $2x = 3y = 4z$. [*Allahadbad*, 1961]

9. Find the equation of the plane through the line $\frac{x-2}{3} = \frac{y}{-1} = \frac{z+3}{2}$ and parallel to the line $\frac{x-1}{4} = \frac{y+2}{2} = \frac{z}{3}$.

10. Find the equation of the plane which contains the line $\frac{x-1}{2} = \frac{y+1}{-1} = \frac{z-3}{4}$ and is perpendicular to the plane.

$$x + 2y + z = . 12.$$ [*A.M.I.E.*, 1977]

11. Find the equation of the two plane through the origin, which are parallel to the line

$$\frac{x-1}{2} = \frac{y+3}{-1} = \frac{z+1}{-2}$$ [Roorkee, 1995]

and distant 5/3 from it. Show that the two planes are perpendicular.

12. Find the equation of a plane through the line of intersection of the planes $x + y + z = 1$ and $2x + 3y + 4z = 5$ which is perpendicular to the planes $x - y + z = 0$. [*Punjabi*, 1969]

13. Find the equation of the plane through the line $3x - y + 2z - 4 = 0 = x + y + z - 2$ and passing through the point (2, 2, 1).

14. Find the equation of the plane through the intersection of the planes $r \cdot n = 4$ and $r \cdot m = 6$ and parallel to the *z*-axis when $n = 2i + 3j - 5k$ and $m = 3i - 4j + 5k$.

15. A plane is drawn through the line $x + y = 1$, $z = 0$ to make an angle $\sin^{-1}\left(\frac{1}{3}\right)$ with the plane $x + y + z = 0$. Prove that two planes can be drawn and find their equations. Prove also that the angle between the plane is $\cos^{-1}(7/9)$. [*Banaras*, 1975]

3.24 INTERSECTION OF THREE PLANES

Any two planes intersect in a line, and this line will intersect any third plane in a point. Therefore, in general, three planes intersect in a point. However, in special cases, the line of intersection of the first two planes could be in the third plane, or be parallel to it. In the former case the three planes have a common line of intersection. In latter case the three planes form a prism, the line of intersection of any two planes being parallel to the third. We now consider these cases analytically.

Let the three planes be represented by the equations

$$a_1x + b_1y + c_1z + d_1 = 0, \quad ...(1)$$

$$a_2x + b_2y + c_2z + d_2 = 0, \quad ...(2)$$

and $$a_3x + b_3y + c_3z + d_3 = 0. \quad ...(3)$$

Solving these, the point of intersection is given by

$$\frac{x}{\begin{vmatrix} b_1 & c_1 & d_1 \\ b_2 & c_2 & d_2 \\ b_3 & c_3 & d_3 \end{vmatrix}} = \frac{-y}{\begin{vmatrix} a_1 & c_1 & d_1 \\ a_2 & c_2 & d_2 \\ a_3 & c_3 & d_3 \end{vmatrix}} = \frac{z}{\begin{vmatrix} a_1 & b_1 & d_1 \\ a_2 & b_2 & d_2 \\ a_3 & b_3 & d_3 \end{vmatrix}} = \frac{-1}{\begin{vmatrix} a_1 & b_1 & c_1 \\ a_2 & b_2 & c_2 \\ a_3 & b_3 & c_3 \end{vmatrix}} \quad ...(4)$$

Three cases arise:

(*i*) If the last determinant is not zero, the planes have a point of intersection at a finite distance from the origin.

(*ii*) If the last determinant is zero, but not any others, we do not get a point of intersection with finite coordinates. In this case the three planes form a prism, and the line of intersection of any two of them is parallel to the third.

(*iii*) If the last determinant and any one of the others are zero, then the other two determinants are also zero. In this case the three planes intersect in a common line.

To prove (*ii*) and (*iii*), we see that the direction ratios l, m, n of the line of intersection of two of the planes, say (2) and (3) are given by

$$\frac{l}{b_2c_3 - b_3c_2} = \frac{-m}{a_2c_3 - a_3c_2} = \frac{n}{a_2b_3 - a_3b_2}.$$

If this line in the plane (1), or is parallel to it, then $a_1 l + b_1 m + c_1 n = 0$, or

$a_1(b_2c_3 - b_3c_2) - b_1(a_2c_3 - a_3c_2) + c_1(a_2b_3 - a_3b_2) = 0,$

i.e., the last determinant in (4) vanishes.

Similarly, the vanishing of the last but one determinant in (4), can be interpreted as the condition that the point $(x_1, y_1, 0)$ lying on the intersection of planes (2) and (3), lies on plane (1) also.

3.25 SHORTEST DISTANCE

Two straight lines which intersect (or are parallel) lio in one plane and are called *coplanar*. Two straight lines which do not line in one plane are called *skew* lines. They neither intersect nor are parallel. Such lines posses a common perpendicular which is the *shortest distance* between the lines. The distance of any point on one line from a point on the other line will always be greater than this shortest distance, except when the two points coincide with the two feet of the common perpendicular to the lines.

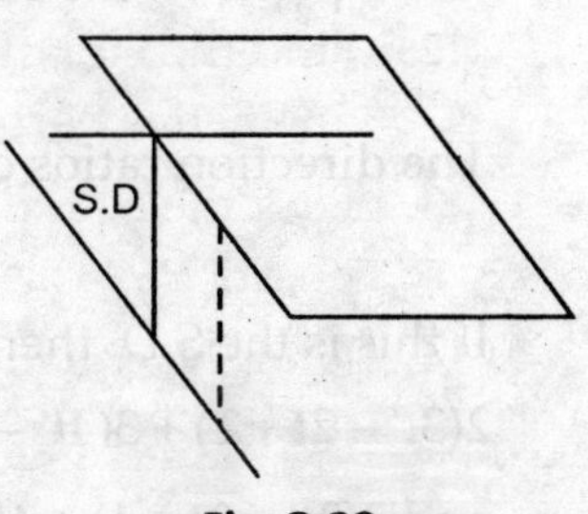

Fig. 3.22

To find the shortest distance between two skew lines, we can first find the plane which passes through the first line (I in the fig.) and is parallel to the second. Then every point on the second line is equidistant from this plane. Hence the shortest distance (S.D.) is equal to the length of the perpendicular to the plane from any point on the second line.

If the length of the perpendicular from a point on the second line to this plane is zero, then that point (and every other point on the second line) lies on the plane. In this case two lines are coplanar.

If the equation of the shortest distance is required we can first find the coordinates of the feet of the common perpendicular to the two lines. This is illustrated in Example 3 below.

***Example* 1.** Show that the lines

$$\frac{x-1}{1} = \frac{y+2}{2} = \frac{z-3}{4} \text{ and } \frac{x+6}{4} = \frac{y-4}{-2} = \frac{z-5}{1}$$

are coplanar; and find the equation of the plane containing them.

The plane containing the first time and parallel to the second is

$$a(x-1) + b(y+2) + c(z-3) = 0,$$

where $\quad 1 . a + 2 . b + 4 . c = 0,$

and $\quad 4 . a - 2 . b + 1 . c = 0$

The last two equations give $a : b : c :: 2 : 3 : -2$. So the above plane becomes

$$2(x-1) + 3(y+2) - 2(z-3) = 0, \text{ or } 2x + 3y - 2z + 10 = 0.$$

The point $(-6, 4, 5)$ on the second line also satisfies this equation. Hence the two lines are coplanar, and the plane containing them is $2x + 3y - 2z + 10 = 0$.

***Example* 2.** Show that the shortest distance between the lines

$$\frac{x-1}{2} = \frac{y-2}{3} = \frac{z-3}{4} \text{ and } \frac{x-2}{3} = \frac{y-4}{4} = \frac{z-5}{5} \text{ is } 1/\sqrt{6}.$$ [*Bhopal*, 1978]

The plane passing through the first line and parallel to the second is

$$a(x-1)+b(y+2)+c(z-3) = 0,$$

where $\quad 2 \,.\, a + 3 \,.\, b + 4 \,.\, c = 0,$

and $\quad 3 \,.\, a + 4 \,.\, b + 5 \,.\, c = 0.$

The last two equations give $a : b : c :: 1 : -2 : 1$. So the above plane becomes

$(x-1)-2(y-2)+(z-3)=0$, or $x-2y+z = 0$.

The shortest distance = perpendicular on this plane from any point on the second line, say from (2, 4, 5),

$$= \frac{2-2.4+5}{\sqrt{(1+2^2+1)}} = \frac{1}{\sqrt{6}}, \text{ taking the numerical value.}$$

Example 3. Find the equation of the shortest distance in Ex. 2. [*Allahabad*, 1969]

Any point on the first line is

$$x = 2t+1,\ y = 3t+2,\ z = 4t+3;$$

and a point on the second line is

$$x = 3t'+2,\ y = 4t'+4,\ z = 5t'+5.$$

The direction ratios of the line joining them are

$$3t'-2t+1,\ 4t'-3t+2,\ 5t'-4t+2.$$

If this is the S.D. then it is perpendicular to both the lines; therefore

$2(3t'-2t+1)+3(4t'-3t+2)+4(5t'-4t+1)=0$

and $2(3t'-2t+1)+4(4t'-3t+2)+5(5t'-4t+1)=0$

or $\quad 38t'-29t'+16 = 0,$

and $\quad 50t'-38t+21 = 0.$

The solution of these equations is $t = \frac{1}{3}$, $t' = -\frac{1}{6}$; so that the feet of S.D. are $\left(\frac{5}{3}, 3, \frac{13}{3}\right)$ and $\left(\frac{3}{2}, \frac{10}{3}, \frac{25}{6}\right)$.

Hence the equation of S.D, is

$$\frac{x-\frac{5}{3}}{\frac{3}{2}-\frac{5}{3}} = \frac{y-3}{\frac{10}{3}-3} = \frac{z-\frac{13}{3}}{\frac{25}{6}-\frac{13}{3}}, \text{ or } \frac{x-\frac{5}{3}}{1} = \frac{y-3}{-2} = \frac{z-\frac{13}{3}}{1}.$$

Note that if we also require the length of S.D., we need not proceed as in Example 2. but find it as the distance between the two feet of S.D.

Example. 4. Show that the lines $r = r_1 + mt$ and $r = r_2 + ns$ intersect if

$$(r_1 - r_2) \,.\, (m \times n) = 0, \text{ but } m \times n \neq 0$$

Find the point of intersection.

The given lines will intersect if some point of the first line lies on the second. The condition for this is

$$r_1 + mt = r_2 + ns, \quad \text{...(1)}$$

for some s and t. To eliminate s, t multiply this by $m \times n$ scalarly, then

$$r_1 \,.\, (m \times n) + m \,.\, (m \times n)t = r_2 \,.\, (m \times n) + n \,.\, (m \times n)s,$$

or $\quad (r_1 - r_2) \,.\, (m \times n) = 0.$

Since for intersection, the two lines must not be parallel, therefore $m \times n \neq 0$.

Multiply (1) scalarly by $r_2 \times n$, then

$$r_1 \,.\, (r_2 \times n) + m \,.\, (r_2 \times n)t = 0.$$

This gives t for the point of intersection. Hence the point of intersection is $r_1 - \frac{\mathbf{r_1} \,.\, (\mathbf{r_2} \times \mathbf{n})}{\mathbf{m} \,.\, (\mathbf{r_2} \times \mathbf{n})} m.$

Examples:

1. Prove that the three planes

$$2x - 3y - 7z = 0, 3x - 14y - 13z = 0, 8x - 31y - 33z = 0$$ [*Roorkee*, 1972]

pass through one line. Find the equations of the line in symmetric form.

2. Examine the nature of intersection of the planes

$$2x - 5y + z = 5, x + y + 4z = 5, x + 3y + 6z = 1.$$ [*Roorkee*, 72]

3. Show that the planes

$$2x - 4y + 2z = 5, 5x - y - z = 8, x + y - z = 7$$

have no common point of intersection, and that they form a prism.

4. Do the planes

$$3x - 4y + 5z = 10, 2x - y + z = 3, x - 3y + 2z = 1$$

intersect in a point? If yes, find its coordinates.

5. Show that the lines

$$\frac{x-1}{2} = \frac{y-2}{3} = \frac{z-3}{4} \text{ and } \frac{x-2}{3} = \frac{y-3}{4} = \frac{z-5}{5}$$ [*Allahabad*, 1971]

are coplanar, and find the equation of the plane containing them.

6. Prove that the lines

$$\frac{x-4}{1} = \frac{y+3}{-4} = \frac{z+1}{7}, \frac{x-1}{2} = \frac{y+1}{-3} = \frac{z+10}{8}$$ [*Kurukshetra*, 1969]

intersect and find the coordinates of their point of intersection.

7. Do the lines $\frac{x+1}{2} = \frac{y+1}{3} = \frac{z+1}{4}$ and $x + 2y + 3z - 14 = 0 = 3x + 4y + 5z - 26$ intersect? If yes, find the point of intersection. [*Punjabi*, 1967]

8. Show that the lines

$$x - 3y + 2z + 4 = 0 = 2x + y + 4z + 1$$

and

$$3x + 2y + 5z - 1 = 0 = 2y + z$$ [*Ranchi*, 1976]

are coplanar.

9. Find the equations of the two straight lines through the origin which intersect the line $\frac{1}{2}(x - 3) = y - 3 = z$ at angles of 60°. [*Delhi*, 1976]

10. A line with direction ratios 2, 7, – 5 is drawn to intersect the lines

$$\frac{x-5}{3} = \frac{y-7}{-1} = \frac{z+2}{7} \text{ and } \frac{x+3}{-3} = \frac{y-3}{2} = \frac{z-6}{4}.$$ [*Gorakhpur*, 1975]

Find the coordinates of the points of intersection and the length intercept on it.

11. Find the shortest distance between the lines

$$\frac{x-5}{3} = \frac{y-7}{-16} = \frac{z-3}{7} \text{ and } \frac{x-9}{3} = \frac{y-13}{8} = \frac{z-15}{-5}$$

12. Find the points on the lines

$$\frac{x-6}{3} = \frac{y-7}{-1} = \frac{z-4}{1} \text{ and } \frac{x}{-3} = \frac{y+9}{2} = \frac{z-2}{4}.$$

which are nearest to each other. Hence find the shortest distance between the lines and its equations. [*Roorkee*, 1995]

13. Find the distance between the parallel lines

$$\frac{x}{6} = \frac{y}{-2} = \frac{z}{1} \text{ and } \frac{x-7}{6} = \frac{y}{-2} = \frac{z+1}{1}.$$

14. Prove that the lines

$$\frac{x-x_1}{l_1} = \frac{y-y_1}{m_1} = \frac{z-z_1}{n_1} \text{ and } \frac{x-x_2}{l_2} = \frac{y-y_2}{m_2} = \frac{z-z_2}{n_2}.$$

are coplanar if

$$\begin{vmatrix} x_1-x_2 & y_1-y_2 & z_1-z_2 \\ l_1 & m_1 & n_1 \\ l_2 & m_2 & n_2 \end{vmatrix} = 0.$$

15. Prove that the lines

$$\frac{x}{l_1} = \frac{y}{m_1} = \frac{z}{n_1}, \frac{x}{l_2} = \frac{y}{m_2} = \frac{z}{n_2}, \frac{x}{l_3} = \frac{y}{m_3} = \frac{z}{n_3}$$

are coplanar if

$$\begin{vmatrix} l_1 & m_1 & n_1 \\ l_2 & m_2 & n_2 \\ l_3 & m_3 & n_3 \end{vmatrix} = 0.$$

16. If $2d$ be the shortest distance between the lines

$$y/b + z/c = 1, x = 0 \text{ and } x/a - z/c = 1, y = 0$$

prove that

$$1/d^2 = 1/a^2 + 1/b^2 + 1/c^2.$$ [*Allahabad*, 1972]

17. Find the condition that the lines

$$(\mathbf{r} - \mathbf{a}) \times \mathbf{b} = 0 = \mathbf{r} \cdot \mathbf{n}_1 - d_1 = 0 = \mathbf{r} \cdot \mathbf{n}_2 - d_2$$

are coplanar. Show that the equation of the plane containing them is

$$\mathbf{r} \cdot [(\mathbf{n}_2 \cdot \mathbf{b})\mathbf{n}_1 - (\mathbf{n}_1 \cdot \mathbf{b})\mathbf{n}_2] = (\mathbf{n}_2 \cdot \mathbf{b})d_1 - (\mathbf{n}_1 \cdot \mathbf{b})d_2.$$ [*Roorkee*, 1979]

18. Prove that the shortest distance between the lines $\mathbf{r} = \mathbf{r}_1 + \mathbf{m}t$ and $\mathbf{r} = \mathbf{r}_2 + \mathbf{n}s$, where t and s are parameters, is $(\mathbf{r}_1 - \mathbf{r}_2) \cdot (\mathbf{m} \times \mathbf{n})/|\mathbf{m} \times \mathbf{n}|$; and that its equation is

$$(\mathbf{r} - \mathbf{r}_1) \cdot \{\mathbf{m} \times (\mathbf{m} \times \mathbf{n})\} = 0 = (\mathbf{r} - \mathbf{r}_2) \cdot \{\mathbf{n} \times (\mathbf{m} \times \mathbf{n})\}.$$

3.26 EQUATION OF A SPHERE

A sphere is the locus of points equidistant from a fixed point. This property gives the equation of a sphere.

Let a sphere with centre at (a, b, c) and radius R be given and let (x, y, z) be any point on it, then we have

$$(x-a)^2 + (y-b)^2 + (z-c)^2 = R^2. \quad \text{...(1)}$$

This is the equation of the sphere. In particular, a sphere with centre at the origin is

$$x^2 + y^2 + z^2 = R^2.$$

An equation of type

$$x^2 + y^2 + z^2 + 2ux + 2vy + 2wz + d = 0$$

represents a sphere, for it may be written as

$$(x+u)^2 + (y+v)^2 + (z+w)^2 = u^2 + v^2 + w^2 - d.$$

Its centre is $(-u, -v, -w)$ and radius $\sqrt{(u^2 + v^2 + w^2 - d)}$.

Vector equation. Let **d** be the vector of the centre C of the sphere and **r** that of any point P on it. Then $CP = R$, the radius of the sphere. So $\mathbf{CP} \cdot \mathbf{CP} = R^2$,

i.e. $$(\mathbf{r} - \mathbf{d})^2 = R^2.$$

This is the vector equation of the sphere. This reduces to (1) on putting $\mathbf{r} = x\mathbf{i} + y\mathbf{j} + z\mathbf{k}$ and $\mathbf{d} = a\mathbf{i} + b\mathbf{j} + c\mathbf{k}$.

***Example* 1.** Find the equation of the sphere having its centre at (1, 2, 3) and passing through the point (4, – 3, 5).

The sphere with centre at (1, 2, 3) and radius R is

$$(x-1)^2 + (y-2)^2 + (z-3)^2 = R^2. \quad \text{...(2)}$$

If this passes through the point (4, – 3, 5) then

$$(4-1)^2 + (-3-2)^2 + (5-3)^2 = R^2 \text{ or } R^2 = 38.$$

Substituting in (2) we get the required equation:

$$x^2 + y^2 + z^2 - 2x - 4y - 6z - 24 = 0.$$

Example **2.** Find the equation of the sphere in the positive octant touching the coordinate planes and the plane $2x + 3y + 6z - 24 = 0$.

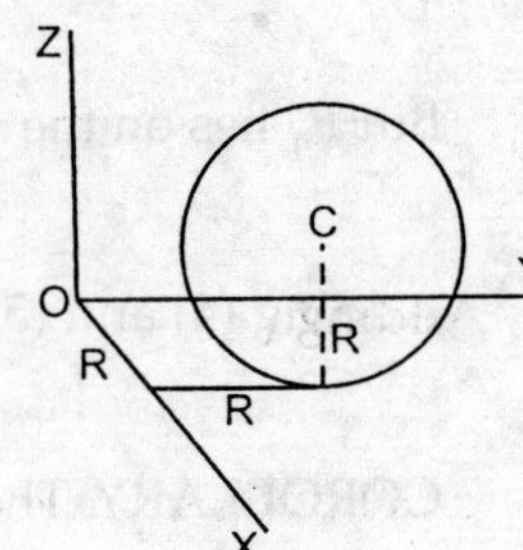

Fig. 3.23

A sphere of radius R, which touches the coordinates planes, will have its centre at ($\pm R$, $\pm R$, $\pm R$). We shall take the + sign all through, as the sphere is in the positive octant. Since it touches a fourth plane, the distance of the centre from this plane also will be R. Therefore

$$\frac{2R + 3R + 6R - 24}{\pm\sqrt{(4+9+36)}} = R,$$

or $$11R - 24 = \pm 7R,$$

i.e. $$R = 24 \div (11 \pm 7) = 6 \text{ or } \frac{4}{3}.$$

Hence there are two such spheres. Their equations are

$$(x-6)^2 + (y-6)^2 + (z-6)^2 = 36,$$

and $$\left(x - \frac{4}{3}\right)^2 + \left(y - \frac{4}{3}\right)^2 + \left(z - \frac{4}{3}\right)^2 = \left(\frac{4}{3}\right)^2;$$

i.e. $$x^2 + y^2 + z^2 - 12(x + y + z) + 72 = 0,$$

and $$9(x^2 + y^2 + z^2) - 24(x + y + z) + 32 = 0.$$

Example 3. Find the equation of the sphere with (x_1, y_1, z_1) and (x_2, y_2, z_2) as the extremities of a diameter.

[*Roorkee, Arch.*, 1968]

Let P_1 and P_2 be the two extremities and P (x, y, z) and point on the sphere. Then P_1P and P_2P are at right angle. Now the direction ratios of these lines are $x - x_1, y - y_1, z - z_1$ and $x - x_2, y - y_2, z - z_2$.

Hence $(x - x_1)(x - x_2) + (y - y_1)(y - y_2) + (z - z_1) = 0$.

This is the required equation.

3.27 TANGENT PLANE

A plane touching a sphere at a given point on it, is perpendicular to the radius drawn to that point. This property gives the equation of a tangent plane.

Let the equation of a sphere be

$$x^2 + y^2 + z^2 + 2ux + 2vy + 2wz + d = 0, \quad \text{...(1)}$$

and let P_1 (x_1, y_1, z_1) be a point on it. Then

$$x_1^2 + y_1^2 + z_1^2 + 2ux_1 + 2vy_1 + 2wz_1 + d = 0. \quad \text{...(2)}$$

The centre C of the sphere has the coordinates $(-u, -v, -w)$, so the direction ratios of CP_1 are $x_1 + u, y_1 + v, z_1 + w$.

Hence the plane through P_1 perpendicular to CP_1 is

$(x_1 + u)(x - x_1) + (y_1 + v)(y - y_1) + (z_1 + w)(z - z_1) = 0.$

or $x_1x + y_1y + z_1z + ux + vy + wz - x_1^2 - y_1^2 - z_1^2 - ux_1 - vy_1 - wz_1 = 0.$

Adding (2) to this equation, it simplifies to

$x_1x + y_1y + z_1z + u(x + x_1) + v(y - y_1) + w(z + z_1) + d = 0.$

This is the equation of the tangent plane to (1) at (x_1, y_1, z_1).

Vector equation. Let the sphere be

$$(r - d)^2 = R^2, \quad ...(3)$$

and let r_1 be a point P_1 on it. If C be the centre of (3), then $CP_1 = r_1 - d$. The tangent plane at P_1 is normal to CP_1 and passes through P_1. Therefore its equation is

$$(r - r_1) . (r_1 - d) = 0. \quad ...(4)$$

But P_1 lies on the sphere (3), therefore

$$(r_1 - d) : (r_1 - d) = R^2. \quad ...(5)$$

Adding (4) and (5), we can write the equation of the tangent plane also as

$$(r - d) . (r_1 - d) = R^2.$$

COROLLARY. The equation of the normal at P_1 is

$$\frac{x - x_1}{x_1 + u} = \frac{y - y_1}{y_1 + v} = \frac{z - z_1}{z_1 + w}.$$

Example. Find the equation of the tangent plane to the sphere at the point (4, – 3, 5).

$$x^2 + y^2 + z^2 - 2x - 4y - 6z - 24 = 0,$$

The tangent plane is

$4x - 3y + 5z - (x + 4) - 2(y - 3) - 3(z + 5) - 24 = 0.$

or $\qquad 3x - 5y + 2z - 37 = 0.$

3.28 SECTION OF A SPHERE

Let

$$S = x^2 + y^2 + z^2 + 2ux + 2vy + 2wz + d = 0 \quad ...(1)$$

be the equation of a sphere, and

$$P = ax + by + cz + \delta = 0. \quad ...(2)$$

be the equation of a plane. If the plane has a real intersection with the sphere, it will cut the sphere in a circle.

I. *To find the centre and radius of the circle $S = 0$, $P = 0$.*

Let C be the centre of the sphere $S = 0$, P_1 a point on the circle, and N the foot of the perpendicular from C on the plane $P = 0$. Then N is the centre of the circle and NP_1 its radius.

We can calculate NP_1 from the relation $NP_1^2 = CP_1^2 - CN^2$

$=$ (radius of sphere)^2 – (perpendicular from C on plane)^2.

The coordinates of N can be found by solving between the equation of the plane and the equations of the line through C perpendicular to the plane. The example below explains the full working.

Fig. 3.24

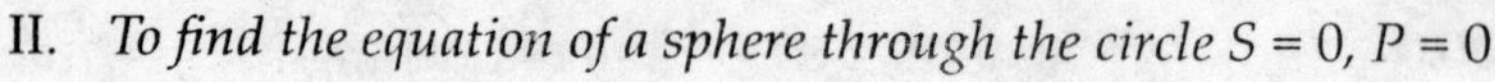

II. *To find the equation of a sphere through the circle $S = 0$, $P = 0$.*

Consider the equation

$$S + \lambda P = 0.$$

This represents a sphere, since the coefficients of x^2, y^2, z^2 are equal and the products xy, yz, zx are absent in this equation. Moreover, the coordinates of any point which satisfy both the equations $S = 0, P = 0$ satisfy $S + \lambda P = 0$ also. Hence it is the equation of a sphere passing through the circle of intersection of $S = 0$ and $P = 0$. Here γ is arbitrary and can be determined from one additional condition.

III. *The points common to the spheres $S = 0, S' = 0$ lie on the plane $S = S'$.*

The common points satisfy both the equations $S = 0, S' = 0$ and so satisfy $S = S'$ also. Moreover, the latter equation represents a plan as all the second degree terms cancel out.

Note that the common points will lie on a circle.

IV. $S + \lambda S' = 0$ *represents a sphere through the circle of intersection of the spheres* $S = 0$ and $S' = 0$

The proof is similar to that for II.

Example **1.** Find the centre and the radius of the circle

$$x^2 + y^2 + z^2 - 2y - 4z = 11, x + 2y + 2z = 15. \qquad \text{[Allahabad, 1969]}$$

The centre of the sphere given by the first equation is (0, 1, 2) and its radius is $\sqrt{(11 + 1 + 4)} = 4$. The perpendicular from (0, 1, 2) on the plane represented by the second equation is

$$\frac{1.0 + 2.1 + 2.2 - 1.5}{\sqrt{(1 + 4 + 4)}} = -\frac{9}{3} = -3.$$

Hence, as in the figure on page 123,

$$NP_1^2 = CP_1^2 - CN^2 = 16 - 9 = 7.$$

Thus the radius of the circle is $\sqrt{7}$.

The centre of the circle is the point of intersection of Any point on the lines is $(t, 2t + 1, 2t + 2)$. If this lies on the plane then

$$\frac{x}{1} = \frac{y-1}{2} = \frac{z-2}{2} \text{ and } x + 2y + 2z = 15.$$

$$t + 2(2t + 1) + 2(2t + 2) = 15,$$

or $\quad 9t = 9, t = 1.$

Hence the centre is (1, 3, 4).

Example **2.** Find the plane in which the circle of intersection of the spheres lies.

$$x^2 + y^2 + z^2 - 5 = 0 \text{ and } x^2 + y^2 + z^2 + 2x + 4y + 6z - 11 = 0$$

Find also the sphere through this circle and passing through the origin.

Subtracting the first equation from the second, we get

$$2x + 4y - 6z - 6 = 0, \text{ or } x + 2y + 3z - 3 = 0.$$

This is the plane in which the circle of intersection lies.

A sphere through this circle is

$$x^2 + y^2 + z^2 + 2x + 4y + 6z - 11 + \lambda(x^2 + y^2 + z^2 - 5) = 0.$$

If this passes through the origin,

$$-11 - 5\lambda = 0, \text{ or } \lambda = -11/5.$$

Substituting in the above equation and simplifying, we get

$$3(x^2 + y^2 + z^2) - 5x - 10y - 15z = 0.$$

3.29 LENGTH OF THE TANGENT

Let be a sphere with centre C. Let $P(x_1, y_1, z_1)$ be an external point and let PT be a line touching the sphere at T. Then the length PT of the tangent line is given by

$$x^2 + y^2 + z^2 + 2ux + 2vy + 2wz + d = 0 \qquad \text{...(1)}$$

$$PT^2 = PC^2 - CT^2$$
$$= \{(x_1 + u)^2 + (y_1 + v)^2 + (z_1 + w)^2\} - (u^2 + v^2 + w^2 - d)$$

on substituting for the centre and radius of the sphere (1). This gives

$$PT^2 = x_1^2 + y_1^2 + z_1^2 + 2ux_1 + 2vy_1 + 2wz_1 + d. \quad ...(2)$$

T
P
C

Fig. 3.25

RADIAL PLANE: The locus of all points from which equal tangents can be drawn to two spheres $S = 0$ and $S' = 0$, is called the *radical plane of the two spheres.* It is evident from the expression (2), that the radical plane of $S = 0$ and $S' = 0$ will be

$$S = S'.$$

We note that the radical plane of two spheres is same as their common plane. However the radical plane has a physical significance even when the spheres do not actually intersect.

Examples :

1. Find the equation of the spheres having its centre at $(-2, 1, -3)$ and
   (*i*) radius 5; (*ii*) passing through the point $(3, -2, 4)$;
   (*iii*) touching the plane XOY; (*iv*) touching the plane $2x - y - 2z = 4$.
2. Find the centre and radius of the following spheres
   (*i*) $x^2 + y^2 + z^2 - 2x + 5z - 5 = 0$; (*ii*) $6(x^2 + y^2 + z^2) - 8x + 10y - 18z - 13 = 0$.
3. Find the equation of the sphere which passes through the four points $(1, 2, 3)$, $(0, -2, 4)$, $(4, -4, 2)$ and $(3, 1, 4)$. Find also the centre and the radius of this sphere.
4. Find the equation of the sphere having the points $(-4, 5, 1)$ and $(4, 1, 7)$ as ends of a diameter. [*Roorkee*, 1963]
5. If the position vectors of two points P, Q and r_1, r_2, show that the sphere on P, Q as diameter is
$$(r - r_1) \,.\, (r - r_2) = 0.$$
6. Show that the centre and radius of the sphere $mr^2 + 2u \,.\, r + d = 0$ are given by $-\underline{u}/m$ and $\sqrt{(\mathbf{u}^2/m^2 - d/m)}$.
7. Find the equation of the sphere having the centre c and (i) passing through a point a, (ii) touching the plane $r \,.\, n + d = 0$.
8. Find the equation of the tangent plane at the point $(a \sin\theta \cos\phi, a \sin\theta \sin\phi, a \cos\theta)$ to the sphere $x^2 + y^2 + z^2 = a^2$.
9. Find the equations of the tangent plane and the normal at the point $(-2, 1, 3)$ to the sphere
$$x^2 + y^2 + z^2 + 12x - 2y - 6z + 30 = 0.$$
10. Find the tangent planes to the sphere
$$x^2 + y^2 + z^2 - 4x - 2y - 6z + 5 = 0$$
which are parallel to the plane $x + 4y + 8z = 0$. Find their points of contact. [*Banaras*, 1976]
11. Show that the plane $2x - 2y + z + 12 = 0$ touches the sphere
$$x^2 + y^2 + z^2 - 2x - 4y + 2z = 3.$$
Find the point of contact.
12. Find the equations of the tangent planes to the sphere $x^2 + y^2 + z^2 = 9$ which pass through the line
$$x + y = 6, x - 2z - 3 = 0.$$
[*Roorkee*, 1975]
13. Find the equation of the sphere which touches the sphere
$$x^2 + y^2 + z^2 - x + 3y + 2z - 3 = 0$$
[*Roorkee*, 1969]
at $(1, 1, -1)$ and passes through the origin.

14. Find the equation of the sphere which passes through the circle

$$x^2 + y^2 + z^2 - 2x + 2y + 4z - 3 = 0,\ 2x + y + z = 4$$ [*Allahabad*, 1973]

and through the point (1, 2, – 1).

15. Find the equations of the spheres which pass through the circle

$$x^2 + y^2 + z^2 = 5,\ x + 2y + 3z = 3$$

and touch the plane $4x + 3y = 15$.

16. Find the equation of the sphere which passes through the point (1, 3, – 2) and the circle

$$y^2 + z^2 = 25,\ x = 0.$$ [*Roorkee*, 1972]

17. Find the equation in which the circle of intersection of the spheres $x^2 + y^2 + z^2 - 2x + 4y - 6z + 12 = 0$ and $x^2 + y^2 + z^2 + 6x - 7y - z = 12$ lies. Find also the equation of the sphere through this circle and passing through (1, 1, 1).

18. Find the equation of the sphere circumscribing the tetrahedron whose faces are

$$x = 0,\ y = 0,\ z = 0,\ x/a + y/b + z/c = 1.$$ [*Banaras*, 1971]

19. Show that the plane $lx + my + nz = p$ touches the sphere $x^2 + y^2 + z^2 + 2ux + 2vy + 2wz + d = 0$, if $(ul + vm + wn + p)^2 = (l^2 + m^2 + n^2)(u^2 + v^2 + w^2 - d)$. [*Gorakhpur*, 1972]

20. Find the equation of the sphere having the circle

$$x^2 + y^2 + z^2 = 16,\ x - 2y + 2z = 3$$ [*Gorakhpur*, '74]

for a great circle. Determine its centre and radius.

21. Find the equation of the spheres, described on the line joining (2, – 1, 4) and (– 2, 2, – 2) as diameter. Also find the area of the circle in which this sphere is intersected by the plane $2x + y - z = 3$. [*Delhi*, 1976]

22. Determine the centre and the radius of the section of the sphere $x^2 + y^2 + z^2 = 16$, by the plane $2x + y + 2z = 9$. Also find the area of the projection of this circle on the xy-plane. [Roorkee, 1971]

[Hint. The projection of an area A on a plane making an angle θ with A is $A \cos \theta$.]

23. Find the centre and radius of the circle in which the sphere $x^2 + y^2 + z^2 - 4x + 2y = 0$ is cut by the plane $x + z = 1$. [*Gorakhpur*, 1975]

24. Find the points of intersection of the line

$$\frac{x-1}{2} = \frac{y+3}{4} = \frac{z-1}{2}$$ [*Banaras*, 1967]

with the sphere $x^2 + y^2 + z^2 = 19$.

25. Show that the sphere

$$x^2 + y^2 + z^2 - 2x + 6y + 14z + 3 = 0$$

divides the line joining the points (2, – 1, – 4) and (5, 5, 5) internally and externally in the ratio 1 : 2. [*Roorkee*, 1971]

26. Find the radical plane of the spheres

$$x^2 + y^2 + z^2 - 2x + 4y - 6z = 10 \text{ and } x^2 + y^2 + z^2 + 6x - 7y - z = 12.$$

27. Show that the spheres

$$x^2 + y^2 + z^2 = 25 \text{ and } x^2 + y^2 + z^2 - 24x - 40y - 18z + 225 = 0$$ [*Delhi*, 1968]

touch externally and find their point of contact.

28. Show that the spheres

$$x^2 + y^2 + z^2 + 6y + 2z + 8 = 0 \quad \text{[Aligarh, 1971]}$$

and $x^2 + y^2 + z^2 + 6x + 8y + 4z + 20 = 0,$

intersect at right angles. Find their plane of intersection.

[Hint. If C_1 and C_2 are the two centres and P a point of intersection, then for orthogonal intersection (draw a figure): $C_1P^2 + C_2P^2 = C_1C_2^2$, i.e. sum of squares of the two radii = square of the distance between the centres.]

29. Show that the spheres intersect orthogonally if

$$x^2 + y^2 + z^2 + 2ux + 2vy + 2wz + d = 0$$

and, $x^2 + y^2 + z^2 + 2u'x + 2v'y + 2w'z + d' = 0$

$$2uu' + 2vv' + 2ww' = d + d'.$$

30. Show that the spheres cut orthogonally if

$$r^2 + 2u_1 . r + d_1 = 0 \text{ and } r^2 + 2u_2 . r + d_2 = 0$$

$$2u_1 . u_2 = d_1 + d_2.$$

31. Find the equation of the sphere which touches the plane $3x + 2y - z + 2 = 0$ at the point $(1, -2, 1)$ and also cuts orthogonally the sphere

$$x^2 + y^2 + z^2 - 4x + 6y + 4 = 0. \quad \text{[Roorkee, 1994]}$$

32. A plane passes through a fixed point (a, b, c) and cuts the axes in A, B, C. Show that the locus of the centre of the sphere $OABC$ is $a/x + b/y + c/z = 2$. [*Bhopal*, 1978]

3.30 EQUATION OF A CYLINDER

A cylinder is a surface generated by a straight line which moves parallel to a given line and passes through a given curve. The moving straight line is called the generator of the cylinder. When the generator is parallel to one of then coordinate axes, we get a very simple form of the equation of the cylinder.

(x, y, z)

(x, y, 0)

Fig. 3.26

For example, let the generator move parallel to the z-axis and pass through the curve $f(x, y) = 0$ in the xy-plane. Then if $(x, y, 0)$ be a point lying on this curve, the point (x, y, z) lies on the cylinder for every value of z, for its lies on the generator through $(x, y, 0)$. Thus in the three dimensional coordinate system the equation.

$$f(x, y) = 0$$

represents a cylinder whose generators are parallel to the z-axis and whose base curve is $f(x, y) = 0, z = 0$.

Similarly, $f(y, z) = 0$ will represent a cylinder with generators parallel to x-axis, and $f(x, z) = 0$ a cylinder with generators parallel to y-axis.

The equation $x^2 + y^2 = a^2$ represents a right circular cylinder since the base curve is a circle. Similarly, $x^2/a^2 + y^2/b^2 = 1$ represents an elliptic cylinder, $y^2 = 4ax$ a parabolic cylinder and $x^2/a^2 - y^2/b^2 = 1$ a hyperbolic cylinder.

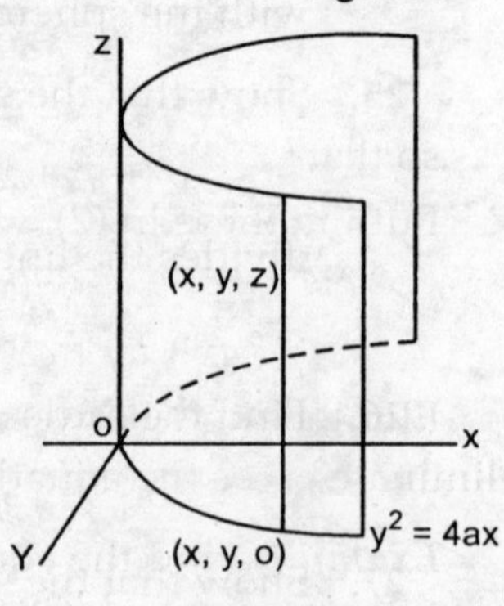

Fig. 3.27

***Example* 1.** Find the equation of the cylinder whose generators are parallel to z-axis and which passes through the curve to intersection of

$$x^2 + y^2 + z^2 = 1 \text{ and } x + y + z = 1. \quad \text{...(1)}$$

To obtain the equation of the cylinder, we eliminate z between the given equations. Thus

$$x^2 + y^2 + (1 - x - y)^2 = 1, \quad ...(2)$$

or $$x^2 + y^2 + xy - x - y = 0, \quad ...(3)$$

being of the form $f(x, y) = 0$, represents a cylinder with generators parallel to z-axis. Moreover, the coordinates of any point which satisfy equations (1), also satisfy (2). So the cylinder passes through the curve (1). Hence (3) given the required equation.

Example 2. Find the equation of the right circular cylinder of radius 3 and axis

$$\frac{x-1}{2} = \frac{y-3}{2} = \frac{z-5}{-1}. \quad \text{[Delhi, 1976]}$$

Every point on the right circular cylinder will be at the same distance 3 from the axis. Since $\sqrt{(2^2 + 2^2 + 1)} = 3$, the direction cosines the axis are 2/3, 2/3, – 1/3. Let $P(x, y, z)$ be a point on the cylinder, A the point (1, 3, 5) and N the foot of perpendicular from P on the axis (see fig. on page 79). Then

$$3^2 = PN^2 = AP^2 - AN^2$$

$$= \{(x-1)^2 + (y-3)^2 + (z-5)\} - \left\{(x-1)\frac{2}{3} + (y-3)\frac{2}{3} - (z-5)\frac{1}{3}\right\}^2$$

or $$81 = 5(x-1)^2 + 5(y-3)^2 + 8(z-5)^2 - 8(x-1)(y-3) + 4(x-1)(z-5) + 4(y-3)(z-5),$$

or $$5x^2 + 5y^2 + 8z^2 - 8xy + 4yz - 6x - 42y - 96z + 225 = 0.$$

3.31 OBLIQUE CYLINDER

To find the equation of the cylinder through the curve of intersection of

$$f(x, y, z) = 0, \; ax + by + cz + d = 0 \quad ...(1)$$

when the generators are parallel to $x/l = y/m = z/n$.

Let (x_1, y_1, z_1) be any point on the curve (1), then

$$f(x_1, y_1, z_1) = 0, \; ax_1 + by_1 + cz_1 + d = 0. \quad ...(2)$$

A generator of the required cylinder is

$$\frac{x - x_1}{l} = \frac{y - y_1}{m} = \frac{z - z_1}{n} = t,$$

so that $$x - lt = x_1, \; y - mt = y_1, \; z - nt = z_1.$$

Putting these in (2), we get

$$f(x - lt, y - mt, z - nt) = 0,$$

$$ax + by + cz + d - (al + bm + cn)t = 0.$$

Eliminating t between these two equations, we get the locus of any point (x, y, z) on a generator. Thus the eliminate gives the equation of the cylinder.

Example. Find the equation of the right circular cylinder through the circle of intersection of

$$x^2 + y^2 + z^2 = 1 \text{ and } x + y + z = 1. \quad ...(3)$$

The circle lies in the plane $x + y + z = 1$. So a generator of the cylinder will be perpendicular to this plane. Its equation is therefore

$$x - x_1 = y - y_1 = z - z_1 = t,$$

where (x_1, y_1, z_1) is a point on the circle (3). Substituting for (x_1, y_1, z_1) in (3), we have

$$(x - t)^2 + (y - t)^2 + (z - t)^2 = 1,$$

and $$x + y + z - 3t = 1.$$

Eliminating t from these, we get

$$(2x - y - z - 1)^2 + (2y - x - z - 1)^2 + (2z - x - y - 1)^2 = 9,$$

or $$6(x^2 + y^2 + z^2) - 6(xy + yz + zx) = 6,$$

or $$x^2 + y^2 + z^2 - yz - zx - xy = 1.$$

3.32 EQUATION OF A CONE

A cone is a surface generated by a straight line which moves so as to intersect a given curve while always passing through a fixed point. The fixed point is called the vertex and the moving line the generator of the cone.

Any cone whose vertex is at the origin, will be represented by a homogeneous equation in x, y, z (i.e. all the terms will be of the same degree).

The reason is that every generator will pass through the origin, and its equation will be of the form So the point (lt, mt, nt) will satisfy the equation of the cone for every value of t. This is possible only if the equation is homogeneous. Thus $x^2 + y^2 = z^2$, $2x^2 - 3y^2 + 4z^2 = 0$, $y^2 = xz$, all represent cones.

$$\frac{x}{l} = \frac{y}{m} = \frac{z}{n} = t \text{ or } x = lt, y = mt, z = nt.$$

Example. Find the equation of the cone with vertex at origin and passing through the curve of intersection of

$$x^2 + y^2 + z^2 = 1 \text{ and } x + y + z = 1. \quad ...(1)$$

To obtain the equation of the cone, we make the first equation homogeneous with the help of the second. Thus

$$x^2 + y^2 + z^2 = (x + y + z)^2. \quad ...(2)$$

being homogeneous, represents a cone with vertex at origin. Moreover, the coordinates of any point which satisfy equations (1) also satisfy (2). So the cone passes through the curve (1). Hence (3) gives the required equation.

or $$xy + yz + zx = 0, \quad ...(3)$$

3.33 RIGHT CIRCULAR CONE

I. *To find the equation of a right circular cone of semi-vertical angle α, having z-axis as its axis and origin as vertex.*

Let (x, y, z) be the coordinates of a point P on the cone. Let PN be the perpendicular from P on OZ. Then $ON = z$, and

$$PN = \sqrt{\{(x - 0)^2 + (y - 0)^2 + (z - z)^2\}}.$$

But from the triangle OPN, $PN = ON \tan \alpha$.

Therefore, $$\sqrt{(x^2 + y^2)} = z, \tan \alpha,$$

or $$x^2 + y^2 = z^2 \tan^2 \alpha.$$

This is the equation of the cone.

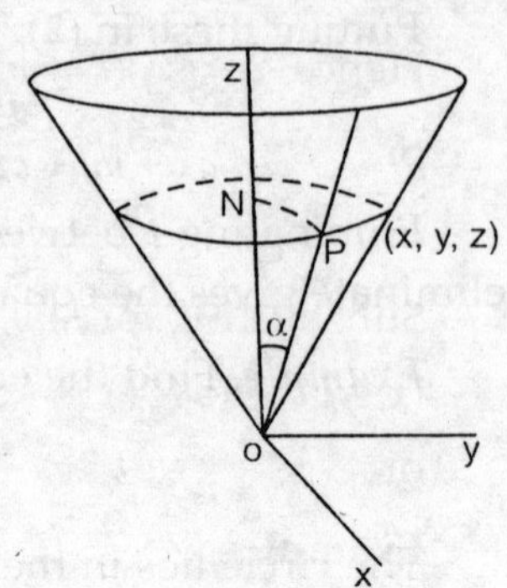

Fig. 3.28

II. *To find the equation of a right circular cone of semi-vertical angle θ, having vertex at A (a, b, c) and axis*

$$\frac{x-a}{\cos\alpha} = \frac{y-b}{\cos\beta} = \frac{z-c}{\cos\gamma}.$$

Let $P(x, y, z)$ be a point on the cone and PN the perpendicular from P on the axis (see fig. on page 79), then $AN = AP \cos\theta$.

Putting
$$AN = (x-a)\cos\alpha + (y-b)\cos\beta + (z-c)\cos\gamma,$$
$$AP = \sqrt{\{(x-a)^2 + (y-b)^2 + (z-c)^2\}}$$

and squaring, we get the required equation.

3.34 CONE THROUGH A GIVEN CURVE

To find the equation of a cone with vertex at (a, b, c) and passing through the curve of intersection of

$$f(x, y, z) = 0 \text{ and } Ax + By + Cz + D = 0.$$

Let the equation of a generator be

$$\frac{x-a}{l} = \frac{y-b}{m} = \frac{z-c}{n} = t, \qquad ...(2)$$

or
$$x = a + lt,\ y = b + mt,\ z = c + nt.$$

Since the generator passes through the curve (1), the point (3) lies on (1) for some value of t; therefore

$$f(a + lt, b + mt, c + nt) = 0,$$
$$Aa + Ba + Cc + Dd + (Al + Bm + Cn)t = 0.$$

Elimination of t between these two equations gives a homogeneous relation of the form $\phi(l, m, n) = 0$ between the unknowns l, m, n. Substituting from (2) we obtain,

$$\phi(x-a, b-y, z-c) = 0$$

as the locus of all generators. Therefore this is the equation of the required cone.

Example. Find the equation of the cone with vertex at [1,1,1] and passing through the curve of intersection of

$$x^2 + y^2 + z^2 = 1 \text{ and } x + y + z = 1 \qquad ...(1)$$

Let a generator be

$$\frac{x-1}{l} = \frac{y-1}{m} = \frac{z-1}{n} = t, \qquad ...(2)$$

Then the point of $1 + lt, 1 + mt, 1 + nt$, lies on (1) for some t; therefore

$$(1+lt)^2 + (1+mt)^2 + (1+nt)^2 = 1,$$
$$3 + (l + m + n)t = 1.$$

From the second equation $t = -2/(l + m + n)$, while the first gives

$$2 + 2(l + m + n)t + (l^2 + m^2 + n^2)t^2 = 0.$$

Hence, $2 - 4 + (l^2 + m^2 + n^2) . 4/(l + m + n)^2 = 0,$

or
$$2(l^2 + m^2 + n^2) - (l + m + n)^2 = 0,$$

or
$$l^2 + m^2 + n^2 - 2mn - 2l - 2lm = 0.$$

Substituting for l, m, n from (2), this gives

$(x-1)^2 + (y-1)^2 + (z-1)^2 - 2(y-1)(z-1) - 2(z-1)(x-1) - 2(x-1)(y-1) = 0,$

or $x^2 + y^2 + z^2 - 2yz - 2zx - 2xy + 2x + 2y + 2z - 3 = 0.$

Examples:

1. Find the equation of the cylinder through the circle of intersection of the sphere $x^2 + y^2 + z^2 = 9$ with the plane $x + 2y - 2z = 6$ where generations are parallel to the z-axis. [*Banaras*, 1978]

2. Find the equation of the cylinder whose generators are parallel to the line

$$\frac{x}{1} = \frac{y}{-2} = \frac{z}{3},$$ [Roorkee, 1971]

and guiding curve is the ellipse $x^2 + 2y^2 = 1, z = 3$.

3. Find the equation of the right circular cylinder of radius 5 and axis

$$\frac{x-2}{2} = \frac{y-3}{-1} = \frac{z+1}{3}.$$

4. Find the equation of the right circular cylinder of radius 2 whose axis passes through (1, 2, 3) and has direction cosines proportional to 2, – 3 and 6. [Bhopal, 1978]

5. Find the equation of the right circular cylinder whose guiding circle is

$$x^2 + y^2 + z^2 = 9, x - y + z = 3.$$

6. Find the equation of the right circular cone having vertex at origin, semi-vertical angle 30° and the straight line $x/1 = y/2 = z/3$ as its axis. [Delhi, 1967]

7. Find the equation of the right circular cone with vertex at origin, semi-vertical angle 60° and the axis $x/3 = y/(-4) = z/5$. [Roorkee, 1976]

8. Find the equation of the right circular cone with vertex at the origin, and passing through the point (1, 1, 2). The axis of the cone is the line $x/2 = -y/4 = z/3$. [Roorkee, 1977]

9. Find the equation of cone whose vertex is (1, 1, 1) and base the circle

$$x^2 + y^2 = 4, z = 2.$$ [Gorakhpur, 1975]

10. Find the equation of the cone whose vertex is the point (1, 2, 3) and whose generating line passes through the conic

$$\frac{x^2}{4} + \frac{y^2}{9} = 1, z = 0$$ [Roorkee, Arch., 1994]

11. Find the equation of the cone whose vertex is (1, 2, 3) and guiding curve the circle

$$x^2 + y^2 + z^2 = 4, x + y + z = 1.$$ [Roorkee, 1979]

12. Find the equation of the cone with its vertex at the origin and base the circle $y^2 + z^2 = b^2, x = a$. Show that the section of this cone by a plane $z = c$ is a hyperbola. [Roorkee, 1974]

3.35 QUADRATIC SURFACES

From what we have seen so far we find that an equation in x, y, z represents a surface in space. A linear equation in x, y, z represents a plane. A quadratic equation in x, y, z will represent a surface known as a *quadratic surface* or a *conicoid*. Spheres, cylinders and cones are a few examples of quadratic surfaces. A pair of planes can also be classified as a quadratic surface. The remaining quadric surfaces are of five types. We consider below their equations in the simplest forms.

I. **Ellipsoid.** The equation

$$\frac{x^2}{a^2} + \frac{y^2}{b^2} + \frac{z^2}{c^2} = 1 \qquad ...(1)$$

represents an ellipsoid. To obtain its shape we consider the section of the ellipsoid (1) by the plane $z = k$. This will be an ellipse given by the equations

$$z = k,$$

$$\frac{x^2}{a^2} + \frac{y^2}{b^2} = 1 - \frac{k^2}{c^2}. \qquad ...(2)$$

The semi-axes of this ellipse are parallel to OX and OY, and are of lengths

$$a\sqrt{(1 - k^2/c^2)}, \; b\sqrt{(1 - k^2/c^2)}.$$

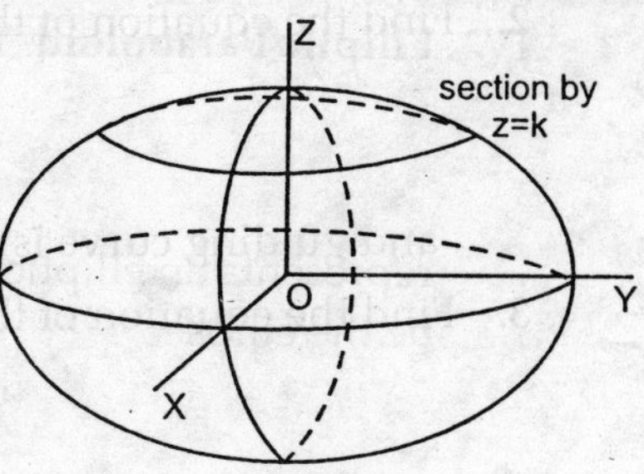

Fig. 3.29

We notice that the section by the xy-plane is the ellipse $z = 0$, $x^2/a^2 + y^2/b^2 = 1$, of semi-axes a and b. As k increases the semi-axes decreases in length. For $k = c$, section (2) reduces to a point ellipse. For $k > c$, we do not real sections. The surface below xy-plane can be obtained by symmetry, for $z = \pm k$ give equal ellipses. In fact the surface is symmetrical about all the three coordinate planes. The shape of the surface is as shown in the diagram above. If necessary, we can actually trace the ellipses (2) for a new particular values of k.

We notice from equation (1) that if the point (x, y, z) lies on the ellipsoid, $(-x, -y, -z)$ also lies on it. The line joining these points passes through the origin and is bisected there. For this reason origin is called the *centre* of the ellipsoid and a chord through the origin is called a *diameter*. This property, as well as that of symmetry about the coordinate planes, holds for hyperboloids also (II and III below). The three mutually perpendicular diameters of lengths $2a$, $2b$, $2c$ are called the axes of the ellipsoid.

If two of the semi-axes of the ellipsoid, say b and c are equal we get an ellipsoid of revolution (called a spheroid). If all the three semi-axes are equal we get the sphere.

II. **Hyperboloid of one sheet.** The equation

$$\frac{x^2}{a^2} + \frac{y^2}{b^2} - \frac{z^2}{c^2} = 1 \qquad \text{...(3)}$$

represents a hyperbola of one sheet. Its section by the plane $z = k$ is the ellipse

$$z = k, \frac{x^2}{a^2} + \frac{y^2}{b^2} = 1 + \frac{k^2}{c^2}.$$

The semi-axes of this ellipse, viz.

$$a\sqrt{(1 + k^2/c^2)}, \ b\sqrt{(1 + k^2/c^2)}.$$

go on increasing as k increases. The smallest section is by the xy-plane. It is seen that the surface is not closed in the z-direction.

The sections by planes $x = k$ or $y = k$ will be hyperbolas. In particular, the section by $x = 0$ is

$$x = 0, y^2/b^2 - z^2/c^2 = 1.$$

If $a = b$, we get a hyperboloid of revolution.

III. **Hyperboloid of two sheets.** The equation

$$-\frac{x^2}{a^2} - \frac{y^2}{b^2} + \frac{z^2}{c^2} = 1 \qquad \text{...(4)}$$

represents a hyperboloid of two sheets. Its section by the plane $z = k$ is the ellipse

$$z = k, \frac{x^2}{a^2} + \frac{y^2}{b^2} = \frac{k^2}{c^2} - 1. \qquad \text{...(5)}$$

Its semi-axes are

$$\frac{a}{c}\sqrt{(k^2 - c^2)}, \ \frac{b}{c}\sqrt{(k^2 - c^2)}$$

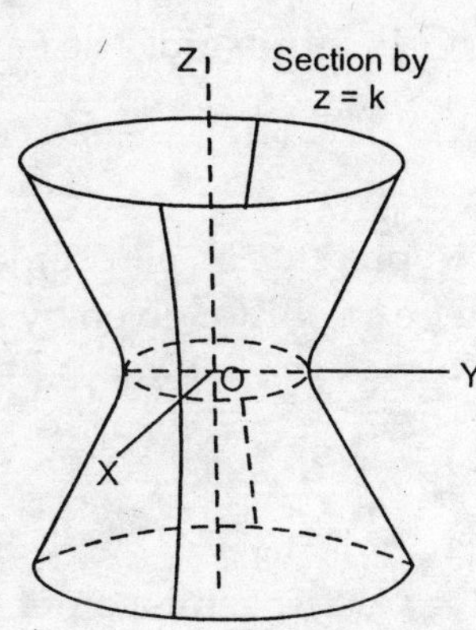

Fig. 3.30

We, therefore, see that the sections (5) are imaginary for $|k| < c$, For $|k| < c$ they are point ellipses. For $|k| < c$ they are real and increase in size as $|k|$ increases. Thus the surface is in two parts, and does not extend between $z = \pm c$. The sections by planes parallel to yz – and xz planes are hyperbolas. Its shape is as shown in the figure.

IV. **Elliptic Paraboloid.** The equation

$$\frac{x^2}{a^2} + \frac{y^2}{b^2} = \frac{z}{c} \qquad ...(6)$$

represents an elliptic paraboloid. To obtain its shape we consider its section by the plane $x = k$. This will be the ellipse

$$z = k, \frac{x^2}{a^2} + \frac{y^2}{b^2} = \frac{k}{c},$$

with semi-axes parallel to OX and OY, and of lengths

$$a\sqrt{(k/c)}, b\sqrt{(k/c)}.$$

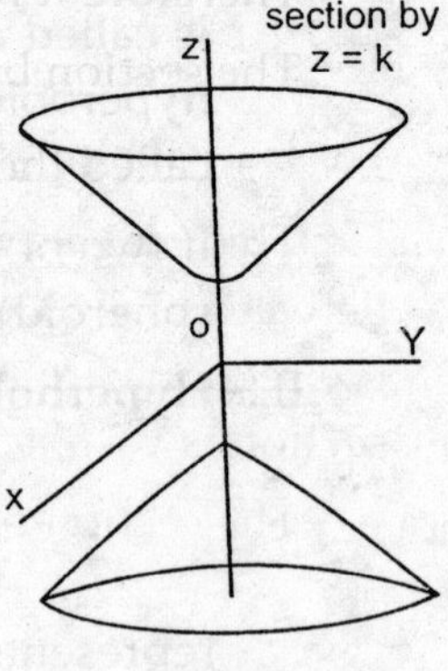

Fig. 3.31

The section by the plane $z = 0$ is a point ellipse and the surface touches the xy-plane there. As k increases, the section increases in size. Assuming c positive, we see that the sections for the negative k will be imaginary. So no part of the surface lies below the xy-plane.

The section of surface (6) by the plane $x = k$ or $y = k$ will be a parabola. This is the reason for calling the surface a parabola. OZ is called the *axis* of the paraboloid and O its *vertex*. The paraboloids do not have a centre. The surface is as shown in the figure.

If c is negative then the paraboloid lies below the xy-plane. If $a = b$, we get a paraboloid of revolution.

V. **Hyperbolic Paraboloid.** The equation

$$\frac{x^2}{a^2} - \frac{y^2}{b^2} = \frac{z}{c} \qquad ...(7)$$

represents a hyperbolic paraboloid. It section by the plane $z = k$ is the hyperbola

$$z = k, \frac{x^2}{a^2} - \frac{y^2}{b^2} = \frac{k}{c}. \qquad ...(8)$$

The section by the xy-plane is the pair of lines

$$z = 0, x^2/a^2 - y^2/b^2 = 0.$$

The asymptotes of all the hyperbolas (8) are parallel to these lines. As k increases the hyperbolas get farther and farther away from their asymptotes. As k changes sign, the hyperbolas also change their location from one pair of opposite corners between the asymptotes to the other pair.

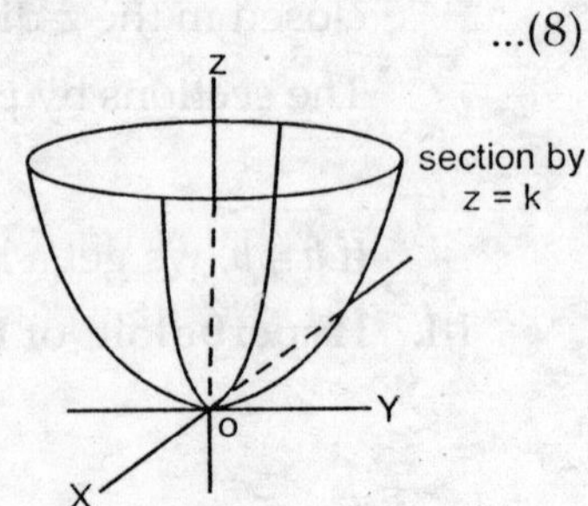

Fig. 3.32

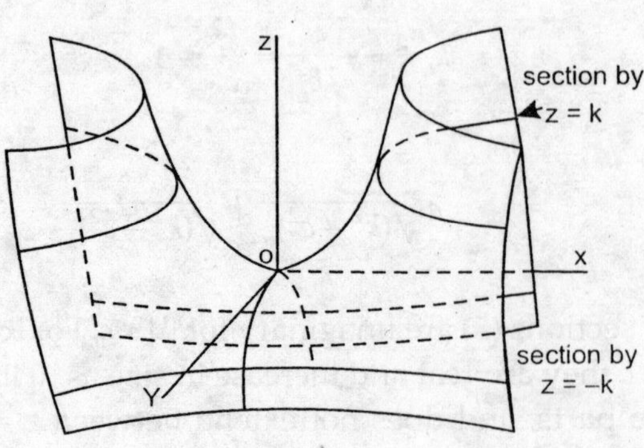

Fig. 3.33

The sections by planes $x = k$, $y = k$ are parabolas. The shape of the surface is as shown in the figure.

If $a = b$, the sections are rectangular hyperbolas. In this case, by a proper choice of axes, the equation of the paraboloid may be also written as $xy = cz$.

Example. What surface is represented by $4x^2 - y^2 - 16z^2 = 36$? [*Punjab*, 1980]

Draw a diagram to show its shape.

The given equation may be written as

$$\frac{x^2}{9} - \frac{y^2}{36} - \frac{z^2}{\frac{9}{4}} = 1.$$

Therefore it represents a hyperboloid of two sheets, with its real axis along OX.

The section by the plane $x = k$ is

$$x = k, y^2 + 16z^2 = 4k^2 - 36.$$

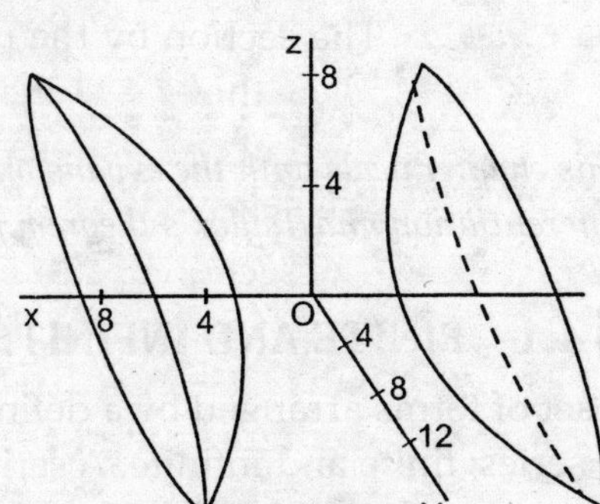

Fig. 3.34

This gives a real ellipse only for $k^2 > 9$. Hence no part of the surface lies between $x = \pm 3$. The surface touches these planes at a point, and for larger k the sections are real ellipses which increase in size as k increases. For $k = 5$, the section is $x = 5$, $y^2/8^2 + z^2/2^2 = 1$.

The section of the surface by the planes $y = 0$ and $z = 0$ are

$$y = 0, x^2/3^2 - z^2/\left(\frac{3}{2}\right)^2 = 1 \text{ and } z = 0, x^2/3^2 - y^2/6^2 = 1$$

respectively. The shape is as shown in the figure.

Examples:

1. Show that $2x^2 + 3y^2 - 5z^2 = 0$ represents a cone with its vertex at the origin. What relation must the direction cosines of the generations of this cone satisfy? [*Aligarh*, 1967]
2. Find the equation of the cylinder through the circle of intersection of the sphere $x^2 + y^2 + z^2 = 9$ with the plane $x + 2y - 2z = 6$ whose generations are parallel to the z-axis. [*Roorkee, Arch.*, '71]
3. Find the equation of the cone with vertex at the origin and having its curve.
$$x^2 - 2y^2 + z^2 = 4x + 9, x - y + z = 7.$$ [*Roorkee, Arch.*, '72]

What surfaces are represented by the following equations?

Draw diagrams to show their shapes.

4. $x^2/4 + y^2/16 + z^2/9 = 1/$
5. $x^2/2 - y^2/3 = z.$ [*Roorkee, Arch.*, 1971]
6. $4x^2 + y^2 - 9z^2 = 16.$ [*Banaras*, 1967]
7. $x^2 + y^2 = 9z^2.$
8. $4x^2 + 9y^2 = 36z.$ [*Roorkee, Arch.*, 1973]
9. $x^2/a^2 - y^2/b^2 = 1.$
10. $z^2 = 4(1 + x^2 + y^2).$ [*Roorkee*, 1978]
11. $9(x^2 + y^2) = 16z.$
12. $x^2 + y^2 = 16.$
13. (*i*) $2x^2 + y^2 = 3 - 2y$; (*ii*) $2x^2 + y^2 = 3 - 2z$.
14. Which of the above surfaces are surfaces of revolution?

Expansion of Functions (Taylor's and Maclaurin's Series)

This chapter deals with the expansion of various function by using Taylor's and Maclaurin's series, Leibnitz's rule for successive differentiation and Taylor's theorem for expansion of function of two variables.

4.1 FINITE AND INFINITE SERIES

A set of terms arranged by a definite law and connected by the sign + ve or – ve is called a series. There are two kinds of series; finite and infinite. A series is said to be finite when the number of terms it contains is definite *i.e.* countable. For example, 1 + 2 + 3 + 4 + 5, is a finite series with 5 terms.

A series is said to be infinite when the number of terms it contains is infinite, *i.e.* uncountable. For example $1 + \frac{1}{2} + \frac{1}{3} + \frac{1}{4} + \frac{1}{5} + \frac{1}{6} + ...$, is an infinite series with infinite terms.

4.2 TAYLOR'S INFINITE SERIES

If a function $f(x)$ possesses derivatives of all orders in the interval $[a, a+h]$ the for every integer h, however small, there corresponds a taylor's expansion with Lagrange's form of remainder *i.e.*,

$$f(a+h) = f(a) + hf'(a) + \frac{h^2}{2!} f''(a) + \frac{h^3}{3!} f'''(a) + ... + \frac{n^{h-1}}{(n-1)!} f^{n-1}(a) + \frac{h^n}{n!} f^n(a+ph) \text{ where } 0 < p < 1$$

or $$f(a+h) = S_n + R_n$$

where $$S_n = f(a) = hf(a) + \frac{h^2}{2!} f'(a) + \frac{h^3}{3!} f''(a) + ...+ \frac{h^{n-1}}{(n-1)!} f^{n-1}(a)$$

and $$R_n = \frac{h^n}{n!} f^n(a+ph)$$

Considering $R_n \to 0$, as $n \to \infty$ then

$$\lim_{n \to \infty} S_n = f(a+h)$$

Hence $$f(a+h) = f(a) + hf(a) + \frac{h^2}{2!} f'(a) + \frac{h^{n-1}}{(n-1} f^{n-1}(a) + ... \text{ is convergent}$$

Thus, if

(*i*) $f(x)$ possesses derivative of all orders in $[a, a+h]$

(*ii*) $R_n = \frac{h^n}{n!} f^n(a+ph)$ where $0 < p < 1$ tends to zero as $n \to \infty$

Then $$f(a+h) \neq f(a)\, hf'(a) = \frac{h^2}{2!} f'(a) + \frac{h^3}{3!} f'(a) + \ldots + \frac{h^{n-1}}{(n-1)!} f^n(a) + \frac{h^n}{n!} f'(a) + \ldots \infty \quad \ldots(i)$$

Hence this is an expansion of $f(a+h)$ as infinite series in ascending integral powers of h and is called Taylor's infinite series.

Note : To express $f(x)$ in ascending integral power of $(x-a)$, change $(a+h)$ to x, *i.e.*, $h = x - a$, in equation(i). Then

$$f(x) = f[a + (x-a)], \text{ we get}$$

$$f(x) = f(a) + (x-a) f'(a) + \frac{(x-a)^2}{2!} f'(a) + \frac{(x-a)^3}{3!} f'(a) + \ldots \ldots + \frac{(x-a)^n}{n!} f^n(a) + \ldots$$

4.3 WORKING METHOD FOR EXPANSION AS TAYLOR'S INFINITE SERIES

To expand $f(x+h)$

Step I. Let $f(x+h)$ be a given function.

Step II. Put $h = 0$ and find $f(x)$.

Step III. Differentiate $f(x)$ a number of times and obtain $f''(x), f'(x), f'(x)\ldots\ldots$

Step IV. Substitute the value of $f(x), f(x), f'(x), f'''(x)$ in

$$f(x+h) = f(x) + hf(x) + \frac{h^2}{2!} f'(x) + \frac{h^3}{3!} f''(x) + \ldots. + \frac{h^n}{n!} f^n(x) + \ldots\ldots.$$

4.4 FAILURE OF TAYLOR'S SERIES

Taylor's series fails to expand $f(x+h)$ as an infinite series in the following situations:

(a) If any of function $f(x), f(x), f''(x)$ ......becomes infinite or does not exist, *i.e.*, indeterminate form for any value of x in the interval under consideration.

(b) If R_n, *i.e*, nth term does not tend to zero as $n \to \infty$

Example **4.1.** Using Taylor's series, prove that

$$e^{x+h} = e^x \left[1 + h + \frac{h^2}{2!} + \frac{h^3}{3!} + \ldots\ldots.\right]$$

Solution : $f(x+h) = e^{x+h}$

Putting $h = 0$, we get $f(x) = e^x$

and $f(x) = e^x$; $f'(x) = e^x$; $f'''(x) = e^x$

$f^{iv}(x) = e^x$; $f^v(x) = e^x$

We know that the Taylor's expansion is given as

$$f(x+h) = f(x) + hf'(x) + \frac{h^2}{2!} f''(x) + \frac{h^3}{3!} f'''(x) + \ldots. \quad \ldots(i)$$

Putting all the above values in (i), we get

$$e^{x+h} = e^x + he^x + \frac{h^2}{2!} e^x\; \frac{h^3}{3!} e^x + \ldots\ldots$$

$$e^{x+h} = e^x \left[1 + h + \frac{h^2}{2!} + \frac{h^3}{3!}\right]$$

Example 4.2. Expand $\sin^{-1}(x+h)$ in power of h.

Solution : $f(x+h) = \sin^{-1}(x+h)$

Putting $h = 0$, we get $f(x) = \sin^{-1} x$

and $$f'(x) = \frac{1}{\sqrt{1-x^2}}$$

$$f''(x) = \frac{x}{(1-x^2)^{3/2}}$$

$$f'''(x) = \frac{1+2x^2}{(1-x^2)^{5/2}} \text{ and so on}$$

We know that the Taylor's series is given by

$$f(x+h) = f(x) + hf'(x) + \frac{h^2}{2!}f''(x) + \frac{h^3}{3!}f'''(x) = \frac{h^4}{4!}f^{IV}(x) = \ldots\ldots$$

$$\therefore \quad \sin^{-1}(x+h) = \sin^{-1}x + \frac{h}{\sqrt{1-x^2}} + \frac{h^2}{2!}\frac{x}{(1-x^2)} + \frac{(1+2x^2)}{(1-x^2)^{5/2}} \times \frac{h^3}{3!} + \ldots\ldots$$

***Example* 4.3.** Apply Taylor's theorem to find the expansions of log sin $(x+h)$.

Solution :

$$f(x+h) = \log\sin(x+h)$$

Put

$$h = 0 \therefore f(x) = \log\sin x$$

and

$$f(x) = \frac{\cos x}{\sin x} = \cot x$$

$$f'(x) = -\text{cosec}^2 x$$

$$f''(x) = -2\,\text{cosec}\,x\,(-\text{cosec}\,x\cot x) = 2\,\text{cosec}^2 x\cot x.$$

We know that the Taylor's theorem is given by

$$f(x+h) = f(x) + hf(x) + \frac{h^2}{2!}f''(x) + \frac{h^3}{3!}f'''(x) + \ldots. \quad \ldots(i)$$

Putting all the above value in equation (*i*) we get

$$\log\sin(x+h) = \log\sin x + h\cot x - \frac{h^2}{2!}\text{cosec}^2 x = \frac{h^3}{3!}(2\,\text{cosec}^2 x\cot x)$$

or

$$\log\sin(x+h) = \log\sin x = h\cot x - \frac{h^2}{2}\text{cosec}^2 x + \frac{h^3}{3}\text{cosec}^2 x\cot x + \ldots\ldots\ldots$$

Example 4.4. Expand sin $(x+h)$ in the powers of h and deduce that

$$\sin(x+h) = \sin x\cos h + \cos x\sin h$$

Solution :

$$f(x+h) = \sin(x+h)$$

Put

$$h = 0 \quad \therefore f(x) = \sin x$$

and

$$f(x) = \cos x;\ f'(x) = -\sin x$$

$$f''(x) = -\cos x;\ f^{iv}_{(x)} = \sin x$$

By Taylor's series

$$f(x+h) = f(x) + hf(x) + \frac{h^2}{2!}f'(x) + \frac{h^4}{4!}f'''(x) + \frac{h^4}{4!}f^{IV}(x) + \ldots\ldots\ldots \quad \ldots(i)$$

Putting the above values in equation (*i*), we get

$$\sin(x+h) = \sin x + h\cos x + \frac{h^2}{2!}(-\sin x) + \frac{h^3}{3!}(-\cos x) + \frac{h^4}{4!}\sin x + \ldots..$$

$$\sin(x+h) = \sin x + h\cos x - \frac{h^2}{2!}\sin x - \frac{h^3}{3!}\cos x + \frac{h^4}{4!}\sin x + \ldots..$$

$$= \sin x\left[1 - \frac{h^2}{2!} + \frac{h^4}{4!} + \ldots\ldots\right] + \cos x\left[h - \frac{h^3}{3!} + \frac{h^5}{5!} + \ldots\ldots\right]$$

$$= \sin x\cos h + \cos x\sin h$$

***Example* 4.5.** Use Taylor's theorem to express the polynomial $2x^3 + 7x^2 + x - 6$ in the power of $(x - 2)$.

Solution :
$$f(x) = 2x^2 + 7x^2 + x - 6$$

it can be written as

$$2x^3 + 7x^2 + x - 6 = f(x) = f[2 + (x - 2)], \text{ where } h = x - 2$$

On expanding $f[2 + (x - 2)]$ by Taylor's series

$$f(x) = f(2) + (x - 2)f'(2) + \frac{(x-2)^2}{2!} f''(2) + \frac{(x-3)^3}{3!} f'''(2) + \ldots\ldots \quad \ldots(i)$$

and
$$f(x) = 2x^3 + 7x^2 + x - 6$$

$\therefore$
$$f(2) = 40$$
$$f'(x) = 6x^2 + 14x + 1$$

$\therefore$
$$f'(2) = 53$$
$$f''(x) = 12x + 14$$

$\therefore$
$$f''(2) = 38$$
$$f''(x) = 12$$

$\therefore$
$$f'''(2) = 12$$

From (i)

$$2x^3 + 7x^2 + x - 6 = 40 + (x - 2)\,53 + \frac{(x-2)^2}{2!}\,38 + \frac{(x-3)^3}{3!}\,12$$
$$= 40 + 53(x - 2) + 19(x - 2)^2 + 2(x - 3)^3$$

***Example* 4.6.** If $f(x) = x^3 + 8x^2 = 15x - 24$, calculate the value of $f\left(\frac{11}{10}\right)$ by the application of Taylor's series

Solution : By Taylor's series

$$f(x + h) = f(x) + hf'(x) + \frac{h^2}{2!} f''(x) + \frac{h^3}{3!} f'''(x) + \ldots\ldots$$

Here
$$f\left(\frac{11}{10}\right) = f\left(1 + \frac{1}{10}\right) \equiv f(x + h) \quad \ldots(i)$$

On comparing

$$x = 1,\ h = \frac{1}{10}$$

On expanding (i) by Taylor's series

$$f\left(\frac{11}{10}\right) = f\left(1 + \frac{1}{10}\right) \equiv f(1) + \frac{1}{10} f(1) + \frac{1}{10^2}\frac{1}{2!} f''(1) + \frac{1}{10^3}\frac{1}{3!} f'''(1) + \ldots\ldots \quad \ldots(A)$$

But
$$f(x) = x^3 + 8x^2 = 15x - 24$$

$\therefore$
$$f(1) = 1 + 8 + 15 - 24 = 0$$
$$f'(x) = 3x^2 + 16x + 15$$

$\therefore$
$$f'(1) = 3 + 16 + 15 = 34$$
$$f''(x) = 6x = 16$$

$\therefore$
$$f''(1) = 6 + 16 = 22$$
$$f'''(x) = 6$$

$\therefore$
$$f'''(1) = 6$$

Putting all the above values in (A), we get

$$f\left(1+\frac{1}{10}\right) = 0+\frac{1}{10}(34)+\frac{1}{10^2}\frac{22}{2!}+\frac{1}{10^3}\frac{6}{3!}$$

$$f\left(1+\frac{1}{10}\right) = 3.4+\frac{11}{100}+\frac{1}{100}=3.4+0.1+0.001$$

$$f\left(\frac{11}{10}\right) = 3.511$$

***Example* 4.7.** Calculate the approximate values of $\sqrt{17}$ to four decimal places by taking the first three terms of an appropriate Taylor's series.

Solution : Let $f(x+h) = \sqrt{x+h}$

Put $h = 0$

$\therefore$ $f(x) = \sqrt{x}$

$$f'(x) = \frac{1}{2}x^{-\frac{1}{2}} = \frac{1}{2\sqrt{x}}$$

$$f''(x) = -\frac{1}{4}x^{-\frac{3}{2}} = \frac{-1}{4x\sqrt{x}}$$

$$f'''(x) = \frac{3}{8}x^{-\frac{5}{2}} = \frac{2}{8x^2\sqrt{x}}$$

But $\sqrt{x+h} = f(x+h) = \sqrt{16+1}$

On comparing $x = 16$

$h = 1$

On expanding by Taylor's series

$$f(x+h) = f(x)+hf'(x)+\frac{h^2}{2!}f''(x)+\frac{h^3}{3!}f'''(x)+\ldots\ldots$$

$$\sqrt{x+h} = \sqrt{x}\times\frac{h}{2\sqrt{x}}-\frac{h^2}{8x\sqrt{x}}+\frac{h^3}{48x^2\sqrt{x}}+\ldots\ldots$$

Putting $x = 16, h = 1$, we get

$$\sqrt{17} = \sqrt{16}+\frac{1}{2\sqrt{16}}-\frac{1}{8\times16\sqrt{16}}+\frac{1}{48(16)^2\sqrt{16}}$$

$$= 4+\frac{1}{8}-\frac{1}{48\times4\times16\times16} = 4.1012$$

***Example* 4.8.** Compute to four decimal place, the value of cos 32° by use of Taylor's series.

Solution : Let $f(x+h) = \cos(x+h)$

Put $h = 0$

$\therefore$ $f(x) = \cos x$

Here $\cos(x+h) = \cos(30° + 2°)$

On comparing, we get $x = 30°, h = 2° = 2° \times \frac{\pi}{180}$ radians and so on.

or $h = 0.03490$

Now $f'(x) = -\sin x\,;\ f''(x) = -\cos x$

$f'''(x) = \sin x\,;\ f^{iv}_{(x)} = \cos x$

Now by Taylor's series

$$\cos(x+h) = f(x+h)$$

$$= f(x) + hf'(x) + \frac{h^2}{2!} f''(x) + \frac{h^3}{3!} f'''(x) + \frac{h^4}{4!} f^{IV}(x) + \ldots\ldots.$$

$$= \cos x + h(-\sin x) + \frac{h^2}{2!}(-\cos x) + \frac{h^3}{3!}\sin x + \frac{h^4}{4!}\cos x + \ldots\ldots.$$

On Putting $x = 30°$ and $h = 0.03490$, we get

$$\cos(30° + 2°) = \cos 30° - (0.03490)\sin 30° - \frac{(0.03490)^2}{2!}\cos 30° + \frac{(0.03490)^3}{3!}\sin 30° + \frac{(0.03490)^4}{4!}\cos 30°$$

$$= \frac{\sqrt{3}}{2} - 0.03490 \times \frac{1}{2} - 0.0060 \times \frac{\sqrt{3}}{2} + 0.0000 \times \frac{1}{2} + \ldots\ldots$$

$$= 0.86602 - 0.1754 - 0.00060 \times 0.86602$$

$$= 0.84857 - 0.00051 = 0.84806 = 0.8461$$

Hence $\cos 32° = 0.8461$

Example **4.9.** Calculate approximately $\log_{10} 404$, if $\log_{10} 4 = 0.6021$.

Solution : Let $f(x+h) = \log_e(x+h)$

put $h = 0 \therefore f(x) = \log_e x$

and

$$f(x) = \frac{1}{x}$$

$$f'(x) = -\frac{1}{x^2}$$

$$f''(x) = \frac{2}{x^3}$$

and so on.

But $f(x+h) = \log_{10}(400+4)$

On comparing $x = 400, h = 4$

Here $\log_e(x+h) = f(x = h)$, then the Taylor's series

$$= f(x) + hf(x) + \frac{h^2}{2!} f'(x) + \frac{h^3}{3!} f''(x) + \ldots.$$

$$\log_e(x+h) = \log_e x + \frac{h}{x} - \frac{h^2}{2x^2} + \frac{2h^3}{6x^3} + \ldots\ldots \quad \ldots(i)$$

We know that

$$\log_e \times \log_a e = \log_a x$$

On multiplying $\log_{10} e$ on both side of (*i*), we get

$$\log_e(x+h) \times \log_{10} e = \log_e x \times \log_{10} e + \log_{10} e \left[\frac{h}{x} - \frac{h^2}{2x^2} + \frac{2h^3}{6x^3} + \ldots\ldots\right]$$

$$\log_{10}(x+h) = \log_{10} x + \log_{10} e \left[\frac{h}{x} - \frac{h^2}{2x^2} + \frac{h^3}{3x^3}\right]$$

On putting $x = 400, h = 4$, we get

$$\log_{10}(404) = \log_{10} 400 + \log_{10} e\left[\frac{4}{400} - \frac{1}{2}\left(\frac{4}{400}\right)^2 + \frac{1}{3}\left(\frac{4}{400}\right)^3\right]$$

$$= \log_{10}(4 \times 10^2) + 0.4343\left[0.01 - \frac{1}{2}(0.0001) + \frac{1}{3}(0.000001)....\right]$$

$$= \log_{10} 4 + 2\log_{10} 10 + 0.4343\,(0.01 - 0.0005 + 0.00000\,.....)$$

$$= 0.6021 + 2 + 0.4343 \times 0.00995$$

$$= 2.6021 + 0.00432 = 2.6064$$

Hence $\log_{10} 404 = 2.6064$

Example **4.10.** Expand $\sin x$ in ascending power of $\left(x - \frac{\pi}{2}\right)$ and hence find the value of sin 91° correct to 4 decimal places.

Solution : Let $f(x) = \sin x$

$$f(x) = f\left[\frac{\pi}{2} + \left(x - \frac{x}{2}\right)\right]$$

[We have done this step to get as $\left(x - \frac{\pi}{2}\right)$

$$= f(a+h) \text{ where } a = \frac{\pi}{2},\ h = \left(x - \frac{\pi}{2}\right)$$

Expand by Taylor's series, then

$$f(x) = f(a) + hf'(a) = \frac{h^2}{2!}f''(a) + \frac{h^3}{3!}f'''(a) + ... \quad ...(i)$$

Here $f(x) = \sin x$

$\therefore$ $f\left(\frac{\pi}{2}\right) = 1$

and $f'(x) = \cos x$

$\therefore$ $f'\left(\frac{\pi}{2}\right) = 0$

$f''(x) = \cos x$

$\therefore$ $f''\left(\frac{\pi}{2}\right) = -1$

$f'''(x) = \cos x$

$\therefore$ $f'''\left(\frac{\pi}{2}\right) = 0$

$f^{iv}(x) = \cos x$

$\therefore$ $f^{iv}\left(\frac{\pi}{2}\right) = 1$

and so on.

Putting all these values in (i), then we get

$$\sin x = f\left(\frac{\pi}{2}\right) + \left(x - \frac{\pi}{2}\right)f\left(\frac{\pi}{2}\right) + \frac{\left(x - \frac{\pi}{2}\right)^2}{2}f'\left(\frac{\pi}{2}\right) + \frac{\left(-\frac{\pi}{2}\right)^3}{2}f'''\left(\frac{\pi}{2}\right) + \frac{\left(-\frac{\pi}{2}\right)^4}{24}f^{iv}\left(\frac{\pi}{2}\right) + ...$$

On putting the value of $f\left(\frac{\pi}{2}\right), f\left(\frac{\pi}{2}\right), f'\left(\frac{\pi}{2}\right)$ and so on in above equation

$$\sin x = 1 + \left(x - \frac{\pi}{2}\right)(0) + \frac{\left(x - \frac{\pi}{2}\right)^2}{2}(-1) + \frac{\left(x - \frac{\pi}{2}\right)^3}{6}(0) + \frac{\left(x - \frac{\pi}{2}\right)^4}{24}(1) + ...$$

or $$\sin x = 1 - \frac{1}{2!}\left(x - \frac{\pi}{2}\right) + \frac{1}{4!}\left(x - \frac{\pi}{2}\right)^4 + \ldots.$$

But $$\sin 91° = \sin(90 + 1°) \therefore 1° = \frac{\pi}{180}$$

$$= \sin\left(\frac{\pi}{2} + \frac{\pi}{180}\right)$$

$$a = \frac{\pi}{2}, \; x - \frac{\pi}{2} = \frac{\pi}{180}$$

Putting these values in (*i*), we get

$$\sin 91° = 1 + \frac{\pi}{180}(0) + \frac{\pi^2}{2!\,(180)^2}(-1) + \frac{\pi^3}{3!\,(180)^3}(0) + \frac{\pi^4}{4!\,(180)^4}(1)$$

$$\sin 91° = 1 - \frac{\pi^2}{2(180)^2} + \frac{\pi^4}{24 \times (180)^4} = 0.0953$$

***Example* 4.11.** Expand log sin x in ascending power of $(x - 3)$

Solution : Let $f(x) = \log \sin x$

$$\therefore \quad \log \sin x = f(x) \Rightarrow f[3 + (x - 3)]$$

$$= f(3 + h)$$ [we have done this step to find h]

Expand by Taylor's series

$$\log \sin x = f(3) + hf'(3) + \frac{h^2}{2!} f''(3) + \frac{h^2}{3!} f'''(3) + \ldots. \qquad \ldots(i)$$

Here $$f(x) = \log \sin x$$

$$\therefore \quad f(3) = \log \sin (3)$$

$$f'(x) = \frac{\cos x}{\sin x} = \cot x$$

$$\therefore \quad f'(3) = \cot (3)$$

$$f''(x) = \operatorname{cosec}^2 x$$

$$\therefore \quad f'(3) = \cot (3)$$

$$f'''(x) = -\operatorname{cosec} x\,(-\operatorname{cosec} x \cot x)$$

$$= 2 \operatorname{cosec}^2 x \cot x$$

$$\therefore \quad f''(3) = 2 \operatorname{cosec}^2 (3) \cot (3)$$

and so on.

Putting all these value in (*i*)

$$\log \sin x = \log \sin (3) + (x - 3) \cot (3) - \frac{(x-3)^2}{2} \operatorname{cosec}^2 (3) + \frac{(x-3)^3}{3} \operatorname{cosec}^3 (3) \cot (3) + \ldots.$$

EXERCISE 4.1

1. Using Taylor's series, prove that

(*a*) $a^{x+h} = a^x \left[1 + h \log a + \frac{h^2}{2!}(\log a)^2 + \frac{h^3}{3!}(\log a)^3 + \ldots\ldots\right]$

(*b*) $\frac{1}{x+h} = \frac{1}{x} - \frac{h}{x^2} + \frac{h^2}{x^3} - \frac{h^3}{x^4} + \ldots\ldots$

(c) $\sec^{-1}(x+h) = \sec^{-1} x + \frac{h}{x\sqrt{x^2-1}} - \frac{2x^2-1}{x^2(x^2-1)^{3/2}}\frac{h^2}{2!}$

(d) $\tan^{-1}(x+h) = \tan^{-1} x + \frac{h}{1+x^2} - \frac{xh^2}{(1+x^2)^2} + ...$

2. Expand the following functions by Taylor's series and prove that

(a) $\log\cos(x+h) = \log\cos x - h\tan x - \frac{h^2}{2}\sec^2 x - \frac{h^3}{3}\sec^2 x \tan x + ...$

(b) $\tan(x+h) = \tan x + h\sec^2 x + h^2 \sec^2 x \tan x + \frac{h^3}{3}\sec^2 x(1+3\tan^2 x) + ...$

(c) $f\left(\frac{x^2}{1+x}\right) = f(x) - \frac{x}{1+x}f'(x) + \left(\frac{x}{1+x}\right)^2 \frac{1}{2!}f''(x) - \left(\frac{x}{1+x}\right)^3 \frac{1}{3!}f'''(x) + ...$

$\left[\text{Hint}: \frac{x^2}{1+x} = x - \frac{x}{1+x} \equiv x+h, \text{ where } h = -\frac{x}{1+x}\right]$

3. Ussing Taylor's series

(a) If $f(x) = x^3 - 6x^2 + 7$; find the value of $\left(\frac{21}{20}\right)$. **Ans.** 1.5426

(b) Calculate the approximate value of $\sqrt{10}$ and $\sqrt{26}$ to four decimal places by taking first three terms of an appropriate Taylor's expansion. **Ans.** 2.1623, 5.099

(c) If $f(x) = x^3 + 2x^2 - 5x = 11$, calculate the value of $f\left(\frac{9}{10}\right)$ by the application for Taylor's series for $f(x+h)$. **Ans.** 8.849

4. Using Taylor's series

(a) Expand e^x in power of $(x-2)$ **Ans.** $e^2\left[1 + (x-2) + \frac{(x-2)^2}{2!} - \frac{(x-2)^3}{3!} + ...\right]$

(b) Expand $\log_e x$ in power of $(x-1)$. Hence evaluate loge 1.1 correct to 4 decimal places.

Ans. $(x-1) - \frac{(x-1)^2}{2} + \frac{(x-1)^3}{3} - \frac{(x-1)^4}{4} + ...$; 0.0953

(c) Compute the value of sin 31° to four decimal places. **Ans.** 0.5151

(d) Expand $\tan x$ in the power of $\left(x - \frac{\pi}{4}\right)$ upto first four terms.

Ans. $1 + 2\left(x-\frac{\pi}{4}\right) + 2\left(x-\frac{\pi}{4}\right)^2 + \frac{8}{3}\left(x-\frac{\pi}{4}\right)^3 + ...$

4.5 MACLAURIN'S INFINITE SERIES

If $a = 0$ and $h = x$ in Taylor's infinite series such that:

(i) $f(x)$ possesses derivatives of all order in the interval $[0, x]$ and

(ii) Maclaurin's remainder $R_n = \frac{x^n}{n!} f^n(a + ph)$ tends to zero as $n \to \infty$, then

$$f(x) = f(0) + xf'(0) + \frac{h^3}{3!}f''(0) + \frac{h^3}{3!}f'''(0) + ... + \frac{x^n}{n!}f^n(0) + ...$$

which is known as Maclaurin's infinite series.

Note : If the function $f(x)$ is denote by y, then Maclaurin infinite series may be written as

$$y = y(0) + xy_1(0) + \frac{x^2}{2!}y_2(0) + \frac{x^3}{3!}y_3(0) + ... + \frac{x^n}{n!}y_n(0) + ...$$

where $y(0)$, $y_1(0)$, $y_2(0)$, $y_3(0)$....$y_n(0)$ etc., denote the value of y, y_1, y_2, y_3, yn, respectively for $x = 0$.

4.6 WORKING METHOD OF EXPANSION AS MACLAURIN'S SERIES

Step I. Put $f(x)$ = given function.

Step II. Differentiate $f(x)$, a number of times *i.e.*, $f(x), f'(x), f''(x), f'''(x)$.... and so on.

Step III. Put $x = 0$ in the result of step II and find $f(0), f'(0), f''(0), f'''(0)$..... and so on.

Strep IV. On substituting the value of $f(0), f'(0), f''(0)$....in Maclaurin's series

we get $$f(x) = f(0) + xf'(0) + \frac{x^2}{2!} f''(0) + \frac{x^3}{3!} f'''(0) + ... \frac{x^n}{n!} f^n(0) + ...$$

4.7 FAILURE OF MACLAURIN'S SERIES

Maclaurin's series fails to expand $f(x)$ in an infinite series in the following situations.

(*a*) If any of $f(x), f'(x), f''(x)$.....becomes infinite or does not exist in the closed interval $[0, x]$

(*b*) If R_n does not tend to zero as $n \to \infty$.

Example **4.12.** Expand a^x in the power of x, by Maclaurin's series.

Solution : $f(x) = a^x$

$\therefore$ $f(0) = a^0 = 1$

and $f'(x) = a^x \log a$

$\therefore$ $f'(0) = a^0 \log a = \log a$

$f''(x) = a^x (\log a)^2$

$\therefore$ $f''(0) = a^0 (\log a)^2 = (\log a)^2$

$f'''(x) = a^x (\log a)^3$

$\therefore$ $f'''(0) = a^0 (\log a)^3 = (\log a)^3$

and so on.

Maclaurin's series is given by

$$f(x) = f(0) + xf'(0) + \frac{x^2}{2!} f''(0) + \frac{x^3}{3!} f'''(0) + ...$$

$$a^x = 1 + x(\log a) + \frac{x^2}{2!} (\log a)^2 + \frac{x^3}{3!} (\log a)^3 \; ....$$

Example **4.13.** Expand $\log(1 + x)$ in the power of x.

Solution : Let $f(x) = \log(1 + x)$

$\therefore$ $f(0) = \log(1 + 0) = 0$

and $f'(x) = \dfrac{1}{1+x}$

$\therefore$ $f'(0) = \dfrac{1}{1+0} = 1$

$f''(x) = -\dfrac{1}{(1+x)^2}$

$\therefore$ $f''(0) = -\dfrac{1}{(1+0)^2} = -1$

$f'''(x) = \dfrac{2}{(1+x)^3}$

$\therefore \qquad f'''(0) = \frac{2}{(1+0)^3} = 2$

$$f^{iv}(x) = -\frac{6}{(1+x)^4}$$

$\therefore \qquad f^{iv}(x) = -\frac{2}{(1+0)^4} = -6$

and so on.

Maclaurin's series is given by

$$f(x) = f(0) + xf'(0) + \frac{x^2}{2!}f''(0) + \frac{x^3}{3!}f'''(0) + \ldots\ldots$$

$$\log(1+x) = 0 + x - \frac{x^2}{2} + \frac{x^3}{3} - \frac{x^4}{4} + \ldots\ldots$$

or $\qquad \log(1+x) = x - \frac{x^2}{2} + \frac{x^3}{3} - \frac{x^4}{4} + \ldots\ldots$

Example 4.14. Expand $\sin^{-1} x$ up to four terms.

Solution : Let $\qquad f(x) = \sin^{-1} x$

$\therefore \qquad f(0) = 0$

and $\qquad f'(x) = \frac{1}{\sqrt{1-x^2}}$

$\therefore \qquad f'(0) = \frac{1}{1+0}$

$$f''(x) = -\frac{1}{2(1-x^2)^{3/2}}(-2x)$$

$$= \frac{x}{(1-x^2)^{3/2}}$$

$\therefore \qquad f''(0) = 0$

$$f'''(x) = \frac{(1-x^2)^{\frac{3}{2}} - x \times \frac{3}{2}(1-x^2)^{\frac{1}{2}}(-2x)}{\left[(1-x^2)^{\frac{3}{2}}\right]}$$

$$= \frac{1+2x^2}{(1-x^2)^{\frac{5}{2}}}$$

$\therefore \qquad f'''(0) = 1$

$$f^{iv}(x) = \frac{(1-x^2)^{\frac{5}{2}}(+4x) - (1+2x^2) \times \frac{5}{2}(1-x^2)^{\frac{3}{2}}(-2x)}{\left[(1-x^2)^{\frac{5}{2}}\right]^2}$$

$$= \frac{(1-x^2)^{\frac{3}{2}}\left[(1-x^2)(+4x) + 5x(1+2x^2)\right]}{(1-x^2)^5}$$

$$f^{iv}(x) = \frac{x-6x^3}{1-x^2} \therefore f^{iv}_{(0)} = 0$$

and so on.

Putting these values in Maclaurin's series

$$f(x) = f(0) + xf'(0) + \frac{x^2}{2!} f''(0) + \frac{x^3}{3!} f'''(0) + \frac{x^4}{4!} f^{iv}(0) + ...$$

$$\sin^{-1} x = 0 + x(1) + \frac{x^2}{2!}(0) + \frac{x^3}{3!}(1) + \frac{x^4}{4!}(0) + \frac{x^5}{5!}(1) + ...$$

$$\sin^{-1} x = x + \frac{x^3}{3!} + \frac{x^5}{5!} + ...$$

***Example* 4.15.** Expand by Maclaurin's series $\frac{e^x}{1+e^x}$ as far as the term x^3.

Solution : Let

$$y = \frac{e^x}{1+e^x} = \frac{e^x+1-1}{1+ex} = 1 - \frac{1}{1+e^x} \quad \therefore y(0) = \frac{1}{2}$$

and

$$y_1 = \frac{e^x}{(1+e^x)^2} = \frac{e^x}{(1+e^x)(1+e^x)} = y(1-y)$$

$$y_1(0) = y(0) - y^2(0) = \frac{1}{2} - \frac{1}{4} = \frac{1}{4}$$

Again

$$y_2 = y_1 - 2yy_1$$

$\therefore$

$$y_2(0) = y_1(0) - 2y(0)\, y_1(0) = \frac{1}{4} - 2\left(\frac{1}{2}\right)\left(\frac{1}{4}\right) = 0$$

and

$$y_3 = y_2 - 2{y_1}^2 - 2yy_2$$

$$y_3(0) = y_2(0) - 2{y_1}^2(0) - 2y(0)\, y_2(0)$$

or

$$y_3(0) = 0 - 2\left(\frac{1}{4}\right)^2 - 2\left(\frac{1}{2}\right)0$$

$$y_3(0) = 0 - 2 \times \frac{1}{16} = -\frac{1}{8}$$

and so on

By Maclaurin's series in terms of y.

$$y = y(0) + xy_1(0) + \frac{x^2}{2!} y_2(0) + \frac{x^3}{3!} y_3(0) + ... \qquad ...(i)$$

Putting all the above value in (*i*), we get

$$\frac{e^x}{1=e^x} = \frac{1}{2} + x\left(\frac{1}{4}\right) + \frac{x^2}{2!}(0) + \frac{x^3}{3!}\left(-\frac{1}{8}\right) + ...$$

$$\frac{e^x}{1+e^x} = \frac{1}{2} + \frac{x}{4} - \frac{1}{48} x^3 + ...$$

***Example* 4.16.** Use Maclaurin's formula to show that

$$e^x \sec x = 1 + x + \frac{2x^2}{2!} + \frac{4x^3}{3!} + ... \qquad ...(i)$$

Solution : Let

$$y = e^x \sec x$$

and

$$y_1 = e^x \sec x + e^x \sec x \tan x$$

$$y_1 = y + y \tan x \qquad ...(ii)$$

Again

$$y_2 = y_1 + y_1 \tan x + y \sec^2 x \qquad ...(iii)$$

$$y_3 = y_2 + y_2 \tan x = y_1 \sec^2 x + y_1 \sec^2 x + y\,(2 \sec^2 x \tan x)$$

$$y_3 = y_2 + y_2 \tan x + 2y_1 \sec^2 x + 2y \sec^2 x \tan x \qquad ...(iv)$$

and so on

Putting $x = 0$ in the above expressions, we get

$$y(0) = 1$$

$$y_1(0) = y(0) + y(0) \tan(0) = 1$$

$$y_2(0) = y_1(0) + y_1(0)\,(0) + y(0) \sec^2 (0) = 1 + 1\,(0) + 1(1) = 2$$

$$y_3(0) = y_2(0) + y_2(0) \times 0 + 2\,y_1(0)\,(1) + 2\,y(0)\,(0)$$

$$= 2 + 0 + 2 + 0$$

$$\therefore \quad y_3(0) = 4$$

and so on

By Maclaurin's series, we get

$$y = y(0) + xy_1(0) + \frac{x^2}{2!} y_2(0) + \frac{x^3}{3!} y_3(0) + \ldots\ldots$$

or

$$e^x \sec x = 1 + x(1) + \frac{x^2}{2!}(2) + \frac{x^3}{3!}(4) + \ldots$$

$$e^x \sec x = 1 + x + \frac{2x^2}{2!} + \frac{4x^3}{3!} + \ldots$$

Example **4.17.** Using Maclurin's series, prove that

$$\log(\sec x + \tan x) = x + \frac{x^3}{6} + \frac{x^5}{24} + \ldots\ldots$$

Solution : Let

$$y = \log(\sec x + \tan x) \qquad ...(i)$$

and

$$y_1 = \frac{(\sec x \tan x + \sec^2 x)}{\sec x = \tan x} = \sec x \qquad ...(ii)$$

$$y_2 = \sec x \tan x$$

$$= y_1 \tan x \qquad ...(iii)$$

Again

$$y_3 = y_2 \tan x + y_1 \sec^2 x$$

$$= y_2 \tan x + {y_1}^3 \qquad ...(iv)$$

$$y_4 = y_3 \tan x + y_2 \times {y_1}^2 + 3{y_1}^2 y_2$$

$$= y_3 \tan x + 4{y_1}^2 y_2 \qquad ...(v)$$

$$y_5 = y_4 \tan x + y^3 \sec^2 x + 8y_1{y_2}^2 + 4{y_1}^2 y_3$$

$$= y_4 \tan x + y_3{y_1}^2 + 8y_1{y_2}^2 + 4y_3{y_1}^2$$

$$= y_4 \tan x + 5{y_1}^2 y_3 + 8y_1{y_2}^2 \qquad ...(vi)$$

Putting $x = 0$ in above equations, we get

$$y(0) = \log(\sec 0 + \tan 0) = 0$$

$$y_1(0) = 1$$

$$y_2(0) = y_1(0) \times 0 = 0$$

$$y_3(0) = 0 + 1 = 1$$

$$y_4(0) = 0 + 4 \times 1 \times 0 = 0$$

$$y_5(0) = 0 + 5 \times 12 \times 1 + 8 \times 1 \times 0 = 5$$

and so on

Maclaurins series is given by

$$y = y(0) + xy_1(0) + \frac{x^2}{2!} y_2(0) + \frac{x^3}{3!} y_3(0) + \frac{x^4}{4!} y_4(0) + \frac{x^5}{5!} y_5(0) + ...$$

$$\log(\sec x + \tan x) = x + \frac{x^3}{6} + \frac{x^5}{24} + ...$$

4.8 SOME USEFUL EXPANSIONS

$$(1+x)^n = 1 + nx + \frac{n(n-1)}{2!}x^2 + \frac{n(n-1)(n-2)}{3!}x^3 + ... \qquad (x < 1)$$

$$e^x = 1 + x + \frac{x^2}{2!} + \frac{x^3}{3!} + \frac{x^4}{4!} + ...$$

$$a^x = 1 + x(\log a) + \frac{x^2}{2!}(\log a)^2 + \frac{x^3}{3!}(\log a)^3 + ... \qquad (a > 0)$$

$$\log(1+x) = x - \frac{x^2}{2} + \frac{x^3}{3} - \frac{x^4}{4} + ...$$

$$\sin x = x - \frac{x^3}{3!} + \frac{x^5}{5!} + \frac{x^7}{7!} + ...$$

$$\cos x = 1 - \frac{x^2}{2!} + \frac{x^4}{4!} - \frac{x^6}{6} + ...$$

$$\tan x = x + \frac{x^3}{3!} + \frac{2x^5}{15} + ...$$

$$\sinh x = x + \frac{x^3}{3!} + \frac{x^5}{5!} + \frac{x^7}{7!} + ...$$

$$\cosh x = 1 + \frac{x^2}{2!} + \frac{x^4}{4!} + \frac{x^6}{6!} + ...$$

$$\sin^{-1} x = x + \frac{1}{2} \times \frac{x^3}{3} \times \frac{1}{2} \times \frac{3}{4} \times \frac{x^5}{5} + \frac{1}{2} \times \frac{3}{4} \times \frac{5}{6} \times \frac{x^7}{7} + ...$$

$$\cos^{-1} x = 1 + x + \frac{x^3}{6} + \frac{3x^5}{40} + ...$$

$$\tan^{-1} x = x - \frac{x^3}{3} + \frac{x^5}{5} + \frac{x^7}{7} + ...$$

Example 4.18. Prove that

$$\frac{x}{\sin x} = 1 + \frac{x^2}{6} + \frac{7x^4}{360} + ...$$

Solution : L.H.S

$$\frac{x}{\sin x} = x(\sin x)^{-1}$$

$$= x\left[x - \frac{x^3}{3!} + \frac{x^5}{5!} - ....\right]^{-1} \qquad \left[\because \sin x = x - \frac{x^3}{3!} + \frac{x^5}{5!} - ......\right]$$

$$= xx^{-1}\left[1-\frac{x^2}{3!}+\frac{x^4}{5!}-\ldots\ldots\right]^{-1}$$

$$= \left[1-\frac{x^2}{3!}+\frac{x^4}{5!}-\ldots\ldots\right]^{-1}$$

Putting $z = -\frac{x^2}{3!}+\frac{x^4}{5!}-\ldots$, we get

L.H.S. $= (1+z)^{-1}$

Expanding by binomial theorem, we get

$$= 1-z+z^2-z^3+\ldots\ldots$$

Putting the value of

$$z = 1-\left(-\frac{x^2}{3!}+\frac{x^4}{5!}-\ldots\ldots\right)+\left(-\frac{x^2}{3!}+\frac{x^4}{5!}-\ldots\ldots\right)^2-\left(-\frac{x^2}{3!}+\frac{x^4}{5!}-\ldots\right)^3+\ldots\ldots$$

$$= 1+\frac{x^2}{6}-\frac{x^4}{120}+\ldots\ldots+\frac{x^4}{36}+\frac{x^8}{(120)^2}+\ldots\ldots$$

$$= 1+\frac{x^2}{6}+x^4\left(-\frac{1}{120}+\frac{1}{36}\right)+\ldots\ldots$$

$$= 1+\frac{x^2}{6}+\frac{7x^4}{360}+\ldots\ldots$$

Hence $\frac{x}{\sin x} = 1+\frac{x^2}{6}+\frac{7x^4}{360}+\ldots\ldots$

***Example* 4.19.** Expand $\log(1+\sin^2 x)$ in power of x as far as the term x^6.

Solution : We know that

$$\sin x = x-\frac{x^3}{3!}+\frac{x^5}{5!}-\frac{x^7}{7!}+\ldots\ldots$$

$$\therefore \quad (\sin x)^2 = \left(x-\frac{x^3}{3!}+\frac{x^5}{5!}+\frac{x^7}{7!}+\ldots\ldots\right)^2$$

$$= \left[x-\left(\frac{x^3}{6}-\frac{x^5}{120}+\ldots\right)\right]$$

$$= x^2-2x\left(\frac{x^3}{6}-\frac{x^5}{120}+\ldots\right)+\left(\frac{x^3}{6}-\frac{x^5}{120}\ldots\right)^2$$

$$= x^2-\frac{x^4}{3}+\frac{x^6}{60}+\frac{x^6}{36}+\ldots\ldots$$

$$= x^2-\frac{x^4}{3}+\frac{2x^6}{45}+\ldots\ldots = z \text{ (say)}$$

$$\therefore \quad \log(1+\sin x) = \log(1+z)$$

$$= z-\frac{z^2}{2}+\frac{z^2}{3}+\ldots$$

$$= \left(x^2-\frac{x^4}{3}+\frac{2x^6}{45}+\ldots\right)-\frac{1}{2}\left(x^2\,\frac{x^4}{3}+\frac{2x^6}{45}+\ldots\right)^2+\frac{1}{3}\left(x^2\,\frac{x^4}{3}+\frac{2x^6}{45}+\ldots\right)^3-\ldots\ldots$$

$$= \left(x^2-\frac{x^4}{3}+\frac{2x^6}{45}+\ldots\right)-\frac{1}{2}\left(x^4-\frac{2x^6}{3}-\ldots\right)+\frac{1}{3}\left(x^6+\ldots\right)\ldots$$

$$= x^2 - \frac{x^4}{3} - \frac{x^4}{2} + \frac{2x^6}{45} + \frac{x^6}{3} + \frac{x^6}{3} + ...$$

$$= x^2 + \frac{5x^4}{6} + \frac{32x^6}{45} - .....$$

Hence $$\log(1+\sin^2 x) = x^2 - \frac{5x^4}{6} + \frac{32x^6}{45} ....$$

Example **4.20.** Prove that

$$\log\frac{\tan x}{x} = \frac{x^2}{3} + \frac{7}{90}x^4 + ...$$

Solution : $$\text{L.H.S.} = \log\frac{\tan x}{x}$$

We know that

$$\tan x = x + \frac{x^3}{3} + \frac{2}{15}x^5 + ...$$

$$\therefore \quad \log\frac{\tan x}{x} = \log\frac{\left(x + \frac{x^3}{3} + \frac{2x^5}{15} ......\right)}{x} = \log\left(1 + \frac{x^2}{3} + \frac{2x^4}{15} + ...\right)$$

$$= \log(1+z), \text{ where } z = \frac{x^2}{3} + \frac{2x^4}{15} + ...$$

$$= z + \frac{z^2}{2} + \frac{z^3}{3} ......$$

$$= \left(\frac{x^2}{3} + \frac{2x^4}{15} + ...\right) - \frac{1}{2}\left(\frac{x^2}{3} + \frac{2x^4}{15} + ...\right)^2 + \frac{1}{3}\left(\frac{x^2}{3} + \frac{2x^4}{15} + ...\right)^3 ....$$

$$= \frac{x^2}{3} + x^4\left(\frac{2}{15} - \frac{1}{18}\right) + ...$$

$$= \frac{x^2}{3} + \frac{7x^4}{90} + ...$$

Hence $$\log\frac{\tan x}{x} = \frac{x^2}{3} + \frac{7x^4}{90} + ...$$

Example **4.21** Prove that

$$\tan^{-1}\frac{2x^2}{1-x^2} = 2\left(x - \frac{x^3}{3} + \frac{x^5}{5} - .....\right)$$

Solution : Let $$y = \tan^{-1}\frac{2x}{1-x^2}$$

Putting $x = \tan\theta$, $\theta = \tan^{-1} x$, then we get

$$y = \tan^{-1}\left(\frac{2\tan\theta}{1-\tan^2\theta}\right)$$

$$= \tan^{-1}\tan 2\theta$$

$$= 2\theta$$

$$= 2\tan^{-1} x \qquad ...(i)$$

and $$y_1 = \frac{2}{1+x^2} \quad ...(ii)$$

$$y_2 = -\frac{4x}{(1+x^2)^2} \quad ...(iii)$$

$$y_3 = -\frac{4(1-3x^2)}{(1+x^2)^3} \quad ...(iv)$$

$$y_4 = \frac{48x-48x^3}{(1+x^2)^4} \quad ...(v)$$

$$y_5 = \frac{48-480x^2+280x^4}{(1+x^2)^5} \quad ...(vi)$$

and so on

Putting $x = 0$ in above equations

$$y(0) = 0$$
$$y_1(0) = 2$$
$$y_2(0) = 0$$
$$y_3(0) = -4$$
$$y_4(0) = 0$$
$$y_5(0) = 48$$

From Maclaurin's series

$$y = y(0) + xy_1(0) + \frac{x^2}{2!}y_2(0) + \frac{x^3}{3!}y_3(0) + \frac{x^4}{4!}y_4(0) + \frac{x^5}{5!}y_5(0) + ...$$

or $$y = 0 + x(2) + \frac{x^2}{2!}(0) + \frac{x^3}{3!}(-4) + \frac{x^4}{4!}(0) + \frac{x^5}{5!}(48) + ...$$

$\therefore$ $$\tan^{-1}\left(\frac{2x}{1-x^2}\right) = 2x - \frac{4x^3}{3!} + \frac{48x^5}{5!} + ...$$

or $$\tan^{-1}\left(\frac{2x}{1-x^2}\right) = 2\left(x - \frac{x^3}{3} + \frac{x^5}{5} - .....\right)$$

4.9 SUCCESSIVE DIFFERENTIATION

Let $y = f(x)$ be a function of x, whish is differentiable. Then $\frac{dy}{dx}$ is known as first differential coefficient of y w.r. t x which may be again differentiated. The term $\frac{d}{dx}\left(\frac{dy}{dx}\right)$ is called the second differential coefficient of y w.r. t x and is denoted by $\frac{d^2y}{dx^2}$ and so on. In general, the nth differential coefficient of y w.r.t x is denoted by $\frac{d^ny}{dx^n}$. Usually the differential operator $\frac{d}{dx}$ is represented by the symbol D such that $\frac{dy}{dx} = \text{D}y$. Similarly $\frac{d^ny}{dx^n} = \text{D}^ny$.

Note : If $y = f(x)$,. then y_r is the rth differentiation of y. The nth differential coefficient of y_r is $(n + r)$th differential coefficient of y *i.e.,* $D^{n+r}y = y_{n+r}$ and $D^n y_2 = D^{n+2}y = y_{n+2}$.

4.10 LEIBNITZ'S RULE

This rule help us of find nth differential coefficient of the product of two function. *i.e.,* If u and v are any two function of x such that all their desired differential coefficient exist, then the nth differential coefficient of their product is given by

$$D^n(u \times v) = (D^n u)\, v + nC_1\, D^{n-1}\, u(Dv) + nC_2\, D^{n-2} u\, (D^2 v) + \ldots\ldots\ \ldots.. + {}^nC_r D^{n-r} u\, (D^r v) + \ldots\ldots + u\, (D^n v)$$

Note : (*i*) Differentiation of $(1 - x^2)\, y_2$ for n times by Leibnitz's rule is given by

$$\begin{aligned} D^n\, [y_2\, (1 - x^2)] &= D^n y_2\, (1 - x^2) + nC_1\, D^{n-1} y\, D(1 - x^2) + {}^nC_2\, D^{n-2} y_2\, D_2\, (1 - x^2) \\ &= y_{n+2}\, (1 - x^2) + {}^nC_1 y_{n+1}\, (-2x) + {}^nC_2 y_n\, (-2) \end{aligned}$$

Note : (*ii*) Differentiation of xy_1 for n times by Leibnitz's rule is given by

$$\begin{aligned} Dn\, [y_1\, x] &= D^n y_1\, (x) + {}^nC_1\, y_1\, (Dx) \\ &= y\, n + 1\, (x) + {}^nC_1\, y_n\, (1) \end{aligned}$$

***Example* 4.22.** If $y = (\sin^{-1} x)^2$, prove that

(*i*) $$(1 - x^2)\, \frac{d^2y}{dx^2} - x\frac{dy}{dx} - 2 = 0$$

(*ii*) $$(1 - x^2)\, y_{n+2} - (2n + 1)\, x\, y_{n+1} - n^2 y = 0$$

Also find the expansion upto x^6 term

Solution : Let $y = (\sin^{-1} x)^2$ ...(*i*)

Differentiating (i) w.r.t. '*x*'. we get

$$y_1 = 2\, (\sin^{-1} x) \times \frac{1}{\sqrt{1 - x^2}} \quad ...(ii)$$

Cross multiplying and squaring on both sides

$$(1 - x^2)\, {y_1}^2 = 4(\sin^{-1} x)^2 = 4y$$

Differentiating again,

$$2(1 - x^2)\, y_1 y_2 - 2x{y_1}^2 = 4y_1$$

Dividing by $2y_1$ on both sides, we get

$$(1 - x^2)\, y^2 - xy_1 - 2 = 0 \quad ...(iii)$$

or $$(1 - x^2)\, \frac{d^2y}{dx^2} - x\frac{dy}{dx} - 2 = 0$$

Again, differentiating (*iii*) n times by Leibnits's rule

$$\begin{aligned} D^n[(1 - x^2)\, y_2 - xy_1 - 2] &= 0 \\ D^n\, (1 - x^2)\, y^2 - D^n xy_1 - D^n(2) &= 0 \end{aligned}$$

$$\{(1 - x^2)\, y_{n+2} + {}^nC_1 y_{n+1}\, (-2x) + {}^nC_2 y_n\, (-2)\} - \{xy_{n+1} + {}^nC_1\, y_n\, (1)\} = 0$$

or $$(1 - x^2)\, y_{n+2} - x(2n + 1)\, y_{n+1} - n^2\, y_n = 0 \quad ...(iv)$$

To find expansion up to x^6 term

Put $x = 0$ in above equations

$$\begin{aligned} y(0) &= 0 \\ y_1(0) &= 0 \\ y_2(0) &= 2 \\ y_{n+2} &= n^2\, y_n(0) \end{aligned}$$

Putting $n = 1, 2, 3, 4$ ... in (*v*), we get ...(*v*)

$$n = 1,$$
$$y_3(0) = 1^2 y_1(0) = 0$$
$$n = 2'$$
$$y_4(0) = 2^2 y2(0) = 2^2 \times 2$$
$$n = 3,$$
$$y_5(0) = 3^2\, y_3(0) = 0$$
$$n = 4,$$
$$y_6(0) = 4^2 y_4(0) = 4^2 \times 2^2 \times 2$$

and so on

By Maclaurin's series

$$y = y(0) + xy_1(0) + \frac{x^2}{2!}\, y_2(0) + \frac{x^3}{3!}\, y_3(0) + \frac{x^4}{4!}\, y_4(0) + \frac{x^5}{5!}\, y_5(0) \times \frac{x^6}{6!}\, y_6(0)$$

$$(\sin^{-1} x)^2 = 0 + x(0) + \frac{x^2}{2!}\,(2) + \frac{x^3}{3!}\,(0) + \frac{x^4}{4!}\, 2^2 \times 2 + \frac{x^5}{5!}\,(0) + \frac{x^6}{6!}\, 4^2 \times 2^2 \times 2 + \ldots\ldots$$

$$(\sin^{-1} x)^2 = \frac{2x^2}{2!} + 2 \times 2^2\, \frac{x^4}{4!} + 2 \times 2^2 \times 4^2\, \frac{x^6}{6!} + \ldots\ldots$$

***Example* 4.23.** By forming a differential equation, prove that

$$\sin(m \sin^{-1} x) = mx + \frac{m(1^2 - m^2)}{3!}\, x^3 + \frac{m(1^2 - m^2)(3^2 - m^2)}{5!}\, x^5 + \ldots\ldots$$

Solution : Let

$$y = \sin(m \sin^{-1} x) \qquad \ldots(i)$$

$$\therefore \quad y_1 = \cos(m \sin^{-1} x)\, \frac{m}{\sqrt{1 - x^2}}$$

Cross multiplying and squaring on both sides

$$(1 - x^2)\, y_1{}^2 = m^2 \cos^2(m \sin^{-1} x) = m^2\, [1 - \sin^2(m \sin^{-1} x)]$$

$$\therefore \quad (1 - x^2)\, y_1{}^2 = m^2\,(1 - y^2) \qquad \ldots(ii)$$

Differentiating the above expression,

$$(1 - x^2)\, 2y_1 y_2 - 2xy_1{}^2 = m^2\,(-2yy_1)$$

Dividing by $2y_1$ on both sides, we get

$$(1 - x^2) y_2 - xy_1 = -m^2 y$$

or

$$(1 - x^2) y_2 - xy_1 + m^2 y = 0 \qquad \ldots(iii)$$

or

$$(1 - x^2)\, \frac{d^2 y}{dx^2} - x\, \frac{dy}{dx} + m^2\, y = 0$$

which is a differential equation.

Again differentiating (*iii*) n times by Leibinitz's rules

$$D_n\,(1 - x^2) y^2 - xy_1 + m^2 y] = 0$$

or

$$D_n\,(1 - x^2) y_2 - D^n\,(xy_1) + D^n\,(m^2 y) = 0$$

$$\{y_{n+2}\,(1 - x_2) + n\mathrm{C}_1 y_{n+1}\,(-2x) + {}^n\mathrm{C}_2 y_n\,(-2)\} - \{y_{n+1}\, x + {}^n\mathrm{C}_1 y_n\,(1)\} + m^2 y_n = 0$$

$$(1 - x^2) y_{n+2} - (2n + 1) xy_{n+1} - (n^2 - m^2) y_n = 0 \qquad \ldots(iv)$$

Putting $x = 0$ in above equations, we get

$$y(0) = \sin[m \sin^{-1}(0)] = 0$$

$$y_1(0) = m$$

$$y_2(0) = -m^2 y(0) = 0$$

$$y_{n+2}(0) = (n^2 - m^2)\, y_n(0) \qquad ...(v)$$

Putting $n = 1, 2, 3, 4....$ in (v), we obtain

$n = 1,$

$$y_3(0) = (1^2 - m^2)\, y_1(0) = m(1^2 - m^2)$$

$n = 2,$

$$y_4(0) = (2^2 - m^2)\, y_2(0) = 0$$

$n = 3,$

$$y_5(0) = (3^2 = m^2)\, y_3(0) = m(1^2 - m^2)\,(3^2 - m^2)$$

and so on

By Maclaurin's sereis

$$y = y(0) + xy_1(0) + \frac{x^2}{2!} y_2(0) + \frac{x^3}{3!} y_3(0) + \frac{x^4}{4!} y_4(0) + \frac{x^5}{5!} y_5(0) + ...$$

$$\sin(m \sin^{-1} x) = 0 + mx + \frac{x^2}{2!}(0) + \frac{x^3}{3!} m(1^2 - m^2) + \frac{x^4}{4!}(0) + \frac{x^5}{5!} m(1^2 - m^2)(3^2 - m^2)$$

$$\sin(m \sin^{-1} x) = mx + m(1^2 - m^2)\frac{x^3}{3!} + m(1^2 - m^2)(3^2 - m^2)\frac{x^5}{5!} + ...$$

***Example* 4.24.** By forming a differential equation, expand $e^{m \cos^{-1} x}$ upto x^5 term.

Solutin : Let

$$y = e^{m \cos^{-1} x} \qquad ...(i)$$

$$\therefore \quad y_1 = e^{m \cos^{-1} x} \times \frac{-m}{\sqrt{1-x^2}} = \frac{-my}{\sqrt{1-x^2}} \qquad ...(ii)$$

Cross multiplying and squaring on both sides

$$(1 - x^2) {y_1}^2 = m^2 y^2$$

Differentiating again, we get

$$(1 - x^2) 2y_1 y_2 - 2x{y_1}^2 = 2m^2 y y_1$$

Dividing by $2y_1$ on both sides

$$(1 - x^2)\, y_2 - xy_1 = m^2 y$$

or

$$(1 - x^2) y_2 - xy_1 - m^2 y = 0 \qquad ...(iii)$$

or

$$(1 - x^2)\frac{d^2 y}{dx^2} - x\frac{dy}{dy} - m^2 y = 0$$

which is a required differential equation of order two.

Again differentiating (iii) n times by Leibnitz's rule

$$(1 - x^2) y_{n+2} - 2nxy_{n+1} - n(n-1)\, y_n - xy_{n+1} - ny_n - m^2 y\, n = 0$$

or $(1 - x^2)\, y_{n+2} - (2n + 1)\, xy_{n+1} - (n^2 + m^2)\, y_n = 0$ $\qquad ...(iv)$

On putting $x = 0$ in above equations, we get

$$y(0) = e^{m\cos^{-1}(0)} = e^{m\frac{\pi}{2}}$$

$$y_1(0) = \frac{-m\,y(0)}{\sqrt{1-0^2}} = -m\,y(0) = -m\,e^{m\frac{\pi}{2}}$$

$$y_2(0) = m^2 y(0) = m^2 e^{m\frac{\pi}{2}}$$

$$y_{n+2}(0) = (n_2 + m_2)\,y_n(0) \qquad ...(v)$$

Putting $n = 1, 2, 3,$ in (v), we get

$$n = 1,$$

$$y_3(0) = (1^2 + m^2)\,y_1(0) = -m(1^2 + m^2)\,e^{m\frac{\pi}{2}}$$

$$n = 2,$$

$$y_4(0) = (2^2 + m^2)y_2(0) = m^2(2^2 + m^2)\,e^{m\frac{\pi}{2}}$$

$$n = 3,$$

$$y_5(0) = (3^2 + m^2)\,y^3(0) = -m^2(1^2 + m^2)(3^2 + m^2)\,e^{m\frac{\pi}{2}}$$

and so on

Maclaurin's series is given by

$$y = y(0) + xy_1(0) + \frac{x^2}{2!}y_2(0) + \frac{x^3}{3!}y_3(0) + \frac{x^4}{4!}y_4(0) + \frac{x^5}{5!}y_5(0) ....$$

or

$$e^{m\cos^{-1}x} = e^{m\frac{\pi}{2}} + x\{-me^{m\frac{\pi}{2}}\} + \frac{x^2}{2!}\{m^2 e^{m\frac{\pi}{2}}\} + \frac{x^3}{3!}\{-m(m^2+1^2)\,e^{m\frac{\pi}{2}}\}$$

$$\frac{x^4}{4!}\{m^2(2^2+m^2)2^{m\frac{\pi}{2}}\} + \frac{x^5}{5!}\{-m^2(1^2+m^2)(3^2 = m^2)\,e^{m\frac{\pi}{2}}\} + ...$$

$$e^{m\cos^{-1}}x = e^{m\frac{\pi}{2}}\left[1 - mx + m^2\frac{x^2}{2!} - m(1^2+m^2)\frac{x^3}{3!} + m^2(2^2+m^2)\frac{x^4}{4!} - m^2(1^2+m^2(3^2+m^2\frac{x^5}{5!} + ...\right]$$

***Example* 4.25.** Expand $\log[x + \sqrt{1+x^2}]$, by forming a differential equation.

Solution : Let $y = \log[x + \sqrt{1 = x^2}]$ ...(i)

$$\therefore \quad y_1 = \frac{1}{[x+\sqrt{1+x^2}]}\left[1 + \frac{1}{2}\times\frac{1}{\sqrt{1+x^2}}\times 2x\right] + \frac{1}{\sqrt{1+x^2}}$$

Cross multiplying and squaring on both sides

$$(1 + x_2)\,y_1{}^2 - 1 = 0 \qquad ...(ii)$$

Differentiating (ii) again w.r. t x we get

$$(1+x^2)\,2y_1y_2 + 2xy_1{}^2 = 0 \qquad \text{since } y_1 \neq 0$$

$$2y_1\,[(1+x^2)y_2 + xy_1] = 0$$

$$\therefore \quad (1+x^2)y_2 + xy_1 = 0 \qquad ...(iii)$$

which is differential equation.

or

$$(1+x^2)\frac{d^2y}{dx^2} + x\frac{dy}{dx} = 0$$

Differentiating (*iii*) n times by Leibnitz's rule, we get

$$\{(1+x^2)\,y_{n+2} + {}^nC_1\,(2x)\,y_{n+1} + {}^nC_2\,(2)\,y_n\} + \{xy_{n+1} + {}^nC_1\,(1)\,y_n\} = 0$$

$$(1+x^2)y_{n+2} + (2n+1)\,xy_{n+1} + n^2y_n = 0 \qquad \text{...(iv)}$$

Putting $\qquad x = 0$ in above equations

$$y(0) = \log\,[0 + \sqrt{1+0^2}] = 0$$

$$y_1(0) = 1$$

$$y_2(0) = 0$$

$$y_{n+2}(0) + 0 + n^2y_n(0) = 0$$

$$y_{n+2}(0) = -\,n^2y_n(0) \qquad \text{...(v)}$$

Putting $n = 1, 2, 3, 4,$ in (v), we get

$$n = 1,\; y_3(0) = -\,1^2y_1(0) = -\,1^2 \times 1$$

$$n = 2,\; y_4(0) = -\,2^2y_2(0) = -\,2^2 \times 0 = 0$$

$$n = 3,\; y_5(0) = -\,3^2y_3(0) = -\,3^2(-1^2 \times 1) = 1 \times 1^2 \times 3^2$$

and so on

Maclaurin's series is given by

$$y = y(0) + xy_1(0) + \frac{x^2}{2!}\,y_2\,(0) + \frac{x^3}{3!}\,y_3(0) + \frac{x^4}{4!}\,y_4(0) + \frac{x^5}{5!}\,y_5(0)\;....$$

or

$$\log\,[x + \sqrt{1+x^2}] = 0 + x(1) + \frac{x^2}{2!}\,(0) + \frac{x^3}{3!}\,(-\,1\,.\,1^2) + \frac{x^4}{4!}\,(0) + \frac{x^5}{5!}\,(1\,.\,1^2\,.\,3^2) + ...$$

$$= x - \frac{1}{2} \times \frac{x^3}{3} + \frac{1\times 3}{2\times 4} \times \frac{x^5}{5}$$

EXERCISE 4.2

1. Using Maclaurin's series find expansion of

(*a*) e^x **Ans.** $1 = x + \frac{x^2}{2!} + \frac{x^3}{3!} + \frac{x^4}{4!} + ...$

(*b*) $\log\,(1-x)$ **Ans.** $-\,x\;\frac{x^2}{2} - \frac{x^3}{3} + \frac{x^4}{4}\;....$

(*c*) $\cos x$ **Ans.** $1 - \frac{x^2}{2!} + \frac{x^4}{4!} - \frac{x^6}{6!} + ...$

(*d*) $\tan x$ **Ans.** $x + \frac{x^3}{6} + \frac{2}{15}x^5 + ...$

(*e*) $\cos^{-1} x$ **Ans.** $1 + x + \frac{x^3}{6} + \frac{3x^5}{40} + ...$

(*f*) $\tan^{-1} x$ **Ans.** $x - \frac{x^3}{3} + \frac{x^5}{5} + \frac{x^7}{7} + ...$

2. By using Maclaurin's series prove that

(*a*) $e^{\sin x} = 1 + x + \frac{x^2}{2} - \frac{x^4}{8} + ...$

(*b*) $\frac{e^x}{\cos x} = 1 + x + \frac{2x^2}{2!} + \frac{4x^3}{3!} + ...$

(c) $\log \sec x = \frac{x^2}{2} + \frac{x^4}{12} + \frac{x^6}{45} + ...$

(d) $\log (1 + e^x) = \log 2 + \frac{x}{2} + \frac{x^2}{8} - \frac{x^4}{192} + ...$

(e) $\log \frac{\sin x}{x} = -\left(\frac{x^2}{6} + \frac{x^4}{180} + \frac{x^6}{2835} + ...\right)$

3. Prove that

(a) $\sin^{-1}\left(\frac{2x^2}{1+x^4}\right) = 2\left(x^2 - \frac{x^6}{3} + \frac{x^{10}}{5} - .....\right)$

(b) $\tan^{-1}\left(\frac{x}{\sqrt{1-x^2}}\right) = x + \frac{1^2}{3}x^3 + \frac{1^2 \times 3^2}{5!}x^5 + ...$

(c) $\sin^{-1}\left(\frac{2x}{1+x^2}\right) = 2\left(x - \frac{x^3}{3} + \frac{x^5}{5} + \frac{x^7}{7} + ...\right)$

(d) $\tan^{-1}\frac{\sqrt{1+x^2-1}}{x} = \frac{1}{2}\left(x - \frac{x^3}{3} + \frac{x^5}{5} + \frac{x^7}{7} + ...\right)$

4. By forming a second order differential equation, prove that

(a) $\cos(m \sin^{-1} x) = 1 - \frac{m^2}{2!}x^2 - \frac{m^2(2^2 - m^2)}{4!}x^4 - \frac{m^2(2^2 - m^2)(4^2 - m^2)}{6!}x^6 + ...$

(b) $e^{m \sin^{-1} x} = 1 + mx + \frac{m^2x^2}{2!} + \frac{m(1^2\ m^2)}{3!}x^3 + \frac{m^2(2^2 + m^2)}{4!}x^4 + ...$

(c) $[x + \sqrt{1+x^2}]m = 1 + mx + \frac{m^2x^2}{2!} + \frac{m^2(m^2 - 1^2)}{3!}x^3 + \frac{m^2(m^2 - 2^2)}{4!}x^4 + ...$

4.11 TAYLOR'S SERIES FOR A FUNCTION OF TWO VARIABLES

When the function is given in two variables and its partial derivatives upto nth order are finite and continuous for all (x, y) where $a < x < a + h$, $b < y < b + k$, then

$$f(a+h, b+k) = f(a, b) + \left(h\frac{\partial}{\partial x} + k\frac{\partial}{\partial y}\right)f + \frac{1}{2}\left(h\frac{\partial}{\partial x} + \frac{\partial}{\partial y}\right)^2 f + \frac{1}{3!}\left(h\frac{\partial}{\partial x} + \frac{\partial}{\partial y}\right)^3 f + ...$$

Proof : Let us suppose that $f(x + h, y + k)$ is a function of one variable only *i.e.*, x (say) where y is assumed as constant

On expanding by Taylor's series for one variable

$$f(x+h, y+k) = f(x, y+k) + h\frac{\partial}{\partial x}f(x, y+k) + \frac{h^2}{2!}\frac{\partial^2}{\partial x^2}f(x, y+k) + + \frac{h^3}{3!}\frac{\partial^3}{\partial x^3}f(x, y+k) \quad ...(i)$$

Now expanding $f(x, y + k)$ as a function of y only *i.e.*,

$$f(x, y+k) = f(x, y) + k\frac{\partial}{\partial y}f(x, y) + \frac{k^2}{2!}\frac{\partial^2}{\partial y^2}f(x, y) + \frac{k^3}{3!}\frac{\partial^3}{\partial y^3}f(x, y) + \frac{k^4}{4!}\frac{\partial^4}{\partial y^4}f(x, y) \quad ...(ii)$$

Using (*ii*) and (*i*), we get

$$f(x+h, y+k) = f(x, y) + k\frac{\partial}{\partial y}f(x, y) + \frac{k^2}{2!}\frac{\partial^2}{\partial y^2}f(x, y) + \frac{k^3}{3!}\frac{\partial^3}{\partial y^3}f(x, y) + \frac{k^4}{4!}\frac{\partial^4}{\partial y^4}f(x, y)$$

$$+ h\frac{\partial}{\partial x}\left\{f(x, y) + k\frac{\partial}{\partial y}f(x\ y) + \frac{k^2}{2!}\frac{\partial^2}{\partial y^2}f(x, y) + \frac{k^3}{3!}\frac{\partial^3}{\partial y^3}f(x, y) + \frac{k^4}{4!}\frac{\partial^4}{\partial y^4}f(x, y)\right\}$$

$$+\frac{h^2}{2!}\frac{\partial}{\partial x^2}\left\{f(x,y)+k\frac{\partial}{\partial y}f(x\ y)+\frac{k^2}{2!}\frac{\partial^2}{\partial y^2}f(x,y)+\frac{k^3}{3!}\frac{\partial^3}{\partial y^3}f(x,y)+\frac{k^4}{4!}\frac{\partial^4}{\partial y^4}f(x,y)\right\}+\ldots.$$

$$= f(x,y)+h\frac{\partial}{\partial x}f(x,y)+k\frac{\partial}{\partial y}f(x,y)+$$

$$\frac{1}{2!}\left(h^2\frac{\partial^2 f(x,y)}{\partial x^2}+k^2\frac{\partial^2 f(x,y)}{\partial y^2}+2hk\frac{\partial^2 f(x,y)}{\partial x\partial y}\right)$$

$$+\frac{1}{3!}\left(h^3\frac{\partial^3 f(x,y)}{\partial x^3}+3h^2k\frac{\partial^3 f(x,y)}{\partial x^2\partial y}+3hk^2\frac{\partial^3 f(x,y)}{\partial x\partial y^2}\right)+\ldots$$

$$= f(x,y)+\left(h\frac{\partial}{\partial x}+k\frac{\partial}{\partial y}\right)f(x,y)+\frac{1}{2!}\left(h\frac{\partial}{\partial x}+k\frac{\partial}{\partial y}\right)^2 f(x,y)+$$

$$\frac{1}{3!}\left(h\frac{\partial}{\partial x}+k\frac{\partial}{\partial y}\right)^3 f(x,y)+\ldots$$

Hence

$$f(x+h,y+k) = f(x,y)+\left(h\frac{\partial}{\partial x}+k\frac{\partial}{\partial y}\right)f(x,y)+\frac{1}{2!}\left(\frac{\partial}{\partial x}+\frac{\partial}{\partial y}\right)^2 f(x,y)$$

$$+\frac{1}{3!}\left(h\frac{\partial}{\partial x}+k\frac{\partial}{\partial y}\right)^3 f(x,y)+\ldots \qquad \ldots(iii)$$

This is the expansion for Taylor's theorem for a function of two variables in term of h, k at the point x and y.

Note (1) : Expansions at point a, b in the powers of h and k is obtained by

Putting $x = a$ and $y = b$ in equation (*iii*), we get

$$f(a+h,b+k)=f(a,b)+[hf_x(a,b)+kf_y(a,b)+\frac{1}{2!}[h^2 f_{xx}(a,b)+2hk\,f_{xy}(a,b)+k^2 f_{yy}(a,b)]$$

$$+\frac{1}{3!}[h^3 f_{xxx}(a,b)+k^3 f_{yyy}(a,b)+3h^2k f_{xxy}(a,b)+3hk^2 f_{xyy}(a,b)]+ \qquad \ldots(iv)$$

Note : (2) Expansion at (0, 0) in the power of x and y is obtained by

Putting $x = 0$, $y = 0$ and replacing h by x and k by y in equation (*iii*), we get which is called Maclausin's expansion of $f(x, y)$.

$$f(x,y)=f(0,0)+[xf_x(0,0)+yf_y(0,0)]+\frac{1}{2!}[x^2 f_{xx}(0,0)+2xy_{xy}(0,0)+y^2 f_{yy}(0,0)]$$

$$+\frac{1}{3!}[x^3 f_{xx}(0,0)+y^3(f_{yyy}(0,0)+3x^2y f_{xxy}(0,0)+3xy^2 f_{xyy}(0,0)]+ \qquad \ldots(v)$$

***Example* 4.26.** Expand $e^x \log(1+y)$ in power of x and y upto term of third degree.

Solution : Let $f(x, y) = e^x \log(1 + y)$

Taylor's series in the powers of x and y

$$f(x,y) = f(0,0)+x\left(\frac{\partial f}{\partial x}\right)_{(0,0)}+y\left(\frac{\partial f}{\partial y}\right)_{(0,0)}+\frac{x^2}{2!}\left(\frac{\partial^2 f}{\partial x^2}\right)_{(0,0)}+\frac{y^2}{2!}\left(\frac{\partial^2 y}{\partial y^2}\right)_{(0,0)}$$

$$+\frac{2xy}{2!}\left(\frac{\partial^2 f}{\partial x\,\partial y}\right)_{(0,0)}+\frac{x^3}{3!}\left(\frac{\partial^3 f}{\partial x^3}\right)_{(0,0)}+\frac{y^3}{3!}\left(\frac{\partial^3 f}{\partial y^3}\right)_{(0,0)}+$$

$$\frac{3x^2y}{3!}\left(\frac{\partial^3 f}{\partial x^2\,\partial y}\right)_{(0,0)}+\frac{3xy^2}{3!}\left(\frac{\partial^3 f}{\partial x\,\partial y^2}\right)_{(0,0)}+\ldots\ldots \quad \ldots.(A)$$

Now $\qquad f(x, y) = e^x \log(1 + y),$

$\therefore \qquad f(0, 0) = 0 \qquad \text{...}(i)$

$$\frac{\partial f}{\partial x} = e^x \log(1 + y),$$

$\therefore \qquad \left(\frac{\partial f}{\partial x}\right)(0, 0) = 0 \qquad \text{...}(ii)$

$$\frac{\partial f}{\partial x} = \frac{e^x}{1 + y},$$

$\therefore \qquad \left(\frac{\partial f}{\partial x}\right)(0, 0) = 1 \qquad \text{...}(iii)$

Again $\qquad \frac{\partial^2 f}{\partial x^2} = e^x \log(1 + y)$

$\therefore \qquad \left(\frac{\partial^2 f}{\partial x^2}\right)_{(0, 0)} = 0 \qquad \text{...}(iv)$

$$\frac{\partial^2 f}{\partial y^2} = -\frac{e^x}{(1 + y)^2}$$

$\therefore \qquad \left(\frac{\partial^2 f}{\partial y^2}\right)_{(0, 0)} = -1 \qquad \text{...}(v)$

$$\frac{\partial^2 f}{\partial x\, \partial y} = \frac{e^x}{1 + y}$$

$\therefore \qquad \left(\frac{\partial^2 y}{\partial x\, \partial y}\right)_{(0, 0)} = 1 \qquad \text{...}(vi)$

and $\qquad \frac{\partial^3 f}{\partial x^3} = e^x \log(1 = y)$

$\therefore \qquad \left(\frac{\partial^3 f}{\partial x^3}\right)_{(0, 0)} = 0 \qquad \text{...}(vii)$

$$\left(\frac{\partial^3 f}{\partial y^3}\right) = \frac{2e^x}{(1 + y)^3}$$

$\therefore \qquad \left(\frac{\partial^2 f}{\partial y^2}\right)_{(0, 0)} = 2 \qquad \text{...}(viii)$

$$\frac{\partial^3 f}{\partial x^2 \partial y} = \frac{e^x}{1 + y}$$

$\therefore \qquad \left(\frac{\partial^3 f}{\partial x^2\, \partial y}\right)_{(0, 0)} = 1 \qquad \text{...}(ix)$

$$\frac{\partial^3 f}{\partial x\, \partial y^2} = -\frac{e^x}{(1 + y)^2}$$

$\therefore \qquad \left(\frac{\partial^3 f}{\partial x\, \partial y^2}\right)_{(0, 0)} = -1 \qquad \text{...}(x)$

On putting all these values in equation (A), we get

$$e^x \log(1+y) = 0 + x(0) + y(1) + \frac{1}{2!}[x^2(0) + 2xy(1) + y^2(-1)] + \frac{1}{3!}[x^3(0) + 3x^2y(1) + 3xy^2(-1) + y^3(2)] + \ldots\ldots$$

or
$$e^x \log(1+y) = y + \frac{1}{2!}(2xy - y^2) + \frac{1}{3!}[3x^2y - 3xy^2 + 2y^3] + \ldots\ldots$$

Hence
$$e^x \log(1+y) = y + xy - \frac{y^2}{2} + \frac{1}{2}(x^2y - xy^2) + \frac{y^3}{3} + \ldots\ldots$$

Example **4.27.** Expand $e^x \sin y$ in power of x and y as far as term of third degree.

Solution : Let
$$f(x, y) = e^x \sin y$$

Taylor's series in the powers of x and y

$$f(x, y) = f(0,0) + xf_x(0,0) + yf_y(0,0) + \frac{x^2}{2!}f_{xx}(0,0) + \frac{y^2}{2!}f_{yy}(0,0)$$

$$+ \frac{2xy}{2!}f_{xy}(0,0) + \frac{x^3}{3!}f_{xxx}(0,0) + \frac{y^3}{3!}f_{yyy}(0,0) + \frac{3x^3y}{3!}f_{xxy}(0,0) + \frac{3xy^2}{3!}f_{xyy}(0,0) + \ldots\ldots \quad \ldots\ldots(A)$$

Now

$$f(x, y) = e^x \sin y$$
$$\therefore \quad f_x(0,0) = 0 \quad \ldots(i)$$
$$f_x(x, y) = e^x \sin y$$
$$\therefore \quad f_x(0,0) = 0 \quad \ldots(ii)$$
$$f_y(x, y) = e_x \cos y$$
$$\therefore \quad f_y(0,0) = 1 \quad \ldots(iii)$$
$$f_{xx}(x, y) = e_x \sin y$$
$$\therefore \quad f_{xx}(0,0) = 0 \quad \ldots(iv)$$
$$f_{xy}(x, y) = e^x \cos y$$
$$\therefore \quad f_{xy}(0,0) = 1 \quad \ldots(v)$$
$$f_{yy}(x, y) = \sin y$$
$$\therefore \quad f_{yy}(0,0) = -1 \quad \ldots(vi)$$
$$f_{xxx}(x, y) = e_x \sin y$$
$$\therefore \quad f_{xxx}(0,0) = 1 \quad \ldots(vii)$$
$$f_{xxy}(x, y) = e^x \cos y$$
$$\therefore \quad f_{xxy}(0,0) = 1 \quad \ldots(viii)$$
$$f_{xyy}(x, y) = -e^x \sin y$$
$$\therefore \quad f_{xyy}(0,0) = 0 \quad \ldots(ix)$$
$$f_{yyy}(x, y) = -e^x \cos y$$
$$\therefore \quad f_{yyy}(0,0) = -1 \quad \ldots(x)$$

Putting all these values in equation (A), we get

$$f(x, y) = e^x \sin y + 0 + x(0) + y(1) + \frac{x^2}{2!}(0) + xy(1) + \frac{y^2}{2!}(-1) = \frac{x^3}{3!}(0)$$

$$+ \frac{3x^2y}{3!}(1) + \frac{3xy^2}{3!}(0)\ \frac{y^3}{3!}(-1) + \ldots$$

Hence $$e^x \sin y = y + xy - \frac{y^2}{z} + \frac{x^2 y}{z} + \frac{y^3}{6} + \ldots$$

Example. **4.28.** Expand $e^x \cos y$ near the point $\left(1, \frac{\pi}{4}\right)$ by Taylor's series.

Solution : Let $$f(x, y) = e^x \sin y$$

$$\therefore \quad f\left(1, \frac{\pi}{4}\right) = \frac{e}{\sqrt{2}} \qquad \ldots(i)$$

Here $f(x, y) = f\left\{1 + (x-1), \frac{\pi}{4} + \left(y - \frac{\pi}{4}\right)\right\}$ *i.e.,* Taylor's series in the powers of $(x - 1)$ and $\left(y - \frac{\pi}{4}\right)$ at point $\left(1, \frac{\pi}{4}\right)$, therefore

$$f(x, y) = f\left(1, \frac{\pi}{4}\right) + (x-1) fx\left(1, \frac{\pi}{4}\right) + \left(y - \frac{\pi}{4}\right) fy\left(1, \frac{\pi}{4}\right) + \frac{(x-1)^2}{2!} f_{xx}\left(1, \frac{\pi}{4}\right)$$

$$+ \frac{\left(y - \frac{\pi}{4}\right)^2}{2!} fyy\left(1, \frac{\pi}{4}\right) + \frac{2(x-1)\left(y - \frac{\pi}{4}\right)}{2!} f_{xy}\left(1, \frac{\pi}{4}\right)$$

$$+ \frac{(x-1)^3}{3!} f_{xxx}\left(1, \frac{\pi}{4}\right) + \frac{\left(y - \frac{\pi}{4}\right)^3}{3!} f_{yyy}\left(1, \frac{\pi}{4}\right) + \frac{3(x-1)^2\left(y - \frac{\pi}{4}\right)}{3!} f_{xxy}\left(1, \frac{\pi}{4}\right)$$

$$+ \frac{3(x - 1\left(y - \frac{\pi}{4}\right)^2}{3!} fxyy\left(1, \frac{\pi}{4}\right)^2 + \ldots \qquad \ldots(A)$$

From (*i*)

$f_x(x, y) = e^x \cos y$

$$\therefore \quad f_x\left(1, \frac{\pi}{4}\right) = \frac{e}{\sqrt{2}} \qquad \ldots(ii)$$

$$f_y(x, y) = -e^x \sin y$$

$$\therefore \quad f_y\left(1, \frac{\pi}{4}\right) = \frac{e}{\sqrt{2}} \qquad \ldots(iii)$$

$$f_{xx}(x, y) = e^x \cos y$$

$$\therefore \quad f_{xx}\left(1, \frac{\pi}{4}\right) = \frac{e}{\sqrt{2}} \qquad \ldots(iv)$$

$$f_{yy}(x, y) = -e^x \cos y$$

$$\therefore \quad f_{yy}\left(1, \frac{\pi}{4}\right) = \frac{e}{\sqrt{2}} \qquad \ldots(v)$$

$$f_{xy}(x, y) = -e^x \sin y$$

$$\therefore \quad f_{xy}\left(1, \frac{\pi}{4}\right) = \frac{e}{\sqrt{2}} \qquad \ldots(vi)$$

$$f_{xxx}(x, y) = e^x \cos y$$

$$\therefore \quad f_{xxx}\left(1, \frac{\pi}{4}\right) = \frac{e}{\sqrt{2}} \qquad ...(vii)$$

$$f_{yyy}(x, y) = e^x \sin y$$

$$\therefore \quad f_{yyy}\left(1, \frac{\pi}{4}\right) = \frac{e}{\sqrt{2}} \qquad ...(viii)$$

$$f_{xxy}(x, y) = -e^x \sin y$$

$$\therefore \quad f_{xxy}\left(1, \frac{\pi}{4}\right) = \frac{e}{\sqrt{2}} \qquad ...(ix)$$

$$f_{xyy}(x, y) = -e^x \cos y$$

$$\therefore \quad f_{xyy}\left(1, \frac{\pi}{4}\right) = \frac{e}{\sqrt{2}} \qquad ...(x)$$

Putting these values in equation (A), we get

$$e^x \cos y = \frac{e}{\sqrt{2}}\left[(x-1)\frac{e}{\sqrt{2}} + \left(y-\frac{\pi}{4}\right)\left(\frac{-e}{\sqrt{2}}\right)\right] + \frac{1}{2!}\left[(x-1)^2\frac{e}{\sqrt{2}} + 2(x-1)\left(y-\frac{\pi}{4}\right)\left(-\frac{e}{\sqrt{2}}\right) + \left(y-\frac{\pi}{4}\right)^2\left(-\frac{e}{\sqrt{2}}\right)\right]$$

$$+ \frac{1}{3!}\left[(x-1)^3\frac{e}{\sqrt{2}} + \left(y-\frac{\pi}{4}\right)^3\frac{e}{\sqrt{2}} + {}^3(x-1)^2\left(y-\frac{\pi}{4}\right)\left(-\frac{e}{\sqrt{2}}\right) + {}^3(x-1)\left(y-\frac{\pi}{4}\right)^2\left(-\frac{e}{\sqrt{2}}\right)\right] + ...$$

$$= \frac{e}{\sqrt{2}}\left[\begin{matrix} 1 + (x-1) - \left(y-\frac{\pi}{4}\right) + \frac{(x-1)^2}{2} - 2(x-1)\left(y-\frac{\pi}{4}\right) - \left(y-\frac{\pi}{4}\right)^2 \\ + \frac{(x-1)^3}{6} + \frac{\left(y-\frac{\pi}{4}\right)^3}{6} - \frac{(x-1)^2\left(y-\frac{\pi}{4}\right)}{2} - \frac{3(x-1)\left(y-\frac{\pi}{4}\right)^2}{2} + ... \end{matrix}\right]$$

***Example* 4.29.** If $f(x, y) = \tan^{-1}(xy)$, compute an approximate value of $f(0.9, 1.2)$.

Solution : $f(x, y) = \tan^{-1}(xy) \therefore f(1, -1) = -\frac{\pi}{4}$ ...(*i*)

Let us expand $f(x, y)$ near the point $(1, -1)$.

$$f(0.9), -1.2) = f(1-0.1, -1-0.2)$$

$$= f(1,-1) + [(-0.1) f_x(1,-1) + (-0.2) f_y(1,-1)] + \frac{1}{2!}[(-0.1)^2$$

$$f_{xx}(1,-1) + (-0.2)^2 f_{yy}(1-1) + 2(-0.1)(-0.2) f_{xy}(1,-1)] + ... \qquad ...(A)$$

From equation (1),

$$f_x(x, y) = \frac{y}{1+x^2y^2}$$

$$\therefore \quad f_x(1,-1) = -\frac{1}{2} \qquad ...(ii)$$

$$f_y(x, y) = \frac{x}{1+x^2y^2}$$

$$\therefore \quad f_y(1,-1) = \frac{1}{2} \qquad ...(iii)$$

$$f_{xx}(x, y) = -\frac{2xy^3}{(1+x^2y^2)^2}$$

$$\therefore \quad f_{xx}(1, -1) = \frac{1}{2} \quad ...(iv)$$

$$f_{yy}(x, y) = \frac{-2xy^3}{(1 + x^2y^2)^2}$$

$$\therefore \quad f_{xy}(1, -1) = \frac{1}{2} \quad ...(v)$$

$$f_{xy}(x, y) = \frac{1-2xy^3}{(1 + x^2y^2)^2}$$

$$\therefore \quad f_{yy}(1, -1) = 0 \quad ...(iv)$$

Putting all these values in equation (A), we get

$$f(0.9, -1.2) = -\frac{\pi}{4} + (-0.1)\left(-\frac{1}{2}\right) + (-0.2)$$

$$\left(\frac{1}{2}\right) + \frac{1}{2}\left[(-0.1)^2\left(\frac{1}{2}\right) + 0.1)(-0.2)(0) + (-0.2)^2\left(\frac{1}{2}\right)\right] + ...$$

$$= -\frac{\pi}{4} + 0.05 - 0.1 + \frac{1}{2}(0.005 + 0.04)$$

***Example* 4.30.** Expand $x^2y + 3y - 2x$ in powers of $x - 1$ and $y + 2$ using Taylor's series.

Solution : Let $f(x, y) = f[1 + (x - 1), -2 + (y + 2)]$

Expand by Taylor's series in power of $x - 1$ and $y + 2$

$$f(x, y) = f(1, -2) + (x - 1) f_x(1, -2) + (y + 2) f_y(1, -2)$$

$$+ \frac{1}{2!}[x + 1)^2 f_{xx}(1, -2) + (y + 2)^2 f_{yy}(1, -2) + 2(x - 1)(y + 2) f_{xy}(1, -2)]$$

$$+ \frac{1}{3!}[x + 1)^3 f_{xx}(1, -2) + (y + 2)^3 f_{yy}(1, -2) + 3(x - 1)^2 (y + 2) f_{xy}(1, -2)]$$

$$+ 3(x - 1)(y + 2)^2 f_{xyy}(1, -2)] + ... \quad ...(A)$$

Here

$$f(x, y) = x^2y + 3y - 2x$$

$$\therefore \quad f(1, -2) = -10 \quad ...(i)$$

$$f_x(x, y) = 2xy - 2$$

$$\therefore \quad f_x(1, -2) = -6 \quad ...(ii)$$

$$f_y(x, y)\, x^2 + 3$$

$$\therefore \quad f_y(1, -2) = 4 \quad ...(iii)$$

$$f_{xx}(x, y) = 2y$$

$$\therefore \quad f_{xx}(1, -2) = 4 \quad ...(iv)$$

$$f_{yy}(x, y) = 0$$

$$\therefore \quad f_{yy}(1, -2) = 0 \quad ...(v)$$

$$f_{xy}(x, y) = 2x$$

$$\therefore \quad f_{xy}(1, -2) = 0 \quad ...(vi)$$

$$f_{xxx}(x, y) = 0$$

$$\therefore \quad f_{xxx}(1,-2) = 0 \quad ...(vii)$$

$$f_{xxy}(x, y) = 2$$

$$\therefore \quad f_{xxy}(1,-2) = 2 \quad ...(viii)$$

$$f_{xyy}(x, y) = 0$$

$$\therefore \quad f_{xxy}(1,-2) = 0 \quad ...(ix)$$

$$f_{yyy}(x, y) = 0$$

$$\therefore \quad f_{yyy}(1,-2) = 0 \quad ...(x)$$

Putting all these values in equation (A), we get

$$x^2y + 3y - 2x = -10 + [(x - 1(-6) + (y+2)(4)] +$$

$$\frac{1}{2!}[(x-1)^2(-4) + 2(x-1)(y+2)(2) + (y+2)^2(0)] + \frac{1}{3!}[(x-1)^3(0) + 3(x-1)^2(y+2)(2) + 3(x-1)(y+2)^2(0)]$$

$$(y+2)^3(0)]$$

$$= -10 - 6(x-1) + 4(y+2) - 2(x-1)^2 + 2(x-1)(y+2) + (x-1)^2(y+2)$$

EXERCISE 4.3

1. Expand $e^x \cos y$ at (0, 0) upto third degree term in x, y

Ans. $1 = x + \frac{1}{2}(x^2 - y^2) + \frac{1}{3!}(x^3 - xy^2) + ...$

2. Expand $\cos x \cos y$ in power of x and y upto fourth order term

Ans. $1 - \frac{x^2}{2} - \frac{y^2}{2} + \frac{x^4}{24} + \frac{y^4}{24} + \frac{x^2y^2}{4} + ...$

3. Expand $f(x, y) = \tan^{-1}\left(\frac{y}{x}\right)$ in the neighbourhood of (1, 1) upto third degree terms. Hence compute $f(1.1, 0{,}9)$ approximately

Ans. $\frac{\pi}{4} - \frac{1}{2}[(x-1) - (y-1) + \frac{1}{4}[(x-1) - (y-1)^2] - \frac{1}{12}[(x-3)^2 + 3(x-1)^2(y-1) - 3(x-1)(y-1)^2 - (y-1)^3] + ..$ and value is 0.6857

4. Expand $\sin xy$ in power of $(x-1)$ and $\left(y - \frac{\pi}{2}\right)$ upto the second degree terms.

Ans. $1 - \frac{1}{8}\pi^2(x-1)^2 - \frac{1}{2}\pi(x-1)\left(y - \frac{\pi}{2}\right) - \frac{1}{2}\left(y - \frac{\pi}{2}\right)^2 + ...$

5. Expand e^{xy} at (1, 1) upto these terms.

Ans. $e\left[1 + (x-1) + (y-1) + \frac{1}{2!}(x-1)^2 + 4(x-1)(y-1) + (y-1)^2 + ...\right]$

ASYMPTOTES AND CURVATURE

This chapter is devoted to the study of asymptotes and its types, in respect of both polar and cartesian co-ordinates. The chapter further deals with curvature and other topics associated with it.

4.12 BRANCH OF A CURVE

Suppose that a curve is prescribed by the equation $y = f(x)$ such that y has two or more values for every value of x. Corresponding to these distinct values of y we shall get different branches of the curve

Thus if $x^2 + y^2 = a^2$, be a curve then $y^2 = a^2 - x^2$

or $y = \pm\sqrt{a^2 - x^2}$

Hence, corresponding to $y = \sqrt{a^2 - x^2}$ and $y = -\sqrt{a^2 - x^2}$ there are two branches of the curve *i.e.*, circle $x^2 + y^2 = a^2$. The branches are referred to as the upper half and lower half of the circle.

4.13 ASYMPTOTE

An asymptote of a curve is a straight line at a finite distance from the origin, to which to tangent to the curve tends as the point of contact recedes to infinity. Hence when a point P on the infinite branch of the curve tends to infinite, its perpendicular distance from the asymptotes tends to zero.

Note: Asymptotes may be parallel either to x axis or to y-axis and accordingly they are called horizontal and vertical asymptotes. If an asymptotes is not parallel to either of the co-ordinate axes, then it is called an oblique asymptote.

4.14 ASYMPTOTES PARALLEL TO COORDINATE AXIS

(*a*) **Asymptotes parallel to y-axis :** Let $y = f(x)$ be any curve and let $x = c$ be asymptotes of the curve. For estimating the value of parameter c, consider a point $A(x, y)$ on the curve AB and let δ be its distance from the line $x = c$ *i.e.*,

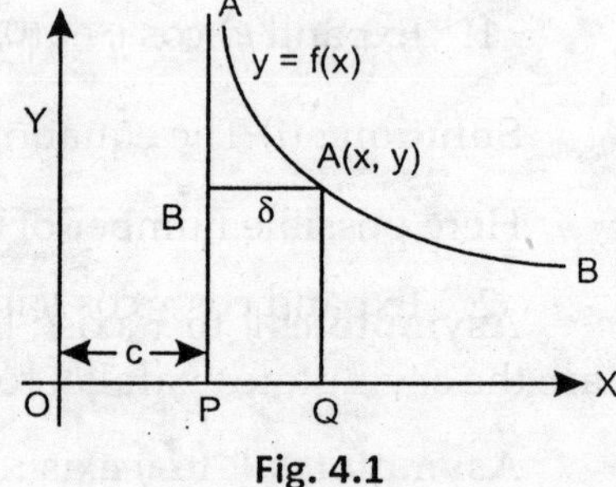

Fig. 4.1

$$\therefore \quad \delta = |x - c|$$

Then by definition of an asymptote $\delta \to 0$ as $A(x, y) \to \infty$ along the curve AB.

Therefore as $A(x, y) \to \infty$ or as $y \to \infty$ along an infinite branch of the curve $y = f(x)$ then $\delta \to 0$

$$\therefore \quad \delta = |x - c| \to 0 \text{ as } y \to \infty$$

or $\quad x \to c$ as $y \to \infty$

$$\therefore \quad \underset{y \to \infty}{Lt}\ x = c, \text{ then } x = c \text{ is the asymptote } \parallel \text{ to } y\text{-axis.}$$

(*b*) **Asymptotes parallel to x-axis :** Similarly to obtain the asymptote parallel to x-axis, find form the given equation of the curve the value 'd' of which y tends as $x \to \infty$, then $y = d$ is an asymptotes parallel to x-axis.

4.15 ASYMPTOTES PARALLEL TO CO-ORDINATE AXES FOR *f*(*x*, *y*) = 0 *i.e.*, ALGEBRAIC CURVE

Case I : Asymptotes parallel to *y*-axis :

Let an algebraic equation $f(x, y) = 0$ of nth order of the curve be given by

$$y^n Q_0(x) + y^{n-1} Q_1(x) + y^{n-2} Q_2(x) + ... + Q_n(x) = 0 \quad ...(i)$$

Where $Q_0(x), Q_1(x), Q_2(x) ... Q_n(x)$ are polynomials in x

Dividing (*i*) by y^n on both side, we get

$$Q_0(x) + \frac{1}{y} Q_1(x) + \frac{1}{y^2} Q_2(x) + ...+ \frac{1}{y^n} Q_n(x) = 0$$

If $x = k$ is an asymptotes of the curve parallel to y-axis of equation (*i*) then $k = \underset{y \to \infty}{Lt}\ x$. Therefore taking $y \to \infty$, and $x \to k$

$$\underset{y \to \infty}{Lt} \; Q_0(x) \Rightarrow Q_0(k) = 0$$

$\therefore$ k is a roof of the equation $Q_0(x) = 0$ ...(*ii*)

If $k_1, k_2 \ldots . k_r$ etc be the roots of he equation $Q_0(x) = 0$. Then $x = k_1, x = k_2 \ldots ., x = k_r$ etc are the asymptotes parallel to y-axis.

Since $Q_0(x) = 0$ {from (*ii*)} has a root k, therefore $x - k$ will be a factor of $Q_0(x)$

Also $Q_0(x)$ is the coefficient of y^n in the equation of the curve, we have the result.

The asymptotes parallel to x-axis are obtained by equating to zero, the coefficient of y^n.

Note : If the coefficient of the highest power of y is a constant, then there is no asymptotes parallel to y-axis.

Case II: Asymptotes parallel to x-axis:

In case (*i*) interchanging y and x and proceed as before then a working method for finding asymptotes parallel to x axis is given by equating to zero the coefficient to the highest power of x present in the given equation of the curve.

Note : If the coefficient of the highest power of x is a constant, then there is no asymptotes parallel to x axis.

Example 4.30. Find the asymptotes, parallel to the axes for the curve.

(*i*) $x^2y^2 = a^2(x^2 + y^2)$ (*ii*) $\dfrac{a^2}{x^2} - \dfrac{b^2}{y^2} = 1$

(*iii*) $x^2y^2 - x^2y - xy^2 + x + y - 1 = 0$ (*iv*) $xy^3 - x^3 - ax^2 - ay^2 = 0$

Solution : (*i*) The equation of the curve is $x^2y^2 - a^2x^2 - a^2y^2 = 0$

Here possible number of asymptotes is 4, since the equation is of fourth degree.

Asymptotes ‖ to x axis : Equating to zero the coefficient of highest power of x *i.e.,* $y^2 - a^2 = 0 \Rightarrow y = \pm a$, which are the asymptotes parallel to x axis.

Asymptotes ‖ to y axis : Equating to zero the coefficient of highest power of y *i.e,* $x^2 - a^2 = 0 \Rightarrow x = \pm a$, which are the asymptotes parallel to y axis.

(*ii*) The equation of curve is $\dfrac{a^2}{x^2} - \dfrac{b^2}{y^2} = 1$

or $a^2y^2 - b^2x^2 = x^2y^2$

or $x^2y^2 - a^2y^2 + b^2x^2 = 0$

Here possible asymptotes are 4.

Asymptotes ‖ to x axis : Equating o zero the coefficient of highest power of x *i.e,* $y^2 + b^2 = 0 \Rightarrow y^2 = -b^2 \Rightarrow y = \pm ib$ which hive imaginary values of y, so there is no asymptotes parallel to x axis.

Asymptotes ‖ to y axis : Equating to zero the coefficient of highest power of y *i.e.,* $x^2 - a^2 = 0 = \Rightarrow x = \pm a$, which are the asymptotes ‖ to y axis.

Hence $x = \pm a$ are the only real asymptotes.

(*iii*) The equation of curve is $x^2y^2 - x^2y - xy^2 + x + y - 1 = 0$

Here possible asymptotes are 4.

Asymptotes ‖ to x axis : Equating to zero the coefficient of highest power of x *i.e.,* $y^2 - y = 0$

or $y(y - 1) = 0$

or $y = 0$ and $y = 1$ are the asymptotes ‖ to x-axis.

Asymptotes || to y axis : Equating to zero the coefficient to highest power of y *i.e.,*

$$x^2 - x = 0$$

$$x\,(x-1) = 0$$

$\Rightarrow$ $x = 0$ and $x = 1$ are the asymptotes || to y- axis.

Hence the asymptotes are $x = 0, x = 1$ and $y = 0, y = 1$.

(iv) The equation of curve is given by $xy^3 - x^3 - ax^3 - ay^2 = 0$

Here possible asymptotes are 4.

Asymptotes || to x axis : The coefficient of highest power of x is 1 which is constant. So there is no asymptotes || to x-axis.

Asymptotes || to y axis : Equating to zero the coefficient of highest power y *i.e.,* $x = 0$

$\therefore$ y-axis itself is the asymptote.

Example 4.31. Show that the asymptotes of the curve $x^2y^2 = a^2\,(x^2 + y^2)$ from a square of side $2a$.

Solution : Here the equation of curve is $x^2y^2 - a^2x^2 - a^2y^2 = 0$.

The required Asymptotes are same as example 4.1 *(i) i.e.,* $x = \pm\, a$, $y = \pm\, a$.

Hence $AB = BC = CD = DA = 2a$

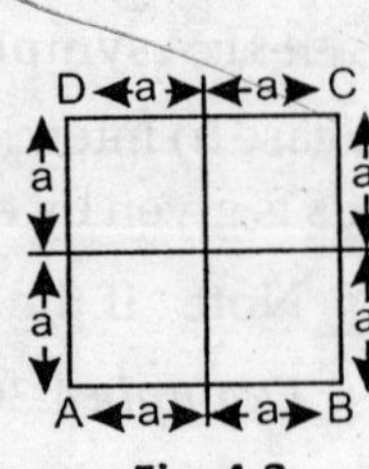

Fig. 4.2

EXERCISE 4.4

Find the asymptotes parallel to axis of the following curves.

1. $x^2y^2 - y^2 - 2 = 0$ **Ans.** $y = 0\ x = \pm\, 1$
2. $y^3 - xy^2 = x^2 + 1$ **Ans.** None
3. $y^3 - xy^2 - x^2 + 1$ **Ans.** None
4. $xy^3 - x^3 = a\,(x^2 + y^2)$ **Ans.** $x = 0$
5. $x^2y^2 - a^2\,(x^2 + y^2) - a^2\,(x + y)\,a^4 = 0$ **Ans.** $x = \pm\, a$, $y = \pm\, a$.

4.16 OBLIQUE ASYMPTOTES

The asymptotes which is not parallel to either of the co-ordinate axes is called an oblique asymptotes, which is given by the equation $y = mx + c$, where m and c are finite constants.

If $y = mx + c$ is an asymptote to the curve $f(x, y) = 0$, prove that

$$m = \underset{x \to \infty}{Lt}\left(\frac{y}{x}\right) \text{ and } c = \underset{x \to \infty}{Lt}\,(y - mx).$$

Suppose $p(x, y)$ be any point of the infinite branch of the curve $f(x, y) = 0$ and $_p = PM$ be the $\perp$ (perpendicular) distance of the point $p(x, y)$ from the line $y = mx + c$ then

$$= \frac{y - mx - c}{\sqrt{1 + m^2}} \qquad \ldots(i)$$

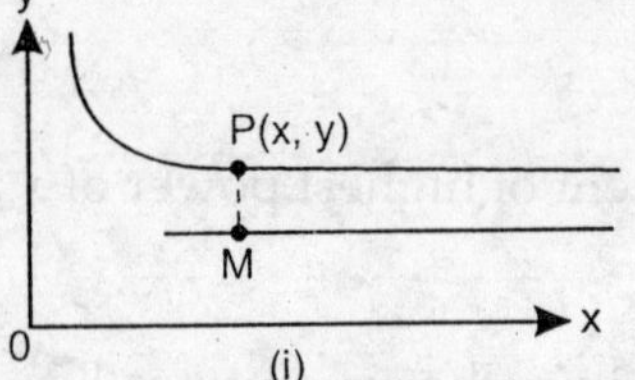

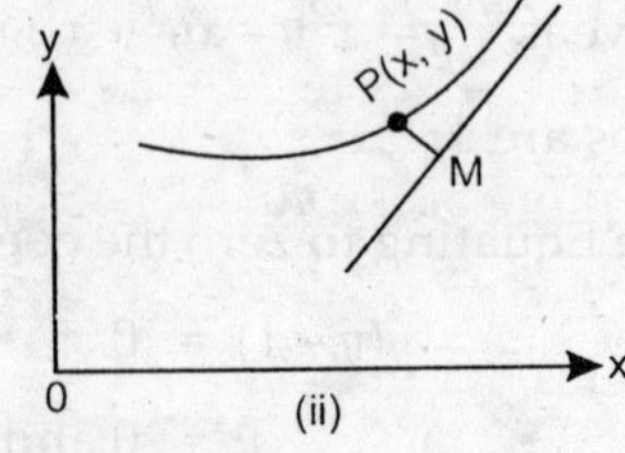

Fig. 4.3

In the first figure the abscissa x tends to infinity as $p(x, y) \to \infty$ along the infinite branch of the curve.

Equation (*i*) can be written as

$$y - mx = c + p\sqrt{1+m^2}$$

Dividing by x on both sides

$$\frac{y}{x} = m + \frac{c}{x} + \frac{p}{x}\sqrt{1+m^2}$$

Since $y = mx + c$ is an asymptotes, so by definition, $p \to 0$ as $x \to \infty$, then equation (*ii*) becomes

$$\underset{x \to \infty}{Lt}\left(\frac{y}{x}\right) = \underset{y \to \infty}{Lt}\left(m + \frac{c}{x} + \frac{P}{x}\sqrt{1+m^2}\right) = m$$

Hence $$\underset{x \to \infty}{Lt}\left(\frac{y}{x}\right) = m.$$

Again form (*i*)

$$y - mx = c + p\sqrt{1+m^2}$$

$$\therefore \quad \underset{x \to \infty}{Lt}(y - mx) = \underset{x \to \infty}{Lt}\left(c\ p\sqrt{1+m^2}\right)\ \underset{x \to \infty}{Lt}\ c + \underset{x \to \infty}{\lim}\ p\sqrt{1+m^2}$$

$$\underset{x \to \infty}{Lt}(y - mx) = c + 0$$

$$\because \quad \underset{x \to \infty}{Lt}\ p = 0$$

Hence $$\underset{x \to \infty}{Lt}(y - mx) = c$$

Therefore Asymptotes is

$$y = mx + c$$

where $$m = \underset{x \to \infty}{Lt}\left(\frac{y}{x}\right) \text{ and } c = \underset{x \to \infty}{Lt}(y - mx)$$

4.17 TO FIND THE OBLIQUE ASYMPTOTES OF THE GENERAL ALGEBRAIC CURVE

Suppose that the general equation of nth degree of an algebraic curve be

$$(a_0 x^n + a_1 x^{n-1} y + a_2 x^{n-2} y^2 + \ldots + a_n y^n) + (b_0 x^{n-1} + b_1 x^{n-2} y + b_2 x^{n-3} y^2 + \ldots b_{n-1} y^{n-1})$$
$$+ (c_0 x^{n-2} + c_1 x^{n-3} y + \ldots + c_{n-2} y^{n-2}) + \ldots + (p_0 x + p_1 y) + k = 0$$

The above equation can be written as

$$x^n Q_n\left(\frac{y}{x}\right) + x^{n-1} Q_{n-1}\left(\frac{y}{x}\right) + x^{n-2} Q_{n-2}\left(\frac{y}{x}\right) + \ldots + x\, Q_1\left(\frac{y}{x}\right) + Q_0\left(\frac{y}{x}\right) = 0 \quad \ldots(i)$$

Where $Q_n\left(\frac{y}{x}\right), Q_{n-1}\left(\frac{y}{x}\right), Q_{n-2}\left(\frac{y}{x}\right) \ldots Q_0\left(\frac{y}{x}\right)$ are the polynomials in $\frac{y}{x}$ of degree $n, n-1, n-2 \ldots 0$, at the most

Let $y = mx + c$ be the asymptote of the curve, here m and c are finite and to be found out.

To find m

Divide (*i*) by x^n on both sides, we get

$$Q_n\left(\frac{y}{x}\right) + \frac{1}{x} Q_{n-1}\left(\frac{y}{x}\right) + \frac{1}{x^2} Q_{n-2}\left(\frac{y}{x}\right) + \ldots + \frac{1}{x^{n-1}} Q_1\left(\frac{y}{x}\right) + \frac{1}{x^n} Q_0\left(\frac{y}{x}\right) = 0 \qquad \ldots(ii)$$

Taking limit as $x \to \infty$ such that $\underset{x \to \infty}{Lt} = m$, we get

$$Q_n(m) + 0 + 0 + 0 + \ldots + 0 + 0 = 0$$

$\therefore$ $Qn(m) = 0$ which gives n values of m say m_1, m_2, m_3 etc.

To find c corresponding to the value of m

Let us take $y - mx = p$ so that as $x \to \infty, p \to c$ because. $\underset{x \to \infty}{Lt}(y - mx) = c$

Now $\qquad y = mx + p$

$\therefore$ $\qquad \frac{y}{x} = m + \frac{p}{x}$

On putting $\frac{y}{x} = m + \frac{p}{x}$ in (*i*), we get

$$x^n, Q_n\left(m + \frac{p}{x}\right) + x^{n-1} Q_{n-1}\left(m + \frac{p}{x}\right) + x_{n-2} Q_{n-2}\left(m + \frac{p}{x}\right) + \ldots + = 0 \qquad \ldots(iii)$$

On expanding each term by Taylor's series, we get

$$x^n\left[Q_n(m) + \frac{p}{x}Q'_n + \frac{p^2}{2!x^2}Q''_n(m) + \ldots\right] + x^{n-1}\left[Q_{n-1}(m) + \frac{p}{x}Q'_{n-1}(m) + \ldots\right] + x^{n-2}\left[Q_{n-2}(m) + \frac{p}{x}Q'_{n-2}(m) + \ldots\right] + \ldots = 0$$

On putting $Q_n(m) = 0$ and arranging the rest of the terms in descending power of x *i.e.*

$$x^{n-1}[pQ'_n(m) + Q_{n-1}(m)] + x^{n-2}\left[\frac{p^2}{2!}Q''_n(m) + pQ'_{n-1}(m) + Q_{n-2}(m)\right] + \ldots = 0 \qquad \ldots(iv)$$

Dividing by x^{n-1} on both sides, we get

$$[pQ'_n(m) + Q_{n-1}(m)] + \frac{1}{x}\left[\frac{p^2}{2!}\right]Q''_n(m) + pQ'_{n-1}(m) + \ldots + \ldots = 0 \qquad \ldots(v)$$

Taking $x \to \infty$. such that $p \to c$, then

$$cQ'_n(m) = Q_{n-1}(m) = 0 \qquad \ldots(vi)$$

If $Q'_n(m) \neq 0$ and $Q_n(m) = 0$ has real and distinct values of m, then c is given by

$$c = -\frac{Q_{n-1}(m)}{Q'_n(m)} \qquad \ldots(vii)$$

Put $\qquad m = m_1, m_2, m_3 \ldots$ then we get $c = c_1, c_2, c_3 \ldots$ and the asymptotes are

$$y = m_1 x + c_1, y = m_2 x + c^2, \ldots$$

Note :

(*i*) When $Q'_n(m) = 0$ and $Q_{n-1}(m) \neq 0$ then from equation (*vii*), c is infinite so there is no asymptotes.

(*ii*) When $Q_n(m) = 0$ and m has two real and equal values *i.e.*, $m = m_1, m_1$ then $Q'_n(m) = 0$, and (*vii*) becomes $c.\ 0 + 0 = 0$ *i.e.*, then it is not possible to find c.

$[Q'_n(m) = 0 \because Q_n(m) = 0$ has two equal values of $m]$

Case of Parallel Asymptotes :

When $Q'_n(m)$ and $Q_{n-1}(m) = 0$. Then equation (*iv*) reduces to $x^{n-2}\left[\frac{p^2}{2}Q''_n(m) + pQ'_{n-1}(m) + Q_{n-2}(m)\right]$

$$+x^{n-3}\left[\frac{p^3}{3!}Q''_n(m) + \frac{p^2}{2}Q''_{n-1}(m) + pQ'_{n-2}(m) + Q_{n-3}(m)\right] + \ldots = 0$$

On Dividing by x^{n-2} on both sides, we get

$$\frac{p^2}{2!}Q''_n(m) + pQ'_{n-1}(m) + Q_{n-2}(m) + \frac{1}{x}\left[\frac{p^3}{3!}Q''_n(m) + \frac{p^2}{2}Q''_{n-1}(m) + pQ'_{n-2}(m) + Q_{n-3}(m)\right] + \ldots = 0.$$

Taking limit as $x \to \infty$, $p \to c$, we get

or $$\frac{c^2}{2!}Q''_n(m) + cQ'_{n-1}(m) + Q_{n-2}(m) = 0$$

If $Q''_n(m) \neq 0$, then we get two value of c *i.e.,* c_1 and c_2 corresponding to the two equal value of m.

i.e., m_1, m_1 and therefore asymptotes are given by

$$y = m_1x + c_1 \text{ and } y = m_1x + c_2$$

When $Q''_n(m) = 0$, $Q'_{n-1}(m) = 0$ and $Q_{n-2}(m) = 0$ *i.e.,* when m has three real and equal values then equation (*iv*) reduce to

$$x^{n-3}\left[\frac{p^3}{3!}Q''_n(m) + \frac{p^2}{2!}Q''_{n-1}(m) + pQ_{n-2}(m) + Q_{n-3}(m)\right] + x_{n-4}[\ldots] + \ldots = 0$$

On dividing by x^{n-3} on both sides we get

$$\frac{p^3}{3!}Q''_n(m) + \frac{p^2}{2!}Q''_{n-1}(m) + pQ'_{n-2}(m) + Q_{n-3}(m) + \frac{1}{x}[\ \ ] + \ldots = 0$$

Taking limit as $x \to 0$ $p \to c$ then we get

or $$\frac{c^3}{3!}Q''_n(m) + \frac{c^2}{2!}Q''_{n-1}(m) + C\,Q'_{n-2}(m) + Q_{n-3}(m) = 0$$

If $Q''_n(m) \neq 0$ then we get three value of c *i.e.,* c_1, c_2 and c_3 corresponding to the three equal values of m *i.e.* m_1, m_1, m_1 and therefore asymptotes are given by

$$y = m_1x + c_1,\ y = m_1x = c_2,\ y = m_1x + c_3$$

Note : Total number of asymptotes of a curve can not exceed the degree of a curve.

4.18 WORKING RULE FOR FINDING OBLIQUE ASYMPTOTES OF AN ALGEBRAIC CURVE OF –*n*th DEGREE

1. **To find Q_n(m) :** It can be obtained by putting $x = 1$ and $y = m$ in the highest degree term of the given equation of the curve.
2. **To find the value of m :** Putting $Q_n(m) = 0$, Solve for m. Let $m = m_1, m_2, m_3, \ldots$, be the roots.
3. **To find $Q_{n-1}(m)$:** It can be obtained by putting $x = 1$ and $y = m$ in the next lower degree terms of the given equation of the curve.

Case I : When the real and distinct value *i.e.,* $m = m_1, m_2, m_3, \ldots$ then the values of c corresponding to the values of m's are given by

$$c = \frac{Q_{n-1}(m)}{Q'_n(m)} \text{ provided } Q'_n(m) \neq 0$$

Then required asymptote are given by

$$y = m_1x + c_1 ; y = m_2x + c_2 ; ...$$

Case II : When m has two real and equal values then the value of c corresponding to m's are given by

$\frac{c^2}{2!} Q''_n(m) + c\, Q'_{n-1}(m) + Q_{n-2}(m) = 0$ provided $Q''_n(m) \neq 0$

Now the term $Q_{n-2}(m)$ can be found out by putting $x = 1$, $y = m$ in next lower degree term *i.e.* Q_{n-2} (m) then required asymptotes are given by

$$y = m_1x + c_1 ; y = m_2x + c_2 ;....$$

Case III : When m has three real and equal values then the value of c corresponding to m's are given by $\frac{c^3}{3!} Q''_n$ $(m) + \frac{c^2}{2!} Q''_{n-1}(m) + CQ'_{n-2}(m) + Q_{n-3}(m) = 0$, provided $Q'''n(m) \neq 0$

Now the term Q_{n-3} (m) can be found out by putting $x = 1$, $y = m$ in next lower degree term *i.e.* Q_{n-3} (m) then required asymptotes are given by

$$y = m_1x + c_1 ; y = m_2x + c_2 ; y = m_1x + c_3 ...$$

Example 4.32. Find all the asymptotes of the following curves.

(a) $x^3 + 2x^2y - xy^2 - 2y^3 + 3xy + 3y^2 + x + 1 = 0$

(b) $x^3 + 4x^2y + 4xy^2 + 5x^2 + 15xy + 10y^2 - 2y + 1 = 0$

Solution : (*a*) The equation of the curve is $x^3 + 2x^2y - xy^2 - 2y^3 + 3xy + 3y^2 + x + 1 = 0$

Here number of possible asymptotes is 3.

(*i*) **Asymptotes ‖ to x-axis :** The coefficient of highest power of x *i.e.* of x^3 is constant, so there is no asymptotes parallel to x-axis.

(*ii*) **Asymptotes ‖ to y-axis :** The coefficient of highest power of y *i.e.* of y^3 is constant, so there is no asymptotes parallel to y-axis.

Hence there is no asymptotes parallel to either axis.

(*iii*) **Oblique Asymptotes :** Put $x = 1$, $y = m$ is highest degree term, *i.e.* third degree term we get

$$Q_3(m) = 1 = 2m - m^2 - 2m^3$$

The slop of asymptotes (*i.e.* value of m) are the roots of $Q_n(m)$ *i.e.* $Q_3(m) = 0$

or $$1 + 2m - m^2 - 2m^3 = 0,$$

or $$2m^3 + m^2 - 2m - 1 = 0$$

On solving, we get $m = -1, 1 - \frac{1}{2}$, *i.e.* real and distinct roots. Now next lower degree term is $Q_2(m)$

Put $x = 1$, $y = m$ is second degree terms of the given equation

$$Q_2(m) = 3m + 3m^2$$

Now c is given by

$$c\, Q_3(m) + Q_2(m) = 0$$

or $$c(2 - 2m - 6m^2) + (3m^2 + 3m) = 0$$

$$c = -\frac{3m^2 + 3m}{2 - 2m - 6m^2}$$

$$c = \frac{3(m^2 + m)}{(6m^2 + 2m - 2} \quad ...(i)$$

when $m = -1$, then form (i), $c = \dfrac{3(1-1)}{6\times 1+2\times -1-2} = 0$

when $m = 1$, then form (i), $c = \dfrac{3(1+1)}{6\times 1\times 2\times 1-2} = 1$

when $m = -\dfrac{1}{2}$, then from (i), $c = \dfrac{3\left(\frac{1}{4}-\frac{1}{2}\right)}{6\times\frac{1}{4}-2\times\frac{1}{2}-2} = \dfrac{1}{2}$

Putting these value of m and c in $y = x + c$ the required asymptotes are

$$y = -x+0,\ y = x = 1,\ y = -\frac{1}{2}x+\frac{1}{2}$$

or $$x + y = 0,\ x - y = 1 = 0,\ x + 2y - 1 = 0$$

(*b*) The equation of curve is

$$x^3 = 4x^2y + 4xy^2 = 5x^2 = 15xy + 10y^2 - 2y + 1 = 0$$

Here number of possible asymptotes is three.

(*i*) **Asymptotes parallel to *x*-axis :** The coefficient of highest power of x *i.e.* of x^3 is constant so there is no asymptote parallel to x- axis.

(*ii*) **Asymptotes parallel to y-axis :** Equating to zero the coefficient of highest power of y *i.e.,* of y^3 the asymptotes parallel to y-axis is given by

$$4x = 10 = 0 \text{ or } 2x = 5 = 0$$

(*iii*) **Oblique Asymptotes :** Put $x = 1$, $y = m$ in highest degree term and net lower degree terms *i.e.* third and second degree term are

$$Q_3(m) = 1 + 4m + 4m^2$$
$$Q_2(m) = 5 + 15m + 10m^2$$

slop of asymptotes (*i.e,* value of m) are given by

$$Q_n(m) = 0 \text{ i.e., } Q_3(m) + 0$$
$$4m^2 + 4m + 1 = 0$$
$$(2m = 1)^2 = 0 \Rightarrow m = -\frac{1}{2}, -\frac{1}{2} \text{ are two real and equal roots.}$$

Now c is given by

$$\frac{c^2}{2!}Q''_3(m) + cQ'_2(m) + Q_1(m) = 0 \qquad \text{...}(i)$$

$$Q'_3(m) = 8m + 4,\ Q'_2(m) = 20m + 15,$$
$$Q''_3(m) = 8,\ Q'_1(m) = -2m$$

Putting these values in (i)

$$\frac{c^2}{2!}(8) + c(20m + 15) + (-2m) = 0$$

Put $m = -\dfrac{1}{2}$, it becomes

$$\frac{1}{2!}c^2(8) + c\,[-10 = 15] + 1 = 0$$

$$4c^2 + 5c + 1 = 0$$

or $$(4c+1)(c+1) = 0$$

$$c = -1, c = -\frac{1}{4}$$

Hence required asymptotes are given by

$$y = -\frac{1}{2}x - 1$$

and $$y = -\frac{1}{2}x - \frac{1}{4}$$

or $$x + 2y + 2 = 0$$

and $$2x + 4y + 1 = 0$$

Example 4.33. Find all the asymptotes of the curve

(*a*) $y^2(x-2a) = x^3 - a^3$ (*b*) $x^3 + y^3 - 3axy = 0$

(*c*) $y^2 = 4ax$ (*d*) $6x^2 + xy - 2y^2 + x + 2y + 1 = 0$

Show the parabola $y^2 = 4\,ax$ has no asymptotes.

Solution : (*a*) The equation of curve is $xy^2 - 2ay^2 - x^3 + a^3 = 0 +$ or $x^3 - xy^2 + 2ay^2 - a^3 = 0$

Here number of possible asymptotes is three.

Now we find parallel asymptotes.

(*i*) **Asymptotes parallel to *x*-axis :** The coefficient of highest power of x(*i.e.* of x^3) is constant so there is no asymptotes parallel to *x*-axis.

(*ii*) **Asymptotes parallel to y-axis :** equating to zero the coefficient of highest power of y (*i.e.* of y^2) *i.e.*, $2a - x = 0$ so the asymptotes parallel to *y*-axis is given by

$$x = 2a$$

(*iii*) **Oblique Asymptotes :** Put $x = 1$, $y = m$ in highest and next lower degree term *i.e.* third and second degree terms *i.e.*, slopes of asymptotes are given by $Q_3(m) = 0$ *i.e.*

$$Q_3(m) = 1 - m^2, \text{ and } Q_2(m) = 2am^2$$

$1 - m^2 = 0$ or $m = \pm 1$ the roots are real and distinct.

Now c is given by

$$c\,Q'_3(m) + Q_2(m) = 0$$

$$c \times (-2m) + 2am^2 = 0 \qquad ...(i)$$

Put $$m = 1 \text{ in } (i)\text{, then}$$

$$-2c + 2a = 0$$

$$c = a$$

Again put $$m = -1 \text{ in } (i)\text{, then}$$

$$2c = 2a = 0$$

$$c = -a$$

Hence, the required asymptotes are given by putting the value of m and c in $y = mx + c$ *i.e.*

$$y = x + a \text{ and } y = -x - a$$

or $$x - y + a = 0 \text{ and } y + x + a = 0$$

(*b*) The equation of curve is $x^3 + y^3 - 3axy = 0$ ...(*ii*)

Here number of possible asymptotes is three

Now we find parallel asymptotes.

(i) **Asymptotes || to *x*-axis :** The coefficient of highest power of x (*i.e.* of x^3) is constant so there is no asymptotes || to x-axis.

(ii) **Asymptotes || to *y*-axis :** The coefficient of highest power of y (*i.e.* y^2) is constant, so there is no asymptotes || to y-axis.

(iii) **Oblique Asymptotes :** Put $x = 1$, $y = m$ in highest and next lower degree term *i.e.* third and second degree term

$$Q_3(m) = 1 + m^3, \text{ and } Q_2(m) = -3am$$

slops of asymptotes (value of m) are given by

$$Q_3(m) = 0 \text{ i.e. } m^3 + 1 = 0$$

$$(m+1)(m^2 + 1 - m) = 0$$

$$\text{Either } m = -1 \text{ or } m^2 - m + 1 = 0$$

$$m = \frac{1 \pm \sqrt{1-4}}{2}$$

$$m = \frac{1 \pm i\sqrt{3}}{2} \text{ (Imaginary)}$$

Hence only real value of m is –1.

Now c is given by

$$cQ'_3(m) = Q_2(m) = 0$$

$$c\,(3m^2) = (-3am) = 0 \qquad ...(iii)$$

Put $m = 1$ in *(ii)*, then

$$c(3) + 3a = 0$$

$$c = -a$$

Hence the required asymptotes is given by putting the value of m and c in $y = mx + c$ *i.e.*

$$y = -x - a$$

or

$$x + y + a = 0$$

(c) The equation of curve is

$$y^2 = 4ax$$

Here number of possible asymptotes is two.

Now we find parallel asymptotes

(i) **Asymptotes parallel to x-axis :** The coefficient of highest power of x(*i.e.* of x) is $4a$ *i.e.*, constant so no asymptotes there is parallel to x-axis.

(ii) **Asymptotes parallel to y-axis :** The coefficient of highest power of y(*i.e.* of y^2) is constant so there is no asymptotes parallel yo y-axis.

(iii) **Oblique asymptotes :** Put $x = 1$, $y = m$ in highest and next lower degree term *i.e.*, second first degree term, *i.e.*,

$$Q_2(m) = m^2, \text{ and } Q_1(m) = -4a$$

slopes of asymptotes (*i.e.*, values of m) is given by.

$$Q_2(m) = 0 \text{ i.e., } m^2 = 0,\ m = 0, 0$$

Also we know that asymptotes (if any) corresponding to $m = 0$ are parallel to x-axis. But it has been proved above that there are no asymptotes parallel to coordinate axes.

Hence the curve have no oblique asymptotes

Since the parabola $y^2 = 4ax$ neither oblique asymptotes nor parallel asymptotes so it has no asymptotes.

(*d*) The equation of curve is

$$6x^2 + xy - 2y^2 + x = 2y + 1 = 0$$

Here number of possible asymptotes is two.

To find parallel asymptotes

(*i*) **Asymptotes parallel to x-axis :** Equating to zero the coefficient of highest power of x (*i.e.* of x^2) and coefficient is 6, so no asymptote parallel to x-axis

(*ii*) **Asymptotes parallel to y-axis :** The coefficient of highest power of y (*i.e.* of y^2), and coefficient is (–2), so no asymptotes is parallel to y-axis.

(*iii*) **Obliques asymptotes :** Put $x = 1$, $y = m$ in highest and next lower degree terms *i.e.* second and firs degree terms *i.e.*

$$Q_2(m) = 6 + m - 2m^2$$

$$Q_1(m) = 1 = 2m$$

Slopes of asymptotes (*i,e.* value of m) are given by

$$Q_2(m) = 0 \text{ i.e. } 2m^2 - 6 = 0$$

or $$m = 2, -\frac{3}{2} \text{ i.e. real and distinct roots and } c \text{ is given by}$$

$$c = -\frac{Q_1(m)}{Q'_2(m)}$$

or $$c = -\frac{1+2m}{4m-1}$$

when $$m = 2, c = -\frac{5}{7}$$

when $$m = -\frac{3}{2}, c = -\frac{2}{7}$$

Hence the required asymptotes are given by putting the value of m and c in $y = mx + c$ *i.e.*

$$y = 2x - \frac{5}{7} ; y = \frac{-3}{2}x - \frac{2}{7}$$

EXERCISE 4.5

1. Find all asymptotes, of the following curves

   (*a*) $xy^2 + 4a^2 (2a - x)$ **Ans.** $x = 0$, *i.e.*, y-axis

   (*b*) $y^2 (x - 2a) = x^3 - a^2$ **Ans.** $x + y + a = 0, x - y + a = 0$

   (*c*) $x^3 - 6x^2y = 11xy^2 - 6y^3 + x + y + 1 = 0$ **Ans.** $x - y = 0, x - 2y = 0, x - 3y = 0$

   (*d*) $x(y - x)^2 = x (y - x) + 2$ **Ans.** $x = 0, y = x, x + 1 = y$.

2. Find all asymptotes of the following curves

   (*a*) $(x - y)^2 (x + 2y - 1) = 3x + y - 7$ **Ans.** $x + 2y - 1 = 0, y = x \pm \frac{2}{\sqrt{3}}$

   (*b*) $x^3 + 3x^2y - 4y^3 - x + y + 3 = 0$ **Ans.** $y - x = 0, 2y + x - 1 = 0, 2y + x + 1 = 0$

(c) $(x+y)^2 (x+y+2) = x+9y-2$ **Ans.** $x+y-2=0, x+y+2 \pm \sqrt{5}=0$

(d) $y^3 - 2xy^2 - x^2y + 2x^3 = 3y^2 - 7xy + 2x^2 + 2y + 2x + 1 = 0$ **Ans.** $2x - y = 0, x - y - 1 = 0, x + y + 2 = 0.$

3. Find all asymptotes of the following curves

(a) $x^3 - x^2y - xy^2 \; y^3 + 2x^2 - 4y^2 + 2xy + x + y + 1 = 0$

Ans. $x + y - 1 = 0, 2x - 3y + 3 + \sqrt{5} = 0$ and $2x - 2y + 3 - \sqrt{5} = 0$

(b) $x^2(x-y)^2 - a^2 (x^2 + y^2) = 0$ **Ans.** $x = \pm a, x - y = \pm \sqrt{2}\, a$

(c) $x^2y^2(x^2-y^2)^2 - (x^2+y^2)^3 = 0$ **Ans.** $x = \pm 1, y = \pm 1, x - y \pm \sqrt{2} = 0, x + y \pm \sqrt{2} = 0$

(d) $y^3 - 3x^2y + xy^2 - 3x^3 + 2y^2 + 2xy + 4x + 5y + 6 = 0$ **Ans.** $x + y = 0, \sqrt{3}\, x - y - 1 = 0, \sqrt{3}\, x + y + 1 = 0$

4.19 INTERSECTION OF A CURVE AND ITS ASYMPTOTES

Let $y = mx + c$ be an asymptote of the curve ...(*i*)

$$x^nQ_n\left(\frac{y}{x}\right) + x^{n-1}Q_{n-1}\left(\frac{y}{x}\right) + x^{n-1}Q_{n-2}\left(\frac{y}{x}\right) + \ldots = 0 \text{ of } n^{\text{th}} \text{ degree} \quad \ldots(ii)$$

The points of intersection of the curve and the asymptotes are given by putting the value of y from (*i*) in (*ii*)

$$x^nQ_n\left(m+\frac{c}{x}\right) + x^{n-1}Q_{n-1}\left(m+\frac{c}{x}\right) + x^{n-1}Q_{n-2}\left(m+\frac{c}{x}\right) + \ldots = 0$$

or

$$x^nQ_n(m) + \{CQ^1{}_n(m) + Q_{n-1}(m)\}\, x^{n-1} + \left[\frac{1}{2!}c^2 Q''_n(m) + cQ'_{n-1}(m) + Q_{n-2}(m)\right]x^{n-2} + \ldots = 0 \quad \ldots(iii)$$

Since $y = mx + c$ is the asymptote of the equation (*i*), then

$$Q_n(m) = 0, cQ'_n(m) + Q_{n-1}(m) = 0$$

The equation (*iii*) reduces to

$$\left[\frac{1}{2!}c^2 Q''_n(m) + cQ'_{n-1}(m) + Q_{n-2}(m)\right]x^{n-2} + \ldots = 0 \quad \ldots(iv)$$

the coefficients of x^{n-2} will be non-zero provided there are no parallel asymptotes.

Since the equation (*iv*) is of degree $n - 2$ in x, there will be $n - 2$ points on intersection.

Now for one asymptotes we have $n - 2$ points of intersection. But there are at most n asymptotes for the equation of n degree. Hence the will be at the most $n(n - 2)$ points of intersection.

Asymptotes by Inspection

If the equation of the curve be put in the form

$F_n + F_{n-2} = 0$ where F_n is of degree n[*i.e.* contains terms of degree n and may also contain lower degree terms] and F_{n-2} is of degree $(n - 2)$ at the most, then $F_n = 0$ will give the asymptotes and $F_{n-2} = 0$ will give the curve on which the points of intersection of the curve and asymptotes lie.

***Example* 4.34.** Find the asymptotes of the curve $x^2y - xy^2 + xy + y^2 + x - y = 0$ and show that they cut the curve in 3 points which lie on the straight line $x + y = 0$.

Solution : The curve is

$$x^2y - xy^2 \; y^2 + xy + x - y = 0, \text{ Let } f(x, y) = x^2y - xy^2 + y^2 + xy + x - y = 0$$

(*i*) The asymptotes ∥ to x-axis is $y = 0$

(*ii*) The asymptotes ∥ to y-axis is $x - 1 = 0$

Now let us find oblique asymptotes

$$Q_3(m) = m - m^2$$
$$Q_2(m) = m + m^2$$

The slopes of the asymptotes are given by

$$m - m^2 = 0,\ m(1-m) = 0 \Rightarrow m = 0, 1$$

Now c is given by

$$cQ'_n(m) + Q_{n-1}(m) = 0$$

$$c = -\frac{Q_{n-1}(m)}{Q'_n(m)} = -\frac{Q_2(m)}{Q'_3(m)} = -\frac{(m+m^2)}{(1-2m)}$$

for $m = 0, c = 0$

$$m = 1\ c = -\frac{2}{-1} = 2$$

The asymptotes are $y = 0, y = x + 2$

There are $3(3-2) = 3$ points of intersection of the asymptotes and the curve

The points of intersection lie on the curve

$$-F_{n-2} = F_n - (x^2y - xy^2 + xy + y^2 + x - y) = 0$$
$$-F_1 = F_3 - (x^2y - xy^2 + xy + y^2 + x - 3) = 0$$
$$-F_1 = y(x-1)(x-y+2) - (x^2y - xy^2 + xy + y^2 + x - y = 0$$
$$= x^2y - xy - xy^2 + y^2 + 2xy - 2y - x^2y + xy^2 - xy - y - y^2 - x + y = 0$$
$$-F_1 = -x - y = 0 \text{ or } x + y = 0$$

∴ The points of intersection lie on $x + y = 0$

Example **4.35.** Find the equation of the hyperbola with $x + y - 1 = 0$, $x + y - 2 = 0$ as its asymptotes and pass through the origin.

Solution : Joint equation of the asymptotes is

$$F_2 = (x+y-1)(x+y-2) = 0$$

The equation of the curve is

$$F_n + F_{n-2} = 0$$

Here $n = 2$

$$F_2 + F_0 = 0 \qquad (F_0 \text{ is constant})$$

$$(x+y-1)(x+y-2) + F_0 = 0$$

It pass through (0, 0)

$$2 + F_0 = 0$$
$$F_0 = -2$$

The equation of the hyperbola is

$$(x+y-1)(x+y-2) - 2 = 0$$

Example **4.36.** Find the equation of the cubic which has the same asymptotes as the curve $x^3 - 6x^2y + 11xy^2 - 6y^3 + x + y + 1 = 0$ and which passes through the points (0, 0). (1, 0).

Solution : The cubic equation of the curve can be put as

$$F_3 + F_1 = 0$$

∴ $F_3 = 0$ gives the asymptotes, *i.e.*

$$x^3 - 6x^2y + 11xy^2 - 6y^3 = 0 \text{ will give the asymptotes.}$$

Now the curve which has these asymptotes is

$x^3 - 6x^2y + 11xy^2 - 6y^3 + ax + by + c = 0$, $F_1 = ax + by + c$

it passes through (0, 0), (1, 0), (0, 1)

$$c = 0$$

$$1 + a = 0 \text{ or, } a = -1$$

$$-6 + b = 0 \text{ or, } b = 6$$

The equation of the curve $x^3 - 6x^2y + 11xy^2 - 6y^3 - x + 6y = 0$

Example 4.37. Show that four asymptotes of the curve $xy(x^2 - y^2) + 25y^2 + ax^2 - 144 = 0$ cut it again in 8 points on an ellipse whose eccentricity is $\frac{4}{5}$.

Solution : The given curve can be put in the form

$$F_4 + F_2 = 0$$

where
$$F_4 = xy\,(x^2 - y^2)$$

$$F_2 = 25y^2 + 9x^2 - 144$$

$F_4 = 0$ or $xy(x^2 - y^2) = 0$ gives the asymptotes, therefore the asymptotes are $x = 0$, $y = 0$, $y = \pm x$.

There are $4(4 - 2) = 4 \times 2 = 8$ points of intersection of the curve and the asymptotes and these 8 points lie on the ellipse.

$$9x^2 + 25y^2 = 144$$

or
$$\frac{x^2}{16} + \frac{y^2}{\frac{144}{25}} = 1$$

$$\frac{x^2}{(4)^2} + \frac{y^2}{\left(\frac{12}{5}\right)^2} = 1$$

$$a = 4,\ b = \frac{12}{5}$$

Now
$$b^2 = a^2(1 - e^2)$$

$\therefore$
$$\left(\frac{12}{5}\right)^2 = 16(1 - e^2)$$

$$\frac{144}{25} = 16(1 - e^2)$$

$$\frac{9}{25} = 1 - e^2$$

$$e^2 = 1 - \frac{9}{25} = \frac{16}{25}$$

$$e = \frac{4}{5}$$

EXERCISE 4.6

1. Find all asymptotes of the curve $xy(x^2 - y^2) + 2x^2 + 2y^2 + 1 = 0$ **Ans.** $x = 0, y = 0$ and $x + y = 0$
2. Find the equation of the cubic which has the same asymptotes as the curve $x^3 - 6x^2y + 11xy - 6y^3 + 4x + 5y + 7 = 0$ and which passes through the points (0, 0), (0, 2) and (2, 0). **Ans.** $x^3 - 6x^2y + 11xy - 6y^3 - 4x + 24y = 0$
3. Find the equation of the be-quadratic curve which was $x = 0, y = -x$ for asymptotes and which passes through (a, b) and which cuts the asymptotes in 8 points that lie on the curcle $x^2 + y^2 = a^2$.
   Ans. $bxy(y^2 - x^2) + a(a^2 - b^2)(x^2 + y^2 - a^2) = 0$
4. Find the asymptotes of the curve $x^3 + 3x^2y - 4y^3 - x + y + 3 = 0$ and show that they cut the curve in 8 points which lie on the line $x - y - z = 0$. **Ans.** $y - x = 0, 2y + x - 1 = 0$ $2y - x + 1 = 0$

4.20 ASYMPTOTES IN POLAR COORDINATE

When a curve is given in the form of r and θ *i.e.* $r = f(\theta)$, it is called a polar curve.

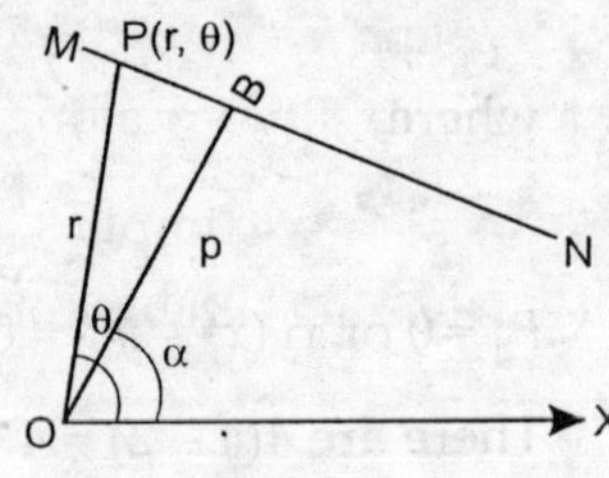

Fig. 4.4

***Theorem* 1.** The polar equation of any line is $p = r\cos(\theta - \alpha)$ where p is the length of perpendicular from the pole to the line and α is the angle which the perpendicular makes with the initial line (x-axis).

Proof : Let O be the pole and OX be nth initial line.

Let $P(r, \theta)$ be any point on the given line such that $OP = r$ and $\angle XOP = \theta$

Let $OB = p$ be the length of perpendicular from pole to the given line MN and $\angle XOB = \alpha$

$$\therefore \quad \angle BOP = \angle XOP - \angle XOB = \theta - \alpha$$

Now in ΔOPB

$$\cos(\theta - \alpha) = \frac{OB}{OP} = \frac{p}{r}$$

$$\therefore \quad p = r\cos(\theta - \alpha) \text{ which is the required equation of the line.}$$

***Theorem* 2.** The length of the perpendicular from a point $p(\Upsilon_1, \theta_1)$ to the line $p = r\cos(\theta - \alpha)$ is $\pm\{p - r_1\cos(\theta_1 - \alpha)\}$.

Proof : Let $p = r\cos(\theta - \alpha)$ be the equation of the line AB.

$$\therefore \quad OM = p \text{ and } \angle XOM = \alpha$$

Now through $p(r_1, \theta_1)$ draw a line $A'B'$ || to AB and Let $PQ = d$ be the length of perpendicular from $A'B'$ to the line AB.

$$\therefore \quad ON = OM + MN = OM + QP = p + d$$

$$\text{and} \quad \angle XON = \alpha$$

$$\therefore \text{ Equation of line } A'B' \text{ is } p + d = r\cos(\theta - \alpha)$$

It passes through $p(r_1, \theta_2)$

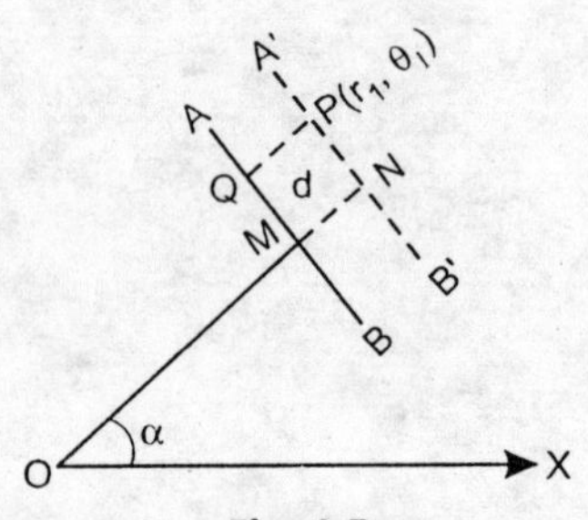

Fig. 4.5

$$\therefore \quad p + d = r_1\cos(\theta_1 - \alpha)$$

$$\text{or} \quad d = -p + r_1\cos(\theta 1 - \alpha)$$

$$= -[p - r_1\cos(\theta_1 - \alpha)$$

Similarly by taking p below, we can see that

$$d = \{p - r_1\cos(\theta_1 - \alpha)\}$$

On combining the two; d, the perpendicular distance of $p(r_1, \theta_1)$ from the line $p = r\cos(\theta - \alpha)$ is given by $d = \pm \{p - r_1 \cos(\theta_1 - \alpha)\}$

***Theorem* 3.** Polar equation of an asymptote to the curve $\frac{1}{r} = f(\theta)$ is $p = r\sin(\theta_1 - \theta)$ where $p = \underset{\theta \to \theta_1}{Lt}\left(\frac{d\theta}{du}\right)$, $u = \frac{1}{r}$ and θ_1 is a root of the equation obtained by putting $u = \frac{1}{r} = 0$, *i.e.*, $f(\theta) = 0$.

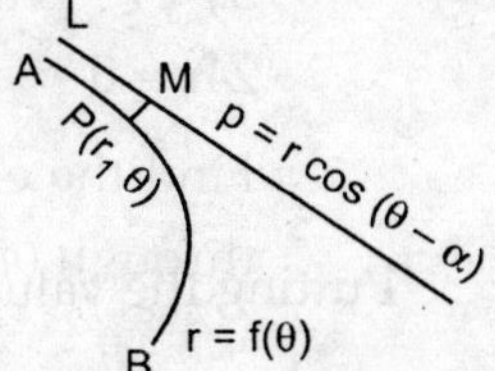

Fig. 4.6

Proof : Let $\quad p = r\cos(\theta - \alpha) \quad$ ...(*i*)

be an asymptotes to the curve $\quad r = f(\theta) \quad$ ...(*ii*)

we are to determine p and α.

Let $p(r, \theta)$ be any point on the curve $r = f(\theta)$.

PM, the $\perp$ distance of $p(r, \theta)$ from the line $p = r(\cos(\phi - \alpha)$ is

$$PM = p - r\cos(\theta - \alpha) \quad \text{...(iii)}$$

(from theorem)

$\therefore$ line $p = r\cos(\theta - \alpha)$ is an asymptotes to the curve $r = f(\theta)$, therefore $PM \to 0$ as $p(r, \theta) \to \infty$ along the curve.

As $p(r, \theta) \to \infty$ along the curve, $r \to \infty$ and let θ ends to a finite value (say θ_1).

Also $r \to \infty$ implies $u \to 0$ $\quad \left[\because r = \frac{1}{\mu}\right]$

Dividing (*iii*) by r on both sides, we get

$$\frac{PM}{r} = \frac{p}{r} - \cos(\theta_1 - \alpha)$$

Taking limit as $r \to \infty$, $\theta \to \theta_1$ and $PM \to 0$, so that

$$0 = 0 - \cos(\theta_1 - \alpha)$$

or $\quad \cos(\theta_1 - \alpha) = 0$

or $\quad \theta_1 - \alpha = \frac{\pi}{2}$

$\therefore \quad \alpha = \theta_1 - \frac{\pi}{2} \quad$ ...(*iv*)

Again taking limit in (*iii*) as $PM \to 0$, $r \to \infty$ and $\theta \to \theta_1$

$$0 = p - \underset{\substack{r \to \infty \\ \theta \to \theta_1}}{Lt} r\cos(\theta - \alpha)$$

or $\quad p = \underset{\substack{r \to \infty \\ \theta \to \theta_1}}{Lt} r\cos(\theta - \alpha)$

From (*iv*) $\quad p = \underset{\substack{r \to \infty \\ \theta \to \theta_1}}{Lt} r\cos\left(\theta - \theta_1 + \frac{\pi}{2}\right)$

or $\quad p = \underset{\substack{r \to \infty \\ \theta \to \theta_1}}{Lt} -r\cos(\theta - \theta_1) \quad$ (It is of the form $\infty \times 0$)

or $\quad p = \underset{\substack{r \to \infty \\ \theta \to \theta_1}}{Lt} \left(\frac{-\sin(\theta - \theta_1)}{\left(\frac{1}{r}\right)}\right) \quad \left(Form, \frac{0}{0}\right)$

or $\quad p = \underset{\substack{r \to \infty \\ \theta \to \theta_1}}{Lt} - \frac{\sin(\theta - \theta_{1u})}{} \quad \left(\frac{0}{0}\right)$

or $$p = \underset{\substack{r \to \infty \\ \theta \to \theta_1}}{Lt} - \frac{\cos(\theta - \theta_1)}{\frac{du}{d\theta}}$$ (By L' Hospital's Rule)

$$= \underset{u \to 0}{Lt} \frac{1}{\frac{du}{d\theta}} \underset{\theta \to \theta_1}{Lt} [-\cos(\theta - \theta_1)]$$

$$p = \underset{u \to 0}{Lt}\left(-\frac{d\theta}{du}\right) \qquad ...(iv) \left(\because \underset{\theta \to \theta_1}{Lt} \cos(\theta - \theta_1) = 1\right)$$

Putting the value of α in (i) then the equation of asymptotes is

$$p = r\cos\left(\theta - \theta_1 + \frac{\pi}{2}\right) = r\cos\left[\frac{\pi}{2} - (\theta_1 - \theta)\right]$$

or $$p = r\sin(\theta_1 - \theta)$$

where $$p = \underset{u \to 0}{Lt}\left(-\frac{d\theta}{du}\right) = -\frac{1}{f'(\theta)}$$

$$p = -\frac{1}{f'(\theta)}$$

$\therefore$ Asymptote is $$r\sin(\theta_1 - \theta) = \underset{u \to 0}{Lt}\left(-\frac{\sin d\theta}{du}\right) = -\frac{1}{f'(\theta)}$$

$$\therefore \quad \left\{\begin{array}{l} \frac{1}{r} = f(\theta) \Rightarrow u = f(\theta) \\ \frac{du}{d\theta} = f'(\theta) \Rightarrow \frac{du}{d\theta} = \frac{1}{f'(\theta)} \end{array}\right.$$

4.21 WORKING METHOD FOR FINDING POLAR ASYMPTOTES

If α be the root of the equation $f(\theta) = 0$, then $r\sin(\theta - \alpha) = \frac{1}{f'(\alpha)}$ is called the asymptotes of a polar curve $\frac{1}{r} = f(\theta)$. We adopt the following step to find the above equation.

Step I : Find the value of $\frac{1}{r}$ from the equation of curve and let it be $f(\theta)$. Also change all t-ratios if any into $\sin\theta$ and $\cos\theta$.

Step II : Put $f(\theta) = 0$ and solve for θ and let $\theta = \alpha, \beta, \gamma$... be the roots.

Step III : Find $f'(\theta)$ *i.e.* differentiate $f(\theta)$ with respect to 'θ' and find $f'(\theta)$ at $\theta = \alpha, \beta, \gamma, ...$

Step IV : The asymptotes corresponding to $\theta = \alpha$ is

$$r\sin(\theta - \alpha) = \frac{1}{f'(\alpha)}, \text{ where } f'(\alpha) \neq 0$$

Note : If $\frac{1}{f'(\alpha)}$ at $\theta = \alpha$ is infinite, then the curve has no asymptote corresponding to $\theta = \alpha$.

Remarks : Some helpful results of trigonometry are as follows:

1. If $\sin\theta = 0$, then $\theta = n\pi$, where n is any integers *i.e* $n \in I$
2. If $\cos\theta = 0$, then $\theta = (2n + 1)\pi/2$, $n \in I$
3. If $\sin\theta = \sin\alpha$, then $\theta = n\pi + (-1)n\ \alpha$, $n \in I$
4. If $\cos\theta = \cos\alpha$, then $\theta = 2n\pi \pm \alpha$, $n \in I$
5. If $\tan\theta = \tan\alpha$, then $\theta = n\pi + \alpha$, $n \in I$

6. $\frac{1}{(-1)^n} = (-1)^n, n \in I$
7. $\sin(n\pi + \theta) = (-1)^n \cos\theta$
8. $\cos(n\pi + \theta) = (-1)^n \cos\theta$
9. $\sin(n\pi - \theta) = -(-1)^n \sin\theta$

***Example* 4.38.** Find the asymptotes of the curve $r\,\theta = a$

Step I : Find the value of $\frac{1}{r}$ *i.e.* $\frac{\theta}{a} = f(\theta)$ (say)

Step II : Put $f(\theta) = 0$ *i.e.* $\frac{\theta}{a} = 0$

$$\theta = 0 = \alpha \text{ (say)}$$

Step III : Find $f'(\theta)$ $\therefore f'(\theta) = \frac{1}{a}$

Step IV : Find the value of $f'(\alpha)$ where $\alpha = 0$

$$f'(\alpha) = \frac{1}{\alpha}$$

Step V : Find the value of $\frac{1}{f'(\alpha)}$ *i.e., a.*

The required asymptotes is given by

$$r \sin(\theta - \alpha) = \frac{1}{f'(\alpha)} \text{ i.e. } r\sin(\theta - \alpha) = a$$

$$r \sin\theta = \alpha.$$

***Example* 4.39.** Find the asymptotes of the curve $r = \frac{a}{1-\cos\theta}$

Solution : The given equation of the curve is

$$r = \frac{a}{1-\cos\theta}$$

Step I : Find the value of $\frac{1}{r} = f(\theta)$ (say) where k is any integers.

$$\therefore \quad \frac{1}{r} = \frac{1-\cos\theta}{a} = f(\theta) \text{ (say)}$$

Step II :

$$\text{Put } f(\theta) = 0$$

$$\therefore \quad 1 - \cos\theta = 0$$

$$\text{or} \quad \cos\theta = 1 \Rightarrow \theta = 0$$

$$\theta = 2k\,\pi + 0 = 2k\,\pi = \theta_1 \text{ (say)}$$

Step III : Find $f'(\theta)$ *i.e.*

$$\therefore \quad f'(\theta) = \frac{\sin\theta}{a}$$

Step IV : Find $f'(\theta_1)$ *i.e.* $f'(2k\,\pi) = \frac{\sin(2k\,\pi}{a} = 0$, because k is any integer. Also find $\frac{1}{f'(\theta_1)} = \frac{1}{f'(2k\pi)} = \infty$

Hence there is no asymptotes the given curve.

***Example* 4.40.** Find the asymptotes of the curve $r \sin n\,\theta = a$

Solution : The given equation of the curve is $r \sin\theta = a$

Step I : Find the value of $\frac{1}{r}$ *i.e.* $f(\theta)$ (say)

$$\therefore \quad \frac{1}{r} = \frac{\sin n\theta}{a} = f(\theta) \text{ (say)}$$

Step II : Put

$$f(\theta) = 0$$

$$\therefore \quad \frac{\sin n\theta}{a} = 0$$

or

$$\sin n\theta = 0$$

or

$$\sin n\theta = \sin m\pi$$

$$n\theta = m\pi$$

where m is any integers.

$$\theta = \frac{m\pi}{n} = \theta_1 \text{ (say)}$$

Step III : Find $f'(\theta)$, *i.e.*

$$f'(\theta) = \frac{n \cos n\theta}{a}$$

$$\therefore \quad f'(\theta_1) = f'\left(\frac{m\pi}{n}\right)$$

$$= \frac{n}{a} \cos\left(\frac{m\pi . n}{n}\right) = \frac{n}{a} \cos m\pi = \frac{n}{a} (-1)^m$$

Step IV : Find $\frac{1}{f'(\theta_1)}$ *i.e*

$$\frac{1}{f'(\theta_1)} = \frac{a}{n(-1)^m}$$

$$= \frac{a}{n} (-1)^m$$

Hence asymptotes is given by

$$r \sin(\theta - \theta_1) = \frac{1}{f'(\theta_1)}$$

$$r \sin\left(\theta - \frac{m\pi}{n}\right) = \frac{a(-1)^m}{n}$$

where m is any integer

Example 4.41. Find the asymptotes of the curve $r = a \operatorname{cosec} \theta + b$

Solution : The equation of the curve is

$$r = a \operatorname{cosec} \theta + b$$

Step I : Find the value of $\frac{1}{r} = f(\theta)$ (say)

$$\therefore \quad \frac{1}{r} = \frac{\sin\theta}{a + b\sin\theta} = f(\theta) \text{ (say)}$$

Step II : Put $f(\theta) = 0$

i.e.

$$\frac{\sin\theta}{a + b\sin\theta} = 0$$

or

$$\sin\theta = 0$$

or $\sin\theta = \sin n\pi$

or $\theta = n\pi = \theta_1$ (say)

Step III : Find $f(\theta)$ *i.e.*

$$\therefore \quad f(\theta) = \frac{(a + b\sin\theta)\cos\theta - \sin\theta \times b\cos\theta}{(a+b\sin\theta)^2}$$

$$\therefore \quad f(\theta) = \frac{a\cos\theta}{(a+b\sin\theta)^2}$$

Step IV : Find the value of $f(\theta_1)$ *i.e.*

$$f(\theta_1) = f(n\pi)$$

$$= \frac{a\cos n\pi}{(a+b\sin n\pi)^2}$$

$$= \frac{(-1)^n}{a}$$

Step V : Find the value of $\frac{1}{f(\theta_1)}$

$$\frac{1}{f(n\pi)} = \frac{a}{(-1)^n}$$

Hence asymptotes is given by

$$r\sin(\theta - \theta_1) = \frac{1}{f(\alpha)}$$

$$r\sin(\theta - n\pi) = \frac{1}{f(n\pi)}$$

or $-r\sin(n\pi - \theta) = \frac{a}{(-1)^n}$

or $-(-1)^n r\sin(-\theta) = \frac{a}{(-1)^n}$

or $r\sin\theta = \frac{a}{(-1)^{2n}}$

$\therefore \quad r\sin\theta = a$

EXERCISE 4.7

Find the asymptotes of the following polar curves.

1. $r = \frac{a\theta}{\theta - 1}$ **Ans.** $r\sin(\theta - 1) = a$
2. $r = a\tan\theta$ **Ans.** $r\cos\theta = \pm a$
3. $r = a(\sec\theta + \tan\theta)$ **Ans.** $r\cos\theta = 0,\ r\cos\theta = 2a$
4. $r\sin\theta = 2\cos 2\theta$ **Ans.** $r\sin\theta = 2$
5. $r = a\sec\theta + b\tan\theta$ **Ans.** $r\cos\theta\ a \pm b$
6. $r = \frac{2a}{1 - 2\cos\theta}$ **Ans.** $r\sin\left(\frac{\pi}{3} \pm \theta\right) + \frac{2a}{\sqrt{3}} = 0$
7. $r\cos 2\theta = a\sin 3\theta$ **Ans.** $\frac{a}{2}\sin\left(\frac{n\pi}{2} + \frac{3\pi}{4}\right) = r\sin\left[(2n+1)\frac{\pi}{4} - \theta\right]$

CURVATURE

The given curve AMB is bent more sharply at the point M then at the point N. The bending of a curve at a particular point is called curvature of the curve at that point. So the curvature at M is more than the curvature at N. In this chapter a mathematical expression for the curvature of a curve at a given point has been discussed. It will give a definite numerical measure of bending of the curve at the point.

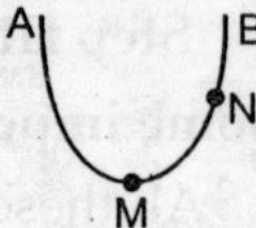

Fig. 4.7

4.22 MATHEMATICAL DEFINITION OF A CURVATURE

Let AB be a curve and P, Q be two neighbouring points on this curve. Let an arc $AP = s$ and the arc $AQ = s + \delta s$. A is a fixed point on the curve from which arcs length are measured. Let the tangents at P and Q make angle Ψ and $\Psi + \delta\Psi$ respectively with a fixed line, *i.e.*, x-axis, then.

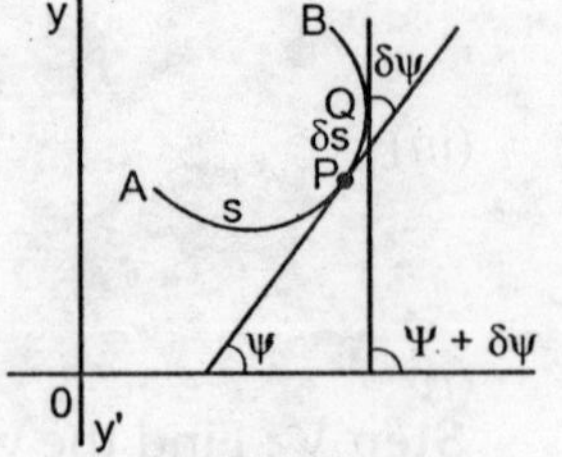

Fig. 4.8

(*i*) The angle $\delta\Psi$ through which the tangent turns as its point of contact travels along the arc PQ is called the total bending or total curvature of arc PQ.

(*ii*) The ratio $\frac{\delta\psi}{\delta s}$ is called the mean or average curvature of the are PQ.

(*iii*) The limiting value of the mean curvature when $Q \to P$ is called the curvature of the curve at the point P.

Thus the curvature (K) at point P is $\underset{Q \to P}{Lt} \frac{\delta\psi}{\delta s} = \underset{\delta s \to 0}{Lt} \frac{\delta\psi}{\delta s} = \frac{d\psi}{ds}$ and

(*iv*) The reciprocal of the curvature of the curve at P, provided this curvature is not zero, is called the radius of curvature of the curve at P and denoted by ρ *i.e.*

$$\rho = \frac{1}{K} = \frac{ds}{d\psi}$$

4.23 CURVATURE OF A CIRCLE

Statement : To show that the curvature of a circle is constant and equal to the reciprocal of its radius.

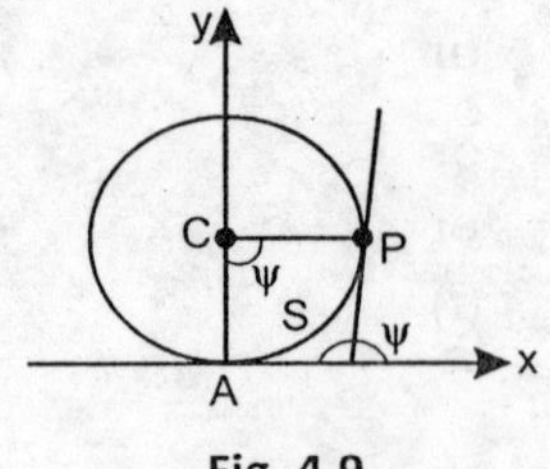

Fig. 4.9

Proof : Consider a circle with centre C and radius r.

Let A be a fixed point on the circle and tangent at A is x-axis.

Let P be any point on the circle such that

arc $AP = s$ and tangent at P make an angle Ψ with x-axis.

$$\therefore \quad \angle ACP = \Psi$$

[$\because$ Angle between two lines = Angle between their perpendicular]

$$\therefore \quad s = r\Psi \qquad \text{[From trigonometry } l = r \times \theta \text{ for a circle]}$$

Differentiating w.r.t Ψ,

$$\frac{ds}{d\psi} = r$$

$\therefore$ Curvature $= \frac{d\psi}{ds} = \frac{1}{r}$ *i.e.*, curvature at every point of the circle is reciprocal of it's radius and hence is constant.

Note:

(*i*) $\because$ Curvature at any point of the circle $= \frac{1}{r}$

As r increase curvature decreases.

Hence curvature becomes smaller than as the radius of circle becomes larger and larger.

(*ii*) Curvature = $\frac{1}{r}$. So radius of curvature = r which is the radius of circle.

(*iii*) The intrinsic equation of the curve is obtained by the relation between the length of arc of a curve measured from a given fixed point on the curve and the angle Ψ between the tangents at its extremities

"Some important result of the derivatives of arcs"

All these result can be obtained from the following y-ratios of Ψ and ϕ.

(*i*) $$\sin \Psi = \frac{dy}{ds}$$

(*ii*) $$\cos \Psi = \frac{dy}{ds}$$

(*iii*) $$\tan \Psi = \frac{dy}{ds}$$

or

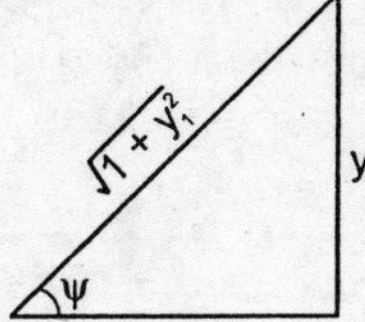

(*i*) $$\sin \Psi = \frac{y_1}{\sqrt{1+y_1^2}}$$

(*ii*) $$\cos \Psi = \frac{y_1}{\sqrt{1+y_1^2}}$$

(*iii*) $$\tan \Psi = \frac{y_1}{1}$$

and

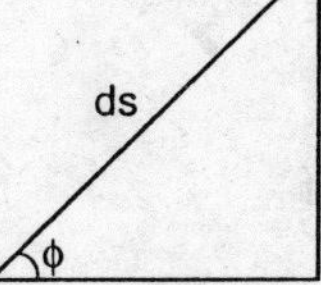

Fig. 4.10.

(*i*) $$\sin \phi = r\frac{d\theta}{ds}$$

(*ii*) $$\cos \phi = \frac{dr}{ds}$$

(*iii*) $$\tan \phi = r\frac{d\theta}{dr} = \frac{r}{r_1}$$

or

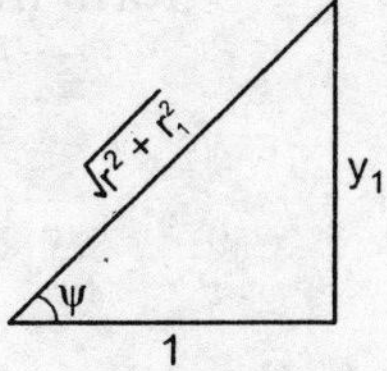

Fig. 4.11.

(*i*) $$\sin \phi = \frac{r}{\sqrt{r^2+r_1^2}}$$

(*ii*) $$\cos \phi = \frac{r_1}{\sqrt{r^2+r_1^2}}$$

(*iii*) $$\tan \phi = \frac{r}{\sqrt{r^2+r_1^2}}$$

4.24 RADIUS OF CURVATURE OF CARTESIAN CURVE

(*a*) **When equation of curve is given in the explicit form $y = f(x)$:**

Derivation : equating of curve is $y = f(x)$. Let Ψ be the angel which tangent at any point $P(x, y)$ makes with the x-axis. Then slope of tangent

$$\tan \Psi = \frac{dy}{ds}$$

differentiating both sides with respect to 's'.

$$\sec^2\Psi \times \frac{d\psi}{ds} = \frac{d}{ds}\left(\frac{dy}{dx}\right)$$

or $$\sec^2\Psi \times \frac{1}{\rho} = \frac{d^2y}{dx^2} \times \frac{dx}{ds}$$

or $$\sec^2\Psi \times \frac{1}{\rho} = \frac{d^2y}{dx^2} \times \cos\Psi \qquad \left[\because \cos\psi = \frac{dx}{ds}\right]$$

$$\rho = \frac{\sec^3\psi}{\dfrac{d^2y}{dx^2}}$$

$$\rho = \frac{(1+\tan^2\psi)^{3/2}}{\dfrac{d^2y}{dx^2}}$$

$$\rho = \frac{\left[1+\left(\dfrac{dy}{ds}\right)^2\right]^{3/2}}{\dfrac{d^2y}{dx^2}} \qquad ...(i)$$

If $$\frac{dy}{dx} = y_1 \text{ and } \frac{d^2y}{dx^2} = y_2, \text{ then } (i) \text{ becomes}$$

$$\rho = \frac{(1+y^2{}_1)^{3/2}}{y_2}$$

Corollary : When equation of curve is given in the form $x = f(y)$. Then by interchanging x and y is above form the ρ is.

$$\rho = \frac{\left[1+\left(\dfrac{dx}{dy}\right)^2\right]^{3/2}}{\dfrac{d^2x}{dy^2}}$$

(*b*) When the equation of the curve is given in the implicit form, *i.e.* $f(x, y) = 0$

Derivation : Here the equation of curve is $f(x, y) = 0$ and we know that

$$\frac{dy}{dx} = -\frac{f_x}{f_y}\text{. Provided } f_y \neq 0$$

Again $$\frac{d^2y}{dx^2} = \frac{f_{xx}(f_y)^2 - 2f_{xy}f_xf_y + f_{yy}(f_x)^2}{(f_y)^3}$$

Substituting the value of $\frac{dy}{dx}$ and $\frac{d^2y}{dx^2}$ in the equation (*i*) then we get

$$\rho = \frac{\left[1+\left(-\dfrac{f_x}{f_y}\right)^2\right]^{3/2}(f_y)^3}{f_{xx}(f_y)^2 - 2f_{xy}f_xf_y + f_{yy}(f_x)^2}$$

or $$\rho = \frac{[(f_x)^2+(f_y)^2]^{3/2}}{f_{xx}(f_y)^2 - 2f_{xy}f_xf_y + f_{yy}(f_x)^2} \text{ (in magnitude)}$$

(c) When the equation of curve is given in parametric form :

Derivation : Let the parametric equation of the curve be

$$x = f(t),\ y = \phi(t),\ t \text{ being a parameter. Then}$$

$$\frac{dx}{dt} = x', \text{ and } \frac{dy}{dt} = y', \text{ by chain rule}$$

$$\frac{dy}{dt} = \frac{y'}{x'}$$

Again

$$\frac{d}{dx}\left(\frac{dy}{dx}\right) = \frac{d}{dt}\left(\frac{y'}{x'}\right)\times\frac{dt}{dx}$$

$$= \frac{x'y'' - y'x''}{(x')^2}\times\frac{1}{x'}$$

$$\frac{d^2y}{dx^2} = \frac{x'y'' - y'x''}{(x')^3}$$

On substituting the values of $\frac{dy}{dx}$ and $\frac{d^2y}{dx^2}$ in $\rho = \dfrac{\left[1+\left(\frac{dy}{dx}\right)^2\right]^{3/2}}{\frac{d^2y}{dx^2}}$

$$= \frac{\left[1+\frac{y'^2}{x'^2}\right]^{3/2}\times(x')^3}{x'y'' - y'x''} = \frac{[x'^2 + y'^2]^{3/2}}{x'y'' - y'x''}$$

***Example* 4.42.** Find the radius of curvature of the curve $y = e^x$ at the point where it crosses the y-axis

Solution : The equation of curve is

$$y = e^x \qquad ...(i)$$

The curve crosses the y-axis where $x = 0$

$\therefore$ $y = 1$, hence points is (0, 1)

For radius of curvature

$$\frac{dy}{dx} = e^x \therefore \left(\frac{dy}{dx}\right)(0, 1) = 1$$

and

$$\frac{d^2y}{dx^2} = e^x \therefore \left(\frac{dy}{dx}\right)(0, 1) = 1$$

The radius of curvature of the curve at point (0, 1) is given by

$$\rho = \frac{\left[1+\left(\frac{dy}{dx}\right)^2\right]^{3/2}}{\frac{d^2y}{dx^2}}$$

$$\therefore \quad P_{at}(0, 1) = \frac{(1+1)^{3/2}}{1}$$

$$P_{at}(0, 1) = 2\sqrt{2} \textbf{ Ans.}$$

Example 4.43. Find the radius of curvature at the given point of the following curves

(i) $x^3 + y^3 = 3axy$ **at** $\left(\frac{3a}{2}, \frac{3a}{2}\right)$

(ii) $\sqrt{x} + \sqrt{y} = \sqrt{a}$ **at** $\left(\frac{a}{4}, \frac{a}{4}\right)$

Solution : (*i*) The equation of curve is

$$x^3 + y^3 = 3axy \quad ...(i)$$

Differentiating (*i*) with respect to '*x*', then

$$3x^2 + 3y^2\frac{dy}{dx} = 3a\left(x\frac{dy}{dx} + y\right)$$

or

$$(y^2 - ax)\frac{dy}{dx} = ay - x^2 \quad ...(ii)$$

At point $\left(\frac{3a}{2}, \frac{3a}{2}\right)$

$$\left(\frac{9a^2}{4} - \frac{3a^2}{2}\right)\frac{dy}{dx} = \frac{3a^2}{2} - \frac{9a^2}{4}$$

$$\therefore \quad \frac{dy}{dx} = -1$$

Differentiating (*ii*) with respect to '*x*', we get

$$(y^2 - ax)\frac{d^2y}{dx^2} + \left(2y\frac{dy}{dx} - a\right)\frac{dy}{dx} = a\frac{dy}{dx} - 2x$$

At point $\left(\frac{3a}{2}, \frac{3a}{2}\right)$

$$\left(\frac{9a^2}{4} - \frac{3a^2}{2}\right)\frac{dy}{dx} + [3a(-1) - a](-1) = a(-1) - 3a \quad \left(x = \frac{3a}{2}\right)$$

or

$$\frac{3a^2}{4} \times \frac{d^2y}{dx^2} = -8a$$

$$\therefore \quad \frac{d^2y}{dx^2} = -\frac{32}{3a}$$

Hence ρ at

$$\left(\frac{3a}{2}, \frac{3a}{2}\right) = \frac{\left[1 + \left(\frac{dy}{dx}\right)^2\right]^{3/2}}{\frac{d^2y}{dx^2}} = \frac{(1+1)^{3/2}}{\frac{32}{3a}}$$

$$\rho = \frac{3a}{8\sqrt{2}} \text{ (in magnitude)}$$

(*ii*) The equation of the curve is

$$\sqrt{x} + \sqrt{y} = \sqrt{a} \quad ...(i)$$

Differentiation (*i*) with respect to '*x*' we get

$$\frac{1}{2}x^{-\frac{1}{2}} + \frac{1}{2}y^{-\frac{1}{2}}\frac{dy}{dx} = 0$$

$$\therefore \quad \frac{dy}{dx} = -\frac{x^{\frac{1}{2}}}{y^{\frac{1}{2}}} = -\sqrt{\frac{y}{x}}$$

Again differentiating w.r.t 'x', we get

$$\frac{d^2y}{dx^2} = -\frac{x^{\frac{1}{2}} \times \frac{1}{2} y^{\frac{1}{2}} \frac{dy}{dx} - y^{\frac{1}{2}} \times \frac{1}{2} x^{\frac{1}{2}}}{x}$$

$$= -\frac{\sqrt{\frac{x}{y}}\left(-\frac{y}{x}\right) - \sqrt{\frac{y}{x}}}{2x} = \frac{1+\sqrt{\frac{y}{x}}}{2x} = \frac{\sqrt{x}+\sqrt{y}}{2x\sqrt{x}}$$

$$\frac{d^2y}{dx^2} = \frac{\sqrt{a}}{2x\sqrt{x}} \text{ from equation } (i)$$

At point $\left(\frac{a}{4}, \frac{a}{4}\right)$, $\frac{dy}{dx} = -1$

and $\frac{d^2y}{dx^2} = \frac{\sqrt{a}}{2 \times \frac{a}{4} \times \frac{\sqrt{a}}{2}} = \frac{4}{a}$

$$\therefore \quad \rho\left(\frac{a}{4}, \frac{a}{4}\right) = \frac{\left[1+\left(\frac{dy}{dx}\right)^2\right]^{3/2}}{\frac{d^2y}{dx^2}} = \frac{(1+1)^{3/2}}{\frac{4}{a}}$$

Hence $\rho = \frac{a}{\sqrt{2}}$

***Example* 4.44.** Prove that the radius of curvature for the catenary $y = c \cosh \frac{x}{c}$ is equal to the portion of the normal intercepted between the curve and the x-axis and that it caries as the square of the ordinate.

Solution : The equation of curve is

$$y = c \cosh \frac{x}{c} \qquad \text{...}(i)$$

Differentiating (i) w.r.t 'x', then we get

$$\frac{dy}{dx} = c \sinh \frac{x}{c} \times \frac{1}{c} \qquad \left[\because \frac{d(\cos h\theta)}{d\theta} = \sinh\theta\right]$$

or $\frac{dy}{dx} = \sinh \frac{x}{c}$

Again differentiating w.r.t 'x', then we get

or $\frac{d^2y}{dx^2} = \cosh \frac{x}{c} \times \frac{1}{c}$

$$\frac{d^2y}{dx^2} = \frac{1}{c} \cosh \frac{x}{c}$$

$\therefore$ Radius of curvature
$$\rho = \frac{\left[1+\left(\frac{dy}{dx}\right)^2\right]^{3/2}}{\frac{d^2y}{dx^2}} = \frac{\left(1+\sinh^2 \frac{x}{c}\right)^{3/2}}{\frac{1}{c}\cosh \frac{x}{c}}$$

$$\text{or} \qquad = \frac{\left(\cosh^2 \frac{x}{c}\right)^{3/2}}{\frac{1}{c}\cosh \frac{x}{c}} \qquad \left[\begin{array}{l} \because 1 = \cosh^2 x - \sinh^2 \theta \\ \because \cosh^2 \theta = 1 + \sinh^2 \theta \end{array}\right]$$

$$\text{or} \qquad = \frac{c \times \cosh^3 \frac{x}{c}}{\cosh \frac{x}{c}}$$

Hence $\rho = c \times \cosh^2 \frac{x}{c}$ ...(*ii*)

Now portion of the normal intercepter between the curve and the x-axis

$$= \text{Length of normal}$$

$$= y\sqrt{1 + \left(\frac{dy}{dx}\right)^2}$$

$$= c \text{ cosn } \frac{x}{c}\sqrt{1 = \sinh^2 \frac{x}{c}}$$

$$= c \cosh \frac{x}{c} \times \cosh \frac{x}{c}$$

$$= c \cosh^2 \frac{x}{c} \qquad ...(iii)$$

From (*ii*) and (*iii*)

$$\rho = \text{Length of normal}$$

Again, $\dfrac{\rho}{(\text{ordinate})^2} = \dfrac{\rho}{y^2} = \dfrac{c \cosh^2 \frac{x}{c}}{c^2 \cosh^2 \frac{x}{c}} = \dfrac{1}{c} = \text{constant}$

$\therefore$ ρ varies as y^2 *i.e.* as square of the ordinate.

***Example* 4.45.** Show that for the rectangular hyperbola $xy = c^2$,

$$\rho = \frac{(x^2 + y^2)^{3/2}}{2c^2}$$

Solution : The equation of rectangular hyperbola is

$$xy = c^2$$

or $y = \dfrac{c^2}{x}$ ...(*i*)

Differentiating (*i*) w.r.t 'x', then we get

$$\frac{dy}{dx} = -\frac{c^2}{x^2}$$

or $y_1 = -\dfrac{c^2}{x^2}$

Again, differentiating w.r.t 'x', then $\dfrac{d^2y}{dx^2} = \dfrac{2c^3}{x^3}$

or $y_2 = \dfrac{2c^2}{x^3}$

Radius of curvature at any point (x, y) for explicit from is

$$\rho = \frac{(1+y_1^2)^{3/2}}{y_2}$$

$$= \frac{\left[1+\frac{c^4}{x^4}\right]^{3/2}}{\frac{2c^2}{x^3}}$$

$$= \frac{(x^4+c^4)^{3/2}}{(x^4)^{3/2}} \times \frac{x^3}{2c^2} = \frac{(x^4+c^4)^{3/2}}{2c^2x^3}$$

$$= \frac{(x^4+y^2x^2)^{3/2}}{2c^2x^3} \qquad (\because xy = c^2)$$

$$\rho = \frac{(x^2+y^2)^{3/2}}{2c^2}$$

Important Note:

(i) Parabola

- Equation of parabola is $y^2 = 4ax$
- $0(0, 0)$ is the vertex of this parabola.
- x-axis is the axis of parabola.
- y-axis is the tangent to the parabola is the length of Latus rectum of the parabola.
- The point $(a, 0)$ on x-axis is called the focus of parabola.
- The line $x = -a$, $\parallel$ to y-axis is called directrix of the parabola.
- Any point (x, y) on the parabola $y^2 = 4ax$ is $(at^2, 2at)$ satisfies the equation of the parabola $y^2 = 4ax$, $x = at^2$, $y = 2at$ are parametric equations of the parabola.
- Any chord of the parabola passing through the focus is called a focal chord.
- If t_1 and t_2 are the extremities of a focal chord of the parabola $y^2 = 4ax$. Then $t_1 t_2 = -1$.
- Equation of tangent at 't' is $ty = x + at^2$

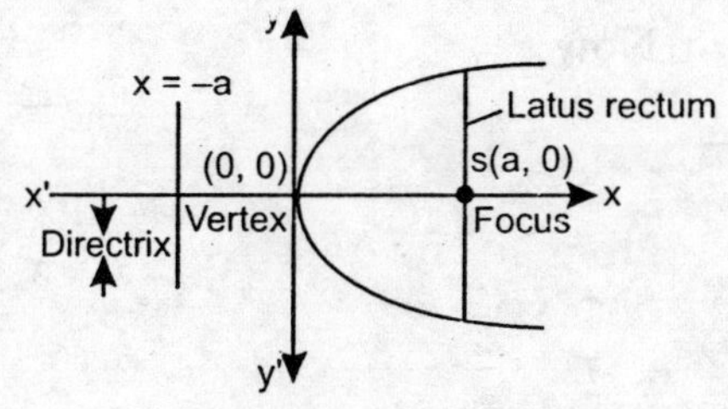

Fig. 4.12

(ii) Ellipse

- Equation of ellipse is $\frac{x^2}{a^2} + \frac{y^2}{b^2} = 1$
- $C(0, 0)$ is called centre of the ellipse.
- x-axis is called the major axis of the ellipse and length of major axis as '$2a$'
- y-axis is called the minor axis of the ellipse and length of minor axis is '$2b$'.
- Latus Rectum of the ellipse $= \frac{2b^2}{a}$
- Semi Latus Rectum of the ellipse $= \frac{b^2}{a}$.
- Any point $P(x, y)$ of the ellipse $\frac{x^2}{a^2} + \frac{y^2}{b^2} = 1$ is $(a \cos \theta, b \sin \theta)$.

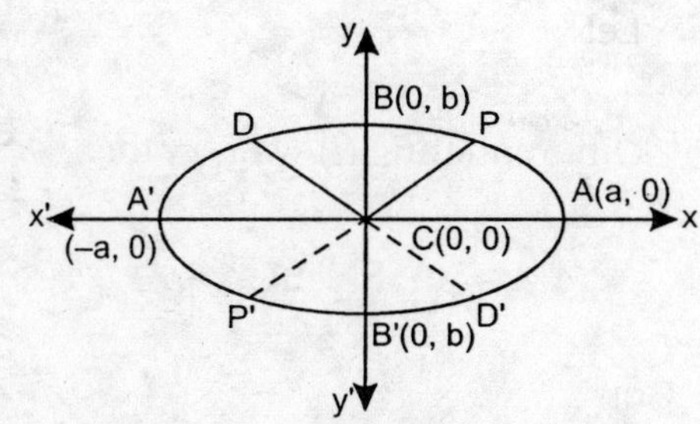

Fig. 4.13

- Eccentricity e of the ellipse is given by

$$b^2 = a^2(1-e^2),\ 0<e<1$$

- Equation of tangent at$(a\cos\theta, b\sin\theta)$ is $\dfrac{x\cos\theta}{a}+\dfrac{y\sin\theta}{b}=1.$
- Conjugate diameter of the ellipse.

If CP and CD are two semi-conjugate diameter of an ellipse and P is the point 'θ' *i.e.* $(a\cos\theta, b\sin\theta)$ then the point D is $(\pi/2+\theta)$ *i.e.*,

$$\left[a\cos\left(\frac{\pi}{2}+\theta\right), b\left(\frac{\pi}{2}+\theta\right)\right]$$

$\therefore$ D is $(-a\sin\theta, b\cos\theta)$

Example **4.46.** The tangents at two points P, Q on the cycloid $x=a(\theta-\sin\theta)$, $y=a(1-\cos\theta)$ are at right angles; show that if ρ_1, ρ_2 be the radii of curvature at these points, then ${\rho_1}^2+{\rho^2}_2=16a^2$.

Solution : The point on the cycloid is

$$x = (\theta-\sin\theta),\quad y=a(1-\cos\theta)$$

and
$$x' = a(1-\cos\theta),\ y'=a\sin\theta$$

$$x'' = a\sin\theta,\ y''=\cos\theta$$

Now
$$x^2+y^2 = a^2(1-2\cos\theta+\cos^2\theta+\sin^2\theta)$$

$$= 2a^2(1-\cos\theta)=4a^2\sin^2\frac{\theta}{2}$$

$$x'y''-y'x'' = a^2\cos\theta(1-\cos\theta)-a^2\sin^2\theta$$

$$= a^2(\cos\theta-\cos^2\theta-\sin^2\theta)$$

$$= a^2(\cos\theta-1)=-a^2(1-\cos\theta)=-2a^2\sin^2\frac{\theta}{2}$$

$\therefore$ Radius of curvature ρ at

$$'\theta' = \frac{(x'^2+y'^2)^{3/2}}{x'y''-y'x''}=\frac{\left(4a^2\sin^2\frac{\theta}{2}\right)^{3/2}}{-2a^2\sin^2\frac{\theta}{2}}$$

$$= \frac{8a^2\sin^3\frac{\theta}{2}}{-2a^2\sin^2\frac{\theta}{2}}$$

$$= 4a\sin\frac{\theta}{2}\ \text{(in magnitude)}$$

Let $\theta=\theta_1$ and $\theta=\theta_2$ be the points corresponding to P and Q then

$$\rho_1 = 4a\sin\frac{\theta_1}{2},$$

$$\rho_2 = 4a\sin\frac{\theta_2}{2}$$

Now by chain rule

$$\frac{dy}{dx} = \frac{\frac{dy}{d\theta}}{\frac{dx}{d\theta}}=\frac{y'}{x'}=\frac{\sin\theta}{1-\cos\theta}$$

$$= \frac{2\sin\theta/2\cos\theta/2}{2\sin^2\theta/2}=\cot\frac{\theta}{2}$$

which is the slope of tangent at 'θ'

$\therefore$ Slopes of tangents at p and Q are $\cot \frac{\theta_1}{2}$ and $\frac{\theta_2}{2}$

It is given that the tangents at P and Q are perpendicular

$$\therefore \quad \cot\frac{\theta_1}{2}\cot\frac{\theta_2}{2} = -1 \qquad \because m_1 m_2 = -1$$

$$\Rightarrow \quad \cot\frac{\theta_2}{2} = -\tan\frac{\theta_1}{2} = \cot\left(\frac{\pi}{2}+\frac{\theta_1}{2}\right)$$

$$\Rightarrow \quad \frac{\theta_2}{2} = \frac{\pi}{2}+\frac{\theta_1}{2} \Rightarrow \theta_2 = \pi + \theta_1$$

$$\therefore \quad \rho_2 = 4a\sin\left(\frac{\pi}{2}+\frac{\theta_1}{2}\right) = 4a\cos\frac{\theta_1}{2}$$

Hence
$$\rho_1^2 + \rho_2^2 = 16a^2\left(\sin^2\frac{\theta_1}{2}+\cos^2\frac{\theta_1}{2}\right) = 16a^2$$

***Example* 4.47.** Prove that the radius of curvature at any point of the astroid $x^{2/3} + y^{2/3} = a^{2/3}$ is three times the lengths of the perpendicular from the origin to the tangent at that point.

Solution : The given equation of the curve (astroid) is $x^{2/3} + y^{2/3} = a^{2/3}$

$\therefore$ The parametric equation of the curve are $x = a\cos^3 t$, $y = a\sin^3 t$

and
$$x' = -3a\cos^2 t\sin t,$$
$$y' = 3a\sin^2 t\cos t$$
$$x'' = -3a(\cos^3 t - 2\cos t\sin^2 t) = 3a\cos t(2\sin^2 t - \cos^2 t)$$
$$y'' = 3a(2\sin t\cos^2 t - \sin^3 t) = 3a\sin t(2\cos^2 t - \sin^2 t)$$

$$\therefore \quad x'^2 + y'^2 = 9a^2(\cos^4 t\sin^2 t + \sin^4 t\cos^2 t) = 9a^2\cos^2 t\sin^2 t$$

and
$$x'y'' - y'x'' = -9a^2\sin^2 t\cos^2 t(2\cos^2 t - \sin^2 t) - 9a^2\sin^2\cos^2 t(2\sin^2 t - \cos^2 t)$$
$$= -9a^2\sin^2 t\cos^2 t(\cos^2 t + \sin^2 t) = -9a^2\sin^2 t\cos^2 t$$

$\therefore$ Radius of curvature ρ at any point 't' *i.e.*, $(a\cos^3 t, a\sin^3 t)$

$$= \frac{(x'^2+y'^2)^{3/2}}{x'y''-y'x''} = \frac{(9a^2\cos^2 t\sin^2 t)^{3/2}}{-9a^2\sin^2 t\cos^2 t}$$

$$= \frac{27a^3\cos^3 t\sin^3 t}{-9a^2\sin^2 t\cos^2 t} = 3a\sin t\cos t \qquad \text{(Neglect negative sign)}$$

Now
$$\frac{dy}{dx} = \frac{y'}{x'} = \frac{3a\sin^2 t\cos t}{-3a\cos^2 t\sin t} = -\tan t$$

$\therefore$ The equation of tangent at $(a\cos^3 t, a\sin^3 t)$ is

$$y - a\sin^3 t = -\tan t(x - a\cos^3 t)$$

or
$$y - a\sin^3 t = -\frac{\sin t}{\cos t}(x - a\cos^3 t)$$

or
$$y\cos t - a\sin^3 t\cos t = -x\sin t + a\sin t\cos^3 t$$

or
$$x\sin t + y\cos t - a\sin t\cos t(\sin^2 t + \cos^2 t) = 0$$

or
$$x\sin t + y\cos t - a\sin\cos t = 0$$

p = length of $\perp$ from the origin (0, 0) to the tangent at 't'

$$= \frac{|0 + 0 - a \sin t \cos t|}{\sqrt{\sin^2 t + \cos^2}} = a \sin t \cos t \text{ (Neglecting negative sign)}$$

Hence, $\rho = 3p$

***Example* 4.48.** Show that for the parabola $y^2 = 4ax$, ρ^2 varies as $(SP)^3$, where ρ is the radius of curvature at any point p of the parabola and S is the focus of the parabola.

Solution : The parametric equation of parabola $y^2 = 4ax$ are at^2, $y = 2at$

Any point P on parabola is $(at^2, 2at)$

Denoting the differentiation w.r.t parameter 't' by dashes, we have

p(at², 2at)

S(a, 0)

Fig. 4.14

$$x' = 2at, \quad y' = 2a$$

$$x'' = 2a, \quad y'' = 0$$

$$\therefore \text{ Radius of curvature } \rho \text{ at } P\,(at^2, 2at) = \frac{(x'^2 + y'^2)^{3/2}}{x'\,y'' - y'x''} = \frac{(4a^2 t^2 + 4a^2)^{3/2}}{0 - 4a^2}$$

$$= \frac{(4a^2)^{3/2}\,(t^2+1)^{3/2}}{-4a^2} = -\frac{8a^3(1+t^2)^{3/2}}{4a^2} = 2a\,(1+t^2)^{3/2}$$

[Neglecting negative sign]

Also SP = distance between $S(a, 0)$ and $P(at^2, 2at)$

$$= a\sqrt{(at^2 - a)^2 + (2at - 0)^2} = \sqrt{a^2\,(t^2-1)^2 + 4a^2 t^2}$$

$$= a\sqrt{(t^2+1)^2} = a(t^2+1)$$

and $(SP)^3 = a^3\,(1+t^2)^3$

Also
$$\rho^2 = 4a^2(1+t^2)^3$$

$$= \frac{4}{a} \times a^3\,(1+t^2)^3$$

$$= \frac{4}{a}\,(SP)^3$$

Hence $\rho^2 \propto (SP)^3$

***Example* 4.49.** Prove that for the ellipse $\frac{x^2}{a^2} + \frac{y^2}{b^2} = 1$ $\rho = \frac{a^2b^2}{p^3}$ where p is the perpendicular from the centre upon the tangent at (x, y).

Solution : The equation of ellipse is $\frac{x^2}{a^2} + \frac{y^2}{b^2} = 1$

Let $(a \cos \theta, b \sin \theta)$ be any point on ellipse then the Parametric equation of the ellipse are

$$x = a \cos \theta,$$

$$y = b \sin \theta$$

and
$$x' = -a \sin \theta,$$

$$y' = b \cos \theta$$

$$x'' = -a \cos \theta,$$

$$y'' = -b \sin \theta$$

$$\therefore \text{ Radius of curvature } \rho \text{ at '}\theta\text{'} = \frac{(x'^2 + y'^2)^{3/2}}{x'y'' - y'x''} = \frac{(a^2 \sin^2 \theta + b^2 \cos^2 \theta)^{3/2}}{ab \sin^2 \theta + ab \cos^2 \theta}$$

$$= \frac{(a^2 \sin^2 \theta + b^2 \cos^2 \theta)^{3/2}}{ab} \quad ...(i)$$

Equating of tangent at $(a \cos \theta, b \sin \theta)$ is $\frac{x}{a} \cos \theta + \frac{y}{b} \sin \theta = 1$

$p =$ length of $\perp$ form centre (0, 0) on tangent

$$= \frac{1}{\sqrt{\frac{\cos^2\theta}{a^2} + \frac{\sin^2\theta}{b^2}}} = \frac{ab}{\sqrt{a^2 \sin^2 \theta = b^2 \cos^2 \theta}}$$

$$\therefore \quad p^3 = \frac{a^3 b^3}{(a^2 \sin^2 \theta + b^2 \cos^2 \theta)^{3/2}} \quad ...(ii)$$

Multiplying (*i*) and (*ii*), $\rho p^3 = a^2b^2$

Hence $\rho = \dfrac{a^2b^2}{p^3}$

Example 4.50. In the ellipse $\frac{x^2}{a^2} + \frac{y^2}{b^3} = 1$, show that the radius of curvature at an end of the major axis is equal to the semi-latus rectum of the ellipse.

Solution : The equation of ellipse is

$$\frac{x^2}{a^2} + \frac{y^2}{b^3} = 1 \quad ...(i)$$

Differentiating (i) w.r.t '*x*' then we get

$$\frac{2x}{a^2} + \frac{2y}{b^2}\frac{dy}{dx} = 0$$

or

$$\frac{dy}{dx} = -\frac{b^2}{a^2}\left(\frac{x}{y}\right)$$

Again Differentiating w.r.t '*x*' we get

$$\frac{d^2y}{dx^2} = -\frac{b^2}{a^2}\left(\frac{y \times 1 - x \times \frac{dy}{dx}}{y^2}\right)$$

$$= -\frac{b^2}{a^2y^2}\left[y - x\left(-\frac{b^2x}{a^2y}\right)\right] = -\frac{b^2}{a^2y^2}\left[\frac{a^2y^2 + b^2x^2}{a^2y}\right.$$

$$= -\frac{b^2y^2}{a^2y^2}\left(\frac{a^2b^2}{a^2y}\right) \quad [\because \text{ from (1) } a^2y^2 + b^2x^2 = a^2b^2]$$

$$= -\frac{b^4}{a^2y^3}$$

$$\therefore \quad \rho \text{ at } (x, y) = \frac{\left[1 + \left(\frac{dy}{dx}\right)^2\right]^{3/2}}{\frac{d^2y}{dx^2}} = \frac{\left[1 + \left(\frac{b^2x}{a^2y}\right)^2\right]^{3/2}}{-b^4 / (a^2y^3)}$$

$$= \frac{[a^4y^2 + b^4x^2]^{3/2}}{a^4b^4} \text{ (In magnitude)}$$

Now the co-ordinate of end of major axis are (0, 0)

$$\therefore \qquad \rho \text{ at } (a, 0) = \frac{[a^4 \times 0 + b^4 a^2]^{3/2}}{a^4 b^4} = \frac{b^6 a^3}{a^4 b^4}$$

$$= \frac{b^2}{a} \text{ which is semi latus rectum of the ellipse.}$$

Example **4.51.** For the curve $y = \dfrac{ax}{a+x}$, if ρ is the radius of curvature at any point (x, y), show that $\left(\dfrac{2\rho}{a}\right)^{2/3} = \left(\dfrac{y}{x}\right)^2 + \left(\dfrac{x}{y}\right)^2$

Solution : The equation of the curve is

$$y = \frac{ax}{a+x} \qquad \ldots(i)$$

Differentiating (i) with respect to 'x', we get

$$\frac{dy}{dx} = a\,\frac{(a+x)\times 1 - x}{(a+x)^2} = \frac{a^2}{(a+x)^2}$$

$$= a^2\,(a+x)^{-2}$$

Again differentiating w.r.t 'x', we get

$$\frac{d^2y}{dx^2} = \frac{d}{dx}\left(\frac{dy}{dx}\right) = -\,2a^2\,(a+x)^{-3} = \frac{-2a^2}{(a+x)^3}$$

$$\frac{d^2y}{dx^2} = \frac{-2y^3}{ax^3}$$

Now $$1 + \left(\frac{dy}{dx}\right)^2 = 1 + \frac{a^4}{(a+x)^4} = 1 + \frac{a4}{\left(\dfrac{ax}{y}\right)^4} = 1 = \frac{y^4}{x^4} = \frac{x^4 + y^4}{x^4}$$

$$\therefore \text{ Radius of curvature } p \text{ at } (x, y) = \frac{\left[1 + \left(\dfrac{dy}{dx}\right)2\right]^{\frac{3}{4}}}{\dfrac{d^2y}{dx^2}} = \frac{\left(\dfrac{x^4 + y^4}{x^4}\right)^{\frac{3}{4}}}{\dfrac{2y^3}{ax^3}}$$

$$= \frac{a}{2}\,\frac{(x^4 + y^4)^{\frac{3}{2}}}{x^3\,y^3} \qquad \text{(In magnitude)}$$

$$\therefore \qquad \left(\frac{2\rho}{a}\right)^{2/3} = \frac{x^4 + y^4}{x^2\,y^2}$$

$$= \frac{x^2}{y^2} + \frac{y^2}{x^2}$$

$$= \left(\frac{x}{y}\right)^2 + \left(\frac{y}{x}\right)^2$$

Example **4.52.** If ρ_1 **and** ρ_2 be the radii of curvature at the extremities of a focal chord of the parabola $y^2 = 4ax$, Prove that

$$(\rho_1)^{-\frac{2}{3}} + (\rho_2)^{-\frac{2}{3}} = (2a)^{-\frac{2}{3}}$$

Solution : Let the extremities of a focal chord be the point $p(at^2, 2at_1)$ and $Q\,(at_2{}^2, 2at_2)$. Then from coordinate geometry, $t_1 \times t_2 = -1$

Also $x = at^2$, $y = 2$ at are the parametric equation of the parabola $y^2 = 4ax$

$$\therefore \quad x' = \frac{dx}{dt} = 2at$$

$$y' = \frac{dy}{dx} = 2a$$

and
$$x'' = \frac{d^2x}{dt^2} = 2a$$

$$y'' = \frac{d^2y}{dt^2} = 0$$

$\therefore$ Radius of curvature ρ at the point 't' $= \dfrac{[(x'^2) + (y'^2)]^{\frac{3}{2}}}{x'y'' - y'x''}$

$$= \frac{[4a^2t^2 + 4a^2]^{\frac{3}{2}}}{2at \times 0 - 2a \times 2a} = \frac{8a^3(1+t^2)^{3/2}}{-4a}$$

$$= 2a(1+t)^{\frac{3}{2}} \qquad \text{(Neglected negative sign)}$$

If ρ_1 be radius of curvature at $P(at_1^2, 2at_1)$, the

$$\rho_1 = 2a(1 + ta_1^2)^{\frac{3}{2}}$$

and ρ_2 be the radius of curvature at $Q(at_2^2, 2at_2)$, then

$$\rho_2 = 2a(1+t_2^2)^{\frac{3}{2}}$$

$$\therefore \quad (\rho_1)^{\frac{-2}{3}} + (\rho_2)^{\frac{-2}{3}} = (2a)^{\frac{-2}{3}}\left[\frac{1}{1+t_1^2} + \frac{1}{1+t_2^2}\right]$$

$$= (2a)^{\frac{-2}{3}}\left[\frac{t_1^2 + t_2^2 + 2}{t_1^2 + t_2^2 + 1 + 1}\right] \qquad [\because t_1^2 + t_2^2 = 1]$$

$$= (2a)^{\frac{-2}{3}}$$

Example **4.53.** Find the radius of curvature at any point (x, y) of the curve $x^{\frac{2}{3}} + y^{\frac{2}{3}} = a^{\frac{2}{3}}$.

Solution : The equation of curve is

$$x^{\frac{2}{3}} + y^{\frac{2}{3}} = a^{\frac{2}{3}} \qquad ...(i)$$

Differentiating both sides of (i) w.r.t 'x', we get

$$\frac{2}{3}x^{-\frac{1}{3}} + \frac{2}{3}y^{-\frac{1}{3}}\frac{dy}{dx} = 0$$

or
$$\frac{dy}{dx} = -\frac{x^{-\frac{1}{3}}}{y^{-\frac{1}{3}}} = -y^{\frac{1}{3}}x^{\frac{1}{3}}$$

Again Differentiating w.r.t 'x', we get

$$\frac{d^2y}{dx^2} = -\left(\frac{1}{3}y^{-\frac{2}{3}}\frac{dy}{dx}\right)x^{-\frac{1}{3}} + y^{-\frac{1}{3}} \times \frac{1}{3}x^{-\frac{4}{3}}$$

$$= -\frac{1}{3}y^{-\frac{2}{3}}x^{-\frac{1}{3}}\left(-y^{\frac{1}{3}}x^{-\frac{1}{3}}\right) + \frac{1}{3}y^{\frac{1}{3}}x^{\frac{4}{3}}$$

$$= \frac{1}{3}y^{-\frac{1}{3}}x^{-\frac{2}{3}} + \frac{1}{3}y^{\frac{1}{3}}x^{-\frac{4}{3}}$$

$$= \frac{1}{3}x^{-\frac{4}{3}}y^{-\frac{1}{3}}\left(x^{\frac{2}{3}}+y^{\frac{2}{3}}\right) = \frac{1}{3}x^{-\frac{4}{3}}y^{-\frac{1}{3}}a^{\frac{2}{3}}$$

$\therefore$ ρ at the point (x, y)

$$= \frac{\left[1+\left(\frac{dy}{dx}\right)^2\right]^{\frac{3}{2}}}{\frac{d^2y}{dx^2}} = \frac{3\left(x^{\frac{2}{3}}+y^{\frac{2}{3}}\right)^{\frac{3}{2}}}{\left(x^{\frac{2}{3}}\right)^{\frac{3}{2}}x^{-\frac{4}{3}}y^{-\frac{1}{3}}a^{\frac{2}{3}}}$$

$$= \frac{3\left(a^{\frac{2}{3}}\right)^{\frac{3}{2}}}{xx^{-\frac{4}{3}}y^{-\frac{1}{3}}a^{\frac{2}{3}}} = \frac{3a}{x^{-\frac{1}{3}}y^{-\frac{1}{3}}a^{2/3}} = 3a^{\frac{1}{3}}x^{\frac{1}{3}}y^{\frac{1}{3}}$$

Hence ρ at (x, y) is $3a^{\frac{1}{3}}x^{\frac{1}{3}}y^{\frac{1}{3}}$

EXERCISE 4.8

1. Find radius of curvature at any point (x, y) of the curves

   (*i*) $ay^2 = x^3$ (*ii*) $xy = c^2$ (*iii*) $y^2 = 4ax$. **Ans.** (*i*) $\rho = \frac{x^{\frac{1}{2}}(4a+9x)^{\frac{3}{2}}}{6a}$ (*ii*) $\frac{(x^2+y^2)^{\frac{3}{2}}}{2c^2}$ (*iii*) $\rho = \frac{2}{\sqrt{a}}(x+a)^{\frac{3}{2}}$

2. Find the of curvature at point $\left(\frac{1}{4}, \frac{1}{4}\right)$ of the curve if the curve $\sqrt{x} + \sqrt{y} = 1$. **Ans.** $\rho = \frac{1}{\sqrt{2}}$

3. (*a*) Find the radius of curvature at any point $(0, c)$ of the catenary $y = c \cosh \frac{x}{c}$ **Ans.** $\rho = c$

   (*b*) Find the points on the parabola $y^2 = 8x$ at which the radius of curvature is $7\frac{13}{16}$

   Ans. Points are $\left(\frac{9}{8}, 3\right)$ and $\left(\frac{9}{8}, 3\right)$

4. Show that the radius of curvature ρ at any point (x, y) of the curve $x^{\frac{2}{3}} + y^{\frac{2}{3}} = a^{\frac{2}{3}}$ satisfies $\rho^3 = 27axy$.

5. Find the radius of curvature at any point (r, θ) of the curve **Ans.** $\rho = 3a \sin\theta \cos\theta$

   (*a*) $x = a\cos^3\theta$, $y = a\sin^3\theta$ **Ans.** $\rho = 3a\sin\theta\cos\theta$

   (*b*) $x = a\cos\theta$, $y = b\sin\theta$ **Ans.** $\rho = (a^2\sin^2\theta + b^2\cos 2\theta)\,\frac{3}{2}$

   (*c*) $x = a(\theta + \sin\theta)$ $y = a(1 - \cos\theta)$ **Ans.** $\rho = 4a\cos\frac{\theta}{2}$

6. If ρ_1 and ρ_2 be the radii of curvature at the extremities of two conjugate semi-diameters of the ellipse $\frac{x^2}{a^2} + \frac{y^2}{b^2} = 1$, prove that $\left(\rho_1^{2/3} + \rho_2^{2/3}\right)(ab)\frac{2}{3} = a^2 + b^2$.

4.25 RADIUS OF CURVATURE OF POLAR CURVES

When equation of the curve is given in r and θ *i.e.*, $r = f(\theta)$ or $f(r, \theta) = 0$ is called polar equation of the curve.

Statement : If ρ be the radius of curvature of the curve $r = f(\theta)$ or $f(r, \theta) = 0$ then,

$$\rho = \frac{(r^2 + r_1^2)^{\frac{3}{2}}}{r^2 - rr_2^2 + 2r_1^2}$$

Where $r_1 = \frac{dr}{d\theta}$; $r_2 = \frac{d^2r}{d\theta^2}$

Proof : Let the tangent at any point $P(r, \theta)$ on the curve AB make an angle Ψ with the initial line OX.

then
$$\Psi = \theta + \phi$$

$$\therefore \quad \frac{d\psi}{ds} = \frac{d\theta}{ds} + \frac{d\phi}{ds}$$

$$= \frac{d\theta}{ds} + \frac{d\phi}{d\theta} \times \frac{d\theta}{ds}$$

or
$$\frac{1}{\rho} = \frac{d\theta}{ds}\left[1 + \frac{d\phi}{d\theta}\right]$$

$$\frac{1}{\rho} = \frac{\sin\phi}{r}\left[1 + \frac{d\phi}{d\theta}\right] \quad ...(i) \left[\because \sin\phi = r\frac{d\theta}{ds}\right]$$

But tan
$$\phi = r \times \frac{d\theta}{dr} = \frac{r}{\frac{dr}{d\theta}} = \frac{r}{r_1} \quad ...(ii)$$

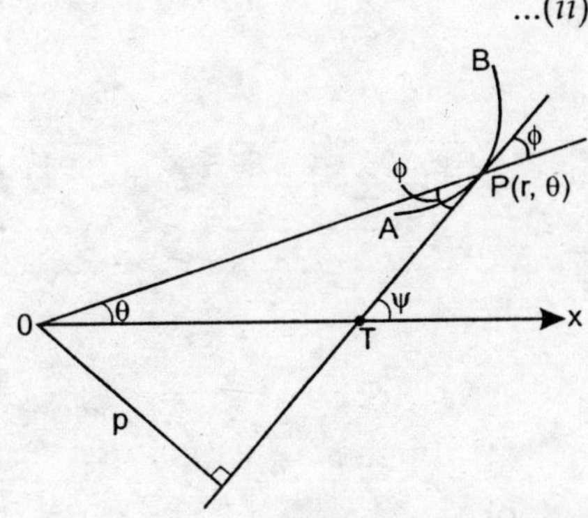

Fig. 4.15

To find $\frac{d\phi}{d\theta}$ differentiating *(ii)* with respect to 'θ', then we get

$$\sec^2\phi \frac{d\phi}{d\theta} = \frac{r_1 \times r_1 - r r_2}{r_1^2}$$

$$= \frac{r_1^2 - r r_2}{r_1^2}$$

$$\therefore \quad \frac{d\phi}{d\theta} = \frac{r_1^2 - r r^2}{r_1^2 \sec^2\phi} = \frac{r_1^2 - r r_2}{r_1^2\left(1 + \frac{r_2}{r_1^2}\right)}$$

$$= \frac{r_1^2 - r r_2}{r_1^2 + r^2} \quad \left(\because \tan\phi = \frac{r}{r_1}\right)$$

Also
$$\tan\phi = \frac{r}{r_1} \text{ gives}$$

$$\sin\phi = \frac{r}{\sqrt{r^2 + r_1^2}}$$

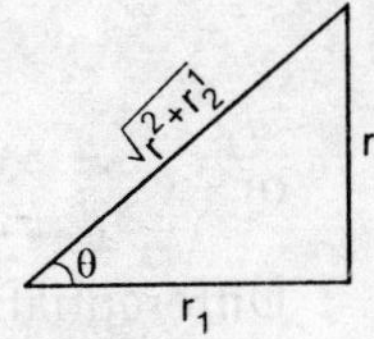

Fig. 4.16

Substituting the value of $\frac{d\phi}{d\theta}$ and $\sin\phi$ in *(i)*, then we get

$$\frac{1}{\rho} = \frac{1}{\sqrt{r^2 + r_1^2}}\left[1 + \frac{r_1^2 - r r_2}{r_1^2 + r^2}\right]$$

or
$$\frac{1}{\rho} = \frac{r_2 + 2r_1^2 - r r_2}{(r^2 + r_1^2)^{\frac{3}{2}}}$$

Hence,
$$\rho = \frac{(r^2 + r_1^2)^{\frac{3}{2}}}{r^2 + 2r_1^2 - r r_2}$$

4.26 RADIUS OF CURVATURE OF PEDAL EQUATIONS

When an equation of curve if given in the form of p and r *i.e.*, $p = f(r)$ then it is called pedal equation.

Proof : Let the pedal equation of the curve be $p = f(r)$

From he previous article,

$$\Psi = \theta + \phi$$

$$\therefore \quad \frac{d\psi}{ds} = \frac{d\theta}{ds} + \frac{d\phi}{ds}$$

$$\frac{1}{\rho} = \frac{d\theta}{ds} + \frac{d\phi}{ds} \quad ...(i)$$

But we know that

$$p = r\sin\phi \quad ...(ii)$$

Differentiating (*ii*) w.r.t 'r', we get

$$\frac{dp}{ds} = \sin\phi + r\cos\phi\,\frac{d\phi}{dr}$$

$$= r\frac{d\theta}{ds} + r\frac{dr}{ds}\frac{d\phi}{dr} \qquad \left[\because \sin\phi = r\frac{d\theta}{ds},\ \cos\phi = \frac{dr}{ds}\right]$$

$$= r\left[\frac{d\theta}{ds} + \frac{d\phi}{ds}\right]$$

$$= r \times \frac{1}{\rho}$$

or

$$\frac{dp}{dr} = r \times \frac{1}{\rho}$$

$$\therefore \quad \rho = \frac{r}{\frac{dp}{dr}}$$

Hence

$$\rho = r\frac{dr}{dp}$$

Example **4.54.** Find the radius of curvature at any point (p, r) of the curve $p^2 = ar$.

Solution : The equation of curve is

$$p^2 = a\,r \quad ...(i)$$

or

$$p = (a\,r)^{\frac{1}{2}}$$

Differentiating (*i*) with respect to 'r', we get

$$\frac{dp}{dr} = \frac{1}{2}(a\,r)^{-\frac{1}{2}} \times a$$

$$\frac{dp}{dr} = \frac{1}{2}\,\frac{a}{a^{1/2}\,r^{1/2}}$$

$$= \frac{1}{2}\,\frac{a^{1/2}}{r^{1/2}}$$

Now radius of curvature in pedal form is

$$r\frac{dr}{dp} = r \times \frac{2r^{1/2}}{a^{1/2}}$$

$$\therefore \quad p = r\frac{dr}{dp} = 2\,2\frac{r^{3/2}}{a^{1/2}}$$

Example **4.55.** For the cardioid $r = a\,(1 + \cos\theta)$, show that radius of curvature at (r, θ) is $\frac{2}{3}\sqrt{2ar}$ and prove that $\frac{\rho^2}{r}$ is constant.

Solution : The equation of cardioid is $r = a\,(1 + \cos\theta)$

$$\therefore \quad r_1 = \frac{dr}{d\theta} = -a\sin\theta$$

and

$$r_2 = \frac{d^2r}{d\theta^2} = -a\cos\theta$$

$\therefore$ Radius of curvature

$$\rho \text{ at } (r, \theta) = \frac{(r^2 + r_1^2)^{\frac{3}{2}}}{r^2 - r\,r_2 + 2r_1^2}$$

$$= \frac{[a^2(1+\cos\theta)^2 + a^2\sin^2\theta]^{\frac{3}{2}}}{a^2(1+\cos\theta)^2 - a(1+\cos\theta)(-a\cos\theta) + 2a^2\sin^2\theta}$$

$$= \frac{[a^3(1+\cos\theta)^2 + (1-\cos^2\theta)]^{\frac{3}{2}}}{a^2(1+\cos\theta)^2 + \cos\theta(1+\cos\theta) + 2(1-\cos^2\theta)}$$

$$= \frac{a[1+\cos\theta)^{\frac{3}{2}}1 + \cos\theta + 1 - \cos\theta)]^{\frac{3}{2}}}{(1+\cos\theta)(1+\cos\theta+\cos\theta+2-2\cos\theta)}$$

$$= \frac{a(1+\cos\theta)^{\frac{3}{2}} \times 2^{\frac{3}{2}}}{3(1+\cos\theta)}$$

$$= \frac{a(1+\cos\theta)^{\frac{1}{2}} \times 2\sqrt{2}}{3}$$

$$= \frac{a}{3} \times \sqrt{\frac{r}{a}} \times 2\sqrt{2} = \frac{2}{3}\sqrt{2\,a\,r}$$

$$\therefore \quad \rho = \frac{2}{3}\sqrt{2\,a\,r}$$

Now

$$\rho^2 = \frac{8a\,r}{9}$$

$$\therefore \quad \frac{\rho^2}{r} = \frac{8a}{9} \text{ which is constant.}$$

Example **4.56.** Find the radius of curvature at any point of the curve $r^n = a^n \cos n\theta$

Solution : The equating of curve is $r^n = a^n \cos n\theta$ ...(*i*)

$$n\log r = n\log a + \log\cos n\theta \qquad ...(ii)$$

Differentiating (*ii*) w.r.t 'θ', we get

$$\frac{n}{r}\frac{dr}{d\theta} = 0 + \frac{1}{\cos n\theta}(-n\sin n\theta)$$

$$\therefore \quad \frac{dr}{d\theta} = r_1 = -r \tan n\theta$$

We know that tan $\phi = r \dfrac{d\theta}{dr}$

$$= r \times \frac{1}{-r \tan n\theta} = -\cot n\theta$$

$$\tan\phi = \tan\left(\frac{\pi}{2} + n\theta\right)$$

$$\therefore \quad \phi = \frac{\pi}{2} + n\theta$$

Also $p = r\sin\phi$

$$= r \sin\left(\frac{\pi}{2} + n\theta\right)$$

$$= r\cos n\theta = r \times \frac{r^n}{a^n} \qquad (\because \text{From } (i))$$

$$\therefore \quad p\,a^n = r^{n+1} \qquad ...(iii)$$

which is a pedal equation

Differentiating (iii) w.r.t 'r', we get

$$a^n \frac{dp}{dr} = (n+1)\, r^n$$

$\therefore$ Radius of curvature in pedal form, $\rho = r\dfrac{dr}{dp}$

$$= r \times \frac{a^n}{(n+1)r^n} = \frac{a^n}{(n+1)r^{n-1}}$$

$$\therefore \quad \rho = \frac{a^n}{(n+1)\,r^{n-1}}$$

Example **4.57.** Find the radius of curvature of the curve $\frac{2a}{r} = 1 + \cos\theta$, hence show that the square of the radius of curvature varies as the cube of the focal distance.

Solution : The equation of curve is

$$\frac{2a}{r} = 1 + \cos\theta \qquad ...(i)$$

which is a parabola with focus as pole

Taking logarithm on both sides

$$\log 2a - \log r = \log(1 + \cos\theta)$$

Differentiating equation (*ii*) w.r.t 'θ', we get

$$\frac{1}{r}\frac{dr}{d\theta} = \frac{-\sin\theta}{\cdot 1 + \cos\theta} = -\frac{2\sin\frac{\theta}{2}\cos\frac{\theta}{2}}{2\cos^2\frac{\theta}{2}}$$

$$-\frac{1}{2}\frac{dr}{d\theta} = -\tan\frac{\theta}{2}$$

$$\therefore \quad \frac{1}{2}\frac{dr}{d\theta} = \tan\frac{\theta}{2} \qquad ...(iii)$$

$\because$ $$\tan\phi = r\frac{d\theta}{dr}$$

$\therefore$ $$\cot\phi = \frac{1}{r}\frac{dr}{d\theta} = \tan\frac{\theta}{2} \qquad \text{[From (iii)]}$$

$$\cot\phi = \cot\left(\frac{\pi}{2}-\frac{\theta}{2}\right)$$

and $$\phi = \frac{\pi}{2}-\frac{\theta}{2}$$

Also $$p = r\sin\phi$$

$$= r\sin\left(\frac{\pi}{2}-\frac{\theta}{2}\right)$$

$$p = r\cos\frac{\theta}{2} \qquad \ldots(iv)$$

Eliminating θ between (*i*) and (*iv*), we get

$$\frac{2a}{r} = 1+\cos\theta$$

$$= 2\cos^2\frac{\theta}{2}$$

$$= 2\frac{p^2}{r^2}$$

or $$p^2 = ar \qquad \ldots(v)$$

Differentiating (*v*) w.r.t '*p*', we get

$$2p = a\frac{dr}{dp}$$

$\therefore$ $$\frac{dr}{dp} = \frac{2p}{a} \qquad \ldots(vi)$$

Now, radius of curvature in pedal form, $\rho = r\frac{dr}{dp} = r\left(\frac{2p}{a}\right)$ [Form (*vi*)]

$$= \left(\frac{2r}{a}\right)p$$

$$= \left(\frac{2r}{a}\right)\times\sqrt{ar} \qquad \text{[From (v)]}$$

$$\rho = \frac{2}{\sqrt{a}}r^{\frac{3}{2}}$$

$\therefore$ $$\rho^2 = \frac{4}{a}r^3$$

Thus $\rho^2 \propto r^3$, where r is the distance of the point form the pole (*i.e.*, focus)

Hence ρ^2 varies as the cube of the focal distance.

***Example* 4.58.** If ρ_1, ρ_2 be the radii of curvature at the extremities of any chord through the pole of the cardiode $r = a(1+\cos\theta)$; show that $\rho_1{}^2+\rho_2{}^2 = \frac{16a^2}{9}$.

Solution : The equation of cardiode is $r = 1\ (1 = \cos\theta)$ ...(*i*)

Let ρ_1 and ρ_2 be the radii of curvature at the extremities p and Q of the chord POQ of the curve.

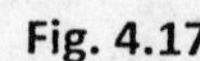

Fig. 4.17

Differentiating (*i*) w.r.t 'θ', we get

$$r_1 = \frac{dr}{d\theta} = -a\sin\theta \qquad ...(ii)$$

Again, differentiating (*ii*) w.r.t 'θ', we get

$$r_2 = \frac{d^2r}{d\theta^2} = -a\cos\theta \qquad ...(iii)$$

∴ Radius of curvature ρ at $(r, \theta) = \dfrac{(r^2 + r_1^2)^{\frac{3}{2}}}{r^2 + 2r_1^2 - r\, r_2}$

$$= \frac{[a^2(1+\cos\theta)^2 + a^2\sin^2\theta)^{\frac{3}{2}}}{a^2(1+\cos\theta)^2 + 2a^2\sin^2\theta + a^2\cos\theta\,(1+\cos\theta)}$$

$$= \frac{a^3[1+\cos^2\theta + 2\cos\theta + \sin^2\theta]^{\frac{3}{2}}}{a^2[1+\cos^2\theta + 2\cos\theta + 2\sin^2\theta + \cos\theta + \cos^2\theta]}$$

$$= \frac{a[2+2\cos\theta]^{\frac{3}{2}}}{[3+3\cos\theta]} = \frac{a \times 2^{\frac{3}{2}}(1+\cos\theta)^{\frac{3}{2}}}{3(1+\cos\theta)}$$

$$= \frac{a \times 2^{\frac{3}{2}}(1+\cos\theta)}{3} = \frac{a \times 2^{\frac{3}{2}}\left(2\cos^2\frac{\theta}{2}\right)^{\frac{1}{2}}}{3}$$

$$= \frac{a \times 2^{\frac{3}{2}} \times 2^{\frac{1}{2}}\cos\frac{\theta}{2}}{3} = \frac{4a}{3}\cos\frac{\theta}{2}$$

Let $\angle XOP = \theta \; \therefore \; \angle XOQ = \pi + \theta$

∴ Radius of curvature ρ_1 at $p = \dfrac{4a}{3}\cos\dfrac{\theta}{2}$ ...(*A*)

Changing θ to π + θ in (*A*) then radius of curvature

$$\rho_2 \text{ at } Q = \frac{4a}{3}\cos\left(\frac{\pi+\theta}{2}\right)$$

$$= \frac{4a}{3}\cos\left(\frac{\pi}{2} + \frac{\theta}{2}\right)$$

$$= -\frac{4a}{3}\sin\frac{\theta}{2} \qquad ...(B)$$

$$= \frac{4a}{3}\sin\frac{\theta}{2}$$ (Neglect Negative sign)

Squaring add adding (*A*) and (*B*), then we get

$$\rho_1^2 + \rho_2^2 = \frac{16a^2}{9}\left(\cos 2\frac{\theta}{2} + \sin^2\frac{\theta}{2}\right)$$

$$\rho_1^2 + \rho_2^2 = \frac{16a^2}{9}$$

EXERCISE 4.9

1. Find the radius of curvature at the point (p, r) of the following curve

(i) $r^3 = a^2 p$ (ii) $2ap^2 = r^3$ **Ans.** (i) $\rho = \frac{a^2}{3r}$, (ii) $\frac{2}{3}\sqrt{2ar}$

2. Find the radius of curvature at the point (p, r) of the curve $r^3 = 2ap^2$. **Ans.** $\frac{2}{3}\sqrt{2ar}$

3. Find the radius of curvature at point (r, θ) of the curve

(a) $r = a(1 - \cos\theta)$ **Ans.** $\frac{2}{3}\sqrt{2ar}$

(a) $r = a \sin n\theta$ **Ans.** $\frac{[r^2 + n^2a^2 - n^2r^2]^{\frac{3}{2}}}{[r^2 + 2a^2n^2 - n^2r^2}$

4. Find the radius of curvature at point (r, θ) of the curve $\frac{1}{r} = (1 + e\cos\theta)$ **Ans.** $\frac{l[1 + e^2 + 2e\cos\theta]^{\frac{3}{2}}}{(1 + e\cos\theta)^3}$

4.27 RADIUS OF CURVATURE AT THE ORIGIN

When the curve passes through the origin, following technique may be adopted for finding the radius of curvature at origin.

(i) Method of Direct Substitution :

Calculate the value of $\frac{dy}{dx}$ and $\frac{d^2y}{dx^2}$ at origin and substituting these values direct in the formula

$$\rho = \frac{\left[1 + \left(\frac{dy}{dx}\right)^2\right]^{3/2}}{\frac{d^2y}{dx^2}}$$

(ii) Method of Expansion :

When the above method fails or becomes very laborious, we obtain the value of $\frac{dy}{dx}$ and $\frac{d^2y}{dx^2}$ at origin by the following method.

Let $y = f(x)$ be the equation of the curve. Since it passes through the origin, therefore

$$f(0) = 0$$

$\therefore$ By Maclaurin's expansion,

$$y = f(0) + x f(0) + \frac{x^2}{2!} f'(0) + \frac{x^3}{3!} f''(0) + ...$$

Put $\quad f(0) = 0$

i.e. $$y = x f(0) + \frac{x^2}{2!} f''(0) + \frac{x^3}{3!} f'''(0) + ...$$

i.e. $$y = px + \frac{1}{2!} qx^2 + \frac{1}{3!} rx^3 + ... \qquad ...(i)$$

where
$$p = f(0) = y_1(0)$$
$$q = f'(0) = y_2(0) \text{ etc.}$$

Differentiating (i) w.r.t 'x', the we get

$$y_1 = p + \frac{2qx}{2!} + \frac{3rx^2}{3!} + \ldots \qquad \ldots(ii)$$

$$y_2 = \frac{2q}{2!} + \frac{6rx}{3!} + \ldots \qquad \ldots(iii)$$

Putting $x = 0$, in (*ii*) and (*iii*), at origin

$$y_1(0) = p,$$

$$y_2(0) = \frac{2q}{2!} = q$$

Hence putting these value of y_1 and y_2 in $\rho = \frac{(1+y_1^2)^{\frac{3}{2}}}{y_2}$ we have

$$\rho = \frac{(1+p^2)^{3/2}}{q}$$

(*iii*) Newton's method

(*a*) If a curve passes through the origin and axis of x is the tangent at origin. Then, radius of curvature ρ at the origin $\underset{\substack{x \to 0 \\ y \to 0}}{Lt} \frac{x^2}{2y}$

Derivation : Since x-axis is the tangent at the origin, therefore, its slope is

$$y_1(0) = \left(\frac{dy}{dx}\right)_{0,0} = 0$$

Now $\underset{\substack{x \to 0 \\ y \to 0}}{Lt} \frac{x^2}{2y}$ is of the form $\left(\frac{0}{0}\right)$

$\therefore$ By L' Hospital Rule

$$\underset{\substack{x \to 0 \\ y \to 0}}{Lt} \frac{x^2}{2y} = \underset{\substack{x \to 0 \\ y \to 0}}{Lt} \frac{2x}{2y_1} = \underset{\substack{x \to 0 \\ y \to 0}}{Lt} \frac{x}{y1} \qquad \left(\frac{0}{0} \text{ from}\right)$$

$$\because \quad y_1(0) = 0, x = 0$$

$$\therefore \quad \underset{\substack{x \to 0 \\ y \to 0}}{Lt} \frac{x^2}{2y} = \underset{\substack{x \to 0 \\ y \to 0}}{Lt} \frac{1}{y_2} = \frac{1}{y_2(0)} \qquad \ldots(i)$$

But ρ at origin $= \frac{[1+y_1^2(0)]^{\frac{3}{2}}}{y_2(0)} = \frac{1}{y_2(0)}$

$$\therefore \quad \rho = \underset{\substack{x \to 0 \\ y \to 0}}{Lt} \frac{x^2}{2y}$$

(*b*) If a curve passes through the origin and axis of y is the tangent at origin, then radius of curvature ρ at origin $= \underset{\substack{x \to 0 \\ y \to 0}}{Lt} \frac{y^2}{2x}$

Derivation : Proof is similar as (*a*) interchanges the letter x and y in this derivation

(*c*) Curvature at origin when Polar equation of the curve is given.

Derivation : If the initial line is the tangent at origin (Pole), then

$$\rho \text{ at the origin} = \underset{\substack{x\to 0\\ y\to 0}}{Lt} \frac{x^2}{2y}$$

$$= \underset{\substack{x\to 0\\ y\to 0}}{Lt} \frac{r^2\cos^2\theta}{2r\sin\theta}$$

$$= \underset{\substack{x\to 0\\ y\to 0}}{Lt} \frac{r}{2} \times \frac{1}{\sin\theta} \times \cos^2\theta$$

$$= \underset{\substack{x\to 0\\ y\to 0}}{Lt} \frac{r}{2\theta} \times \frac{\theta}{\sin\theta} \times \cos^2\theta \qquad \left\{ \because \begin{array}{l} \dfrac{\theta}{\sin\theta} \to 1 \\ \cos\theta \to 1 \\ \text{when } \theta \to 0 \end{array} \right.$$

$$= \underset{\substack{x\to 0\\ y\to 0}}{Lt} \left(\frac{r}{2\theta}\right)\left(\frac{0}{0}\right)$$

$$= \underset{\substack{x\to 0\\ y\to 0}}{Lt} \frac{\frac{dr}{d\theta}}{2}$$

$$\rho(\text{at pole}) = \underset{\substack{x\to 0\\ y\to 0}}{Lt} \left(\frac{1}{2}\frac{dr}{d\theta}\right)$$

***Note* :** The length to an algebraic curve at origin are obtained to zero to lowest degree term.

***Example* 4.59.** Find the radius of curvature at the origin for the curve

$$x^3 - 2xy^2y + 3xy^2 - 4y^3 + 5x^2 - 6xy + 7y^2 - 8y = 0$$

Solution : The given curve passes through origin therefore tangent at origin is given by equating to zero the lowest degree term in the curve *i.e.,*

$$y = 0 \text{ i.e., } x\text{-axis is the tangent at the origin}$$

$\therefore$ By Newton's formula ρ at (0, 0)

$$= \underset{\substack{x\to 0\\ y\to 0}}{Lt} \frac{x^2}{2y}$$

Dividing the equation of the curve by $2y$, then taking limit as x and y both tend to zero, then

$$\underset{\substack{x\to 0\\ y\to 0}}{Lt} \left[x\frac{x^2}{2y} - x^2\frac{2y}{2y} + \frac{3}{2}xy - 2y^2 + 5\frac{x^2}{2y} - 3x + \frac{7}{2}y - 4\right] = 0$$

$$5\underset{\substack{x\to 0\\ y\to 0}}{Lt} \frac{x^2}{2y} - 4 = 0, \text{ Other term become zero}$$

$$\therefore \qquad 5\rho - 4 = 0$$

$$\rho = \frac{4}{5}$$

***Example* 4.60.** Find the radius of curvature at origin for the curve $x^3 + y^2 - 2x^2 + 6y = 0$

Solution : The equation of the curve is

$$x^3 + y^2 - 2x^2 + 6y = 0$$

The curve passes through origin, so tangent at origin is given by equating to zero the lowest degree term in the curve, *i.e.,*

$\therefore$ $$6y = 0$$
$$y = 0 \text{ i.e., } x\text{-axis is the tangent at the origin.}$$

$\therefore$ ρ at (origin) $\underset{\substack{x \to 0 \\ y \to 0}}{Lt} \left(\dfrac{x^2}{2y}\right)$

Dividing (*i*) by $2y$, then we get

$$\underset{\substack{x \to 0 \\ y \to 0}}{Lt} \left(x\frac{x^2}{2y} + \frac{y}{2} - 2\frac{x^2}{2y} + \frac{6}{2}\right) = 0$$

$$\underset{\substack{x \to 0 \\ y \to 0}}{Lt} -2\frac{x^2}{2y} + 3 = 0$$

or $$-2\rho + 3 = 0$$

$\therefore$ $$\rho = \frac{3}{2}$$

Example **4.61.** Show that the radius of curvature of the curve $y^2 = x^2 \dfrac{(a+x)}{(a-x)}$ at origin is $a\sqrt{2}$

Solution : The equation of the curve is

$$y^2 = x^2 \frac{(a+x)}{(a-x)}$$

or $$(a-x)\,y^2 = x^2\,(a+x) \qquad ...(i)$$

The curve passes through origin, *i.e.*, x-axis is the tangent at origin, therefore tangent at origin is given by equating to zero the lowest degree term in the curve

$$a(x^2 - y^2) = 0$$

therefore $y = \pm ax$ are the tangent at origin therefore Newton's method is not applicable.

Now by the method of expansion $y = px + \dfrac{qx^2}{2!} + ...$

Putting the value of this y in (*i*)

$$\left(px + q\frac{x^2}{2} + ...\right)^2 (a-x) = x^2\,(a+x) \qquad ...(ii)$$

Equating the coefficient of x^2 on both sides, then

$$p^2 a = a$$
$$p = \pm 1$$

Equating the coefficient of x^3 on both sides, then

$$-p^2 + apq = 1 \qquad ...(iii)$$

when $p = 1$, then form (*iii*)

$$-1 + aq = 1 \Rightarrow q = \frac{2}{a}$$

$\therefore$ Radius of curvature ρ at origin $= \dfrac{(1+p^2)^{3/2}}{q} = \dfrac{(1+1)^{3/2}}{2/a}$

$$= \frac{2\sqrt{2}}{2} \times a = a\sqrt{2}$$

When = – 1, then form (*iii*)

$$-1 - aq = 1$$

$$-aq = 2$$

$$q = \frac{-2}{a}$$

$$\therefore \quad \rho \text{ at origin} = \frac{(1+p^2)^{3/2}}{q} = \frac{(1+1)^{3/2}}{+2/a}$$

$$= \frac{2\sqrt{2}}{-2} \times a = -a\sqrt{2}$$

$$= a\sqrt{2} \text{ (in magnitude)}$$

Hence ρ at origin is $a\sqrt{2}$

Example **4.62.** Apply Newton's formula to find the radius of curvature at the origin for the cycloid.

$$x = a(\theta + \sin\theta)$$

$$y = a(1 - \cos\theta)$$

Solution : When $\theta = 0$, $x = 0$ and $y = 0$ therefore origin corresponding to $\theta = 0$.

Also, by chain rule, $\dfrac{dy}{dx} = \dfrac{\frac{dy}{d\theta}}{\frac{dx}{d\theta}} = \dfrac{a\sin\theta}{a(1+\cos\theta)} = \dfrac{\sin\theta}{1+\cos\theta}$

At $\theta = 0$ *i.e.*, origin

$$\frac{dy}{dx} = \frac{\sin\theta}{1+\cos\theta} = 0, \text{ i.e., tangent at origin is } x\text{-axis}$$

$\therefore \rho$ at origin

$$\underset{\substack{x\to 0\\ y\to 0}}{Lt} \frac{x^2}{2y} = \underset{\substack{\theta\to 0\\ r\to 0}}{Lt} \frac{a^2(\theta+\sin\theta)^2}{2a(1-\cos\theta)}, \left(\frac{0}{0}\right) \text{ form}$$

$$= \underset{\theta\to 0}{Lt} \frac{2a(\theta+\sin\theta)(1+\cos\theta)}{2\sin\theta}, \left(\frac{0}{0} \text{ form}\right)$$

$$= \underset{\theta\to 0}{Lt} \frac{a(\theta+\sin\theta)(-\sin\theta)+a(1+\cos\theta)^2}{\cos\theta} = 4a$$

$$\therefore \quad \rho = 4a$$

Example **4.63.** Find the radius of curvature of the curve $r = a\sin n\theta$ at the pole.

Solution : The equation of the curve is

$$r = a\sin n\theta \qquad \text{...(i)}$$

Differentiating (*i*) with respect to 'θ', then

$$\frac{dr}{d\theta} = na\cos n\theta$$

$$\tan\phi = r\frac{d\theta}{dr}$$

$$\therefore \quad \tan\phi = \frac{1}{n}\tan n\theta \qquad \text{...(ii)}$$

When $\theta = 0$, then from (*i*), $r = 0$

Therefore the curve passes through the pole *i.e.*, origin

Also from (*ii*), $\tan\phi = 0$, when $\theta = 0$

$$\therefore \quad \phi = 0$$

Therefore initial line is the tangent to the curve at pole

Now by Newton's method for polar curve

$$\rho \text{ at pole} = \underset{\substack{r\to 0\\ \theta\to 0}}{Lt}\left(\frac{1}{2}\times\frac{dr}{d\theta}\right)$$

$$= \underset{\substack{r\to 0\\ \theta\to 0}}{Lt}\left(\frac{1}{2}\times na\cos n\theta\right)$$

$$= \frac{na}{2}$$

Hence radius of curvature, ρ at pole is $= \dfrac{na}{2}$

EXERCISE 4.10

Find the radius of curvature at origin for the curves

1. $2x^4 + 4x^3y + xy^2 + 6y^3 - 3x^2 - 2xy + y^2 - 4x = 0$ **Ans.** $\rho = 2$
2. $x^3 + y^3 = 3ax\,y$ **Ans.** $\rho = 2$
3. $a(y^2 - x^2) = x^3$ **Ans.** $\rho = 2a\sqrt{2}$

4.28 CENTRE OF CURVATURE, CIRCLE OF CURVATURE AND EVOLUTE

(1) **Centre of Curvature :** Centre of curvature for any point P of a curve AB is the point on the positive direction of the normal at p(*i.e.* the direction on the concave side of curve) at a distance ρ from it.

Let PD be the normal to the curve at P and C be a point on it such that $PC = \rho$, then C is the centre of curvature of the curve at P. Hence centre of curvature at any point $P(x, y)$ on the curve $y = f(x)$ is give by

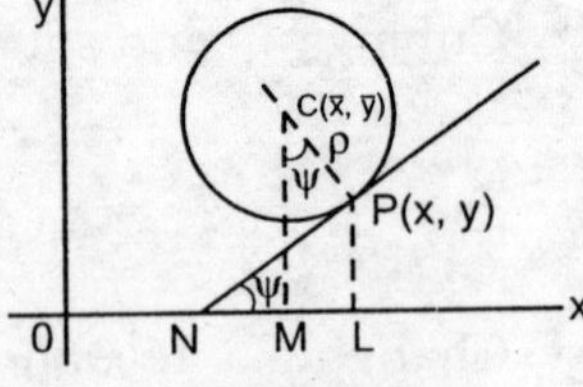

Fig. 4.18

$$\bar{x} = x - \rho\sin\Psi = x - \frac{y_1(1+y_1^2)}{y_2}$$

$$\bar{y} = y + \rho\cos\Psi = y + \frac{1+y_1^2}{y_2},\; y_1 = \frac{dy}{dx} \text{ and } y_2 = \frac{d^2y}{dx^2}$$

(2) **Evolute of a Curve :** The locus of the centre of curvature of the given curve is called the evolute of the curve and the curve it self is called involute.

(3) The circle with it's centre at the centre of curvature C and radius equal to ρ is called the circle of curvature of the curve at the point P.

Let ρ be the radius of curvature and $(\bar{x}, \bar{y})$ the centre of curvature at a given point, then the equation of the circle of curvature is given by $(x-\bar{x})^2 + (y-\bar{y})^2 = \rho^2$.

Example **4.64.** Find the coordinate of the centre of curvature at $(at^2, 2at)$ on the parabola $y^2 = 4ax$. Also find the equation of evolute.

Solution : The parametric equation of the curve $y^2 = 4ax$ is

$$x = at^2 \; ; \; y = 2at$$

$$\frac{dx}{dt} = 2at \; ; \; \frac{dy}{dt} = 2a$$

By chain rule of differentiation

$$y_1 = \frac{dy}{dx} = \frac{dy}{dt} \times \frac{dt}{dx} = \frac{\frac{dy}{dt}}{\frac{dx}{dt}} = \frac{1}{t}$$

and
$$y_2 = \frac{d^2y}{dx^2} = -\frac{1}{t^2} \times \frac{dt}{dx} = -\frac{1}{2at^3}$$

If $(\bar{x}, \bar{y})$ is the centre of curvature at 't', then

$$\bar{x} = x - \frac{y_1(1 + y_1^2)}{y_2} = 3at^2 + 2a \qquad \text{...(i)}$$

$$\bar{y} = y + \frac{(1 + y_1^2)}{y_2} = -2at^3 \qquad \text{...(ii)}$$

The coordinate of $(\bar{x}, \bar{y})$ are $(2a + 3at^2, -2at^3)$

Now evolute of a curve is the locus of its centre of curvature

To eliminate the parameter 't', we have

From (i)
$$t^2 = \frac{\bar{x} - 2a}{3a} \qquad \text{...(iii)}$$

From (ii)
$$t^3 = -\frac{\bar{y}}{2a} \qquad \text{...(iv)}$$

Cubing (iii) and squaring (iv), we get

$$\left(\frac{\bar{x} - 2a}{3a}\right)^3 = \frac{\bar{y}^2}{4a^2} \text{ or } 27\, a\, \bar{y}^2 = 4\,(\bar{x} - 2a)^3$$

Changing $\bar{x}$ to x and $\bar{y}$ to y, then we get the locus of $(\bar{x}, \bar{y})$ or the evolute of parabola as $27ay^2 = 4(x - 2a)^3$.

***Example* 4.65.** Find the co-ordinate of the centre of curvature of the rectangular hyperbola $xy = a^2$.

Solution : The equation of curve is

$$xy = a^2 \qquad \text{...(i)}$$

$$\therefore \quad y_1 = \frac{dy}{dx} = -a^2 x^{-2} \text{ and } y_2 = \frac{d^2y}{dx^2} = 2a^2 x^{-3}$$

Let $(\bar{x}, \bar{y})$ be the co-ordinate of centre of curvature at (x, y), then we get

$$\bar{x} = x - \frac{\frac{dy}{dx}\left(1 + \left(\frac{dy}{dx}\right)^2\right)}{\frac{d^2y}{dx^2}} = x - \frac{\left(-\frac{a^2}{x^2}\right)\left(1 + \frac{a^4}{x^4}\right)}{\frac{2a^2}{x^3}}$$

$$\bar{x} = \frac{3x}{2} + \frac{y^2}{2x} \qquad (\because a^4 = x^2 y^2)$$

and
$$\bar{y} = y + \frac{\left[1 + \left(\frac{dy}{dx}\right)^2\right]}{\frac{d^2y}{dx^2}}$$

$$= y + \frac{\left[1+\left(\frac{a^4}{x^4}\right)\right]}{2\frac{a^2}{x^3}} = y + \frac{x^4 + a^4}{2a^2 x}$$

$$\bar{y} = y + \frac{x^4 + x^2y^2}{2x^2 y} = \frac{3y}{2} + \frac{x^2}{2y}$$

Hence required centre of curvature is $\left(\frac{3x}{2} + \frac{y^2}{2x}, \frac{3y}{2} + \frac{x^2}{2y}\right)$

Example **4.66.** Show that the centre of curvature and equation of circle at the point $\left[\frac{a}{4}, \frac{a}{4}\right]$ on the curve $\sqrt{x} + \sqrt{y} = \sqrt{a}$ is $\left(\frac{3a}{4}, \frac{3a}{4}\right)$, and $\left(x - \frac{3a}{4}\right)^2 + \left(y - \frac{3a}{4}\right)^2 = \frac{a^2}{2}$.

Solution : Here the equation of curve is

$$\sqrt{x} + \sqrt{y} = \sqrt{a} \qquad \text{...}(i)$$

Differentiating (*i*) w.r.t '*x*' we get

$$\frac{1}{2} x^{-1/2} + \frac{1}{2} y^{-1/2} \times y_1 = 0 \qquad \text{...}(ii)$$

Again differentiating (*ii*) w.r.t '*x*' we get

$$-\frac{1}{4} x^{-3/2} - \frac{1}{4} y^{-3/2} \times y_1 + \frac{1}{2} y^{-1/2} y_2 = 0 \qquad \text{...}(iii)$$

At the point $\left[\frac{a}{4}, \frac{a}{4}\right]$

From (*ii*),

$$\frac{1}{2} \times \frac{2}{\sqrt{a}} + \frac{1}{2} \times \frac{2}{\sqrt{a}} y_1 = 0$$

$$\therefore \quad y_1 = -1$$

From (*iii*), $-\frac{1}{4} \times \frac{4}{a} \frac{2}{\sqrt{a}} - \frac{1}{4} \times \frac{2}{\sqrt{a}} (-1)^2 + \frac{1}{2} \times \frac{2}{\sqrt{a}} y_2 = 0$

or

$$-\frac{4}{a\sqrt{a}} + \frac{1}{\sqrt{a}} y_2 = 0$$

$$\therefore \quad y_2 = \frac{4}{a}$$

Now

$$\rho\left(\frac{a}{4}, \frac{a}{4}\right) = \frac{(1 + y_1^2)^{3/2}}{y_2}$$

$$= \frac{(1+1)^{3/2}}{4/a} = 2\sqrt{2} \times \frac{a}{4} = \frac{a}{\sqrt{2}}$$

Let $(\bar{x}, \bar{y})$ be the centre of curvature at $\left[\frac{a}{4}, \frac{a}{4}\right]$ then

$$\bar{x} = x - \frac{y_1(1 + y_1^2)}{y_2}$$

$$= \frac{a}{4} - \frac{-(1+1)}{\frac{4}{a}}$$

$$= \frac{a}{4} + \frac{a}{2}$$

$$= \frac{3a}{4}$$

$$\bar{y} = y + \frac{1+y_1^2}{y_2}$$

$$= \frac{a}{4} + \frac{1+1}{\frac{4}{a}}$$

$$= \frac{a}{4} + \frac{2a}{4} = \frac{3a}{4}$$

Equation of circle of curvature is

$$(x-\bar{x})^2 + (y-\bar{y})^2 = \rho^2 \text{ i.e.}$$

$$\left(x - \frac{3a}{4}\right)^2 + \left(y - \frac{3a}{4}\right)^2 = \frac{a^2}{2}$$

***Example* 4.67.** Show that the evolute of the tractrix $x = c\cos t + c\log\tan\frac{t}{2}$, $y = C$ sint is the catenary $y = c\cosh\frac{x}{c}$

Solution : The equation of curve is $x = c\cos t + c\log\tan\frac{t}{2}$, $y = c\sin t$ ...(i)

$$\frac{dx}{dt} = -c\sin t + \frac{c}{\tan\frac{t}{2}} \times \frac{1}{2}\sec^2\frac{t}{2} = -c\sin t + \frac{c\cos\frac{t}{2}}{\sin\frac{t}{2}} \times \frac{1}{2\cos^2\frac{t}{2}}$$

$$= -c\sin t + \frac{c}{2\sin\frac{t}{2}\cos\frac{t}{2}} = -c\sin t + \frac{c}{\sin t}$$

$$= \frac{c(1-\sin^2 t)}{\sin t}$$

$$= \frac{c\cos^2 t}{\sin t}$$

$$\frac{dy}{dt} = c\cos t$$

∴ By chain rule of differentiation,

$$y_1 = \frac{dy}{dt} = \frac{\frac{dy}{dt}}{\frac{dx}{dt}} = c\,\text{cost} \times \frac{\sin t}{c\cos^2 t} = \tan t$$

$$y_2 = \sec^2 t \times \frac{dt}{dx} = \frac{1}{\cos^2 t} \times \frac{\sin t}{c\cos^2 t} = \frac{\sin t}{c\cos^4 t}$$

If $(\bar{x}, \bar{y})$ is the centre of curvature at any point on the curve, then

$$\bar{x} = x - \frac{y_1(1+y_1^2)}{y_2} = c\cos t + c\log\tan\frac{t}{2} - \frac{\tan t(1+\tan 2\, t)}{\frac{\sin t}{c\cos^4 t}}$$

$$= c\cos t + c\log\tan\frac{t}{2} - \frac{\cos^4 t}{\sin t}\times\frac{\sin t}{\cos t}\times\frac{1}{\cos^2 t}$$

or $$\bar{x} = c\cos t + c\log\tan\frac{t}{2} - c\cos t$$

or $$\bar{x} = c\log\tan\frac{t}{2} \qquad \ldots(ii)$$

and $$\bar{y} = y_{\cdot} + \frac{1+y_1^2}{y_2} = c\sin t + \frac{\frac{1+\tan^2 t}{\sin t}}{c\cos^4 t}$$

$$= c\sin t + \frac{c\cos^4 t}{\sin t}\times\sec^2 t = c\sin t + \frac{c\cos^2 t}{\sin t}$$

$$= \frac{c(\sin^2 t + \cos^2 t)}{\sin t}$$

or $$\bar{y} = \frac{c}{\sin t} \qquad \ldots(iii)$$

Evolute of the give curve is the locus of $(\bar{x}\ \bar{y})$. Let us eliminate t between (*ii*) and (*iii*), then

CURVE TRACING

The present chapter deals with tracing of curve in Cartesian, polar and parametric coordinate systems for determining approximate shape of a curve. This is helpful in the application of integration in finding the length, area and volume of the region bounded by the curves.

4.29 CURVE AND ITS FORMS

To determine the approximate shapes of curves from their Cartesian, polar or parametric equations without plotting a large number of points lying on them. We shall consider mainly those curves whose equations can be solved for y, x or r. Basically curve tracing is an analytical method of drawing an approximate shape of a curve by the study of some of its important characteristic such as symmetry, intercepts, asymptotes, tangents, multiple points and region of existence. Curve tracing is very useful in the application of integration in finding lengths, area volume etc.

4.30 PROCEDURE OF CURVE TRACING FOR CARTESIAN CURVES

I. Symmetry

Firstly, we find symmetry of the cures. In this regard the following points are helpful.

(*a*) **Symmetry about x-axis :** If we replace y by $-y$ and there is no change in the equation of the curve *i.e.*, the power of y is even, then the curve is symmetrical about x-axis.

Example : The parabola $y^2 = 4ax$ is symmetrical about the x-axis.

(*b*) **Symmetry about the y-axis :** If we replace x by $-x$ and there is no change in the equation of the curve *i.e.*, the power of x is even, then the curve is symmetrical about the y-axis.

Example : The parabola $x^2 = 4ay$ is symmetrical about y-axis.

(*c*) **Symmetry in opposite quadrants :** If we replace x by $-x$ and y by $-y$ and there is no change in the equation of the curve *i.e.*, the powers of x and y or xy are even, then the curve is symmetrical in opposite quadrants.

Example : The curve $xy = c^2$ is symmetrical in opposite quadrants.

(*d*) **Symmetry about the line $y = x$:** If x and y are interchanged (*i.e.*, x is changed by y and y by x) and there is no change in the equation of the curve, the curve is symmetrical about the line $y = x$.
Example : The curve $x^3 + y^3 = 3axy$ is symmetrical about the line $y = x$.

II. Nature of the origin :

(*a*) Firstly find whether the curve passes through the origin or not. If (0, 0) satisfies the given equation of the curve, then the curve passes through origin.

(*b*) If the curve passes through the origin, find the equation of tangent at origin by equation to zero the lowest degree terms in the equation of the curve.

(*c*) If there are two tangents at the origin, then origin will be a double points. Find the nature of double point.

Note : If the two tangents are real and distinct, then origin is a node.

If the two tangent are real and coincident, then origin is cusp.

If the two tangents are imaginary, then origin is a conjugate points or an isolated point.

III. Points of intersection of the curve with the co-ordinate axes:

Find the points where the curve cuts the coordinate axes.

To find the points where the curve cuts the x-axis, put $y = 0$ in the equation of the curve and solve the resulting equation for x.

Similarly the points of intersection with the y-axis are obtained by putting $x = 0$ and solving the resulting equation for y. Find also the tangents at these points, if necessary.

IV. Asymptotes:

Find all the asymptotes (if any) *i.e.*, parallel or oblique.

V. Special points (or Singularities) :

Find $\frac{dy}{dx}$ *i.e.*, the slope of the tangents to the curve at the point (x, y)

Find the points where

(*i*) $\frac{dy}{dx} = 0$, *i.e.*, tangents are $\parallel$ to x-axis.

(*ii*) $\frac{dy}{dx} = \infty$, *i.e.*, tangents are $\parallel$ to y-axis

(*iii*) $\frac{dy}{dx}$ is positive, *i.e.*, the function is increasing *i.e.*, the curve is rising.

(*iv*) $\frac{dy}{dx}$ is negative *i.e.*, the curve is falling.

Find $\frac{d^2y}{dx^2}$ (if not tedious) and if

(*i*) $\frac{d^2y}{dx^2}$ is positive, *i.e.*, the curve is concave upwards.

(*ii*) $\frac{d^2y}{dx^2}$ is negative, *i.e.*, the curve is concave downward.

(*iii*) Also find points of inflection, *i.e.*, at the point of inflection; $\frac{d^2y}{dx^2} = 0$ and $\frac{d^3y}{dx^3} \neq 0$.

VI. Region where the curve does not exist :

Find the region of x-y-plane where the curve does not exist. Such aregion is obtained on solving the given equation for one variable in terms of the other variables, *i.e.,*

$$x^2 = 4ay$$

or
$$y = \frac{x^2}{4a}$$

Putting the values of x (positive & negative) and get the value of y.

Note **: Sometimes inconvenient step may be omitted without any disadvantages.**

Example **4.68.** Trace the curve $ay^2 = x^3$ (semi-cubical parabola)

Solution : The equation of curve is $ay^2 = x^3$ ...(*i*)

1. **Symmetry :** Since in the equation of the curve the power if y are even therefore, the curve is symmetrical about the axis of x.
2. **Origin :** The curve passes through the origin because (0, 0) satisfies the equation of the curve.
3. **Tangent at origin :** Equating to zero the lowest degree term in equation of the curve, *i.e.,* $ay^2 = 0$ *i.e.,* $y = 0$, $y = 0$ thus the origin is double points and it is a cusp since there are two coincident tangent at the origin.
4. **Points of intersection :** The curve does not intersect the co-ordinate axes any where else except at the origin.
5. **Asymptotes :** Obviously the curve has no asymptotes.
6. **Region :** The equation of curve is

$$ay^2 = x^3$$

$$\therefore \quad y^2 = \frac{x^3}{a}$$

when
$$x = 0, y^2 = 0$$

when $x > 0$, y^2 is positive *i.e.,* y is real so the curve exist in the region $x > 0$

As x increases, y^2 also increases and when $x \to \infty$, $y^2 \to \infty$.

when $x < 0$, y^2 is negative *i.e.* y is imaginary, therefore the curve does not exist in the region $x < 0$.

Taking all these facts in to consideration, the shape of the curve is shown in the figure 4.19.

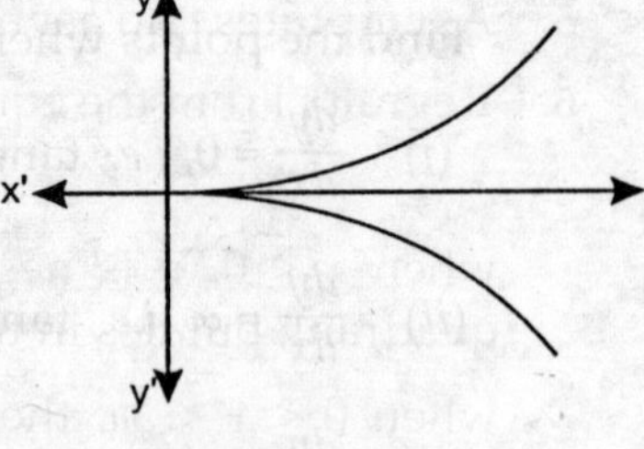

Fig. 4.19

Example **4.69.** Trace the curve $y^2 (2a - x) = x^3$

Solution : The equation of the curve is

$$y^2 (2a - x) = x^3 \qquad ...(i)$$

1. **Symmetry :** The curve is symmetrical about x-axis, since the power of y is even.
2. **Origin :** The curve passes through origin because (0, 0) satisfies the equation of the curve.
3. **Tangent at origin :** Equating to zero the lowest degree term in the equation of the curve, *i.e.,* $2ay^2 = 0$ *i.e.,* $y = 0$, $y = 0$ are two coincident tangents at the origin, therefore the origin is a cusp.
4. **Points of intersection :** The curve meets the co-ordinate axes only at the origin.
5. **Asymptotes :** There are three possible asymptotes *i.e.,* $x = 2a$ is the asymptotes $\|$ to y-axis and the remaining asymptotes are oblique which are found out as follows:

   Put $x = 1$, and $y = m$ in third degree term *i.e.,* $m^2 + 1 = 0$ or $m = \pm i$ which are imaginary roots. Hence $x = 2a$ is only real asymptotes.

6. **Region :** Solving the equation of the curve of y, we get

$$y^2 = \frac{x^3}{(2a - x)}$$

when $x = 0$, $y^2 = 0$

when $x \to 2a$, $y^2 \to \infty$, therefore, $x = 2a$ is an asymptotes of the curve.

when $0 < x < 2a$, y^2 is positive *i.e.* y is real so the curve exists in this region.

when $x > 2a$, y^2 negative *i.e.* y is imaginary, therefore the curve does not exist in the region $x > 2a$. When $x < 0$, y^2 is negative so the curve does not exist in the region $x < 0$.

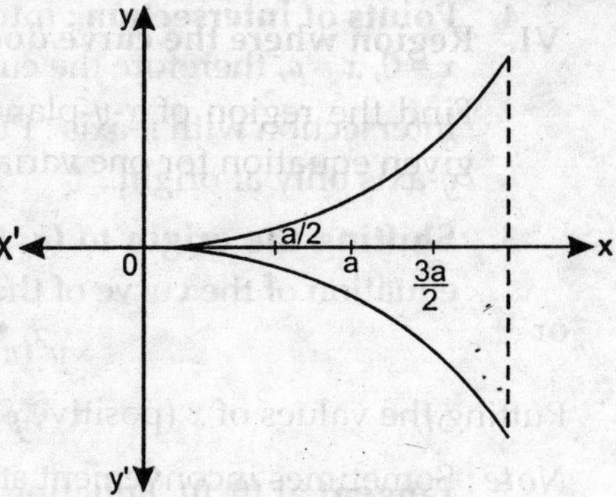

Fig. 4.20

Combining all these facts, the shape of the curve is as shown in Figure 4.20.

Example 4.70. Trace the curve $y = x^3 - 3ax^2$

Solution : The given curve is

$$y = x^3 - 3ax^2$$

or

$$y = x^2(x - 3a) \quad \text{...}(i)$$

1. **Symmetry :** No symmetry.
2. **Origin :** The curve passes through origin because (0, 0) satisfies the equation of curve.
3. **Tangent at origin :** Equating to zero the lowest degree terms in the equation of the curve, *i.e.*, $y = 0$ *i.e.* x-axis is tangent at origin.
4. **Points of intersection :** Intersection with x-axis: Put $y = 0$ in the equation of the curve *i.e.* $x^3 - 3ax^2 = 0$ $\therefore$ $x = 0$ or $x = 3a$ therefore the points of intersection with x-axis is (0, 0) and (3a, 0).

   Intersection with y-axis :

   Put $x = 0$ in the equation of the curve *i.e.*, $y = 0$, therefore (0, 0) is only points of intersection with y-axis.
5. **Asymptotes :** no asymptotes.
6. **Region :** From the equation of curve

$$y = x^2\,(x - 3a)$$

when $x < 0$, y is negative, therefore the curve does not lie in second quadrant, but lies in third quadrant.

when $0 < x < 3a$, then y is negative, therefore the curve lies in fourth quadrant for value of x such that $0 < x < 3a$.

When $x > 3a$, y is positive, therefore the curve lies in the first quadrant for value of $x > 3a$.

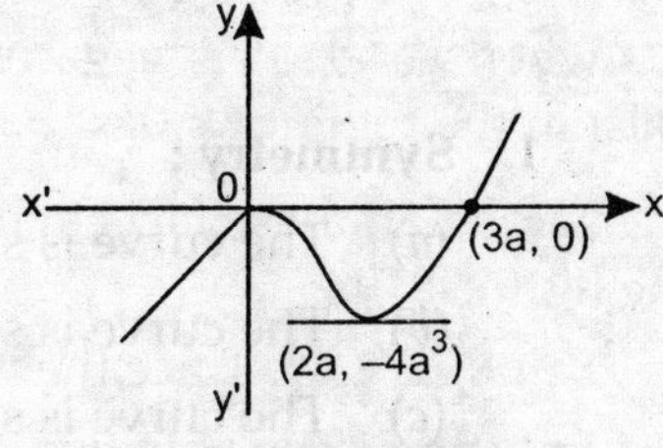

Fig. 4.21

The shape of the curve is shown in the figure 4.21.

Example 4.71. Trace the curve $y^2(a + x) = x^2(a - x)$

Solution : The equation of curve is

$$y^2(a + x) = x^2(a - x) \quad \text{...}(i).$$

1. **Symmetry :** The curve is symmetrical about the x-axis.
2. **Origin :** The curve passes through the origin.
3. **Tangent at origin :** Equating to zero the lowest degree term in the equation of the curve, *i.e.* $a(y^2 - x^2) = 0$ *i.e.* $y = \pm x$, since there are two real and distinct tangent at the origin, therefore, the origin is a node on the curve.

4. **Points of intersection :** Intersection with x-axis : Put $y = 0$ in the equation of the curve, *i.e.*, $x^2(a - x) = 0$ or $x = 0, x = a$, therefore the curve intersects the x-axis at (0, 0).

   Intersection with y-axis : Put $x = 0$ in the equation of the curve *i.e.* $y = 0$, therefore, the curve intersects the y-axis only at origin.

5. **Shifting the origin to $(a, 0)$:** Replace x by $x + a$ and y by $y + 0$ in the equation of the curve then the equation of the curve of the curve becomes

$$y^2(a + x + a) = (x + a)^2 (a - x - a)$$

or
$$y^2(2a + x) = -(x^2 + 2ax + a^2)\,x$$

Tangent at (0, 0). Equating to zero the lowest degree term *i.e.* $x = 0$ (*i.e.* the new y-axis) is the tangent at the new origin.

6. **Asymptotes :** Here $x = -a$ is the only real asymptotes to the curve.

7. **Region :** Solving the equation of the curve for y *i.e.*

$$y^2 = \frac{x^2(a - x)}{x + a}$$

when $x = 0, y^2 = 0$

when $x = a, y^2 = 0$

when $0 < x < a$, y^2 is positive, therefore, the curve exist in this region.

when $x > a$, y^2 is negative, therefore, the curve does not exist in the region $x > a$.

when $x \to -a$, $y^2 = \infty$, therefore, $x = -a$ is an asymptotes of the curve.

when $-a < x < 0$, y^2 is positive, therefore, the curve lies in this region.

when $x < -a$, y^2 is negative, therefore the curve does not exist in the region $x < -a$.

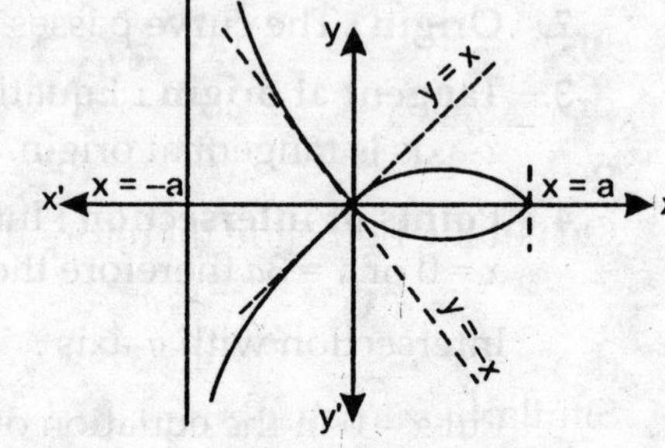

Fig. 4.22

Hence the shape of the curve is as shown in Figure 4.22.

Example 4.72. Trace the curve $x^{2/3} + y^{2/3} = a^{2/3}$ (Astroid).

Solution : The equation of curve is

$$x^{2/3} + y^{2/3} = a^{2/3} \quad ...(i)$$

1. **Symmetry :**
   (*a*) The curve is symmetrical about the x-axis.
   (*b*) The curve is symmetrical about the y-axis.
   (*c*) The curve is symmetrical about the line $y = x$.
   (*d*) The curve is symmetrical about the line $y = -x$
   (*e*) The curve is also symmetrical in opposite quadrants.
2. **Origin :** The curve does not passes through origin.
3. **Points of intersection :** (*a*) Intersection with x-axis :

   Put $y = 0$ in the equation of curve, *i.e.* $x^{2/3} = a^{2/3}$

   On cubing both side $x^2 = a^2$ or $x = \pm a$ therefore intersection with x-axis are the points $(a, 0)$ and $(-a, 0)$

   (*b*) Intersection with y-axis. Put $x = 0$ in the equation of the curve $y^{2/3} = a^{2/3}$

   On cubing both sides $y^2 = a^2$ or $y = \pm a$, therefore intersection with y-axis are $(0, a)$, $(0, -a)$

   (*c*) Intersection with the line $y = x$.

Putting $y = x$ in the equation of the curve *i.e.*,

$$x^{2/3} + x^{2/3} = a^{2/3}$$

or $$2x^{2/3} = a^{2/3}$$

On cubing both sides $$8x^2 = a^2$$

$$x = \pm \frac{a}{2\sqrt{2}}$$

therefore intersection with the line $y = x$ are $\left(\pm\frac{a}{2\sqrt{2}}, \pm\frac{a}{2\sqrt{2}}\right)$

(*d*) Intersection with the line $y = -x$ are

$\left(\pm\frac{a}{2\sqrt{2}}, \pm\frac{a}{2\sqrt{2}}\right)$

4. **Asymptotes :** The curve has no asymptotes.
5. **Region :** Solving the equation of the curve for y, *i.e.*,

$$y^{2/3} = a^{2/3} - x^{2/3}$$

$$y = (a^{2/3} - x^{2/3})^{3/2}$$

For y to be real,

$$(a^{2/3} - x^{2/3})^{3/2} > 0$$

or $$(a^{2/3} - x^{2/3}) > 0$$

or $$a^{2/3} > x^{2/3}$$

or $$x^{2/3} < a^{2/3}$$

On cubing both side therefore the curve lies between the line $x = -a$ and $x = a$.

$$x^2 < a^2$$

$\therefore$ $$-a < x < a$$

Similarly interchanging x and y we see that the curve lies between the lines

$$y = -a \text{ and } y = a$$

6. **Special Points :** The equation of curve is

$$x^{2/3} + y^{2/3} = a^{2/3} \quad \text{...(i)}$$

Differentiating (1) with respect to 'x' then we get

$$\frac{2}{3}x^{-1/3} + \frac{2}{3}y^{-1/3}\frac{dy}{dx} = 0$$

$\therefore$ $$\frac{dy}{dx} = -\frac{y^{-\frac{1}{3}}}{x^{-\frac{1}{3}}} \quad \text{...(ii)}$$

If $$\frac{dy}{dx} = 0, \text{ then } y^{1/3} = 0, \text{ or } y = 0$$

Putting $y = 0$ in the equation of the curve we get, $x = \pm a$, therefore tangent are || to x-axis at point $(\pm a, 0)$ *i.e.* x-axis is a tangent at the point $(\pm a, 0)$.

Similarly, $\frac{dy}{dx} \to \infty$ when $x = 0$

Putting $x = 0$ in the equation of the curve we get, $y = \pm a$, therefore tangent are || to y-axis at points $(0, \pm a)$.

Again from equation (2), $\frac{dy}{dx}$ is negative when x and y both are positive, hence y decreases as x increases in the first quadrant.

The shape of the curve is shown in the Figure 4.24.

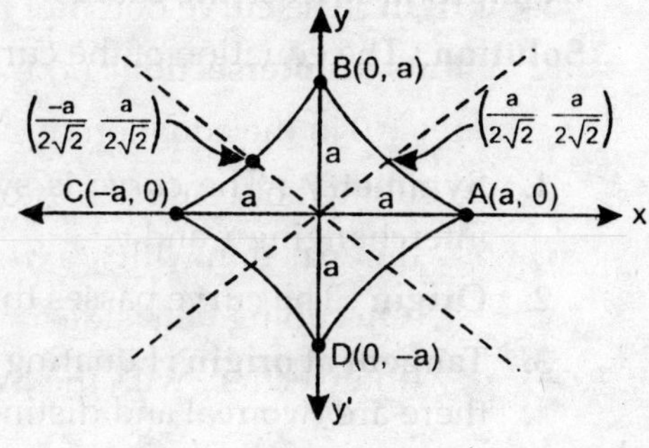

Fig. 4.23

Note : The point where the line $y = x$ meets the astroid $x^{2/3} + y^{2/3} = a^{2/3}$ *i.e.* the point $\left(\frac{a}{2\sqrt{2}}, \frac{a}{2\sqrt{2}}\right)$ is called the vertex of the astroid.

Example 4.73. Trace the curve $y^2(x^2 + y^2) + a^2(x^2 - y^2) = 0$

Solution : The equation of the curve is

$$y^2(x^2 + y^2) + a^2(x^2 - y^2) = 0 \qquad ...(i)$$

1. **Symmetry :** The curve is symmetrical about both axis.
2. **Origin :** The curve passes through origin.
3. **Tangent at origin :** Equating to zero the lowest degree term in the equation of the curve, *i.e.* $a^2(x^2 - y^2) = 0$ or $y = \pm x$ are the two tangents at the origin, therefore, origin is a node.
4. **Points of intersections :** Intersection with x-axis : Putting $y = 0$ in the equating of the curve, *i.e.* $x = 0$, therefore the curve intersects the x-axis at origin.

   Intersection with y-axis : Putting $x = 0$ in the equation of the curve *i.e.*,

   $$y^2(0 + y^2) + a^2(0 - y^2) = 0$$

   or $$y^2(y^2 - a^2) = 0$$

   $y = 0$, $y = \pm a$, therefore $(0, 0)$, $(0, a)$ and $(0, -a)$ the points on intersection with y-axis.
5. **Shifting the origin :** Replace x by $x + 0$ and y by $y + a$ in the equation of the curve, then equation of curve becomes

   $$(y + a)^2 \{x^2 + (y + a)^2\} + a^2 \{x^2 - (y + a)^2\} = 0$$

   or $$(y^2 + 2ay + a^2)(x^2 + y^2 + 2ay + a^2) + a^2 \{x^2 - y^2 - 2ay - a^2\} = 0$$

   Equating to zero the lowest degree term *i.e.* $2a^3y = 0$ or $y = 0$ is the tangent at the ne origin. Thus new x-axis is tangent at the new origin.

   We need not find the tangent at $(0, -a)$ as the curve is (symmetrical about x-axis).
6. **Asymptotes :** All four asymptotes are imaginary.
7. **Region :** Solving the equation of the curve for x *i.e.*

   $$x^2 = \frac{y^2(a^2 - y^2)}{(a^2 + y^2)}$$

   when $y = 0$, $x^2 = 0$

   when $y = a$, $x^2 = 0$

   when $0 < y < a$, x^2 is positive therefore, the curve exist in the region $0 < y < a$

   when $y > a$, x^2 is negative therefore the curve does not exist in the region $y > a$.

   Hence the shape of the curve is shown in Figure 4.25.

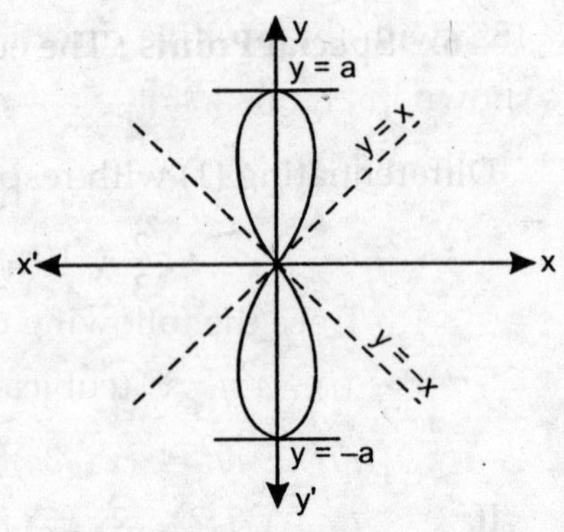

Fig. 4.24

Example 4.74. Trace the curve $x^3 + y^3 = 3axy$

Solution : The equation of the curve is

$$x^3 + y^3 = 3axy \qquad ...(i)$$

1. **Symmetry :** The curve is symmetrical about the line $y = x$ since its equation remains unchanged on interchanging x and y.
2. **Origin :** The curve passes through origin.
3. **Tangent at origin :** Equating to zero the lowest degree term in the curve, *i.e.*, $3axy = 0$ *i.e.*, $x = 0$, $y = 0$. Since there are two real and distinct tangent at the origin, therefore, the origin is a node on the curve.

4. **Points of intersection :** Intersection with x-axis :

   Put $y = 0$ then $x = 0$, therefore (0, 0) is point of intersection with the x-axis. Similarly origin is only point of intersection with y-axis.

5. **Asymptotes :** There are no || asymptotes to both axes.

   Find oblique asymptotes, putting $x = 1$, $y = m$ in third degree term them we get

$$\phi_3(m) = m^3 + 1$$

put
$$\phi_3(m) = 0 \text{ i.e.}$$
$$m^3 + 1 = 0$$
$$m = -1, \text{ is only real root}$$

Also
$$\phi_2(m) = -3am \text{ and } c \text{ is given by}$$
$$c\,\phi'(m) + \phi_2(m) = 0$$
$$c\,(3m^2) - (-3am) = 0$$

when $m = -1$, $c = -a$, therefore, the required asymptotes is given by putting the value of m and c in $y = mx + c$ in

$$y = mx + c$$
$$y = -x - a \text{ is the only real asymptotes of the curve.}$$

6. **Region :** If we replace x by $-x$ and y by $-y$ then R.H.S of equation is positive while L.H.S. become both negative but x and y bot be both negative therefore the curve does not exist in the third quadrant.

7. **Special Points :** The curve meets the line $y = x$ at the point $\left(\frac{3a}{2}, \frac{3a}{2}\right)$.

   Also form the equation of the curve

$$\frac{dy}{dx} = -\frac{3x^2 - 3ay}{3y^2 - 3ax}$$

$\therefore \frac{dy}{dx}$ at $\left(\frac{3a}{2}, \frac{3a}{2}\right)$ is -1, therefore the tangent at this point makes an angle 135° with the positive direction of the x-axis. Hence the shape of the curve is shown in Figure 4.25.

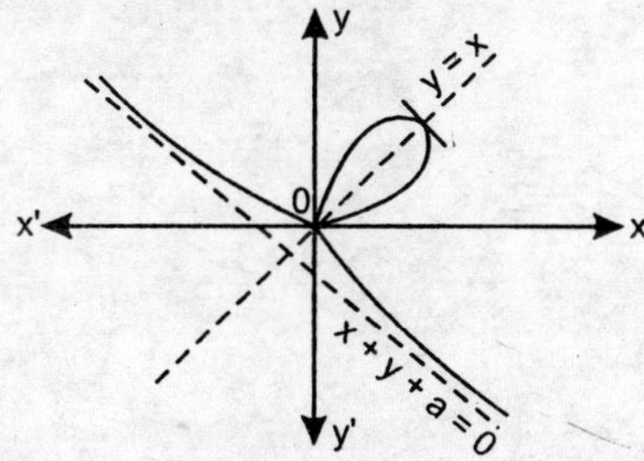

Fig. 4.25

EXERCISE 4.11

1. Trace the following curves :

   (i) $a^2y = x^3$ (cubical parabola) (ii) $y^2 = 4ax$

   (iii) $9ay^2 = x(x - 3a)^2$ (iv) $y^2(a + x) = x^2(a - x)$

   (v) $y^2(x^2 + y^2) + a^2(x^2 - y^2) = 0$

2. Trace the following curves :

   (i) $xy^2 = a^2(a - x)$ (ii) $y(x^2 + 4a^2) = 8a^3$

   (iii) $a^2y^2 = x^2(a^2 - x^2)$ (iv) $y^2(1 - x^2) = x^2(1 + x^2)$

Answers:

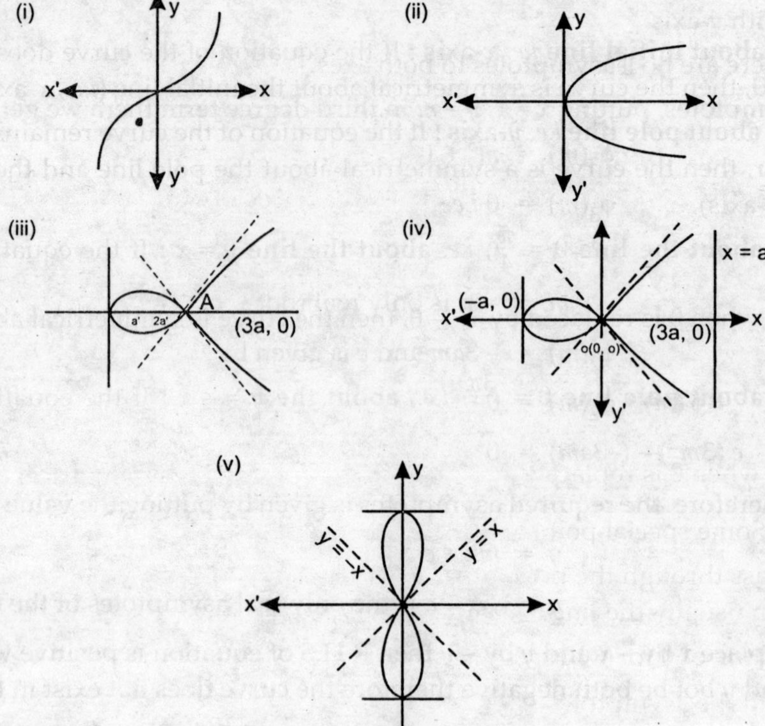

Fig. 4.26

2.

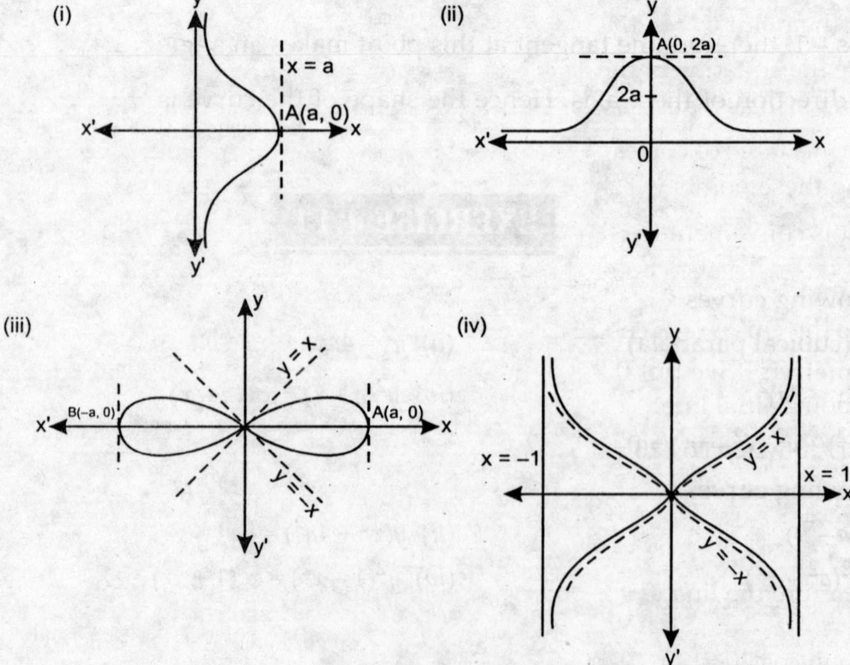

Fig. 4.27

4.31 PROCEDURE OF CURVE TRACING FOR POLAR CURVES

1. **Symmetry :**

(*a*) **Symmetry about initial line** *i.e.* **x-axis :** If the equation of the curve does not change by changing the sign of θ, then the curve is symmetrical about the initial line (*i.e.*, x-axis).

(*b*) **Symmetry about pole line** *i.e.* **y-axis :** If the equation of the curve remains unchanged by changing the sign of r, then the curve is a symmetrical about the pole line and the pole is the centre of the curve (*i.e.* y-axis).

(*c*) **Symmetry about the line $\theta = \frac{\pi}{4}$,** *i.e.* **about the line $y = x$:** If the equation of the curve remains unchanged when θ is replaced by $\frac{\pi}{2} - \theta$, then the curve is symmetrical about the line $\theta = \frac{\pi}{4}$.

(*d*) **Symmetry about pole line $\theta = \frac{3\pi}{4}$** *i.e.,* **about the $y = -x$:** If the equation of the curve remains unchanged when θ is replaced by $\frac{3\pi}{4} - \theta$, then the curve is symmetrical about the line $\theta = \frac{3\pi}{4}$.

2. **Pole or origin :** (Some special points on the curve)

The curve will pass through the pole if for some value of θ, the value of r comes out to be zero, *i.e.* If $r = 0$ when $\theta = \alpha$ then usually the line $\theta = \alpha$ will be a tangent to the curve at the pole.

Hence find the value of θ for which r is maxima or minima or $r \to \infty$.

3. **Asymptotes :** Find the asymptotes if the curve possesses as infinite branch. If $r \to \infty$ as $\theta \to \alpha$ one should not assume that $\theta = \alpha$ is an asymptote. The asymptotes may be $\parallel$ to the line $\theta = \alpha$ or even may not exist at all. Find the asymptotes by the method which is given in the chapter on asymptotes.

4. **Find $\tan \phi$** *i.e.* $r \frac{d\theta}{dr}$ which will indicate the direction of the tangent at any point. If for $\theta = \alpha$, ϕ comes out to be zero, then the line $\theta = \alpha$ will be a tangent to the curve at the point $\theta = \alpha$. If for $\theta = \alpha$, ϕ comes out to be $\frac{\pi}{2}$, then at the point $\theta = \alpha$, the tangent will be $\perp$ to the radius vector $\theta = \alpha$.

5. **Region :** Solve the equation of the curve for r by giving the values to θ, and find how varies as θ increases from 0 to ∞ and also as θ decreases from 0 to $-\infty$.

We should form a table of corresponding value of r and θ which would give us a number of points of the curve. Plotting these points we shall find the shape of the curve.

Note : Sometimes it is convenient to change the equation from the polar to Cartesian form by the relation.

$$x = r \cos, \theta, y = r \sin \theta$$

***Example* 4.75.** Trace the curve $r = a \cos 2\theta$.

Solution : (*i*) Symmetry : If we put $\theta = -\theta$ then there is no change in the given equation of the curve so the curve is symmetrical about initial line.

(*ii*) Poles : Here, $r = 0$ when $\cos 2\theta = 0$

or $$2\theta = \pm \frac{\pi}{2}$$

or $\theta = \pm \frac{\pi}{4}$, therefore the line $\theta = \pm \frac{\pi}{4}$ are tangent to the curve at the pole

Again, $r = a$ when $\cos 2\theta = 1$ or $2\theta = 0$, $\theta = 0$

(*iii*) Asymptotes : Let $\frac{1}{r} = \frac{1}{a \cos 2\theta} = f(\theta)$

Put $$f(\theta) = 0$$

Hence there is no asymptotes to the curve

(iv) Find tan ϕ : *i.e.* $r\,\dfrac{d\theta}{dr}$

Here
$$\tan\phi = r\frac{d\theta}{dr} = \frac{a\cos 2\theta}{-2a\sin 2\theta} = -\frac{1}{2}\cot 2\theta$$

$$\Rightarrow \quad \tan\phi = -\frac{1}{2}\cot 2\theta$$

Here $\phi = 90°$ when $2\theta = 0$ *i.e.,* $\theta = 0$, therefore at $\theta = 0$, the tangent is $\perp$ to the radius vector $\theta = 0$.

(v) Region : The following table gives the corresponding values of θ and r.

θ	0	$\frac{\pi}{6}$	$\frac{\pi}{4}$	$\frac{\pi}{3}$	$\frac{\pi}{2}$	$\frac{2\pi}{3}$	$\frac{3\pi}{4}$	$\frac{5\pi}{4}$	π
r	a	$\frac{a}{2}$	0	$-\frac{a}{2}$	$-a$	$-\frac{a}{2}$	0	$-\frac{a}{2}$	$-a$

The variation of θ from π to 2π need not be considered because of symmetry about the initial line

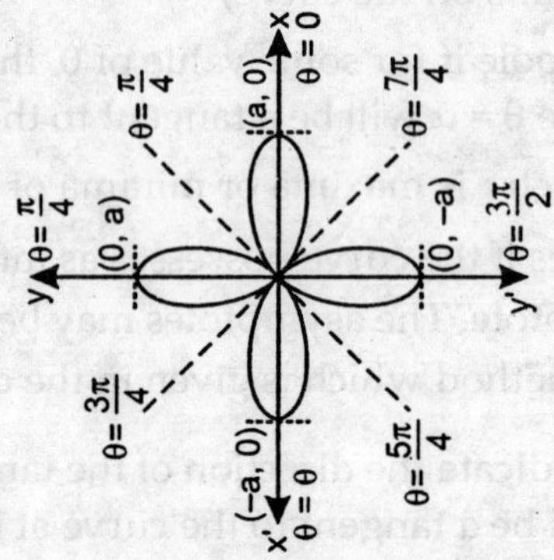

Fig. 4.28

Important Note : The above example is a particular case of the curve of the type $r = a\cos n\theta$ which has n loops when n is odd, and $2n$ loops when n is even.

Example 4.76. Trace the curve $r = a\sin 3\theta$. (Three leaved rose)

Solution : The equation of the curve is $r = a\sin 3\theta$

(1) Symmetry : The curve remains unchanged when θ is replaced by $(\pi - \theta)$, so the curve is symmetrical about the line $\theta = \pi/2$.

(2) Pole or origin : Here, $r = 0$, when $\sin 3\theta = 0$, when gives

$$3\theta = 0, \pi, 2\pi, 3\pi \ldots$$

or
$$\theta = 0, \frac{\pi}{3}, \frac{2\pi}{3}, \pi \ldots$$

Therefore the curve passes through the pole and tangents at the pole are $\theta = 0$, $\theta = \dfrac{\pi}{3}$ and $\theta = \dfrac{2\pi}{3}$ the other values of θ give the same tangents.

$$r = a, \text{ when } \theta = \frac{\pi}{6} \text{ i.e. the curve cute line at } \left(a, \frac{\pi}{6}\right)$$

(3) Asymptotes : There are no asymptotes, because for any finite values of θ, r does not tend to infinity.

(4) Region and Special Points:

(*i*) Here $r = a\sin 3\theta$

$\therefore$ $|\sin 3\theta| \le 1$ therefore r is never greater than a therefore the curve lies within the circle $r = a$.

(*ii*) Points :

θ	0	$\frac{\pi}{6}$	$\frac{\pi}{3}$	$\frac{\pi}{2}$	$\frac{2\pi}{3}$	$\frac{5\pi}{6}$	π
r	0	a	0	$-a$	0	a	0

Here it is concluded that when θ increases from 0 to $\frac{\pi}{6}$, r is positive and increases from 0 to a then θ increases from $\frac{\pi}{6}$ to $\frac{\pi}{3}$, r is positive and decreases from a to 0 therefore one loop is complete between the line θ = 0 and θ = $\frac{\pi}{3}$

Again θ increases from $\frac{\pi}{3}$ to $\frac{\pi}{2}$, r is negative and increase from 0 to – a. As θ increase from $\frac{\pi}{2}$ to $\frac{2\pi}{3}$, r is negative and decreases from – a to 0, thus second loop is complete between the line $\theta = \frac{\pi}{3}$ and $\theta = \frac{2\pi}{3}$ in the opposite direction. As θ increases from $\frac{2\pi}{3}$ to $\frac{5\pi}{6}$,r remains positive and increase from 0 to a. As θ increases from $\frac{5\pi}{6}$ to π, r remains positive and decrease from a to 0. Thus third loop is complete between the line $\theta = \frac{2\pi}{3}$ and θ = π.

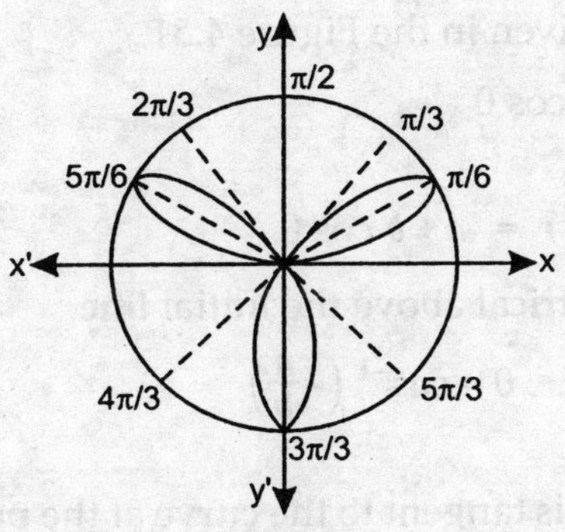

Fig. 4.29

When θ varies form π to 2π, the same loops are repeated as r is periodic the shape of the curve is given above in the Figure 4.30.

Note : The curve $r = a \sin n\theta$ or $r = a \cos n\theta$ consists with n or $2n$ loop according as n is odd or even.

Example 4.77. Trace the curve $r = a(1 + \cos\theta)$ (Cardiod)

Solution : The equation of the curve is $r = a(1 + \cos\theta)$.

(1) Symmetry : The equation of the curve remains unchanged when θ is replaced (–θ) therefore the curve is symmetrical about initial line.

(2) Pole or origin : Here $r = 0$ when θ = π, Pole lies on the curve and θ = π is the tangent at pole. $r = 2a$, when θ = 0 *i.e.* the curve cuts the initial line θ = 0 at (2*a*, 0) and he lines $\theta = \pm\frac{\pi}{2}$ ατ $\left(a, \frac{\pi}{2}\right), \left(a, \frac{\pi}{2}\right)$.

(3) Asymptotes : There are no asymptotes, because for any finite value of θ, r does not tend to infinity.

(4) Find tan ϕ : *i.e.* $r\frac{d\theta}{dr}$

$$\tan\phi = -\cot\frac{\theta}{2}$$

$$\tan\phi = \tan\left(\frac{\pi}{2}, \frac{\theta}{2}\right)$$

Which implies that $\phi = \frac{\pi}{2} + \frac{\theta}{2}$

$\therefore \phi = \frac{\pi}{2}$ when $\theta = 0$ and $r = 2a$ *i.e.,* at $(2a, 0)$ the tangent is perpendicular to initial line.

(5) Region : The corresponding values of θ and r are given below

θ	0	$\frac{\pi}{4}$	$\frac{\pi}{2}$	π
r	$2a$	$a\left(1+\frac{1}{\sqrt{2}}\right)$	a	0

The variation of θ from π to 2π need not be considered because of symmetry about initial line.

Note : (*i*) r is greater than $2a$, so no portion of the curve lies to the right of the tangent at $(2a, 0)$

(*ii*) Here $|\cos\theta| \le 1$, so $r \le 2a$ *i.e.* curve lies with in circle $r = 2a$.

(*iii*) When θ increases from 0 to π, r remains positive and decreases from $2a$ to 0.

When θ increases from 0 to 2π, r remains positive and increase from 0 to 2a.

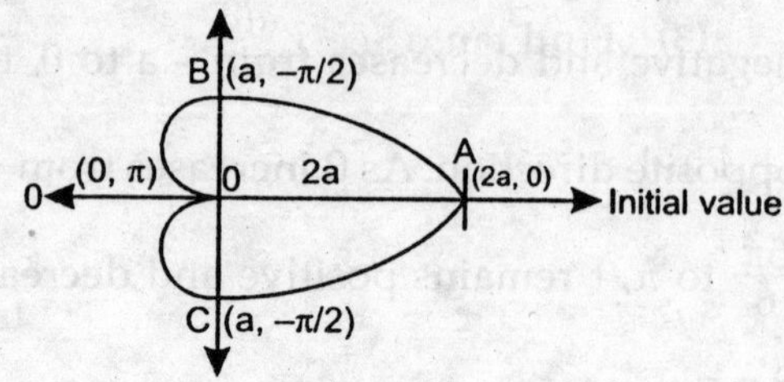

Fig. 4.30

The shape of the curve is given in the Figure 4.31.

Example **4.75.** Trace the curve $r = a + b\cos\theta$ (Limacon) when $a < b$

Solution : The equation of the curve is

$$r = a + b\cos\theta$$

(1) Symmetry : The curve is symmetrical above the initial line.

(2) Pole : $r = 0$, when $a + b\cos\theta = 0$, *i.e.* $\theta = \cos^{-1}\left(-\frac{a}{b}\right)$

Since $\frac{a}{b} < 1$, therefore $\cos^{-1}\left(-\frac{a}{b}\right)$ is tangent to the curve at the pole r is maximum when $\cos\theta = 1$ *i.e.,* $\theta = 0$, then $r = a + b$.

Also r is minimum when $\cos\theta = -1$ *i.e.,* $\theta = \pi$, then $r = a - b$ which is negative.

(3) Find tan ϕ : *i.e.* $r\frac{d\theta}{dr}$

Here $$\frac{d\theta}{dr} = -b\sin\theta$$

$\therefore$ $$\tan\phi = r\frac{d\theta}{dr} = -\frac{r}{b\sin\theta}$$

$$\tan\phi = -\frac{(a + b\cos\theta)}{b\sin\theta}$$

when $\theta = 0$, $\phi = 90°$ and when $\theta = \pi$, again $\phi = 90°$, therefore at the points $\theta = 0$ and $\theta = \pi$, the tangent is perpendicular to the radius vector.

(4) Region : The corresponding values of r and θ is given by the following table :

θ	0	$\frac{\pi}{2}$	$\cos^{-1}\left(-\frac{a}{b}\right)$	cos–1 $\left(-\frac{a}{b}\right) < \theta < \pi$	π
r	$a + b$	a	c	r is negative	$a - b$

Note : The variation of θ from π to 2π need not be considered because the curve is symmetrical about the initial line. Hence the shape of the curve is given in the Figure 4.31.

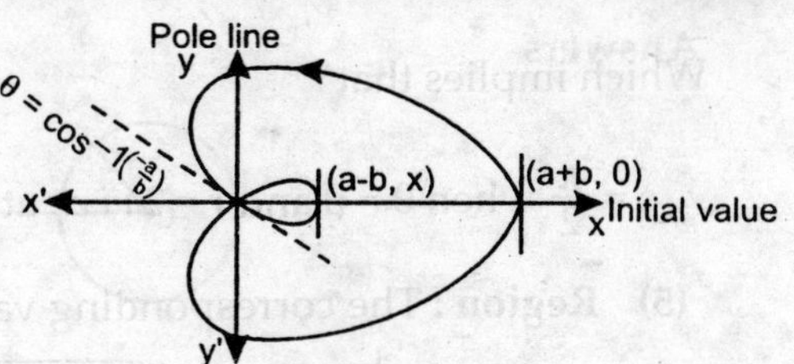

Fig. 4.31

Example **4.78.** Trace the curve $0 = a\,e^{m\theta}$ (Equiangular spiral)

Solution : The equation of the curve is

$$r = a\,e^{m\theta}$$

(1) Symmetry : The curve is not symmetrical about the initial lines as well as the line $\perp$ to the initial line.

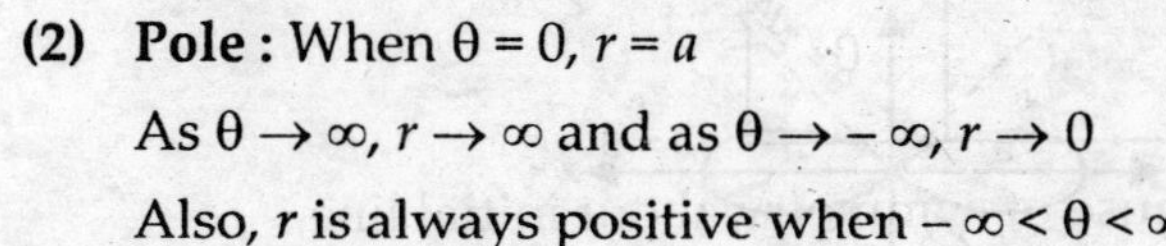

(2) Pole : When $\theta = 0$, $r = a$

As $\theta \to \infty$, $r \to \infty$ and as $\theta \to -\infty$, $r \to 0$

Also, r is always positive when $-\infty < \theta < \infty$

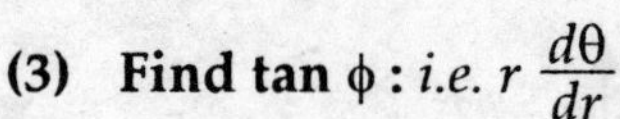

(3) Find tan ϕ : *i.e.* $r\dfrac{d\theta}{dr}$

Find $$\frac{dr}{d\theta} = a\,m\,e^{m\theta}$$

$\therefore$ $$\tan\phi = r\frac{d\theta}{dr}$$

$$= \frac{r}{a\,m\,e^{m\theta}} = \frac{a\,e^{m\theta}}{a\,m\,e^{m\theta}} = \frac{1}{m}$$

$$\tan\phi = \frac{1}{m}$$

$\therefore$ $$\phi = \tan^{-1}\left(\frac{1}{m}\right) = \text{constant } i.e. \text{ radius vector is constant}$$

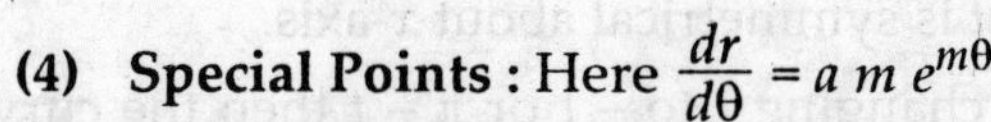

(4) Special Points : Here $\dfrac{dr}{d\theta} = a\,m\,e^{m\theta}$

When $-\infty < \theta < \infty$, r is positive throughout, therefore r increases continuously as θ increases from $-\infty$ to ∞.

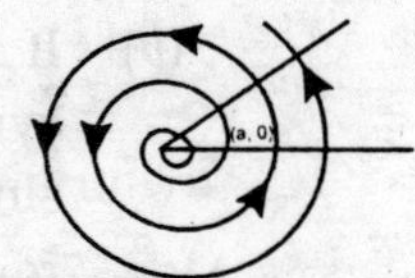

Fig. 4.32

Hence it is concluded that the angle between the radius vector and tangent always remains constant. The shape of curve is given in Figure 4.32.

EXERCISE 4.12

Trace the following curves :

1. $r = a$
2. $r = 2a\cos\theta$
3. $r = a(1 + \sin\theta)$
4. $r = a(1 - \cos\theta)$
5. $r^2 = a^2\cos 2\theta$
6. $r^2 = a^2\sin 2\theta$
7. $r = a + b\cos\theta$, when $a > b$

Answers

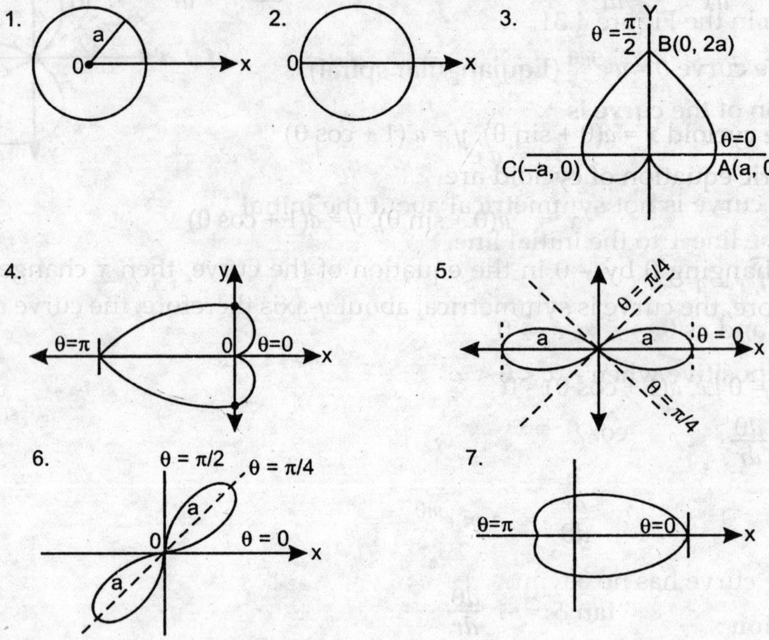

Fig. 4.33

Parametric curve : If the equation of a curve if given in the following form *i.e.*, $x = f(t)$, $y = \phi(t)$, then it is called parametric form of the curve. In some cases the curve can be easily traced by eliminating the parameter. But if it is not convenient to eliminate t, then the following procedure is followed.

4.32 PROCEDURE OF CURVE TRACING FOR PARAMETRIC CURVES

1. **Symmetry :** (*a*) If $x = f(t)$ is even and $y = \phi(t)$ is odd function of t, changing t to $-t$ or t to $\pi - t$ then the curve is symmetrical about x-axis *i.e.* the parabola $x = at^2$, $2at$ is symmetrical about x-axis.

   (*b*) If $x = f(t)$ is odd and $y = \phi(t)$ is even function of t on changing t to $-t$ or $\pi - t$ then the curve is symmetrical about y-axis, *i.e*, the cycloid $x = a(t + \sin t)$, $y = a(1 - \cos t)$ and the ellipse $x = a \cos t$, $y = b \sin t$ is symmetrical about y-axis.

   (c) If $x = f(t)$ and $y = \phi(t)$ both are odd function of t on changing t to $-t$ or $t = \pi - t$, then the curve is symmetrical in opposite quadrants *i.e.* the rectangular hyperbolla $x = ct$, $y = \frac{c}{t}$ is symmetrical in opposite quadrants.

2. **Origin :** Determine t for which $x = 0$ and $y = 0$, then the curve passes through the origin or putting both x and y equal to zero and find the value of t if there is any common value of t, then the curve passes through the origin.

3. **Asymptotes :** Find asymptotes || to x-axis and y-axis.

4. **Point of Intersection :**

   (*i*) The point of intersection of the curve with the x-axis are given by the roots of the equation $\phi(t) = 0$.

   (*ii*) The point of intersection of the curve with the y-axis are given by the roots of the equation $f(t) = 0$.

5. **Region :** If possible, find the greatest and least value of x and y therefore, the line parallel to the axes between which the curve lies or does not lie.

   For example : If the curve is $x = a \cos \theta$, $y = b \sin \theta$; then the greatest value of x is a and the least value is – a therefore the curve lies entirely between the lines $x = \pm a$

   ($\because$ greatest and least value of $\cos \theta$ are 1 and – 1)

   Similarly, the curve lies entirely between the lines $y = \pm b$.

6. **Special Point :** Find $\frac{dy}{dx}$ and $\frac{dy}{dt}$ and then consider the signs of $\frac{dx}{dt}$ and $\frac{dy}{dt}$ is suitable interval of t.

Find $\frac{dy}{dx} = \frac{dy/dt}{dx/dt}$

Example 4.79. Trace the cycloid $x = a(\theta + \sin\theta)$, $y = a(1 + \cos\theta)$

Solution : The parametric equation of cycloid are

$$x = a(\theta + \sin\theta),\ y = a(1 + \cos\theta)$$

(1) **Symmetry :** On changing θ by $-\theta$ in the equation of the curve, then x changes by $-x$ and y remains unchanged therefore, the curve is symmetrical about y-axis therefore, the curve does not passes through the origin.

(2) **Origin :** putting $y = 0$ *i.e.* $a(1 + \cos\theta) = 0$

$\therefore$ $$\cos\theta = -1$$

or $$\theta = \pi$$

For $$\theta = \pi,\ x = a\pi \neq 0$$

(3) **Asymptotes :** The curve has no asymptotes.

(4) **Point of intersection :**

Intersection with x-axis : Putting $y = 0$ in the equation of the curve then $\theta = \pi$ which gives $x = a\pi$, therefore intersection with the x-axis is $(a\pi, 0)$.

Intersection with y-axis : Putting $x = 0$ in the equation of the curve *i.e.*, $\theta + \sin\theta = 0$ which is satisfied by only $\theta = 0$

For $\theta = 0$, $y = 2a$, therefore, intersection with the y-axis is $(0, 2a)$

(5) **Region :** We know that

$$-1 \le \cos\theta \le 1$$

$\therefore$ $$1 - 1 \le 1 + \cos\theta \le 1 + 1$$

or $$0 \le (1 + \cos\theta) \le 2$$

$$0 \le a(1 + \cos\theta) \le 2a$$

or $$0 \le y \le 2a$$

therefore the curve lies entirely between the line $y = 0$ and $y = 2a$.

(6) **Special Points :** From the equation of the curve

$$\frac{dx}{d\theta} = a(1 + \cos\theta) \text{ and } \frac{dy}{dx} = -a\sin\theta$$

$\therefore$ $$\frac{dy}{dx} = \frac{dy/d\theta}{dx/d\theta} = -\frac{a\sin\theta}{a(1+\cos\theta)}$$

$$\frac{dy}{dx} = -\tan\frac{\theta}{2}$$

Corresponding value of x, y and $\frac{dy}{dx}$ for different values of θ are

$\theta =$	$-\pi$	$-\frac{\pi}{2}$	0	$\frac{\pi}{2}$	π
$x =$	$-a\pi$	$-a\left(\frac{\pi}{2}+1\right)$	0	$-a\left(\frac{\pi}{2}+1\right)$	$a\pi$
$y =$	0	a	$2a$	a	0
$\frac{dy}{dx}$	∞	1	0	-1	$-\infty$

Therefore, points on the curve are $(-a\pi, 0)$, $\left[-a\left(\frac{\pi}{2}+1\right), a\right]$, $[0, 2a]$, $\left[a\left(\frac{\pi}{2}+1\right), a\right]$, $(a\pi, 0)$ and the tangent at the points $(-a\pi, 0)$ and $(a\pi, 0)$ are parallel to y-axis and at a $(0, 2a)$ the tangent || to the x-axis

$$\left(\because \frac{dy}{dx} = 0\right)\left(\because \frac{dy}{dx} = \infty \text{ or } -\infty\right)$$

Again $$\frac{dy}{dx} = -\tan\frac{\theta}{2}$$

Again differentiating w.r.t, 'x', we get

$$\frac{d^2y}{dx^2} = -\left(\sec^2\frac{\theta}{2}\right)\frac{1}{2}\frac{d\theta}{dx}$$

$$\frac{d^2y}{dx^2} = -\frac{1}{2}\left(\sec^2\frac{\theta}{2}\right)\frac{1}{2a}\sec^2\frac{\theta}{2}\left[\because \frac{d\theta}{dx} = \frac{1}{a(1+\cos\theta)} = \frac{1}{2a\cos^2\frac{\theta}{2}} = \frac{1}{2a}\sec^2\frac{\theta}{2}\right]$$

$$\frac{d^2y}{dx^2} = -\frac{1}{4a}\sec^4\frac{\theta}{2}$$

therefore, $\frac{d^2y}{dx^2}$ is negative for all values of θ.

For value of $\theta > \pi$ or $< -\pi$, the same types of branches of the curve will be obtained.

The curve consists of congruent breaches on both sides of y-axis which extend of infinity.

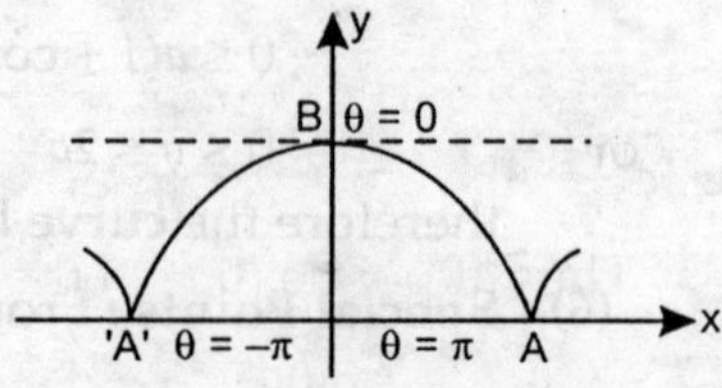

Fig. 4.34

Hence the shape of the curve is shown in the Figure 4.34.

Note : *B* is called vertex of the cycloid and line *A'A i.e.* the line joining the two end points is called base of the cycloid and a is called the radius of the generating circle.

***Example* 4.80.** Trace the curve $x = a\left[\cos t + \frac{1}{2}\log\tan^2\frac{t}{2}\right]$, $y = a\sin t$ (Tractrix)

Solution : The equation of curve are

$$x = a\cos t + \frac{a}{2}\log\tan^2\frac{t}{2},\ y = a\sin t \qquad ...(i)$$

(1) Symmetry :

(*a*) On change t to $-t$ in the equation of the curve (*i*), x remains unchanged and y changes to $-y$, therefore, the curve is symmetrical about x-axis.

(*b*) On changing t to $\pi - t$ in equation of the curve (*i*), y is unchanged and x changes to $-x$ therefore, for curve is symmetrical about y-axis.

(2) Origin : Putting $y = 0$ *i.e.* $a \sin t = 0$ or $t = 0$ and then $x = a + \frac{a}{2} \log 0 = -\infty$, therefore, the curve does not pass through the origin.

(3) Asymptotes : When $t = 0$, $x \to -\infty$ and $y = 0$ thus $y = 0$ *i.e.*, x-axis is an asymptotes to the curve.

(4) Points of Intersection : The curve does not meet x-axis. It meets y-axis (putting $x = 0$) where $x = a \cos t + \frac{a}{2} \log \tan^2 \frac{t}{2} = 0$ which is satisfied by $t = \pm \frac{\pi}{2}$ and then $y = a \sin\left(\pm \frac{p}{2}\right) = \pm a$

Thus the curve meets y-axis at $(0, \pm a)$

(5) Region : We know that

$$-1 \le \sin t \le 1$$

$$\therefore \quad -a \le a \sin t \le a$$

or $$-a \le y \le a$$

Thus the curve lies entirely between he lines $y = \pm a$.

(6) Special Points : From the equation of curves (i)

$$\frac{dx}{dt} = -a \sin t + \frac{a}{2} \frac{1}{\tan^2 t/2} 2 \tan \frac{t}{2} \sec^2 \frac{t}{2} \frac{t}{2}$$

$$= a \left\{ -\sin t + \frac{1}{2 \sin \frac{t}{2} \cos \frac{t}{2}} \right\}$$

$$= a \left\{ -\sin t + \frac{1}{\sin t} \right\}$$

$$= a \left\{ \frac{\cos^2 t}{\sin t} \right\}$$

and $$\frac{dy}{dt} = a \cos t$$

$$\therefore \quad \frac{dy}{dx} = \frac{dy/dt}{dx/dt} = \frac{a \cos t \sin t}{a \cos^2 t} = \tan t \quad \ldots(ii)$$

When $t = \pm \frac{\pi}{2} \frac{\pi}{2}$, $\frac{dy}{dx} = \pm \infty$ and then $x = 0$, $y = \pm a$.

Thus at points $(0, \pm a)$ the tangents is || to y-axis *i.e.* y axis itself is a tangent at the point $(0, \pm a)$

When $t = 0$, $\frac{dy}{dx} = 0$ and then $x \to \infty$, $y = 0$ which implies that $y = 0$ x axis is an asymptote.

Again $$\frac{dy}{dx} = \tan t$$

Differentiating w.r.t 'x' on both sides

$$\frac{d^2y}{dx^2} = \sec^3 t \frac{dt}{dx}$$

$$= \frac{\sin t}{a \cos^4 t}$$

therefore, $\frac{d^2y}{dx^2}$ is positive for value of t in $(0, \pi)$

The shape of curve in Figure 4.35.

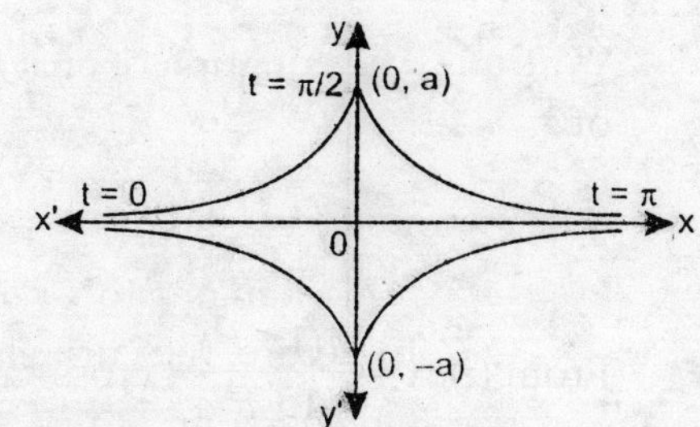

Fig. 4.35

Example **4.81.** Trace the curve $x^{2/3} + y^{2/3} = a^{2/3}$ (Asteroids)

Solution : The parametric equation of the curve are

$$x = a\cos^3 t,\ y = a\sin^3 t$$

Here
$$\frac{dx}{dt} = -3a\cos^2 t\sin t,\ \frac{dy}{dt} = 3a\sin^2 t\cos t$$

$$\therefore \quad \frac{dx}{dt} = \frac{dy/dt}{dx/dt} = \frac{3a\sin^2 t\cos t}{sa\cos^2 t\sin t} = -\tan t$$

Also the equation of the curve can be written as

$$\left(\frac{x^2}{a^2}\right)^{\frac{1}{3}} + \left(\frac{y^2}{a^2}\right)^{\frac{1}{3}} = 1 \quad ...(i)$$

we observe the following facts about the curve

(*i*) The curve is symmetrical about both axis. It is also symmetrical about the $y = \pm x$.

(*ii*) The curve does not passes thought origin.

(*iii*) Point of intersection. Intersection with the x-axis, putting $y = 0$ in the equation of the curve *i.e.*, $\left(\frac{x^2}{a^2}\right)^{\frac{1}{3}} = 1$ or $x^2 = a^2$ or $x = \pm a$, therefore the curve cuts the x-axis at $(a, 0)$ and $(-a, 0)$

Similarly the curve crosses the y-axis at $(0, a)$ and $(0, -a)$.

(*iv*) At the point $(a, 0)$ we have $x = a$, therefore, $\cos^2 t = 1$ or $t = 0$.

When $t = 0$, $\frac{dy}{dx} = 0$, hence at the point $(a, 0)$ the x-axis is tangent to the curve.

Again at the point $(0, a)$ we have $y = a$ therefore $\sin^3 t = 1$ or $t = \frac{\pi}{2}$, hence at the point $(0, a)$ the y-axis is tangent to the curve.

(*v*) value of $\sin t$ and $\cos t$ can not numerically exceed 1. Therefore, in this curve the values of x and y are bounded by the lines $x = a$, $x = -a$, $y = a$ and $y = -a$, Hence the shape of the curve is as shown in the Figure 4.36.

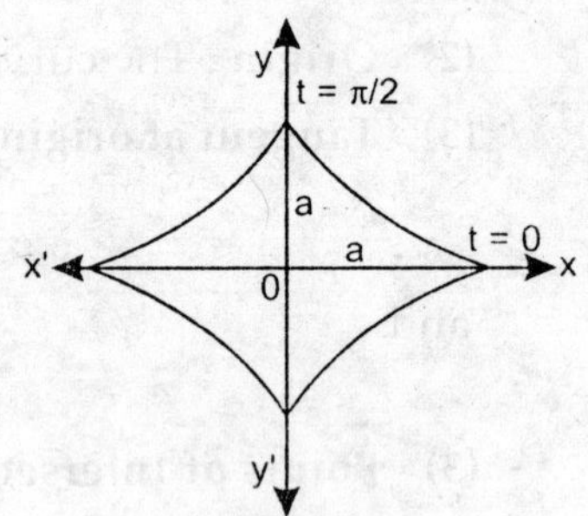

Fig. 4.36

Example **4.80.** Trace the curve $x = \frac{a(t+t^3)}{1+t^4}$, $y = \frac{1(t-t^3)}{1+t^4}$

Solution : The equation of the curve are

$$x = \frac{a(t+t^3)}{1+t^4},\ y = \frac{a(t-t^3)}{1+t^4} \quad ...(i)$$

Eliminating 't' between x and y and form a Cartesian equation of the curve

Now
$$\frac{x}{y} = \frac{1+t^2}{1-t^2}$$

or
$$y + t^2 y = x - t^2 x$$

or
$$t^2(x = y) = x - y$$

$$\therefore \quad t^2 = \frac{x-y}{x+y} \quad ...(ii)$$

From (*i*), $x = \frac{a(1+t^3)}{1+t^4}$, cross multiplying and squaring

$x^2(1+t^4)^2 = a^2 t^2 (1+t^2)^2$

Putting the value of t^2 from (*ii*),

$$x^2 \left\{1 + \frac{(x-y)^2}{(x+y)^2}\right\}^2 = \frac{a^2(x-y)}{x+y}\left\{1 + \frac{x-y}{x+y}\right\}^2$$

or $$x^2 \frac{\{(x+y)^2 + (x-y)^2\}^2}{(x+y)^4} = \frac{a^2(x-y)}{(x\ y)}\left\{\frac{2x}{x+y}\right\}^2$$

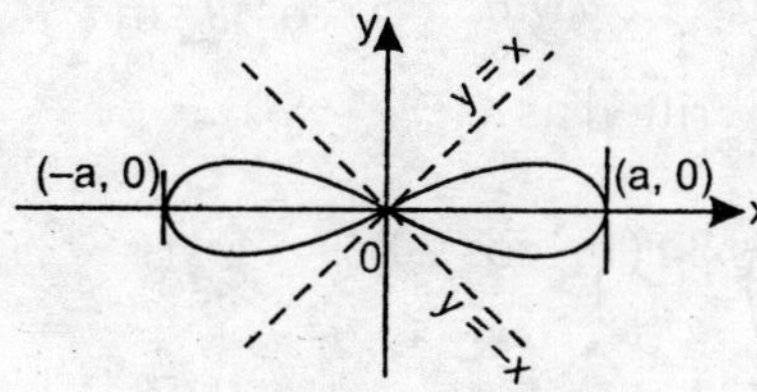

Fig. 4.37

or $$\frac{4x^2(x^2+y^2)^2}{(x+y)^4} = \frac{4a^2x^2(x-y)}{(x+y)^3}.$$

or $$\frac{(x^2+y^2)^2}{x+y} = a^2(x-y)$$

or $$(x^2+y^2)^2 = a^2(x^2-y^2) \qquad ...(A)$$

(1) Symmetry : The curve (*A*) is symmetrical about both axis *i.e.* *a* and *y* axis.

(2) Origin : The curve (*A*) passes through the origin.

(3) Tangent at origin : Equating to zero the lowest degree term in the equation of the curve *i.e.*,

$$a^2(x^2-y^2) = 0$$

$$\therefore \quad x^2 - y^2 = 0$$

$y = \pm x$, therefore $y = \pm x$ is the tangents at origin.

(3) Points of Intersection :

Intersection with x-axis ; Putting $y = 0$ in the equation of then curve (*A*), the $x^4 = a^2 x^2$

or $$x^2(x^2 - a^2) = 0 \text{ i.e.,}$$

Either $$x^2 = 0 \text{ i.e, } x = 0$$

or $$x^2 = a^2 \text{ i.e. } x = \pm a$$

therefore intersection the curve (*A*) with x-axis at $(-a, 0)$ $(0, 0)$ and $(a, 0)$

Intersection with y-axis : Putting $x = 0$ in the equation of the curve (*A*), then

$$y^4 = -a^2y^2$$

or $$y^2(y^2 + a^2) = 0$$

$\therefore$ Either $$y^2 = 0 \text{ i.e. } y = 0 \text{ or } y = \pm i\,a$$

Hence only point of intersection of curve (*A*) with y-axis is (0, 0)

Thus the shape of the curve is given in Figure 4.37.

EXERCISE 4.13

Trace the following curves:

1. $x = a(\theta + \sin\theta),\ y = a(1 - \cos\theta)$
2. $x = a(\theta - \sin\theta),\ y = a(1 - \cos\theta)$
3. $x = a\{\cos\theta - \log(1 + \cos\theta)\},\ y = a\sin\theta$
4. $x = \dfrac{a(1-t^2)}{1+t^2},\ y = \dfrac{at(1-t^2)}{1+t^2}$
5. $x = a\sin 2\theta\,(1 + \cos 2\theta),\ y = a\cos 2\theta\,(1 - \cos 2\theta)$

Answers

1.

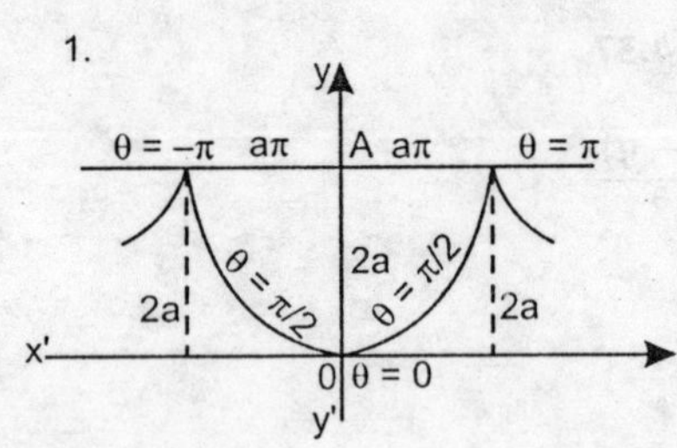

2.

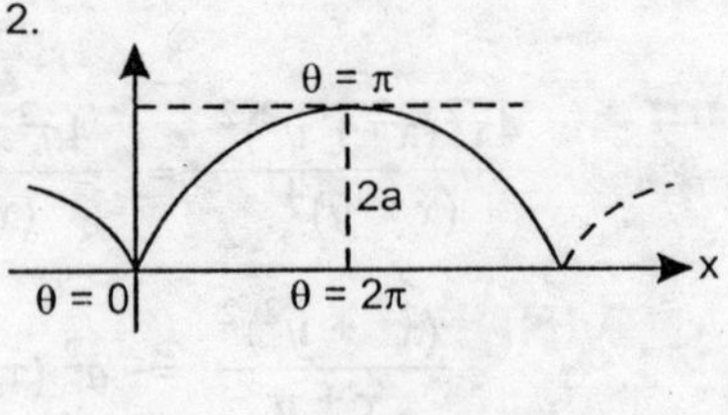

3.

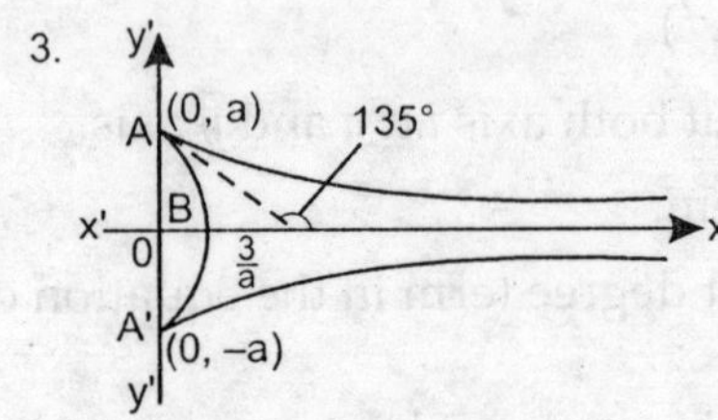

4.

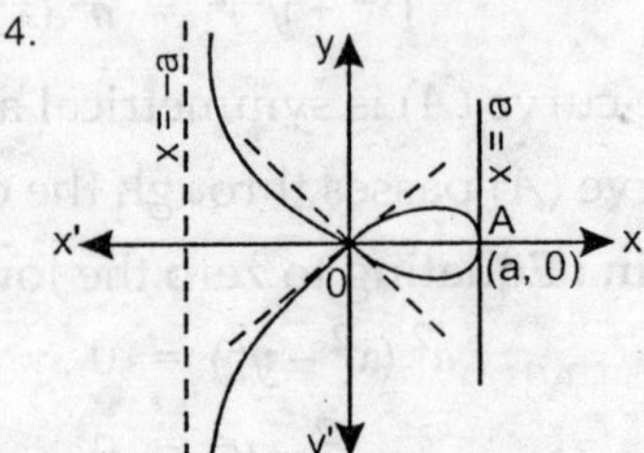

5.

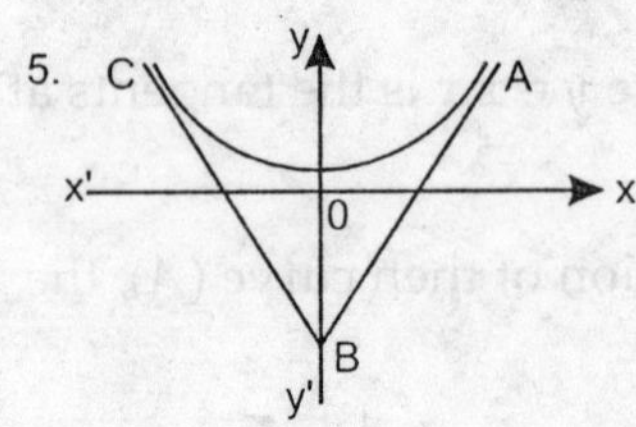

Fig. 4.38

Partial Differentiation

This chapter deals with partial derivatives, homogeneous functions, Euler's theorem for homogeneous functions, composite function total derivatives, jacobians and it's properties.

5.1 DEPENDENT AND INDEPENDENT VARIABLES

The area of a rectangle depends upon its length and breadth and according it may be stated that area is the function of two variables, *i.e.*, length and breadth.

If z be the area of rectangle and x and y be the length and breadth respectively, then z is called a function of two variable x and y. Symbolically, it is written as

$$z = f(x, y)$$

The variable x and y are called **independent variable,** while z is called the **dependent variable.** Similarly, z can be defined as function of more than two variables.

5.2 PARTIAL DERIVATIVES

Let $z = f(x, y)$ be a function of two independent variables x and y, If y is taken as constant and x changes, then z becomes a function of x only. The derivative of z with respect to a x taking y as constant is called partial derivative of z with respect to x and denoted by

$$\frac{\partial z}{\partial x} \text{ or } \frac{\partial f}{\partial x} \text{ or } f_x(x, y)$$

and
$$\frac{\partial z}{\partial x} = \lim_{\delta x \to 0} \frac{f(x + \delta x, y) - f(x, y)}{\delta x}$$

Similarly if x is kept constant and y changes, then x become a function of y only. The derivative of z with respect to y taking x as constant called parlayed of x with respect y and is denoted by

$$\frac{\partial z}{\partial y} \text{ or } \frac{\partial f}{\partial y} \text{ or } f_y(x, y)$$

and
$$\frac{\partial z}{\partial y} = \lim_{\delta y \to 0} \frac{f(x, y - \delta y) - f(x, y)}{\delta y}$$

In general, $\frac{\partial z}{\partial x}$ or $fx(x, y)$ and $\frac{\partial z}{\partial y}$ or $f_y(x, y)$ are also functions of x and y and so these can be differentiated again partially with respect to x and y. That is

$$\frac{\partial}{\partial x}\left(\frac{\partial z}{\partial x}\right) = \frac{\partial^2 z}{\partial x^2} \text{ or } \frac{\partial^2 f}{\partial x^2} \text{ or } f_{xx}(x, y)$$

$$\frac{\partial}{\partial y}\left(\frac{\partial z}{\partial y}\right) = \frac{\partial^2 z}{\partial y^2} \text{ or } \frac{\partial^2 f}{\partial y^2} \text{ or } f_{yy}(x, y)$$

$$\frac{\partial}{\partial x}\left(\frac{\partial z}{\partial y}\right) = \frac{\partial^2 x}{\partial x\,\partial y} \text{ or } \frac{\partial^2 f}{\partial x\,\partial y} \text{ or } f_{xy}(x, y)$$

$$\frac{\partial}{\partial y}\left(\frac{\partial z}{\partial x}\right) = \frac{\partial^2 z}{\partial y\partial x} \text{ or } \frac{\partial^2 f}{\partial y\partial x} \text{ or } f_{xy}(x, y)$$

Corrollary : In general

$$\frac{\partial^2 z}{\partial y\partial x} = \frac{\partial^2 z}{\partial x\partial y}$$

Taking an example: $z = ax^2 + 2hxy + by^2$ and differentiating it partially with respect to x and y, one obtains

$$\frac{\partial z}{\partial x} = 2ax + 2hy \qquad \text{...(i)}$$

and
$$\frac{\partial z}{\partial y} = 2hx + 2by \qquad \text{...(ii)}$$

Again differentiating (*i*) with respect to y and (*ii*) with respect to x, we get

$$\frac{\partial}{\partial y}\left(\frac{\partial z}{\partial x}\right) = 2h \text{ or } \frac{\partial^2 z}{\partial y\,\partial x} = 2h \qquad \text{...(A)}$$

and
$$\frac{\partial}{\partial x}\left(\frac{\partial z}{\partial y}\right) = 2h \text{ or } \frac{\partial^2 z}{\partial x\,\partial y} \qquad \text{...(B)}$$

From (*A*) and (*B*), it is observed that

$$\frac{\partial^2 z}{\partial y\,\partial x} = \frac{\partial^2 z}{\partial x\,\partial y}$$

Example 5.1. Evaluate $\frac{\partial z}{\partial x}$ and $\frac{\partial z}{\partial y}$, if

(*i*) $z = x^2 y - x \sin xy$

(*ii*) $z = \log(x^2 + y^2)$

Solution : (*i*) $z = x^2 y - x \sin xy$

$$\begin{aligned}\frac{\partial z}{\partial x} &= 2xy - x\frac{\partial}{\partial x}(\sin xy) - \sin xy \times \frac{\partial}{\partial x}(x)\\ &= 2xy - x\cos xy \times y - \sin xy \times 1\\ &= 2xy - xy\cos xy - \sin xy\end{aligned}$$

and
$$\begin{aligned}\frac{\partial z}{\partial y} &= \frac{\partial}{\partial y}(x^2 y) - \frac{\partial}{\partial y}(x \sin xy)\\ &= x^2 - x\cos xy \times x\\ &= x^2 - x^2\cos xy\\ &= x^2(1 - \cos xy)\end{aligned}$$

(*ii*)
$$z = \log(x^2 + y^2)$$

$$\frac{\partial z}{\partial x} = \frac{1}{x^2+y^2}(2x) = \frac{2x}{x^2+y^2}$$

Example **5.2.** If $u = \sin^{-1}\left(\frac{x}{y}\right) + \tan^{-1}\left(\frac{y}{x}\right)$, then find the value of $x\frac{\partial u}{\partial x} + y\frac{\partial u}{\partial y}$

Solution: $u = \sin^{-1}\left(\frac{x}{y}\right) + \tan^{-1}\left(\frac{y}{x}\right)$

$$\therefore \quad \frac{\partial u}{\partial x} = \frac{1}{\sqrt{1-\left(\frac{x}{y}\right)^2}}\left(\frac{1}{y}\right) + \frac{1}{1+\left(\frac{y}{x}\right)^2}\left(\frac{-y}{x^2}\right) = \frac{1}{\sqrt{y^2-x^2}} - \frac{y}{x^2+y^2}$$

or
$$x\frac{\partial u}{\partial x} = \frac{x}{\sqrt{y^2-x^2}} - \frac{xy}{x^2+y^2} \quad ...(i)$$

Similarly
$$\frac{\partial u}{\partial y} = \frac{1}{\sqrt{1-\left(\frac{x}{y}\right)^2}}\left(\frac{-x}{2}\right) + \frac{1}{1+\left(\frac{y}{2}\right)^2}\frac{1}{x} = -\frac{x}{y\sqrt{y^2-x^2}} + \frac{x}{x^2+y^2}$$

or
$$y\frac{\partial u}{\partial y} = -\frac{x}{\sqrt{y^2-x^2}} + \frac{xy}{x^2+y^2} \quad ...(ii)$$

On adding (*i*) and (*ii*), we get

$$x\frac{\partial u}{\partial x} + y\frac{\partial u}{\partial y} = 0$$

Example **5.3.** If $z = e^{ax+by}(ax - by)$, prove that

Solution : $z = a^{ax+by} f(ax - by)$

$$\frac{\partial z}{\partial x} = a\,e^{ax+by} f(ax-by) + e^{ax+by} f'(ax-by)\,a$$

$$b\frac{\partial z}{\partial x} = ab\,e^{ax+by}\{f(ax-by) + f'(ax-by)\} \quad ...(i)$$

Similarly
$$\frac{\partial z}{\partial y} = be^{ax+by} f(ax-by) + e^{ax+by} f'(ax-by)(-b)$$

$$a\frac{\partial z}{\partial y} = ab\,e^{ax+by}\{f(ax-by) - f'(ax-by)\} \quad ...(ii)$$

On adding (*i*) and (*ii*) we get

$$b\frac{\partial z}{\partial x} + a\frac{\partial z}{\partial y} = 2ab\,e^{ax+by} f(ax-by)$$

$$b\frac{\partial z}{\partial x} + a\frac{\partial z}{\partial y} = 2abz$$

Example **5.4.** If $u = \tan^{-1} a \times \log(x^2+y^2) + b\tan^{-1}\left(\frac{y}{x}\right)$, prove that

$$\frac{\partial^2 u}{\partial x^2} + \frac{\partial^2 u}{\partial y^2} = 0$$

Solution: $u = \tan^{-1} a \times \log(x^2+y^2) + b\tan^{-1}\left(\frac{y}{z}\right)$

$$\frac{\partial u}{\partial x} = \frac{2x\tan^{-1}}{(x^2+y^2)} + \frac{b}{y^2}\left(\frac{-y}{x^2}\right) = \frac{2x\tan^{-1} a}{x^2+y^2} - \frac{by}{x^2+y^2}$$

Again
$$\frac{\partial^2 u}{\partial x^2} = \tan^{-1} a \frac{[(x^2+y^2)\times 2 - 2x \times 2x]}{(x^2+\ y^2)} + \frac{by \times 2x}{(x^2+y^2)^2}$$

$$= \tan^{-1} a\left[\frac{2x^2+2y^2-4x^2}{(x^2+y^2)^2}\right] + \frac{2xyb}{(x^2+y^2)^2}$$

$$= 2\tan^{-1} a \frac{(y^2-x^2)}{(x^2-y^2)} + \frac{2xyb}{(x^2+y^2)} \quad ...(i)$$

Similarly
$$\frac{\partial u}{\partial y} = \tan^{-1} a \frac{2y}{x^2+y^2} + \frac{b}{1+\frac{y^2}{x^2}}\left(\frac{1}{x}\right)$$

$$= \frac{2y \tan^{-1} a}{x^2+y^2} \frac{bx}{x^2+y^2}$$

Again
$$\frac{\partial^2 u}{\partial y^2} = \tan^{-1} a \frac{[(x^2+y^2)\times 2 - 2y \times 2y]}{(x^2+y^2)^2} - \frac{bx \times 2y}{(x^2+y^2)^2}$$

$$= 2\tan^{-1} a \frac{(x^2-y^2)}{(x^2+y^2)^2} - \frac{2bxy}{(x^2+y^2)} \quad ...(ii)$$

On adding (*i*) and (*ii*), we get

$$\frac{\partial^2 x}{\partial x^2} + \frac{\partial^2 u}{\partial y^2} = 0$$

Example **5.5.** If $u = (1-2xy+y^2)^{-\frac{1}{2}}$, show that

$$\frac{\partial}{\partial x}\left\{(1-x^2)\frac{\partial u}{\partial x}\right\} + \frac{\partial}{\partial y}\left(y^2 \frac{\partial u}{\partial y}\right) = 0$$

Solution :
$$u = (1-2xy+y^2)^{-\frac{1}{2}} \quad ...(i)$$

Differentiating equation (*i*) partially with respect to x, we get

$$\frac{\partial u}{\partial x} = \frac{-1}{2}(1-2xy+y^2)^{-\frac{3}{2}}(-2y) = \frac{y}{(1-2xy+y^2)^{3/2}}$$

$$\therefore \quad (1-x^2)\frac{\partial u}{\partial x} = \frac{(1-x^2)y}{(1-2xy+y^2)^{3/2}}$$

Again differentiating with respect to x, we get

$$\frac{\partial}{\partial x}\left\{1-x^2\frac{\partial u}{\partial x}\right\} = \frac{(1-2xy+y^2)^{\frac{3}{2}}(-2xy) - (y-yx^2)\times\frac{3}{2}(1-2xy+y^2)^{\frac{1}{2}}(-2y)}{[(1-2xy+y^2)^{\frac{3}{2}}]^2}$$

$$= \frac{(1-2xy+y^2)^{\frac{1}{2}}[1-2xy+y^2](-2xy) + 3(y^2-x^2y^2)}{(1-2xy+y^2)^3}$$

$$\frac{x^2y^2 - 2xy^3 - 2xy + 3y^2}{(1-2xy+y^2)^{5/2}} \quad ...(ii)$$

Similarly differentiating (*i*) partially with respect to *y*, we get

$$\frac{\partial u}{\partial y} = \frac{-1}{2}(1-2xy=y^2)^{-\frac{3}{2}}(-2x+2y) = \frac{(x-y)}{(1-2xy+y^2)^{3/2}}$$

$$\therefore \quad y^2\frac{\partial u}{\partial y} = \frac{xy^2-y^3}{(1-2xy+y^2)^{3/2}}$$

Again differentiating partially with respect to *y*, we get

$$\frac{\partial}{\partial y}\left(y^2\frac{\partial u}{\partial y}\right) = \frac{(1-2xy+y^2)^{3/2}(2xy-3y^2)-(xy^2-y^3)\times\frac{3}{2}(1-2xy+y^2)^{1/2}(-2+2y)}{(1-2xy+y^2)^3}$$

$$= (1-2xy+y^2)^{\frac{1}{2}}\frac{[1-2xy+y^2(2xy-3y^2)+3(3xy^2+y^3)(x-y)]}{(1-2xy^2)^3}$$

$$= \frac{-x^2y^2+2xy^3+2xy-3y^2}{(1-2xy+y^2)^{5/2}} \quad ...(iii)$$

On adding (*ii*) and (*iii*), we get

$$\frac{\partial}{\partial x}\left\{(1-x^2)\frac{\partial u}{\partial x}\right\}+\frac{\partial}{xy}\left(y^2\frac{\partial u}{\partial y}\right) = \frac{x^2y^2-2xy^3-2xy+3y^2}{(1-2xy+y^2)^{5/2}}+\frac{-x^2y^2+2xy^3+2xy-3y^2}{(1-2xy+y^2)^{5/2}}$$

$$= \frac{0}{(1-2xy+y^2)^{5/2}} = 0$$

***Example* 5.6.** If $u = \frac{1}{\sqrt{t}}e^{-\frac{x^2}{4a^2t}}$, prove that

$$\frac{\partial u}{\partial t} = a^2\frac{\partial^2 u}{\partial x^2}$$

Solution:

$$u = \frac{1}{\sqrt{t}}e^{-\frac{x^2}{4a^2t}}$$

$$\frac{\partial u}{\partial t} = -\frac{1}{2}t^{-\frac{3}{2}}e^{-\frac{x^2}{4a^2t}}+\frac{1}{\sqrt{t}}e^{-\frac{x^2}{4a^2t}}\left[-\frac{x^2}{4a^2}(-1)t^{-2}\right]$$

$$= e^{-\frac{x^2}{4a^2t}}\left[-\frac{t^{-\frac{3}{2}}}{2}+\frac{1}{\sqrt{t}}\frac{x^2}{4t^2a^2}\right]$$

$$= e^{\frac{x^2}{4a^2t}}\left[-\frac{t^{-\frac{3}{2}}}{2}+\frac{x^2}{4a^2t^{\frac{5}{2}}}\right] \quad ...(i)$$

Similarly

$$\frac{\partial u}{\partial x} = \frac{1}{\sqrt{t}}\times e^{-\frac{x^2}{4a^2t}}\left(-\frac{2x}{4a^2t}\right)$$

Again

$$\frac{\partial^2 u}{\partial x^2} = e^{-\frac{x^2}{4a^2t}}\left[-\frac{t^{\frac{-3}{2}}}{2a^2}+\frac{x^2}{4a^2t^{\frac{5}{2}}}\right]$$

$$a^2\frac{\partial^2 u}{\partial x^2} = e^{-\frac{x^2}{4a^2t}}\left[-\frac{t^{\frac{-3}{2}}}{2}+\frac{x^2}{4a^2t^{\frac{5}{2}}}\right] \quad ...(ii)$$

From (*i*) and (*ii*), we get

Example **5.7.** If $\frac{x^2}{a^2+u}+\frac{y^2}{b^2+u}+\frac{z^2}{c^2+u}=1$ prove that

$$\left(\frac{\partial u}{\partial x}\right)^2+\left(\frac{\partial u}{\partial y}\right)^2+\left(\frac{\partial u}{\partial z}\right)^2 = 2\left(x\frac{\partial u}{\partial x}+y\frac{\partial u}{\partial y}+z\frac{\partial u}{\partial z}\right)$$

Solution : $\frac{x^2}{a^2+u}+\frac{y^2}{b^2+u}+\frac{z^2}{c^2+u}=1$

or $x^2(a^2+u)^{-1}+y^2(b^2+u)^{-1}+z^2(c^2+u)^{-1}= 1$ ...(A)

Let u be the function of x, y and z, differentiating (A) partially with respect to x, y and z, we get

$$2x(a^2+u)^{-1}-x^2(a^2+u)^{-2}\frac{\partial u}{\partial x}-y^2(b^2+u)^{-2}\frac{\partial u}{\partial x}-z^2(c^2+u)^{-2}\frac{\partial u}{\partial x}=0$$

or $\frac{2x}{a^2+u}-\frac{\partial u}{\partial x}\left\{\frac{x^2}{(a^2+u)^2}+\frac{y^2}{(b^2+u)^2}+\frac{z^2}{(c^2+u)^2}\right\}=0$

$$\frac{2x}{a^2+u} = \frac{\partial u}{\partial x}\Sigma\frac{x^2}{(a^2+u)^2}$$

Note : $\left\{\Sigma\frac{x^2}{(a^2+u)^2}=\frac{x^2}{(a^2+u)^2}+\frac{y^2}{(b^2+u)^2}+\frac{z^2}{(c^2+u)^2}\right\}$

$\therefore$ $$\frac{\partial u}{\partial x} = \frac{\frac{2x}{(a^2+u)}}{\Sigma\frac{x^2}{(a^2+u)^2}} \quad ...(i)$$

Similarly $$\frac{\partial u}{\partial y} = \frac{\frac{2y}{(b^2+u)}}{\Sigma\frac{x^2}{(a^2+u)^2}} \quad ...(ii)$$

and $$\frac{\partial u}{\partial z} = \frac{\frac{2z}{(c^2+u)}}{\Sigma\frac{x^2}{(a^2+u)^2}} \quad ...(iii)$$

L.H.S. $$\left(\frac{\partial u}{\partial x}\right)^2+\left(\frac{\partial u}{\partial y}\right)^2+\left(\frac{\partial u}{\partial z}\right)^2 = \frac{4\left[\frac{x^2}{(a^2+u)^2}+\frac{y^2}{(b^2+u)^2}+\frac{z^2}{(c^2+u)^2}\right]}{\left[\sum\frac{x^2}{(a^2+u)^2}\right]^2}$$

$$= \frac{4}{\sum\frac{x^2}{(a^2+u)^2}} \quad ...(iv)$$

R.H.S

$$2\left(x\frac{\partial u}{\partial x}+y\frac{\partial u}{\partial y}+z\frac{\partial u}{\partial z}\right) = \frac{4\left[\frac{x^2}{a^2+u}+\frac{y^2}{b^2+u}+\frac{z^2}{c^2+u}\right]}{\sum\frac{x^2}{(a^2+u)^2}}$$

$$= \frac{4 \times (1)}{\sum \frac{x^2}{(a^2+u)^2}} = \frac{4}{\sum \frac{x^2}{(a^2+u)^2}} \quad \text{(From given function)} \quad ...(v)$$

From (*iv*) and (*v*)

$$\left(\frac{\partial u}{\partial x}\right)^2 + \left(\frac{\partial u}{\partial y}\right)^2 + \left(\frac{\partial u}{\partial z}\right)^2 = 2\left(x\frac{\partial u}{\partial x} + y\frac{\partial u}{\partial y} + z\frac{\partial u}{\partial z}\right)$$

***Example* 5.8.** If $x^x y^y z^z = c$, show that at $x = y = z$

$$\frac{\partial^2 z}{\partial x \partial y} = (-x \log ex)^{-1}$$

Solution: Given function is $x^x y^y z^z = c$

Taking logarithm on both sides,

$$x \log x + y \log y + z \log z = \log c \quad ...(A)$$

Let z be the function of x, y. Differentiating (A) partially with respect to x and y, we get

$$\frac{x}{y} + \log x + \frac{z}{z}\frac{\partial z}{\partial x} + \log z \frac{\partial z}{\partial x} = 0$$

$$(1 + \log x) + \frac{\partial z}{\partial x}(1 + \log z) = 0$$

$$\therefore \quad \frac{\partial z}{\partial x} = -\frac{1 + \log x}{1 + \log z} \quad ...(i)$$

Similarly

$$\frac{\partial z}{\partial y} = -\frac{1 + \log y}{1 + \log z} \quad ...(ii)$$

Again differentiating (*i*) partially with respect to y, we get

$$\frac{\partial}{\partial y}\left(\frac{\partial z}{\partial x}\right) = \frac{(1 + \log x)\left(\frac{1}{z}\frac{\partial z}{\partial y}\right)}{(1 + \log z)^2}$$

$$\frac{\partial^2 z}{\partial y \partial x} = \frac{(1 + \log x)}{z(1 + \log z)^2}\frac{\partial z}{\partial y}$$

$$\frac{\partial^2 z}{\partial y\, \partial x} = \frac{-(1 + \log x)(1 + \log y)}{z(1 + \log z)^3} \quad \text{[From (ii)]}$$

At $x = y = z$

$$\left(\frac{\partial^2 z}{\partial y\, \partial x}\right)_{at\ x=y=z} = \frac{-(1 + \log x)^2}{x(1 + \log x)^3} = -\frac{1}{x(\log e + \log x)}$$

Replacing 1 in the denominator by log e, we get

$$\left(\frac{\partial^2 z}{\partial y\, \partial x}\right)_{at\ x=y=z} = -\frac{1}{x[\log e + \log x]} = -\frac{1}{x \log ex} = -(x \log ex)^{-1}$$

Hence

$$\left(\frac{\partial^2 z}{\partial y\, \partial x}\right)_{at\ x=y=z} = -(x \log ex)^{-1}$$

Example 5.9. If $u = \log(x^3 + y^3 + z^3 - 3xyz)$, show that $\left(\frac{\partial}{\partial x} + \frac{\partial}{\partial y} + \frac{\partial}{\partial z}\right)^2 u = -\frac{9}{(x+y+z)^2}$

Solution : $u = \log(x^3 + y^3 + z^3 - 3xyz)$ ...(A)

Differentiating (A) partially with respect to x, we get

$$\frac{\partial u}{\partial x} = \frac{3x^2 - 3yz}{x^3 + y^3 + z^3 - 3xyz} \quad ...(i)$$

Similarly
$$\frac{\partial u}{\partial y} = \frac{3y^2 - 3zx}{x^2 + y^3 + z^3 - 3xyz} \quad ...(ii)$$

$$\frac{\partial u}{\partial z} = \frac{3z^2 - 3xy}{x^3 + y^3 + z^3 - 3xyz} \quad ...(iii)$$

On adding (*i*), (*ii*) and (*iii*) ; we get

$$\frac{\partial u}{\partial x} + \frac{\partial u}{\partial y} + \frac{\partial u}{\partial z} = \frac{3(x^2 + y^2 + z^2 - xy - yz - zx)}{x^3 + y^3 + z^3 - 3xyz}$$

Note : $a^3 + b^3 + c^3 - 3abc = (a + b + c)(a^2 + b^2 + c^2 - c^2 - ab - ca)$

$$= \frac{3}{x + y + z}$$

L.H.S
$$\left(\frac{\partial}{\partial x} + \frac{\partial}{\partial y} + \frac{\partial}{\partial z}\right)^2 u$$

$$= \left(\frac{\partial}{\partial x} + \frac{\partial}{\partial y} + \frac{\partial}{\partial z}\right)\left(\frac{\partial u}{\partial x} + \frac{\partial u}{\partial y} + \frac{\partial u}{\partial z}\right)$$

$$= \left(\frac{\partial}{\partial x} + \frac{\partial}{\partial y} + \frac{\partial}{\partial z}\right)\left(\frac{3}{x + y + z}\right)$$

$$= \frac{-3}{(x + y + z)^2} + \frac{-3}{(x + y + z)^2} + \frac{-3}{(x + y + z)^2}$$

$$= -\frac{9}{(x + y + z)^2}$$

EXERCISE 5.1

1. Evaluate $\frac{\partial z}{\partial x}$ and $\frac{\partial z}{\partial y}$; if

(*i*) $x + y + z = \log z$

Ans. $\frac{\partial z}{\partial x} = \frac{\partial z}{\partial y} = \frac{z}{1 - z}$

(*ii*) $z = \sin^{-1}\left(\frac{y}{x}\right)$

Ans. $\frac{\partial z}{\partial x} = -\frac{y}{x\sqrt{x_2 - y_2}}, \frac{\partial z}{\partial y} = \frac{1}{\sqrt{x^2 - y^2}}$

(*iii*) $z = \log x \times \tan^{-1}(x^2 + y^2)$

Ans. $\frac{\partial z}{\partial x} = \frac{\tan^{-1}(x^2 - y^2)}{x} + \frac{2x \log x}{1 + (x^2 + y^2)^2}, \frac{\partial z}{\partial y} = \frac{2y \log x}{1 + (x^2 + y^2)^2}$

2. If $z = \tan(y + ax) - (y - ax)^{3/2,}$ prove that $\frac{\partial^2 z}{6x^2} = a^2 \frac{\partial^2 z}{\partial y^2}$

3. If $z(x+y) = x^2 + y^2$, prove that $\left(\frac{\partial z}{\partial x} - \frac{\partial z}{\partial y}\right)^2 = 4\left(1 - \frac{\partial z}{\partial x} - \frac{\partial z}{\partial y}\right)$

4. Given that $v = e^{r\cos\theta}\sin(r\sin\theta)$, $u = e^{r\cos\theta}\cos(r\sin\theta)$

Prove that $\frac{\partial u}{\partial r} = \frac{1}{r}\frac{\partial v}{\partial \theta}$ and $\frac{\partial v}{\partial r} = -\frac{1}{r}\frac{\partial u}{\partial \theta}$.

5. If $u = (\log x)\tan^{-1}(x^2 + y^2)$, show that $\frac{\partial^2 u}{\partial x\,\partial y} = \frac{\partial^2 u}{\partial y\,\partial x}$

6. If $z = x^2\tan^{-1}\left(\frac{y}{x}\right) - y^2\tan^{-1}\left(\frac{y}{x}\right)$ prove that $\frac{\partial^2 z}{\partial y\,\partial x} = \frac{x^2 - y^2}{x^2 + y^2}$

7. If $u = (x^2 + y^2 + z^2)^{-\frac{1}{2}}$, show that $\frac{\partial^2 u}{\partial x^2} + \frac{\partial^2 u}{\partial y^2} + \frac{\partial^2 u}{\partial z^2} = 0$

8. If $u = \frac{y}{z} + \frac{z}{x}$ show that $x\frac{\partial u}{\partial x} + y\frac{\partial u}{\partial y} + z\frac{\partial u}{\partial z} = 0$

9. If $\theta = t^n e^{\frac{-r^2}{4t}}$ find what value of n will make

$$\frac{1}{r^2}\frac{\partial}{\partial r}\left(r^2\frac{\partial \theta}{\partial r}\right) = \frac{\partial \theta}{\partial t}$$

Ans. $n = \frac{-3}{2}$

10. The equation $\frac{\partial z}{\partial t} = \mu\frac{\partial^2 z}{\partial x^2}$ refers to the conduction of heat along a bar without radiation. The function z is prescribed as: where A, g and n are positive constants. Establish the relation : $g = \sqrt{\frac{n}{2\mu}}$

$$z = Ae^{-gx}\sin(nt - gx)$$

11. If $u = \log(x^2 + y^2 + z^2)$, prove that

$$(x^2 + y^2 + z^2)\left(\frac{\partial^2 u}{\partial x^2} + \frac{\partial^2 u}{\partial y^2} + \frac{\partial^2 u}{\partial z^2}\right) = 2$$

12. Find the value of n so that the equation $v = r^n(3\cos^2\theta - 1)$ satisfies the relation

$$\frac{\partial}{\partial r}\left(r^2\frac{\partial v}{\partial r}\right) + \frac{1}{\sin\theta}\frac{\partial}{\partial \theta}\left(\sin\theta\frac{\partial v}{\partial \theta}\right) = 0$$

Ans. $n = 2, -3$

13. If $u = r^m$ and $r^2 = x^2 + y^2 + z^2$, prove that

$$u_{xx} + u_{yy} + u_{zz} = m(m+1)\,r^{m-2}$$

14. If $u = \log(\tan x + \tan y)$, prove that

$$\sin 2x\frac{\partial u}{\partial x} + \sin 2y\frac{\partial u}{\partial y} = 2$$

15. If $u = \tan^{-1}\frac{xy}{\sqrt{1 + x^2 + y^2}}$, show that

$$\frac{\partial^2 u}{\partial x\,\partial y} = (1 + x^2 + y^2)^{-\frac{3}{2}}$$

16. If $u = \log(e^x + e^y)$, show that $rt - s^2 = 0$ where

$$r = \frac{\partial^2 z}{\partial x^2}, t = \frac{\partial^2 z}{\partial y^2}, s = \frac{\partial^2 z}{\partial x^2 y}$$

17. If $u = e^{xyz}$, find the value of $\frac{\partial^3 u}{\partial x\, \partial y\, \partial z}$ **Ans.** $\frac{\partial^3 u}{\partial x\, \partial y\, \partial z} = e^{xyz}(1 + 3xyz + x^2y^2z^2)$

5.3 WHICH VARIABLE IS TO BE TREATED AS CONSTANT

For changing cartesian co-ordinates into polar co-ordinates, put

$$x = r\cos\theta, \text{ and } y = r\sin\theta$$

To Find $\frac{\partial r}{\partial x}$, relation between r and x is required, so above relation can be written as

$$r = x\sec\theta, \text{ and} \qquad \ldots(i)$$

$$r^2 = x^2 + y^2 \qquad \ldots(ii)$$

Differentiating (*i*) with respect to x heating θ as constant

$$\frac{\partial r}{\partial x} = \sec\theta \qquad \ldots(iii)$$

Differentiating (*ii*) with respect to x keeping y as constant

$$2r\frac{\partial r}{\partial x} = 2x \text{ or } \frac{\partial r}{\partial x} = \frac{x}{r} = \cos\theta \qquad \ldots(iv)$$

From (*iii*) $\frac{\partial r}{\partial x} = \sec\theta$ and from (*iv*) $\frac{\partial r}{\partial x} = \cos\theta$. These two values of $\frac{\partial r}{\partial x}$ cause confusion. This can be avoided by adopting the following notations :

(*i*) $\left(\frac{\partial r}{\partial x}\right)_\theta$ means the partial derivative of r with respect to x keeping θ as constant

Then : $\left(\frac{\partial r}{\partial x}\right)_\theta = \sec\theta$

(*ii*) $\left(\frac{\partial r}{\partial x}\right)_y$ means the partial derivatives of r with respect to x keeping y as constant.

Then : $\left(\frac{\partial r}{\partial x}\right)_y = \cos\theta$

Example **5.10.** If $x = r\cos\theta$, $y = r\sin\theta$, find

(*i*) $\left(\frac{\partial x}{\partial r}\right)_\theta$ (*ii*) $\left(\frac{\partial y}{\partial \theta}\right)_r$ (*iii*) $\left(\frac{\partial r}{\partial x}\right)_r$ (*iv*) $\left(\frac{\partial \theta}{\partial y}\right)_x$

Solution : (*i*) $\left(\frac{\partial x}{\partial r}\right)_\theta$ means the partial derivative of x with respect to r, keeping θ as constant.

Here $x = r\cos\theta$

$\therefore$ $\left(\frac{\partial x}{\partial r}\right)_\theta = \cos\theta$

(*ii*) $\left(\frac{\partial y}{\partial \theta}\right)_r$ means the partial derivative of y with respect to θ keeping r as constant

Here $y = r\sin\theta$

$\therefore$ $\left(\frac{\partial y}{\partial \theta}\right)_r = r\cos\theta$

(*iii*) $\left(\frac{\partial r}{\partial x}\right)_y$ = means the partial derivative of r with respect to x, keeping y as constant. Now r can be expressed as a function of x and y, *i.e.*,

$$r = \sqrt{x^2 + y^2}$$

$$\left(\frac{\partial r}{\partial x}\right)_y = \frac{1}{2}(x^2 + y^2)^{-\frac{1}{2}}(2x) = \frac{x}{\sqrt{x^2 + y^2}}$$

(*iv*) $\left(\frac{\partial \theta}{\partial y}\right)_x$ means the partial derivative of θ with respect to y keeping x as constant. Now θ can be expressed as a function of x and y, *i.e.*,

$$\theta = \tan^{-1}\frac{y}{x}$$

$$\left(\frac{\partial \theta}{\partial y}\right)_x = \frac{1}{1 + \frac{y^2}{x^2}}\left(\frac{1}{x}\right) = \frac{x}{x^2 + y^2}$$

***Example* 5.11.** If $u = f(r)$, where $r^2 = x^2 + y^2$; prove that

$$\frac{\partial^2 u}{\partial x^2} + \frac{\partial^2 u}{\partial y^2} = f''(r) + \frac{1}{2} f(r)$$

Solution :

$$u = f(r) \qquad ...(i)$$

and

$$r^2 = x^2 + y^2 \qquad ...(ii)$$

By chain rule,

$$\frac{\partial u}{\partial x} = \frac{\partial u}{\partial r} \times \frac{\partial r}{\partial x} = f(r)\,\frac{\partial r}{\partial x}$$

Again,

$$\frac{\partial^2 u}{\partial x^2} = \frac{\partial^2 u}{\partial r^2}\left(\frac{\partial r}{\partial x}\right)^2 + \frac{\partial u}{\partial r}\frac{\partial^2 r}{\partial x^2} \qquad ...(iii)$$

Similarly

$$\frac{\partial^2 u}{\partial y^2} = \frac{\partial^2 u}{\partial r^2}\left(\frac{\partial r}{\partial y}\right)^2 + \frac{\partial u}{\partial r}\frac{\partial^2 r}{\partial y^2}$$

$$= f''(r)\left(\frac{\partial r}{\partial y}\right)^2 + f(r)\frac{\partial^2 r}{\partial y^2} \qquad ...(iv)$$

From equation (*ii*) $2r\frac{\partial r}{\partial x} = 2x$

or

$$\frac{\partial r}{\partial x} = \frac{x}{y'}\,\frac{\partial r}{\partial y} = \frac{y}{r}$$

and

$$\frac{\partial^2 r}{\partial x^2} = \frac{r^2 - x^2}{r^3},\ \frac{\partial^2 r}{\partial y^2} = \frac{r^2 - y^2}{r^3}$$

Equations (*iii*) and (*iv*) become

$$\frac{\partial^2 u}{\partial x^2} = f''(r)\frac{x^2}{r^2} + f(r)\left(\frac{r^2 - x^2}{r^3}\right) \qquad ...(v)$$

and

$$\frac{\partial^2 u}{\partial y^2} = f''(r)\ \frac{y^2}{r^2} + f(r)\frac{r^2 - y^2}{r^3} \qquad ...(vi)$$

Adding (*v*) and (*vi*) we get

$$\frac{\partial^2 u}{\partial r^2} + \frac{\partial^2 u}{\partial y^2} = f''(r) + \frac{1}{r} f(r)$$

Example **5.12.** If $u = lx + my$, $v = mx - ly$, show that

(*i*) $$\left(\frac{\partial u}{\partial x}\right)_y \left(\frac{\partial x}{\partial u}\right)_v = \frac{l^2}{l^2 + m^2}$$

(*ii*) $$\left(\frac{\partial y}{\partial v}\right)_x \left(\frac{\partial v}{\partial y}\right)_u = \frac{l^2 + m^2}{l^2}$$

Solution : $$u = lx + my \quad \ldots(i)$$
and $$v = mx - ly \quad \ldots(ii)$$

(*i*) $\left(\frac{\partial u}{\partial x}\right)_y \left(\frac{\partial x}{\partial u}\right)_v$ $\left(\frac{\partial u}{\partial x}\right)_y$ implies the partial derivative of u with respect to x, keeping y as constant therefore a relation is needed for expressing u as a function of x and y only

From equation (*i*) $\left(\frac{\partial u}{\partial x}\right)_y = l$

Again $\left(\frac{\partial x}{\partial u}\right)_v$ implies the partial derivative of x, with respect to u, keeping v as constant. Therefore a relation is needed for expressing x as a function of u and v only. Eliminating y between (*i*) and (*ii*) by multiplying (*i*) by l (*ii*) by m and then adding

$$lu + mv = (l^2 + m^2)\,x$$

$$x = \frac{l}{l^2 + m^2}u + \frac{m}{l^2 + m^2}v$$

$$\therefore \quad \left(\frac{\partial x}{\partial u}\right)_v = \frac{l}{l^2 + m^2}$$

Hence $$\left(\frac{\partial u}{\partial x}\right)_y \left(\frac{\partial x}{\partial u}\right)_v = \frac{l^2}{l^2 + m^2}$$

(*ii*) $\left(\frac{\partial y}{\partial v}\right)_x \left(\frac{\partial u}{\partial y}\right)_u$

Now $\left(\frac{\partial y}{\partial v}\right)x$ means the partial derivative of y, with respect to v, keeping x as constant.

There a relation is needed for expressing y as function of v and x

From (*ii*) $$y = \frac{mx}{l} - \frac{1}{l}v$$

$$\therefore \quad \left(\frac{\partial y}{\partial v}\right)_x = \frac{-1}{l}$$

Also $\left(\frac{\partial v}{\partial y}\right)_u$ means the partial derivative of v with respect to y, keeping u as constant.

Therefore a relation is needed for expressing v as a function of y and u. Eliminating x between (*i*) and (*ii*)

$$v = \frac{mu}{l} - \left(\frac{l^2 + m^2}{l}\right)y$$

$$\left(\frac{\partial v}{\partial y}\right)_u = \frac{l^2 + m^2}{l}$$

Hence $$\left(\frac{\partial y}{\partial v}\right)_x \left(\frac{\partial v}{\partial y}\right)_u = \left(\frac{-1}{l}\right)\left(-\frac{l^2 + m^2}{l}\right) = \frac{l^2 + m^2}{l^2}$$

EXERCISE 5.2

1. If $x = r\cos\theta$, $y = r\sin\theta$, prove that

$$(i)\ \frac{\partial r}{\partial x} = \frac{dx}{dr} \quad (ii)\ r\frac{\partial\theta}{\partial x} = \frac{1}{r}\frac{\partial x}{\partial\theta} \quad (iii)\ \frac{\partial^2\theta}{\partial x^2} + \frac{\partial^2\theta}{\partial x^2} + \frac{\partial^2\theta}{\partial y^2} = 0$$

2. If $x = r\cos\theta$, $y = r\sin q$, prove that

$$(i)\ \frac{\partial^2 r}{\partial x^2} + \frac{\partial^2 r}{\partial y^2} = \frac{1}{r}\left[\left(\frac{\partial r}{\partial x}\right)^2 + \left(\frac{\partial r}{\partial y}\right)^2\right] \text{and } (ii)\ \frac{\partial^2 r}{\partial x^2}\frac{\partial^2 r}{\partial y^2} = \left(\frac{\partial^2 r}{\partial x\partial y}\right)^2$$

3. If $x^2 = au + bv$, $y^2 = au - bv$, prove that

$$\left(\frac{\partial u}{\partial x}\right)_y\left(\frac{\partial x}{\partial u}\right)_v = \left(\frac{\partial v}{\partial y}\right)_x\left(\frac{\partial y}{\partial v}\right)_u = \frac{1}{2}$$

4. If $u = f(r)$ and $r^2 = x^2 + y^2 + z^2$, prove that

$$u_{xx} + u_{yy} + u_{zz} = f''(r) + \frac{2}{r}f(r)$$

5.4 HOMOGENEOUS FUNCTION

A function $f(x, y)$ is said to be homogeneous function of x, y if the power of each term x, y is the same e.g. $f(x, y) = x^3y + xy^3$.

A function $f(x, y)$ is a homogenous function of order n, if the degree of each term in x, y is equal to n, *i.e*,

$$a_0x^n + a_1x^{n-1}y + a_2x^{n-2}y^2 + \dots + a_{n-1}xy^{n-1} + a_ny^n \qquad \text{...(A)}$$

is a homogenous function of order n.

The expression (A) can be written as

$$x^n\left[a_0 + a_1\left(\frac{y}{x}\right) + a_2\left(\frac{y}{x}\right)^2 + \dots + a_{n-1}\left(\frac{y}{x}\right)^{n-1} + a_n\left(\frac{y}{x}\right)^n\right] = xn\,f\left(\frac{y}{x}\right) \qquad \text{...(B)}$$

Hence any function $f(x, y)$ which can be expressed in the form $x^n\,\phi\left(\frac{y}{x}\right)$ is called homogeneous function of degree n in x and y.

For example :

(*i*) $x^3\sin\left(\frac{y}{x}\right)$ is homogeneous function of degree 3 in x and y.

In general, a function $f(x, y, z, t.....)$ is said to be homogeneous function of degree n in $x, y, z, t....$ if it can be expressed in the form $x^n\,\phi\left(\frac{y}{x}, \frac{z}{x}, \frac{t}{x}...\right)$.

(*ii*) $\cos^{-1}\dfrac{x+y}{\sqrt{x}+\sqrt{y}}$ is not a homogenous function because it can not be written as expression (B).

5.5 EULER'S THEOREM ON HOMOGENEOUS FUNCTION OF TWO VARIABLES

Statement : It u be a homogenous function in x, y of order n, then $x\frac{\partial u}{\partial x} + y\frac{\partial u}{\partial y} = nu$.

Proof : Since u is a homogenous function in x, y of degree n, so u can be written as

$$u = x^n\,\phi\left(\frac{y}{x}\right) \qquad \text{...}(i)$$

Differentiating (*i*) partially with respect to to x, we get

$$\frac{\partial u}{\partial x} = n\, x^{n-1}\, \phi\left(\frac{y}{x}\right) + n^{x}\, f'\left(\frac{y}{x}\right)\left(\frac{-y}{x^2}\right)$$

$$= n\, x^{n-1}\, \phi\left(\frac{y}{x}\right) - x^{n-2}\, y\, \phi'\left(\frac{y}{x}\right)$$

Multiplying both sides by x^n

$$x\frac{\partial u}{\partial x} = n\, x^{n}\, \phi\left(\frac{y}{x}\right) - x^{n-1}\, y\, \phi'\left(\frac{y}{x}\right) \qquad ...(ii)$$

Again differentiating (*i*) partially with respect to y, we get

$$\frac{\partial u}{\partial y} = x^{n}\, \phi\left(\frac{y}{x}\right) \times \frac{1}{x}$$

$$= x^{n-1}\, \phi'\left(\frac{y}{x}\right)$$

Multiplying both sides by y

$$y\frac{\partial u}{\partial y} = x^{n-1}\, y\, \phi'\left(\frac{y}{x}\right) \qquad ...(iii)$$

Adding (*ii*) and (*iii*), we get

$$x\frac{\partial u}{\partial x} + y\frac{\partial u}{\partial y} = n\, x^{n}\, \phi\left(\frac{y}{x}\right)$$

or

$$x\frac{\partial u}{\partial x} + y\frac{\partial u}{\partial y} = nu$$

Euler's theorem on homogeneous function of three and more than three variables:

(*a*) A function $u(x, y, z)$ is said to be homogenous function of degree n in x, y, z, if it can be expressed as

$$u = x^{n}\, \phi\left(\frac{y}{x}, \frac{z}{x}\right)$$

Euler's theorem may then be written as

$$x\frac{\partial u}{\partial x} + y\frac{\partial u}{\partial y} + z\frac{\partial u}{\partial z} = nu$$

(*b*) If u be a homogenous function of degree n in x, y, z, t, p and so on, then

$$u = x^{n}\, \phi\left(\frac{y}{x}, \frac{z}{x}, \frac{t}{x} ....\right)$$ and then Euler's theorem becomes

$$x\frac{\partial u}{\partial x} + y\frac{\partial u}{\partial y} + z\frac{\partial u}{\partial z} + t\frac{\partial u}{\partial t} = nu$$

Deduction : If u is a homogeneous function in x, y of degree n and $u = f(z)$, then

$$x\frac{\partial z}{\partial x} + y\frac{\partial z}{\partial y} = \frac{f(z)}{f(z)}$$

Proof : Hence u is homogenous function in x and y of degree n. Then by Euler's theorem

$$x\frac{\partial u}{\partial x} + y\frac{\partial u}{\partial y} = nu \qquad ...(i)$$

Now $u = f(z)$ give, then $\frac{\partial u}{\partial x} = f(z)\,\frac{\partial z}{\partial x}$ and $\frac{\partial u}{\partial y} = f(z)\,\frac{\partial z}{\partial y}$

Substitution the value of $\frac{\partial u}{\partial x}$ and $\frac{\partial u}{\partial y}$ in (*i*), then

$$\left.\begin{aligned} x\,f(z)\frac{\partial z}{\partial x} + y\,f(z)\frac{\partial z}{\partial y} &= n(u) \\ f(z)\left[x\frac{\partial z}{\partial x} + y\frac{\partial z}{\partial y}\right] &= nf(z) \\ f(z)\,x\frac{\partial z}{\partial x} + y\frac{\partial z}{\partial y} &= \frac{n\,f(z)}{f(z)} \end{aligned}\right\}$$

***Example* 5.13.** Show that $f(x, y, z) = \dfrac{4x^3 + 2y^2 z}{x + 2y + 3z}$ is homogenous. What is the degree of the function ?

Solution : The given function $f(x, y, z) = \dfrac{4x^3 + 2y^2 z}{x + 2y + 3x}$ is homogeneous because each term of numerator (Nr) and denominator (Dr) has same power. Therefore

Degree of function = Power of Nr minus the power of Dr, i.e., $n = 3 - 1 = 2$

Hence it is second degree function.

***Example* 5.14.** Verify Euler's theorem, when

(*i*) $f(x, y) = ax^2 + 2hxy + by^2$

(*ii*) $f(x, y) = \sin^{-1}\left(\frac{x}{y}\right) + \tan^{-1}\left(\frac{y}{x}\right)$

Solution : (*i*) $f(x, y) = ax^2 + 2hxy + by^2$ ...(*i*)

which is homogeneous function of degree 2. Therefore by Euler's theorem

$$x\frac{\partial f}{\partial x} + x\frac{\partial f}{\partial y} = 2\,f(x, y) \quad \text{...(ii)}$$

Differentiating (*i*) partially x and y, then we get

$$\frac{\partial f}{\partial x} = 2ax + 2hy$$

and $$x\frac{\partial f}{\partial x} = 2ax^2 + 2hxy \quad \text{...(iii)}$$

Similarly $$\frac{\partial f}{\partial y} = 2hx + 2by$$

and $$y\frac{\partial f}{\partial y} = 2hxy + 2by^2 \quad \text{...(iv)}$$

Additing (*iii*) and (*iv*), we get

$$x\frac{\partial f}{\partial x} + y\frac{\partial f}{\partial y} = 2[ax^2 + 2hxy + by^2]$$

Hence $$x\frac{\partial f}{\partial x} + y\frac{\partial f}{\partial y} = 2\,f(x, y), \text{ from equation } (i)$$

(*ii*) $f(x, y) = \sin^{-1}\left(\frac{x}{y}\right) + \tan^{-1}\left(\frac{y}{x}\right)$ which is *a*

$$x\frac{\partial f}{\partial x} + y\frac{\partial f}{\partial y} = 0 \quad \text{...(ii)}$$

Alternatively : Differentiating (*i*) partially with respect to *x*, then

$$\frac{\partial f}{\partial x} = \frac{1}{\sqrt{1-\frac{x^2}{y^2}}}\left(\frac{1}{y}\right)+\frac{1}{1+\frac{y^2}{x^2}}\left(\frac{-y}{x^2}\right)=\frac{1}{\sqrt{y^2-x^2}}-\frac{y}{x^2+y^2}$$

$$\therefore \quad x\frac{\partial f}{\partial x} = \frac{x}{\sqrt{y^2-x^2}}-\frac{xy}{x^2-y^2} \qquad ...(iii)$$

Again differentiating (*i*) partially with respect to *y*, then we get

$$\frac{\partial f}{\partial y} = \frac{1}{\sqrt{1-\frac{x^2}{y^2}}}\left(-\frac{x}{y^2}\right)+\frac{1}{1+\frac{y^2}{x^2}}\times\frac{1}{x}=\frac{-x}{y\sqrt{y2-x2}}+\frac{x}{x^2+y^2}$$

$$\therefore \quad y\frac{\partial f}{\partial y} = \frac{-x}{\sqrt{y^2-x^2}}+\frac{xy}{x^2+y^2} \qquad ...(iv)$$

Adding (*iii*) and (*iv*), we get

$$x\frac{\partial f}{\partial x}+y\frac{\partial f}{\partial y} = 0$$

Hence from (*ii*) and (*v*), the theorem is verified

***Example* 5.15.** If $u=\sin^{-1}\frac{x^2+y^2}{x+y}$ show that $x\frac{\partial u}{\partial x}+y\frac{\partial u}{\partial y}=\tan u$

Solution : $u=\sin^{-1}\frac{x^2+y^2}{x+y}$

Here *u* is not homogenous function, Let $v=\frac{x^2+y^2}{x+y}$,

$$u = \sin^{-1} v$$

or

$$v = \sin u = \frac{x^2+y^2}{x+y} \qquad ...(i)$$

Now *v* is homogenous function in *x*, *y* of degree one. Therefore by Euler's theorem

$$x\frac{\partial v}{\partial x}+y\frac{\partial v}{\partial y} = nv$$

or

$$x\frac{\partial v}{\partial x}+y\frac{\partial v}{\partial y} = \sin u \qquad ...(ii)$$

Diferentiating (*i*) partially with respect to *x* and *y*, then we get

$$\frac{\partial v}{\partial x} = \cos u\,\frac{\partial u}{\partial x}$$

or

$$x\frac{\partial v}{\partial x} = \cos u\, x\frac{\partial u}{\partial x} \qquad ...(iii)$$

Similarly

$$y\frac{\partial v}{\partial y} = \cos u\, y\,\frac{\partial u}{\partial y} \qquad ...(iv)$$

Adding (*iii*) and (*iv*), we get

$$x\frac{\partial v}{\partial x} + y\frac{\partial v}{\partial y} = \cos u\left[x\frac{\partial u}{\partial x} + y\frac{\partial u}{\partial y}\right]$$

From (*ii*)

$$\sin u = \cos u\left[x\frac{\partial u}{\partial x} + y\frac{\partial u}{\partial y}\right]$$

or

$$x\frac{\partial u}{\partial x} + y\frac{\partial u}{\partial y} = \tan u.$$

Example **5.16.** Show that $x\,\dfrac{\partial u}{\partial x} + y\dfrac{\partial u}{\partial y} + z\dfrac{\partial u}{\partial z} = 2\tan u$, where $u = \sin^{-1}\left(\dfrac{x^3+y^3+x^3}{ax+by+cz}\right)$

Solution :

$$u = \sin^{-1}\frac{x^3+y^3+z^3}{ax+by+cz}$$

Hence u is not a homogeneous function. Let $v = \left(\dfrac{x^3+y^3+z^3}{ax+by+cz}\right)$, then

$$u = \sin^{-1} v$$

or

$$v = \sin u = \frac{x^3+y^3+z^3}{ax+by+cz} \quad ...(i)$$

Now v is a homogenous function in x, y and x of degree 2. Therefore by Euler's theorem

$$x\frac{\partial v}{\partial x} + y\frac{\partial v}{\partial y} + z\frac{\partial v}{\partial z} = 2v$$

or

$$x\frac{\partial v}{\partial x} + y\frac{\partial v}{\partial y} + z\frac{\partial v}{\partial z} = 2\sin u \quad ...(ii)$$

Differentiating (*i*) partially with respect to x, y and z, then

$$\frac{\partial v}{\partial x} = \cos u\,\frac{\partial u}{\partial x}$$

or

$$x\frac{\partial v}{\partial x} = \cos u \times x\frac{\partial u}{\partial x} \quad ...(iii)$$

and

$$\frac{\partial v}{\partial y} = \cos u\,\frac{\partial u}{\partial y}$$

or

$$y\frac{\partial v}{\partial y} = \cos u \times y\,\frac{\partial u}{\partial y} \quad ...(iv)$$

Similarly

$$\frac{\partial v}{\partial z} = \cos u\,\frac{\partial u}{\partial z}$$

or

$$z\frac{\partial v}{\partial z} = \cos u \times z\,\frac{\partial u}{\partial z} \quad ...(v)$$

Adding (*iii*), (*iv*) and (*v*), we get

$$x\frac{\partial v}{\partial x} + y\frac{\partial v}{\partial y} + z\frac{\partial v}{\partial z} = \cos u\left[x\frac{\partial u}{\partial x} + y\frac{\partial u}{\partial y} + z\frac{\partial u}{\partial z}\right]$$

From (*ii*)

$$2\sin u = \cos u\left(x\frac{\partial u}{\partial x} + y\frac{\partial u}{\partial y} + z\frac{\partial u}{\partial z}\right)$$

or $$x\frac{\partial u}{\partial x}+y\frac{\partial u}{\partial y}+z\frac{\partial u}{\partial z}=\frac{2\sin u}{\cos u} = 2\tan u$$

Example 5.17. If $u=\sin^{-1}\dfrac{x+2y+3z}{\sqrt{x^8+y^8+z^3}}$, show that $x\dfrac{\partial u}{\partial x}+y\dfrac{\partial u}{\partial y}+z\dfrac{\partial u}{\partial z}=-3\tan u$

Solution : $u=\sin^{-1}\dfrac{x+2y+3z}{\sqrt{x^8+y^8+z^8}}$

Here u is not a homogenous function

Let $$v(x,y,z) = \frac{x+2y+3z}{\sqrt{x^8+y^8+z^8}}$$

or $$v(tx,ty,tz) = \frac{tx+2ty+3tz}{\sqrt{t^8x^8+t^8y^8+t^8z^8}}=\frac{t(x+2y+3z)}{t^4\sqrt{x^8+y^8+z^8}}$$

or $$v(tx,ty,tz) = t^{-3}v(x,y,z)$$

$\therefore$ $$\sin u = v(tx,ty,tz)=t^{-3}v(x,t,z) \quad ...(i)$$

which implies that sin u is homogeneous function of degree –3 in x, y, z

$\therefore$ By Euler's theorem for (*i*)

$$x\frac{\partial v}{\partial x}+y\frac{\partial v}{\partial z}+z\frac{\partial v}{\partial z} = -3\sin u \quad ...(ii)$$

Differentiating (*i*) partially with respect to x, y and x, then

$$\frac{\partial v}{\partial x} = \cos u\frac{\partial u}{\partial x}$$

or $$x\frac{\partial v}{\partial x} = \cos u\times x\frac{\partial u}{\partial x} \quad ...(iii)$$

and $$\frac{\partial u}{\partial y} = \cos u\frac{\partial u}{\partial y}$$

or $$y\frac{\partial v}{\partial y} = \cos u\times y\frac{\partial u}{\partial y} \quad ...(iv)$$

Similarly $$\frac{\partial v}{\partial z} = \cos u\frac{\partial u}{\partial z}$$

or $$y\frac{\partial v}{\partial z} = \cos u\times z\frac{\partial u}{\partial z} \quad ...(v)$$

Adding (*iii*), (*iv*) and (*v*), we get

$$x\frac{\partial v}{\partial x}\,y\frac{\partial v}{\partial y}+z\frac{\partial v}{\partial z} = \cos u\left(x\frac{\partial u}{\partial x}+y\frac{\partial u}{\partial y}+z\frac{\partial u}{\partial z}\right)$$

From (*ii*) $$-3\sin u = \cos u\left(x\frac{\partial u}{\partial x}+y\frac{\partial u}{\partial y}+z\frac{\partial u}{\partial z}\right)$$

or $$\frac{-3\sin u}{\cos u} = x\frac{\partial u}{\partial x}+y\frac{\partial u}{\partial y}+z\frac{\partial u}{\partial z}$$

Hence $$x\frac{\partial u}{\partial x}+y\frac{\partial u}{\partial y}+z\frac{\partial u}{\partial z} = -3\tan u$$

Example **5.18.** If $u = \log \dfrac{x^4 + y^4}{x + y}$, show that

$$x\frac{\partial u}{\partial x} + y\frac{\partial u}{\partial y} = 3.$$

Solution : $u = \log \dfrac{x^4 + y^4}{x + y}$

Here u is not a homogenous function

Let $e^u = \dfrac{x^4 + y^4}{x + y}$, which is a homogenous function of degree 3 in x, y

$\therefore$ By Euler's theorem

$$x\frac{\partial(e^u)}{\partial x} + y\frac{\partial(e^u)}{\partial y} = n\, e^u$$

$$x\, e^u \frac{\partial u}{\partial x} + ye^u \frac{\partial u}{\partial y} = 3e^u$$

or $$x\frac{\partial u}{\partial x} + y\frac{\partial u}{\partial y} = 3$$

Examples **5.19.** If $u = \tan^{-1}\left(\dfrac{x^3 + y^3}{x - y}\right)$, prove that $x\dfrac{\partial u}{\partial x} + y\dfrac{\partial u}{\partial x} + y\dfrac{\partial u}{\partial y} = \sin 2u$ and $x^2 \dfrac{\partial u^2}{\partial x^2} + 2xy \dfrac{\partial^2 u}{\partial y^2} + y^2 \dfrac{\partial^2 u}{\partial y^2} =$ $2 \cos 3u \sin u$

Solution : $u = \tan^{-1}\left(\dfrac{x^3 - y^3}{x - y}\right)$

Here u is not a homogenous function. Let $v = \dfrac{x^3 + y^3}{x - y}$, then

$$u = \tan^{-1} v$$

or $$v = \tan u = \frac{x^3 + y^3}{x - y} \qquad ...(i)$$

Now v is a homogenous function of degree 2 in x, y. Therefore by Euler's theorem for equation (*i*)

$$x\frac{\partial v}{\partial x} + y\frac{\partial v}{\partial y} = nv$$

or $$x\frac{\partial v}{\partial x} + y\frac{\partial v}{\partial y} = 2 \tan u \qquad ...(ii)$$

Differentiating (*i*) partially with respect to x and y, then

$$\frac{\partial v}{\partial x} = \sec^2 u \frac{\partial u}{\partial x}$$

or $$x\frac{\partial v}{\partial x} = \sec^2 u \times x \frac{\partial u}{\partial x} \qquad ...(iii)$$

Similarly $$\frac{\partial v}{\partial y} = \sec^2 u \frac{\partial u}{\partial y}$$

or $$y\frac{\partial v}{\partial y} = \sec^2 u \times y \frac{\partial u}{\partial y} \qquad ...(iv)$$

Adding (*iii*) and (*iv*), we get

$$x\frac{\partial v}{\partial x} + y\frac{\partial v}{\partial y} = \sec^2 u \left(x\frac{\partial u}{\partial y} + y\frac{\partial u}{\partial y}\right)$$

From (*ii*)
$$2\tan u = \sec^2 u\left(x\frac{\partial u}{\partial y} + y\frac{\partial u}{\partial y}\right)$$

or
$$\frac{2\tan u}{\sec^2 u} = \left(x\frac{\partial u}{\partial y} + y\frac{\partial u}{\partial y}\right)$$

or
$$x\frac{\partial u}{\partial x} + y\frac{\partial u}{\partial y} = \sin 2u \qquad ...(v)$$

Again differentiating (v) partially with respect to x and y, then

$$\text{or } x\frac{\partial^2 u}{\partial x^2} + \frac{\partial u}{\partial x} + y\frac{\partial^2 u}{\partial x\,\partial y} = 2\cos 2u\,\frac{\partial u}{\partial x}\;\frac{\partial^2 u}{\partial x^2} + x\frac{\partial u}{\partial x} + yx\frac{\partial^2 u}{\partial x\,\partial y} = 2\cos 2u \times x\frac{\partial u}{\partial x} \qquad ...(iv)$$

Similarly
$$y^2\frac{\partial^2 u}{\partial y^2} + y\frac{\partial u}{\partial y} + xy\frac{\partial^2 u}{\partial x\,\partial y} = 2\cos 2u \times y\,\frac{\partial u}{\partial y} \qquad ...(viii)$$

Adding (vi) and (vii), we get

$$x^2\frac{\partial^2 u}{\partial x^2} + 2xy\frac{\partial^2 u}{\partial x\,\partial y} + y^2\frac{\partial^2 u}{\partial y^2} + \left(x\frac{\partial u}{\partial y} + \frac{\partial u}{\partial y}\right) = 2\cos 2u\left(x\frac{\partial u}{\partial y} + y\frac{\partial u}{\partial y}\right)$$

From (v)

$$x^2\frac{\partial^2 u}{\partial x^2} + 2xy\frac{\partial^2 u}{\partial x\,\partial y} + y^2\frac{\partial^2 u}{\partial y^2} + \sin 2u = 2\cos 2u\sin 2u$$

or
$$\begin{aligned}
x^2\frac{\partial^2 u}{\partial x^2} + 2xy\frac{\partial^2 u}{\partial x\,\partial y} + y^2\frac{\partial^2 u}{\partial y^2} &= 2\cos 2u\sin 2u - \sin 2u\\
&= 2(2\cos^2 u - 1)\,2\sin u\cos u - 2\sin u\cos u\\
&= 2\sin u\,(4\cos^3 u - 2\cos u - \cos u)\\
&= 2\sin u\,(4\cos^3 u - 3\cos u)\\
&= 2\cos 3u\sin \text{u}
\end{aligned}$$

***Example* 5.20.** If $u = \sin^{-1}\dfrac{x+y}{\sqrt{x}-\sqrt{y}}$, prove that $x^2\dfrac{\partial^2 u}{\partial x^2} + 2xy\dfrac{\partial^2 u}{\partial x\,\partial y} + y^2\dfrac{\partial^2 u}{\partial y^2} = = -\dfrac{\sin u\cos 2u}{4\cos^3 u}$

Solution: $u = \sin^{-1}\dfrac{x+y}{\sqrt{x}-\sqrt{y}}$

Here u is not a homogeneous function so let $\dfrac{x+y}{\sqrt{x}-\sqrt{y}} = v$, then

$$u = \sin^{-1}$$

or
$$v = \sin u = \frac{x+y}{\sqrt{x}-\sqrt{y}} \qquad ...(i)$$

Now v is homogenous function of degree $\dfrac{1}{2}$ in x, y

∴ By Euler's theorem for (i)

$$x\frac{\partial v}{\partial x} + y\frac{\partial v}{\partial y} = nv$$

or
$$x\frac{\partial v}{\partial x} + y\frac{\partial v}{\partial y} = \frac{1}{2}\sin u \qquad ...(ii)$$

Differentiating (i) partially with respect to x and y, then

$$\frac{\partial v}{\partial x} = \cos u \frac{\partial u}{\partial x}$$

or
$$x\frac{\partial v}{\partial x} = \cos u \times x\frac{\partial u}{\partial x} \quad ...(iii)$$

Similarly
$$\frac{\partial v}{\partial y} = \cos u \frac{\partial u}{\partial y}$$

or
$$y\frac{\partial v}{\partial y} = \cos u \times y\frac{\partial u}{\partial y} \quad ...(iv)$$

$$x\frac{\partial v}{\partial x} + y\frac{\partial v}{\partial y} = \cos u \left(x\frac{\partial u}{\partial y} + y\frac{\partial u}{\partial x}\right)$$

From (ii)

$$\frac{1}{2}\sin u = \cos u\left(x\frac{\partial u}{\partial x} + y\frac{\partial u}{\partial y}\right)$$

or
$$x\frac{\partial u}{\partial x} + y\frac{\partial u}{\partial y} = \frac{1}{2}\tan u \quad ...(v)$$

Again differentiating (v) partially with respect to x and y, then

$$x^2\frac{\partial^2 u}{\partial x^2} + \frac{\partial u}{\partial x} + y\frac{\partial^2 y}{\partial x\,\partial y} = \frac{1}{2}\sec^2 u\frac{\partial u}{\partial x}$$

or
$$x^2\frac{\partial^2 u}{\partial x^2} + x\frac{\partial u}{\partial x} + xy\frac{\partial^2 u}{\partial x\,\partial y} = \frac{1}{2}\sec^2 u \times x\frac{\partial u}{\partial u} \quad ...(vi)$$

Similarly
$$y^2\frac{\partial^2 u}{\partial y^2} + y\frac{\partial u}{\partial y} + xy\frac{\partial^2 u}{\partial x\,\partial y} = \frac{1}{2}\sec^2 u \times y\frac{\partial u}{\partial y} \quad ...(vii)$$

Additing (vi) and (vii), we get $x^2\frac{\partial^2 u}{\partial x^2} + 2xy\frac{\partial^2 u}{\partial x\,\partial y} + y^2\frac{\partial^2 u}{\partial y^2} + \left(x\frac{\partial u}{\partial x} + y\frac{\partial u}{\partial y}\right) = \frac{1}{2}\sec^2 u\left(x\frac{\partial u}{\partial x} + y\frac{\partial u}{\partial y}\right)$

From (v)

$$x^2\frac{\partial^2 u}{\partial x^2} + 2xy\frac{\partial^2 u}{\partial x\,\partial y} + y^2\frac{\partial^2 u}{\partial y^2} = \frac{1}{2}\sec^2 u\left(\frac{1}{2}\tan u\right) - \frac{1}{2}\tan u$$

$$= \frac{1}{4}\frac{\sin u}{\cos^3 u} - \frac{1}{2}\frac{\sin u}{\cos u} = \frac{1}{2}\frac{\sin u}{\cos u}\left(\frac{1-2\cos^2 u}{2\cos^2 u}\right)$$

$$= -\frac{\sin u\cos^2 u}{4\cos^3 u}$$

***Example* 5.21.** If $z = x\,\phi\left(\frac{y}{x}\right) + \psi\left(\frac{y}{x}\right)$, prove that $x^2\frac{\partial^2 u}{\partial x^2} + 2xy\frac{\partial^2 z}{\partial x\,\partial y} + y^2\frac{\partial^2 z}{\partial y^2} = 0$

Solution : $z = x\phi\left(\frac{y}{x}\right) + \psi\left(\frac{y}{x}\right)$ ...(i)

Let
$$u = x\phi\left(\frac{y}{x}\right) \text{ and } v = \psi\left(\frac{y}{x}\right)$$

Here u is homogeneous function of degree 1 in x, y

∴ By Euler's theorem $x\frac{\partial u}{\partial x} + y\frac{\partial u}{\partial y} = u$ ...(A)

Similarly v is a homogeneous function of degree 0 in x, y.

$\therefore$ By Euler's theorem

$$x\frac{\partial v}{x} + y\frac{\partial v}{\partial y} = 0 \quad ...(B)$$

Now (*i*) becomes

$$z = u + v \quad ...(ii)$$

and

$$x\frac{\partial z}{\partial x} + y\frac{\partial z}{\partial y} = \left(x\frac{\partial u}{\partial x} + y\frac{\partial u}{\partial y}\right) + \left(x\frac{\partial v}{\partial x} + y\frac{\partial v}{\partial y}\right).$$

From (*A*) and (*B*)

$$x\frac{\partial z}{\partial x} + y\frac{\partial z}{\partial y} = u \quad ...(iii)$$

Again differentiate (*iii*) partially with respect to x and y. Then

$$x\frac{\partial^2 z}{\partial x^2} + \frac{\partial z}{\partial x} + y\frac{\partial^2 z}{\partial x\,\partial y} = \frac{\partial u}{\partial x}$$

or

$$x^2\frac{\partial^2 z}{\partial x^2} + x\frac{\partial z}{\partial y} + xy\frac{\partial^2 z}{\partial x\,\partial y} = x\frac{\partial u}{\partial x} \quad ...(iv)$$

Similarly,

$$y^2\frac{\partial^2 z}{\partial y^2} + y\frac{\partial z}{\partial y} + xy\frac{\partial^2 z}{\partial x\,\partial y} = y\frac{\partial u}{\partial y} \quad ...(v)$$

Add (*iv*) and (*v*), we get

$$x^2\frac{\partial^2 z}{\partial x^2} + 2xy\frac{\partial^2 z}{\partial x\,\partial y} + y^2\frac{\partial^2 z}{\partial y^2} + \left(x\frac{\partial u}{\partial x} + y\frac{\partial z}{\partial y}\right) = \left(x\frac{\partial u}{\partial x} + y\frac{\partial u}{\partial y}\right)$$

From equation (A) and (B)

$$x^2\frac{\partial^2 z}{\partial x^2} + 2xy\frac{\partial^2 z}{\partial x\,\partial y} + y^2\frac{\partial^2 z}{\partial y^2} = u$$

or

$$x^2\frac{\partial^2 z}{\partial x^2} + 2xy\frac{\partial^2 z}{\partial x\,\partial y} + y^2\frac{\partial^2 z}{\partial y^2} = 0$$

***Example* 5.22.** If $z = x^n f_1\left(\frac{y}{x}\right) + y^{-n} f_2\left(\frac{x}{y}\right)$, prove that $x^2\frac{\partial^2 x}{\partial x^2} + 2xy\frac{\partial^2 z}{\partial x\,\partial y} + y^2\frac{\partial^2 z}{\partial y^2} + x\frac{\partial z}{\partial x} + y\frac{\partial z}{\partial y} = n^2 z$

Solution: $x^n f_1\left(\frac{y}{x}\right) = y^{-n} f_2\left(\frac{x}{y}\right)$ ...(A)

Let

$$u = x^n f_1\left(\frac{y}{x}\right), \text{ then by Euler's theorem}$$

$$x\frac{\partial u}{\partial x} + y\frac{\partial u}{\partial y} = nu \quad ...(i)$$

and

$$v = y^{-n} f_2\left(\frac{x}{y}\right), \text{ then by Euler's Theorem}$$

$$x\frac{\partial v}{\partial y} + y\frac{\partial v}{\partial y} = -nv \quad ...(ii)$$

From (*A*)

$$z = u + v \text{ and} \quad ...(B)$$

$$x\frac{\partial z}{\partial x}+y\frac{\partial z}{\partial y} = \left(x\frac{\partial u}{\partial x}+y\frac{\partial u}{\partial y}\right)+\left(x\frac{\partial v}{\partial x}+y\frac{\partial v}{\partial y}\right)$$

From (*i*) and (*ii*)

$$x\frac{\partial z}{\partial y}+y\frac{\partial z}{\partial y} = nu - nv = n(n-v) \quad ...(C)$$

Differentiating (C) partially with respect to x and y, then

$$x\frac{\partial^2 z}{\partial x^2}+\frac{\partial z}{\partial x}+y\frac{\partial^2 z}{\partial x\partial y} = n\left(\frac{\partial u}{\partial x}-\frac{\partial v}{\partial x}\right)$$

or
$$x^2\frac{\partial^2 z}{\partial x^2}+x\frac{\partial z}{\partial x}+xy\frac{\partial^2 z}{\partial x\,\partial y} = n\left(x\frac{\partial u}{\partial x}-x\frac{\partial v}{\partial x}\right)$$

Similarly,
$$y^2\frac{\partial^2 z}{\partial y^2}+y\frac{\partial z}{\partial y}+xy\frac{\partial^2 z}{\partial y\,\partial x} = n\left(y\frac{\partial u}{\partial y}-y\frac{\partial v}{\partial y}\right) \quad ...(E)$$

Adding (D) and (E), we get

$$x^2\frac{\partial^2 z}{\partial x^2}+2xy\frac{\partial^2 z}{\partial x\,\partial y}+y^2\frac{\partial^2 z}{\partial y^2}+x\frac{\partial z}{\partial x}+y\frac{\partial z}{\partial y} = n\left(x\frac{\partial u}{\partial x}+y\frac{\partial u}{\partial y}\right)-n\left(x\frac{\partial u}{\partial x}+y\frac{\partial u}{\partial y}\right)$$

$$= n[(nu)-(-nv)]$$

$$= n^2(n+v) = n^2 z$$

***Example* 5.23.** If $u = \text{cosec}^{-1}\left(\dfrac{x^{\frac{1}{2}}+y^{\frac{1}{2}}}{x^{\frac{1}{3}}+y^{\frac{1}{3}}}\right)^{\frac{1}{2}}$, prove that

$$x^2\frac{\partial^2 u}{\partial x^2}+2xy\frac{\partial^2 u}{\partial x\,\partial y}+y^2\frac{\partial^2 u}{\partial y^2} = \frac{\tan u}{144}(13+\tan^2 u)$$

Solution:
$$u = \text{cosec}^{-1}\left(\frac{x^{\frac{1}{2}}+y^{\frac{1}{2}}}{x^{\frac{1}{3}}+y^{\frac{1}{3}}}\right)^{\frac{1}{2}}$$

Here u is not homogeneous function

Let
$$v = \left(\frac{x^{\frac{1}{2}}+y^{\frac{1}{2}}}{x^{\frac{1}{3}}+y^{\frac{1}{3}}}\right)^{\frac{1}{2}}, \text{ then } v = \text{cosec } u = \left(\frac{x^{\frac{1}{2}}+y^{\frac{1}{2}}}{x^{\frac{1}{3}}+y^{\frac{1}{3}}}\right)^{\frac{1}{2}} \quad ...(i)$$

Now v is a homogeneous function of degree $\frac{1}{12}$ in x, y

∴ By Euler's theorem

$$x\frac{\partial v}{\partial y}+y\frac{\partial v}{\partial y} = nv = \frac{1}{12}\text{cosec } u \quad ...(ii)$$

Differentiating (*i*) partially with respect to x and y, then

$$\frac{\partial v}{\partial x} = -\text{cosec } u \cot u\frac{\partial u}{\partial x}$$

or
$$x\frac{\partial v}{\partial x} = -\text{cosec } u \cot u \times x\frac{\partial u}{\partial x} \quad ...(iii)$$

Similarly
$$y\frac{\partial v}{\partial y} = -\text{cosec } u \cot u \times y\frac{\partial v}{\partial y} \quad ...(iv)$$

Adding (*iii*) and (*iv*), we get

$$x\frac{\partial v}{\partial x} + y\frac{\partial v}{\partial y} = -\text{cosec } u \cot u\left(x\frac{\partial u}{\partial x} + y\frac{\partial u}{\partial y}\right)$$

From (*ii*)

or $$\frac{1}{12}\text{ coses } u = -\text{cosec } u \cot u\left(x\frac{\partial u}{\partial x} + \frac{\partial u}{\partial y}\right)$$

$$x\frac{\partial u}{\partial x} + y\frac{\partial u}{\partial y} = -\frac{\tan u}{12} \quad \text{...(v)}$$

Again differentiating (*v*) partially with respect *x* and *y*, then

$$x^2\frac{\partial^2 u}{\partial x^2} + \frac{\partial u}{\partial x} + y\frac{\partial^2 u}{\partial x\, \partial y} = \frac{1}{12}\sec^2 u\frac{\partial u}{\partial x}$$

or $$x^2\frac{\partial^2 u}{\partial x^2} + x\frac{\partial u}{\partial y} + xy\frac{\partial^2 u}{\partial x\, \partial y} = -\frac{1}{12}\sec^2 u \times x\frac{\partial u}{\partial x} \quad \text{...(vi)}$$

Similarly $$y^2\frac{\partial^2 u}{\partial y^2} + y\frac{\partial u}{\partial y} + xy\frac{\partial^2 u}{\partial x\, \partial y} = -\frac{1}{12}\sec^2 u \times y\frac{\partial u}{\partial y} \quad \text{...(vii)}$$

Adding (*vi*) and (*vii*), we get

$$x^2\frac{\partial^2 u}{\partial x^2} + y^2\frac{\partial^2 u}{\partial y^2} + 2xy\frac{\partial^2 u}{\partial x\, \partial y} = -\frac{1}{12}\sec^2 u\left(x\frac{\partial u}{\partial x} + y\frac{\partial u}{\partial y}\right) - \left(x\frac{\partial u}{\partial x} + y\frac{\partial u}{\partial y}\right)$$

From equation (*v*)

$$x^2\frac{\partial^2 u}{\partial x^2} + y^2\frac{\partial^2 u}{\partial y^2} + 2xy\frac{\partial^2 u}{\partial x\, \partial y} = -\frac{1}{12}\sec^2 u\left(-\frac{1}{12}\tan u\right) - \left(-\frac{1}{12}\tan u\right)$$

$$= \frac{1}{12}\tan u\left(\frac{1}{12}\sec^2 u + 1\right) = \frac{\tan u}{144}(\sec^2 u + 12)$$

$$= \frac{\tan u}{144}(1 + \tan^2 u + 12) = \frac{\tan u}{144}(13 + \tan^2 u)$$

EXERCISE 5.3

1. Show that $f(x, y, z) = \dfrac{5x^5 + 3z^2y^3}{x^2 + 3y^2 + 2x^2}$ is homogeneous. What is the degree of the function? **Ans. Yes, 3**

2. Verifyt Euler's theorem, when

   (*i*) $u = \dfrac{x^2(x^2 - y^2)^3}{(x^2 + y^2)^3}$ (*ii*) $u = \dfrac{x^{\frac{1}{4}} + y^{\frac{1}{4}}}{x^{\frac{1}{5}} + y^{\frac{1}{5}}}$

3. If $u = \cos^{-1}\dfrac{x + y}{\sqrt{x} + \sqrt{y}}$, show that $x\dfrac{\partial u}{\partial x} + y\dfrac{\partial u}{\partial y} = \dfrac{-1}{2}\cos u$

4. (*i*) If $u = \sin^{-1}\left(\dfrac{\sqrt{x} - \sqrt{y}}{\sqrt{x} + \sqrt{y}}\right)$, show that $\dfrac{\partial u}{\partial x} = -\dfrac{y}{x}\dfrac{\partial u}{\partial y}$

(*ii*) If $\sin u = \dfrac{x^2+y^2}{x+y}$, *show that* $x\dfrac{\partial u}{\partial x} + y\dfrac{\partial u}{\partial y} = \tan u$

(*iii*) If $u = \cos^{-1}\left(\dfrac{xy+yz+zx}{x^2+y^2+z^2}\right)$, show that $x\dfrac{\partial u}{\partial x} + y\dfrac{\partial u}{\partial y} + z\dfrac{\partial u}{\partial z} = 0$

(*iv*) If $u = \sec^{-1}\left(\dfrac{x^2-y^3}{x+y}\right)$, show that $x\dfrac{\partial u}{\partial x} + y\dfrac{\partial u}{\partial y} = 2\cot u$

5. If $u = \tan^{-1}\left(\dfrac{x^3+y^3}{x-y}\right)$, prove that $x\dfrac{\partial u}{\partial x} + y\dfrac{\partial u}{\partial y} = \sin 2u$

6. (*i*) If $u = \log\left(\dfrac{x^5+y^5+z^5}{x^2+y^2+z^2}\right)$ show that $x\dfrac{\partial u}{\partial x} + y\dfrac{\partial u}{\partial y} + z\dfrac{\partial u}{\partial z} = 3$

(*ii*) If $u = \log\left(\dfrac{x^3-y^3}{x+y}\right)$, show that $x\dfrac{\partial u}{\partial x} + y\dfrac{\partial u}{\partial y} = 2$

7. (*i*) If $u = \tan^{-1}\left(\dfrac{x^2-y^2}{x-y}\right)$, prove that

$$x\frac{\partial u}{\partial x} + y\frac{\partial u}{\partial y} = \frac{1}{2}\sin 2u$$

and $$x^2\frac{\partial^2 u}{\partial x^2} + 2xy\frac{\partial^2 u}{\partial y^2} = \frac{1}{4}(\sin 4u - 2\sin 2u)$$

(*ii*) If $u = \sin^{-1}(x^3+y^3)^{\frac{2}{5}}$, prove that

$$x^2\frac{\partial^2 u}{\partial x^2} + 2xy\frac{\partial^2 u}{\partial x\partial y} + y^2\frac{\partial^2 u}{\partial y^2} = \frac{6}{5}\tan u\left(\frac{6}{5}\sin^2 a - 1\right)$$

(*iii*) If $u = \sin^{-1}\left(\dfrac{x+y}{\sqrt{x}+\sqrt{y}}\right)$, prove that

$$2x\frac{\partial u}{\partial x} + 2y\frac{\partial u}{\partial y} = \tan\left[\sin^{-1}\left(\frac{x+y}{\sqrt{x}+\sqrt{y}}\right)\right] \text{ and } x^2\frac{\partial^2 u}{\partial x^2} + 2xy\frac{\partial^2 u}{\partial x\partial y} + y^2\frac{\partial^2 u}{\partial y^2} = \frac{1}{2}\tan u\left(\frac{1}{2}\sec^2 u - 1\right)$$

8. (*i*) If $z = x^2\tan^{-1}\left(\dfrac{y}{x}\right) - y^2\tan^{-1}\left(\dfrac{x}{y}\right)$, prove that

$$x^2\frac{\partial^2 z}{\partial x^2} + 2xy\frac{\partial^2 z}{\partial x\,\partial y} + y^2\frac{\partial^2 z}{\partial y^2} = 2z$$

(*ii*) If $z = \log\dfrac{x+y}{\sqrt{x^2+y^2}} + \sin^{-1}\dfrac{x+y}{\sqrt{x}+\sqrt{y}}$, evaluate $x^2\dfrac{\partial^2 z}{\partial x^2} + 2xy\dfrac{\partial^2 z}{\partial x\,\partial y} + y^2\dfrac{\partial^2 z}{\partial y^2}$

Ans. $-\dfrac{\sin v\cos 2v}{4\cos^3 v}$, where $\sin v = \dfrac{x+y}{\sqrt{x}+\sqrt{y}}$

(*iii*) If $u = \tan^{-1}\left(\frac{y^2}{x}\right)$, show that

$$x^2 \frac{\partial^2 u}{\partial x^2} + 2xy \frac{\partial^2 u}{\partial x\, \partial y} + y^2 \frac{\partial^2 u}{\partial y^2} = \sin 2u \sin^2 u.$$

9. If $u = \sin^{-1}\left[\frac{x^{\frac{1}{4}} + y^{\frac{1}{4}}}{x^{\frac{1}{6}} + y^{\frac{1}{6}}}\right]$

$$x^2 \frac{\partial^2 u}{\partial x^2} + 2xy \frac{\partial^2 u}{\partial x\, \partial y} + y^2 \frac{\partial^2 u}{\partial y^2} = \frac{\tan u}{144}[\tan 2u - 11]$$

5.6 TOTAL DERIVATIVES

If

$$u = f(x, y), \text{ then}$$

$$du = \frac{\partial f}{\partial x} dx + \frac{\partial f}{\partial y} dy \qquad ...(i)$$

This is know as total derivative of u

5.7 COMPOSITE FUNCTIONS

(*i*) If $u = f(x, y)$ where $x = f(t)$ and $y = y(t)$, then u is called a composite function of t (single variable) and $\frac{du}{dt}$ can be found as

$$\frac{du}{dt} = \frac{\partial u}{\partial x}\frac{dx}{dt} + \frac{\partial u}{\partial y}\frac{dy}{dt}$$

or

$$\frac{du}{dt} = \frac{\partial f}{\partial x}\frac{dx}{dt} + \frac{\partial f}{\partial y}\frac{dy}{dt}$$

(*ii*) If $u = f(x, y)$ where $x = \phi\,(s, t)$ and $y = \psi(s, t)$, then u is called a composite function of (two variable) s and t, there $\frac{\partial u}{\partial s}$ and $\frac{\partial u}{\partial t}$ is defined as

$$\frac{\partial u}{\partial s} = \frac{\partial u}{\partial x}\frac{\partial x}{\partial s} + \frac{\partial u}{\partial y}\frac{\partial y}{\partial s}$$

and

$$\frac{\partial u}{\partial t} = \frac{\partial u}{\partial x}\frac{\partial x}{\partial t} + \frac{\partial u}{\partial y}\frac{\partial y}{\partial t}$$

5.8 DIFFERENTIATION OF COMPOSITE FUNCTIONS

Statement : If $u = f(x, y)$ where $x = f(t)$, $y = f(t)$, i.e., u is a composite function of t then $\frac{du}{dt} = \frac{\partial u}{\partial x}\frac{dx}{dt} + \frac{\partial u}{\partial y}\frac{dy}{dt}$

Proof : $u = f(x, y)$ ...(*i*)

Let δu, δx, δy by increments in u, x and y corresponding to an increment δt in t, then

$$u + \delta u = f(x + \delta x, y + \delta y) \qquad ...(ii)$$

Subtracting (*i*) from (*ii*), we get

$$\delta u = f(x + \delta x, y + \delta y) - f(x, y)$$

$$= f(x + \delta x, y + \delta y) - f(x, y + \delta y) + f(x, y + \delta y) - f(x, y)$$

$$\frac{\delta u}{\delta t} = \frac{f(x + dx, y + sy) - f(x, y + dy)}{\delta t} + \frac{f(x, y + dy) - f(x, y)}{\delta t}$$

$$\frac{\delta u}{\delta t} = \frac{f(x+\delta x, y+sy) - f(x, y+\delta y)}{\delta x} \times \frac{\delta x}{\delta t} + \frac{f(x, y+\delta y) - f(x, y)}{\delta y} \times \frac{\delta y}{\delta t}$$

As $\delta t \to 0$, both δx and δy $\to$ 0. Then the above equations becomes.

$$\lim_{dt \to 0} \frac{\delta u}{\delta t} = \lim_{dx \to 0} \frac{f(x+\delta x, y+dy) - f(x, y+\delta y)}{\delta x} \times \lim_{\delta t \to 0} \frac{\delta x}{\delta t} + \lim_{\delta y \to 0} \frac{f(x, y+\delta y) - f(x, y)}{\delta y} \times \lim_{\delta t \to 0} \frac{\delta y}{\delta t} \quad \text{...(iii)}$$

By definition of derivative

$$\lim_{\delta t \to 0} \frac{\delta u}{\delta t} = \frac{du}{dt}, \lim_{\delta t \to 0} \frac{\delta x}{\delta t} = \frac{dx}{dt}, \lim_{\delta t \to 0} \frac{\delta y}{\delta t} = \frac{dy}{dt}$$

and
$$\lim_{\delta y \to 0} \frac{f(x+\delta x, y+\delta y)}{\delta x} = \frac{\partial u}{\partial y}$$

Similarly
$$\lim_{\delta x \to 0} \frac{f(x, y+\delta y) - f(x, y) - f(x, y) + 8y}{\delta y} = \frac{\partial u}{\partial y}$$

Hence (*iii*) becomes

$$\frac{du}{dt} = \frac{\partial u}{\partial x}\frac{dx}{dt} + \frac{\partial u}{\partial y}\frac{dy}{dt}$$

Note : 1. If $u = f(x, y, z)$ and x, y, z are function of t, then u be a composite function of t and

$$\frac{du}{dt} = \frac{\partial u}{\partial x}\frac{dx}{dt} + \frac{\partial u}{\partial y}\frac{dy}{dt} + \frac{\partial u}{\partial z}\frac{dz}{dt}$$

2. If $u = f(x, y)$ where $y = \phi(x)$ then since $x = \phi(x)$, u is a composite function of x

$$\therefore \quad \frac{du}{dx} = \frac{\partial u}{\partial x} + \frac{\partial u}{\partial y}\frac{dy}{dx}$$

If $u = f(x, y)$ where $x = \psi(y)$ then since $y = \psi(y)$, u is a composite function of y then

Similarly
$$\frac{du}{dy} = \frac{\partial u}{\partial x}\frac{dx}{dy} + \frac{\partial u}{\partial y}$$

5.9 DIFFERENTIATION OF IMPLICIT FUNCTIONS

If $u\ (x, y) = c$ be an implicit relation between x and y which defines y as a differentiable function of x then

$$\frac{du}{dx} = 0 = \frac{\partial u}{\partial x} + \frac{\partial u}{\partial y}\frac{dy}{dx},$$ which gives an important relation

$$\frac{dy}{dx} = -\frac{\frac{\partial u}{\partial x}}{\frac{\partial u}{\partial y}}$$

Example **5.24.** If $u = \sin\left(\frac{x}{y}\right)$, $x = e^t$ and $y = t^2$ find $\frac{du}{dt}$ as a function of t, verify your result by direct substitution.

Solution : $u = \sin\left(\frac{x}{y}\right)$ and $x = e^t$, $y = t^2$

Here u is composite function of t. There

$$\frac{du}{dt} = \frac{\partial u}{\partial x}\frac{dx}{dt} + \frac{\partial u}{\partial y}\frac{dy}{dt}$$

$$= \cos\left(\frac{x}{y}\right)\left(\frac{1}{y}\right)e^t + \cos\left(\frac{x}{y}\right)\left(\frac{-x}{y^2}\right)2t$$

$$= \cos\left(\frac{e^t}{t^2}\right)\frac{e^t}{t^2} - \cos\left(\frac{e^t}{t^2}\right)\left(\frac{e^t}{t^4}\right)2t$$

$$= \left(\frac{1-2}{t^3}\right)e^t \cos\left(\frac{e^t}{t^2}\right) \quad ...(A)$$

Also $$u = \sin\left(\frac{x}{y}\right) = \sin\left(\frac{e^t}{t^2}\right)$$

$$\frac{du}{dt} = \cos\left(\frac{e^t}{t^2}\right)\frac{t^2e^t - e^t\,2t}{t^4}$$

$$= \left(\frac{t-2}{t^3}\right)e^t \cos\left(\frac{e^t}{t^2}\right) \quad ...(B)$$

Hence (A) and (B) are same.

***Example* 5.25.** If $u = \sin^{-1}(x - y)$, $x = 3t$, $y = 4t^3$, show that $\frac{du}{dt} = \frac{3}{\sqrt{1-t^2}}$

Solution : $u = \sin^{-1}(x - y)$, $x = 3t$, $y = 4t^3$

Here u is composite function of t. Therefore

$$\frac{du}{dt} = \frac{\partial u}{\partial x}\frac{dx}{dt} + \frac{\partial u}{\partial y}\frac{dy}{dt}$$

$$= \frac{1}{\sqrt{1-(x-y)^2}} \times 3 + \frac{1}{\sqrt{1-(x-y)^2}}(-1) \times 12t^2$$

$$= \frac{3(1-4t^2)}{\sqrt{1-(x-y)^2}} = \frac{3(1-4t^2)}{1-(3t-4t^3)^2}$$

$$= \frac{3(1-4t^2)}{\sqrt{1-9t^2+24t^4-16t^6}} = \frac{3(1-4t^2)}{\sqrt{(1-t^2)(1-4t^2)^2}}$$

$$\frac{du}{dt} = \frac{3}{\sqrt{1-t^2}}$$

***Example* 5.26.** If $u = \sin(x^2 + y^2)$,. where $a^2x^2 + b^2y^2 = c^2$, find $\frac{du}{dx}$.

Solution : $u = \sin(x^2 + y^2)$ and $\phi(x, y) = c$ *i.e.*, (Implicit form)

Let u be a composite function of x. Then

$$\frac{du}{dx} = \frac{\partial u}{\partial x}\frac{dx}{dx} + \frac{\partial u}{\partial y}\frac{dy}{dx}$$

$$= \frac{\partial u}{\partial x} + \frac{\partial u}{\partial y}\frac{dy}{dx} \quad ...(A)$$

Here $$\frac{\partial u}{\partial x} = 2x\cos(x^2 + y^2)$$

and $$\frac{\partial u}{\partial y} = 2y\cos(x^2 + y^2)$$

Also differentiating $a^2x^2 + b^2y^2 = c^2$ with respect to x, we obtain

$$2a^2x + 2b^2y\frac{dy}{dx} = 0$$

$$\frac{dy}{dx} = -\frac{a^2x}{b^2y}$$

From (A)

$$\frac{du}{dx} = 2x\cos(x^2+y^2) + 2y\cos(x^2+y^2)\left(-\frac{a^2x}{b^2y}\right)$$

$$\frac{du}{dx} = 2\left(x - \frac{a^2}{b^2}x\right)\cos(x^2+y^2)$$

$$\frac{du}{dx} = \frac{2(b^2-a^2)}{b^2}\,x \times \cos(x^2+y^2)$$

***Example* 5.27.** Find $\frac{dy}{dx}$ when $(\cos x)^y = (\mathbf{sin}\ y)^x$

Solution : $(\cos x)^y - (\sin y)^x = 0$

Let $f(x, y) = (\cos x)^y - (\sin y)^x$

$$\frac{\partial f}{\partial x} = y(\cos x)^{y-1}(-\sin x) - (\sin y)^x \log(\sin y)$$

$$\frac{\partial f}{\partial x} = -[y\sin x(\cos x)^{y-1} + (\sin y)^x \log(\sin y)]$$

Similarly $\frac{\partial f}{\partial y} = (\cos x)^y \log(\cos x) - x(\sin y)^{x-1}\cos y$

For implicit form $\frac{dy}{dx} = -\frac{\frac{\partial f}{\partial x}}{\frac{\partial f}{\partial y}}$ ...(*i*)

$$\frac{dy}{dx} = \frac{y\sin x(\cos c)^{y-1} + (\sin x)^y \log(\sin y)}{(\cos x)^y \log(\cos x) - x(\sin y)^{x-1}(\cos y)}$$

Put $(\cos x)^y = (\sin y)^x$, then we get

$$\frac{dy}{dx} = \frac{(\cos x)^y(y\tan x + \log\sin y)}{(\cos x)^y(\log\cos x - x\cot y)}$$

$$\frac{dy}{dx} = \frac{y\tan x + \log\sin y}{\log\cos x - x\cot y}$$

***Example* 5.28.** If $f(x, y) = 0$ and $p = \frac{\partial f}{\partial x}, q = \frac{\partial f}{\partial y}, r = \frac{\partial^2 f}{\partial x^2}, t = \frac{\partial^2 f}{\partial y^2}, s = \frac{\partial^2 f}{\partial x\,\partial y}$, then show that

(*i*) $\frac{d^2y}{dx^2} = -\frac{q^2r - 2qps + p^2t}{q^3}$

(*ii*) $\left(\frac{\partial f}{\partial y}\right)^3 \frac{d^2y}{dx^2} = 2\frac{\partial f}{\partial x}\frac{\partial f}{\partial y}\frac{\partial^2 f}{\partial x\,\partial y} - \left(\frac{\partial f}{\partial y}\right)^2\frac{\partial^2 f}{\partial y^2} - \left(\frac{\partial f}{\partial x}\right)^2\frac{\partial^2 f}{\partial x^2}$

Solution: (*i*) $f(x, y) = 0$ is an implicit function, then

$$\frac{dy}{dx} = -\frac{\dfrac{\partial f}{\partial x}}{\dfrac{\partial f}{\partial y}} = -\frac{p}{q} \quad \text{...(i)}$$

Again differentiate (*i*) with respect to x on both side, we obtain

$$\frac{d^2y}{dx^2} = -\frac{q\dfrac{dp}{dx} - q\dfrac{dq}{dx}}{q^2} \quad \text{...(ii)}$$

But

$$\frac{dp}{dx} = \frac{\partial p}{\partial x}\frac{dx}{dx} + \frac{\partial p}{\partial y}\frac{dy}{dx}$$

$$\frac{dp}{dx} = \frac{\partial p}{\partial x} + \frac{\partial p}{\partial y}\frac{dy}{dx}$$

$$\frac{dp}{dx} = \frac{\partial^2 f}{\partial x^2} + \frac{\partial^2 f}{\partial x\,\partial y}\frac{dy}{dx} \qquad \left(\because p\frac{\partial f}{\partial x}\right)$$

$$\frac{dp}{dx} = r + s\left(\frac{-p}{q}\right) = \frac{rq - sq}{q}$$

Also

$$\frac{dq}{dx} = \frac{\partial q}{\partial x}\frac{dx}{dx} + \frac{\partial q}{\partial y}\frac{dy}{dx}$$

$$\frac{dq}{dx} = \frac{\partial^2 f}{\partial y\,\partial x} + \frac{\partial^2 f}{\partial y^2}\left(\frac{dy}{dx}\right) \qquad \left(\because q = \frac{\partial f}{\partial y}\right)$$

$$\frac{dq}{dx} = s + t\left(\frac{-q}{q}\right) = \frac{sq - pt}{q}$$

Substituting $\frac{dq}{dx}$ and $\frac{dp}{dx}$ in equation (*ii*), then we get

$$\frac{d^2y}{dx^2} = -\frac{q\dfrac{(rq - sp)}{q} - p\dfrac{(sq - tp)}{q}}{q^2}$$

$$\frac{d^2y}{dx^2} = -\frac{q^2r - 2pqs + tp^2}{q^3} \quad \text{...(iii)}$$

(*ii*) Put $= \frac{\partial f}{\partial x}, q = \frac{\partial f}{\partial y}, r = \frac{\partial^2 f}{\partial y^2}, t = \frac{\partial^2 f}{\partial y^2}$ and $s = \frac{\partial^2 f}{\partial x\,\partial y}$ in then

$$\frac{d^2y}{dx^2} = \frac{\left(\dfrac{\partial f}{\partial y}\right)^2\dfrac{\partial^2 f}{\partial x^2} - 2\left(\dfrac{\partial f}{\partial x}\right)\left(\dfrac{\partial f}{\partial y}\right)\left(\dfrac{\partial^2 f}{\partial x\,\partial y}\right) + \dfrac{\partial^2 f}{\partial y^2}\left(\dfrac{\partial f}{\partial x}\right)^2}{\left(\dfrac{\partial f}{\partial y}\right)^2}$$

or

$$\left(\frac{\partial f}{\partial y}\right)^3\frac{d^2y}{dx^2} = -\left(\frac{\partial f}{\partial y}\right)^2\frac{\partial^2 f}{\partial x^2} + 2\frac{\partial f}{\partial x}\frac{\partial f}{\partial y}\frac{\partial^2 f}{\partial x\,\partial y} - \frac{\partial^2 f}{\partial y^2}\left(\frac{\partial f}{\partial x}\right)^3$$

***Example* 5.29.** If $y^3 - 3ax^2 + x^3 = 0$, then prove that

$$\frac{d^2y}{dx^2} = -\frac{2a^2x^2}{y^5}$$

Solution : $f(x, y) = y^3 - 3ax^2 + x^3$

$$p = \frac{\partial f}{\partial x} = -6ax + 3x^2$$

$$q = \frac{\partial f}{\partial y} = 3y^2$$

$$r = \frac{\partial^2 f}{\partial x^2} = -6a + 6x$$

$$t = \frac{\partial^2 f}{\partial y^2} = 6y$$

$$s = \frac{\partial^2 f}{\partial x\, \partial y} = 0$$

Equation (i) is an implicit function. Then

$$\frac{d^2y}{dx^2} = \frac{q^2r - 2pqs + p^2t}{q^3}$$

$$= -\frac{(3y^2)^2(-6a + 6x) - 2(-6ax + 3x^2)\,3y^2\,(0) + (-6ax + 3x^2)^2(6y)}{(3y^2)^3}$$

$$= -\frac{54y^4(-a + x) + 54(-2ax + x^2)^2\,y}{27\,y^6}$$

$$= -\frac{2y^3(-a + x) + 2(4a^2x^2 + x^4 - ax^3)}{y^5}$$

But $\quad y^3 = 3ax^2 - x^3$, then

$$\frac{d^2y}{dx^2} = \frac{2(3ax^2 - x^2)(x - a) + 2(4a^2x^2 + x^4 - 4ax^3)}{y^5}$$

$$= -\frac{6ax^3 - 6a^2x^2 - 2x^4 + 2ax^3 + 8a^2x^2 + 2x^4 - 8ax^3}{y^5}$$

$$= --\frac{2a^2x^3}{y^5}$$

***Example* 5.30.** If $f(x, y) = 0$ and $\phi(y, z) = 0$, show that $\frac{\partial f}{\partial y}\frac{\partial \phi}{\partial z}\frac{dz}{dx} = \frac{\partial f}{\partial x}\frac{\partial \phi}{\partial y}$

Solution : $f(x, y) = 0$ ...(i)

and $\phi(y, z) = 0$ ...(ii)

Differentiating (i) with respect to x, we get

$$\frac{\partial f}{\partial x} = 0 = \frac{\partial f}{\partial x}\frac{dx}{dx} + \frac{\partial f}{\partial y}\frac{dy}{dx}$$

or
$$\frac{dy}{dx} = -\frac{\frac{\partial f}{\partial x}}{\frac{\partial f}{\partial y}} \quad ...(iii)$$

Differentiating (*ii*) with respect to y, we get

$$\frac{\partial \phi}{\partial y} = 0 = \frac{\partial \phi}{\partial y}\frac{dy}{dy} + \frac{\partial \phi}{\partial z}\frac{dz}{dy}$$

or
$$\frac{dz}{dy} = -\frac{\frac{\partial \phi}{\partial y}}{\frac{\partial \phi}{\partial z}}$$

Differentiating (*ii*) with respect to y, we get

$$\frac{\partial \phi}{\partial y} = 0 = \frac{\partial \phi}{\partial y}\frac{dy}{dy} + \frac{\partial \phi}{\partial z}\frac{dz}{dy}$$

or
$$\frac{dz}{dy} = -\frac{\frac{\partial \phi}{\partial y}}{\frac{\partial \phi}{\partial z}} \quad ...(iv)$$

Multiplying (*iii*) and (*iv*), we get

$$\frac{dy}{dx} \times \frac{dz}{dy} = \left(-\frac{\frac{\partial f}{\partial x}}{\frac{\partial f}{\partial y}}\right)\left(-\frac{\frac{\partial \phi}{\partial y}}{\frac{\partial \phi}{\partial z}}\right)$$

$$\frac{dz}{dx} = \frac{\left(\frac{\partial f}{\partial x}\right) \times \left(\frac{\partial \phi}{\partial y}\right)}{\left(\frac{\partial f}{\partial y}\right) \times \left(\frac{\partial \phi}{\partial z}\right)}$$

$$\frac{\partial f}{\partial y}\frac{\partial \phi}{\partial z}\frac{dz}{dx} = \frac{\partial f}{\partial x}\frac{\partial \phi}{\partial y}$$

***Example* 5.31.** If the curves $f(x, y) = 0$ and $\phi(y, z) = 0$, touch,
Show that at the point of contact.

$$\frac{\partial f}{\partial x}\frac{\partial \phi}{\partial y} = \frac{\partial f}{\partial y}\frac{\partial \phi}{\partial z}.$$

Solution :
$$f(x, y) = 0, \text{ then } \frac{dy}{dx} = -\frac{\frac{\partial f}{\partial x}}{\frac{\partial f}{\partial y}} \quad ...(i)$$

and
$$\phi(y, z) = 0, \text{ then } \frac{dz}{dy} = -\frac{\frac{\partial \phi}{\partial y}}{\frac{\partial \phi}{\partial z}} \quad ...(ii)$$

At the point of contact slope of both curve must be equal *i.e.*, $m_1 = m_2$

$$\frac{dy}{dx} = \frac{dz}{dy} \Rightarrow \frac{\frac{\partial f}{\partial x}}{\frac{\partial f}{\partial y}} = \frac{\frac{\partial \phi}{\partial y}}{\frac{\partial \phi}{\partial z}}$$

$$\frac{\partial f}{\partial z}\frac{\partial \phi}{\partial z} = \frac{\partial \phi}{\partial y}\frac{\partial f}{\partial y}$$

***Example* 5.32.** If $\phi(x, y, z) = 0$, show that $\left(\frac{\partial y}{\partial z}\right)_x\left(\frac{\partial z}{\partial x}\right)_y\left(\frac{\partial x}{\partial y}\right)_z = -1$

Solution : The given relation is implicit function and let y be a function of x and z. Then

$$\left(\frac{\partial y}{\partial z}\right)_x = -\frac{\frac{\partial \phi}{\partial z}}{\frac{\partial \phi}{\partial y}} \quad ...(i)$$

Similarly let z be a function of y. Then

$$\left(\frac{\partial z}{\partial x}\right)_y = -\frac{\frac{\partial \phi}{\partial x}}{\frac{\partial \phi}{\partial z}} \quad ...(ii)$$

and let x be a function y and z, then

$$\left(\frac{\partial x}{\partial y}\right)_z = -\frac{\frac{\partial \phi}{\partial y}}{\frac{\partial \phi}{\partial x}} \quad ...(iii)$$

Multiplying (*i*), (*ii*) and (*iii*) desired result is obtained.

***Example* 5.33.** If $u = 2xy^2 - 3x^2y$ and if x increases at the rate of 2 cm/sec and it passes through the value $x = 3$. Show that if y passing through the value $y = 1$ cm, y must be decreasing at the rate of $2\frac{2}{15}$ per sec. In order that u shall remain constant.

Solution : $u = 2xy^2 - 3x^2y$ and $\frac{dx}{dt} = 2$ cm/sec. Here $x = 3$ cm, $y = 1$ cm

Also it is given that u shall remain constant. *i.e.* $\frac{du}{dt} = 0$

By total derivative, $\frac{du}{dt} = \frac{\partial u}{\partial x}\frac{dx}{dt} + \frac{\partial u}{\partial y}\frac{dy}{dt}$

But $\frac{du}{dt} = 0$

$\therefore$ $0 = (2y^2 - 6xy)2 + (4xy - 3x^2)\frac{dy}{dt}$

Point are $x = 3$ and $y = 1$, then

$$0 = [2(1)2 - 6 \times 3 \times 1]\,2 + (4 \times 3 \times 1 - 3 \times 9)\frac{dy}{dx}$$

$$0 = -32 - 15\frac{dy}{dt} = -\frac{32}{15} = -2\frac{2}{15}$$

Hence y decrease at the rate of $2\frac{2}{15}$ cm/sec.

Example **5.34.** At a given instant the side of rectangle are 4 fit and 3 fit respectively and they are increasing at the rate of 1.5 ft/sec and 0.5 ft/sec respectively. Find the rate at which the area in increasing at that instant.

Solution : Let x and y are the sides of rectangle and A be the area of rectangle then

$$A = xy \quad \text{...}(i)$$

Here $$x = 4 \text{ ft}, y = 3ft \text{ and } \frac{dx}{dt} = 1.5 \text{ ft/sec}, \frac{dy}{dt} = 0.5 \text{ ft/sec}$$

Also A is the composite function of x and y. So the change in area

$$\frac{\partial A}{\partial t} = \frac{\partial A}{\partial x}\frac{dx}{dt} + \frac{\partial A}{\partial y}\frac{dy}{dt}$$

$$= y(1.5) + x(0.5) \quad (\therefore x = 4, y = 3)$$

$$= 3 \times 1.5 + 4 \times 0.5 = 6ft^2/\text{secl},$$

Hence increasing rate in area is 6.5 ft^2/sec.

Example **5.35.** If $z = \log(u^2 + v)$ and $u = e^{x^2y^2}$, $v = x^2$, $v = x^2 + y$ find $\frac{\partial z}{\partial x}$ and $\frac{\partial z}{\partial y}$

Solution : Here z is the function of u and v and u, v are the function of x and y then by change of variable

$$\frac{\partial z}{\partial x} = \frac{\partial z}{\partial u}\frac{\partial u}{\partial x} + \frac{\partial z}{\partial v}\frac{\partial v}{x}$$

$$= \frac{2u}{u^2+v}(2xe^{x^2} + y^2) + \frac{2x}{u^2+v} = \frac{4x\,ue^{x^2+y^2} + 2x}{u^2+v}$$

$$= \frac{2x(2u^2+1)}{u^2+v}$$

Similarly $$\frac{\partial z}{\partial y} = \frac{\partial z}{\partial u}\frac{\partial u}{\partial y} + \frac{\partial z}{\partial v}\frac{\partial v}{\partial y} = \frac{2u(2y)\,2^{x^2+y^2}}{u^2+v} + \frac{1}{u^2+v}$$

$$= \frac{4yu^2+1}{u^2+v}$$

Example **5.36.** If x is function of x and y and if $x = e^u + e^{-v}$ and $y = e^{-u} - e^v$ then prove that $\frac{\partial z}{\partial u} - \frac{\partial z}{\partial v} = x\frac{\partial z}{\partial x} - y\frac{\partial z}{\partial y}$

Solution : Let $z = f(x, y)$

$$x = e^u + e^{-v}, y = e^{-u} - e^v \text{ (given)} \quad \text{...}(i)$$

By change of variables,

$$\frac{\partial z}{\partial u} = \frac{\partial z}{\partial x}\frac{\partial x}{\partial u} + \frac{\partial z}{\partial y}\frac{\partial y}{\partial u}$$

$$= \frac{\partial z}{\partial x}e^u + \frac{\partial z}{\partial y}(-e^{-v})$$

$$= \frac{\partial z}{\partial x}e^u - \frac{\partial z}{\partial y}e^{-v} \quad \text{...}(ii)$$

Similarly $$\frac{\partial z}{\partial v} = -\frac{\partial z}{\partial x}e^{-v} - \frac{\partial z}{\partial y}e^v \quad \text{...}(iii)$$

Subtracting (*iii*) from (*ii*), we get

$$\frac{\partial z}{\partial u}-\frac{\partial z}{\partial v} = \frac{\partial z}{\partial x}(e^{u}+e^{-v})-\frac{\partial z}{\partial y}(e^{-u}-e^{v})$$

From (*i*)
$$\frac{\partial z}{\partial u}-\frac{\partial z}{\partial v} = \frac{\partial z}{\partial x}\times\frac{\partial z}{\partial y}y$$

Hence
$$\frac{\partial z}{\partial u}-\frac{\partial z}{\partial v} = x\frac{\partial z}{\partial x}-y\frac{\partial z}{\partial y}.$$

Example 5.37. If $u = f(r, s)$, $r = x + at$, $s = y + bt$ and x, y, t, are independent variables, show that

$$\frac{\partial u}{\partial t} = a\frac{\partial u}{\partial x}+b\frac{\partial u}{\partial y}$$

Solution : By change of variables

$$\frac{\partial u}{\partial t} = \frac{\partial u}{\partial r}\frac{\partial r}{\partial t}+\frac{\partial u}{\partial s}\frac{\partial s}{\partial t}$$

$$= \frac{\partial u}{\partial r}a+\frac{\partial u}{\partial s}b$$

$$= a\frac{\partial u}{\partial r}+b\frac{\partial u}{\partial s} \quad \text{...(i)}$$

Similarly
$$\frac{\partial u}{\partial x} = \frac{\partial u}{\partial r}\frac{\partial r}{\partial x}+\frac{\partial u}{\partial s}\frac{\partial s}{\partial x}$$

$$= \frac{\partial u}{\partial r}\times 1+\frac{\partial u}{\partial s}\times 0$$

$$= \frac{\partial u}{\partial r} \quad \text{...(ii)}$$

and
$$\frac{\partial u}{\partial y} = \frac{\partial u}{\partial r}\frac{\partial r}{\partial y}+\frac{\partial u}{\partial s}\frac{\partial s}{\partial y}$$

$$= \frac{\partial u}{\partial r}\times 0+\frac{\partial u}{\partial s}\times 1 \quad \text{...(iii)}$$

$$= \frac{\partial u}{\partial s}$$

Substituting the results of expression (*ii*) and (*iii*), the identity (*i*) can be rewritten as $\frac{\partial u}{\partial t} = a\frac{\partial u}{\partial x}+b\frac{\partial u}{\partial y}$

Example 5.38. If $u = f(x, y)$ and $x = r\cos\theta$, $y = r\sin\theta$, prove that $\left(\frac{\partial u}{\partial x}\right)^2+\left(\frac{\partial u}{\partial y}\right)^2 = \left(\frac{\partial u}{\partial r}\right)^2+\frac{1}{r^2}\left(\frac{\partial u}{\partial \theta}\right)^2$

Solution : $u = f(x, y)$

and
$$x = r\cos\theta$$

$$y = r\sin\theta$$

From the above relations, $x^2+y^2 = r^2$ and $\theta = \tan^{-1}\left(\frac{y}{x}\right)$

By change of variables

$$\frac{\partial u}{\partial r} = \frac{\partial u}{\partial x}\frac{\partial x}{\partial r}+\frac{\partial u}{\partial y}\frac{\partial y}{\partial r}$$

$$= \frac{\partial u}{\partial x}\cos\theta+\frac{\partial u}{\partial y}\sin\theta \quad \text{...(i)}$$

Similarly $$\frac{\partial u}{\partial \theta} = \frac{\partial u}{\partial x}\frac{\partial x}{\partial \theta} + \frac{\partial u}{\partial y}\frac{\partial y}{\partial \theta}$$

$$\frac{\partial u}{\partial \theta} = \frac{\partial u}{\partial x}(-r\sin\theta) + \frac{\partial u}{\partial y}(r\cos\theta)$$

$$= -r\sin\theta\frac{\partial u}{\partial x} + r\cos\theta\frac{\partial u}{\partial y}$$

$$\frac{1}{r}\frac{\partial u}{\partial \theta} = -\sin\theta\frac{\partial u}{\partial x} + \cos\theta\frac{\partial u}{\partial y} \quad ...(ii)$$

Squaring and adding (*i*) and (*ii*), we get

$$\left(\frac{\partial u}{\partial r}\right)^2 + \frac{1}{r^2}\left(\frac{\partial u}{\partial x}\right)^2 = \left(\frac{\partial u}{\partial x}\right)^2 + \left(\frac{\partial u}{\partial y}\right)^2 + 2\sin\theta\cos\theta\frac{\partial u}{\partial x}\frac{\partial u}{\partial y} - 2\sin\theta\cos\theta\frac{\partial u}{\partial x}\frac{\partial u}{\partial y}$$

or $$\left(\frac{\partial u}{\partial x}\right)^2 + \left(\frac{\partial u}{\partial y}\right)^2 = \left(\frac{\partial u}{\partial r}\right)^2 + \frac{1}{r^2}\left(\frac{\partial u}{\partial \theta}\right)^2$$

***Example* 5.39.** If $u = f(2x - 3y, 3y - 4z - 2x)$, prove that $\frac{1}{2}\frac{\partial u}{\partial x} + \frac{1}{3}\frac{\partial u}{\partial y} + \frac{1}{4}\frac{\partial u}{\partial z} = 0$

Solution: $u = f(2x - 3y, 3y - 4z, 4z - 2x)$

Let $2x - 3y = r$, $3y - 4z - 2x = t$, then

$$u = f(r, s, t) \text{ and} \quad ...(i)$$

$$r = 2x - 3y \quad ...(ii)$$

$$s = 3y - 4z \quad ...(iii)$$

$$t = 4z - 2x \quad ...(iv)$$

By change of variable and using (*ii*), (*iii*) and (*iv*), we get

$$\frac{\partial u}{\partial x} = \frac{\partial u}{\partial r}\frac{\partial r}{\partial x} + \frac{\partial u}{\partial s}\frac{\partial s}{\partial x} + \frac{\partial u}{\partial t}\frac{\partial t}{\partial x}$$

$$= \frac{\partial u}{\partial r}(2) + \frac{\partial u}{\partial s}(0) + \frac{\partial u}{\partial t}(-2)$$

$$= 2\frac{\partial u}{\partial r} - 2\frac{\partial u}{\partial t}$$

Similarly $$\frac{\partial u}{\partial y} = \frac{\partial u}{\partial r}\frac{\partial r}{\partial y} + \frac{\partial u}{\partial s}\frac{\partial s}{\partial y} + \frac{\partial u}{\partial t}\frac{\partial t}{\partial y}$$

$$= -3\frac{\partial u}{\partial r} + 3\frac{\partial u}{\partial s} \quad ...(vi)$$

and $$\frac{\partial u}{\partial z} = \frac{\partial u}{\partial r}\frac{\partial r}{\partial z} + \frac{\partial u}{\partial s}\frac{\partial s}{\partial z} + \frac{\partial u}{\partial t}\frac{\partial t}{\partial z}$$

$$= \frac{\partial u}{\partial r}(0) + \frac{\partial u}{\partial s}(-4) + \frac{\partial u}{\partial t}(4)$$

$$= -4\frac{\partial u}{\partial s} + 4\frac{\partial u}{\partial t} \quad ...(vii)$$

From (*v*), *vi* and (*vii*) may be written as

$$\frac{1}{2}\frac{\partial u}{\partial x} = \frac{\partial u}{\partial r} - \frac{\partial u}{\partial t}$$

$$\frac{1}{3}\left(\frac{\partial u}{\partial y}\right) = -\frac{\partial u}{\partial r} + \frac{\partial u}{\partial s}$$

$$\frac{1}{4}\frac{\partial u}{\partial x} = -\frac{\partial u}{\partial s}+\frac{\partial u}{\partial t}$$

Adding all these equations.

$$\frac{1}{2}\frac{\partial u}{\partial x}\frac{1}{3}\frac{\partial u}{\partial y}+\frac{1}{4}\frac{\partial u}{\partial x} = 0$$

***Example* 5.40.** If $x = u + v + w$, $y = vw + uv + wu$ and $z = uvw$ and f is *a* function of x, y, z show that

$$u\frac{\partial f}{\partial u}+v\frac{\partial f}{\partial v}+w\frac{\partial f}{\partial w} = x\frac{\partial f}{\partial x}+2y\frac{\partial f}{\partial y}+3z\frac{\partial f}{\partial z}$$

Solution: Here f is a function of x, y, z *i.e.* $f(x, y, z)$

and $$x = u+v+w \quad ...(i)$$

$$y = uv+uv+wu \quad ...(ii)$$

$$z = uvw \quad ...(iv)$$

By change of variable and using (*ii*), (*iii*) and (*iv*), we get

$$\frac{\partial f}{\partial u} = \frac{\partial f}{\partial x}\frac{\partial x}{\partial u}+\frac{\partial f}{\partial y}\frac{\partial y}{\partial u}+\frac{\partial f}{\partial z}\frac{\partial z}{\partial u}$$

$$= \frac{\partial f}{\partial x}(1)+\frac{\partial f}{\partial y}(v+w)+\frac{\partial f}{\partial z}vw$$

$$= \frac{\partial f}{\partial x}+(v+w)\frac{\partial f}{\partial y}+\frac{\partial f}{\partial z}vw \quad ...(v)$$

Similarly $$\frac{\partial f}{\partial v} = \frac{\partial f}{\partial x}\frac{\partial x}{\partial v}+\frac{\partial f}{\partial y}\frac{\partial y}{\partial v}+\frac{\partial f}{\partial z}\frac{\partial z}{\partial v}$$

$$= \frac{\partial f}{\partial x}(1)+\frac{\partial f}{\partial y}(u+w)+\frac{\partial f}{\partial z}uw$$

$$= \frac{\partial f}{\partial x}+(u+w)\frac{\partial f}{\partial y}+uw\frac{\partial f}{\partial z} \quad ...(vi)$$

and $$\frac{\partial f}{\partial w} = \frac{\partial f}{\partial x}\frac{\partial x}{\partial w}+\frac{\partial f}{\partial y}\frac{\partial y}{\partial w}+\frac{\partial f}{\partial z}\frac{\partial z}{\partial w}$$

$$= \frac{\partial f}{\partial x}(1)+\frac{\partial f}{\partial y}(v+u)+\frac{\partial f}{\partial z}uv$$

$$= \frac{\partial f}{\partial x}+(v+u)\frac{\partial f}{\partial y}+uv\frac{\partial f}{\partial z} \quad ...(vii)$$

Multiplying (v) by u, (*vi*) by v and (*vii*) by w and adding, we get

$$u\frac{\partial f}{\partial u}+v\frac{\partial f}{v}+w\frac{\partial f}{\partial w} = \frac{\partial f}{\partial x}(u+v+w)+\frac{\partial f}{\partial y}(2uv = 2vw+2wu)+\frac{\partial f}{\partial z}(3uvw)$$

$$u\frac{\partial f}{\partial u}+v\frac{\partial f}{\partial v}+w\frac{\partial f}{\partial} = x\frac{\partial f}{\partial x}+2y\frac{\partial f}{\partial y}+3z\frac{\partial f}{\partial z}$$

***Example* 5.41.** If $u = x^2 - y^2$, $v = 2xy$ and $f(x, y) = \theta(u, v)$, show by substitution that

$$\frac{\partial^2 f}{\partial x^2} + \frac{\partial^2 f}{\partial y^2} = 4(x^2 + y^2)\left(\frac{\partial^2 \theta}{\partial u^2} + \frac{\partial^2 \theta}{\partial v^2}\right)$$

Solution : $f(x, y) = \theta(u, v)$ ...(*i*)

and $u = x^2 - y^2, v = 2xy$ ...(*ii*)

By change of variable and using (*ii*), we get

$$\frac{\partial f}{\partial x} = \frac{\partial \theta}{\partial u}\frac{\partial u}{\partial x} + \frac{\partial \theta}{\partial v}\frac{\partial v}{\partial x}$$

$$= \frac{\partial \theta}{\partial u}(2x) + \frac{\partial \theta}{\partial v}(2y) \quad ...(iii)$$

Similarly $$\frac{\partial f}{\partial y} = \frac{\partial \theta}{\partial u}\frac{\partial u}{\partial y} + \frac{\partial \theta}{\partial v}\frac{\partial v}{\partial y}$$

$$= \frac{\partial \theta}{\partial u}(-2y) + \frac{\partial \theta}{\partial v}(2x) \quad ...(iv)$$

Again differentiating (*iii*) and (*iv*) partially with respect to x and y, we get

From (*iii*)

$$\frac{\partial^2 f}{\partial x^2} = \frac{\partial}{\partial x}\left(2x\frac{\partial \theta}{\partial u}\right) + \frac{\partial}{\partial x}\left(2y\frac{\partial \theta}{\partial v}\right)$$

$$= 2\frac{\partial \theta}{\partial u} + 2x\frac{\partial^2 \theta}{\partial u^2}\frac{\partial u}{\partial x} + 2y\frac{\partial^2 \theta}{\partial v^2}\frac{\partial v}{\partial x}$$

$$= 2\frac{\partial \theta}{\partial u} + 2x\frac{\partial^2 \theta}{\partial u^2}(2x) + 2y\frac{\partial^2 \theta}{\partial v^2}(2y)$$

$$= 2\frac{\partial \theta}{\partial u} + 4x^2\frac{\partial^2 \theta}{\partial u^2} + 4y^2\frac{\partial^2 \theta}{\partial v^2} \quad ...(v)$$

Similarly, from (*iv*)

$$\frac{\partial^2 f}{\partial y^2} = 4y^2\frac{\partial^2 \theta}{\partial u^2} + 4x^2\frac{\partial^2 \theta}{\partial v^2} - 2\frac{\partial \theta}{\partial u} \quad ...(vi)$$

Adding (*v*) and (*vi*) than, we get

$$\frac{\partial^2 f}{\partial x^2} + \frac{\partial^2 f}{\partial y^2} = 4(x^2 + y^2)\frac{\partial^2 \theta}{\partial u^2} + 4(x^2 + y^2)\frac{\partial^2 \theta}{\partial v^2} + 0$$

$$\frac{\partial^2 f}{\partial x^2} + \frac{\partial^2 f}{\partial y^2} = 4(x^2 + y^2)\left(\frac{\partial^2 \theta}{\partial u^2} + \frac{\partial^2 \theta}{\partial v^2}\right)$$

***Example* 5.42.** If $x + y = 2e^{\theta}\cos\varphi$ and $x - y = 2ie^{\theta}\sin\varphi$, then show that $\frac{\partial^2 u}{\partial \theta^2} + \frac{\partial^2 u}{\partial \varphi^2} = 4xy\frac{\partial^2 u}{\partial x\,\partial y}$

Solution : $x + y = 2e^{\theta}\cos\varphi$ ...(*i*)

$$x - y = 2ie^{\theta}\sin\varphi \quad ...(ii)$$

Adding (*i*) and (*ii*)

$$x = e^{\theta + i\varphi} \quad ...(iii)$$

Subtracting (*ii*) from (*i*)

$$y = e^{\theta - i\varphi} \quad ...(iv)$$

Let u be a composite function of θ and φ, such that

$$u = f(x, y) \quad ...(A)$$

and

$$x = e^{\theta} \times e^{i\varphi}$$

$$y = e^{\theta} \times e^{-i\varphi} \quad ...(B)$$

By change of variables, using (A) and (B), we get

$$\frac{\partial u}{\partial \theta} = \frac{\partial u}{\partial x}\frac{\partial x}{\partial \theta} + \frac{\partial u}{\partial y}\frac{\partial y}{\partial \theta}$$

$$= \frac{\partial u}{\partial x}(e^{\theta} e^{i\phi}) + \frac{\partial u}{\partial y}(e^{\theta} e^{-i\phi}) \quad ...(v)$$

Similarly

$$\frac{\partial u}{\partial \varphi} = i\frac{\partial u}{\partial x}(e^{\theta} e^{i\varphi}) - i\frac{\partial u}{\partial y}(e^{\theta} e^{-i\varphi}) \quad ...(vi)$$

Again differentiating (*v*) and (*vi*) partially with respect to θ and φ, we get From (*v*)
From (*v*)

$$\frac{\partial^2 u}{\partial \theta^2} = \frac{\partial}{\partial \theta}\left[\frac{\partial u}{\partial x} e^{\theta} e^{i\phi}\right] + \frac{\partial}{\partial \theta}\left[\frac{\partial u}{\partial y} e^{\theta} e^{-i\varphi}\right]$$

$$= \frac{\partial u}{\partial x} e^{\theta} e^{i\varphi} + e^{\theta} e^{i\varphi} \frac{\partial}{\partial \theta}\left(\frac{\partial u}{\partial x}\right) + \frac{\partial u}{\partial y} e^{\theta} e^{-i\varphi} + e^{\theta} e^{-i\varphi} \frac{\partial}{\partial \theta}\left(\frac{\partial u}{\partial y}\right)$$

$$= \frac{\partial u}{\partial x} e^{\theta} e^{i\varphi} + \frac{\partial u}{\partial y} e^{\theta} e^{-i\varphi} + e^{\theta} e^{i\varphi} \frac{\partial}{\partial x}\left(\frac{\partial u}{\partial \theta}\right) + e^{\theta} e^{-i\varphi} \frac{\partial}{\partial y}\left(\frac{\partial u}{\partial \theta}\right)$$

$$\left[\therefore \frac{\partial}{\partial x}\left(\frac{\partial}{\partial y}\right) = \frac{\partial}{\partial y}\left(\frac{\partial}{\partial x}\right)\right]$$

$$\frac{\partial^2 u}{\partial \theta^2} = \frac{\partial u}{\partial x} e^{\theta} e^{i\varphi} + \frac{\partial u}{\partial y} e^{\theta} e^{-i\varphi} + e^{\theta} e^{i\varphi} \frac{\partial}{\partial x}\left[\frac{\partial u}{\partial x} e^{\theta} e^{i\varphi} + \frac{\partial u}{\partial y} e^{\theta} e^{-i\varphi}\right] +$$

$$e^{\theta} e^{-i\varphi} \frac{\partial}{\partial y}\left[\frac{\partial u}{\partial x} e^{\theta} e^{i\varphi} + \frac{\partial u}{\partial y} e^{\theta} e^{-i\varphi}\right]$$

$$\frac{\partial^2 u}{\partial \theta^2} = \frac{\partial u}{\partial x} e^{\theta} e^{i\varphi} + \frac{\partial u}{\partial y} e^{\theta} e^{-i\varphi} + e^{2\theta} e^{2i\varphi} \frac{\partial^2 u}{\partial x^2} + e^{2\theta} \frac{\partial^2 u}{\partial x\, \partial y} + e^{2\theta} e^{-2i\varphi} \frac{\partial^2 u}{\partial y^2} \quad ...(vii)$$

Similarly from (*vi*)

$$\frac{\partial^2 u}{\partial \varphi^2} = -\frac{\partial u}{\partial x} e^{\theta} e^{i\varphi} - \frac{\partial u}{\partial y} e^{\theta} e^{-i\varphi} - \frac{\partial^2 u}{\partial x^2} e^{2\theta} e^{2i\varphi} + \frac{\partial^2 u}{\partial x\, \partial y} e^{2\theta} - \frac{\partial^2 u}{\partial y^2} e^{2\theta} e^{-2i\varphi} \quad ...(viii)$$

Adding (*vii*) and (*vii*), we get

$$\frac{\partial^2 u}{\partial \theta^2} + \frac{\partial^2 u}{\partial \varphi^2} = 4e^{2\theta} \frac{\partial^2 u}{\partial x\, \partial y}$$

or

$$\frac{\partial^2 u}{\partial \theta^2} + \frac{\partial^2 u}{\partial \varphi^2} = 4xy \frac{\partial^2 u}{\partial x\, \partial y}$$

EXERCISE 5.4

1. (*i*) If $u = x^2 + y^2 + z^2$ and $x = e^{2t}$, $y = e^{2t}$, cos $3t$, $z = e^{2t}$ sin $3t$, find $\frac{du}{dt}$ as a total derivatives and verify the result by direct substitution. **Ans.** $\frac{du}{dt} = 8e^{4t}$

   (*ii*) If $u = x^3 + y^3$ where $x = a$ cost, $y = b$ sin, t, find $\frac{du}{dt}$ and verify the result.

   Ans. $\frac{du}{dt} = 3 \text{ sin}t \text{ cos}t \, (b^3 \text{ sin}t - a^3 \text{ cos}t)$

2. (*i*) Find the total derivative coefficient of x^2y with respect to x when x and y connected by the relation $x^2 + xy + y^2 = 1$.

   Ans. $2xy - \frac{x^2(2x+y)}{x+2y}$

   (*ii*) If $x^3 + 3x^2y + 6xy^2 + y^3 = 1$, find $\frac{dy}{dx}$

   Ans. $\frac{dy}{dx} = -\frac{x^2 + 2xy + 2y^2}{x^2 - 4xy + y^2}$

   (*iii*) If $u = x \log xy$ where $x^3 + y^3 + 3xy = 1$, find $\frac{du}{dx}$

   Ans. $\frac{du}{dx} = \log xy + 1 - \frac{x(x^2+y)}{y(x+y^2)}$

3. Find $\frac{dy}{dx}$, when $x^y = y^x$.

   Ans. $\frac{dy}{dx} = \frac{y(y - x\log)}{x(x - y\log x)}$

4. (*i*) If $x^4 + y^4 = 4a^2xy$, show that $(a^2x - y^3)^3 \frac{d^2y}{dx^2} = 2a^2xy\,(3a^4 + x^2y^2)$

   (*ii*) If $x^3 + y^3 - 3axy = 0$, show that $\frac{d^2y}{dx^2} = \frac{2a^3xy}{(ax - y^2)^3}$

5. If x increases at the rate of 2 cm/sec at the instant when $x = 3$ cm and $y = 1$ cm, at what rate must y be changing in order that the function $2xy - 3x^2y$ shall be neither increasing nor decreasing ?

   Ans. $\frac{dy}{dt} = \frac{-32}{15}$ cm/sec

6. (*i*) If $u = f(x - y, y - z, z - x)$, prove that $\frac{\partial u}{\partial x} + \frac{\partial u}{\partial y} + \frac{\partial u}{\partial z} = 0$

   (*ii*) If u = f (r, s, t), where $r = \frac{x}{y}$; $s = \frac{y}{x}$; and $x = \frac{z}{x}$ then prove that $x\frac{\partial u}{\partial x} + y\frac{\partial u}{\partial y} + z\frac{\partial u}{\partial z} = 0$

7. (*i*) If $z^3 + xy - y^2z = 6$, obtain the expression for $\frac{\partial y}{\partial x}$ and $\frac{\partial z}{\partial x}$ in term of x, y, z and find their value at the point (0, 1, 2,).

   $\left[\text{Hint: If } f(x, y, z) = c, \text{ then } \frac{\partial y}{\partial x} = -\frac{\frac{\partial f}{\partial x}}{\frac{\partial f}{\partial y}} \text{ and } \frac{\partial z}{\partial x} = -\frac{\frac{\partial f}{\partial x}}{\frac{\partial f}{\partial z}}\right]$

   Ans. $\left(\frac{\partial y}{\partial x}\right) at\ (0, 1, 2) = \frac{1}{4}$, $\left(\frac{\partial z}{\partial x}\right) at\ (0, 1, 2) = -\frac{1}{11}$

8. Give that $u = f(x^2 + y^2 + z^2)$ where $x = r \cos\theta \cos\phi$, $y = r\,\phi \cos\phi$ and $z = r \sin\theta$. Find $\frac{\partial u}{\partial \theta}$ *and* $\frac{\partial u}{\partial \phi}$

   Ans. $\frac{\partial u}{\partial \theta} = \frac{\partial u}{\partial \phi} = 0$

9. If z is function of x and u and y be other variable such that $u = lx + my$, $v = ly - mx$, show that

$$\frac{\partial^2 z}{\partial x^2} + \frac{\partial^2 z}{\partial y^2} = (l^2 + m^2)\left(\frac{\partial^2 z}{\partial u^2} + \frac{\partial^2 z}{\partial v^2}\right)$$

5.10 CONCEPT OF A JACOBIAN

If u and v are function of two independent variable x and y, then the determinant $\begin{vmatrix} \frac{\partial u}{\partial x} & \frac{\partial u}{\partial y} \\ \frac{\partial v}{\partial x} & \frac{\partial v}{\partial y} \end{vmatrix}$ is called the Jacobian of u, v with respect to x, y and is written as $\frac{\partial(u, v)}{\partial(x, y)}$ *or* $J\left(\frac{u, v}{x, y}\right)$

Similarly, the Jacobian of u, v, w with respect to x, y, z is called to x, y, z is defined as

$$\frac{\partial(u, v, w)}{\partial(x, y, z)} = \begin{pmatrix} \frac{\partial u}{\partial x} & \frac{\partial u}{\partial y} & \frac{\partial u}{\partial z} \\ \frac{\partial v}{\partial x} & \frac{\partial v}{\partial y} & \frac{\partial v}{\partial z} \\ \frac{\partial w}{\partial x} & \frac{\partial w}{\partial y} & \frac{\partial z}{\partial y} \end{pmatrix}$$

In similarly way Jacobians of four or more variable can be defined. The application of Jacobians is in connection with the change of variable in multiple integrals.

5.11 PROPERTIES OF JACOBIANS

There are two properties of Jacobians

(*a*) If $J = \frac{\partial(u, v)}{\partial(x, y)}$ and $J' = \frac{\partial(x, y)}{\partial(u, v)}$, then prove that $JJ' = 1$

Proof : Let $u = f(x, y)$ and $v = g(x, y)$ ...(*i*)

and let $x = f(u, v)$ and $y = \psi(u, v)$ ...(*ii*)

By change of variables

$$\frac{\partial u}{\partial u} = \frac{\partial u}{\partial x} \times \frac{\partial x}{\partial x} + \frac{\partial u}{\partial y} \times \frac{\partial y}{\partial u} = 1 \qquad \left[\because \frac{\partial u}{\partial u} = 1\right]$$

$$\frac{\partial u}{\partial v} = \frac{\partial u}{\partial x} \times \frac{\partial x}{\partial v} + \frac{\partial x}{\partial v} \times \frac{\partial y}{\partial v} = 0 \qquad \left[\because \frac{\partial u}{\partial v} = 0\right]$$

and

$$\frac{\partial v}{\partial u} = \frac{\partial v}{\partial x} \times \frac{\partial x}{\partial u} + \frac{\partial v}{\partial y} \times \frac{\partial y}{\partial u} = 0 \qquad \left[\because \frac{\partial v}{\partial u} = 0\right]$$

$$\frac{\partial v}{\partial v} = \frac{\partial v}{\partial x} \times \frac{\partial x}{\partial v} + \frac{\partial v}{\partial y} \times \frac{\partial y}{\partial v} = 1 \qquad \left[\because \frac{\partial v}{\partial v} = 1\right]$$

Here

$$J = \frac{\partial(u, v)}{\partial(x, y)} = \begin{vmatrix} \frac{\partial u}{\partial x} & \frac{\partial u}{\partial y} \\ \frac{\partial v}{\partial x} & \frac{\partial v}{\partial y} \end{vmatrix}$$

and

$$J' = \frac{\partial(x, y)}{\partial(u, v)} = \begin{vmatrix} \frac{\partial x}{\partial u} & \frac{\partial x}{\partial v} \\ \frac{\partial y}{\partial u} & \frac{\partial y}{\partial v} \end{vmatrix}$$

so $$JJ' = \begin{vmatrix} \frac{\partial u}{\partial x} & \frac{\partial u}{\partial y} \\ \frac{\partial v}{\partial x} & \frac{\partial v}{\partial y} \end{vmatrix} \begin{vmatrix} \frac{\partial x}{\partial u} & \frac{\partial x}{\partial v} \\ \frac{\partial y}{\partial u} & \frac{\partial y}{\partial v} \end{vmatrix}$$

$$JJ' = \begin{vmatrix} \frac{\partial u}{\partial x}\frac{\partial x}{\partial u} + \frac{\partial u}{\partial y}\frac{\partial y}{\partial u} & \frac{\partial u}{\partial x}\frac{\partial x}{\partial v} + \frac{\partial u}{\partial y}\frac{\partial y}{\partial v} \\ \frac{\partial v}{\partial x}\frac{\partial x}{\partial u} + \frac{\partial v}{\partial y}\frac{\partial y}{\partial u} & \frac{\partial v}{\partial x}\frac{\partial v}{\partial v} + \frac{\partial v}{\partial y}\frac{\partial y}{\partial v} \end{vmatrix}$$

From above relations

$$JJ' = \begin{vmatrix} 1 & 0 \\ 0 & 1 \end{vmatrix} = 1$$

Hence $$JJ' = 1$$

(*b*) If u, v are function of r, s and r, s function of x, y, then $\frac{\partial(u, v)}{\partial(x, y)} = \frac{\partial(u, v)}{\partial(r, s)} \times \frac{\partial(r, s)}{\partial(x, y)}$

Prof : Let $= \left.\begin{matrix} u = f(r, s) \\ v = g(r, s) \end{matrix}\right]$ ...(*i*)

and $= \left.\begin{matrix} r = \phi(x, y) \\ s = \varphi(x, y) \end{matrix}\right]$ ...(*ii*)

$$\text{R.H.S} = \frac{\partial(u, v)}{\partial(r, s)} \times \frac{\partial(r, s)}{\partial(x, y)}$$

By definition of Jacobians

$$\frac{\partial(u, v)}{\partial(r, s)} \times \frac{\partial(r, s)}{\partial(x, y)} = \begin{vmatrix} \frac{\partial u}{\partial r} & \frac{\partial u}{\partial s} \\ \frac{\partial v}{\partial r} & \frac{\partial v}{\partial s} \end{vmatrix} \begin{vmatrix} \frac{\partial r}{\partial x} & \frac{\partial r}{\partial y} \\ \frac{\partial s}{\partial x} & \frac{\partial s}{\partial y} \end{vmatrix} = \begin{vmatrix} \frac{\partial u}{\partial r}\frac{\partial r}{\partial x} + \frac{\partial u}{\partial s}\frac{\partial s}{\partial x} & \frac{\partial u}{\partial r}\frac{\partial r}{\partial y} + \frac{\partial u}{\partial s}\frac{\partial s}{\partial y} \\ \frac{\partial v}{\partial r}\frac{\partial r}{\partial s} + \frac{\partial v}{\partial s}\frac{\partial s}{\partial x} & \frac{\partial v}{\partial r}\frac{\partial r}{\partial y} + \frac{\partial v}{\partial s}\frac{\partial s}{\partial y} \end{vmatrix}$$

By total derivatives

$$= \begin{vmatrix} \frac{\partial u}{\partial x} & \frac{\partial u}{\partial y} \\ \frac{\partial v}{\partial x} & \frac{\partial v}{\partial y} \end{vmatrix} = \frac{\partial(u, v)}{\partial(x, y)}$$

***Example* 5.43.** If $x = r\cos\theta$, $y = r\sin\theta$, evaluate $\frac{\partial(r, \theta)}{\partial(x, y)}, \frac{\partial(x, y)}{\partial(r, \theta)}$ and prove that $\frac{\partial(r, \theta)}{\partial(x, y)} \times \frac{\partial(x, y)}{\partial(r, \theta)} = 1$

Solution : By definition of Jacobians

$$\frac{\partial(r, \theta)}{\partial(x, y)} = \begin{vmatrix} \frac{\partial r}{\partial x} & \frac{\partial r}{\partial y} \\ \frac{\partial \theta}{\partial x} & \frac{\partial \theta}{\partial y} \end{vmatrix} \quad \text{...(A)}$$

and $$\frac{\partial(x, y)}{\partial(r, \theta)} = \begin{vmatrix} \frac{\partial x}{\partial r} & \frac{\partial x}{\partial \theta} \\ \frac{\partial y}{\partial r} & \frac{\partial y}{\partial \theta} \end{vmatrix} \quad \text{...(B)}$$

Here $$x = r\cos\theta \quad \text{...(i)}$$

$$y = r\sin\theta \quad \text{...(ii)}$$

and $$x^2 + y^2 = r^2 \quad ...(iii)$$

$$\tan^{-1}\frac{y}{x} = \theta \quad ...(iv)$$

From (*iii*)

$$2r\frac{\partial r}{\partial x} = 2x, \quad \therefore \frac{\partial r}{\partial x} = \frac{x}{r}$$

$$2r\frac{\partial r}{\partial y} = 2y, \quad \therefore \frac{\partial r}{\partial y} = \frac{y}{r}$$

From (*iv*)

$$\frac{\partial\theta}{\partial x} = \frac{1}{1+\frac{y^2}{x^2}}\left(-\frac{y}{x^2}\right) = \frac{-y}{x^2+y^2}$$

$$\frac{\partial\theta}{\partial x} = \frac{-y}{r^2}$$

$$\frac{\partial\theta}{\partial y} = \frac{1}{1+\frac{y^2}{x^2}}\frac{1}{x} = \frac{x}{x^2+y^2}$$

$$\frac{\partial\theta}{\partial y} = \frac{x}{r^2}$$

From (A)

$$\frac{\partial(r,\theta)}{\partial(x,y)} = \begin{vmatrix} \frac{\partial r}{\partial x} & \frac{\partial r}{\partial y} \\ \frac{\partial\theta}{\partial x} & \frac{\partial\theta}{\partial y} \end{vmatrix} = \begin{vmatrix} \frac{x}{r} & \frac{y}{r} \\ \frac{-y}{r^2} & \frac{x}{r^2} \end{vmatrix}$$

$$= \frac{x^2}{r^3} + \frac{y^2}{r^3} = \frac{x^2+y^2}{r^3} = \frac{1}{r} \quad ...(C)$$

Now, from (*i*) $$\frac{\partial x}{\partial r} = \cos, \theta, \frac{\partial x}{\partial\theta} = -r\cos\theta$$

and from (*ii*) $$\frac{\partial y}{\partial r} = \sin\theta, \frac{\partial y}{\partial\theta} = r\cos\theta$$

From (B), $$\frac{\partial(x,y)}{\partial(r,\theta)} = \begin{vmatrix} \frac{\partial x}{\partial r} & \frac{\partial x}{\partial\theta} \\ \frac{\partial y}{\partial r} & \frac{\partial y}{\partial\theta} \end{vmatrix} = \begin{vmatrix} \cos\theta & -r\sin\theta \\ \sin\theta & r\sin\theta \end{vmatrix}$$

$$r\cos^2\theta + r\sin^2\theta = r \quad ...(D)$$

Again $$\frac{\partial(r,\theta)}{\partial(x,y)} \times \frac{\partial(x,y)}{\partial(r,\theta)} = \frac{1}{r}\times r = 1$$ From (C) and (D)

Hence $$\frac{\partial(r,\theta)}{\partial(x,y)} \times \frac{\partial(x,y)}{\partial(r,\theta)} = 1$$

***Example* 5.44.** If $x = u(1 - v)$ and $y = uv$, Prove that $JJ' = 1$

Solution : Let $J = \frac{\partial(x, y)}{\partial(u, v)}$ *and* $J = \frac{\partial(u, v)}{\partial(x, y)}$

By definition of Jacobians

$$\frac{\partial(x, y)}{\partial(u, v)} = \begin{vmatrix} \frac{\partial x}{\partial u} & \frac{\partial x}{\partial v} \\ \frac{\partial y}{\partial u} & \frac{\partial y}{\partial v} \end{vmatrix}$$

Here $x = u(1 - v)$ and $y = uv$, then

$$\frac{\partial x}{\partial u} = 1 - v$$

$$\frac{\partial x}{\partial v} = -u$$

$$\frac{\partial y}{\partial u} = v$$

$$\frac{\partial y}{\partial v} = u$$

$$\therefore \quad \frac{\partial(x, y)}{\partial(u, v)} = \begin{vmatrix} 1 - v & -u \\ v & u \end{vmatrix} = u(1 - v) + uv = 0 \qquad \text{...(A)}$$

Hence $\frac{\partial(x, y)}{\partial(u, v)} = u$

Again $J' = \frac{\partial(u, v)}{\partial(x, y)}$

By definition of jacobians

$$\frac{\partial(u, v)}{\partial(x, y)} = \begin{vmatrix} \frac{\partial u}{\partial x} & \frac{\partial u}{\partial y} \\ \frac{\partial v}{\partial x} & \frac{\partial v}{\partial y} \end{vmatrix}$$

Here $x = u - uv$ and $y = uv \therefore x = u - y$

or $u = x + y \therefore \frac{\partial u}{\partial x} = 1, \frac{\partial u}{\partial y} = 1$

Again $v = \frac{y}{u}$

or $v = \frac{y}{x + y)} \therefore \frac{\partial v}{\partial x} = \frac{-1}{(x + y)^2}$ *and* $\frac{\partial v}{\partial y} = \frac{x}{(x + y)^2}$

Now $\frac{\partial(u, v)}{\partial(x, y)} = \frac{x}{(x + y)^2} + \frac{y}{(x + y)^2} = \frac{(x + y)}{(x + y)^2}$

$$\frac{\partial(u, v)}{\partial(x, y)} = \frac{1}{(x + y)} = \frac{1}{u} \qquad \text{...(B)}$$

From (A) and (B), we get

$$JJ' = u \times \frac{1}{u} = 1$$

***Example* 5.45.** If $x = a \cosh \xi \cos \eta$, $y = a \sinh \xi \sin \eta$ show that $\frac{\partial(x, y)}{\partial(\xi, \eta)} = \frac{1}{2} a^2 (\cosh 2\xi - \cos 2\eta)$

Solution : $x = a \cosh \xi \cos \eta$

and $\quad y = a \sinh \xi \sin \xi$

by definition of Jacobians

$$\frac{\partial(x, y)}{6(\xi, \eta)} = \begin{vmatrix} \frac{\partial x}{\partial \xi} & \frac{\partial x}{\partial \eta} \\ \frac{\partial y}{\partial \xi} & \frac{\partial y}{\partial \eta} \end{vmatrix} \qquad \text{...(A)}$$

From the given relations

$$\frac{\partial x}{\partial \xi} = a \sinh \xi \cos \eta$$

$$\frac{\partial x}{\partial \eta} = -a \cosh \xi \sin \eta$$

$$\frac{\partial y}{\partial \xi} = a \cosh \xi \sinh \eta$$

$$\frac{\partial y}{\partial \xi} = a \cosh \xi \sinh \eta$$

$$\frac{\partial y}{\partial \eta} = a \text{ sing } \xi \cos \eta$$

From (A)

$$\frac{\partial(x, y)}{\partial(\xi, \eta)} = \begin{vmatrix} a \sinh \xi \cos \eta & -a \cosh \xi \sin \eta \\ a \cosh \xi \sin \eta & a \sinh \xi \cos \eta \end{vmatrix}$$

$$= a^2 \sinh^2 \xi \cos^2 \eta + a^2 \cosh^2 \xi \sin^2 h$$

From hyperbolic function

$$\cosh^2\xi - \sinh^2\xi = 1$$

and

$$\cosh^2 + \sinh^2 \xi = \cosh 2\xi$$

$\therefore$

$$\frac{\partial(x, y)}{\partial(\xi, \eta)} = a^2 [\sinh^2 \xi (1 - \sin^2\eta) + (1 + \sinh^2\xi] \sin^2\eta]$$

$$= a^2 [\sinh^2 \xi + \sin^2 \eta]$$

$$= a^2 \left[\frac{\cosh 2\xi - 1}{2} + \frac{1 - \cos 2\eta}{2}\right] = \frac{a^2}{2} = \frac{a^2}{2} [\cosh 2\xi - \cos^2\eta]$$

Hence

$$\frac{\partial(x, y)}{\partial(\xi, \eta)} = \frac{a^2}{2} [\cosh 2\xi - \cos 2\eta]$$

***Example* 5.46.** If $u = \frac{x + y}{1 - xy}$ and $v = \tan^{-1} x + \tan^{-1} y$, find $\frac{\partial(u, v)}{\partial(x, y)}$

Solution : By definition of Jacobians

$$\frac{\partial(u, v)}{\partial(x, y)} = \begin{vmatrix} \frac{\partial u}{\partial x} & \frac{\partial u}{\partial y} \\ \frac{\partial v}{\partial x} & \frac{\partial v}{\partial y} \end{vmatrix} \qquad \text{...(A)}$$

Here $u = \frac{x + y}{1 - xy}$ and $v = \tan^{-1} x + \tan^{-1} y$. Therefore $\frac{\partial u}{\partial x} = \frac{(1 - xy) - (x + y)(-y)}{(1 - xy)^2} = \frac{1 - xy + xy + y^2}{(1 + xy)^2}$

$$\frac{\partial u}{\partial x} = \frac{1+y^2}{(1-xy)^2}$$

and
$$\frac{\partial u}{\partial y} = \frac{(1-xy)-(x+y)(-x)}{(1-xy)^2} = \frac{1-xy+x^2+xy}{(1-xy)^2}$$

$$\frac{\partial u}{\partial y} = \frac{1+x^2}{(1-xy)^2}$$

Similarly
$$\frac{\partial v}{\partial x} = \frac{1}{1+x^2} \text{ and } \frac{\partial v}{\partial y} = \frac{1}{1+y^2}$$

From (A)
$$\frac{\partial(u,v)}{\partial(x,y)} = \begin{vmatrix} \dfrac{1+y^2}{(1-xy)^2} & \dfrac{1+x^2}{(1-xy)^2} \\ \dfrac{1}{1+x^2} & \dfrac{1}{1+y^2} \end{vmatrix}$$

$\therefore$
$$\frac{\partial(u,v)}{\partial(x,y)} = 0$$

***Example* 5.47.** If $u = \frac{yz}{x}$, $v = \frac{zx}{y}$ and $w = \frac{xy}{z}$, show that $\frac{\partial(u,v,w)}{\partial(x,y,z)} = 4$

Solution : By definition of Jacobians

$$\frac{\partial(u,v,w)}{\partial(x,y,z)} = \begin{vmatrix} \dfrac{\partial u}{\partial x} & \dfrac{\partial u}{\partial y} & \dfrac{\partial u}{\partial z} \\ \dfrac{\partial v}{\partial x} & \dfrac{\partial v}{\partial y} & \dfrac{\partial v}{\partial z} \\ \dfrac{\partial w}{\partial x} & \dfrac{\partial w}{\partial y} & \dfrac{\partial w}{\partial z} \end{vmatrix}$$

Here
$$u = \frac{yz}{x},\ v = \frac{zx}{y} \text{ and } w = \frac{xy}{z}. \text{ Then}$$

$$\frac{\partial u}{\partial x} = \frac{-yz}{x^2},\ \frac{\partial u}{\partial y} = \frac{z}{x},\ \frac{\partial u}{\partial z} = \frac{y}{x}$$

and
$$\frac{\partial v}{\partial x} = \frac{z}{y},\ \frac{\partial v}{\partial y} = \frac{-zx}{y^2},\ \frac{\partial v}{\partial z} = \frac{x}{y}$$

Similarly,
$$\frac{\partial w}{\partial x} = \frac{y}{z},\ \frac{\partial w}{\partial y} = \frac{x}{z},\ \frac{\partial w}{\partial z} = \frac{-xy}{z^2}$$

$$\frac{\partial(u,v,w)}{\partial(x,y,z)} = \begin{vmatrix} \dfrac{-yz}{x^2} & \dfrac{z}{x} & \dfrac{y}{x} \\ \dfrac{z}{y} & \dfrac{-zx}{y^2} & \dfrac{x}{y} \\ \dfrac{y}{z} & \dfrac{x}{z} & \dfrac{-xy}{z^2} \end{vmatrix}$$

Taking $\frac{1}{x}, \frac{1}{y}$ and $\frac{1}{z}$ as common from rows R_1, R_2 and R_3 respectively

$$= \frac{1}{xyz}\begin{vmatrix} \frac{-yz}{x} & z & y \\ z & \frac{-zx}{y} & x \\ y & x & \frac{-xy}{z} \end{vmatrix}$$

$$= \frac{1}{xyz}\left[-\frac{yz}{x}\left(\frac{x^2yz}{yz} - x^2\right) - z(-xy + xy) + y(zx + zx)\right]$$

$$= \frac{1}{xyz}[0 + 2xyz + 2xyz] = 4$$

Hence $\qquad \frac{\partial(u, v, w)}{\partial(x, y, z)} = 4$

Example **5.48.** If $F = xu + v - y,\ G = u^2 + vy + w,\ H = zu - v + vw,$ compute $\frac{\partial(F, G, H)}{\partial(u, v, w)}$.

Solution : By definition of Jacobians

$$\frac{\partial(F, G, H)}{\partial(u, v, w)} = \begin{vmatrix} \frac{\partial F}{\partial u} & \frac{\partial F}{\partial v} & \frac{\partial F}{\partial w} \\ \frac{\partial G}{\partial u} & \frac{\partial G}{\partial v} & \frac{\partial G}{\partial w} \\ \frac{\partial H}{\partial u} & \frac{\partial H}{\partial v} & \frac{\partial H}{\partial w} \end{vmatrix} \qquad ...(A)$$

Here $\qquad F = xu + v - y, G = u^2 + vy + w$

and $\qquad H = zu - v + vw$

so, $\qquad \frac{\partial F}{\partial u} = x,\ \frac{\partial F}{\partial v} = 1,\ \frac{\partial F}{\partial w} = 0$

$$\frac{\partial G}{\partial u} = 2u,\ \frac{\partial G}{\partial v} = y,\ \frac{\partial G}{\partial w} = 1$$

and $\qquad \frac{\partial H}{\partial u} = z,\ \frac{\partial H}{\partial v} = -1 + w,\ \frac{\partial H}{\partial w} = v$

Substituting these values in (A)

$$\frac{\partial(F, G, H)}{\partial(u, v, w)} = \begin{vmatrix} x & 1 & 0 \\ 2u & y & 1 \\ z & -1 + w & v \end{vmatrix}$$

$$= x[yv + 1 - w] - 1[2uv - z] = xyv + x - xw - 2uw + z$$

Hence $\qquad \frac{\partial(F, G, H)}{\partial(u, v, w)} = xy\underline{z} + x - wx - 2uv + z$

Example **5.49.** If $u = x + y + z,\ uv = y + z,\ uvw = z,$ show that $\frac{\partial(x, y, z)}{\partial(u, v, w)} = u^2v$

Solution : By definition of Jacobians

$$\frac{\partial(x, y, z)}{\partial(u, v, w)} = \begin{vmatrix} \frac{\partial x}{\partial u} & \frac{\partial x}{\partial v} & \frac{\partial x}{\partial w} \\ \frac{\partial y}{\partial u} & \frac{\partial y}{\partial v} & \frac{\partial y}{\partial w} \\ \frac{\partial z}{\partial u} & \frac{\partial z}{\partial v} & \frac{\partial z}{\partial w} \end{vmatrix} \qquad ...(A)$$

Here $u = x + y + z; \quad uv = y + z$

$\therefore$ $u = x + uv$ or $x = u - uv$

Also $y = uv - z = uv - uvw$ ...(*ii*)

and $z = uvw$ ...(*iii*)

From (*i*), (*ii*) and (*iii*),

So,

$$\frac{\partial x}{\partial u} = 1 - v, \frac{\partial x}{\partial v} = -u, \frac{\partial x}{\partial w} = 0$$

$$\frac{\partial y}{\partial u} = v - vw, \frac{\partial y}{\partial v} = u - uw, \frac{\partial y}{\partial w} = -uv$$

$$\frac{\partial z}{\partial u} = vw, \frac{\partial z}{\partial v} = uw, \frac{\partial z}{\partial w} = uv$$

Substituting these values in (A)

$$\frac{\partial(x, y, z)}{\partial(u, v, w)} = \begin{vmatrix} 1-v & -u & 0 \\ v - vw & u-uw & -uv \\ vw & uw & uv \end{vmatrix}$$

$$= (1 - v)\,[u^2v - u^2wv + u^2vw] + u\,[uv^2 - uv^2w + uv^2w]$$

$$= (1 - v)\,(u^2v) + u\,(uv^2) = u^2v - u^2v^2 + u^2v^2 = u^2v$$

Hence $$\frac{\partial(x, y, z)}{\partial(u, v, w)} = u^2v$$

EXERCISE 5.5

1. (*i*) If $r = \sqrt{x^2 + y^2}$ and $\theta = \tan^{-1}\left(\frac{y}{x}\right)$, evaluate $\frac{\partial(r, \theta)}{\partial(x, y)}$ **Ans.** $\frac{1}{\sqrt{x^2 + y^2}}$

   (*ii*) If $u = x^2 - 2y$, $v = x + y$, prove that $\frac{\partial(u, v)}{\partial(x, y)} = 2x + 2$

   (*iii*) If $x = r\cos$, $y = r\sin\theta$, $z = z$, Evaluate $\frac{\partial(x, y, z)}{\partial(\theta, q, z)}$ **Ans.** r

2. (*i*) If $u = \frac{x}{y - z}$, $v = \frac{y}{z - x}$, $w \frac{z}{x - y}$ show that $\frac{\partial(u, v, w)}{\partial(x, y, z)} = 0$

   (*ii*) If $u = \frac{y^2}{2x}$, $v \frac{x^2 + y^2}{2x}$, find $\frac{(u, v)}{\partial(x, y)}$ **Ans.** $\frac{-y}{2x}$

3. (*i*) If $u = x^2 - 2y$, $v = x + z$, $w = x - 2y + 3z$, find $\frac{\partial(u, v, w)}{\partial(x, y, z)}$ **Ans.** $10x + 4$

   (*ii*) If $u = (1 - r)^{-\frac{1}{2}}$, $v = y(1 - r^2)^{-\frac{1}{2}}$, $w = z(1 - r^2)^{-\frac{1}{2}}$ where $r^2 = x^2 + y^2 + z^2$, show that

$$\frac{\partial(u, v, w)}{\partial(x, y, z)} = (1 - r^2)^{-\frac{5}{2}}$$

4. If $u = e^x \sin y$, $v = e^x \cos y$ find the value of $\frac{\partial(u, v)}{\partial(x, y)}$ **Ans.** $e^{x+y}.\sin(x + y)$

5. If $u = a\cosh x \cos y$ and

   $v = a\sinh x \sinh y$, show that

$$\frac{\partial(u, v)}{\partial(x, y)} = \frac{a^2}{2}(\cos 2x - \cos 2y)$$

6. If

$$x = r \sin\theta \cos\theta$$
$$y = r \sin\theta \sin\theta$$
$$z = r \cos\theta, \text{ then}$$

$$\frac{\partial(r, q, f)}{}$$

7.

$$x = r \cos\theta$$
$$y = r \sin\theta, \text{ then}$$

$$\frac{\partial(x, y)}{\partial(r, \theta)} = r$$

5.12 CONCEPT OF MAXIMA AND MINIMA

Let $f(x, y)$ be any function of two independent variables x and y, supposed to be continuous for all values of these variables in the neighbourhood of their values a and b respectively. Then $f(a, b)$ is said to be a maximum or minimum value of $f(x, y)$ according as $f(a + h, b + k)$ is lesser then or greater than $f(a, b)$ for all sufficiently small values of h and k, (positive or negative) provided both of them are not equal to zero.

5.13 STATIONARY AND EXTREME POINTS

A point (a, b) is called a stationary point, if both the first order partial derivatives of the function $f(x, y)$ vanish at this point, A stationary point which is either a point of maxima or a minima is called an extreme point and the value of this function at this points is called an extreme value. A stationary point is not necessary an extreme point.

5.14 NECESSARY CONDITION FOR EXISTENCE OF A MAXIMUM OR MINIMA OF $f(x, y)$ at $x = a$, $y = b$

By definition it is obvious that we shall have a maximum or a minimum of $f(x, y)$ at $x = a$, $y = b$ if the expression $f(a + h, b + k) - f(a, b)$ is of invariable sign for all sufficiently small independent value of h and k, provided both of them are not equal to zero. If the sign of $f(a + h, b + k) - f(a, b)$ is negative, we shall have maximum of $f(x, y)$ at $x = a$, $y = b$. If it is positive, $f(x, y)$ has a minimum a $x = a$, $y = b$.

By Taylor's theorem for two variable, we have

$$f(a + h, b + k) = f(a, b) + \left(h\frac{\partial f}{\partial x} + k\frac{\partial f}{\partial y}\right)_{\substack{x=a\\x=b}} + \frac{1}{2!}\left(h^2\frac{\partial^2 f}{\partial x^2} + k^2\frac{\partial^2 f}{\partial y^2} + 2hk\frac{\partial^2 f}{\partial x\,\partial y}\right)_{\substack{x=a\\y=b}} + \ldots$$

$$\therefore \quad f(a + h, b + k) - f(a, b) = h\left(h\frac{\partial f}{\partial x}\right)_{\substack{x=a\\y=b}} + k\left(\frac{\partial f}{\partial y}\right)_{\substack{x=a\\y=a}} + \text{term of the second and higher order in } h \text{ and } k \quad \ldots(i)$$

By taking h and k, sufficiently small, the first degree terms in h and k can be made to govern the sign of the R.H.S and therefore of the L.S.H. of equation (i)

$$\text{Thus, the sign of } [f(a + h, b + k) - f(a, b)] = \text{the sign of } \left[h\left(\frac{\partial f}{\partial x}\right)_{\substack{x=a\\y=b}} + k\left(\frac{\partial f}{\partial y}\right)_{\substack{x=a\\y=b}}\right] \quad \ldots(ii)$$

Taking $k = 0$, we find that the right hand side of (ii) change sign when h changes sign. Therefore $f(x, y)$ can not have a maximum or minimum at $x = a$, $y = b$ if $\left(\frac{\partial f}{\partial x}\right)_{\substack{x=a\\y=b}} \neq 0$

Similarly taking $h = 0$, it is observed that $f(x, y)$ can not have maximum or a minimum at $x = a$, $y = b$ if $\left(\frac{\partial f}{\partial y}\right)_{\substack{x=a\\y=b}} \neq 0$

Thus a set necessary condition $f(x, y)$ should have maximum or a minimum values at $x = a$, $y = b$ is that

$$\left(\frac{\partial f}{\partial x}\right)_{\substack{x=a\\y=b}} = 0 \text{ and } \left(\frac{\partial f}{\partial y}\right)_{\substack{x=a\\y=b}} = 0$$

These condition are necessary but not sufficient for the existence of maxima or minima.

5.15 SUFFICIENT CONDITION FOR MAXIMA OR MINIMA

Let
$$r = \left(\frac{\partial^2 f}{\partial x^2}\right)_{\substack{x=a\\y=b}}, s = b\left(\frac{\partial^2 f}{\partial x\, \partial y}\right)_{\substack{x=a\\y=b}} \text{ and } t = \left(\frac{\partial^2 f}{\partial y^2}\right)_{\substack{x=a\\y=b}}$$

If $\left(\frac{\partial f}{\partial x}\right)_{\substack{x=a\\y=b}} = 0$ and $\left(\frac{\partial f}{\partial y}\right)_{\substack{x=a\\y=b}} = 0$, i.e., if the necessary condition for the existence of maxima or minima are satisfied, the new have

$$f(a+h, b+k) - f(a, b) = \frac{1}{2!}(rh^2 + 2hks + tk^2) + P_3$$

where P_3, consists of terms of third and higher orders in h and k.

or
$$\Delta f = \frac{1}{2r}[h^2 r^2 + 2kh\, rs + k^2 \text{ rt}] + \ldots\ldots$$

$$= \frac{1}{2r}[(h^2 r^2 + 2hkrs + k^2 s^2) + (k^2 rt - k^2 s^2)] + \ldots\ldots$$

$$= \frac{1}{2r}[(h\, r + ks^2) + k^2(rt - s^2)] + \ldots \quad \ldots(iii)$$

From equation (*iii*)

$(h\, r + ks)^2$ is always positive and $k^2(rt - s^2)$ will positive if $rt - s^2 > 0$. In this case Δf will have the same sign as that of r for all values of h and k.

Hence, if $rt - s^2 > 0$ then $f(x, y)$ has a maximum or a minimum value at (a, b) accordingly as $r < 0$ or $r > 0$.

If $rt - s^2 < 0$, then Δf changes sign with h and k. Hence the function has neither a maximum nor a minimum value at (a, b), the point (a, b) is saddle point in this case.

If $rt - s^2 = 0$. no conclusion can be drawn about a maximum value at (a, b) and hence further investigation is required.

5.16 WORKING RULES FOR MAXIMA AND MINIMA

Suppose $z = f(x, y)$ be function of x and y, then

Step I Find $\frac{\partial z}{\partial x}$ and $\frac{\partial z}{\partial y}$

Step II Put $\frac{\partial z}{\partial x} = 0$ and $\frac{\partial z}{\partial y}$ and solve for x and y. Let (a, b), (c, d), ...be the values of (x, y), these point are called stationary points.

Step III For each stationary points, find $r = \frac{\partial^2 z}{\partial x^2}$, $s = \frac{\partial^2 z}{\partial x \partial y}$, $t = \frac{\partial^2 z}{\partial y^2}$

Step IV (*a*) If $rt - s^2 > 0$ and $r < 0$ for a point (a, b) for step (*iii*), then z has a maximum value at (a, b).

(*b*) If $rt - s^2 > 0$ and $r > 0$ for a particular point (a, b) of step (*iii*) then z has a minimum value at (a, b).

(*c*) If $rt - s^2 < 0$ for a particular point (a, b) of step (*iii*) then z has no extreme (maxima or minima) value at (a, b).

(*d*) If $rt - s^2 = 0$, the case is doubtful and required further investigation.

***Example* 5.50.** Find the extreme values of $xy(a - x - y)$

Solution : Let $u = xy(a - x - y)$, then

$$\frac{\partial u}{\partial x} = ay - 2xy - y^2 \text{ and}$$

$$\frac{\partial u}{\partial y} = ax - axy - x^2$$

Also
$$r = \frac{\partial^2 u}{\partial x^2} = -2y, \; s = \frac{\partial^2 u}{\partial x\, \partial y} = a - 2x - 2y$$

and
$$t = \frac{\partial^2 u}{\partial y^2} = -2x$$

For maximum or minima of u

$$\text{Put } \frac{\partial u}{\partial x} = 0 \text{ and } \frac{\partial u}{\partial y} = 0$$

Thus, we have

$$ay - 2xy - y^2 = 0$$

$$ax - x^2 - 2xy = 0$$

These equations can be written as

$$y(a - ax - y) = 0, \; x(a - x - 2y) = 0$$

So that we have to consider the four pairs of equation, *i.e.*,

$$y = 0, x = 0; \; a - 2x - y = 0, x = 0;$$

$$y = 0, a - x - 2y = 0; \; a - 2x - y = 0, \; a - x - 2y = 0$$

Solving these, we get the following pairs of the values of x and y which make the function stationary $(0, 0)$, $(0, a)$, $(a, 0)$ $\left(\frac{a}{3}, \frac{a}{3}\right)$

For (0, 0)

$$r = 0, s = a, t = 0$$

so that $rt - s^2$ is negative.

Therefore we have neither a maximum nor a minimum of u at $(0, 0)$

For (0, *a*),

$$r = -2a, \quad s = -a, t = 0$$

So that $rt - s^2$ in negative.

Therefore u has not an extreme value at $(0, a)$. Similarly it can be shown that u has not an extreme values at $(a, 0)$

For $\left(\frac{a}{3}, \frac{a}{3}\right)$

$$r = -\frac{2}{3}a, \quad s = -\frac{a}{3}, \quad t = -\frac{2}{3}a \text{ and } rt - s^2 \text{ is positive.}$$

Therefore u has an extreme value at $\left(\frac{a}{3}, \frac{a}{3}\right)$ and will be maximum or minimum according as r is negative or positive *i.e.*, according as a positive or negative.

$\therefore$ The extreme value of $u = \frac{a^3}{27}$

Example **5.51.** Examine the following function for extreme values $u = x^4 + y^4 - 2x^2 + 4xy - 2y^2$

Solution : Here $u = x^4 + y^4 - 2x^2 + 4xy - 2y^2$ then

$$\frac{\partial u}{\partial x} = 4x^3 - 4x + 4y \text{ and} \quad ...(i)$$

$$\frac{\partial u}{\partial y} = 4y^3 + 4x - 4y \quad ...(ii)$$

Also

$$r = \frac{\partial^2 u}{\partial x^2} = 12x^2 - 4$$

$$t = \frac{\partial^2 u}{\partial y^2} = 12y^2 - 4$$

$$s = \frac{\partial^2 u}{\partial x \partial y} = 4$$

For maxima or minima

$$\frac{\partial u}{\partial x} = \frac{\partial u}{\partial y} = 0$$

$$\therefore \quad 4x^3 - 4x + 4y = 0 \quad ...(i)$$

and

$$4y^3 + 4x - 4y = 0 \quad ...(ii)$$

On solving these equation, we get

$x + y = 0$ and $x^2 + y^2 - xy = 0$, which gives imaginary values of x or y

or

$$y = -x$$

From (ii)

$$4y^3 - 4y - 4y = 0$$

or

$$4y^3 - 8y = 0$$

or

$$4y\,(y^2 - 2) = 0$$

or

$$y = 0, y = \pm\sqrt{2}$$

Hence possible extreme points are $(0, 0), (-\sqrt{2}, \sqrt{2}), (\sqrt{2}, -\sqrt{2})$

For (0, 0)

$$(r)_{(0,0)} = -4, (t)(0, 0) = -4, (s)(0, 0) = 4.$$

Now

$$rt - s^2 = -4 \times -4 - 16 = 0$$

Further Investigation in needed. Hence $u(x, y)$ has not extreme value at (0, 0) and so $rt - s^2 = 384 > 0$

For $(-\sqrt{2}, \sqrt{2})$

$$r(-\sqrt{2}, \sqrt{2}) = 20$$

$$t(-\sqrt{2}, \sqrt{2}) = 20$$

$$s(-\sqrt{2}, \sqrt{2}) = 4$$

and r is positive. Hence $u(x, y)$ at $(-\sqrt{2}, \sqrt{2})$

For $(\sqrt{2}, -\sqrt{2})$

$$r(\sqrt{2}, -\sqrt{2}) = 20$$

$$t(\sqrt{2}, -\sqrt{2}) = 20$$

$$s(\sqrt{2}, -\sqrt{2}) = 4$$

and so $rt - s^2 = 384 > 0$ and r is also positive. Hence $u(x, y)$ is maximum at $(\sqrt{2}, -\sqrt{2})$

***Examples* 5.52.** Find the maximum and minimum value of

(*i*) $x^3 + y^3 - 3axy$

(*ii*) $xy + \frac{a^3}{x} + \frac{a^3}{y}$

Solution : (*i*) Let $f(x, y) = x^3 + y^3 - 3axy$

and $$\frac{\partial f}{\partial x} = 3x^2 - 3ay, \frac{\partial f}{\partial y} = 3y^2 - 3ax$$

For maximum or minima

$$\frac{\partial f}{\partial x} = 0, \frac{\partial f}{\partial y} = 0$$

$\therefore$ $$3x^2 - 3ay = 0, 3y^2 - 3ax = 0$$

On solving these equations, we get

$$x^2 = ay \text{ and } y^2 = ax$$

or $$x = \frac{y^2}{a}$$

or $$\left(\frac{y^2}{a}\right)^2 = ay \Rightarrow y^4 - a^3y = 0$$

$$y(y^3 - a^3) = 0$$

or $$y = 0 \; y = a \Rightarrow x = 0, x = a.$$

Hence possible extreme points are (0, 0), (a, a)

Now find r, s, t

$$r = \frac{\partial^2 f}{\partial x^2} = 6x, t = \frac{\partial^2 f}{\partial y^2} = 6y, s = \frac{\partial^2 f}{\partial x \partial y} = -3a$$

For (0, 0)

$$(r)_{(0,0)} = 0, (t)(0, 0) = 0, (s)_{(0,0)} = -3a.$$

and $$rt - s^2 = 0 - (-3a)^2 = -9a^2 < 0 \text{ for any value of a}$$

Hence at (0, 0) the given function os neither maximum nor minimum

For (a, a)

$$(r)_{(a,a)} = 6a, \; (t)_{(a,a)} = 6a, \quad (s)_{(a,a)} = -3a$$

and $$rt - s^2 = 6a \times 6a - (-3a)^2$$

$$= 36a^2 - 9a^2 = 27a^2 > 0 \text{ for any value of } a$$

Hence at (a, a) the function $f(x, y)$ is minimum because r is positive if $a > 0$.

(*ii*) Let
$$f(x, y) = xy + \frac{a^3}{x} + \frac{a^3}{y}$$

and
$$\frac{\partial f}{\partial x} = y - \frac{a^3}{x^2}, \frac{\partial f}{\partial y} = x - \frac{a^3}{y^2}$$

Also
$$r = \frac{\partial^2 f}{\partial x^2} = \frac{2a^3}{x^3}, t = \frac{\partial^2 f}{\partial y^2} \text{ and } s = \frac{\partial^2 f}{\partial x \partial y} = 1$$

For maxima or minima,
$$\frac{\partial f}{\partial x} = \frac{\partial f}{\partial y} = 0$$

Hence
$$y - \frac{a^3}{x^2} = 0, x - \frac{a^3}{y^2} = 0$$

or
$$y = \frac{a^3}{x^2} \text{ and } x = \frac{a^3}{y^2}$$

or $y = \frac{a^3}{a^6} y^4$ or $y(a^6 - a^3 y^3) = 0$ either $y = 0$, or $(a^3 - y^3) = 0 \Rightarrow y = a$

when $y = 0, x = \infty$

when $y = a, x = a$

Hence only (a, a) is possible extreme point

For (a, a)

$$(r)_{(a, a)} = 2, \ (t)_{(a, a)} = 2. \ (s)_{(a, a)} = 1.$$

and
$$rt = s^2 = 2 \times 2 - (1)^2 = 4 - 1 = 3 > 0$$

Here r is also positive so the given function $f(x, y)$ is minimum at (a, a).

Example **5.53.** Find the maximum and minimum value of

(*i*) $x^3 + 3xy^2 - 15x^2 - 15y^2 + 72x$

(*ii*) $2(x^2 - y^2) - x^4 + y^4$

Solution : (*i*) Let $f(x, y) = x^3 + 3xy^2 - 15x^2 - 15y^2 + 72x$

$\therefore$
$$\frac{\partial f}{\partial x} = 3x^2 + 3y^2 - 30x + 72, \frac{\partial^2 f}{\partial x^2} = 6x - 30$$

$$\frac{\partial f}{\partial y} = 6xy - 30y, \frac{\partial^2 f}{\partial y^2} = 6x - 30, \text{ and } \frac{\partial^2 f}{\partial x \partial y} = 6y$$

From maxima or minima,
$$\frac{\partial f}{\partial x} = 0, \frac{\partial f}{\partial y} = 0, \text{ i.e.,}$$

$$3x^2 + 3y^2 - 30x + 72 = 0$$

or
$$x^2 + y^2 - 10x + 24 = 0 \quad ...(i)$$

and
$$6xy - 30y = 0 \quad ...(ii)$$

From (*ii*), $6y(x - 5) = 0$, either $y = 0$ or $x = 5$

When $y = 0$, from (*i*)

$$x^2 - 10x + 24 = 0 \text{ or } x^2 - 6x - 4x + 24 = 0$$

$$(x - 6) - 4(x - 6) = 0$$

$$x = 4, 6$$

Again, when $x = 5$ from (*i*) $y = \pm 1$

Hence possible extreme points are (4, 0), (6, 0), (5, 1), (5, –1)

For (4, 0)

$$r(4, 0) = -6,$$

$$t(4, 0) = -30 \text{ and}$$

$$s(4, 0) = 0$$

and $rt - s^2 = -6 \times 30 - (0)^2 = 180 > 0$ and negative, therefore the function $f(x, y)$ is maximum at (4, 0)

For (6, 0)

$$r(6, 0) = 6$$

$$t(6, 0) = -30$$

$$s(6, 0) = 0$$

and $rt - s^2 = -180 < 0$, at (6, 0) the given function $f(x, y)$ is neither maximum nor minima

For (5, 1)

$$r(5, 1) = 0$$

$$t(5, 1) = 0$$

$$s(5, 1) = 6$$

and $rt - s^2 = -36 < 0$ the given function $f(x, y)$ is neither maximum nor minimum at (5, 1) and (5, –1)

(*ii*) Let

$$f(x, y) = 2x^2 - 2y^2 - x^4 + y^4$$

and
$$\frac{\partial f}{\partial x} = 4x - 4x^3, \frac{\partial f}{\partial y} = -4y + 4y^3$$

Also
$$r = \frac{\partial^2 f}{\partial x^2} = 4 - 12x^2, t = \frac{\partial^2 f}{\partial y^2} = -4 + 12y^2$$

and
$$s = \frac{\partial^2 f}{\partial x \partial y} = 0$$

For maxima or minima,

$$\frac{\partial f}{\partial x} = \frac{\partial f}{\partial y} = 0, \text{ i.e.,}$$

$$4x - 4x^3 = 0, \quad -4y + 4y^3 = 0$$

or
$$4x(1 - x^2) = 0, \quad -4y(1 - y^2) = 0$$

either
$$x = 0, \text{ or } x \pm 1, \text{ either } y = 0 \text{ or } y = \pm 1$$

Hence possible extreme points are (0, 0), (1, 1), (–1, –1), (1, 0), (–1, 0), (0, 1), (0, –1)

For (0, 0)

$$r_{(0,0)} = 4, (t)_{(0,0)} = -4, \text{ and } (s)_{(0,0)} = 0$$

$$rt - s^2 = -16 < 0$$

Hence the function is neither maxima nor minima at (0, 0)

For (1, 1)

$$(r)_{(1,1)} = -8,\ (t)_{(1,1)} = 8, \text{ and } (s)_{(1,1)} = 0$$

and $$rt - s^2 = -64 > 0$$

Hence the given function is neither maxima noir minima at (1, 1)

For (1, 0)

$$(r)_{(1,0)} = -8,\ (t)_{(1,0)} = -4,\ (s)_{(1,0)} = 0$$

and $$rt - s^2 = 32 > 0$$

Hence the function is maxima at (1, 0) because r is negative

For (–1, 0)

$$(r)_{(-1,0)} = -8,\ (t)(-1, 0) = -4, \text{ and } (s)_{(1,0)} = 0$$

$\therefore$ $$rt - s^2 = 32 > 0$$

Hence the given function is maxima at (–1, 0)

For (0, 1)

$$(r)_{(0,1)} = 4,\ (t)_{(0,1)} = 8 \text{ and } (s)_{(0,1)} = 0$$

$\therefore$ $$rt - s^2 = 32 > 0 \text{ and r is positive}$$

So the given function at (0, 1) is minima.

For (0, –1)

$$(r)_{(0,-1)} = 4,\ (t)_{(0,-1)} = 8, \text{ and } (s)(0, -1) = 0$$

$\therefore$ $$rt - s^2 = 32 > 0 \text{ and } r \text{ is positive}$$

Hence the given function at (0, –1) is minima.

For (–1, 1)

$$(r)_{(-1,-1)} = -8,\ (t)_{(-1,-1)} = 8, \text{ and } (s)_{(-1,-1)} = 0$$

$\therefore$ $$rt - s^2 = -64 - 0 \Rightarrow -64 < 0$$

therefore the given function is neither maxima nor minima at (–1, –1).

Hence At ($\pm$ 1, 0), the function is maxima

At (0, $\pm$ 1), the function is minima

***Example* 5.54.** Examine maxima or minima of the function sin x sin y sin $(x + y)$

Solution : Let $f(x, y) = \sin x \sin y \sin (x + y)$

and $$\frac{\partial f}{\partial x} = \sin y\,[\sin x \cos (x + y) + \sin (x + y) \cos x]$$

or $$\frac{\partial f}{\partial x} = \sin y\,[\sin (x + y + x)] \text{ or } [\sin (2x + y)].\ \sin y]$$

Similarly $\frac{\partial f}{\partial y} = \sin x\,[\cos y \sin (x + y) + \sin y \cos (x + y)]$

or $$\frac{\partial f}{\partial y} = \sin x \sin (x + y + y) \text{ or } \sin x \sin (x, + 2y)$$

For maxima or minima

$$\frac{\partial f}{\partial x} = \frac{\partial f}{\partial y} = 0 \textit{ i.e.,}$$

Sin y sin $(2x + y) = 0$ either sin $y = 0$ $y = 0$ or sin$(2x + y) = 0$

sin x sin $(x + 2) = 0$ either sin $x = 0$ $x = 0$ or sin $(x + 2y) = 0$

Hence, $x = 0$, and $y = 0$

and sin$(2x + y)$ = sin π or $2x + y = \pi$

sin$(x + 2)$ = sin π or $x + 2y = \pi$

On simplification

$$y = \frac{\pi}{3}, x = \frac{\pi}{3}$$

Hence possible extreme are $(0, 0)$ $\left(\frac{\pi}{3}, \frac{\pi}{3}\right)$

Also
$$r = \frac{\partial^2 f}{\partial x^2} = 2 \sin y \cos(2x + y)$$

$$t = \frac{\partial^2 f}{\partial y^2} = 2 \sin x \cos(x + 2y)$$

$$x = \frac{\partial^2 f}{\partial x \partial y} = \cos x \sin(x + 2y) + \sin x \cos(x + 2y)$$

$$= \sin(x + 2y + x)$$

$$= \sin(2x + 2y)$$

At (0, 0)

$$(r)_{(0, 0)} = 2 \sin 0 \cos(0 + 0) = 0$$

$$(t)_{(0, 0)} = 2 \sin 0 \cos(0 + 0) = 0$$

and
$$s_{(0, 0)} = \sin(0 + 0) = 0$$

$\therefore$
$$rt - s^2 = 0 \times 0 - 0 = 0$$

Hence further investigation is required

At $\left(\frac{\pi}{3}, \frac{\pi}{3}\right)$

$$(r)_{\left(\frac{\pi}{3}, \frac{\pi}{3}\right)} = 2 \sin \frac{\pi}{3} \cos\left(2 \times \frac{\pi}{3} + \frac{\pi}{3}\right)$$

$$= 2 \times \frac{\sqrt{3}}{2} \times \cos \pi$$

$$= -\sqrt{3}$$

$$(t)_{\left(\frac{\pi}{3}, \frac{\pi}{3}\right)} = 2 \times \sin \frac{\pi}{3} \cos\left(\frac{\pi}{3} + 2\frac{\pi}{3}\right)$$

$$= 2 \frac{\sqrt{3}}{2} \times \cos \pi$$

$$= -\sqrt{3}$$

$$(s)_{\left(\frac{\pi}{3}, \frac{\pi}{3}\right)} = \sin\left(2 \times \frac{\pi}{3} + 2\frac{\pi}{3}\right) = \sin\left(\frac{2\pi + 2\pi}{3}\right) = \sin \frac{4\pi}{3}$$

$$= \sin\left(\pi + \frac{\pi}{3}\right) = -\sin\frac{\pi}{3} = -\frac{\sqrt{3}}{2}$$

$$\therefore \quad rt - s^2 = -\sqrt{3} \times -\sqrt{3} - \left(-\frac{\sqrt{3}}{2}\right)^2$$

$$3 - \frac{3}{4} = \frac{9}{4} > 0 \text{ and } r \text{ is negative}$$

Hence the given function is maximum at $\left(\frac{\pi}{3}, \frac{\pi}{3}\right)$.

Example **5.55.** Examine for minimum and maximum values of $\sin x + \sin y + \sin(x + y)$

Solution. Let $f(x, y) = \sin x + \sin y + \sin + (x + y)$

and $\quad f_x = \text{c}\llcorner\ x + \cos(x + y),$

$$f_y = \cos y + \cos(x + y)$$

Also $\quad r = f_{xx} = -\sin x - \sin(x + y),$

$$t = f_{yy} = -\sin y - \sin(x + y)$$

and $\quad s = f_{xy} = -\sin(x + y)$

For maxima or minima

$$f_x = f_y = 0 \text{ i.e.,}$$

$$\cos x + \cos(x + y) = 0 \qquad \text{...}(i)$$

and $\quad \cos y + \cos(x + y) = 0 \qquad \text{...}(ii)$

Substracting (*ii*) from (*i*), we get

$$\cos x - \cos y = 0$$

or $\quad \cos x = \cos y$

$$\therefore \quad x = y$$

From (*i*)

$$\cos x + \cos 2x = 0$$

$$\cos 2x = -\cos x$$

$$\cos 2x = \cos(\pi - x)$$

or $\quad 2x = \pi - x$

$$3x = \pi$$

$$x = \frac{\pi}{3} \quad \text{and } y = \frac{\pi}{3}$$

At $\left(\frac{\pi}{3}, \frac{\pi}{3}\right)$

$$r = -\frac{\sqrt{3}}{2} - \frac{\sqrt{3}}{2} = -\sqrt{3}$$

$$s = \frac{\sqrt{3}}{2}, t = \frac{\sqrt{3}}{2} - \frac{\sqrt{3}}{2} = -\sqrt{3}$$

$$\therefore \quad rt - s^2 = 3 - \frac{3}{4} = \frac{9}{4} > 0 \text{ and } r > 0$$

Hence $f(x, y)$ has maximum value at $\left(\frac{\pi}{3}, \frac{\pi}{3}\right)$ and maximum value at $\left(\frac{\pi}{3}, \frac{\pi}{3}\right)$ is

$$= f\left(\frac{\pi}{3}, \frac{\pi}{3}\right) = \sin\frac{\pi}{3} + \sin\frac{\pi}{3} + \sin\frac{2\pi}{3}$$

$$= \frac{\sqrt{3}}{2} + \frac{\sqrt{3}}{2} + \frac{\sqrt{3}}{2} = \frac{3\sqrt{3}}{2}$$

***Example* 5.56.** Show that the rectangular solid of maximum volume that can be inscribed in a given sphere is a cube.

Solution : Suppose x, y, z the length, breath and height of the rectangular solid. If V is the volume of the solid, then

$$V = xyz \qquad ...(i)$$

Since each diagonal of solid passes through the centre of the sphere then

Each diagonal = diameter of sphere = D (say)

$$\therefore \quad \sqrt{x^2 + y^2 + z^2} = D$$

$$\text{or} \quad D^2 = x^2 + y^2 + z^2$$

$$\text{or} \quad z = \sqrt{D^2 - x^2 - y^2}$$

Now from (i)

$$V = x\,y\,z = xy\sqrt{D^2 - x^2 - y^2}$$

$$\text{or} \quad V^2 = x^2y^2(D^2 - x^2 - y^2) = f(x, y) \text{ (say)}$$

$$\text{or} \quad f(x, y) = x^2y^2D^2 - x^4y^2 - x^2y^4 \qquad ...(ii)$$

$$\therefore \quad f_x = 2xy^2D^2 - 4x^3y^2 - 2xy^4 = 2xy^2(D^2 - 2x^2 - y^2)$$

$$\text{and} \quad f_y = 2x^2yD^2 - 2x^4y - 4x^2y^3 = 2x^2y(D^2 - x^2 - 2y^2)$$

For maxima or minima $\quad f_x = f_y = 0$ i.e.,

$$2xy^2(D^2 - 2x^2 - y^2) = 0$$

$$\text{and} \quad 2x^2y(D^2 - x^2 - 2y^2) = 0$$

$$D^2 - 2x^2 - y^2 = 0 \qquad ...(iii)$$

$$D^2 - x^2 - 2y^2 = 0 \qquad ...(iv)$$

$$\therefore \quad x \neq 0 \quad y \neq 0$$

Subtracting (iv) from (iii)

$$-x^2 + y^2 = 0$$

$$x^2 = y^2 \Rightarrow x = y$$

From (iii) $D^2 - 2x^2 - x^2 = 0$

$$D^2 = 3x^2$$

$$x = y = \frac{D}{\sqrt{3}}$$

and $$z = \sqrt{D^2 - x^2 - y^2}$$

$$z = \sqrt{D^2 - \frac{D^2}{3} - \frac{D^2}{3}} = \frac{D}{\sqrt{3}}$$

Also $$r = f_{xx} = 2D^2 y^2 - 12x^2 y^2 - 2y^4$$

$$s = f_{xy} = 4D^2 xy - 8x^3 y - 8xy^3$$

$$t = f_{yy} = 2D^2 x^2 - 2x^4 - 12x^2 y^2$$

At $$x = y = \frac{D}{\sqrt{3}}$$

$$r = 2D^2\left(\frac{D}{\sqrt{3}}\right)^2 - 12\left(\frac{D^2}{\sqrt{3}}\right)\left(\frac{D^2}{3}\right) - \frac{2D^4}{9} = -\frac{8D^4}{9}$$

$$s = 4D^2\left(\frac{D}{3}\right)^2 - 8\left(\frac{D^3}{3\sqrt{3}}\right)\left(\frac{D}{\sqrt{3}}\right) - 8\left(\frac{D}{\sqrt{3}}\right)\left(\frac{D^3}{3\sqrt{3}}\right) = -\frac{4D^4}{9}$$

and $$t = 2D^2\left(\frac{D^2}{3}\right) - 2\left(\frac{D^4}{9}\right) - 12\left(\frac{D^2}{3}\right)\left(\frac{D^2}{3}\right) = -\frac{8D^4}{9}$$

$$\therefore \quad rt - s^2 = \frac{64D^8}{81} - \frac{16D^8}{81} = \frac{16D^8}{27} > 0.$$ for all value of D, and r is negative

Therefore, V^2 and hence V is maximum when $x = y = z$ i.e., when the rectangular solid is a cube.

***Example* 5.57.** A rectangular box, open at the top, is to have a given capacity. Find the dimension of the box requiring least material for its construction.

Solution : Suppose x, y and z be the length, breadth and height of the rectangular box respectively. Let V be the given capacity *i.e.*, volume and S is the surface then

$$V = x\,y\,z \text{ which is given, so } V \text{ is constant, } \therefore z = \frac{V}{xy} \qquad ...(i)$$

and the surface is $S = xy + 2xz + 2yz$

$$S = xy + 2x\frac{V}{xy} + 2y\frac{V}{xy}$$

Let $$S = f(x, y) = xy + \frac{2V}{y} + \frac{2V}{x}$$

$$\therefore \quad f_x = y - \frac{2V}{x^2}, \; f_y = x - \frac{2V}{y^2}$$

For maxima or minima $f_x = f_y = 0$ i.e.,

$$y = \frac{2V}{x^2} \text{ and } x = \frac{2V}{y^2} \qquad ...(iii)$$

Using (*ii*) and (*iii*), we get

$$x = \frac{2V}{4V^2} \times x^4$$

$$x = \frac{x^4}{2V} \Rightarrow x^3 = 2V \quad \because x \neq 0$$

$$x = (2V)^{1/3}$$

and $$y = \frac{2V}{x^2} \Rightarrow \frac{2V}{(2V)^{2/3}} \Rightarrow (2V)^{1/3}$$

$\therefore$ $$x = y = (2V)^{1/3} \text{ is a stationary point.}$$

Now $$r = f_{xx} = \frac{4V}{x^3},\ s = f_{xy} = 1,\ t = f_{yy} = \frac{4V}{x^3}$$

At $$x = y = (2V)^{1/3}$$

$$r = \frac{4V}{2V} = 2,\ s = 1,\ t = \frac{4V}{2V} = 2$$

and $$rt - s^2 = 4 - 1 = 3 > 0 \text{ and } r > 0$$

The given function S is minimum when $x = y = (2V)^{1/3}$

Also $$z = \frac{V}{xy} = \frac{V}{(2V)^{2/3}} = \frac{V^{1/3}}{2^{2/4}} = \frac{(2V)^{1/3}}{2} = \frac{y}{2}$$

Hence S is minimum when $x = y = 2z = (2V)^{1/3}$

***Example* 5.58.** Find the volume of largest rectangular Parallelpiped that can be inscribed in the ellipsoid $\frac{x^2}{a^2} + \frac{y^2}{b^2} + \frac{z^2}{c^2} = 1.$

Solution : Let (x, y, z) be the vertex of the parallelepiped and suppose it lies on the ellipsoid $\frac{x^2}{a^2} + \frac{y^2}{b^2} + \frac{z^2}{c^2} = 1$ ...(*i*)

Let its dimensions be $2x$, $2y$ and $2z$ therefore the volume V is given by

$$V = 2x \times 2y \times 2z = 8xyz$$

or $$V^2 = 64\, x^2 y^2 z^2$$

$$= 64\, x^2 y^2 c^2 \left(1 - \frac{x^2}{a^2} - \frac{y^2}{b^2}\right)$$

From (*i*)

or $$= 64\, c^2 \left[x^2 y^2 - \frac{x^4 y^2}{a^2} - \frac{x^2 y^4}{b^2}\right] = f(x, y) \quad \text{(say)}$$

$\therefore$ $$f_x = 64\, c^2 \left[2xy^2 - \frac{4x^3 y^2}{a^2} - \frac{2xy^4}{b^2}\right]$$

$$f_y = 64\, c^2 \left[2x^2 y - \frac{2x^4 y}{a^2} - \frac{4x^2 y^3}{b^2}\right]$$

For maxima or minima

$$f_x = f_y = 0 \text{ i.e.,}$$

$$128c^2\, xy^2 \left[1 - \frac{2x^2}{a^2} - \frac{y^2}{b^2}\right] = 0$$

and $$128c^2\, x^2 y \left(1 - \frac{x^2}{a^2} - \frac{2y^2}{b^2}\right) = 0$$

$\therefore$
$$1-\frac{2x^2}{a^2}-\frac{y^2}{b^2}=0 = 0 \quad \therefore x \neq 0, y \neq 0 \qquad \text{...}(ii)$$

and
$$1-\frac{x^2}{a^2}-\frac{2y^2}{b^2}=0 = 0 \qquad \text{...}(iii)$$

Subtracting (*iii*) from (*ii*), we get

$$-\frac{x^2}{a^2}+\frac{y^2}{b^2} = 0 \quad \text{or} \quad y=\frac{bx}{a}$$

From (*ii*) on putting the value of y

$$1-\frac{2x^2}{a^2}-\frac{x^2}{a^2} = 0$$

$\therefore$
$$x^2 = \frac{a^2}{3} \quad \text{or} \quad x=\frac{a}{\sqrt{3}}$$

and
$$y = \frac{b}{a}\times\frac{a}{\sqrt{3}}=\frac{b}{\sqrt{3}}$$

From (*i*)
$$z^2 = c^2\left[1-\frac{x^2}{a^2}-\frac{y^2}{b^2}\right]=c^2\left[1-\frac{1}{3}-\frac{1}{3}\right]=\frac{c^2}{3}$$

$$z = \frac{c}{\sqrt{3}}$$

Hence
$$x = \frac{a}{\sqrt{3}},\ y=\frac{b}{\sqrt{3}} \text{ and } z=\frac{c}{\sqrt{3}} \text{ is a stationary point}$$

Also
$$r = f_{xx}=64c^2\left(2y^2-\frac{12x^2y^2}{a^2}-\frac{2y^4}{b^2}\right)$$

$$s = f_{xy}=64c^2\left[4xy-\frac{8x^3y}{a^2}-\frac{8xy^3}{b^2}\right]$$

$$t = f_{yy}=64c^2\left[2x^2-\frac{2x^4}{a^2}-\frac{12x^2y^2}{b^2}\right]$$

At point $\left(\frac{a}{\sqrt{3}},\frac{b}{\sqrt{3}}\frac{c}{\sqrt{3}}\right)$

$$r = 64c^2\left[\frac{ab^2}{3}-\frac{12}{a^2}\left(\frac{b^3}{3}\right)\left(\frac{a^2}{3}\right)-\frac{2}{b^2}\left(\frac{b^4}{9}\right)\right]=\frac{-512}{9}b^2c^2$$

$$s = 64c^2\left[4\left(\frac{a}{\sqrt{3}}\right)\left(\frac{b}{\sqrt{3}}\right)-\frac{8}{a^2}\left(\frac{a^3}{3\sqrt{3}}\right)\left(\frac{b}{\sqrt{3}}\right)-\frac{8}{b^2}\left(\frac{a}{\sqrt{3}}\right)\left(\frac{b^3}{3\sqrt{3}}\right)\right]=-\frac{256}{9}abc^2$$

$$t = 64c^2\left[a\left(\frac{a^2}{3}\right)-\frac{2}{a^2}\left(\frac{a^4}{9}\right)-\frac{12}{b^2}\left(\frac{a^2}{3}\right)\left(\frac{b^2}{3}\right)\right]=-\frac{512}{9}a^2c^2$$

$\therefore$
$$rt-s^2 = \left(\frac{512}{9}\right)^2 a^2b^2c^4-\left(-\frac{256}{9}\right)^2 a^2b^2c^4$$

$$= \left(\frac{256}{9}\right)^2 a^2b^2c^4\,(4-1)=3\,a^2b^2c^4\left(\frac{256}{9}\right)^2>0 \text{ and } r<0$$

$\therefore$ V^2 and hence V is maximum at

$$x = \frac{a}{\sqrt{3}}, \quad y = \frac{b}{\sqrt{3}} \quad \text{and} \quad z = \frac{c}{\sqrt{3}}$$

Maximum value $= 8 \times \frac{a}{\sqrt{3}} \times \frac{b}{\sqrt{3}} \times \frac{c}{\sqrt{3}} = \frac{8abc}{\sqrt{3}}$ **Ans.**

EXERCISE 5.6

1. Discus the maxima and minima of the following functions
   (*a*) $x^3 y^2 (1 - x - y)$ **Ans.** Maximum value $= \frac{1}{432}$ at $\left(\frac{1}{2}, \frac{1}{3}\right)$
   (*b*) $x^2 + y^2 + 6x + 12$ **Ans.** Minimum value = 3 at (–3, 0)
   (*c*) $x^3 + y^3 + 3xy$ **Ans.** Maximum value = 1 (–1, –1)
   (*d*) $x^3 y^2 (12 - x - y)$ **Ans.** Maximum at (6, 4)
2. Examine for extreme values
   (*a*) $\cos x \cos y \cos (x + y)$ **Ans.** Minimum value $-\frac{3}{2}$ at $\left(\frac{2\pi}{3}, \frac{2\pi}{3}\right)$
   (*b*) $\cos x\, t \cos y + \cos (x + y)$ **Ans.** Maximum value $\frac{1}{8}$ at $\left(\frac{\pi}{3}, \frac{\pi}{3}\right)$
3. In a plane triangle *ABC*, find the maximum value of cos *A* cos *B* cos *C*
4. Show that the all triangles inscribed in a circle, the one with minimum are is equilateral.
   [**Hint :** Let ABC be a triangle inscribed in a circle of radius *R* let *x, y, z* be the angles of triangle ABC so that $0 < x < \pi$, $0 < y < \pi$, $0 < z < \pi$ and $x + y + z = \pi$, if *s* be the area of ΔABC then $s = \frac{1}{2} ab \sin z$ where *a*, *b* denote the length of the side containing the angle *z*. Now by the relation of side and angle

$$R = \frac{a}{2 \sin x} = \frac{b}{2 \sin y} = \frac{c}{2 \sin z}$$

$$\therefore \quad a = 2R \sin x, \quad b = 2R \sin y, \quad c = 2R \sin z$$

and

$$S = \frac{1}{2} 2R \sin x \times 2R \sin y \times \sin [\pi - (x + y)]$$

$$S = 2R^2 \sin x \sin y \times \sin (x + y)$$

$$\frac{S}{2R^2} = \sin x \sin y \sin (x + y) = f(x, y) \text{ (say)}$$

Now we have to maximize $\frac{S}{2R^2}$

5.17 LAGRANGE'S METHOD OF UNDETERMINED MULTIPLIERS

When a function contains more than two variable *i.e.*, three, four or more and these variable are not independent but are connected by some given relation then we use Lagrange's method for maxima or minima.

Let $f(x, y, z)$ be a function of *x, y, z* which is to be examined for maximum or minimum value and suppose that the variables *x, y, z* are connected by the relation.

$$\phi(x, y, z) = 0$$

Consider Lagrange's function

$$F(x, y, z) = f(x, y, z) + \lambda\, \phi(x, y, z) \quad \text{...(A)}$$

where λ is parameter

For maxima or minimum of $F(x, y, z)$

$$\frac{\partial F}{\partial x} = 0 \Rightarrow \frac{\partial f}{\partial x} + \lambda \frac{\partial \phi}{\partial x} = 0 \quad ...(i)$$

$$\frac{\partial F}{\partial y} = 0 \Rightarrow \frac{\partial f}{\partial y} + \lambda \frac{\partial \phi}{\partial y} = 0 \quad ...(ii)$$

$$\frac{\partial F}{\partial z} = 0 \Rightarrow \frac{\partial f}{\partial z} + \lambda \frac{\partial \phi}{\partial z} = 0 \quad ...(iii)$$

On solving the (*i*), (*ii*) and (*iii*), we get the value of x, y, z and λ for a maximum or minima $F(x, y, z)$.

Note : Lagrange's method does not enable us to find whether there is a maximum or minimum. This fact is determined from the physical consideration of the problem.

***Example* 5.59.** Find the minimum value of $x^2 + y^2 + z^2$, given that

(*i*) $x\,y\,z = a^3$

(*ii*) $ax + by + cz = p$

Solution : (*i*) Let $u = x^2 + y^2 + z^2$ ...(*i*)

and $f(x, y, z) = x\,y\,z - a^3$ ...(*ii*)

By Lagragian's method

$$F(x, y, z) = u(x, y, z) + \lambda(x, y, z) \quad ...(iii)$$

For maxima or minima of (*iii*)

$$\frac{\partial F}{\partial x} = \frac{\partial u}{\partial x} + \lambda \frac{\partial \phi}{\partial x} = 0 \text{ or } 2x + \lambda yz = 0 \quad ...(A)$$

$$\frac{\partial F}{\partial y} \equiv \frac{\partial u}{\partial y} + \lambda \frac{\partial \phi}{\partial y} = 0 \text{ or } 2y + \lambda xz = 0 \quad ...(B)$$

$$\frac{\partial F}{\partial z} = \frac{\partial u}{\partial z} + \lambda \frac{\partial \phi}{\partial z} = 0 \text{ or } 2z + \lambda\, xy = 0 \quad ...(C)$$

Now multiple (*A*) by x and (*B*) by y and subtracting (*B*) from (*A*), we get

$$2(x^2 - y^2) = 0 \quad \text{or} \quad y = x$$

Similarly from (B) and (C)

$$2(z^2 - y^2) = 0 \quad \text{or} \quad z = y$$

Hence $x = y = z$

Now from (*ii*) *i.e.*, relation

$$xyz = a^3$$

at $x = y = z,\ x^3 = a^3$

$$x = a$$

$$y = a$$

$$z = a$$

From (*i*) $x^2 + y^2 + z^2$

or $a^2 + a^2 + a^2 = 3a^2$

Which is the minimum value of the function.

(*ii*) Let $u = x^2 + y^2 + z^2$ ...(*i*)

and $$\phi = ax + by + cz - p \quad ...(ii)$$

By Lagrange's method

$$F(x, y, z) = u(x, y, z) = \lambda\, \phi(x, y, z) = 0 \quad ...(iii)$$

For maxima or minima (*iii*)

$$\frac{\partial F}{\partial x} = \frac{\partial u}{\partial x} + \lambda \frac{\partial \phi}{\partial x} = 0$$

or $$x = -\frac{\lambda a}{2}$$

$$\frac{\partial F}{\partial y} = \frac{\partial u}{\partial y} + \lambda \frac{\partial \phi}{\partial y} = 0$$

or $$2y + \lambda\, b = 0$$

or $$y = -\frac{\lambda b}{a}$$

$$\frac{\partial F}{\partial z} = \frac{\partial u}{\partial z} + \lambda \frac{\partial \phi}{\partial z} = 0$$

or $$2z + \lambda\, c = 0$$

or $$z = -\frac{\lambda c}{2}$$

Now putting the value of x, y, z in (*ii*),

$$ax + by + cz = p$$

$$-\frac{\lambda a^2}{2} - \frac{\lambda b^2}{2} - \frac{\lambda c^2}{2} = p$$

$$-\lambda(a^2 + b^2 + c^2) = 2p$$

or $$\lambda = \frac{-2p}{a^2 + b^2 + c^2}$$

$\therefore$ $$x = \frac{2ap}{2(a^2 + b^2 + c^2)}$$

or $$x = \frac{ap}{a^2 + b^2 + c^2}$$

Similarly $$y = \frac{2bp}{2(a^2 + b^2 + c^2)}$$

or $$y = \frac{bp}{a^2 + b^2 + c^2}$$

and $$z = \frac{2cp}{a^2 + b^2 + c^2}$$

Hence the minimum value of u is

$$x^2 + y^2 + z^2 = \frac{a^2 p^2}{(a^2 + b^2 + c^2)^2} + \frac{b^2 p^2}{(a^2 + b^2 + c^2)^2} + \frac{c^2 p^2}{(a^2 + b^2 + c^2)^2}$$

or $$\frac{p^2 (a^2 + b^2 + c^2)}{(a^2 + b^2 + c^2)^2} = \frac{p^2}{a^2 + b^2 + c^2}$$

***Example* 5.60.** Find the maximum and minimum distance of the point (3, 4, 12) from the sphere $x^2 + y^2 + z^2 = 1$.

Solution : Let A (A, x, z) be any point on the sphere and suppose that B (3, 4, 12) be given point, then distance AB is given by

$$AB = \sqrt{(x-3)^2 + (y-4)^2 + (z-12)^2}$$

$$AB^2 = (x-3) + (y-4)^2 + (z-12)^2 = f(x, y, z) \text{ (say)} \quad ....(i)$$

and x, y, z and connected by the relation.

$$f(x, y, z) = x^2 + y^2 + z^2 - 1 \quad ...(ii)$$

Consider Lagrange's function

$$F(x, y, z) = f(x, y, z) + \lambda\phi\,(x\;y\;z) \quad ...(A)$$

For maxima or minima of (A)

$$\frac{\partial F}{\partial x} = 0 \Rightarrow \frac{\partial f}{\partial x} + \lambda\frac{\partial \phi}{\partial x} = 0$$

or $$2(x-3) + \lambda\, 2x = 0 \quad ...(iii)$$

$$\frac{\partial F}{\partial y} = 0 \Rightarrow \frac{\partial f}{\partial y} + \lambda\frac{\partial \phi}{\partial y} = 0$$

or $$2(y-4) + \lambda\, 2y = 0 \quad ...(iv)$$

$$\frac{\partial F}{\partial z} = 0 \Rightarrow \frac{\partial f}{\partial z} + \lambda\frac{\partial \phi}{\partial z} = 0$$

or $$2(z-12) + \lambda\, 2z = 0 \quad ...(v)$$

On multiplying (*iii*) by x, (*iv*) by y and (*v*) by z and adding then

$$2(x^2 + y^2 + z^2) - 6x - 8y - 24z + 2\lambda\,(x^2 + y^2 + z^2) = 0$$

or $2(1) - 6x - 8y - 24z + 2y = 0$

From (*ii*)

$$3x + 4y + 12\,z = \lambda + 1 \quad ...(vi)$$

From (*iii*), (*iv*) and (*v*), we get

$$x = \frac{3}{1+\lambda}\; y = \frac{4}{1+\lambda},\; z = \frac{12}{1+\lambda}$$

Putting the value of x, y, z in (*vi*), we get,

$$\frac{9}{1+\lambda} + \frac{16}{1+\lambda} + \frac{144}{1+\lambda} = 1 + \lambda$$

or $$(1+\lambda)^2 = 169$$

or $$1 + \lambda = \pm 13$$

or $$\lambda = 12, \text{ or } \lambda = -14$$

When $$\lambda = 12, x = \frac{3}{13}, \quad y = \frac{4}{13}, \quad z = \frac{12}{13}$$

When $$\lambda = -14, x = -\frac{3}{13}, \; y = -\frac{4}{13}, z = -\frac{12}{13}$$

Thus we get two stationary points $P\left(\frac{3}{13}, \frac{4}{13}, \frac{12}{13}\right)$ and $Q\left(\frac{-3}{13}, \frac{-4}{13}, \frac{-12}{13}\right)$ on the sphere which are at a minimum or minimum distance from the given point A.

Now $$AP = \sqrt{\left(3-\frac{3}{13}\right)^2+\left(4-\frac{4}{13}\right)^2+\left(12-\frac{12}{13}\right)^2} = 12$$

and $$AQ = \sqrt{\left(3+\frac{3}{13}\right)^2+\left(4+\frac{4}{13}\right)^2+\left(12+\frac{12}{13}\right)^2} = 14$$

$\therefore P\left(\frac{3}{13}, \frac{4}{13}, \frac{12}{13}\right)$ is at a minimum distance from A and the minimum distance = 12 and Q $\left(\frac{-3}{13}, \frac{-4}{13}, \frac{-12}{13}\right)$ is at a maximum distance from A and the maximum distance = 14.

***Example* 5.61.** Find the dimensions of the rectangular box, open at the top of maximum capacity whose surface is 432 square cm.

Solution : Let x, y, z be the length breadth and height of the box then volume V is

$$V = xyz$$

and surface of box whose tope is open

$$S = xy + 2yz + 2zx = 432$$

Let $$V = xyz \qquad ...(i)$$

and $$f(x, y, z) = xy + 2yz - 432 \qquad ...(ii)$$

By Lagrange's method

$$F(x, y, z) = V(x, y, z) + \lambda\, \phi(x, y, z) \qquad ...(A)$$

For maxima or minima

$$\therefore \quad \frac{\partial F}{\partial x} = \frac{\partial V}{\partial x} + \lambda \frac{\partial \phi}{\partial x} = 0 \text{ or } \lambda\,(y + 2z) + yz = 0 \qquad ...(iii)$$

$$\frac{\partial F}{\partial y} = \frac{\partial V}{\partial y} + \lambda \frac{\partial \phi}{\partial y} = 0 \text{ or } \lambda\,(x + 2z) + xy = 0 \qquad ...(iv)$$

$$\frac{\partial F}{\partial z} = \frac{\partial V}{\partial z} + \lambda \frac{\partial \phi}{\partial z} = 0 \text{ or } \lambda\;(2y + 2x) + xy = 0 \qquad ...(v)$$

On multiplying (*iii*) by x, (*iv*) by y and by z

$$\lambda\,(xy + 2xz) + xyz = 0 \qquad ...(vi)$$

$$\lambda\,(xy + 2yz) + xyz = 0 \qquad ...(vii)$$

$$\lambda\,(2yz + 2zx) + xyz = 0 \qquad ...(viii)$$

From (*vi*) and (*vii*)

$$2xz = 2zy \quad \text{or} \quad x = y$$

Similarly from (*vii*) and (*viii*), we get

$$xy = 2zx$$

or $$z = \frac{y}{2}$$

From (*ii*),

$$xy + 2yz + 2zx = 432$$

$$x^2 + x^2 + x^2 = 432$$

$$3x^2 = 432 \Rightarrow x = 12$$

$$y = 12$$

$$z = 6$$

Dimension are 12 cm, 12 cm, 6 cm.

***Examples* 5.62.** If $x\,y\,z = 8$ find the values of x, y and x for which $u = \dfrac{5xyz}{(x + 2y + 4z)}$ is a maximum.

Solution : Let $u = \dfrac{5xyz}{(x + 2y + 4z)}$ ...(*i*)

where the variable x, y, z are connected by the relation

$$x\,y\,z = 8$$

Equation (*ii*) can be taken as

$$f(x, y, z) = x\,y\,z - 8 \qquad ...(iii)$$

By Lagrangian method,

$$F(x, y, z) = u(x, y, z) + \lambda\,\phi\,(x, y, z) \qquad ...(A)$$

For maxima or minima

$$\frac{\partial F}{\partial x} = \frac{\partial F}{\partial y} = \frac{\partial F}{\partial z} = 0$$

$\therefore$

$$\frac{\partial u}{\partial x} + \lambda\frac{\partial \phi}{\partial x} = 0 \equiv \frac{(x + 2y + 4z) \times 5yz - 5xyz}{(x + 2y + 4z)^2} + \lambda yz = 0$$

$$= \frac{10y^2z + 20yz^2}{(z + 2y + 4z)^2} + \lambda yz = 0 \qquad ...(iv)$$

$$\frac{\partial u}{\partial y} + \lambda\frac{\partial \phi}{\partial y} = 0 = \frac{(x + 2y + 4z) \times 5xz - 5xyz\,(2)}{(x + 2y + 4z)^2} + \lambda xz = 0$$

or

$$= \frac{5x^2z + 20\,xz^2}{(x + 2y + 4z)^2} + \lambda\,xz = 0 \qquad ...(v)$$

$$\frac{\partial u}{\partial z} + \lambda\frac{\partial \phi}{\partial z} = 0 = \frac{(x + 2y + 4z) \times 5xy - 5xyz\,(4)}{(x + 2y + 4z)^2} + \lambda xy = 0$$

or

$$\frac{5x^2y - 10xy^2}{(x + 2y + 4z)^2} + \lambda\,xz = 0 \qquad ...(vi)$$

On multiplying (*iv*) by x and (*v*) by y and subtracting (*iv*) from (*v*)

$$\frac{5x^2yz - 10xy^2z}{(x + 2y + 4z)^2} = 0 \text{ or } 5xyz\,(x - 2y) = 0 \qquad \because \; xyz \neq 0$$

$\therefore$

$$x = 2y$$

Again multiplying (*v*) by y and (*vi*) by z and subtracting (*v*) from (*vi*)

$$\frac{10xy^2z - 20xyz^2}{(x + 2y + 4z)^2} = 0$$

or

$$10xyz\,(y - 2z) = 0$$

$\therefore$

$$xyz \neq 0 \quad \because \quad y = 2z$$

Hence

$$x = 2y, \quad z = \frac{y}{2}$$

From (*ii*)

$$xyz = 8$$

$$2y \times y \times \frac{y}{2} = 8 \quad \text{or} \quad y^3 = 23 \quad \text{or} \quad y = 2$$

Hence $x = 2y = 2 \times 2 = 4$

$$z = \frac{y}{2} = \frac{2}{1} = 1$$

Hence value of x, y, z

$$x = 4$$
$$y = 2$$
$$z = 1$$

Example **5.63.** Prove that the rectangular solid of maximum volume which can be inscribed in sphere is a cube.

Solution : Let $2x$, $2y$ and $2z$ be length-breath and height of the rectangular solid and suppose that R be the radius of the sphere then volume of solid is

$$V = 8xyz \quad \text{...(i)}$$

Again let $A(x, y, z)$ be any point on the sphere, then distance from origin to this point (origin being the control) is

$$(x-0)^2 + (y-0)^2 + (z-0)^2 = R^2$$

or $$x^2 + y^2 + z^2 = R^2 \quad \text{...(ii)}$$

Let $$\phi(x, y, z) = x^2 + y^2 + z^2 - R^2$$

By Lagrange's method

$$F(x, y, z) = V(x, y, z) + \lambda\, \phi(x, y, z) \quad \text{...(iii)}$$

For maximum or minimum of (*iii*)

$$\frac{\partial F}{\partial x} = \frac{\partial V}{\partial x} + \lambda \frac{\partial \phi}{\partial x} = 0$$

$$\frac{\partial F}{\partial y} = \frac{\partial V}{\partial y} + \lambda \frac{\partial \phi}{\partial y} = 0$$

$$\frac{\partial F}{\partial z} = \frac{\partial V}{\partial z} + \lambda \frac{\partial \phi}{\partial z} = 0$$

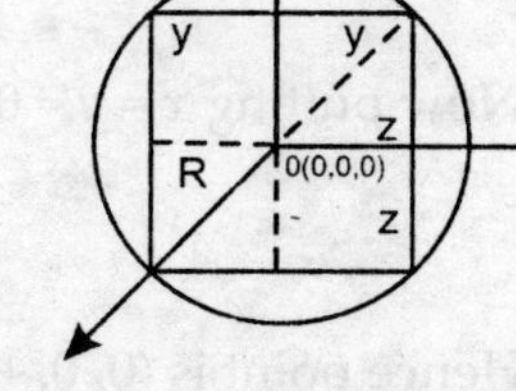

Fig. 5.1

Hence from (*i*) and (*ii*), we get

$$\frac{\partial F}{\partial x} = 8yz + \lambda(2x) = 0 \quad \text{...(iv)}$$

$$\frac{\partial F}{\partial y} = 8xz + \lambda(2y) = 0 \quad \text{...(v)}$$

and $$\frac{\partial F}{\partial z} = 8xy + \lambda(2z) = 0 \quad \text{...(vi)}$$

On multiplying (*iv*) by x, (*v*) by y and (*vi*) by z, we get

$$8xyz = -2\lambda\, x^2,\ 8xyz = -2\lambda y^2 \text{ and } 8xyz = -2\lambda z^2$$

$$\therefore \quad x^2 = y^2 = z^2 \text{ or } x = y = z$$

Hence rectangular solid is a cube.

Example **5.64.** Find the point on the surface $z^2 = xy + 1$ nearest to the origin.

Solution : Let $P(x, y, z)$ be any point on the surface and $0(0, 0, 0)$ be the origin then the distance from origin to $P(x, y, z)$ is given by

$$D^2 = (x-0)^2 + (y-0)^2 + (z-0)^2$$

Let $D^2 = u,$

$\therefore$ $u = x^2 + y^2 + z^2$ ...(*i*)

and $f(x, y, z) = z^2 - xy - 1 = 0$ ...(*ii*)

By Lagrange's method

$$F(x, y, z) = u(x, y, z) + \lambda\phi(x, y, z) \quad \text{...(A)}$$

For maxima or minima of (A),

$$\frac{\partial F}{\partial x} = \frac{\partial u}{\partial x} + \lambda\frac{\partial \phi}{\partial x} = 0$$

$$2x + \lambda(-y) = 0 \quad \text{...(iii)}$$

$$\frac{\partial F}{\partial y} = \frac{\partial u}{\partial y} + \lambda\frac{\partial \phi}{\partial y} = 0$$

$$2y + \lambda(-x) = 0 \quad \text{...(iv)}$$

$$\frac{\partial F}{\partial z} = \frac{\partial u}{\partial z} + \lambda\frac{\partial \phi}{\partial z} = 0$$

$$2z + \lambda(2z) = 0 \quad \text{...(v)}$$

From (*v*) $2z(\lambda + 1) = 0 \Rightarrow 2z \neq 0$

$\therefore$ $\lambda = -1$

From (*i*) and (*ii*), we get

$$2x + y = 0 \quad \text{...(vi)}$$

$$2y + x = 0 \quad \text{...(vii)}$$

On solving for x and y

$$x = 0 \text{ and } y = 0$$

Now putting $x = y = 0$, in (*ii*)

$$z^2 = xy + 1$$

$$z^2 = 0 + 1 \Rightarrow z^2 = 1 \Rightarrow z = \pm 1.$$

Hence point is $(0, 0, \pm 1)$.

***Examples* 5.65.** The temperature T at any point (x, y, z) in space is $T = 400\, xyz^2$. Find the highest temperature on the surface of the unit sphere $x^2 + y^2 + z^2 = 1$.

Solution : Hence $T = 400\, xy^2$ ...(*i*)

and $\phi(x, y, z) = x^2 + y^2 + z^2 - 1$

By Lagrange's method,

$$F(x, y, z) = T(x, y, z) + \lambda\phi(x, y, z) \quad \text{...(ii)}$$

For maxima or minima or $F(x, y, z)$

$$\frac{\partial F}{\partial x} = \frac{\partial T}{\partial x} + \lambda\frac{\partial \phi}{\partial x} = 0$$

$$\frac{\partial F}{\partial y} = \frac{\partial T}{\partial y} + \lambda\frac{\partial \phi}{\partial y} = 0$$

$$\frac{\partial F}{\partial z} = \frac{\partial T}{\partial z} = \lambda\frac{\partial \phi}{\partial z} = 0$$

From (*i*) and (*ii*), we get

$$400\,yz^2 + \lambda\,(2x) = 0 \quad ...(iii)$$

$$400\,xz^2 + \lambda\,(2y) = 0 \quad ...(iv)$$

$$800\,xyz + \lambda\,(2z) = 0 \quad ...(v)$$

On multiplying (*iii*) by x (*iv*) and (*v*) by z and adding, we get $1600xyz^2 + 2\lambda\,(x^2 + y^2 + z^2) = 0$

From (*v*)

$$1600xyz^2 + 2\lambda(1) = 0 \quad \text{or} \quad \lambda = -\,800xyz^2$$

Putting the value of λ in (*iii*)

$$400\,yz^2 + 2x\,(-800x\,y\,z^2) = 0$$

or $$400yz^2\,(1 - 4x^2) = 0$$

$\therefore$ $$400yz^2 \neq 0$$

$\therefore$ $$4x^2 = 1$$

$$x = \pm\frac{1}{2}$$

Similarly from (*iv*)

$$y = \pm\frac{1}{2}$$

and from (*v*) $$z = \pm\frac{1}{2}$$

$\therefore$ From (*i*) $$T = 400\,xyz^2$$

$$= 400\,\frac{1}{2} \times \frac{1}{2} \times \frac{1}{2}$$

$$T = 50$$

***Example* 5.66.** Divide 24 into three parts such that he continued product of the first, square of the second and the cube of the third may be maximum.

Solution : Let three parts be x, y and z such and z such that

$$x + y + z = 24$$

or $$\phi(x, y, z) = x + y + z - 24 \quad ...(i)$$

and the product of the first, square of second and cube of third is $x\,y^2\,z^3$

Let $$u = xy^2\,z^3 \quad ...(ii)$$

By Lagrange's method

$$F(x, y, z) = u(x, y, z) + \lambda\phi\,(x, y, z) \quad ...(A)$$

For maxima or minima of (A)

$$\frac{\partial F}{\partial x} = \frac{\partial u}{\partial x} + \lambda\frac{\partial \phi}{\partial x} = 0 \quad ...(iii)$$

or $$y^2\,z^3 + \lambda = 0$$

$$\frac{\partial F}{\partial y} = \frac{\partial u}{\partial y} + \lambda\frac{\partial \phi}{\partial y} = 0 \quad ...(iv)$$

or $$2xyz^3 + \lambda = 0$$

$$\frac{\partial F}{\partial z} = \frac{\partial u}{\partial z} + \lambda\,\frac{\partial \phi}{\partial z} = 0$$

or $$3xy^2z^2 + \lambda = 0$$

From (*iii*) and (*iv*) ...(*v*)

$$y^2 z^3 = 2xyz^3$$
$$y = 2x$$
$\therefore$ $$x \neq 0,\ yh \neq 0,\ z \neq 0$$
or $$x = \frac{y}{2}$$

Similarly from (*iv*) and (*v*)

$$2xyz^3 = 3xy^2z^2 \quad \text{or} \quad 2z = 3y$$
$$y = \frac{2z}{3}$$

From (*i*), we get

$$x + y + z = 24$$
$$\frac{z}{3} + \frac{2z}{3} + z = 24$$
or $$\frac{\partial z}{3} = 24 \Rightarrow 2z = 24, \quad z = 12$$
$\therefore$ $$x = \frac{z}{3} = \frac{12}{3} \quad \text{or} \quad x = 4$$
and $$y = \frac{2z}{3} = \frac{2}{3} \times 12 \quad \text{or} \quad y = 8$$
Hence $$x = 4, \quad y = 8 \text{ and } \quad z = 12$$

***Example* 5.67.** If $= a^3 x^2 + b^3y^2 + c^3 z^2$ where $\frac{1}{x} + \frac{1}{y} + \frac{1}{z} = 1$. Show that stationary value of u is given by $x = \frac{\Sigma a}{a}$, $y = \frac{\Sigma a}{b}$ and $c = \frac{\Sigma a}{c}$.

Solution : Let $$u = a^3 x^2 + b^3 y^2 + c^3 z^2 \quad \text{...(i)}$$
and $$\phi(x, y, z) = \frac{1}{x} + \frac{1}{y} + \frac{1}{z} - 1 \quad \text{...(ii)}$$

By Lagrange's method

$$F(x, y, z) = u(x, y, z) + \lambda\phi(x, y, z) \quad \text{...(A)}$$

For maxima or minima of (A).

$$\frac{\partial F}{\partial x} = \frac{\partial u}{\partial x} + \lambda\frac{\partial \phi}{\partial x} = 0 \quad \text{or} \quad 2a^3x + \lambda\left(-\frac{1}{x^2}\right) = 0 \quad \text{...(iii)}$$
or $$2a^3 x^3 - \lambda = 0$$
$$\frac{\partial F}{\partial y} = \frac{\partial u}{\partial y} + \lambda\frac{\partial \phi}{\partial y} = 0 \quad \text{or} \quad 2c^3y + \lambda\left(-\frac{1}{y^2}\right) = 0 \quad \text{...(iv)}$$
$$2b^3y^3 - \lambda = 0$$
$$\frac{\partial F}{\partial z} = \frac{\partial u}{\partial z} + \lambda\frac{\partial \phi}{\partial z} = 0 \quad \text{or} \quad 2c^3z + \lambda\left(-\frac{1}{z^2}\right) = 0$$
or $$2c^3 z^3 - \lambda = 0 \quad \text{...(v)}$$

From (*iii*) and (*iv*), we get

$$2a^3x^3 = 2b^3y^3 \quad \text{or} \quad ax = by$$

Similarly from (*iv*) and (*v*), we get

$$2b^3y^3 = 2c^3z^3 \quad \text{or} \quad by = cz$$

$$\therefore \quad x = \frac{b}{a}y, \quad z = \frac{b}{c}y$$

From (*ii*)

$$\frac{1}{x} + \frac{1}{y} + \frac{1}{z} = 1$$

or

$$\frac{a}{by} + \frac{1}{y} + \frac{c}{by} = 1$$

or

$$\frac{a+b+c}{by} = 1$$

or

$$y = \frac{a+b+c}{b}$$

or

$$y = \frac{\Sigma a}{b}$$

Similarly

$$x = \frac{\Sigma a}{a} \quad \text{and} \quad z = \frac{\Sigma a}{c}$$

Example **5.68.** Find the stationary values of $x^2 + y^2 = z^2$ subject to $ax^2 + by^2 + cz^2 = 1$ and $lx + my + nz = 0$

Solution : Let

$$u(x, y, z) = x^2 + y^2 + z^2 \qquad \text{...}(i)$$

and

$$\phi(x, y, z) = ax^2 + by^2 + cz^2 - 1 \qquad \text{...}(ii)$$

$$\psi(x, y, z) = lx = my + nz \qquad \text{...}(iii)$$

By Lagrange's method

$$F(x, y, z) = u(x, y, z) + \lambda_1 \phi(x, y, z) + \lambda_2 \psi(x, y, z) \qquad \text{...(A)}$$

For maxima or minima of (A)

$$\frac{\partial F}{\partial x} = \frac{\partial u}{\partial x} + \lambda_2 \frac{\partial \phi}{\partial x} + \lambda_2 \frac{\partial \psi}{\partial x} = 0$$

$$\frac{\partial F}{\partial y} = \frac{\partial u}{\partial y} + \lambda_1 \frac{\partial \phi}{\partial y} + \lambda_2 \frac{\partial \psi}{\partial y} = 0$$

$$\frac{\partial F}{\partial z} = \frac{\partial u}{\partial z} + \lambda_1 \frac{\partial \phi}{\partial z} + \lambda_2 \frac{\partial \psi}{\partial z} = 0$$

Using (*i*), (*ii*) and (*iii*), we get

$$2x + \lambda_1 (2ax) + \lambda_2 l = 0$$

or

$$2x + 2ax\,\lambda_1 + l\,\lambda_2 = 0 \qquad \text{...}(iv)$$

$$2y + \lambda_1 (2by) + m\,\lambda_2 = 0$$

or

$$2y + 2by\,\lambda_1 + m\,\lambda_2 = 0 \qquad \text{...}(v)$$

and

$$2z + \lambda_1 (2cz) + \lambda_2 (n) = 0$$

or

$$2z + 2cz\,\lambda_1 + n\,\lambda_2 = 0 \qquad \text{...}(vi)$$

On multiplying (*iv*) by x, (*v*) by (*vi*) by z and adding them

$$(2x^2 + 2y^2 + 2z^2) + (2ax^2 + 2by^2 + 2cz^2)\,\lambda_1 + (l_x + my + nz)\,\lambda_2 = 0$$

From (*i*), (*ii*) and (*iii*), we get

$$2u + 2\,\lambda_1(1) + 0 = 0$$

$$\therefore \quad \lambda_1 = -u$$

On putting $\lambda_1 = -u$ in (iv), (v) and (vi) and (vi), we get

$$2x - 2axu + \lambda_2 l = 0$$

$$2x(1 - au) = -\lambda_2 l$$

$$\therefore \quad x = \frac{-\lambda_2 l}{2(1-au)}$$

and $$2y - 2byu + \lambda_2 m = 0$$

$$2y(1 - bu) = -\lambda_2 \text{m}$$

$$\therefore \quad y = \frac{-y_2 m}{2(1-bu)}$$

Similarly

$$2z - 2czu + \lambda_2 n = 0$$

$$2z(1 - cu) = -\lambda_2 n$$

$$z = \frac{-\lambda_2 n}{2(1-cu)}$$

On putting the value of x, y, z in (iii), we get

$$lx + my + nz = 0$$

or $$\frac{-\lambda_2 l^2}{2(1-au)} + \frac{-\lambda_2 m^2}{2(l-bu)} + \frac{-\lambda_2 n^2}{2(1-cu)} = 0$$

or $$-\frac{\lambda_2}{2}\left[\frac{l^2}{1-au} + \frac{m^2}{1-bu} + \frac{n^2}{1-cu}\right] = 0$$

Hence $$\frac{l^2}{1-au} + \frac{m^2}{1-bu} + \frac{n^2}{1-cu} = 0$$

Example **5.69.** Prove that the stationary values of $u = \frac{x^2}{a^4} + \frac{y^2}{b^4} + \frac{z^2}{c^4}$ where $lx + \text{my} + nz = 0$ and $\frac{x^2}{a^2} + \frac{y^2}{b^2} + \frac{z^2}{c^2} = 1$ are the roots of the equation $\frac{l^2 a^4}{1-a^2 u} + \frac{m^2 b^4}{1-b^2 u} + \frac{n^2 c^4}{1-c^2 u} = 0$.

Solution : Let $$u(x, y, z) = \frac{x^2}{a^4} + \frac{y^2}{b^4} + \frac{z^2}{c^4} \quad \text{...(i)}$$

and $$f(x, y, z) = lx + my + xz \quad \text{...(ii)}$$

$$\psi(x, y, z) = \frac{x^2}{a^2} + \frac{y^2}{b^2} + \frac{z^2}{c^2} - 1 \quad \text{...(iii)}$$

Consider Lagrange's function

$$F(x, y, z) = \left(\frac{x^2}{a^4} + \frac{y^2}{b^4} + \frac{z^2}{c^4}\right) + \lambda(lx + my + nz) + \mu\left(\frac{x^2}{a^2} + \frac{y^2}{b^2} + \frac{z^2}{c^2} - 1\right) \quad \text{...(A)}$$

For maxima or minima or (A),

$$\frac{\partial F}{\partial x} = 0 \Rightarrow \frac{2x}{a^4} + \lambda l + \frac{2\mu x}{a^2} = 0 \quad \text{...(i)}$$

$$\frac{\partial F}{\partial y} = 0 \Rightarrow \frac{2y}{b^4} + \lambda m + \frac{2\mu y}{b^2} = 0 \quad \text{...(ii)}$$

and $$\frac{\partial F}{\partial z} = 0 \Rightarrow \frac{2z}{c^4} + \lambda n + \frac{2z\mu}{c^2} = 0 \quad ...(iii)$$

Multiplying (*i*) by x, (*ii*) by y and by z and adding them, we get

$$2\left(\frac{x^2}{a^4} + \frac{y^2}{b^4} + \frac{z^2}{c^4}\right) + \lambda(lx + my + nz) + 2\mu\left(\frac{x^2}{a^2} + \frac{y^2}{b^2} + \frac{z^2}{c^2}\right) = 0$$

or $$2u + \mu(0) + 2\mu(1) = 0 \quad \text{[From given relations and function]}$$

$$\therefore \quad \mu = -u$$

From (*i*) $$\frac{2x}{a^4} + \lambda l - \frac{2ux}{a^2} = 0$$

$$\frac{2x}{a^4}(1 - a^2 u) = -\lambda l$$

$$\therefore \quad x = -\frac{\lambda l a^4}{2(1 - a^2 u)}$$

From (*ii*) $$y = -\frac{\lambda m b^4}{2(1 - b^2 u)}$$

and from (*iii*) $$z = -\frac{\lambda n c^4}{2(1 - c^2 u)}$$

To eliminate λ between x, y, z, multiplying the value of x, y, z by l, m, n respectively and adding them

$$lx + my + nz = -\frac{\lambda}{2}\left[\frac{l^2 a^4}{1 - a^2 u} + \frac{m^2 b^4}{1 - b^2 u} + \frac{n^2 c^4}{1 - c^2 u}\right]$$

But $$lx + my + nx = 0$$

$$\therefore \quad \frac{l^2 a^4}{1 - a^2 u} + \frac{m^2 b^4}{1 - b^2 u} + \frac{n^2 c^4}{1 - c^2 u} = 0 \quad \because \; l \neq 0$$

EXERCISE 5.7

1. Find the minimum value of $x^2 + y^2 + z^2$ when

   (*a*) $x + y + z = 3a$ **Ans.** Minimum value is $3a^2$

   (*b*) $ax + by + cz = a + b + c$ **Ans.** $\frac{(a + b + c)^2}{a^2 + b^2 + c^2}$

   (*c*) $xy + yz + zx = 3a^2$ **Ans.** $3a^2$

2. (*a*) Find the minimum value of the function $x + y + z$ subject to the condition $\frac{a}{z} + \frac{b}{y} + \frac{c}{z} = 1$.

   Ans. Stationary point is $\left\{a^{\frac{1}{2}}\left(a^{\frac{1}{2}} + b^{\frac{1}{2}} + c^{\frac{1}{2}}\right), b^{\frac{1}{2}}(a^{\frac{1}{2}} + b^{\frac{1}{2}} + c^{\frac{1}{2}}), c^{\frac{1}{2}}(a^{\frac{1}{2}} + b^{\frac{1}{2}} + c^{\frac{1}{2}})\right\}$ and minimum value is $(\sqrt{a} + \sqrt{b} + \sqrt{c})^2$

   (*b*) Find the minimum value of the function $ax + by + cz = 1$ subject to the condition $\frac{a^2}{x^2} + \frac{b^2}{y^2} + \frac{c^2}{z^2} = 1$.

   Ans. Stationary point is $\left\{a^{\frac{1}{2}}\left(a^{\frac{3}{2}} + b^{\frac{3}{2}} + c^{\frac{3}{2}}\right),\right\}\left\{b^{\frac{1}{2}}(a^{\frac{3}{2}} + b^{\frac{3}{2}} + c^{\frac{3}{2}})\right\}, \left\{c^{\frac{1}{2}}(a^{\frac{3}{2}} + b^{\frac{3}{2}} + c^{\frac{3}{2}})\right\}$ and minimum values of function is $(a^{\frac{3}{2}} + b^{\frac{3}{2}} + c^{\frac{3}{2}})^2$

3. (*a*) Find the shortest and the longest distance from the point (1, 2, –1) to the sphere $x^2 + y^2 + z^2 = 24$

Ans. Shortest distance is $\sqrt{6}$, Longest distance $3\sqrt{6}$.

(*b*) Find the minimum and maximum distance from the origin to the curve

$5x^2 + 6xy + 5y^2 - 8 = 0$ **Ans.** Minimum distance is 1, Maximum distance is 4.

4. (*a*) A rectangular box, open at the top, is to have a volume of 32 c.c. Find the dimensions of the box requiring least material for it construction.

Ans. $x = 4, y = 4, z = 4$.

(*b*) Find the dimension of the rectangular box, open at the top with maximum capacity, whose surface are is 256 sqm/cm.

Ans. $x = 8$ cm, $y = 8$ cm, $z = 4$ cm.

5. Given that $x + y + z = a$. Find the maximum value of $x^m y^n z^p$.

Ans. Maximum value $a^{m+n+p} \dfrac{m^n n^n p^p}{(m+n+p)^{m+n+p}}$

5.18 DIFFERENTIATION UNDER THE INTEGRAL SIGN

Suppose $F(\alpha)$ is the value of a define integral $\int_a^b f(x, \alpha)\,dx$, where α is parameter. To find $F(\alpha)$, first we have to evaluate the integral $\int_a^b f(x, \alpha)\,dx$, and then differentiate $F(\alpha)$ w.r.t. α. However it is not always possible the integral and then to find the derivative, such problems are solved by reversing the order of the integration and differentiation i.e., first differentiate $F(x, a)$ partially w.r.t. "α" and then integrate it. Thus is general $\int_a^b f(x, \alpha)\,dx = F(\alpha)$

5.19 LEIBNITZ'S RULE FOR DIFFERENTIATION UNDER THE INTEGRAL SIGN

(*a*) when the limits of integration are constant *i.e.* independent of parameter

Statement : If $f(x, \alpha)$ and $\frac{\partial}{\partial \alpha} f(x, \alpha)$ be continuous function of x and α then

$$\frac{d}{d\alpha}\left[\int_a^b f(x, \alpha)\,dx\right] = \int_a^b \frac{\partial}{\partial \alpha} f(x, \alpha)\,dx$$

Proof : Let $\int_a^b f(x, \alpha)\,dx = F(\alpha)$ ...(*i*)

where α is parameter

Suppose $\delta\alpha$ is increment in α then (*i*) becomes

$$F(\alpha + \delta\alpha) = \int_a^b f(x, \alpha + \delta\alpha)\,dx \quad ...(ii)$$

Substraction (*i*) from (*ii*) then, we get

$$F(\alpha + \delta\alpha) - F(\alpha) = \int_a^b f(x, \alpha + \delta\alpha)\,dx - \int_a^b f(x, \alpha)\,dx$$

$$= \int_a^b [f(x, \alpha, \delta\alpha) - f(x, \alpha)]\,dx$$

Dividing by $\delta\alpha$ on both sides, we get

$$\frac{F(\alpha+\delta\alpha)-F(\alpha)}{\delta\alpha} = \int_a^b \frac{f(x,\alpha,+\delta\alpha)-f(x+\alpha)}{\delta\alpha}dx$$

Taking limits on both sides as $\delta\alpha \to 0$, then

$$\lim_{da\to 0}\frac{F(\alpha+\delta\alpha)-F(\alpha)}{\delta\alpha} = \lim_{\delta\alpha\to 0}\int_a^b \frac{f(x,\alpha+d\alpha)-F(x,\alpha)}{\delta\alpha}dx$$

$$\therefore \quad \frac{dF}{d\alpha} = \int_a^b \frac{\partial}{\partial\alpha}f(x,\alpha)\,dx$$

This formula is useful for evaluating definite integrals which are otherwise impossible to evaluate.

Example **5.70.** Prove that $\int_a^\infty \frac{\tan^{-1} ax}{x(1+x^2)}dx = \frac{\pi}{2}\log(1+a)\ a \ge 0.$

Solution : Let $F(a) = \int_0^\infty \frac{\tan^{-1} ax}{x(1+x^2)}dx$ ...(A)

Here a is parameter

Differentiating (A) w.r.t a on both sides, we get

$$F(a) = \int_0^\infty \frac{x\,dx}{x(1+x^2)(1+a^2x^2)}$$

$$= \int_0^\infty \frac{dx}{(1+x^2)(1+a^2x^2)} \qquad \text{...(B)}$$

Now $$\frac{1}{(1+x^2)(1+a^2x^2)} = \frac{Ax+B}{1+x^2}+\frac{Cx+D}{1+a^2x^2} \qquad \text{...(C)}$$

$$\therefore \quad 1 = (Ax+B)(1+a^2x^2)+(Cx+D)(1+x^2)$$

Comparing the coefficient of various like powers on both side

Coefficient of x^3

$$0 = a^2A+C$$

Coefficient of x^2 ...(*i*)

$$0 = a^2B+D$$

Coefficient of x ...(*ii*)

$$0 = A+C$$

Constant term ...(*iii*)

$$1 = B+D \qquad \text{...(iv)}$$

From (*iii*) $A = -C$

From (*i*) $a^2A = C = 0$

$\therefore$ $-a^2C+C = 0$

or $C(1-a^2) = 0$

or $C = 0$ and $A = 0$

From (*iv*)

$$B = 1-D$$

From (*ii*)

$$Ba^2 + D = 0$$

or $$(1-D)a^2 + D = 0$$

or $$-a^2 = D(1-a^2)$$

$\therefore$ $$D = \frac{a^2}{a^2-1}$$

and $$B = 1 - D = 1 - \frac{a^2}{a^2-1} = \frac{-1}{a^2-1}$$

$\therefore$ $$B = \frac{1}{1-a^2}$$

Putting these values of partial fraction in (C), we get

$$F'(a) = \int_0^\infty \left[\frac{Ax+B}{1+x^2} + \frac{Cx+D}{1+a^2x^2}\right]dx$$

$\therefore$ $$F'(a) = \int_0^\infty \frac{dx}{(1-a^2)(1+x^2)} + \int_0^\infty \frac{a^2}{(a^2-1)}\frac{dx}{(1+a^2x^2)}$$

$$= \frac{1}{(1-a^2)}\left[\tan^{-1}x\right]_0^\infty + \frac{a^2}{(a^2-1)}\int_0^\infty \frac{dx}{a^2\left(x^2+\frac{1}{a^2}\right)}$$

$$= \frac{\pi}{2(1-a^2)} + \frac{a}{a^2-1}\frac{\pi}{2}$$

$\therefore$ $$F'(a) = \frac{\pi}{2(1-a^2)} + \frac{\pi^2}{2(a^2-1)} \qquad ...(D)$$

Integrating (D) w.r.t. on both sides, we get

$$F(a) = \frac{\pi}{2}\int \frac{da}{1-a^2} + \frac{\pi}{2}\int \frac{a}{a^2-1}da$$

$$F(a) = \frac{\pi}{2}\left[\int\left\{\frac{A}{(1+a)} + \frac{B}{1-a}\right\}da\right] - \frac{\pi}{2}\times\frac{1}{2}\int\frac{-2a}{1-a^2}da + C$$

$$F(a) = \frac{\pi}{2}\left[\int\left\{\frac{A}{1+a} + \frac{B}{1-a}\right\}da\right] - \frac{\pi}{2}\times\frac{1}{2}\int\frac{-2a}{1-a^2}da + C$$

$\therefore$ $$F(a) = \frac{\pi}{2}\left[\frac{1}{2}\left\{\int\frac{da}{1+a} - \frac{da}{1-a}\right\}\right] - \frac{\pi}{2}\times\frac{1}{2}\log(1-a^2) = C$$

$\therefore$ $$F(a) = \frac{\pi}{2}\left[\log(1+a)^{\frac{1}{2}} - \log(1-a)^{\frac{1}{2}}\right] - \frac{\pi}{2}\log(1-a^2)^{\frac{1}{2}} + C \qquad ...(E)$$

When $a = 0$ then from equation (*A*)

$$F(0) = 0$$

Again when a = 0 then from (*E*)

$$F(0) = 0 = \frac{\pi}{2}\times 0 - \frac{\pi}{2}\times 0 + c$$

$\therefore$ $$C = 0$$

Putting C = 0 in (E) then

$$F(a) = \frac{\pi}{2}\left[\log\left(\frac{1+a}{1-a}\right)^{\frac{1}{2}} - \log(1-a^2)^{1/2}\right] + 0$$

$$F(a) = \frac{\pi}{2}\log(1+a)$$

***Example* 5.71.** Evaluate $\int_0^1 \frac{x^\alpha - 1}{\log x}\, dx \quad a \geq 0$

Solution: Let $$F(\alpha) = \int_0^1 \frac{x^\alpha - 1}{\log x}\, dx \quad ...(i)$$

Here a is parameter. Differentiate (*i*) with respect to α on both side

$$F'(\alpha) = \frac{d}{d\alpha}\int_0^1 \frac{x^\alpha - 1}{\log x} dx$$

$$= \int_0^1 \frac{\partial}{\partial\alpha}\frac{x^\alpha - 1}{\log x} dx = \int_0^1 \frac{x^\alpha \log x}{\log x}\, dx$$

$$F'(\alpha) = \int_0^1 x^\alpha\, dx$$

Integration w.r.t x on R.H.S, we get

$$F(\alpha) = \frac{1}{\alpha+1} \quad ...(ii)$$

On integrating (*ii*) w.r.t. α then we get

$$F(\alpha) = \int \frac{d\alpha}{1+\alpha}$$

$$\therefore \quad F(\alpha) = \log(1+\alpha) + c$$

When α = 0 from (i), $F(0) = 0$ ...(*iii*)

Similarly, when $a = 0$ from (*iii*)

$$F(0) = \log(1+0) + c$$

$$0 = c$$

Putting the value of c in (*iii*), we get

$$F(\alpha) = \log(1+\alpha) + 0$$

$$\therefore \quad F(\alpha) = \log(1+\alpha)$$

***Example* 5.72.** Evaluate $\int_0^\pi$ log (1 + a cos x) dx, using the method of differentiation under the sign of integration.

Solution : Let $$F(a) = \int_0^\pi \log(1 + a\cos x)\, dx \quad ...(i)$$

where a is parameter

Differentiating (*i*) w.r.t a on both sides, we get

$$F'(a) = \frac{d}{da}\int_0^\pi \log(1 + a\cos x)\, dx$$

$$F'(a) = \int_0^\pi \frac{\partial}{\partial a}\log(1 + a\cos x)\, dx$$

$$F'(a) = \int_0^\pi \frac{\cos x}{1 + a\cos x}\, dx \quad ...(ii)$$

$$= \frac{1}{a}\int_0^{\pi} \frac{1 + a\cos x - 1}{1 + a\cos x}dx$$

or $$F'(a) = \int_0^{\pi}\left\{\frac{1}{2} + \frac{-\frac{1}{a}}{1 + a\cos x}\right\}dx$$

On integrating R.H.S w.r.t 'x' we get

$$F'(a) = \frac{\pi}{2} - \frac{1}{a}\int_0^{\pi}\frac{dx}{1 + a\cos x}$$

From trigonometry, $$\cos x = \frac{1 - \tan^2\frac{x}{2}}{1 + \tan^2\frac{x}{2}}$$

$\therefore$ $$F'(a) = \frac{\pi}{a} - \frac{1}{2}\int_0^{\pi}\frac{\sec^2 x/2}{\left(1 + \tan^2\frac{x}{2}\right) + a\left(1 - \tan^2\frac{x}{2}\right)}dx$$

Put $$\tan\frac{x}{2} = t$$

$\therefore$ $$\frac{1}{2}\sec^2\frac{x}{2}\,dx = dt$$

when $$x = 0,\ t = 0,\ \text{when } x = \pi,\ t = \infty$$

$\therefore$ $$F'(a) = \frac{\pi}{a} - \frac{1}{a}\int_0^{\infty}\frac{2\,dt}{(1 + t^2) + a(l - t^2)}$$

$$= \frac{\pi}{a} - \frac{1}{a}\int_0^{\infty}\frac{2\,dt}{(1 + a) + (1 - a)t^2}$$

$\therefore$ $$F'(a) = \frac{\pi}{a} - \frac{2}{a(1 - a)}\int_0^{\infty}\frac{dt}{\left\{\sqrt{\frac{1 + a}{1 - a}}\right\}^2 + t^2}$$

$$= \frac{\pi}{a} - \frac{2}{a(1 - a)}\sqrt{\frac{1 - a}{1 + a}}\left[\tan^{-1}\frac{t}{\sqrt{\frac{1 + a}{1 - a}}}\right]_0^{\infty}$$

$$= \frac{\pi}{a} - \frac{2}{a\sqrt{1 - a^2}}\frac{\pi}{2}$$

$$F'(a) = \frac{\pi}{a} - \frac{\pi}{a\sqrt{1 - a^2}} \qquad \text{...}(iii)$$

On integrating (*iii*) w.r.t. *a*, we get

$$F(a) = \pi\int\frac{da}{a} - \pi\int\frac{da}{a\sqrt{1 - a^2}} + c$$

$$= \pi\log a - \pi\int\frac{da}{a\sqrt{1 - a^2}} + c$$

Put $$a = \cos t$$

$$da = -\sin t\,dt$$

$$\therefore \quad F(a) = \pi \log a - \pi \int \frac{-\sin t\, dt}{\cos t \sqrt{1-\cos^2 t}} + c$$

$$F(a) = \pi \log a + \pi \int \frac{\sin t}{\cos t \sin t} dt + c$$

$$F(a) = \pi \log a + \pi \int \sec t\, dt + c$$

$$F(a) = \pi \log a + \pi \log \left(\frac{1}{a} + \frac{\sqrt{1-a^2}}{2}\right) + c$$

$$F(a) = \pi \log a + \pi \log \{1 + \sqrt{1-a^2}\} - \pi \log a + c$$

$$F(a) = \pi \log \{1 + \sqrt{1-a^2}\} + c \qquad \text{...(iv)}$$

Initially, when $a = 0$ then from, (*i*),

$$F(0) = 0$$

Again from (*iv*), when $a = 0$, then

$$F(0) = \pi \log \{1 + \sqrt{1-0}\} + c$$

or

$$0 = \pi \log 2 + c$$

$$\therefore \quad c = -\pi \log 2$$

Putting the value of c in equation (*iv*) then we get

$$F(a) = \pi \log [1 + \sqrt{1-a^2}] - \pi \log 2$$

$$F(a) = \pi \log \frac{1 + \sqrt{1-a^2}}{2}$$

Example **5.73.** Differentiating $\int_0^x \frac{dx}{x^2+a^2} = \frac{1}{a} \tan^{-1} \frac{x}{2}$ under the integral sign. Find the value of $\int_0^x \frac{dx}{(x^2+a^2)^2}$.

Solution : $\int_0^x \frac{dx}{(x^2+a^2)} = \frac{1}{a} \tan^{-1} \frac{x}{2}$ ...(*i*)

where a is parameter

Differentiating (*i*) w.r.t a on both sides, then we get

$$\frac{d}{da} \int_0^x \frac{dx}{(x^2+a^2)} = \frac{d}{da}\left[\frac{1}{2} \tan^{-1}\left(\frac{x}{a}\right)\right]$$

$$\int_0^x \frac{-2a}{(x^2+a^2)^2} dx = -\frac{1}{a^2} \tan{-1} \frac{x}{a} + \frac{1}{a}\left(\frac{1}{1+\frac{x^2}{a^2}}\right)\left(-\frac{x}{a^2}\right)$$

On simplification

$$\int_0^x \frac{dx}{(a^2+x^2)^2} = \frac{1}{2a^3} \tan^{-1}\left(\frac{x}{a}\right) + \frac{x}{2a^2 (a^2+x^2)^2}$$

Example **5.74.** By successive differentiation $\int_0^1 x^m\, dx = \frac{1}{m+1}$ w.r.t 'm'. Evaluate $\int_0^1 x^m (\log x)^n\, dx$.

Solution : $\int_x^1 x^m\, dx = \frac{1}{m+1}$ ...(*i*)

Differentiating (*i*) w.r.t. 'm' on both sides we get

$$\frac{d}{dm} \int_0^1 x^m\, dx = \frac{d}{dm}\left(\frac{1}{m+1}\right)$$

or $$\int_0^1 x^m \log x\,dx = \frac{(-1)}{(m+1)^2} \qquad ...(ii)$$

Again differentiating (*ii*) w.r.t. m on both sides

$$\int_0^1 x^m (\log x)^2 dx = \frac{(-1)(-2)}{(m+1)^3}$$

$$\vdots \quad \vdots \qquad \vdots$$

$$\int_0^1 x^m (\log x)^n dx = \left(\frac{(-1)(-2)\dots(-n)}{(m+1)^{n+1}}\right)$$

Hence $$\int_0^1 x^m (\log x)^n\, dx = \frac{(-1)^n\, n!}{(m+1)^{n+1}}$$

***Example* 5.75.** Prove that $\int_0^{\pi/2} \frac{\log(1+y\sin^2 x)}{\sin^2 x}\,dx = \pi\,[\sqrt{1+y}-1]$ where y is parameter.

Solution : Let $F(y) = \int_0^{\pi/2} \frac{\log(1+y\sin^2 x)}{\sin^2 x}\,dx \qquad ...(i)$

Differentiating (*i*) w.r.t. y on both sides

$$F(y) = \int_0^{\pi/2} \frac{\sin^2 x}{\sin^2 x\,(1+y\sin^2 x)}\,dx$$

$$= \int_0^{\pi/2} \frac{dx}{1+y\sin^2 x}$$

$$= \int_0^{\pi/2} \frac{dx}{\cos^2 + \sin^2 x + y\sin^2 x} \qquad [\because I = \cos^2 x + \sin^2 x]$$

or $$F'(y) = \int_0^{\pi/2} \frac{dx}{\cos^2 x + (1+y)\sin^2 x}$$

$$= \int_0^{\pi/2} \frac{dx\sec^2 x}{1+(1+y)\tan^2 x}$$

Put $$\sqrt{1+y}\,\tan x = t$$

$$\sqrt{1+y}\,\sec^2 x\,dx = dt$$

$$\sec^2 = \frac{dt}{\sqrt{1+y}}$$

when $x = \frac{\pi}{2}$ then $t = \infty$

when $x = 0$, then $t = 0$

$\therefore$ $$F'(y) = \int_0^{\infty} \frac{dt}{\sqrt{1+y}\,[1+t^2]}$$

$$F'(y) = \frac{1}{\sqrt{1+y}}\,[\tan^{-1} t\,]_0^{\infty}$$

$$F'(y) = \frac{\pi}{2\sqrt{1+y}} \qquad ...(ii)$$

Integrating (*ii*) w.r.t. '*y*' on both sides we get

$$F(y) = \frac{\pi}{2}\int (1+y)^{-\frac{1}{2}} dy = \frac{\pi}{2}\left[\frac{(1+y)^{\frac{1}{2}}}{\frac{1}{2}}\right] + c$$

$$F(y) = \pi\sqrt{1+y} + c \qquad ...(iii)$$

Initially when $y = 0$ from (*i*)

$$F(0) = 0$$

Again from (*iii*)

$$F(0) = \pi\sqrt{1+0} + c$$

$$\therefore \quad c = -\pi$$

Hence $$\int_0^{\pi/2} \log \frac{(1+y\sin^2 x)}{\sin^2 x} dx = \pi[\sqrt{1+y} - 1]$$

***Example* 5.76.** Prove that $\int_0^{\infty} e^{-x^2} \cos \alpha x \, dx = \frac{\sqrt{\pi}}{2} e^{-\frac{a^2}{4}}$ were α is parameter.

Solution : Let $F(a) = \int_0^{\infty} e^{-x^2} \cos \alpha x \, dx$

Differentiating (*i*) w.r.t. α on both sides, we get

$$F'(\alpha) = \int_0^{\infty} e^{-x^2} (-x^2 \sin \alpha x) \, dx$$

$$= \frac{1}{2}\int_0^{\infty} \sin \alpha x \, (-2xe^{-x^2}) dx$$

On integrating by parts

$$= \frac{1}{2}\left[\sin \alpha x \, e^{-x^2}\Big|_0^{\infty} - \int_0^{\infty} \alpha \cos \alpha x e^{-x^2} dx\right]$$

$$= -\frac{1}{2}\left[(0-0) - \int_0^{\infty} \alpha \cos \alpha \, x \, e^{-x^2} dx\right]$$

$$= -\frac{\alpha}{2}\int_0^{\infty} e^{-x^2} \cos \alpha x \, dx$$

$$F'(\alpha) = -\frac{\alpha}{2} F(\alpha) \qquad ...(ii)$$

Separating the variable

$$\frac{F'(\alpha)}{F(\alpha)} = -\frac{\alpha}{2}$$

Integrating w.r.t 'α' then, we get.

$$\log F(\alpha) = -\frac{\alpha^2}{4} + c \qquad ...(iii)$$

Initially, when $\alpha = 0$ From (*i*)

$$F(0) = \int_0^{\infty} e^{-x^2} dx$$

Put $$x^2 = z$$

$$2x\,dx = dz$$

when $x = \infty$ then $z = \infty$

when $x = \infty$ then $z = \infty$

$$dx = \frac{dz}{2\sqrt{z}}$$

$$F(0) = \int_0^\infty e^{-z} \times \frac{dz}{2} \times z^{-\frac{1}{2}}$$

$$= \frac{1}{2}\int_0^\infty e^{-z} z^{\frac{-1}{2}} dz$$

$$= \frac{1}{2}\int_0^\infty e^{-z} z^{\frac{1}{2}-1}\, dz$$

$$= \frac{1}{2}\Gamma\frac{1}{2}$$ ∴ By gama function

∴ $$\Gamma\frac{1}{2} = \sqrt{\pi}$$

$$= \frac{1}{2}\frac{\pi}{2}$$

∴ $$F(0) = \frac{\sqrt{\pi}}{2}$$

Again, when $\alpha = 0$ From (*iii*)

$$\log F(0) = -\frac{0}{4} + c$$

$$\log\frac{\sqrt{\pi}}{2} = +c$$

∴ $$c = \log\frac{\sqrt{\pi}}{2}$$

Putting the value of c in (*iii*) then

$$\log F(\alpha) = -\frac{\alpha^2}{4} + \log\frac{\sqrt{\pi}}{2}$$

or $$\log F(\alpha) - \log\frac{\sqrt{\pi}}{2} = -\frac{\alpha^2}{4}$$

or $$\log\frac{F(\alpha)}{\frac{\sqrt{\pi}}{2}} = -\frac{\alpha^2}{4}$$

or $$\frac{F(\alpha)}{\sqrt{\frac{\pi}{2}}} = e^{-\frac{\alpha^2}{4}}$$

$$F(\alpha) = \frac{\sqrt{\pi}}{2} \times e^{\frac{\alpha^2}{4}}$$

Hence $$\int_0^\infty e^{-x^2} \cos ax\, dx = \frac{\sqrt{\pi}}{2} e^{-\frac{\alpha^2}{4}}$$

Example **5.77.** Prove that $\int_0^\infty xe^{-x2} \sin \alpha x\, dx = \frac{\sqrt{\pi}}{4} \alpha\, e^{-\frac{\alpha^2}{4}}$.

Solution : From the result example *i.e.* $\int_0^\infty e^{-x^2} \cos \alpha x\, dx = \frac{\sqrt{\pi}}{2} e^{\frac{\alpha^2}{4}}$, α is a parameter

Differentiation (*i*) with respect to α on both sides, we get

$$\int_0^\infty -xe^{-x^2} \sin ax\, dx = \frac{\sqrt{\pi}}{2}\, e^{\frac{\alpha^2}{4}} \left(-\frac{2\alpha}{4}\right)$$

$$\therefore \qquad \int_0^\infty xe^{-x^2} \sin ax\, dx = \frac{\sqrt{\pi}}{4}\, \alpha\, e^{\frac{\alpha^2}{4}}$$

Hence

$$\int_0^\infty xe^{-x^2} \sin \alpha x\, dx = \frac{\sqrt{\pi}}{4} \alpha\, e^{-\frac{\alpha^2}{4}}$$

Example **5.78.** Evaluate the integral $\int_0^\infty \frac{e^{-x} \sin bx}{x} dx$

Solution : Let $I(b) = \int_0^\infty \frac{e^{-x} \sin bx}{x} dx$, where b is parameter

Differentiating (*i*) with respect to '*b*' on both sides, we get

$$\frac{dI}{db} = \int_0^\infty \frac{\partial}{\partial b}\left(\frac{e^{-x} \sin bx}{x}\right) dx$$

$$= \int_0^\infty x \frac{e^{-x} \cos bx}{x} dx$$

$$= \int_0^\infty e^{-x} \cos bx\, dx$$

$$\therefore \qquad \frac{dI}{db} = \left[\frac{e^{-x}}{1+b^2}(-\cos bx + b \sin bx)\right]_0^\infty$$

$$\left\{\because \int e^{ax} \cos bx\, dx = \frac{e^{ax}}{a^2+b^2}(a \cos bx + b \sin bx)\right\}$$

$$\frac{dI}{db} = \frac{1}{1+b^2} \qquad \text{...(ii)}$$

Integrating both sides of (*ii*) with respect to '*b*' we have

$$I(b) = \tan^{-1} b + c \qquad \text{...(iii)}$$

At initially when $b = 0$, from (*i*)

$$I(0) = 0$$

Similarly when $b = 0$ from (*iii*)

$$I(0) = \tan^{-1} 0 + c$$

$$0 = c$$

Hence (*iii*) becomes

$$I = \tan^{-1} b$$

or

$$\int_0^\infty \frac{e^{-x} \sin bx}{x} dx = \tan^{-1} b$$

(*b*) When the limits of integration are given in the form of a parameter

If $f(x, \alpha)$, $\frac{\partial f(x,\alpha)}{\partial \alpha}$ be continuous function of x and α then

$$\frac{d}{d\alpha}\left[\int_{\phi(\alpha)}^{\psi(\alpha)} f(x,a)\,dx\right] = \int_{\phi(\alpha)}^{\psi(\alpha)} \frac{\partial}{\partial\alpha} f(x,\alpha)\,dx + \frac{d\psi}{d\alpha}\, f[\psi(\alpha),\alpha] - \frac{d\phi}{d\alpha} f[f(\alpha),\alpha]$$

Provided $\phi(\alpha)$ and $\psi(\alpha)$ possess continuous first order order derivatives w.r.t α.

***Example* 5.79.** Verify the rule of differentiation under the sign of interrogation for $\int_0^{a^2} \tan^{-1}\left(\frac{x}{a}\right) dx$, where a is parameter.

Solution : Let $I(a) = \int_0^{a^2} \tan^{-1}\frac{x}{a}\,dx$ ...(*i*)

Differentiating (*i*) with respect to (*a*) on both sides, we get

$$\frac{dI(a)}{da} = \int_0^{a^2} \frac{\partial}{\partial a}\left(\tan^{-1}\frac{x}{2}\right) dx + 2a\tan^{-1}\frac{a^2}{a} - 0$$

$$= \int_0^{a^2} \frac{1}{1+\frac{x^2}{a^2}}\left(-\frac{x}{a^2}\right)dx + 2z\tan^{-1}\alpha$$

$$= \int_0^{a^2} \frac{-x}{a^2+x^2}\,dx + 2a\tan^{-1}a$$

$$= \left[-\frac{1}{2}\log(a^2+x^2)\right]_0^{a^2} + 2a\tan^{-1}a$$

$$= -\frac{1}{2}\log(a^2+a^4) + \frac{1}{2}\log a^2 + 2a\tan^{-1}a.$$

$$= -\frac{1}{2}\log\frac{a^2+a^4}{a^2} + 2a\tan^{-1}a$$

$$\frac{dI(a)}{da} = \frac{1}{2}\log(1+a^2) + 2a\tan^{-1}a \qquad ...(ii)$$

Now, on integrating (*i*) by parts w.r.t '*x*' we have

$$I(a) = \int_0^{a^2} 1.\tan^{-1}\frac{x}{a}\,dx$$

$$= \left\{\tan^{-1}\left(\frac{x}{a}\right)x\right\}_0^{a^2} - \int_0^{a^2} \frac{1}{1+\frac{x^2}{a^2}}\frac{1}{2}x\,dx$$

$$= \left(a^2\tan^{-1}\frac{a^2}{a} - 0\right) - \int_0^{a^2} \frac{ax}{a^2+x^2}\,dx$$

$$= a^2\tan^{-1}a - \frac{a}{2}\left[\log(x^2+a^2)\right]_0^{a^2}$$

$$= a^2\tan^{-1}a - \frac{a}{2}\log(a^2+a^4)\ \frac{a}{2}\log a^2$$

$$\therefore \quad I(a) = a^2\tan^{-1}a - \frac{a}{2}\log(a^2+1)$$

Differentiating with respect to 'a' on both sides, we get

$$\frac{d}{da} I(a) = a^2 \frac{1}{1+a^2} + 2a \tan^{-1} a - \frac{a}{2} \frac{2a}{1+a^2} - \frac{1}{2} \log (a^2+1)$$

$$= \frac{a^2}{1+a^2} + 2a \tan^{-1} a - \frac{a^2}{1+a^2} - \frac{1}{2} \log (a^2+1) \qquad ...(iii)$$

$$\therefore \quad \frac{DI(a)}{da} = -\frac{1}{2} \log (a^2+1) + 2a \tan^{-1} a$$

Hence from equation (*ii*) and (*iii*), the rule is verified.

***Example* 7.80.** Evaluate $\int_0^{\alpha} \frac{\log(1+\alpha x)}{1+x^2} dx$ and hence show that $\int_0^1 \frac{\log(1+x)}{1+x^2} dx = \log 2.$

Solution : Let $F(\alpha) = \int_0^{\alpha} \frac{\log(1+\alpha x)}{1+x^2} dx$, hence α is parameter. ...(*i*)

Differentiating (*i*) w.r.t. on both sides, we get

$$F'(\alpha) = \int_0^{\alpha} \frac{\partial}{\partial \alpha} \frac{\log(1+\alpha x)}{1+x^2} dx - \frac{d(0)}{d\alpha}[0] + \frac{d\alpha}{d\alpha}\left[\frac{\log(1+\alpha^2)}{1+\alpha^2}\right]$$

$$= \int_0^{\alpha} \frac{x}{(1+\alpha x)(1+x^2)} dx + \frac{\log(1+\alpha^2)}{1+\alpha^2} \qquad ...(ii)$$

Taking $\frac{x}{1(1+\alpha x)(1+x^2)} = \frac{A}{(1+\alpha x)} + \frac{Bx+C}{1+x^2}$

$$x = A(1+x^2) + (Bx+C)(1+\alpha x) \qquad ...(A)$$

Put $1+\alpha x = 0 \quad \therefore = -\frac{1}{\alpha}$

$$-\frac{1}{\alpha} = A\left(1+\frac{1}{\alpha^2}\right) + \left[B \times -\frac{1}{\alpha^2} + C\right](1-1)$$

$$-\frac{1}{\alpha} = A\left(\frac{\alpha^2+1}{\alpha^2}\right) \qquad \therefore A = \frac{-\alpha}{\alpha^2+1}$$

Again comparing the coefficient of like power on both sides of (*A*)

Coeff of x^2

$$0 = A + B\alpha \qquad ...(B)$$

Coeff of x

$$1 = C\alpha + B \qquad ...(C)$$

Constant term

$$\hat{} = A + C \qquad ...(D)$$

$$\therefore \quad A = -C$$

From equation (*B*), (*C*) and (*D*), on simplification

$$B = \frac{1}{1+\alpha^2}$$

and $$C = \frac{\alpha}{1+\alpha^2}$$

From (*ii*)

$$\therefore \quad F'(a) = -\frac{\alpha}{\alpha^2+1}\int_0^\alpha \frac{dx}{1+\alpha x} + \frac{1}{1+\alpha^2}\int_0^\alpha \frac{x}{1+x^2}dx + \frac{\alpha}{1+\alpha^2}\int_0^\alpha \frac{1}{1+x^2}dx + \frac{\log(1+\alpha^2)}{1+\alpha^2}$$

or $$F'(\alpha) = -\frac{a}{1+\alpha^2}\int_0^\alpha \frac{dx}{1+ax} + \frac{1}{2(1+\alpha^2)}\int_0^\alpha \frac{2x}{1+x^2}dx + \frac{\alpha}{1+\alpha^2} + \frac{\log(1+\alpha^2)}{1+\alpha^2}$$

$$= \frac{\alpha}{1+\alpha^2}\left\{\frac{1}{\alpha}\log(1+\alpha x)\right\}_0^\alpha + \frac{1}{2(1+\alpha^2)}\{\log(1+x^2)\}_0^\alpha + \frac{\alpha}{1+\alpha^2}[\tan^{-1}x]_0^\alpha + \frac{\log(1+\alpha^2)}{(1+\alpha^2)}$$

$$F'(\alpha) = \frac{1}{(1+\alpha^2)}\log(1+\alpha^2) + \frac{\log(1+\alpha^2)}{2(1+\alpha^2)} + \frac{\alpha\tan^{-1}\alpha}{2(1+\alpha^2)} + \frac{\alpha\tan^{-1}\alpha}{1+\alpha^2} + \frac{\log(1+\alpha^2)}{(1+\alpha^2)}$$

$$\therefore \quad F'(\alpha) = \frac{\log(1+\alpha^2)}{2(1+\alpha^2)} + \frac{\alpha\tan^{-1}\alpha}{1+\alpha^2} \qquad ...(iii)$$

On integrating (*iii*) both sides w.r.t tp 'α' then

$$F(\alpha) = \frac{1}{2}\int \frac{\log(1+\alpha^2)}{(1+\alpha^2)}da + \int \frac{\alpha\tan^{-1}\alpha}{(1+\alpha^2)}d\alpha$$

Integrating by parts, taking $\log(1+\alpha^2)$ as first function

$$= \frac{1}{2}\log(1+\alpha^2)\tan^{-1}\alpha - \frac{1}{2}\int \frac{2\alpha}{(1+\alpha^2)}\tan^{-1}\alpha\, d\alpha + \int \frac{a\tan^{-1}\alpha}{1+\alpha^2}d\alpha + c$$

$$F(\alpha) = \frac{1}{2}\log(1+\alpha^2)\tan^{-1}a + c \qquad ...(iv)$$

Initially, when $\alpha = 0$, from (*i*)

$$F(0) = 0$$

Again, when $\alpha = 0$ from (*iv*)

$$F(0) = \frac{1}{2}\log(1+0)\tan^{-1}0 + c$$

$$0 = \frac{1}{2}\times 0 + c$$

$$\therefore \quad c = 0$$

Hence $$F(\alpha) = \frac{1}{2}\log(1+\alpha)\tan^{-1}\alpha$$

$$= \int_0^\alpha \frac{\log(1+\alpha x)}{1+x^2} = \frac{1}{2}\log^2(1+\alpha)\tan^{-1}\alpha \qquad ...(v)$$

To show that $$\int_0^1 \frac{\log(1+x)}{1+x^2}dx = \frac{\pi}{8}\log_e 2$$

Put $\alpha = 1$ in (*iv*), then

$$F(\alpha) = \int_0^1 \frac{\log(1+x)}{1+x^2}dx = \frac{1}{2}\log 2\tan^{-1}(1)$$

$$= \frac{1}{2}\log 2 \times \frac{\pi}{4} = \frac{\pi}{8}\log_e 2$$

Example 5.81. If $y = \int_0^x f(t)\sin[k(x-t)]\,dt$. Prove that y satisfy the differential equation $\frac{d^2y}{dx^2} + k^2 y = k f(x)$.

Solution : Let $y(x) = \int_0^x f(t)\sin k(x-t)\,dt$, here x is parameter ...(*i*)

Differentiating (*i*) w.r.t. x on both sides, we get

$$\frac{dy}{dx} = \frac{d}{dx}\int_0^x f(t)\,k(x-t)\,dt + \frac{dx}{dx} \times f(x)\sin k(x-x) - 0$$

$$= \int_0^x f(t)\frac{\partial}{\partial x}\{\sin k(x-t)\}\,dt$$

{∴ Lebnitz rule for differentiation under the sign of integration}

$$= \int_0^x f(t)\,k\cos k(x-t)\,dt$$

$$\frac{dy}{dx} = k\int_0^x f(t)\cos k(x-t)\,dt \qquad ...(ii)$$

Again, differentiating (*ii*) w.r.t. x on both sides then, we get

$$\frac{d^2y}{dx^2} = k\int_0^x f(t)\,\{-\sin k(x-t)\,\mathrm{d}t + x - t\}\,dt + k\frac{dx}{dx}f(x)\cos k(x-x) - 0$$

$$\frac{d^2y}{dx^2} = -k^2\int_0^x f(t)\,k(x-t)\,dt + k f(x)$$

Hence $$\frac{d^2y}{dx^2} = k^2\,y(x) = k f(x)$$

EXERCISE 5.8

1. Prove that

$$\int_0^\infty \frac{e^{-ax}\sin x}{x} = \cot^{-1}\alpha \text{ and hence deduce that } \int_0^\infty \frac{\sin}{x}\,dx = \frac{\pi}{2}$$

2. $$\int_0^\infty \frac{1-e^{-ax}}{x}e^{-x}\,dx = \log(1+a),\ (a > -1)$$

3. (*a*) $$\int_0^{\pi/2} \frac{\log(1+(\sin\alpha\cos x)}{\cos x}\,dx = \pi\,\alpha \text{ where } \alpha \text{ is parameter.}$$

(*b*) $$\int_0^{\pi/2} \frac{\log(1+\cos\alpha\cos x)}{\cos x}\,dx = \frac{1}{2}\left\{\frac{\pi^2}{4} - \alpha^2\right\}$$

4. If $|a| < 1$, Evaluate $\int_0^\pi \frac{\log(1+a\cos x)}{\cos x}\,dx$. **Ans.** $\pi\sin^{-1} a$

5. Show that $\int_{\frac{\pi}{6a}}^{\frac{\pi}{2a}} \frac{\sin ax}{x}\,dx =$ constant.

Reduction Formulae: Definite Integrals

6.1 REDUCTION FORMULAE

A formula which connects an integral with another in which the integrand is of the same type but of a lower degree or order is called a *reduction formula.* Usually the reduction formula has to be used repeatedly in order to evaluate the given integral. This method of integration is called integration by successive reduction.

Reduction formulae are generally obtained by integration by parts. They are useful when the integral cannot be directly obtained otherwise.

Consider for example $\int (x^2 + k)^n \, dx$, where $n > 1$. Taking unity as the second factor and integrating by parts, we have

$$\int (x^2 + k)^n \,.\, 1 \, dx = (x^2 + k)^n x - \int n(x^2 + k)^{n-1} \,.\, 2x \,.\, x \, dx$$

$$= x(x^2 + k)^n - 2n \int (x^2 + k)^{n-1} (x^2 + k - k) \, dx$$

$$= x(x^2 + k)^n - 2n \int (x^2 + k)^n \, dx + 2nk \int (x^2 + k)^{n-1} \, dx.$$

The first integral on the right is the same as the one we started with. Transposing it to the left and dividing by $2n + 1$, we obtain

$$\int (x^2 + k)^n \, dx = \frac{x(x^2 + k)^n}{2n + 1} + \frac{2nk}{2n + 1} \int (x^2 + k)^{n-1} \, dx.$$

This is a reduction formula since it connects two integrals of the same type but with indices n and $n - 1$.

Putting $n = - m + 1$, this formula becomes

$$\int \frac{dx}{(x^2 + k)^{m-1}} = -\frac{x}{(2m - 3)(x^2 + k)^{m-1}} + \frac{(2m - 2)k}{(2m - 3)} \int \frac{dx}{(x^2 + k)^m}.$$

Dividing by $(2m - 2)k/(2m - 3)$ and transposing one term, this gives

$$\int \frac{dx}{(x^2 + k)^m} = \frac{x}{(2m - 2)k(x^2 + k)^{m-1}} + \frac{(2m - 3)}{(2m - 2)k} \int \frac{dx}{(x^2 + k)^{m-1}}.$$

This reduction formula is useful when $m > 1$.

***Example* 6.1.** Evaluate $\int x^4 \sqrt{(a^2 - x^2)} \, dx$.

Integrating by parts, we have

$$\int x^n \sqrt{(a^2 - x^2)} \, dx = -\frac{1}{2} \int x^{n-1} (a^2 - x^2)^{1/2} (-2x) \, dx$$

$$= -\frac{1}{2} x^{n-1} \frac{(a^2 - x^2)^{3/2}}{3/2} + \frac{1}{2} \int (n - 1) x^{n-2} \,.\, \frac{(a^2 - x^2)^{3/2}}{3/2} dx$$

$$= -\frac{1}{3}x^{n-1}(a^2-x^2)^{3/2} + \frac{n-1}{3}\int x^{n-2}(a^2-x^2)\sqrt{(a^2-x^2)}\,dx$$

$$= \frac{1}{3}x^{n-1}(a^2-x^2)^{3/2} + \frac{n-1}{3}a^2\int x^{n-2}\sqrt{(a^2-x^2)}\,dx - \frac{n-1}{3}\int x^n\sqrt{(a^2-x^2)}\,dx$$

Transposing the last term to the left and dividing by $(n+2)/3$, we obtain the reduction formula:

$$\int x^n\sqrt{(a^2-x^2)}\,dx = -\frac{x^{n-1}(a^2-x^2)^{3/2}}{n+2} + \frac{n-1}{n+2}a^2\int x^{n-2}\sqrt{(a^2-x^2)}\,dx.$$

Applying this formula successively, we have

$$\int x^4\sqrt{(a^2-x^2)}\,dx = -\frac{x^3(a^2-x^2)^{3/2}}{6} + \frac{3}{6}a^2\int x^2\sqrt{(a^2-x^2)}\,dx$$

$$= -\frac{1}{6}x^3(a^2-x^2)^{3/2} + \frac{1}{2}a^2\left\{-\frac{1}{4}x(a^2-x^2)^{3/2} + \frac{1}{4}a^2\int\sqrt{(a^2-x^2)}\,dx\right\}$$

$$= C - \frac{1}{6}\left(x^3 + \frac{3}{4}a^2x\right)(a^2-x^2)^{3/2} + \frac{1}{8}a^4\left\{\frac{1}{2}x\sqrt{(a^2-x^2)} + \frac{1}{2}a^2\sin^{-1}(x/a)\right\}$$

Examples 6.2.

Integrate

1. $1/(x^2+1)^3$.
2. $(x^2+a^2)^{5/2}$. [*Allahabad*, 1969]
3. Prove that

$$\int (a^2+x^2)^{n/2}\,dx = \frac{x(a^2+x^2)^{n/2}}{n+1} + \frac{na^2}{n+1}\int (a^2+x^2)^{n/2-1}\,dx.$$

4. If $I_n = \int_0^a (a^2-x^2)^n\,dx$, prove that

$$I_n = \frac{2na^2}{2n+1}I_{n-1} \text{ for } n \geq 2.$$ [*Gorakhpur*, 1997]

5. If $I_n = \int x^n\sqrt{(a-x)}\,dx$, prove that

$$(2n+3)\,I_n = 2anI_{n-1} - 2x^n(a-x)^{3/2}.$$ [*Roorkee, Arch.*, 1977]

6. If $I_n = \int x^n(a^2-x^2)^{1/2}\,dx$, prove that

$$(n+2)\,I_n = (n-1)a^2\,I_{n-2} - x^{n-1}(a^2-x^2)^{3/2}.$$ [*Roorkee, Arch.*, 1994]

6.2 INTEGRATION OF $\sin^m x \cos^n x$

When m and n are positive integers (or zero)

$$\int \sin^m x\cos^n x\,dx.$$

can be evaluated by successive reduction or by expressing $\sin^m x\cos^n x$ as the sum of sines or cosines of multiples of x. But in the following two cases, substitution can be profitably employed.

I. *m or n is an odd positive integer*. If $m = 2r+1$, then

$$\sin^m x\cos^n x = \sin^{2r+1} x\cos^n x = (1-\cos^2 x)^r\cos^n x\sin x.$$

So the substitution $\cos x = t$ will reduce the integrand to a sum of powers of t.

Similarly, if $n = 2r + 1$. we can put $\sin x = t$.

II. *$m + n$ is an even negative integer.* Let $m + n = -2r$, then

$$\sin^m x \cos^n x = \tan^m x \cos^{m+n} x = \tan^m x \sec^{2r} x.$$

$$= \tan^m x (1 + \tan^2 x)^{r-1} \sec^2 x.$$

So the substitution $\tan x = t$ will reduce the integrand to a sum of powers of t.

***Example* 6.3.** Integrate $\sin^7 x$.

$$\int \sin^7 x\, dx = \int (1 - \cos^2 x)^3 \sin x\, dx = -\int (1 - t^2)^3\, dt, \text{ where } t = \cos x,$$

$$= -\int (1 - 3t^2 + 3t^4 - t^6)\, dt = -t + t^3 - \frac{3}{5} t^5 + \frac{1}{7} t^7 + C$$

$$= C - \cos x + \cos^3 x - \frac{3}{5} \cos^5 x + \frac{1}{7} \cos^7 x.$$

***Example* 6.4.** Integrate $\sin^{5/6} x \cos^3 x$.

$$\int \sin^{5/6} x \cos^3 x\, dx = \int \sin^{5/6} x (1 - \sin^2 x) \cos x\, dx$$

$$= \int t^{5/6}(1 - t^2)\, dt, \text{ where } t = \sin x,$$

$$= \int (t^{5/6} - t^{17/6})\, dt = \frac{6}{11} t^{11/6} - \frac{6}{23} t^{23/6} + C$$

$$= \frac{6}{11} \sin^{11/6} x - \frac{6}{23} \sin^{23/6} x + C.$$

***Example* 6.5.** Integrate $1/\sin^3 x \cos^5 x$.

$$\int \frac{dx}{\sin^3 x \cos^5 x} = \int \frac{\sec^8 x}{\tan^3 x} dx = \int \frac{(1 + \tan^2 x)^3 \sec^2 x}{\tan^3 x} dx$$

$$= \int \frac{(1 + t^2)^3\, dt}{t^3}, \text{ where } t = \text{tax } x,$$

$$= \int \left(\frac{1}{t^3} + \frac{3}{t} + 3t + t^3\right) dt = -\frac{1}{2} t^{-2} + 3 \log t + \frac{3}{2} t^2 + \frac{1}{4} t^4 + C$$

$$= C - \frac{1}{2} \cot^2 x + 3 \log \tan x + \frac{3}{2} \tan^2 x + \frac{1}{4} \tan^4 x.$$

***Example* 6.6.** Integrate $\sin^2 x \cos^4 x$.

Let $\cos x + i \sin x = z$; then $\cos x - i \sin x = z^{-1}$.

Therefore $2 \cos x = z + z^{-1}$, $2i \sin x = z - z^{-1}$.

Also by De Moivre's Theorem, $2 \cos px = z^p + z^{-p}$,

$$2i \sin px = z^p - z^{-p}.$$

Therefore,

$$2^2 2^4 i^2 \sin^2 x \cos^4 x = (z - z^{-1})^2 (z + z^{-1})^4$$

$$= (z^6 + z^{-6}) + 2(z^4 + z^{-4}) - (z^2 + z^{-2}) - 4.$$

$$= 2 \cos 6x + 4 \cos 4x - 2 \cos 2x - 4.$$

Hence

$$\int \sin^2 x \cos^4 x\, dx = C - 2^{-6} \left(\frac{1}{3} \sin 6x + \sin 4x - \sin 2x - 4x\right).$$

6.3 REDUCTION FORMULA FOR $\int \sin^m x \cos^n x\, dx$

$$\int \sin^m x \cos^n x\, dx = \int \sin^m x \cos x \cos^{n-1} x\, dx$$

$$= \frac{\sin^{m+1} x \cos^{n-1} x}{m+1} + \frac{n-1}{m+1}\int \sin^{m+1} x \cos^{n-2} x \sin x\, dx, \text{ on integration by parts,}$$

$$= \frac{\sin^{m+1} x \cos^{n-1} x}{m+1} + \frac{n-1}{m+1}\int \sin^m x \cos^{n-2} x\,(1-\cos^2 x)\, dx$$

$$= \frac{\sin^{m+1} x \cos^{n-1} x}{m+1} + \frac{n-1}{m+1}\int \sin^m x \cos^{n-2} x\, dx - \frac{n-1}{m+1}\int \sin^m x \cos^n x\, dx.$$

Transposing the last term to the left and dividing by $1 + (n-1)/(m+1)$, *i.e.* by $(m+n)/(m+1)$, we get the reduction formula

$$\int \sin^m x \cos^n x\, dx = \frac{\sin^{m+1} x \cos^{n-1} x}{m+n} + \frac{n-1}{m+n}\int \sin^m x \cos^{n-2} x\, dx. \quad ...(1)$$

By writing

$\int \sin^m x \cos^n x\, dx$ as $\int \sin^{m-1} x \cos^n x \sin x\, dx$ and integrating by parts we can obtain the reduction formula

$$\int \sin^m x \cos^n x\, dx = -\frac{\sin^{m-1} x \cos^{n+1} x}{m+n} + \frac{m-1}{m+n}\int \sin^{m-2} x \cos^n x\, dx, \quad ...(2)$$

which diminishes the power of $\sin x$ instead of that of $\cos x$.

This could also be obtained from (1) by replacing x by $\frac{1}{2}\pi - x$, dx by $-dx$, and then interchanging m and n.

Putting $m = 0$ in (1) and $n = 0$ in (2) we obtain the reduction formulae

$$\int \cos^n x\, dx = \frac{\cos^{n-1} x \sin x}{n} + \frac{n-1}{n}\int \cos^{n-2} x\, dx; \quad ...(3)$$

$$\int \sin^m x\, dx = -\frac{\sin^{m-1} x \cos x}{m} + \frac{m-1}{m}\int \sin^{m-2} x\, dx. \quad ...(4)$$

This could also be obtained independently, by proceeding in manner similar to the above.

Example 6.7. Integrate $\sin^4 x \cos^2 x$.

$$\int \sin^4 x \cos^2 x\, dx = \frac{1}{6}\sin^5 x \cos x + \frac{1}{6}\int \sin^4 x\, dx, \text{ by (1),}$$

$$= \frac{1}{6}\sin^5 x \cos x + \frac{1}{6}\left\{-\frac{1}{4}\sin^3 x \cos x + \frac{3}{4}\int \sin^2 x\, dx\right\}, \text{ by (4),}$$

$$= \frac{1}{6}\sin^5 x \cos x - \frac{1}{24}\sin^3 x \cos x + \frac{1}{8}\left\{-\frac{1}{2}\sin x \cos x + \frac{1}{2}\int dx\right\}, \text{ applying (4) again,}$$

$$= \frac{1}{48}(8\sin^5 x - 2\sin^3 x - 3\sin x)\cos x + \frac{1}{16}x + C.$$

6.4 THE INTEGRAL $\int_0^{\pi/2} \sin^m x \cos^n x\, dx$

Putting the limits 0 and $\frac{1}{2}\pi$ in the reduction formula (1) of the last article, we get

$$\int_0^{\pi/2} \sin^m \cos^n x\, dx = \frac{n-1}{m+n}\int_0^{\pi/2} \sin^m x \cos^{n-2} x\, dx \quad ...(5)$$

since the integrand part $(\sin^{m+1} x \cos^{n-1} x)/(m+n)$ is zero both at $\frac{1}{2}\pi$ and 0.

Similarly, the formula (4) gives

$$\int_0^{\pi/2} \sin^m x\,dx = \frac{m-1}{m}\int_0^{\pi/2} \sin^{m-2} x\,dx \qquad ...(6)$$

Case I. Let m and n be even positive integers. Then, applying formula (5) repeatedly till the power of $\cos x$ becomes zero, and after that applying formula (6) repeatedly, we have

$$\int_0^{\pi/2} \sin^m x \cos^n x\,dx = \frac{n-1}{m+n}\cdot\frac{n-3}{m+n-2}\cdot\frac{n-5}{m+n-4}\cdot\ldots\frac{1}{m+2}\int_0^{\pi/2}\sin^m x\,dx$$

$$= \frac{(n-1)(n-3)(n-5)\ldots 1}{(m+n)(m+n-2)\ldots(m+2)}\cdot\frac{m-1}{m}\cdot\frac{m-3}{m-2}\cdot\ldots\frac{1}{2}\int_0^{\pi/2} dx$$

$$= \frac{(m-1)(m-3)\ldots 1\times(n-1)(n-3)\ldots 1}{(m+n)(m+n-2)(m+n-4)\ldots 2}\left(\frac{1}{2}\pi\right)$$

Case II. Let n be an even positive integer and m an odd positive integer. Proceeding as in Case I, we find

$$\int_0^{\pi/2} \sin^m x \cos^n x\,dx = \frac{(n-1)(n-3)(n-5)\ldots 1}{(m+n)(m+n-2)\ldots(m+2)}\times\frac{m-1}{m}\cdot\frac{m-3}{m-2}\cdot\ldots\frac{2}{3}\int_0^{\pi/2}\sin x\,dx$$

$$= \frac{(m-1)(m-3)\ldots 2\times(n-1)(n-3)\ldots 1}{(m+n)(m+n-2)(m+n-4)\ldots 3.1}\times 1.$$

Case III. Let n be an odd positive integer. Proceeding as in Case I, we have

$$\int_0^{\pi/2} \sin^m x \cos^n x\,dx = \frac{n-1}{m+n}\cdot\frac{n-3}{m+n-2}\ldots\frac{2}{m+3}\int_0^{\pi/2}\sin^m x\cos x\,dx$$

$$= \frac{(n-1)(n-3)\ldots 2}{(m+n)(m+n-2)\ldots(m+3)}\left[\frac{\sin^{m+1} x}{m+1}\right]_0^{\pi/2}$$

$$= \frac{(n-1)(n-3)\ldots 2}{(m+n)(m+n-2)\ldots(m+3)(m+1)}$$

$$= \frac{(m-1)(m-3)\ldots\times(n-1)(n-3)\ldots 2}{(m+n)(m+n-2)\ldots(m+1)(m-1)(m-3)\ldots}\times 1,$$

on multiplying the numerator and denominator by the factors $(m-1)(m-3)\ldots$. The result is true whether m is odd or even.

Thus in every case

$$\int_0^{\pi/2} \sin^m x\cos^n x\,dx = \frac{(m-1)(m-3)(m-5)\ldots\times(n-1)(n-3)\ldots}{(m+n)(m+n-2)(m+n-4)\ldots}\times k, \qquad ...(7)$$

where $k = 1$, except when m and n are both even integers in which case $k = \frac{1}{2}\pi$. The last factor in a product like $(m-1)(m-3)\ldots$ is 2 or 1, according as the other factors are even or odd.

The formula holds also when m or n is zero, provided we omit all negative factors in the numerator, and regard 0 as an even number in determining the value of k.

The above result can also be stated in terms of the Gamma Function : $\Gamma(x)$ The properties of this function which we need are

$$\Gamma(p+1) = p\,\Gamma(p),$$

$$\Gamma(1) = 1, \Gamma\left(\frac{1}{2}\right) = \sqrt{\pi}.$$

Dividing the numerator and denominator of (7) by a suitable power of 2, we can write it as

$$\int_0^{\pi/2} \sin^m x \cos^n x\, dx = \frac{\Gamma\left(\frac{m+1}{2}\right)\Gamma\left(\frac{n+1}{2}\right)}{2\Gamma\left(\frac{m+n+2}{2}\right)}$$

***Example* 6.8.** Evaluate $\int_0^{\pi/2} \sin^4 x \cos^2 x\, dx$.

$$\int_0^{\pi/2} \sin^4 x \cos^2 x\, dx = \frac{3\times 1\times 1}{6\times 4\times 2}\left(\frac{1}{2}\pi\right), \text{ by (7)}, = \frac{1}{32}\pi.$$

***Example* 6.9.** Evaluate $\int_0^a x^2(a^2-x^2)^{3/2} dx$.

Put $x = a\sin\theta,\ dx = a\cos\theta\, d\theta.$

Then $\theta = 0$ when $x = 0$, and $\theta = \frac{1}{2}\pi$ when $x = a$.

Hence
$$\int_0^a x^2(a^2-x^2)^{3/2} dx = a^6 \int_0^{\pi/2} \sin^2\theta \cos^4\theta\, d\theta$$

$$= a^6 \frac{\Gamma\left(\frac{3}{2}\right)\Gamma\left(\frac{5}{2}\right)}{2\Gamma(4)} = a^6 \frac{\frac{1}{2}\sqrt{\pi}\cdot\frac{3}{2}\cdot\frac{1}{2}\sqrt{\pi}}{2.3.2.1} = \frac{1}{32}\pi a^6.$$

***Example* 6.10.**

Integrate

1. $\cos^5 x$.
2. $\cos^2 x \sin^3 x$.
3. $\cos^{3/4} x \sin^5 x$.
4. $1/\sin x \cos^3 x$.
5. $1/\sqrt{(\cos^3 x \sin^5 x)}$.
6. $\sec x \tan^3 x$.
7. $\sin^2 x$.
8. $\sin^2 x \cos^6 x$.
9. $\sin^6 x$.
10. $\sin^2 x \cos^2 x$.

Evaluate

11. $\int_0^{\pi/2} \cos^3 x \sin^4 x\, dx$. [*Allahabad*, 1970]
12. $\int_0^{\pi/4} \sin^6 x\, dx$.
13. $\int_0^{\pi/4} \sin^5 x \cos^2 x\, dx$.

14. $\int_0^{\pi/8} \cos^3 4x \, dx$.

15. $\int_0^{\pi/4} x^4 (1 - x^2)^{3/2} \, dx$.

16. $\int_0^a x^4 \sqrt{(a^2 - x^2)} \, dx$.

17. $\int_0^a \frac{x^4}{\sqrt{(a^2 - x^2)}} dx$.

18. $\int_0^1 x^{3/2} \sqrt{(1 - x)} \, dx$.

19. $\int_0^a \frac{x^4 \, dx}{(a^2 + x^2)^4}$.

20. $\int \cos x \cos 2x \cos 3x \, dx$.

6.5 INTEGRATION OF $1/(a + b \cos x)$

To integrate $1/(a + b \cos x)$, we write $a + b \cos x$ as

$$a\left(\cos^2 \frac{1}{z} x + \sin^2 \frac{1}{2} x\right) + b\left(\cos^2 \frac{1}{z} x - \sin^2 \frac{1}{z} x\right),$$

divide the numerator and denominator by $\cos^2 \frac{1}{2} x$, and then we put $\tan \frac{1}{2} x = t$. This gives an integral of the form $\int dt/(t^2 \pm c^2)$, which can be directly integrated.

Since $\sin x = 2 \sin \frac{1}{2} x \cos \frac{1}{2} x$, the integral of $1/(a + b \sin x)$ or $1/(a + b \cos x + c \sin x)$ can be evaluated in a similar fashion.

***Example* 6.11.** Integrate $1/(4 + 5 \cos x)$.

$$\int \frac{dx}{4 + 5\cos x} = \int \frac{dx}{4\left(\cos^2 \frac{1}{2}x + \sin^2 \frac{1}{2}x\right) + 5\left(\cos^2 \frac{1}{2}x - \sin^2 \frac{1}{2}x\right)}$$

$$= \int \frac{dx}{(4+5)\cos^2 \frac{1}{2}x + (4-5)\sin^2 \frac{1}{2}x}$$

$$= 2\int \frac{\frac{1}{2}\sec^2 \frac{1}{2}x \, dx}{9 - \tan^2 \frac{1}{2}x} = 2\int \frac{dt}{9 - t^2}, \text{ where } t = \tan \frac{1}{2} x,$$

$$= -\frac{2}{6} \log \frac{t-3}{t+3} + C = \frac{1}{3} \log \frac{\tan \frac{1}{2}x + 3}{\tan \frac{1}{2}x - 3} + C$$

***Example* 6.12.** Integrate $1/(4 + 5 \sin x)$.

$$\int \frac{dx}{4 + 5\sin x} = \int \frac{dx}{4\left(\cos^2 \frac{1}{2}x + \sin^2 \frac{1}{2}x\right) + 10 \sin \frac{1}{2}x \cos \frac{1}{2}x}$$

$$= \int \frac{\frac{1}{2}\sec^2 \frac{1}{2}x \, dx}{2 + 2\tan^2 \frac{1}{2}x + 5\tan \frac{1}{2}x} = \frac{1}{2}\int \frac{dt}{t^2 + \frac{5}{2}t + 1}, \text{ putting } \tan \frac{1}{2} x = t,$$

$$= \frac{1}{2}\int \frac{dt}{\left(t+\frac{5}{4}\right)^2 - \frac{25}{16}+1} = \frac{1}{2}\cdot\frac{1}{2}\cdot\frac{4}{3}\cdot\log\frac{t+\frac{5}{4}-\frac{3}{4}}{t+\frac{5}{4}+\frac{3}{4}} + C$$

$$= \frac{1}{3}\log\frac{t+\frac{1}{2}}{t+2} + C = \frac{1}{3}\log\frac{2\tan\frac{1}{2}x+1}{\tan\frac{1}{2}x+2} + C.$$

6.6 FUNCTIONS OF sin x and cos x

Function of $\sin x$ and $\cos x$ can be integrated by a variety of substitutions. Two easily integrable cases are :

(*i*) When the integrand can be expressed as $f(\sin x) \cdot \cos x$ [or $f(\cos x) \cdot \sin x$].

(*ii*) $1/(a + b\cos^2 x + c\sin^2 x)$. This can be integrated by dividing the numerator and denominator by $\cos^2 x$ and putting $\tan x = t$.

The same substitution is applicable for integrating $1/(a\cos x + b\sin x)^2$.

***Example* 6.13.** Integrate $1/(\sin x + \sin 2x)$

$$\int \frac{dx}{\sin x + \sin 2x} = \int \frac{dx}{\sin x + 2\sin x\cos x} = \int \frac{\sin x\, dx}{\sin^2 x(1+2\cos x)}$$

$$= \int \frac{-dt}{(1-t^2)(1+2t)}, \text{ on putting } \cos x = t,$$

$$= \int \frac{dt}{(t-1)(t+1)(2t+1)}$$

$$= \int \left\{\frac{1}{(t-1)2.3} + \frac{1}{(t+1)(-2)(-1)} + \frac{1}{\left(-\frac{3}{2}\right)\left(\frac{1}{2}\right)(2t+1)}\right\} dt$$

$$= \frac{1}{6}\log(t-1) + \frac{1}{2}\log(t+1) - \frac{4}{3}\cdot\frac{1}{2}\log(2t+1) + C$$

$$= \frac{1}{6}\log(1-\cos x) + \frac{1}{2}\log(1+\cos x) - \frac{2}{3}\log(1+2\cos x) + C'.$$

***Example* 6.14.** Integrate $1/(a^2 - b^2\cos^2 x)$, $a > b$.

$$\int \frac{dx}{a^2 - b^2\cos^2 x} = \int \frac{\sec^2 x\, dx}{a^2\sec^2 x - b^2} = \int \frac{\sec^2 x\, dx}{a^2(1+\tan^2 x) - b^2}$$

$$= \int \frac{dt}{a^2 - b^2 + a^2t^2}, \text{ where } \tan x = t,$$

$$= \frac{1}{\sqrt{(a^2-b^2)}}\cdot\frac{1}{a}\tan^{-1}\frac{at}{\sqrt{(a^2-b^2)}} + C$$

$$= \frac{1}{a\sqrt{(a^2-b^2)}}\tan{-1}\frac{a\tan x}{\sqrt{(a^2-b^2)}} + C.$$

***Example* 6.15.**

Integrate

1. $1/(3 + 4\cos x)$. [*Allahabad*, 1970]

2. $1/(5 + 4\sin x)$. [*Roorkee*, 1978]

Evaluate

3. $\int_0^{\pi/2} \frac{dx}{2+3\cos x}$. [*Gorakhpur*, 1972]

4. $\int_0^{\pi} \frac{d\theta}{5+3\cos\theta}$.

5. $\int_0^{\pi/2} \frac{\cos x\, dx}{(1+\sin x)(2+\sin x)}$. [*Kurukshetra*, 1973]

Integrate

6. $1/(3+2\cos x)\sin x$.
7. $\tan x/(a+b\tan^2 x)$.
8. $1/(2\sin x+\cos x)^2$.
9. $1/(a^2\cos^2+b^2\sin^2 x)$. [*Roorkee*, 1972]
10. $1/(a\sin x+b\cos x)$.
11. $\sin x/\sin(x-\alpha)$. [Hint. Put $x-\alpha=t$]
12. $1/(a+b\tan x)$. [*Allahabad*, 1970]
13. $\sqrt{(1+\sin x)}$.
14. $\frac{x-\sin x}{1-\cos x}$.
15. $1/\sqrt{\{\sin^3 x\sin(x+\alpha)\}}$. [*AMIE*, 1978]
16. Evaluate $\int_0^{\pi/4} \frac{\sin 2\theta\, d\theta}{\sin^4\theta+\cos^4\theta}$.

6.7 OTHER TRANSCENDENTAL FUNCTIONS

There are no general rules which will gives us the integral of every transcendental function; but quite often a suitable substitution will reduce a given function to some easily integrable form. Some typical cases are given below:

(*i*) *A rational function of* e^{ax}. Putting $e^{ax}=t$, this can be reduced to a rational function of t (see Example. 6.15).

(*ii*) *Logarithmic or inverse trigonometric function multiplied by an easily integrable rational function (or unity).* Integration by parts will convert it to integration of an algebraic function, as in Example. 6.16.

(*iii*) *Functions involving* $\sin^{-1} x$, $\cos^{-1} x$ or $\tan^{-1} x$. The substitution $x=\sin\theta$, $\cos\theta$ or $\tan\theta$ may give an integrable form (Example. 6.17).

(*iv*) *The functions* $e^{ax}\sin bx$, $e^{ax}\cos bx$, $\cosh ax\cos bx$, etc. We can get the result as in Example. 6.18. by integrating by parts twice.

(*v*) *Integration in series.* In some cases a function of x may be expanded in powers of x and integrated term by term.*

(*vi*) *Successive reduction.* This will be considered in the next section.

Example **6.16.** Integrate $1/(1+e^x-2e^{2x})$.

Putting $e^x = t$ and $e^x\,dx=dt$, i.e. $dx=dt/t$, we have

$$\int\frac{dx}{1+e^x-2e^{2x}} = \int\frac{dt}{t(1+t-2t^2)} = \int\frac{dt}{t(1-t)(1+2t)}$$

$$= \int\left\{\frac{1}{t}+\frac{1}{3(1-t)}+\frac{1}{\left(-\frac{1}{2}\right)\left(\frac{3}{2}\right)(1+2t)}\right\}dt$$

* This process is justifiable only under certain conditions.

$$= \log t - \frac{1}{3}\log(t-1) - \frac{4}{3}\cdot\frac{1}{2}\log(1+2t) + C$$

$$= \frac{1}{3}\{3\log e^x - \log(e^x - 1) - 2\log(1+2e^x)\} + C$$

$$= \frac{1}{3}\log\{e^{3x}/(e^x-1)(1+2e^x)^2\} + C.$$

***Example* 6.17.** Integrate $\{\log(1+x)\}/x^2$.

Integrating by parts.

$$\int \frac{\log(1+x)}{x^2}dx = -\frac{\log(1+x)}{x} + \int \frac{1}{1+x}\cdot\frac{1}{x}dx$$

$$= -\frac{\log(1+x)}{x} + \int\left(\frac{1}{x} - \frac{1}{1+x}\right)dx$$

$$= C - (1/x)\log(1+x) + \log x - \log(1+x).$$

***Example* 6.18.** Integrate $(\sin^{-1}x)/(1-x^2)^{3/2}$.

Putting $x = \sin\theta$ and $dx = \cos\theta\, d\theta$, we have

$$\int \frac{\sin^{-1}x}{(1-x^2)^{3/2}}dx = \int \frac{\theta\,.\cos\theta\, d\theta}{(1-\sin^2\theta)^{3/2}} = \int \theta \sec^2\theta\, d\theta$$

$= \theta\tan\theta - \int \tan\theta\, d\theta$, on integrating by parts,

$$= \theta\tan\theta + \log\cos\theta + C$$

$$= \frac{x\sin^{-1}x}{\sqrt{(1-x^2)}} + \log\sqrt{(1-x^2)} + C.$$

***Example* 6.19.** Integrate $e^{ax}\sin bx$ on

$$\int e^{ax}\sin bx\, dx = e^{ax}\left(-\frac{\cos bx}{b}\right) - \int ae^{ax}\left(-\frac{\cos bx}{b}\right)dx$$

$$= -\frac{1}{b}e^{ax}\cos bx + \frac{a}{b}\left[e^{ax}\frac{(\sin bx)}{b} - \int ae^{ax}\frac{\sin bx}{b}dx\right],$$

on integrating by parts twice. Transposing the last term to the left we have

$$\left(1+\frac{a^2}{b^2}\right)\int e^{ax}\sin bx\, dx = \frac{1}{b^2}e^{ax}(a\sin bx - b\cos bx),$$

or

$$\int e^{ax}\sin bx\, dx = \frac{e^{ax}}{a^2+b^2}(a\sin bx - b\cos bx).$$

***Example* 6.20.**

Integrate

1. $1/(e^x - 1)$. [*Allahabad*, 1969]

2. $1/(e^x - 1)^2$. [*Bhopal*, 1977]

3. $1/(1+e^x)(1+e^{-x})$.

4. $1/(1+3e^x+2e^{2x})$.

5. $\log(1+x^2)$.
6. $\log\{x+\sqrt{(x^2-a^2)}\}$.
7. $x^2(\log x)^2$.
8. $x\sin^{-1}x$.
9. $(\tan^{-1}x)/(1+x)^2$.
10. $\sin^{-1}\sqrt{\{x/(a+x)\}}$.
11. $\dfrac{x\tan^{-1}x}{(1+x^2)^{3/2}}$. [Roorkee, Arch., 1977]
12. $\dfrac{x\sin^{-1}x}{\sqrt{(1-x^2)}}$.
13. $(\sin^{-1}x)^2$.
14. $\sin x\sinh x$.
15. $e^{m}\cos^{-1}x$.
16. $\sqrt{(\tan\theta)}$.
    [Hint : Put $\theta=t^2$, $\theta=\tan^{-1}t^2$.]
17. $e^x\dfrac{1+\sin x}{1+\cos x}$. [Roorkee, Arch., 1978]

Integrate by expanding the integrand

18. e^{-x^2}.
19. $\sin(1/x)$.
    Evaluate
20. $\int_0^\infty e^{-ax}\sin bx\,dx,\ a>0$.
21. $\int\dfrac{xe^x}{(x+1)^2}dx$. [AMIE, 1978]
    [Hint. Integrate by parts taking xe^x as the first function.]

6.8 REDUCTION FORMULAE

In a number of cases the integral can be obtained by successive reduction. Some of the reduction formulae are derived below. Except in (i), these are obtained by integrating by parts.

(*i*) $\tan^n x$ and $\cot^n x$.

$$\begin{aligned}\int\tan^n x &= \int\tan^{n-2}x\tan^2 x\,dx\\ &= \int\tan^{n-2}x(\sec^2 x-1)\,dx\\ &= \int\tan^{n-2}x\sec^2 x\,dx-\int\tan^{n-2}x\,dx\\ &= \frac{\tan^{n-1}x}{n-1}-\int\tan^{n-2}x\,dx.\end{aligned}$$

Similarly, or by putting $x+\frac{1}{2}\pi$ for x in the above,

$$\int\cot^n x\,dx = \frac{\cot^{n-1}x}{n-1}-\int\cot^{n-2}x\,dx.$$

(*ii*) $\sec^n x$ and $\operatorname{cosec}^n x$.

$$\begin{aligned}\int \sec^n x\,dx &= \int \sec^{n-2} x \,.\, \sec^2 x\,dx \\ &= \sec^{n-2} x \tan x - (n-2)\int \sec^{n-3} x \,.\, \sec x \tan x \,.\, \tan x\,dx \\ &= \sec^{n-2} x \tan x - (n-2)\int \sec^{n-2} x\,(\sec^2 x - 1)\,dx \\ &= \sec^{n-2} x \tan x + (n-2)\int \sec^{n-2} x\,dx - (n-2)\int \sec^n x\,dx.\end{aligned}$$

Transposing the last term to the left and dividing by $n - 1$, we get

$$\int \sec^n x\,dx = \frac{\sec^{n-2} x \tan x}{n-1} + \frac{n-2}{n-1}\int \sec^{n-2} x\,dx.$$

Similarly, or by putting $x + \frac{1}{2}\pi$ for x in the above,

$$\int \operatorname{cosec}^n x\,dx = -\frac{\operatorname{cosec}^{n-2} x \cot x}{n-1} + \frac{n-2}{n-1}\int \operatorname{cosec}^{n-2} x\,dx.$$

(*iii*) $x^n \sin mx$ or $x^n \cos m^x$. Interchanging twice by parts

$$\begin{aligned}\int x^n \sin mx\,dx &= -\frac{x^n \cos mx}{m} + \frac{n}{m}\int x^{n-1} \cos mx\,dx \\ &= -\frac{x^n \cos mx}{m} + \frac{n}{m}\left\{\frac{x^{n-1}\sin mx}{m} - \frac{n-1}{m}\int x^{n-2}\sin mx\,dx\right\};\end{aligned}$$

i.e.
$$\int x^n \sin mx\,dx = -\frac{x^n \cos mx}{m} + \frac{nx^{n-1}\sin mx}{m^2} - \frac{n(n-1)}{m^2}\int x^{n-2}\sin mx\,dx.$$

The formula for $\int x^n \cos mx$ can be obtained similarly.

***Example* 6.21.**

Integrate

1. $\tan^4 x$.
2. $\tan^5 x$. [*Roorkee*, 1979]
3. $\cot^3 x$.
4. $\sec^3 x$. [*Roorkee*, 1972]
5. $\operatorname{cosec}^4 x$.
6. $\sec^5 x$.
7. Evaluate $\int_0^{\pi/4} \frac{dx}{\cos^3 x}$. [*Gorakhpur*, 1972]
8. If $U_n = \int_0^{\pi/4} \tan^n x\,dx$, prove that

$$U_n + U_{n-2} = 1/(n-1).$$ [Allahabad, 1970]

Integrate

9. $x^2 \sin 2x$.
10. $x \sin^2 x$.
11. $e^{2x} \cos^3 x$.
12. If $U_n = \int_0^{\pi/2} x^n \sin x\,dx$ and $n > 1$, prove that

$$U_n + n(n-1)U_{n-2} = n(\pi/2)^{n-1}.$$

Hence evaluate $\int_0^{\pi/2} x^4 \sin x \, dx$. [*Roorkee*, 1995]

13. Integrating by parts twice, obtain a reduction formula for

$$I_m = \int_0^{\infty} e^{-x} \sin^m x \, dx,$$

where $m \geq 2$, in the form

$$(1 + m^2)I_m = m(m-1)I_{m-2}.$$

Hence evaluate I_4. [*Roorkee*, 1996]

14. If $I_{(m,n)} = \int_0^{\pi/2} \cos^m x \cos nx \, dx$, prove that

$$I_{(m,n)} = \left\{\frac{m(m-1)}{m^2 - n^2}\right\} I_{(m-2,n)}.$$ [*Allahabad*, 1969]

15. Find a reduction formula for, $\int x^m (\log x)^n \, dx$. [*Banaras*, 1979]

6.9 DEFINITE INTEGRALS

As already defined

$$\int_a^b f(x)\,dx = [F(x)]_a^b = F(b) - F(a),$$

where $dF(x)/dx = f(x)$. As shown earlier, the definite integral does not depend on the constant of integration C. The following properties of the definite integrals are found in evaluating them.

(*i*) $\int_a^b \mathbf{f(x)\,dx} = \int_a^b \mathbf{f(t)\,dt}$.

For both sides are equal to $F(b) - F(a)$.

(*ii*) $\int_a^b \mathbf{f(x)\,dx} = -\int_a^b \mathbf{f(x)\,dx}$.

For $F(b) - F(a) = -\{F(a) - F(b)\}$.

(*iii*) $\int_a^b \mathbf{f(x)\,dx} = \int_a^c \mathbf{f(x)\,dx} + \int_c^b \mathbf{f(x)\,dx}$

For the right-hand side is equal to
$F(c) - F(a) + F(b) - F(c)$,
which is equal to $F(b) - F(a)$.

(*iv*) $\int_0^a \mathbf{f(x)\,dx} = \int_0^a \mathbf{f(a-x)\,dx}$.

For, on putting $a - x = t$ and $-dx = dt$,

$$\int_0^a f(a-x)\,dx = -\int_a^0 f(t) = \int_0^a f(t)\,dt \text{ by (ii)}, = \int_0^a f(x)\,dx, \text{ by (i)}.$$

(*v*) $\int_a^a \mathbf{f(x)\,dx} = 2\int_0^a \mathbf{f(x)\,dx}$, if $f(-x), = f(x)$; and $= 0$, if $f(-x) = -f(x)$.

PROOF. $\int_a^a f(x)\,dx = \int_a^0 f(x) + \int_0^a f(x)\,dx$. ...(1)

Putting $x = -t$, in the first integral on the right,

$$\int_{-a}^{0} f(x)\,dx = -\int_{a}^{0} f(-t)\,dt = \int_{0}^{a} f(-t)\,dt = \int_{0}^{a} f(-x)\,dx$$

Substituting this in (1), we get

$$\int_{-a}^{a} f(x)\,dx = \int_{0}^{a} \{f(x) + f(-x)\}\,dx$$

from which the result follows at once.

Functions like x^4, $\cos x$, etc. for which $f(-x) = f(x)$ are called *even* functions. Functions like x^3, $\sin x$, etc. for which $f(-x) = -f(x)$, are called *odd* functions.

(*vi*) $$\int_{0}^{2a} f(x)\,dx = 2\int_{0}^{a} f(x)\,dx, \text{ if } f(2a-x) = f(x),$$

and $$= 0, \quad \text{if } f(2a-x) = -f(x).$$

This can be proved in a manner similar to (*v*).

***Example* 6.22.** Evaluate $\int_{0}^{\pi} \frac{x\sin x}{1+\cos^2 x}\,dx$. [*Bhopal*, 1978]

Let $$I = \int_{0}^{\pi} \frac{x\sin x}{1+\cos^2 x}\,dx.$$

Then $$I = \int_{0}^{\pi} \frac{(\pi - x)\sin x}{1+\cos^2 x}\,dx, \text{ by (iv)}.$$

Adding the two values of I we get

$$2I = \int_{0}^{\pi} \frac{(\pi - x + x)\sin x}{1+\cos^2 x}\,dx = \pi\int_{0}^{\pi} \frac{\sin x\,dx}{1+\cos^2 x}\,dx$$

$$= -\pi\,[\tan^{-1}\cos x]_{0}^{\pi} = -\pi\left(-\frac{1}{4}\pi - \frac{1}{4}\pi\right).$$

Hence $$I = \frac{1}{4}\pi^2.$$

***Example* 6.23.** Evaluate $\int_{0}^{\pi} \sin^4 x\,dx$.

$$\int_{0}^{\pi} \sin^4 x\,dx = 2\int_{0}^{\pi/2} \sin^4 x\,dx, \text{ by (vi), since } \sin^4(\pi - x) = \sin^4 x,$$

$$= 2 \cdot \frac{3.1}{4.2} \cdot \frac{\pi}{2} = \frac{3}{8}\pi.$$

***Example* 6.24.** Evaluate $\int_{0}^{\pi/2} \log\sin x\,dx$. [*Roorkee*, 1971]

Let $$I = \int_{0}^{\pi/2} \log\sin x\,dx.$$

Then $$I = \int_{0}^{\pi/2} \log\cos x\,dx, \text{ by } (iv).$$

Adding the two values of I, we get

$$2I = \int_{0}^{\pi/2} (\log\sin x + \log\cos x)\,dx$$

$$= \int_0^{\pi/2} \{\log(2\sin x \cos x) - \log 2\}\, dx$$

$$= \int_0^{\pi/2} \log \sin 2x\, dx - \frac{1}{2}\pi \log 2$$

$$= \frac{1}{2}\int_0^{\pi} \log \sin u\, du - \frac{1}{2}\pi \log 2, \text{ where } x = \frac{1}{2}u,$$

$$= \int_0^{\pi/2} \log \sin u\, du - \frac{1}{2}\pi \log 2, \text{ by } (vi),$$

$$= I - \frac{1}{2}\pi \log 2.$$

Therefore $$I = -\frac{1}{2}\pi \log 2,$$

i.e. $$\int_0^{\pi/2} \log \sin x\, dx = -\frac{1}{2}\pi \log_e 2.$$

Example 6.25.

Evaluate

1. $\int_0^{\pi} \sin^3 x\, dx.$

2. $\int_0^{\pi} x \sin^3 x\, dx.$ [*Roorkee, Arch.*, 1973]

3. $\int_0^{\pi} \sin^3 \theta (1 + 2\cos\theta)(1 + \cos\theta)^2 d\theta.$

4. $\int_0^{\pi/2} \frac{\sqrt{(\sin x)}\, dx}{\sqrt{(\sin x)} + \sqrt{(\cos x)}}.$ [*Banaras*, 1979]

5. $\int_0^{\pi/2} \frac{\sin^2 x}{\sin x + \cos x} dx$ [*Roorkee*, 1979]

6. $\int_0^{\pi/2} \frac{d\theta}{1 + \tan\theta}.$ [*Roorkee, Arch.*, 1978]

7. $\int_0^{\pi} \frac{x\, dx}{a^2 \cos^2 x + b^2 \sin^2 x}$

8. $\int_0^{\pi} \frac{x \tan x\, dx}{\sec x + \tan x}.$

9. $\int_0^{\pi/2} \log \tan x\, dx.$

10. $\int_0^{\pi/4} \log(1 + \tan\theta)\, d\theta.$ [*Allahabad*, 1973]

11. $\int_0^{\pi} \log(1 + \cos x)\, dx.$ [*Roorkee*, 1973]

12. $\int_0^{\pi/2} x \cot x\, dx.$

13. $\int_0^{\pi/2} x^2 \operatorname{cosec}^2 x\, dx.$ [*Gorakhpur*, 96]

14. $\int_0^\infty \frac{\log(x+1/x)}{1+x^2}dx$.

15. $\int_0^1 \frac{\log(1+x)}{1+x^2}dx$.

[Hint. Put $x = \tan\theta$.]

16. If $f(2a-x) = f(x)$, prove that

$\int_0^{2a} f(x)\,dx = 2\int_0^a f(x)\,dx$.

What is the value of the integral on the left is $f(2a-x) = -f(x)$?

6.10 THE INTEGRAL AS A SUM

We have so far looked upon integration as the operation which is the inverse of differentiation; and we defined the integral of $f(x)$ as that function which on differentiation gives $f(x)$.

It is also possible to define the integral directly as the limit of the sum of a series of numbers. In fact this is how integration originated.

Let f be a function defined for all values of x form a to b $(b > a)$. Divide the interval from a to b into n parts of lengths $\delta x_1, \delta x_2, ..., \delta x_n$; and let $x_1, x_2, ... x_n$ be any points lying in the subintervals $\delta x_1, \delta x_2, ..., \delta x_n$ respectively. Then the integral

$$\int_a^b f(x)\,dx \quad ...(1)$$

$$f(x_1)\delta x_1 + f(x_2)\delta x_2 + ... + f(x_n)\delta x_n,$$

is defined as the limit of the sum

when $n \to \infty$ and each of the lengths $\delta x_1, \delta x_2, ...,$ tends to zero. ...(2)

The integral (1) exists only when the sum (2) has a definite limit. It can be shown that the integral always exists when the function f is finite and continuous in the interval $a \le x \le b$. However, it may exist even for certain functions which are discontinuous or unbounded.

We prove in the next section the equivalence of the two definitions of the integral. This theorem is known as the *Fundamental Theorem of the Integral Calculus*. For convenience of proof we shall divide the interval $b - a$ into n subintervals of equal length h, and take the points $x_1, x_2, x_3, ...,$ at the beginning of each subinterval, i.e. at $a, a + h, a + 2h, ...$. With these restrictions

$$\int_a^b f(x)\,dx = \lim_{n\to\infty} h[f(a) + f(a+h) + f(a+2h) + ... + f\{a + (n-1)h\}]$$

6.11 FUNDAMENTAL THEOREM OF INTEGRAL CALCULUS

If

$$\int f(x)dx = F(x) + C, \quad ...(1)$$

then $\lim_{n\to\infty} h[f(a) + f(a+h) + f(a+2h) + ... + f\{a+(n-1)h\}] = F(b) - F(a)$, ...(2)

where $h = (b-a)/n$ *and* $f(x)$ *is continuous for* $a \le x \le b$, *a and b being fixed numbers.*

PROOF : Equation (1) implies that

$$\lim_{h\to\infty} \frac{F(x+h) - F(x)}{h} = f(x).$$

Therefore $$\frac{F(x+h)-F(x)}{h} = f(x)+\varepsilon,$$

where $\varepsilon \to 0$ as $h \to 0$.

Giving to x the values $a, a+h, a+2h, \ldots$ in succession, we get

$$F(a+h)-F(a) = hf(a)+h\varepsilon_1,$$
$$F(a+2h)-F(a+h) = hf(a+h)+h\varepsilon_2,$$
$$F(a+3h)-F(a+2h) = hf(a+2h)+h\varepsilon_3,$$
$$\ldots \qquad \ldots$$
$$F(a+nh)-F\{a+(n-1)h\} = hf\{a+(n-1)h\}+h\varepsilon_n.$$

Since $a+nh=b$, we get on addition

$$F(b)-F(a) = h[f(a)+f(a+h)+\ldots+f\{a+(n-1)h\}]+h(\varepsilon_1+\varepsilon_2+\ldots+\varepsilon_n). \quad \ldots(3)$$

Now suppose ε_r is the term in $\varepsilon_1+\varepsilon_2+\ldots+\varepsilon_n$ which has the largest numerical value. Then

$h|\varepsilon_1+\varepsilon_2+\ldots+\varepsilon_n| \le hn\,|\varepsilon_r| \le (b-a)\,|\varepsilon_r|$.

This tends to zero as $h \to 0$, since $\varepsilon_r \to 0$. Hence the theorem follows on taking the limit of (3), when $n \to \infty$ and $h \to 0$.

***Example* 6.26.** Evaluate $\int_a^b x^2\,dx$ directly from the definition of the integral as the limit of a sum.

$$\int_a^b x^2dx = \lim_{n\to\infty} h[a^2+(a+h)^2+(a+2h)^2+\ldots\{a+(n-1)h\}^2], \text{ where } b-a=nh,$$
$$= \lim_{n\to\infty} h[na^2+\{1+2+3+\ldots+(n-1)\}2ah+\{1^2+2^2+\ldots+(n-1)^2\}h^2]$$
$$= \lim_{n\to\infty} h\left[na^2+\frac{1}{2}n(n-1)2ah+\frac{1}{6}(n-1)(2n-1)nh^2\right]$$
$$= \lim_{n\to\infty}\left[nh\,.\,a^2+nh\,.\,(nh-h)\,.\,a+\frac{1}{3}\,.\,(nh-h)\,.\left(nh-\frac{1}{2}h\right).\,nh\right]$$
$$= (b-a)a^2+(b-a)^2a+\frac{1}{3}(b-a)^3$$
$$= \frac{1}{3}(b-a)\{3a^2+3(b-a)a+b^2-2ab+a^2\}$$
$$= \frac{1}{3}(b-a)(a^2+ab+b^2) = \frac{1}{3}b^3-\frac{1}{3}a^3.$$

6.12 THE INTEGRAL AS AN AREA

The area enclosed by the curve $y=f(x)$, the axis of x and the ordinates at $x=a$ and $x=b$ is given by $\int_a^b f(x)\,dx$

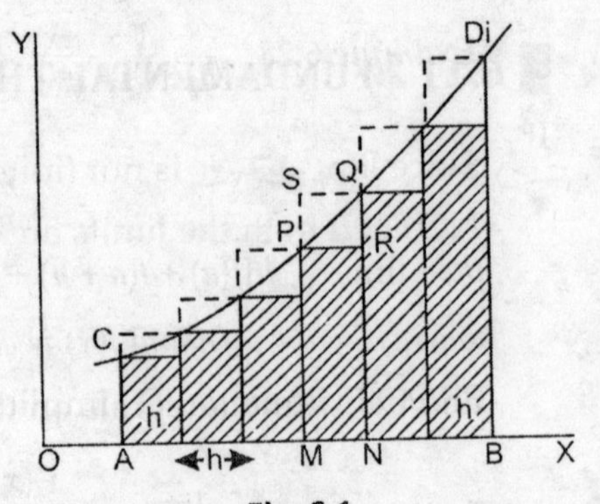

Fig. 6.1

PROOF. The required area is $ABCD$. Divide AB into n parts of equal length h [= $(b-a)/n$], and construct ordinates at the points of subdivision. Suppose, for the sake of convenience, that $f(x)$ increases continuously as we go form A to B.

We notice that the area $MNQP$ lies between the two rectangles $MNRP$ and $MNQS$. Therefore the whole area $ABDC$ lies between the sum of the smaller rectangular areas (shaded in the figure) and the sum of the larger ones (topped by the dotted lines). Thus the area $ABDC$ lies between

$h[f(a) + f(a + h) + ... + f\{a + (n - 1)h\}]$, ...(1)

and $h[f(a + h) + f(a + 2h) + ... + f(a + nh)]$. ...(2)

The limit of the sum (1), as $n \to \infty$ and $h \to 0$, is

$$\int_a^b f(x)\,dx \quad ...(3)$$

The limit of the sum (2) is the same, since the difference of (1) and (2) is $h[f(a + nh) - f(a)]$; *i.e.* $h[f(b) - f(a)]$, which tends to zero as $h \to 0$. Hence the area *ABDC*, which lies between (1) and (2), is also equal to (3).

REMARKS

1. If $f(x)$ decreases continuously as we go from *A* to *B*, the proof is similar. If $f(x)$ increases from *A* to *E* ($x = c$) and then decreases from *E* to *D*, the area *ABDC*

$$= \text{area } AEFC + \text{area } EBDF$$

$$= \int_a^c f(x)\,dx + \int_c^b f(x)\,dx$$

$$= \int_a^b f(x)\,dx.$$

Fig. 6.2

The proof can be extended to any finite number of turning points.

2. If x_1 be any point between a and $a + h$, x_2 any point between $a + h$ and $a + 2h$, and so on, then the sum

$$h[f(x_1) + f(x_2) + ... + f(x_n)] \quad ...(4)$$

lies in magnitude between the sums (1) and (2). Hence as $h \to 0$, the limit of sum (4) also is equal to (3)..

We shall often use this result in the form

$$\lim_{n\to\infty} \sum_{r=1}^{n} f(x_r)\,h = \int_a^b f(x)\,dx$$

6.13 INTEGRALS WITH INFINITE LIMITS

We define an integral with one of its limits infinite as follows :

$$\int_a^\infty f(x)\,dx = \lim_{b\to\infty} \int_a^b f(x)\,dx,$$

Provided the limit is a definite number.

$$\int_{-\infty}^b f(x)\,dx = \lim_{a\to-\infty} \int_a^b f(x)\,dx,$$

Provided the limit is a definite number.

***Example* 6.27.** $\int_1^\infty \frac{dx}{x^3} = \lim_{b\to\infty} \int_1^b \frac{dx}{x^3} = \lim_{b\to\infty} \frac{1}{2}\left(1 - \frac{1}{b^2}\right) = \frac{1}{2}.$

***Example* 6.28.** $\int_1^\infty \frac{dx}{\sqrt{x}} = \lim_{b\to\infty} \int_1^b \frac{dx}{\sqrt{x}} = \lim_{b\to\infty} 2(\sqrt{b} - 1).$

Since $\lim_{b\to\infty} \sqrt{b}$ is not finite, the integral under consideration does no exist.

NOTE. If both the limits are infinite, we can write

$$\int_{-\infty}^\infty f(x)\,dx = \int_{-\infty}^c f(x)\,dx + \int_c^\infty f(x)\,dx,$$

and define the integrals on the right in the above manner.

6.14 IMPROPER INTEGRALS

Let $f(x)$ be continuous for all values of x from a to b $(b > a)$, except that $f(x) \to \infty$ as $x \to b$; then we define $\int_a^b f(x)\,dx$ to mean $\lim_{\varepsilon\to 0} \int_a^{b-\varepsilon} f(x)\,dx$, provided that the limit is a definite number.

Similarly, if $f(x) \to \infty$ as $x \to a$ and $f(x)$ is otherwise continuous, we define $\int_a^b f(x)\,dx$ to mean $\lim_{\varepsilon\to 0}$, provided that the limit is a definite number.

Again, if $f(x)$ is continuous for all values of x from a to b, except that $f(x) \to \infty$ as $x \to c$, when c lies between a and b, we define $\int_a^b f(x)\,dx$ to mean $\lim_{\varepsilon\to 0} \int_a^{c-\varepsilon} f(x)\,dx + \lim_{\varepsilon\to 0} \int_{c+\varepsilon}^b f(x)\,dx$, provided that each limit is a definite number.

The above definitions hold also when the limit of $f(x)$ at $x = a$, b or c is $-\infty$, or when the limits on the right and on the left at $x = c$ are infinite, but of different signs.

***Example* 6.29.**

$$\int_0^1 \frac{dx}{\sqrt{(1-x)}} = \lim_{\varepsilon\to 0} \int_0^{1-\varepsilon} \frac{dx}{\sqrt{(1-x)}}$$

$$= \lim_{\varepsilon\to 0} [-2\sqrt{(1-x)}]_0^{1-\varepsilon} = \lim_{\varepsilon\to 0} [-2\sqrt{\varepsilon} + 2] = 2.$$

***Example* 6.30.**

$$\int_0^1 \frac{dx}{x^3} = \lim_{\varepsilon\to 0} \int_\varepsilon^1 \frac{dx}{x^3} = \lim_{\varepsilon\to 0} \left[-\frac{1}{2x^2}\right]_\varepsilon^1$$

$$= \lim_{\varepsilon\to 0} \left[-\frac{1}{2} + \frac{1}{2\varepsilon^2}\right].$$

Since $\lim_{\varepsilon\to 0} (1/\varepsilon^2) = \infty$, the integral does not exist.

***Example* 6.31.**

$$\int_{-1}^1 \frac{dx}{x^{2/3}} = \lim_{\varepsilon\to 0} \int_{-1}^{-\varepsilon} \frac{dx}{x^{2/3}} + \lim_{\varepsilon\to 0} \int_\varepsilon^1 \frac{dx}{x^{2/3}}$$

$$= \lim_{\varepsilon\to 0} [3x^{1/3}]_{-1}^{-\varepsilon} + \lim_{\varepsilon\to 0} [3x^{1/3}]_\varepsilon^1$$

$$= \lim_{\varepsilon\to 0} [-3\varepsilon^{1/3} + 3] + \lim_{\varepsilon\to 0} [3 - 3(\varepsilon')^{1/3}]$$

$$= 6.$$

6.15 DIFFERENTIATION UNDER SIGN OF INTEGRATION

Let $f(x, t)$ be a function of two variables x and t. Then the definite integral

$$\int_A^B f(x, t)\,dt \qquad \text{...(1)}$$

where A and B are constants, is a function of x say $\phi(x)$. The following theorem is useful for obtaining its differential coefficient $\phi(x)$.

If $f(x, t)$ and $\partial f/\partial x$ are continuous in the domain $a \le x \le b$ and $A \le t \le B$, then for $a \le x \le b$

$$\frac{d}{dx} \int_A^B f(x, t)\,dt = \int_A^B \frac{\partial f(x, t)}{dx}\,dt \qquad \text{...(2)}$$

The left hand side is equal to $\lim_{h\to 0} \frac{1}{h} \int_A^B f(x+h, t)\,dt - \int_A^B f(x, t)\,dt$

$$= \lim_{h\to 0} \int_A^B \frac{f(x+h, t) - f(x, t)}{h}\,dt = \int_A^B \frac{\partial f}{\partial x}\,dt$$

The last step, while looking obvious, needs uniform continuity of $\partial f/\partial x$ in the given domain. When A and B are finite, continuity implies uniform continuity. In case (1) is an improper integral the required conditions are ensured if $\left|\frac{\partial f}{\partial x}\right| \le M(t)$ and $\int_A^B M(t)\,dt$ exists.

When the limits A and B are not constants but are functions of x, the result (2) gets modified.

Put
$$f(x, t) = \frac{\partial F(x, t)}{\partial t} \qquad ...(3)$$

Then
$$\int_{A(x)}^{B(x)} f(x, t)\, dt = [F(x, t)]_{A(x)}^{B(x)} = F(x, B) - F(x, A).$$

Therefore
$$\frac{d}{dx}\int_{A(x)}^{B(x)} f(x, t)\, dt = \frac{d}{dx}[F(x, B) - F(x, A)]$$

$$= \frac{\partial F(x, B)}{\partial x} + \frac{\partial F(x, B)}{\partial B}\frac{dB}{dx} - \frac{\partial F(x, A)}{\partial x} - \frac{\partial F(x, A)}{\partial A}\frac{dA}{dx}$$

$$= \frac{\partial}{\partial x}[F(x, B) - F(x, A)] + f(x, B)\frac{dB}{dx} - f(x, A)\frac{dA}{dx} \text{ by (3)}$$

$$= \int_{A(x)}^{B(x)} \frac{\partial f(x, t)\, dt}{\partial x} + f(x, B)\frac{dB}{dx} - f(x, A)\frac{dA}{dx}.$$

This formula, and its particular case (2), is known as *Leibnitz's rule*. This rule can be utilised for evaluating certain integrals which may be difficult to evaluate otherwise. For example, the result

$$\int_0^\infty e^{-ax}\, dx = \left[\frac{e^{-ax}}{-a}\right]_0^\infty = a^{-1}, a > 0.$$

gives on differentiating both sides with respect to a, n times,

$$\int_0^\infty (-x)^n e^{-ax}\, dx = (-1)(-2) \dots (-n)a^{-n-1}$$

Therefore
$$\int_0^\infty x^n e^{-ax}\, dx = n!a^{-n-1}.$$

Example **6.32.** By Leibnitz's rule

$$\frac{d}{dx}\int_1^{x^2} e^{-xt^2}\, dt = \int_1^{x^2} \frac{\partial}{dx}(-e^{-xt^2})dt + e^{-x.x^2}\frac{dx^2}{dx} - 0$$

$$= 2xe^{-x^3} - \int_1^{x^2} t^2 e^{-xt^2}\, dt.$$

Example **6.33.** Evaluate
$$\phi(x) = \int_0^\infty e^{-t^2} \sin(2xt)\, \text{dt}. \qquad ...(4)$$

By Leibnitz's rule

$$\frac{d\phi}{dx} = -2\int_0^\infty t\, e^{-t^2} \sin(2xt)\, dt. \qquad ...(5)$$

[Here we have an improper integral for which $|t\, e^{-t^2} \sin(2xt)| \le te^{-t^2}$ and $\int_0^\infty te^{-t^2} dt$ exists.]

Integrating by parts, (5), gives

$$\frac{d\phi}{dx} = [e^{-t^2} \sin(2xt)]_0^\infty - 2x\int_0^\infty e^{-t2} \cos(2xt)\, dt$$

$$= 0 - 2x\, \phi(x).$$

Hence ϕ satisfies the differential equation

$$\frac{d\phi}{dx} = -2x\phi, \text{ or } \frac{1}{\phi}\frac{d\phi}{dx} = -2x,$$

whose solution is

$$\log \phi = \log c - x^2, \text{ or } \phi = ce^{-x^2}. \quad ...(6)$$

But by (4), $$\phi(0) = \int_0^\infty e^{-t^2} dt = \frac{1}{2}\sqrt{\pi}.$$

This gives $$c = \frac{1}{2}\sqrt{\pi}. \text{ Therefore,}$$

$$\int_0^\infty e^{-t^2} \cos(2xt)\, dt = \frac{1}{2}\sqrt{\pi}\, e^{-x^2}.$$

On putting $t = au$ $(a > 0)$ and $x = b/2a$, we get a more useful result

$$\int_0^\infty e^{-a^2u^2} \cos bu\, du = \frac{\sqrt{\pi}}{2a} e^{-b^2/4a^2}.$$

Example **6.34.**

Evaluate the following integrals from the definition of integral as the limit of a sum :

1. $\int_1^2 x\, dx.$
2. $\int_a^b \frac{dx}{x^2}.$
3. $\int_a^b e^x dx.$
4. $\int_a^b \cos x\, dx.$

Obtain the following differential coefficients

5. $\frac{d}{dx}\int_{\pi/2}^{\pi} \frac{\cos(xt)}{t} dt.$
6. $\frac{d}{du}\int_1^2 \log(ux)\, dx.$
7. $\frac{d}{dx}\int_{x^2}^2 \log(1+t^2)\, dt.$
8. $\frac{d}{dx}\int_x^{\tan x} e^{-t^2} dt.$
9. $\frac{d}{dx}\int_{\sqrt{x}}^{x} \frac{dt}{x^2+t+1}.$
10. Evaluate $\int_0^1 x^n \log x\, dx$ by differentiating both sides of
$$\int_0^1 x^n dx = \frac{1}{n+1}$$
with respect to n $(n > -1)$.
11. Evaluate $\int_0^\infty \frac{dx}{(x^2+a^2)^n}$ by repeated differentiation of $\int_0^\infty \frac{dx}{(x^2+a^2)}$
12. If $\phi(a) = \int_0^\infty \frac{e^{-ax}\sin x}{x} dx$, where $a > 0$, show that $\phi'(a) = -1/(a^2+1)$. Hence prove that
$$\int_0^\infty \frac{e^{-ax}\sin x}{x} dx = \frac{1}{2}\pi - \tan^{-1} a.$$

13. Use the above result to show that $\int_0^\infty \frac{\sin x}{x}\,dx = \frac{1}{2}\pi$. [*Poona*, 1976]

14. Given $\int_{-\infty}^{\infty} e^{-x^2}dx = \sqrt{\pi}$, evaluate $\int_{-\infty}^{\infty} x^2 e^{-x^2}dx$.

[Hint. Put $x = au$ in the first integral, divide by a, and then differentiate with respect to a.]

INTEGRAL CALCULUS-I

This chapter is devoted to the study of single integral, volume and surface area of revolutions, multiple integral and its applicable, double integral, area and volume as double integral.

6.16 SOLID OF REVOLUTION AND SURFACE OF REVOLUTION

If a plane area is revolved about a fixed line in its own plane, then the body so generated by the plane area is called the solid of revolution and the surface so generated is called the surface of revolution and the fixed line about which the solid revolves is called the axis of revolution.

6.17 (A) VOLUME OF SOLID OF REVOLUTION (FOR CERTAIN CURVES)

(*a*) **Revolution about x-axis**

The volume of solid generated by the revolution about x-axis of the area bounded by the curve $y = f(x)$, the x-axis and the ordinates, $x = a$, $x = b$ is $\int_a^b \pi y^2\,dx$

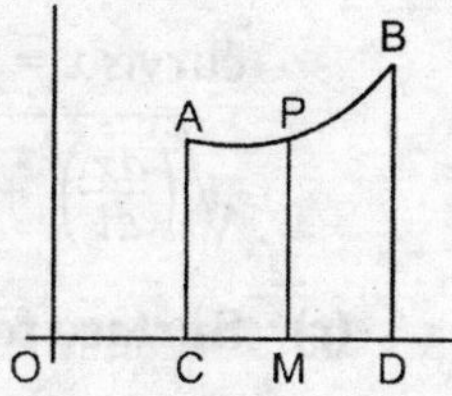

Fig. 6.3

(*b*) **Revolution about y-axis**

The volume of solid generated by the revolution about y-axis of the area bounded by the curve $x = f(y)$, the y-axis and the abscissae $y = a$, $y = b$ is $\int_a^b \pi x^2\,dx$

(*c*) **Revolution about any axis**

The volume of the solid generated by the revolution about any axis CD of the area bounded by the curve AB, the axis CD and the perpendiculars AC, BD on the axis is $\int_{OC}^{OD} \pi (PM)^2 d(OM)$ where O is a fixed point on the axis CD and PM is perpendicular from any point P of the curve AB on CD.

(B) VOLUME OF SOLID OF REVOLUTION (FOR PARAMETRIC CURVES)

(*a*) The volume of the solid generated by the revolution about x-axis, of the area bounded by the curve $x = f(t)$, $y = \phi(t)$, the x-axis and the ordinates where $t = a$, $t = b$ is $\int_a^b \pi y^2 \frac{dx}{dt}\,dt$

(*b*) The volume of the solid generated by the revolution about y-axis, of the area bounded by the curve $x = f(t)$, $y = \phi(t)$, the y-axis and the abscissae at the points $t = a$, $t = b$ is $\int_a^b \pi x^2 \frac{dy}{dt}\,dt$

(*c*) Volume of solid revolution between two solids. The volume of the solid generated by the revolution about the x-axis of the area bounded by the curves $y = f(x)$, $y = \phi(x)$ and the ordinates $x = a$, $x = b$ is $\int_a^b \pi (y_1^2 - y_2^2)\,dx$ where y_1 is the ordinates of the upper curve and y_2 that of the lower curve.

(C) VOLUME OF THE SOLID OF REVOLUTION (FOR THE POLAR CURVES)

The volume of the solid generated by the revolution of the area bounded by the curve $r = f(\theta)$ and the radii vectors $\theta = \alpha, \theta = \beta$

(*a*) about the initial line OX $(\theta = 0)$ is $\int_{\alpha}^{\beta} \frac{2}{3}\pi r^3 \sin\theta \, d\theta$

(*b*) about the line OY $\left(\theta = \frac{\pi}{2}\right)$ is $\int_{\alpha}^{\beta} \frac{2}{3}\pi r^3 \cos\theta \, d\theta$

(D) SURFACE AREA OF SOLID OF REVOLUTION

(*a*) Surface formula for Cartesian equation

The curved surface of the solid generated by the revolution about x-axis, of the area bounded by the curve $y = f(x)$, the x-axis and the ordinates $x = a$, $x = b$ is $\int_a^b 2\pi y \frac{ds}{dx} dx$ where $\frac{ds}{dx} = \sqrt{1 + \left(\frac{dy}{dx}\right)^2}$

(*b*) Surface formula for parametric equation

The curved surface of the solid generated by the revolution about x-axis, of the area bounded by the curve $x = f(t)$, $y = \phi(t)$ the x-axis and the ordinates at the points $t = a$, $t = b$ is $\int_a^b 2\pi y \frac{ds}{dt} dt$ where $\frac{ds}{dx} = \sqrt{\left(\frac{dx}{dt}\right)^2 + \left(\frac{dy}{dt}\right)^2}$

(*c*) Surface formula for Polar equations

The curved surface of the solid generated by the revolution, about the initial line of the area bounded by the curve $r = f(\theta)$ and the radii vectors $\theta = \alpha$, $\theta = \beta$ is

$$\int_{\theta=\alpha}^{\theta=\beta} 2\pi y \frac{ds}{d\theta} d\theta \qquad \text{where } \frac{ds}{d\theta} = \sqrt{r^2 + \left(\frac{dr}{d\theta}\right)^2} \text{ and } y = r \sin\theta.$$

Example **6.35.** Find the volume of the solid generated by revolving the ellipse $\frac{x^2}{a^2} + \frac{y^2}{b^2} = 1$ about the major axis.

Solution. The equation of the ellipse is $\frac{x^2}{a^2} + \frac{y^2}{b^2} = 1$

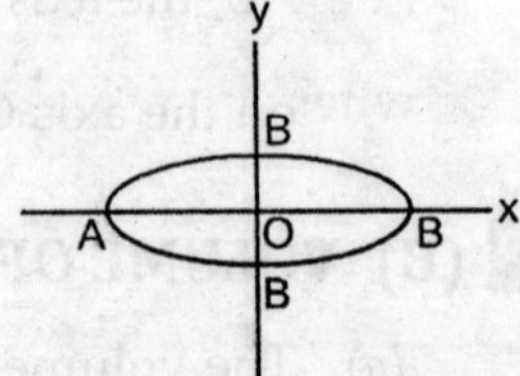

Fig. 6.4

$$\therefore \quad \frac{y^2}{b^2} = 1 - \frac{x^2}{a^2} = \frac{a^2 - x^2}{a^2}$$

$$y^2 = \frac{b^2}{a^2}(a^2 - x^2)$$

The ellipse is symmetrical about y-axis.

Required volume generated by the ellipse about x-axis

$$= 2\int_0^a \pi y^2 dx = 2\pi \int_0^a \frac{b^2}{a^2}(a^2 - x^2)\, dx$$

$$= 2\pi \frac{b^2}{a^2} \int_0^a (a^2 - x^2)\, dx = 2\pi \frac{b^2}{a^2}\left[a^2 x - \frac{x^3}{3}\right]_0^a = \frac{4}{3}\pi a b^2$$

***Example* 6.36.** Find the volume generated by the revolution about the initial line of the limacon $r = a + b\cos\theta$ $a > b$.

Solution. We shall revolve only the shaded region above the initial line

$$V = \frac{2}{3}\pi\int_0^{\pi}(a+b\cos\theta)^2\sin\theta\,d\theta$$

$$= -\frac{2}{3}\frac{\pi}{b}\int_{a+b}^{a-b}t^3\,dt \qquad a+b\cos\theta = t \quad b\sin\theta\,d\theta = -dt$$

$$= -\frac{2}{3}\frac{\pi}{b}\left[\frac{(a-b)^4-(a+b)^4}{4}\right] = -\frac{2\pi}{3b}\left[\frac{-8a^3b-8ab^3}{4}\right] = \frac{4}{3}\pi a(a^2+b^2)$$

Fig. 6.5

***Example* 6.37.** Find the volume of the solid generated by revolving the finite region bounded by the curves $y = x^2+1$, $y = 5$ about the line $x = 3$.

Solution. The required region is given in the figure. The volume is given by

$$V = \pi\int_{y=1}^{5}(x_1^2 - x_2^2)\,dy$$

$$= \pi\int_1^5[(3+\sqrt{y-1})^2-(3-\sqrt{y-1})^2]\,dy \qquad \left\{\because \begin{array}{l} x^2 = y-1 \\ x = \pm\sqrt{y-1}\end{array}\right.$$

$$= 12\pi\int_1^5\sqrt{y-1}\,dy = 12\pi\left[\left(\frac{2}{3}\right)(y-1)^{3/2}\right]_1^5 = 8\pi\,(8) = 64\pi$$

Fig. 6.6

***Example* 6.38.** Find the volume of the solid generated by the revolution of the loop of the curve $(a-x)y^2 = (a+x)x^2$ about the x-axis

Solution. The shape of the curve is shown in the figure

The curve is symmetrical about x-axis

$\therefore$ Required volume $\qquad V = \int_{-a}^{0}\pi y^2\,dx = \int_{-a}^{0}\pi\frac{a+x}{a-x}x^2\,dx$

Let $\quad a - x = z \quad dx = -dz$

when $\quad x = -a \quad z = 2a$

$\qquad x = 0 \quad z = a$

Fig. 6.7

$$V = \pi\int_{2a}^{a}\frac{a+a-z}{z}(a-z)^2(-dz) = \pi\int_a^{2a}\frac{2a-z}{z}(a-z)^2\,dz$$

$$= \pi\int_a^{2a}\frac{(2a-z)}{z}(a^2+z^2-2az)\,dz$$

$$= \pi\int_a^{2a}\left[\frac{2a^3}{z}+2az-4a^2-a^2-z^2+2az\right]dz$$

$$= \pi\int_a^{2a}\left[\frac{2a^3}{z}+4az-z^2-5a^2\right]dz$$

$$= \pi \left[2a^3 \log z + 2az^2 - \frac{z^3}{3} - 5a^2 z \right]_a^{2a}$$

$$= \pi \left[\left\{ 2a^3 \log 2a + 2a(2a)^2 - \frac{(2a)^3}{3} - 5a^2(2a) \right\} - \left\{ 2a^3 \log a + 2a^3 - \frac{a^3}{3} - 5a^3 \right\} \right]$$

$$= 2\pi a^3 \left[\log 2 - \frac{2}{3} \right]$$

Example 6.39. Find the volume generated by the revolution of the area under one complete arc of the cycloid $x = a(\theta - \sin\theta)$, $y = a(1 - \cos\theta)$, the axis of revolution being the x-axis.

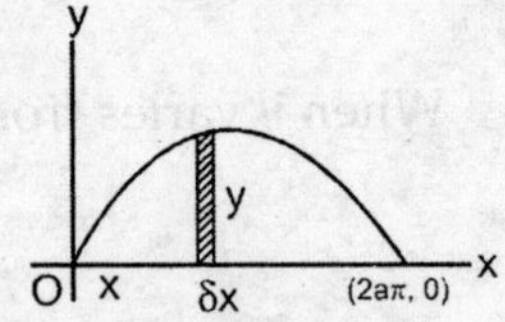

Fig. 6.8

Solution. Let the area under arc be divided into n strips of width dx by lines || to the y-axis and y be the height of the typical strip at a distance x from y-axis.

$$\text{Volume} = \int_0^{2\pi a} \pi y^2 dx = \int_0^{2\pi} \pi a^2 (1 - \cos\theta)^2 a(1 - \cos\theta)\, d\theta$$

$$= \int_0^{2\pi} \pi a^3 (1 - \cos\theta)^3 d\theta$$

$$= \int_0^{2\pi} \pi a^3 \times \left(2\sin^2 \frac{\theta}{2}\right)^3 d\theta$$

$$= 8\pi a^3 \int_0^{2\pi} \sin^6\left(\frac{\theta}{2}\right) d\theta = 16\pi a^3 \int_0^{\pi} \sin^6 \phi\, d\phi \qquad (\theta = 2\phi)$$

$$= 32\pi a^3 \int_0^{\pi/2} \sin^6 \phi\, d\phi \qquad (\sin(\pi - \phi) = \sin\phi)$$

$$= 32\pi a^3 \frac{5 \times 3 \times 1}{6 \times 4 \times 2} \frac{\pi}{2} = 5\pi^2 a^3$$

Example 6.40. Find the volume of the surface generated by revolving the Cardioid $r = a(1 + \cos\theta)$ about initial line.

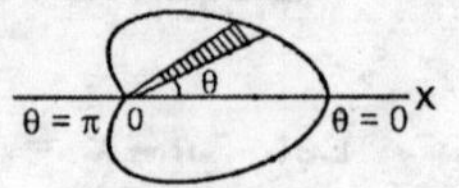

Fig. 6.9

Solution. Volume generated $= \frac{2}{3}\pi \int_0^{\pi} r^3 \sin\theta\, d\theta$

$$= \frac{2}{3}\pi \int_0^{\pi} a^3 (1 + \cos\theta)^3 \sin\theta\, d\theta = -\frac{2}{3}\pi a^3 \int_0^{\pi} (1 + \cos\theta)^3 (-\sin\theta)\, d\theta$$

$$= -\frac{2}{3} a^3 \left[\frac{(1 + \cos\theta)^4}{4} \right]_0^{\pi} = \frac{2}{3}\pi a^3 \left[\frac{(1+1)^4}{4} \right] = \frac{8}{3}\pi a^3$$

Example 6.41. Show that the volume of the solid formed by the revolution of the Cissoid $y^2(2a - x) = x^3$ about its asymptotes is $2\pi^2 a^3$

Solution. The asymptote of this curve is $x = 2a$.

$$\text{Required volume} \qquad V = 2\pi \int_0^{\infty} (2a - x)^2 dy$$

From the equation of the curve $\quad y = \dfrac{x^{3/2}}{\sqrt{2a - x}}$

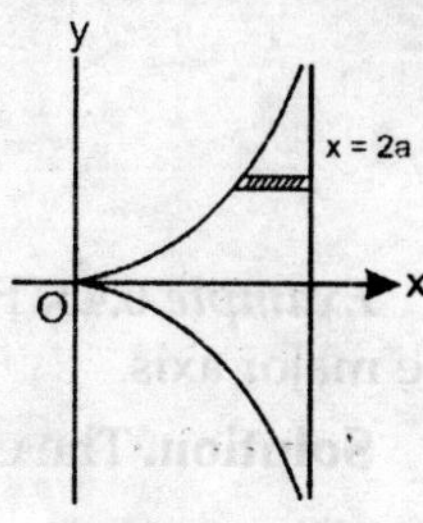

Fig. 6.10

$$\frac{dy}{dx} = \frac{(3a - x)\sqrt{x}}{(2a - x)^{3/2}}$$

$$dy = \frac{(3a - x)\sqrt{x}}{(2a - x)^{3/2}}\,dx$$

When y varies from 0 to ∞, x varies from 0 to $2a$

$$\therefore \quad V = 2\pi \int_0^{\infty} (2a - x)^2\, dy = 2\pi \int_0^{2a} (2a - x)^2 \frac{(3a - x)\sqrt{x}}{(2a - x)^{3/2}}\,dx$$

$$= 2\pi \int_0^{2a} (a - x)\sqrt{2ax - x^2}\,dx + 4\pi a \int_0^{2a} \sqrt{2ax - x^2}\,dx$$

Put $\quad x = 2a\sin^2\theta$, we get $\displaystyle\int_0^{2a} \sqrt{2ax - x^2}\,dx = \int_0^{\frac{\pi}{2}} 2a \times 4a \sin^2\theta \cos^2\theta\, d\theta = \pi\,\frac{a^2}{2}$

$$\therefore \quad V = \pi \left[\frac{(2ax - x^2)^{3/2}}{3/2}\right]_0^{2a} + 4\pi\, a\, \frac{\pi a^2}{2} = 0 + 2\pi^2 a^3 = 2\pi^2 a^3$$

***Example* 6.42.** The area cut off from the right parabola $y^2 = 4ax$ by the chord joining the vertex to an end of the latus rectum through four right-angles about the chord. Find the volume of the solid generated.

Solution. The equation of the parabola is

$$y^2 = 4ax$$

The co-ordinates of B are $(a, 2a)$

Hence the equation of OB

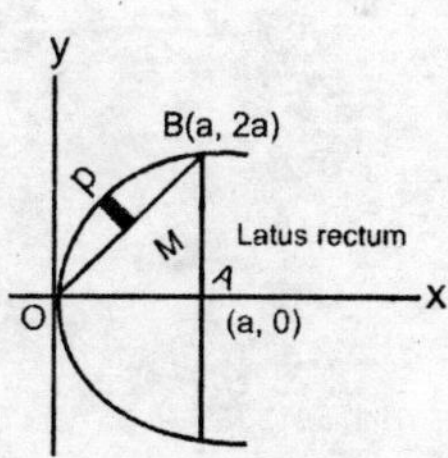

Fig. 6.11

$$y - 0 = \frac{2a - 0}{a - 0}\,(x - 0)$$

$$2x - y = 0$$

Let $P(at^2, 2at)$ be a point on the arc OB and PM the perpendicular from P to OB.

$$PM = \frac{2at^2 - 2at}{\sqrt{4 + 1}} = \frac{2at^2(t - 1)}{\sqrt{5}}$$

$$OP = \sqrt{(at^2 - 0)^2 + (2at - 0)^2} = at\sqrt{t^2 + 4}$$

$$OM^2 = OP^2 - PM^2 = a^2 t^2 (t^2 + 4) - \frac{4a^2 t^2 (t - 1)^2}{5}$$

$$OM^2 = \frac{a^2 t^2 (t + 4)^2}{5}$$

$$OM = \frac{at(t + 4)}{\sqrt{5}}$$

Required volume

$$= \int^{t} \pi (PM)^2\, d(OM) = \int^{1} \frac{4a^2 t^2 (t - 1)^2}{\ldots}\, d\left[\frac{at(t + 4)}{\ldots}\right]$$

$$= \frac{4\pi a^3}{5\sqrt{5}}\int_0^1 t^2(t^2-2t+1)(2t+4)\,dt = \frac{4\pi a^3}{5\sqrt{5}}\int_0^1 (2t^5-6t^3+4t^2)\,dt = \frac{2\pi a^3}{15\sqrt{5}}$$

***Example* 6.43.** Find the surface area of the solid generated by the revolution of the ellipse $x^2 + 4y^2 = 16$ about the major axis,

Solution. The equation of the ellipse is

$$x^2 + 4y^2 = 16$$

$$4y^2 = 16 - x^2 \quad \therefore \quad y = \frac{\sqrt{16-x^2}}{2}$$

$$\frac{dy}{dx} = \frac{1}{2}\cdot\frac{1}{2}\cdot\frac{1}{\sqrt{16-x^2}}\times(-2x) = -\frac{x}{2\sqrt{16-x^2}}$$

$$dy = -\frac{x}{2\sqrt{16-x^2}}\,dx$$

$$\frac{ds}{dx} = \sqrt{1+\left(\frac{dy}{dx}\right)^2} = \sqrt{1+\frac{x^2}{4(16-x^2)}} = \sqrt{\frac{64-3x^2}{4(16-x^2)}}$$

The ellipse $x^2 + 4y^2 = 16$ meets x-axis where $y = 0$,

$$\therefore \qquad x^2 = 16 \quad x = \pm 4$$

For the upper half of the ellipse in first quadrant x varies from 0 to 4.

The ellipse is symmetrical about y-axis

$$\therefore \text{ Required surface} = 2\times\int_0^4 2\pi y\frac{ds}{dx}dx = 4\pi\int_0^4 \frac{\sqrt{16-x^2}}{2}\times\sqrt{\frac{64-3x^2}{4(16-x^2)}}\,dx$$

$$= \pi\int_0^4 \sqrt{64-3x^2}\,dx = \sqrt{3}\,\pi\int_0^4\sqrt{\frac{64}{3}-x^2}\,dx$$

$$= \sqrt{3}\,\pi\left[\frac{x\sqrt{\frac{64}{3}-x^2}}{2} + \frac{64}{3\times 2}\sin^{-1}\frac{x}{8}\right]_0^4 = \sqrt{3}\,\pi\left[2\sqrt{\frac{64}{3}-16} + \frac{32}{3}\sin^{-1}\frac{\sqrt{3}}{2}\right]$$

$$= \sqrt{3}\,\pi\left[2\times\frac{4}{\sqrt{3}} + \frac{32}{3}\frac{\pi}{3}\right] = 8\pi\left[1+\frac{4\pi}{3\sqrt{3}}\right]$$

***Example* 6.44.** Find the area of the surface of revolution generated by revolving one arc of the curve $y = \sin x$ about the x-axis.

Solution. One arc of the curve $y = \sin x$ lies in $(0, \pi)$, Further

$$\sqrt{1+\left(\frac{dy}{dx}\right)^2} = \sqrt{1+\cos^2 x}$$

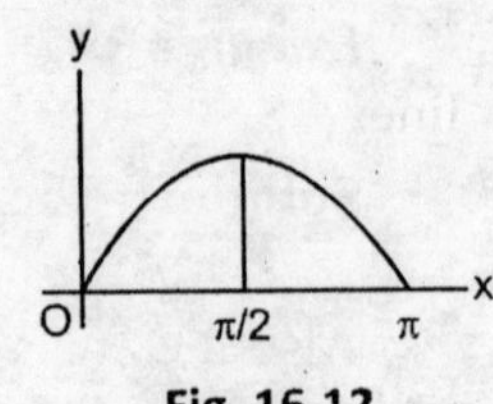

Fig. 16.12

Required surface area $$= 2\pi\int_0^\pi y\sqrt{1+\left(\frac{dy}{dx}\right)^2}\,dx$$

$$= 2\pi\int_0^\pi \sin x\sqrt{1+\cos^2 x}\,dx$$ Put $t = \cos x$ $\therefore$ $dt = -\sin x\,dx$

$$= -2\pi \int_{1}^{-1} \sqrt{1+t^2}\, dt = 2\pi \int_{-1}^{1} \sqrt{1+t^2}\, dt = 2\pi \left[\frac{t\sqrt{1+t^2}}{2} + \frac{1}{2}\sinh^{-1}(t)\right]_{-1}^{1}$$

$$= \pi\, \{[\sqrt{2} + \sinh^{-1}(1)] - [-\sqrt{2} - \sinh^{-1}(1)]\} = 2\pi\, [\sqrt{2} + \sinh^{-1}(1)]$$

***Example* 6.45.** Find the surface of the solid generated by revolving the astroid $x = a\cos^3 t$, $y = a\sin^3 t$ about the axis of x.

Solution. Let the shaded portion OAB be revolved above x-axis.

$$\frac{ds}{dt} = \sqrt{\left(\frac{dx}{dt}\right)^2 + \left(\frac{dy}{dt}\right)^2}$$

$$= \sqrt{(-3a\cos^2 t\sin t)^2 + (3a\sin^2 t\cos t)^2}$$

$$= \sqrt{9a^2\cos^2 t\sin^2 t \times (\cos^2 t + \sin^2 t)}$$

$$= 3a\cos t\sin t.$$

Fig. 6.13

Surface area
$$S = 2\int_0^{\pi/2} 2\pi\, y\frac{ds}{dt}\, dt = 4\pi \int_0^{\pi/2} a\sin^3 t \times 3a\cos t\sin t\, dt$$

$$= 12\pi a^2 \int_0^{\pi/2} \sin^4 t\cos t\, dt = 12\pi a^2 \left[\frac{\sin^5 t}{5}\right]_0^{\pi/2} = \frac{12}{5}\pi a^2$$

***Example* 6.46.** A quadrant of a circle of radius a bounded by the tangents at its extremities revolves about one of the tangents. Show that the surfaces area so generated is $\pi(\pi - 2)\, a^2$.

Solution. Let the equation of the circle be

$$x^2 + y^2 = a^2 \text{ or } x = a\cos t,\ y = a\sin t \text{ and } P(x, y) \text{ be any point on it}$$

$$\text{PM} = \text{NA} = a - x = a - a\cos t = a\,(1 - \cos t)$$

$$\frac{ds}{dt} = \sqrt{\left(\frac{dx}{dt}\right)^2 + \left(\frac{dy}{dt}\right)^2} = \sqrt{a^2\sin^2 t + a^2\cos^2 t} = a$$

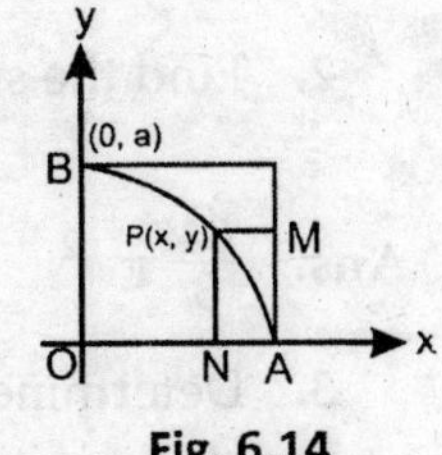

Fig. 6.14

Surface area
$$S = \int_0^{\pi/2} 2\pi\,(PM)^2 \times \frac{ds}{dt} \times dt = 2\pi a^2 \int_0^{\pi/2} (1 - \cos t)\, dt$$

$$= 2\pi a^2 \int_0^{\pi/2} (1 - \cos t)\, dts = 2\pi a^2\, [t - \sin t]_0^{\pi/2}$$

$$= 2\pi a^2 \left[\frac{\pi}{2} - 1\right] = \pi a^2\, (\pi - 2)$$

***Example* 6.47.** Find the surface of the solid formed by revolving the cardiod $r = a(1 + \cos\theta)$ about the initial line.

Solution. The cardiod is symmetrical about the initial line and for its upper half, θ varies from 0 to π.

Also
$$\frac{ds}{d\theta} = \sqrt{r^2 + \left(\frac{dr}{d\theta}\right)^2} = \sqrt{a^2(1+\cos\theta)^2 + a^2\sin^2\theta}$$

$$= a\sqrt{2(1+\cos\theta)} = a\sqrt{4\cos^2\theta/2} = 2a\cos\frac{\theta}{2}$$

$\therefore$ Required surface
$$= \int_0^{\pi} 2\pi\, y\frac{ds}{d\theta}\, d\theta = 2\pi \int_0^{\pi} r\sin\theta \times 2a\cos\frac{\theta}{2}\, d\theta \qquad \begin{cases} x = r\cos\theta \\ y = r\sin\theta \end{cases}$$

$$= 4\pi a \int_0^{\pi} a(1+\cos\theta)\sin\theta\cos\frac{\theta}{2}\,d\theta$$

$$= 4\pi a^2 \int_0^{\pi} 2\cos^2\frac{\theta}{2} \times 2\sin\frac{\theta}{2}\cos\frac{\theta}{2}\cos\frac{\theta}{2}\,d\theta$$

$$= 16\pi a^2 \int_0^{\pi} \cos^4\frac{\theta}{2}\sin\frac{\theta}{2}\,d\theta = 16\pi a^2(-2)\int_0^{\pi}\cos^4\frac{\theta}{2}\left(-\sin\frac{\theta}{2}\times\frac{1}{2}\right)d\theta$$

$$= -32\pi a^2\left[\frac{\cos^5\theta/2}{5}\right]_0^{\pi} = -\frac{32\pi a^2}{5}(0-1) = \frac{32\pi a^2}{5}$$

***Example* 6.48.** Find the surface of the sphere of radius 'a', equation of the circle being $r = a$.

Solution. Equation of the circle is $r = a$. Let the shaded portion be revolved about x-axis.

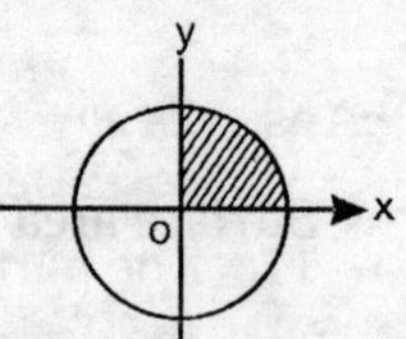

Fig. 6.15

$$\frac{ds}{d\theta} = \sqrt{r^2 + \left(\frac{dr}{d\theta}\right)^2} = \sqrt{a^2+0} = a$$

$$S = 2\int_0^{\pi/2} 2\pi r\sin\theta\frac{ds}{d\theta}\,d\theta = 4\pi\int_0^{\pi/2} a\sin\theta\times a\,d\theta$$

$$= 4\pi\int_0^{\pi/2} a^2\sin\theta\,d\theta = 4\pi a^2\,[-\cos\theta]_0^{\pi/2} = 4\pi a^2$$

EXERCISE 6.1

1. Find the surfaces of the solid generated by the revolution of astroid $x^{2/3} + y^{2/3}$ about x-axis.

Ans. $\frac{12}{5}\pi a^2$

2. Find the surface area of the solid generated by revolving the cycloid
$x = a(\theta - \sin\theta),\ y = a(1-\cos\theta)$ about x-axis.

Ans. $\frac{64}{3}\pi a^2$

3. Determine the area of the surface generated by the revolution of the loops of the curve $r^2 = a^2\cos 2\theta$ about initial line.

Ans. $2\pi a^2(2-\sqrt{2})$

4. Find the volume of the solid formed by the revolution of the curve $y^2(a+x) = x^2(a-x)$ about x-axis.

Ans. $2\pi a^3\left[\log 2 - \frac{2}{3}\right]$

5. The arc lying between $\theta = -\frac{\pi}{2}$ and $\theta = \frac{\pi}{2}$ of the curve $x = a\cos^3\theta,\ y = a\sin^3\theta$ rotates about the axis of x. Find the volume of the solid so generated.

Ans. $\frac{16\pi a^2}{105}$

6. Find the volume generated by revolving the area bounded by the curve $\sqrt{x} + \sqrt{y} = \sqrt{a},\ x = 0,\ y = 0$ about x-axis.

Ans. $\frac{\pi a^3}{12}$

7. Find the volume formed by the revolution of the curve $27ay^2 = 4(x - 2a)^2$ about x-axis.

Ans. $48\pi a^3$

8. Area bounded by x-axis, $y^2 = 4ax$ and the ordinate $x = 3a$ is revolved about x-axis. Find the volume generated.

Ans. $18\pi a^3$

9. Show that the volume of the solid of generated by revolving the area included between the curves $y^2 = x^3$, $x^2 = y^3$ about x-axis is $\frac{5\pi}{28}$.

10. Find the surface and volume of the ellipsoid formed by the revolution of ellipse $\frac{x^2}{a^2} + \frac{y^2}{b^2} = 1$ round its major axis respectively.

Ans. $2\pi ab = \sqrt{1-e^2} + e^{-1}\sin^{-1} e$ and $\frac{4}{3}\pi ab^2$.

11. The part of the parabola $y^2 = 4ax$ cut off by the latus rectum revolves about the tangent at the vertex. Find the volume of the reel thus generated.

Ans. $\frac{4}{5}\pi a^3$

12. The part of the parabola $y^2 = 4ax$ cut off by the latus rectum revolves about the tangent at the vertex. Find the area of the surface so generated.

Ans. $\pi a^2 [3\sqrt{2} + \log(\sqrt{2} - 1)]$

13. For the curve $r^2 = a^2 \cos 2\theta$, prove that the volume of revolution about the initial line is $\frac{\pi a^2}{6\sqrt{2}}[3\log(\sqrt{2}+1) - \sqrt{2}]$

14. Find the volume of the solid generated by revolving the area included between the curve $\frac{y+8}{x} = x - 2$ and x-axis about the line $x + 5 = 0$.

Ans. 432π

15. Find the surface area of the solid formed by the revolution of the loop of the curve given by $3ay^2 = x(x-a)^2$.

Ans. $\frac{4a}{\sqrt{3}}$

16. Find the volume and surface of the solid generated by the revolution of the loop of the curve $x = t^2$, $y = \frac{t^3}{3}$ about x-axis.

Ans. $\frac{3\pi}{4}$, 3π

17. A quadrant of a circle of radius 'a' revolves about its chord. Show that the volume of the spindle generated is $\frac{\pi a^3}{6\sqrt{2}}(10 - 3\pi)$.

6.18 MULTIPLE INTEGRALS AND THEIR APPLICATIONS

DOUBLE INTEGRALS: As already known, the definite integral $\int_a^b f(x)$ is defined as

$$\int_a^b f(x)dx = \underset{\substack{x\to\infty\\ \delta x\to 0}}{Lt}\ [f(x_1)\delta x_1 + f(x_2)\delta x_2 + f(x_3)\delta x_3 + \ldots + f(x_n)\delta x_n]$$

Consider a function $f(x, y)$ of the independent variables x and y defined at each point of the region S of xy plane. Divide the region into elementary areas

$\delta A_1, \delta A_2, \delta A_3, \ldots \delta A_n$

Thus
$$\iint_S f(x, y)dA = \underset{\substack{x\to\infty\\ \delta x\to 0}}{Lt}\ [f(x_1, y_1)\delta A_1 + f(x_2, y_2)\delta A_2 + f(x_3, y_3)\delta A_3 + \ldots + f(x_n, y_n)\delta A_n]$$

6.19 EVALUATION OF DOUBLE INTEGRALS

There are two other methods of evaluating double integrals.

(1) If A is described as $f_1(x) \le y < f_2(x) (y_1 \le y \le y_2)$ and $a \le x \le b$.

Then
$$\iint_A f(x, y)dA = \int_a^b \boxed{\int_{y_1}^{y_2} f(x, y)\,dy\,dx}$$

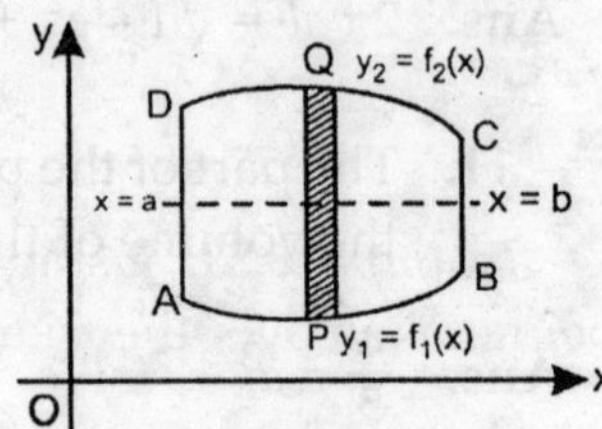

Fig. 6.16

$f(x, y)$ is first integrated with respect to y (assuming x as constant) between the limits y_1 and y_2 and integrating the result w.r.t. x between the limits a and b.

In the region we consider a strip of area $\delta x \delta y$. Then integrating w.r.t. y from y_1 to y_2 converts small rectangle $\delta x\ \delta y$ to a strip $PQ(y\ \delta x)$, while the integration of the result w.r.t. x. corresponds to sliding the strip PQ from AD to BC covering the whole region $ABCD$.

Second Method

$$\iint_A f(x, y)dA = \int_c^d \boxed{\int_{x_1}^{x_2} f(x, y)\,dx\,dy}$$

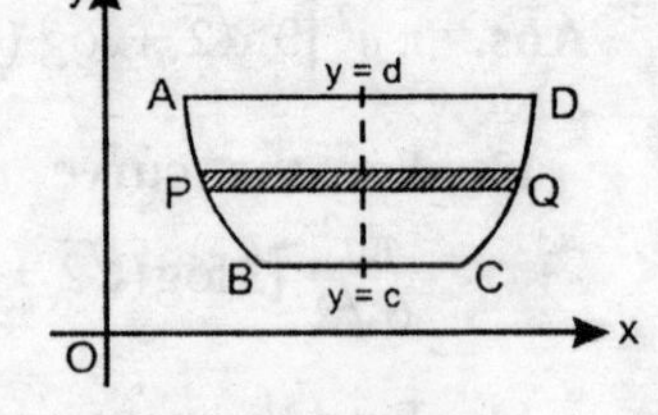

Fig. 6.17

Here $f(x, y)$ is first integrated w.r.t. from x_1 to x_2 and then sliding the strip PQ from BC to AD covering the whole area $ABCD$.

***Example* 6.49.** Evaluate $\iint xy\ dx\ dy$ over the region in the positive quadrant for which $x + y \le 1$.

Solution. $x + y = 1$ represents a line AB, and $x + y < 1$ represents a plane OAB.

Consider a strip parallel to y-axis, Q lies on x-axis and P lies on the line AB. The limits for y are 0 to $1 - x$. The strip PQ is then slided || to y-axis from $x = 0$ to $x = 1$ cover the whole region

Fig. 6.18

$$\text{Required integral} = \int_0^1 x\,dx \int_0^{1-x} y\,dy = \int_0^1 x\,dx\left(\frac{y^2}{2}\right)_0^{1-x}$$

$$= \frac{1}{2}\int_0^1 x(1-x)^2 dx$$

$$= \frac{1}{2}\int_0^1 (x - 2x^2 + x^3)dx = \frac{1}{2}\left[\frac{x^2}{2} - \frac{2x^3}{3} + \frac{x^4}{4}\right]_0^1 = \frac{1}{24}$$

***Example* 6.50.** Evaluate $\iint_S xy\,dx\,dy$ where S is the domain bounded by x-axis, ordinate $x = 2a$ and the curve $x^2 = 4ay$

Solution. The area of integration is OAB which is bounded by x-axis, ordinate $x = 2a$ and the curve $x^2 = 4ay$.

Consider a strip $PQ \parallel$ to y-axis meeting x-axis at P and the curve at Q. Integrating first w.r.t. y, from the limits $y = 0$ to $y = \frac{x^2}{4a}$ and then sliding $PQ \parallel$ to y-axis from $x = 0$ to $x = 2a$ to cover the whole region.

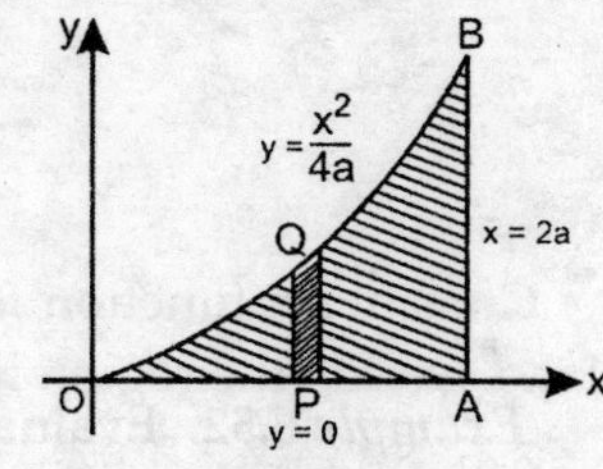

Fig. 6.19

Hence
$$\iint_S xy\,dx\,dy = \int_0^{2a}\int_0^{x^2/4a} xy\,dx\,dy = \int_0^{2a} x\left(\frac{y^2}{2}\right)_0^{\frac{x^2}{4a}} dx$$

$$= \int_0^{2a} x\frac{x^4}{16a^2}\times\frac{1}{2}dx = \int_0^{2a}\frac{x^5}{32a^2}dx$$

$$= \frac{1}{32a^2}\left(\frac{x^6}{6}\right)_0^{2a} = \frac{64a^6}{192a^2} = \frac{a^4}{3}$$

***Example* 6.51.** Using transformation $x + y = u;\ y = uv$, show that $\iint [xy(1 - x - y)]^{\frac{1}{2}}\,dx\,dy = \frac{2\pi}{105}$, the integration being taken over the triangle $x = 0,\ y = 0,\ x + y = 1$.

Solution. Integral $= [xy(1 - x - y)]^{\frac{1}{2}}\,dx\,dy$

We have $x + y = u \quad y = uv \quad \therefore x = u - y = u - uv = u(1 - v)$

Now
$$dx\,dy = \frac{\partial(x, y)}{\partial(u, v)}du\,dv = \begin{vmatrix}\frac{\partial x}{\partial u} & \frac{\partial x}{\partial v}\\ \frac{\partial y}{\partial u} & \frac{\partial y}{\partial v}\end{vmatrix} du\,dv$$

$$dx\,dy = \begin{vmatrix}1 - v & -u\\ v & u\end{vmatrix} du\,dv = u du\,dv$$

$$u = x + y = 1$$

$$u = 1,\ x + uv = u$$

$$\therefore \quad x = u(1 - v) \quad y = uv, \quad x = 0, \quad y = 0$$

$$\Rightarrow \quad 0 = u(1 - v) \quad uv = 0$$

$$u = 0,\ v = 1 \quad u = 1, \quad v = 0$$

Hence u varies from 0 to 1 and v varies from 0 and 1

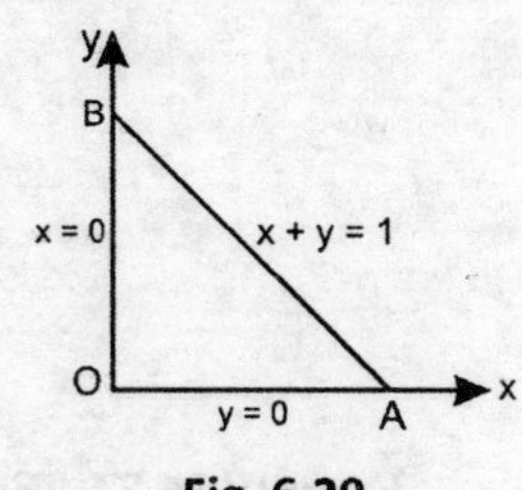

Fig. 6.20

Now
$$\iint [xy(1 - x - y)]^{\frac{1}{2}}\,dx\,dy = \iint (u - uv)^{\frac{1}{2}}(uv)^{\frac{1}{2}}(1 - u)^{\frac{1}{2}}\,u\,du\,dv$$

$$= \int_0^1 u^2(1 - u)^{\frac{1}{2}}du\int_0^1 v^{\frac{1}{2}}(1 - v)^{\frac{1}{2}}dv$$

$$= \frac{\Gamma 3\,\Gamma\frac{3}{2}}{\Gamma\frac{9}{2}}\cdot\frac{\Gamma\frac{3}{2}\Gamma\frac{3}{2}}{\Gamma 3}\left\{\int_0^1 x^{m-1}(1 - x)^{n-1} = \frac{\Gamma m\Gamma n}{\Gamma m + n}\right\}$$

$$= \frac{2\Gamma\frac{3}{2}}{\frac{7}{2}\times\frac{5}{2}\times\frac{3}{2}\Gamma\frac{3}{2}} \times \frac{\frac{1}{2}\Gamma\frac{1}{2}\times\frac{1}{2}\Gamma\frac{1}{2}}{2} \qquad \Gamma 3 = 2! = 2$$

$$= \frac{16}{7\times5\times3} \times \frac{\frac{1}{2}\times\sqrt{\pi}\times\frac{1}{2}\sqrt{\pi}}{2} = \frac{2}{7\times5\times3}\,\pi = \frac{2\pi}{105}$$

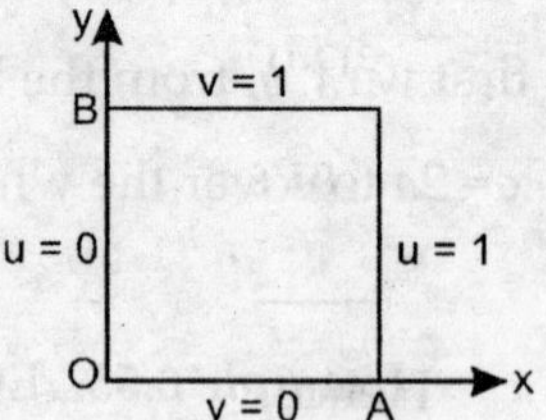

Fig. 6.21

Example **6.52.** Evaluate $\int_0^1 \int_0^{\sqrt{1+x^2}} \frac{dydx}{1+x^2+y^2}$

Solution.

$$\int_0^1 \int_0^{\sqrt{1+x^2}} \frac{dydx}{(1+x^2)+y^2} = \int_0^1 \left[\frac{1}{\sqrt{1+x^2}}\tan^{-1}\frac{y}{\sqrt{1+x^2}}\right]_0^{\sqrt{1+x^2}}$$

$$= \int_0^1 \frac{1}{\sqrt{1+x^2}}\{\tan^{-1}1 - \tan^{-1}0\}\,dx = \int_0^1 \frac{1}{\sqrt{1+x^2}} \times \frac{\pi}{4}dx$$

$$= \frac{\pi}{4}\int_0^1 \frac{1}{\sqrt{1+x^2}}dx = \frac{\pi}{4}\,[\log(x+\sqrt{1+x^2})]_0^1$$

$$= \frac{\pi}{4}\,[\log(1+\sqrt{2}) - \log 1] = \frac{\pi}{4}\,[\log(\sqrt{2}+1)]$$

Example **6.53.** Evaluate $\iint_R y\,dx\,dy$ where R is the region bounded by the parabola $y^2 = 4x$ and $x^2 = 4y$.

Solution. Solving $y^2 = 4x$, $x^2 = 4y$ we get

$$x = 0, 4$$

$$y = 0, 4$$

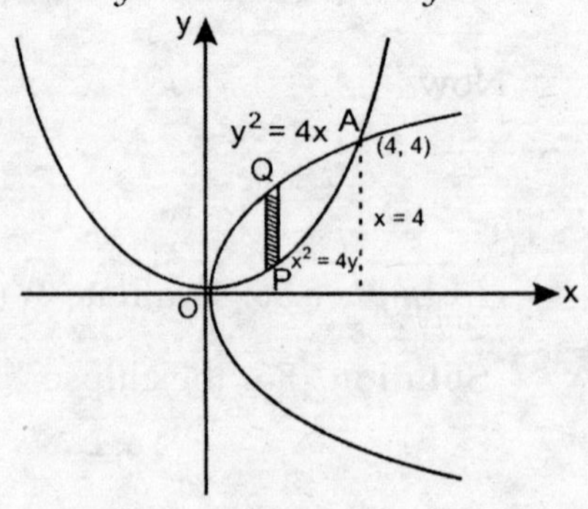

Fig. 6.22

Co-ordinates of A are (4, 4)

The region R is expressed as $0 \le x \le 4,\ \frac{x^2}{4} \le y \le 2\sqrt{x}$

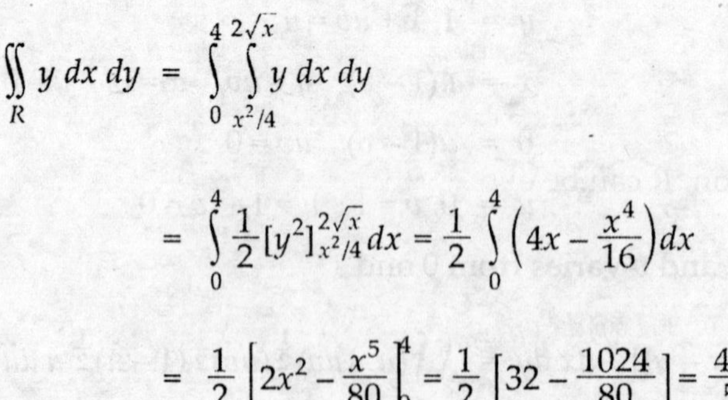

$$\therefore \qquad \iint_R y\,dx\,dy = \int_0^4 \int_{x^2/4}^{2\sqrt{x}} y\,dx\,dy$$

$$= \int_0^4 \frac{1}{2}[y^2]_{x^2/4}^{2\sqrt{x}}dx = \frac{1}{2}\int_0^4 \left(4x - \frac{x^4}{16}\right)dx$$

$$= \frac{1}{2}\left[2x^2 - \frac{x^5}{80}\right]_0^4 = \frac{1}{2}\left[32 - \frac{1024}{80}\right] = \frac{48}{5}$$

Example **6.54.** Evaluate $\int_0^1 \int_0^{x^2} e^{y/x}\,dy\,dx$

Solution.

$$\int_0^1 \int_0^{x^2} e^{y/x}\,dy\,dx = \int_0^1 \left[\frac{e^{y/x}}{1/x}\right]_0^{x^2} dx$$

$$= \int_0^1 (xe^{y/x})_0^{x^2} dx = \int_0^1 (x e^x - x)\,dx = \int_0^1 xe^x - \int_0^1 x\,dx$$

$$= (xe^x)_0^1 - \int_0^1 e^x dx - \int_0^1 x\,dx = (e-0) - (e^x)_0^1 - \left(\frac{x^2}{2}\right)_0^1$$

$$= e-(e-1) - \frac{1}{2} = 1 - \frac{1}{2} = \frac{1}{2}$$

***Example* 6.55.** Evaluate $\iint_S \sqrt{xy - y^2}\, dx\, dy$ where S is the triangle with vertices (0, 0), (10, 1) and (1, 1)

Solution. Equation of OB is $y = x$

Equation of OA is $\quad y = \frac{1}{10}x$

Taking a strip $PQ \parallel$ to x-axis when P lies on $y = x$ and Q lies on $y = \frac{1}{10}x$. The region of triangle OAB is given by $y \le x \le 10y$ and $0 \le y \le 1$

Fig. 6.23

$$\therefore \quad \iint_S \sqrt{xy - y^2}\, dx\, dy = \int_0^1 dy \int_y^{10y} (xy - y^2)^{1/2} dx$$

$$= \int_0^1 dy \left[\frac{2}{3} \times \frac{1}{y}(xy - y^2)^{\frac{3}{2}}\right]_y^{10y}$$

$$= \int_0^1 \frac{2}{3} \times \frac{1}{y}(9y^2)^{\frac{3}{2}} dy = 18 \int_0^1 y^2 dy = 18\left[\frac{y^3}{3}\right]_0^1 = \frac{18}{3} = 6$$

***Example* 6.56.** Evaluate $\iint (x+y)^2 dx\, dy$ over the area bounded by the ellipse $\frac{x^2}{a^2} + \frac{y^2}{b^2} = 1$.

Solution. For the ellipse

$$\frac{y}{b} = \pm\sqrt{1 - \frac{x^2}{a^2}} \text{ or } y = \pm \frac{b}{a}\sqrt{a^2 - x^2}$$

$\therefore$ The region of integration, R can be expressed as $-a \le x \le a$, $-\frac{b}{a}\sqrt{a^2 - x^2} \le y \le \frac{b}{a}\sqrt{a^2 - x^2}$

$$\iint_R (x+y)^2 dx\, dy = \iint_R (x^2 + y^2 + 2xy)\, dx\, dy = \int_{-a}^{a} \int_{-b/a\sqrt{a^2-x^2}}^{b/a\sqrt{a^2-x^2}} (x^2 + y^2 + 2xy)\, dx\, dy$$

$$= \int_{-a}^{a} \int_{-b/a\sqrt{a^2-x^2}}^{b/a\sqrt{a^2-x^2}} (x^2 + y^2)\, dy\, dx + \int_{-a}^{a} \int_{-b/a\sqrt{a^2-x^2}}^{b/a\sqrt{a^2-x^2}} 2xy\, dx\, dy$$

$$= \int_{-a}^{a} \int_{0}^{b/a\sqrt{a^2-x^2}} 2(x^2 + y^2)\, dy\, dx + 0$$

Since $x^2 + y^2$ is an even function of y, $2xy$ is an odd function of y

$$= \int_{-a}^{a}\left[2\left(x^2 y + \frac{y^3}{3}\right)\right]_0^{b/a\sqrt{a^2-x^2}} dx$$

$$= 2\int_{-a}^{a}\left[x^2 \frac{b}{a}\sqrt{a^2-x^2} + \frac{1}{3}\frac{b^3}{a^3}(a^2-x^2)^{3/2}\right]dx$$

$$= 4\int_{0}^{a}\left[\frac{b}{a}x^2\sqrt{a^2-x^2} + \frac{b^3}{3a^3}(a^2-x^2)^{3/2}\right]dx \text{ Put } x = a\sin\theta \quad dx = a\cos\theta\, d\theta$$

$$= 4\int_{0}^{\pi/2}\left(\frac{b}{a}a^2\sin^2\theta \times a\cos\theta + \frac{b^3}{3a^3} \times a^3\cos^3\theta\right) \times a\cos\theta\, d\theta$$

$$= 4\int_{0}^{\pi/2}\left[a^3 b\sin^2\theta\cos^2\theta + \frac{ab^3}{3}\cos^4\theta\right]d\theta$$

$$= 4\left[a^3 b \times \frac{1\times 1}{4\times 2} \times \frac{\pi}{2} + \frac{ab^3}{3}\frac{3\times 1}{4\times 2} \times \frac{\pi}{2}\right] = \frac{\pi}{4}[a^3 b + ab^3] = \frac{\pi}{4}ab[a^2+b^2]$$

Example **6.57.** Evaluate $\int_0^{2a}\int_0^{\sqrt{2ax-x^2}}(x^2+y^2)\,dx\,dy$

Solution. Here $y = 0$, $y = \sqrt{2ax - x^2}$ or $x^2 + y^2 - 2ax = 0$

$$I = \int_0^{2a}\int_0^{\sqrt{2ax-x^2}}(x^2+y^2)\,dx\,dy$$

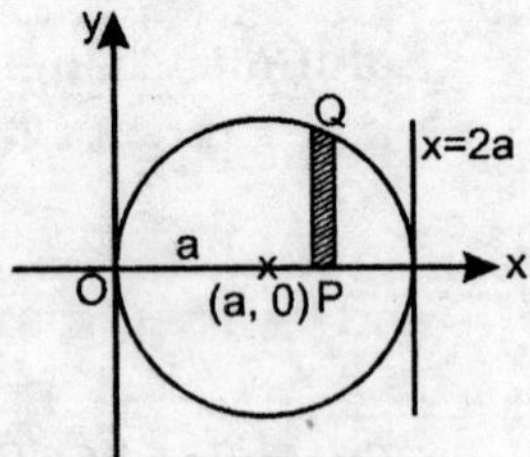

Fig. 6.24

Region of integration is upper half of the circle $x^2 + y^2 - 2ax = 0$

Let us convert Cartesian to polar co-ordinates by putting $x = r\cos\theta$, $y = r\sin\theta$

$x^2 + y^2 - 2ax = 0 \quad r^2 - 2\,ra\cos\theta \quad r = 2\,a\cos\theta$

Limits of r are from 0 to $2a\cos\theta$

Limits of θ are from 0 to $\pi/2$

$$I = \int_0^{2a}\int_0^{\sqrt{2ax-x^2}}(x^2+y^2)\,dy\,dx$$

$$= \int_0^{\pi/2}\int_0^{2a\cos\theta} r^2 \times r\,d\theta\,dr$$

$$= \int_0^{\pi/2} d\theta \int_0^{2a\cos\theta} r^3 dr = \int_0^{\pi/2} d\theta\left[\frac{r^4}{4}\right]_0^{2a\cos\theta}$$

$$= 4\int_0^{\pi/2} a^4\cos^4\theta\,d\theta = 4\,\frac{3\times 1\times\pi}{4\times 2\times 2}a^4 = \frac{3\pi}{4}a^4$$

Fig. 6.25

Example **6.58.** Evaluate the integral $\iint xy(x^2+y^2)^{n/2}\,dx\,dy$ over the positive quadrant of $x^2+y^2=4$ (changing to polar co-ordinates).

Solution. The given integral $I=\iint xy(x^2+y^2)^{n/2}\,dx\,dy$ ...(1)

On putting $x=r\cos\theta$, $y=r\sin\theta$, $x^2+y^2=r^2$ and $dx\,dy=r\,dr\,d\theta$ in (1), we get

$$I = \iint (r\cos\theta)(r\sin\theta)r^n \times r\,dr\,d\theta$$

$$= \int_0^{\pi/2}\int_0^{2} r^{n+3}\cos\theta\sin\theta\,d\theta\,dr$$

$$= \int_0^{2} r^{n+3}dr \int_0^{\pi/2}\cos\theta\sin\theta\,d\theta$$

$$= \left[\left(\frac{r^{n+4}}{n+4}\right)\right]_0^2 \int_0^{\pi/2}\frac{\sin 2\theta}{2}\,d\theta$$

$$= \left(\frac{2^{n+4}}{n+4}\right)\left(-\frac{\cos 2\theta}{4}\right)_0^{\pi/2} = \frac{2^{n+4}}{n+4}\left(-\frac{\cos\pi-1}{4}\right) = \frac{2^{n+4}}{n+4}\,\frac{1}{2}.$$

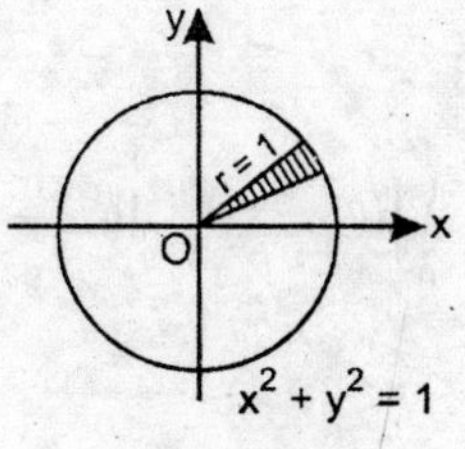

Fig. 6.26

Example **6.59.** Evaluate the integral $\iint \frac{1-x^2-y^2}{1+x^2+y^2}\,dx\,dy$ over the + *ve* quadrant of the circle $x^2+y^2=1$.

Solution. Changing to polar coordinate by putting $x=r\cos\theta$, $y=r\sin\theta$.

The circle $x^2+y^2=1$ changes to

$$r^2 = 1 \quad r=1$$

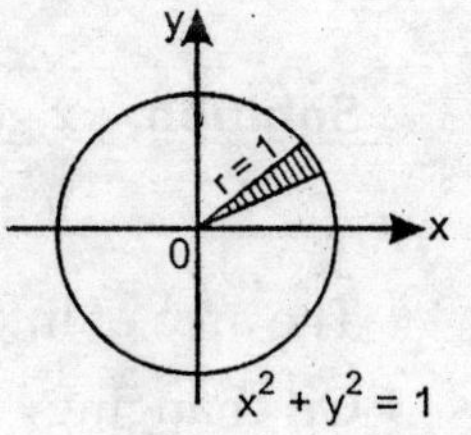

and θ varies from 0 to $\pi/2$ in the + ve quadrants of the circle, the integral becomes

$$= \int_0^{\pi/2}\int_0^1 \frac{1-r^2}{1+r^2}r\,dr\,d\theta = \int_0^{\pi/2}\int_0^1 \frac{(1+r^2)-2r^2}{1+r^2}r\,dr\,d\theta$$

$$= \int_0^{\pi/2}\int_0^1\left(1-\frac{2r^2}{1+r^2}\right)r\,dr\,d\theta = \int_0^{\pi/2}\int_0^1\left(-1+\frac{2}{r^2+1}\right)r\,dr\,d\theta$$

$$= \int_0^{\pi/2}\left[-\frac{r^2}{2}+\log(r^2+1)\right]_0^1 d\theta$$

$$= \int_0^{\pi/2}\left(-\frac{1}{2}+\log 2\right)d\theta = \left(-\frac{1}{2}+\log 2\right)\frac{\pi}{2} = -\frac{\pi}{4}+\frac{\pi}{2}\log 2$$

Fig. 6.27

Example **6.60.** Evaluate $\iint \sqrt{(a^2-x^2-y^2)}\,dx\,dy$ over the semi-circle $x^2+y^2=ax$ in the + ve quadrant.

Solution. Let us use polar co-ordinate $x=r\cos\theta$; $y=r\sin\theta$

The circle $x^2+y^2=ax$ becomes

$$r^2 = ar\cos\theta$$

$$r = a\cos\theta$$

$\therefore$ r goes from 0 to $a \cos \theta$ goes from 0 to $\frac{\pi}{2}$

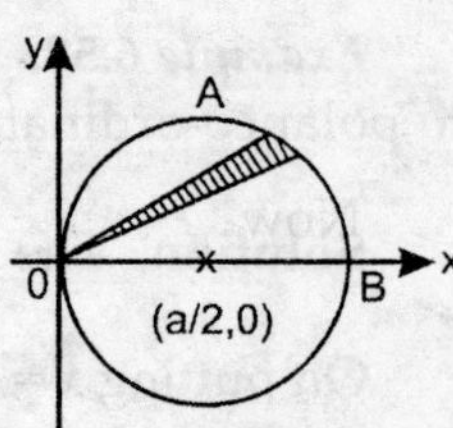

Fig. 6.28

The integral becomes

$$= \int_0^{\pi/2} \int_0^{a\cos\theta} \sqrt{a^2 - r^2}\, r\, dr\, d\theta$$

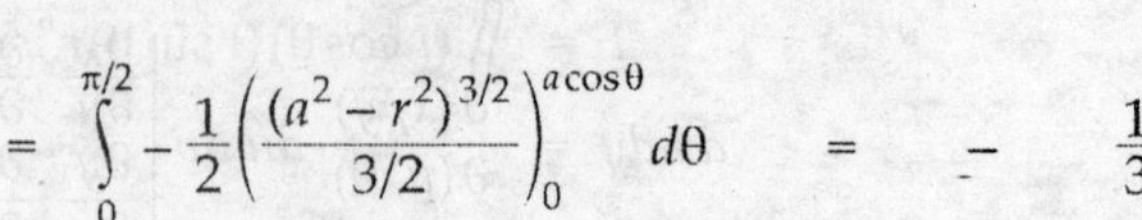

$$= \int_0^{\pi/2} -\frac{1}{2}\left(\frac{(a^2-r^2)^{3/2}}{3/2}\right)_0^{a\cos\theta} d\theta \quad = \quad -\frac{1}{3}$$

$$\int_0^{\pi/2} [(a^2 - r^2)^{3/2}]_0^{a\cos\theta}$$

$$= -\frac{1}{3}\int_0^{\pi/2} [(a^2 - a^2\cos^2\theta)^{3/2} - a^3]\, d\theta$$

$$= -\frac{1}{3}\, a^3 \int_0^{\pi/2} [(1-\cos^2\theta)^{3/2} - 1]\, d\theta = -3\,\frac{1}{3}\, a^3 \int_0^{\pi/2} (\sin^3\theta - 1)\, d\theta$$

$$= -\frac{a^3}{3}\left[\frac{2}{3} - \frac{\pi}{2}\right] = -\frac{a^3}{3}\left[\frac{2}{3} - \frac{\pi}{2}\right]$$

Example **6.61.** Evaluate $\int_0^{\infty}\int_0^{\infty} e^{-(x^2+y^2)}\, dx\, dy$ by changing to polar coordinates.

Solution. x goes from $x = 0$ to $x = \infty$

y goes from $y = 0$ to $y = \infty$

Fig. 6.29

The area of integration is the whole area bounded by x-axis and y-axis

On changing

θ goes from $\quad \theta = 0$ to $\theta = \frac{\pi}{2}$

r goes from $\quad r = 0$ to $r = \infty$

$$\int_0^{\pi/2}\int_0^{\infty} e^{-r^2} r\, dr\, d\theta = -\frac{1}{2}\int_0^{\pi/2}\int_0^{\infty} (-2re^{-r^2})\, d\theta$$

$$= -\frac{1}{2}\int_0^{\pi/2} (e^{-r^2})_0^{\infty} = -\frac{1}{2}\int_0^{\pi/2} (0-1)\, d\theta = \frac{1}{2}\,(\theta)_0^{\pi/2} = \frac{\pi}{4}$$

Example **6.62.** Evaluate the integral $\iint_R (x+y)^2\, dx\, dy$ where R is the parallelogram in x-y plane using the transformation $u = x + y$, $v = x - 2y$.

Solution. Changing the co-ordinate to u and v

$$A(1, 0) \rightarrow A'(1, 1)$$

$$B(3, 1) \rightarrow B'(4, 1)$$

$$C(2, 2) \to C'(4, -2)$$

$$D(0, 1) \to D'(1, -2)$$

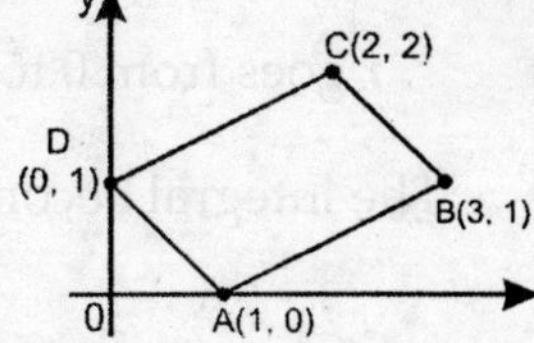

Fig. 6.30

Now $\quad 3y = u - v, \quad 3x = 2u + v$

$\therefore \quad y = \dfrac{u - v}{3}, x = \dfrac{2u + v}{3}$

Then
$$dx\,dy = \frac{\partial(x, y)}{\partial(u, v)} du\,dv \begin{vmatrix} \frac{\partial x}{\partial u} & \frac{\partial x}{\partial v} \\ \frac{\partial y}{\partial u} & \frac{\partial y}{\partial v} \end{vmatrix} du\,dv$$

$$= \begin{vmatrix} \frac{2}{3} & \frac{1}{3} \\ \frac{1}{3} & -\frac{1}{3} \end{vmatrix} du\,dv = -\frac{1}{3}\,du\,dv$$

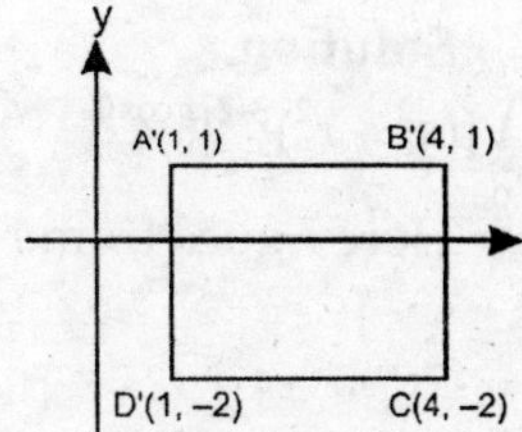

Fig. 6.31

u goes from 1 to 4

v goes from – 2 to 1

Required integral
$$= \int_{-2}^{1}\int_{1}^{4} (x + y)^2 \left(-\frac{1}{3} du\,dv\right)$$

$$= \int_{-2}^{1}\int_{1}^{4} u^2 \left(-\frac{1}{3} du\,dv\right)$$

$$= -\frac{1}{3}\int_{-2}^{1} \left(\frac{u^3}{3}\right)_1^4 dv = -\frac{1}{9}\int_{-2}^{1} (64 - 1)\,dv$$

$$= -\frac{1}{9} \times 63\,(v)_{-2}^{1} = -\frac{1}{9} \times 63 \times 3 = -21$$

6.20 DOUBLE INTEGRATION IN POLAR CO-ORDINATES

To evaluate $\int_{\theta_1}^{\theta_2}\int_{r_1}^{r_2} f(r, \theta)\,dr\,d\theta$, we first integrate w.r.t. r between the limits $r = r_1$ to $r = r_2$ by keeping θ constant and then integrating the result w.r.t. θ from θ_1 to θ_2. In this integral r_1 and r_2 are the function of θ and θ_1, θ_2 are constants.

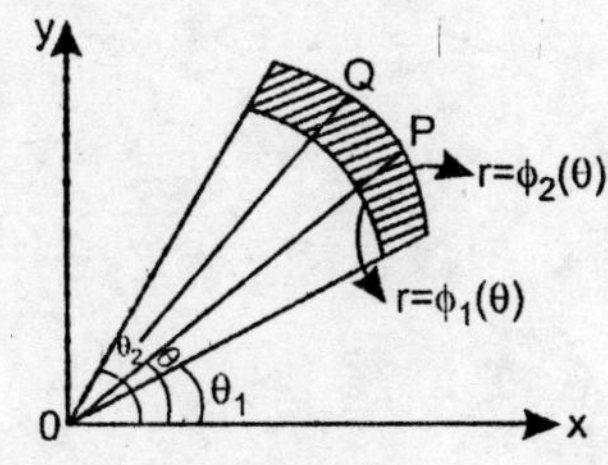

Fig. 6.32

Example **6.63.** Evaluate $\displaystyle\int_{-\pi/2}^{\pi/2}\int_{0}^{2\cos\theta} r^2\,dr\,d\theta$

Solution.
$$I = \int_{-\pi/2}^{\pi/2}\int_{0}^{2\cos\theta} r^2\,dr\,d\theta = \int_{-\pi/2}^{\pi/2} \left(\frac{r^3}{3}\right)_0^{2\cos\theta}$$

$$= \frac{1}{3}\int_{-\pi/2}^{\pi/2} (2\cos)^3\,d\theta = \frac{8}{3}\int_{-\pi/2}^{\pi/2} \cos^3\theta\,d\theta$$

$$= 2 \times \frac{8}{3} \int_0^{\pi/2} \cos^3\theta \, d\theta = \frac{16}{3} \left[\sin\theta - \frac{1}{3}\sin^3\theta\right]_0^{\pi/2}$$

$$= \frac{16}{3}\left[\sin\frac{\pi}{2} - \frac{1}{3}\left(\sin\frac{\pi}{2}\right)^3\right] = \frac{16}{3}\left[1 - \frac{1}{3}\right] = \frac{32}{9}$$

***Example* 6.64.** Evaluate $\iint_R r^2 \sin\theta \, dr \, d\theta$ where R is the semi-circle $r = 2a \cos\theta$ above the initial line.

Solution. $$I = \iint_R r^2 \sin\theta \, dr \, d\theta$$

Here r goes from $r = 0$ to $r = 2a \cos\theta$ and θ goes from $\theta = 0$ to $\theta = \frac{\pi}{2}$

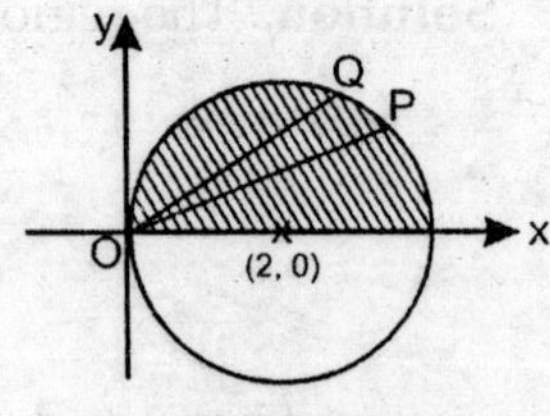

Fig. 6.33

$$I = \int_0^{\pi/2} \int_0^{2a\cos\theta} r^2 \sin\theta \, dr \, d\theta$$

$$\int_0^{\pi/2} \sin\theta \, d\theta \int_0^{2a\cos\theta} r^2 dr = \int_0^{\pi/2} \sin\theta \, d\theta \left(\frac{r^3}{3}\right)_0^{2a\cos\theta}$$

$$= \frac{1}{3}\int_0^{\pi/2} (8a^3\cos^3\theta) \sin\theta \, d\theta$$

$$= \frac{8a^3}{3}\int_0^{\pi/2} \cos^3\theta \sin\theta \, d\theta = \frac{8a^3}{3} \frac{2 \times 1}{4 \times 2} = \frac{2a^3}{3}$$

***Example* 6.65.** Evaluate $\iint_R \frac{r \, dr \, d\theta}{\sqrt{a^2 + r^2}}$ over one loop of the lemniscate $r^2 = a^2 \cos 2\theta$.

Solution. The region of integration is the area bounded by the curve

$$r = 0 \quad r = a\sqrt{\cos 2\theta} \text{ and } \theta = -\frac{\pi}{4}; \theta = \frac{\pi}{4}$$

$$I = \iint_R \frac{r \, dr \, d\theta}{\sqrt{a^2 + r^2}} = \int_{-\pi/4}^{\pi/4} \int_0^{a\sqrt{\cos 2\theta}} \frac{r \, dr \, d\theta}{\sqrt{a^2 + r^2}}$$

Fig. 6.34

$$= \int_{-\pi/4}^{\pi/4} \int_0^{a\sqrt{\cos 2\theta}} \frac{1}{2} \cdot \frac{2r \, dr \, d\theta}{\sqrt{a^2 + r^2}}$$

$$= \int_{-\pi/4}^{\pi/4} \left[\frac{1}{2} \frac{(a^2 + r^2)^{1/2}}{1/2}\right]_0^{a\sqrt{\cos 2\theta}} d\theta$$

$$= \int_{-\pi/4}^{\pi/4} \{(a^2 + a^2\cos 2\theta)^{1/2} - a\} \, d\theta$$

$$= a \int_{-\pi/4}^{\pi/4} \{[1 + \cos 2\theta]^{1/2} - 1\} \, d\theta = a \int_{-\pi/4}^{\pi/4} (\sqrt{2}\cos\theta - 1)$$

$$= 2a \int_0^{\pi/4} (\sqrt{2}\cos\theta - 1)\, d\theta = 2a\,[\sqrt{2}\sin\theta - \theta]_0^{\pi/4}$$

$$= 2a\left[\sqrt{2}\frac{1}{\sqrt{2}} - \frac{\pi}{4}\right] = 2a\left[1 - \frac{\pi}{4}\right]$$

***Example* 6.66.** Evaluate $\iint r^3\, dr\, d\theta$ over the area bounded by the curves

$$r = 2\cos\theta, \quad r = 4\cos\theta, \quad \theta = -\frac{\pi}{2}, \quad \theta = \frac{\pi}{2}$$

Solution. The region of integration is the area bounded by the curves

$$r = 2\cos\theta, \quad r = 4\cos\theta, \text{ and}$$

$$\theta = -\frac{\pi}{2}, \theta = \frac{\pi}{2}$$

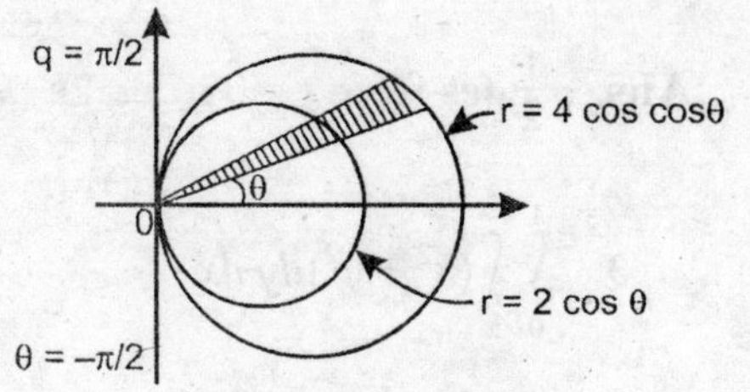

Fig. 6.35

$$I = \iint r^2\, dr\, d\theta = \int_{-\pi/2}^{\pi/2} d\theta \int_{2\cos\theta}^{4\cos\theta} r^3\, dr$$

$$= \int_{-\pi/2}^{\pi/2} \left[\frac{r^4}{4}\right]_{2\cos\theta}^{4\cos\theta} d\theta$$

$$= \frac{1}{4}\int_{-\pi/2}^{\pi/2} (256\cos^4\theta - 16\cos^4\theta)\, d\theta$$

$$= \frac{1}{4} \times 240 \int_{-\pi/2}^{\pi/2} \cos^4\theta\, d\theta = 60 \times 2 \int_0^{\pi/2} \cos^4\theta\, d\theta$$

$$= \frac{3 \times 1}{4 \times 2} \times \frac{\pi}{2} \times 120 = \frac{3\pi}{16} \times 120 = \frac{45\pi}{2}$$

***Example* 6.67.** Find the whole area enclosed by the lamieon $r = a + b\cos\theta\ a > b$.

Solution. The given curve is symmetrical about the initial line

$$\text{Required area} = 2\int_0^{\pi}\int_0^{a+b\cos\theta} r\, dr\, d\theta = 2\int_0^{\pi}\left[\frac{r^2}{2}\right]_0^{a+b\cos\theta} d\theta$$

$$= \int_0^{\pi} (a + b\cos\theta)^2\, d\theta$$

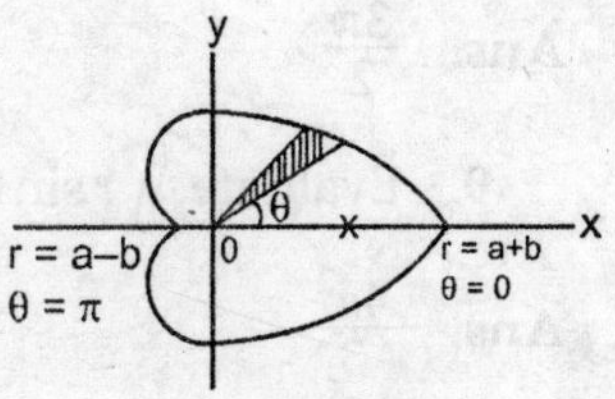

$$= \int_0^{\pi} (a^2 + b^2\cos^2\theta + 2ab\cos\theta)\, d\theta$$

Fig. 6.36

$$= [a^2\theta + 2ab\sin\theta]_0^{\pi} + \frac{b^2}{2}\left[\theta + \frac{\sin 2\theta}{2}\right]_0^{\pi}$$

$$= a^2\pi + \frac{b^2}{2}\left[\pi + \frac{1}{2}\sin 2\pi\right] = a^2\pi + \frac{b^2\pi}{2} = \pi\left(a^2 + \frac{b^2}{2}\right)$$

EXERCISE 6.2

1. $\int_1^2\int_0^x \frac{dxdy}{x^2+y^2}$

Ans. $\frac{\pi}{4}\log 2$

2. $\int_0^1 dx \int_0^x e^{y/x}\, dy$

Ans. $\frac{1}{2}(e-1)$

3. $\int_0^1\int_x^{\sqrt{x}} (x^2+y^2)\, dy\, dx$

Ans. $\frac{3}{35}$

4. Evaluate $\iint (x^2+y^2)\, dx\, dy$ over the region in the positive quadrant for which $x+y \le 1$.

Ans. $\frac{1}{6}$

5. Evaluate $\iint y\, dx\, dy$ where R is the region bounded by the ellipse $\frac{x^2}{a^2}+\frac{y^2}{b^2}=1$

Ans. $\frac{ab^2}{3}$

6. Evaluate $\iint xy(x+y)\, dx\, dy$ over the area between $y=x^2$ and $y=x$.

Ans. $\frac{3}{36}$

7. Find by double integration the smaller of the area bounded by the circle $x^2+y^2=9$ and the region $x+y=3$.

Ans. $\frac{9}{4}(\pi-2)$

8. Find by double integration the area of the positive quadrant of the ellipse $4x^2+ay^2=36$.

Ans. $\frac{3\pi}{2}$

9. Evaluate $\iint r\sin\theta\, dr\, d\theta$ over the cardiod $r=a(1+\cos\theta)$ about the initial line.

Ans. $\frac{4a^2}{3}$

10. Evaluate $\iint_R r^2\sin\theta\, dr\, d\theta$ where R is the region bounded by the semi-circle $r=2a\cos\theta$, above the initial line.

Ans. $\frac{2a^2}{3}$

11. Find the area that is inside the circle $r = a \cos \theta$ and out side the cardiod $r = a(1 - \cos \theta)$.

Ans. $\frac{a^2}{3}[3\sqrt{3} - \pi]$

6.21 CHANGE OF ORDER OF INTEGRATION

In case of double integral with constant limits, it is immaterial whether the limits are changed or not i.e., the order of integration is immaterial. But if it is not so, that is the limits are variable, the order of integration has got a meaning whole evaluating a double integral. It sometimes facilities the evaluation of the integral. On changing the order of integration, the limits are also changed.

Example **6.68.** By changing the order of integration evaluate $\int_0^4 \int_y^4 \frac{x\,dx\,dy}{x^2 + y^2}$

Solution. The region of integration for the given integral is the area bounded by

$$x = y, \quad x = 4, \quad y = 0, \quad y = 4$$

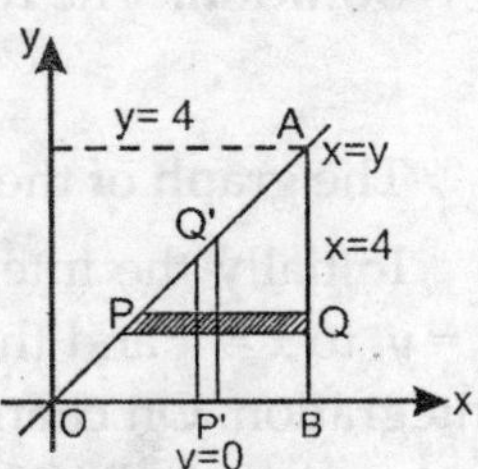

Fig. 6.37

The graph of the region is OAB

Initially the integration is first performed w.r.t. x along a strip PQ || to x-axis from $x = y$, to $x = 4$ and then slided || to x-axis from $y = 0$ to $y = 4$ to cover the whole region of integration. On changing the order of integration the integration is first performed w.r.t. y along a strip $P'Q'$ || to y-axis from $y = 0$ to $y = x$ and then slided || to y-axis from $x = 0$ to $x = 4$.

Thus
$$I = \int_0^4 \int_0^x \frac{x\,dx\,dy}{x^2 + y^2} = \int_0^4 dx \int_0^x \frac{x}{x^2 + y^2}\,dy$$

$$= \int_0^4 x\left[\frac{1}{x}\tan^{-1}\frac{y}{x}\right]_0^y dx = \int_0^4 \frac{\pi}{4}\,dx = \frac{\pi}{4}\int_0^4 dx = \frac{\pi}{4}[x]_0^4 = \pi$$

Example **6.69.** Change the order of integration in the integral and evaluate $\int_0^{4a} \int_{x^2/4a}^{2\sqrt{ax}} dy\,dx$

Solution. The region of integration is bounded by the curves

$$y = \frac{x^2}{4a}, \quad y = 2\sqrt{ax} \quad \text{or} \quad y^2 = 4ax$$

and
$$x = 0, \quad x = 4a$$

The graph of the region of integration is OAO. The point of intersection of the two curve is $A(4a, 4a)$. Initially the integration is first performed w.r.t. y and then w.r.t. x along a strip PQ || to y-axis from $y = \frac{x^2}{4a}$ to $y = 2\sqrt{ax}$ and then slided from $x = 0$ to $x = 4$ || to y-axis.

Fig. 6.38

On changing the order of integration, the integration is first performed along a horizontal strip $P'Q'$ from $x = \frac{y^2}{4a}$ to $x = 2\sqrt{ay}$ and then strip is slided from $y = 0$ to $y = 4a$ || to x-axis to cover the region of integration

Thus
$$I = \int_0^{4a} \int_{y^2/4a}^{2\sqrt{ay}} dy\,dx = \int_0^{4a} [x]_{y^2/4a}^{2\sqrt{ay}} dy = \int_0^{4a} \left(2\sqrt{ay} - \frac{y^2}{4a}\right) dy$$

11. Find the area that is inside the circle $r = a\cos\theta$ and out side the cardiod $r = a(1 - \cos\theta)$.

Ans. $\frac{a^2}{3}[3\sqrt{3} - \pi]$

6.21 CHANGE OF ORDER OF INTEGRATION

In case of double integral with constant limits, it is immaterial whether the limits are changed or not i.e., the order of integration is immaterial. But if it is not so, that is the limits are variable, the order of integration has got a meaning whole evaluating a double integral. It sometimes facilities the evaluation of the integral. On changing the order of integration, the limits are also changed.

Example **6.68.** By changing the order of integration evaluate $\int_0^4 \int_y^4 \frac{x\,dx\,dy}{x^2 + y^2}$

Solution. The region of integration for the given integral is the area bounded by

$$x = y, \quad x = 4, \quad y = 0, \quad y = 4$$

The graph of the region is OAB

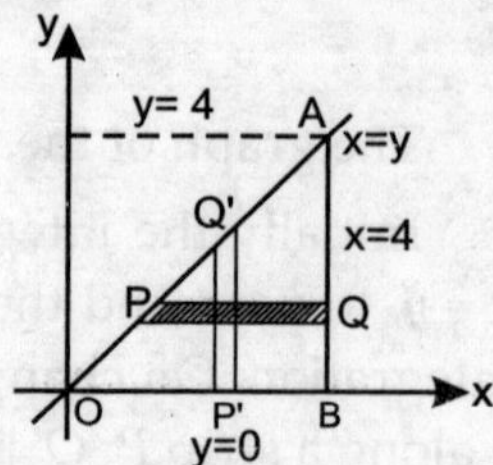

Fig. 6.37.

Initially the integration is first performed w.r.t. x along a strip $PQ \parallel$ to x-axis from $x = y$, to $x = 4$ and then slided $\parallel$ to x-axis from $y = 0$ to $y = 4$ to cover the whole region of integration. On changing the order of integration the integration is first performed w.r.t. y along a strip $P'Q' \parallel$ to y-axis from $y = 0$ to $y = x$ and then slided $\parallel$ to y-axis from $x = 0$ to $x = 4$.

Thus
$$I = \int_0^4 \int_0^x \frac{x\,dx\,dy}{x^2 + y^2} = \int_0^4 dx \int_0^x \frac{x}{x^2 + y^2}\,dy$$

$$= \int_0^4 x\left[\frac{1}{x}\tan^{-1}\frac{y}{x}\right]_0^y dx = \int_0^4 \frac{\pi}{4}\,dx = \frac{\pi}{4}\int_0^4 dx = \frac{\pi}{4}[x]_0^4 = \pi$$

Example **6.69.** Change the order of integration in the integral and evaluate $\int_0^{4a} \int_{x^2/4a}^{2\sqrt{ax}} dy\,dx$

Solution. The region of integration is bounded by the curves

$$y = \frac{x^2}{4a}, \quad y = 2\sqrt{ax} \quad \text{or} \quad y^2 = 4ax$$

and
$$x = 0, \quad x = 4a$$

The graph of the region of integration is OAO. The point of intersection of the two curve is $A(4a, 4a)$. Initially the integration is first performed w.r.t. y and then w.r.t. x along a strip $PQ \parallel$ to y-axis from $y = \frac{x^2}{4a}$ to $y = 2\sqrt{ax}$ and then slided from $x = 0$ to $x = 4 \parallel$ to y-axis.

On changing the order of integration, the integration is first performed along a horizontal strip $P'Q'$ from $x = \frac{y^2}{4a}$ to $x = 2\sqrt{ay}$ and then strip is slided from $y = 0$ to $y = 4a \parallel$ to x-axis to cover the region of integration

Fig. 6.38.

Thus
$$I = \int_0^{4a} \int_{y^2/4a}^{2\sqrt{ay}} dy\,dx = \int_0^{4a} [x]_{y^2/4a}^{2\sqrt{ay}}\,dy = \int_0^{4a}\left(2\sqrt{ay} - \frac{y^2}{4a}\right)dy$$

On changing the order of integration it is first performed w.r.t. y along a vertical strip PQ from $y=0$ to $y=4-x^2$ and then strip if slided || to y-axis from $x=1$ to $x=2$ to cover the whole region.

Thus $$\int_0^1\int_1^{\sqrt{4-v}}(x+y)\,dx\,dy = \int_1^2\int_0^{4-x^2}(x+y)\,dx\,dy$$

$$= \int_1^2 dx\left[xy+\frac{y^2}{2}\right]_0^{4-x^2}$$

$$= \int_1^2 dx\left[x(4-x^2)+\frac{(4-x^2)^2}{2}\right] = \int_1^2\left[(4x-x^3)+\frac{1}{2}(16+x^4-8x^2)\right]dx$$

$$= \left[4\frac{x^2}{2}-\frac{x^4}{4}+8x+\frac{x^5}{10}-\frac{4x^3}{3}\right]_1^2 = \left[2x^2-\frac{x^4}{4}+8x+\frac{x^5}{10}-\frac{4}{3}x^3\right]_1^2$$

$$= \left(8-4+16+\frac{32}{10}-\frac{32}{3}\right)-\left(2-\frac{1}{4}+8+\frac{1}{10}-\frac{4}{3}\right) = \frac{241}{60}$$

Example 6.72. Change the order of integration in the integral $\int_0^b\int_0^{a/b\sqrt{b^2-y^2}} xy\,dx\,dy$

Solution. The region of integration is bounded by the curves

$$x = 0,\quad x=\frac{a}{b}\sqrt{b^2-y^2}\quad\text{or}\quad\left(\frac{x}{a}\right)^2=\frac{b^2-y^2}{b^2}$$

or $$\frac{x^2}{a^2}+\frac{y^2}{b^2} = 1$$

and $$y = 0,\quad y=b$$

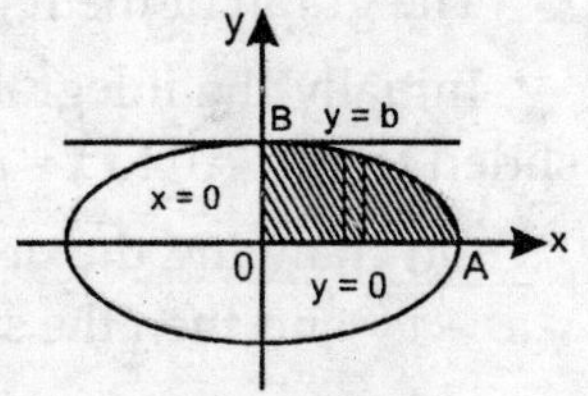

Fig. 6.41

The graph of the region of integration is AOB. Initially the integration is first performed w.r.t. x and then w.r.t. y. On changing the order of integration the integration is first performed w.r.t. y along a strip PQ from $y=0$ $y=b\sqrt{1-\frac{x^2}{a^2}}=\frac{b}{a}\sqrt{a^2-x^2}$ and then the strip PQ is slided || to y-axis from $x=0$ to $x=a$ to cover the whole region of integration.

Thus $$I = \int_{x=0}^{a}\int_{y=0}^{\frac{b}{a}\sqrt{a^2-x^2}} xy\,dx\,dy$$

$$= \int_0^a x\left[\frac{y^2}{2}\right]_0^{\frac{b}{a}\sqrt{a^2-x^2}} dx = \frac{1}{2}\int_0^a x\left[\frac{b^2}{a^2}(a^2-x^2)\right]dx$$

$$= \frac{1}{2}\frac{b^2}{a^2}\left[\frac{a^2x^2}{2}-\frac{x^4}{4}\right]_0^a = \frac{1}{2}\frac{b^2}{a^2}\left[\frac{a^4}{2}-\frac{a^4}{4}\right] = \frac{1}{2}\frac{b^2}{a^2}\left[\frac{a^4}{4}\right] = \frac{1}{8}a^2b^2$$

Example **6.73.** Change the order of integration in $\int_0^\infty \int_x^\infty \frac{e^{-y}}{y} dy\, dx$ and hence solve it.

Solution. The region of integration is bounded by the curves

$$y = x, \quad y = \infty$$

and

$$x = 0, \quad x = \infty$$

The region of integration is divided by its vertical strips like PQ.

Initially the integration is first performed w.r.t. y along the strip PQ from $y = x$ and $y = \infty$ and then slided || to y-axis from $x = 0$ to $x = \infty$.

On changing the order of integration, the integration is first integrated w.r.t. x from $x = 0$ to $x = y$ and then strip is slided from $y = 0$ to $y = \infty$ to cover the whole region of integration.

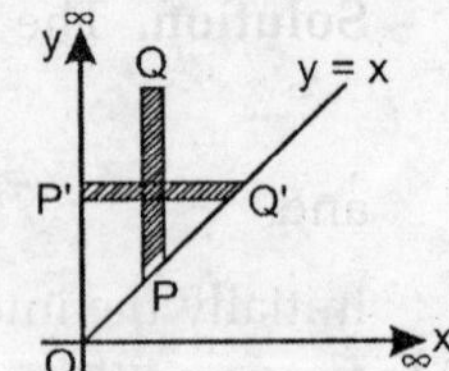

Fig. 6.42

Thus
$$I = \int_0^\infty \int_x^\infty \frac{e^{-y}}{y} dy\, dx = \int_0^\infty \frac{e^{-y}}{y} [x]_0^y dy$$

$$= \int_0^\alpha \frac{e^{-y}}{y} \times y\, dy = \int_0^\infty e^{-y} dy = \left[\frac{e^{-y}}{-1}\right]_0^\infty = 1$$

Example **6.74.** Change the order of integration in the integral $\int_0^a \int_0^{\sqrt{a^2-x^2}} f(x, y)\, dy\, dx$.

Solution. The region of integration is bounded by the curves

$$y = 0, \quad y = \sqrt{a^2 - x^2} \quad \text{or} \quad x^2 + y^2 = a^2$$

and

$$x = 0, \quad x = a$$

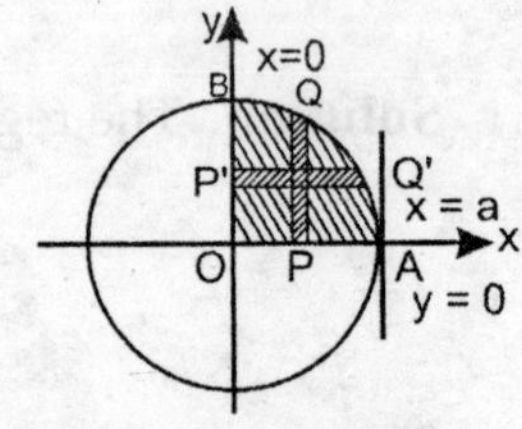

Fig. 6.43

The graph of the region of integration is OAB

Initially the integration is first performed w.r.t. y along a PQ || to y-axis from $y = 0$ to $y = \sqrt{a^2 - x^2}$ and then slided from $x = 0$ to $x = a$ to cover the whole region.

On changing the order of integration is first performed w.r.t. x along a strip $P'Q'$ || to x-axis from $x = 0$ to $x = \sqrt{a^2 - y^2}$ and then the strip is slided || to x-axis from $y = 0$ to $y = a$ for convering the region integration.

Thus
$$I = \int_0^a \int_0^{\sqrt{a^2-y^2}} f(x, y)\, dx\, dy$$

Example **6.75.** Change the order of integration and evaluate $\int_0^4 \int_y^4 \frac{x\, dx\, dy}{x^2 + y^2}$.

Solution. The region of integration is bounded by the curves $x = y$, $x = 4$ and $y = 0$, $y = 4$. The graph of the region of integration is OAB. Co-ordinates of B are $x = 4$, $y = 4$.

Initially the integration is first performed w.r.t. x along a strip PQ || to x-axis from $x = y$ to $x = 4$ and then slided || to x-axis from $y = 0$, $y = 4$.

On changing the order of integration, the integration is first performed w.r.t. y along a strip $P'Q'$ || to y-axis from $y = 0$ to $y = x$ and then slided from $x = 0$, $x = 4$ || to y-axis to cover the whole region of integration.

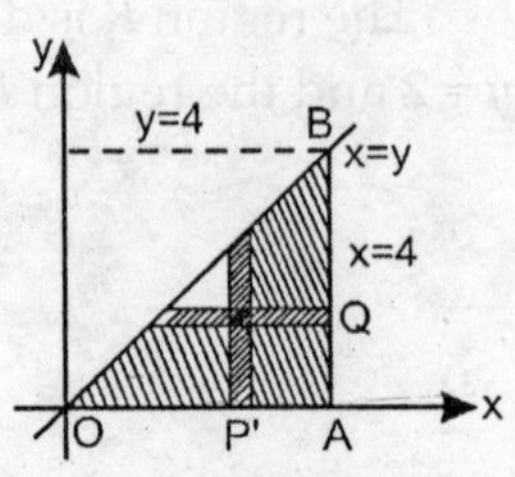

Fig. 6.44

Thus $$I = \int_{x=0}^{4}\int_{y=0}^{x} \frac{x\,dx\,dy}{x^2+y^2} = \int_0^4 x\left[\left(\tan^{-1}\frac{y}{x}\right)\frac{1}{x}\right]_1^0 dx$$

$$= \int_0^4 (\tan^{-1}1 - 0)\,dx = \int_0^4 \frac{\pi}{4}dx = \pi$$

***Example* 6.76.** Change the order of integration $\int_0^{\infty}\int_0^{x} xe^{\frac{x^2}{y}}\,dy\,dx$

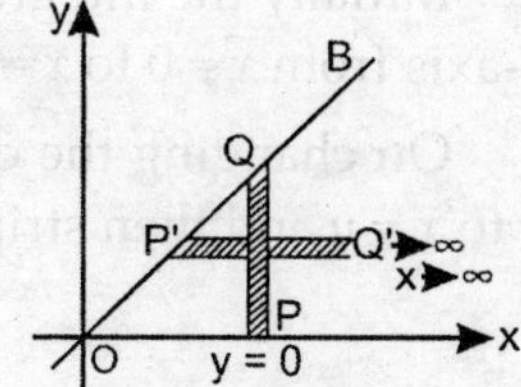

Fig. 6.45

Solution. The region of integration is bounded by the curves

$$y = 0, \quad y = x$$

and $$x = 0, \quad x = \infty$$

Initially the integration is first integrated w.r.t. y along a strip PQ || to y-axis from $y = 0$ to $y = x$ and then slided from $x = 0$ to $x = \infty$ to cover the region of integrate.

On changing the order of integration, the integration is first performed w.r.t. x along a strip $P'\,Q'$ || to x-axis from $x = y$ to $x = \infty$ and the slided || to x-axis from $y = 0$ to $y = \infty$

Thus $$I = \int_{y=0}^{\infty}\int_{x=y}^{\infty} xe^{\frac{x^2}{y}}\,dx\,dy = \int_{y=0}^{\infty} -\frac{y}{2}\int_{x=y}^{\infty} -\frac{2x}{y}e^{\frac{x^3}{y}}\,dx\,dy$$

$$= \int_{y=0}^{\infty} -\frac{y}{2}\left[e^{\frac{x^2}{y}}\right]_y^{\infty} dy = \int_{y=0}^{\infty} -\frac{y}{2}(0 - e^{-y})\,dy = +\frac{1}{2}\int_0^{\infty} ye^{-y}dy$$

$$= \frac{1}{2}\left\{[-ye^{-y}]_0^{\infty} + \int e^{-y}dy\right\} = \frac{1}{2}\,[0+1] = \frac{1}{2}$$

***Example* 6.77.** Evaluate $\iint_R x^2\,dx\,dy$ where R is the region in the first quadrant bounded by the hyperbola $xy = 16$ and the lines $y = x$, $y = 0$, $x = 8$.

Solution. The region of integration is the area OABC

Given by $$xy = 16 \quad ...(i)$$

$$x = 8 \quad ...(ii)$$

$$y = x \quad ...(iii)$$

From (*i*) and (*ii*) we get $y = 2$, $x = 8$

The co-ordinates of point B are (8, 2).

From (*i*) and (*iii*), we get $x^2 = 16$ $x = 4$, $y = 4$ is the coordinate of C are (4, 4)

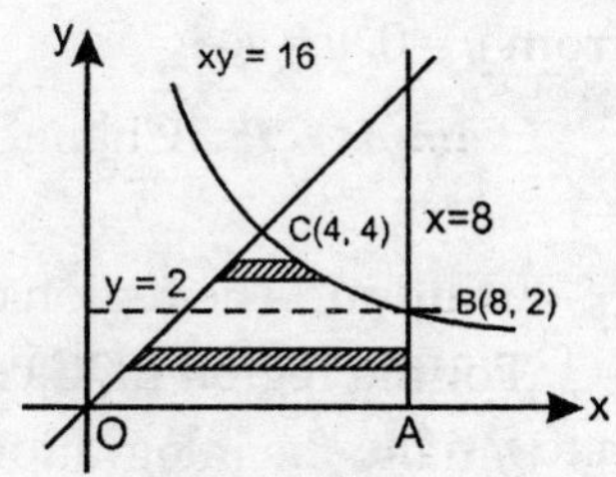

Fig. 6.46

The region R is divided into two parts R_1 and R_2, R_1 is the region below the line $y = 2$ and the region R_2 is the region above $y = 2$.

$$I_1 = \int_0^2\int_y^8 x^2\,dx\,dy$$

$$I_2 = \int_2^4\int_y^{16/y} x^2\,dx\,dy$$

$$I = I_1 + I_2 = \int_0^2\int_y^8 x^2\,dx\,dy + \int_2^4\int_y^{16/y} x^2\,dx\,dy$$

$$= \int_0^2 dy\left[\frac{x^3}{3}\right]_y^8 + \int_2^4 dy\left(\frac{x^3}{3}\right)_y^{16y}$$

$$= \frac{1}{3}\int_0^2 (8^3 - y^3)\,dy + \frac{1}{3}\int_2^4 \left(\frac{16^3}{y^3} - y^3\right)dy$$

$$= \frac{1}{3}\left(8^3 y - \frac{y^4}{4}\right)_0^2 + \frac{1}{3}\left[\frac{16^3}{-2y^2} - \frac{y^4}{4}\right]_2^4$$

$$= \frac{1}{3}\left[8^3 \times 2 - \frac{2^4}{4}\right] + \frac{1}{3}\left[\frac{-16^3}{2\times 4^2} - \frac{4^4}{4} + \frac{16^3}{2\times 2^2} + \frac{2^4}{4}\right] = 274$$

Example **6.78.** Change the order of integration in the integral $\int_0^{2a}\int_{\sqrt{2ax-x^2}}^{\sqrt{2ax}} f(x, y)\,dx\,dy.$

Solution. The region of integration is bounded by

$$y = \sqrt{2ax - x^2} \quad \text{or} \quad y^2 = 2ax - x^2 \text{ or } x^2 + y^2 - 2ax = 0$$

$$y = \sqrt{2ax}\ y^2 = 2ax$$

and $\qquad x = 0,\ x = 2a.$

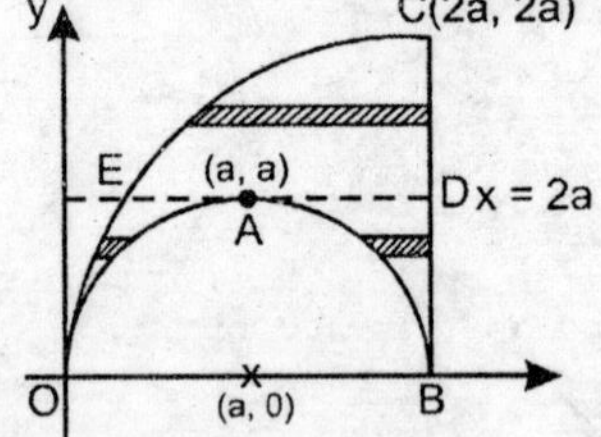

Fig. 6.47

The region of integration is *OABCO*

Initially the integration is performed w.r.t. y and then w.r.t. to x. On changing the order of integration, the integration is first performed w.r.t. to x and w.r.t. y. For this we divide the region in three sub-region.

EOA, ABD and *EDC.*

For the region *EOA* consider the strip || to x-axis which extends from $x = \dfrac{y^2}{2a}$ to $x =$

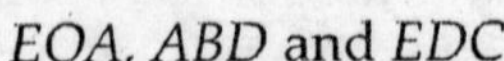

$a - \sqrt{a^2 - y^2}$ and then slides from $y = 0$ $y = a$.

$$I_1 = \int_0^a \int_{\frac{y^2}{2a}}^{a-\sqrt{a^2-y^2}} f(x, y)\,dx\,dy$$

For the region *ABD* consider the strip || to x-axis extending from $x = a + \sqrt{a^2 - y^2}$ to $x = 2a$ and the strip slides from $y = 0$, $y = a$.

$$I_2 = \int_0^a \int_{a+\sqrt{a^2-y^2}}^{2a} f(x, y)\,dx\,dy$$

For the region *EDC* consider the strip || to x-axis extending from $x = \dfrac{y^2}{2a}$ to $y = 2a$ and the strip slides from $y = a$ to $y = 2a$.

$$I_3 = \int_a^{2a}\int_{y^2/2a}^{2a} f(x, y)\,dx\,dy$$

$$I = I_1 + I_2 + I_3 = \int_0^a \int_{\frac{y^2}{2a}}^{a-\sqrt{a^2-y^2}} f(x, y)\,dx\,dy + \int_0^a \int_{a+\sqrt{a^2-y^2}}^{2a} f(x, y)\,dx\,dy + \int_0^{2a}\int_{\frac{y^2}{2a}}^{2a} f(x, y)\,dx\,dy$$

Example **6.79.** Change the order of integration of $\int_{a}^{ae^{\pi/4}} \int_{2\log\left(\frac{r}{a}\right)}^{\pi/2} f(r, \theta)\, r\, dr\, d\theta$

Solution.
$$I = \int_{a}^{ae^{\pi/4}} \int_{2\log\left(\frac{r}{a}\right)}^{\pi/2} f(r, \theta)\, r\, dr\, d\theta$$

The region of integration is bounded by the curves

(*i*) $\theta = 2 \log (r/a)$ *i.e.* $r = a\, e^{\theta/2}$ (equi-angular spiral)

(*ii*) $\theta = \pi/2$ (a line through the pole perpendicular to the line *OA*)

(*iii*) $r = a$ (a circle of radius '*a*' with centre at the pole 0)

(*iv*) $r = a\, e^{\pi/4}$ (a circle of radius $a\, e^{\frac{\pi}{4}}$ with centre at pole 0)

The graph of region of integration is to *BML* (shaded)

Initially the integration is firstly w.r.t. 'θ' and then w.r.t. '*r*'

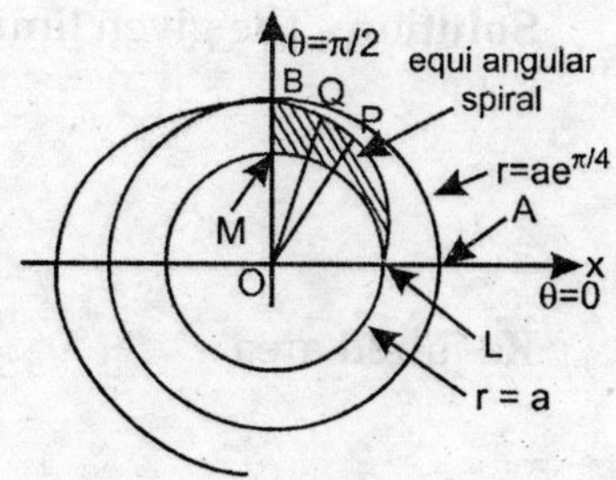

Fig. 6.48

On changing the order of integration, the integration is first performed w.r.t. '*r*' and then w.r.t. 'θ'. To do so we take an elementary triangular strip *OPQ*. On this strip θ is constant but *r* varies from $r = a$ to $r = a\, e^{\theta/2}$.

To cover the whole region of integration θ varies from θ = 0 to θ = π/2. Thus we have

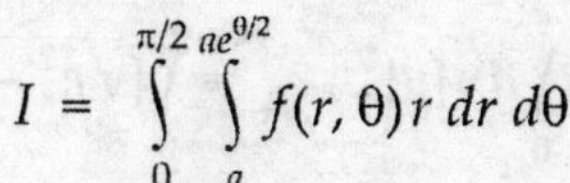
$$I = \int_{0}^{\pi/2} \int_{a}^{ae^{\theta/2}} f(r, \theta)\, r\, dr\, d\theta$$

EXERCISE 6.3

Change the order of the integration and hence evaluate the following:

1. $\int_{0}^{1} \int_{x^2}^{x} (x^2 + y^2)^{-1/2} dy\, dx$ **Ans.** $\int_{0}^{1} dy \int_{y}^{\sqrt{y}} (x^2 + y^2)^{-3/2} dx$

2. $\int_{0}^{a} dy \int_{\frac{x^2}{a}}^{2a-x} dy\, dx\, dy$ **Ans.** $\frac{3a^4}{8}$

3. $\int_{0}^{1} \int_{x}^{2-x} \frac{x}{y} dx\, dy$ **Ans.** $\log \frac{4}{e}$

4. $\int_{0}^{1} \int_{x}^{\sqrt{2-x^2}} \frac{x\, dy\, dx}{\sqrt{x^2 + y^2}}$ **Ans.** $1 - \frac{1}{\sqrt{2}}$

5. $\int_{0}^{a} \int_{x/a}^{\sqrt{x/a}} (x^2 + y^2)\, dx\, dy$ **Ans.** $\frac{a^2}{28} + \frac{a}{20}$

6. $\int_0^a \int_{\sqrt{ax}}^0 \frac{y^2\,dydx}{\sqrt{y^2 - a^2x^2}}$ **Ans.** $\frac{\pi a^2}{6}$

7. $\int_a^{a/\sqrt{2}} \int_y^{\sqrt{a^2-y^2}} \log(x^2 + y^2)\,dx\,dy$ **Ans.** $\frac{\pi a^2}{4}\left[\log a - \frac{1}{2}\right]$

8. $\int_0^1 \int_{4y}^4 e^{x^2}\,dy\,dx$ **Ans.** $\frac{1}{8}(e^{16} - 1)$

9. $\int_0^1 \int_0^{1-y} (x^2 + y^2)\,dx\,dy$ **Ans.** $\frac{1}{6}$

Example **6.80.** Find the area between the curves $x^2 + y^2 = a^2$ and $x + y = a$ as a double integral and evaluate

Solution. The given limits are

$$y = a - x, \quad y = \sqrt{a^2 - x^2}$$

$$x = 0, \quad x = a$$

Required area

$$= \iint dxdy = \int_0^a \int_{a-x}^{\sqrt{a^2-x^2}} dx\,dy$$

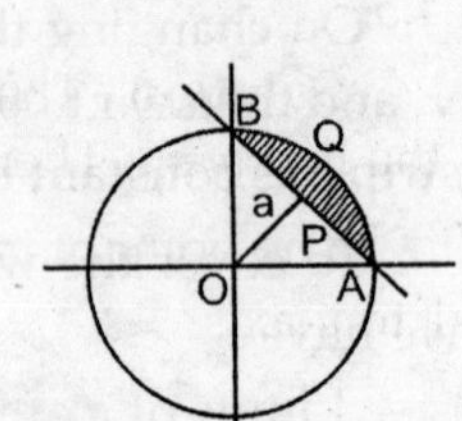

Fig. 6.49

$$= \int_0^a dx\,[y]_{a-x}^{\sqrt{a^2-x^2}} = \int_0^a [\sqrt{a^2 - x^2} - (a - x)]\,dx$$

$$= \left[\frac{x}{2}\sqrt{a^2 - x^2} + \frac{a^2}{2}\sin^{-1}\frac{x}{a}\right]_0^a - \left[ax - \frac{x^2}{2}\right]_0^a$$

$$= 0 + \frac{a^2}{2} \times \frac{\pi}{2} - \frac{a^2}{2} = \frac{\pi a^2}{4} - \frac{a^2}{2} = \frac{\pi a^2}{4} - \frac{a^2}{2}$$

Example **6.81.** Find the area of a plate in the form of a quadrant of the ellipse $\frac{x^2}{a^2} + \frac{y^2}{b^2} = 1$

Solution. Dividing the area into vertical strips of width δx, y-varies from $y = 0$ to $y = b\sqrt{1 - x^2/a^2}$ and x varies from 0 to a

The required area

$$= \int_0^a dx \int_0^{b\sqrt{1-x^2/a^2}} dy$$

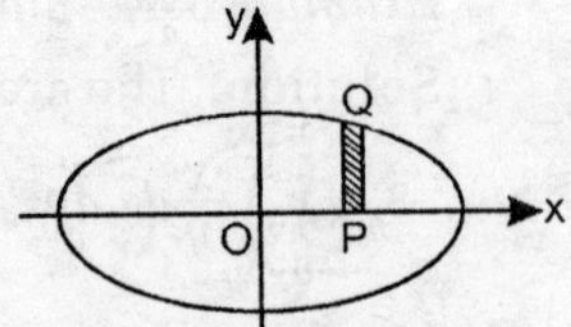

Fig. 6.50

$$= \int_0^a [y]_0^{b\sqrt{1-x^2/a^2}}\,dx = \frac{b}{a}\int_0^a \sqrt{(a^2 - x^2)}\,dx$$

$$= \frac{b}{a}\left[\int_0^a \frac{x\sqrt{(a^2 - x^2)}}{2} + \frac{a^2}{2}\sin^{-1}\frac{x}{a}\right]_0^a = \frac{b}{a}\frac{a^2}{2} \times \frac{\pi}{2} = \frac{\pi ab}{4}$$

***Example* 6.82. Calculate the area included between the curve $r = a(\sec\theta + \cos\theta)$ and its asymptote.**

Solution. The curve is symmetrical about the initial line and has an asymptote $r = a \sec\theta$

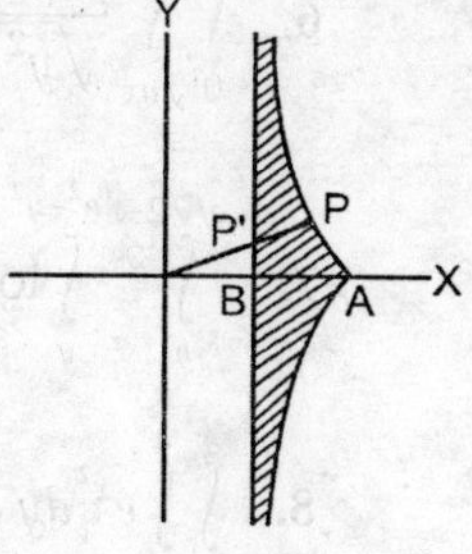

Fig. 6.51

Draw any line cutting the curve at P and its asymptote at P'. Along this line, θ is constant and r varies from $a\sec\theta$ at P' to $a(\sec\theta = \cos\theta)$ at P. Then to get upper half of the area, θ varies from to $\pi/2$.

$$\text{The required area} = 2\int_0^{\pi/2}\int_{a\sec\theta}^{a(\sec\theta+\cos\theta)} r\,dr\,d\theta$$

$$= 2\int_0^{\pi/2}\left(\frac{r^2}{2}\right)_{a\sec\theta}^{a(\sec\theta+\cos\theta)} d\theta$$

$$= a^2\int_0^{\pi/2}(2+\cos^2\theta)\,d\theta = \frac{5\pi a^2}{4}$$

***Example* 6.83.** By double integration, show that the area between the parabola $y = 4x - x^2$ and the line $y = x$ is 4.5.

Solution. The given parabola is $y = 4x - x^2$ and the line is $y = x$ or $y = x(4 - x)$ and the line $y = x$. Then the region is bounded by $y = x$ and $y = 4x - x^2$.

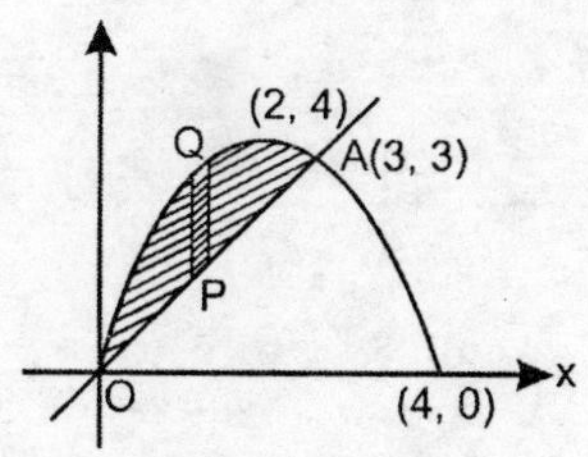

Fig. 6.52

The point of intersection of the parabola and the line is $A(3, 3)$. Consider a strip PQ || to y-axis.

Limits of y are $y = x$ at P and $y = x(4 - x)$ at Q and on sliding the strip, x varies from $x = 0$ to $x = 3$.

Thus the integral

$$\int_0^3\int_x^{4x-x^2} dx\,dy = \int_0^3 [y]_x^{4x-x^2}\,dx$$

$$= \int_0^3 (4x - x^2 - x)\,dx = \int_0^3 (3x - x^2)\,dx$$

$$= \left[\frac{3x^2}{2} - \frac{x^3}{3}\right]_0^3 = 3\times\frac{9}{2} - \frac{27}{3} = \frac{27}{2} - 9 = \frac{27-18}{2} = \frac{9}{2} = 4.5.$$

***Example* 6.84.** Find, by double integral, the area enclosed by the curve $y = \dfrac{3x}{x^2+2}$ and $y = x^2$.

Solution. The area of integration is the region bounded by

$$y = \frac{3x}{x^2+2} \text{ and } y = x^2$$

The point of intersection of the two curves is $A(1, 1)$. Therefore, OAO is the required area. Let us take a strip PQ || to y-axis.

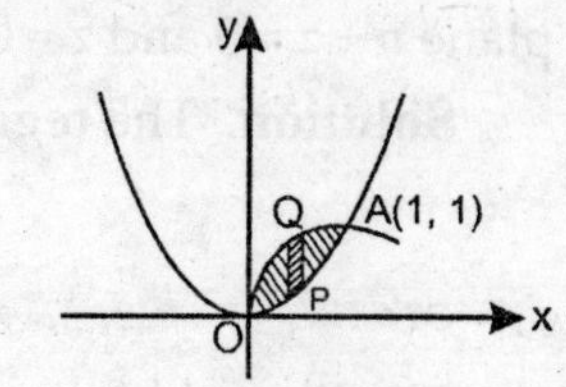

Fig. 6.53

The limits of integration are form $\quad y = x^2$ to $y = \dfrac{3x}{x^2+2}$

and $\quad x = 0$ to $x = 1$

$$\text{Area} = \int_0^1\int_{x^2}^{\frac{3x}{x^2+2}} dx\,dy$$

$$= \int_0^1 [y]_{x^2}^{\frac{3x}{x^2+2}} dx = \int_0^1 \left(\frac{3x}{x^2+2} - x^2\right) dx$$

$$= \int_0^1 \left(\frac{3}{2}\left(\frac{2x}{x^2+2}\right) - x^2\right) dx = \left[\frac{3}{2}\log(x^2+2) - \frac{x^3}{3}\right]_0^1$$

$$= \frac{3}{2}\log 3 - \frac{1}{3} - \frac{3}{2}\log 2 = \frac{2}{3}\log\frac{3}{2} - \frac{1}{3}$$

Example 6.85. Find by double integration the area inside the circle $r = a \sin\theta$ and outside the cardioid $r = a(1 - \cos\theta)$

Sol. The point of intersection of the circle and cardioid is given by $a \sin\theta = a(1 - \cos\theta)$

$$2\sin\frac{\theta}{2}\cos\frac{\theta}{2} = 2\sin^2\frac{\theta}{2}$$

$$\tan\frac{\theta}{2} = 1, \frac{\theta}{2} = \frac{\pi}{2}, \theta = \frac{\pi}{2}$$

$$\text{Required area} = \int_0^{\pi/2} \int_{a(1-\cos\theta)}^{a\sin\theta} r\, dr\, d\theta$$

$$= \int_0^{\pi/2} \left[\frac{r^2}{2}\right]_{a(1-\cos\theta)}^{a\sin\theta} d\theta$$

$$= \int_0^{\pi/2} \frac{1}{2}[a^2\sin^2\theta - a^2(1-\cos\theta)^2] d\theta$$

$$= \frac{a^2}{2}\int_0^{\pi/2} [\sin^2\theta - 1 - \cos^2\theta + 2\cos\theta]$$

$$= \frac{a^2}{2}\left[\frac{1}{2}\times\frac{\pi}{2} - \frac{\pi}{2} - \frac{1}{2}\times\frac{\pi}{2} + 2\times 1\right] = \frac{a^2}{2}\left[\frac{\pi}{4} - \frac{\pi}{2} - \frac{\pi}{4} + 2\right]$$

$$= \frac{a^2}{2}\left[2 - \frac{\pi}{2}\right] = a^2\left[1 - \frac{\pi}{4}\right]$$

Fig. 6.54

6.22 VOLUME AS A DOUBLE INTEGRAL

Volume $V = \iint_R z\, dx\, dy$ where R is the region of integration.

Example 6.86. Find the volume bounded by the cylinder $x^2 + y^2 = 4$, the plane $y + z = 3$ and $z = 0$, using double integral.

Solution. The region of integration is bounded by the curves

$$x^2 + y^2 = 4, \quad y + z = 3, \quad z = 0$$

or
$$z = 3 - y, \quad z = 0$$

The limits of integration are from

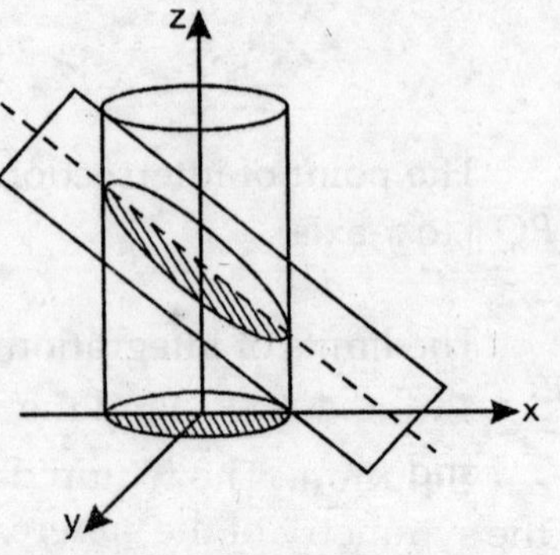

Fig. 6.55

$$x = -\sqrt{4-y^2} \text{ to } \sqrt{4-y^2}$$

and $$y = -2 \text{ to } 2$$

$$V = \int_{-2}^{2}\int_{-\sqrt{4-y^2}}^{\sqrt{4-y^2}} (3-y)\,dx\,dy$$

$$= \int_{-2}^{2} (3-y)\,[x]_{-\sqrt{4-y^2}}^{\sqrt{4-y^2}}\,dy$$

$$= \int_{-2}^{2} (3-y)[\sqrt{4-y^2}+\sqrt{4-y^2}]\,dy = 2\int_{-2}^{2} (3-y)(\sqrt{4-y^2})\,dy$$

$$= 2\int_{-2}^{2} 3\sqrt{4-y^2}\,dy - 2\int_{-2}^{2} y(4-y^2)^{1/2}\,dy$$

$$= 6\int_{-2}^{2} (4-y^2)^{1/2}\,dy - \int_{-2}^{2} 2y\,(4-y^2)^{1/2}\,dy$$

$$= 6\left[\frac{y\sqrt{4-y^2}}{2}+\frac{4}{2}\sin^{-1}\frac{y}{2}\right]_{-2}^{2} + \left[\frac{(4-y^2)^{3/2}}{3/2}\right]_{-2}^{2}$$

$$= 6\left[\left(\frac{2\pi}{2}\right)+\left(\frac{2\pi}{2}\right)\right]+\frac{2}{3}(0) = 6\times 2\pi = 12\pi$$

Example **6.87**. Find the volume common to the cylinders $x^2+y^2=a^2$ and $x^2+z^2=a^2$.

Solution. The region of integration is bounded by $x^2+y^2=a^2$ and $x^2+z^2=a^2$.

Half of the volume lies above xy-plane and half below it. Hence the section of the cylinder $x^2+y^2=a^2$ is the circle $x^2+y^2=a^2$ in the xy-plane.

Required volume $$(V) = 2\iint_A \sqrt{a^2-x^2}\,dy\,dx$$

where A is the area inside the circle $x^2+y^2=a^2$ in x-y plane. If A denotes the area of the quadrant OPQ of this circle, then by symmetry

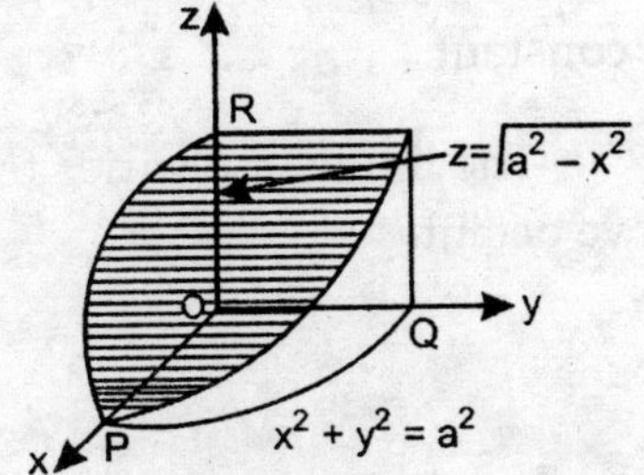

Fig. 6.56

$$= 4\times 2\iint \sqrt{a^2-x^2}\,dy\,dx$$

$$= 8\int_0^a\int_0^{\sqrt{a^2-x^2}} \sqrt{a^2-x^2}\,dy\,dx$$

$$= \int_0^a \sqrt{a^2-x^2}\,[y]_0^{\sqrt{a^2-x^2}}\,dx = 8\int_0^a (a^2-x^2)\,dx = 8\left[a^2x-\frac{x^3}{3}\right]_0^a = \frac{16a^3}{3}$$

Example **6.88**. Find the volume common to the sphere $x^2+y^2+z^2=a^2$ and the cylinder $x^2+y^2=ax$.

Solution. The required volume is the part of sphere $x^2+y^2+z^2=a^2$ lying within the cylinder. On account of the symmetry of the sphere, half of it lies above the plane xy and half below it.

The required volume is easily found by changing to cylindrical to co-ordinates (r, ϕ, z). We therefore have

$$x = r\cos\phi$$
$$y = r\sin\phi$$
$$z = z$$

and
$$J = \frac{\partial(x, y, z)}{\partial(r, \phi, z)} = r$$

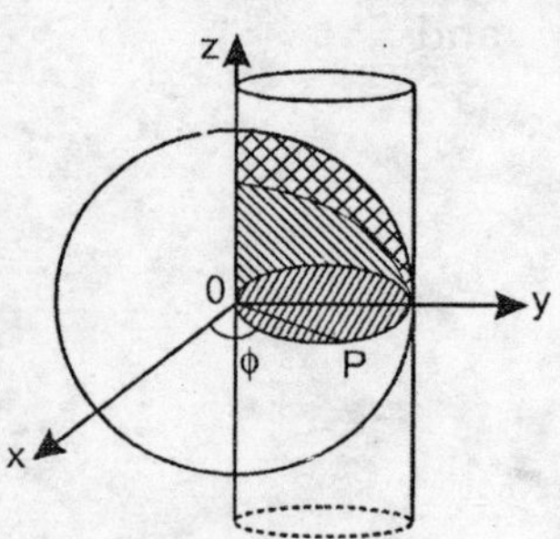

Fig. 6.57

Then the equation of the sphere becomes $r^2 + z^2 = a^2$ and that of cylinder $r = a\cos\phi$

The volume inside the cylinder is bounded by the sphere and equals twice the volume shown shaded in figure for which z varies from 0 to $\sqrt{a^2 - r^2}$; r varies from 0 to $a\cos\phi$ and ϕ varies from 0 to π.

Required volume
$$= 2\int_0^{\pi}\int_0^{a\cos\phi} r\,z\,dr\,d\phi$$

$$= 2\int_0^{\pi}\int_0^{a\cos\phi} r\sqrt{a^2 - r^2}\,dr\,d\phi = 2\int_0^{\pi} -\frac{1}{2}\left[\frac{(a^2 - r^2)^{3/2}}{3/2}\right]_0^{a\cos\phi} d\phi$$

$$= -\frac{2}{3}\int_0^{\pi}[(a^2 - r^2)^{3/2}]_0^{a\cos\phi}\,d\phi = -\frac{2}{3}\int_0^{\pi}[(a^2 - a^2\cos^2\phi)^{3/2} - a^3]$$

$$= -\frac{2}{3}a^3\int_0^{\pi}(\sin^3\phi - 1)\,d\phi = -\frac{4a^3}{3}\left[\frac{2}{3} - \frac{\pi}{2}\right] = \frac{2a^3}{9}\,[3\pi - 4]$$

Example 6.89. Find the volume of the ellipsoid by double integration, the ellipsoid is $\frac{x^2}{a^2} + \frac{y^2}{b^2} + \frac{z^2}{c^2} = 1.$

Solution. Here $z = c\sqrt{1 - \frac{x^2}{a^2} - \frac{y^2}{b^2}}$

The required volume of the ellipsoid is 8 times the volume of the ellipsoid in the + ve constant.

The limits of y varies from 0 to $b\sqrt{1 - \frac{x^2}{a^2}}$ and limits of x varies from 0 to a in the + ve octant.

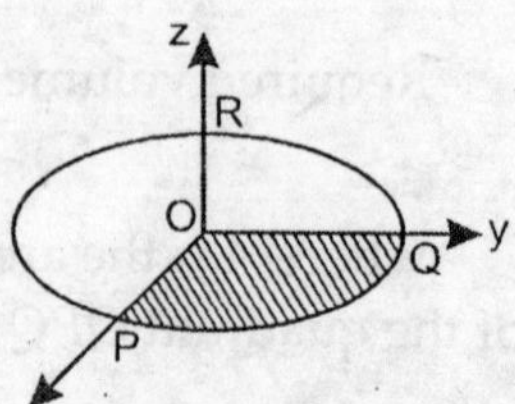

Fig. 6.58

$$\therefore \quad V = 8\int_0^{a}\int_0^{b\sqrt{1-\frac{y^2}{a^2}}} c\sqrt{1 - \frac{x^2}{a^2} - \frac{y^2}{b^2}}\,dy\,dx$$

$$= 8\int_0^{a}\int_0^{b\sqrt{1-\frac{y^2}{a^2}}} c\sqrt{1 - \frac{x^2}{a^2} - \frac{y^2}{b^2}}\,dy\,dx \qquad \text{Put } t = b\sqrt{1 - \frac{x^2}{a^2}}$$

$$= \frac{8c}{b}\int_0^{a}\int_0^{t}\sqrt{t^2 - y^2}\,dx\,dy$$

$$= \frac{8c}{b}\int_0^a \left[y\frac{\sqrt{t^2-y^2}}{2}+\frac{t^2}{2}\sin^{-1}\frac{y}{t}\right]_0^t dx$$

$$= \frac{8c}{b}\int_0^a \left(\frac{t^2}{2}\times\frac{\pi}{2}\right)dx = \frac{8\pi c}{4b}\int_0^a t^2\,dx$$

$$= \frac{2\pi c}{b}\int_0^a b^2\left(1-\frac{x^2}{a^2}\right)dx = 2\pi\, bc\int_0^a\left(1-\frac{x^2}{a^2}\right)dx$$

$$= 2\pi\, bc\left[x-\frac{x^3}{3a^2}\right]_0^a = 2\pi\, bc\left[a-\frac{a}{3}\right]$$

$$= 2\pi\, bc\times\frac{2a}{3} = \frac{4\pi abc}{3}$$

Example **6.90.** Find the volume of the tetrahedron bounded by the co-ordinate planes and the plane $\frac{x}{a}+\frac{y}{b}+\frac{3}{c}=1$. Use double integral approach.

Solution. Here, $z = c\left[1-\frac{x}{a}-\frac{y}{b}\right]$. The limits of y varies from $y = 0$ to $y = b\left(1-\frac{x}{a}\right)$ and x varies from 0 to a.

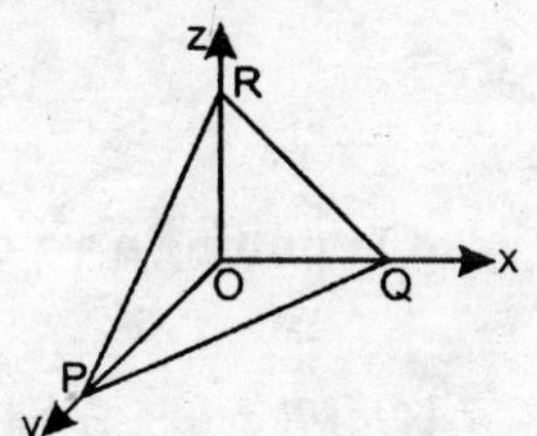

Fig. 6.59

Required volume
$$V = \int_0^a\int_0^{b\left(1-\frac{x}{a}\right)} c\left(1-\frac{x}{a}-\frac{y}{b}\right)dy\,dx$$

$$= c\int_0^a\left[\left(1-\frac{x}{a}\right)y-\frac{y^2}{2b}\right]_0^{b\left(1-\frac{x}{a}\right)} dx$$

$$= c\int_0^a\left[b\left(1-\frac{x}{a}\right)^2-\frac{b^2}{2b}\left(1-\frac{x}{a}\right)^2\right]dx$$

$$= \frac{bc}{2}\int_0^a\left(1-\frac{x}{a}\right)^2dx = \frac{bc}{2}\left[\frac{\left(1-\frac{x}{a}\right)^3}{-\frac{3}{a}}\right]_0^a = \frac{-abc}{6}(0-1) = \frac{abc}{6}$$

Example **6.91.** Find the volume of the solid bounded by the conical surface $(z-2)^2 = \frac{x^3}{3}+\frac{y^2}{2}$ and the plane $z = 0$.

Solution. By symmetry, required volume is given by in xy-plane, the ellipse is

$$V = 4\iint z\,dy\,dx$$

Here
$$z = \sqrt{\frac{x^3}{3}+\frac{y^2}{2}}+2 = 2+\sqrt{\frac{x^3}{3}+\frac{y^2}{2}}$$

in xy-plane, the ellipse is

$$\frac{x^2}{3}+\frac{y^2}{2} = 4$$

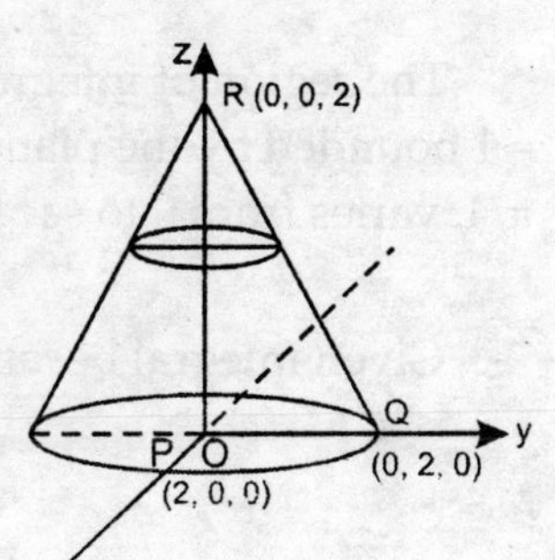

Fig. 6.60

Putting $$x = u\sqrt{3} \quad y = v\sqrt{2}$$

the ellipse reduces to $$\frac{u^2}{4} + \frac{v^2}{4} = 1$$

$$u^2 + v^2 = 4$$

in uv plane.

$$\text{Volume} = 4 \iint z\,dy\,dx = 4\left[2 + \sqrt{\frac{x^3}{3} + \frac{y^2}{2}}\right] dx\,dy$$

$$J = \begin{vmatrix} \frac{\partial x}{\partial u} & \frac{\partial x}{\partial v} \\ \frac{\partial y}{\partial u} & \frac{\partial y}{\partial v} \end{vmatrix} = \sqrt{6}$$

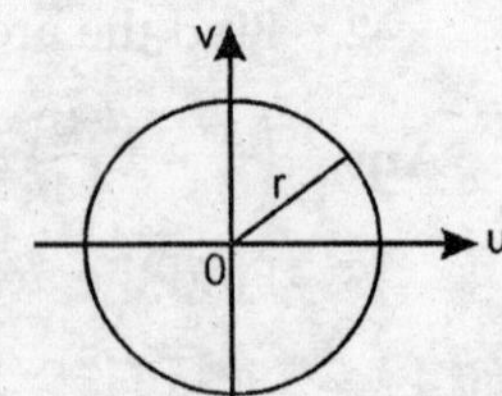

Fig. 6.61

$$= 4\int_0^2 \int_0^{\sqrt{4-u^2}} [2 + \sqrt{u^2 + v^2}]\sqrt{6}\; du\,dv$$

$$= 4\sqrt{6} \int_0^2 \int_0^{\sqrt{4-u^2}} [2 + \sqrt{u^2 + v^2}]$$

$$= 4\sqrt{6} \int_0^{\pi/2} \int_0^2 (2 + r)\, r\, dr\, d\theta$$

On putting $u = r\cos\theta$, $v = r\sin\theta$

$$\text{Volume} = 4\sqrt{6} \int_0^{\pi/2} \left(r^2 + \frac{r^3}{3}\right)_0^2 d\theta = 4\sqrt{6}\,\frac{20}{3} \int_0^{\pi/2} d\theta$$

$$= 4\sqrt{6} \times \frac{20}{3} \times \frac{\pi}{2} = \frac{40\sqrt{6}\,\pi}{3}$$

Example 6.92. Evaluate $\displaystyle\int_0^1 \int_0^{\sqrt{1-x^2}} \int_{\sqrt{x^2+y^2}}^1 \frac{dx\,dy\,dz}{\sqrt{x^2+y^2+z^2}}$.

Solution. Changing to spherical coordinates

$$x = r\sin\theta\cos\phi,$$

$$y = r\sin\theta\cos\phi, \quad z = r\cos\theta$$

$$|J| = r^2\sin\theta \text{ and } x^2 + y^2 + z^2 = r^2$$

The region of integration is common to the cone $z^2 = x^2 + y^2$ and the cylinder $x^2 + y^2 = 1$ bounded by the plane $z = 1$ in the + ve quadrant as in figure. Hence θ varies from 0 to $\pi/4$, varies from 0 to $\sec\theta$ and ϕ varies from 0 to $\pi/2$.

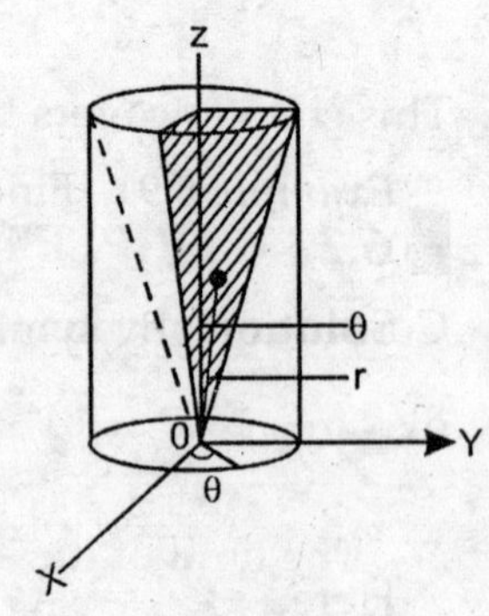

Fig. 6.62

Given integral becomes $\displaystyle\int_0^{\pi/2} \int_0^{\pi/4} \int_0^{\sec\theta} \frac{1}{r} \times r^2 \sin\theta\, dr\, d\theta\, d\phi$

$$= \int_0^{\pi/2} d\phi \int_0^{\pi/4} \left[\frac{r^2}{2}\right]_0^{\sec\theta} \sin\theta\, d\theta = \frac{\pi}{2} \int_0^{\pi/4} \frac{\sec^2\theta}{2} \sin\theta\, d\theta$$

$$= \frac{\pi}{4}\int_0^{\pi/4} \sec\theta \tan\theta \, d\theta = \frac{\pi}{4}(\sec\theta)_0^{\pi/4} = \frac{(\sqrt{2}-1)\pi}{4}$$

EXERCISE 6.4

1. Find the area included between the curves $y^2 = 4a(x + a)$ and $y^2 = 4b(b - x)$.

Ans. $\frac{8}{3}\sqrt{ab}\ (a + b)$

2. Find the area common to the circles $x^2 + y^2 = a^2$ and $x^2 + y^2 = 2ax$.

Ans. $\left(\frac{\pi}{3} - \frac{\sqrt{3}}{4}\right) a^2$

3. Find the area that lies inside the cardiod $r = a(1 + \cos\theta)$ and outside the circle $r = a$.

Ans. $a^2 \frac{(\pi + 8)}{4}$

4. Find the area enclosed by the lemniscate $r^2 = a^2 \cos 2\theta$.

Ans. a^2

5. Show that the area of the region included between the cardiods $r = a(1 + \cos\theta)$ and $r = a(1 - \cos\theta)$ is $\frac{a^2}{2}$ $(3\pi - 8)$.

6. Find by double integration the volume of the sphere $x^2 + y^2 + z^2 = 9$.

Ans. 36π

7. Find the volume bounded by x-y plane, the paraboloid $2z = x^2 + y^2$ and the cylinder $x^2 + y^2 = 4$.

Ans. 4π

8. Find the volume enclosed by the cylinders $x^2 + y^2 = 2ax$ and $z^2 = 2ax$.

Ans. $\frac{128a^3}{15}$

9. Find the volume bounded by surface $1 = x^2 + y^2 - 2$ and the cylinder $x^2 + y^2 = 4$.

Ans. $4\sqrt{3}\,\pi$

10. Find the area bounded by the circles $r = 2\sin\theta$ and $r = 4\sin\theta$.

Ans. 3π

INTEGRAL CALCULUS-II

This chapter covers the triple integral, change of variables, volume as triple integral and beta-gamma functions.

6.23 TRIPLE INTEGRAL

Consider a function $f(x, y, z)$ defined at every point of the 3-dimensional region V. Divide V into n elementary volumes $\delta V_1, \delta V_2, \delta V_3, ..., \delta V_n$. Let (x_r, y_r, z_r) be any point within rth sub-division δV_r. Consider the sum $\sum_{r=1}^{n} f(x_r, y_r, z_r)\delta V_r$

The limit of the sum, if it exists as $n \to \infty$ and $\delta V_r \to 0$ is called the triple integral of $f(x, y, z)$ over the region V and is defined by $\iiint f(x, y, z)\,dV$

For the purpose of evaluation it can be expressed as the integral $\int_{x_1}^{x_2}\int_{y_1}^{y_2}\int_{z_1}^{z_2} f(x, y, z)\,dx\,dy\,dz$ First (x, y, z) is integrated w.r.t. z, then the result w.r.t. y and finally w.r.t. x

Thus
$$I = \int_{x_1}^{x_2}\int_{y_1(x)}^{y_2(x)}\int_{z_1(x,y)}^{z_2(x,y)} f(x, y, z)\,dx\,dy\,dz$$

***Example* 6.93.** Evaluate $\displaystyle\int_0^a \int_0^{\sqrt{a^2-x^2}} \int_0^{\sqrt{a^2-x^2-y^2}} \frac{1}{\sqrt{a^2-x^2-y^2-z^2}}$.

Solution. $\displaystyle I = \int_0^a \int_0^{\sqrt{a^2-x^2}} \int_0^{\sqrt{a^2-x^2-y^2}} \frac{1}{\sqrt{a^2-x^2-y^2-z^2}}$

$$= \int_0^a \int_0^{\sqrt{a^2-x^2}} \left[\sin^{-1}\frac{z}{\sqrt{a^2-x^2-y^2}}\right]_0^{\sqrt{a^2-x^2-y^2}} dy\,dx$$

$$= \int_0^a \int_0^{\sqrt{a^2-x^2}} \{\sin^{-1}1 - \sin^{-1}0\}\,dy\,dx = \int_0^a \int_0^{\sqrt{a^2-x^2}} \frac{\pi}{2}\,dy\,dx$$

$$= \frac{\pi}{2}\int_0^a [y]_0^{\sqrt{a^2-x^2}}\,dx = \frac{\pi}{2}\int_0^a \sqrt{a^2-x^2}\,dx$$

$$= \frac{\pi}{2}\left[\frac{x\sqrt{a^2-x^2}}{2} + \frac{a^2}{2}\sin^{-1}\frac{x}{a}\right]_0^a = \frac{\pi}{4}a^2 \cdot \frac{\pi}{4} = \frac{a^2\pi^2}{8}$$

***Example*. 6.94.** Evaluate $\displaystyle\int_0^1 \int_0^{\sqrt{1-x^2}} \int_0^{\sqrt{1-x^2-y^2}} \boldsymbol{xyz\,dx\,dy\,dz}$.

Solution.
$$I = \int_0^1 \int_0^{\sqrt{1-x^2}} \int_0^{\sqrt{1-x^2-y^2}} xyz\,dx\,dy\,dz = \int_0^1 \int_0^{\sqrt{1-x^2}} \left[\frac{xyz^2}{2}\right]_0^{\sqrt{1-x^2-y^2}} dy\,dx$$

$$= \int_0^1 \int_0^{\sqrt{1-x^2}} \frac{xy}{2}(1-x^2-y^2)\,dy\,dx = \frac{1}{2}\int_0^1 \int_0^{\sqrt{1-x^2}} (xy - x^3y - xy^3)\,dy\,dx$$

$$= \frac{1}{2}\int_0^1 \left[\frac{xy^2}{2} - \frac{x^3y^2}{2} - \frac{xy^4}{4}\right]_0^{\sqrt{1-x^2}} dx$$

$$= \frac{1}{2}\,\frac{1}{2}\int_0^1 x\left[1 - x^2 - x^2(1-x^2) - \frac{(1-x^2)^2}{2}\right]dx$$

$$= \frac{1}{4}\int_0^1 \left[x - 2x^3 + x^5 - \frac{x+5-2x^3}{2}\right]dx$$

$$= \frac{1}{8}\int_0^1 [2x - 4x^3 + 2x^5 - x - x^5 + 2x^3]\,dx$$

$$= \frac{1}{8}\int_0^1 (x - 2x^3 + x^5)\,dx$$

$$= \frac{1}{8}\left[\frac{x^2}{2} - \frac{2x^4}{4} + \frac{x^6}{6}\right]_0^1 = \frac{1}{8}\left[\frac{1}{2} - \frac{1}{2} + \frac{1}{6}\right] = \frac{1}{8} \times \frac{1}{6} = \frac{1}{48}$$

Example. 6.95. Evaluate $\int_1^e \int_1^{\log y} \int_1^{e^x} \log z \, dz\, dx\, dy$.

Solution.

$$I = \int_1^e \int_1^{\log y} \int_1^{e^x} \log z\, dz\, dx\, dy = \int_1^e \int_1^{\log y} [z \log z - z]_1^{e^x} dx\, dy$$

$$= \int_1^e \int_1^{\log y} [e^x \log e^x - e^x + 1]\, dx\, dy$$

$$= \int_1^e \int_1^{\log y} (xe^x - e^x + 1)\, dx\, dy = \int_1^e [xe^x - e^x - e^x + x]_1^{\log y} dy$$

$$= \int_1^e [\log y\, e^{\log y} - e^{\log y} - e^{\log y} + e - 1]\, dy$$

$$= \left[\frac{y^2}{2}\log y - \frac{y^2}{4} - y^2 + y\log y - y + ey - y\right]_1^e$$

$$= \left[\frac{e^2}{2} - \frac{e^2}{4} - e^2 + e - e + e^2 - e\right] - \left[-\frac{1}{4} - 1 - 1 + e - 1\right] = \frac{1}{4}\,[e^2 - 8e + 13]$$

Example. 6.96. Evaluate $\int_0^3 \int_0^4 \int_{x=\frac{y}{2}}^{x=\frac{y}{2}+1} \left(\frac{2x-y}{2} + \frac{z}{3}\right) dx\, dy\, dz$ by applying the transformation.

$$u = \frac{2x-y}{2},\ v = \frac{y}{2},\ w = \frac{z}{3}$$

Solution. Here $u = \frac{2x-y}{2} = x - \frac{y}{2} = x - v \;\Rightarrow\; x = u + v$

The limits of integration of x are from $x = \frac{y}{2} = v$ to $x = \frac{x}{2} + 1 = v + 1$

i.e. from $u + v = v$ to $u + v = v + 1$

$\Rightarrow$ $u = 0$ to $u = 1$

Similarly limits of integration of y are from $y = 0$ to $y = 4$

i.e. $v = 0$ to $v = 2$

Lastly the limits of z are from z = 0 to z = 3

i.e., $w = 0$ to $w = 1$. Also we have

$$du\,dv\,dw = |J|\,dx\,dy\,dz$$

where
$$|J| = \frac{\partial(u,v,w)}{\partial(x,y,z)} = \begin{vmatrix} \frac{\partial u}{\partial x} & \frac{\partial u}{\partial y} & \frac{\partial u}{\partial z} \\ \frac{\partial v}{\partial x} & \frac{\partial v}{\partial y} & \frac{\partial v}{\partial z} \\ \frac{\partial w}{\partial x} & \frac{\partial w}{\partial y} & \frac{\partial w}{\partial z} \end{vmatrix} = \begin{vmatrix} 1 & -\frac{1}{2} & 0 \\ 0 & \frac{1}{2} & 0 \\ 0 & 0 & \frac{1}{3} \end{vmatrix} = \frac{1}{6}$$

$\therefore$
$$du\,dv\,dw = \frac{1}{6}\,dx\,dy\,dz$$

$\Rightarrow$
$$dx\,dy\,dz = 6\,du\,dv\,dz$$

$$I = 6\int_0^1\int_0^2\int_0^1 (u+w)\,du\,dv\,dw = \int_0^1\int_0^2\left(\frac{u^2}{2}+uw\right)_0^1 dv\,dw$$

$$= 6\int_0^1\int_0^2\left(\frac{1}{2}+w\right)dv\,dw = \int_0^1\left(\frac{v}{2}+wv\right)_0^2 dw$$

$$= 6\int_0^1 (1+2w)\,dw = 6\,(w+w^2)_0^1 = 6(1+1) = 12$$

EXERCISE 6.5

1. Evaluate $\int_0^1\int_0^{\sqrt{1-x^2}}\int_0^{\sqrt{1-x^2-y^2}} xyz\,dx\,dy\,dz$

Ans. $\frac{1}{48}$

2. Evaluate $\int_8^x\int_0^x\int_0^{x+y} e^{x+y+z}\,dx\,dy\,dz$

Ans. $\frac{1}{8}e^{4a} - \frac{3}{4}e^{2a} + e^{a} - \frac{3}{4}$

3. Evaluate $\int_0^1\int_{y^2}^1\int_0^{1-x} z\,dz\,dx\,dy$

Ans. $\frac{4}{35}$

4. Evaluate $\int_0^{\pi/2}\int_{y^2}^{a\cos\theta}\int_0^{\sqrt{a^2-r^2}} r\,dz\,dr\,d\theta$

Ans. $\frac{a^3}{3}\left(\frac{\pi}{2}-\frac{2}{3}\right)$

5. Evaluate $\int_{-1}^{1}\int_{0}^{2}\int_{x-3}^{x+2}(x+y+z)\,dy\,dz\,dz$

Ans. 0

6. Evaluate $\int_{1}^{3}\int_{1/x}^{1}\int_{2}^{\sqrt{xy}} xyz\,dz\,dy\,dz$

Ans. $\frac{13}{9}-\frac{1}{6}\log 3$

7. Evaluate $\int_{0}^{\pi/2} d\theta \int_{a}^{a\sin\theta} dv \int_{0}^{a^2-r^2} r\,dz$

Ans. $\frac{5a^3\pi}{64}$

8. Evaluate $\frac{dxdydz}{(x+y+z+1)^3}$ over the integrated row bounded by the co-ordinate planes and the plane $x+y+z=1$.

Ans. $\frac{1}{2}\log 2-\frac{5}{16}$

6.24 CHANGE OF VARIABLES

(*i*) If $\iint_R f(x, y)\,dx\,dy$ is a double integral and if the variables be changed to u and v by the substitution $x=\phi(u, v)$, $y=\psi(u, v)$, then the double integral is changed to $\iint f[\phi(u, v), \psi(u, v)]\,|J|\,du\,dv$ where

$$|J| = \frac{\partial(x, y)}{\partial(u, v)} = \begin{vmatrix} \frac{\partial x}{\partial u} & \frac{\partial x}{\partial v} \\ \frac{\partial y}{\partial u} & \frac{\partial y}{\partial v} \end{vmatrix}$$

(*ii*) To change Cartesian co-ordinate to polar co-ordinates by the transformation $x = r\cos\theta$, $y = r\sin\theta$ $\iint f(x, y)\,dx\,dy = \iint f(r\cos\theta, r\sin\theta)\,|J|\,dr\,d\theta$

where
$$|J| = \frac{\partial(x, y)}{\partial(r, \theta)} = \begin{vmatrix} \frac{\partial x}{\partial r} & \frac{\partial x}{\partial \theta} \\ \frac{\partial y}{\partial r} & \frac{\partial y}{\partial \theta} \end{vmatrix} = \begin{vmatrix} \cos\theta & -r\sin\theta \\ \sin\theta & r\cos\theta \end{vmatrix}$$

$$= r[\cos^2\theta + \sin^2\theta] = r$$

$$\iint f(x, y)\,dx\,dy = \iint f(r\cos\theta, r\sin\theta)\,r\,dr\,d\theta$$

(*iii*) To change Cartesian co-ordinates to cylindrical co-ordinates

$$x = r\cos\phi,\ y = r\sin\phi \quad z = z$$

$$|J| = \frac{\partial(x, y, z)}{\partial(r, \phi, z)} = \begin{vmatrix} \frac{\partial x}{\partial r} & \frac{\partial x}{\partial \phi} & \frac{\partial x}{\partial z} \\ \frac{\partial y}{\partial r} & \frac{\partial y}{\partial \phi} & \frac{\partial y}{\partial z} \\ \frac{\partial z}{\partial r} & \frac{\partial z}{\partial \phi} & \frac{\partial z}{\partial z} \end{vmatrix} = \begin{vmatrix} \cos\phi & -r\sin\phi & 0 \\ \sin\phi & r\cos\phi & 0 \\ 0 & 0 & 1 \end{vmatrix} = r$$

$$\iint f(x, y, z)\,dx\,dy\,dz = r\iint f(r\cos\phi, r\sin\phi, z)\,r\,dr\,d\phi\,dz$$

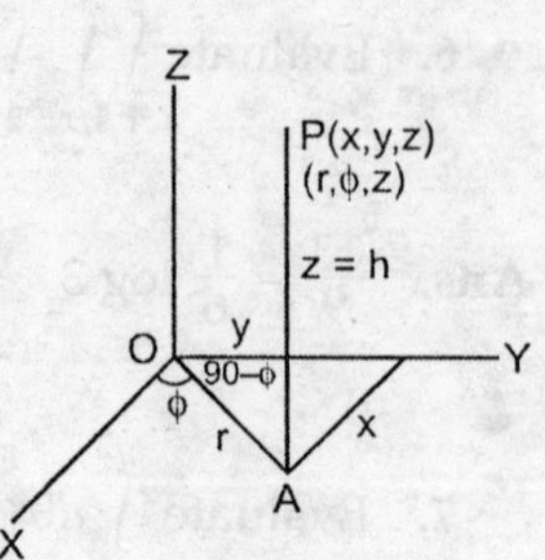

Fig. 6.63

Note: For the cylinder: $x^2 + y^2 = a^2$, $z = 0$, $z = h$

$0 \le \phi \le 2\pi, \quad 0 \le r \le a$

$0 \le z \le h$

(*iv*) To change Cartesian co-ordinates (x, y, z) to spherical co-ordinates (r, θ, ϕ) by the transformation

$$x = r\sin\theta\cos\phi$$

$$y = r\sin\theta\sin\phi$$

$$z = r\cos\theta$$

So that $x^2 + y^2 + z^2 = r^2$

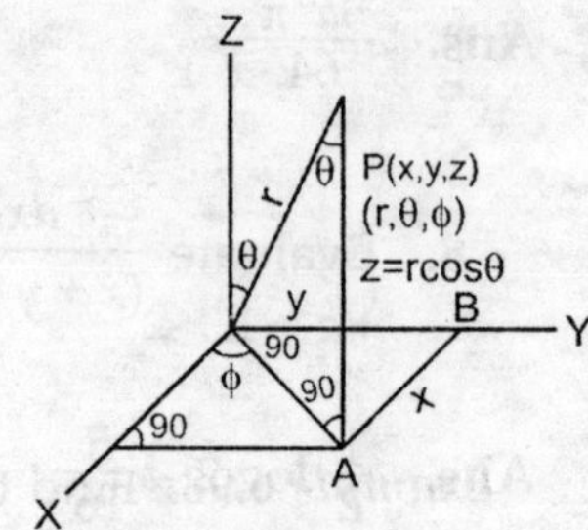

Fig. 6.64

$$|J| = \frac{\partial(x, y, z)}{\partial(r, \theta, \phi)}$$

$$= \begin{vmatrix} \frac{\partial x}{\partial r} & \frac{\partial x}{\partial \theta} & \frac{\partial x}{\partial \phi} \\ \frac{\partial y}{\partial r} & \frac{\partial y}{\partial \theta} & \frac{\partial y}{\partial \phi} \\ \frac{\partial z}{\partial r} & \frac{\partial z}{\partial \theta} & \frac{\partial z}{\partial \phi} \end{vmatrix}$$

$$= \begin{vmatrix} \sin\theta\cos\phi & r\cos\theta\sin\phi & -r\sin\theta\sin\phi \\ \sin\theta\cos\phi & r\cos\theta\sin\phi & r\sin\theta\cos\phi \\ \cos\theta & -r\sin\theta & 0 \end{vmatrix} = r^2\sin\theta$$

$$\iiint_V f(x, y, z)\,dx\,dy\,dz = \iiint_V f(r\sin\theta\cos\phi, r\sin\theta\sin\phi, r\cos\theta)\,r^2\sin\theta\,dr\,d\theta\,d\phi$$

Note: For the sphere $x^2 + y^2 + z^2 = a^2$, $\quad r = a$

If the region of integrating is the whole sphere

$0 \le r \le a \quad 0 \le \theta \le \pi \quad 0 \le \phi \le 2\pi$

6.25 VOLUME AS TRIPLE INTEGRAL

$V = \iiint dx\,dy\,dz$ with appropriate limits of integration.

***Example* 6.97.** Find the volume of the ellipsoid.

$$\frac{x^2}{a^2} + \frac{y^2}{b^2} + \frac{z^2}{c^2} = 1$$

Solution. Let $OABC$ be the positive octant of the given ellipsoid which is bounded by the planes $OAB(z = 0)$; $OBC(x = 0)$ $OCA(y = 0)$ and the surface ABC.

i.e.,
$$\frac{x^2}{a^2} + \frac{y^2}{b^2} + \frac{z^2}{c^2} = 1$$

Fig. 6.65

$$\text{Volume} = 8 \iiint dx\, dy\, dz \text{ with appropriate limits}$$

$$= 8 \int_0^a \int_0^{b\sqrt{1-x^2/a^2}} \int_0^{c\sqrt{1-x^2/a^2-y^2/b^2}} dx\, dy\, dz$$

$$= 8c \int_0^a \int_0^{b\sqrt{1-x^2/a^2}} c\sqrt{1-x^2/a^2-y^2/b^2}\, dy$$

$$= \frac{8c}{b} \int_0^a dx \int_0^P \sqrt{P^2 - y^2}\, dy \qquad P = b\sqrt{1-x^2/a^2}$$

$$= \frac{8c}{b} \int_0^a dx \left[\frac{y\sqrt{P^2-y^2}}{2} + \frac{P^2}{2}\sin^{-1}\frac{y}{P} \right]_0^P = \frac{8c}{b} \int_0^a \frac{b^2}{2}\left(1 - \frac{x^2}{a^2}\right)\frac{\pi}{2}\, dx$$

$$= \frac{8c}{b} \times \frac{b^2}{2} \times \frac{\pi}{2}\left[x - \frac{x^3}{3a^2}\right]_0^a = 2b\,\pi c\left[a - \frac{a}{3}\right] = 2\pi\, bc \times \frac{2a}{3} = \frac{4}{3}\, abc$$

***Example* 6.98.** Find the volume of the solid bounded by the surface $\left(\frac{x}{a}\right)^{2/3} + \left(\frac{y}{b}\right)^{2/3} + \left(\frac{z}{c}\right)^{2/3} = 1.$

Sol. Put
$$X = \left(\frac{x}{a}\right)^{1/3} \quad X^3 = \frac{x}{a} \quad x = aX^3$$

$$Y = \left(\frac{x}{a}\right)^{1/3} \quad Y^3 = \frac{y}{b} \quad y = bY^3$$

$$Z = \left(\frac{z}{c}\right)^{1/3} \quad Z^3 = \frac{z}{c} \quad z = cZ^3$$

The surface transforms to $X^2 + Y^2 + Z^2 = 1$

by the transformation $x = aX^3$, $y = bY^3$, $z = cZ^3$

Now
$$|J| = \frac{\partial(x, y, z)}{\partial(X, Y, Z)} = \begin{vmatrix} \frac{\partial x}{\partial X} & \frac{\partial x}{\partial Y} & \frac{\partial x}{\partial Z} \\ \frac{\partial y}{\partial X} & \frac{\partial y}{\partial Y} & \frac{\partial y}{\partial Z} \\ \frac{\partial z}{\partial X} & \frac{\partial z}{\partial Y} & \frac{\partial z}{\partial Z} \end{vmatrix} = \begin{vmatrix} 3aX^2 & 0 & 0 \\ 0 & 3bY^2 & 0 \\ 0 & 0 & 3cZ^2 \end{vmatrix} = 27\, X^2\, Y^2\, Z^2\, abc$$

$$\text{Volume} = \iiint dx\, dy\, dz = \iiint |J|\, dX\, dY\, dZ = 27 \iiint abc\, X^2 Y^2 Z^2\, dX\, dY\, dZ$$

Now the surface changes to sphere $X^2 + Y^2 + Z^2 = 1$

Changing (X, Y, Z) to spherical co-ordinates

$$X = r \sin\theta \cos\phi$$

$$Y = r \sin\theta \sin\phi$$

$$Z = r\cos\theta$$

We have $X^2 + Y^2 + Z^2 = r^2$

$\therefore$ $1 = r^2 \quad r = 1$

$$|J| = r^2 \sin\theta$$

$\therefore$ $$\text{Volume} = 27\,abc \iiint (r^2\sin^2\theta\cos^2\phi \times r^2\sin^2\theta\sin^2\phi \times r^2\cos^2\theta) \times r^2\sin\theta\, dr\, d\theta\, d\phi$$

$$= 27\,abc \int_{\phi=0}^{2\pi}\int_{\theta=0}^{\pi}\int_{r=0}^{1} r^8 \sin^5\theta\cos^2\theta\cos^2\phi\sin^2\phi\, dr\, d\theta\, d\phi$$

$$= 8 \times 27\,abc \int_0^{\pi/2}\int_0^{\pi/2}\int_0^{1} r^8 \sin^5\theta\cos^2\theta\sin^2\phi\cos^2\phi\, dr\, d\theta\, d\phi$$

$$= 216\,abc \int_0^{\pi/2}\int_0^{\pi/2} \left(\frac{r^9}{9}\right)_0^1 \sin^5\theta\cos^2\theta\sin^2\phi\cos^2\phi\, d\theta\, d\phi$$

$$= \frac{216abc}{9}\int_0^{\pi/2}\left(\frac{4\times2\times1}{7\times5\times3\times1}\right)\sin^2\phi\cos^2\phi\, d\phi$$

$$= \frac{216abc}{9} \times \frac{8}{105}\int_0^{\pi/2}\sin^2\phi\cos^2\phi\, d\phi$$

$$= \frac{8}{105} \times 24abc\left[\frac{1\times1}{4\times2}\times\frac{\pi}{2}\right] = \frac{4}{35}\pi abc$$

Example 6.99. Find by triple integration the volume of the region bounded by the paraboloid $az = x^2 + y^2$ and the cylinder $x^2 + y^2 = R^2$.

Solution. Changing to cylindrical co-ordinates $x = r\cos\theta$; $y = r\sin\theta$; $z = z$ the parabola becomes $az = r^2$, $z = \frac{r^2}{a}$

and the equation of cylinder becomes $r^2 = R^2$, $r = R$. Thus in the common region z varies from z = 0 to z =, r varies from $r = 0$ to $r = R$ and θ varies from 0 to $\frac{\pi}{2}$. Since the required volume is four times the volume in the positive octant.

$$\text{Therefore the required volume} = 4\int_0^{\pi/2}\int_0^{R}\int_0^{r^2/a} r\, dz\, dr\, d\theta = 4\int_0^{\pi/2}\int_0^{R} r\left[\frac{r^2}{a}\right] dr\, d\theta$$

$$= 2\int_0^{\pi/2}\left(\frac{r^4}{4a}\right)_0^R d\theta = \frac{1}{a}\int_0^{\pi/2} R^4\, d\theta = \frac{R^4}{a}\times\frac{\pi}{2} = \frac{\pi R^4}{2a}$$

Example 6.100. Find the volume common to the cylinders $x^2 + y^2 = a^2$ and $x^2 + z^2 = a^2$

Solution. The limits of z varies from $-\sqrt{a^2 - x^2}$ to $\sqrt{a^2 - x^2}$ and that of y varies from $-\sqrt{a^2 - x^2}$ to $\sqrt{a^2 - x^2}$ and x-varies from $-a$ to a

$\therefore$ $$V = \int_{-a}^{a}\int_{-\sqrt{a^2-x^2}}^{\sqrt{a^2-x^2}}\int_{-\sqrt{a^2-z^2}}^{\sqrt{a^2-x^2}} dx\, dy\, dz = \int_{-a}^{a}\int_{-\sqrt{a^2-x^2}}^{\sqrt{a^2-x^2}} [z]_{-\sqrt{a^2-x^2}}^{\sqrt{a^2-x^2}}\, dx\, dy$$

$$= \int_{-a}^{a} \int_{-\sqrt{a^2-x^2}}^{\sqrt{a^2-x^2}} 2\sqrt{a^2-x^2}\, dy\, dx = 2 \int_{-a}^{a} \int_{-(\sqrt{a^2-x^2})}^{\sqrt{a^2-x^2}} \sqrt{a^2-x^2}\, dy\, dx$$

$$= 2 \int_{-a}^{a} \sqrt{a^2-x^2}\, [y]_{-\sqrt{a^2-x^2}}^{\sqrt{a^2-x^2}}\, dx = 2 \int_{-a}^{a} 2(a^2-x^2)\, dx = 4\left[a^2 x - \frac{x^3}{3}\right]_{-a}^{a} = \frac{16}{3} a^3$$

Example 6.101. Find the volume bounded by the xy-plane, the cylinder $x^2 + y^2 = 1$ and the plane $x + y + z = 3$.

Solution. Required volume $\iiint_V dx\, dy\, dz$

The limits of z are 0 to $3 - y - z$

The limit of y are $-\sqrt{1-x^2}$ to $\sqrt{1-x^2}$ and the limits of x are -1 to 1

$$V = \int_{-1}^{1} \int_{-\sqrt{1-x^2}}^{\sqrt{1-x^2}} \int_{0}^{3-x-y} dx\, dy\, dz = \int_{-1}^{1} \int_{-\sqrt{1-x^2}}^{\sqrt{1-x^2}} (3 - x - y)\, dx\, dy$$

$$= \int_{-1}^{1} \left[3y - 3xy - \frac{y^2}{2}\right]_{-\sqrt{1-x^2}}^{\sqrt{1-x^2}} dx = \int_{-1}^{1} \left[y\left(3 - 3x - \frac{y}{2}\right)\right]_{-\sqrt{1-x^2}}^{\sqrt{1-x^2}} dx$$

$$= \left[\sqrt{1-x^2}\left(3 - 3x - \frac{\sqrt{1-x^2}}{2}\right) + \sqrt{1-x^2}\left(3 - 3x + \frac{\sqrt{1-x^2}}{2}\right) dx\right]$$

$$= \int_{-1}^{1} 2\sqrt{1-x^2}\, [3 - 3x]\, dx = 6 \int_{-1}^{1} \sqrt{1-x^2}\, (1 - x)\, dx$$

$$= 6 \int_{-1}^{1} (\sqrt{1-x^2} - x\sqrt{1-x^2})\, dx$$

$$= 6\left[\frac{x\sqrt{1-x^2}}{2} + \frac{1}{2}\sin^{-1} x + \frac{1}{2}\frac{(1-x^2)^{\frac{3}{2}}}{\frac{3}{2}}\right]_{-1}^{1}$$

$$= 6\left[\frac{1}{2} \times \frac{\pi}{2} + \frac{1}{2} \times \frac{\pi}{2}\right] = 3(\pi) = 3\pi$$

Example 6.102. Find volume of the tetrahedron bounded by co-ordinate planes and the plane $\frac{x}{a} + \frac{y}{b} + \frac{z}{c} = 1$.

Solution. Required volume $= \iiint dx\, dy\, dz$

Now $\quad \frac{x}{a} + \frac{y}{b} + \frac{z}{c} = 1$

put $\quad X = \frac{x}{a} \quad x = aX$

$\quad Y = \frac{y}{b} \quad y = bY$

$$Z = \frac{z}{c} \quad z = cZ$$

The plane $X + Y + Z = 1$ ...(*i*)

Now $$\iiint dx\,dy\,dz = \iiint |J|\,dX\,dY\,dZ$$

where $$|J| = \begin{vmatrix} \frac{\partial x}{\partial X} & \frac{\partial x}{\partial Y} & \frac{\partial x}{\partial Z} \\ \frac{\partial y}{\partial X} & \frac{\partial y}{\partial Y} & \frac{\partial y}{\partial Z} \\ \frac{\partial z}{\partial X} & \frac{\partial z}{\partial Y} & \frac{\partial z}{\partial Z} \end{vmatrix} = \begin{vmatrix} a & 0 & 0 \\ 0 & b & 0 \\ 0 & 0 & c \end{vmatrix} = abc$$

$\therefore$ $$V = \iiint dx\,dy\,dz = abc \iiint dX\,dY\,dZ$$

$$= abc \int_0^1 \int_0^{1-X} \int_0^{1-X-Y} dX\,dY\,dZ \text{ from } (i)$$

$$= abc \int_0^1 \int_0^{1-X} (1 - X - Y)\,dX\,dY$$

$$= abc \int_0^1 \left[\frac{(1 - X - Y)}{-2}\right]_0^{1-X} dX = \int_0^1 (1 - X)^2\,dX = \frac{abc}{6}$$

Example **6.103**. Find the volume cut off from the paraboloid $x^2 + \frac{y^2}{4} + z = 1$ by the plane $z = 0$.

Solution. Required volume is four times the volume in the first octant. Since the volume in the first octant is bounded from above by the surface

$$z = 1 - x^2 + \frac{y^2}{4}$$

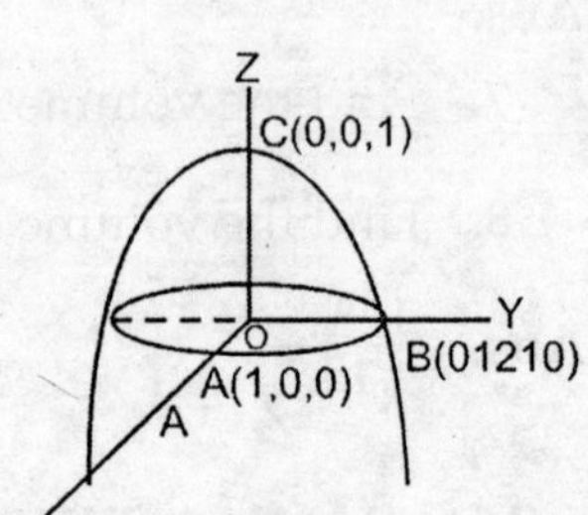

Fig. 6.66

and the base as the quadrant of the ellipse $x^2 + \frac{y^2}{4} = 1$ in xy-plane

$\therefore$ Volume $$(V) = 4\int_0^1 \int_0^{2\sqrt{1-x^2}} \int_0^{1-x^2-\frac{y^2}{4}} dz\,dy\,dx$$

$$= 4\int_0^1 \int_0^{2\sqrt{1-x^2}} [z]_0^{1-x^2-\frac{y^2}{4}}\,dy\,dx$$

$$= 4\int_0^1 \int_0^{2\sqrt{1-x^2}} \left(1 - x^2 - \frac{y^2}{4}\right)dy\,dx$$

$$= 4\int_0^1 \left[2(1 - x^2)^{3/2} - \frac{8}{12}(1 - x^2)^{3/2}\right]dx$$

$$= 4\int_0^1 \frac{4}{3}(1-x^2)^{3/2}\,dx = \frac{16}{3}\int_0^1 (1-x^2)^{3/2}\,dx$$

Let $x = \sin\theta,\ dx = \cos\theta,\ d\theta$

$$V = \frac{16}{3}\int_0^{\pi/2} \cos^3\theta\cos\theta\,d\theta = \frac{16}{3}\int_0^{\pi/2}\cos^4\theta\,d\theta = \frac{16}{3}\times\frac{3\times 1}{4\times 2}\times\frac{\pi}{2} = \pi$$

EXERCISE 6.6

1. Find by triple integration the volume of the paraboloid of revolution $x^2 + y^2 = 4z$ cut off by the plane $z = 4$.

Ans. 32π

2. Find the volume of the solid surfaces $x^2 + y^2 = a^2$ and $x^2 + z^2 = a^2$

Ans. $\frac{16a^2}{3}$

3. Find the volume of the portion of the sphere lying inside the cylinder $x^2 + y^2 = a^2$

Ans. $\frac{2a^3}{9}(3\pi - 4)$

4. Find the volume cut from the sphere $x^2 + y^2 + z^2 = a^2$ by the cone $x^2 + y^2 = z^2$

Ans. $\frac{\pi a^3}{9}(2-\sqrt{2})$

5. Find the volume bounded by the cylinder $x^2 + y^2 = 4$ and the hyperboloid $x^2 + y^2 - z^2 = 1$.

Ans. $4\sqrt{3}\,\pi$

6. Find the volume of the cylinder $x^2 + y^2 - 2ax = 0$ intercepted between the paraboloids $x^2 + y^2 = 2az$ and x-y plane.

Ans. $3\pi a^3$

7. Find the volume of the region bounded by $z = x^2 + y^2$, $z = 0$, $x = -a$, $x = a$ and $y = -b$, $y = a$.

8. Find the volume enclosed by the cylinders $x^2 + y^2 = 2ax$ and $z^2 = 2ax$.

Ans. $\frac{128a^3}{15}$

6.26 GAMMA FUNCTION

1. If n is positive, then the definite integral $\int_0^\infty e^{-x}x^{n-1}\,dx$ which is a function of x is called the Gamma function and is denoted by $\Gamma\nu$ Thus

$$\Gamma n = \int_0^\infty e^{-x}x^{n-1}\,dx$$

In particular

$$\Gamma 1 = \int_0^\infty e^{-x}\,dx = \left[\frac{e^{-x}}{-1}\right]_0^\infty = 1$$

6.27 REDUCTION FORMULA FOR Γ(n)

2. $\Gamma(n+1) = \int_0^\infty e^{-x} x^n\, dx = \int_0^\infty x^n e^{-x}\, dx$

$$= \left[\frac{x^n e^{-x}}{-1}\right]_0^\infty + \int_0^\infty x n^{n-1} e^{-x}\, dx = n\Gamma n$$

$$\Gamma n + 1 = n\Gamma n \text{ is the reduction formula of } \Gamma n$$

$$\Gamma n + 1 = n\Gamma n\; n(n-1)\,\Gamma n - 1$$

$$= n(n-1)(n-2)\,\Gamma n - 2$$

$$= n(n-1)(n-2) \ldots 3\,.\,2\,.\,1\,\Gamma 1$$

$$= n!\quad \Gamma 1 = 1$$

where n is positive

3. $\Gamma n + 1 = n\Gamma n$

$$\Gamma n = \frac{\Gamma n + 1}{n}$$

$$= \frac{(n+1)\Gamma n + 1}{n(n+1)} = \frac{\Gamma n + 2}{n(n+1)} \qquad n \neq 0, -1$$

$$= \frac{(n+2)\Gamma n + 2}{n(n+1)(n+2)} = \frac{\Gamma n + 3}{n(n+1)(n+2)} \qquad n \neq 0, -1, -2$$

$$= \frac{\Gamma n + k + 1}{n(n+1)(n+2) \ldots (n+k)} \qquad n \neq 0, -1, -2, -3, \ldots, -k$$

4. Value of $\Gamma \frac{1}{2}$

We have

$$\Gamma \frac{1}{2} = \int_0^\infty e^{-x}\, x^{\frac{1}{2}}\, dx = 2\int_0^\infty e^{-y^3}\, dy = 2\int_0^\infty e^{-x^2}\, dx \qquad \left\{\begin{array}{l}\text{Putting } x = y^2 \\ dx = 2y\, dy\end{array}\right.$$

$$\left[\Gamma \frac{1}{2}\right]^2 = 4\int_0^\infty e^{-y^2}\, dy \times \int_0^\infty e^{-x^2}\, dx = 4\int_0^\infty\int_0^\infty e^{(x^2+y^2)}\, dx\, dy$$

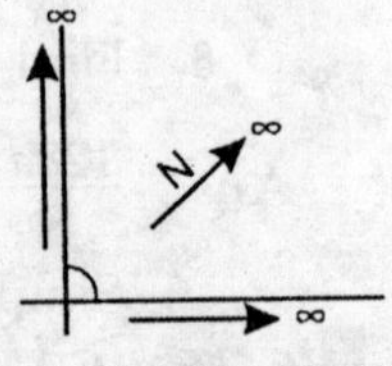

Fig. 6.67

Changing to polar coordinates

$$x = r\cos\theta$$

$$y = r\sin\theta$$

$$\therefore \quad \left[\Gamma \frac{1}{2}\right]^2 = 4\int_0^{\pi/2}\int_0^\infty e^{-r^2}\, r\, dr\, d\theta$$

$$= 4 \times \frac{\pi}{2}\int_0^\infty e^{-r^2}\, r\, dr = 2\pi\left(-\frac{1}{2}e^{-r^2}\right)_0^\infty = \pi$$

$$\Gamma \frac{1}{2} = \sqrt{\pi}$$

6.28 BETA FUNCTION

If m, n are positive then the definite integral $\int_0^1 x^{m-1}(1-x)^{n-1}\,dx$ is called the Beta function and is denoted by $\beta(m, n)$.

Thus $$\beta(m, n) = \int_0^1 x^{m-1}(1-x)^{n-1}\,dx \quad m > 0, n > 0$$

6.29 SOME IMPORTANT DEDUCTIONS

1. $\beta(m, n) = \beta(n, m)$

Since $$\beta(m, n) = \int_0^1 x^{1}(1-x)^{n-1}\,dx$$

Put $$x = 1-y$$

$$dx = -dy$$

$$= -\int_1^0 (1-y)^{m-1} y^{n-1}\,dy$$

$$= \int_0^1 y^{n-1}(1-y)^{m-1}\,dy = \beta(n, m)$$

$$\beta(m, n) = \beta(n, m)$$

2. Relation between Beta and Gamma function

i.e. $$\beta(m, n) = \frac{\Gamma m \Gamma n}{\Gamma m+n}$$

We know that $$\Gamma m = \int_0^\infty e^{-z} z^{m-1}\,dz$$

Putting $$z = x^2 \; dz = 2x\,dx$$

$$\Gamma m = 2\int_0^\infty e^{-x^2} x^{2m-1}\,dx$$

Similarly, $$\Gamma n = 2\int_0^\infty 2^{-y^2} y^{2n-1}\,dy$$

$$\Gamma m\,\Gamma n = 4\int_0^\infty\int_0^\infty e^{(-x^2+y^2)} x^{2m-1} \times y^{2n-1}\,dx\,dy$$

Fig. 6.68

Changing to polar co-ordinates, we have

$$\Gamma m\,\Gamma n = 4\int_0^{\pi/2}\int_0^\infty e^{-r^2} r^{2(m+n)-1} \cos^{2m-1}\theta \sin^{2n-1}\theta\,dr\,d\theta$$

$$= 4\int_0^\infty e^{-r^2} r^{2(m+n)-1}\,dr \int_0^{\pi/2} \cos^{2n-1}\theta \sin^{2n-1}\theta\,d\theta$$

Now 2 $$\int_0^{\infty} e^{-r^2} r^{2(m+n)-1} dr = \Gamma(m+n)$$

On putting $\sin^2\theta = z$, $2\sin\theta\cos\theta\, d\theta = dz$

$$\therefore \quad 2\int_0^{\pi/2} \cos^{2n-1}\theta \sin^{2n-1}\theta\, d\theta = \int_0^1 (1-z)^{m-1} z^{n-1}\, dz$$

$$= \int_0^1 z^{n-1}(1-z)^{m-1} dz = \beta(n, m) = \beta(m, n)$$

$\therefore$ From (1) $$\Gamma m\,\Gamma n = \Gamma m + n \times \beta(m, n)$$

$$\beta(m, n) = \frac{\Gamma m \Gamma n}{\Gamma m + n}$$

3. $$\beta(m, n) = 2\int_0^{\pi/2} \sin^{2m-1}\theta \cos^{2n-1}\theta\, d\theta$$

Now $$\beta(m, n) = \int_0^1 x^{m-1}(1-x)^{n-1} dx$$

Put $$x = 2\sin^2\theta$$

$$dx = 2\sin\theta\cos\theta\, d\theta$$

$$\therefore \quad \beta(m, n) = \int_0^{\pi/2} \sin^{2m-2}\theta \cos^{2n-2}\theta 2\sin\theta\cos\theta\, 2\theta$$

$$= 2\int_0^{\pi/2} \sin^{2m-1}\theta \cos^{2n-1}\theta\, d\theta$$

Cor. $$\Gamma\frac{1}{2} = \sqrt{\pi}$$

We know that

$$\beta(m, n) = 2\int_0^{\pi/2} \sin^{2m-1}\theta \cos^{2n-1}\theta\, d\theta$$

Put $$m = \frac{1}{2},\ n = \frac{1}{2}$$

$$\beta\left(\frac{1}{2}, \frac{1}{2}\right) = 2\int_0^{\pi/2} \sin^{2\times 1/2-1}\theta \cos^{2\times 1/2-1}\theta\, d\theta = 2 \times \frac{\pi}{2} \quad \text{...}(i)$$

We know that

$$\beta(m, n) = \frac{\Gamma m \Gamma n}{\Gamma m + n}$$

$$\beta\left(\frac{1}{2}, \frac{1}{2}\right) = \frac{\left(\Gamma\frac{1}{2}\right)^2}{\Gamma 1} = \left(\Gamma\frac{1}{2}\right)^2 \qquad ...(ii)$$

From (*i*) and (*ii*)

$$\Gamma\frac{1}{2} = \sqrt{\pi}$$

4. $$\beta(m, n) = \int_0^\infty \frac{y^{n-1}}{(1+y)^{m+n}} = \int_0^\infty \frac{x^{m-1}+x^{n-1}}{(1+x)^{m+n}}\,dx$$

We have $$\beta(m, n) = \int_0^\infty x^{m-1}(1-x)^{n-1}\,dx$$

$$\text{Put } x = \frac{1}{1+y} \Rightarrow 1-x = 1-\frac{1}{1+y} = \frac{y}{1+y}\,dx = \frac{-dy}{(1+y)^2}$$

$$= \int_\infty^0 \left(\frac{1}{1+y}\right)^{m-1}\left(\frac{y}{1+y}\right)^{n-1}\frac{-1}{(1+y)^2}\,dy$$

$$= \int_0^\infty \frac{y^{n-1}}{(1+y)^{m+n}}\,dy = \int_0^\infty \frac{y^{n-1}}{(1+y)^{m+n}}\,dy$$

$$\beta(m, n) = \int_0^\infty \frac{y^{n-1}}{(1+y)^{m+n}}\,dy = \int_0^1 \frac{y^{n-1}}{(1+y)^{m+n}}\,dy + \int_1^\infty \frac{y^{n-1}}{(1+y)^{m+n}}\,dy$$

Putting $y = \frac{1}{z}$ in the second integral we have $dy = -\frac{1}{z^2}$

$$\therefore \int_1^\infty \frac{y^{n-1}}{(1+y)^{m+n}}\,dy = \int_1^0 \frac{\left(\frac{1}{z}\right)^{n-1}}{\left(1+\frac{1}{z}\right)^{m+n}}\left(-\frac{1}{z^2}\right)dz = \int_0^1 \frac{y^{n-1}}{(1+y)^{m+n}}\,dz$$

$$\therefore \quad \beta(m, n) = \int_1^0 \frac{y^{n-1}}{(1+y)^{m+n}}\,dy + \int_0^1 \frac{z^{m-1}}{(1+z)^{m+n}}\,dz = \int_0^1 \frac{x^{n-1}+x^{m-1}}{(1+x)^{m+n}}\,dx$$

5. $$\int_0^{\pi/2} \sin^p x\cos^q dx = \frac{1}{2}\beta\left(\frac{p+1}{2}, \frac{q+1}{2}\right) = \frac{\Gamma\frac{p+1}{2}\Gamma\frac{q+1}{2}}{2\Gamma\frac{p+q+2}{2}}$$

We know that

$$\beta(m, n) = 2\int_0^{\pi/2} \sin^{2m-1}\theta\cos^{2n-1}\theta\,d\theta$$

$$\beta\left(\frac{p+1}{2}, \frac{q+1}{2}\right) = 2\int_0^{\pi/2} \sin^p x\cos^q x\,dx$$

$$\therefore \quad \int_0^{\pi/2} \sin^p x \cos^q x\, dx = \frac{1}{2}\beta\left(\frac{p+1}{2}, \frac{q+1}{2}\right) = \frac{\Gamma\frac{p+1}{2}\Gamma\frac{q+1}{2}}{2\Gamma\frac{p+q+2}{2}}$$

6.

(*a*) $$\int_0^{\pi/2} \sin^p x\, dx = \frac{\Gamma\frac{p+1}{2}}{\Gamma\frac{p+2}{2}} \times \Gamma\frac{\pi}{2}$$

(*b*) $$\int_0^{\pi/2} \cos^\theta x\, dx = \frac{\Gamma\frac{q+1}{2}}{\Gamma\frac{q+2}{2}} \times \Gamma\frac{\pi}{2}$$

(*a*) We know that $$\int_0^{\pi/2} \sin^p x \cos^q x\, dx = \frac{1}{2}\beta\left(\frac{p+1}{2}, \frac{q+1}{2}\right) = \frac{\Gamma\frac{p+1}{2}\Gamma\frac{q+1}{2}}{2\Gamma\frac{p+q+2}{2}} \quad ...(1)$$

Putting $q = 0$ in (1) we get

$$\int_0^{\pi/2} \sin^p x\, dx = \frac{\Gamma\frac{p+1}{2}\Gamma\frac{1}{2}}{2\Gamma\frac{p+2}{2}} = \frac{\Gamma\frac{p+1}{2}\sqrt{\pi}}{2\Gamma\frac{p+2}{2}} = \frac{\Gamma\frac{p+1}{2}}{\Gamma\frac{p+2}{2}} \times \frac{\sqrt{\pi}}{2}$$

(*b*) Putting $p = 0$ in (1) we get

$$\int_0^{\pi/2} \cos^q x\, dx = \frac{\Gamma\frac{q+1}{2}}{\Gamma\frac{q+2}{2}} \times \frac{\sqrt{\pi}}{2}$$

***Example* 6.104.** Evaluate $\int_0^1 x^m \left(\log\frac{1}{x}\right)^{n-1} dx$ in terms of t-functions and hence find $\int_0^1 x^y \left(\log\frac{1}{x}\right)^3 dx$

Solution. Given $\int_0^1 x^m \left(\log\frac{1}{x}\right)^{n-1} dx$

Put $$\log\frac{1}{x} = y$$

$$\frac{1}{x} = e^y \Rightarrow x^m = e^{-my}$$

$$-\frac{1}{x^2}dx = e^y dy$$

$$dx = x^2 e^y\, dy = \frac{1}{e^{2y}} e^y\, dy = -e^{-y}\, dy$$

$$\int_0^1 e^{-my}\left(\log\frac{1}{x}\right)^{n-1} dx = -\int_\infty^0 e^{-my} \times (y)^{n-1} e^{-y}\, dy$$

$$= -\int_{\infty}^{0} e^{-my-y}\, y^{n-1}\, dy = -\int_{\infty}^{0} e^{-(m+1)y} \times y^{n-1} dy$$

Put $(m+1)\, y = t$

$$dy = \frac{1}{m+1}\, dt$$

$$\int_0^1 x^n \left[\log\left(\frac{t}{n}\right)\right]^{n-1} dx = \int_0^{\infty} e^{-1}\left(\frac{t}{m+1}\right)^{n-1} \frac{1}{m+1} dt$$

$$= \frac{1}{(m+1)^n} \int_0^{\infty} e^{-1} t^{n-1} dt = \frac{1}{(m+1)^n}\, \Gamma n$$

Example **6.105**. Show that $\int_0^{\pi/2} \sqrt{\cot\theta}\, d\theta = \frac{1}{2}\Gamma\frac{1}{4}\Gamma\frac{3}{4}$.

Solution.

$$\int_0^{\pi/2} \sqrt{\cot\theta}\, d\theta = \int_0^{\pi/2} \cos^{\frac{1}{2}}\theta \sin^{-\frac{1}{2}}\theta\, d\theta$$

$$= \int_0^{\pi/2} \sin^{-\frac{1}{2}}\theta \cos^{\frac{1}{2}}\theta\, d\theta = \frac{1}{2}\, \beta\left(\frac{1}{4}, \frac{3}{4}\right) = \frac{1}{2} \left|\!\underline{\frac{1}{4}} \left|\!\underline{\frac{3}{4}}\right.\right.$$

Example **6.106**. Evaluate $\int_0^1 \frac{dx}{\sqrt{1-x^4}}$.

Solution.

$$\text{Put } x^2 = \sin\theta$$

$$2x\, dx = \cos\theta\, d\theta$$

$$dx = \frac{\cos\theta}{2\sqrt{\sin\theta}} d\theta$$

$$\int_0^1 \frac{dx}{\sqrt{1-x^4}} = \int_0^1 \frac{\cos\theta}{2\sqrt{\sin\theta}} \times \frac{1}{\sqrt{1-\sin^2\theta}} d\theta$$

$$= \frac{1}{2}\int_0^{\pi/2} \sin^{-1/2}\theta\, d\theta = \frac{\Gamma\left(\frac{-\frac{1}{2}+1}{2}\right)\frac{\sqrt{\pi}}{2}}{2\Gamma\left(\frac{-\frac{1}{2}+2}{2}\right)} = \frac{\Gamma\frac{1}{4}}{\Gamma\frac{3}{4}}\frac{\sqrt{\pi}}{4}$$

Example **6.107**. Evaluate $\int_0^{\pi/2} \sqrt{\tan\theta}\, d\theta$

Solution.

$$\int_0^{\pi/2} \sqrt{\tan\theta}\, d\theta = \int_0^{\pi/2} \sin^{\frac{1}{2}}\theta \cos^{\frac{1}{2}}\theta\, d\theta = \frac{\Gamma\left(\frac{\frac{1}{2}+1}{2}\right)\Gamma\left(\frac{-\frac{1}{2}+1}{2}\right)}{2\Gamma\left(\frac{\frac{1}{2}-\frac{1}{2}+2}{2}\right)}$$

$$= \frac{\Gamma\frac{3}{4}\Gamma\frac{1}{4}}{2\Gamma 1} = \frac{1}{2}\Gamma\frac{3}{4}\Gamma\frac{1}{4}$$

***Example* 6.108.** Evaluate.

(*a*) $\overline{|3.5}$ (*b*) $\beta\left(\frac{9}{2}, \frac{7}{2}\right)$

Solution.

(*a*) $\Gamma\, 3.5 = \Gamma\,\frac{7}{2} = \frac{5}{2} \times \frac{3}{2} \times \frac{1}{2}\Gamma\frac{1}{2}$

$$= \frac{15}{8} \times \sqrt{\pi}$$

(*b*) $\beta\left(\frac{9}{2}, \frac{7}{2}\right) = \dfrac{\Gamma\frac{9}{2}\Gamma\frac{7}{2}}{2\Gamma\left(\frac{9}{2}+\frac{7}{2}\right)}$

$$= \frac{\frac{7}{2}\times\frac{5}{2}\times\frac{3}{2}\times\frac{1}{2}\Gamma\frac{1}{2}\times\frac{5}{2}\times\frac{3}{2}\times\frac{1}{2}\Gamma\frac{1}{2}}{2\Gamma 8} = \frac{7\times5\times3\times5\times3\times1}{128\times 7!}\frac{(\sqrt{\pi})^2}{2}$$

***Example* 6.109.** Evaluate $\int_0^\infty \frac{x^c}{c^x}dx$

Solution.

$$\int_0^\infty x^c c^{-x}\,dx = \int_0^\infty x^c e^{\log e^{-x}dx}$$

$$= \int_0^\infty x^c e^{-x\log c}\,dx = \int_0^\infty e^{-x\log c} x^c\,dx$$

Put

$$x\log c = t$$

$$dx = \frac{dt}{\log c}$$

$$\int_0^\infty x^c c^{-x}\,dx = \int_0^\infty e^{-1}\left(\frac{t}{\log c}\right)^c \frac{dt}{\log c}\ \frac{1}{(\log c)^{c+1}}\int_0^\infty e^{-t}t^c\,dt$$

$$= \frac{1}{(\log c)^{c+1}}\int_0^\infty e^{-t}t^{(c+1)-1}\,dt = \frac{1}{(\log c)^{c+1}}\,\Gamma(c+1)$$

***Example* 6.110.** Prove that

(i) $\beta\left(m, \frac{1}{2}\right) = 2^{2m-1}\,\beta(m, n)$

(ii) $\Gamma m\,\Gamma\left(m+\frac{1}{2}\right) = \frac{\sqrt{\pi}}{2^{2m-1}}\,\Gamma(2m)$

Solution.

(*i*) We know that

$$\beta(m, n) = 2\int_0^{\pi/2} \sin^{2m-1}\theta \cos^{2n-1}\theta \, d\theta \quad ...(i)$$

Putting $n = \frac{1}{2}$ in (*i*) we have

$$\beta\left(m, \frac{1}{2}\right) = 2\int_0^{\pi/2} \sin^{2m-1}\theta \, d\theta$$

Putting $n = m$ in (*i*), we have

$$\beta(m, m) = 2\int_0^{\pi/2} (\sin\theta\cos\theta)^{2m-1} d\theta$$

$$= \frac{2}{2^{2m-1}}\int_0^{\pi/2} \sin^{2m-1}(2\theta)\, d\theta$$

Putting $2\theta = \phi$

$$d\theta = \frac{1}{2}\, d\phi$$

$$\beta(m, n) = \frac{1}{2^{2m-1}}\int_0^{\pi} \sin^{2m-1}\phi \, d\phi = \frac{2}{2^{2m-1}} \times 2\int_0^{\frac{\pi}{2}} \sin^{2m-1}\phi\, d\phi$$

$$2^{2m-1}\beta(m, n) = 2\int_0^{\pi/2} \sin^{2n-1}\theta \, d\theta = \beta\left(m, \frac{1}{2}\right)$$

(*ii*) Now $2^{2m-1}\beta(m, m) = \frac{2^{2m-1}\Gamma m \Gamma m}{\Gamma 2m} = \frac{\Gamma m \Gamma \frac{1}{2}}{\Gamma\left(m + \frac{1}{2}\right)} \quad \left[\because 2^{2m-1}\beta(m, m) = \beta\left(m, \frac{1}{2}\right) = \frac{\Gamma m \Gamma \frac{1}{2}}{\Gamma\left(m + \frac{1}{2}\right)}\right]$

$$\Gamma m\, \Gamma\left(m + \frac{1}{2}\right) = \frac{\sqrt{\pi}\,\Gamma 2m}{2^{2m-1}}$$

***Example* 6.111.** Prove that $\beta(m, n) = \beta(m + 1, n) + \beta(m, n + 1)$.

Solution. $\beta(m + 1, n) + \beta(m, n + 1)$

$$= \int_0^1 x^m (1 - x)^{n-1} dx + \int_0^1 x^{m-1}(1 - x)^n$$

$$= \int_0^1 x^{m-1}\{x + 1 - x\}(1 - x)^{n-1} dx = \int_0^1 x^{m-1}(1 - x)^{n-1} dx = \beta(m, n)$$

***Example.* 6.77.** Prove that $\Gamma\left(n + \frac{1}{2}\right) = \frac{\sqrt{\pi}\,\Gamma(2n+1)}{2^{2n}\Gamma(n+1)}$

Solution. $\beta(m, n) = \frac{\Gamma m \Gamma n}{\Gamma m + n}$

$$\beta\left(n+\frac{1}{2}, n+\frac{1}{2}\right) = \frac{\left[\Gamma n+\frac{1}{2}\right]^2}{\Gamma 2n+1} \quad ...(i)$$

Now
$$\beta\left(n+\frac{1}{2}, n+\frac{1}{2}\right) = \int_0^1 x^{n-\frac{1}{2}}(1-x)^{n-\frac{1}{2}}$$

$$\text{Putting } x = \sin^2\theta,\ dx = 2\sin\theta\cos\theta\, d\theta$$

$$\beta\left(n+\frac{1}{2}, n+\frac{1}{2}\right) = \int_0^{\pi/2} \sin^{2n-1}\theta\cos^{2n-1}\theta\sin\theta\cos\theta\, d\theta$$

$$= 2\int_0^{\pi/2} \sin^{2n}\theta\cos^{2n}\theta\, d\theta = 2\int_0^{\pi/2}\left(\frac{\sin 2\theta}{2}\right)^{2n} d\theta$$

$$= \frac{1}{2^{2n-1}}\int_0^{\pi/2} \sin^{2n}2\theta\, d\theta = \frac{1}{2^{2n-1}} \times \frac{1}{2}\int_0^{\pi} \sin^{2n}\phi\, d\theta \qquad \text{Putting } \phi = 2\theta$$

$$= \frac{1}{2^{2n}} \times 2\int_0^{\pi/2} \sin^{2n}\phi\, d\phi \quad \left\{\int_0^{2a} f(x)\,dx = 2\int_0^{a} f(x)\,dx \text{ if } f(2a-x) = f(x)\right.$$

$$\beta\left(n+\frac{1}{2}, n+\frac{1}{2}\right) = \frac{1}{2^{2n-1}}\,\frac{\sqrt{\pi}}{2}\,\frac{\Gamma\frac{2n+1}{2}}{\Gamma\frac{2n+2}{2}} = \frac{\sqrt{\pi}\,\Gamma n+\frac{1}{2}}{2^{2n}\,\Gamma n+1}$$

$$= \frac{\sqrt{\pi}\left(\Gamma n+\frac{1}{2}\right)}{2^{2n}\,\Gamma n+1} = \frac{\left[\Gamma n+\frac{1}{2}\right]^2}{\Gamma\, 2n+1} \text{ from } (i)$$

$$\therefore \quad \Gamma\left(n+\frac{1}{2}\right) = \frac{\sqrt{\pi}\,\Gamma\, 2n+1}{2^{2n}\,\Gamma\, n+1}$$

Cor. Putting $n = \frac{1}{4}$, we have

$$\Gamma\frac{3}{4} = \frac{\sqrt{\pi}\,\Gamma\frac{3}{2}}{\sqrt{2}\,\Gamma\frac{5}{4}} = \frac{\sqrt{\pi}\,\frac{1}{2}\,\Gamma\left(\frac{1}{2}\right)}{\sqrt{2}\,\frac{1}{4}\,\Gamma\frac{1}{4}}$$

$$\Rightarrow \quad \Gamma\frac{1}{4}\,\Gamma\frac{3}{4} = \pi\sqrt{2}$$

Example 6.112. Show that $\iint x^{m-1}y^{n-1}\,dx\,dy = \frac{a^m b^n}{2n}\left[\beta\left(\frac{m}{2}, \frac{n}{2}+1\right)\right]$ where the positive quadrant of the ellipse $\frac{x^2}{a^2} + \frac{y^2}{b^2} = 1$ is $x \geq 0 \quad y \geq 0$.

Solution. Put $\frac{x}{a} = X \quad x = aX$

$$\frac{y}{b} = Y \quad y = bY$$

and the ellipse changes to $X^2 + Y^2 = 1 \quad X \geq 0, Y \geq 0$

$$\therefore \quad \iint x^{m-1} y^{n-1} dx\, dy = \iint a^{m-1} X^{m-1} b^{n-1} Y^{n-1} ab\, dX\, dY$$

$$= a^m b^n \iint X^{m-1} Y^{n-1} dX\, dY$$

$$= a^m b^n \int_0^1 \int_0^{\sqrt{1-X^2}} X^{m-1} Y^{n-1} dX\, dY$$

$$= a^m b^n \int_0^1 X^{m-1} \left[\frac{Y^n}{n}\right]_0^{\sqrt{1-X^2}} dX$$

$$= \frac{a^m b^n}{n} \int_0^1 X^{m-1} (1-X^2)^{\frac{n}{2}} dX$$

Fig. 6.69

put $\qquad X^2 = t, X = t^{1/2}, dX = \frac{1}{2\sqrt{t}} dt$

$$= \frac{a^m b^n}{2n} \int_0^1 t^{\frac{m-1}{2} - \frac{1}{2}} (1-t)^{n/2} dt$$

$$= \frac{a^m b^n}{2n} \int_0^1 t^{\frac{m}{2}-1} (1-t)^{(n/2+1)-1} dt = \frac{a^m b^n}{2n} \left[\beta\left(\frac{m}{2}, \frac{n}{2}+1\right)\right]$$

***Example* 6.113.** Prove that

(a) $\iint_D x^{l-1} y^{m-1} dx\, dy = \frac{\Gamma(l)\Gamma(m)}{\Gamma(l+m+1)} h^{l+m}$ where D is the domain $x \geq 0, y \geq 0$ and $x + y \leq h$

(b) $\iiint_V x^{l-1} y^{m-1} z^{n-1} dx\, dy\, dz = \frac{\Gamma(l)\Gamma(m)\Gamma(n)}{\Gamma(l+m+n)} \quad x \geq 0\ y \geq 0\ z \geq 0 \quad x + y + z \leq 1$

Solution. Putting $\frac{x}{h} = X, \quad \frac{y}{h} = Y$ we have $X + Y \leq 1$

(*a*)
$$= \iint_D (hX)^{l-1} (hY)^{m-1} h^2\, dX\, dY$$

$$= h^{l+m} \iint_D X^{l-1} Y^{m-1} dX\, dY$$

$$= h^{l+m} \int_0^1 X^{l-1} \left[\frac{Y^m}{m}\right]_0^{1-X} dX$$

$$= \frac{h^{l+m}}{m} \int_0^1 X^{l-1} (1-X)^m dX = \frac{h^{l+m}}{m} \beta[l, m+1]$$

$$= \frac{h^{l+m}}{m} \frac{\Gamma l \times \Gamma(m+1)}{\Gamma(l+m+1)} = \frac{h^{m+1}}{m} \frac{\Gamma l \times m \Gamma m}{\Gamma(l+m+1)}$$

$$= \frac{h^{l+m}\Gamma l\,\Gamma m}{\Gamma(l+m+1)} \qquad ...(i)$$

(b) Taking $y + z \le 1 - x$ (= h, say), the triple integral

$$= \int_0^1 \int_0^{1-x} \int_0^{1-x-y} x^{l-1} y^{m-1} z^{n-1}\, dz\, dy\, dx = \int_0^1 x^{l-1} \left[\int_0^h \int_0^{h-y} y^{m-1} z^{n-1}\, dz\, dy \right]$$

$$= x^{l-1} \frac{\Gamma m \Gamma n}{\Gamma(m+n+1)} h^{m+n}\, dx \text{ by } (i)$$

$$= \frac{\Gamma m \Gamma n}{\Gamma(m+n+1)} \int_0^1 x^{l-1}(1-x)^{m+n}\, dx = \frac{\Gamma m \Gamma n}{\Gamma(m+n+1)}\, \beta(l, m+n+1)$$

$$= \frac{\Gamma m \Gamma n}{\Gamma(m+n+1)} \times \frac{\Gamma l \Gamma(l+m+n+1)}{\Gamma(l+m+n+1)}$$

$$= \frac{\Gamma(l)\Gamma(m)\Gamma(n)}{\Gamma(l+m+n+1)}$$ The result is known as dirichlet's integral

Example **6.114**. Prove that $\Gamma n \Gamma 1 - n = \dfrac{\pi}{\sin n\pi}$

Solution.

$$\beta(m, n) = \frac{\Gamma m \Gamma n}{\Gamma(m+n)}$$

$$\Rightarrow \quad \int_0^\infty \frac{x^{n-1}}{(1+x)^{m+n}}\, dx = \frac{\Gamma m \Gamma n}{\Gamma(m+n)} \qquad \text{(already proved)}$$

$$\Gamma m \Gamma n = \Gamma(m+n) \int_0^\infty \frac{x^{n-1}}{(1+x)^{m+n}}\, dx$$

Putting $m + n = 1$ we get

$$\Gamma n \Gamma 1 - n = \Gamma 1 \int_0^\infty \frac{x^{n-1}}{(1+x)}\, dx$$

$$\Rightarrow \quad \Gamma n \Gamma 1 - n = \frac{\pi}{\sin n\pi} \text{ (remember)}$$

Example **6.115**. Show that $\int_0^\infty e^{-x^2}\, dx = \frac{1}{2}\sqrt{\pi}$

Solution.

$$\text{Put } x^2 = u$$

$$2x\, dx = du \Rightarrow 2\sqrt{u}\, dx = du$$

or

$$dx = \frac{du}{2\sqrt{u}}$$

when $x = 0$, we have $u = 0$ when $n \to \infty$, $u \to \infty$

$$\int_0^\infty e^{-x^2}\, dx = \int_0^\infty e^{-u} \frac{du}{2\sqrt{u}} = \frac{1}{2} \int_0^\infty e^{-u} u^{-\frac{1}{2}}\, du$$

$$= \frac{1}{2}\int_0^\infty e^{-u}u^{+\frac{1}{2}-1} = \frac{1}{2}\Gamma\frac{1}{2} = \frac{1}{2}\sqrt{\pi}$$

Cor. show that $\int_0^\infty e^{-x^2}dx = \frac{\sqrt{\pi}}{2}$

Example 6.1ı5. Evaluate : $\int_0^\infty \frac{x^4(1+x^5)}{(1+x)^{15}}$

Solution.

$$\int_0^\infty \frac{x^4(1+x^5)}{(1+x)^{15}}dx = \int_0^\infty \frac{x^4}{(1+x)^{15}}dx + \int_0^\infty \frac{x^9}{(1+x)^{15}}$$

$$= \int_0^\infty \frac{x^{5-1}}{(1+x)^{5+10}}dx + \int_0^\infty \frac{x^{10-1}}{(1+x)^{10+5}}dx$$

$$= \beta(5, 10) + \beta(10, 5) = \beta(5, 10) + \beta(5, 10)$$

$$= 2\beta(5, 10) = \frac{2\Gamma5\Gamma10}{\Gamma5+10} = 2\frac{\Gamma5\Gamma10}{\Gamma15} = \frac{2\times 4!\times 9!}{14!} = \frac{1}{5005}$$

Example 6.117. Show that $\int_0^1 \frac{1}{(1-x^n)}dx = \frac{\sqrt{\pi}\,\Gamma\frac{1}{n}}{n\Gamma\left(\frac{1}{n}+\frac{1}{2}\right)}$

Solution Putting $x^n = \sin^2\theta$, we get

$$x = (\sin\theta)^{2/n}$$

$$dx = \frac{2}{n}(\sin\theta)^{\frac{2}{n}-1}\cos\theta\, d\theta$$

$$\int_0^1 \frac{dx}{\sqrt{1-x^n}}dx = \frac{2}{n}\int_0^{\pi/2}\frac{1}{\sqrt{1-\sin^2\theta}}(\sin\theta)^{\frac{2}{n}-1}\cos\theta\, d\theta$$

$$= \frac{2}{n}\int_0^{\pi/2}\frac{(\sin\theta)^{\frac{2}{n}-1}}{\cos\theta}\cos\theta\, d\theta$$

$$= \frac{2}{n}\int_0^{\pi/2}(\sin\theta)^{2\left(\frac{1}{n}\right)-1}(\cos\theta)^{2\left(\frac{1}{2}\right)-1}d\theta = \frac{2}{n}\,\frac{1}{2}\,\beta\left(\frac{1}{n},\frac{1}{2}\right)$$

$$= \frac{\Gamma\frac{1}{n}\Gamma\frac{1}{2}}{n\Gamma\left(\frac{1}{n}+\frac{1}{2}\right)} = \frac{\sqrt{\pi}\,\Gamma\frac{1}{n}}{n\Gamma\left(\frac{1}{n}+\frac{1}{2}\right)}$$

Example 6.118. Evaluate:

(*i*) $\Gamma\frac{1}{4}\Gamma\frac{3}{4}$ (*ii*) (2.5, 1.5)

Solution.

(i) $\Gamma\frac{1}{4}\Gamma\frac{3}{4} = \frac{\Gamma\frac{1}{4}\Gamma\frac{3}{4}}{1} = \frac{\Gamma\frac{1}{4}\Gamma\frac{3}{4}}{\Gamma 1} = \frac{\Gamma\frac{1}{4}\Gamma\frac{3}{4}}{\Gamma\left(\frac{1}{4}+\frac{3}{4}\right)} = \beta\left(\frac{1}{4}, \frac{3}{4}\right)$

(ii) $\beta(2.5, 1.5) = \beta\left(\frac{5}{2}, \frac{3}{2}\right)$

$$= \frac{\Gamma\frac{5}{2}\times\Gamma\frac{3}{2}}{\Gamma\left(\frac{5}{2}+\frac{3}{2}\right)} = \frac{\frac{3}{2}\times\frac{1}{2}\Gamma\frac{1}{2}\times\frac{1}{2}\Gamma\frac{1}{2}}{\Gamma 4} = \frac{3}{8}\times\frac{\sqrt{\pi}\times\sqrt{\pi}}{3!} = \frac{\pi}{16}$$

***Example* 6.119.** Show that $\int_0^\infty e^{-2x}x^{n-1}\,dx = \frac{\Gamma n}{2^n}$

Solution. Put $2x = t,\ dx = \frac{1}{2}\,dt$

$$\int_0^\infty e^{-2x}x^{n-1}\,dx = \int_0^\infty e^{-t}\left(\frac{t}{2}\right)^{n-1}\times\frac{1}{2}dt = \frac{1}{2^n}\int_0^\infty e^{-t}t^{n-1}\,dt = \frac{1}{2^n}\Gamma n$$

***Example* 6.120.** Show that

(i) $\int_0^\infty e^{-ax}\cos bx\, x^{n-1}\,dx = \frac{\Gamma n}{[\sqrt{(a^2+b^2)}]^n}\cos n\theta.$

(ii) $\int_0^\infty e^{-ax}\sin bx\, x^{n-1}\,dx = \frac{\Gamma n}{[\sqrt{(a^2+b^2)}]^n}\sin n\theta$

Solution. Consider $\int_0^\infty e^{-ax}e^{ibx}x^{n-1}\,dx$

$$= \int_0^\infty e^{-(a-ib)x}x^{n-1}\,dx$$

Put $(a-ib)\,x = t,\ dx = \frac{dt}{a-ib}$

$$\text{Integral} = \int_0^\infty e^{-t}\left(\frac{t}{a-ib}\right)^{n-1}\times\frac{dt}{(a-ib)}$$

$$= \int_0^\infty e^{-t}t^{n-1}\frac{1}{(a-ib)^n}dt = \frac{1}{(a-ib)^n}\int_0^\infty e^{-t}t^{n-1}dt$$

$$= \frac{1}{(a-ib)^n}\Gamma n = \frac{\Gamma n}{[r(\cos\theta - i\sin\theta)]^n}$$

where $a = r\cos\theta,\ b = r\sin\theta$

$$\text{Integral} = \frac{\Gamma n}{r^n e^{-in\theta}} = \frac{\Gamma n}{(\sqrt{a^2+b^2})n}e^{in\theta}$$

Equating real and imaging parts we get

(i) $\int_0^{\infty} e^{-ax} \cos bx\, x^{n-1}\, dx = \frac{\Gamma n}{(a^2+b^2)^{n/2}} \cos n\theta$

(ii) $\int_0^{\infty} e^{-ax} \sin bx\, x^{n-1}\, dx = \frac{\Gamma n}{(a^2+b^2)^{n/2}} \sin n\theta$

Example **6.121**. Show that

(i) $\Gamma\left(-\frac{1}{2}\right) = -2\sqrt{\pi}$ (ii) $\Gamma\left(-\frac{3}{2}\right) = \frac{4\sqrt{\pi}}{3}$

Solution. We know that

$$\Gamma n\, \Gamma(1-n) = \frac{\pi}{\sin n\pi}$$

$$\text{Put } n = \frac{1}{2}$$

$$\Gamma\left(-\frac{1}{2}\right)\Gamma\left(1+\frac{1}{2}\right) = \frac{\pi}{\sin\left(-\frac{1}{2}\pi\right)}$$

$$\Gamma\left(-\frac{1}{2}\right)\Gamma\left(\frac{3}{2}\right) = \frac{\pi}{-\sin\frac{\pi}{2}}$$

$$\Gamma\left(-\frac{1}{2}\right) = \frac{\pi}{-1} \times \frac{1}{\frac{1}{2}\Gamma\frac{1}{2}} = \frac{\pi}{-1} \times \frac{2}{\sqrt{\pi}} = -2\sqrt{\pi}$$

(ii) Put $n = -\frac{3}{2}$

$$\Gamma\left(-\frac{3}{2}\right)\Gamma\left(1+\frac{3}{2}\right) = \frac{-\pi}{\sin\frac{3\pi}{2}} = \frac{-\pi}{-1} = \pi$$

$$\Gamma\left(-\frac{3}{2}\right) = \pi \times \frac{1}{\Gamma\frac{5}{2}} = \pi \times \frac{1}{\frac{3}{2} \times \frac{1}{2} \times \Gamma\frac{1}{2}}$$

$$= \frac{4}{3}\pi \frac{1}{\sqrt{\pi}} = \frac{4}{3}\sqrt{\pi}$$

Example **6.122**. Evaluate $\int_0^{\infty} \sqrt{x}\, e^{-3\sqrt{x}}\, dx$

Solution. Put $x^{\frac{1}{3}} = t \quad x = t^3 \quad dx = 3t^2\, dt$

$$I = \int_0^{\infty} \sqrt{x}\, e^{-3\sqrt{x}}\, dx = \int_0^{\infty} t^{+3/2} e^{-1} \times 3t^2\, dt$$

$$= 3\int_0^{\infty} e^{-t} t^{7/2} dt = 3 \times \Gamma\frac{9}{2} = 3 \times \frac{7}{2} \times \frac{5}{2} \times \frac{3}{2} \times \frac{1}{2} \times \sqrt{\pi}$$

$$= \frac{315}{16}\sqrt{\pi}$$

EXERCISE 6.7

1. Evaluate $\Gamma -\frac{5}{2}$

Ans. $\frac{-8\sqrt{\pi}}{15}$

2. Show that $\frac{\Gamma\frac{1}{3}\Gamma\frac{5}{6}}{\Gamma\frac{2}{3}} = \sqrt{\pi}\, 2^{\frac{1}{3}}$

3. Prove that $\int_0^{\infty} \frac{x}{1+x^6} dx = \frac{\pi}{3\sqrt{3}}$

4. Show that $\int_0^{\infty} 4x^4 - e^{-x^4} dx = \Gamma\frac{5}{4}$

5. Show that $\int_0^{\infty} \frac{e^{-5t}}{\sqrt{t}} dt = \sqrt{\frac{\pi}{5}}\; 5 > 0$

6. Prove that $\int_0^{a} x^{n-1}(a-x)^{n-1} dx = a^{m+n-1}\,\beta(m, n)$

7. Prove that $\int_0^{\pi/2} \sqrt{\sin\theta}\, d\theta \times \int_0^{\pi/2} \frac{d\theta}{\sqrt{\sin\theta}} = \pi$

8. Show that $\int_0^{1} \frac{x^{m-1}(1-x)^{n-1}}{(a+x)^{m+n}} dx = \frac{\Gamma m \Gamma n}{a^n (1+a)^m \Gamma(m+n)}$

9. Show that $\frac{\beta(m+2, n-2)}{\beta(m, n)} = \frac{m(m+1)}{(n-1)(n-2)}$

10. Prove that $\int_0^{\infty}\int_0^{\infty} e^{(ax^2+by^2)} x^{2n-1} y^{2n-1} dx\, dy = \frac{\Gamma m \Gamma n}{4a^m b^n}$

11. The plane $\frac{x}{a} + \frac{y}{b} + \frac{z}{c} = 1$ meets the axes in A, B, C. Apply Dirichlets integral to find the volume of the tetrahedron $OABC$.

Ans. $\frac{abc}{6}$

12. Show that $\iint x^{2l-1} y^{2n-1} dx\, dy = \frac{1}{4l} a^{2(l+m)} \frac{\Gamma(l+1)\Gamma m}{\Gamma l + m + 1}$

Vector Calculus

This chapter deals with scalar products, vector product, gradient, divergence, curl of vector with geometrical interpellation, line integral, surface integral, volume integral green's, Gauss's Divergence and Stoke's theorems.

7.1 SCALAR AND VECTOR QUANTITIES

A quantity which has amplitude as well as direction is called **vector** and a quantity which has only magnitude and no direction is called **scalar**. A vector is represented by a letter with an arrow over it such as $\vec{a}$ and its magnitude by $|\vec{a}|$. In this chapter we give working knowledge of vectors. Some of these have wide application in the field of Science and Engineering.

7.2 SCALAR AND VECTOR PRODUCT, ANGULAR VELOCITIES

(*a*) **Scalar Product.** If $\vec{a}$ and $\vec{b}$ are two vectors and θ is the angle between $\vec{a}$ and $\vec{b}$, then the scalar product of these vectors is defined as $\vec{a}\cdot\vec{b} = |\vec{a}|\,|\vec{b}|\cos\theta$.

(*b*) **Vector Product or Cross Product.** If $\vec{a}$ and $\vec{b}$ are two vectors and θ is the angle between them then the magnitude of cross product of $\vec{a}$ and $\vec{b}$ is defined as $|\vec{a}\times\vec{b}| = |\vec{a}|\,|\vec{b}|\sin\theta$ and its direction is perpendicular to both $\vec{a}$ and $\vec{b}$ and forms a right handed system If $\hat{n}$ be a unit vector perpendicular to both $\vec{a}$ and $\vec{b}$, then

$$\vec{a}\times\vec{b} = |\vec{a}|\,|\vec{b}|\sin\theta\,\hat{n}$$

Fig. 7.1

(*c*) **Angular velocity.** Let a rigid body be rotating about the axis *OA* with the angular velocity $\vec{w}$ which is a vector.

Let *P* be any point on the body such that

$$\vec{OP} = \vec{r}$$

$$AOP = \theta$$

and $$AP \perp OA$$

Let *V* be the velocity of *P* and $\hat{n}$ be a unit vector ⊥ to $\vec{w}$ and $\vec{r}$ then $\vec{w}\times\vec{r} = (wr\sin\theta)\,\hat{n} = (w\,AP)\,\hat{n}$

$= $ (speed of *P*) $\hat{n}$ = velocity of *P* ⊥ to $\vec{w}$ and $\vec{r}$

$$\therefore \quad \vec{V} = \vec{w}\times\vec{r}$$

Fig. 7.2

7.3 DIFFERENTIATION OF VECTORS

Let O be the origin and P be the position of a moving particle at time 't'.

Let $$\overrightarrow{OP} = \vec{r}$$

and Q be the position of the particle at the time $t + \delta t$ such that

$$\overrightarrow{OQ} = \vec{r} + \delta\vec{r}$$

$$\therefore \quad P\vec{Q} = \overrightarrow{OQ} - \overrightarrow{OP} = (\vec{r} + \delta r) - \vec{r} = \delta\vec{r}$$

$\frac{\delta\vec{r}}{\delta t}$ is a vector. As $\delta t \to 0$, Q tends to P and the chord PQ becomes tangent at P

We define $$\frac{d\vec{r}}{dt} = \lim_{\delta t \to 0} \frac{\delta\vec{r}}{dt}, \text{ then}$$

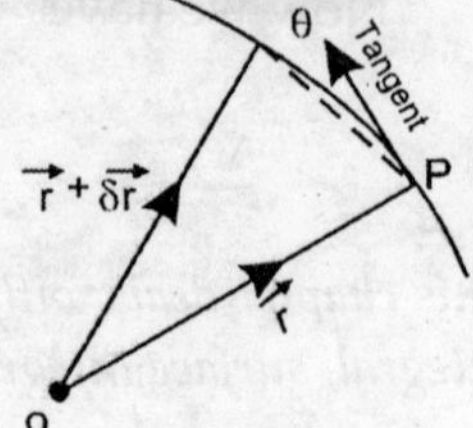

Fig. 7.3

$\frac{d\vec{r}}{dt}$ is the vector in the direction of the tangent at P

$\frac{d\vec{r}}{dt}$ is also called the differential coefficients of $\vec{r}$ w.r.t. 't'.

7.4 FORMULAE OF DIFFERENTIATION

(*i*) $$\frac{d}{dt}(\vec{F} + \vec{G}) = \frac{d\vec{F}}{dt} + \frac{d\vec{G}}{dt}$$

(*ii*) $$\frac{d}{dt}(\vec{F}\,\phi) = \frac{d\vec{F}}{dt}\phi + \vec{F}\frac{d\phi}{dt}$$

(*iii*) $$\frac{d}{dt}(\vec{F}\cdot\vec{G}) = \vec{F}\cdot\frac{d\vec{G}}{dt} + \frac{d\vec{F}}{dt}\cdot\vec{G}$$

(*iv*) $$\frac{d}{dt}(\vec{F}\times\vec{G}) = \vec{F}\times\frac{d\vec{G}}{dt} + \frac{d\vec{F}}{dt}\times\vec{G}$$

7.5 SCALAR AND VECTOR POINT FUNCTION

(*i*) **Field:** If a function is defined in any region of space for every point of the region, then this region is known as field.

(*ii*) **Scalar point function:** A function $\phi(x, y, z)$ is called a scalar point function if it associates a scalar with every point in space *e.g.* temperature distribution in a heated body, density of a body ,speed of body are some examples of a scalar point function.

(*iii*) **Vector point function:** If a function $F(x, y, z)$ defines a vector at every point of a region, then $F(x, y, z)$ is called a vector point function. The velocity of a moving fluid, gravitational force, are examples of a vector point function.

7.6 VECTOR DIFFERENTIAL OPERATOR DEL *i.e.* ∇

The operator ∇ is defined as

$$\nabla = i\frac{\partial}{\partial x} + j\frac{\partial}{\partial y} + k\frac{\partial}{\partial z}$$

(*ii*) **Gradient:** The gradient of a scalar point function ϕ is defined as $\nabla\phi$, and is written as gradient ϕ.

$$\text{Gradient } \phi = \nabla\phi = \left(i\frac{\partial}{\partial x} + j\frac{\partial}{\partial y} + k\frac{\partial}{\partial z}\right)\phi = i\frac{\partial\phi}{\partial x} + j\frac{\partial\phi}{\partial y} + k\frac{\partial\phi}{\partial z}$$

Gradient ϕ is a vector quantity.

7.7 GEOMETRICAL INTERPRETATION OF GRADIENT

$$\nabla\phi = i\frac{\partial\phi}{\partial x} + j\frac{\partial\phi}{\partial y} + k\frac{\partial\phi}{\partial z}$$

Let a level surface (the surface on which the value of the function remains the same) $\phi(x, y, z) = k$ passes through a point P whose position vector is $\vec{R}$.

Let another level surface $\phi = \delta\phi = k'$ be a neighbouring surface which passes through a Q whose position vector is $\vec{R} + \delta\vec{R}$

$$\therefore \qquad \overrightarrow{PQ} = \delta\vec{R}$$

Now we have

$$\vec{\nabla}\phi \cdot \delta\vec{R} = \left(\hat{i}\frac{\partial\phi}{\partial x} + \hat{j}\frac{\partial\phi}{\partial y} + k\frac{\partial\phi}{\partial z}\right)(\hat{i}\delta x + \hat{j}\delta y + \hat{k}\delta y) \qquad \because \vec{R} = \hat{i}x + \hat{j}x + \hat{k}z$$

$$= \frac{\partial\phi}{\partial x}\delta x + \frac{\partial\phi}{\partial y}\delta y + \frac{\partial\phi}{\partial z}\delta z$$

$$= \delta\phi$$

$$\vec{\nabla}\phi \cdot \delta\vec{R} = \delta\phi$$

Fig. 7.4

Now there are two cases

Case I: When Q lies on the surface on which P lies, then $\delta\phi = 0$

$$\nabla\phi \cdot \delta\vec{R} = 0$$

$\therefore$ $\nabla\phi$ is $\perp$ to be every $\delta\vec{R}$ lying on the surface

$\nabla\phi$ is normal to the surface at P

Let $\hat{n}$ be unit normal at P to the surface

$$\nabla\phi = |\nabla\phi|\,\hat{n}$$

Hence $\nabla\phi$ is a vector normal to the surface at P and the magnitude is $|\nabla\phi|$

Case II: When P and Q lie on two different level surfaces

then
$$\frac{\partial\phi}{\partial n} = \underset{\delta n \to 0}{Lt}\frac{\delta\phi}{\delta n} = \underset{\delta n \to 0}{Lt}\frac{\nabla\phi \cdot \delta\vec{R}}{\delta n} = |\nabla\phi| \underset{\delta n \to 0}{Lt}\frac{|\delta\vec{R}|\cos\theta}{\delta n}$$

Now $|\delta\vec{R}|\cos\theta = PQ\cos\theta = \delta n$

$$\frac{\partial\phi}{\partial n} = |\nabla\phi| \underset{\delta n \to 0}{Lt}\frac{|\delta\vec{R}|\cos\theta}{\delta n} = \underset{\delta n \to 0}{Lt}|\nabla\phi|\frac{\delta n}{\delta n} = |\nabla\phi|$$

Hence the magnitude of $\nabla\phi$ is the rate of charge of ϕ in the direction of the normal to the surface.

7.8 DIRECTIONAL DERIVATIVE OF 'ϕ' IN THE DIRECTION OF $\overrightarrow{PQ}$

If $PQ = |\delta\vec{R}|\,\delta r$ and n' is a unit vector normal in the direction of PQ, then the limiting value of $\frac{\delta\phi}{\delta r}$ as $\delta r \to 0$ is known as the directional derivative of ϕ at P along the direction of PQ

Since
$$\delta r = \frac{\delta n}{\cos\theta} = \frac{\delta n}{\hat{n}.n'} \quad \text{and } \delta n = PP'$$

$$\frac{\partial\phi}{\partial r} = \underset{\delta r \to 0}{Lt}\frac{\delta\phi}{\delta r} = \underset{\delta r \to 0}{Lt}\frac{\nabla\phi \cdot \delta\vec{R}}{\delta r}$$

$$= \underset{\delta r \to 0}{Lt} \frac{|\nabla\phi||\delta \vec{R}|\cos\theta}{\delta r} = \underset{\delta r \to 0}{Lt} \frac{|\nabla\phi|\delta r\cos\theta}{\delta r}$$

$$= \underset{\delta r \to 0}{Lt} |\nabla\phi|\cos\theta \qquad \because \delta r = |\delta \vec{R}|$$

$$= \underset{\delta r \to 0}{Lt} |\nabla\phi|\hat{n}.\hat{n}' = \nabla\phi.\hat{n}'$$

which is the directional derivative of ϕ in the direction of $\hat{n}'$. Thus the directional derivative of ϕ in the direction of $\hat{n}'$ is the resolved part of ϕ in the direction of $\hat{n}'$.

Since $\nabla\phi . \hat{n}' = |\nabla\phi| \cos\theta \le |\nabla\phi|$

It follows that $|\nabla\phi|$ is maximum in the direction of ∇f

***Example* 7.1.** Find the directional derivative of $f = x^2 yz + 4xz^2$ in the direction $2\hat{i} - \hat{j} - 2\hat{k}$ at the point $(1, -2, -1)$

Solution. Unit vector $\hat{n}' = \frac{2\hat{i} - \hat{j} - 2\hat{k}}{\sqrt{4+1+4}} = \frac{[2\hat{i} - \hat{j} - 2\hat{k}]}{3}$ in the direction $2\hat{i} - \hat{j} - 2\hat{k}$

$$\nabla f = \hat{i}\frac{\partial}{\partial x}[x^2yz + 4xz^2] + \hat{j}\frac{\partial}{\partial y}[x^2 yz + 4xz^2] + \hat{k}\frac{\partial}{\partial z}[x^2 yz + 4x z^2]$$

$$= \hat{i}\,[2xyz + 4z^2] + \hat{j}\,[x^2z] + \hat{k}\,[x^2y + 8\,xz]$$

The directional derivative of f in the direction $2\hat{i} - \hat{j} - 2\hat{k}$ at $(1, -2, -1)$ is $\nabla f . \hat{n}'$

$$= \frac{1}{3}[2\,(2xyz) + 8z^2 + (-x^2z) - 2x^2y - 16\,xz]$$

$$= \frac{1}{3}[4xyz + 8z^2 - x^2z - 2x^2y - 16\,xz]$$

$$= \frac{1}{3}[8 + 8 + 1 + 4 + 16] = \frac{1}{3}[37]$$

***Example* 7.2.** The temperature at any point in space is given by $T = xy + yz + zx$. Determine the directional derivative of T in the direction of the vector $3\hat{i} - 4\hat{k}$ at any point (1, 1, 1).

Solution.

$$T = xy + yz + zx$$

$$\nabla T = \left(\hat{i}\frac{\partial}{\partial x} + \hat{j}\frac{\partial}{\partial y} + \hat{k}\frac{\partial}{\partial z}\right)T$$

$$= \left(\hat{i}\frac{\partial}{\partial x} + \hat{j}\frac{\partial}{\partial y} + \hat{k}\frac{\partial}{\partial z}\right)(xy + yz + zx)$$

$$= \hat{i}\,(y + z) + \hat{j}\,(x + z) + \hat{k}\,(y + x)$$

$$\nabla T \text{ at } (1, 1, 1) = 2\hat{i} + 2\hat{j} + 2\hat{k}$$

Directional derivative of T at (1, 1, 1) in the direction $3\hat{i} - 4\hat{k}$

$$= (2\hat{i} + 2\hat{j} + 2\hat{k}) . \frac{3\hat{i} - 4\hat{k}}{\sqrt{9+16}} = \frac{1}{5}(6 - 8) = -\frac{2}{5}$$

***Example* 7.3.** Find the directional derivative of $f(x, y, z) = x^2 y^2 z^2$ at the point $(1, 1, -1)$ in the direction of the tangent to the curve $x = e^t$, $y = 2 \sin t + 1$ and $z = t - \cos t$ at $t = 0$.

Solution.

$$f(x, y, z) = x^2 y^2 z^2$$

$$\nabla f = 2xy^2 z^2 i + 2x^2 yz^2 j + 2x^2 y^2 z k$$

$$\text{At } t = 0 \quad x = 1, y = 1, z = 0 - 1 = -1 \text{ i.e., } (1, 1, -1)$$

Vector normal to the surface $f(x, y, z) = x^2 y^2 z^2$ at $(1, 1, -1)$ is $2i + 2j - 2k$

Now $\vec{r} = xi + yj + zk$

$$\vec{r} = e^t i + (2 \sin t + 1) j + (t - \cos t) k$$

$$\text{Tangent} = \frac{d\vec{r}}{dt} = i e^t + 2j \cos t + k(1 + \sin t)$$

tangent at $(1, 1, -1)$ *i.e.*, at $t = 0$ is $i + 2j + k$

Directional derivative in the direction of the tangent

$$= (2i + 2j - 2k) \cdot \frac{i + 2j + k}{\sqrt{6}} = \frac{2 + 4 - 2}{\sqrt{6}} = \frac{2\sqrt{6}}{3}$$

***Example* 7.4.** Find the constants a and b so that the surface $ax^2 - byz = (a + 2)x$ is orthogonal to the surface $4x^2 y + z^3 = 4$ at the point $(1, -1, 2)$.

Solution. Let $\phi_1 = ax^2 - byz - (a + 2)x$

$$\phi_2 = 4x^2 y + z^3 - 4$$

Let n_1, n_2 be the normal vectors to the surfaces

$$\phi_1 = 0, \quad \phi_2 = 0$$

$$\vec{n}_1 = \nabla\phi_1 = \hat{i}\,[2ax - (a + 2)] + \hat{j}\,[-bz] + \hat{k}\,[-by]$$

and $\vec{n}_2 = \nabla\phi_2 = \hat{i}\,[8xy] + \hat{j}\,[4x^2] + \hat{k}\,[3z^2]$

since the given surfaces are orthogonal at $(1, -1, 2)$ so the angle between the normals $\vec{n}_1$ and $\vec{n}_2$ is 90°.

Therefore $\vec{n}_1 \cdot \vec{n}_2 = 0$

$$[(a - 2)\,\hat{i} - 2b\,\hat{j} + bk] \cdot [-8i + 4j + 12k] = 0$$

$$-8(a - 2) - 2b \times 4 + 12b = 0$$

$$-8a + 16 + 4b = 0$$

$$-2a + b + 4 = 0$$

$$2a - b = 4$$

Since the point $(1, -1, 2)$ lies on the surfaces.

$$ax^2 - byz = (a + x)x$$

We get $a + 2b = a + 2$

$$b = 1$$

Putting $b = 1$ in (1) we get

$$2a - 1 = 4$$

$$2a = 5$$

$$a = 2.5$$

Hence $a = 2.5$ and $b = 1$.

***Example* 7.5.** What is the greatest rate of increase of $u = x^2 + yz^2$ at the point $(1, -1, 3)$.

Solution. $u = x^2 + zy^2$

$$\nabla u = \hat{i}\frac{\partial}{\partial x}(x^2 + yz^2) + \hat{j}\frac{\partial}{\partial y}(x^2 + yz^2) + \hat{k}\frac{\partial}{\partial z}(x^2 + yz^2)$$

$$= 2x\hat{i} + z^2\,\hat{j} + 2yz\,\hat{k} = 2\hat{i} + 9\hat{j} - 6\hat{k}$$

Hence the greatest rate of increase of u at $(1, -1, 1) = |\nabla u|$

$$= |2\hat{i} + 9\hat{j} - 6\hat{k}| = \sqrt{4 + 81 + 36} = 11$$

***Example* 7.6.** What is the directional derivative of $f = xy^2 + yz^3$ at the point $(2, -1, 1)$ in the discretion of normal to the surface $x \log z - y^2 = -4$ at $(-1, 2, 1)$.

Solution. Let $\phi(x, y, z) = x \log z - y^2$

Normal to $\phi = \nabla\phi = \frac{\partial\phi}{\partial x}\hat{i} + \frac{\partial\phi}{\partial y}\hat{j} + \frac{\partial\phi}{\partial z}\hat{k}$

$$= \hat{i}\,\log z - 2y\,\hat{j} + \frac{x}{z}\hat{k} = -4\,\hat{j} - \hat{k} \text{ at } (-1, 2, 1)$$

Also $f = xy^2 + yz^3$

$$\nabla f = \hat{i}\frac{\partial f}{\partial x} + \hat{j}\frac{\partial f}{\partial y} + \hat{k}\frac{\partial f}{\partial z} = \hat{i}\,y^2 + \hat{j}\,(2xy + z^3) + \hat{k}\,(3yz^2)$$

$$= \hat{i} - 3\hat{j} - 3\hat{k} \text{ at } (2, -1, 1)$$

Unit vector in the direction $-4\,\hat{j} - \hat{k} = \frac{-4\hat{j} - \hat{k}}{\sqrt{17}}$

The required directional derivative of f at $(2, -1, 1)$ in the direction of $\frac{-4\hat{j} - \hat{k}}{\sqrt{17}}$ is

$$= (\hat{i} - 3\hat{j} - 3\hat{k}) \cdot \frac{-4\hat{j} - \hat{k}}{\sqrt{17}} = \frac{15}{\sqrt{17}}$$

***Example* 7.7.** Calculate the angle between the tangent planes to the surface $x \log z = y^2 - 1$, $x^2y = 2 - z$ at the point $(1, 1, 1)$.

Solution. The surfaces are

$$\phi_1 = x \log z - y^2 + 1$$

$$\phi_2 = x^2y - 2 + z$$

Normal to the surface $\phi_1 = 0$

is $n_1 = \nabla\phi_1 = \hat{i}\,\log z - 2y\,\hat{j} + \frac{x}{z}\,\hat{k}$

Normal to the surface $\phi_2 = 0$

is $n_2 = \nabla\phi_2 = \hat{i}2xy + \hat{j}x^2 + \hat{k}$

$$n_1 \text{ at } (1, 1, 1) = \hat{i}\,\log 1 - 2\,\hat{j} + \hat{k} = -2\,\hat{j} + \hat{k}$$

$$n_2 \text{ at } (1, 1, 1) = 2\hat{i} + \hat{j} + \hat{k}$$

Angle between n_1 and n_2 is

$$\theta = \cos^{-1}\left[\frac{-2+1}{\sqrt{4+1}\,\sqrt{4+1+1}}\right] = \cos^{-1}\left[\frac{-1}{\sqrt{30}}\right]$$

***Example* 7.8.** Find the angle between the surfaces $x^2 + y^2 + z^2 = 9$ and $z = x^2 + y^2 - 3$ at $(2, -1, 2)$

Solution. Let $\phi_1 = x^2 + y^2 + z^2 - 9$

$$\phi_2 = z - x^2 - y^2 + 3$$

The normal to the surface $\phi_1 = 0$

is $n_1 = \nabla\phi_1 = 2x\hat{i} + 2y\hat{j} + 2z\hat{k}$

The normal to the surface $\phi_2 = 0$

is $n_2 = \nabla\phi_2 = -2x\hat{i} - 2y\hat{j} + \hat{k}$

$$n_1 \text{ at } (2, -1, 2) = 4\hat{i} - 2\hat{j} + 4\hat{k}$$

$$n_2 \text{ at } (2, -1, 2) = -4\hat{i} + 2\hat{j} + \hat{k}$$

Angle between n_1 and n_2 is

$$\theta = \cos^{-1}\left[\frac{-16-4+4}{\sqrt{16+4+16}\sqrt{16+4+1}}\right] = \cos^{-1}\left(\frac{-16}{6\times\sqrt{21}}\right)$$

Example 7.9. If f and $\vec{G}$ are point functions, prove that the components of the latter, normal and tangential to the surface $f = 0$ are $\frac{(\vec{G}.\nabla f)\nabla f}{(\nabla f)^2}$ and $\frac{\nabla f \times (\vec{G}\times\nabla f)}{(\nabla f)^2}$

Solution. Let OP represent the magnitude of the vector $\vec{G}$

i.e., $OP = |\vec{G}|$

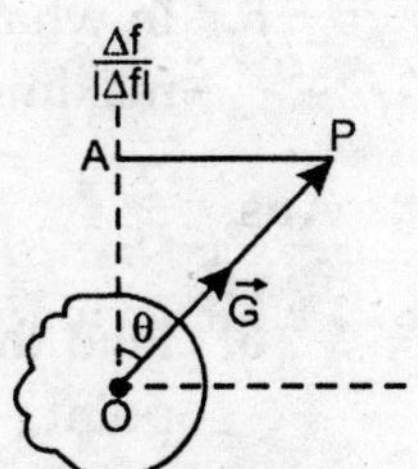

Fig. 7.5

Let θ be the angle between OP and normal to the surface $f = 0$

Now unit vector normal to the surface $f = 0$ is $\frac{\nabla f}{|\nabla f|}$

The magnitude of the components of $\vec{G}$ normal to the surface

$$OP\cos\theta = OA$$

Now $\vec{G}.\nabla f = |\vec{G}||\nabla f|\cos\theta$

$$|\vec{G}|\cos\theta = \frac{\vec{G}.\nabla f}{|\nabla f|}$$

Normal of the component of $\vec{G}$ normal to the surface $= \frac{\vec{G}.\nabla f}{|\nabla f|}$

$\therefore$ Vector component of $\vec{G}$ normal both surface $= \frac{(\vec{G}.\nabla f)}{|\nabla f|^2}\nabla f$

Tangential vector component of $\vec{G}$

$$= \vec{G} - \frac{(\vec{G}.\nabla f)\nabla f}{|\nabla f|^2} = \frac{\vec{G}|\nabla f|^2 - (\vec{G}.\nabla f)\nabla f}{|\nabla f|^2}$$

$$\frac{(\nabla f.\nabla f)\vec{G} - (\vec{G}\times\nabla f)\nabla f}{|\nabla f|^2} = \frac{\nabla f\times(\vec{G}\times\nabla f)}{(\nabla f)^2}$$

EXERCISE 7.1

1. Find a unit vector to the surface $x^3 + y^3 + 3xyz = 3$ at the point (1, 2, – 1).

Ans. $\dfrac{-\hat{i} + 3\hat{j} + 2\hat{k}}{\sqrt{4}}$

2. In what direction from (3, 1, – 2) is the directional derivative of $\phi = x^2\, y^2\, z^4$ maximum. Find also this magnitude of maximum.

Ans. $96\,(i + 3\hat{j} - 3\hat{k})96\sqrt{19}$

3. Calculate the angle between the normals to the surface $xy = z^2$ at the points (4, 1, 2) and (3, 3, – 3).

Ans. $\cos^{-1}\left(-\dfrac{1}{\sqrt{30}}\right)$

4. Find the directions derivative of the function $\phi = x^2 - y^2 + 2z^2$ at the point (1, 2, 3) in the direction of the line PQ when Q is the point (5, 0, 4).

Ans. $\left(-\dfrac{28}{\sqrt{21}}\right)$

5. In what direction from (1, 1, – 1) is the directional derivative of $\phi\ (x^2 - xy^2 + yz^2)$ maximum. Also find maximum magnitude.

Ans. $\dfrac{2}{3}$

6. Find the directional derivative of $\phi = 3e^{2x-y+z}$ at the point A (1, 1, 1) in the direction $\overline{AB}$ where B is the point (– 3, 5, 6).

Ans. $\dfrac{-5}{3}$

7. If the directional derivative of $\phi = ax^2y + by^2z + cz^2x$ at the point (1, 1, 1) has maximum magnitude 15 in the direction parallel to the line.

$$\frac{x-1}{2} = \frac{y-3}{-2} = \frac{Z}{1}, \text{ find the value of } a, b, c.$$

Ans. $a = 4\,k,\ b = -\,11\,k,\ c = 10\,k$ where $k = \pm\,\dfrac{5}{9}$

7.9 DEL APPLIED TO VECTOR POINT FUNCTIONS

(*i*) **Divergence:**

Let $$\vec{F} = F_1\,(x, y, z)\,\hat{i} + F_2\,(x, y, z)\,\hat{j} + F_3\,(x, y, z)\,\hat{k}$$

define a vector field, then divergence of a vector field denoted by div $\vec{F}$ is defined as

$$\text{div}\ \vec{F} = \nabla\,.\,\vec{F} = \hat{i}\,.\,\frac{\partial \vec{F}}{\partial x} + \hat{j}\,.\,\frac{\partial \vec{F}}{\partial y} + \hat{k}\,.\,\frac{\partial \vec{F}}{\partial z}$$

If $$\vec{F} = \hat{i}F_1 + \hat{j}F_2 + \hat{k}F_3$$

then $$\text{div}\ \vec{F} = \left(\hat{i}\frac{\partial}{\partial x} + \hat{j}\frac{\partial}{\partial y} + \hat{k}\frac{\partial}{\partial j}\right).\,(\hat{i}F_1 + \hat{j}F_2 + \hat{k}F_3) = \frac{\partial F_1}{\partial x} + \frac{\partial F_2}{\partial y} + \frac{\partial F_3}{\partial j}$$

(*ii*) **Curl:** The curl of a vector point function $\vec{F}$ is defined as

$$\text{curl } \vec{F} = \vec{\nabla} \times \vec{F} = \hat{i} \times \frac{\partial \vec{F}}{\partial x} + \hat{j} \times \frac{\partial \vec{F}}{\partial y} + \hat{k} \times \frac{\partial \vec{F}}{\partial z}$$

If $\vec{F} = \hat{i}F_1 + \hat{j}F_2 + \hat{k}F_3$, then

$$\text{curl } \vec{F} = \vec{\nabla} \times \vec{F} = \left(i\frac{\partial}{\partial x} + j\frac{\partial}{\partial y} + k\frac{\partial}{\partial z}\right) \times (iF_1 + jF_2 + kF_3)$$

$$= i\left(\frac{\partial F_3}{\partial y} - \frac{\partial F_2}{\partial z}\right) + \hat{j}\left(\frac{\partial F_1}{\partial z} - \frac{\partial F_3}{\partial x}\right) + \hat{k}\left(\frac{\partial F_2}{\partial x} - \frac{\partial F_1}{\partial y}\right)$$

$$= \begin{vmatrix} \hat{i} & \hat{j} & \hat{k} \\ \frac{\partial}{\partial x} & \frac{\partial}{\partial y} & \frac{\partial}{\partial z} \\ F_1 & F_2 & F_3 \end{vmatrix}$$

Note: Curl $\vec{F}$ can also be written as

$$\text{curl } \vec{F} = \sum\left(\frac{\partial F_3}{\partial y} - \frac{\partial F_2}{\partial z}\right)\hat{i}$$

7.10 PHYSICAL INTERPRETATION OF DIVERGENCE

Consider the motion of the fluid having velocity $\vec{V} = \hat{i}V_x + \hat{j}V_y + \hat{k}V_z$ at a point P (x, y, z). Consider a small parallelopiped with edge $\delta x, \delta y, \delta z$ parallel to the axes in the mass of the fluid with one of its corner at P.

Now the amount of fluid entering the face $P\,B'$ in the unit time $= V_y\, \delta z\, \delta x$ and the amount of fluid leaving the face $P'\,B$ in unit time $= V_{y+\delta y}\, \delta z\, \delta x$

But
$$V_{y+\delta y} = \left[V_y + \frac{\partial V_y}{\partial y}\delta y\right] \text{nearly}$$

Amount of fluid leaving the face

$$P'\,B \text{ in unit time} = \left(V_y + \frac{\partial V_y}{\partial y}\delta y\right) \delta z\, \delta x$$

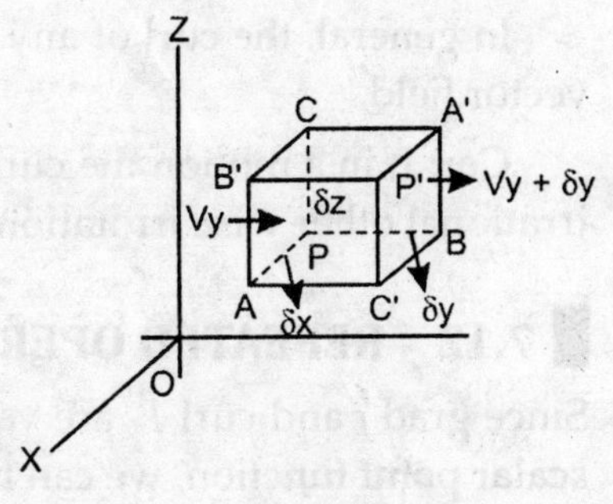

Fig. 7.6

The net decrease of the amount of fluid due to flow across these two faces

$$= \frac{\partial V_y}{\partial y}\, \delta x\, \delta y\, \delta z$$

Hence the total decrease of fluid inside the parallelopiped in unit time

$$= \left(\frac{\partial V_x}{\partial x} + \frac{\partial V_y}{\partial y} + \frac{\partial V_z}{\partial z}\right) \delta x\, \delta y\, \delta z$$

Thus the rate of loss of fluid per unit volume

$$= \frac{\partial V_x}{\partial x} + \frac{\partial V_y}{\partial y} + \frac{\partial V_z}{\partial z} = \text{div } \vec{V}$$

Therefore div $\vec{V}$ gives the rate at such fluid is originating at a point per unit volume.

Similarly if $\vec{V}$ represents an electric flux, div $\vec{V}$ is the amount of flux which diverges per unit volume in unit time. If $\vec{V}$ represents heat flux, div $\vec{V}$ is the rate at which heat is issuing from a point per unit volume. In general if $\vec{V}$ represents any physical quantity, the div $\vec{V}$ gives at each point the rate per unit volume at which the physical quantity is issuing at that point. This explains the justification for the name divergence of a vector point function.

Cor: if div $\vec{V} = 0$, every where, then such a point function $\vec{V}$ is called a solenoidal vector function.

7.11 PHYSICAL INTERPRETATION OF CURL

Consider the rotation of a rigid body about a fixed axis through O. If ω be its angular velocity, then the velocity $\vec{V}$ of any particle $P(\vec{R})$ of the body is given by,

$$\vec{V} = \vec{\omega} \times \vec{R}$$

If $$\vec{\omega} = \omega_1 \hat{i} + \omega_2 j + \omega_3 k$$

and $$\vec{R} = x\hat{i} + y\hat{j} + z\hat{k}$$

then $$\vec{V} = \omega + \vec{R} = \begin{vmatrix} \hat{i} & \hat{j} & \hat{k} \\ \omega_1 & \omega_2 & \omega_3 \\ x & y & z \end{vmatrix}$$

$$= \hat{i}\,(\omega_2 z - \omega_3 y) + \hat{j}\,(\omega_3 x - \omega_1 z) + \hat{k}\,(\omega_1 y - \omega_2 x)$$

Fig. 7.7

$$\text{curl } \vec{V} = \begin{vmatrix} \hat{i} & \hat{j} & \hat{k} \\ \dfrac{\partial}{\partial x} & \dfrac{\partial}{\partial y} & \dfrac{\partial}{\partial z} \\ \omega_2 z - \omega_3 y & \omega_3 x - \omega_1 z & \omega_1 y - \omega_2 x \end{vmatrix}$$

$$= \hat{i}\,(\omega_1 + \omega_1) + \hat{j}\,(\omega_2 + \omega_2) + \hat{k}\,(\omega_3 + \omega_3)$$

$$= 2\,(\omega_1 \hat{i} + \omega_2 \hat{j} + \omega_3 \hat{k}) = 2\omega$$

$$\vec{\omega} = \frac{1}{2} \text{ curl } \vec{V}$$

Then the angular velocity of rotation at any point is equal to half the curl of its velocity vector point function.

In general, the curl of any vector point function gives the measure of the angular velocity at any point of the vector field.

Cor: If in a motion the curl of the velocity vector point function $\vec{V}$ is zero *i.e.*, curl $\vec{V} = 0$ then $\vec{V}$ is said to be irrational other wise irrotational.

7.12 REPEATED OPERATION BY ∇

Since grad f and curl $\vec{F}$ are vector point function we can find their divergence and curl, where as div $\vec{F}$ being a scalar point function, we can find its gradient only.

Thus we have the following identities

1. div grad $f = \nabla \cdot \nabla f = \nabla^2 f = \dfrac{\partial^2 f}{\partial x^2} + \dfrac{\partial^2 f}{\partial y^2} + \dfrac{\partial^2 f}{\partial z^2}$
2. curl grad $f = \nabla \times \nabla f = 0$
3. div curl $\vec{F} = \nabla \cdot \nabla \times \vec{F} = 0$
4. curl curl $\vec{F}$ = grad div $\vec{F} - \nabla^2 \vec{F}$ *i.e.* $\nabla \times \nabla \times \vec{F} = \nabla(\nabla \cdot \vec{F}) - \nabla^2 \vec{F}$
5. grad div $\vec{F}$ = curl curl $\vec{F} + \nabla^2 \vec{F}$ *i.e.* $\nabla(\nabla \cdot \vec{F}) = \nabla \times (\nabla \times \vec{F}) + \nabla^2 \vec{F}$

Proofs

1. div grad $f = \nabla \cdot (\nabla f) = \nabla^2 f = \dfrac{\partial^2 f}{\partial x^2} + \dfrac{\partial^2 f}{\partial y^2} + \dfrac{\partial^2 f}{\partial z^2}$

Now $$\nabla . \nabla f = \left(\hat{i}\frac{\partial}{\partial x}+\hat{j}\frac{\partial}{\partial y}+\hat{k}\frac{\partial}{\partial z}\right).\left(\hat{i}\frac{\partial f}{\partial x}+\hat{j}\frac{\partial f}{\partial y}+\hat{k}\frac{\partial}{\partial z}\right)$$

$$= \frac{\partial^2 f}{\partial x^2}+\frac{\partial^2 f}{\partial y^2}+\frac{\partial^2 f}{\partial z^2} = \left(\frac{\partial^2}{\partial x^2}+\frac{\partial^2}{\partial y^2}+\frac{\partial^2}{\partial z^2}\right) f = \nabla^2 f$$

2. curl grad $f = \nabla \times (\nabla f) = \nabla \times \left(\hat{i}\frac{\partial f}{\partial x}+\hat{j}\frac{\partial f}{\partial y}+\hat{k}\frac{\partial f}{\partial z}\right)$

$$= \begin{vmatrix} \hat{i} & \hat{j} & \hat{k} \\ \frac{\partial}{\partial x} & \frac{\partial}{\partial y} & \frac{\partial}{\partial z} \\ \frac{\partial f}{\partial x} & \frac{\partial f}{\partial y} & \frac{\partial f}{\partial z} \end{vmatrix} = \sum \hat{i}\left(\frac{\partial^2 f}{\partial y \partial z}-\frac{\partial^2 f}{\partial z \partial y}\right) = 0$$

3. div curl $\vec{F} = \nabla . \nabla \times F = \left(\hat{i}\frac{\partial}{\partial x}+\hat{j}\frac{\partial}{\partial y}+\hat{k}\frac{\partial}{\partial z}\right).\left(\hat{i}\frac{\partial \vec{F}}{\partial x}+\hat{j}\frac{\partial \vec{F}}{\partial y}+\hat{k}\frac{\partial \vec{F}}{\partial z}\right)$

$$= \sum\left[\hat{i}.\left(\hat{i}\times\frac{\partial^2 \vec{F}}{\partial x^2}\right)+\hat{i}.\left(\hat{j}\times\frac{\partial^2 \vec{F}}{\partial x \partial y}\right)+\hat{i}.\left(\hat{k}\times\frac{\partial^2 \vec{F}}{\partial x \partial z}\right)\right]$$

$$= \sum\left[\hat{i}\times i.\frac{\partial^2 \vec{F}}{\partial x^2}+i\times j.\frac{\partial^2 \vec{F}}{\partial x \partial y}+i\times k.\frac{\partial^2 F}{\partial x \partial z}\right]$$

$$= \sum\left(\hat{k}\frac{\partial^2 \vec{F}}{\partial x \partial y}-\hat{j}\frac{\partial^2 \vec{F}}{\partial x \partial z}\right) = 0$$

4. curl curl $\vec{F} = \nabla \times (\times \vec{F}) = \sum i\frac{\partial}{\partial x}\times\left[\hat{i}\times\frac{\partial \vec{F}}{\partial x}+\hat{j}\times\frac{\partial \vec{F}}{\partial y}+\hat{k}\times\frac{\partial \vec{F}}{\partial z}\right]$

$$= \sum\left[\hat{i}\times\hat{i}\times\frac{\partial^2 F}{\partial x^2}+\hat{i}\times j\times\frac{\partial^2 F}{\partial x \partial y}+\hat{i}\times\hat{k}\times\frac{\partial^2 F}{\partial x \partial z}\right]$$

$$= \sum\left[\left(i.\frac{\partial^2 F}{\partial x^2}\right)i-(i.i)\frac{\partial^2 F}{\partial x^2}+\left(i.\frac{\partial^2 F}{\partial x \partial y}\right)j-(\hat{i}.j)\frac{\partial^2 F}{\partial x \partial y}+\left(i.\frac{\partial^2 F}{\partial x \partial z}\right)k-(i.k)\frac{\partial^2 F}{\partial x \partial z}\right]$$

$$= \sum i\frac{\partial}{\partial x}\left[\hat{i}.\frac{\partial \vec{F}}{\partial x}+\hat{j}\frac{\partial \vec{F}}{\partial y}+\hat{k}.\frac{\partial \vec{F}}{\partial z}\right]-\sum\frac{\partial^2 F}{\partial^2 x}$$

$$= \sum i\frac{\partial}{\partial x}\left[\hat{i}.\frac{\partial \vec{F}}{\partial x}+\hat{j}.\frac{\partial \vec{F}}{\partial y}+\hat{k}.\frac{\partial \vec{F}}{\partial z}\right]-\nabla^2 F$$

$$= \nabla(\nabla . F)-\nabla^2 F = \text{grad}(\text{div}\,\vec{F})-\nabla^2\vec{F}$$

5. In the previous problem we have proved that curl curl $\vec{F} = \nabla(\nabla . \vec{F})-\nabla^2\vec{F}$

$\Rightarrow$ $$\nabla(\nabla . \vec{F}) = \text{curl curl}\,\vec{F}+\nabla^2\vec{F} = \nabla\times(\nabla\times\vec{F})+\nabla^2\vec{F}$$

7.13 PROPERTIES OF DIVERGENCE AND CURL

1. $\nabla(fg) = f\nabla g + g\nabla f$

2. $\nabla (f\vec{G}) = \nabla f . \vec{G} + f \nabla . \vec{G}$

3. $\nabla \times (f\vec{G}) = \nabla f \times \vec{G} + f\nabla \times \vec{G}$

4. $\nabla \times (\vec{F} . \vec{G}) = (\vec{F} . \nabla)\vec{G} + (\vec{G} . \nabla)\vec{F} + \vec{F} \times (\nabla \times \vec{G}) + \vec{G} \times (\nabla \times \vec{F})$

5. $\nabla . (\vec{F} \times \vec{G}) = \vec{G} . (\nabla \times \vec{F}) - \vec{F} . (\nabla \times \vec{G})$

6. $\nabla \times (\vec{F} \times \vec{G}) = F(\nabla . \vec{G}) - G(\nabla . \vec{F}) + \vec{G} . \nabla \vec{F} - \vec{F} . \vec{G}$

Proofs:

1. $\nabla (fg) = \sum i \frac{\partial}{\partial x}(fg) = \sum i \left\{\frac{\partial f}{\partial x} g + f \frac{\partial g}{\partial x}\right\}$

$$= \sum i \frac{\partial f}{\partial x} g + \sum i\, f \frac{\partial g}{\partial x} = g \sum i \frac{\partial f}{\partial x} + f \sum i \frac{\partial g}{\partial x} = g\nabla f + f\nabla g$$

2. $\nabla . (f\vec{G}) = \sum i \frac{\partial}{\partial x}(f\vec{G}) = \sum i . \left[\frac{\partial f}{\partial x}\vec{G} + f\frac{\partial \vec{G}}{\partial x}\right]$

$$= \sum \left(i \frac{\partial f}{\partial x}\right) . \vec{G} + \sum f\left(\Sigma i . \frac{\partial \vec{G}}{\partial x}\right) = \vec{G} . \nabla f + f\nabla . \vec{G}$$

3. $\nabla \times (f\vec{G}) = \sum i \times \frac{\partial}{\partial x}(f\vec{G}) = \sum i \times \left(\frac{\partial f}{\partial x}\vec{G} + f\frac{\partial \vec{G}}{\partial x}\right)$

$$= \sum i \frac{\partial f}{\partial x} \times \vec{G} + \sum i \times \frac{\partial \vec{G}}{\partial x} f = \nabla f \times G + f\nabla \times \vec{G}$$

4. $\nabla (\vec{F} . \vec{G}) = \sum i \frac{\partial}{\partial x}(\vec{F} . \vec{G}) = \sum i \frac{\partial \vec{F}}{\partial x}\vec{G} + \sum i\, F . \frac{\partial \vec{G}}{\partial x}$

Let us consider

$$\vec{F} \times (\nabla \times \vec{G}) = \vec{F} \times \sum i \frac{\partial}{\partial x} \times \vec{G} = \vec{F} \times \sum i \times \frac{\partial \vec{G}}{\partial x}$$

$$= \sum \vec{F} \times i \times \frac{\partial \vec{G}}{\partial x} = \sum \left[\left(\vec{F} . \frac{\partial \vec{T}}{\partial x}\right) i - (\vec{F} . i)\frac{\partial \vec{G}}{\partial x}\right]$$

$$\sum i\vec{F} . \frac{\partial \vec{G}}{\partial x} = \vec{F} \times (\nabla \times \vec{G}) + \sum (\vec{F} . i)\frac{\partial \vec{G}}{\partial x}$$

Similarly

$$\sum i \frac{\partial \vec{F}}{\partial x} . \vec{G} = \vec{G} \times (\nabla \times \vec{F}) + \sum (\vec{G} . i)\frac{\partial \vec{F}}{\partial x}$$

$\therefore$

$$\nabla (\vec{F} . \vec{G}) = \vec{F} \times (\nabla \times \vec{G}) + \vec{G} \times (\nabla \times \vec{F}) + \sum (\vec{F} . i)\frac{\partial \vec{G}}{\partial x} + \sum (G . i)\frac{\partial \vec{F}}{\partial x}$$

$$= \vec{F} \times (\nabla \times \vec{G}) + \vec{G} \times (\nabla \times \vec{F}) + (\vec{F} . \nabla)\vec{G} + (\vec{G} . \nabla)\vec{F}$$

5.

$$\nabla . (\vec{F} \times \vec{G}) = \vec{G} . (\nabla \times \vec{F}) + \vec{F}(\nabla \times \vec{G})$$

$$\nabla . (\vec{F} \times \vec{G}) = \sum i . \frac{\partial \vec{F}}{\partial x}(\vec{F} \times \vec{G})$$

$$= \sum i.\left(\frac{\partial \vec{F}}{\partial x}\times\vec{G}\right)+\sum i.\left(F\times\frac{\partial \vec{G}}{\partial x}\right)$$

$$= \sum i.\left(\frac{\partial \vec{F}}{\partial x}\times\vec{G}\right)-\sum i.\left(\frac{\partial \vec{G}}{\partial x}\times\vec{F}\right)$$

$$= \sum\left(i\times\frac{\partial \vec{F}}{\partial x}\right).\vec{G}-\sum\left(i\times\frac{\partial \vec{G}}{\partial x}\right).\vec{F}$$

$$= \vec{G}.(\nabla\times\vec{F})-\vec{F}.(\nabla\times\vec{G})$$

6.

$$\nabla\times(\vec{F}\times\vec{G}) = \sum i\times\frac{\partial}{\partial x}(\vec{F}\times\vec{G})$$

$$= \sum i\times\frac{\partial \vec{F}}{\partial x}\times\vec{G}+\sum i\times\vec{F}\times\frac{\partial \vec{G}}{\partial x}$$

$$= \sum (i.G)\frac{\partial \vec{F}}{\partial x}-\left(i.\frac{\partial \vec{F}}{\partial x}\right)\vec{G}+\left(i.\frac{\partial \vec{G}}{\partial x}\right)\vec{F}-(i.F)\frac{\partial \vec{G}}{\partial x}$$

$$= \nabla\vec{F}.\vec{G}-(\nabla.\vec{F})\vec{G}+(\nabla.\vec{G})\vec{F}-\nabla\vec{G}.\vec{F}$$

$$= \vec{F}(\nabla.\vec{G})-\vec{G}(\nabla.\vec{F})+\vec{G}.(\nabla\vec{F})-\vec{F}.(\nabla\vec{G})$$

***Example* 7.10.** Prove that $\nabla r^n = nr^{n-2}\vec{R}$, when $r^2 = x^2+y^2+z^2$ $R = \hat{i}x+\hat{j}y+\hat{k}z$

Solution.

$$\nabla r^n = \nabla(x^2+y^2+z^2)^{n/2} = \sum i\frac{\partial}{\partial x}[x^2+y^2+z^2]^{n/2}$$

$$= \sum i\frac{n}{2}(x^2+y^2+z^2)^{\frac{n}{2}-1}\times 2x = \sum n(x^2+y^2+z^2)^{\frac{n-2}{2}}ix$$

$$= \Sigma n\, r^{n-2}\, ix = nr^{n-2}\,\Sigma\, ix = nr^{n-2}\,\vec{R}$$

***Example* 7.11.** If $\vec{R} = \hat{i}x+\hat{j}y+\hat{k}z$

$r^2 = x^2+y^2+z^2$

Prove that

(*i*) $\nabla\left(\frac{1}{r^2}\right) = -\frac{2\vec{R}}{r^4}$

(*ii*) $\text{div}\,(r^n\vec{R}) = (3+n)r^n$

(*iii*) $\nabla\left(\frac{\nabla.\text{R}}{\text{r}}\right) = -\frac{3}{\text{r}^3}\text{R}$

Solution. (*i*)

$$\nabla\left(\frac{1}{r^2}\right) = \sum i\frac{\partial}{\partial x}\left(\frac{1}{r^2}\right) = \sum i\frac{-2}{r^3}.\frac{\partial r}{\partial x}$$

$$= -2\sum\hat{i}\frac{1}{r^3}.\frac{\partial r}{\partial x} = -2\sum\hat{i}\frac{1}{r^3}\frac{x}{r} = -2\sum\hat{i}\frac{x}{r^4}$$

$$= -2\frac{2}{r^4}\sum(ix) = -\frac{2}{r^4}\vec{R}$$

(*ii*)

$$\text{div}\,(r^n.\vec{R}) = r^n\,\nabla.\vec{R} + \nabla r^n.\vec{R}$$

$$= r^n \sum i.\frac{\partial}{\partial x}(\hat{i}x+\hat{j}y+\hat{k}z) + \nabla r^n.\vec{R}$$

$$= r^n \sum i.i + \nabla r^n \vec{R} = r^n \sum 1 + (\nabla r^n)\vec{R} = 3r^n + (\nabla r^n).\vec{R}$$

Now

$$\nabla r^n = \sum i\left(\frac{\partial}{\partial x}\right)(r^n) = \sum inr^{n-1}\frac{\partial r}{\partial x}$$

$$= \sum inr^{n-1}\frac{x}{r} = \sum nr^{n-2}ix = nr^{n-2}\sum ix = nr^{n-2}\vec{R}$$

$\therefore$

$$\text{div}\,(r^n.\vec{R}) = 3r^n + nr^{n-2}\vec{R}.\vec{R} = 3r^n + n.r^n = (3+n)r^n$$

(*iii*)

$$\nabla\left(\frac{\nabla.R}{r}\right) = \nabla\left[\frac{\sum i\frac{\partial}{\partial x}.(\hat{i}x+\hat{j}y+\hat{k}z)}{r}\right]$$

$$= \nabla\left[\frac{3}{r}\right] = 3\nabla\left(\frac{1}{r}\right) = 3\left[\sum i\frac{\partial}{\partial x}\left(\frac{1}{r}\right)\right]$$

$$= 3\sum i\left(-\frac{1}{r^2}\right)\frac{\partial r}{\partial x} = -\frac{3}{r^2}\sum\frac{ix}{r} = -\frac{3}{r^3}\sum ix = -\frac{3}{r^3}\vec{R}$$

***Example* 7.12.** If $r^2 = x^2+y^2+z^2$ $\vec{R} = \hat{i}x+\hat{j}y+\hat{k}z$ then show that $\nabla^2 f(r) = f''(r) + \frac{2}{r}f'(r)$

Solution. Now

$$\nabla^2 f(r) = \frac{\partial^2 f(r)}{\partial x^2} + \frac{\partial^2 f(r)}{\partial y^2} + \frac{\partial^2 f(r)}{\partial z^2}$$

$$\frac{\partial f(r)}{\partial x} = \frac{\partial}{\partial r}\{f(r)\}\frac{\partial r}{\partial x} = f'(r)\frac{x}{r} = \frac{f'(r)}{r}x$$

$$\frac{\partial^2 f(r)}{\partial x^2} = \frac{\partial}{\partial x}\left[\frac{f'(r)}{r}.x\right] = \frac{f'(r)}{r} + x\frac{rf''(r)-f'(r)}{r^2}\frac{\partial r}{\partial x}$$

$$= \frac{f'(r)}{r} + x\frac{rf''(r)-f'(r)}{r^2}.\frac{x}{r} = \frac{f'(r)}{r} + \frac{x^2}{r^3}[rf''(r)-f'(r)]$$

Similarly

$$\frac{\partial^2 f(r)}{\partial y^2} = \frac{f'(r)}{r} + \frac{y^2}{r^3}[rf''(r)-f'(r)]$$

$$\frac{\partial^2 f(r)}{\partial z^2} = \frac{f'(r)}{r} + \frac{z^2}{r^3}[rf''(r)-f'(r)]$$

$$\frac{\partial^2 f(r)}{\partial x^2} + \frac{\partial^2 f(r)}{\partial y^2} + \frac{\partial^2 f(r)}{\partial z^2} = \frac{3f'(r)}{r} + \frac{x^2+y^2+z^2}{r^3}[rf''(r)-f'(r)]$$

$$= \frac{3f'(r)}{r} + \frac{1}{r}[rf''(r)-f'(r)]$$

$$= \frac{3f'(r)}{r} + f''(r) - \frac{f'(r)}{r} = f''(r) + \frac{2f'(r)}{r}$$

***Example* 7.13.** Prove that div (grad r^n) = $n(n+1)r^{n-2}$

Solution. We have grad $r^n = \nabla r^n = n\,r^{n-2}\vec{R}$

$$= n\,[(n-2)(r^{n-4})\,\vec{R}\,.\,\vec{R} + r^{n-2}\,\nabla\,.\,R]$$

$$= n\,(n-2)(r^{n-4}\,.\,r^2) + n\,r^{n-2}\,.\,3$$

$$= n\,r^{n-2}\,(n-2+3) = n\,(n+1)\,r^{n-2}.$$

***Example* 7.14.** Prove that $\vec{A}\,.\,\nabla\left(\vec{B}\,.\,\nabla\frac{1}{r}\right) = \frac{3(\vec{A}\,.\,\vec{R})(\vec{B}\,.\,\vec{R})}{r^3} - \frac{A\,.\,B}{r^3}$ where A and B are constant vectors and $\vec{R} = x\hat{i} + y\hat{j} + j\hat{k}$

Solution. We have

$$\nabla\frac{1}{r} = \frac{1}{r^3}\vec{R}$$

$$\vec{B}\,.\,\nabla\frac{1}{r} = -\frac{\vec{B}\,.\,\vec{R}}{r^3}$$

$$\nabla\left(B\,.\,\nabla\frac{1}{r}\right) = -\nabla\left[\frac{1}{r^3}(\vec{B}\,.\,\vec{R})\right] = -\left[\sum i\frac{\partial}{\partial x}\left(\frac{1}{r^3}(\vec{B}\,.\,\vec{R})\right)\right]$$

$$= -\sum i\left[\frac{\partial}{\partial x}\left(\frac{1}{r^3}\right)\vec{B}\,.\,\vec{R} + \frac{1}{r^3}\frac{\partial}{\partial x}(\vec{B}\,.\,\vec{R})\right] \qquad \left\{\frac{\partial}{\partial x}(\vec{B}\,.\,\vec{R}) = \frac{\partial\vec{B}}{\partial x}.\vec{B}\,.\vec{R}\,.\frac{\partial}{\partial x}\vec{R}\right.$$

$$= \sum i\left[\left(-\frac{3}{r^4}\frac{\partial r}{\partial x}\right)(\vec{B}\,.\,\vec{R}) + \frac{1}{r^3}(\vec{B}\,.\,i)\right] \qquad = \vec{B}\,.\,i$$

$$= -\sum i\left[\frac{x}{r}\left(-\frac{3}{r^4}\right)(\vec{B}\,.\,\vec{R}) + \frac{1}{r^3}(\vec{B}\,.\,i)\right] \qquad (\vec{B}\text{ being constant})$$

$$= \frac{3}{r^5}\sum i\,x(\vec{B}\,.\,\vec{R}) - \frac{1}{r^3}\sum(\vec{B}\,.\,i)i = \frac{3}{r^5}\vec{R}\,(\vec{B}\,.\,\vec{R}) - \frac{1}{r^3}\sum i\,(\vec{B}\,.\,i)$$

$$= \frac{3}{r^5}\,R\,(\vec{B}\,.\,\vec{R}) - \frac{1}{r^3}\vec{B} = \frac{3}{r^5}\,R\,(\vec{B}\,.\,\vec{R}) - \frac{\vec{B}}{r^3}$$

$$A\left[\nabla\left(\vec{B}\,.\,\frac{1}{r}\right)\right] = \frac{3}{r^5}\,(\vec{A}\,.\,\vec{R})(\vec{B}\,.\,\vec{R}) - \frac{\vec{A}\,.\,\vec{B}}{r^3}$$

***Example* 7.15.** If r is the distance of a point (x, y, z) from the origin. Prove that curl $\left[\hat{k}\times\text{grad}\frac{1}{r}\right]$ + grad $\left[\hat{k}\,.\,\text{grad}\frac{1}{r}\right] = 0$ where $\hat{k}$ is a unit vector in the direction of z-axis.

Solution. Since r is the distance of a point (x, y, z) from the origin $r^2 = x^2 + y^2 + z^2$

Now
$$\text{grad}\frac{1}{r} = \nabla\left(\frac{1}{r}\right) = \left(i\frac{\partial}{\partial x} + j\frac{\partial}{\partial y} + k\frac{\partial}{\partial z}\right)\left(\frac{1}{r}\right) = \sum i\left(-\frac{1}{r^2}\frac{\partial r}{\partial x}\right)$$

$$= -\sum i\left(\frac{1}{r^2}\right).\frac{x}{r} = -\sum i\frac{x}{r^3} = -\frac{\vec{R}}{r^3}$$

$$k\times\text{grad}\,\frac{1}{r} = -\frac{\hat{k}\times\vec{R}}{r^3} = -\frac{\hat{k}\times[\hat{i}x + \hat{j}y + \hat{k}z]}{r^3} = -\frac{[\hat{j}x - \hat{i}y]}{r^3} = \frac{\hat{i}y - \hat{j}x}{r^3}$$

$$\text{curl}\left[k\times\text{grad}\frac{1}{r}\right] = \nabla\times\left(\frac{iy - jx}{r^3}\right) = \nabla\times\left[\frac{1}{r^3}(iy - jx)\right]$$

$$= \frac{1}{r^3}\,[\nabla\times(iy - jx)] + \nabla\left(\frac{1}{r^3}\right)\times(iy - xj)$$

But $$\nabla \times (iy - jx) = \begin{vmatrix} \hat{i} & \hat{j} & \hat{k} \\ \frac{\partial}{\partial x} & \frac{\partial}{\partial y} & \frac{\partial}{\partial z} \\ y & -x & 0 \end{vmatrix} = \hat{i}\ (0+0) + \hat{j}\ (0-0) + \hat{k}\ (-1-1) = -2\ \hat{k}$$

$$\text{curl } \left(k \times \text{grad}\frac{1}{r}\right) = \frac{2\hat{k}}{r^3} + \left(-\frac{3}{r^5}\vec{R}\right) \times (\hat{i}y - x\hat{j})$$

$$= -\frac{2k}{r^3} - \frac{3}{r^5}\ (\hat{i}x + \hat{j}y + \hat{k}z) \times (\hat{i}y - x\hat{j})$$

$$= -\frac{2k}{r^3} - \frac{3}{r^5}\ [-\hat{k}x^2 - ky^2 + \hat{j}yz + ixz]$$

$$= \left[\frac{-2\hat{k}(x^2+y^2+z^2) + 3kx^2 + 3ky^2 - 3jyz - 3ixz}{r^5}\right]$$

$$= \left[\frac{kx^2 + ky^2 - 2\hat{k}z^2 - 3\hat{j}yz - 3\hat{i}zx}{r^5}\right]$$

$$= \left[\frac{kx^2 + ky^2 - 2\hat{k}z^2 - 3\hat{j}yz - 3\hat{i}zx}{r^5}\right] \quad \text{... } (i)$$

$$\text{grad}\left(\frac{1}{r}\right) = \left(-\frac{1}{r^3}\vec{R}\right)$$

$$\text{k. grad}\left(\frac{1}{r}\right) = \frac{\hat{k}\,.\,\vec{R}}{r^3} = -\frac{k\,.\,[\hat{i}x + \hat{j}y + \hat{k}z]}{r^3} = -\frac{z}{r^3}$$

$$\text{grad } \left[k\,.\,\text{grad}\left(\frac{1}{r}\right)\right] = \nabla\left(-\frac{z}{r^3}\right) = -\nabla\left(\frac{1}{r^3}\right) z - \frac{1}{r^3}\nabla z$$

$$= \frac{3}{r^5}\ \vec{R}\ z - \frac{1}{r^3}\ [\nabla z] = \frac{3}{r^5}\ [\hat{i}x + \hat{j}y + \hat{k}z]\ z - \frac{1}{r^3}\left[i\frac{\partial}{\partial x} + j\frac{\partial}{\partial y} + k\frac{\partial}{\partial z}\right]z = \frac{3}{r^5}\ [ixz + jyz + kz^2] - \frac{\hat{k}}{r^3}$$

$$= \frac{3ixz + 3jyz + 3kz^2 - \hat{k}(x^2+y^2+z^2)}{r^5}$$

$$= \left[\frac{kx^2 + ky^2 - 2kz^2 - 3ixz - 3jzy}{r^5}\right] \quad \text{...}(ii)$$

$$\text{grad}\left(\hat{k}\,.\,\text{grad}\frac{1}{r}\right) + \text{curl}\left[k \times \text{grad}\frac{1}{r}\right] = 0, \text{ from } (i) \text{ and } (ii).$$

***Example* 7.16.** Find curl curl curl $\vec{F}$, where $\vec{F}$ is a solenoidal vector $|\nabla\,.\,\vec{F} = 0|$

Solution. Let $$F = \hat{i}f_1 + \hat{j}f_2 + \hat{k}f_3$$

$$\text{curl } \vec{F} = \begin{vmatrix} \hat{i} & \hat{j} & \hat{k} \\ \frac{\partial}{\partial x} & \frac{\partial}{\partial y} & \frac{\partial}{\partial z} \\ f_1 & f_2 & f_3 \end{vmatrix} = \sum i\left(\frac{\partial f_3}{\partial y} - \frac{\partial f_2}{\partial z}\right)$$

$$\text{curl curl } \vec{F} = \begin{vmatrix} \hat{i} & \hat{j} & \hat{k} \\ \frac{\partial}{\partial x} & \frac{\partial}{\partial y} & \frac{\partial}{\partial z} \\ \frac{\partial f_3}{\partial y} - \frac{\partial f_2}{\partial z} & \frac{\partial f_1}{\partial z} - \frac{\partial f_3}{\partial x} & \frac{\partial f_2}{\partial x} - \frac{\partial f_1}{\partial y} \end{vmatrix}$$

$$= \sum i\left[\frac{\partial}{\partial y}\left(\frac{\partial f_2}{\partial x} - \frac{\partial f_1}{\partial y}\right) - \frac{\partial}{\partial z}\left(\frac{\partial f_1}{\partial z} - \frac{\partial f_3}{\partial x}\right)\right]$$

$$= \sum i\left[\frac{\partial^2 f_2}{\partial y \partial x} - \frac{\partial^2 f1}{\partial y^2} - \frac{\partial^2 f_1}{\partial z^2} + \frac{\partial^2 f_3}{\partial x \partial z}\right]$$

$$= \sum i\left[\frac{\partial^2 f_2}{\partial y \partial x} + \frac{\partial^2 f_3}{\partial x \partial z} + \frac{\partial^2 f_1}{\partial x^2} - \left(\frac{\partial^2 f_1}{\partial x^2} + \frac{\partial^2 f_1}{\partial y^2} + \frac{\partial^2 f_1}{\partial z^2}\right)\right]$$

$$= \sum i \frac{\partial}{\partial x}\left[\frac{\partial f_2}{\partial y} + \frac{\partial f_3}{\partial z} + \frac{\partial f_1}{\partial x}\right] - \sum i\left(\frac{\partial^2 f_1}{\partial x^2} + \frac{\partial^2 f_1}{\partial y^2} + \frac{\partial^2 f_1}{\partial z^2}\right)$$

$$= \sum i \frac{\partial}{\partial x}(\nabla . \vec{F}) - \sum i \nabla^2 f_1$$

$$= \nabla (\nabla . \vec{F}) - [i\nabla^2 f_1 + j\nabla^2 f_2 + k\nabla^2 f_3]$$

$$= \nabla [\nabla . F] - \nabla^2 F = - \nabla^2 \vec{F}$$ $\vec{F}$ is solenoidal $\therefore$ $\boxed{\nabla . \vec{F} = 0}$

Now curl curl curl curl $\vec{F}$

$$= \text{curl curl } [- \nabla^2 F] = \nabla [\nabla . (-\nabla 2 F)] - \nabla^2 (- \nabla^2 \vec{F})$$

$$= - \nabla [\nabla . \nabla^2 F] + \nabla^4 F$$

$$= -\nabla\left[i\frac{\partial}{\partial x} + j\frac{\partial}{\partial y} + k\frac{\partial}{\partial z}\right].\left\{\left(\frac{\partial^2 f_1}{\partial x^2} + \frac{\partial^2 f_1}{\partial y^2} + \frac{\partial^2 f_1}{\partial z^2}\right) + \left(\frac{\partial^2 f_2}{\partial x^2} + \frac{\partial^2 f_2}{\partial y^2} + \frac{\partial^2 f_2}{\partial z^2}\right) + \left(\frac{\partial^2 f_32}{\partial x^2} + \frac{\partial^2 f_3}{\partial y^2} + \frac{\partial^2 f_3}{\partial z^2}\right)\right\} + \nabla^4 \vec{F}$$

$$= - \nabla \left[\sum i . \frac{\partial}{\partial x}\left(\frac{\partial^2 f_1}{\partial x^2} + \frac{\partial^2 f_1}{\partial y^2} + \frac{\partial^2 f_1}{\partial z^2}\right)\right] + \nabla^4 \vec{F}$$

$$= - \nabla \left[\sum \frac{\partial^2}{\partial x^2}\left(i\frac{\partial f_1}{\partial x} + j\frac{\partial f_2}{\partial y} + k\frac{\partial f_3}{\partial z}\right)\right] + \nabla^4 \vec{F}$$

$$= - \nabla \left[\sum \frac{\partial^2}{\partial x^2}[\nabla . f1 + \nabla . f2 + \nabla . f_3]\right] + \nabla^4 F$$

$$= - \nabla \left[\sum \frac{\partial^2}{\partial x^2}[\nabla . F]\right] + \nabla^4 \vec{F} = \nabla^4 \vec{F}$$ $\boxed{\nabla . F = 0}$

or curl curl curl curl F $= \nabla [\nabla (- \nabla^2 \vec{F})] = - \nabla [\nabla^2 (\nabla . \vec{F})] + \nabla^4 \vec{F} + \nabla^4 \vec{F}$

$$= 0 + \nabla^4 \vec{F}$$ $\boxed{\nabla . \vec{F} = 0}$

$$= \nabla^4 \vec{F}$$

EXERCISE 7.2

1. Find the divergence and curl of the vectors

   (i) $\vec{V} = (x\,y\,z)\,\hat{i} + (3x^2y)\,\hat{j} + (xz^2 + y^2z)\,\hat{k}$

   (ii) $\vec{R} = (x^2 + yz)\,\hat{i} + (y^2 + zx)\,\hat{j} + (z^2 + xy)\,\hat{k}$

Ans. $yz + 3x^2 + 2zx - y^2 - 2yz\,\hat{i} + (xy - z^2)\,\hat{j} + x\,(6y - z)\,\hat{k}$

2. Find curl $\vec{F}$ where $\vec{F} = \text{grad}\,(x^2 + y^3 + z^3 - 3xyz)$

Ans. 0

3. Show that each of the vectors

   (i) $(x + 3y)\,\hat{i} + (y - 3z)\,\hat{j} + (x - 2z)\,\hat{k}$

   (ii) $3x^4 z^2\,\hat{i} + 4x^2 z^2\,\hat{j} + 3x^2 y^2\,\hat{k}$ are solenoidal

4. If $\vec{A}$ is a constant vector and $\vec{R} = x\hat{i} + y\hat{j} + z\hat{k}$

   Prove that

   (i) grad $(\vec{A}\,.\,\vec{R}) = \vec{A}$

   (ii) curl $(\vec{A} \times \vec{R}) = 2\vec{A}$

5. Prove that

   (i) $\nabla \vec{A}^2 = 2\,(\vec{A}\,.\,\vec{V}) = 2\vec{A} \times (\vec{V}\,.\,\vec{A})$ where $\vec{A}$ is a constant vector.

   (ii) curl $\{(\vec{A}\,.\,\vec{R})\,\vec{R}\} = \vec{A}\,.\,\vec{R}$

6. If r and $\vec{R}$ have their usual meanings and $\vec{A}$ is a constant vector then prove that $\nabla \times \dfrac{(\vec{A}\,.\,\vec{R})}{r^n} = \dfrac{2-n}{r^n}\vec{A} + \dfrac{n(\vec{A}\,.\,\vec{R})}{r^{n+2}}\vec{R}$

7. If V_1 and V_2 be vectors joining the fixed points (x_1, y_1, z_1) and (x_2, y_2, z_2) prove that

   (i) div $(V_1 \times V_2) = 0 - 4$, grad $(V_1\,.\,V_2) = V_1 + V_2$

   (ii) curl $(V_1 \times V_2) = 2(V_1 - V_2)$

8. If $u\,\vec{F} = \nabla v$ where u, v are scalar fields and $\vec{F}$ is a vector field, show that $\vec{F}\,.\,\text{curl}\,\vec{F} = 0$

   Remember

   (i) $\nabla\,.\,\vec{R} = 3$ (ii) $(\vec{A}\,.\,\nabla)\,\vec{R} = A$

   (iii) $\nabla \times \nabla = 0$ (iv) $\vec{R}\,.\,\vec{R} = r^2$

   (v) $\nabla\,.\,\nabla \times \vec{F} = 0$ (vi) $\nabla\,.\,\nabla = \nabla^2$

7.14 SOME BASIC CONCEPTS

1. **Smooth curve:** A curve C which possess a unique tangent at every point of it is called a smooth curve and a curve C which is composed of a finite number of smooth curves is called a piece wise smooth curve. In figure (2) the curve consists of a number of smooth curves AB, BC, CD, DE. Such a curve is called piece wise smooth curve.

Fig. 7.8

2. **Smooth surface:** A surface S which has a unique normal at each of its point and this normal depends continuously on the points of S, is called smooth surface.

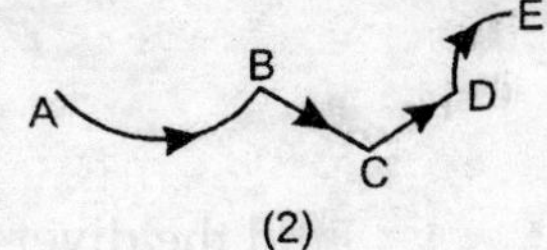

(2)

Fig. 7.9

3. **Connected region**

(*a*) **Simply connected region:** A region S in which each closed curve can be reduced a point without passing out of the region is called a simply connected region. The region in term to a circle is simply connected.

(*b*) **Multi-connected region:** A region which is not simply connected is called multi-connected. The region between two concentrated circles is called multi-connected region see fig. (3).

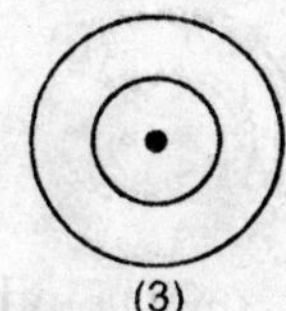
(3)

Fig. 7.10

7.15 A LINE INTEGRAL

Consider a continuous vector function $F(\vec{R})$ which is defined at each point of a curve C in space. Divide C into n parts at the points $A = P_0, P_1, P_2, \dots, P_n$. Let their position vectors be $R_0, R_1, R_2, \dots, R_n$.

Let $\vec{u}_i$ be the position vector of any point on the curve.

Now consider the sum $S = \sum_{i=0}^{n} F(u_i)\,\delta R_i$ where $\delta Ri = R_i - R_{i-1}$. The limit of the sum S as $n \to \infty$ in such a way that $|\delta R_i| \to 0$, provided it exists is called the tangential the line integral of $F(R)$ along C and is written as

$$\int_c \vec{F}(\vec{R})\,d\vec{R} = \int_c F\cdot\frac{d\vec{R}}{dt}dt$$

P_n=B, C, P_2, P_1, P_0=A

(4)

Fig. 7.11

When the path of integration is a closed curve, the symbol of integration is used as $\oint$ in place of $\int$

If $$\vec{F}(\vec{R}) = \hat{i}\,f(x, y, z) + \hat{j}\,\phi(x, y, z) + \hat{k}\,\psi(x, y, z)$$

and $$d\vec{R} = \hat{i}dx + \hat{j}dy + \hat{k}dz$$

then $$\int_c \vec{F}(\vec{R})\,d\vec{R} = \int_c (fdx + \phi dy + \psi dz)$$

Case 1. Circulation: If $\vec{F}$ represents the velocity of a fixed particle then the line integral $\int_c \vec{F}\cdot d\vec{R}$ is called the circulation of F around the curve. When the circulation $\vec{F}$ around any closed curve in a region vanishes, then $\vec{F}$ is called irrotational in the region.

Case 2. Work: If $\vec{F}$ represents the force acting on a particle moving along any arc AB, then the work done by the force $\vec{F}$ during the displacement from A to $B = \int_A^B \vec{F}\cdot d\vec{R}$.

Example. 7.17. Evaluate the integral $\int_c (x^2 + x)\,dx + (x + y^2)\,dy$ where C is the square formed by the lines $y = \pm 1$, $x = \pm 1$.

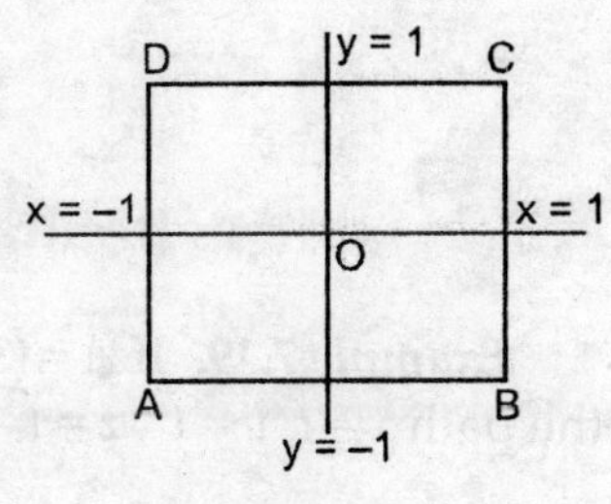

Fig. 7.12

Solution.

(*a*) Along the line $AB = -1$, $dy = 0$ x varies from -1 to 1

Given integral $$I = \int_{-1}^{1} (x^2 + x)\,dx$$

$$= 2\int_0^1 x^2\,dx = \left[\frac{x^3}{3}\right]_0^1 = \frac{2}{3} \qquad \left(\int_{-1}^{1} x\,dx = 0, \text{ function being odd}\right)$$

(*b*) Now along *BC*. $x = 1 \quad dx = 0$

y varies from – 1 to 1.

$$I = \int_{-1}^{1}(1+y^2)\,dy = \left[y + \frac{y^3}{3}\right] = 2 + \frac{2}{3} = \frac{8}{3}$$

(*c*) Along *CD* $y = 1 \quad dy = 0$

x varies from 1 to – 1.

$$I = \int_{1}^{-1}(x^2 + x)\,dx = \left[\frac{x^2}{3} + \frac{x^2}{2}\right]_1^{-1} = -2\left[\frac{1}{3}\right] = \frac{-2}{3}$$

(*d*) Along *DA* $x = -1 \quad dx = 0$

y varies from 1 to – 1.

$$I = \int_{1}^{-1}(1+y^2)\,dy = -\frac{8}{3}$$

Hence along the closed curve *ABCD*, $I = 0$

Example. **7.18.** A vector field is given by $\vec{F} = (\sin y)\,\hat{i} + x\,(1 + \cos y)\,\hat{j}$.

Evaluate the line integral over a circular path given by $x^2 + y^2 = a^2$ and $z = 0$.

Solution. Line integral $I = \int_c \vec{F}\,.\,d\vec{R}$

$$= \int_c [(\sin y)\hat{i} + x(1+\cos y)\hat{j}]\,.\,[\hat{i}dx + \hat{j}dy]$$

$$= \int_c \sin y\,dx + x(1+\cos y)\,dy$$

$$= \int_c (\sin y\,dx + x\cos y\,dy) + x\,dy = \int_c d[x\sin y] + x\,dy$$

Parametric equations of the circle $x^2 + y^2 = a^2$, $z = 0$ are $x = a\cos\theta$, $y = a\sin\theta$

$$I = \int_0^{2\pi} d[a\cos\theta\sin(a\sin\theta)] + (a\cos\theta)(a\cos\theta\,d\theta)$$

$$= [a\cos\theta\sin(a\sin\theta)]_0^{2\pi} + a^2\int_0^{2\pi}\cos^2\theta\,d\theta$$

$$= 0 + \frac{a^2}{2}\left[\theta + \frac{\sin 2\theta}{2}\right]_0^{2\pi} = 0 + \frac{a^2}{2} \times 2\pi = \pi a^2$$

Example **7.19.** If $A = (3x^2 + 6y)\,\hat{i} - 14\,yzx\,\hat{j} + 20\,x\,z^2\,\hat{k}$, evaluate $\int \vec{A}\,.\,d\vec{R}$ from the point (0, 0, 0) to (1, 1, 1) along the path $x = t$, $y = t^2$, $z = t^3$.

Solution. $I = \int \vec{A}\,.\,d\vec{R} = \int (3x^2 + 6y)\,dx + (-14yzx)\,dy + 20xz^2\,dz \quad [d\vec{R} = \hat{i}\,dx + \hat{j}\,dy + \hat{k}\,dz]$

$$x = 0, \quad y = 0, \quad z = 0$$

$$\Rightarrow \qquad t = 0 \text{ and}$$

$$\Rightarrow \qquad x = 1, \quad y = 1, \quad z = 1$$

$$t = 1$$

$$I = \int_0^1 9t^2 dt - 14t^5 \times 2t dt + 20t^7 \times 3t^2 dt$$

$$= \int_0^1 (9t^2 - 28t^7 + 60t^9) dt = \left[\frac{9t^3}{3} - \frac{28t^8}{8} + \frac{60t^{10}}{10}\right]_0^1$$

$$= \frac{9}{3} - \frac{28}{8} + \frac{60}{10} = 3 - \frac{7}{2} + 6 = \frac{6 - 7 + 12}{2} = \frac{11}{2}$$

***Example.* 7.20.** If c is a simple closed curve in x-y plane (a) not enclosing the origin, (b) enclosing the origin, evaluate $\int_c \vec{F} \cdot d\vec{R}$ where $\vec{F} = \dfrac{y\hat{i} - x\hat{j}}{x^2 + y^2}$.

Solution.

$$\vec{F} \cdot d\vec{R} = \frac{(y\hat{i} - x\hat{j})}{x^2 + y^2} \cdot (\hat{i}dx + \hat{j}dy)$$

taking $x = r\cos\theta$

$y = r\sin\theta$

$$= \frac{y}{x^2 + y^2} dx - \frac{x}{x^2 + y^2} dy$$

$dx = -r\sin\theta \, d\theta + \cos\theta \, dr$

$dy = r\cos\theta \, d\theta + \sin\theta \, dr$

$$\frac{r\sin\theta}{r^2}(-r\sin\theta \, d\theta + \cos\theta \, dr)$$

$$- \frac{r\cos\theta}{r^2}(r\cos\theta \, d\theta + \sin\theta \, dr) = -\frac{r^2}{r^2} d\theta = -d\theta$$

Fig. 7.13

(a) $$\int_c \vec{F} \cdot d\vec{R} = \int_c -d\theta = -[\theta]_{\theta_1}^{\theta_2} = 0 \text{ (not enclosing the origin)}$$

(b) $$\int_c \vec{F} \cdot d\vec{R} = \int_c -d\theta = -[\theta]_0^{2\pi} = -2\pi \text{ (enclosing the origin)}$$

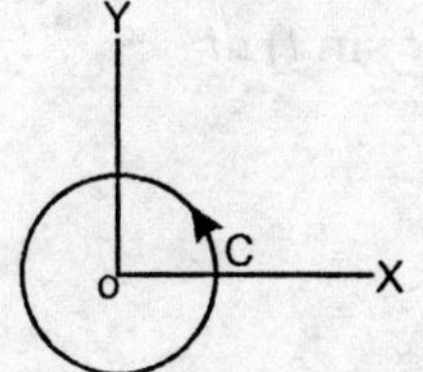

Fig. 7.14

***Example* 7.21.** Evaluate $\int_c (y - \sin x)\,dx + \cos x\,dy$ where c is the plane triangle enclosed by the lines $y = 0$, $x = \dfrac{\pi}{2}$ and $y = \dfrac{2}{\pi}x$.

Solution. The co-ordinates of B, the point of intersection of the line $x = \dfrac{\pi}{2}$ and $y = \dfrac{2}{\pi}x$ are $x = \dfrac{2}{\pi}$, $y = 1$.

Now along OA, $\qquad y = 0, dy = 0,$

x varies from 0 to $\dfrac{\pi}{2}$

$$I_1 = \int (y - \sin x)dx + \cos x \, dy = \int_0^{\pi/2} -\sin x \, dx = [\cos x]_0^{\pi/2} = -1.$$

Along AB: $x = \frac{\pi}{2}, dx = 0$

y varies from 0 to 1 $I_2 = \int_0^1 (y-1)\times 0 + \cos\frac{\pi}{2}dy = 0$

Along BO: $y = \frac{2}{\pi}x,\ dy = \frac{2}{\pi}dx$

Fig. 7.15

x varies from $\frac{\pi}{2}$ to 0

$$I_3 = -\int_0^{\frac{p}{2}}\left[\left(\frac{2}{p}x - \sin x\right)dx + \cos x\left(\frac{2}{p}dx\right)\right]$$

$$= -\left[\frac{x^2}{\pi} + \cos x + \frac{2}{\pi}\sin x\right]_0^{\frac{\pi}{2}} = -\left(\frac{\pi^2}{4\pi} - 1 + \frac{2}{\pi}\right) = -\left[\frac{\pi^2}{4\pi} - 1 + \frac{2}{\pi}\right]$$

Hence required result $= I_1 + I_2 + I_3$

$$= -1 - \left[\frac{\pi^2}{4\pi} - 1 + \frac{2}{\pi}\right] = -\frac{\pi}{4} - \frac{2}{\pi}$$

Example 7.22. If $\vec{F} = 2y\hat{i} - z\hat{j} + x\hat{k}$, $d\vec{R} = \hat{i}dx + \hat{j}dy + \hat{k}dz$, evaluate $\int \vec{F} \times d\vec{R}$ along a curve $x = \cos t$, $y = \sin t$, $z = 2\cos t$ from $t = 0$ to $t = \frac{\pi}{2}$

Solution.

$$\vec{F} \times d\vec{R} = \begin{vmatrix} \hat{i} & \hat{j} & \hat{k} \\ 2y & -z & x \\ dx & dy & dz \end{vmatrix}$$

$$= \hat{i}\,(-z\,dz - x\,dy) + \hat{j}\,(x\,dx - 2y\,dz) + \hat{k}\,(2ydy + z\,dx)$$

$\vec{F}$ □ $d\vec{R} = \hat{i}$ $[(-2\cos t)\times(-2\sin t) - \cos t \times \cos t]\,dt + \hat{j}\,(-\cos t\times\sin t + 2\sin t\times 2\sin t)\,dt + \hat{k}\,(2\cos t\sin t - 2\cos t\sin t)\,dt$

$$\vec{F} \times d\vec{R} = \int_0^{\pi/2}(4\cos t\sin t - \cos^2 t)\hat{i}\,dt + (4\sin^2 t - \sin t\cos t)\hat{j}\,dt$$

$$= \left(4\times\frac{1}{2} - \frac{1}{2}\frac{\pi}{2}\right)\hat{i} + \left(4\frac{1}{2}\times\frac{\pi}{2} - \frac{1}{2}\right)\hat{j} = \left(2 - \frac{\pi}{4}\right)\hat{i} + \left(\pi - \frac{1}{2}\right)\hat{j}$$

Example 7.23. Find the work done in moving a particle in the force field $\vec{F} = 3x^2\,\hat{i} + (2xz - y)\,\hat{j} + z\,\hat{k}$ along (*a*) the straight line from (0, 0, 0) to (2, 1, 3) (*b*) the curve defined by $x^2 = 4y = 3y^3 = 4z$ from $x = 0$ to $x = 2$.

Solution.

(*a*) the equations of the straight line from (0, 0, 0) (2, 1, 3) are $\frac{x}{2} = \frac{y}{1} = \frac{z}{3} = \lambda$ (say)*i*

$x = 2\lambda \quad y = \lambda \quad z = 3\lambda$ are the parametric equations.

The point (0, 0, 0) and (2, 1, 3) corresponds to $\lambda = 0$, and $\lambda = 1$ respectively.

$\therefore$ Work done $= \int_c \vec{F}\,.\,d\vec{R} = \int_c \{3x^2\hat{i} + (2xz - y)\hat{j} + z\hat{k}\}\,.\,\{\hat{i}dx + \hat{j}dy + \hat{k}dz\}$

$$= \int_c [3x^2 dx + (2xz - y)dy + zdz]$$

$$= \int_0^1 2(2\lambda)^2 (2d\lambda) + [4\lambda(3\lambda) - \lambda]d\lambda + [3\lambda][3d\lambda]$$

$$= \int_0^1 24\lambda^2 d\lambda + (12\lambda^2 - \lambda)d\lambda + 9\lambda d\lambda$$

$$= \int_0^1 (36\lambda^2 + 8\lambda)d\lambda = \left(\frac{36\lambda^3}{3} + \frac{8\lambda^2}{2}\right)_0^1 = 12 + 4 = 16$$

(*b*) Let $x = \lambda$ in $x^2 = 4y$, $3x^3 = 8z$. Then the parametric equations are $x = \lambda$, $y = \frac{\lambda^2}{4}$, $z = \frac{3\lambda^3}{8}$ and λ varies from 0 to 2.

$$\therefore \quad \text{Work done} = \int_c \vec{F} \cdot d\vec{R} = \left[3\lambda^2 d\lambda + \left(2\lambda\frac{3\lambda^3}{8} - \frac{\lambda^2}{4}\right)d\left(\frac{\lambda^2}{4}\right) + \frac{3\lambda^3}{8}d\left(3\frac{\lambda^2}{8}\right)\right]$$

$$= \int_c \vec{F} \cdot d\vec{R} = \int_0^2 \left(3\lambda^2 - \frac{\lambda^3}{8} + \frac{51}{64}\lambda^5\right)d\lambda = \left|\lambda^3 - \frac{\lambda^4}{32} + \frac{17}{128}\lambda^2\right|_0^2 = 16.$$

***Example* 7.24.** If $\vec{F} = (2x + y)\,\hat{i} + (3y - x)\,\hat{j}$, evaluate $\int_c \vec{F} \cdot d\vec{R}$ where c is the curve in xy plane consisting of the straight lines from (0, 0) to (2, 0) and then to (3, 2).

Solution. The path of integration c has been shown in the figure. It consists of the straight line OA and AB.

We have
$$\int_c \vec{F} \cdot d\vec{R} = \int_c [(2x + y)\hat{i} + (3y - x)\hat{j} \cdot (dx\hat{i} + dy\hat{j})]$$

$$= \int_c (2x + y)dx + (3y - x)dy$$

Now along the straight line OA, $y = 0$, $dy = 0$, x varies from 0 to 2. The equation of the straight line AB is

$$y - 0 = \frac{2-0}{3-2}(x - 2) \text{ i.e. } y = 2x - 4.$$

Along AB $y = 2x - 4$, $dy = 2\,dx$, x varies from 2 to 3

$$\int_c \vec{F} \cdot d\vec{R} = \int_0^2 [(2x + 0)dx + 0] + \int_2^3 (2x + 2x - 4)dx + (6x - 12 - x)dx$$

$$= \left[\frac{x^2}{2} \times 2\right]_0^2 + \int_2^3 (14x - 28)dx$$

$$= 4 + 14\int_2^3 (x - 2)dx = 4 + 14\left[\frac{(x-2)^2}{2}\right]_2^3 = 4 + 7 = 11.$$

Fig. 7.16

EXERCISE 7.3

(LINE INTEGRAL)

1. Evaluate $\int_C \vec{F} \cdot d\vec{R}$ where $\vec{F} = x^2 y^2 \hat{i} + y\hat{j}$ and c is $y^2 = 4x$ in the xy plane from (0, 0) to (4, 4).

Ans. -264

2. Evaluate $\int_c \vec{F} \cdot d\vec{R}$ where $\vec{F} = z\hat{i} + x\hat{j} + y\hat{k}$ and c is the arc of the curve $\vec{R} = \cos\lambda\ \hat{i} + \sin\lambda\ \hat{j} + \lambda\ \hat{k}$ from $\lambda = 0$ to $\lambda = 2\pi$.

Ans. 3π

3. Evaluate $\int_c \vec{F} \cdot d\vec{R}$ where $\vec{F} = xy\hat{i} + yz\hat{j} + zx\hat{k}$ and the curve c is $\vec{R} = \lambda\hat{i} + \lambda^2\hat{j} + \lambda^3\hat{k}$ λ varying from -1 to 1.

Ans. $\dfrac{10}{7}$

4. Evaluate $\int_c \vec{F} \cdot d\vec{R}$ where $\vec{F} = xy\hat{i} + (x^2 + y^2)\hat{j}$ and c is the rectangle in x-y plane bounded by the lines $y = 2$, $x = 4$, $y = 10$ and $x = 1$.

Ans. 60

5. Evaluate $\int_c -\dfrac{y^3\hat{i} + x^3\hat{j}}{(x^2 + y^2)^2} \cdot d\vec{R}$ where c is the boundary of the square $x = \pm a$, $y = a \pm$ in counter clockwise sense.

Ans. $\pi + 2$

6. Using the line integral, compute the work done by if force $\vec{F}$: $(2y + z)\ \hat{i} + xz\ \hat{j} + (yz - x)\ \hat{k}$ when it moves a particle from the point (0, 0, 0) to (2, 1, 1) along the curve $x = 2\lambda^2 y = \lambda$, $z = \lambda^2$.

Ans. $8\dfrac{8}{35}$

7. If $\vec{F} = 3\,xy\ \hat{i} - y^2\ \hat{j}$, evaluate $\int_c \vec{F} \cdot d\vec{R}$ where c is the curve in xy plane $y = 2x^2$ from (0, 0) to (1, 2).

Ans. $-\dfrac{7}{6}$

8. Evaluate $\int_c (xy + z^2)\,dS$ where c is the arc of the helix $x = \cos\lambda$, $\sin\lambda$, $y = \sin\lambda$, $z = \lambda$ which joins the points (1, 0, 0) and $(-1, 0, \pi)$.

Ans. $\dfrac{\pi^3\sqrt{2}}{3}$

7.16 SURFACE INTEGRAL

Any integral which is to be evaluated over a surface is called a surface integral.

(*a*) **Surface Integral:** Let $\vec{F}$ $(\vec{R})$ be a continuous vector function and S be a surface. Divide S into a finite number of sub-surfaces. Let the surface element surrounding any point P $(\vec{R})$ be $\delta\ \vec{S}$ which can be regarded as vector, its magnitude being the area and its direction that of the outward normal to the element.

From the sum $\Sigma\ \vec{F}\ (\vec{R}) \cdot \delta\ \vec{S}$ where the summation extends over all the sub-surfaces.

Then the limit of the sum as the number of sub-surfaces tends to infinity in such a way that the area to each sub-surface tends to zero is called the normal surface integral of $\vec{F}\ (\vec{R})$ over S and is denoted by $\int_S \vec{F}\,.\,d\vec{S}$ or $\int_S \vec{F}\,.\,\hat{n}\,ds$ where $\hat{n}$ is a outward unit normal at P to S.

(*b*) **Flux across a surface:** If $\vec{F}$ represents the velocity of a fluid particle then the total outward flux of $\vec{F}$ across a closed surfaces S is the surface integral $\iint_S \vec{F}\,.\,d\vec{S}$. When the flux of $\vec{F}$ across any surface S in a region vanishes, $\vec{F}$ is said to be solenoidal vector point function in the region.

Note: In order to evaluate surface integral, it is convenient to express them as double integral over the orthogonal projection of the surface 'S' on one of the co-ordinate planes.

Consider a point P on the surfaces S: $z = f(x, y)$. Let its projection on x-y plane be the region R. Divide into area elements by drawing lines parallel to the axis of x and y.

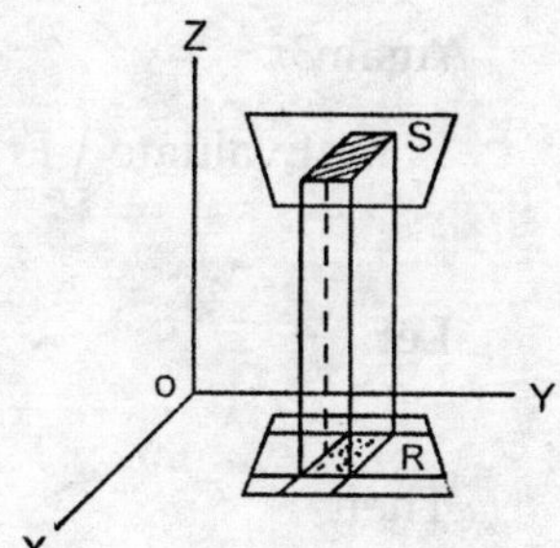

Fig. 7.17

On the element $\delta x\ \delta y$ as base erect a cylinder having generators parallel to OZ and meeting the surface in an element δS

As $\delta x\ \delta y$ in the projection of δS on the xy-plane.

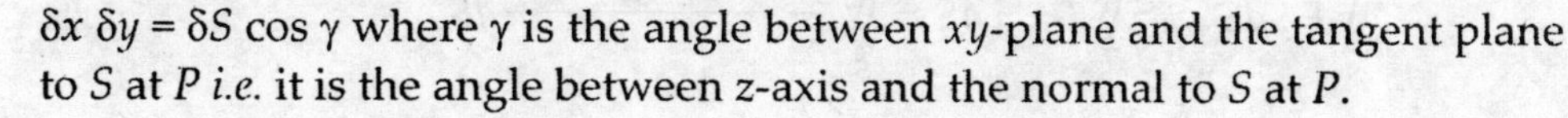

$\therefore$ $\delta x\ \delta y = \delta S \cos \gamma$ where γ is the angle between xy-plane and the tangent plane to S at P *i.e.* it is the angle between z-axis and the normal to S at P.

$$dS = \frac{dx\ dy}{\cos\gamma} = \frac{dx\ dy}{|\bar{n}\,.\,\bar{k}|}$$

where k is the unit vector along z-axis and $\hat{n}$ is a unit vector normal to the surface at P.

Hence $$\iint_S \vec{F}\,.\,d\vec{S} = \iint_R F\,.\,\hat{n}\,\frac{dx\ dy}{|\bar{n}\,.\,\bar{k}|}$$

Thus the surface integral can be evaluated with the help of the double integral over R.

7.17 VOLUME INTEGRAL

Let S be a surface and $\vec{F}\ (\vec{R})$ be a continuous vector functions. Let S be a surface enclosing the region E. Divide E into finite number of sub-intervals $E_1, E_2, \ldots E_n$. Let δv_i be the volume of the sub-region E_i enclosing any point whose position vector is R_i.

Consider the sum $$V = \sum_{i=1}^{n} \vec{F}(\vec{R}_i)\,\delta v_i$$

The limit of this sum as $n \to \infty$ in such a way that $\delta v_i \to 0$, is called the volume integral of $\vec{F}\ (\vec{R})$ over E and is symbolically written as $\int_E F dv$.

If $$\vec{F} = f(x, y, z)\ \hat{i} + \phi(x, y, z)\ \hat{j} + \psi(x, y, z)\ \hat{k}$$

so that $$\delta v = \delta x\ \delta y\ \delta z \text{ then}$$

$$\int_E F dv = \hat{i} \iiint_E f\ dx\ dy\ dz + \hat{j} \iiint_E \phi\ dx\ dy\ dz + \hat{k} \iiint_E \psi\ dx\ dy\ dz$$

***Example* 7.25.** If $\vec{r}$ denotes the position vector of the point (x, y, z) measured from origin 0 and $r = |\vec{r}|$, evaluate $\iint_S \frac{\vec{r}}{r^3}\,.\,d\vec{S}$ where S denotes the sphere of radius 'a' with centre at the origin.

Solution. The equation of the sphere S is $x^2 + y^2 + z^2 = a^2$

A vector normal to the sphere S at the point (x, y, z) is given by $\nabla(x^2 + y^2 + z^2) = 2x\hat{i} + 2y\hat{j} + 2z\hat{k}$. If $\hat{n}$ denotes the unit vector along the outward drwn normal to the sphere S at the point (x, y, z) then

$$\hat{n} = \frac{2x\hat{i} + 2y\hat{j} + 2z\hat{k}}{\sqrt{4x^2 + 4y^2 + 4z^2}}$$

$$= \frac{x\hat{i} + y\hat{j} + z\hat{k}}{a} \quad \text{Since } x^2 + y^2 + z^2 = a2$$

Again $\vec{r} = x\hat{i} + y\hat{j} + z\hat{k}$

$$r = |\vec{r}| = \sqrt{x^2 + y^2 + z^2}$$

Let $\vec{F} = \dfrac{\vec{r}}{r^3}$

Then $\iint_S \vec{F} \cdot d\vec{S} = \iint_S \vec{F} \cdot \hat{n}\, dS = \iint_S \dfrac{\vec{r}}{r^3} \cdot \dfrac{(x\hat{i} + y\hat{j} + z\hat{k})}{a} dS$

$$\iint_S \vec{F} \cdot d\vec{S} = \iint_S \frac{\vec{r}}{r^3} \cdot \frac{(x\hat{i} + y\hat{j} + z\hat{k})}{a} dS$$

$$= \iint \frac{\vec{r} \cdot \vec{r}}{a(x^2 + y^2 + z^2)^{3/2}} dS = \iint \frac{x^2 + y^2 + z^2}{a(x^2 + y^2 + z^2)^{3/2}}$$

$$= \iint \frac{1}{a(x^2 + y^2 + z^2)^{1/2}} dS = \iint \frac{1}{a^2} dS = \frac{1}{a^2} \iint dS = \frac{1}{a^2} 4\pi a^2 = 4\pi$$

***Example* 7.26.** The velocity is $\vec{F} = y\hat{i} + 2\hat{j} + xz\hat{k}$, show that the flux of water through the parabolic cylinder $y = x^2$, $0 \le x \le 3$, $0 \le z \le 2$, is 69 m^3/sec.

Solution. A vector normal to the surface $y = x^2$ is

$$\nabla(x^2 - y) = \hat{i}\frac{\partial}{\partial x}(x^2 - y) + \hat{j}\frac{\partial}{\partial y}(x^2 - y) + \hat{k}\frac{\partial}{\partial z}(x^2 - y)$$

$$= 2x\,\hat{i} - \hat{j}$$

Thus the unit vector $\hat{n}$ normal to the surface at any point (x, y, z) on $y = x^2$

$$\hat{n} = \frac{2x\hat{i} - \hat{j}}{\sqrt{4x^2 + 1}}$$

$$\iint_S F \cdot \hat{n} dS = \iint_S (y\hat{i} + 2\hat{j} + xz\hat{k}) \cdot \frac{2xi - j}{\sqrt{4x^2 + 1}} = \iint_S \frac{2xy - 2}{\sqrt{4x^2 + 1}} \cdot dS$$

Now taking projection of the surface on zx plane.

$$= \iint_S \frac{2xy - 2}{\sqrt{4x^2 + 1}} \cdot \frac{dx\,dz}{|\hat{n} \cdot \hat{j}|} = \iint_S \frac{2xy - 2}{\sqrt{4x^2 + 1}} \cdot \frac{dx\,dz}{\left|\dfrac{2x\hat{i} - \hat{j}}{\sqrt{4x^2 + 1}} \cdot j\right|}$$

$$= \iint_S (2x^3 - 2)\frac{dx\,dz}{1} = \int_0^2\int_0^3 [(2x^3 - 2)\,dx]\,dz \qquad (y = x^2)$$

$$= \int_0^2 \left(\frac{2x^4}{4} - 2x\right)_0^3 dz = \int_0^2 2\left(\frac{81}{4} - 3\right)dz$$

$$= 4\left(\frac{81}{4} - 3\right) = 4 \times \frac{69}{4} = 69 \text{ m}^3/\text{sec.}$$

Example. **7.27.** Evaluate $\int_S \vec{F}\,.\,\hat{n}dS$ where $\vec{F} = 2x^2y\,\hat{i} - y^2\,\hat{j} + 4xz^2\,\hat{k}$ and S in the closed surface of the region in the first octaint bounded by the cylinder $y^2 + z^2 = 9$ and the planes $x = 0$, $x = 2$, $y = 0$ and $z = 0$.

Solution. The given surface is piece-wise smooth and is bounded by

(*a*) S_1 – the rectangular surface *OBCD* in *xz* plane.

(*b*) S_2 – the rectangular surface *AODE* is *xy* plane.

(*c*) S_3 – the circular quadrant *OAB* in *yz*-plane.

(*d*) S_4 – the circular quadrant *CDE*.

(*e*) S_5 – the curved surface *ABCE* in the first octant.

Now
$$\int_S \vec{F}\,.\,\hat{n}dS = \int_{S_1} \vec{F}\,.\,\hat{n}dS + \int_{S_2} \vec{F}\,.\,\hat{n}dS + \int_{S_3} \vec{F}\,.\,\hat{n}dS + \int_{S_4} \vec{F}\,.\,\hat{n}dS + \int_{S_5} \vec{F}\,.\,\hat{n}dS$$

Now for the surface S_1

$$y = 0, \quad \hat{n} = -\text{j}$$

$$\int_{S_1} \vec{F}\,.\,\hat{n}dS = \int_{S_1} (2x^2y\hat{i} - y^2\hat{j}\,.\,4xz^2\hat{k})\,.\,(-\,j)\,dS = \int_{S_1} y^2 dS = 0$$

For the surface S_2
$$z = 0, \quad \hat{n} = -\,\hat{k}$$

$$\int_{S_2} \vec{F}\,.\,\hat{n}dS = \int_{S_2} (2x^2y\hat{i} - y^2\hat{j} + 4xz^2\hat{k})\,.\,(-\,k)\,dS = -\int_{S_2} 4xz^2 dS = 0$$

For the surface S_3
$$x = 0 \quad \hat{n} = -\,i$$

$$\int_{S_3} \vec{F}\,.\,\hat{n}dS = \int_{S_3} (2x^2y\hat{i} - y^2\hat{j} + 4xz^2\hat{k})\,.\,(-\,i)\,dS$$

$$= \int_{S_3} -\,2x^2y dS = 0$$

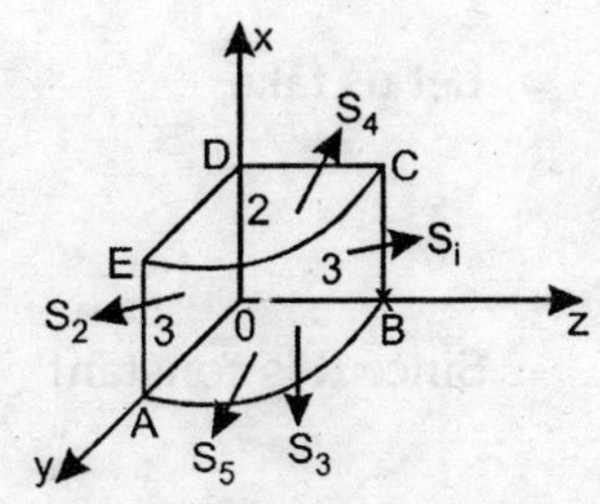

Fig. 7.18

For the surface S_4
$$x = 2\,.\,n = i$$

$$\int_{S_4} \vec{F}\,.\,\hat{n}dS = \int_{S_4} (2x^2y\hat{i} - y^2\hat{j} + 4xz^2\hat{k})\,.\,\hat{i}\;dS$$

$$= \int_{S_4} 2x^2y dS = 8\int_{S_4} y dS = 8\iint y\;dy\;dz$$

$$= 8\int_0^3\int_0^{\sqrt{9-z^2}} y\;dy\;dz = 8\int_0^3 \left[\frac{y^2}{2}\right]_0^{\sqrt{9-z^2}} dz = \frac{8}{2}\int_0^3 (9 - z^2)\,dz = 72$$

For the surface S_5

$$\nabla (y^2 + z^2) = 2y\hat{j} + 2z\hat{k}$$

$$\hat{n} = \frac{2y\hat{i} + 2z\hat{k}}{\sqrt{4y^2 + 4z^2}} = \frac{y\hat{i} + z\hat{k}}{3} \qquad \boxed{y^2 + z^2 = 9}$$

$$\int_{S_5} \vec{F} \cdot \hat{n} dS = \int_{S_5} (2x^2 y\hat{i} - y^2\hat{j} + 4xz^2\hat{k}) \cdot \frac{(y\hat{j} + z\hat{k})}{3} dS$$

$$= \int_{S_5} \frac{-y^3 + 4xz^3}{3} dS = \int_{S_5} \frac{(-y^3 + 4xz^3)\, dxdy}{3 \cdot \frac{z}{3}} \qquad \boxed{dS = \frac{dxdy}{|\hat{x} \cdot \hat{k}|}}$$

$$= \int_{S_5} \frac{-y^3 + 4xz^3}{z} dxdy = \int_0^2 \int_0^3 \left(-\frac{y^3}{z} + 4xz^2\right) dy\, dx$$

Put

$$y = 3 \sin \theta$$

$$dy = 3 \cos \theta\, d\theta \qquad \boxed{\text{Since } y^2 + z^2 = 9}$$

$$z = 3 \cos \theta$$

$$\int_0^2 \int_0^{\frac{\pi}{2}} \left[-\frac{27 \sin^3\theta}{3\cos\theta} + 4x(9\cos^2\theta)\right] 3 \cos\theta\, d\theta\, dx = \int_0^2 \left[-27 \times \frac{2}{3} + 108x \times \frac{2}{3}\right] dx = 108$$

Hence $\int \vec{F} \cdot \hat{n} dS = 0 + 0 + 0 + 72 + 108 = 180$

***Example* 7.28.** Evaluate $\iint_S (x^3\hat{i} + y^3\hat{j} + z^3\hat{k}) \cdot \hat{n} dS$ where S is the surface of the sphere of radius R and $\hat{\boldsymbol{n}}$ is the unit normal outward.

Solution. Equation of the sphere is $x^2 + y^2 + z^2 = R^2$

$$\nabla(x^2 + y^2 + z^2) = 2x\hat{i} + 2y\hat{j} + 2z\hat{k}$$

$$\hat{n} = \frac{2x\hat{i} + 2y\hat{j} + 2z\hat{k}}{2\sqrt{x^2 + y^2 + z^2}} = \frac{x\hat{i} + y\hat{j} + z\hat{k}}{R} = \frac{x}{R}\hat{i} + \frac{y}{R}\hat{j} + \frac{z}{R}\hat{k}$$

Let us take

$$x = R \sin \theta \cos \phi$$

$$y = R \sin \theta \sin \phi$$

$$z = R \cos \theta$$

Since R is constant

$$dS = R^2 \sin \theta\, d\theta\, d\phi$$

$$= \iint_S (x^3\hat{i} + y^3\hat{j} + z^3\hat{k}) \cdot \left(\frac{x}{R}\hat{i} + \frac{y}{R}\hat{j} + \frac{z}{R}\hat{k}\right) dS = \iint_S \frac{x^4 + y^4 + z^4}{R} dS$$

$$\text{Integral} = \frac{8}{R} \int_0^{\frac{\pi}{2}} \int_0^{\frac{\pi}{2}} R^4 (\sin^4\theta \cos^4\phi + \sin^4\theta \sin^4\phi + \cos^4\theta) R^2 \sin\theta\, d\theta\, d\phi$$

$$= 8R^5 \int_0^{\frac{\pi}{2}}\int_0^{\frac{\pi}{2}} (\sin^5\theta\cos^4\phi + \sin^5\theta\sin^4\phi + \cos^4\theta\sin\theta)\,d\theta d\phi$$

$$= 8R^5 \int_0^{\frac{\pi}{2}} \left(\frac{4}{5\times 3}\cos^4\phi + \frac{4}{5\times 3}\sin^4\phi + \frac{3\times 1}{5\times 3}\right)d\phi$$

$$= 8R^5 \int_0^{\frac{\pi}{2}} \left[\frac{4}{15}(\sin^4\phi + \cos^4\phi) + \frac{1}{5}\right]d\theta$$

$$= 8\,R^5\left[\frac{4}{15}\left(\frac{3}{4.2}\frac{\pi}{2} + \frac{3}{4.2}\frac{\pi}{2}\right) + \frac{1}{5}\frac{\pi}{2}\right] = 8\,R^5\left[\frac{\pi}{10} + \frac{\pi}{10}\right] = 8\,R^5 \times \frac{2\pi}{10} = \frac{8\pi\,R^5}{5}$$

Example 7.29. Evaluate $\int_S \vec{F}\cdot\hat{n}dS$ where $\vec{F} = 6z\hat{i} - 4\hat{j} + y\hat{k}$ and S is the portion of the plane $2x + 3y + 6z = 12$ in the first octant.

Fig. 7.19

Solution. The equation of the plane is $2x + 3y + 6z = 12$

$$\nabla(2x + 3y + 6z) = 2\hat{i} + 3\hat{j} + 6\hat{k}$$

$$\hat{n} = \frac{2\hat{i} + 3\hat{j} + 6\hat{k}}{\sqrt{4+9+36}} = \frac{2\hat{i} + 3\hat{j} + 6\hat{k}}{7}$$

$$\int_S \vec{F}\cdot\hat{n}dS = \int (6z\hat{i} - 4\hat{j} + y\hat{k})\cdot\frac{2\hat{i} + 3\hat{j} + 6\hat{k}}{7}\cdot\frac{dxdy}{\frac{6}{7}} \qquad \left\{dS = \frac{dxdy}{|\hat{n}\cdot\hat{k}|}\right\}$$

$$= \int_S \frac{12z - 12 + 6y}{7}\cdot\frac{dxdy}{\frac{6}{7}} = \int_S (2z - 2 + y)\,dxdy$$

Now $\qquad 2x + 3y + 6z = 12, \quad 2z = \frac{1}{3}[12 - 2x - 3y]$

$$\int_S F\cdot\hat{n}dS = \int_S \left[\frac{1}{3}(12 - 2x - 3y) - 2 + y\right]dx\,dy = \int_0^6 \int_{y=0}^{\frac{12-2x}{3}} \left(\frac{6-2x}{3}\right)dx\,dy$$

$$= \int_0^6 \left(\frac{6-2x}{3}y\right)_0^{\frac{12-2x}{3}} dx = \frac{1}{9}\left[72x - \frac{36x^2}{2} + \frac{4x^3}{3}\right]_0^6$$

$$= \frac{1}{9}\left[72\times 6 - 18\times 36 + \frac{4}{3}\times 6\times 6\times 6\right] = \frac{1}{9}(72) = 8$$

Example 7.30. Evaluate $\iint_S \vec{A}\cdot\hat{n}dS$ where $\vec{A} = z\hat{i} + x\hat{j} - 3y^2\hat{k}$, S is the surface of the cylinder $x^2 + y^2 = 16$ included in the first octant between $z = 4$, $z = 5$.

Solution. $\qquad \nabla(x^2 + y^2) = 2x\hat{i} + 2y\hat{j}$

$$\hat{n} = \frac{2(x\hat{i} + y\hat{j})}{\sqrt{4x^2 + 4y^2}} \qquad x^2 + y^2 = 16$$

$$= \frac{x\hat{i} + y\hat{j}}{4}$$

$$\vec{A} \cdot \hat{n} = (z\hat{i} + x\hat{j} - 3y^2\hat{k}) \cdot \left(\frac{x\hat{i} + y\hat{j}}{4}\right)$$

$$= \frac{xz + xy}{4}$$

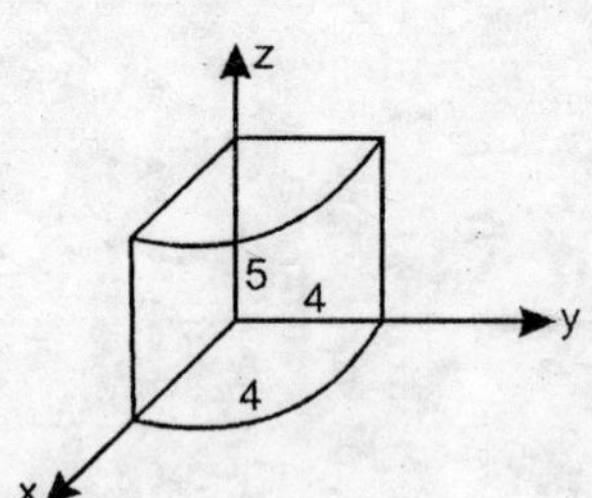

Fig. 7.20

Taking projection in yz plane.

$$dS = \frac{dy\,dz}{|n \cdot i|} = \frac{dydz}{\frac{x}{4}}$$

$$\iint_S \vec{A} \cdot \hat{n}dS = \iint_S \frac{x(y+z)}{4} \cdot \frac{dy\,dz}{\frac{x}{4}} = \int_0^5 (8 + 4z)\,dz = \left(8z + 4\frac{z^2}{2}\right)_0^5$$

$$= 8 \times 5 + 2 \times 25 = 40 + 50 = 90$$

Example **7.31.** Evaluate $\iint_S \vec{F} \cdot \hat{n}dS$ where $\vec{F} = 4x\hat{i} - 2y^2\hat{j} + z^2\hat{k}$ and S is the closed surface consisting of the cylinder $x^2 + y^2 = 4$ and circular discs $z = 0$, $z = 3$.

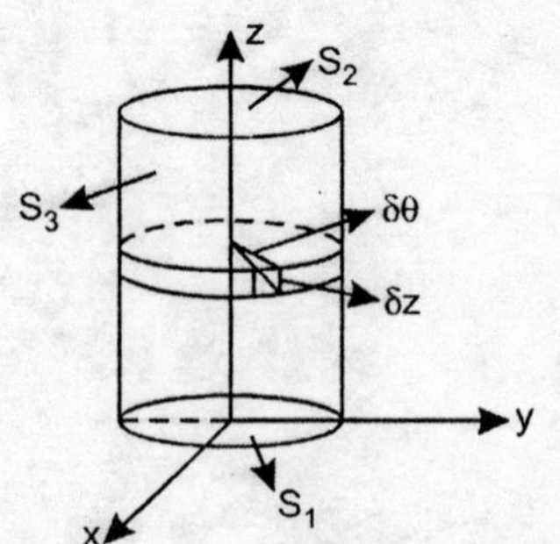

Fig. 7.21

Solution. Here the surface S consists of there surfaces

(*i*) the surface S_1 of the base is the plane $z = 0$

(*ii*) the surface S_2 of the top is the plane $z = 5$

(*iii*) the surface S_3 of the convex portion of the cylinder.

For the Surface S_1 $\qquad z = 0, \ \hat{n} = -\hat{k}$

$$\iint_{S_1} \vec{F} \cdot d\vec{S} = \iint_{S_1} (4x\hat{i} - 2y^2\hat{j} + z^2\hat{k}) \cdot (-k)\,dS$$

$$= -\iint z^2 dS = 0$$

For the Surface S_2 $\qquad z = 3 \quad \hat{n} = k$

$$\iint_{S_2} \vec{F} \cdot d\vec{S} = \iint z^2 dS = 9 \iint dS = 9 \times \pi \times 4 = 36\,\pi$$

For the Convex Surface S_3

$$\nabla(x^2 + y^2) = 2x\hat{i} + 2y\hat{j}$$

$$\hat{n} = \frac{2x\hat{i} + 2y\hat{j}}{\sqrt{4x^2 + 4y^2}} = \frac{x\hat{i} + y\hat{j}}{2}$$

$$\iint_{S_3} \vec{F} \cdot d\vec{S} = \iint_{S_3} (4x\hat{i} - 2y^2\hat{j} + z^2\hat{k}) \cdot \frac{x\hat{i} + y\hat{j}}{2} dS = \iint_{S_3} (2x^2 - y^3)\,dS$$

put $\qquad x = 2\cos\theta,\quad y = 2\sin\theta \qquad x^2 + y^2 = 4$

$$\iint_{S_3} \vec{F}\,.\,dS = \int_{z=0}^{3}\int_{\theta=0}^{2\pi} (8\cos^2\theta - 8\sin^3\theta)\,2\,d\theta\,dz \qquad dS = 2\,d\theta\,dz$$

$$= 2\int_0^{2\pi} 8(\cos^2\theta - \sin^3\theta)[z]_{z=0}^{3}d\theta = 48\left[\int_0^{2\pi}\cos^2\theta\,d\theta - \int_0^{2\pi}\sin^3\theta\,d\theta\right]$$

$$= 48\left[4\int_0^{\frac{\pi}{2}}\cos^2\theta\,d\theta\right] = 192\,.\,\frac{1}{2}\,.\,\frac{\pi}{2} = 48\,\pi$$

Hence the required surface integral

$$= 0 + 36\,\pi + 48\,\pi = 84\,\pi$$

***Example.* 7.32.** If $\vec{F} = (2x^2 - 3z)\,\hat{i} - 2xy\,\hat{j} - 4x\,\hat{k}$, then evaluate $\iiint_V \vec{\nabla}\,.\,\vec{F}\,dV$ where V is the closed region bounded by the plane $x = 0$, $y = 0$, $z = 0$ and $2x + 2y + z = 4$.

Solution. We have $\qquad \vec{F} = (2x^2 - 3z)\,\hat{i} - 2xy\,\hat{j} - 4x\,\hat{k}$

$$\vec{\nabla}\,.\,\vec{F} = \left(\hat{i}\frac{\partial}{\partial x} + \hat{j}\frac{\partial}{\partial y} + \hat{k}\frac{\partial}{\partial z}\right).\,[(2x^2 - 3z)\,\hat{i} - 2xy\,\hat{j} - 4x\,\hat{k}]$$

$$= \frac{\partial}{\partial x}(2x^2 - 3z) - \frac{\partial}{\partial y}(2\,xy) - \frac{\partial}{\partial z}(4\,x)$$

$$= 4x - 2x = 2x$$

$$\iiint_V \nabla\,.\,\vec{F}dV = \iiint_V 2x\,dx\,dy\,dz = 2\int_{x=0}^{2}\int_{y=0}^{2-x}\int_{z=0}^{4-2x-2y} x\,dx\,dy\,dz$$

$$= 2\int_{x=0}^{2}\int_{y=0}^{2-x} x(z)_{z=0}^{4-2x-2y}dx\,dy = 2\int_{x=0}^{2}\int_{y=0}^{2-x} x(4 - 2x - 2y)\,dx\,dy$$

$$= 2\int_{x=0}^{2}[(4xy - 2x^2y - xy^2)]_{y=0}^{2-x}dx$$

$$= 2\int_{x=0}^{2}[4x(2 - x) - 2x^2(2 - x) - x(2 - x)^2]dx$$

$$= 2\int_0^2 (x^3 - 4x^2 + 4x)\,dx$$

$$= 2\left[\frac{1}{4}x^4 - \frac{4}{3}x^3 + 2x^2\right]_0^2 = 2\left[4 - \frac{32}{3} + 8\right] = \frac{8}{3}$$

***Example* 7.33.** Evaluate $\iiint_V (2x + y)\,dV$ where V is the closed region bounded by the cylinder $z = 4 - x^2$ and the planes $x = 0$, $y = 2$, $z = 0$, $y = 0$

Solution.
$$\iiint_V (2x + y)\,dV = \int_0^2\int_0^2\int_0^{4-x^2} (2x + y)\,dx\,dy\,dz = \int_0^2\int_0^2 (2x + y)[z]_0^{4-x^2}\,dx\,dy$$

$$= \int_0^2\int_0^2 (2x + y)(4 - x^2)\,dy\,dx$$

$$= \int_0^2\int_0^2 (8x - 2x^3 + 4y - x^2y)\,dy\,dx = \int_0^2 \left[8xy - 2x^3y + 2y^2 - \frac{x^2y^2}{2}\right]_0^2$$

$$= \int_1^2 (16x - 4x^3 + 8 - 2x^2)\,dx = \left[\frac{16x^2}{2} - \frac{4x^4}{4} + 8x - \frac{2x^3}{3}\right]_0^2$$

$$= 8 \times 4 - 16 + 16 - \frac{16}{3} = 32 - 16 + 16 - \frac{16}{3}$$

$$= 32 - \frac{16}{3} = \frac{96 - 16}{3} = \frac{80}{3}$$

***Example* 7.34.** If $\vec{F} = 2z\hat{i} - z\hat{j} + z\hat{k}$, evaluate $\iiint_V \vec{F}\,.\,dV$ where V is the region bounded by the surface $x = 0$, $y = 0$, $x = 2$, $y = 4$, $z = x^2$, $z = 2$.

Solution.
$$\iiint_V \vec{F}\,.\,dV = \iiint (2x\hat{i} - x\hat{j} + y\hat{k})\,dx\,dy\,dz$$

$$= \int_0^2\int_0^4\int_{x^2}^2 (2z\hat{i} - x\hat{j} + y\hat{k})\,dz\,dy\,dz$$

$$= \int_0^2\int_0^4 \left(2\frac{z^2}{2}\hat{i} - xz\hat{j} + yz\hat{k}\right)_{x^2}^2 dy\,dx$$

$$= \int_0^2\int_0^4 (4\hat{i} - 2x\hat{j} + 2y\hat{k}) - (x^4\hat{i} - x^3\hat{j} + yx^2\hat{k})\,dy\,dx$$

$$= \int_0^2 \left[4y\hat{i} - 2yx\hat{j} + y^2\hat{k} - x^4y\hat{i} + x^3\hat{j} - \frac{y^2}{2}x^2k\right]_0^4 dx$$

$$= \int_0^2 (16\hat{i} - 8x\hat{j} + 16\hat{k} - 4x^4\hat{i} + 4x^3\hat{j} - 8x^2\hat{k})\,dx$$

$$= \left[16x\hat{i} - 4x^2\hat{j} + 16x\hat{k} - 4\frac{x^5}{5}\hat{i} + 4\frac{x^4}{4}\hat{j} - 8\frac{x^3}{3}\hat{k}\right]_0^2$$

$$= 32\hat{i} - 16\hat{j} + 32\hat{k} - \frac{128}{5}\hat{i} + 16\hat{j} - \frac{64}{3}\hat{k}$$

$$= \frac{32\hat{i}}{5} + \frac{32\hat{k}}{3} = \frac{32}{15}[3\hat{i} + 5\hat{k}]$$

EXERCISE 7.4

1. Evaluate $\iint_S \vec{F} \cdot \hat{n} dS$ where $f = yz\hat{i} + zx\hat{j} + xz\hat{k}$ and S is the part of the surface of the sphere $x^2 + y^2 + z^2 = 1$ where lies in the first quadrant.

Ans. $\frac{3}{8}$

2. If $\vec{F} = 4xz\hat{i} - y^2\hat{j} + yz\hat{k}$, Evaluate $\iint_S \vec{F} \cdot \hat{n} dS$ where S in the surface of the cube bounded by $n = 0, y = 0, z = 0, x = 1, y = 1, z = 1$.

Ans. $\frac{3}{2}$

3. Evaluate $\iint \vec{F} \cdot \hat{n} dS$ where $\vec{F} = 4x\hat{i} - 2y^2\hat{j} + z^2\hat{k}$ where S is the surfaces of the sphere cylinder $x^2 + y^2 = 4$ included in the first octant between $z = 0, z = 3$.

Ans. -32

4. If $\vec{F} = (2x^3 - 3z)\,\hat{i} - 2xy\hat{j} - 4x\hat{k}$, then evaluate $\iiint_V \vec{V} \cdot \vec{F} dV$ where V is the closed region bounded by the planes $x = 0, y = 0, z = 0$ and $2x + 2y + z = 4$.

Ans. $\frac{8}{3}(\hat{j} + \hat{k})$

5. Evaluate $\iiint_V \phi\, dV$ where $\phi = 45x^2y$ and V is the region bounded by the plane $4x + 2y + z = 8, x = 0, y = 0, z = 0$.

Ans. 128

7.18 GREEN'S THEOREM

Green's Theorem in plane.

(Relation between the line integral and double integral).

Statement. Let S be a closed bounded region in x-y plane whose boundary C consists of finitely many smooths curves. If $M(x, y)$ and $N(x, y)$ be continuous function of x, y having continuous partial derivatives $\frac{\partial M}{\partial y}, \frac{\partial N}{\partial x}$ then $\int_C (M\,dx + N\,dy) = \iint \left(\frac{\partial N}{\partial x} - \frac{\partial M}{\partial y}\right) dx\,dy$

***Example.* 7.35.** Verify Green's theorem in a plane for $\int_C [(x^2 - 2xy)\,dx + (x^2y + 3)\,dy]$ where C is the boundary of the region defined by $y^2 = 8x$ and $x = 2$.

Solution. By Green's theorem we have

$$\iint_S \left(\frac{\partial N}{\partial x} - \frac{\partial M}{\partial y}\right) dx\,dy = \int_C (M\,dx + N\,dy)$$

Here $M = x^2 - 2xy,\ N = x^2y + 3$

$$\frac{\partial N}{\partial x} = 2\,xy,\ \frac{\partial M}{\partial y} = -\,2x$$

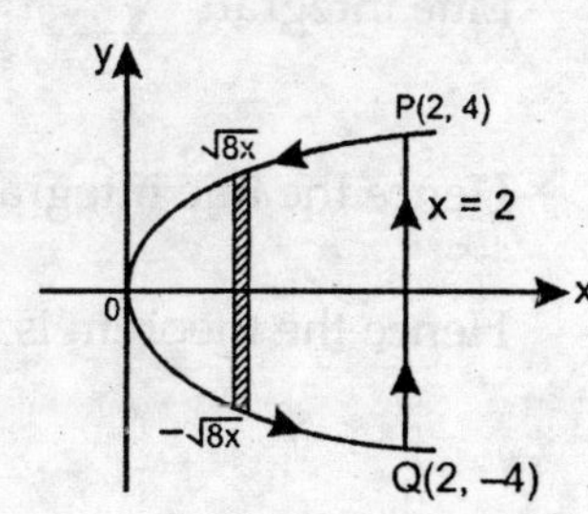

Fig. 7.22

$$\iint_S \left(\frac{\partial N}{\partial x} - \frac{\partial M}{\partial y}\right) dx\,dy = \iint_S (2xy + 2x)\,dx\,dy$$

Now the points of intersection of $y^2 = 8x$ and $x = 2$ are $P(2, 4)$ and $Q(2, -4)$

$$= \iint_S (2xy + 2x)\,dx\,dy$$

$$= \int_0^2 \int_{-\sqrt{8x}}^{\sqrt{8x}} (2xy + 2x)\,dx\,dy = \int_0^2 \left(2x\frac{y^2}{2} + 2xy\right)_{-\sqrt{8x}}^{\sqrt{8x}} dx$$

$$= \int_0^2 [(x \times 8x + 2x\sqrt{8x}) - (x \times 8x - 2x\sqrt{8x})]\,dx$$

$$= \int_0^2 4x\sqrt{8x}\,dx = 8\sqrt{2}\int_0^2 x^{3/2}dx = 8\sqrt{2} \times \left[\frac{x^{5/2}}{\frac{5}{2}}\right]_0^2$$

$$= \frac{16\sqrt{2}}{5}[2^{5/2} - 0] = \frac{16}{5} \times 2^3 = \frac{128}{5}$$

Now let us calculate the line integral along C, and the integral along line $x = 2$

Along $y^2 = 8x$, $x = \frac{y^2}{8}$, $dx = \frac{2y}{8}\,dy$, y varies from -4 to 4

Line integral along C $= \int_{-4}^{4}\left(\frac{y^4}{64} - 2\frac{y^2}{8}y\right)\frac{1}{4}y\,dy + \left(\frac{y^4}{64}y + 3\right)dy$

$$= \int_{-4}^{4}\left(\frac{y^5}{256} + \frac{y^5}{64} - \frac{1}{16}y^4 + 3\right)dy = \int_{-4}^{4}\left(\frac{y^5}{256} + \frac{y^5}{64} - \frac{1}{16}y^4 + 3\right)dy$$

$$= -2\int_0^4\left(-\frac{1}{16}y^4 + 3\right)dy$$

$\int_{-a}^{a} f(x)\,dx$ 0 or $2\int_0^a f(x)\,dx$ according as $\begin{cases} f(-x) = -f(x) \\ f(-x) = f(x) \end{cases}$

$$= -2\left[-\frac{y^5}{80} + 3y\right]_0^4 = -2\left[-\frac{1}{80}(4)^5 + 3 \times 4\right] = \frac{128}{5} - 24$$

Along the line $x = 2$, we have $x = 2$, $dx = 0$, y varies from -4 to 4.

Line integral $= \int_{-4}^{4}[0 + (2^2y + 3)]\,dy = \int_{-4}^{4}(4y + 3)\,dy = 24$

Hence the line integral $= \frac{128}{5} - 24 + 24 = \frac{128}{5}$

Hence the theorem is verified.

Example 7.36. Verify Green's theorem in plane for $\int_C (xy + y^2)\,dx + x^2\,dy$ when C is the closed curve of the region bounded by $y = x$ and $y = x^2$.

Solution. The points of intersection of $y = x$ and $y = x^2$ are $x = 0, y = 0$ and $x = 1, y = 1$.

i.e., $O\,(0, 0)$, $P\,(1, 1)$

Now
$$\int_S \left(\frac{\partial N}{\partial x} - \frac{\partial M}{\partial y}\right) dx\,dy = \iint_S \left[\frac{\partial}{\partial x}(x^2) - \frac{\partial}{\partial y}(xy + y^2)\right] dx\,dy$$

$$= \iint_S (2x - x - 2y)\,dx\,dy = \iint_S (x - 2y)\,dx\,dy$$

$$= \int_0^1 \int_{x^2}^{x} (x - 2y)\,dy\,dx = \int_0^1 (xy - y^2)_{x^2}^{x}\,dx$$

$$= \int_0^1 [(x^2 - x^2) - (x^3 - x^4)]\,dx$$

$$= \int_0^1 (x^4 - x^3)\,dx = \left[\frac{x^5}{5} - \frac{x^4}{4}\right]_0^1 = \frac{1}{5} - \frac{1}{4} = -\frac{1}{20}$$

Fig. 7.23

Let us evaluate the integral along the curve C . $y = x^2$ along the line $y = x$.

Along $y = x^2$, the curve OP $dy = 2x\,dx$, therefore along the curve $y = x^2$ the line integral $= \int_0^1 (x^3 + x^4)\,dx + x^2\,(2x)$ $dx = \int_0^1 (3x^3 + x^4)\,dx = \frac{19}{20}$

Along $y = x$, the line PO $dy = dx$ therefore along the line $y = x$ the line integral $= \int_1^0 [(x)(x) + x^2 + x^2]\,dx = \int_1^0 3x^2\,dn$ $= \left(\frac{3x^3}{3}\right)_1^0 = 0 - 1 = -1$ the required integral $= \frac{19}{20} - 1 = -\frac{1}{20}$

Hence the theorem is verified.

Example. 7.37. Verify Green's theorem for $\int_C (3x^2 - 8y^2)\,dx + (4y - 6\,xy)\,dy$ where C is the boundary of the region bounded by $x = 0$, $y = 0$ and $x + y = 1$.

Solution. By Green's theorem

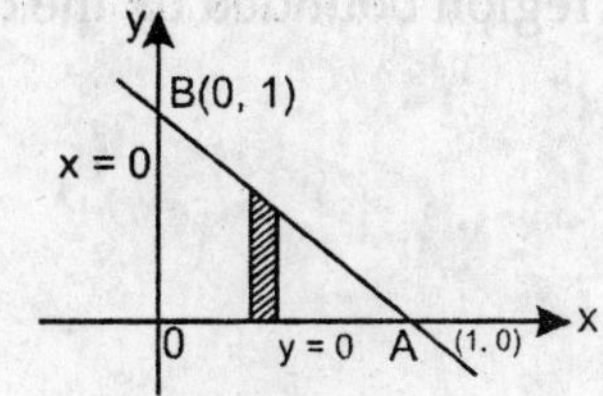

Fig. 7.24

$$\int_C M dx + N dy = \iint_S \left(\frac{\partial N}{\partial x} - \frac{\partial M}{\partial y}\right) dx\,dy$$

Region S is bounded by $x = 0$, $y = 0$ and $x + y = 1$

$$\therefore \quad \int_C M dx + N dy = \int_{OA} M dx + N dy + \int_{AB} M dx + N dy + \int_{BO} M dx + N dy$$

Now along OA $\quad y = 0\ dy = 0$, x goes from 0 to 1

$$\therefore \quad \int_{OA} M dx + N dy = \int_0^1 3x^2\,dx = \left[3\frac{x^3}{3}\right]_0^1 = 1$$

Along AB $x + y = 1, x = 1 - y, dx = -dy$

$$\int_{AB} Mdx + Ndy = \int_0^1 [(3(1-y)^2 - 8y^2)(-dy) + (4y - 6(1-y)y)dy]$$

$$= \int_0^1 [-3(1 + y^2 - 2y) + 8y^2 + 4y - 6y + 6y^2]\,dy$$

$$= \int_0^1 (-3 + 11y^2 + 4y)\,dy = \int_0^1 (11y^2 + 4y - 3)\,dy$$

$$= \left[\frac{11y^3}{3} + 2y^2 - 3y\right]_0^1 = \left(\frac{11}{3} + 2 - 3\right) = \frac{8}{3}$$

Along BO $\qquad x = 0, dx = 0, y$ varies from 1 to 0

$$\int_{BO} Mdx + Ndy = \int_1^0 (4y - 6 \times 0)\,dy = \int_1^0 4y\,dy = \left[\frac{4y^2}{2}\right]_1^0 = -2$$

$\therefore$ **Line integral** $\qquad = 1 + \frac{8}{3} - 2 = \frac{5}{3}$

Now $$\iint_S \left(\frac{\partial N}{\partial n} - \frac{\partial M}{\partial y}\right) dx\,dy = \iint_S (-6y + 16y)\,dx\,dy = \int_0^1 \int_0^{1-x} 10y\,dy\,dx$$

$$= \int_0^1 10\left[\frac{y^2}{2}\right]_0^{1-x} dx = \int_0^1 5(1-x)^2\,dx = 5\left[\frac{(1-x)^3}{-3}\right]_0^1 = \frac{5}{3}$$

Hence the theorem is verified.

***Example* 7.38.** If C is a simple closed curve in $x - y$ plane not enclosing the origin, show that $\int_c \vec{F} \cdot d\vec{R} = 0$ where $\vec{F} = \dfrac{y\hat{i} - x\hat{j}}{x^2 + y^2}$.

Solution. Since the origin lies out side the curve C, the function $\vec{F}$ is continuous and differentiable in the region bounded by the curve C.

$$\int_C \vec{F} \cdot d\vec{R} = \int_C \frac{y\hat{i} - x\hat{j}}{x^2 + y^2} \cdot (\hat{i}dx + \hat{j}dy)$$

$$= \int_C \left(\frac{y}{x^2 + y^2}\right) dx + \left(-\frac{x}{x^2 + y^2}\right) dy$$

Now put $\qquad M = \dfrac{y}{x^2 + y^2}$

$\qquad N = \dfrac{x}{x^2 + y^2}$ and apply Green's theorem

$$\int_c \left(\frac{y}{x^2 + y^2}\right) dx + \left(-\frac{x}{x^2 + y^2}\right) dy = \int_C Mdx + Ndy$$

$$= \iint_S \left(\frac{\partial N}{\partial x} - \frac{\partial M}{\partial y}\right) dx\, dy = \iint_S \left(\frac{y^2 - x^2}{(x^2+y^2)} - \frac{y^2 - x^2}{x^2+y^2}\right) dx\, dy = 0$$

***Example* 7.39.** Apply Green's theorem to evaluate $\int_C (y - \sin x)\, dx + \cos x\, dy$ where C is the plane triangle enclosed by lines $y = 0,\ x = \frac{\pi}{2},\ y = \frac{2}{\pi}x$.

Solution. The co-ordinates of A are $\left(\frac{\pi}{2}, 0\right)$, B are $\left(\frac{\pi}{2}, 1\right)$

y = 2/π x
B(π/2, 1)
x = π/2
P
Q
0
y = 0
A (π/2, 0)

Fig. 7.25

Now $$\int_C (y - \sin x)\, dx + \cos x\, dy = \int_C M dx + N dy$$

where $$N = y - \sin x$$
$$N = \cos x$$

Now by Green's theorem

$$\int_C M dx + N dy = \iint_S \left(\frac{\partial N}{\partial x} - \frac{\partial M}{\partial y}\right) dx\, dy$$

M, N and $\frac{\partial N}{\partial x}, \frac{\partial M}{\partial y}$ are continuous functions of x and y

$$\therefore \quad \int_C (y - \sin x)\, dx + \cos x\, dy = \iint_S (-\sin x - 1)\, dx\, dy$$

$$= -\iint_S (1 + \sin x)\, dx\, dy = \int_0^1 \int_{\frac{\pi}{2}y}^{\frac{\pi}{2}} (1 + \sin x)\, dx\, dy$$

$$= \int_0^1 \left[(x - \cos x)\right]_{\frac{\pi}{2}y}^{\frac{\pi}{2}} dy = -\int_0^1 \left[\frac{\pi}{2} - \left(\frac{\pi}{2}y - \cos\frac{\pi}{2}y\right)\right] dy$$

$$= -\int_0^1 \left[\frac{\pi}{2}y - \frac{\pi}{2}\frac{y^2}{2} + \sin\frac{\frac{\pi y}{2}}{\frac{\pi}{2}}\right]_0^1 = -\left(\frac{\pi}{2} - \frac{\pi}{4} + \frac{2}{\pi}\right) + 0 = -\left[\frac{\pi}{4} + \frac{2}{\pi}\right]$$

***Example* 7.40.** Apply Green's theorem to prove that the area enclosed by a plane curve is $\frac{1}{2}\int_C (\boldsymbol{x\, dy - y\, dx})$. Hence find the area of an ellipse whose major and minor axes are of length a and b.

Solution. Consider $\int_C x\, dy - y\, dx$

Here $$M = -y,\ N = x$$

Now M, N are continuous and differentiable of x and y. Also $\frac{\partial M}{\partial y}, \frac{\partial N}{\partial x}$ are continuous and differentiable in the closed region bounded by C.

By Green's theorem

$$\int_C M dx + N dy = \iint_S \left(\frac{\partial N}{\partial x} - \frac{\partial M}{\partial x}\right) dx\, dy$$

$$\oint_C x\,dy - y\,dx = \iint_S (1+1)\,dx\,dy = 2\iint_S dx\,dy$$

$$\iint_S dx\,dy = \frac{1}{2}\int x\,dy - y\,dx$$

Area enclosed by the closed curve $= \frac{1}{2}\int (x\,dy - y\,dx)$.

Now any point of the ellipse is given by $x = a\cos\theta$, $y = b\sin\theta$, θ **being the parameter**

Hence the area of the ellipse

$$= \frac{1}{2}\int x\,dy - y\,dx = \frac{1}{2}\int_0^{2\pi} a\cos\theta b\cos\theta\,d\theta - b\sin\theta(-a\sin\theta)\,d\theta$$

$$= \frac{1}{2}\text{ ab}\int_0^{2\pi}(\cos^2\theta + \sin^2\theta)\,d\theta = \frac{1}{2}\text{ ab}\int_0^{2\pi} d\theta = \frac{1}{2}ab\,[\theta]_0^{2\pi} = \pi\,ab.$$

7.19 STOKE'S THEOREM

(Relation between line and Surface Integral)

If S **be an open surface bounded by a closed curve C and** $\vec{F} = f_1\hat{i} + f_2\hat{j} + f_3\hat{k}$ **be any continuous and differentiable vector point function, then**

$$\int_C \vec{F}\,.\,d\vec{R} = \iint_S \text{curl}\,\vec{F}\,.\,\hat{n}dS = \iint_S (\nabla\times\vec{F})\,.\,\hat{n}dS$$

where $\hat{n} = \cos\alpha\,\hat{i} + \cos\beta\,\hat{j} + \cos\gamma\,\hat{k}$

is a unit vector along outward normal at any point of S.

Now

$$\nabla\times\vec{F} = \begin{vmatrix} \hat{i} & \hat{j} & \hat{k} \\ \frac{\partial}{\partial x} & \frac{\partial}{\partial y} & \frac{\partial}{\partial z} \\ f_1 & f_2 & f_3 \end{vmatrix}$$

$$= \hat{i}\left(\frac{\partial f_3}{\partial y} - \frac{\partial f_2}{\partial z}\right) + \hat{j}\left(\frac{\partial f_1}{\partial z} - \frac{\partial f_3}{\partial x}\right) + \hat{k}\left(\frac{\partial f_2}{\partial x} - \frac{\partial f_1}{\partial y}\right)$$

$$(\nabla\times\vec{F})\,.\,\hat{n} = \sum\left(\frac{\partial f_3}{\partial y} - \frac{\partial f_2}{\partial z}\right)\cos\alpha$$

Writing $d\vec{R} = \hat{i}dx + \hat{j}dy + \hat{k}dz$

We get

$$(\vec{F}\,.\,d\vec{R}) = (f_1\hat{i} + f_2\hat{j} + f_3\hat{k})\,.\,(\hat{i}dx + \hat{j}dy + \hat{k}dz)$$

$$= f_1\,dx + f_2\,dy + f_3\,dz$$

Stoke's theorem becomes

$$\int_C (f_1dx + f_2dy + f_3dz) = \iint_S \left(\frac{\partial f_3}{\partial y} - \frac{\partial f2}{\partial z}\right)\cos\alpha + \left(\frac{\partial f_1}{\partial z} - \frac{\partial f_3}{\partial x}\right)\cos\beta + \left(\frac{\partial f_2}{\partial x} - \frac{\partial f_1}{\partial y}\right)\cos\gamma$$

***Example* 7.41.** Verify Stokes theorem for $\vec{F} = (2x - y)\,\hat{i} - yz^2\hat{j} - y^2z\hat{k}$ where S is the upper half of the surface $x^2 + y^2 + z^2 = 1$ and C is its boundary.

Solution. We have to prove that

$$\int_C \vec{F} \cdot d\vec{R} = \iint_S \text{curl } \vec{F} \cdot \hat{n} dS$$

$$\text{L.H.S.} = \int_C \vec{F} \cdot d\vec{R}$$

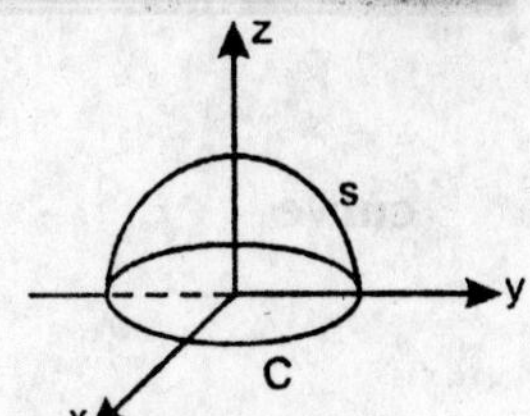

Fig. 7.26

Now C is the circle $x^2 + y^2 = 1$ in x-y plane where $z = 0$. Its parametric equations are therefore $x = \cos\theta$, and $y = \sin\theta$

$$\int_C \vec{F} \cdot d\vec{R} = \int_C ((2x - y)\hat{i} - yz^2\hat{j} - y^2 z\hat{k}) \cdot (idx + jdy)$$

$$= \int_C (2x - y)\,dx - yz\,dy = \int_C (2x - y)\,dx \ (z = 0)$$

$$= \int_0^{2\pi} (2\cos\theta - \sin\theta)\,d(\cos\theta) = \int_0^{2\pi} (2\cos\theta - \sin\theta)(-\sin\theta)\,d\theta$$

$$= -\int_0^{2\pi} (2\cos\theta\sin\theta - \sin^2\theta)\,d\theta = \pi$$

$$\text{R.H.S.} = \iint_S \text{curl } \vec{F} \cdot \hat{n} dS$$

Now curl

$$\vec{F} = \nabla \times \vec{F} = \begin{vmatrix} \hat{i} & \hat{j} & \hat{k} \\ \frac{\partial}{\partial x} & \frac{\partial}{\partial y} & \frac{\partial}{\partial z} \\ 2x - y & -yz^2 & -y^2 z \end{vmatrix}$$

$$= \hat{i}\,(-2yz + 2yz) + \hat{j}\,(0 + 0) + \hat{k}\,(1) = \hat{k}$$

curl

$$\vec{F} \cdot \hat{n} = \hat{k} \cdot \hat{n}$$

$$\therefore \quad \int_S curl\, \vec{F} \cdot \hat{n} dS = \iint_S \hat{k} \cdot \hat{n} dS$$

$$= \iint_S \hat{k} \cdot \hat{n} \frac{dx\,dy}{|\hat{n} \cdot \hat{k}|} = \iint_S dx\,dy = \pi r^2 \cdot 1^2 = \pi$$

r = radius of the sphere

= radius of the circle 'C'

Hence the theorem to verified.

Example **7.42.** Evaluate by Stoke's theorem $\int_C yz\,dx + zx + dy + xy\,dz$ where C is the curve $x^2 + y^2 = 4$ and $z = y^2$.

Solution. $\int_C (yz\,dx + zx\,dy + xy\,dz)$

$$= \int_C (\hat{i}yz + \hat{j}zx + \hat{k}xy) \cdot (\hat{i}dx + \hat{j}dy + \hat{k}dz) = \int_C \vec{F} \cdot d\vec{R}$$

i.e.

$$\vec{F} = \hat{i}yz + \hat{j}zx + \hat{k}xy \text{ and } d\vec{R} = \hat{i}dx + \hat{j}dy + \hat{k}dz$$

curve $\vec{F} = \begin{vmatrix} \hat{i} & \hat{j} & \hat{k} \\ \frac{\partial}{\partial x} & \frac{\partial}{\partial y} & \frac{\partial}{\partial z} \\ yz & zx & xy \end{vmatrix} = \hat{i}\,(x-x) + \hat{j}\,(y-y) + \hat{k}\,(z-z) = 0$

$$\int_C \vec{F} . d\vec{R} = \int_S \text{curl}\,\vec{F} . \hat{n} dS = 0$$

Example **7.43.** Verify Stoke's theorem for the function $\vec{F} = (x^2 + y^2)\,\hat{i} - 2xy\hat{j}$ taken round the rectangle bounded by $x = \pm a$, $y = 0$, $y = b$.

Solution. $\vec{F} = (x^2 + y^2)\,\hat{i} - 2xy\hat{j}$ and $d\vec{R} = \hat{i}dx + \hat{j}dy$

Fig. 7.27

$$\int_C \vec{F} . d\vec{R} = \int_C (x^2 + y^2)\,dx - 2xy\,dy$$

$$= \int_{DA} (x^2 + y^2)\,dx - 2xy\,dy + \int_{AB} (x^2 + y^2)\,dx - 2xy\,dy$$

$$+ \int_{BC} (x^2 + y^2)\,dx - 2xy\,dy + \int_{CD} (x^2 + y^2)\,dx - 2xy\,dy$$

Along *DA* $x = -a$ $dx = 0$ y varies from b to 0

$$\int_{DA} (x^2 + y^2)\,dx - 2xy\,dy = \int_b^0 + 2\,ay\,dy = 2a\left[\frac{y^2}{2}\right]_b^0 = 2a\left[0 - \frac{b^2}{2}\right] = -\,ab^2$$

Along *AB* $y = 0$, $dy = 0$, x varies from $-a$ to a

$$\int_{AB} (x^2 + y^2)\,dx - 2xy\,dy = \int_{-a}^{a} x^2 dx = \left[\frac{x^3}{3}\right]_{-a}^{a} = \frac{1}{3}\,[a^3 + a^3] = \frac{2a^3}{3}$$

Along *CD* $y = b$, $by = 0$, x varies from a to $-a$

$$\int_{CD} (x^2 + b^2)\,dx - 2xy\,dy = \int_a^{-a} (x^2 + b^2)\,dx = \left[\frac{x^3}{3} + b^2 x\right]_a^{-a} = \left(-\frac{a^3}{3} - ab^2\right) - \left(\frac{a^3}{3} + ab^2\right) = -\,\frac{2a^3}{3} - 2ab^2$$

Along *BC* $x = a$, $dx = 0$ y varies from 0 to b

$$\int_{BC} (x^2 + y^2)\,dx - 2xy\,dy = \int_0^b -\,2ay\,dy = -\,2a\left(\frac{y^2}{2}\right)_0^b = -\,\frac{2ab^2}{2}$$

$$\therefore \quad \int_C \vec{F} . d\vec{R} = -\,ab^2 + \frac{2a^3}{3} - \frac{2a^3}{3} - 2ab^2 - ab^2 = 4\,ab^2$$

Also curve $\vec{F} = \begin{vmatrix} \hat{i} & \hat{j} & \hat{k} \\ \frac{\partial}{\partial x} & \frac{\partial}{\partial y} & \frac{\partial}{\partial z} \\ x^2 + y^2 & -\,2xy & 0 \end{vmatrix}$

$$= \hat{i}\,(0-0)+\hat{j}\,(0+0)+\hat{k}\,(-2y-2y) = -4y\,\hat{k}$$

$$\text{curl}\,\vec{F}\,.\,\hat{n} = (4y\,\hat{k})\,.\,\hat{k} = -4y$$

$$\iint_S \text{curl}\,\vec{F}\,.\,\hat{n}dS = \int_{-a}^{a}\int_{0}^{b} -4y\,dx\,dy = -\int_{-a}^{a}\left[4\frac{y^2}{2}\right]_0^b dx$$

$$= -2\int_{-a}^{a} b^2\,dx = -2\,b^2\,[x]_{-a}^{a}$$

$$= -2\,b^2\,[a+a] = -4\,ab^2.$$

Stoke's theorem is verified.

***Example* 7.44.** If S be the surface of the sphere $x^2+y^2+z^2=9$. Prove that $\iint_S \textbf{curl}\,\vec{F}\,.\,d\vec{S} = 0$ or $\iint_S \vec{F}\,.\,\hat{n}dS = 0$.

Solution. Let S be the surface of the sphere $x^2+y^2+z^2=9$ cut the surface S by x-y plane passing through the origin. Let S_1 and S_2 denote its upper and lower portions. Let C be the common curve bounding both these options.

$$\iint_S \text{curl}\,\vec{F}\,.\,\hat{n}\,dS = \iint_{S_1} \text{curl}\,\vec{F}\,.\,\hat{n}\,dS + \iint_{S_2} \text{curl}\,\vec{F}\,.\,\hat{n}\,dS$$

$$= \int_C \vec{F}\,.\,d\vec{R} - \int_C \vec{F}\,.\,d\vec{R} = 0$$

Using Stoke's theorem.

The second integral is – ve because the curve C is traversed in a direction opposite to that of the first.

***Example* 7.45.** Prove that curl grad $V = 0$ by Stoke's theorem.

Solution. Let C be a simple closed curve and S be the surface enclosed by V.

Then by Stoke's theorem, we have

$$\int_C \text{grad}\,V\,.\,d\vec{R} = \iint_S \text{curl}\,(\text{grad}\,V)\,.\,\hat{n}\,dS$$

But grad

$$\vec{V}\,.\,d\vec{R} = \left(\hat{i}\frac{\partial V}{\partial x}+\hat{j}\frac{\partial V}{\partial y}+\hat{k}\frac{\partial V}{\partial z}\right).\,(\hat{i}dx+\hat{j}dy+\hat{k}dz)$$

$$= \frac{\partial V}{\partial x}dx+\frac{\partial V}{\partial y}dy+\frac{\partial V}{\partial z}dz = dV$$

$$\int_C \text{grad}\,V\,.\,d\vec{R} = \int_C dV = [V]_A^A = 0 \text{ where A is any position on C.}$$

$$= \iint_S \hat{n}\,.\,\text{curl}\,(\text{grad}\,\vec{V})\,dS$$

Since it is true for all surfaces S.

$$\text{curl}\,(\text{grad}\,\vec{V}) = 0$$

***Example* 7.46.** Prove that $\int_C \vec{A}\times\vec{R}\,.\,d\vec{R} = 2\text{A}\int_S d\vec{S}$ where A is any constant vector and deduce that $\int_C \vec{R}\times d\vec{R}$ is twice the vector area of the surface enclosed by C.

Solution. We have $\int_C A \times R \,.\, dR = \iint_S \text{curl}\,(A \times R) d\vec{S}$ (By Stoke's theorem)

Let

$$\vec{A} = \hat{i}a_1 + \hat{j}a_2 + \hat{k}a_3$$

$$\therefore \quad \vec{A} \,\square\, \vec{R} = \begin{vmatrix} \hat{i} & \hat{j} & \hat{k} \\ a_1 & a_2 & a_3 \\ x & y & z \end{vmatrix} = \hat{i}\,(a_2 z - a_3 y) + \hat{j}\,(a_3 x - a_1 z) + \hat{k}\,(a_1 y - a_2 x)$$

$$\text{curl}\,(\vec{A} \times \vec{R}) = \nabla \times (\vec{A} \times \vec{R})$$

$$= \begin{vmatrix} \hat{i} & \hat{j} & \hat{k} \\ \frac{\partial}{\partial x} & \frac{\partial}{\partial y} & \frac{\partial}{\partial z} \\ a_2 z - a_3 y & a_3 x - a_1 z & a_1 y - a_2 x \end{vmatrix}$$

$$= \hat{i}\,(a_1 + a_1) + \hat{j}\,(a_2 + a_2) + \hat{k}\,(a_3 + a_3)$$

$$= 2\,[\hat{i}a_1 + \hat{j}a_2 + \hat{k}a_3] = 2\vec{A}$$

Then

$$\int_C \vec{A} \times \vec{R} \,.\, d\vec{R} = \iint_S \text{curl}\,(\vec{A} \times \vec{R}) \,.\, d\vec{S}$$

$$= \iint 2\vec{A} \,.\, d\vec{S} = 2\vec{A} \,.\, \iint d\vec{S} \;\backslash\; \int_C \vec{A} \,.\, \vec{R} \times d\vec{R} = 24 \iint d\vec{S}$$

also

$$\vec{A} \,.\, \int_C \vec{R} \times d\vec{R} = 2A \,.\, \iint d\vec{S}$$

$$\therefore \quad \int_C \vec{R} \times d\vec{R} = 2 \iint d\vec{S}$$

Example 7.47. Evaluate by Stoke's theorem. $\oint_C (\sin z\, dx - \cos x\, dy + \sin y\, dz)$ where C is the boundary of the rectangle.

$0 \le x \le \pi$, $0 \le y \le 1$, $z = 3$.

Solution.

$$\oint_C (\sin z\, dx - \cos x\, dy + \sin y\, dz)$$

$$= \oint_C (\hat{i}\sin z - \hat{j}\cos x + \hat{k}\sin y) \,.\, (\hat{i}dx + \hat{j}dy + \hat{k}dx)$$

$$= \int_C \vec{F} \,.\, dR \quad \text{where } F = \hat{i}\,\sin z - \hat{j}\,\cos x + \hat{k}\,\sin y$$

$$= \iint_S \text{curl}\,\vec{F} \,.\, \hat{n}\, dS \text{ by Stoke's theorem.}$$

Now curl

$$\vec{F} = \nabla \times \vec{F} = \begin{vmatrix} \hat{i} & \hat{j} & \hat{k} \\ \frac{\partial}{\partial x} & \frac{\partial}{\partial y} & \frac{\partial}{\partial z} \\ \sin z & -\cos x & \sin y \end{vmatrix}$$

$$= \cos y\,\hat{i} + \cos z\,\hat{j} + \sin x\,\hat{k}$$

Rectangle lies in the plane $z = -3$ [|| to x-y plane]

$$\therefore \quad \hat{n} = \hat{k}$$

$$\text{curl } \vec{F} \cdot \hat{n} = (\cos y\, \hat{i} + \cos 2\, \hat{j} + \sin x\, \hat{k}) \cdot \hat{k} = \sin x$$

Given integral

$$= \iint_S \text{curl } \vec{F} \cdot \hat{n}\, dS = \iint_S \sin x\, dx\, dy$$

$$= \int_{x=0}^{\pi} \int_0^1 \sin x\, dy\, dx = \int_{x=0}^{\pi} \sin x [y]_0^1\, dx = \int_0^{\pi} \sin x\, dx = 2$$

***Example* 7.48.** Apply Stoke's theorem to find the value of $\int_C (y\, dx + z dy + x\, dz)$ where C is the curve of intersection of sphere $x^2 + y^2 + z^2 = a^2$ and $x + z = a$.

Solution. The points of intersection of the sphere $x^2 + y^2 + z^2 = a^2$ and $x + z = a$ are $A\ (a, 0, 0)$, $C\ (0, 0, a)$.

The curve C is evidently a circle lying in the plane $x + z = a$ and having $A\ (a, 0, 0)$ and $C\ (0, 0, a)$ as the extremities of the diameter.

i.e. $\int_C (y\, dx + z dy + x\, dz)$

$$= \int_C (y\hat{i} + z\hat{j} + x\hat{k}) \cdot d\vec{R}$$

where

$$\vec{F} = y\hat{i} + z\hat{j} + x\hat{k}$$

$$d\vec{R} = \hat{i} dx + \hat{j} dy + \hat{k} dz$$

$$= \iint_S \text{curl}\, (y\hat{i} + z\hat{j} + x\hat{k}) \cdot \hat{n} dS$$

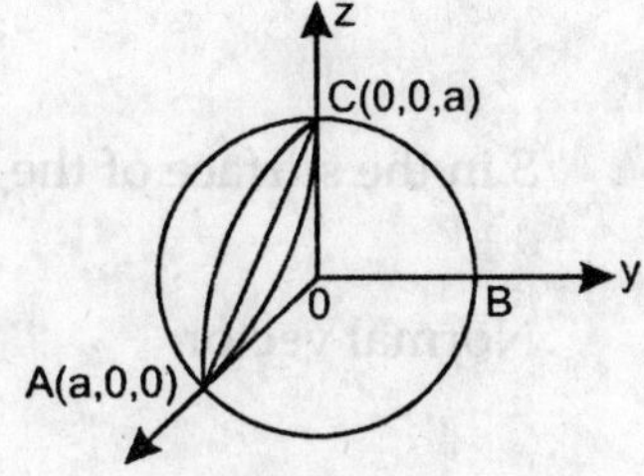

Fig. 7.28

where S is the circle on AC as diameter and curl $(y\hat{i} + z\hat{j} + x\hat{k}) = \begin{vmatrix} i & j & k \\ \frac{\partial}{\partial x} & \frac{\partial}{\partial y} & \frac{\partial}{\partial z} \\ y & z & x \end{vmatrix} = -(i + j + k)$

$$\int_C y dx + z dy + x dz = \iint_S -(\hat{i} + \hat{j} + \hat{k}) \cdot \left(\frac{1}{\sqrt{2}}\hat{i} + \frac{1}{\sqrt{2}}\hat{k}\right) dS \quad \left\{\begin{matrix} \because \nabla (x + z) = \hat{i} + \hat{k} \\ \hat{n} = \frac{1}{\sqrt{2}}\hat{i} + \frac{1}{\sqrt{2}}\hat{k} \end{matrix}\right\}$$

$$= \frac{-2}{\sqrt{2}} \iint_S dS = -\frac{2}{\sqrt{2}} \cdot \pi \cdot \left(\frac{a}{\sqrt{2}}\right)^2 = -\frac{\pi a^2}{\sqrt{2}} \quad \left(\iint_S dS = \text{Area of circle } S\right)$$

***Example* 7.49.** Show that $\int_C \vec{R} \cdot d\vec{R} = 0$, $\vec{R} = \hat{i}x + \hat{j}y + \hat{k}z$

Solution. Now by Stoke's theorem

$$\int_C \vec{R} \cdot d\vec{R} = \int_S \text{curl } \vec{R} \cdot \hat{n} dS$$

Now

$$\text{curl } \vec{R} = \begin{vmatrix} \hat{i} & \hat{j} & \hat{k} \\ \frac{\partial}{\partial x} & \frac{\partial}{\partial y} & \frac{\partial}{\partial z} \\ x & y & z \end{vmatrix} = 0$$

$$\int \vec{R} \cdot d\vec{R} = 0$$

***Example* 7.50.** Use the Stoke's theorem to evaluate $\int_C [(x+2y)\,dx + (x-z)\,dy + (y-z)\,dz]$ where C is the boundary of the triangle with vertices (2, 0, 0) (0, 3, 0) and (0, 0, 6) oriented in the anti-clockwise direction.

Solution. Here $\int_C [(x+2y)\,dx + (x-z)\,dy + (y-z)\,dz]$

$$= \int [\hat{i}(x+2y) + \hat{j}(x-z) + \hat{k}(y-z)] \cdot (\hat{i}dx + \hat{j}dy + \hat{k}dz)$$

$$\int_C \vec{F} \cdot d\vec{R} = \iint_S curl\,\vec{F} \cdot \hat{n}\,dS \text{ where } \vec{F} = (x+2y)\hat{i} + (x-2)\hat{j} + (y-z)\hat{k}$$

$$\text{curl } \vec{F} = \begin{vmatrix} \hat{i} & \hat{j} & \hat{k} \\ \frac{\partial}{\partial x} & \frac{\partial}{\partial y} & \frac{\partial}{\partial z} \\ x+2y & x-z & y-z \end{vmatrix}$$

$$= (1+1)\,\hat{i} - (0+0)\,\hat{j} + (1-2)\,\hat{k}$$

$$= 2\hat{i} - \hat{k}$$

Fig. 7.29

S in the surface of the plane $\frac{x}{2} + \frac{y}{3} + \frac{z}{6} = 1$ (of ∇ABC) $\hat{n}$ is the unit normal to the plane ABC.

Normal vector $= \nabla\left[\frac{x}{2} + \frac{y}{3} + \frac{z}{6}\right] = \frac{\hat{i}}{2} + \frac{\hat{j}}{3} + \frac{\hat{k}}{6} = \frac{1}{6}[3\hat{i} + 2\hat{j} + \hat{k}]$

$$\hat{n} = \frac{\frac{1}{6}[3\hat{i} + 2\hat{j} + \hat{k}]}{\frac{1}{6}\sqrt{9+4+1}} = \frac{1}{\sqrt{14}}(3\hat{i} + 2\hat{j} + \hat{k})$$

$$\text{curl } \vec{F} \cdot \hat{n} = (2\hat{i} - \hat{k}) \cdot \frac{1}{\sqrt{14}}(3\hat{i} + 2\hat{j} + \hat{k}) = \frac{1}{\sqrt{14}}(6-1) = \frac{5}{\sqrt{14}}$$

$$\int_C \vec{F} \cdot d\vec{R} = \iint_C curl\,\vec{F} \cdot \hat{n}\,dS = \iint_S \frac{5}{\sqrt{14}}dS = \frac{5}{\sqrt{14}}\iint_S dS$$

$$= \frac{5}{\sqrt{14}}\iint_R \frac{dx\,dy}{|\hat{n} \cdot \hat{k}|} = \frac{5}{\sqrt{14}}\iint_R \frac{dx\,dy}{\frac{1}{\sqrt{14}}} = 5\iint_R dx\,dy$$

where R is the projection of S on x-y plane.

$$= 5\,(\text{Area of the } \nabla OAB) = 5 \cdot \frac{2 \times 3}{2} = 15$$

***Example* 7.51.** Verify Stoke's theorem for vector field $\vec{F} = (x^2 - y^2)\,\hat{i} + 2yx\hat{j}$ over the box bounded by the plane $x = 0, x = a, y = 0, y = b, z = 0, z = c$, the face $z = 0$ is cut.

Solution. Here C is the boundary $OABC$

$$\int_C \vec{F} \cdot d\vec{R} = \int_C (x^2 + y^2)dx + 2xy\,dy$$

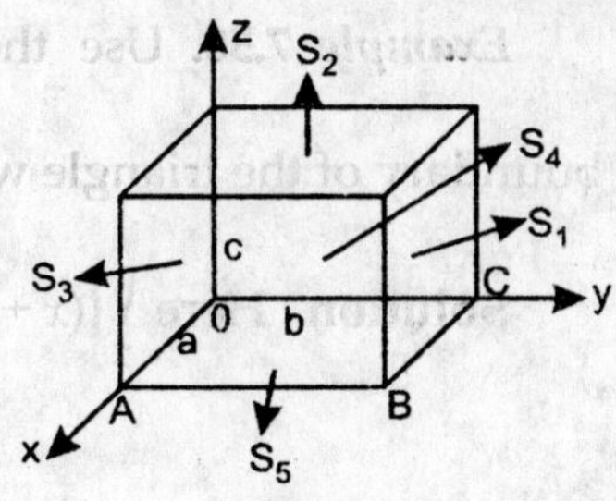

Fig. 7.30

Along *AB* $z = 0, x = a, y$ varies from 0 to b

$$dx = 0$$

$$\int_{AB} = \int_0^b 2xy\,dy = a\,[y^2]_0^b = ab^2$$

Along *BC* $z = 0, y = b, x$ varies from a to 0.

$$dy = 0$$

$$\int_{BC} = \int_a^0 (x^2 - y^2)\,dx = \left(\frac{x^2}{3} - b^2 x\right)_a^0 = -\frac{a^3}{3} + ab^2$$

Along *CO* $z = 0, x = 0, y$ varies from b to 0

$$dx = 0$$

$$\int_{CO} = \int_b^0 2xy\,dy = \left[2x \cdot \frac{y^2}{2}\right]_b^0 = [x\,y^2]_b^0 = 0$$

Along *OA* $z = 0, y = 0, x$ varies from 0 to a.

$$dy = 0$$

$$\int_{OA} = \int_0^a x^2 dx = \left[\frac{x^3}{3}\right]_0^a = \frac{a^3}{3}$$

So Line integral $= ab^2 - \dfrac{a^3}{3} + ab^2 + 0 + \dfrac{a^3}{3} = 2ab^2$

Now $$\iint_S \text{curl}\,\vec{F}\cdot\hat{n}\,dS = \iint_{S_1} \text{curl}\,\vec{F}\cdot\hat{n}\,dS + \iint_{S_2} \text{curl}\,\vec{F}\cdot\hat{n}\,dS$$

$$+ \iint_{S_3} \text{curl}\,\vec{F}\cdot\hat{n}\,dS + \iint_{S_4} \text{curl}\,\vec{F}\cdot\hat{n}\,dS + \iint_{S_5} \text{curl}\,\vec{F}\cdot\hat{n}\,dS$$

For S_1 $\quad y = b \quad \hat{n} = \hat{j} \quad dS = dz\,dx$

$$\iint_{S_1} \text{curl}\,\vec{F}\cdot\hat{n}\,dS = \iint_{S_1} \nabla\times\vec{F}\cdot\hat{n}\,dS$$

Now $$\nabla\times\vec{F} = \begin{vmatrix} \hat{i} & \hat{j} & \hat{k} \\ \dfrac{\partial}{\partial x} & \dfrac{\partial}{\partial y} & \dfrac{\partial}{\partial z} \\ x^2 - y^2 & 2xy & 0 \end{vmatrix} = (2x + 2y)\,\hat{k} = 4y\,\hat{k}$$

$$\iint_{S_1} \text{curl}\,\vec{F}\cdot\hat{n}\,dS = \iint_{S_1} 4y\,\hat{k}\cdot\hat{j} = 0$$

For S_2 $\quad z = c \quad y$ varies from 0 to $b \quad \hat{n} = k$

x varies from 0 to a

$$\iint_{S_2} \text{curl}\,\vec{F}\cdot\hat{n}\,dS = \iint_{S_2} 4y\hat{k}\cdot\hat{k}\,dS = \iint_{S_2} dy\,dS$$

$$= \int_0^a \int_0^b 4y\, dx\, dy = \int_0^a (2y^2)_0^b\, dx = 2\, b^2 \times a = 2\, ab^2$$

For S_3 $\qquad \hat{n} = -\hat{j} \quad y = 0$

$$\iint_{S_3} 4y\hat{k} \,.\, (-\hat{j})\, dS = 0$$

For S_4 $\qquad \hat{n} = -\hat{i} \quad x = 0$

$$\iint_{S_4} \text{curl}\, \vec{F} \,.\, \hat{n}\, dS = \iint_{S_4} (4yk) \,.\, (-\hat{i})\, dS = 0$$

For S_5 $\qquad \hat{n} = \hat{i} \quad x = a$

$$\iint_{S_5} \text{curl}\, \vec{F} \,.\, \hat{n}\, dS = \iint_{S_5} 4y\hat{k} \,.\, (\hat{i})\, dS = 0$$

Now
$$\iint_{S_5} \text{curl}\, \vec{F} \,.\, \hat{n}\, dS = 0 + 2ab^2 + 0 + 0 + 0 = 2ab^2$$

Hence Proved

Example 7.52. Apply Stoke's theorem to calculate $\int_C 4y\, dx + 2zdy + 6y\, dz$ where C is the curve if intersection of $x^2 + y^2 + z^2 = 6z$ and $z = x + 3$

Solution. Now $\int_C 4y\, dx + 2zdy + 6ydz = \int_C (4y\hat{i} + 2z\hat{j} + 6y\hat{k}) \,.\, (\hat{i}dx + \hat{j}dy + \hat{k}dz) = \int_C \vec{F} \,.\, d\vec{R}$

where
$$\vec{F} = 4y\hat{i} + 2z\hat{j} + 6y\hat{k}$$

$$d\vec{R} = \hat{i}dx + \hat{j}dy + \hat{k}dz$$

Now curl
$$\vec{F} = \begin{vmatrix} \hat{i} & \hat{j} & \hat{k} \\ \dfrac{\partial}{\partial x} & \dfrac{\partial}{\partial y} & \dfrac{\partial}{\partial z} \\ 4y & 2z & 6y \end{vmatrix}$$

$$= (6 - 2)\, \hat{i} - (0 - 0)\, \hat{j} + (0 - 4)\, \hat{k} = 4\hat{i} - 4\hat{k}$$

S is the surface of the circle

$$x^2 + y^2 + z^2 = 6z, \quad z = x + 3$$

$\hat{n}$ is the unit normal to the plane

$$x - z + 3 = 0$$

$$\hat{n} = \frac{\nabla (x - z + 3)}{|\nabla (x - z + 3)|} = \frac{\hat{i} - \hat{k}}{\sqrt{1 + 1}} = \frac{\hat{i} - \hat{k}}{\sqrt{2}}$$

$$\text{curl}\, \vec{F} \,.\, n = \frac{\hat{i} - \hat{k}}{\sqrt{2}} \,.\, (4\hat{i} - 4\hat{k}) = \frac{4 + 4}{\sqrt{2}} = 4\sqrt{2}$$

$$\int_C \vec{F} \,.\, d\vec{R} = \int_S \text{curl}\, \vec{F} \,.\, \hat{n}\, dS = \iint_S 4\sqrt{2}\, (dx\, dz) = 4\sqrt{2} \text{ (area of the circle)}$$

Centre of the sphere $x^2 + y^2 + z^2 = 6z$ or $x^2 + y^2 + (z-3)^2 = 9$ lies on the plane $z = x + 3$, it means that the given circle is the greatest circle of the sphere, whose radius is equal to the radius of the sphere.

Radius of the circle = 3 Area = π (3)2 = 9π

$$\iint_S \text{curl } \vec{F} \,.\, \hat{n}\, dS = 4\sqrt{2}\,(9\pi) = 36\sqrt{2}\,\pi$$

7.20 GAUSS'S DIVERGENCE THEOREM

(Relation between Surface and Volume Integral)

Statement: The volume integral of the divergence of a vector point function $\vec{F}$ taken over the volume V enclosed by a surface S is equal to the surface integral of the normal component of $\vec{F}$ taken over the closed surface S.

i.e.
$$\iiint_V \text{div}\,.\,\vec{F}\, dV = \iiint_V \nabla\,.\,\vec{F}\, dV = \int \vec{F}\,.\,\hat{n} dS$$

If
$$\vec{F} = f_1\hat{i} + f_2\hat{j} + f_3\hat{k}, \text{ then}$$

$$\nabla\,.\,\vec{F} = \text{div } \vec{F} = \frac{\partial f_1}{\partial x} + \frac{\partial f_2}{\partial y} + \frac{\partial f_3}{\partial z}$$

Suppose α, β, γ are the angles made by the outward drawn unit normal with the positive direction of the axes, then

$$\hat{n} = \cos\alpha\,\hat{i} + \cos\beta\,\hat{j} + \cos\gamma\,\hat{k}$$

$$\vec{F}\,.\,\hat{n} = f_1\cos\alpha + f_2\cos\beta + f_3\cos\gamma$$

Thus the divergence theorem is

$$\iiint_V \left(\frac{\partial f_1}{\partial x} + \frac{\partial f_2}{\partial y} + \frac{\partial f_3}{\partial z}\right) dx\, dy\, dz = \iint_S (f_1\cos\alpha + f_2\cos\beta + f_3\cos\gamma)\, dS$$

$$= \iint_S (f_1\, dy\, dz + f_2\, dz\, dx + f_3\, dx\, dy)$$

i.e. $\cos\alpha$.
$$dS = dy\, dz \text{ etc.}$$

***Example* 7.53.** Evaluate $\iint \vec{r}\,.\,\hat{n}dS$ when S is a closed surface.

Solution. By Gauss Divergence theorem

$$\iint \vec{r}\,.\,\hat{n}.dS = \iiint_V \nabla\,.\,\vec{r}\, dV$$

$$= \iiint_V \left(\hat{i}\frac{\partial}{\partial x} + \hat{j}\frac{\partial}{\partial y} + \hat{k}\frac{\partial}{\partial z}\right).(x\hat{i} + y\hat{j} + z\hat{k})\, dV$$

$$= \iiint_V \left(\frac{\partial x}{\partial x} + \frac{\partial y}{\partial y} + \frac{\partial z}{\partial z}\right) dV = 3\iiint_V dV = 3\,V.$$

V is the volume enclosed by surface S.

***Example* 7.54.** Show that $\int_S \hat{n} dS = 0$ for any closed surface S.

Solution. Let $\vec{a}$ be any constant vector

$$\vec{a} \cdot \int_S \hat{n} dS = \int_S \vec{a} \cdot \hat{n} dS$$

$$= \iiint_V \text{div } \vec{a} \, dV \qquad \text{(By Divergence theorem)}$$

$$= 0 \qquad [\text{div } \vec{a} = 0, \text{ as } \vec{a} \text{ is a constant vector}]$$

$$\int_S \hat{n} dS = 0$$

***Example* 7.55.** Verify divergence theorem for $\vec{F} = x^2 i + z\hat{j} + yz\hat{k}$, taken over the cube bounded by $x = 0, x = 1, y = 0, y = 1$.

Solution. Divergence theorem states that $\int_V \text{div } \vec{F} \, . dV = \int_S \vec{F} \, . \hat{n} dS$

$$\text{div } \vec{F} = \frac{\partial}{\partial x}(x^2) + \frac{\partial}{\partial y}(z) + \frac{\partial}{\partial z} yz = 2x + y$$

$$\int_V div \, \vec{F} \, . dV = \int_0^1 \int_0^1 \int_0^1 () dx \, dy \, dz = \int_0^1 \int_0^1 \int_0^1 (2x + y) dx \, dy \, dz$$

$$= \int_0^1 \int_0^1 (x^2 + xy)_0^1 \, dy \, dz = \int_0^1 \int_0^1 (1 + y) dy \, dz$$

$$= \int_0^1 \left(y + \frac{y^2}{2} \right)_0^1 dz = \int_0^1 \frac{3}{2} dz = \frac{3}{2} \qquad \text{...(1)}$$

Now let

S_1 = face $OABC$ ∴ $z = 0, \hat{n} = -\hat{k}$

S_2 = face $DEFG$ ∴ $z = 1, \hat{n} = \hat{k}$

S_3 = face $BCDG$ ∴ $y = 1, \hat{n} = \hat{j}$

S_4 = face $OEFA$ ∴ $y = 0, \hat{n} = -\hat{j}$

S_5 = face $OCDE$ ∴ $x = 0, \hat{n} = -\hat{i}$

S_6 = face $ABGF$ ∴ $x = 1, \hat{n} = \hat{i}$

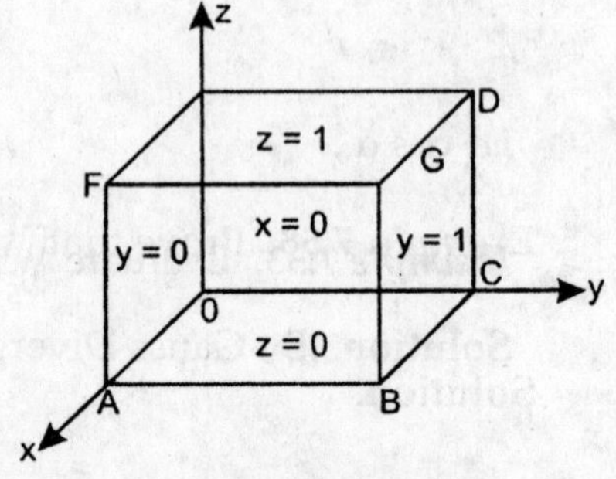

Fig. 7.31

$$\int_S \vec{F} \, . \hat{n} dS = \int_{S_1} \vec{F} \, . \hat{n} dS + \int_{S_2} \vec{F} \, . \hat{n} dS + \int_{S_3} \vec{F} \, . \hat{n} dS + \int_{S_4} \vec{F} \, . \hat{n} dS + \int_{S_5} \vec{F} \, . \hat{n} dS + \int_{S_6} \vec{F} \, . \hat{n} dS$$

Now $\int_{S_1} \vec{F} \, . \hat{n} dS = \int_{S_1} (x^2 \hat{i} + z\hat{j} + yz\hat{k}) . (-k) dx \, dy = 0$

$$\int_{S_2} \vec{F} \, . \hat{n} dS = \int_0^1 \int_0^1 y \, dx \, dy = \frac{1}{2}, \int_{S_3} \vec{F} \, . \hat{n} dS = \int_0^1 \int_0^1 z \, dx \, dz = \frac{1}{2}$$

$$\int_{S_4} \vec{F}\,.\,\hat{n}dS = \int_0^1\int_0^1 -z\,dx\,dz = -\frac{1}{2},\ \int_{S_5} \vec{F}\,.\,\hat{n}dS = 0$$

$$\int_{S_6} \vec{F}\,.\,\hat{n}dS = \int_0^1\int_0^1 dy\,dz = 1$$

$$\int_{S} \vec{F}\,.\,\hat{n}dS = 0 + \frac{1}{2} + \frac{1}{2} - \frac{1}{2} + 0 + 1 = \frac{3}{2} \quad ...(2)$$

From (1) and (2) Divergence theorem is verified.

***Example* 7.56.** Find the surface integral $\int \vec{F}\,.\,\hat{n}dS$ where $\vec{F} = (2x+3z)\,\hat{i} - (xz+y)\,\hat{j} + (y^2+2z)\,\hat{k}$ and S is the surface of the sphere having centre $(3-1, 2)$ and radius 3.

Solution. Equation of the sphere is

$$(x-3)^2 + (y+1)^2 + (z-2)^2 = 3^2 \quad ...(1)$$

Now by Gauss's Divergence theorem

$$\iiint \vec{F}\,.\,\hat{n}dS = \iiint_V \nabla\,.\,F\,dV = \iiint_V (2-1+2)dV = \iiint 3dV = 3\,V$$

where V is the volume of the sphere (1).

$$V = \frac{4}{3}\pi r^3 = \frac{4}{3}\pi (3)^3 = \frac{4}{3}\pi \times 27 = 36\,\pi$$

$$\iint \vec{F}\,.\,\hat{n}dS = 3V = 3 \times 36\,\pi = 108\,\pi$$

***Example* 7.57.** Using divergence theorem prove that $\iint_S \vec{R}\,.\,d\vec{S} = 3\,V$.

Solution. According to Gauss Divergence theorem

$$\iint_S \vec{R}\,.\,d\vec{S} = \iiint_V \nabla\,.\,\vec{R}\,dV = \iiint_V 3\,dV = 3V$$

***Example.* 7.58.** Prove that $\iint_S \nabla r^2 d\vec{S} = 6\,V$.

Solution.

$$\text{L.H.S.} = \iint \nabla r^2 d\vec{S} = \iiint_V \nabla\,.\,(\nabla r^2)dV$$

$$= \iiint_V \nabla\,.\left[\Sigma i\frac{\partial}{\partial x}(x^2+y^2+z^2)\right]dV$$

$$= \iiint_V \nabla\,.\,(\Sigma 2\,xi)\,dV = \iiint_V \nabla\,.\,2\vec{R}\,dV = 2\iiint_V 3\,dV = 6\,V$$

***Example* 7.59.** If S is any closed surface enclosing a volume V and $\vec{F} = ax\hat{i} + by\hat{j} + cz\hat{k}$, Prove that $\iint_S \vec{F}\,.\,\hat{n}dS = (a+b+c)V$.

Solution. By Gauss Divergence theorem

$$\iint \vec{F}\,.\,\hat{n}dS = \iiint_V \nabla\,.\,(ax\hat{i} + by\hat{j} + cz\hat{k})dV = \iiint_V (a+b+c)\,dV = (a+b+c)V$$

Example 7.60. Using divergence theorem evaluate $\iint_S \vec{F}\,.\,\hat{n}dS$ where $\vec{F} = x^3\hat{i} + y^3\hat{j} + z^3\hat{k}$ and S is the surface of the sphere $x^2 + y^2 + z^2 = a^2$.

Solution. $$\iint \vec{F}\,.\,\hat{n}dS = \iiint_V \nabla\,.\,\vec{F}\,dV = \iiint_V 3(x^2 + y^2 + z^2)dV$$

Using spherical co-ordinates

$$x = r\sin\theta\cos\phi$$

$$y = r\sin\theta\sin\phi$$

$$z = r\cos\theta$$

We have

$$\iint \vec{F}\,.\,\hat{n}dS = 3\int_{\phi=0}^{2\pi}\int_{\theta=0}^{\pi}\int_{z=0}^{a} r^2\,.\,r^2\sin\theta\,dr\,d\theta\,d\phi$$

$$= 3\int_{\phi=0}^{2\pi}\int_{0}^{\pi}\left(\frac{r^5}{5}\right)_0^a \sin\theta\,d\theta\,d\phi = 3\int_0^{2\pi}\int_0^{\pi}\frac{a^5}{5}\sin\theta\,d\theta\,d\phi$$

$$= \frac{3}{5}a^5\int_0^{2\pi} -(\cos\theta)_0^{\pi}\,d\phi$$

$$= \frac{3}{5}a^5\int_0^{2\pi}(1+1)\,d\phi = \frac{6}{5}a^5\int_0^{2\pi}d\phi = \frac{12a^5}{5}\pi$$

Example 7.61. Verify Gauss divergence theorem for the function $\vec{F} = y\hat{i} + x\hat{j} + z^2\hat{k}$ over the cylindrical region $x^2 + y^2 = 9$, $z = 0$ and $z = 2$.

Solution. Here $\vec{F} = y\hat{i} + x\hat{j} + z^2\hat{k}$

The given surface consists of three surfaces (*i*) bottom surfaces S_1 (*ii*) Upper surfaces S_2 and (*iii*) the curved surfaces S_3.

Now we have

$$\iint_S \vec{F}\,.\,\hat{n}dS = \iint_{S_1}\vec{F}\,.\,\hat{n}dS + \iint_{S_2}\vec{F}\,.\,\hat{n}dS + \iint_{S_3}\vec{F}\,.\,\hat{n}dS$$

Now Surface S_1 $\quad \hat{n} = -\hat{k} \quad z = 0$

$$\iint_{S_1}\vec{F}\,.\,\hat{n}dS = \iint_{S_1}(y\hat{i} + x\hat{j} + z^2\hat{k})\,.\,(-k)\,.\,dS$$

$$= \iint_{S_1} -z^2 dS$$

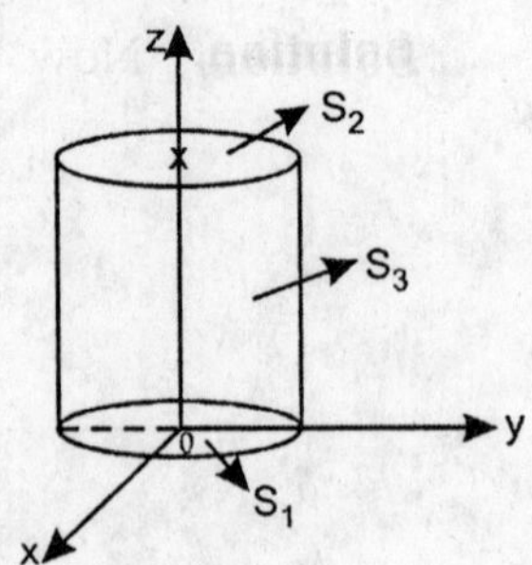

Fig. 7.32

For Surface S_2 $\quad \hat{n} = \hat{k} \quad z = 2$

$$\iint_{S_2}\vec{F}\,.\,\hat{n}dS = \iint_{S_2} +z^2 dS$$

$$= 4\iint dS = 4 \times 9\pi = 36\pi \qquad [\text{Area} = \pi r^2 = 9\pi]$$

For Surface S_3 (curved surface)

$$\hat{n} = \frac{2x\hat{i} + 2y\hat{j}}{\sqrt{4x^2+4y^2}} = \frac{x\hat{i}+y\hat{j}}{\sqrt{x^2+y^2}} = \frac{x\hat{i}+y\hat{j}}{3}$$

$$dS = \frac{dz\,dx}{|\hat{n}.\hat{j}|} = \frac{dz\,dx}{\frac{y}{3}} = 3\,\frac{dz\,dx}{y} \qquad \text{[Taking projection on } z\text{-}x \text{ plane]}$$

$$\vec{F}.\hat{n} = \frac{xy}{3} + \frac{xy}{3} = \frac{2xy}{3}$$

$$\iint \vec{F}.\hat{n}dS = \iint_S \frac{2xy}{3} \times 3\,\frac{dz\,dx}{y} = \iint_S 2x\,dx\,dz$$

$$= 2\int_0^2\int_{-3}^3 x\,dx\,dz = \int_0^2 \left(\frac{x^2}{2}\right)_{-3}^3 dz = 0$$

Hence $$\iint_S \vec{F}.\hat{n}dS = 0 + 36\pi + 0 = 36\pi$$

Now $$\iiint_V \nabla.F\,dV = \iiint_V 2z\,dV = 2\iiint_V z\,dV$$

$$= 2\int_{-3}^3\int_{-\sqrt{9-x^2}}^{\sqrt{9-x^2}}\int_0^2 z\,dx\,dy\,dz = 2\int_{-3}^3\int_{-\sqrt{9-x^2}}^{\sqrt{9-x^2}}\left(\frac{z^2}{2}\right)_0^2 dx\,dy$$

$$= 8\int_{-3}^3\int_0^{\sqrt{9-x^2}} dx\,dy = 8\int_{-3}^3 \sqrt{9-x^2}\,dx$$

$$= 8\times 2\left[\frac{x\sqrt{9-x^2}}{2} + \frac{9}{2}\sin^{-1}\frac{x}{3}\right]_0^3 = 16\times\frac{9}{2}\times\frac{\pi}{2} = 36\text{ p}$$

Hence the theorem is verified.

Example* 7.62.** Evaluate $\iint_a$ ***x dy dz + y dz dx + z dx dy over the sphere of radius '*a*'.

Solution. Now $\iint_S x\,dy\,dz + y\,dz\,dx + z\,dx\,dy$ [sphere $x^2+y^2+z^2=a^2$]

$$= \iiint_V \left(\frac{\partial x}{\partial x} + \frac{\partial y}{\partial y} + \frac{\partial z}{\partial z}\right)dV = \iiint_V (1+1+1)\,dV = 3V$$

$$= 3\times\frac{4}{3}\pi a^3 \qquad \left[V = \frac{4}{3}\pi r^3\right]$$

$$= 4\pi a^3$$

***Example* 7.63.** For the sphere $x^2 + y^2 + z^2 = 9$, prove that $\iint_S \vec{F}\,.\,\hat{n}dS = 36\,\pi(l + m + n)$ where $\vec{F} = \hat{i}lx + \hat{j}my + \hat{k}nz$.

Solution.
$$\iint_S \vec{F}\,.\,\hat{n}dS = \iiint_V (\nabla\,.\,\vec{F})\,dV = \iiint_V (l + m + n)\,dV$$
$$= (l + m + n)V = (l + m + n)\,\frac{4}{3}\,\pi \times 3^3$$
$$= 36\,\pi\,(l + m + n)\ \textbf{Ans.}$$

***Example* 7.64.** Apply divergence theorem to evaluate $\iint_S (lx^2 + my^2 + nz^2)\,dS$ taken over the sphere $(x - a)^2 + (y - b)^2 + (z - i) = r^2$ where l, m, n are direction cosines.

Solution. By Gauss divergence theorem
$$\iint \vec{F}\,.\,\hat{n}dS = \iiint_V (\nabla\,.\,\vec{F})\,dV$$

Now
$$\hat{n} = \frac{2(x-a)\hat{i} + 2(y-b)\hat{j} + 2(z-c)\hat{k}}{\sqrt{4(x-a)^2 + 4(y-b)^2 + 4(z-c)^2}}$$
$$= \frac{(x-a)\hat{i} + (y-b)\hat{j} + (z-c)\hat{k}}{\sqrt{r^2}}$$
$$= \frac{(x-a)\hat{i} + (y-b)\hat{j} + (z-c)\hat{k}}{r}$$

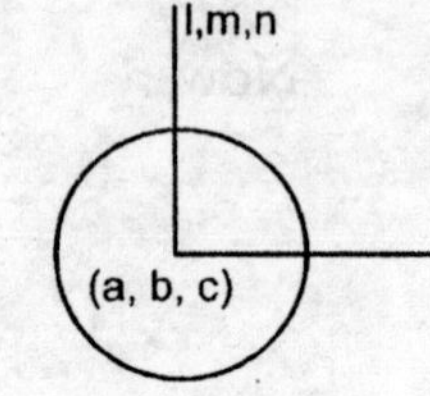

Fig. 7.33

Now
$$l = \frac{x-a}{\sqrt{(x-a)^2 + (y-b)^2 + (z-c)^2}} = \frac{x-a}{r}$$
$$m = \frac{y-b}{z}$$
$$n = \frac{z-c}{z}$$
$$\iint_S (lx^2 + my^2 + nz^2)\,dS = \iint_S (l\hat{i} + m\hat{j} + n\hat{k})\,.\,(\hat{i}x^2 + \hat{j}y^2 + \hat{k}z^2)\,dS$$
$$= \iint_S \hat{n}\,.\,\vec{F}\ dS = \iint_S \vec{F}\,.\,\hat{n}dS$$

where
$$\hat{n} = \hat{i}l + \hat{j}m + \hat{k}n$$
$$\vec{F} = \hat{i}x^2 + \hat{j}y^2 + \hat{k}z^2$$

So by divergence theorem
$$\iint_S (lx^2 + my^2 + nz^2)\,dS = \iiint_V \nabla\,.\,F\,dV$$
$$= \iiint_V 2(x + y + z)\,dV \qquad \ldots(1)$$

Given $(x-a)^2+(y-b)^2+(z-c)^2 = r^2$

Put
$$x-a = R\sin\theta\cos\phi \qquad R^2=r^2$$
$$y-b = R\sin\theta\sin\phi \qquad R=r$$
$$z-c = R\cos\theta$$

$$\iint_S (lx^2+my^2+nz^2)\,dS = \iiint_V 2(x+y+z)\,dV$$

$$= \iiint_V [(a+b+c)+R(\sin\theta\cos\phi+\sin\theta\sin\phi+\cos\theta)\times R^2\sin\theta\,dR\,d\theta\,d\phi$$

$$8\times 2\int_0^{\pi/2}\int_0^{\pi/2}\int_0^{r}[(a+b+c)+R(\sin\theta\cos\phi+\sin\theta\sin\phi+\cos\theta)]\times R^2\sin\theta\,dR\,d\theta\,d\phi$$

$$= 8\times 2\int_0^{\pi/2}\int_0^{\pi/2}\left[(a+b+c)\frac{R^3}{3}+\frac{R^4}{4}(\sin\theta\cos\phi+\sin\theta\sin\phi+\cos\theta)\right]_0^r\sin\theta\,d\theta\,d\phi$$

$$= 8\times 2\int_0^{\pi/2}\int_0^{\pi/2}\left[(a+b+c)\frac{r^3}{3}\sin\theta+\frac{r^4}{4}(\sin^2\theta\cos\phi+\sin^2\phi+\cos\theta\sin\theta\right]d\theta\,d\phi$$

$$= 16\int_0^{\frac{\pi}{2}}\left[(a+b+c)\frac{r^3}{3}+\frac{r^4}{4}\left(\frac{1}{2}.\frac{\pi}{2}\cos\phi+\frac{1}{2}.\frac{\pi}{2}\sin\phi+\frac{1}{2}\right)\right]d\phi$$

$$= 16\left[(a+b+c)\frac{r^3}{3}\times\phi+\frac{r^4}{4}\left(\frac{\pi}{4}(\sin\phi)+\frac{\pi}{4}(-\cos\phi)+\frac{1}{2}\phi\right)\right]_0^{\frac{\pi}{2}}$$

$$= \frac{8\pi}{3}(a+b+c)r^3+4r^4\left(\frac{\pi}{4}+\frac{\pi}{4}+\frac{\pi}{4}\right)=\frac{8\pi}{3}[(a+b+c)]r^3+3\pi r^4$$

Example 7.65. Evaluate $\iint\frac{dS}{\sqrt{a^2x^2+b^2y^2+c^2z^2}}$ over the closed surface of the ellipsoid $ax^2+by^2+cz^2=1$

Solution. Now
$$\hat{n} = \frac{2\hat{i}ax+2\hat{j}by+2\hat{k}iz}{2\sqrt{a^2x^2+b^2y^2+c^2z^2}}=\frac{\hat{i}ax+\hat{j}by+\hat{k}cz}{\sqrt{a^2x^2+b^2y^2+x^2z^2}}$$

$$\iint_S\frac{dS}{\sqrt{a^2x^2+b^2y^2+c^2z^2}} = \iint_S(\hat{i}x+\hat{j}y+\hat{k}z).\frac{(\hat{i}ax+\hat{j}by+\hat{k}cz)}{\sqrt{a^2x^2+b^2y^2+x^2z^2}}dS$$

$$[(\hat{i}x+\hat{j}y+\hat{k}z).(\hat{i}ax+\hat{j}by+\hat{k}cz)=1]$$

$$= \iint_S\vec{F}.n\,dS=\iiint_V\nabla.\vec{F}\,dV=\iiint_V(1+1+1)\,dV$$

$$\vec{F} = \hat{i}x+\hat{j}y+\hat{k}y$$

$$= 3\iiint_V dV=3V=3\times\frac{4}{3}\frac{1}{\sqrt{a}}.\frac{1}{\sqrt{b}}.\frac{1}{\sqrt{c}}=\frac{4\pi}{\sqrt{abc}}$$

Hint: Ellipsoid

is $\left(\frac{x}{1/a}\right)^2 + \left(\frac{y}{1/b}\right)^2 + \left(\frac{z}{1/c}\right)^2 = 1$

Put $A = \frac{1}{\sqrt{a}}$ Ellipsoid becomes

$B = \frac{1}{\sqrt{b}}$ $Ax^2 + By^2 + Cz^2 = 1$

$C = \frac{1}{\sqrt{c}}$ Volume $V = \frac{4}{3}ABC\pi = \frac{4}{3}\frac{\pi}{\sqrt{abc}}$

***Example* 7.66.** Using divergence theorem to evaluate $\iint_S xdy\,dz + ydz\,dx + z\,dx\,dy$ where S is the portion of the plane of the plane $x + 2y + 3z = 6$ which lie in the first quadrant.

Solution. $\int_S (xdy\,dz + ydz\,dx + z\,dx\,dy) = \iiint_V \left(\frac{\partial x}{\partial x} + \frac{\partial y}{\partial y} + \frac{\partial z}{\partial z}\right) dx\,dy\,dz$

$$= \iiint_V (1+1+1)\,dx\,dy\,dz$$

$$= \iiint_V 3\,dx\,dy\,dz$$

V = Volume of the tetrahedron

$$= \frac{\text{base} \times \text{height}}{3}$$

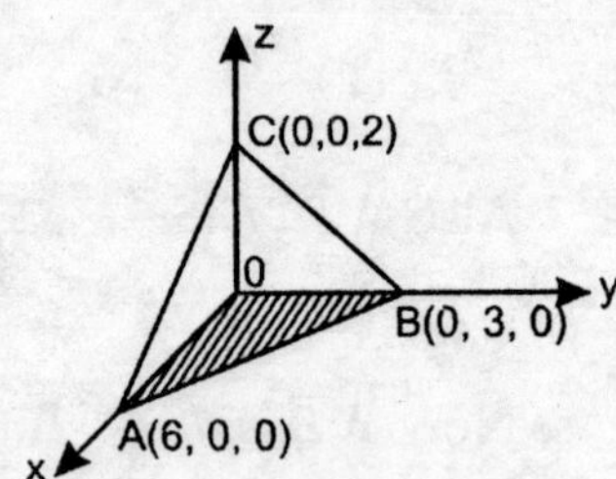

Fig. 7.34

In order to find the area of the $\Delta\,AOB$, put $z = 0$.

The triangle $\Delta\,OAB$ lies in x-y plane, its vertices are A (6, 0), B (0, 3), O (0, 0)

$$\text{Area of } \Delta\,OAB = \frac{1}{2}\begin{vmatrix} 0 & 0 \\ 6 & 0 \\ 0 & 3 \\ 0 & 0 \end{vmatrix} = 9$$

$$\text{Volume of the tetrahedron} = \frac{9 \times 2}{3} = 6$$

$$\iint_S (x\,dy\,dz + y\,dz\,dx + z\,dx\,dy) = \iiint_V 3\,dV = 3\,V = 3 \times 6 = 18$$

***Example* 7.67.** Verify the divergence theorem for the function $\vec{F} = 2x^2y\hat{i} - y^2\hat{j} + 4xz^2\hat{k}$ taken over the region in the first octant bounded by $y^2 + z^2 = 9$, $x = 2$.

Solution. We have $\iiint_V \nabla . F\,dV$

$$= \iiint (4xy - 2y + 8xz)\,dV$$

$$= \int_{x=0}^{2}\int_{y=0}^{3}\int_{z=0}^{\sqrt{9-y^2}} (4xy - 2y + 8xz)\,dx\,dy\,dz$$

$$= \int_0^2\int_0^3 (4xyz - 2yz + 4xz^2)_0^{\sqrt{9-y^2}}\; dy\, dx$$

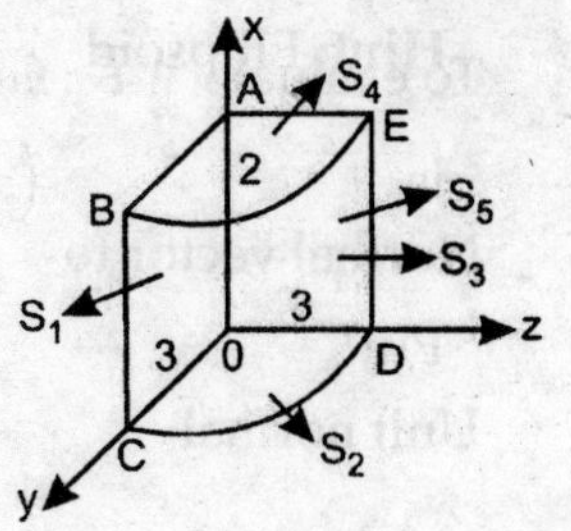

$$= \int_0^2\int_0^2 \left[4xy(\sqrt{9-y^2}) - 2y\sqrt{9-y^2} + 4x(9-y^2)\right] dy\, dx$$

Fig. 7.35

$$= \int_0^2\int_0^2 \left[4x\left(\frac{-2y}{-2}\right)\sqrt{9-y^2} + (-2y)\sqrt{9-y^2} + (36x + 4xy^2)\right] dy\, dx$$

$$= \int_0^2 \left(-2x\frac{(9-y^2)^{3/2}}{3/2} + \frac{(9-y^2)^{3/2}}{3/2} + 36xy - 4x\frac{y^3}{3}\right)_0^3 dx$$

$$= \int_0^2 \left[(36xy - 4xy^3) - \left(\frac{2x(9-y^2)^{3/2}}{3/2}\right) + \frac{(9-y^2)^{3/2}}{3/2}\right]_0^3 dx$$

$$= \int_0^2 \left[\left(108x - \frac{4}{3}x \times 27\right) - \left(-\frac{4}{3}x \times 27\right) + \frac{2}{3} \times 27\right] dx$$

$$= \left(108\frac{x^2}{2} - 18x\right)_0^2 = 54 \times 4 - 18 \times 2 = 216 - 36 = 180 \qquad ...(1)$$

Also $\iint_S \vec{F}\,.\,\hat{n}dS = \int_{S_1 = OB} \vec{F}\,.\,\hat{n}dS + \iint_{S_2 = OCD} \vec{F}\,.\,\hat{n}dS + \iint_{S_3 = AD} \vec{F}\,.\,\hat{n}dS + \iint_{S_4 = ABE} \vec{F}\,.\,\hat{n}dS + \iint_{S_5 = \text{Curved surface } BCDE} \vec{F}\,.\,\hat{n}dS$

Now $\iint_{S_1} \vec{F}\,.\,\hat{n}dS = \iint_{OABC} (2x^2y\hat{i} - y\hat{j} + 4xz^2\hat{k})\,.\,(-k)\,dS = \iint_{S_1} 4xz^2\,dS = 0$ (Since $z = 0$, in S_1)

$$\iint_{S_2} \vec{F}\,.\,\hat{n}dS = \iint_{OCD} (2x^2z\hat{i} - y^2\hat{j} + 4xz^3\hat{k})\,.\,(-\hat{i})\,dS = \iint_{S_2} 2x^2z\;dS = 0 \text{ (Since } x = 0, \text{ in } S_2)$$

$$\iint_{S_3} \vec{F}\,.\,\hat{n}dS = \iint_{AODE} (2x^2y\hat{i} - y^2\hat{j} + 4xz^3\hat{k})\,.\,(-\hat{k})\,dS = \iint_{S_3} -4y^2\,dS = 0 \text{(Since } y = 0, \text{ in } S_3)$$

$$\iint_{S_4} \vec{F}\,.\,\hat{n}dS = \iint_{ABE} (2x^2y\hat{i} - y2\hat{j} + 4xz^3\hat{k})\,.\,(i)\,dS = \iint_{S_4} 2x^2y\;dS$$

$$= \iint_{S_4} 8y\;dS \qquad \text{(Since } x = 2 \text{ in } S_4)$$

$$= 8\int_0^3\int_{z=0}^{\sqrt{9-y^2}} y\;dy\;dz = 8\int_0^3 \left[\frac{y^2}{2}\right]_0^{\sqrt{9-z^2}} dz$$

$$= 8\int_0^3 \frac{9-z^2}{2}dz = 4\int_0^3 (9-z^2)dz = 4\left[9z - \frac{z^3}{3}\right]_0^3 = 4[27-9] = 72$$

To evaluate $\iint_{S_5} \vec{F}\,.\,\hat{n}dS$, we proceed as below.

Normal vector to $S_5 = \nabla[y^2 + z^2] = 2y\hat{j} + 2z\hat{k}$

Unit normal $\hat{n} = \dfrac{2y\hat{j} + 2z\hat{k}}{\sqrt{4y^2 + 4z^2}} = \dfrac{y\hat{j} + z\hat{k}}{\sqrt{y^2 + z^2}} = \dfrac{y\hat{j} + z\hat{k}}{3}$

$$\iint_{S_5} \vec{F}\,.\,\hat{n}dS = \iint_{BCDE} (2x^2 y\hat{j} - y^2\hat{j} + 4xz^2\hat{k})\,.\,\frac{y\hat{i} + z\hat{k}}{3} dS$$

$$= \iint \frac{-y^3 + 4xz^3}{3} dS$$

$$= \iint \frac{-y^3 + 4xz^3}{3} \times \frac{dy\,dx}{|\hat{n}\,.\,\hat{k}|} = \iint \frac{-y^3 + 4xz^3}{3} \times \frac{dx\,dy}{z/3}$$

$$= \iint \left(-\frac{y^3}{z} + 4xz^2\right) dx\,dy \qquad \text{(Taking projection on } x\text{-}y \text{ plane)}$$

Now $y^2 + z^2 = 9$, its parametric equations are $y = 3 \sin \theta$, $z = 3 \cos \theta$

$$= \int_0^2 \int_0^{\frac{\pi}{2}} \left(-\frac{27 \sin^3 \theta}{3 \cos \theta} \times 4x3^2 \cos^2 \theta\right) \times 3 \cos \theta \, d\theta \, dx$$

$$= \int_0^2 \int_0^{\pi/2} (-27 \sin^3 \theta + 4x \times 27 \cos^3 \theta) \, d\theta \, dx$$

$$= \int_0^2 \left(-27 \times \frac{2}{3} + 108x \times \frac{2}{3}\right) = \int_0^2 (-18 + 72x) \, dx$$

$$= \left[-18x + 72\frac{x^2}{2}\right]_0^2 = (-18x + 36x^2)_0^2 = 108$$

$$\therefore \qquad \iint_S \vec{F}\,.\,\hat{n}dS = 0 + 0 + 6 + 72 + 108 = 180$$

Hence theorem is verified.

EXERCISE 7.4

1. Verified Green's theorem in plane for $\oint_C (2xy - x^2)\,dx + (x^2 z)\,dy$ when C is the boundary of the given bounded by $y = x^2$ and $y^2 = x$.

2. Verify Green's theorem in xy plane for $\int_C (xy^2 - 2xy)\,dz = (x^2 y + z)dy$

3. Verify Green's theorem for the function $\vec{F} = 4xz\hat{i} - y^2\hat{j} + yz\hat{k}$ over the surface S of the cube bounded by $x = 0$, $x = 1$, $y = 0$, $y = 1$ and $z = 0$, $z = 1$.

4. Verify Stoke's theorem for $\vec{F} = -y^2\hat{i} + x^2\hat{j}$ in the region $x^2 + y^2 = 1, z = 0$.

5. Verify Stoke's theorem for $\vec{F} = (y - z + 2)\,\hat{i} - (yz + 4)\,\hat{j} - 2z\,\hat{k}$, where S is the surface of the cuboid $x = 0, y = 0, z = 0\ x = 2, y = 2, z = 2$ above the x-y plane.

6. Apply Stoke's theorem to evaluate $\int_C ydx + zdy + xdz$ where C is the curve of intersection of $x^2 + y^2 + z^2 = a^2 x + z = a$.

Ans. $-\dfrac{\pi a^2}{\sqrt{2}}$

7. Verify Stoke's theorem for $\vec{F} = -y\hat{i} + 2yz\hat{j} + y^2\hat{k}$ over the surface S in the upper half of the sphere $x^2 + y^2 + z^2 = 1$.

8. Verify Stoke's theorem for $\vec{F}\ . = (x^2 + y^2)\,\hat{i} + 2xy\,\hat{j}$ in the rectangular region in xy-plane, the rectangle being given by $(0, 0)$, $(a, 0)$, $(0, b)$ and (a, b).

9. Verify Gauss divergence theorem for the function $\vec{F} = y\hat{i} + x\hat{j} + z^2\hat{k}$ over the cylindrical region bounded by $x^2 + y^2 = 9$, $z = 0$ and $z = 2$.

10. Using divergence theorem, evaluate $\iint \vec{R}\,.\,\hat{n}dS$ where S is the surface of the sphere $x^2 + y^2 + z^2 = 4$

Ans. $108\,\pi$

11. Evaluate $\iint x\,dy\,dz + y\,dz\,dx + z\,dx\,dy$ over the surface of the sphere of radius 'a'.

Ans. $4\pi a^2$

12. Evaluate $\iint (x\hat{i} + y\hat{j} + z^2\hat{k})\,\hat{n}\,d\vec{S}$ where S is the closed surface bounded by the cone $x^2 + y^2 = z^2$ and the plane $z = 1$.

Ans. $\dfrac{7}{6}\pi$

13. Verify divergence theorem for $\vec{F} = (x^3 - yz)\,\hat{i} - 2x^2y\,\hat{j} + 2\,\hat{k}$ taken over the cube bounded by the planes $x = 1, x = 0, y = 1, y = 0, z = 1, z = 0$.

Ans. 84π

14. Verify divergence theorem for $\vec{F} = (x^2 - yz)\,\hat{i} + (y^2 - zx)\,\hat{j} + (z^2 - xy)\,\hat{k}$ taken over the rectangular parallelo piped $0 \le x \le a$, $0 \le y \le b$, $0 \le z \le c$.

15. Show that $\int_S (ax\hat{i} + by\hat{j} + cz\hat{k})\,.\,\hat{n}dS = \dfrac{4}{3}\,\pi\,(a + b + c)$ where S is the surface of the sphere $x^2 + y^2 + z^2 = 1$.

Infinite Series

This chapter covers the topics on sequence, series, convergence and divergence, properties of infinite series, series of positive terms. Comparison test. D' Alembert's ratio test, Raabe's test. Logarithmic test, Gauss's test, Cauchy root test, Integral test, alternating series: Lebnitiz's rule, absolutely convergent and conditionally convergent series

8.1 SEQUENCE

A sequence is a succession of number or terms formed according to some mathematical rule. The nth term of a sequence is denoted by u_n. It is customary to denote the sequence by the symbol $<u_n>$ or by (u_n).

For example, if $$u_n = 3n + 1$$

By putting different values of n in the u_n, different terms of the sequence are obtained, *i.e.*, when $n = 1$, $u_1 = 4$, when $n = 2$, $u_2 = 7$, similarly when $n = 3$ $u_3 = 10$ .... so on.

A sequence having infinite number of terms is known as an infinite sequence.

8.2 LIMIT

If a sequence tends to a limit k, then we write $\underset{n \to \infty}{Lt}\ u_n = k$ or simply $u_n \to k$ as $n - \infty$.

8.3 REAL SEQUENCE

A real sequence is a function whose domain is the set N of all natural numbers and the range is a subset R of real numbers.

Symbolically, $f: N \to R$ is a real sequence.

8.4 RANGE OF SEQUENCE

The set of all distinct terms of a sequence is called its range. Therefore, the range of a sequence $<u_n>$ is the set $\{u_1, u_2, u_3, ...\}$

***Example* 8.1.** The range of the sequence $<u_n> = <(-1)^n>$ is $\{-1, 1\}$, a finite set and the range of function is $(-1, 1)$. a finite set and the range of function is $\{-1, 1\}$.

***Example* 8.2.** The range of the sequence $<u_n> = \left\langle \frac{1}{n} \right\rangle = \left\langle \frac{1}{n} : n \in N \right\rangle$, an infinite set.

8.5 CONSTANT SEQUENCE

A sequence in which all the terms are equal is called a constant sequence.

The sequence $<u_n> = <a, a, a, a ....>$ is a constant sequence and its range is the singleton set (a).

8.6 BOUNDED AND UNBOUNDED SEQUENCE

(*i*) **Bounded below sequence :** A sequence $<u_n>$ if said to be bounded below is ∃ there exists a real number k such that $u_n \geq k\ \forall\ n \in N$, the set of all natural number *i.e.*, the sequence $<u_n> = <1,2,3,4, ...>$ is bounded below as there exist a number 1 which is less then each terms of the sequence i.e., $u_n > 1\ \forall\ n \in$ N.

(*ii*) **Bounded above sequence :** A sequence $<u_n>$ is said to be bounded above if ∃ (there exists) a real number I such that $u_n \leq I\ \forall\ n \in N$, the set of all natural number *i.e.*, the sequence $<u_n> = <-1, -2, -3, -4, ...>$ is bounded above as ∃ (there exists) a number –1 which is greater than each term of the sequence *i.e.*, $u_n \leq \{-1\}\ \forall\ n \in$ N.

(*iii*) **Bounded sequence :** A sequence $<u_n>$ is said to be bounded both below and above *i.e.*, a sequence $<u_n>$ is bounded if ∃ (there exists) two real number k and l $(k \leq l)$ such that $k \leq u_n \leq l\ \forall\ n \in N$, the set of all natural numbers.

If we choose M = maximum of $\{|k|, |l|\}$, then the sequence $<u_n>$ is defined as a bounded sequence provided $|un| \leq M\ \forall\ n \in$ N.

Example : If $u_n = \frac{1}{n}$ then $<u_n> \left\{\frac{1}{2}, \frac{1}{2}, \frac{1}{3}, \frac{1}{4}, \frac{1}{5} ....\right\}$

Here $u_n \geq 0$ and $u_n \leq 1\ \forall\ n \in N$ *i.e.*, $0 < u_n \leq 1\ \forall\ n \in N$.

Hence the sequence $<u_n>$ is bounded.

(*iv*) **Unbounded sequence :** If ∃ (there exists) a real number M such that $|u_n| \geq M\ \forall\ n \in N$ then the sequence $<u_n>$ is defined to be an unbounded sequence.

Example : Let the nth term of the sequence $<u_n>$ defined by $u_n = 3^{n-1}$, *i.e.*,

$$<u_n> = \{1, 3, 3^2 ....\}$$

Here $u_n > 1\ \forall\ n \in N$ but there exist no real number l such that $u_n < l$. Hence the sequence $<u_n>$ *i.e.* 3^{n-1} is unbounded above.

8.7 CONVERGENT, DIVERGENT AND OSCILLATORY SEQUENCE

(*i*) **Convergent sequence :** A sequence $<u_n>$ is said to be a convergent sequence if $\underset{n \to \infty}{Lt}\ u_n$ is finite.

Example **8.3 :** The $<u_n> = \left\langle 1, \frac{1}{4}, \frac{1}{9}, \frac{1}{16}, ... \frac{1}{n^2} ...\right\rangle$ is a convergent, since

$$u_n = \frac{1}{n^2} \text{ and } \underset{n \to \infty}{Lt}\ u_n = 0, \text{ which is finite.}$$

(*ii*) **Divergent sequence :** A sequence $<u_n>$ is said to be a divergent sequence if $\underset{n \to \infty}{Lt}\ u_n$ is not finite *i.e.*, $\underset{n \to \infty}{Lt}\ u_n = \infty$ or $-\infty$.

Example **8.4 :** The sequence $<u_n> = <3, 5, 7, ...,(2n + 1) ....>$ is a divergent sequence, since

$$u_n = 2n + 1 \quad \text{and} \quad \underset{n \to \infty}{Lt}\ u_n = \infty$$

Similarly, the sequence $<u_n> = <-2^2, -2^3, ...2^n ....>$ is also divergent sequence as

$$u_n = -2^n \quad \text{and} \quad \underset{n \to \infty}{Lt}\ u_n = -\infty$$

(*iii*) **Oscillatory sequence :** A sequence $<u_n>$ is said to be oscillatory if it is neither a convergent sequence nor a divergent. Oscillotory sequence are of two types.

(*a*) A bounded sequence which is not convergent is said to oscillate finitely.

Here $u_n = (-1)^n$ and this sequence is bounded.

Also $$\underset{n \to \infty}{Lt}\ u_{2n} = \underset{n \to \infty}{Lt}\ (-1)^{2n} = 1$$

and $$\underset{n\to\infty}{Lt}\ u_{an+1} = \underset{n\to\infty}{Lt}\ (-1)^{2n+1} = -1$$

Since $\underset{n\to\infty}{Lt}\ u_{2n} \neq \underset{n\to\infty}{Lt}\ u_{2n+1}$, therefore $\underset{n\to\infty}{Lt}\ u_n$ does not exist which implies that the sequence $<u_n>$ is not convergent. Hence the sequence $<u_n>$ oscillates finitely in this case.

(*b*) An unbounded sequence which not divergent is said to oscillate infinitely.

Example : Suppose that the sequence be $<u_n>$

$$\therefore \quad u_n = n(-1)^n \text{ and the sequence is unbounded.}$$

Also $$\underset{n\to\infty}{Lt} = \underset{n\to\infty}{Lt}\ 2n(-1)^{2n} = \infty.$$

and $$\underset{n\to\infty}{Lt}\ u_{2n+1} = \underset{n\to\infty}{Lt}\ (2n+1)(-1)^{2n+1} = -\infty$$

Thus the sequence $<u_n>$ is not divergent and hence this sequence oscillates infinitely.

Note : If we say that $\underset{n\to\infty}{Lt}\ u_n = l$, then it means

$$\underset{n\to\infty}{Lt}\ u_{2n} = \underset{n\to\infty}{Lt}\ u_{2n+1} = l$$

Similarly, $\underset{n\to\infty}{Lt}\ u_n = \infty$, the it means

$$\underset{n\to\infty}{Lt}\ u_{2n} = \underset{n\to\infty}{Lt}\ u_{2n+1} = \infty$$

8.8 MONOTONIC SEQUENCE

When the sequence $<u_n>$ is either increasing or decreasing, such a sequence is called monotonic sequence When the sequence.

(*i*) 3, 5, 7, 9 .... is monotonic sequence.

(*ii*) $1, \frac{1}{2}, \frac{1}{3}, \frac{1}{4}, \ldots\ldots$ is also a monotonic sequence.

(*iii*) 1, –1, 1, –1, ... is not a monotonic sequence.

Note : A sequence which is monotonic and bounded is a convergent sequence.

Example **8.5.** Determine the general term of the following sequences. Prove that the following sequence are convergent.

(*a*) $\frac{1}{2}, \frac{1}{4}, \frac{1}{8}, \frac{1}{16} \ldots$ (*b*) $\frac{1^2}{1!}, \frac{2^2}{2!}, \frac{3^2}{3!}, \frac{4^2}{4!}, \ldots$

Solution : (*a*) $\frac{1}{2}, \frac{1}{4}, \frac{1}{8}, \frac{1}{16} \ldots$

The general term of the sequence is $\frac{1}{2^n}$ *i.e.* $u_n = \frac{1}{2^n}$ and $\underset{n\to\infty}{Lt}\ u_n = 0$. which is finite.

Hence the sequence is convergent.

Solution : (*b*) $\frac{1^2}{1!}, \frac{2^2}{2!}, \frac{3^2}{3!}, \frac{4^2}{4!}, \frac{5^2}{5!} \ldots$

The general term of the sequence is $\frac{n^2}{n!}$ *i.e*, $u_n = \frac{n^2}{n!}$

and $$\underset{n\to\infty}{Lt}\ u_n = \underset{n\to\infty}{Lt}\ \frac{n^2}{n(n-1)(n-2)!}$$

or $$\underset{n\to\infty}{Lt}\ u_n = \underset{n\to\infty}{Lt}\ \frac{1}{\left(1-\frac{1}{n}\right)(n-2)!}$$

$$= 0, \text{ which is finite}$$

Hence the sequence is convergent.

Example **8.6.** Which of the following sequence are convergent ? Given the terms as

(*a*) $u_n = \frac{n+1}{n}$ (*b*) $u_n = n^2$ (*c*) $u_n = \frac{1}{n}$

Solution : (*a*) $u_n = \frac{n+1}{n}$ which is finite. There the sequence is convergent.

$$u_n = \frac{n\left(1+\frac{1}{n}\right)}{n} \text{ and } \underset{n\to\infty}{Lt}\ u_n = \underset{n\to\infty}{Lt}\ 1+\frac{1}{n} = 1.$$

(*b*) $$u_n = n^2$$

and $$\underset{n\to\infty}{Lt}\ u_n = \underset{n\to\infty}{Lt}\ n^2 = \infty \text{ (Infinite)}$$

Hence the given sequence is divergent.

(*c*) $$u_n = \frac{1}{n}$$

and $$\underset{n\to\infty}{Lt}\ u_n = \underset{n\to\infty}{Lt}\ \frac{1}{n} = 0 \text{ (finite)}$$

Hence the given sequence is convergent.

EXERCISE 8.1

Determine the general term of each of the following sequence and check the convergence or divergence of the sequence.

1. 1, –1, 1, –1, ... **Ans.** $u_n = (-1)^{n-1}$, Convergent

2. $\frac{1}{2}, \frac{2}{3}, \frac{3}{4}, \cdot \frac{4}{5}$ **Ans.** $u_n = \frac{n}{n+1}$, Convergent

3. 3, 6, 9, 12 .... **Ans.** $u_n = 3n$, Divergent

4. $\frac{2}{1}, \frac{3}{2}, \frac{4}{5}, \frac{4}{3} \ldots$ **Ans.** $u_n = \frac{n+1}{n}$, Convergent

5. $1+\frac{1}{1}, 1+\frac{1}{2}, 1+\frac{1}{3}, 1+\frac{1}{4} \ldots$ **Ans.** $u_n = 1+\frac{1}{n}$, Convergent

8.9 SOME STANDARD LIMITS

I. $\underset{n \to \infty}{Lt}\ x^n = 0$ if $x < 1$ and $\underset{n \to \infty}{Lt}\ n^x = \infty$ if $x > 1$

II. $\underset{n \to \infty}{Lt}\ \frac{x^n}{n!} = 0$ for all values of x

III. $\underset{n \to \infty}{Lt}\ \frac{\log n}{n} = 0$

IV. $\underset{n \to \infty}{Lt}\ n^h = \infty$

V. $\underset{n \to \infty}{Lt}\ \frac{1}{n^h} = 0$

VI. $\underset{n \to \infty}{Lt}\left[\frac{m^n - 1}{n}\right] = \log m$ or $\underset{n \to 0}{Lt}\left[\frac{m^{\frac{1}{n}} - 1}{\frac{1}{n}}\right] = \log$

VII. $\underset{n \to \infty}{Lt}\ (n)^{\frac{1}{n}} = 1$

VIII. $\underset{n \to \infty}{Lt}\left(1 + \frac{1}{n}\right)^n$

IX. $\underset{n \to \infty}{Lt}\ (n!)^{\frac{1}{n}} = \infty$

X. $\underset{n \to \infty}{Lt}\left(\frac{n!}{n}\right)^{\frac{1}{n}} = \frac{1}{e}$

XI. $\underset{n \to \infty}{Lt}\ nx^n = 0$ if $x < 0$

XII. $\underset{n \to \infty}{Lt}\ \frac{\sin n}{n} = 1$ *or* $\underset{n \to \infty}{Lt}\ \frac{n}{\sin n} = 1$

XIII. $\underset{n \to \infty}{Lt}\ \frac{\tan n}{n} = 1$ or $\underset{n \to \infty}{Lt}\ \frac{n}{\tan n} = 1$

XIV. $\underset{n \to \infty}{Lt}\ \cos n = 1$ or $\underset{n \to \infty}{Lt}\ \frac{1}{\cos n} = 1$

XV. $\underset{n \to \infty}{Lt}\left(1 + \frac{x}{n}\right)^n = e^x$

8.10 SERIES

Let $\{u_n\}$ be a sequence of real numbers, then the expression of the form $u_1 + u_2 + u_3 + \ldots + u_n + \ldots$ in which every term in followed by another to some definite law is called a series. In other world a series is the sun of sequence.

For example $1 + 3 + 5 + 7 + 9 + \ldots\ldots$ is a series.

If the number of term of a series is finite, the series is called finite series.

If the number of terms of a series is infinite, then it is called an infinite series, *i.e*, $u_1 + u_2 + u_3 + u_4 + ... + u_n +$ ... ∞, is an infinite series and it denoted by $\sum_{n=1}^{\infty} u_n$ or Σu_n. The sum of the first n term of a series is denoted by S_n.

8.11 CONVERGENT, DIVERGENT AND OSCILLATORY SERIES

Consider the infinite series

$$\Sigma u_n = u_1 + u_2 + u_3 + ... + u_n + ... \infty \text{ and let the sum of first } n \text{ terms be}$$

$$S_n = u_1 + u_2 + u_3 + ... + u_n$$

(*a*) If S_n tends to a finite number as $n \to \infty$, then the series Σu_n is said to be a convergent series.

(*b*) If S_n tends to infinity as $n \to \infty$, thin the series Σu_n is said to a be a divergent series.

(*c*) If S_n does not tends to a unique limit, finite or infinite, then the series Σu_n is called oscillatory series.

8.12 PROPERTIES OF INFINITE SERIES

The nature of the series remains unchanged if

(*a*) the sign of all the terms are changed.

(*b*) the addition or removal of a finite number of terms is done.

(*c*) each term of the series in multiplied or divided by the same fixed number m which is not zero.

Note 1: (*a*) If Σu_n and Σv_n are convergent series then $\Sigma(u_n + v_n)$ is also convergent.

(*b*) If Σu_n and Σv_n are two divergent series having all terms positive then $\Sigma(u_n + v_n)$ also divergent.

Note 2: (*a*) A.P. Se Series : a, $(a + b)$, $(a + ad)$, $(a + 3d)$, ... is an A.P. series.

$$n^{\text{th}} \text{ term is } u_n = a + (n - 1)\, d$$

where, a is first term and d is common difference i.e., $d = u_2 - u_1 = u_3 - u_2$

Sum of first n terms is $S_n = \frac{n}{2}\, [2a + (n - 1)]$

$$S_n = \frac{n}{2}\, (a + l) \text{ when last is given.}$$

(*b*) G.P. Series : $a,\ ar,\ ar^2,\ ar^3 + ...$ is a G.P. series.

n^{th} term is $u_n = ar^{n-1}$ where, a is first term and r is common ratio i.e.,

$$r = \frac{ar}{a} = \frac{ar^2}{ar}$$

Sum of first n term is $S_n = \frac{a(r^n - 1)}{r - 1}$, when $r > 1$

$$S_n = \frac{a(1 - r^n)}{1 - r} \text{ when } r < 1$$

and $$S_\infty = \frac{a}{1 - r}, \text{ when number of terms is infinite.}$$

8.13 PARTIAL SUMS

If $\Sigma u_n = u_1 + u_2 + u_3 + ... u_n + u_{n+1} + ...$ is an infinite series, where the terms may be positive or negative, then $S_n = u_1 + u_2 + u_3 + + ... u_4 + u_n$ is called the nth partial sum of Σu_n. Hence the nth partial sum of an infinite series is the sum of its first n terms.

Example 8.7. Examine the nature of the following series

(*a*) $1+2+3+4 \,.....\, + n + ... \infty$ (*b*) $1+1+\frac{1}{2}+\frac{1}{4}+\frac{1}{8}+ ....$

(*c*) $1+4+7+10+ .....$ (*d*) $1+3+5+7= ...... \infty$

(*e*) $\frac{1}{1.2}+\frac{1}{2.3}+\frac{1}{3.4} .... \infty$ (*f*) $1-\frac{1}{3}+\frac{1}{3^2}+\frac{1}{3^3}+\frac{1}{3^4}+ ... \infty$

(*g*) $6-10+4+6-10+6-10+4+ .... \infty$

Solution : (*a*) $1+2+3+4+ ...... + n + ... \infty$ is an A.P. series

$$S_n = \frac{n(n+1)}{2}$$

$$\underset{n\to\infty}{Lt}\ S_n = \underset{n\to\infty}{Lt}\ \frac{n(n+1)}{2} = \infty$$

Hence, the series is divergent.

Solution : (*b*) $1+\frac{1}{2}+\frac{1}{4}+\frac{1}{8}+ .... \infty$ is a G.P. series

$$\therefore \qquad S_n = \frac{1}{1-\frac{1}{2}} = 2$$

and $$\underset{n\to\infty}{Lt}\ S_n = 2$$

Hence the series is convergent.

Solution : (*c*) $1+4+7+10+ ..... \infty$ is an A.P. series

$$\therefore \qquad S_n = \frac{n}{2}\,[2a+(n-1)d]$$

$$= \frac{n}{2}[2\times 1+(n-1)\times 3] = \frac{n}{2}[2+3n-3] = \frac{\pi}{2}[3n-1]$$

$$\underset{n\to\infty}{Lt}\ S_n = \underset{n\to\infty}{Lt}\ \frac{n}{2}\,(3n-1) = \infty$$

Hence the series is divergent.

Solution : (*d*) $1+3+5+7+ ..... \infty$ is an A.P. series

$$\therefore \qquad S_n = \frac{n}{2}\,[2a+(n+1)\,d]$$

$$= \frac{n}{2}[2\times 1+(n-1)\times 2] = \frac{n}{2}[2+2n-2]$$

$$S_n = \frac{n}{2}(2n)$$

$$S_n = n^2$$

$$\underset{n\to\infty}{Lt}\ S_n = \underset{n\to\infty}{Lt}\ n^2 = \infty$$

Hence the series is divergent.

Solution : (*e*) $\frac{1}{1.2}+\frac{1}{2.3}+\frac{1}{3.4}+\infty$ is an infinite series which can be written as

$$\therefore \quad S_n = \frac{1}{1.2}+\frac{1}{2.3}+\frac{1}{3.4}+....+\frac{1}{n(n+1)}$$

$$S_n = \left(1-\frac{1}{2}\right)+\left(\frac{1}{2}-\frac{1}{3}\right)+\left(\frac{1}{3}-\frac{1}{4}\right)+.....+\left(\frac{1}{n}-\frac{1}{n+1}\right)$$

$$S_n = \left(1+\frac{1}{2}+\frac{1}{3}+....\frac{1}{n}\right)-\left(\frac{1}{2}+\frac{1}{3}+\frac{1}{4}+....+\frac{1}{n+1}\right)=1-\frac{1}{n+1}$$

Hence the series is $\underset{n\to\infty}{Lt} S_n = 1$ convergent.

Solution : (*f*) $1-\frac{1}{3}+\frac{1}{3^2}-\frac{1}{3^3}+\frac{1}{3}$ .... ∞ is an infinite G.P.

Here
$$S_n = \frac{1-\left(-\frac{1}{3}\right)^n}{1+\frac{1}{3}}$$

and
$$\underset{n\to\infty}{Lt} S_n = \frac{3}{4}$$

Hence the series is convergent.

Solution : (*g*) $6-10+4+6-10+4+6-10+4+.....\infty$.

$$S = 6-10+4+6-10+4+6-4+.....\infty.$$

$S_{3m} = 0$, when the number of terms is $3m$.

$S_{3m+1} = 6$, when the number of terms is $3m+1$

$S_{3m+2} = -4$, when the number of terms is $3m+2$.

where m is an *y* positive integers.

therefore as the series tends to infinity, the sum of the series varies as shown above.

Hence the given series is an oscillatory series.

EXERCISE 8.2

Discuss the nature of the following series

1. $1+4+7+10+.....\infty$. **Ans.** Divergent
2. $6-5-1+6-5-1+6-5-1+....\infty$. **Ans.** Oscillatory
3. $\frac{1}{1.3}+\frac{1}{1.5}+\frac{1}{5.7}$ ... ∞. **Ans.** Convergent
4. $\log 3+\log\frac{4}{3}+\frac{5}{4}+....\infty$. **Ans.** Divergent
5. $7-4-3+7-4-3+7-4-3+....\infty$ **Ans.** Oscillatory

8.14 GEOMETRIC SERIES

The series $1 + p + p^2 + p^3 + p^4 + ... \infty$ is

(a) convergent if $|p| < 1$

(b) divergent if $p \geq 1$

(c) oscillatory if $p \leq -1$.

Proof : $S_n = 1 + p + p^2 + p^3 + p^4 + ... + p^{n-1}$

$$= \frac{1-p^n}{1-p}$$

Case I. When $|p| < 1$. $\underset{n \to \infty}{Lt}\ p^n = 0$

$$\therefore \quad \underset{n \to \infty}{Lt}\ S_n = \underset{n \to \infty}{Lt}\ \frac{1-p^n}{1-p}$$

$$= \frac{1-0}{1-p} = \frac{1}{1-p}$$

Hence, the series is convergent.

Case II. (i) When $p > 1$ $\underset{n \to \infty}{Lt}\ p^n = \infty$

$$\therefore \quad \underset{n \to \infty}{Lt}\ S_n = \underset{n \to \infty}{Lt}\ \frac{1-p^n}{1-p} = \infty$$

Hence, the series is divergent.

(ii) When $p = 1$, the series becomes $1 + 1 + 1 + 1 .... + n$ terms

and $$S_n = 1 + 1 + 1 + ... = n$$

$$\therefore \quad \underset{n \to \infty}{Lt}\ S_n = \underset{n \to \infty}{Lt}\ n = \infty$$

Hence the series is divergent.

Case III. (i) when $p = -1$, the series becomes $1 - 1 + 1 - 1 + 1 - 1 + ... \infty$

and $$S_n = \begin{cases} 0 \text{ if } n \text{ is even} \\ 1 \text{ if } n \text{ is odd} \end{cases}$$

Hence, the series is oscillatory.

(ii) when $p > -1$, let $p = -k$ therefore $k > 1$, then

$$p^n = (-k)n = (-1)^n k^n$$

$$\therefore \quad \underset{n \to \infty}{Lt}\ S_n = \underset{n \to \infty}{Lt}\ \frac{1-p^n}{1-p}$$

$$= \underset{n \to \infty}{Lt}\ \frac{1-(-1)^n k^n}{1-(-k)}$$

$$= \begin{cases} \infty & \text{if } n \text{ is odd} \\ -\infty & \text{if } n \text{ is even} \end{cases}$$

Hence, the series is oscillatory.

8.15 POSITIVE TERM SERIES

(*i*) If all the terms of the series Sun are positive *i.e.*, $u_n > 0 \ \forall \ n$.

(*ii*) If all therm after few negative terms in an infinite series are positive, such series is a positive terms series e.g., –10 –6 –1 + 5 + 12 + 20 + ... is a positive term series, the nature of a positive term series remains unchanged.

8.16 NECESSARY CONDITION FOR CONVERGENT SERIES

Statement : Let Σu_n be an infinite series then for every convergent series

$$\underset{n \to \infty}{Lt}\ u_n = 0$$

Proof : Let $S_n = u_1 + u_2 + u_3 + u_4 + ... u_n$

and $\underset{n \to \infty}{Lt}\ S_n = l$ (a finite quantity) because the series is convergent.

Also $S_n = S_{n-1} + u_n$

$\therefore$ $S_n = S_{n-1} + u_n$

or $u_n = S_n - S_{n-1}$

$$\therefore \quad \underset{n \to \infty}{Lt} = \underset{n \to \infty}{Lt}\ (S_n - S_{n-1})$$

$$= \underset{n \to \infty}{Lt}\ S_n - \underset{n \to \infty}{Lt}\ S_{n-1}$$

$$= l - l$$

$$= 0$$

Hence $\underset{n \to \infty}{Lt}\ u_n = 0$

But converse of the theorem is not true.

Taking an example.

$\frac{1}{\sqrt{1}} + \frac{1}{\sqrt{2}} + \frac{1}{\sqrt{3}} + \frac{1}{\sqrt{4}} + ... + \frac{1}{\sqrt{n}} + ... \infty$ is divergent

$$\therefore \quad S_n = 1 + \frac{1}{\sqrt{2}} + \frac{1}{\sqrt{3}} + \frac{1}{\sqrt{4}} + ... + \frac{1}{\sqrt{n}} > \frac{1}{\sqrt{n}} + \frac{1}{\sqrt{n}} + \frac{1}{\sqrt{n}} + \frac{1}{\sqrt{n}} + ... + \frac{1}{\sqrt{n}}$$

$$> \frac{n}{\sqrt{n}}$$

$$> \sqrt{n}$$

$$\therefore \quad \underset{n \to \infty}{Lt}\ S_n = \underset{n \to \infty}{Lt}\ \sqrt{n} = \infty$$

Thus, the series is divergent, although $\underset{n \to \infty}{Lt}\ U_n = \underset{n \to \infty}{Lt}\ \frac{1}{\sqrt{n}} = 0$.

Hence $\underset{n \to \infty}{Lt}\ u_n = 0$ is a necessary condition but not a sufficient condition for convergence.

8.17 CAUCHY'S FUNDAMENTAL TEST FOR DIVERGENCE

Let Σu_n be an infinite series, then divergence $\underset{n \to \infty}{Lt}\ u_n \neq 0$, which is called Cauchy's fundamental test for divergence.

***Example* 8.8.** Test for convergence/divergence of following series.

(*a*) $1+\frac{3}{5}+\frac{8}{10}+\frac{15}{17}+......+\frac{2^n-1}{2^n+1}+...\infty$

(*b*) $1+\frac{2}{3}+\frac{3}{4}+\frac{4}{5}+\frac{5}{6}....+\frac{n}{n+1}...\infty$

Solution : (*a*)
$$u_n = \frac{2^n-1}{2^n-2}$$

$\therefore$
$$\underset{n\to\infty}{Lt} = \underset{n\to\infty}{Lt}\frac{2^n-1}{2^n+1}$$

$$= \underset{n\to\infty}{Lt}\frac{1-\frac{1}{2^n}}{1+\frac{1}{2^n}}$$

$$= 1 \neq 0$$

Hence by Cauchy's fundamental test for divergence, the series is divergent.

(*b*)
$$u_n = \frac{n}{n+1}$$

$\therefore$
$$\underset{n\to\infty}{Lt}\ u_n = \underset{n\to\infty}{Lt}$$

$$= \underset{n\to\infty}{Lt}\frac{1}{1+\frac{1}{n}}$$

$$= 1 \neq 0$$

Hence, by Cauchy fundamental test for divergence the series is divergent.

EXERCISE 8.3

Examine the series for convergence or divergence

1. $\frac{1}{1+2^{-1}}+\frac{2}{1+2^{-2}}+\frac{3}{1+2^{-3}}+....$ **Ans. Divergent**

2. $\frac{1}{2}+\frac{2}{3}+\frac{3}{4}+\frac{4}{5}+....$ **Ans. Divergent**

3. $\Sigma\sqrt{\frac{n}{n+1}}$ **Ans. Divergent**

4. $\Sigma(6-n^2)$ **Ans. Divergent**

5. $\Sigma\cos\frac{1}{n}$ **Ans. Divergent**

8.18 HYPER HARMONIC SERIES TEST, P - SERIES TEST FOR $\Sigma\frac{1}{n^p}$

This series $\Sigma\frac{1}{n^p}=\frac{1}{1^p}+\frac{1}{2^p}+\frac{1}{3^p}+....+\frac{1}{n^p}+....\infty$ is (*i*) Convergent if $p > 1$

(*ii*) Divergent of $p \leq 1$

Proof : Case I. When $p > 1$

Let the series $\Sigma\frac{1}{n^p}$ be defined as

$$\Sigma\frac{1}{n^p} = \frac{1}{1^p}+\frac{1}{2^p}+\frac{1}{3^p}+\frac{1}{4^p}+\frac{1}{5^p}+\frac{1}{6^p}+\frac{1}{7^p}+\frac{1}{8^p}+\frac{1}{9^p}+\frac{1}{10^p}+....+\infty$$

the above series can be grouped as

$$\frac{1}{1^p}+\left(\frac{1}{2^p}+\frac{1}{3^p}\right)+\left(\frac{1}{4^p}+\frac{1}{5^p}+\frac{1}{6^p}+\frac{1}{7^p}\right)+\left(\frac{1}{8^p}+\frac{1}{9^p}+\frac{1}{10^p}+\frac{1}{11^p}+\frac{1}{11^p}+\frac{1}{12^p}+\frac{1}{14^p}+\frac{1}{15^p}\right)+....$$

Now $$\frac{1}{1^p} = 1 \qquad ...(i)$$

and $$\frac{1}{2^p}+\frac{1}{3^p}<\frac{1}{2^p}+\frac{1}{2^p}=\frac{2}{2^p} \qquad ...(ii)\left(\because\ \frac{1}{3^p}<\frac{1}{2^p}\right)$$

Similarly $$\frac{1}{4^p}+\frac{1}{5^p}+\frac{1}{6^p}+\frac{1}{7^p}<\frac{1}{4^p}+\frac{1}{4^p}+\frac{1}{4^p}+\frac{1}{4^p}=\frac{4}{4^p} \qquad ...(iii)$$

and $$\frac{1}{8^p}=\frac{1}{9^p}+\frac{1}{10^p}+...+\frac{1}{15^p}<\frac{1}{8^p}+\frac{1}{8^p}+\frac{1}{8^p}+....\frac{1}{8^p}=\frac{8}{8^p} \qquad ...(iv)$$

On adding (*i*), (*ii*), (*iii*) and (*iv*), we get

$$\frac{1}{1^p}+\left(\frac{1}{2^p}+\frac{1}{3^p}\right)+\left(\frac{1}{4^p}+\frac{1}{5^p}+\frac{1}{6^p}+\frac{1}{7^p}\right)+\left(\frac{1}{8^p}+\frac{1}{9^p}+...+\frac{1}{15^p}\right)+...<\frac{1}{1^p}+\frac{2}{2^p}+\frac{4}{4^p}+\frac{8}{8^p}+....$$

$$= <1+\left(\frac{1}{2}\right)^{p-1}+\left(\frac{1}{2}\right)^{2p-2}+\left(\frac{1}{2}\right)^{3p-3}+...$$

$$= <\frac{1}{1-\left(\frac{1}{2}\right)^{p-1}} \left[\text{For } G.P. \text{ series, } r=\left(\frac{1}{2}\right)^{p-1} \text{ and } S_\infty=\frac{1}{1-r}\right]$$

Hence the given series is convergent when $p > 1$

Case II. When $p = 1$

The given series becomes $1+\frac{1}{2}+\left(\frac{1}{3}+\frac{1}{4}\right)+\left(\frac{1}{5}+\frac{1}{6}+\frac{1}{7}+\frac{1}{8}\right)+\left(\frac{1}{9}+\frac{1}{10}+...+\frac{1}{16}\right)+...$

Now $$1+\frac{1}{2} = 1+\frac{1}{2} \qquad ...(i)$$

$$\frac{1}{3}+\frac{1}{4}>\frac{1}{4}+\frac{1}{4}=\frac{1}{2} \qquad ...(ii)\left(\because\ \frac{1}{3}>\frac{1}{4}\right)$$

$$\frac{1}{5}+\frac{1}{6}+\frac{1}{7}+\frac{1}{8}>\frac{1}{8}+\frac{1}{8}+\frac{1}{8}+\frac{1}{8}=\frac{1}{2} \qquad ...(iii)$$

$$\frac{1}{9}+\frac{1}{10}+\frac{1}{11}+...+\frac{1}{16}>\frac{1}{16}+\frac{1}{16}+\frac{1}{16}+...\frac{1}{16}=\frac{8}{16}=\frac{1}{2} \qquad ...(iv)$$

On adding (*i*), (*ii*), (*iii*) and (*iv*), we get

$$1+\frac{1}{2}+\left(\frac{1}{3}+\frac{1}{4}\right)+\left(\frac{1}{5}+\frac{1}{6}+\frac{1}{7}+\frac{1}{8}\right)+\left(\frac{1}{9}+\frac{1}{10}+\ldots+\frac{1}{16}\right)+\ldots$$

$$>1+\frac{1}{2}+\frac{1}{2}+\frac{1}{2}+\ldots$$

$$>1+1+\frac{n}{2}$$

$$>\infty \text{ as, } n\to\infty$$

Hence, the given series id divergent when $p=1$

Case III. When $p<1$

$$\therefore \quad \frac{1}{2^p}>\frac{1}{2} \qquad \ldots(i)$$

$$\frac{1}{3^p}>\frac{1}{3} \qquad \ldots(ii)$$

$$\frac{1}{4^p}>\frac{1}{4} \qquad \ldots(iii)$$

and so on

On adding (*i*), (*ii*), (*iii*), we get $\frac{1}{2^p}+\frac{1}{3^p}+\frac{1}{4^p}+\frac{1}{5^p}+\ldots.>\frac{1}{2}+\frac{1}{3}+\frac{1}{4}+\frac{1}{5}+\ldots$

or $\quad 1+\frac{1}{2^p}+\frac{1}{3^p}+\frac{1}{4^p}+\frac{1}{5^p}+\ldots.>1+\frac{1}{2}+\frac{1}{3}+\frac{1}{4}+\frac{1}{5}+\ldots$

> divergent series as $p=1$

Hence, the given series is divergent when $p<1$.

8.19 COMPARISON TEST

Let Σu_n and Σv_n be two position terms such that $\underset{n\to\infty}{Lt}\frac{u_n}{v_n}=k$ (finite number), then both series converges or diverges together.

Proof. By the definition of limt. $\exists$ a positive number $\in$, however small such that $\left|\frac{u_n}{v_n}-k\right|<\in$ for $n>p$

i.e., $\quad -\in<\frac{u_n}{v_n}-k<+\in$

$$k-\in<\frac{u_n}{v_n}<k+\in \text{ for } n>p$$

Ignoring the first p terms on both sides, we have

$$k-\in<\frac{u_n}{v_n}<k+\in \text{ for all } n \qquad \ldots(i)$$

Case I : When Σv_n is convergent, then

$\underset{n\to\infty}{Lt}(v_1+v_2+v_3+v_n)=q$ (say), then q is a finite number

$$=(k+\in)\,q$$

Hence, Σu_n is also convergent

Case II : When Σv_n is divergent, then

$$\underset{n\to\infty}{Lt}\ (v_1 + v_2 + v_3 + \ldots.. v_n) \to \infty \qquad \ldots(ii)$$

From (*i*),

$$k - \in < \frac{u_n}{v_n}$$

or $u_n > (k - \in)\, v_n$ for all n

$$\underset{n\to\infty}{Lt}\ (u_1 + u_2 + u_3 + \ldots. + u_n) > (k - \in) \times \underset{n\to\infty}{Lt}\ (v_1 + v_2 + \ldots. + v_n)$$

From (*ii*), $\underset{n\to\infty}{Lt}\ (v_1 + v_2 + v_3 + \ldots. + v_n)\infty$

Hence, Σu_n is also divergent.

Note : Comparison test is very useful for testing the convergence or divergence of a series, In this test, we choose Σv_n or (p – series) in such a way that $\underset{n\to\infty}{Lt}\ \frac{u_n}{v_n}$ = finite number, then the nature of both the series is the same, the nature of Σv_n is already, known, so the nature of Σu_n can also be known.

***Example* 8.9.** Examine the following series

(*a*) $\Sigma\sqrt{n^4+1} - \sqrt{n^4-1}$ (*b*) $\Sigma\frac{1}{2n\,(n+1)}$

Solution : (*a*)

$$u_n = \sqrt{n^4+1} - \sqrt{n^4-1}$$

or

$$u_n = \frac{\{\sqrt{n^4+1} - \sqrt{n^4-1}\} \times \{\sqrt{n^4+1} + \sqrt{n^4-1}\}}{\sqrt{n^4+1} + \sqrt{n^4-1}}$$

$$= \frac{(n^4+1) - (n^4-1)}{\sqrt{n^4+1} + \sqrt{n^4-1}} = \frac{2}{n^2\left(\sqrt{1+\frac{1}{n^4}} + \sqrt{1-1\frac{1}{n^4}}\right)}$$

Let

$$v_n = \frac{1}{n^2}$$

$$\frac{v_n}{v_n} = \frac{2}{\left(\sqrt{1+\frac{1}{n^4}} + \sqrt{1-1\frac{1}{n^4}}\right)}$$

and $\underset{n\to\infty}{Lt}\ \frac{v_n}{v_n} = \frac{2}{\sqrt{1}+\sqrt{1}} = 1$, which is finite and non zero therefore, by comparison test both series converges or diverges together.

$$\text{Since } \Sigma v_n = \Sigma\frac{1}{n^p} = \Sigma\frac{1}{n^2}$$

Here $p = 2 > 1$, therefore Σv_n is convergent

Hence Σu_n series is convergent.

Solution : (*b*) $$u_n = \frac{1}{2n(2n-1)} = \frac{1}{2n^2\left(2-\frac{1}{n}\right)}$$

Let $$v_n = \frac{1}{n^2}$$

$$\therefore \quad \frac{u_n}{v_n} = \frac{1}{2\left(2-\frac{1}{n}\right)}$$

and $$\underset{n\to\infty}{Lt}\frac{u_n}{v_n} = \frac{1}{2(2)} = \frac{1}{4},$$ which is finite and non-zero

Also $\Sigma v_n = \Sigma \frac{1}{n^2}$ therefore Σv_n is convergent. Hence the series Σu_n is convergent

***Example* 8.10.** Test for convergence or divergence of the series

(*a*) $\frac{1}{1\times2\times3}+\frac{3}{2\times3\times4}+\frac{5}{3\times4\times5}\ldots\ldots\infty$

(*b*) $\frac{1}{\sqrt{1}+\sqrt{2}}+\frac{1}{\sqrt{2}+\sqrt{3}}+\frac{1}{\sqrt{3}\,\sqrt{4}}+\ldots\ldots\infty$

Solution : (*a*) $\frac{1}{1\times2\times3}+\frac{3}{2\times3\times4}+\frac{5}{3\times4\times5}+\ldots\ldots\infty$

$$\therefore \quad u_n = \frac{2n-1}{n(n+1)(n+2)}$$

$$= \frac{\left(2-\frac{1}{n}\right)}{n^2\left(1+\frac{1}{n}\right)\left(1+\frac{2}{n}\right)}$$

Let $$\Sigma v_n = \Sigma\frac{1}{n^2}$$

$$\therefore \quad \underset{n\to\infty}{Lt}\frac{u_n}{v_n} = \underset{n\to\infty}{Lt}\frac{2-\frac{1}{n}}{\left(1+\frac{1}{n}\right)\left(1+\frac{2}{n}\right)}$$

$$= 2, \text{ finite and non zero}$$

therefore by comparison test, both series Σu_n, Σv_n converge or diverge together, but Σv_n is convergent as $p = 2$. Hence Σu_n is also convergent.

(*b*) $\frac{1}{\sqrt{1}+\sqrt{2}}+\frac{1}{\sqrt{2}+\sqrt{3}}+\frac{1}{\sqrt{3}+\sqrt{4}}+\ldots\ldots\infty$

$$\therefore \quad u_n = \frac{1}{\sqrt{n}+\sqrt{n+1}}$$

$$= \frac{1}{\sqrt{n}\left[1+\sqrt{1+\frac{1}{n}}\right]}$$

Let $$\Sigma v_n = \Sigma \frac{1}{\sqrt{n}}$$

$$\therefore \quad \underset{n \to \infty}{Lt} \frac{u_n}{v_n} = \underset{n \to \infty}{Lt} \frac{1}{\sqrt{n}\left[1+\sqrt{1+\frac{1}{n}}\right]} \times \sqrt{n}$$

$$= \underset{n \to \infty}{Lt} \frac{1}{1+\sqrt{1+\frac{1}{n}}}$$

$= \frac{1}{1+1} = \frac{1}{2}$, which is finite and non zero number therefore, by comparison test, Σu_n and Σv_n converges or diverges together.

Since $$v_n = \frac{1}{n^{\frac{1}{2}}}, p = \frac{1}{2} < 1$$

Hence Σv_n is divergent, so Σu_n is also divergent.

Example **8.11.** Test the convergence of the series $\Sigma u_n = \Sigma \frac{2n^3+5}{4n^2+1}$

Solution : $$u_n = \frac{n^3\left(2+\frac{5}{n^3}\right)}{n^5\left(4+\frac{1}{n^5}\right)} = \frac{1}{n^2} \frac{2+\frac{5}{n^3}}{n+4+\frac{1}{n^5}}$$

Let $$v_n = \frac{1}{n^2}$$

$\therefore \underset{n \to \infty}{Lt} \frac{u_n}{v_n} = \frac{2}{4} = \frac{1}{2}$ which is finite and non zero, therefore by comparison test, Σu_n and Σv_n convergence or diverges together

Since $$v_n = \frac{1}{n^p} = \frac{1}{n^2}$$

Here $P = 2 > 1$, there Σv_n is convergent

Example **8.12.** Test the following series for convergence

(*a*) $\frac{2}{1^p} + \frac{3}{2^p} + \frac{4}{3^p} + \frac{5}{4^p} + ...\infty$

(*b*) $1 + \frac{1}{2^2} + \frac{1}{2^2} + \frac{1}{3^2} + \frac{1}{4^4} + \frac{1}{5^5} + ...\infty$

Solution : (*a*) $\frac{2}{1^p} + \frac{3}{2^p} + \frac{4}{3^p} + \frac{5}{4^p} + ...\infty$

$$u_n = \frac{n+1}{n^p}$$

$$= \frac{\left(1+\frac{1}{n}\right)}{n^{p-1}}$$

Let $$v_n = \frac{1}{n^{p-1}}$$

By comparison test $\quad \underset{n\to\infty}{Lt} \frac{u_n}{v_n} = \underset{n\to\infty}{Lt} \left(1+\frac{1}{n}\right)$

$= 1$, which finite and non zero

therefore by Comparison test, both the series are either convergent of divergent

Since $\quad v_n = \frac{1}{n^{p-1}} + \frac{1}{n^p}$

where $\quad p = p-1$

if $\quad p > 1$ *i.e.*, $p-1 > 1$

or $\quad p > 2$, then the series Σv_n is convergent

if $\quad p \leq 1$ *i.e.*, $p-1 \leq 1$

or $\quad p \leq 2$, then the series Σv_n is divergent

Therefore, the given series is convergent if $p > 2$ and divergent if $p \leq 2$.

(*b*) $1+\frac{1}{2^2}+\frac{2^2}{3^3}+\frac{3^3}{4^4}+\frac{4^4}{5^5}+ \ldots\ldots \infty$

Omitting the first term the series reduce to

$$\frac{1}{2^2}+\frac{2^2}{3^3}+\frac{3^3}{4^4}+\frac{4^4}{5^5}+ \ldots \infty$$

$$u_n = \frac{n^n}{(n+1)^{n+1}}$$

$$= \frac{n^n}{(n+1)^n (n+1)}$$

$$= \frac{n^n}{n^n\, n\left(1+\frac{1}{n}\right)^n \left(1+\frac{1}{n}\right)}$$

$$= \frac{1}{n\left(1+\frac{1}{n}\right)^n \left(1+\frac{1}{n}\right)}$$

Let $\quad v_n = \frac{1}{n}$

By comparison test, $\quad \underset{n\to\infty}{Lt} \frac{u_n}{v_n} = \underset{n\to\infty}{Lt} \frac{1}{n\left(1+\frac{1}{n}\right)^n \left(1+\frac{1}{n}\right)} \times n$

$= \frac{1}{e}$, which is finite and non zero

therefore, by Comparison test, both series are either convergent or divergent together.

Since $v_n = \frac{1}{n}$, $p = 1$, therefore, Σv_n is divergent. Hence the given series Σu_n is also divergent.

Example **8.13.** Test for convergence or divergence of the series whose n^{th} terms are

(*a*) $\sin\frac{1}{n}$ (*b*) $\tan^{-1}\frac{1}{n}$ (*c*) $\frac{1}{2^n+3^n}$

Solution : (*a*) $u_n = \sin\frac{1}{n}$

$$\because \quad \sin\frac{1}{n} = \frac{1}{n} - \frac{1}{3!}\frac{1}{n^3} + \frac{1}{5!}\frac{1}{n^5} - \frac{1}{7!}\frac{1}{n^7} + \ldots\ldots$$

$$\therefore \quad u_n = \frac{1}{n}\left[1 - \frac{1}{3!n^2} + \frac{1}{5!n^4} - \frac{1}{7!n^6} + \ldots\ldots\right]$$

Let $$v_n = \frac{1}{n}$$

By comparison test, $$\underset{n\to\infty}{Lt}\frac{u_n}{u_n} = \underset{n\to\infty}{Lt}\left(1 - \frac{1}{3!n^2} + \frac{1}{5!n^4} - \frac{1}{5!n^4} - \frac{1}{7!n^6} + \ldots\ldots\right)$$

$$= 1, \text{ which is finite and non zero}$$

therefore by comparison test both series are either convergent or divergent together

Since $v_n = \frac{1}{n}$ here $p = 1$, therefore Σv_n is divergent. Hence Σu_n is also divergent.

(*b*) $$u_n = \tan^{-1}\left(\frac{1}{n}\right)$$

$$\therefore \quad \tan^{-1}\left(\frac{1}{n}\right) = \frac{1}{n} - \frac{1}{3n^3} + \frac{1}{5n^5} - \ldots\ldots \infty$$

$$= \frac{1}{n}\left(1 - \frac{1}{3n^2} + \frac{1}{5n^4} + \ldots\right)$$

Let $$v_n = \frac{1}{n}$$

$$\therefore \quad \underset{n\to\infty}{Lt}\frac{u_n}{n_n} = \underset{n\to\infty}{Lt}\left(1 - \frac{1}{3n^2} + \frac{1}{5n^4} + \ldots\right)$$

$$= 1, \text{ which is finite and non zero}$$

$\therefore$ By comparison test, both series are either convergent or divergent together

Since $v_n = \frac{1}{n}$, $p = 1$, therefore Σv_n is divergent. Hence, Σu_n is also divergent

(*c*) $$u_n = \frac{1}{2^n + 3^n} = \frac{1}{3^n\left[1 + \left(\frac{2}{3}\right)^n\right]}$$

Let $$v_n = \frac{1}{3^n}$$

$$\therefore \quad \underset{n\to\infty}{Lt}\frac{u_n}{v_n} = \underset{n\to\infty}{Lt}\frac{1}{\left[1 + \left(\frac{2}{3}\right)^n\right]} \times 3n$$

$$= 1, \text{ which is finite and non zero}$$

therefore by comparison test, both series are either convergent or divergent together

Since $v_n = \frac{1}{3^n}$, which is a geometric series with common ratio $\frac{1}{3} < 1$, therefore Σu_n is convergent.

Hence the given series Σu_n is also convergent.

***Example* 8.14.** Test the series Σu_n for convergence, where $u_n = \sqrt{n^2+1} - n$

Solution :
$$u_n = \frac{(n^2+1)-n^2}{\sqrt{n^2+1}+n} = \frac{1}{n\left(\sqrt{1+\frac{1}{n^2}}+1\right)}$$

Let
$$v_n = \frac{1}{n}$$

By comparison test, $\underset{n\to\infty}{Lt} \frac{u_n}{v_n} = \frac{1}{\sqrt{1+1}} = \frac{1}{2}$, which is finite and non zero

Since
$$v_n = \frac{1}{n^p} = \frac{1}{n}$$

Here $p = 1$ is Σv_n is divergent

Hence the series Σu_n is divergent.

EXERCISE 8.4

Discuss the nature of the following serices.

1. $\frac{2}{1^p} + \frac{3}{2^p} + \frac{4}{3^p} + \frac{5}{4^p}$ ..... ∞ **Ans.** The given services is convergent if $P > 2$ and

2. $1\frac{2^2}{2!} + \frac{3^2}{3!} + \frac{4^2}{4!} + ...$ **Ans.** Convergent

3. $1 + \frac{1\times 2}{1\times 3} + \frac{1\times 2\times 3}{1\times 3\times 5} + \frac{1\times 2\times 3\times 4}{1\times 3\times 5\times 7} + \infty$ **Ans.** Convergent

4. $\frac{1}{1+2} + \frac{2}{1+2^2} + \frac{3}{1+2^3} + ...\infty$ **Ans.** Convergent

5. $\sum_{n=1}^{\infty} \frac{1}{\sqrt{n}+\sqrt{n+1}}$ **Ans.** Divergent

6. $\sum_{n=1}^{\infty} \{\sqrt[3]{n^3+1} - n\}$ **Ans.** Convergent

7. $\sum_{n=1}^{\infty} \frac{n^n}{n!}$ **Ans.** Convergent

8. $\sum_{n=1}^{\infty} \frac{x^n}{1+x^{2n}}, (x > 0)$ **Ans.** Convergent if $x > 1$ and divergent of $x \le 1$

9. $\sum_{n=1}^{\infty} \sin\frac{1}{n}$ **Ans.** Divergent

10. $\sum_{n=1}^{\infty} \frac{n^2}{e^n}$ **Ans.** Divergent

8.20 D'ALEMBERT'S RATIO TEST

Statement : If Σu_n is a positive term series such that $\underset{n\to\infty}{Lt} \dfrac{u_{n+1}}{u_n} = k$ then

(*i*) The series is convergent if $k < 1$

(*ii*) The series is divergent if $k > 1$ and

(*iii*) The test fails to decide the nature of the series if $k = 1$

Solution : Case I : When $\underset{n\to\infty}{Lt} \dfrac{u_{n+1}}{u_n} = k < 1$

By definition of limit, we can find a number $r < 1$ such that $\dfrac{u_{n+1}}{u_n} < r$ for $n \geq m$ $\left[\dfrac{u_2}{u_1} < r, \dfrac{u_3}{u_2} < r, \dfrac{u_4}{u_3} < r \ldots\right]$

Omitting the first m terms, let the series the $u_1 + u_2 + u_3 + u_4 + \ldots.. \infty$

$$= u_1\left(1 + \frac{u_2}{u_1} + \frac{u_3}{u_1} + \frac{u_3}{u_1} + \ldots. \infty\right)$$

$$= u_1\left(1 + \frac{u_2}{u_1} + \frac{u_3}{u_2} \times \frac{u_2}{u_1} + \frac{u_4}{u_3} \times \frac{u_3}{u_2} \times \frac{u_2}{u_1} + \ldots\ldots \infty\right)$$

$$< u_1 (1 + r + r^2 + r^3 + \ldots.. \infty) \qquad (\because r < 1)$$

$$= \frac{u_1}{1-r}. \text{ which is a finite number}$$

Hence, Σu_n is convergent.

Case II. When $\underset{n\to\infty}{Lt} \dfrac{u_{n+1}}{u_n} = k > 1$

By definition of limit, we can find a number such that $\dfrac{u_n + 1}{u_n} > 1$ for $n \geq m$

$\dfrac{u_2}{u_1} > 1, \dfrac{u_3}{u_2} > 1 \dfrac{u_4}{u_3} > 1$

Omitting the first terms, let the series be

$$u_1 + u_2 + u_3 + u_4 + \ldots.. \infty$$

$$= u_1\left(1 + \frac{u_2}{u_1} + \frac{u_3}{u_1} + \frac{u_4}{u_1} + \ldots \infty\right)$$

$$= u_1\left(1 + \frac{u_2}{u_1} + \frac{u_3}{u_1} \times \frac{u_2}{u_1} + \frac{u_4}{u_3} \times \frac{u_3}{u_2} \times \frac{u_2}{u_1} + \ldots. \infty\right)$$

$$\geq u_1 (1 + 1 + 1 + 1 + \ldots. \text{ to } n \text{ terms})$$

$$= n\, u_1$$

$$\underset{n\to\infty}{Lt} (u_1 + u_2 + u_3 + \ldots\ldots + u_n) > \underset{n\to\infty}{Lt} nu_1$$

Let S_n be the sum of n terms then

$$\underset{n\to\infty}{Lt} S_n \geq \underset{n\to\infty}{Lt} nu_1 = \infty$$

Hence, Σu_n is divergent.

Case III. When $\frac{u_{n+1}}{u_n} = k = 1$, when then D' Alembert ratio test fails

Consider the series whose nth therm is $\frac{1}{n^2}$ *i.e.,*

$$u_n = \frac{1}{n^2} \text{ and } u_{n+1} = \frac{1}{(n+1)^2}$$

Here
$$\frac{u_{n+1}}{u_n} = \frac{\frac{1}{(n+1)^2}}{\frac{1}{n^2}}$$

$$= \frac{n^2}{(n+1)^2} = \frac{1}{\left(1+\frac{1}{n}\right)^2}$$

$$\therefore \quad \underset{n\to\infty}{Lt} \frac{u_{n+1}}{u_n} = \underset{n\to\infty}{Lt} \frac{1}{\left(1+\frac{1}{n}\right)^2} = 1 \quad \ldots(i)$$

Again consider another series whose nth terms is $\frac{1}{n}$ *i.e.,*

$$u_n = \frac{1}{n} \quad \text{and} \quad u_{n+1} = \frac{1}{n+1}$$

$$\therefore \quad \underset{n\to\infty}{Lt} \frac{u_{n+1}}{u_n} = \underset{n\to\infty}{Lt} \frac{\frac{1}{n+1}}{\frac{1}{n}}$$

$$= \underset{n\to\infty}{Lt} \frac{n}{n+1} = \underset{n\to\infty}{Lt} \frac{1}{1+\frac{1}{n}} \quad \ldots(ii)$$

$$= 1$$

Thus from (*i*) and (*ii*) in both cases $\underset{n\to\infty}{Lt} \frac{u_{n+1}}{u_n} = 1$

But the first series is convergent as p = 2, and the second series is divergent as $p = 1$

Hence, when $\underset{n\to\infty}{Lt} \frac{u_{n+1}}{u_n} = 1$, the series may be convergent or divergent.

Thus D' Alembert Ratio test fails when $k = 1$.

Note : The D' Alembert ratio test can also be stated in the following form

If Σu_n is a positive term series such that $\underset{n\to\infty}{Lt} \frac{u_n}{u_{n+1}} = k$, then

(*i*) The series is convergent if $k > 1$.

(*ii*) The series is divergent if $k < 1$ and

(*iii*) The test fails to decide the nature of the series if $k = 1$

Usually, we use this test in the above from because it is more convenient for further investigation when this test fails.

***Example* 8.15.** Discuss the convergence and divergence of the following series

(*a*) $1+\frac{2^P}{2!}+\frac{3^P}{3!}+\frac{4^P}{4!}+\ldots\ldots\infty, (p>0)$

(*b*) $\frac{1^2 2^2}{1!}+\frac{2^2 3^2}{2!}+\frac{3^2 4^2}{3!}+\ldots\ldots\infty$

Solution : (*a*) Omitting the first term

$\frac{1^p}{1!}+\frac{2^p}{2!}+\frac{3^p}{4!}+\ldots..\infty$ $\left[\because \frac{1^P}{1!}=1\right]$

$$\therefore \qquad u_n = \frac{n^p}{n!}$$

and
$$u_{n+1} = \frac{(n+1)^p}{(n+1)!}$$

Now
$$\frac{u_n}{u_n+1} = \frac{n^p}{n!}\times\frac{(n+1)!}{(n+1)^P}$$

$$= \frac{n^p\,(n+1)n!}{n!\,(n+1)^p} = \frac{n^p}{(n+1)^{p-1}}$$

$$= \frac{n}{\left(1+\frac{1}{n}\right)^{p-1}}$$

$$\therefore \qquad \underset{n\to\infty}{Lt}\frac{u_n}{u_{n+1}} = \underset{n\to\infty}{Lt}\frac{n}{\left(1+\frac{1}{n}\right)^{P-1}}$$

$$= \infty > 1$$

Therefore, by D' Alembert's Ratio Test, Σu_n is convergent.

(*b*) $\frac{1^2 2^2}{1!}+\frac{2^2 3^2}{2!}+\frac{3^2 4^2}{3!}+\ldots\ldots\infty$

$$\therefore \qquad u_n = \frac{n^2\,(n+1)^2}{n!}$$

and
$$u_{n+1} = \frac{(n+1)^2\,(n+2)^2}{(n+1)!}$$

Now
$$\frac{u_n}{un+1} = \frac{n^2\,(n+1)^2}{n!}\times\frac{(n+1)!}{(n+1)^2\,(n+2)^2}$$

$$= \frac{n\left(1+\frac{1}{n}\right)}{\left(1+\frac{2}{n}\right)^2}$$

$$\therefore \qquad \underset{n\to\infty}{Lt}\frac{u_n}{u_{n+1}} = \underset{n\to\infty}{Lt}\frac{n\left(1+\frac{1}{n}\right)}{\left(1+\frac{1}{n}\right)^2}$$

$$= \infty > 1$$

Therefore by D' Alembert's ratio test Σu_n is convergent.

Example **8.16.** Discuss the convergence or divergence of the following series when nth term

(*a*) $u_n = \frac{n! \times 2^n}{n^n}$ (*b*) $u_n = \frac{n^2}{2^n} + \frac{1}{n^2}$ (*c*) $u_n = \frac{n}{1+2^n}$

Solution : (*a*) $u_n = \frac{n! \times 2^n}{n^n}$ and $u_{n+1} = \frac{(n+1)! \times 2^{n+1}}{(n+1)^{n+1}}$

Now
$$\frac{u_n}{u_{n+1}} = \frac{n! \times 2^n}{n^n} \times \frac{(n+1)(n+1)^n}{(n+1)n! \times 2^n}$$

$$= \frac{1}{2}\frac{(n+1)^n}{n^n} = \frac{1}{2}\left(1+\frac{1}{n}\right)^n$$

$$= \frac{1}{2}\frac{(n+1)^n}{n^n} = \frac{1}{2}\left(1+\frac{1}{n}\right)^n$$

$\therefore$
$$\underset{n\to\infty}{Lt} \frac{u_n}{u_{n+1}} = \frac{1}{2} \times e \qquad \left[\lim_{n\to\infty}\left(1+\frac{1}{n}\right)^n = e\right]$$

$$= \frac{e}{2} > 1$$

Therefore, the given series Σu_n is convergent.

(*b*)
$$u_n = \frac{n^2}{2^n} + \frac{1}{n^2} = P_n + Q_n \text{ (say)}$$

where,
$$P_n = \frac{n^2}{2^n}$$

Here
$$P_{n+1} = \frac{(n+1)^2}{2n+1}$$

and
$$Q_n = \frac{1}{n^2}$$

$\therefore$
$$Q_{n+1} = \frac{1}{(n+1)^2}$$

Now
$$\frac{P_n}{P_{n+1}} = \frac{n^2}{2^n} \times \frac{2^n \times 2}{(n+1)^2} = \frac{2}{\left(1+\frac{1}{n}\right)^2}$$

$\therefore$
$$\underset{n\to\infty}{Lt} \frac{P_n}{P_n+1} = 2$$

Hence, ΣP_n is convergent by D' Alembert's Ratio Test,

Again
$$\frac{Q_n}{Q_{n+1}} = \left(\frac{n+1}{n}\right)^2 = \frac{1}{\left(1+\frac{1}{n}\right)^2}$$

$$\underset{n\to\infty}{Lt}\ \frac{Q_n}{Q_{n+1}} = 1, \text{ the test fails}$$

Also $$\Sigma Q_n = \Sigma \frac{1}{n^2}\ P = 2 > 1$$

Hence, ΣQ_n is convergent

therefore, $\Sigma u_n = \Sigma P_n + \Sigma Q_n$ is convergent

(c) $$u_n = \frac{n}{1+2^n} \quad \text{and} \quad u_{n+1} = \frac{n+1}{1+2^{n+1}}$$

$$\frac{u_n}{u_{n+1}} = \frac{n}{1+2^n} \times \frac{1+2^n \times 2}{n+1}$$

$$= \left(\frac{1}{1+\frac{1}{n}}\right)\left(\frac{2+\frac{1}{2^n}}{1+\frac{1}{2^n}}\right)$$

$$\underset{n\to\infty}{Lt}\ \frac{u_n}{u_{n+1}} = \underset{n\to\infty}{Lt}\left(\frac{1}{1+\frac{1}{n}}\right)\left(\frac{2+\frac{2}{2^n}}{1+\frac{1}{2^n}}\right)$$

$$= 2 > 1$$

Hence the series is convergent by D' Alembert's test.

***Example* 8.17.** Examine the nature of the following series

(a) $1 + 1 + \frac{\alpha+1}{\beta+1} + \frac{(\alpha+1)(2\alpha+1)}{(\beta+1)(2\beta+1)} + \frac{(\alpha+1)(2\alpha+1)(3\alpha+1)}{(\beta+1)(2\beta+1)(3\beta+1)} + \ldots\ldots \infty$

(b) $\frac{1}{3} + \frac{1\times 2}{3\times 5} + \frac{1\times 2\times 3}{3\times 5\times 7} + \frac{1\times 2\times 3\times 4}{3\times 5\times 7\times 9} + \ldots\ \infty$

Solution : (a) Omitting the first term

Here $$u_n = \frac{(\alpha+1)(2\alpha+1)(3\alpha+1)\ldots(n\alpha+1)}{(\beta+1)(2\beta+1)(3\beta+1)\ldots(n\beta+1)}$$

Now $$\frac{u_n}{u_{n+1}} = \frac{(n+1)\beta+1}{(n+1)\alpha+1}$$

$\therefore$ $$\underset{n\to\infty}{Lt}\ \frac{u_n}{u_{n+1}} = \underset{n\to\infty}{Lt}\ \frac{\beta+\frac{1}{n+1}}{\alpha+\frac{1}{n+1}} = \frac{\beta}{\alpha}$$

Hence by D' Alember's ratio test,

(i) the series in convergent if $\frac{\beta}{\alpha} >$ *i.e.*, if $\beta > \alpha > 0$

(ii) the series is divergent if $\frac{\beta}{\alpha} < 1$ *i.e.* if $\alpha > \beta > 0$

(*iii*) the test fail to decide the nature of the series if $\beta = \alpha$ or $\frac{\beta}{\alpha} = 1$

When $\beta = \alpha$, then the series becomes

$1 + 1 + 1 + 1 + \ldots$

$\therefore S_n$ = the sum of n term of this series = n

Since $\underset{n \to \infty}{Lt} S_n = \infty$, hence the series in divergent.

Thus the given series in convergent in $\beta > \alpha > 0$ and divergent of $\alpha > \beta > 0$.

Solution : (*b*)
$$u_n = \frac{1 \times 2 \times 3 \times 4 \times \ldots . xn}{3 \times 5 \times 7 \times 9 \times \ldots . 2n + 1}$$

and
$$u_n + 1 = \frac{1 \times 2 \times 3 \times 4 \times \ldots . \times n(n+1)}{3 \times 5 \times 7 \times 9 \times \ldots . \times (2n+1) \times (2n+3)}$$

$$\frac{u_n}{n_{n+1}} = \frac{2n+3}{n+1}$$

$$\therefore \quad \underset{n \to \infty}{Lt} \frac{u_n}{u_{n+1}} = \underset{n \to \infty}{Lt} \frac{2 + \frac{3}{n}}{1 + \frac{1}{n}} = 2 \text{ which is} > 1$$

Hence by D' Alemert ratio test the given series is convergent.

EXERCISE 8.5

Test for convergence the following series :

1. $\left(\frac{1}{3}\right)^2 + \left(\frac{1 \times 2}{3 \times 5}\right)^2 + \left(\frac{1 \times 2 \times 3}{3 \times 5 \times 7}\right)^2 + \ldots \infty$ **Ans.** Convergent

2. $\frac{2!}{3} + \frac{3!}{3^2} + \frac{4!}{3^3} + \ldots \infty$ **Ans.** Divergent

3. $\sum \frac{n^2 (n+1)^2}{n!}$ **Ans.** Convergent

4. $\sum \frac{n}{n^2 + 1} x^n, x > 0$ **Ans.** Convergent if $x < 1$, Divergent if $x \geq 1$

5. $\frac{x}{1.3} + \frac{x^2}{3.5} + \frac{x^3}{5.7} + \ldots \infty$ **Ans.** Convergent if $x < 1$, Divergent if $x > 1$

6. $1 + \frac{x}{2} + \frac{x^2}{5} + \frac{x^3}{10} + \ldots . + \frac{x^n}{n^2 + 1} + \ldots . \infty$ **Ans.** Convergent for $x < 1$, Divergent for $x > 1$

7. $\frac{x}{2\sqrt{3}} + \frac{x^2}{3\sqrt{4}} + \frac{x^3}{4\sqrt{5}} + \ldots . \infty$ **Ans.** Convergent for $x < 1$, Divergent for $x > 1$

8. $\sum_{n+1}^{\infty} \frac{n^2}{3^n}$ **Ans.** Convergent

9. $\sum_{n=1}^{\infty} \frac{n \times x^n}{(x+1)(n+2)}$ **Ans.** Convergent for $x < 1$, Divergent for $x \geq 1$

10. $\sum_{n=1}^{\infty} \frac{x^n}{(2n)!}$ **Ans.** Convergent

8.21 RAABE'S TEST (HIGHER RATIO TEST)

Statement : If Sun is a positive term series such that

$$\underset{n \to \infty}{Lt}\ n\left(\frac{u_n}{u_{n+1}} - 1\right) = k, \text{ then}$$

(*i*) The series is convergent if $k > 1$

(*ii*) The series is divergent if $k < 1$ and

(*iii*) The test fails to decide the nature of the series if $k = 1$.

Solution : Cases I : When $k > 1$

Let p be a number such that $1 < p < k$. Compare the given series Σu_n with $\Sigma \frac{1}{n^p}$ which is convergent as $p > 1$.

Now $\frac{u_n}{u_{n+1}} > \frac{(n+1)^p}{n^p}$

or $\frac{u_n}{u_{n+1}} > \left(1 + \frac{1}{n}\right)^p$

$$> 1 + \frac{p}{n} + \frac{p(p-1)}{2!} \frac{1}{n^2} + \ldots.$$

or $\frac{u_n}{u_{n+1}} - 1 > \frac{p}{n} + \frac{p(p-1)}{2!} \frac{1}{n^2} + \ldots.$

or $\left(\frac{u_n}{u_{n+1}} - 1\right) > p + \frac{p(p-1)}{2} \frac{1}{n} + \ldots.$

If $\underset{n \to \infty}{Lt} \left(\frac{u_n}{u_{n+1}} - 1\right) > p$ and $p < k$ which is true as $1 < p < k$ therefore, Σu_n is convergent when $k > 1$.

Case II. When $k < 1$, then series diverges.

Proceed as in case I

Case III. When $k = 1$, Raabe's test fails if $k = 1$

Note **:** Raab's test is applied when D' Alembert's Ratio test fails.

Example **8.18.** Test the convergence for the series

$$\frac{x}{1 \times 2} + \frac{x^2}{3 \times 4} + \frac{x^3}{5 \times 6} + \frac{x^4}{7 \times 8} + \ldots.. \infty$$

Solution : $\frac{x}{1 \times 2} + \frac{x^2}{3 \times 4} + \frac{x^3}{5 \times 6} + \frac{x^4}{7 \times 8} + \ldots..\infty$

$$u_n = \frac{x^n}{(2n-1)} \quad \text{and} \quad u_{n+1} = \frac{x^{n+1}}{(zn+1)(2n+2)}$$

Now, $$\frac{u_n}{u_{n+1}} = \frac{x^n}{(2n-1)2n} \times \frac{(2n+1)(2n+2)}{x^n \times x}$$

$$\frac{u_n}{u_{n+1}} = \frac{2n\left(1+\frac{1}{2n}\right)2n\left(1+\frac{1}{n}\right)}{2n\left(1-\frac{1}{2n}\right)}\frac{1}{x}$$

$$\therefore \quad \underset{n\to\infty}{Lt}\frac{u_n}{u_{n+1}} = \underset{n\to\infty}{Lt}\frac{\left(1+\frac{1}{2n}\right)\left(1+\frac{1}{n}\right)}{\left(1-\frac{1}{2n}\right)}\frac{1}{x} = \frac{1}{x}$$

Hence, we conclude that

(*i*) If $\frac{1}{x} > 1$ *i.e.*, $x < 1$, then Σu_n is convergent

(*ii*) If $\frac{1}{x} < 1$ *i.e.*, $x > 1$, then Σu_n is divergent

(*iii*) If $\frac{1}{x} = 1$ i.e., $x = 1$, then D' Alembert ratio test fails

when $$x = 1, = \frac{u_n}{u_{n+1}} = \frac{\left(1+\frac{1}{2n}\right)\left(1+\frac{1}{n}\right)}{\left(1-\frac{1}{2n}\right)}$$

Now apply Raabe's test for the test for the series

$$n\left(\frac{un}{un+1}-1\right) = n\left[\frac{\left(1+\frac{1}{2n}\left(1+\frac{1}{n}\right)\right)}{1-\frac{1}{2n}}-1\right]$$

$$\therefore \quad \underset{n\to\infty}{Lt}\, n\left(\frac{u_n}{u_{n+1}}-1\right) = \underset{n\to\infty}{Lt}\left[\frac{(2n+1)(2n+2)-2n(2n-1)}{2n(2n-1)}\right]$$

$$= \underset{n\to\infty}{Lt}\left[\frac{8n+2}{2n(2n-1)}\right]$$

$$= \underset{n\to\infty}{Lt}\frac{2+\frac{2}{4n}}{\left(1-\frac{1}{2n}\right)} = 2 > 1, \text{ therefore the series is convergent}$$

Hence, the given is convergent if $x > 1$ and divergent if $x > 1$.

***Example* 8.19.** Test the following series for convergence or divergence.

(*a*) $$u_n = \frac{\sqrt{n}}{\sqrt{n^2+1}}x^n \; (x > 0)$$

(*b*) $$u_n = \frac{1\times3\times5\times7\times9 \,.....\, (4n-3)}{2\times4\times6\times8\times10 \,....\, (4n-2)}\times\frac{x^{2n}}{4n}$$

Solution : (*a*) $u_n = \frac{\sqrt{n}}{\sqrt{n^2+1}}x^n$ and $u_{n+1} = \frac{\sqrt{n+1}}{\sqrt{(n+1)^2+1}}x^{n+1}$

Now $$\frac{u_n}{u_{n+1}} = \frac{\sqrt{n}}{\sqrt{n+1}} \cdot \frac{\sqrt{(n+1)^2+1}}{\sqrt{n^2+1}} \times \frac{x^n}{x^{n+1}}$$

$$= \frac{\sqrt{n} \times (n+1)\sqrt{1+\frac{1}{(n+1)^2}}}{\sqrt{n+1} \times n\sqrt{1+\frac{1}{n^2}}} \times \frac{1}{x} = \frac{\sqrt{n+1}}{\sqrt{n}} \times \frac{\sqrt{1+\frac{1}{(n+1)^2}}}{\sqrt{1+\frac{1}{n^2}}} \times \frac{1}{x}$$

$$= \frac{\sqrt{1+\frac{1}{n}}\sqrt{1+\frac{1}{(n+1)^2}}}{\sqrt{1+\frac{1}{n^2}}} \times \frac{1}{x}$$

$$\therefore \qquad \underset{n\to\infty}{Lt} \frac{u_n}{u_{n+1}} = \frac{1}{x}$$

By D'Alembert ratio test, we conclude that

(*i*) If $\frac{1}{x} > 1$ *i.e.* $x < 1$; the series is convergent

(*ii*) If $\frac{1}{x} < 1$ *i.e.,* $x > 1$; the series is divergent

(*iii*) If $\frac{1}{x} = 1$ *i.e.,* $x = 1$; then test fails

when $$x = 1 \text{ then } u_n = \frac{1}{\sqrt{n}} \cdot \frac{1}{\sqrt{1+\frac{1}{n^2}}}$$

Taking $$v_n = \frac{1}{\sqrt{n}}$$

By comparison Test,

$$\frac{u_n}{v_n} = \frac{1}{\sqrt{1+\frac{1}{n^2}}}$$

$$\therefore \qquad \underset{n\to\infty}{Lt} \frac{u_n}{v_n} = \underset{n\to\infty}{Lt} \frac{1}{\sqrt{1+\frac{1}{n^2}}}$$

$= 1$, which is finite and non zero

Therefore by comparison test both series converge or diverge together

Since $$\Sigma u_n = \frac{1}{\sqrt{n}} = \frac{1}{n^p}$$

Here $p = \frac{1}{2} < 1$, so Σv_n is divergent

Hence the series Σu_n is divergent if $x \geq 1$ and convergent if $x < 1$.

Solution : (*b*) $u_n = \frac{1 \times 3 \times 5 \times 7 \times 9 \ldots\ldots (4n-3)}{2 \times 4 \times 6 \times 8 \times 10 \ldots\ldots (4n-2)} \times \frac{x^{2n}}{4n}$

and $$u_{n+1} = \frac{1\times3\times5\times7\times9\ldots.(4n-3)\times(4n-1)}{2\times4\times6\times8\times10\ldots.(4n-2)\times(4n-2)}\times\frac{x^{2n+2}}{4(n+1)}$$

Now $$\frac{u_n}{u_{n+1}} = \frac{x^{2n}(4n+2)(4n+4)}{4n(4n+1)x^{2n}\times x^2}$$

$$= \frac{4n\left(1+\frac{1}{2n}\right)4n\left(1+\frac{1}{n}\right)}{4n\times4n\left(1+\frac{1}{4n}\right)}\frac{1}{x^2}$$

$$= \frac{\left(1+\frac{1}{2n}\right)\left(1+\frac{1}{n}\right)}{\left(1+\frac{1}{4n}\right)}\times\frac{1}{x^2}$$

$$\underset{n\to\infty}{Lt}\frac{u_n}{u_{n+1}} = \frac{1}{x}$$

Hence by *D'* Alembert ratio test, we conclude that

(*i*) If $\frac{1}{x}>1$ *i.e.,* $x<1$, then the series is convergent

(*ii*) If $\frac{1}{x}<1$ *i.e.,* >1. then the series is divergent

(*iii*) If $\frac{1}{x}=1$, then D' Alembert ratio test fails

When $x=1$, $$\frac{u_n}{u_{n+1}} = \frac{\left(1+\frac{1}{2n}\right)\left(1+\frac{1}{n}\right)}{\left(1+\frac{1}{4n}\right)}$$

$$n\left(\frac{u_n}{u_{n+1}}-1\right) = \left(\frac{(4n+2)\,4(n+1)}{4n(4n+1)}-1\right) = n\left(\frac{(4n+2)(n+1)-n(4x-1)}{n(4n+1)}\right)$$

$$= \frac{5n+2}{4n+1} = \frac{5n\left(1+\frac{1}{5n}\right)}{4n\left(1+\frac{1}{4n}\right)}$$

$$\therefore \qquad \underset{n\to\infty}{Lt}\,n\left(\frac{un}{u_{n+1}}-1\right) = \frac{5}{4}>1$$

Hence, the series in convergent when $x\le1$ and divergent when $x>1$.

***Example* 8.20.** Test the convergence of the following series

(*a*) $\frac{x}{1\times2}+\frac{x^2}{3\times4}+\frac{x^3}{5\times6}+\frac{x^4}{7\times8}+\ldots.\infty$

(*b*) $1+\frac{1}{2}x+\frac{1\times3}{2\times4}x^2+\frac{1\times3\times5}{2\times4\times6}x^3+\ldots\ldots\infty$

Solution : (*a*) $\frac{x}{1\times2}+\frac{x^2}{3\times4}+\frac{x^3}{5\times6}+\frac{x^4}{7\times8}+\ldots..\infty$

$$\therefore \qquad u_n = \frac{x^n}{(2n-1)2n}$$

and
$$u_{n+1} = \frac{x^{n+1}}{(2n+1)\,2\,(n+1)}$$

Now
$$\frac{u_n}{u_{n+1}} = \frac{x^n}{2n(2n-1)} \times \frac{2(2n+1)(n+1)}{x^2+x}$$

$$= \frac{2(2n+1)(n+1)}{2n(2n-1)} \times \frac{1}{x}$$

$$= \frac{n^2\left(2+\frac{1}{n}\right)\left(1+\frac{1}{n}\right)}{n^2\left(2-\frac{1}{n}\right)}\frac{1}{x}$$

$$\underset{n\to\infty}{Lt}\ \frac{u_n}{u_{n+1}} = \frac{(2)(1)}{2} \times \frac{1}{x} = \frac{1}{x}$$

Hence by *D'* Alembert ratio test, we conclude that

(*i*) If $\frac{1}{x} > 1$ *i.e.,* $x < 1$, the series is convergent

(*ii*) If $\frac{1}{x} < 1$ *i.e.,* $x > 1$, then the series is divergent, and

(*iii*) If $\frac{1}{x} = 1$ *i.e.,* $x = 1$, then the test fails to decide to nature of the series.

When
$$x = 1, \frac{u_n}{u_{n+1}} = \frac{\left(2+\frac{1}{n}\right)\left(1+\frac{1}{n}\right)}{2-\frac{1}{n}}$$

$$n\left(\frac{u_n}{u_{n+1}} - 1\right) = n\left(\frac{2(n+1)(2n+1)}{2n(2n-1)} - 1\right)$$

$$= n\,\frac{(n+1)(2n+1) - n(2n-1)}{n(2n-1)}$$

$$= \frac{2n^2+2n+n+1-2n^2+n}{2n-1}$$

$$= \frac{4n+1}{2n-1} = \frac{4+\frac{1}{n}}{2-\frac{1}{n}}$$

$$\underset{n\to\infty}{Lt}\left[\frac{u_n}{u_{n+1}} - 1\right] = \frac{4}{2} = 2 > 1$$

Hence the series is convergent when $x < 1$ and divergent when $x > 1$.

Solution : (b) $1 + \frac{1}{2}x + \frac{1\times 3}{2\times 4}x^2 + \frac{1\times 3\times 5}{2\times 4\times 6}x^3 + \ldots\ \infty$

$$\therefore \qquad u_n = \frac{1\times 3\times 5\ldots(2n-1)}{2\times 4\times 6\ldots 2n}x^n$$

and $$u_{n+1} = \frac{1 \times 3 \times 5 \ldots (2n-1) \times (2n+1)}{2 \times 4 \times 6 \ldots 2n \times 2 \times (n+1)} x^{n+1}$$

Now $$\frac{u_n}{u_{n+1}} = \frac{x^n \, 2\,(n+1)}{(2n+1)\,x^{n+1}} = \frac{2n\left(1+\frac{1}{n}\right)}{n\left(2+\frac{1}{n}\right)} \frac{1}{x}$$

$\therefore$ $$\underset{n \to \infty}{Lt} \frac{u_n}{u_{n+1}} = \frac{1}{x}$$

Hence by D' Alemberts ratio test

(i) If $\frac{1}{x} > 1$ *i.e.*, $x < 1$. the series is convergent

(ii) If $\frac{1}{x} < 1$ *i.e.*, $x > 1$, the series is divergent

(iii) If $\frac{1}{x} = 1$, *i.e.*, $x = 1$, then test fails to decide the nature of the series.

When $x = 1$, then $\frac{u_n}{u_{n+1}} = \frac{2n+2}{2n+1}$

Applying Raabe's test, for this test,

$$n\left(\frac{u_n}{u_{n+1}} - 1\right) = n\left(\frac{2n+2}{2n+1} - 1\right) = n\,\frac{2n+2-2n-1}{2n+1} = \frac{1}{2+\frac{1}{n}}$$

$\therefore$ $$\underset{n \to \infty}{Lt}\; n\left(\frac{u_n}{u_{n+1}} - 1\right) = \frac{1}{2} < 1$$

Hence the series is convergent if $x < 1$ and divergent if $x \geq 1$.

Example 8.21. Test the convergent of the following series.

(a) $\frac{3}{7}x + \frac{3 \times 6}{7 \times 10}x^2 + \frac{3 \times 6 \times 9}{7 \times 10 \times 13}x^3 + \frac{3 \times 6 \times 9 \times 12}{7 \times 10 \times 13 \times 16}x^4 + \ldots \infty$

(b) $\frac{x}{1} + \frac{1}{2} \times \frac{x^3}{3} + \frac{1 \times 3}{2 \times 4} \times \frac{x^5}{5} + \frac{1 \times 3 \times 5}{2 \times 4 \times 6} \times \frac{x^7}{7} + \ldots \infty \; (x > 0)$

Solution : (a) $\frac{3}{7}x + \frac{3 \times 6}{7 \times 10}x^2 + \frac{3 \times 6 \times 9}{7 \times 10 \times 13}x^3 + \frac{3 \times 6 \times 9 \times 12}{7 \times 10 \times 13 \times 16}x^4 + \ldots \infty$

$\therefore$ $$u_n = \frac{3 \times 6 \times 9 \times 12 \ldots 3n}{7 \times 10 \times 13 \times 16 \ldots (3n+4)} x^n$$

and $$u_{n+1} = \frac{3 \times 6 \times 9 \times 12 \ldots 3n \times 3(n+1)}{7 \times 10 \times 13 \times 16 \ldots (3n+4) \times (3n+7)} x^{n+1}$$

Now, $$\frac{u_n}{u_{n+1}} = \frac{3n+7}{3n+3} \times \frac{1}{x} = \left[\frac{1+\frac{7}{3n}}{1+\frac{1}{n}}\right]\frac{1}{x}$$

$\therefore$ $$\underset{n \to \infty}{Lt} \frac{u_n}{u_{n+1}} = \frac{1}{x}$$

Hence by D' Alembert's ratio test,

(i) If $\frac{1}{x} > 1$ *i.e.*, $x < 1$, then the series is convergent (*ii*) If $\frac{1}{x} < 1$ *i.e.*, $x > 1$, the series is divergent, and

(*iii*) If $\frac{1}{x} = 1$ *i.e.*, $x = 1$, then the test fails to decide the nature of the series.

When $x = 1$, then $\frac{u_n}{u_{n+1}} = \frac{3n+7}{3n+3}$

Applying Raabe's test, we get

$$n\left(\frac{u_n}{u_{n+1}} - 1\right) = n\left(\frac{3n+7}{3n+3} - 1\right) = n\,\frac{3n+7-3n-3}{3n+3}$$

$$= \frac{4n}{3(n+1)} = \frac{4}{3}\,\frac{1}{\left(1+\frac{1}{n}\right)}$$

$$\therefore \quad \underset{n\to\infty}{Lt} = n\left(\frac{u_n}{u_{n+1}} - 1\right) = \frac{4}{3} > 1, \; \Sigma u_n \text{ is convergent}$$

Hence the series in convergent if $x < 1$ and divergent if $x > 1$.

Solution : (*b*) $\frac{x}{1} + \frac{1}{2} \times \frac{x^3}{3} + \frac{1\times 3}{2\times 4} \times \frac{x^5}{5} + \frac{1\times 3\times 5}{2\times 4\times 6} \times \frac{x^7}{7} + \ldots .. \infty \; (x > 0)$

Omitting the first therm, then nth term are

$$u_n = \frac{1\times 3\times 5 \ldots . 2n-1}{2\times 4\times 6 \ldots . 2n} \times \frac{x^{2n+1}}{2^{n+1}}$$

and $$u_{n+1} = \frac{1\times 3\times 5 \ldots . 2n-1}{2\times 4\times 6 \ldots . 2n} \times \frac{2n-1}{2n+2} \times \frac{x^{2n+2}}{2n+3}$$

By D' Alembert's ratio test, $$\frac{u_n}{u_{n+1}} = \frac{x^{2n+1}}{2n+1} \times \frac{2n+2}{2n+1} \times \frac{2n+3}{x^{2n+3}}$$

$$= \frac{2n\left(1+\frac{1}{n}\right) 2n\left(1+\frac{3}{2n}\right)}{2n\left(1+\frac{1}{2n}\right) 2n\left(1+\frac{1}{2n}\right)} \frac{1}{x^2}$$

$$\underset{n\to\infty}{Lt} \frac{u_n}{u_{n+1}} = \frac{1}{x^2}$$

Hence by D' Alembert ratio test

(*i*) If $\frac{1}{x^2} > 1$ *i.e.*, $x^2 < 1$ then the series in convergent

(*ii*) If $\frac{1}{x^2} < 1$ *i.e.*, $x^2 > 1$, then the series in divergent and

(*iii*) If $\frac{1}{x^2} < 1$ *i.e.*, $x^2 > 1$, then the test fails to decide to nature of the series.

When $$x^2 = 1, \text{ then } \frac{u_n}{u_{n+1}} = \frac{\left(1+\frac{1}{n}\right)\left(1+\frac{3}{2n}\right)}{\left(1+\frac{1}{2n}\right)\left(1+\frac{1}{2n}\right)}$$

Applying Raabe's test, *i.e.,*

$$n\left(\frac{u_n}{u_{n+1}}-1\right) = n\left(\frac{2\,(n+1)\,2\,(n+3)}{(2n+1)\,(2n+1)}-1\right)$$

$$= n\left(\frac{(2n+2)\,(2n+3)-(2n+1)\,(2n+1)}{(2n+1)^2}\right)$$

$$= \frac{n}{(2n+1)}\,[4n^2+4n+6n+6-4n-1]$$

$$= \frac{n\,(6n+5)}{(2n+1)\,(2n+1)} = \frac{n^2\left(6+\frac{5}{n}\right)}{n^2\left(2+\frac{1}{n}\right)^2}$$

$\therefore \underset{n\to\infty}{Lt}\; n\left(\frac{u_n}{u_{n+1}}-1\right) = \frac{6}{(2)^2} = \frac{3}{2} > 1.$ therefore Σu_n is convergent

Hence the series is convergent when $x^2 < 1$ and divergent when $x^2 > 1$.

8.22 LOGARITHMIC TEST

Statement : If Σu_n is a positive term series such that $\underset{n\to\infty}{Lt}\left(n\log\frac{u_n}{u_{n+1}}\right) = k$, then

(*i*) The series is convergent if $k > 1$

(*ii*) The series is divergent if $k, < 1$, and

(*iii*) The test fails to decide the nature of the series if $k = 1$

Solution : Case I : When $k > 1$

Compare Sun and $\Sigma\frac{1}{n^p}$ if $1 < p < k$, Σu_n of convergent

$$\therefore \quad \frac{u_n}{u_n+1} > \frac{(n+1)^p}{n^p} = \left(1+\frac{1}{n}\right)^p \quad \text{...}(i)$$

Taking log on both sides of (*i*), we get

$$\log\frac{u_n}{u_n+1} > p\log\left(1+\frac{1}{n}\right)$$

$$\text{If } \log\frac{u_n}{u_n+1} > p\left(\frac{1}{n}-\frac{1}{2n^2}+\frac{1}{3n^2}-\frac{1}{4n^4}+\ldots\right)$$

$$\text{or } \log\frac{u_n}{u_n+1} > p\left(1-\frac{1}{2n}+\frac{1}{3n^2}-\frac{1}{4n^3}+\ldots\right)$$

$$\therefore \underset{n\to\infty}{Lt}\; n\left(\frac{u_n}{un+1}\right) > p \text{ i.e., } k > p \qquad \left[\because \underset{n\to\infty}{Lt}\; n\log\frac{u_n}{u_{n+1}} = k\right]$$

Hence, the series Σu_n is convergent.

Case II. When $k > 1$

Similarly, compare Σu_n and $\frac{1}{n^p}$ if $k < p < 1$, Σu_n is divergent

$$\therefore \quad \frac{u_n}{u_{n+1}} < \frac{(n+1)^p}{n^p} < \left(1+\frac{1}{n}\right)^p \qquad \text{...(ii)}$$

Taking log on both side of (*ii*), we have

$$\log \frac{u_n}{u_{n+1}} < p \log\left(1+\frac{1}{n}\right)$$

$$< p\left[\frac{1}{n} - \frac{1}{2n^2} + \frac{1}{3n^3} - \frac{1}{4n^4} + \ldots\right]$$

$$n \log \frac{u_n}{u_{n+1}} < p\left[1 - \frac{1}{2n}\right) + \frac{1}{3n^2} - \frac{1}{4n^3} + \ldots\right]$$

$$\therefore \quad \underset{n\to\infty}{Lt}\ n\left(\log \frac{u_n}{u_{n+1}}\right) < p \text{ i.e., } k > p \qquad \left(\because \underset{n\to\infty}{Lt}\ n \log \frac{u_n}{u_{n+1}} = k\right)$$

Hence Σu_n is divergent

Case III : When $k = 1$, then Logarithmic test fails to decide the nature of the series.

Note : When '*e*' is present in the result of '*D*' Alembert ratio test after taken limit, then we apply Logarithmic test in place of Raabe's test.

***Example* 8.22.** Test the convergence of the series

$$x + \frac{2^2 x^2}{21} + \frac{3^2 x^3}{3!} + \frac{4^4 x^4}{4!} + \ldots \infty$$

Solution :

$$x + \frac{2^2 x^2}{2!} + \frac{3^2 x^3}{3!} + \frac{4^4 x^4}{4!} + \ldots \infty$$

$$u_n = \frac{x^n + n^n}{n!}$$

and
$$u_{n+1} = \frac{(n+1)^{n+1}\, x^{n+1}}{(n+1)!}$$

Now,
$$\frac{u_n}{u_{n+1}} = \frac{x^n + n^n}{n!} \times \frac{(n+1)!}{(n+1)^{n+1}\, x^{n+1}}$$

$$= \frac{n^n\, x^n\, (n+1)\, n!}{n!\,(n+1)\left(1+\frac{1}{n}\right)^n x^n .x}$$

$$= \frac{1}{\left(1+\frac{1}{n}\right)^n} \frac{1}{x}$$

$$\therefore \quad \underset{n\to\infty}{Lt} \frac{u_n}{u_{n+1}} = \frac{1}{ex}$$

Hence by *D' Alembert* Ratio Test

(*i*) If $\frac{1}{ex} > 1$ or $x < \frac{1}{e}$, the series on convergent

(*ii*) If $\frac{1}{ex} < 1$ or $x > \frac{1}{e}$, the series is divergent, and

(*iii*) If $\frac{1}{ex} = 1$ or $x = \frac{1}{e}$, the fails to decide the nature of the series.

When $x = \frac{1}{e}$, then $$\frac{u_n}{u_{n+1}} = \frac{1}{\left(1+\frac{1}{n}\right)^n} \times e$$

Now apply Logarithmic test, for this test

$$\log \frac{u_n}{u_{n+1}} = \log \frac{e}{\left(1+\frac{1}{n}\right)^n}$$

$$= \log e - \log\left(1+\frac{1}{n}\right)^n$$

$$= 1 - n\log\left(1-\frac{1}{n}\right)$$

$$= 1 - n\left(\frac{1}{n} - \frac{1}{2n^2} + \frac{1}{3n^3} + \ldots\right)$$

and $$n\log\frac{u_n}{u_{n+1}} = \frac{1}{2} - \frac{1}{3n} + \ldots$$

$\therefore$ $$\underset{n\to\infty}{Lt}\ n\log\frac{u_n}{u_{n+1}} = \underset{n\to\infty}{Lt}\ \frac{1}{2} + 0$$

$$= \frac{1}{2} < 1$$

therefore, the series Σu_n is divergent

Hence the given series is convergent if $x < \frac{1}{e}$ and divergent if $x > \frac{1}{e}$.

***Example* 8.23.** Discuss the convergence of the series : $\frac{1!}{2}x + \frac{2!}{3^2 x} + \frac{3!}{4^3}x^3 + \frac{4!}{5^4}x^4 + \ldots \infty$

Solution : (*a*) $\frac{1!}{2} \times \frac{2!}{3^2}x^2 + \frac{3!}{4^3}x^3 + \frac{4!}{5^4}x^4 + \ldots \infty$

$$u_n = \frac{n!\,x^n}{(n+1)^n}$$

and $$u_{n+1} = \frac{(n+1)!\,x^{n+1}}{(n+2)^{n+1}}$$

Now, $$\frac{u_n}{u_{n+1}} = \frac{n!\,x^n}{(n+1)^n} \times \frac{(n+2)^{n+1}}{(n+1)n! \times x \times x^n} = \frac{(n+2)^{n+1}}{(n+1)^{n+1}} \times \frac{1}{x}$$

$$= \left(\frac{n+2}{n+1}\right)^{n+1} \frac{1}{x} = \left[\frac{1+\frac{2}{n}}{1+\frac{1}{n}}\right]^{n+1} \times \frac{1}{x}$$

$$= \frac{\left(1+\frac{2}{n}\right)^n \left(1+\frac{1}{n}\right)}{\left(1+\frac{1}{n}\right) \times \left(1+\frac{1}{n}\right)^n} \times \frac{1}{x}$$

$$\therefore \quad \underset{n\to\infty}{Lt} \frac{u_n}{u_{n+1}} = \frac{e^2}{e} \times \frac{1}{x}$$

$$= \frac{e}{x}$$

Hence by D' Alembert ratio test

(i) If $\frac{e}{x} > 1$ *i.e.,* $x < e$; the series is convergent

(ii) If $\frac{e}{x} < 1$ *i.e.,* $x > e$; the series is divergent, and

(iii) If $\frac{e}{x} = 1$ *i.e.,* $x = e$ then the test fails to decide to nature of the series.

When $x = e$, then
$$\frac{u_n}{u_{n+1}} = \frac{\left(1+\frac{2}{n}\right)^n \left(1+\frac{2}{n}\right)}{\left(1+\frac{1}{n}\right)\left(1+\frac{1}{n}\right)^n} \times \frac{1}{e}$$

As $\frac{u_n}{u_{n+1}}$ contains e; therefore Logarithmic test is application, for this test

$$n \log \frac{u_n}{u_{n+1}} = n \log \left\{\left(\frac{n+2}{n+1}\right)^{n+1} \times \frac{1}{e}\right\} = n \log \left(\frac{n+2}{n+1}\right)^{n+1} - n \log e$$

$$= n\,(n+1) \log \left(\frac{n+2}{n+1}\right) - n$$

$$= n\,(n+1) \log \left(1+\frac{1}{n+1}\right) - n$$

$$= n\,(n+1) \left[\frac{1}{n+1} - \frac{1}{2}\left(\frac{1}{n+1}\right)^2 + \frac{1}{3}\left(\frac{1}{n+1}\right)^3 - \frac{1}{4}\left(\frac{1}{n+1}\right)^4 \ldots\right] - n$$

$$= n\left[1 - \frac{1}{2\,(n+1)} + \frac{1}{3}\left(\frac{1}{n+1}\right)^2 + \frac{1}{4}\left(\frac{1}{n+1}\right)^3 + \ldots\ \infty\right] - n$$

$$= \frac{-n}{n+1}\left[\frac{1}{2} - \frac{1}{3}\left(\frac{1}{n+1}\right) - \frac{1}{4}\left(\frac{1}{n+1}\right)^2 - \ldots\ \infty\right]$$

$$= \frac{-1}{1+\frac{1}{n}}\left[\frac{1}{2} - \frac{1}{3}\left(\frac{1}{n+1}\right) - \frac{1}{4}\left(\frac{1}{n+1}\right)^2 + \ldots\ \infty\right]$$

$\therefore \underset{n\to\infty}{Lt}\ n \log \frac{u_n}{u_{n+1}} = -1\left(\frac{1}{2}\right) = \frac{-1}{2} < 1.$ therefore, Σu_n is divergent.

Hence the series is convergent if $x < e$ and divergent if $x \geq e$.

Example **8.24.** Discuss the convergence of the series $\frac{a+x}{1!}+\frac{(a+2x)^2}{2!}+\frac{(a+3x)^3}{3!}+\frac{(a+4x)^4}{4!}+....\infty$

Solution : $\frac{a+x}{1!}+\frac{(2+2x)^2}{2!}+\frac{(a+3x)^3}{3!}+....\infty$

$$\therefore \qquad u_n = \frac{(a+nx)^n}{n!}$$

and
$$u_{n+1} = \frac{(a+(n+1)x)^{n+1}}{(n+1)!}$$

By D' Alembert ratio test,
$$\frac{u_n}{u_{n+1}} = \frac{(a+nx)^n}{n!} \times \frac{(n+1)\,n!}{(a+(n+1)\,x)^{n+1}}$$

$$= \frac{x^n n^n \left(1+\frac{a}{nx}\right)^n}{(n+1)^{n+1} x^{n+1}\left[1+\frac{a}{(n+1)x}\right]^{n+1}} \times (x+1)$$

$$= \frac{n^n\left[1+\frac{a}{nx}\right]^n}{(n+1)^n\left[1+\frac{a}{(n+1)x}\right]^{n+1}} \times \frac{1}{x}$$

$$= \frac{\left[\left(1+\frac{a}{nx}\right)^{\frac{nx}{a}}\right]^{\frac{a}{x}}}{\left(1+\frac{1}{n}\right)^n\left[\left\{1+\frac{a}{(n+1)x}\right\}^{\frac{(n+1)x}{a}}\right]^{\frac{a}{x}}} \times \frac{1}{x}$$

$$\underset{n\to\infty}{Lt} \frac{u_n}{u_{n+1}} = \frac{e^{\frac{a}{x}}}{e \times e^{\frac{a}{x}}} \times \frac{1}{x} = \frac{1}{ex}$$

Hence by D' Alembert' ratio test

(*i*) If $\frac{1}{ex} > 1$ or $x < \frac{1}{e}$, then the series is convergent

(*ii*) If $\frac{1}{ex} < 1$ or $x > \frac{1}{e}$, then the series is divergent

(*iii*) If $\frac{1}{ex} = 1$ or $x = \frac{1}{e}$, then the test fails to decide the nature of the series.

When
$$x = \frac{1}{e} \text{ then } \frac{u_n}{u_{n+1}} = \frac{e\left(1+\frac{ae}{n}\right)^n}{\left(1+\frac{1}{n}\right)^n\left(1+\frac{ae}{(n+1)}\right)^{n+1}}$$

Applying Logarithmic test *i.e.*,

$$n \log \frac{u_n}{u_{n+1}} = n \log\left\{\frac{e\left(1+\frac{ae}{n}\right)^n}{\left(1+\frac{1}{n}\right)^n\left(1+\frac{ae}{n+1}\right)^{n+1}}\right\}$$

$$= n\left\{\log e + n\log\left(1+\frac{ae}{n}\right) - n\log\left(1+\frac{1}{n}\right) - (n+1)\log\left(1+\frac{ae}{n+1}\right)\right\}$$

$$= \left\{1 + n\left(\frac{ae}{n} - \frac{a^2e^2}{2n^2} + \frac{a^3e^3}{3n^3} + ...\right) - n\left(\frac{1}{n} - \frac{1}{2n^2} + \frac{1}{3n^3} - \frac{1}{2n^4} + ...\right) - (n+1)\left(\frac{ae}{n+1} - \frac{a^2e^2}{2(n+1)^2} + \frac{a^2e^3}{3(n+1)^3}....\right)\right\}$$

$$= n\left\{1 + \left(ae - \frac{a^2e^2}{2n} + \frac{a^3e^3}{3n^2}....\right) - \left(1 - \frac{1}{2n} + \frac{1}{3n^2} - \frac{1}{4n^3} + ...\right) - \left(ae - \frac{a^2e^2}{2(n+1)} + \frac{a^3e^3}{3(n+1)^2} - ...\right)\right\}$$

$$= -\frac{a^2e^2}{2} + \frac{a^3e^3}{3n} + \frac{1}{2} - \frac{1}{3n}.... + \frac{a^2e^2}{2\left(1+\frac{1}{n}\right)} - \frac{a^3e^3}{3(n+1)\left(1+\frac{1}{n}\right)} + ...$$

$$\therefore \quad \underset{n\to\infty}{Lt}\ n\log\frac{u_n}{u_{n+1}} = \frac{1}{2} < 1,\ \Sigma u_n \text{ is divergent}$$

Hence by Lograthmic test the series is divergent when $x \geq \frac{1}{e}$ and convergent when $x < \frac{1}{e}$.

***Example* 8.25.** Test for convergence of the series $x^2 (\log 2)^q + x^3 (\log 3) (\log 3)^q + x^4 (\log 4)^q + ... \infty$

Solution : Adding one more then we get $x (\log 1)^q + x^2 (\log 2)^q + x^3 (\log 3)^q + x^4 (\log 4)q + ... \infty$

$$\therefore \quad u_n = x^n (\log n)^q, \text{ and } u_{n+1} = x_{n+1} (\log n + 1)^q$$

Now

$$\frac{u_n}{u_{n+1}} = \frac{(\log)^q}{[\log (n+1)]^q} \times \frac{1}{x}$$

$$= \frac{(\log n)^q}{\left[\log n\left(1+\frac{1}{n}\right)\right]^q} \times \frac{1}{x}$$

$$= \frac{(\log n)^q}{\left[\log n + \log\left(1+\frac{1}{n}\right)\right]^q} \times \frac{1}{x}$$

$$\therefore \quad \underset{n\to\infty}{Lt}\ \frac{u_n}{u_{n+1}} = \frac{1}{x} \qquad \left[\because \underset{n\to\infty}{Lt}\ \frac{1}{\log n} = 0 \text{ and } \underset{n\to\infty}{Lt}\ \frac{1}{n} = 0\right]$$

Hence, by D' Alembert Ratio Test,

(*i*) If $\frac{1}{x} > 1$ *i.e.*, $x <$ the series is convergent

(*ii*) If $\frac{1}{x} < 1$ *i.e.*, $x > 1$ the series is divergent, and

(*iii*) If $\frac{1}{x} = 1$. *i.e.*, then test fails to decide the nature of the series.

When

$$x = 1\ \frac{u_n}{u_{n+1}} = \frac{1}{\left[1 + \frac{1}{\log n}\log\left(1+\frac{1}{n}\right)\right]^q}$$

$\therefore$ Apply Logarithmic test, for this test

$$n \log \frac{u_n}{u_{n+1}} = -q\, n \log \left[1 + \frac{1}{\log n}\left(\frac{1}{n} - \frac{1}{2n^2} + \frac{1}{3n^3} \ldots.\right)\right] \text{on expanding } \log\left(1+\frac{1}{n}\right)$$

$$= -q\, n \log \left[1 + \left(\frac{1}{n \log n} - \frac{1}{2n^2 \log n} + \frac{1}{3n^3 \log n} + \ldots.\right)\right]$$

$$= -q\, n \log (1 + z)$$

where
$$z = \frac{1}{n \log n} - \frac{1}{2n^2 \log n} + \frac{1}{3n^3 \log n} \ldots.$$

$$n \log \frac{u_n}{u_{n+1}} = -q\, n \left[z - \frac{z^2}{2} + \frac{z^3}{3} \ldots.\right]$$

$$= -qn \left[\left(\frac{1}{n \log n} - \frac{1}{2n^2 \log n} + \ldots\right) - \frac{1}{2}\left(\frac{1}{n \log n} - \frac{1}{2n^2 \log n} + \ldots\right)^2 + \ldots\right]$$

$$= \frac{-q}{\log n} + \text{all such terms which contain } n^2 \text{ and so on in Denominator}$$

$$\therefore \quad \underset{n \to}{Lt}\; n \log \frac{u_n}{u_{n+1}} = \underset{n \to \infty}{Lt} \frac{-q}{\log n} = 0 < 1, \quad \therefore \Sigma u_n \text{ is divergent}$$

Hence, the given series in divergent of $x > 1$ and convergent of $x < 1$.

***Example* 8.26.** Test for convergence of the series $x + \frac{2^2 x^2}{2!} + \frac{3^3 x^3}{3!} + \frac{4^4 x^4}{4!} + \frac{5^5 x^5}{5!} + \ldots.$

Solution :
$$u_n = \frac{n^n x^n}{n!} \text{ and } u_{n+1} = \frac{(n+1)^{n+1} x^{n+1}}{(n+1)!}$$

Now
$$\frac{u_n}{u_{n+1}} = \frac{1}{x} \times \frac{n^n}{n!} \times \frac{(n+1)!}{(n+1)^{n+1}}$$

$$= \frac{1}{x} \times \frac{n^n (n+1)}{(n+1)^{n+1}}$$

$$= \frac{1}{x} \times \frac{n^n}{(n+1)^n}$$

$$= \frac{1}{x} \times \frac{1}{\left\{1 + \frac{1}{n}\right\}^n}$$

$$\therefore \quad \underset{n \to \infty}{Lt} \frac{u_n}{u_{n+1}} = \frac{1}{ex} \qquad \left[\because \underset{n \to \infty}{Lt}\left(1 + \frac{1}{n}\right)^n = e\right]$$

Hence by D' Alembert ratio test

(*i*) If $\frac{1}{ex} >$ *i.e.*, $x < \frac{1}{e}$, then the series in convergent.

(*ii*) If $\frac{1}{ex} < 1$ *i.e.*, $x > \frac{1}{e}$, then the series in divergent.

(*iii*) If $\frac{1}{ex} = 1$ *i.e.*, $x = \frac{1}{e}$ then the test fails to decide the nature of the series.

When $x = \frac{1}{e}$, then $\frac{u_n}{u_{n+1}} = \frac{e}{\left(1+\frac{1}{n}\right)^n}$

Now applying log test in this case *i.e.*,

$$n \log \frac{u_n}{u_{n+1}} = n \log \frac{e}{\left(1+\frac{1}{n}\right)^n}$$

$$n \log \frac{u_n}{u_{n+1}} = n\left[\log e - n \log\left(1+\frac{1}{n}\right)\right]$$

$$= n\left[1 - n\left(\frac{1}{n} - \frac{1}{2n^2} + \frac{1}{3n^3} \ldots\right)\right]$$

$$= n\left[1 - 1 + \frac{1}{2n} - \frac{1}{3n^2} + \ldots\right] = \frac{1}{2} - \frac{1}{3n} + \ldots.$$

$\therefore \quad \underset{n\to\infty}{Lt} \; n \log \frac{u_n}{u_{n+1}} = \frac{1}{2} < 1$, therefore the series is divergent. Hence by log test, the series in convergent if $x < \frac{1}{e}$ and divergent of $x < \frac{1}{e}$.

8.23 GAUSS'S TEST

Statement. If Σu_n is positive term series such that

$$\frac{u_n}{u_{n+1}} = \alpha + \frac{\beta}{n} + \frac{\gamma}{n^2} \text{ where } \alpha > 0, \text{ then}$$

(*i*) The series is convergent if $\alpha > 1$

(*ii*) The series is divergent if $\alpha < 1$

whatever β may be

(*iii*) When $\alpha = 1$, then $\begin{cases} \text{if } b > 1 \text{ the series is convergent} \\ \text{if } b < 1, \text{ the series is divergent} \end{cases}$

Note : Gauss test never fails as we known that the given series is divergent for $\beta = 1$.

Moreover, the test is applied after the failure of higher ratio test (Raabe's test) only when it is possible to expand to expand $\frac{u_{n}}{u_{n+1}}$ by Binomial theorem.

Example **8.27.** Test for convergence of the series $\frac{2^2 \times 4^2}{3^2 \times 5^2} + \frac{2^2 + 4^2 \times 6^2}{3^2 \times 5^2 \times 7^2} + \frac{2^2 \times 4^2 \times 6^2 \times 8^2}{3^2 \times 5^2 \times 7^2 \times 9^2} + \ldots..$

Solution : $\frac{2^2 \times 4^2}{3^2 \times 5^2} + \frac{2^2 \times 4^2 \times 6^2}{3^2 \times 5^2 \times 7^2} + \ldots\ldots + \frac{2^2 \times 4^2 \times 6^2 \ldots. (2n+2)^2}{3^2 \times 5^2 \times 7^2 \ldots. (2n+3)^2} + \ldots. \infty$

$$\therefore \quad u_n = \frac{2^2 \times 4^2 \times 6^2 \ldots. (2n+2)^2}{3^2 \times 5^2 \times 7^2 \ldots. (2n+3)^2}$$

and

$$u_{n+1} = \frac{2^2 \times 4^2 \times 6^2 \ldots. (2n+2)^2 (2n+5)^2}{3^2 \times 5^2 \times 7^2 \ldots. (2n+3)^2 (2n+5)^2}$$

Now $$\frac{u_n}{u_n+1} = \frac{(2n+5)^2}{(2n+4)^2}$$

$$= \frac{4n^2+25+20n}{4n^2+16+16n}$$

$$= \frac{4+\frac{20}{n}+\frac{25}{n^2}}{4+\frac{16}{n}+\frac{16}{n^2}}$$

$\therefore \quad \underset{n\to\infty}{Lt} \frac{u_n}{u_{n+1}}$ =1, therefore *D'* Alembert ratio test fails

We apply Raabe's test, *i.e.*,

$$\underset{n\to\infty}{Lt}\ n\left(\frac{u_n}{u_{n+1}}-1\right) = \underset{n\to\infty}{Lt}\ n\left[\frac{4n^2+20\,n+25}{4n^2+16n+16}\right]$$

$$= \underset{n\to\infty}{Lt}\ n\left[\frac{4n+9}{4n^2+16n+16}\right]$$

We apply Raabe's, *i.e.*,

$$\underset{n\to\infty}{Lt}\ n\left(\frac{u_n}{u_{n+1}}-1\right) = \underset{n\to\infty}{Lt}\ n\left[\frac{4n^2+20n+25}{4n^2+16n+16}-1\right]$$

$$= \underset{n\to\infty}{Lt}\ n\left[\frac{4n+9}{4n^2+16n+16}\right]$$

$$= \underset{n\to\infty}{Lt}\left[\frac{4n^2+9n}{4n^2+16n+16}\right]$$

$$= \underset{n\to\infty}{Lt}\left[\frac{4+\frac{9}{n}}{4+\frac{16}{n}+\frac{16}{n^2}}\right]$$

= 1, Raabe's test fails

Now apply Gauss test, for this test

$$\frac{u_u}{u_n+1} = \frac{(2n+5)^2}{(2n+4)^2}$$

$$= \left(1+\frac{5}{2n}\right)^2\left(1+\frac{2}{n}\right)^{-2}$$

$$= \left(1+\frac{5}{2n}\right)^2\left(1-\frac{4}{n}+\frac{-2\times-3}{2!}\times\frac{4}{n^2}+....\right)$$

(On expanding by Binomial theorem)

$$= \left(1+\frac{5}{n}+\frac{25}{4n^2}\right)\left(1+\frac{4}{n}+\frac{12}{n^2}+....\right)$$

$$= 1-\frac{4}{n}+\frac{12}{n^2}+\frac{5}{n}-\frac{20}{n^2}+\frac{25}{4n^2}+....$$

$$= 1 + \frac{1}{n} - \frac{7}{n^2}$$

$$= \alpha + \frac{b}{n} + \frac{\gamma}{n^2}$$

Here $\alpha = 1$, $\beta = 1$, thus the series is divergent

Hence by Gauss test, the series is divergent.

Example **8.28.** Test for convergence of the series :

$$1 + \frac{\alpha \times \alpha}{1 \times r} x + \frac{\alpha(\alpha+1)\beta(\beta+1)}{1 \times 2 \times r(r+1)} x^2 + \frac{\alpha(\alpha+1)(\alpha+2)\beta(\beta+1)(\beta+1)}{1 \times 2 \times 3 \times r(r+1)(r+2)} x^3 + \frac{\alpha(\alpha+1)(\alpha+2)(\alpha+3)\beta(\beta+1)(\beta+2)(\beta+3)}{1 \times 2 \times 3 \times r(r+1)(r+2)(r+3)}$$

Solution : Omitting the first term, then

$$u_n = \frac{\alpha(\alpha+1)(\alpha+2)\ldots(\alpha+n-1)\beta(\beta+1)(\beta+2)\ldots(\beta+n-1)}{1 \times 2 \times 3 \ldots n \times r(r+1)(r+2)(r+3)\ldots(r+n-1)} \times x^n$$

and
$$u_{n+1} = \frac{\alpha(\alpha+1)(\alpha+2)\ldots(\alpha+n-1)(\alpha+n)\beta(\beta+1)(\beta+2)\ldots(\beta+n-1)(\beta+n)}{1 \times 2 \times 3 \ldots n(n+1)r(r+1)(r+2)(r+3)\ldots(r+n-1)(r+n)} \times x^{n+1}$$

Now
$$\frac{u_n}{u_{n+1}} = \frac{(n+1)(r+n)}{(\alpha+n)(\beta+n)} \times \frac{1}{x}$$

$$= \frac{\left(1+\frac{1}{n}\right)\left(1+\frac{r}{n}\right)}{\left(1+\frac{\alpha}{n}\right)\left(1+\frac{\beta}{n}\right)} \times \frac{1}{x}$$

$\therefore$
$$\underset{n \to \infty}{Lt} \frac{u_n}{u_{n+1}} = \frac{1}{x}$$

Hence by D' Alembert ratio test,

(*i*) If $\frac{1}{x} > 1$ *i.e.,* $x < 1$, then the series is convergent

(*ii*) If $\frac{1}{x} < 1$ *i.e.,* $x > 1$ then the series is divergent

(*iii*) and If $\frac{1}{x} = 1$ *i.e.,* $x = 1$, then the test fails to decide the nature of the series.

When $x = 1$, then
$$\frac{u_n}{u_{n+1}} = \frac{\left(1+\frac{1}{n}\right)\left(1+\frac{r}{n}\right)}{\left(1+\frac{\alpha}{n}\right)\left(1+\frac{\beta}{n}\right)}$$

$$= \left(1+\frac{1}{n}\right)\left(1+\frac{r}{n}\right)\left(1+\frac{\alpha}{n}\right)^{-1}\left(1+\frac{\beta}{n}\right)^{-1}$$ (On expanding by binomial theorem)

$$= \left(1+\frac{1}{n}\right)\left(1+\frac{r}{n}\right)\left[1-\frac{\alpha}{n}+\frac{\alpha^2}{n^2}+\ldots\right]\left[1-\frac{\beta}{n}+\frac{\beta^2}{n^2}+\ldots\right]$$

$$= \left(1+\frac{1}{n}+\frac{r}{n}+\frac{r}{n^2}\right)\left(1-\frac{\alpha}{n}-\frac{\beta}{n}-\frac{\alpha\beta}{n^2}+\frac{\alpha^2}{n^2}+\ldots\right)$$

$$= 1 + \frac{1}{n}(2 + r - \alpha - \beta) + O\left(\frac{1}{n^2}\right)$$

Hence by Gauss test

(i) If $1 + r - \alpha - \beta >$ *i.e.*, $r > \alpha + \beta$, then the series is convergent

(ii) If $1 + r - \alpha - \beta < 1$ *i.e.*, $r < b$, then the series is divergent

Thus the given series is convergent if $x < 1$ and divergent if $x > 1$ and if $x = 1$, series is convergent is $r > a + b$ and divergent if $r \le \alpha + \beta$.

EXERCISE 8.6

Discuss the nature of the following series

1. $\left(\frac{1}{3}\right)^2 + \left(\frac{1.2}{3.5}\right)^2 + \left(\frac{1.2.3}{3.5.7}\right)^2 + ... \infty$ **Ans.** Convergent.

2. $\frac{2}{1} + \frac{2.5.8}{1.5.9} + \frac{2.5.8.11}{1.5.9.13} + ... \infty$ **Ans.** Convergent

3. $1 + \frac{3}{7}x + \frac{3.6}{7.10}x^2 + \frac{3.6.9}{7.10.13}x^3 + \frac{3.6.912}{7.10.13.16} + x^4 + ...$ **Ans.** The series is convergent

4. $\frac{x}{1.2} + \frac{x^2}{3.4} + \frac{x^3}{5.6} + \frac{x^4}{7.8} + ... \infty$ **Ans.** if $x \le 1$ convergent, if $x > 1$ divergent

5. $1 + \frac{2^2}{3^2} + \frac{2^2.4^2}{3^2.5^2} + \frac{2^2.4^2.6^2}{3^2.5^2.7^2.9^2} + ... \infty$ **Ans.** Divergent

6. $x^2 + \frac{2^2 x^2}{2!} + \frac{3^3 x^3}{3!} + \frac{4^4 x^4}{4!} + \frac{5^5 x^5}{5!} + ...$ **Ans.** Convergent for $x < \frac{1}{e}$, Divergent for $x \ge \frac{1}{e}$

7. $\frac{a}{b} + \frac{a(a+1)}{b(b+1)} + \frac{a(a+1)(a+2)}{b(b+1)(b+2)} + ... \infty$ **Ans.** Convergent for $b > a + 1$, Divergent for $b < a + 1$

8. $u_n = \frac{1 \times 3 \times 5 .... (2n-1)}{2 \times 4 \times 6 .... 2n} \times \frac{x^{2n}}{2n}, x > 0$ **Ans.** Convergent for $x^2 < 1$, Divergent for $x^2 > 1$

9. $1 + \frac{(1!)^2}{2!}x^2 + \frac{(2!)^2}{4!}x^4 + \frac{(3!)^2}{6!}x^6 + ... \infty \; (x > 0)$ **Ans.** Convergent for $x^2 < 4$,.Divergent for $x^2 \ge 4$

10. $\frac{x^2}{2\log 2} + \frac{x^3}{3\log 3} + \frac{x^4}{4\log 4} + ... \infty$

8.24 CAUCHY'S ROOT TEST

When power are involved in u_n then Cauchy root test is used.

Statement : If Sun is a positive term series such that $\underset{n \to \infty}{Lt} (u_n)^{\frac{1}{n}} = k$, then

(i) If, $k < 1$ the series is convergent

(ii) If $k, > 1$, the series is divergent

(iii) and If $k = 1$, the test fails

Proof : By definition of limit

$$|u_n^{\frac{1}{n}} - k| < e \text{ for } n > p$$

$$k - \in < (u_n)^{\frac{1}{n}} < k + \in \text{ for } n > p$$

Case I. When $k > 1$

$$k + \in < r < 1$$

$$(u_n)^{\frac{1}{n}} < k$$

Taking nth power on both sides, then therefore, so the series is divergent $u_n < k^n$, which implies that therefore, the series is convergent

$$< \frac{1}{1-k} \text{ (which is finite quanity)}$$

Case II : When $k > 1$

$$k - e > 1$$

$$(u_n)^{\frac{1}{n}} > k - \in > 1$$

$$u_n > 1$$

$$S_n = u_1 + u_2 + u_3 + ... + u_n > n$$

$$\therefore \underset{n \to \infty}{Lt} S_n \to \infty$$

Case III. When $k = 1$

Then test fails *i.e.,* no conclusion can be drawn about the converge or divergence of the series.

Working rule : First of all find u_n then find $(u_n)^{\frac{1}{n}}$ and taking limit as $n \to 8$ then.

(*i*) If $\underset{n \to \infty}{Lt} (u_n)^{\frac{1}{n}} < 1$, the series is convergent

(*ii*) If $\underset{n \to \infty}{Lt} (u_n)^{\frac{1}{n}} > 1$. the series is divergent

(*iii*) If $\underset{n \to \infty}{Lt} (u_n)^{\frac{1}{n}} = 1$, then test fails and do decide the nature of the series.

Apply comparison test or *D'* Alembert Ratio test which is suitable.

***Example* 8.29.** Test for convergence of the following series

(*a*) $\sum\left(1+\frac{1}{\sqrt{n}}\right)^{-n^{\frac{3}{2}}}$ or $\frac{1}{\sum\left(1+\frac{1}{\sqrt{n}}\right)^{n^{\frac{3}{2}}}}$

(*b*) $\sum \frac{x^n}{n!}$

(*c*) $\frac{1^3}{3} + \frac{2^3}{3^2} + \frac{3^3}{3^3} + \frac{4^3}{3^4} + ... + \frac{n^3}{3^n} + \infty ....$

Solution : (*a*) $u_n = \dfrac{1}{\sum\left(1+\frac{1}{\sqrt{n}}\right)^{n^{\frac{3}{2}}}}$

Since power is involved in u_n the Cauchy root test is applicable.

$$\therefore \qquad (u_n)^{\frac{1}{4}} = \frac{1}{\sum\left(1+\frac{1}{\sqrt{n}}\right)^{\sqrt{n}}}$$

$$\text{and} \qquad \underset{n\to\infty}{Lt}\,(u_n)^{\frac{1}{4}} = \underset{n\to\infty}{Lt}\,\frac{1}{\sum\left(1+\frac{1}{\sqrt{n}}\right)^{\sqrt{n}}}$$

$$= \frac{1}{e}, \text{ which is} < 1 \qquad \left\{\because \underset{n\to\infty}{Lt}\left(1+\frac{1}{n}\right)^n = e\right\}$$

Hence, by Cauchy root test the given series is convergent.

(*b*)
$$u_n = \frac{x^n}{n!}$$

Since power is involved in u_n the Cauchy root test in applicable

$$\text{Now} \qquad (u_n)^{\frac{1}{n}} = \frac{x}{(n!)^{\frac{1}{n}}}$$

$$\therefore \qquad \underset{n\to\infty}{Lt}\,(u_n)^{\frac{1}{n}} = \underset{n\to\infty}{Lt}\,\frac{x}{(n!)^{\frac{1}{n}}}$$

$$= \underset{n\to\infty}{Lt}\left\{\frac{n}{(n!)^{\frac{1}{n}}}\times\frac{x}{n}\right\}$$

$$= \underset{n\to\infty}{Lt}\left\{\frac{(n^n)^{\frac{1}{n}}}{(n!)^{\frac{1}{n}}}\times\frac{x}{n}\right\}$$

$$= e\times\underset{n\to\infty}{Lt}\,\frac{x}{n} \qquad \left\{\because \underset{n\to\infty}{Lt}\left(\frac{n^n}{n!}\right)^{\frac{1}{n}} = e\right\}$$

$$= 0 < 1, \text{(for any value of } x)$$

Hence by Cauchy's root test the given series is convergent.

(*c*) $\sum \frac{n^3}{3^n}$

$$\therefore \qquad u_n = \frac{n^3}{3^n}$$

Since power is involved in u_n, the Cauchy root test is applicable

$$\text{Now} \qquad (u_n)^{\frac{1}{n}} = \frac{(n)^{\frac{3}{n}}}{3}$$

$$\therefore \qquad \underset{n\to\infty}{Lt}\,(u_n)^{\frac{1}{n}} = \underset{n\to\infty}{Lt}\,\frac{(n)^{\frac{3}{n}}}{3}$$

$$= \frac{1}{3} \underset{n\to\infty}{Lt} \left(n^{\frac{1}{n}}\right)^3 \qquad \left\{\because \underset{n\to\infty}{Lt} (n)^{\frac{1}{n}} = 1\right\}$$

$$= \frac{1}{3} < 1$$

Hence by Cauhy's root test, the given series is convergent.

***Example* 8.30.** Test for convergence of the series $\left(\frac{2^2}{1^2}-\frac{2}{1}\right)^{-1}+\left(\frac{3^3}{2^3}-\frac{3}{2}\right)^{-2}+\left(\frac{4^4}{3^4}-\frac{4}{3}\right)^{-3}+....$

Solution : $\left(\frac{2^2}{1^2}-\frac{2}{1}\right)^{-1}+\left(\frac{3^3}{2^3}-\frac{3}{2}\right)^{-2}+\left(\frac{4^4}{3^4}-\frac{4}{3}\right)^{-3}+.....$

$$\therefore \qquad u_n = \left[\frac{(n+1)^{n+1}}{n^{n+1}}-\frac{n+1}{n}\right]^{-n}$$

Since power is involved in u_n, the Cauchy root test is applicable

Now
$$(u_n)^{\frac{1}{n}} = \left[\frac{(n+1)^{n+1}}{n^{n+1}}-\frac{n+1}{n}\right]^{-1}$$

$$= \left(\frac{n+1}{n}\right)^{-1}\left[\left(\frac{n+1}{n}\right)^{n}-1\right]^{-1}$$

Hence
$$\underset{n\to\infty}{Lt} (u_n)^{\frac{1}{n}} = \underset{n\to\infty}{Lt} \left(1+\frac{1}{n}\right)^{-1}\left[\left(1+\frac{1}{n}\right)^{n}-1\right]^{-1}$$

$$= (1+0)^{-1}(e-1)^{-1}$$

$$= \frac{1}{e-1} < 1 \qquad [\because 2 < e < 3]$$

Hence by Cauchy root test the given series is convergent.

***Example* 8.31.** Test the following series for convergent or divergence

(*a*) $\frac{1}{2}+\frac{2}{3}x+\left(\frac{3}{4}\right)^2 x^2+\left(\frac{4}{5}\right)^3 x^3 .... \infty \; (x > 0)$

(*b*) $\frac{2}{1^2}+\frac{3^2}{2^3}x^2+\frac{4^3}{3^4}x^3+...+\frac{(n+1)^n}{n^{n+1}}x^n+....\infty$

Solution : (*a*) $\frac{1}{2}+\frac{2}{3}x+\left(\frac{3}{4}\right)^2 x^2+\left(\frac{4}{5}\right)^3 x^3+....\infty$

Omitting the first term $\frac{2}{3}x+\left(\frac{3}{4}\right)^2 x^2+\left(\frac{4}{5}\right)^3 x^3+....\infty$

$$u_n = \left[\frac{n+1}{n+2}\right]^n x^n$$

Since power is involved in u_n, Cauchy's root test is applicable

Now
$$(u_n)^{\frac{1}{n}} = \frac{n+1}{n+2}\times x$$

$$= \frac{\left(1+\frac{1}{n}\right)}{\left(1+\frac{2}{n}\right)} \times x$$

Hence $$\underset{n\to\infty}{Lt} (u_n)^{\frac{1}{n}} = \underset{n\to\infty}{Lt} \left(\frac{1+\frac{1}{n}}{1+\frac{2}{n}}\right) + x$$

$$= x$$

Therefore, by Cauchy's root test

(*i*) If $x < 1$, the series is convergent

(*ii*) If $x > 1$, the series is divergent

(*iii*) If $x = 1$, the fails to decide the nature of series

When $x = 1$, then

$$u_n = \left(\frac{n+1}{n+2}\right)^n = \frac{\left(1+\frac{1}{n}\right)^n}{\left(1+\frac{2}{n}\right)^n}$$

$$\underset{n\to\infty}{Lt}\ u_n = \underset{n\to\infty}{Lt} \frac{\left(1+\frac{1}{n}\right)^n}{\left[\left(1+\frac{2}{n}\right)^{\frac{n}{2}}\right]^2} = \frac{e}{e^2} = \frac{1}{e}$$, which is finite and non zero therefore Σu_n is divergent

Hence the given series is convergent if $x > 1$ and divergent if $x \geq 1$.

(*b*) $\frac{2}{1^2} + \frac{3^2}{2^3}x^2 + \frac{4^3}{3^4}x^3 + ... + \frac{(n+1)^n}{n^{n+1}}x^n + ... \infty$

$$\therefore \qquad u_n = \frac{(n+1)^n}{n^{n+1}} \times n^n$$

Since power is involved in u_n, the Cauchy roots test is applicable

$$(u_n)^{\frac{1}{n}} = \frac{(n+1)x}{n} \times \frac{1}{(n)^{\frac{1}{n}}}$$

$$\therefore \qquad \underset{n\to\infty}{Lt} (u_n)^{\frac{1}{n}} = \left[\underset{n\to\infty}{Lt} \left(1+\frac{1}{n}\right)\right] \times \left[\underset{n\to\infty}{Lt} \frac{1}{n^{\frac{1}{n}}}\right] x$$

$$= (1+0)\,x = x$$

$\therefore$ By Cauchy root test

(*i*) If $x < 1$, the series is convergent

(*ii*) If $x > 1$, the series is divergent and

(*iii*) If $x = 1$, test fails to decide the nature of series

When $x = 1$, then

$$u_n = \frac{(n+1)^n}{n^{n+1}}$$

$$= \frac{1}{n}\left(1+\frac{1}{n}\right)^n$$

Let $$v_n = \frac{1}{n}$$

Applying Comparison test, we get

$$\frac{u_n}{v_n} = \frac{1}{n}\left(1+\frac{1}{n}\right)^n \times \frac{n}{1}$$

$$\therefore \quad \underset{n\to\infty}{Lt}\frac{u_n}{v_n} = \underset{n\to\infty}{Lt}\left(1+\frac{1}{n}\right) = e \text{ which is finite and non zero.}$$

therefore, by comparison test Σu_n and Σv_n converge or diverge together

since $$v_n = \frac{1}{n}, p = 1$$

$\therefore$ Σv_n is divergent, therefore, Σu_n is also divergent

Hence Σv_n is convergent if $x < 1$ and divergent of $x \geq 1$.

EXERCISE 8.7

Discuss the convergence/divergence of the following series

1. $\Sigma\frac{1}{n^n}$ **Ans.** Convergent
2. $\Sigma\frac{1}{\left(1+\frac{1}{n}\right)n^2}$ **Ans.** Convergent
3. $\Sigma\left(1+\frac{1}{\sqrt{n}}\right)^{-n\frac{3}{2}}$ **Ans.** Convergent
4. $1+\frac{x}{2}+\frac{x^2}{3^2}+\frac{x^3}{4^3}+ \ldots \infty \; (x>0)$ **Ans.** Convergent
5. $\Sigma\left(\frac{nx}{n+1}\right)^n$ **Ans.** Convergent for $x<1$. divergent for $x \geq 1$
6. Test the convergence of the series $x + 2x^2 + 3x^3 + 4x^3 + \ldots . \infty$. **Ans.** Convergent for $x > 1$, divergent for $x \geq 1$
7. $\sum_{n=2}^{\infty}\frac{1}{(\log n)^n}$ **Ans.** (*i*) Convergent

   $\sum_{n=1}^{\infty}\frac{(1+nx)^n}{n^n}$ **Ans.** (*ii*) Convergent for $x<1$. divergent for $x \geq 1$

8. 25 INTEGRAL TEST

Statement : A positive term series $f(1)+f(2)+f(3)+ \ldots f(n) + \ldots .$ where $f(n)$ decreases monotonically, converges or diverges according to the integral $\int_1^{\infty} f(x)\,dx$ is finite or infinite.

$$f(1) + f(2) + f(3) + \ldots + f(n) > \int_1^{n+1} f(x)\,dx$$

$$> f(2) + f(3) + f(4) + \ldots + f(n+1)$$

$$S_n > \int_1^{n+1} f(x)\,dx > S_{n+1} - f(1)$$

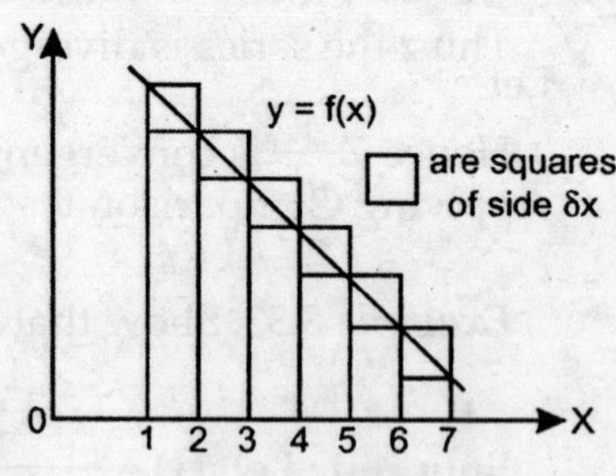

Fig. 8.1

As $n \to \infty$, from second inequality *i.e.*, S_{n+1} the integral has a finite value then $\underset{n\to\infty}{Lt}\ S_{n+1}$ is also finite.

$\therefore\ \Sigma f(n)$ is convergent.

Similarly, if the integral is infinite, then form the first inequality $\underset{n\to\infty}{Lt}\ Sn \to \infty$, is also infinite, so that series $\Sigma f(n)$ is divergent.

***Example* 8.32.** Apply the integral test to determine the convergence of the p-series. $\frac{1}{1^p} + \frac{1}{2^p} + \frac{1}{3^p} + \frac{1}{4^p} + \ldots \infty$

Solution : Let $f(x) = \frac{1}{x^p}$, and limits are 1 to ∞

$$\therefore \qquad \int_1^{\infty} f(x)\,dx = \underset{m\to\infty}{Lt} \int_1^{m} \frac{1}{x^p}\,dx$$

$$= \underset{m\to\infty}{Lt} \left[\frac{x^{1-p}}{1-p}\right]_1^m$$

$$\int_1^{\infty} f(x)\,dx = \underset{m\to\infty}{Lt} \frac{1}{1-p}[m^{1-p} - 1]$$

(*i*) When $p > 1$, then

$$\int_1^{\infty} f(x)\,dx = \underset{m\to\infty}{Lt} \frac{1}{1-p}\left[\frac{1}{m^{p-1}} - 1\right]$$

Taking limit

$$= \frac{1}{p-1}, \text{ which is finite.}$$

By integral test, the series is convergent for $p > 1$

(*ii*) When $p > 1$

$$\int_1^{\infty} f(x)\,dx = \frac{1}{1-p}\left[\underset{m\to\infty}{Lt}\ m^{1-p} - 1\right]$$

$$\frac{1}{1-p}[\infty - 1] = \infty$$

Hence, the series is divergent, for $p < 1$

(*iii*) When $p = 1$

$$\int_1^{\infty} f(x)\,dx = \int_1^{\infty} \frac{1}{x}\,dx = [\log x]_1^{\infty} \to \infty$$

Thus, the series is divergent for $p = 1$.

Hence $\Sigma \frac{1}{n^p}$ is convergent if $p > 1$ and divergent if $p \le 1$.

Example **8.33.** Show that by integral test the series $\sum_{n=2}^{\infty} \frac{1}{n(\log n)^p}$ converges If $p > 1$ and diverges if $0 < p \le 1$.

Solution : Let $f(x) = \frac{1}{x(\log x)^p}$, $p > 0$ and $x \in (2, \infty)$

Then $f(x)$ is positive and monotonically decreasing in $[2, \infty]$

By Integral Test

$$\int_2^{\infty} f(x)\,dx = \int_2^{\infty} \frac{dx}{x(\log x)^p}$$

When $p \ne 1$

$$\therefore \quad \int_2^{\infty} f(x)\,dx = \int_2^{\infty} \frac{(\log x)^{-p}}{x}\,dx = \left[\frac{(\log x)^{-p+1}}{-p+1}\right]_2^{\infty}$$

Case I : When $p > 1$, then

$1 - p < 0$ *i.,e* $p - 1$ is positive

$$\therefore \quad \int_2^{\infty} f(x)\,dx = \frac{-1}{p-1}\left[\frac{1}{(\log x)^{p-1}}\right]_2^{\infty} = \frac{-1}{p-1}\left[0 - \frac{1}{(\log 2)^{p-1}}\right]$$

$$= +\frac{1}{(p-1)(\log 2)^{p-1}}, \text{ which is finite}$$

Therefore by integral test, the series $\sum_{n=1}^{\infty} \frac{1}{n(\log n)^p}$ is convergent for $p > 1$.

Case II : When $p < 1$

$p - 1 < 0$ *i.e.,* $p - 1$ is negative.

$$\therefore \quad \int_2^{\infty} f(x)\,dx = \frac{1}{1-p}\left[(\log x)^{1-p}\right]_2^{\infty}$$

$$= \frac{1}{1-p}\left[(\log \infty)^{1-p} - (\log 2)^{1-p}\right] = \infty$$

Therefore the series $\sum_{n=2}^{\infty} \frac{1}{n(\log n)^p}$ is divergent for $p < 1$.

Case III : When $p = 1$ then

$$\int_2^{\infty} f(x)\,dx = \int_2^{\infty} \frac{dx}{x\log x} = [\log(\log x)]_2^{\infty}$$

$= \infty$, so the series in divergent in this case.

Hence the series $\sum_{n=2}^{\infty} \frac{1}{n(\log n)^p}$ converges if $p > 2$ and diverges if $0 < p < 1$.

***Example* 8.34.** Apply integral test to check the converges of the following series :

(i) $\sum_{n=1}^{\infty} \frac{1}{\sqrt{n}}$

(ii) $\sum_{n=1}^{\infty} \frac{n}{(n^2+1)^2}$

Solution : (i) Let $f(x) = \frac{1}{\sqrt{x}}$ and $x \in [1, \infty]$.

Then $f(x)$ is positive and monotonically decreasing in $[1, \infty]$

Now
$$\int_1^{\infty} f(x)\,dx = \int_1^{\infty} \frac{1}{\sqrt{x}}\,dx$$
$$= 2[\sqrt{x}\,]_1^{\infty}$$
$$= 2\,[\sqrt{x} - 1] = \infty$$

Therefore $\int_1^{\infty} f(x)\,dx$ is divergent.

Hence by integral test the series $\sum_{n=1}^{\infty} \frac{1}{\sqrt{n}}$ is divergent.

Solution : (ii) Let $f(x) = \frac{x}{(x^2+1)^2}$ and $x \in [1, \infty]$

Then $f(x) > 0$ and monotonically decreasing in $[1, \infty]$

$\therefore$
$$\int_1^{\infty} f(x)\,dx = \int_1^{\infty} \frac{x}{(x^2+1)}\,dx = \frac{1}{2}\left[-\frac{1}{x+1}\right]_1^{\infty}$$
$$= \frac{1}{2}\left[0 + \frac{1}{2}\right] = \frac{1}{4},$$ which is finite therefore $\int_1^{\infty} f(x)\,dx$ is converges to $\frac{1}{4}$

Hence by integral test the series $\sum_{n=1}^{\infty} \frac{n}{(n^2+1)^2}$ is convergent.

EXERCISE 8.8

Discuss the convergence of the following series.

1. $\sum_{n=2}^{\infty} \frac{1}{n \log n}$ **Ans. Convergent**

2. $\sum_{n=1}^{\infty} n^2 e^{-n^2}$ **Ans. Convergent**

3. $\sum_{n=1}^{\infty} \frac{1}{n\,(\log n)^2}$ **Ans. Convergent**

4. $\sum_{n=1}^{\infty} \frac{1}{(n+1)^2}$ **Ans.** Convergent

5. $\sum_{n=1}^{\infty} \frac{1}{n\sqrt{(n^2-1)}}$ **Ans.** Convergent

8.26 ALTERNATING SERIES

A series in which the terms are alternately positive and negative is called an alternating series. Thus an alternating series is of the form

$u_1 - u_2 + u_3 - u_4 + ... + (1)^{n-1} u_n + ...$ where $u_n > 0 \ \forall \ n$. It is denoted by $\sum_{n=1}^{\infty} (-1)^{n-1}$

Note : The series obtained from alternating series by taking each term as positive is known as absolute series. Absolute series of $\Sigma(-1)^{n-1}_{u_n}$ is denoted by $\Sigma|u_n|$. For examples, absolute series for $1 - \frac{1}{2} + \frac{1}{3} - \frac{1}{4} + ... + (-1)^n \frac{1}{n} + ...$

$$\Sigma u_n = \sum_{n=1}^{\infty} (-1)^{n-1} \frac{1}{n}$$

$$\therefore \quad \Sigma u_n = \sum_{n=1}^{\infty} \frac{1}{n} = \Sigma|u_n|$$

Alternating Series Test : (Leibnitz's Test)

An infinite series $\Sigma(-1)^{n-1}$ un is convergent in which the terms are

(*i*) Alternative positive and negative

(*ii*) Each term in numerically less the proceeding therm, and

(*iii*) $\underset{n \to \infty}{Lt} u_n = 0$

Proof : The given alternative series is

$$u_1 - u_2 + u_3 + u_4 + ... = \sum_{n=1}^{\infty} (-1)^{n-1} u_n \quad ...(i)$$

Suppose that $u_1 > u_2 > u_3 > u_4 > ..... > u_n > u_{n+1}$ ...(*ii*)

and $\underset{n \to \infty}{Lt} u_n = 0$

Consider the sum of $2n$ terms and it can be written as

$$S_{2n} = (u_1 - u_2) + (u_3 - u_4) + ... + (u_{2n-1} - u_{2n}) \quad ...(iii)$$

Equation (*iii*) can also be written as

$$S_{2n} = u_1 - (u_2 - u_3) - (u_4 - u_5) - (u_6 - u_7) ....u_{2n} \quad ...(iv)$$

Since $u_1 > u_2 > u_3 > u_4 > ..... u_n > u_{n+1}$, then the expression within the brakets in equation (*iii*) and (*iv*) are all positive.

It is clear from equation (*iii*) that S_{2n} is positive and increases with n also from (*iv*) it is obvious that S_{2n} always remains less that u_1

Hence S_{2n} must tend to a finite limit.

Moreover
$$\underset{n\to\infty}{Lt}\ (S_{2n+1}) = \underset{n\to\infty}{Lt}\ (S_{2n} + u_{2n+1})$$

$$\underset{n\to\infty}{Lt}\ (S_{2n+1}) = \underset{n\to\infty}{Lt}\ S_{2n} + 0 \qquad [\because \text{ From } (ii)]$$

Hence $\underset{n\to\infty}{Lt}\ S_n$ tends to the same limit whether n is even or odd, therefore the given series is convergent.

Note : (*i*) An alternating series will not be convergent if any one of two condition (*ii*) and (*iii*) is not satisfied.

(*ii*) If $\underset{n\to\infty}{Lt}\ u_n \neq 0$ then the given series is oscillatory.

Example 8.35. Test for convergence the series $1 - \frac{1}{2} + \frac{1}{3} - \frac{1}{4} + \ldots$

Solution :

(*i*) The terms of the given series are alternative positive and negative

(*ii*) $u_n = \frac{1}{n}$ and $u_{n+1} = \frac{1}{n+1}$, clearly each term of given series is numerically less than its preceding term, and

(*iii*) $u_n = \frac{1}{n}$, taking limit as $n \to s$, then

$$\underset{n\to\infty}{Lt}\ u_n = \underset{n\to\infty}{Lt}\ \frac{1}{n} = 0$$

Hence the given alternative series satisfies all condition of Leibnitz's tests, therefore the given series is convergent.

Example 8.36. Discuss the convergence of the following series.

(*a*) $2 - \frac{3}{2} + \frac{4}{3} - \frac{5}{4} + \ldots \infty$

(*b*) $\frac{1}{6} - \frac{2}{11} + \frac{3}{16} - \frac{4}{21} + \frac{5}{26} \ldots \infty$

(*c*) $1 - \frac{1}{2} + \frac{1}{4} - \frac{1}{8} + \frac{1}{16} - \ldots\infty$

Solution : (*a*) $2 - \frac{3}{2} + \frac{4}{5} - \frac{5}{4} + \ldots\infty$

(*i*) The terms of the given series are alternating positive and negative

(*ii*)
$$u_n = \frac{n+1}{n} \text{ and } u_{n+1} = \frac{n+2}{n+1}$$

$\therefore$
$$u_n - u_{n+1} = \frac{n+1}{n} - \frac{n+2}{n+1}$$

$$= \frac{(n+1)^2 - n(n+2)}{n(n+1)}$$

$$= \frac{n^2 + 1 + 2n - n^2 - 2n}{n(n+1)}$$

$$= \frac{1}{n(n+1)} > 0$$

therefore, $u_n > u_{n+1}$, *i.e.* each term is numerically less than the proceeding term

(iii) and
$$u_n = \frac{n+1}{n}$$

$$\underset{n\to\infty}{Lt}\ u_n = \underset{n\to\infty}{Lt}\ \frac{(n+1)}{n}$$

$$= \underset{n\to\infty}{Lt}\ \left(1+\frac{1}{n}\right) = 1 \neq 0$$

Therefore *(iii)* condition of Leibnitz's test is not satisfied so it an oscillatory series.

(b) $\frac{1}{6} - \frac{2}{11} + \frac{3}{16} - \frac{4}{21} + \frac{5}{26} \ldots$

(i) The terms of the given series are alternating positive and negative

(ii)
$$u_n = \frac{5n}{5n+1}$$

and
$$u_{n+1} = \frac{n+1}{5(n+1)+1}$$

$$= \frac{n+1}{5n+6}$$

$$\therefore \quad u_n - u_{n+1} = \frac{n}{5n+1} - \frac{n+1}{5n+6}$$

$$= \frac{n(5n+6) - (5n+1)(n+1)}{(5n+1)(5n+6)}$$

$$= \frac{5n^2 + 6n - 5n^2 - 5n - n - 1}{(5n+1)(5n+6)}$$

$$= \frac{-1}{(5n+1)(5n+6)} > 0 \text{ (numerically)}$$

therefore, $u_n > u_{n+1}$, *i.e.*, each term is numerically less than its proceeding terms

(iii) and
$$u_n = \frac{n}{5n+1}$$

$$\therefore \quad \underset{n\to\infty}{Lt}\ \frac{n}{5n+1}$$

$$\underset{n\to\infty}{Lt}\ \frac{n}{5n+1} = \frac{1}{5} \neq 0$$

Therefore the series is oscillatory series.

(c) $1 - \frac{1}{2} + \frac{1}{4} - \frac{1}{8} + \frac{1}{16} + \ldots\ldots$

(i) The terms of the given infinite series are alternating positive and negative.

(ii) $u_n = \frac{1}{2^{n-1}}$ and $u_{n+1} = \frac{1}{2^n}$

$$\therefore u_n - u_{n+1} = \frac{1}{2^{n-1}} - \frac{1}{2^n} = \frac{1}{2^n} > 0 \ \forall\ n \quad \therefore \quad u_n > u_{n+1}$$

(*iii*) $u_n = \frac{1}{2^{n-1}}$, $\underset{n\to\infty}{Lt} \quad u_n = \underset{n\to\infty}{\frac{1}{}} \frac{1}{2^{n-1}} = 0$

Hence by Leibnitz's test the given series in convergent

Example **8.37.** Test the convergence of the series

(*i*) $\sum_{n=1}^{\infty} (-1)^{n-1} \frac{x^n}{n^2}$

(*ii*) $\sum_{n=1}^{\infty} (-1)^{n-1} \frac{n}{n^2+1}$

Solution : $\sum_{n=1}^{\infty} (-1)^{n-1} \frac{x^n}{n^2}$

The given $\left(\frac{x}{1^2} - \frac{x^2}{2^2} + \frac{x^3}{3^2} - \frac{x^4}{4^2} + \ldots\right)$ is an alternative series because it has positive and negative terms

Also $\quad |u_n| = \frac{x^n}{n^2}$ and $|u_{n+1}| = \frac{x^{n+1}}{(n+1)^2}$

$$\therefore \quad \underset{n\to\infty}{Lt}\left|\frac{u_n}{u_{n+1}}\right| = \left|\frac{1}{x}\right|$$

Therefore, by D'Alembert ratio test, the series $\Sigma |u_n|$ is convergent if $\left|\frac{1}{x}\right| > 1$ *i.e.,* $|x| < 1$ or $-1 < x < 1$, the series is divergent if $\left|\frac{1}{x}\right| < 1$ *i.e.,* $|x| > 1$ and when $\left|\frac{1}{x}\right| = 1$ or $|x| = 1$ then this test fail.

When $x = 1$, the series becomes

(*i*) $\frac{1}{1^2} - \frac{1}{2^2} + \frac{1}{3^2} - \frac{1}{4^2} + \ldots \infty$ which is an alternating series

(*i*) $u_n > u_{n+1}$

(*iii*) $u_n = \frac{1}{n^2} \quad \therefore \underset{n^2}{Lt}\ u_n = 0$

Hence the series in convergent by Leibnitz's rule

When $x = -1$, the series becomes

$$-\left[\frac{1}{1^2} + \frac{1}{2^2} + \frac{1}{3^2} \ldots \infty\right] = -\sum \frac{1}{n^2}$$

therefore, the series in convergent.

Hence the series is convergent if $|x| < a$ and divergent if $|x| > 1$.

(*ii*) $\sum_{n=1}^{\infty} (-1)^{n-1} \frac{n}{n^2+1}$

The terms of the series are alternately positive and negative.

Here $|u_n| = \frac{n}{n^2+1}$ and $|u_{n+1}| = \frac{n+1}{(n+1)^2+1}$

$$\therefore \qquad |u_n| - |u_{n+1}| = \frac{n}{n^2+1} - \frac{n+1}{(n+1)^2+1}$$

$$= \frac{n(n^2+2n+2)-(n+1)(n^2+1)}{(n^2+1)(n^2+2n+2)}$$

$$= \frac{n^3+2n^2+2n-n^3-n^2-n-1}{(n^2+1)(n^2+2n+2)}$$

$$= \frac{n^2+n-1}{(n^2+1)(n^2+2n+2)} > 0 \qquad (\because n > 0)$$

therefore, $|u_{n+1}| < |u_n|$ *i.e.,* each term is numerically less than its preceding term and $u_n = \frac{n}{n^2+1}$ and $u_n = \frac{n}{n^2+1}$

$$\therefore \qquad \underset{n \to \infty}{Lt} \frac{n}{n^2\left(1+\frac{1}{n^2}\right)} = 0$$

Hence, the series is convergent by Leibunitz's rule.

***Example* 8.38.** Show that the series $\frac{\log 2}{2^2} - \frac{\log 3}{3^2} + \frac{\log 4}{4^2} + \ldots\ldots$ convergent.

Solution : The terms of the given series are alternating positive negative.

$$|u_n| = \frac{\log(n+1)}{(n+1)^2} \quad \text{and} \quad |u_{n+1}| = \frac{\log(n+2)}{(n+2)^2}$$

We prove the $|u_{n+1}| < |u_n| \ \forall \ n$, for this

Let us take $\qquad f(x) = \frac{\log x}{x^2}$ then

$$f'(x) = \frac{x^2 \times \left(\frac{1}{x}\right) - \log x \times 2x}{x^4} = \frac{1-2\log x}{x^2}$$

$$= < 0 \text{ whenever } x < e^{\frac{1}{2}}$$

therefore, the function $f(x)$ is monotonically decreasing for all $x > e^{\frac{1}{2}}$

But $2 < e < 3$ so $1 < e^{\frac{1}{2}} < 2$

Therefore $f(n+2) < f(n+1) \quad \therefore$ for all $n \quad \therefore \quad u_{n+1} < u_n$

also $\qquad u_n = \frac{\log(n+1)}{(n+1)^2}$

$$\therefore \qquad \underset{n \to \infty}{Lt}\ u_n = \underset{n \to \infty}{Lt} \left[\frac{\log(n+1)}{n+1} \times \frac{1}{n+1}\right] = 0$$

$$\therefore \qquad \underset{n \to \infty}{Lt} \frac{\log(n+1)}{n} = 0$$

Hence all condition of Leibnitz's test are satisfied by the given series so the series is convergent.

EXERCISE 8.9

Discuss the convergence of the following series.

1. $1-\frac{1}{2}+\frac{1}{3}-\frac{1}{4}+\frac{1}{5}$ ..... ∞ **Ans.** Convergent

2. $\frac{1}{2^2}-\frac{1}{4^2}+\frac{1}{6^2}-\frac{1}{8^2}+$ ..... ∞ **Ans.** Convergent

3. $1-\frac{1}{\sqrt[2]{2}}+\frac{1}{\sqrt[3]{3}}-\frac{1}{\sqrt[4]{4}}+$ ..... ∞ **Ans.** Convergent

4. $\frac{1}{\log 2}-\frac{1}{\log 3}+\frac{1}{\log 4}-\frac{1}{\log 5}+$ .....∞ **Ans.** Convergent

5. $\frac{1}{1P}-\frac{1}{2P}+\frac{1}{3P}-\frac{1}{4P}+$ ....∞ **Ans.** Convergent

6. $\frac{1}{1P}-\frac{1}{2P}+\frac{1}{3P}-\frac{1}{4P}+$ .... ∞ **Ans.** If $P > 0$, Convergent
If $P < 0$, Oscillatory

7. $\frac{x}{1+x}-\frac{x^2}{1+x^2}+\frac{x^3}{1+x^3}-\frac{x^4}{1+x^4}$ **Ans.** Convergent

8. $\sum_{n=1}^{\infty}(-1)^{n-1}\frac{-n}{2n-1}$ **Ans.** Oscillatory

9. Prove that the series $x-\frac{x^2}{2}+\frac{x^3}{3}-\frac{x^4}{4}+$.... 0. convergent if $-1 < x \le 1$.

10. $\sum_{n=1}^{\infty}\frac{\cos n\pi}{n^2+1}$ **[Hint.** $\cos n\pi = (-1)^n$**]** **Ans.** Convergent

11. $\left(\frac{1}{2}-\frac{1}{\log 2}\right)-\left(\frac{1}{2}-\frac{1}{\log 3}\right)+\left(\frac{1}{2}-\frac{1}{\log 4}\right)-\left(\frac{1}{2}-\frac{1}{\log 5}\right)+$.... ∞ **Ans.** Oscillatory

8.27 ALTERNATING CONVERGENT SERIES

Alternating convergent series are of two types

(*a*) Absolutely convergent series

(*b*) Conditionally convergent series

(*a*) Absolutely Convergent Series. If the series $u_1 + u_2 + u_3 + u_4 + u_5 +$ .... be such that $|u_1| + |u_2| + |u_3| + |u_4| + |u_5| +$ ..... is convergent then $u_1 + u_2 + u_3 + u_5 +$ .... is called absolutely convergent

(*b*) Conditionally Convergent Series. If the series $u_1 + u_2 + u_3 + u_4 + u_5$ .... be convergent but $|u_1| + |u_2| + |u_3| + |u_4| + |u_5| +$ ..... is divergent, then the given series $u_1 + u_2 + u_4 + u_5 +$ ..... is called conditionally convergent or semi convergent.

Note : (*i*) If any alternating series Σu_n is convergent by Leibnitz's test and it is also convergent by Comparison test, D' Alemberts Ratio Test or Cauchy root test, then it is said to be absolutely convergent.

(*ii*) If any alternating series sum is convergent by Lebnitz's test and it is divergent by comparison test, D' Alembert Ratio Test or Cauchy roots test then it said to be conditionally convergent.

Example **8.39.** Test whether the following series are absolutely convergent or conditionally convergent.

(*i*) $1-\frac{1}{2^2}+\frac{1}{3^2}+\frac{1}{3^2}-\frac{1}{4^2}+.....+(-1)^{n-1}\frac{1}{n^2}+....\infty$

(*ii*) $\frac{1}{2}-\frac{2}{5}+\frac{3}{10}-\frac{4}{17}+....(-1)n-1\frac{n}{n^2+1}+....\infty$

Solution :

(*i*) $1-\frac{1}{2^2}+\frac{1}{3^2}-\frac{1}{4^2}+....+(-1)^{n-1}\frac{1}{n^2}+.....\infty$

In this series, terms are alternating positive negating and each term is less than the preceding term numerically, *i.e.*, $u_{n+1}<u_n$

Now $$u_n = \frac{1}{n^2}$$

$\therefore$ $$\underset{n\to\infty}{Lt} = \underset{n\to\infty}{Lt}\frac{1}{n^2}=0$$

Hence the given series satisfies all condition of Leibnitz's test, therefore it is convergent

Again, $\Sigma|u_n|=\sum\frac{1}{n^2}$ which is convergent by p-test ($p=2$)

Hence the given series in absolutely convergent.

(*ii*) $\frac{1}{2}-\frac{2}{5}+\frac{3}{10}-\frac{4}{17}+.....(-1)^{n-1}\frac{n}{n^2+1}+....\text{ to }\infty$

In this series, terms are alternating positive negative and each term is less then the preceding term numerically, *i.e.*, $u_{n+1}<u_n\ \forall\ n$

Here $$u_n = \frac{n}{n^2+1}$$

and $$\underset{n\to\infty}{Lt}\ u_n = \underset{n\to\infty}{Lt}\frac{1}{\left(n+\frac{1}{n}\right)}=0$$

Hence the given series satisfies the all conditions of Leibnitz test therefore it is convergent

Again $$\Sigma|u_n| = \frac{n}{n^2+1}=\frac{1}{1^2+1}+\frac{2}{2^2+1}+\frac{3}{3^2+1}+.....\infty$$

Here $$u_n = \frac{n}{n^2+1}\text{ take } v_n=\frac{1}{n}$$

$\therefore$ $$\frac{u_n}{v_n} = \frac{n}{n^2+1}\times\frac{n}{1}$$

$$= \frac{n^2}{n^2\left(1+\frac{1}{n}\right)}$$

$$= 1+\frac{1}{n}$$

and
$$\underset{n\to\infty}{Lt}\frac{u_n}{u_n} = \underset{n\to\infty}{Lt}\frac{1}{\left(1+\frac{1}{n}\right)}$$
$$= 1 \neq 0, \text{ i.e., finite and non zero.}$$

but $\Sigma v_n = \frac{1}{n}$ is divergent, therefore, $\Sigma |u_n|$ is divergent by comparison test Hence, Σu_n is not absolutely convergent *i.e.*, conditionally convergent.

Example 8.40. (*a*) Show that the series $1-\frac{1}{2}+\frac{1}{3}-\frac{1}{4}+\frac{1}{5}-\frac{1}{6}+\ldots\ldots\infty$ is convergent but not absolutely convergent.

(b) What can you say about the series $\sum_{n=1}^{\infty}(-1)^{n-1}\sin\frac{1}{n}$

Solution : (*a*) $1-\frac{1}{2}+\frac{1}{3}-\frac{1}{4}-\frac{1}{5}-\frac{1}{6}+\ldots\infty$

(*i*) The terms of the series are alternately positive and negative

(*ii*) $u_n = \frac{1}{n}$ and $u_{n+1} = \frac{1}{n+1}$, clearly $u_{n+1} < u_n$

(*iii*) and $\underset{n\to\infty}{Lt}\ u_n \quad u_n = \underset{n\to\infty}{Lt}\frac{1}{n} = 0$

Thus all conditions of Leibnitz's test are satisfied by the given series therefore the series is convergent. But by comparison test, the series $\Sigma|u_n| = 1+\frac{1}{2}+\frac{1}{3}+\frac{1}{4}+\frac{1}{5}+\frac{1}{6}+\ldots\infty$ is divergent $(\because p = 1)$

Hence, the given series is conditionally convergent.

(*b*) $\sum_{n=1}^{\infty}(-1)^{n-1}\sin\frac{1}{n} = \sin\frac{1}{1}-\sin\frac{1}{2}+\sin\frac{1}{3}-\sin\frac{1}{4}+\ldots\infty$

(*i*) The terms are alternately positive and negative

Here
$$u_n = \sin\frac{1}{n} \text{ and } u_{n+1} = \sin\frac{1}{n+1}$$

$\therefore$
$$u_n - u_{n+1} = \sin\frac{1}{n} - \sin\frac{1}{n+1} > 0,\ \forall \geq 1 \text{ i.e.,}$$
$$u_n > u_{n+1}\ \forall\ n$$

and
$$\underset{n\to\infty}{Lt}\ u_n = \underset{n\to\infty}{Lt}\ \sin\frac{1}{n} = 0$$

therefore, the given series is convergent, but when all terms are positive the series becomes $\Sigma|u_n| = \sin\frac{1}{1} + \sin\frac{1}{2}+\sin\frac{1}{3}+\ldots$

Here
$$u_n = \sin\frac{1}{n},\ \text{take } v_n = \frac{1}{n}$$

$\therefore$
$$\frac{u_n}{v_n} = \frac{\sin\frac{1}{2}}{\frac{1}{n}}$$

$$\underset{n\to\infty}{Lt}\frac{u_n}{v_n} = \underset{n\to\infty}{Lt}\frac{\sin\frac{1}{n}}{\frac{1}{n}} = 1 \qquad \left\{\because \underset{n\to\infty}{Lt}\frac{\sin x}{x} = 1\right\}$$

Which is finite and non zero, but $p = 1$ therefore $\Sigma v_n = \sum \frac{1}{n}$ is divergent by p-test

$\therefore \Sigma |u_n|$ is also divergent

Hence the given series is conditionally convergent.

Example **8.41.** Discuss the series for convergence and absolutely convergent.

(*a*) $1 - \frac{1}{3} + \frac{1}{2} - \frac{1}{3^3} + \frac{1}{2^2} - \frac{1}{3^5} + \frac{1}{2^3} - \frac{1}{3^7} + ... \infty$

(*b*) $\frac{1}{1.3} - \frac{1}{2.4} + \frac{1}{3.5} - \frac{1}{4.6} + ... \infty$

(*c*) $1 - \frac{1}{2^3}(1+2) + \frac{1}{3^3}(1+2+3) - \frac{1}{4^3}(1+2+3+4) + ... \infty$

Solution :

(*a*) $1 - \frac{1}{3} + \frac{1}{2} - \frac{1}{3^3} + \frac{1}{2^2} - \frac{1}{3^5} + \frac{1}{2^3} - \frac{1}{3^7} + ... \infty$

The terms of the series are alternating position, negative and the given series can be written

$$\left(1 + \frac{1}{2} + \frac{1}{2^2} + \frac{1}{2^3} + ...\right) - \left(\frac{1}{3} + \frac{1}{3^3} + \frac{1}{5^5} + ...\right)$$

Here
$$S_\infty = \frac{1}{1-\frac{1}{2}} - \frac{\frac{1}{3}}{1-\frac{1}{3^2}}$$

$$= 2 - \frac{3}{8} = \frac{13}{8}$$

$\therefore$
$$\underset{n\to\infty}{Lt} S_n = \frac{13}{8},$$

therefore, the given series is convergent

Again, $\left(1 + \frac{1}{2} + \frac{1}{2^2} + \frac{1}{2^3} + ...\right) + \left(\frac{1}{3} + \frac{1}{3^3} + \frac{1}{3^5} + \frac{1}{3^7} + ...\right)$

$\therefore$
$$\underset{n\to\infty}{Lt} S_n = 2 + \frac{3}{8} = \frac{19}{8}$$

Hence $|u_n|$ is also a convergent therefore, the given series is absolutely convergent.

(*b*) $\frac{1}{1.3} - \frac{1}{2.4} + \frac{1}{3.5} - \frac{1}{3.6} + ...$

The terms if the series are alternating positive, negative

(*i*) The series is alterately position and negative

(*ii*) $$u_n = \frac{1}{n(n+2)},\; u_{n+1} = \frac{1}{(n+1)(n+3)}$$

$\therefore$ $$u_n - u_{n+1} = \frac{1}{n(n+2)} - \frac{1}{(n+1)(n+3)}$$

$$= \frac{2n+3}{n(n+1)(n+2)(n+3)} > 0 \;\forall\; n$$

$\therefore$ $$u_n > u_{n+1} \;\forall\; n$$

(*iii*) Also $$\underset{n\to\infty}{Lt}\, u_n = \underset{n\to\infty}{Lt}\, \frac{1}{n(n+2)} = 0$$

$\therefore$ By Leibnitz's test the series Σu_n is convergent

Again $$|u_n| = \frac{1}{n(n+2)} = \frac{1}{n2\left(1+\frac{2}{n}\right)} \text{ take } |v_n| = \frac{1}{n^2}$$

$\therefore$ $$\underset{n\to\infty}{Lt}\left|\frac{u_n}{v_n}\right| = \underset{n\to\infty}{Lt}\, \frac{1}{1+\frac{2}{n}}$$

$= 1$ which is and finite non zero

$\therefore$ By comparison test, $\Sigma|u_n|$ and $\Sigma|u_v|$ is convergent or divergent together. Since $|v_n| = \frac{1}{n^2}$. Therefore, $\Sigma|u_n|$ and $|v_n|$ both convergent

Hence, the given series Σu_n is absolutely convergent.

(*c*) The given series is $1 - \frac{1}{2^3}(1+3) + \frac{1}{3^3}(1+2+3) - \frac{1}{4^3}(1+2+3+4) + \ldots. \infty$

or $$\frac{1}{1^3}(1) - \frac{1}{2^3}(1+2) + \frac{1}{3^3} + \frac{1}{3^3}(1+2+3) - \frac{1}{4^3}(1+2+3+4) + \ldots..$$

(*i*) the terms of the series are alternating positive and negative.

(*ii*) $$\sum_{n=1}^{\infty} u_n = \sum_{n=1}^{\infty} (-)^{n-1} \frac{(1+2+3+4+\ldots.+n)}{n^3}$$

$$= \sum_{n=1}^{\infty} (-)^{n-1} \frac{\frac{n(n+1)}{2}}{n^3}$$

$$= \sum_{n=1}^{\infty} (-)^{n-1} \frac{(n+1)}{2n^2}$$

$\therefore$ $$u_n = \frac{n+1}{2n^2},\; u_{n+1} = \frac{(n+2)}{2(n+1)^2}$$

$\therefore$ $$u_n - u_{n+1} = \frac{n+1}{2n^2} - \frac{n+2}{2(n+1)^2}$$

$$= \frac{(n+1)^3 - n^2(n+2)}{2n^2(n+1)^2}$$

$$= \frac{n^2+3n+1}{2n^2(n+1)^2} > 0 \;\forall\; n$$

therefore $u_n > u_{n+1} \ \forall\ n$

Also
$$\underset{n\to\infty}{Lt}\ u_n = \underset{n\to\infty}{Lt}\ \frac{n+1}{2n^2}$$
$$= \underset{n\to\infty}{Lt}\ \frac{1}{2}\left(\frac{1}{n}+\frac{1}{n^2}\right) = 0$$

Therefore by Leibnitz's test the series Σu_n is convergent

Again,
$$|u_n| = \frac{n+1}{2n^2} = \frac{\left(1+\frac{1}{n}\right)}{2n}, \text{ take } |v_n| = \frac{1}{n}$$

$\therefore$
$$\underset{n\to\infty}{Lt}\ \frac{u_n}{v_n} = \frac{1}{2}, \text{ which is finite and non zero, therefore}$$

But Comparison test,

$\Sigma|u_n|$ and $\Sigma|v_n|$ are convergent or divergent together

But $|v_n| = \frac{1}{n}$, and $p = 1$, therefore $|v_n|$ is divergent.

Hence the given series Σu_n is conditionally convergent.

***Example* 8.42.** (*a*) Find the interval of convergence for the series $x - \frac{x^2}{\sqrt{2}} + \frac{x^3}{\sqrt{3}} - \frac{x^4}{\sqrt{4}} + \ldots.\ \infty$

(*b*) Prove that the series $\frac{\sin x}{1^3} - \frac{\sin 2x}{2^3} + \frac{\sin 3x}{2^3} - \ldots.\ \infty$ converge absolutely.

Solution :

(*a*) $x - \frac{x^2}{\sqrt{2}} + \frac{x^3}{\sqrt{3}} - \frac{x^4}{\sqrt{4}} + \ldots..\ \infty$

The terms of the series are alternating positive negative.

Here
$$\Sigma u_n = \sum \frac{(-1)^{n-1}\, x^n}{\sqrt{n}}$$

$\therefore$
$$|u_n| = \left|\frac{x^n}{\sqrt{n}}\right| \text{ and } |u_{n+1}| = \left|\frac{x^{n+1}}{\sqrt{n+1}}\right|$$

By D' Alembert ratio rest

$$\left|\frac{u_n}{u_{n+1}}\right| = \left|\frac{x^n}{\sqrt{n}}\right|\left|\frac{\sqrt{n+1}}{x^n \times x}\right| = \left|\frac{\sqrt{1+\frac{1}{n}}}{x}\right|$$

$\therefore$
$$\underset{n\to\infty}{Lt}\ \left|\frac{u_n}{u_{n+1}}\right| = \frac{\sqrt{1+\frac{1}{n}}}{|x|} = \frac{1}{|x|}$$

therefore, by D' Alembert's ratio test,

(*i*) The given series Σu_n is convergent if $\frac{1}{|x|} > 1$ *i.e* $|x| < 1$ or $-1 < x < 1$

(*ii*) The given series sum is convergent if $\frac{1}{|x|} > 1$ *i.e* $|x| > 1$, and (*iii*) The D'Alembert's ratio test fails if $|x| = 1$ *i.e* $x = \pm 1$

When $x = 1$, then the series becomes $1 - \frac{1}{\sqrt{2}} + \frac{1}{\sqrt{2}} - \frac{1}{\sqrt{4}} + \ldots.. \infty$, which is an alternative series and this is convergent by Lebnitz's test

When $x = -1$, then the series becomes

$$-\left(1 + \frac{1}{\sqrt{2}} + \frac{1}{\sqrt{3}} + \frac{1}{\sqrt{4}} + \ldots.. \infty\right) = \sum \frac{1}{\sqrt{n}}$$ which is divergent by Comparison test or *p*-test

Hence the given series is convergent if $-1 < x < 1$.

(*b*) $\frac{\sin x}{1^3} - \frac{\sin 2x}{2^3} + \frac{\sin 3x}{2^3} \ldots\ldots \infty$

(*i*) The terms of the series are alternative positive and negative.

(*ii*) $$\text{Here } \Sigma u_n = \sum_{n=1}^{\infty} (-1)^{n-1} \frac{\sin n x}{n^3}$$

Now $$|u_n| = \left|\frac{\sin n x}{n^3}\right|$$

$$|u_{n+1}| = \left|\frac{\sin (n+1)x}{(n+1)^3}\right|$$

$\therefore |u_n| - u_{n+1} > 0$, because $|\sin x| \leq 1$

(*iii*) $$\underset{n \to \infty}{Lt}\ u_n = \underset{n \to \infty}{Lt} \frac{|\sin n x|}{n^3}$$

$$= 0 \qquad \{\because |\sin n x| \leq 1\}$$

Hence by Leibnitz's test Σu_n is convergent

Also $$\Sigma |u_n| = \frac{|\sin n x|}{n^3} = \frac{1}{n^3}|\sin nx| \quad \text{take } |v_n| = \frac{1}{n^3}$$

By comparison test, $$\left|\frac{u_n}{v_n}\right| = |\sin nx|$$

$$\underset{n \to \infty}{Lt} \left|\frac{u_n}{v_n}\right| = \text{finite and non zero} \qquad \{\because |\sin nx| < 1\}$$

Here $P = 3$, therefore $\Sigma |u_n|$ and $\Sigma |v_n|$ are convergent together

Hence the given series is absolutely convergent.

EXERCISE 8.10

1. Test the following series for absolutely convergent or conditional convergent.

(*a*) $1 - \frac{1}{2\sqrt{2}} + \frac{1}{3\sqrt{3}} - \frac{1}{4\sqrt{4}} + \ldots. \infty$ **Ans.** Absolutely convergent

(*b*) $1 - \frac{1}{2P} + \frac{1}{3P} - \frac{1}{4P} + \ldots. \infty\ (P > 0)$ **Ans.** Absolutely convergent of $p > 1$ conditionally convergent if $0 < p \leq 1$

(c) $1-\frac{1}{3!}+\frac{1}{5!}-\frac{1}{7!}+\ldots.\infty$ **Ans.** Absolutely convergent

2. Find the internal of convergence of the series

(a) $x-\frac{x^3}{3}+\frac{x^5}{5}-\frac{x^7}{7}+\ldots..\infty$ **Ans.** $-1\ x\leq 1$

(b) $x-\frac{x^2}{2^2}+\frac{x^3}{3^2}-\frac{x^4}{4^2}+\ldots..$ **Ans.** $-1<x\leq 1$

(c) $1+\frac{x}{2}+\frac{x^2}{3}+\frac{x^3}{4}+\ldots..\infty$ **Ans.** Divergent

3. Test the absolute convergent or conditionally convergent of the series

(a) $\sum_{n=1}^{\infty}(-1)^n\frac{1}{n}$ **Ans.** Conditionally convergent

(b) $\sum_{n=1}^{\infty}(-1)^n\{\sqrt{(n+1)}-\sqrt{n}\}$ **Ans.** Conditionally convergent

(c) $\sum_{n=1}^{\infty}(-1)^n\frac{\cos x}{n\sqrt{n}}$ **Ans.** Absolutely convergent

(d) $\sum_{n=1}^{\infty}(-1)^n\frac{x^n}{n+1}$ **Ans.** Absolutely convergent series if $0<x<1$

4. Prove that $\frac{1}{a}+\frac{1}{a+1}-\frac{1}{a+2}+\frac{1}{a+3}+\frac{1}{a+4}-\frac{1}{a+5}+\ldots..$ is a divergent series.

Fourier Series

The contents of this chapter are Fourier series, Dirichelt's condition, even-odd functions, Fourier series for discontinuous functions, change of interval, half-range series and Parseval's theorem.

9.1 FOURIER SERIES

Many engineering and physical problems involve the study of fourier series for their solution, for example conduction of heat, electrodynamics and mechanical vibrations. It is necessary to express these functions in a series of sines and cosines. Most of the single valued functions which occur in applied mathematics can be expressed in the form

$$\frac{a_0}{2} + a_1 \cos x + a_2 \cos 2x + \ldots\ldots\ldots\ldots + b_1 \sin x + b_2 \sin 2x + \ldots\ldots\ldots\ldots$$

within a desired range of the values of the variable. Such a series is known as the Fourier Series. Some important definite integrals.

(*a*) $\int_c^{c+2\pi} \sin mx \, dx = 0$

(*b*) $\int_c^{c+2\pi} \cos mx dx = 0$

(*c*) $\int_c^{c+2\pi} \sin mx \cos nx dx = 0, \ m \neq n$

(*d*) $\int_c^{c+2\pi} \cos mx \sin nx dx = 0, \ m \neq n$

(*e*) $\int_c^{c+2\pi} \sin^2 mx dx = \pi, \ m \neq 0$

(*f*) $\int_c^{c+2\pi} \cos^2 mx dx = \pi, \ m \neq 0$

(*g*) $\int_c^{c+2\pi} \cos mx \sin nx dx = 0, \ m \neq n$

(*h*) $\int_c^{c+2\pi} \sin mx \sin nx \, dx = 0, \ m \neq n$

9.2 DIRICHLET'S CONDITIONS FOR FOURIER SERIES

If a function $f(x)$ is defined in the interval $c \le x \le c + 2\pi$ and is

(i) finite, single valued in the interval.

(ii) has a finite number of singularities in the interval.

(iii) is periodic.

(iv) has finite number of maxima and minima in the interval.

then the fourier series for the function $f(x)$ is defined as

$$f(x) = \frac{a_0}{2} + \sum_{n=1}^{\infty} a_n \cos nx + \sum_{n=1}^{\infty} b_n \sin nx$$

where
$$a_0 = \frac{1}{\pi} \int_c^{c+2\pi} f(x)\,dx$$

$$a_n = \frac{1}{\pi} \int_c^{c+2\pi} f(x) \cos nx\,dx$$

$$b_n = \frac{1}{\pi} \int_c^{c+2\pi} f(x) \sin nx\,dx$$

a_0, a_n, b_n are known as Euler's formulae.

The above four conditions are known as Dirichlet's conditions.

Note 1 : It must be noted that the fourier expansion is not valid in every case.

Note 2 : The Euler's formulae a_0, a_n, b_n are obtained as follows, we have

$$f(x) = \frac{a_0}{2} + \sum_{n=1}^{\infty} a_n \cos nx + \sum_{n=1}^{\infty} b_n \sin nx$$

$$\int_c^{c+2\pi} f(x)\,dx = \int_c^{c+2\pi} \frac{a_0}{2}\,dx + \int_c^{c+2\pi} \sum_1^{\infty} a_n \cos nx\,dx + \int_c^{c+2\pi} \sum_1^{\infty} b_n \sin nx\,dx$$

$$\int_c^{c+2\pi} f(x)\,dx = \frac{a_0}{2} \times 2\pi + a_n \times 0 + b_n \times 0$$

$$= a_0 \pi$$

$$a_0 = \frac{1}{\pi} \int_c^{c+2\pi} f(x)\,dx \quad \left\{ \begin{array}{l} \text{Since } \int_c^{c+2\pi} \cos mx dx = 0 \\ \int_c^{c+2\pi} \sin mx\,dx \end{array} \right.$$

Also multiplying (i) by cos nx and integrating between the limits $(c, c + 2\pi)$

$$\int_c^{c+2\pi} f(x) \cos nx\,dx = \int_c^{c+2\pi} \frac{a_0}{2} \cos nx\,dx + \int_c^{c+2\pi} \left(\sum_1^{\infty} a_n \cos nx\right) \cos nx\,dx + \int_c^{c+2\pi} \left(\sum_1^{\infty} b_n \sin nx\right) \cos nx\,dx$$

$$= \frac{a_0}{2} \times 0 + a_n \pi + 0$$

$$a_n = \frac{1}{\pi} \int_c^{c+2\pi} f(x) \cos nx \, dx \quad \left\{ \begin{aligned} & \int_c^{c+2\pi} \left(\sum_1^{\infty} an \cos nx \right) \cos nx \, dx \\ &= \int_c^{c+2\pi} (a_1 \cos x + a_2 \cos x \ldots. + an \cos nx) \cos nx \\ &= a_n \pi \left\{ \text{since} \int_c^{c+2\pi} \cos nx \cos mx dx = 0 \; m \neq n = \pi, m = n \right\} \end{aligned} \right.$$

Similarly, $$b_n = \frac{1}{\pi} \int_c^{c+2\pi} f(x) \sin mx \, dx$$

Further

(*i*) If $c = 0$, then the interval becomes $0 \leq x \leq 2\pi$.

$$f(x) = \frac{a_0}{2} + \sum_1^{\infty} a_n \cos nx + \sum_1^{\infty} b_n \sin nx$$

where $$a_0 = \frac{1}{\pi} \int_0^{2\pi} f(x) dx$$

$$a_n = \frac{1}{\pi} \int_0^{2\pi} f(x) \cos nx \, dx$$

$$b_n = \frac{1}{\pi} \int_0^{2\pi} f(x) \sin nx \, dx$$

(*ii*) If $c = -\pi$, the interval becomes $-\pi \leq x \leq \pi$

$$a_0 = \frac{1}{\pi} \int_{-\pi}^{\pi} f(x) dx$$

$$a_n = \frac{1}{\pi} \int_{-\pi}^{\pi} f(x) \cos nx \, dx$$

$$b_n = \frac{1}{\pi} \int_{-\pi}^{\pi} f(x) \sin nx \, dx$$

***Example* 9.1.** Obtain a Fourier series to represent e^{-ax} from $-\pi$ to x. Hence derive series for $\pi/\sinh \pi$.

Solution. Let $$f(x) = e^{-ax} \text{ and}$$

$$e^{-ax} = f(x) = \frac{a_0}{2} + \sum_{n=1}^{\infty} a_n \cos nx + \sum_{n=1}^{\infty} b_n \sin nx$$

Then, $$a_0 = \frac{1}{\pi} \int_{-\pi}^{\pi} f(x) dx = \frac{1}{\pi} \int_{-\pi}^{\pi} e^{-ax} dx = \frac{1}{\pi} \left[\frac{e^{-ax}}{-a} \right]_{-\pi}^{\pi}$$

$$= \frac{1}{-a\pi}[e^{-ax} - e^{a\pi}] = \frac{2}{a\pi}\sinh a\pi$$

$$a_n = \frac{1}{\pi}\int_{-\pi}^{\pi} f(x)\cos nx\,dx = \frac{1}{\pi}\int_{-\pi}^{\pi} e^{-ax}\cos nx\,dx$$

$$= \frac{1}{\pi}\left[\frac{e^{-ax}}{a^2+n^2}\{-a\cos nx + n\sin nx\}\right]_{-\pi}^{\pi}$$

$$= \frac{1}{\pi(a^2+n^2)}[e^{-ax}\{-a\cos n\pi + 0\} - e^{ax}\{-a\cos(-n\pi) + 0\}]$$

$$= \frac{1}{\pi(a^2+n^2)}[-a(-1)^n\{e^{-a\pi} - e^{-a\pi}\}]$$

$$= \frac{2a(-1)^n}{\pi(a^2+n^2)}\sin h\, a\pi$$

$$b_n = \frac{1}{\pi}\int_{-\pi}^{\pi} f(x)\sin nx\,dx = \frac{1}{\pi}\int_{-\pi}^{\pi} e^{-ax}\sin nx\,dx$$

$$= \frac{1}{\pi}\left\{\frac{e^{-ax}}{a^2+n^2}(-a\sin nx - n\cos nx)\right\}_{-\pi}^{\pi}$$

$$= \frac{1}{\pi\{a^2+n^2\}}\{e^{-ax}(-0 - n\cos n\pi) - e^{ax}(0 - n\cos(-n\pi))\}$$

$$= \frac{1}{\pi(a^2+n^2)}\{(-1)^n n\{e^{ax} - e^{-ax}\}\}$$

$$= \frac{2(-1)^n n}{\pi(a^2+n^2)}\sinh a\pi$$

$$= e^{-ax} = f(x) = \frac{\sinh a\pi}{\pi a} + \sum_{n=1}^{\infty}\frac{2a(-1)^n}{\pi(a^2+n^2)}\sinh a\pi\cos nx + \sum_{n=1}^{\infty}\frac{2n(-1)^n}{\pi(a^2+n^2)}\sinh a\pi\sin nx$$

$$= \frac{\sinh a\pi}{\pi a} - \frac{2a}{\pi(a^2+1^2)}\sin h\, a\pi\cos x + \frac{2a}{\pi(a^2+2^2)}\sinh a\pi\cos 2x$$

$$- \frac{2a}{\pi(a^2+3^2)}\sin h\, a\pi\cos 3x \ldots\ldots\ldots\ldots$$

$$- \frac{2}{\pi(a^2+1^2)}\sin h\, a\pi\sin x + \frac{2\times 2}{\pi(a^2+2^2)}\sin h\, a\pi\sin 2x$$

$$- \frac{2\times 3}{\pi(a^2+3^2)}\sin h\, a\pi\sin 3x \ldots\ldots\ldots\ldots$$

Put $x = 0,\ a = 1$

$$1 = \frac{\sin h\pi}{\pi} + \sum_{n=1}^{\infty} \frac{2(-1)^n}{\pi(a^2+n^2)}$$

$$= \frac{2\sin h\pi}{\pi}\left[\frac{1}{2} + \sum_{n=1}^{\infty} \frac{(-1)^n}{\pi(a^2+n^2)}\right]$$

$$\frac{\pi}{\sin h\pi} = 2\left[\frac{1}{2} - \frac{1}{1^2+1^2} + \frac{1}{1^2+2^2} + \frac{1}{1^2+3^2} \cdots\cdots\right]$$

$$= 2\left[\frac{1}{1^2+2^2} - \frac{1}{1^2+3^2} + \frac{1}{1^2+4^2} \cdots\cdots\right]$$

Example **9.2.** Prove that $x^2 = \frac{\pi^2}{3} + 4\sum_{n=1}^{\infty} (-1)\frac{\cos nx}{n^2}$, $-\pi < x < \pi$

Hence show that

1. $\sum_{n=1}^{\infty} \frac{1}{n^2} = \frac{\pi^2}{6}$

2. $\sum_{n=1}^{\infty} \frac{1}{(2n-1)^2} = \frac{\pi^2}{8}$

3. $\frac{1}{1^2} - \frac{2}{2^2} + \frac{1}{3^2} \cdots\cdots = \frac{\pi^2}{12}$

4. $\sum_{n=1}^{\infty} \frac{1}{n^4} = \frac{\pi^4}{90}$

Solution. Let $f(x) = x^2$ and

$$x^2 = f(x) = \frac{a_0}{2} + \sum_{1}^{\infty} a_n \cos nx + \sum_{1}^{\infty} b_n \sin nx$$

Then,

$$a_0 = \frac{1}{\pi}\int_{-\pi}^{\pi} f(x)\,dx = \frac{1}{\pi}\int_{-\pi}^{\pi} x^2\,dx = \frac{1}{\pi}\left[\frac{x^3}{3}\right]_{-\pi}^{\pi} = \frac{2}{3}\pi^2$$

$$a_n = \frac{1}{\pi}\int_{-\pi}^{\pi} f(x)\cos nx\,dx = \frac{1}{\pi}\int_{-\pi}^{\pi} x^2\cos nx\,dx = \frac{2}{\pi}\int_{0}^{\pi} x^2\cos nx\,dx$$

$$= \frac{2}{\pi}\left[x^2\frac{\sin nx}{n} - 2x\frac{(-\cos nx)}{n^2} + 2\frac{(-\sin nx)}{n^3}\right]_0^{\pi}$$

$$= \frac{2}{\pi}\left[\frac{2x\cos nx}{n^2}\right]_0^{\pi} \quad \text{(since } \sin nx = 0\text{, between the limits 0 and } \pi\text{)}$$

$$= \frac{4}{\pi n^2}\left[\pi\cos n\pi + 0 \times \cos(0)\right]$$

$$= \frac{4}{n^2}(-1)^n$$

$$b_n = \frac{1}{\pi}\int_{-\pi}^{\pi} f(x)\sin nx\,dx = \frac{1}{\pi}\int_{-\pi}^{\pi} x^2 \sin nx\,dx = 0 \text{ (since } x^2 \sin nx \text{ is an odd function)}$$

$$\therefore \quad x^2 = \frac{\pi^2}{3} + \sum_{n=1}^{\infty} 4\frac{(-1)^n}{n^2}\cos nx \qquad ...(1)$$

$$= \frac{\pi^2}{3} + 4\left[\frac{-\cos x}{1^2} + \frac{\cos 2x}{2^2} - \frac{\cos 3x}{3^2} + \frac{\cos 4x}{4^2} \text{......}\right]$$

$$= \frac{\pi^2}{3} - 4\left[\frac{\cos x}{1^2} - \frac{\cos 2x}{2^2} + \frac{\cos 3x}{3^2} - \frac{\cos 4x}{4^2} \text{......}\right]$$

Put $x = \pi$ in (1), $\pi^2 = \dfrac{\pi^2}{3} + 4\sum_{n=1}^{\infty}\left[\dfrac{(-1)^n}{n^2}\cos n\pi\right]$

$$\frac{2\pi^2}{3} = +4\sum_{n=1}^{\infty}\left[\frac{(-1)^n(-1)^n}{n^2}\right]$$

$$= +4\sum_{n=1}^{\infty}\frac{1}{n^2}$$

$$\sum_{n=1}^{\infty}\frac{1}{n^2} = \frac{\pi^2}{6} \qquad ...(i)$$

$$\therefore \quad \frac{\pi^2}{6} = \frac{1}{1^2} + \frac{1}{2^2} + \frac{1}{3^2} + \frac{1}{4^2} \text{......}$$

Put $x = 0$ in (1), $0 = \dfrac{\pi^2}{3} + 4\Sigma\dfrac{(-1)^n}{n^2}$

$$= \frac{\pi^2}{3} - 4\left[\frac{1}{1^2} + \frac{1}{2^2} + \frac{1}{3^2} + \frac{1}{4^2} \text{......}\right]$$

$$\frac{\pi^2}{12} = \frac{1}{1^2} - \frac{1}{2^2} + \frac{1}{3^2} \text{......} \qquad ...(ii)$$

Again (*i*) and (*ii*)

$$\frac{\pi^2}{6} + \frac{\pi^2}{12} = 2\left[\frac{1}{1^2} + \frac{1}{3^2} + \frac{1}{5^2} \text{......}\right]$$

$$\frac{\pi^2}{8} = \frac{1}{(2n-1)^2}$$

We have also $x^2 = \dfrac{\pi^2}{3} + 4\sum_{n=1}^{\infty}(-1)^n\dfrac{\cos nx}{n^2}$

$$\therefore \quad \int_{-\pi}^{\pi} x^4 dx = \int_{-\pi}^{\pi}\frac{\pi^2}{3}x^2 dx + 4\int_{-\pi}^{\pi}\sum_{n=1}^{\infty}(-1)^n\frac{x^2\cos nx}{n^2}dx$$

$$\frac{2\pi^5}{5} = \frac{1}{9}(2\pi^5) + 4\sum_{n=1}^{\infty}\frac{(-1)^n}{n^2}\int_{-\pi}^{\pi} x^2\cos nx\,dx$$

$$= \frac{2}{9}\pi^5 + 4\sum_{n=1}^{\infty}\frac{(-1)^n}{n^2}.2\left[x^2\frac{\sin nx}{n} - 2x\frac{(-\cos nx)}{n^2} + 2\left(\frac{-\sin nx}{n^3}\right)\right]_0^{\pi}$$

$$= \frac{2}{9}\pi^5 + 8\sum_{n=1}^{\infty}\frac{(-1)^n}{n^2}\left(\frac{2\pi\cos n\pi}{n^2}\right)$$

$$= \frac{2}{9}\pi^5 + 16\pi\sum_{n=1}^{\infty}\frac{(-1)^n(-1)^n}{n^4}$$

$$= \frac{2}{9}\pi^5 + 16\pi\sum_{n=1}^{\infty}\frac{1}{n^4}$$

$$16\pi\sum_{n=1}^{\infty}\frac{1}{n^4} = \frac{2\pi^5}{5} - \frac{2}{9}\pi^5 = \frac{2\pi^5}{45}\ [4]$$

$$= \frac{8\pi^5}{45}$$

$$= \frac{\pi^4}{90}$$

Example 9.3. Find the Fourier series for the function $f(x) = x + x^2$ in the interval $-\pi < x < \pi$. Hence deduce that

$$\frac{\pi^2}{16} = 1 + \frac{1}{2^2} + \frac{1}{3^2} \ldots\ldots$$

Solution. Let $x + x^2 = f(x)$ and

$$f(x) = \frac{a_0}{2} + \sum_{n=1}^{\infty} a_n \cos nx + \sum_{n=1}^{\infty} b_n \sin nx$$

Then

$$a_0 = \frac{1}{\pi}\int_{-\pi}^{\pi} f(x)\,dx = \frac{1}{\pi}\int_{-\pi}^{\pi}(x + x^2)\,dx = \frac{2\pi^2}{3}$$

$$a_n = \frac{1}{\pi}\int_{-\pi}^{\pi} f(x)\cos nx\,dx = \frac{1}{\pi}\int_{-\pi}^{\pi}(x + x^2)\cos nx\,dx$$

$$= \frac{1}{\pi}\int_{-\pi}^{\pi}(x\cos nx + x^2\cos nx)\,dx$$

$$= \frac{1}{\pi}\times 2\int_0^{\pi} x^2\cos nx\,dx \qquad \left\{\begin{matrix} x\cos nx \text{ is odd} \\ x^2\cos nx \text{ is even}\end{matrix}\right\}$$

$$= \frac{2}{\pi}\left[x^2\frac{\sin nx}{n} - 2x\frac{(-\cos nx)}{n^2} + 2\frac{(-\sin nx)}{n^3}\right]_0^{\pi}$$

$$= \frac{2}{\pi}\left[\frac{2\pi\cos n\pi}{n^2}\right] = \frac{4(-1)^n}{n^2}$$

$$b_n = \frac{1}{\pi}\int_{-\pi}^{\pi}(x + x^2)\sin nx\,dx$$

$$= \frac{1}{\pi}\int_{-\pi}^{\pi}(x\sin nx + x^2\sin nx)\,dx$$

$$= \frac{2}{\pi}\int_{0}^{\pi} x\sin nx\,dx \qquad \left(\text{as} \int_{-\pi}^{\pi} x^2 \sin nx dx = 0\right)$$

$$= \frac{2}{\pi}\left[x\,\frac{-\cos nx}{n} - \left(\frac{-\sin nx}{n^2}\right)\right]_0^{\pi}$$

$$= \frac{2}{\pi}\left[\frac{-\pi\cos n\pi}{n}\right] = \frac{-2(-1)^n}{n}$$

$$x + x^2 = \frac{\pi^3}{3} + 4\sum_{1}^{\infty}\frac{(-1)^n}{n^2}\cos nx - 2\sum_{1}^{\infty}\frac{(-1)^n}{n}\sin nx$$

Put $x = -\pi$

$$-\pi + \pi^2 = \frac{\pi^2}{3} + 4\sum_{n=1}^{\infty}\frac{1}{n^2}$$

$$= \frac{\pi^2}{3} + 4\left[\frac{1}{1^2} + \frac{1}{2^2} + \frac{1}{3^2} \dots\dots\right] \qquad \dots(i)$$

Put $\pi = x$

$$+\pi + \pi^2 = \frac{\pi^2}{3} + 4\left[\frac{1}{1^2} + \frac{1}{2^2} + \frac{1}{3^2} \dots\dots\right] \qquad \dots(ii)$$

Adding (*i*) and (*ii*)

$$2\pi^2 = 2\,\frac{\pi^2}{3} + 8\left[\frac{1}{1^2} + \frac{1}{2^2} + \frac{1}{3^2} \dots\dots\right]$$

$$\frac{4\pi^2}{3} = 8\left[\frac{1}{1^2} + \frac{1}{2^2} + \frac{1}{3^2} \dots\dots\right]$$

$$\frac{\pi^2}{6} = \frac{1}{1^2} + \frac{1}{2^2} + \frac{1}{3^2} \dots\dots$$

***Example* 9.4.** Evaluate $f(x) = x \sin x$, $0 < x < 2x$ as Fourier series and hence deduce that $\frac{\pi - 2}{4} = \frac{1}{1.3} - \frac{1}{3.5} + \frac{1}{5.7}$

Solution. Let $f(x) = x \sin x$ and

$$x\sin x = f(x) = \frac{a_0}{2} + \sum_{n=1}^{\infty} a_n\cos nx + \sum_{n=1}^{\infty} b_n\sin nx$$

Then

$$a_0 = \frac{1}{\pi}\int_{0}^{2\pi} x\sin x\,dx = \frac{1}{\pi}[x(-\cos x) - (-\sin x)]_0^{2\pi}$$

$$= \frac{1}{\pi}[-2\pi] = -2$$

$$a_n = \frac{1}{\pi}\int_{0}^{2\pi} x\sin x\cos nx\,dx$$

$$= \frac{1}{2\pi}\int_0^{2\pi} x[\sin(n+1)x - \sin(n-1)x]\,dx$$

$$= \frac{1}{2\pi}\left[x\left(\frac{-\cos(n+1)x}{n+1} + \frac{\cos(n-1)x}{n-1}\right) - \left(\frac{-\sin(n+1)x}{(n+1)^2} + \frac{\sin(n-1)x}{(n-1)^2}\right)\right]_0^{2\pi}$$

$$= \frac{1}{2\pi}\left[2\pi\left(-\frac{\cos 2(n+1)\pi}{n+1} + \frac{\cos 2(n-1)\pi}{n-1}\right)\right]$$

$$= \frac{1}{2\pi}\left[2\pi\left(-\frac{1}{n+1} + \frac{1}{n-1}\right)\right]$$

$$= \frac{1}{2\pi}\left[2\pi\left(\frac{+2}{n^2-1}\right)\right], n \neq 1$$

$$= \frac{2}{n^2-1}, n \neq 1$$

When $n = 1, a_1 = \frac{1}{\pi}\int_0^{2\pi} x \sin x \cos x\,dx$

$$= \frac{1}{2\pi}\int_0^{2\pi} x \sin 2x\,dx = \frac{1}{2\pi}\left[x\left(-\frac{\cos 2x}{2}\right) - 1.\left(-\frac{\sin 2x}{4}\right)\right]_0^{2\pi}$$

$$= +\frac{1}{2\pi}[-\pi] = -\frac{1}{2}$$

$$b_n = \frac{1}{\pi}\int_0^{2\pi} x \sin x \sin nx\,dx$$

$$= \frac{1}{2\pi}\int_0^{2\pi} x[\cos(n-1)x - \cos(n+1)x]\,dx$$

$$= \frac{1}{2\pi}\left[x\left(\frac{\sin(n-1)x}{n-1} - \frac{\sin(n+1)x}{n+1}\right) - \left(\frac{-\cos(n-1)x}{(n-1)^2} - \frac{-\cos(n+1)x}{(n+1)^2}\right)\right]_0^{2\pi}$$

$$= \frac{1}{2\pi}\left[\frac{\cos(n-1)\pi}{(n-1)^2} - \frac{\cos 2(n+1)\pi}{(n+1)^2} - \frac{1}{(n-1)^2} + \frac{1}{(n+1)^2}\right]$$

$$= \frac{1}{2\pi}\left[\left(\frac{1}{(n-1)^2} - \frac{1}{(n+1)^2}\right) - \left(\frac{1}{(n-1)^2} - \frac{1}{(n+1)^2}\right)\right]$$

$$= 0, n \neq 1$$

When $n = 1, b_1 = \frac{1}{\pi}\int_0^{2\pi} x \sin^2 x\,dx$

$$= \frac{1}{2\pi}\int_0^{2\pi} x(1 - \cos 2x)\,dx$$

$$= \frac{1}{2\pi}\left[x\left(x - \frac{\sin 2x}{2}\right) - \left(\frac{x^2}{2} + \frac{\cos 2x}{4}\right)\right]_0^{2\pi}$$

$$= \frac{1}{2\pi}\left[x^2 - \frac{x^2}{2} - \frac{\cos 2x}{4} - \frac{x \sin 2x}{2}\right]_0^{2\pi}$$

$$= \frac{1}{2\pi}\left[\left(\frac{4\pi^2}{2} - \frac{1}{4}\right) - \left(-\frac{1}{4}\right)\right]$$

$$= \frac{1}{2\pi} \times 2\pi^2 = \pi$$

$$x \sin x = -1 - \frac{1}{2}\cos x + \sum_{n=2}^{\infty} \frac{2}{n^2 - 1}\cos nx + \pi \sin x + \sum_{n=2}^{\infty}(0)$$

Put $x = \frac{\pi}{2}$

$$\frac{\pi}{2} = -1 + \pi + 2\sum_{n=2}^{\infty} \frac{1}{n^2 - 1}\cos\frac{n\pi}{2}$$

$$= -1 + \pi + 2\left[\frac{1}{2^2 - 1}\cos \pi + \frac{1}{3^2 - 1}\cos\frac{3\pi}{2} + \frac{1}{4^2 - 1}\cos 2\pi \ldots\ldots\right]$$

$$-\pi + \frac{\pi}{2} = -1 - 2\left[\frac{1}{1.3} - \frac{1}{3.5} + \frac{1}{5.7}\ldots\ldots\right]$$

$$1 - \frac{\pi}{2} = -2\left[\frac{1}{1.3} - \frac{1}{3.5} + \frac{1}{5.7}\ldots\ldots\right]$$

$$\frac{\pi - 2}{4} = \frac{1}{1.3} - \frac{1}{3.5} + \frac{1}{5.7}\ldots\ldots$$

***Example* 9.5.** If $f(x) = \left(\frac{\pi - x}{4}\right)^2$ in the range $0 < x < 2\pi$, show that $f(x) = \frac{\pi^2}{48} + \frac{1}{4}\sum_{n=1}^{\infty}\frac{\cos nx}{n^2}$

Solution. Let $f(x) = \left(\frac{\pi - x}{4}\right)^2$ and

$$\left(\frac{\pi - x}{4}\right)^2 = f(x) = \frac{a_0}{2} + \sum_{n=1}^{\infty} a_n \cos nx + \sum_{n=1}^{\infty} b_n \sin nx$$

Then

$$a_0 = \frac{1}{\pi}\int_0^{2\pi} f(x)\,dx = \frac{1}{\pi}\int_0^{2\pi}\left(\frac{\pi - x}{4}\right)^2 dx = \frac{\pi^2}{24}$$

$$a_n = \frac{1}{\pi}\int_0^{2\pi}\left(\frac{\pi - x}{4}\right)^2 \cos nx\,dx = \frac{1}{16\pi}\int_0^{2\pi}(\pi - x)^2 \cos nx\,dx$$

$$= \frac{1}{16\pi}\left[(\pi - x)^2\frac{\sin nx}{n} - 2(\pi - x)(-1)\frac{-\cos nx}{n^2} + 2(-1)^2\left(\frac{-\sin nx}{n^3}\right)\right]_0^{2\pi}$$

$$= \frac{1}{16\pi}\left[(\pi - x)^2\frac{\sin nx}{n} + 2(\pi - x)\left(\frac{-\cos nx}{n^2}\right) - 2\frac{\sin nx}{n^2}\right]_0^{2x}$$

$$= \frac{1}{16\pi}\left[-2(\pi - x)\frac{\cos nx}{n^2}\right]_0^{2\pi}$$

$$= \frac{+1}{16\pi}\left[2\pi\frac{\cos 2n\pi}{n^2} + 2\pi\frac{1}{n^2}\right]$$

$$= \frac{+\pi}{16\pi}\left[\frac{2}{n^2} + \frac{2}{n^2}\right] = \frac{+1}{16}\left[+\frac{4}{n^2}\right] = +\frac{1}{4n^2}$$

$$b_n = \frac{1}{\pi}\int_0^{2\pi} f(x)\sin nx\, dx = \frac{1}{\pi}\int_0^{2\pi}\left(\frac{\pi - x}{4}\right)^2 \sin nx\, dx = 0$$

Hence,
$$\left(\frac{\pi - x}{4}\right)^2 = \frac{\pi^2}{48} + \frac{1}{4}\sum_{n=1}^{\infty}\frac{\cos nx}{n^2}$$

***Example* 9.6.** Prove that in the range $-\pi < x < \pi$

$$\cosh ax = \frac{2a^2}{\pi}\sin a\pi\left\{\frac{1}{2a^2} + \sum_{n=1}^{\infty}\frac{(-1)^n}{n^2 + a^2}\cos nx\right\}$$

Solution. Let $f(x) = \cosh ax$ and

$$\cosh ax = f(x) = \frac{a_0}{2} + \sum_{n=1}^{\infty} a_n \cos nx + \sum_{n=1}^{\infty} b_n \sin nx$$

Then
$$a_0 = \frac{1}{\pi}\int_{-\pi}^{\pi} f(x)\, dx = \frac{1}{\pi}\int_{-\pi}^{\pi} \cos h\, ax\, dx$$

$$= \frac{1}{\pi}\left[\frac{\sin h\, ax}{a}\right]_{-\pi}^{\pi} = \frac{2}{\pi a}\sin h\, a\pi$$

$$a_n = \frac{1}{\pi}\int_{-\pi}^{\pi} f(x)\cos nx\, dx = \frac{1}{\pi}\int_{-\pi}^{\pi}\cos h\, ax \cos nx\, dx$$

$$= \frac{1}{2\pi}\int_{-\pi}^{\pi}(e^{ax}\cos nx + e^{-ax}\cos nx)\, dx$$

$$= \frac{1}{2\pi}\left[\int_{-\pi}^{\pi} e^{ax}\cos nx\, dx + \int_{-\pi}^{\pi} e^{-ax}\cos nx\, dx\right]$$

$$= \frac{1}{2\pi}\left[\frac{e^{ax}}{a^2 + n^2}\{a\cos nx + n\sin nx\} + \frac{e^{-ax}}{a^2 + n^2}\{-a\cos nx + n\sin nx\}\right]_{-\pi}^{\pi}$$

$$= \frac{1}{2\pi(a^2 + n^2)}[a\cos nx\{a^{ax} - e^{-ax}\} + n\sin nx\{e^{ax} + e^{-ax}\}]_{-\pi}^{\pi}$$

$$= \frac{1}{2\pi(a^2 + n^2)}[2a(-1)^n \sin h\, a\pi + 2a(-1)^n \sin h\, a\pi] \qquad [\because \sin n\pi = 0]$$

$$= \frac{2(-1)^n}{2\pi(a^2 + n^2)}[2a \sin h\, a\pi]$$

$$= \frac{2a(-1)^n}{\pi(a^2+n^2)} \sin h\, a\pi$$

$$b_n = \frac{1}{\pi}\int_{-\pi}^{\pi} \sin h\, ax \sin x\, dx$$

$$= \frac{1}{\pi}\frac{e^{ax}-e^{-ax}}{2} \sin nx\, dx$$

$$= \frac{1}{2\pi}\int_{-\pi}^{\pi}\left[\frac{e^{ax}}{a^2+n^2}\{a \sin nx - n\cos nx\}\right]_{-\pi}^{\pi} - \frac{1}{2\pi}\left[\frac{e^{-ax}}{a^2+n^2}\{-a \sin nx - n\cos nx\}\right]_{-\pi}^{\pi}$$

$$= \frac{1}{2\pi(a^2+n^2)}\{e^{a\pi}(0-n\cos n\pi) - e^{-ax}(0-n\cos(-n\pi)) + e^{-ax}(0-n\cos n\pi) - e^{ax}(0-n\cos(-n\pi))\}$$

$$= \frac{1}{2\pi(a^2+n^2)}[0] = 0$$

Hence

$$\cosh ax = \frac{1}{\pi a}\sin h\, a\pi + \frac{2a}{\pi}\sum_{n=1}^{\infty}\frac{(-1)^n}{a^2+n^2}\sin h\, a\pi$$

$$= \frac{2a}{\pi}\sin h\, a\pi\left[\frac{1}{2a^2} + \sum_{n=1}^{\infty}\frac{(-1)^n}{n^2+a^2}\cos nx\right]$$

EXERCISE 9.1

1. Find Fourier series for the function $f(x) = x\cos x,\ 0 < x < 2\pi$

Ans. $x\cos x = -\frac{1}{2}\sin x + 2\sum_{n=2}^{\infty}\frac{(-1)^n}{n^2-1}\sin nx$

2. Expand $f(x) = e^{-ax}$ in $-\pi < x > \pi$ Hence derive for $\frac{\pi}{\sin h\pi}$

Ans. $e^{ax} = \frac{2\sin h\, a\pi}{\pi}\left[\frac{1}{2a} - a\left(\frac{\cos x}{a^2+1^2} - \frac{\cos 2x}{a^2+2^2} + \frac{\cos 3x}{a^2+3^2}\right) - \left(\frac{\sin x}{a^2+1^2} - \frac{2\sin 2x}{a^2+2^2} + \frac{3\sin 3x}{a^2+3^2}\right)......\right]$

3. $f(x) = \begin{cases} -x, & -\pi < x < 0 \\ x, & 0 < x < \pi \end{cases}$

Ans. $f(x) = \frac{-\pi}{2} + \frac{4}{\pi}\left[\frac{\cos x}{1^2} + \frac{\cos 3x}{3^2} + \frac{\cos 5x}{5^2} + ...\right]$

4. Expand $f(x) = \sin ax$ in $-\pi < x < \pi$ as Fourier series.

Ans. $\sin ax = \frac{2\sin a\pi}{\pi}\left[\frac{\sin x}{1^2-a^2} - \frac{2\sin 2x}{2^2-a^2} + \frac{3\sin 3x}{3^2-a^2}\right]$

5. Expand $f(x) = x - x^2$ in $-\pi < x < \pi$ as Fourier series.

Ans. $x - x^2 = \frac{-\pi^2}{3} + 4\left[\frac{\cos x}{1^2} - \frac{\cos 2x}{2^2} + \frac{\cos 3x}{3^2} + \ldots\ldots\right]$

6. Obtain Fourier series for the function $\left.\begin{aligned} f(x) &= 1 + \frac{2x}{\pi} \\ &= 1 - \frac{2\pi}{\pi} \end{aligned}\right\} \begin{aligned} -\pi < x < 0 \\ 0 < x < \pi \end{aligned}$

Ans. $f(x) = \frac{8}{\pi^2}\left[\frac{\cos x}{1^2} + \frac{\cos 3x}{3^2} + \frac{\cos 5x}{5^2} + \ldots\ldots\right]$

7. Find Fourier series expansion for $f(x) = |\sin x|, -\pi < x < \pi$

Ans. $|\sin x| = \frac{2}{\pi} - \frac{4}{\pi}\left[\frac{\cos 2x}{3} + \frac{\cos 4x}{15} + \ldots\ldots\right]$

8. Find Fourier series of the function given $f(x + 2\pi) = f(x)$

$$f(x) = \begin{cases} x + \pi & 0 \le x \le \pi \\ -x + \pi & -\pi \le x \le 2\pi \end{cases}$$

Ans. $f(x) = \frac{\pi}{2} + \frac{4}{\pi}\left[\frac{\cos x}{1^2} + \frac{\cos 3x}{3^2} + \frac{\cos 5x}{5^2} + \ldots\ldots\right]$

9. An alternating-current after passing through a rectifier has the form' where I_0 is the maximum current and the period is 2π. Express i as a Fourier series.

$$\begin{aligned} i &= I_0 \sin x, & 0 \le x \le \pi \\ &= 0, & \pi \le x \le 2\pi \end{aligned}$$

Ans. $i = \frac{I_0}{2} + \frac{I_0}{2}\sin x - \frac{2I_0}{\pi}\left[\frac{\cos 2x}{2^2 - 1} + \frac{\cos 4x}{4^2 - 1} + \frac{\cos 6x}{6^2 - 1} + \ldots\ldots\right]$

10. Find the Fourier series to represent to function $f(x)$ is given by $f(x) = x$ for $0 \le x \le \pi$ and $f(x) = 2 \sin x$ for $\pi \le x \le 2\pi$. Deduce that $\frac{1}{1^2} + \frac{1}{3^2} + \frac{1}{5^2} + \ldots\ldots\ldots\ldots\ldots\ldots = \frac{\pi^2}{8}$

Ans. ?

11. Expand $f(x) = |\cos x|$ in $-\pi < x < \pi$ as Fourier series.

Ans. $|\cos x| = \frac{2}{\pi} + \frac{4}{\pi}\left[\frac{\cos 2x}{3} - \frac{\cos 4x}{15} + \ldots\ldots\right]$

12. Expand $f(x) = \sqrt{1 - \cos x}$ in $(0, 2\pi)$ as Fourier series.

Ans. $\sqrt{1 - \cos x} = \frac{2\sqrt{2}}{\pi} - \frac{4\sqrt{2}}{\pi}\sum_{n=2}^{\infty}\frac{\cos nx}{4n^2 - 1}$

9.3 EVEN AND ODD FUNCTIONS

If $f(-x) = f(x)$ the function $f(x)$ is an even function, Example : x^2, $\cos x$.

If $f(-x) = -f(x)$ the function $f(x)$ is odd function, Example : x, $\sin x$.

When $f(x)$ is even

$$a_0 = \frac{1}{\pi}\int_{-\pi}^{\pi} f(x)\,dx = \frac{2}{\pi}\int_{0}^{\pi} f(x)\,dx$$

$$a_n = \frac{1}{\pi}\int_{-\pi}^{\pi} f(x)\cos nx\,dx = \frac{2}{\pi}\int_{0}^{\pi} f(x)\cos nx\,dx$$

$$b_n = \frac{1}{\pi}\int_{-\pi}^{\pi} f(x)\sin nx\,dx = 0 \quad \text{(even function × odd function) = odd function}$$

When $f(x)$ is odd

$$a_0 = \frac{1}{\pi}\int_{-\pi}^{\pi} f(x)\,dx = 0$$

$$a_n = \frac{1}{\pi}\int_{-\pi}^{\pi} f(x)\cos nx\,dx = 0$$

$$b_n = \frac{1}{\pi}\int_{-\pi}^{\pi} f(x)\sin nx\,dx$$

$$= \frac{2}{\pi}\int_{0}^{\pi} f(x)\sin nx\,dx \quad \text{(odd function × odd function) = even function}$$

9.4 FOURIER SERIES FOR DISCONTINUOUS FUNCTIONS

Let a function $f(x)$ be discontinuous at $x = x_0$

This implies $f(x_0 - 0) < f(x_0)$

$f(x_0 + 0) > f(x_0)$

Both the limit $f(x_0 - 0)$, $f(x_0 + 0)$ exist but are not equal. The value of the function at a point of discontinuity is taken as the mean of $f(x_0 - 0)$, $f(x_0 + 0)$

i.e.,

$$f(x_0) = \frac{f(x_0 - 0) + f(x_0 + 0)}{2}$$

***Example* 9.7.** Find the Fourier expansion for $f(x)$ if

$$f(x) = -\pi, \quad -\pi < x < 0$$
$$= x, \quad 0 < x < \pi$$

deduce that $\frac{1}{1^2} + \frac{1}{3^2} + \frac{1}{5^2} + \dots\dots\dots = \frac{\pi^2}{8}$

Solution. Let $f(x) = \frac{a_0}{2} + \sum_{n=1}^{\infty} a_n \cos nx + \sum_{n=1}^{\infty} b_n \sin nx$

Then

$$a_0 = \frac{1}{\pi}\int_{-\pi}^{\pi} f(x)\,dx = \frac{1}{\pi}\int_{-\pi}^{0} -\pi\,dx + \frac{1}{\pi}\int_{0}^{\pi} x\,dx$$

$$= \frac{1}{\pi}(-\pi x)_{-\pi}^{0} + \frac{1}{\pi}\left(\frac{x^2}{2}\right)_0^{\pi}$$

$$= \frac{1}{\pi}\left[0-\pi^2+\frac{\pi^2}{2}-0\right] = \frac{-\pi^2}{2}\times\frac{1}{\pi}$$

$$= \frac{-\pi^2}{2}\times\frac{1}{\pi} = \frac{-\pi}{2}$$

$$a_n = \frac{1}{\pi}\int_{-\pi}^{\pi}(-\pi)\cos nx\,dx + \frac{1}{\pi}\int_0^{\pi} x\cos nx\,dx$$

$$= \frac{1}{\pi}\left(-\pi\frac{\sin nx}{n}\right)_{-\pi}^{\pi} + \frac{1}{\pi}\left[x\frac{\sin nx}{n} - \left(\frac{-\cos nx}{n^2}\right)\right]_0^{\pi}$$

$$= \frac{1}{\pi}(-\pi\times 0) + \frac{1}{\pi}\left[\frac{\cos nx}{n^2} + \left(\frac{-1}{n^2}\right)\right]$$

$$= \frac{\cos nx - 1}{\pi n^2} = \frac{(-1)^n - 1}{\pi n^2} = \begin{cases} 0, & \text{if } n \text{ is even} \\ \frac{-2}{\pi n^2}, & \text{if } n \text{ is odd} \end{cases}.$$

$$a_1 = -\frac{2}{\pi 1^2},\ a_2 = 0,\ a_3 = -\frac{2}{\pi a^3}\ a_3 = 0,\ a_5 = -\frac{2}{\pi 5^2}\ \ldots\ldots$$

$$b_n = \frac{1}{\pi}\left[\int_{-\pi}^{0}(-\pi)\sin nx\,dx + \int_0^{\pi} x\sin nx\,dx\right]$$

$$= \frac{1}{\pi}\left[(-\pi)\frac{-\cos nx}{n}\right]_{-\pi}^{0} + \frac{1}{\pi}\left[x\frac{(-\cos nx)}{n} - \frac{(-\sin nx)}{n^2}\right]_0^{\pi}$$

$$= \frac{1}{\pi}\left[(+\pi)\left(\frac{\cos nx}{n}\right)\right]_{-\pi}^{0} + \frac{1}{\pi}\left[-x\frac{\cos nx}{n} + \frac{\sin nx}{n^2}\right]_0^{\pi}$$

$$= \frac{1}{\pi}\left[\pi\left(\frac{1}{n}\right) - \pi\left(\frac{\cos n\pi}{n}\right)\right] + \frac{1}{\pi}\left[-\pi\frac{\cos n\pi}{n}\right]$$

$$= \frac{\pi}{\pi}\left[\frac{1}{n} - \frac{2(-1)^n}{n}\right]$$

$$= \frac{\pi}{\pi}\left[\frac{1}{n} - \frac{2(-1)^n}{n}\right]$$

$$= \frac{1}{n} - 2\frac{(-1)^n}{n} = \frac{1}{n}[1-2(-1)^n]$$

$$b_1 = 3,\ b_2 = -\frac{1}{2},\ b_3 = 1,\ b_4 = -\frac{1}{4}$$

$$f(x) = \frac{-\pi}{4} - \frac{2}{\pi}\left[\cos x + \frac{\cos 3x}{3^2} + \frac{\cos 5x}{5^2} + \ldots\ldots\right] + 3\sin x - \frac{\sin 2x}{2} + \frac{3\sin 3x}{3} - \frac{\sin 4x}{4}\ \ldots\ldots$$

Putting $x = 0$ **we obtain**

$$f(0) = \frac{-\pi}{4} - \frac{2}{\pi}\left[1 + \frac{1}{3^2} + \frac{1}{5^2}\ \ldots\ldots\right]$$

Since $x = 0$ is the point of discontinuity

$$f(0) = \frac{f(0-0)+f(0+0)}{2} = \frac{-\pi+0}{2} = \frac{-\pi}{2}$$

$$\frac{-\pi}{2} = \frac{-\pi}{4} - \frac{2}{\pi}\left[1 + \frac{1}{3^2} + \frac{1}{5^2} + \ldots\ldots\right]$$

$$\frac{-\pi}{4} = -\frac{2}{\pi}\left[1 + \frac{1}{3^2} + \frac{1}{5^2} + \ldots\ldots\right]$$

$$\frac{\pi^2}{8} = 1 + \frac{1}{3^2} + \frac{1}{5^2} + \ldots\ldots$$

***Example* 9.8.** If $f(x) = 0$ for $-\pi \le x \le 0$ and $f(x) = \sin x$ for $0 \le x \le \pi$ prove that $f(x) = \frac{1}{x} + \frac{\sin x}{2} - \frac{2}{\pi}\sum_{m=1}^{\infty}\frac{\cos 2mx}{4m^2 - 1}$

Hence show that $\frac{1}{1.3} - \frac{1}{3.5} + \frac{1}{5.7} \ldots\ldots\ldots = \frac{\pi - 2}{4}$

Solution. Let $f(x) = \frac{a_0}{2} + \sum_{n=1}^{\infty} a_n \cos nx + \sum_{n=1}^{\infty} b_n \cos nx$

then
$$a_0 = \frac{1}{\pi}\int_{-\pi}^{\pi} f(x)\,dx = \frac{1}{\pi}\left[\int_{-\pi}^{0} 0\,dx + \int_{0}^{\pi} \sin x\,dx\right]$$

$$= \frac{1}{\pi}[-\cos \pi]_0^{\pi} = \frac{1}{\pi}[1 + 1] = \frac{2}{\pi}$$

$$a_n = \frac{1}{\pi}\int_{-\pi}^{\pi} f(x)\cos nx\,dx = \frac{1}{\pi}\left[\int_{0}^{\pi} \sin x \cos nx\,dx\right]$$

$$= \frac{1}{2\pi}\left[\int_{0}^{\pi}[\sin(1+n)x + \sin(1-n)x]\,dx\right]$$

$$= \frac{1}{2\pi}\int_{0}^{\pi}[\sin(n+1)x - \sin(n-1)x]\,dx$$

$$= \frac{1}{2\pi}\left[-\cos\frac{(n+1)x}{n+1} - \frac{-\cos(n-1)\pi}{n-1}\right]_0^{\pi}, n \ne 1$$

$$= \frac{1}{2\pi}\left[\left(\frac{-\cos(n+1)\pi}{n+1} + \frac{\cos(n-1)\pi}{n-1}\right) - \left(-\frac{1}{n+1} + \frac{1}{n-1}\right)\right]$$

$$= \frac{1}{2\pi}\left[-\frac{(-1)^{n+1}}{n+1} + \frac{(-1)^{n-1}}{n-1} + \frac{1}{n+1} - \frac{1}{n-1}\right]$$

When n is even $n + 1$, $n - 1$ are odd

$$a_n = \frac{1}{2\pi}\left[\frac{1}{n+1} - \frac{1}{n-1} + \frac{1}{n+1} - \frac{1}{n-1}\right]$$

$$= \frac{1}{\pi}\left[\frac{1}{n+1} - \frac{1}{n-1}\right] = \frac{1}{\pi}\left[\frac{-2}{n^2-1}\right] : n \ne 1$$

When n is odd $n + 1$, $n - 1$ are even

$$a_n = \frac{1}{2\pi}\left[-\frac{1}{n+1} + \frac{1}{n-1} + \frac{1}{n+1} - \frac{1}{n-1}\right] = 0 \quad : \quad n \ne 1$$

When $n = 1$

$$a_1 = \frac{1}{\pi}\int_0^{\pi} \sin x \cos x\, dx = \frac{1}{2\pi}\int_0^{2\pi} \sin 2x\, dx$$

$$= \frac{1}{2\pi}\left(\frac{-\cos 2x}{2}\right)_0^{\pi}$$

$$= -\frac{1}{4\pi}[\cos 2\pi - \cos 0]$$

$$= -\frac{1}{4\pi}[1-1] = 0$$

Now

$$b_n = \frac{1}{\pi}\int_{-\pi}^{\pi} f(x)\sin nx\, dx$$

$$= \frac{1}{\pi}\left[\int_{-\pi}^{0} 0 \sin nx\, dx + \int_0^{\pi} \sin x \sin nx\, dx\right]$$

$$= \frac{1}{2\pi}\left[\int_0^{\pi} 2\sin x \sin nx\, dx\right]$$

$$= \frac{1}{2\pi}\left[\int_0^{\pi} [\cos(n-1)x - \cos(n+1)x]\, dx\right]$$

$$= \frac{1}{2\pi}\left[\frac{\sin(n-1)x}{n-1} - \frac{\sin(n+1)x}{n+1}\right]_0^{\pi} = 0,\ n \neq 1$$

When

$$n = 1,\ b_1 = \frac{1}{\pi}\int_0^{\pi} \sin x \sin x\, dx = \frac{1}{\pi}\int_0^{\pi} \sin^2 x dx$$

$$= \frac{1}{\pi}\int_0^{\pi} \frac{1-\cos 2x}{2} dx$$

$$= \frac{1}{2\pi}\left[x - \frac{\sin 2x}{2}\right]_0^{\pi}$$

$$= \frac{1}{2\pi}[\pi] = \frac{1}{2}$$

$$f(x) = \frac{1}{\pi} + \frac{1}{\pi}\sum_{2,4,6}^{\infty} \frac{-2}{n^2-1}\cos nx + \frac{1}{2}\sin x$$

$$f(x) = \frac{1}{\pi} + \frac{1}{2}\sin x - \frac{2}{\pi}\sum \frac{\cos nx}{n^2-1} \quad \text{where } n \text{ is even integer.}$$

Put $n = 2m$

$$f(x) = \frac{1}{\pi} + \frac{1}{2}\sin x - \frac{2}{\pi}\left[\frac{\sum_{m=1}^{\infty}\cos 2mx}{4m^2-1}\right]$$

Put $x = \frac{\pi}{2}$ $\sin\frac{\pi}{2} = \frac{1}{\pi} + \frac{1}{2} - \frac{2}{\pi}\sum\frac{(-1)^m}{4m^2-1}$

$$\frac{1}{2} - \frac{1}{\pi} = \frac{-2}{\pi}\left[\frac{\Sigma(-1)^m}{(2m-1)(2m+1)}\right]$$

$$\frac{\pi-2}{2\pi} = \frac{-2}{\pi}\left[\frac{-1}{1.3} + \frac{1}{3.5} - \frac{1}{5.7}\ldots\ldots\right]$$

$\therefore$ $$\frac{1}{1.5} - \frac{1}{3.5} + \frac{1}{5.7} = \frac{\pi-2}{4}$$

***Example* 9.9.** Find Fourier series of the following $f(x) = x^2$ $0 \le x \le \pi$ and $f(x) = -x^2 - \pi \le x \le 0$

Solution. When $-\pi \le x \le 0, 0 \le -x \le \pi$

$\therefore$ $$f(-x) = (-x)^2 = x^2 = -f(x)$$

When $0 \le x \le \pi, -\pi \le -x \le 0$

$\therefore$ $$f(-x) = -(-x)^2 = -x^2 = -f(x)$$

$\Rightarrow f(x)$ is an odd function of x in $[-\pi, \pi]$. Also the graph of above function is symmetrical about the origin.

$\therefore$ $$a_0 = 0 = a_n$$

Let $$f(x) = \sum_{n=1}^{\infty} b_n \sin nx$$

when $$b_n = \frac{1}{\pi}\left[\int_{-\pi}^{0} -x^2 \sin nx dx + \int_0^{\pi} x^2 \cos nx\, dx\right]$$

$$= \frac{1}{\pi}\left[\frac{-x^2(-\cos nx)}{n} - \frac{(-2x)(-\sin nx)}{n^2} + \frac{(-2)(-\cos nx)}{n^3}\right]_{-\pi}^{0}$$

$$+ \frac{1}{\pi}\left[\frac{x^2(-\cos nx)}{n} - \frac{(2x)(-\sin nx)}{n^2} + \frac{(2)(\cos nx)}{n^3}\right]_0^{\pi}$$

$$= \frac{1}{\pi}\left[\left(\frac{-2}{n^3}\right) - \left\{\frac{-\pi^2(-\cos n\pi)}{n} + \frac{-2\cos n(-\pi)}{n^3}\right\}\right] + \frac{1}{\pi}\left[\left(\frac{-\pi^2\cos n\pi}{n} + \frac{2\cos n\pi}{n^3}\right) - \left(\frac{2}{n^3}\right)\right]$$

$$= \frac{1}{\pi}\left[\frac{-2}{n^3} - \frac{\pi^2(-1)^n}{n} + \frac{2(-1)^n}{n^3} - \frac{\pi^2(-1)^2}{n} + \frac{2(-1)^n}{n^3} - \frac{2}{n^3}\right]$$

$$= \frac{1}{\pi}\left[\frac{-4}{n^3} - \frac{2\pi^2(-1)^n}{n} + \frac{4(-1)^n}{n^3}\right]$$

$$b_1 = \frac{1}{\pi}\left[\frac{-4}{1^3} + \frac{2\pi^2}{1} - \frac{4}{1^3}\right] = \frac{1}{\pi}\left[\frac{2\pi^2}{1} - \frac{8}{1^3}\right]$$

$$b_2 = \frac{1}{\pi}\left[\frac{-4}{2^3} - \frac{2\pi^2}{2} + \frac{4}{2^3}\right]$$

$$b_2 = \frac{1}{\pi}[-\pi^2] = -\pi$$

$$b_3 = \frac{1}{\pi}\left[-\frac{4}{3^3} + \frac{2\pi^2}{3} - \frac{4}{3^3}\right]$$

$$= \frac{1}{\pi}\left[-\frac{8}{3^3} + \frac{2\pi^2}{3}\right] = \frac{1}{\pi}\left[\frac{2\pi^2}{3} - \frac{8}{3^3}\right]$$

$$b_4 = \frac{1}{\pi}\left[-\frac{4}{4^3} - \frac{2\pi^2}{4} + \frac{4}{4^3}\right]$$

$$= \frac{1}{\pi}\left[-\frac{2\pi^2}{4}\right] = \frac{\pi}{2}$$

$$f(x) = \left(\frac{2\pi}{1^3} - \frac{8}{\pi 1^3}\right)\sin x - \pi \sin 2x + \left(\frac{2\pi}{3} - \frac{8}{\pi 3^3}\right)\sin 3x - \frac{\pi}{2}\sin 4x + \ldots\ldots$$

***Example* 9.10.** Show that for $-\pi < x < \pi$

$$\sin ax = \frac{2\sin a\pi}{\pi}\left[\frac{\sin x}{1^2 - a^2} - \frac{2\sin 2x}{2^2 - a^2} + \frac{3\sin 3x}{3^2 - a^2} \ldots\ldots\right]$$

Solution. Let $f(x) = \sin ax$ and

$$\sin ax = f(x) = \frac{a_0}{2} + \sum_{n=1}^{\infty} a_n \cos nx + \sum_{n=1}^{\infty} b_n \sin nx$$

We know that $f(-x) = \sin(-ax) = -\sin ax = -f(x)$

Therefore, function $f(x)$ is an odd function

$\therefore$

$$a_0 = 0,\ a_n = 0$$

and

$$b_n = \frac{1}{\pi}\int_{-\pi}^{\pi} \sin ax \sin nx\, dx$$

$$= \frac{2}{\pi}\int_0^{\pi} \sin ax \sin nx\, dx$$

$$= \frac{1}{\pi}\int_0^{\pi} [\cos(n-a)x - \cos(n+a)x]\, dx$$

$$= \frac{1}{\pi}\left[\frac{\sin(n-a)x}{n-a} - \frac{\sin(n+a)x}{n+a}\right]_0^{\pi}$$

$$= \frac{1}{\pi}\left[\frac{\sin(n-a)x}{n-a} - \frac{\sin(n+a)x}{n+a}\right]$$

$$= \frac{1}{\pi}\left[\frac{(-1)^{n+1}\sin a\pi}{n-a} - \frac{(-1)^n \sin a\pi}{n+a}\right]$$

$$= \frac{(-1)^{n+1}}{\pi}\left[\frac{2n\sin a\pi}{n^2-a^2}\right], n \neq a$$

$$= \frac{2(-1)^{n+1}}{\pi(n^2-a^2)} n \sin a\pi$$

$$f(x) = \sum_{n=1}^{\infty} \frac{2(-1)^{n+1}}{(n^2-a^2)\pi} n \sin a\pi \sin nx$$

$$= \frac{2\sin a\pi}{\pi}\left[\frac{1}{1^2-a^2}\sin x - \frac{2}{2^2-a^2}\sin 2x + \frac{3}{3^2-a^2}\sin 3x \ldots\ldots\right]$$

***Example* 9.11.** For a function $f(x)$ defined by $f(x) = |x| - \pi < x < \pi$, obtain the Fourier series. Also deduce that $\frac{1}{1^2}+\frac{1}{3^2}+\frac{1}{5^2}+\ldots\ldots = \frac{\pi^2}{8}$

Solution. Let $f(x) = |x|$ and

$$|x| = f(x) = \frac{a_0}{2} + \sum_{n=1}^{\infty} a_n \cos nx + \sum_{n=1}^{\infty} b_n \sin nx$$

then
$$a_0 = \frac{1}{\pi}\int_{-\pi}^{\pi} f(x)\,dx = \int_{-\pi}^{\pi} |x|\,dx$$

$$= \frac{1}{\pi}\int_{-\pi}^{0}(-x)\,dx + \frac{1}{\pi}\int_{0}^{\pi} x\,dx$$

$$= \frac{1}{x}\left[\frac{-x^2}{2}\right]_{-\pi}^{0} + \frac{1}{\pi}\left[\frac{x^2}{2}\right]_{0}^{\pi}$$

$$= \frac{1}{2\pi}[(0+\pi^2)+(\pi^2-0)] = 0$$

$$= \frac{2\pi^2}{2\pi} = \pi$$

$$a_n = \frac{1}{\pi}\int_{-\pi}^{\pi} f(x)\cos nx\,dx = \frac{1}{\pi}\int_{-\pi}^{\pi}|x|\cos nx\,dx = \frac{2}{\pi}\int_{0}^{\pi} x\cos nx\,dx$$

$$= \frac{2}{\pi}\left[x\frac{\sin nx}{n} - \left(\frac{-\cos nx}{n^2}\right)\right]_0^{\pi}$$

$$= \frac{2}{\pi}\left[x\frac{\sin nx}{n} + \frac{\cos nx}{n^2}\right]_0^{\pi}$$

$$= \frac{2}{\pi}\left[\frac{\cos nx}{n^2} - \frac{1}{n^2}\right] = \frac{2}{\pi}[(-1)^n - 1]\frac{1}{n^2}$$

Case I When n is even $\quad a_n = 0$

Case II When n is odd
$$a_n = \frac{-4}{\pi n^2}$$

$$b^n = \frac{1}{\pi}\int_{-\pi}^{\pi} f(x)\sin nx dx = \frac{1}{\pi}\int_{-\pi}^{0} -(x)\sin nx dx + \frac{1}{\pi}\int_{0}^{\pi} x\sin nx\, dx$$

$$= \frac{1}{\pi}\left[-x\left(\frac{-\cos nx}{n}\right)-(-1)\left(\frac{\sin nx}{n^2}\right)\right]_{-\pi}^{0} + \frac{1}{\pi}\left[x\left(\frac{-\cos nx}{n}\right)-1\left(\frac{-\sin nx}{n^2}\right)\right]_{0}^{\pi}$$

$$= \frac{1}{\pi}\left[x\frac{\cos nx}{n} - \frac{\sin nx}{n^2}\right]_{-\pi}^{0} + \frac{1}{\pi}\left[-x\frac{\cos nx}{n} + \frac{\sin nx}{n^2}\right]_{0}^{\pi}$$

$$= \frac{1}{\pi}\left[-\left(-\pi\frac{(-1)^n}{n}\right)+\left(-\pi\frac{(-1)^n}{n}\right)\right] = \frac{1}{\pi}[0] = 0$$

$$|x| = f(x) = \frac{\pi}{2} - \frac{4}{\pi}\left[\frac{1}{1^2}\cos x + \frac{1}{3^2}\cos 3x + ...\right]$$

Put
$$x = 0$$

$$0 = \frac{\pi}{2} - \frac{4}{\pi}\left[\frac{1}{1^2}+\frac{1}{3^2}+\frac{1}{5^2}+......\right]$$

$$\therefore \quad \frac{\pi^2}{8} = \frac{1}{1^2}+\frac{1}{3^2}+\frac{1}{5^2}+......$$

Note : Since $|x|$ is an even function, it can be expanded as $|x| \frac{a_0}{2} + \sum_{n=1}^{\infty} a_n \cos nx$

***Example* 9.12.** Obtain Fourier series for the function

$$f(x) = x, \quad 0 \le x \le \pi$$
$$= 2\pi - x, \quad \pi \le x \le x \text{ Deduce that } \frac{1}{1^2}+\frac{1}{3^2}+\frac{1}{5^2}+......... = \frac{\pi^2}{8}$$

Solution. Let $f(x) = \frac{a_0}{2} + \sum_{n=1}^{\infty} a_n \cos nx + \sum_{n=1}^{\infty} b_n \sin nx$

Then
$$a_0 = \frac{1}{\pi}\int_0^{2\pi} f(x)dx = \frac{1}{\pi}\left[\int_0^{\pi} xdx + \int_{\pi}^{2\pi}(2\pi - x)dx\right] = \frac{1}{\pi}\left[\left|\frac{x^2}{2}\right|_0^{\pi} + \left|2\pi x - \frac{x^2}{2}\right|_{\pi}^{2\pi}\right] = \pi$$

$$a_n = \frac{1}{\pi}\left[\int_0^{\pi} x\cos nx\, dx + \int_{\pi}^{2\pi}(2\pi - x)\cos n\, xdx\right]$$

$$= \frac{1}{\pi}\left[x\left(\frac{\sin nx}{n}\right)-1\left(\frac{-\cos nx}{n^2}\right)\right]_0^{\pi} + \frac{1}{\pi}\left[(2\pi - x)\frac{\sin nx}{n} - (-1)\left(\frac{-\cos nx}{n^2}\right)\right]_{\pi}^{2\pi}$$

$$= \frac{1}{\pi}\left[\left(\frac{\cos n\pi}{n^2} - \frac{1}{n^2}\right) + \left(-\frac{\cos 2n\pi}{n^2} + \frac{\cos n\pi}{n^2}\right)\right]$$

$$= \frac{1}{\pi n^2}[(-1)^n - 1 - 1 + (-1)^n] = \frac{2}{\pi n^2}[(-1)^n - 1]$$

$$a_n = \begin{cases} \dfrac{-4}{\pi n^2}, \text{ if } & n \text{ is odd} \\ 0, \text{ if } & n \text{ is even} \end{cases}$$

$$b_n = \frac{1}{\pi}\int_0^{2\pi} f(x)\sin nx\, dx$$

$$= \frac{1}{\pi}\left[\int_0^{\pi} x\sin nx\, dx + \int_{\pi}^{2\pi}(2\pi - x)\sin nx dx\right]$$

$$= \frac{1}{\pi}\left[x\left(\frac{-\cos nx}{n}\right) - \left(-\frac{\sin nx}{n^2}\right)\right]_0^{\pi} + \left[(2\pi - x)\left(-\frac{\cos nx}{n}\right) - (-1)\left(-\frac{\sin nx}{n^2}\right)\right]_{\pi}^{2\pi}$$

$$= \frac{1}{\pi}\left[-\frac{\pi\cos n\pi}{n} + \frac{\pi\cos n\pi}{n}\right] = 0$$

$$\therefore \quad f(x) = \frac{\pi}{2} - \frac{4}{\pi}\left[\frac{\cos x}{1^2} + \frac{\cos 2x}{3^2} + \frac{\cos 5x}{5^2} \ldots\ldots\right]$$

Putting $x = 0$, we get

$$0 = \frac{\pi}{2} - \frac{4}{\pi}\left[\frac{1}{1^2} + \frac{1}{3^2} + \frac{1}{5^2} \ldots\ldots\right] \Rightarrow \frac{\pi^2}{8} = \frac{1}{1^2} + \frac{1}{3^2} + \frac{1}{5^2}$$

***Example* 9.13.** Obtain Fourier series for the function $f(x)$ given by

$$f(x) = 1 + \frac{2x}{\pi}, \quad -\pi \le x \le 0$$

$$= 1 - \frac{2x}{\pi}, \quad 0 \le x \le \pi$$

Hence deduce that $\dfrac{1}{1^2} + \dfrac{1}{3^2} + \dfrac{1}{5^2} \ldots\ldots = \dfrac{\pi^2}{8}$

Solution. When $f(x) = 1 + \dfrac{2x}{\pi}, \quad -\pi \le x \le 0$, then $0 \le -x \le \pi$

$$f(-x) = 1 - \frac{2x}{\pi}, \quad 0 \le x \le \pi = f(x)$$

Also when $f(x) = 1 - \dfrac{2x}{\pi} \quad 0 \le x \le \pi$, then $-\pi \le -x \le 0$

$$f(-x) = 1 - \frac{2(-x)}{\pi} = f(x) \quad -\pi \le x \le 0$$

Hence $f(-x) = f(x)$ in both the cases, therefore the function is an even function in $[-\pi, \pi]$

$$\therefore \quad b_n = 0$$

Hence,
$$f(x) = \frac{a_0}{2} + \sum_{n=1}^{\infty} a_n \cos nx$$

Now
$$a_0 = \frac{2}{\pi}\int_0^{\pi} f(x)dx = \frac{2}{\pi}\int_0^{\pi}\left(1 - \frac{2x}{\pi}\right)dx$$

$$= \frac{2}{\pi}\left[x - \frac{x^2}{\pi}\right]_0^{\pi} = 0$$

$$a_n = \frac{2}{\pi}\int_0^{\pi} f(x)\cos nx dx = \frac{2}{\pi}\int_0^{\pi}\left(1 - \frac{2x}{\pi}\right)\cos nx dx$$

$$= \frac{2}{\pi}\left[\left(1-\frac{2x}{\pi}\right)\frac{\sin nx}{n} - \left(-\frac{2}{\pi}\right)\frac{(-\cos nx)}{n^2}\right]_0^{\pi}$$

$$= \frac{2}{\pi}\left[\left(-\frac{2}{\pi}\frac{\cos nx}{n^2}\right)+\left(\frac{2}{\pi}\frac{1}{n^2}\right)\right]$$

$$= \frac{2}{\pi}\left[-\frac{2}{n}\frac{(-1)^n}{n^2}+\frac{2}{\pi}\frac{1}{n^2}\right]$$

$$= \frac{4}{n^2\pi^2}[-(-1)^n+1] = \frac{4}{\pi^2 n^2}[1-(-1)^n]$$

$\therefore$ $$f(x) = \frac{4}{\pi^2}\left[\frac{2\cos x}{1^2}+\frac{2\cos 3x}{3^2}+\frac{2\cos 5x}{5^2}\ \ldots\ldots\right] [\because 1-(-1)^n = 0 \text{ when } n \text{ is even}]$$

Now, $$f(0) = 1 = \frac{8}{\pi^2}\left[\frac{1}{1^2}+\frac{1}{3^2}+\frac{1}{5^2}\ \ldots\ldots\right]$$

$\therefore$ $$\frac{1}{1^2}+\frac{1}{3^2}+\frac{1}{5^2} = \frac{\pi^2}{8}$$

9.5 CHANGE OF INTERVAL (CHANGE OF SCALE)

We have so far considered only integral of length 2π or *i.e.*, $(-\pi, \pi)$ or $(0, 2\pi)$, but most of the practical problems involve use of function with different periods say '$2l$'. In such case we transform by simple change of variable, the interval of arbitrary length *i.e.* $2l$ into length of 2π, and the determine fourier series.

Let the function be defined in the interval $(-l, l)$ *i.e.*, $-l < x < l$, we have to change it to $-\pi < x < \pi$, for that we put $z = \frac{\pi x}{l}$ or $x = \frac{lz}{\pi}$ $\therefore$ $f(x) = f\left(\frac{lz}{\pi}\right) = F(z)$

as
$$x = -l \Rightarrow z = -\pi$$
$$x = l \Rightarrow z = \pi$$

$F(z)$ is defined in the entered $(-\pi, \pi)$

$$F(z) = \frac{a_0}{2} + \Sigma a_n \cos nz + \Sigma b_n \sin nz$$

where
$$a_0 = \frac{1}{\pi}\int_{-\pi}^{\pi} F(z)\,dz$$

$$a_n = \frac{1}{\pi}\int_{-\pi}^{\pi} F(z)\cos nz\,dz$$

$$b_n = \frac{1}{\pi}\int_{-\pi}^{\pi} F(z)\sin nz\,dz$$

by inverse transformation

$$f(x) = \frac{a_0}{2} + \sum a_n \cos n\frac{\pi x}{l} + \Sigma b_n \sin n\frac{\pi x}{l}$$

where
$$a_0 = \frac{1}{l}\int_{-l}^{l} f(x)\,.\,\frac{\pi}{l}\,dx$$

$$a^n = \frac{1}{l}\int_{-l}^{l} f(x)\cos\frac{n\pi x}{l}dx$$

$$b_n = \frac{1}{l}\int_{-l}^{l} f(x)\sin\frac{n\pi x}{l}dx$$

9.6 HALF RANGE SERIES

Suppose we wish to find a Fourier series for the function $f(x)$ which is defined in half the interval 'say' $0 \le x \le \pi$ or $0 \le x \le l$. This can be done by defining a function $F(x)$ in the interval $-l \le x \le l$ such that $F(x) : f(x)$ in the interval $0 \le x \le l$.

Now if we define $F(x)$ in $-l \le x \le l$ then the series of $F(x)$ can be obtained. $F(x)$ in $-l < x < l$ is chosen in either of the following two ways.

1. **Half range cosine series**

Let $F(x)$ be an even function in $-l < x < l$ such that $F(x) = f(x)$ in $0 \le x \le l$

we have
$$a_0 = \frac{1}{l}\int_{-l}^{l} F(x)dx = \frac{2}{l}\int_0^3 F(x)dx = \frac{2}{l}\int_0^l f(x)dx$$

$$a_n = \frac{1}{l}\int_{-l}^{l} F(x\cos\frac{n\pi x}{l}dx = \frac{2}{l}\int_0^l F(x)\cos\frac{x\pi x}{l}dx$$

$$= \frac{2}{l}\int_0^l f(x)\cos\frac{n\pi x}{l}dx$$

$$b_n = \frac{1}{l}\int_{-l}^{l} F(x)\sin\frac{n\pi x}{l}dx = 0$$

in
$$f(x) = \frac{a_0}{2}\sum_{n=1}^{\infty} b_n \sin\frac{n\pi x}{l}$$

2. **Half range sine series**

Let $F(x)$ be an odd function in $-l \le x \le l$

$$a_0 = 0 \quad a_n = 0$$

$$b_n = \frac{2}{l}\int_0^l F(x)\sin\frac{n\pi x}{l}dx$$

$$= \frac{2}{l}\int_0^l f(x)\sin\frac{n\pi x}{l}dx$$

$$f(x) = \Sigma\, b_n \sin\frac{n\pi x}{l}$$

OR

1. **Half range sine series**

If it required to expand the function $f(x)$ as a sine series in the interval $0 < x < l$ we extend the function to $-l < x < l$ by reflecting it in the origin, so that $f(-x) = -f(x)$. for which $a_0 = a_n = 0$

Then the function $f(x)$ will be odd and we therefore have

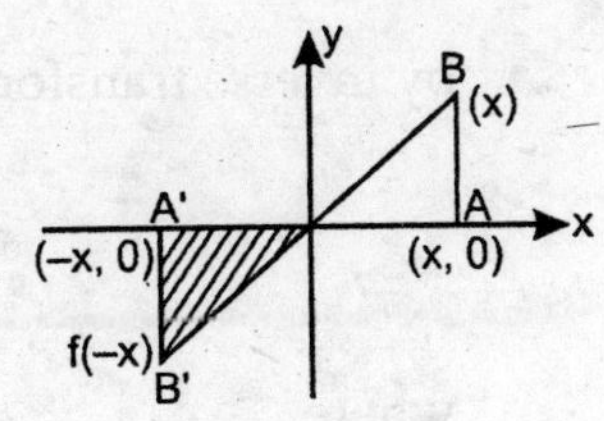

Fig. 9.1

$$f(x) = \sum_{n=1}^{\infty} b_n \sin\frac{n\pi}{l}x$$

where $$b_n = \frac{2}{l}\int_0^l f(x)\sin\frac{n\pi}{l}x\,dx$$

2. Half range cosine series

If it is required to expand $f(x)$ as a cosine series the interval $0 \le x \le l$ we extend the function to $-l \le x \le l$ by reflecting $f(x)$ in y-axis so that $f(-x) = f(x)$ for which $b_n = 0$.

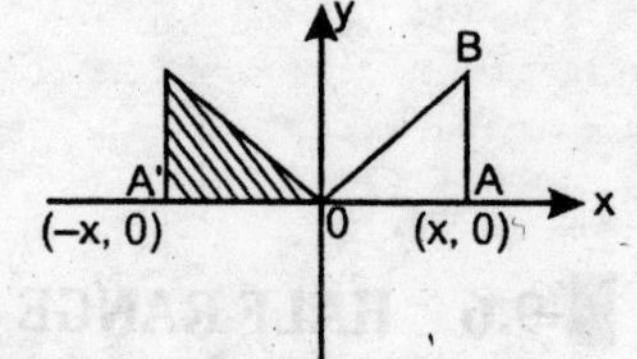

Fig. 9.2

Then $f(x)$ be an even function and the fourier series will be

$$f(x) = \frac{a_0}{2} + \sum_{n=1} a_n \cos\frac{x\pi n}{l}$$

where $$a_0 = \frac{2}{l}\int_0^l f(x)\,dx$$

$$a_n = \frac{2}{l}\int_0^l f(x)\cos\frac{x\pi n}{l}\,dx$$

$$b_n = 0$$

Note : If $l = \pi$, then $0 \le x \le l\,\pi$ and half sine series will be

$$f(x) = \sum_{n=1}^{\infty} b_n \sin nx \qquad a_0 = 0,\ a_n = 0$$

then $$b_n = \frac{2}{\pi}\int_0^\pi f(x)\sin nx\,dx$$

and half range cosine series will be

$$f(x) = \frac{a_0}{2} + \sum_{n=1}^{\infty} a_n \cos nx$$

where $$a_0 = \frac{2}{\pi}\int_0^\pi f(x)\,dx$$

$$a_n = \int_0^\pi f(x)\cos nx\,dx \text{ and } b_n = 0$$

***Example* 9.13.** Expand $f(x) = e^{-x}$ as Fourier series in the interval $(-l, l)$.

Solution. Let $$f(x) = e^{-x} \text{ and}$$

$$e^{-x} = f(x) = \frac{a_0}{2} + \Sigma a_n \cos\frac{n\pi x}{l} + \Sigma b_n \sin\frac{n\pi x}{l}$$

then $$a_0 = \frac{1}{l}\int_{-l}^l e^{-x}\,dx = \frac{1}{l}\,[e^l - e^{-l}] = \frac{2}{l}\sinh t$$

$$a_n = \frac{1}{l}\int_{-l}^l e^{-x}\cos\frac{n\pi x}{l}\,dx$$

$$= \frac{1}{l}\left[\frac{e^{-x}}{1+\frac{n^2\pi^2}{l^2}}\left\{-\cos\frac{n\pi x}{l} + \frac{n\pi}{l}\sin\frac{n\pi x}{l}\right\}\right]_{-l}^{l} \quad \left[\because \int e^{ax}\cos bx\,dx = \frac{e^{ax}}{a^2+b^2}(a\cos bx + b\sin bx)\right]$$

$$= \frac{2l(-1)^n}{l^2+n^2\pi^2}\sinh l$$

$$b_n = \frac{1}{l}\int_{-l}^{l} e^{-x}\sin\frac{n\pi x}{l}dx$$

$$= \frac{1}{l}\left[\frac{e^{-x}}{1+\left(\frac{n^2\pi^2}{l^2}\right)}\left(-\sin\frac{n\pi x}{l}-\frac{n\pi}{l}\cos\frac{n\pi x}{l}\right)\right]_{-l}^{l}$$

$$= \left[\frac{2n\pi(-1)^n\sinh l}{l^2+n^2\pi^2}\right]$$

$$e^{-x} = \frac{\sinh l}{l}+\sum\frac{2l(-1)^n}{l^2+n^2\pi^2}\sinh l\cos\frac{n\pi x}{l}+\sum\frac{2n\pi(-1)^n}{l^2+n^2\pi^2}\sinh l\sin\frac{n\pi x}{l}$$

***Example* 9.14.** Expand as Fourier series the function $f(x) = \pi x \; 0 \le x \le 1 = \pi(2-x) \; 1 \le x \le 2$

Solution. Let

$$f(x) = \frac{a_0}{2}+\sum a_n\cos\frac{n\pi x}{1}+\sum b_n\sin\frac{n\pi x}{1} \quad l = 1$$

then

$$a_0 = \frac{1}{1}\int_0^2 f(x)\,dx = \int_0^1 f(x)\,dx+\int_1^2 f(x)\,dx$$

$$= \int_1^1 \pi x\,dx+\int_1^2 \pi(2-x)\,dx = \pi$$

$$a_n = \int_0^2 f(x)\cos\frac{n\pi x}{1}dx = \int_0^1 \pi x\cos n\pi x dx+\int_1^2 \pi(2-x)\cos n\pi x\,dx$$

$$= \pi\left[x\frac{\sin n\pi x}{n\pi}-\frac{(-\cos n\pi x)}{n^2\pi^2}\right]_0^1+\pi\left[(2-x)\frac{\sin n\pi x}{n\pi}+\left(-\frac{\cos n\pi x}{n^2\pi^2}\right)\right]_1^2$$

$$= \pi\left[\frac{(-1)^n}{n^2\pi^2}-\frac{1}{n^2\pi^2}\right]+\left[\left(\frac{-\cos 2n\pi}{n^2\pi^2}\right)+\frac{\cos n\pi}{n^2\pi^2}\right]$$

$$= \pi\left[\frac{(-1)^n}{n^2\pi^2}-\frac{1}{n^2\pi^2}-\frac{1}{n^2\pi^2}+\frac{(-1)^n}{n^2\pi^2}\right]$$

$$= 2\pi\left[\frac{(-1)^n-1}{n^2\pi^2}\right]$$

$$= \frac{2}{n^2\pi}[(-1)^n-1]$$

$$= \begin{Bmatrix} 0 & \text{when } n \text{ is even} \\ \frac{-4}{n^2\pi} & \text{when } n \text{ is odd} \end{Bmatrix}$$

$$b_n = \int_0^2 f(x)\sin n\pi\, x dx = \int_0^2 \pi x\sin n\pi x\,dx+\int_1^2 \pi(2-x)\sin n\pi x\,dx$$

$$= \pi\left[x\frac{(-\cos n\pi x)}{n\pi} - \frac{(-\sin n\pi x)}{n^2\pi^2}\right]_0^1 + \pi\left[(2-x)\frac{(-\cos n\pi x)}{n\pi} + \frac{(-\sin n\pi x)}{n^2\pi^2}\right]_1^2$$

$$= \pi\left[\frac{-\cos\pi}{n\pi}\right] + \pi\left[\frac{-\sin 2n\pi}{n^2\pi^2} + \frac{\cos n\pi}{n\pi}\right]$$

$$= \pi\left[-\frac{(-1)^n}{n\pi} + \frac{(-1)^n}{n\pi}\right] = 0$$

$$f(x) = \frac{\pi}{2} - \frac{4}{\pi}\left[\frac{\cos\pi x}{1^2} + \frac{\cos 3\pi x}{3^2} + \frac{\cos 5\pi x}{5^2} + \ldots\ldots\right]$$

Put $x = 0\ f(0) = \left[\frac{1}{1^2} + \frac{1}{3^2} + \frac{1}{5^2}\ \ldots\ldots\right]$

$$0 = \frac{\pi}{2} - \frac{4}{\pi}\left[\frac{1}{1^2} + \frac{1}{3^2} + \frac{1}{5^2}\ \ldots\ldots\right]\frac{1}{1^2} + \frac{1}{3^2} + \frac{1}{5^2} + \ldots\ldots = \frac{\pi^2}{8}$$

$$\frac{1}{1^2} + \frac{1}{3^2} + \frac{1}{5^2} + \ldots\ldots = \frac{\pi^2}{8}$$

***Example* 9.15.** Develop $f(x)$ in Fourier series in the interval $(-2, 2)$ if $f(x) = 0,\ -2 < x < 0$

$$= 1,\ 0 < x < 2$$

Solution. Let $f(x) = \frac{a_0}{2} + \Sigma a_n \cos\frac{n\pi x}{2} + \Sigma b_n \sin\frac{n\pi x}{2}$

then $a_0 = \frac{1}{2}\int_{-2}^{2} f(x)\,dx = \frac{1}{2}\left[\int_{-2}^{0} 0\,dx + \int_{0}^{2} 1\,dx\right] = \frac{1}{2} \times 2 = 1$

$$a_n = \frac{1}{2}\int_{-2}^{2} f(x)\cos\frac{n\pi x}{2}\,dx$$

$$= \frac{1}{2}\int_{0}^{2}\cos\frac{n\pi x}{2}\,dx = \frac{1}{2}\left[\frac{\sin\left(\frac{n\pi x}{2}\right)}{\frac{n\pi}{2}}\right]_0^2 = 0$$

$$b_n = \frac{1}{2}\int_{-2}^{2} f(x)\,.\,\sin\frac{n\pi x}{2}\,dx = \frac{1}{2}\int_{0}^{2} 1\,.\,\sin\frac{n\pi x}{2}\,dx$$

$$= \frac{1}{2}\left[-\frac{\cos\frac{n\pi x}{2}}{\frac{n\pi}{2}}\right]_0^2$$

$$= \frac{-1}{n\pi}\left[\cos\frac{n\pi x}{2}\right]_0^2$$

$$= \frac{-1}{n\pi}[\cos n\pi - \cos 0] = \frac{1}{n\pi}[1-(-1)^n]$$

$$b_n = \begin{cases}\frac{1}{n\pi}[1-(-1)^n] = \frac{2}{n\pi} & \text{When } n \text{ is odd}\\ 0 & \text{When } n \text{ is even}\end{cases}$$

$b_1 =$

$$b_2 = 0, \; b_4 = 0$$

$$f(x) = \frac{1}{2} + \frac{2}{\pi}\sin\frac{\pi x}{2} + \frac{2}{3\pi}\sin\frac{3\pi x}{2} + \frac{2}{5\pi}\sin\frac{5\pi x}{2} + ...$$

***Example* 9.16.** A sinusoidal voltage E sin wt is passing through a half wave rectifier which clips the negative portion of the wave. Develop the resulting function.

$$u(t) = \begin{cases} 0, & \frac{-T}{2} < t < 0 \\ E\sin wt, & 0 < t < \frac{T}{2} \end{cases}$$

and $\qquad T = \frac{2\pi}{w}$ in a Fourier series.

Solution. Let

$$u(t) = \sum_{n=1}^{\infty} a_n \cos\frac{n\pi t}{\frac{T}{2}} + \sum_{n=1}^{\infty} b_n \sin\frac{n\pi t}{\frac{T}{2}}$$

Then

$$a_0 = \frac{1}{\frac{T}{2}}\int_0^{\frac{T}{2}} u(t)\,dt = \frac{2}{T}\int_0^{\frac{T}{2}} E\sin wt\,dt \text{ [since } u(t) = 0 \text{ in } \frac{-T}{2} < t < 0]$$

$$= \frac{2E}{T}\int_0^{\frac{T}{2}} \sin wt\,dt = \frac{2E}{T}\left[\frac{-\cos wt}{w}\right]_0^{\frac{T}{2}}$$

$$= \frac{-2E}{Tw}\left[\cos\frac{wT}{2} - \cos 0\right]$$

$$= \frac{2E}{Tw}\left[1 - \cos\frac{wT}{2}\right]$$

But

$$wT = 2\pi$$

$$= \frac{2E}{2\pi}[1 - \cos\pi] = \frac{2E}{2\pi} \times 2 = \frac{2E}{\pi}$$

$$a_n = \frac{1}{\frac{T}{2}}\int_0^{\frac{T}{2}} u(t)\cos\frac{n\pi t}{\frac{T}{2}}\,dt$$

$$= \frac{2E}{T}\int_0^{\frac{T}{2}} \sin wt \cos\frac{2n\pi t}{T}\,dt$$

$$= \frac{E}{T}\int_0^{\frac{T}{2}} \left[\sin\left(wt + \frac{2n\pi t}{T}\right) + \sin\left(wt - \frac{2n\pi t}{T}\right)\right]dt$$

$$= \frac{E}{T}\int_0^{\frac{T}{2}} \left[\sin\left(\frac{2n\pi t}{T} + wt\right) - \sin\left(\frac{2n\pi t}{T} - wt\right)\right]dt$$

$$= \frac{Ew}{2\pi}\left[\int_0^{\frac{\pi}{w}} \sin(nw + w)t - \sin(nw - wt)\right]dt$$

$$= \frac{Ew}{2\pi}\left[\int_0^{\frac{\pi}{w}} \sin(n+1)wt - \sin(n-1)wt\right]dt$$

$$= \frac{Ew}{2\pi}\left[-\frac{\cos(n+1)wt}{(n+1)w} + \frac{\cos(n-1)wt}{(n-1)w}\right]_0^{\frac{\pi}{w}}$$

$$= \frac{Ew}{2\pi w}\left[\left(-\frac{\cos(n+1)\pi}{n+1}+\frac{\cos(n-1)\pi}{n-1}\right)-\left(-\frac{1}{n+1}+\frac{1}{n-1}\right)\right]$$

$$= \frac{E}{2\pi}\left[-\frac{(-1)^{n+1}}{n+1}+\frac{(-1)^{n-1}}{n-1}+\frac{1}{n+1}-\frac{1}{n-1}\right]$$

when n is odd, $n+1$ and $n-1$ are even

$$= \frac{E}{2\pi}\left[-\frac{(+1)}{n+1}+\frac{(+1)}{n-1}+\frac{1}{n+1}-\frac{1}{n-1}\right]=0, \quad n \neq 1$$

when n is even, $n+1$ and $n-1$ is odd

$$= \frac{E}{2\pi}\left[\frac{1}{n+1}-\frac{1}{n-1}+\frac{1}{n+1}-\frac{1}{n-1}\right]=\frac{-2E}{\pi}\,\frac{1}{n^2-1}, \quad n \neq 1$$

when $n = 1$

$$a_1 = \frac{E}{T}\int_0^{\frac{T}{2}} 2\sin wt\cos\frac{2\pi t}{T}\,dt$$

$$= \frac{E}{T}\int_0^{\frac{T}{2}} 2\sin wt\cos wt\,dt$$

$$= \frac{Ew}{2\pi}\int_0^{\frac{2\pi}{w}} \sin 2wt\,dt = \frac{Ew}{2\pi}\left[-\frac{\cos 2wt}{2w}\right]_0^{\frac{2\pi}{w}} = \frac{Ew}{2\pi}\left[-\frac{\cos 4\pi-\cos 0}{2w}\right]=0$$

$$b_n = \frac{2}{T}\int_0^{\frac{T}{2}} E\sin wt\sin\frac{n\pi t}{\frac{T}{2}}\,dt$$

$$= \frac{2E}{T}\int_0^{\frac{T}{2}} \sin wt\sin\frac{2n\pi t}{t}\,dt$$

$$= \frac{E}{T}\int_0^{\frac{T}{2}} \sin wt\sin n\,wt\,dt$$

$$= \frac{Ew}{2\pi}\int_0^{\frac{\pi}{w}} 2\sin wt\sin n\,wt\,dt$$

$$= \frac{Ew}{2\pi}\left[\frac{\sin(n-1)wt}{(n-1)w}-\frac{\sin(n+1)wt}{(n+1)w}\right]_0^{\frac{\pi}{w}}$$

$$= \frac{E}{2\pi}\left[\frac{\sin(n-1)wt}{n-1}-\frac{\sin(n+1)wt}{n+1}\right]_0^{\frac{\pi}{w}}$$

$$= \frac{E}{2\pi}\left[\left(\frac{\sin(n-1)\pi}{n-1}-\frac{\sin(n+1)\pi}{n+1}\right)-(0-0)\right]$$

$$= \frac{E}{2\pi}[0]=0 \text{ if } \quad n\neq 1$$

If $n = 1$ we have

$$b_1 = \frac{Ew}{2\pi}\int_0^{\frac{\pi}{w}} 2\sin^2 wt\,dt$$

$$= \frac{Ew}{2\pi}\int_0^{\frac{\pi}{w}} (1 - \cos 2wt)\, dt = \frac{E}{2}$$

$$u(t) = \frac{E}{\pi} + \frac{E}{2}\sin\frac{2\pi T}{T} - \frac{2E}{\pi}\left[\frac{1}{1.3}\cos\frac{4\pi t}{T} + \frac{1}{3.5}\cos\frac{8\pi t}{T} \ldots\ldots\right]$$

$$= \frac{E}{\pi} + \frac{E}{2}\sin wt - \frac{2E}{\pi}\left[\frac{1}{1.3}\cos 2wt + \frac{1}{3.5}\cos 4wt + \ldots\ldots\right]$$

***Example* 9.17.** Obtain cosine and sine series for $f(x) = x$ in the interval $0 \le x \le \pi$. Hence show that $\frac{1}{1^2} + \frac{1}{3^2} + \frac{1}{5^2}$ $\ldots\ldots = \frac{\pi^2}{8}$.

Solution. For cosine series

Let us extend the interval to $-\pi \le x \le \pi$ by reflecting $f(x)$ in y-axis, such that $f(-x) = f(x)$. Hence $f(x)$ is even,

$\therefore \quad b_n = 0$ and

$$a_0 = \frac{2}{\pi}\int_0^{\pi} f(x)\,dx = \frac{2}{\pi}\int_0^{\pi} x\,dx = \frac{2}{\pi}\left[\frac{x^2}{2}\right]_0^{\pi} = \pi$$

$$a_n = \frac{2}{\pi}\int_0^{\pi} f(x)\cos nx\,dx = \frac{2}{\pi}\int_0^{\pi} x\cos nx\,dx$$

$$= \frac{2}{\pi}\left[x\frac{\sin nx}{n} - \frac{(-\cos nx)}{n^2}\right]_0^{\pi} = \frac{2}{\pi n^2}\left[(-1)^n - 1\right]$$

$$= \begin{cases} \dfrac{-4}{\pi n^2}, & \text{when } n \text{ is odd} \\ 0, & \text{when } n \text{ is even} \end{cases}$$

$$x = \frac{a_0}{2} + \sum_{n=1}^{\infty} a_n \cos nx$$

$$= \frac{\pi}{2} - \frac{4}{\pi}\left[\frac{\cos x}{1^2} + \frac{\cos 3x}{3^2} + \frac{\cos 5x}{5^2} \ldots\ldots\right]$$

$$= \frac{\pi}{2} - \frac{4}{\pi}\left[\frac{1}{1^2} + \frac{1}{3^2} + \frac{1}{5^2} \ldots\ldots\right]$$

$$\therefore \quad \frac{1}{1^2} + \frac{1}{3^2} + \frac{1}{5^2} \ldots\ldots = \frac{\pi^2}{8}$$

For sine series

Let us extend its interval to $-\pi < x < \pi$ by reflecting $f(x)$ in the origin so that $f(-x) = -f(x)$. Hence $f(x)$ is odd.

$$\therefore \quad a_0 = 0 \text{ and } a_n = 0$$

$$b_n = \frac{2}{\pi}\int_0^{\pi} x\sin nx\,dx = \frac{2}{\pi}\left[x\frac{(-\cos nx)}{n} - \left(\frac{-\sin nx}{n^2}\right)\right]_0^{\pi}$$

$$= \frac{2}{\pi}\left[\frac{-\pi\cos n\pi}{n} - (0)\right] = \frac{-2}{n}(-1)^n = \frac{2}{n}(-1)^{n+1}$$

Hence, $$x = \Sigma b_n \sin nx = 2\sum_{n=1}^{\infty}\frac{(-1)^{n+1}}{n}\sin nx$$

***Example* 9.18.** Find the half range cosine series for the function $f(x) = (x-1)^2$ in the interval $0 < x < 1$. Hence show that

$$\pi^2 = 8\left(\frac{1}{1^2} + \frac{1}{3^2} + \frac{1}{5^3} = \ldots\ldots\right)$$

Solution. Extend the interval of $f(x)$ to $-1 < x < 1$ by reflecting $f(x)$ in y-axis such that $f(-x) = f(x)$ and the new function is even function in $(-1, 1)$

$$\therefore \quad b_n = 0 \text{ and } a_0 = \frac{2}{1}\int_0^1 (x-1)^2 dx = \frac{2}{1}\left[\frac{(x-1)^3}{3}\right]_0^1$$

$$= \frac{2}{3}$$

$$a_n = \frac{2}{1}\int_0^1 (x-1)^2 \cos\frac{n\pi x}{1} dx$$

$$= \frac{2}{1}\left[(x-1)^2\frac{\sin n\pi x}{n\pi} - 2(x-1)\frac{(-\cos n\pi x)}{n^2\pi^2} + 2\frac{(-\sin n\pi x)}{n^3\pi^3}\right]_0^1$$

$$= 2\left[0 + \frac{2}{n^2\pi^2}\right] = \frac{4}{n^2\pi^2}$$

Therefore half range cosine series is

$$f(x) = \frac{1}{3} + \frac{4}{\pi^2}\left[\sum_{n=1}^{\infty}\frac{\cos n\pi x}{n^2}\right]$$

$$(x-1)^2 = \frac{1}{3} + \frac{4}{\pi^2}\left[\frac{\cos \pi x}{1^2} + \frac{\cos 2\pi x}{2^2} + \frac{\cos 3\pi x}{3^2} \ldots\ldots\right]$$

put $x = 0$

$$1 = \frac{1}{3} + \frac{4}{\pi^2}\left[\frac{1}{1^2} + \frac{1}{2^2} + \frac{1}{3^2} \ldots\ldots\right]$$

$$\frac{\pi^2}{8} = \frac{1}{1^2} + \frac{1}{2^2} + \frac{1}{3^2} \ldots\ldots$$

***Example* 9.19.** Obtain a Fourier expansion of $x \sin x$ as a cosine series in $(0, \pi)$. Also show that $\frac{1}{1.3} - \frac{1}{3.5} + \frac{1}{5.7} \ldots\ldots = \frac{\pi - 2}{4}$

Solution. For cosine series let us extend the interval of $f(x)$ to $-\pi < x < \pi$ by reflecting $f(x)$ in y-axis such that $f(-x) = f(x)$ then $f(x)$ is even function in the interval $(-\pi, \pi)$, $b_n = 0$

$$a_0 = \frac{2}{\pi}\int_0^{\pi} x \sin x \, dx = \frac{2}{\pi}[x(-\cos x) - 1(-\sin x)]_0^{\pi}$$

$$= \frac{2}{\pi}[-\pi \cos \pi] = 2$$

$$a_n = \frac{2}{\pi}\int_0^{\pi} x \sin x \cos nx \, dx = \frac{1}{\pi}\int_0^{\pi} x(2\cos nx \sin x) dx$$

$$= \frac{1}{\pi}\int_0^{\pi} x[\sin(n+1)x - \sin(n-1)x] dx$$

$$= \frac{1}{\pi}\left[x\left(\frac{-\cos(n+1)x}{n+1}+\frac{\cos(n-1)x}{n-1}\right)-1\left(\frac{-\sin(n+1)x}{(n+1)^2}-\frac{\sin(n-1)x}{(n-1)^2}\right)\right]_0^\pi$$

$$= \frac{1}{\pi}\left[\pi\left(\frac{-\cos(n+1)\pi}{n+1}+\frac{\cos(n-1)\pi}{n-1}\right)\right],\ n \neq 1$$

$$= \left[-\frac{(-1)^{n+1}}{n+1}+\frac{(-1)^{n-1}}{n-1}\right]$$

$$= (-1)^{n+1}\left[\frac{1}{n-1}-\frac{1}{n+1}\right] = \frac{2(-1)^{n+1}}{n^2-1}$$

$$= \frac{2(-1)^{n+1}}{n^2-1},\ n \neq 1$$

when $n = 1$

$$a_1 = \frac{2}{\pi}\int_0^\pi x \sin x \cos x\, dx = \frac{1}{\pi}\int_0^\pi x \sin 2x\, dx = \frac{1}{\pi}\left[x\left(-\frac{\cos 2x}{2}\right)-1.\left(-\frac{\sin 2x}{2^2}\right)\right]_0^\pi$$

$$= -\frac{1}{2}$$

$$x \sin x = 1 - \frac{1}{2}\cos x - 2\left[\frac{\cos 2x}{1.3}-\frac{\cos 3x}{2.4}+\frac{\cos 4x}{3.5}\ldots\ldots\right]$$

Putting

$$x = \frac{\pi}{2}$$

$$\frac{\pi}{2} = 1 - 2\left[-\frac{1}{1.3}+\frac{1}{3.5}-\frac{1}{5.7}\ldots\ldots\right]$$

$$\frac{\pi-2}{4} = \frac{1}{1.3}-\frac{1}{3.5}+\frac{1}{5.7}\ldots\ldots$$

Example **9.20.** If $f(x) = x,\ 0 < x < \frac{\pi}{2}$

$$= \pi - x,\quad \frac{\pi}{2} < x < \pi$$

Show that

(*i*) $f(x) = \frac{4}{\pi}\left(\sin x - \frac{1}{3^2}\sin 3x + \frac{1}{5^2}\sin 5x\right)$

(*ii*) $f(x) = \frac{\pi}{4} - \frac{2}{\pi}\left[\frac{1}{1^2}\cos 2x + \frac{1}{3^2}\cos 6x + \frac{1}{5^2}\sin 10x\right]$

Solution.

(*i*) For half range sine series, let us extend the interval of $f(x)$ to $-\pi < x < \pi$ such that $f(x) = -f(x)$. Hence $f(x)$ is an odd function $a_0 = 0$, $a_n = 0$

$$b_n = \frac{2}{\pi}\int_0^\pi f(x)\sin nx\, dx$$

$$= \frac{2}{\pi}\left[\int_0^{\frac{\pi}{2}} x \sin nx\, dx + \int_{\frac{\pi}{2}}^{\pi} (\pi - x) \sin nx\, dx\right]$$

$$= \frac{2}{\pi} = \left[\left(\frac{x(-\cos nx)}{n} - 1\left(\frac{-\sin nx}{n^2}\right)\right)_0^{\frac{\pi}{2}} + \left((\pi - x)\left(\frac{-\cos nx}{n}\right) - (-1)\frac{(-\sin nx)}{n^2}\right)_{\frac{\pi}{2}}^{\pi}\right]$$

$$= \frac{2}{\pi}\left[-\frac{\pi}{2n}\cos\frac{n\pi}{2} + \frac{1}{n^2}\sin\frac{n\pi}{2} + \frac{\pi}{2n}\cos\frac{n\pi}{2} + \frac{1}{n^2}\sin\frac{n\pi}{2}\right]$$

$$= \frac{4}{n^2\pi}\sin\frac{n\pi}{2}, \text{ which is 0 if } n \text{ is even}$$

Hence half range sine series is

$$f(x) = \frac{4}{\pi}\left[\frac{\sin x}{1^2} - \frac{\sin 3x}{3^2} + \frac{\sin 5x}{5^2} - \ldots\ldots\right]$$

(*ii*) For half range cosine series

$$b_n = 0, \quad \text{since } f(-x) = f(x)$$

$$a_0 = \int_0^{\pi} f(x)\, dx = \frac{2}{\pi}\left[\int_0^{\frac{\pi}{2}} x\, dx + \int_{\frac{\pi}{2}}^{\pi} (\pi - x)\, dx\right] = \frac{\pi}{2}$$

$$a_n = \frac{2}{\pi}\left[\int_0^{\frac{\pi}{2}} x \cos nx\, dx + \int_{\frac{\pi}{2}}^{\pi} (\pi - x)\cos nx\, dx\right]$$

$$= \frac{2}{\pi}\left[\left(\frac{x \sin nx}{n} - 1\left(\frac{-\cos nx}{n^2}\right)\right)_0^{\frac{\pi}{2}} + \left((\pi - x)\frac{\sin nx}{n} - (-1)\left(\frac{-\cos nx}{n^2}\right)\right)_{\frac{\pi}{2}}^{\pi}\right]$$

$$= \frac{2}{\pi}\left[\frac{\pi}{2n}\sin\frac{n\pi}{2} + \frac{1}{n^2}\cos\frac{n\pi}{2} - \frac{1}{n^2} - \frac{\cos n\pi}{n^2} - \frac{\pi}{2n}\sin\frac{n\pi}{2} + \frac{\cos\frac{n\pi}{2}}{n^2}\right]$$

$$= \frac{2}{\pi}\left[\frac{2}{n^2}\cos\frac{n\pi}{2} - \frac{1}{n^2} - \frac{(-1)^n}{n^2}\right] = \frac{2}{\pi n^2}\left[2\cos\frac{n\pi}{2} - (-1)^n - 1\right]$$

$$a_1 = \frac{2}{\pi}[-1 + 1] = 0$$

$$a_2 = \frac{2}{\pi}\left[\frac{2}{2^2}\cos\pi - \frac{1}{2^2} - \frac{1}{2^2}\right] = \frac{-2}{\pi}$$

$$a_3 = 0$$

$$a_4 = \frac{2}{\pi}\left[\frac{1}{8} - \frac{1}{16} - \frac{1}{16}\right] = 0$$

$$a_5 = \frac{2}{\pi}[0] = 0$$

$$a_6 = \frac{2}{\pi}\left[-\frac{1}{18}-\frac{1}{36}-\frac{1}{36}\right]=\frac{-2}{\pi . 3^2}$$

$$a_7 = 0 \quad a_8=0 \quad a_9=0$$

$$a_{10} = \frac{2}{\pi}\left[-\frac{1}{50}-\frac{1}{100}-\frac{1}{100}\right]=-\frac{2}{25\pi}$$

Then,
$$f(x) = \frac{\pi}{4}-\frac{2}{\pi}\left[\frac{\cos 2x}{1^2}+\frac{\cos 6x}{3^2}+\frac{\cos 10x}{5^2}\ldots\ldots\right]$$

***Example* 9.21.** Obtain half range cosine series for $f(x) = Kx,\ 0 \le x \le \frac{l}{2}$

$$= K(l-x),\ \frac{l}{2} \le x \le l$$

Deduce the sum of the series $\frac{1}{1^2}+\frac{1}{3^2}+\frac{1}{5^2}\ldots\ldots$

Solution. Let
$$f(x) = \frac{a_0}{2}+\sum_{n=1}^{\infty} a_n \cos\frac{n\pi x}{l}$$

then
$$a_0 = \frac{2}{l}\int_0^l f(x)\,dx = \frac{2}{l}\left[\int_0^{\frac{l}{2}} Kx\,dx+\int_{\frac{l}{2}}^{l} K(l-x)\,dx\right]$$

$$= \frac{2}{l}\left[\left(K\frac{x^2}{2}\right)_0^{\frac{l}{2}}+K\left((lx)-\frac{x^2}{2}\right)_{\frac{l}{2}}^{l}\right]$$

$$= \frac{2}{l}\left[\left(K\frac{l^2}{8}\right)+K\left(\left(l^2-\frac{l^2}{2}\right)-K\left(\frac{l^2}{2}-\frac{l^2}{8}\right)\right)\right]$$

$$= \frac{2}{l}\left[K\frac{l^2}{4}\right]=\frac{Kl}{2}$$

$$a_n = \frac{2}{l}\left[\int_0^l f(x)\cos\frac{n\pi x}{l}\,dx\right]$$

$$= \frac{2}{l}\left[\int_0^{\frac{l}{2}} K\,x\cos\frac{n\pi x}{l}\,dx+\int_{\frac{l}{2}}^{l} K(l-x)\cos\frac{n\pi x}{l}\,dx\right]$$

$$= \frac{2K}{l}\left\{\left[x\left(\frac{\sin\frac{n\pi x}{2}}{\frac{n\pi}{l}}\right)-\left(\frac{-\cos\frac{n\pi x}{l}}{\frac{n^2\pi^2}{l^2}}\right)\right]_0^{\frac{l}{2}}+\left[(l-x)\frac{\sin\frac{n\pi x}{l}}{\frac{n\pi}{l}}-(-1)-\left(\frac{\cos\frac{n\pi x}{l}}{\frac{n^2\pi^2}{l^2}}\right)\right]_{\frac{l}{2}}^{l}\right\}$$

$$= \frac{2K}{l}\left[\left(\frac{\frac{l}{2}}{\frac{n\pi}{l}}\sin\frac{n\pi}{2}+\frac{\cos\frac{n\pi}{2}}{\frac{n^2\pi^2}{l^2}}-\frac{l}{\frac{n^2\pi^2}{l^2}}\right)+\left(\frac{-l^2}{n^2\pi^2}\cos n\pi-\frac{l^2}{2n\pi}\sin\frac{n\pi}{2}+\frac{l^2}{n^2\pi^2}\cos\frac{n\pi}{2}\right)\right]$$

$$= \frac{2kl}{n^2\pi^2}\left[2\cos\frac{n\pi}{2} - 1 - \cos n\pi\right]$$

$$= \frac{2kl}{n^2\pi^2}\left[2\cos\frac{n\pi}{2} - 1 - (-1)^n\right]$$

When n is odd, $\cos\frac{n\pi}{2} = 0$ and $\cos n\pi = 1$

$$a_1 = a_3 = a_5 = ... = 0$$

When n is even

$$a_n = \frac{2kl}{n^2\pi^2\left[2\cos\frac{n\pi}{2} - 2\right]}$$

$$a_2 = \frac{4kl}{2^2\pi^2}[-2] = \frac{-8kl}{2^2\pi^2}$$

$$a_4 = \frac{4kl}{4^2\pi^2}[\cos 2\pi - 1] = 0$$

$$a_6 = \frac{2kl}{6^2\pi^2}[2\cos 3\pi - 1 - \cos 6\pi] = \frac{-8kl}{6^2\pi^2} \text{ and so on.}$$

$$f(x) = \frac{kl}{4} - \frac{8kl}{\pi^2}\left[\frac{1}{2^2}\cos\frac{2\pi x}{l} + \frac{1}{6^2}\cos\frac{6\pi x}{l} ......\right]$$

Put $x = l, f(l) = 0$

$$0 = \frac{kl}{4} - \frac{8kl}{\pi^2}\left[\frac{1}{2^2} + \frac{1}{6^2} + ...\right]$$

$$\frac{1}{2^2} + \frac{1}{6^2} + ... = \frac{\pi^2}{32}$$

$$\frac{1}{1^2} + \frac{1}{3^2} + ... = \frac{\pi^2}{8}$$

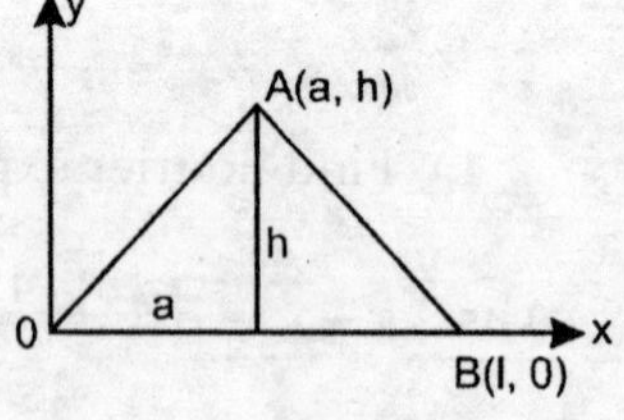

Fig. 9.3

***Example* 9.22.** For the function defined by the graph OAB, find half range sine series.

Solution. Equation of OA $y = \frac{h}{a}x \quad 0 \le x \le a$

Equation of AB $y = \frac{h(x-l)}{a-l} \quad a \le x \le l$

Now $f(x) = \Sigma_{b_n} \sin\frac{n\pi x}{l}$

where $b_n = \frac{2}{l}\int_0^l f(x)\sin\frac{n\pi x}{l}dx = \frac{2}{l}\left[\int_0^a \frac{h}{a}\sin\frac{n\pi x}{l}dx + \int_a^l \frac{h(x-l)}{a-l}\sin\frac{n\pi x}{l}dx\right]$

$$= \frac{2h}{al}\int_0^a x\sin\frac{n\pi x}{l}dx + \frac{2h}{l(a-l)}\int_a^l \left((x-l)\sin\frac{n\pi x}{l}dx\right)$$

$$= \frac{2h}{al}\left[x\left(\frac{-\cos\frac{n\pi x}{l}}{\frac{n\pi}{l}}\right) - \left(\frac{-\sin\frac{n\pi x}{l}}{\frac{n^2\pi^2}{l^2}}\right)\right]_0^a$$

$$+ \frac{2h}{l(a-l)}\left[(x-l)\left(\frac{-\cos\frac{n\pi x}{l}}{\frac{n\pi}{l}}\right) - \left(\frac{-\sin\frac{n\pi x}{l}}{\frac{n^2\pi^2}{l^2}}\right)\right]_a^l$$

$$= \frac{2h}{al}\left[-\frac{al}{n\pi}\cos\frac{n\pi a}{l} + \frac{l^2}{n^2\pi^2}\sin\frac{n\pi a}{l}\right] + \frac{2h}{(a-l)}\left[\frac{l^2}{n^2\pi^2}\sin\frac{n\pi a}{l} + \frac{l(a-l)}{n\pi}\cos\frac{n\pi a}{l} - \frac{l^2}{n^2\pi^2}\sin\frac{n\pi a}{l}\right]$$

$$= \frac{2hl}{n^2\pi^2}\left[\frac{1}{a} - \frac{1}{a-l}\right]\sin\frac{n\pi a}{l}$$

$$= \frac{-2hl^2}{an^2\pi^2(a-l)}\sin\frac{n\pi a}{l}$$

$$f(x) = \sum_{n=1}^{\infty}\frac{2hl^2}{a(l-a)n^2\pi^2}\sin\frac{n\pi a}{l}\sin\frac{n\pi x}{l}$$

$$= \frac{2hl^2}{a(l-a)\pi^2}\left[\sin\frac{\pi a}{l}\sin\frac{\pi n}{l} + \frac{1}{2^2}\sin\frac{2\pi a}{l}\sin\frac{2\pi}{l} + \frac{1}{3^2}\sin\frac{3\pi a}{l}\sin\frac{3\pi n}{l}\ \ldots\ldots\right]$$

EXERCISE 9.2

1. Find Fourier expansion of $f(x) = x - x^2, -1 < x < 1$

Ans. $x - x^2 = -\frac{1}{3} + \frac{4}{\pi^2}\left[\frac{\cos\pi x}{1^2} - \frac{\cos 2\pi x}{2^2} + \frac{\cos 3\pi x}{3^2}\ \ldots\ldots\right] + \frac{2}{\pi}\left[\frac{\sin\pi x}{1}\ \ldots\ldots\right]$

2. Expand as Fourier expansion of $f(x) = e^{-x}$ in $(-1, 1)$

Ans. $e^{-n} = \sin h\, 1\left[\frac{1}{l} - 2l\left[\frac{1}{l^2+\pi^2}\cos\frac{\pi x}{l} - \frac{1}{l^2 2^2\pi^2}\sin\frac{2\pi x}{l}\right] + \frac{1}{l^2+3^3\pi^2}\cos\frac{3\pi x}{l} + \ldots\ldots\right]$

$$-2\pi\left[\frac{1}{e^2+\pi^2}\sin\frac{\pi x}{l} - \frac{2}{3^2+2^2\pi^2}\sin\frac{2\pi x}{l}\ \ldots\ldots\right]$$

3. Expand $f(x) = 0, \quad -2 < x < 0$ and $f(x) = 1, \quad 0 < x < 2$ as Fourier series.

Ans. $f(x) = \frac{1}{4} - \frac{2}{\pi^2}\left[\cos\pi x + \frac{\cos 3\pi x}{3^2} + \frac{\cos 5\pi x}{5^2}\ \ldots\ldots\right]$

4. Expand $f(x) = x, \quad 0 < x < 1$

$= \quad 0, \quad 1 < x < 2$

as Fourier series.

Ans. $f(x) = \frac{1}{2} + \frac{2}{\pi}\left[\sin\frac{\pi x}{2} + \frac{1}{3}\sin\frac{3\pi x}{2} + \frac{1}{5}\sin\frac{5\pi x}{2}\right]$

5. Expand $f(x) = \begin{cases} 0, & -2 < x < -1 \\ k, & -1 < x < 1 \\ 0, & 1 < x < 2 \end{cases}$ as Fourier series.

Ans. $f(x) = \frac{k}{2} + \frac{2k}{\pi}\left[\cos\frac{\pi t}{2} - \frac{1}{3}\cos\frac{3\pi t}{2} + \frac{1}{5}\cos\frac{5\pi}{2}\right]$

6. Obtain Fourier expansion of $f(x) = |\cos x|$ in $(-\pi, \pi)$

Ans. $|\cos x| = \frac{2}{\pi} + \frac{4}{\pi}\left[\frac{1}{3}\cos 2x - \frac{1}{15}\cos 4x + \ldots\ldots\right]$

7. Expand $f(x) \begin{cases} \frac{1}{4} - x & 0 < x < \frac{1}{2} \\ x - \frac{3}{4} & \frac{1}{2} < x < 1 \end{cases}$ as Fourier sine series

Ans. $f(x) = \left(\frac{1}{\pi} - \frac{4}{\pi^2}\right)\sin \pi x + \left(\frac{1}{3\pi} - \frac{4}{3^2\pi^2}\right)\sin 3\pi x - \left(\frac{1}{5\pi} - \frac{4}{5^2\pi^2}\right)\sin 5\pi n - 1$

8. Prove that in $0 < x < l$

$$x = \frac{l}{2} - \frac{4l}{\pi^2}\left\{\frac{\cos \pi x}{l} + \frac{1}{3^2}\cos\frac{3\pi x}{l} + \frac{1}{5^2}\cos\frac{\sin x}{l} + \ldots\ldots\right\}$$

Hence deduce that $\frac{1}{1^4} + \frac{1}{3^4} + \frac{1}{5^4} + \ldots\ldots = \frac{\pi^4}{96}$

9. Show that a constant function 'a' can be expanded as $\frac{4a}{\pi}\left\{\sin x + \frac{\sin 3x}{3} + \frac{\sin 5x}{5} + \ldots\ldots\right\}$ in the range $0 < x < \pi$

10. Show that for $0 < x < \pi$

$x(\pi - x) = \frac{\pi^2}{6} - \left(\frac{\cos 2x}{1^2} + \frac{\cos 4x}{2^2} + \frac{\cos 6x}{3^2}\right)$ and hence evaluate $\sum_{n=1}^{\infty} \frac{1}{n^4}$

11. Obtain for series sine for unity in $0 < x < \pi$ and hence show that $1 + \frac{1}{3^2} + \frac{1}{5^2} + \frac{1}{3^2} \ldots\ldots = \frac{\pi^2}{8}$

Ans. $1 = \frac{4}{\pi}\left[\frac{\sin x}{1} + \frac{\sin 3x}{3} + \frac{\sin 5x}{5} + \ldots\ldots\right]$

12. Prove that in the range $(0, l)$

$x = \frac{l}{2} - \frac{4l}{\pi^2}\left[\sum_{m=1}^{\infty} \frac{1}{(2m-1)^2}\cos(2m-1)\frac{\pi x}{l}\right]$ and hence deduce that

$$\frac{1}{1^4} + \frac{1}{3^4} + \frac{1}{5^4} + \ldots\ldots = \frac{\pi^4}{96}$$

13. Expand as Fourier series

$$f(x) = \begin{cases} \frac{1}{4} - x & 0 < x < \frac{1}{2} \\ x - \frac{1}{4} & \frac{1}{2} < x < 1 \\ 1 & 1 < x < \frac{3}{2} \\ x - 1 & \frac{3}{2} < x < 2 \end{cases}$$

Ans. $f(x) = \frac{7}{16} + \frac{1}{\pi}\left(\frac{1}{\pi} + \frac{1}{4}\right)\cos \pi x \frac{-3}{4}\left(\frac{1}{\pi} + \frac{1}{2}\right)\sin \pi x$

14. Obtain a Fourier series for $f(x) = 1 - t^2$ $\quad -1 \le t \le 1$

Ans. $1 - t^2 = \frac{2}{3} + \frac{4}{\pi^2}\left[\cos \pi t - \frac{\cos 2\pi t}{2^2} + \frac{\cos 3\pi t}{3^2} + \ldots\ldots\right]$

15. Develop $\sin\left(\frac{\pi x}{l}\right)$ as half range cosine series in $0 < x < l$

Ans. $\sin\left(\frac{\pi x}{l}\right) = \frac{2}{\pi} - \frac{4}{\pi}\left[\frac{\frac{\cos 2\pi x}{l}}{1.3} + \frac{\frac{\cos 4\pi x}{l}}{3-5} + \frac{\frac{\cos 6\pi x}{l}}{5.7} + \ldots\ldots\right]$

Parseval's theorem on Fourier series statement

If the fourier series of $f(x)$ over an interval $c < x < c + 2l$ is given as

$$f(x) = \frac{a_0}{2} + \sum_{n=1}^{\infty} a_n \cos\frac{n\pi x}{l} + \sum_{n=1}^{\infty} b_n \sin\frac{n\pi x}{\sin l}$$

then

$$= \frac{1}{2l}\int_c^{c+2l} [f(x)]^2\, dx = \frac{a_0^2}{4} + \frac{1}{2}\sum_{n=1}^{\infty}(a_n^2 + b_n^2)$$

Proof : Now we have in the interval $c < x < c + 2\,l$

$$f(x) = \frac{a_0}{2} + \sum_{n=1}^{\infty} a_n \cos\frac{n\pi x}{l} + \sum_{n=1}^{\infty} b_n \sin\frac{n\pi x}{l} \qquad \ldots(i)$$

where

$$a_0 = \frac{1}{l}\int_c^{c+2l} f(x)\, dx$$

$$a_n = \frac{1}{l}\int_c^{c+2t} f(x)\cos\frac{n\pi x}{l}\, dx \qquad \ldots(ii)$$

$$b_n = \frac{1}{l}\int_c^{c+2e} f(x)\sin\frac{n\pi x}{l}\, dx$$

Multiplying both sides of (*i*) by $f(x)$, we wave

$$[f(x)]^2 = \frac{a_0}{2} f(x) + \sum_{n=1}^{\infty} a_n f(x)\cos\frac{n\pi x}{l} + \sum_{n=1}^{\infty} b_n f(x)\sin\frac{n\pi x}{l}\, dx$$

Integrating both sides w.r.t. x, between the limit c to $c + 2l$, we have

$$\int_c^{c+2l} [f(x)]^2\, dx = \frac{a_0}{2}\int_c^{c+2l} f(x)\, dx + \sum_{n=1}^{\infty} a_n \int_c^{c+2l} f(x)\cos\frac{n\pi x}{l}\, dx + \sum_{n=1}^{\infty} b_n \int_c^{c+2l} f(x)\sin\frac{n\pi x}{l}\, dx$$

$$= \frac{a_0}{2} a_0 l + \sum_{n=1}^{\infty} a_n (l a_n) + \sum_{n=1}^{\infty} b_n (l b_n)$$

$$= \frac{a_0^2}{2} l + \sum_{n=1}^{\infty} l a_n^2 + \sum_{n=1}^{\infty} l b_n^2$$

Case I : If $c = 0$, $\frac{1}{2l}\int_0^{2l} [f(x)]^2 dx = \frac{a_0^2}{4} + \frac{1}{2}\sum_{n=1}^{\infty}(a_n^2 + b_n^2)$

Case II : If $c = -l$, the interval becomes $-l < x < l$

$$\frac{1}{2l}\int_{-l}^{l} [f(x)]^2 dx = \frac{a_0^2}{4} + \frac{1}{2}\sum_{n=1}^{\infty}(a_n^2 + b_n^2)$$

(*i*) If $f(x)$ is even in $(-l, l)$ then $\frac{2}{l}\int_0^l [f(x)^2 dx] = \frac{a_0^2}{2} + \sum_1^{\infty} a_n^2$

(*ii*) If $f(x)$ is odd in $(-l, l)$ then $\frac{2}{l}\int_0^l [f(x)]^2 dx = \sum_{n=1}^{\infty} b_n^2$

***Example* 9.23.** Find the fourier series for unity in $0 < x < \pi$ and hence show that $1 + \frac{1}{3^2} + \frac{1}{5^2} \ldots\ldots = \frac{\pi^2}{8}$

Solution. Since we have half range sine series for half the interval for $f(x) = 1$ in $0 < x < \pi$

$$a_0 = 0, \quad a_n = 0 \text{ and } f(x) = \sum_{n=1}^{\infty} b_n \sin nx\, dx$$

where
$$b_n = \frac{2}{\pi}\int_0^{\pi} f(x) \sin nx\, dx$$

$$= \frac{2}{\pi}\int_0^{\pi} \sin nx\, dx = \frac{2}{\pi}\left[-\frac{\cos nx}{n}\right]_0^{\pi}$$

$$= \frac{-2}{\pi}\left[\frac{(-1)^n}{n} - \frac{1}{n}\right] = \frac{2}{\pi}\left[\frac{1}{n} - \frac{(-1)^n}{n}\right]$$

$$= \begin{cases} 0 & \text{when } n \text{ is even} \\ \frac{4}{\pi n} & \text{when } n \text{ is odd} \end{cases}$$

Now from Parseval's theorem on Fourier constants $\int_c^{c+2l} [f(x)]^2 dx = 2l\left[\frac{a_0^2}{4} + \frac{l}{2}\sum_{n=1}^{\infty}(a_n^2 + b_n^2)\right]$

Here
$$\pi\int_0^{\pi} (1)^2 dx = \frac{\pi}{2}\left[\frac{16}{\pi^2 1^2} + \frac{16}{\pi^2 3^3} + \frac{16}{\pi^2 5^2} \ldots\ldots\right]$$

$$\pi = \frac{8}{\pi}\left[\frac{1}{1^2} + \frac{1}{3^2} + \frac{1}{5^2} \ldots\ldots\right]$$

$$\frac{\pi^2}{8} = \frac{1}{1^2} + \frac{1}{3^2} + \frac{1}{5^2} \ldots\ldots$$

***Example* 9.24.** Prove that in $0 < x < l$

$$x = \frac{l}{2} - \frac{4l}{\pi^2}\left[\cos\frac{\pi x}{l} + \frac{1}{3^2}\cos\frac{3\pi x}{l} + \frac{1}{5^2}\cos\frac{5\pi x}{l}\ ......\right]$$

hence deduce that
$$\frac{\pi^4}{96} = \frac{1}{1^4} + \frac{1}{3^4} + \frac{1}{5^4} + ...$$

Solution. Here $f(x) = x$

We have to expand $f(x)$ in half of interval $0 < x < e$ as cosine series for which

$$x = f(x) = \frac{a_0}{2} + \sum_{n=1}^{\infty} a_n \cos\frac{n\pi x}{l}$$

where
$$a_0 = \frac{2}{l}\int_0^l f(x)\cos\frac{n\pi x}{l}\,dx = \frac{2}{l}\int_0^l x\cos\frac{n\pi x}{l}\,dx = l$$

$$a_n = \frac{2}{l}\int_0^l f(x)\cos\frac{n\pi x}{l}\,dx$$

$$= \frac{2}{l}\int_0^l x\cos\frac{n\pi x}{l}\,dx$$

$$= \frac{2}{l}\left[\frac{x\sin\frac{n\pi x}{l}}{\frac{n\pi}{l}} - \frac{-\cos\frac{n\pi x}{l}}{\left(\frac{n\pi}{l}\right)^2}\right]_0^l$$

$$= \frac{2}{l}\left[\left(\frac{l}{n\pi}\right)^2[(-1)^n - 1]\right]$$

$$= \frac{2l}{n^2\pi^2}[(-1)^n - 1] = 0 \text{ if } n \text{ is even}$$

$$= \frac{-4l}{n^2\pi^2} \text{ if } n \text{ is odd}$$

$\therefore$ from (1)

$$x = \frac{l}{2} - \frac{4l}{\pi^2}\left[\frac{1}{1^2}\cos\frac{n\pi x}{l} + \frac{1}{3^2}\cos\frac{3n\pi}{l} + ...\right]$$

Now $\int_0^l [f(x)]^2 dx = \frac{l}{2}\left[\frac{a_0^2}{2} + a_1^2 + a_2^2 + a_3^2\ ......\right]$

$$\left[\frac{x^3}{3}\right]_0^l = \frac{l}{2}\left\{\frac{a_0^2}{2} + a_1^2 + a_2^2 + a_3^2 + ...\right\}$$

$$= \frac{l}{2}\left[\frac{l^2}{2} + \frac{16l^2}{1^4\pi^4} + \frac{16l^2}{3^4\pi^4} + \frac{16l^2}{5^4\pi^4}\ ......\right]$$

$$\frac{l^3}{3} = \frac{l^3}{2}\left[\frac{l}{2} + \frac{16}{\pi^4}\left\{\frac{1}{1^4} + \frac{1}{3^4} + \frac{1}{5^4}\right\}\right]$$

$$\frac{1}{3} - \frac{1}{4} = \frac{16}{2\pi^4}\left\{\frac{1}{1^4} + \frac{1}{3^4} + \frac{1}{5^4}\ ......\right\}$$

$$\frac{\pi^4}{12} = 8\left[\frac{1}{1^4}+\frac{1}{3^4}+\frac{1}{5^4}\ldots\ldots\right]$$

$$\frac{\pi^4}{96} = \frac{1}{1^4}+\frac{1}{3^4}+\frac{1}{5^4}\ldots\ldots$$

EXERCISE 9.3

1. Find Fourier series of x^2 in $(-\pi, \pi)$. Use Parseval's identify to prove that $\frac{\pi^4}{90} = 1 + \frac{1}{2^4} + \frac{1}{3^4} + \ldots\ldots$

Ans. $x^2 = \frac{\pi^2}{3} + \sum_{n=1}^{\infty} \frac{4(-1)^n}{n^2} \cos nx$

2. If $f(x) = \Sigma b_n \sin \frac{n\pi x}{l}$ in $(0, l)$, then show that $\int_0^l [f(x)]^2 dx = \frac{l}{2} \sum_{n=1}^{\infty} b_n^2$

3. If $\frac{a_0}{2} + \Sigma a_n \cos \frac{n\pi x}{l}$ is a half range cosine series of $f(x)$ of period $2l$ in $(0, l)$, then show that $f(x) = \frac{a_0^2}{2} + \sum_{n=1}^{\infty} a_n^2$ using the result evaluate $1^{-4} + 3^{-4} + 5^{-4} + \ldots\ldots\ldots$ from the half range cosines series of the function defined in (0, 2) by

$$f(x) = \begin{Bmatrix} \pi x & 0 < x < 1 \\ \pi(2-x) & 1 < x < 2 \end{Bmatrix}$$

EXERCISE 9.4

1. Check for even-odd function:

(*i*) $f(x) = \begin{Bmatrix} -x & -4 \le x \le 0 \\ x & 0 \le x \le 4 \end{Bmatrix}$

(*ii*) $f(x) = \begin{Bmatrix} -x^2 & -\pi \le x \le 0 \\ x^2 & 0 \le x \le \pi \end{Bmatrix}$

Solution. (*i*) even (*ii*) odd

2. Name the condition for expansion of a function in Fourier series.

Solution. Dirichelt's condition

3. What is the value of a_0 and a_n if the given function is odd?

Solution. $a_0 = a_n = 0$

4. What is b_n equal if given function is even?

Solution. $b_n = 0$

5. If $f(x)$ is and odd function in $[-c, c]$, the graph of $f(x)$ is symmetrical about which axes?

Solution. Origin

6. What is the period of:_

(*i*) $\tan x$ (*ii*) $\cos x$

Solution. (*i*) π (*ii*) 2π

7. The fourier series of the function $f(x)$ in the interval $-a < x < b$ is given by

$$f(x) = \frac{a_0}{2} + \sum_{n=1}^{\infty}(a_n \cos n\theta + b_n \sin n\theta) \text{ where } \theta = \frac{2\pi x}{a+b}, \text{ find } a_n \text{ and } b_n.$$

Solution. $a_n = \dfrac{2}{(a+b)}\displaystyle\int_{-a}^{b} f(x)\cos n\theta\, dx,\ b_n = \dfrac{2}{(a+b)}\displaystyle\int_{-a}^{b} f(x)\sin n\theta\, dx$

8. Find the Fourier Series for $f(x) = \begin{cases} -1, & -1 < x < 0 \\ 0, & 0 < x < 1 \end{cases}$. Also find the sum of the series at $x = -1$.

Solution. $-\dfrac{1}{2} + \dfrac{2}{\pi}\displaystyle\sum_{n=1}^{\infty}\dfrac{1}{(2n-1)}\sin(2n-1)\pi x,\ -\dfrac{1}{2}$

9. Find the Fourier Series for

$$\begin{aligned} f(x) &= 0, -\pi \le x < -\pi/2 \\ &= x, -\pi/2 < x < \pi/2 \\ &= 0, \pi/2 < x \le \pi \quad \text{where} \\ f(x+2\pi) &= f(x) \end{aligned}$$

Solution. $\dfrac{2}{\pi} = \displaystyle\sum_{n=1}^{\infty}\left[\dfrac{\sin n\pi/2}{n^2} - \dfrac{\pi\cos n\pi/2}{2n}\right]\sin nx$

10. Find the half-range series for the function

$$\begin{aligned} f(x) &= 1, 0 < x < l/2 \\ &= 0, l/2 < x < 1 \end{aligned}$$

Solution. $\dfrac{4}{\pi} = \displaystyle\sum_{m=0}^{\infty}\dfrac{(-1)^{m+1}}{(2m+1)}\cos\dfrac{(2m+1)}{l}\pi x$

Differentiation

10.1 FUNCTIONS

In mathematical analysis we often come across related quantities; for example, the volume of a sphere depends upon its radius, the velocity of a falling particle depends upon the distance through which it has fallen, the square of a number depends upon the number itself. In each case we have a variable quantity depending upon another variable quantity. Such relationship between two variables is given the name of a *function*. We now give its definition.

Let-X and y be two sets, and suppose that to each element x of X there corresponds, by some rule, a single element y of Y. Then the set f of all ordered pairs (x, y) is called a **function** *which maps X into Y.*

The set X is called the *domain* of the function, and Y its *codomain*. The element y, which corresponds to x, is called the image of x or the *value* of the function at x. The set of all images is called the *range* of the function. Since all the elements of Y need not be images, the range is a subset of the codomain. When the range in same as codomain, we say that f maps X onto Y.

A function is best denoted by a single letter, such as f. The value of the function if then denoted by $f(x)$. It is a common practice, however, to say "the function $y = f(x)$" and to use "value of the function" for particular values like $f(a)$, $f(b)$, etc.

The above definition of a function is applicable to all sorts of sets. However, in elementary calculus X and Y are most often the set R of real numbers.* For this reason X and Y are often not specified; only the rule relating y to x is given. Thus rule is quite often an equation. But it could be a table, an operation or a verbal instruction. The only criterion is that we should be able to assign one element of Y to each element of X.

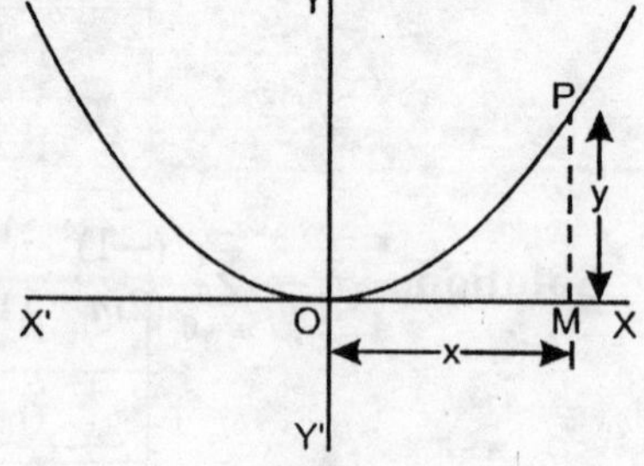

Fig. 10.1

10.2 GRAPHICAL REPRESENTATION

Draw through a point O two straight lines X' OX, Y' OY perpendicular to each other. Take a point M on OX such that $OM = x$, and draw MP of length y parallel to OY. Then x and y are the Cartesian coordinates of the point P.

The coordinates can be written as an ordered pair (x, y). Therefore any ordered pair (x, y) can be represented on this diagram by a point having the coordinates x and y.

A function f, being a set of ordered pairs, will therefore be represented by a set of points. This set of points constitutes the graph of the function. Quite often these points are adjacent to each other and a curve can be drawn through them. The above figure shows the graph of the function defined by the rule $y = x^2$.

10.3 LIMITS

Let f be a function given by the rule $y = f(x)$. Consider a set of positive number $h_1, h_2, h_3, ..., h_n, ...$, which continually decreases, *i.e.*, $h_1 > h_2 > h_3 > ... > h_n > ... > 0$.

---

* Sometimes X or Y is an interval, the set of all real number lying between two given number a and b.

It may then happen that the values

$$f(a+h_1), f(a+h_2), f(a+h_3), \ldots, f(a+h_n), \ldots$$

a-h_n a+h_n

O a-h_1 a a+h_1

Fig. 10.2

Continually approach a number A as h_n gets smaller and smaller. This number A is called the limit of $f(x)$ at a (or as x tends to a). We write

$$\lim_{x \to a} f(x) = A.$$

$x \to a$ is read as "x tends to a".

In fact A is the limit on the right, since we have considered only the values of x greater than a (*i.e.* on the right of a in the figure). If we consider the values

$$f(a-h_1), f(a-h_2), f(a-h_3), \ldots,$$

and find that they continually approach a number B as h_n gets smaller and smaller, we call B the *limit* of $f(x)$ *on the left*. When $A = B$ (*as happen in most cases*), we call A the limit of $f(x)$ at a. The limits on the right and left are respectively denoted by

$$\lim_{x \to a^+} f(x) \text{ and } \lim_{x \to a^-} f(x).$$

Definition: If $|f(x) - A|$ can be made as small as we please for all values of x for which $|x - a|$ is sufficiently small (but not zero), we say that $f(x)$ has the limit A as $x \to a$.

***Example* 10.1.** Find $\lim_{x \to 0} \frac{\sin x}{x}$, where x is in radians.

We find from the tables the following values of x and $(\sin x)/x$.

x	$(\sin x)/x$	x	$(\sin x)/x$
0.3	0.9851	–0.3	0.9851
0.1	0.9983	–0.1	0.9983
0.05	0.9996	–0.05	0.9996
0.02	0.99993	–0.02	0.99993
0.01	0.99998	–0.01	0.99998

We see that $(\sin x)/x$ continually approaches the value 1 as $x \to 0$. Hence

$$\lim_{x \to a} (\sin x)/x = 1.$$

***Example* 10.2.** Find $\lim_{x \to a} x^3$.

We see that $\quad (a+h)^3 = a^3 + 3a^2h + 3ah^2 + h^3.$

Therefore $\quad |x^3 - a^3| = |3a^2h + 3ah^2 + h^3|$

$$= |h| \, . \, |3a^2 + 3ah + h^2|.$$

Since the factor $|3a^2 + 3ah + h^2|$ is finite, we can make the right hand side (and so $|x^3 - a^3|$) as small as we please by taking $|h|$ sufficiently small. Therefore

$$\lim_{x \to a} x^3 = a^3.$$

Note: The student should not that in finding the limit at a we do use $f(a)$, the value of $f(x)$ at a. The limit depends on the values of $f(x)$ at points in the neighbourhood of $x = a$, other than a itself.

10.4 INFINITE LIMITS

Consider the function given by $y = 1/x^3$. When x take the value 0.1, 0.01, 0.001, ..., y takes successively increasing values 103, 106, 109, ... In fact, $1/x^3$ can be made as a large as we please by taking x sufficiently small. We express this by saying that the limit of $1/x^3$ when $x \to 0$ from the right is infinity (denoted by ∞).

Similarly, when $x \to 0$ from the left, $1/x^3$ takes up values which are negative but continually increasing in magnitude. In this case, we say

$$\lim_{x \to 0^-} -(1/x^3) = -\infty.$$

Since the limits on the right and left are different, $\lim_{x \to 0}(1/x^3)$ does not exist. For a function like $1/x^2$ both the limits are $+\infty$, and we write

$$\lim_{x \to 0} (1/x^2) = \infty.$$

The student should note that ∞ is not a number like 10^6 or 10^{20}. It is used above only to indicate that $1/x^2$ can take up a value as large as we please for all sufficiently small values of $|x|$.

10.5 CONTINUITY

If we draw the graph of the curve $y = x^3$, we find that the graph is continuous, *i.e.* it has no breaks in it. The curve $y = 1/x^3$, however has a discontinuity at the points $x = 0$ (see fig 10.3).

For a curve to be continuous at $x = a$, the difference $f(x) - f(a)$ should be small when x is any point near a. This gives us the following definition of continuity

A function f is said to be continuous at $x=a$, if $/f(x)-f(a)|$ can be made as small as we please for all values of x for which $|x-a|$ is sufficiently small.

Comparing this definition with that of the limit, we se that for a function continuous at $x = a$

$$\lim_{x \to a} f(x) = f(a)$$

Fig. 10.3

When $\lim_{x \to a} f(x) \neq f(a)$, or when the limit does not exist or is infinite, the function is discontinuous at $x = a$.

It can be easily verified (or seen from their graphs) that

(*i*) x^n is continuos for all values of x when n is positive.

(*ii*) $1/x^n$ (n positive) is continuous for all values of x except x=0.

(*iii*) e^x, $\sin x$ and $\cos x$ are continuous for all values of x; $\log x$ is continuous for all values of x>0. (For x<0, $\log x$ is not real.)

(*iv*) $\tan x$ is continuous for all values of x except $\pm\frac{1}{2}\pi, \pm\frac{3}{2}\pi, \pm\frac{5}{2}\pi$, etc. As $x \to \frac{1}{2}\pi$ from the right, than $x \to -\infty$. As $x \to \frac{1}{2}\pi$ from the left, $\tan x \to +\infty$. Thus $\lim_{x \to \pi/2} \tan x$ does not exist.

Similar statements can be made for $\cot x$, $\sec x$ and $\operatorname{cosec} x$.

10.6 DIFFERENTIAL COEFFICIENT

If f is a function of x, $\lim_{h \to 0} \dfrac{f(a+h)-f(a)}{h}$

is called the DIFFERENTIAL COEFFICIENT of $f(x)$ at $x = a$. ...(1)

It is denoted by $f'(a)'$ or $\{df(x)/dx\}x=a$. Quite often we write x itself for a in (1), it being understood that in finding the limit when $h \to 0$, x is kept constant. In this case the differential coefficient of $f(x)$ is denoted by $f(x)$

$\frac{df(x)}{dx}$, or $Df(x)$.

When the limit (1) exists, we say that $f(x)$ is differentiable at $x = a$. When this limits does not exist, the differential coefficient of $f(x)$ does not exist at $x=a$, and the function is not differentiable at $x = a$. The process of finding the differential coefficient is called differentiation. The differential coefficient is also called the derivative.

A function $f(x)$ must be continuous at $x=a$ if its differential coefficient at $x=a$ is to exist. But continuity of $f(x)$ does not ensure its differentiability. If $f(x)$ is continuous at $x=a$, $f'(x)$ may exit at $x=a$, or may not.

10.7 DIFFERENTIATION OF SIMPLE FUNCTIONS

(*i*) If $f(x)=x^n$, $f(x+h) = (x+h)^n$. Therefore by definition,

$$\frac{d}{dx}x^n = \lim_{h\to 0}\frac{(x+h)^n - x^n}{h}$$

$$= \lim_{h\to 0} x^n \frac{(1+h/x)^n - 1}{h}, x \neq 0,$$

$$= \lim_{h\to 0} x^n \frac{1 + n\frac{h}{x} + \frac{(n-1)}{1.2}\frac{h^2}{x^2} + \ldots - 1}{h}$$

$$= \lim_{h\to 0}\,[nx^{n-1} + h\{\frac{1}{2}n(n-1)x^{n-2} + \ldots\}]$$

$$= nx^{n-1}$$

$f'(o)$ can be obtained by a direct application of definition.

(*ii*)
$$\frac{d}{dx}\sin x = \lim_{h\to 0}\frac{\sin(x+h) - \sin x}{h}$$

$$= \lim_{h\to 0}\frac{2\cos(x+\frac{1}{2}h)\sin\frac{1}{2}h}{h}$$

$$= \lim_{h\to 0}\cos(x+\tfrac{1}{2}h).\frac{\sin\frac{1}{2}h}{\frac{1}{2}h}$$

$$= \cos x,$$

since $(\sin\frac{1}{2}h)/(\frac{1}{2}h) \to 1$ as $h \to 0$.

We can similarly prove

$$\frac{d}{dx}\cos x = -\sin x.$$

(*iii*)
$$\frac{d}{dx}e^x = \lim_{h\to 0}\frac{e^{x+h} - e^x}{h} = \lim_{h\to 0} e^x\frac{e^h - 1}{h}$$

$$= \lim_{h\to 0} e^x \frac{1 + h + \frac{1}{2}h^2 + (h^3/3!) + \ldots - 1}{h}$$

$$= \lim_{h\to 0} e^x\,[1 + h\{\tfrac{1}{2} + \tfrac{1}{6}h + \ldots\}] = e^x$$

(iv) $$\frac{d}{dx}\log x = \lim_{h\to 0}\frac{\log(x+h)-\log x}{h}$$

$$= \lim_{h\to 0}\frac{\log(1+h/x)}{h}$$

$$= \lim_{h\to 0}\left[\frac{1}{x}+h\left\{-\frac{1}{2x^2}+\frac{h}{3x^3}-\ldots\right\}\right]$$

$$= 1/x.$$

(v) If $f(x) = c$ for all values of x, then

$$\frac{d}{dx}f(x) = \lim_{h\to 0}\frac{f(x+h)-f(x)}{h} = \lim_{h\to 0}\frac{c-c}{h} = 0$$

i.e the differential coefficient of a constant is zero.

10.8 DIFFERENTIATION OF A SUM, PRODUCT OR QUOTIENT

A direct application of definition shows that

$$\frac{d}{dx} - \{f(x) + f_2(x)\} = \frac{d}{dx}f_1(x) + \frac{d}{dx}f_2(x) \tag{1}$$

and $$\frac{d}{dx}\{cf(x)\} = c\,\frac{d}{dx}f(x),$$

where c is a constant, Formula (1) can be extended to a sum (of difference) of several functions.

For product or quotient of two function we have the following rules:

(i) *Differential coefficient of the product of two functions = first functions × diff. coeff. of second + second function × diff. coeff of first*

(ii) *Differential coefficient of the quotient of two function*

$$= \frac{[\textit{diff.coeff. of numerator (denomiator)} -(\textit{numerator})(\textit{diff. coeff. of denomiator})]}{\textit{square of denomiator}}$$

We leave the proof of the first of these as an exercise to the student and prove only the second one here.

Let $$f(x) = f_1(x)/f_2(x),$$

then $$f(x+h) = f_1(x+h)/f_2(x+h).$$

Therefore $$\frac{d}{dx}\left\{\frac{f_1(x)}{f_2(x)}\right\} = \lim_{h\to 0}\left\{\frac{f_1(x+h)}{f_2(x+h)}-\frac{f_2(x)}{f_2(x)}\right\}\div h$$

$$= \lim_{h\to 0}\frac{f_1(x+h)f_2(x)-f_1(x)\,f_2(x+h)}{f_2(x+h)\,f_2(x)\,h}$$

$$= \lim_{h\to 0}\frac{\left\{\frac{f_1(x+h)-f_1(x)}{h}\right\}f_2(x)-f_1(x)\left\{\frac{f_2(x+h)-f_2(x)}{h}\right\}}{f_2(x+h)f_2(x)}$$

$$= \frac{\left\{\frac{df_1(x)}{dx}\right\} f_2(x) - f_1(x)\left\{\frac{df_2(x)}{dx}\right\}}{\{f_2(x)\}^2}$$

Application of this rule gives:

$$\frac{d}{dx} \tan x = \frac{d}{dx}\left(\frac{\sin x}{\cos x}\right) = \frac{d(\sin x)/dx . \cos x - \sin x . d(\cos x)/dx}{\cos^2 x}$$

$$= \frac{\cos x . \cos x - \sin x(-\sin x)}{\cos^2 x}$$

i.e.,
$$\frac{d}{dx} \tan x = \sec^2 x.$$

Similarly,
$$\frac{d}{dx} \cot x = -\operatorname{cosec}^2 x.$$

Again
$$\frac{d}{dx} \sec x = \frac{d}{dx}\left(\frac{1}{\cos x}\right) = \frac{d1/dx . \cos x - 1(\cos x)/dx}{\cos^2 x}$$

$$= \frac{0 + \sin x}{\cos^2 x} = \sec x \tan x.$$

Similarly
$$\frac{d}{dx} \operatorname{cosec} x = -\operatorname{cosec} x \cot x.$$

Example **10.3.** Differentiate $3x^5 e^x = 2x^2$.

$$\frac{d}{dx}(3x^5 e^x + 2x^2) = 3\frac{d}{dx}(x^5 e^x) + 2\frac{d}{dx}(x^2)$$

$$= 3\{x^5 . e^x + 5x^4 . e^x\} + 2. 2x, \text{ by rule } (i)$$

$$= (3x^5 = 15x^4)e^x + 4x.$$

10.9 DIFFERENTIATION OF A FUNCTION OF A FUNCTION

To differentiate $f\{g(x)\}$, when f and g are both continuous functions, put $g(x) = t$ and $g(x + h) = t + k$.

Then as $h \to 0$, $g(x + h) \to g(x)$; so that k also $\to 0$.

Therefore

$$\frac{d}{dx} fg(x)\} = \lim_{h-0} \frac{f\{g(x+h)\} - f\{g(x)\}}{h}$$

$$= \lim_{h-0} \frac{f(t+k) - f(t)}{h}$$

$$= \lim_{h-0} \left\{\frac{f(t+k) - f(t)}{k}, \frac{k}{h}\right\}$$

$$= \frac{df(t)}{dt} . \frac{dt}{dx} = \frac{df\{g(x)\}}{dg(x)} . \frac{dg(x)}{dx}$$

Thus we differentiate $f\{g(x)\}$ first treating $g(x)$ as a single variable, and then multiply it by $dg(x)/dx$.

***Example* 10.4.** Find the differential coefficient of $\sin x^2$.

$$\frac{d}{dx}\sin x^2 = \frac{d}{dx^2}\sin x^2 . \frac{dx^2}{dx} = (\cos x^2)\, 2x.$$

***Example* 10.5.** Find the differential coefficient of $\log \sin e^{5x}$.

$$\frac{d}{dx}\log \sin e^{5x} = \frac{d(\log \sin e^{5x})}{d(\sin e^{5x})} . \frac{d}{dx}\sin e^{5x}$$

$$= \frac{1}{\sin e^{5x}}\frac{d(\sin e^{5x})}{d(e^{5x})}\frac{de^{5x}}{dx}$$

$$= \frac{\cos e^{5x}}{\sin e^{5x}} . \frac{d(e^{5x})}{d(5x)} . \frac{d(e^{5x})}{d(5x)}$$

$$= (\cot e^{5x})e^{5x} . 5.$$

Examples

Writes down the differential coefficients of

1. x^{-4}
2. $x^{2/3}$
3. $5x^6$.
4. $4x^3 + 3\sin x$.
5. $x^n + a^n$.
6. $x^3 \log x$.
7. $e^x \sin x$.
8. $(\cos x)/\log x$.
9. $(\cot x)/(x + ex)$.
10. $\sin x^3$.
11. $\tan^3 x$.
12. e^{3x}.
13. $\log(1 + \sin x)$.
14. $\cos(\cos x)$
15. $e\cot x$
16. $\sqrt{(\log x)}$.
17. $(ax + b)^n$.
18. $e^{\sin}\, x/ \sin^n$.
19. $\sin^2(\log x^2)$.
20. $x \log x . \log \log x$
21. $\log\{\sqrt{(x-a)} + \sqrt{(x-b)}\}$.
22. $(x-1)^2 (x+2)^3 (x^2+4)$.
23. Prove that if $f(x)$ is differentiable at $x = a$, it is continuous there.
24. Draw the graph of the following function:

$$f(x) = \begin{cases} x, & 0 \le x \le 1, \\ 2-x, & 1 \le x \le 2. \end{cases}$$

Discuss the continuity and differentiability at $x = 1$. [*Roorkee* 1973]

25. Show that the function defined by $y = x$ when $x > 0$; $y = 0$ when $x = 0$ and $y = x$ when $x < 0$; is continuous at 0 but not differentiable at $x = 0$. [*AMIE*, 1978]

10.10 INVERSE FUNCTION

Let $y = \sin{-1}\, x$; then $x = \sin y$.

Differentiating both side with respect to x, we get

$$1 = \frac{d}{dx}\sin y = \frac{d}{dy}\sin y . \frac{d}{dy} = \cos y \frac{dy}{dx} \qquad ...(1)$$

Therefore,

$$\frac{dy}{dx} = \frac{1}{\cos y} = \frac{1}{\sqrt{(1-\sin^2 y}} = \frac{1}{\sqrt{(1-x^2)}};$$

i.e., $$\frac{d}{dx}(\sin^{-1} x) = \frac{1}{\sqrt{(1-x^2)}}$$

Proceeding in a similar fashion, we can prove that

$$\frac{d}{dx}(\cos^{-1} x) = \frac{1}{\sqrt{(1-x^2)}};$$

$$\frac{d}{dx}(\tan^{-1} x) = \frac{1}{1+x^2}; \quad \frac{d}{dx}(\cot^{-1} x) = -\frac{1}{1+x^2};$$

$$\frac{d}{dx}(\sec^{-1} x) = \frac{1}{x\sqrt{x^2-1}}; \quad \frac{d}{dx}(\text{cosec}^{-1} x) = -\frac{1}{x\sqrt{x^2-1}}$$

We notice from (1) the result

$$\frac{dy}{dx} \times \frac{dx}{dy} = 1$$

This is true for all cases, and can be obtained from the result of Article 10.9.

10.11 HYPERBOLIC FUNCTIONS

$$\frac{d}{dx}(\sinh x) = \frac{d}{dx}\{\tfrac{1}{2}(e^x - e^{-x})\} = \tfrac{1}{2}e^x - \tfrac{1}{2}\frac{de^{-x}}{d(-x)}\frac{d(-x)}{dx}$$

$$= \tfrac{1}{2}e^x \; \tfrac{1}{2}e^{-x} = \cosh x.$$

Similarly, $$\frac{d}{dx}(\cosh x) = \sinh x.$$

$$\frac{d}{dx}(\tanh x) = \frac{d}{dx}\left(\frac{\sinh x}{\cosh x}\right)$$

$$= \frac{\cosh x.\cosh x - \sinh x.\sinh x}{\cosh^2 x}$$

$$= 1/\cosh^2 x = \text{sech}^2 x.$$

We can similarly prove that

$$\frac{d}{dx}(\coth x) = \text{cosech}^2 x,$$

$$\frac{d}{dx}(\text{sech}\, x) = -\text{sech}\, x \tan x,$$

$$\frac{d}{dx}(\text{cosech}\, x) = -\text{cosech}\, x \coth x.$$

If $$y = \sinh^{-1} x, \text{ then } \sinh y = x.$$

Therefore differentiating with respect to x,

$$\cosh y\frac{dy}{dx} = 1,$$

or $$\frac{dy}{dx} = \frac{1}{\cosh y} = \frac{1}{\sqrt{(1+\sinh^2 y)}},$$

i.e. $$\frac{d}{dx}(\sin^{-1} x) = \frac{1}{\sqrt{(1+x^2)}}.$$

We can similarly prove that

$$\frac{d}{dx}(\cosh^{-1} x) = \frac{1}{\sqrt{x^2-1}} \quad \frac{d}{dx}(\tanh^{-1} x) = \frac{1}{1-x^2}$$

10.12 LOGARITHMIC DIFFERENTIATION

Let

$$y = \{f(x)\}^{g(x)}$$

To differentiate such a function, we first take the logarithm of both sides; then

$$\log y = \log \{f(x)\}g(x).$$
$$= g(x) \log f(x)$$

Differentiation of both the sides will then give dy/dx.

***Example* 10.6.** Differentiate $(\sin x)^{\log x}$.

Let $$y = (\sin x) \log x,$$

then $$\log y = \log x. \log \sin x,$$

Differentiating, $$\frac{1}{y}\frac{dy}{dx} = \log x. \frac{1}{\sin x}. \cos x + \frac{1}{x}\log \sin x.$$

Therefore $$\frac{dy}{dx} = (\sin x)\log x\}\log x. \cot x + (\log \sin x)/x\}.$$

10.13 IMPLICIT FUNCTIONS

If the relation between x and y is given by an equation involving both, and this equation is not immediately solvable for y, then y is called an implicit function of x. On the other hand y is called an explicit function of x, when y is given directly in terms of x.

By differentiating the given equation involving x and y, it is always possible to obtain dy/dx.

***Example* 10.7.** Find dy/dx when $x^5 + y^5 + 5xy = c$.

Differentiating the given equation with respect to x, we get

$$5x^4 + 5y^4\frac{dy}{dx} + 5x\frac{dy}{dx} + 5 = 0.$$

Therefore $$\frac{dy}{dx} = -\frac{x^4 + y}{x + y^4}$$

10.14 PARAMETRIC EQUATIONS

Sometimes x and y are both expressed in terms of a third variables, usually called a *parameter*. We can find dy/dx in such cases by using 1.33.

This if $$x = f(t) \text{ and } y = g(t), \text{ then}$$

$$\frac{dy}{dx} = \frac{dy}{dt}\cdot\frac{dt}{dx} = \frac{dy}{dt} \div \frac{dx}{dt}.$$

***Example* 10.8.** If $x = a \cos \theta$ and $y = b \sin \theta$, find dy/dx

Here $\frac{dy}{dx} = \frac{dy/d\theta}{dx/d\theta} = \frac{b\cos\theta}{-a\sin\theta} = -\frac{b}{a}\cot\theta.$

Examples

Differentiate

1. $\log \sec^{-1} x^4$.
2. $(\sin^{-1} x^4)4$.
3. $\tan^{-1} e^{2x+1}$
4. $\operatorname{cosec}^3 (m \sin^{-1} x)$.
5. $\cot x \coth x$.
6. $\sinh x/ (x + \cosh x)$.
7. $x^{\sin x}$.
8. $(\log x)^{\sin x}$
9. $xe^x \sin x$.
10. $(\cot x)^{\sin x} + (\tan x) \sin x$
11. $\sin^{-1}\left(\frac{2x}{1+x^2}\right)$ [Roorkee, Arch, 1973]
12. $\sin^{-1}\left(\frac{1-x^2}{1+x^2}\right)$ [Roorkee, 1973]

Find $\frac{dy}{dx}$ when

13. $x^3 + 3axy + y^3 = a^3$
14. $y = x^y$. [Banaras, 1979]
15. $y^x + + x^y = c$. [Roorkee, 1971]
16. $x = a(t-\sin t),\ y = a(1-\cos t)$. [Bhopal, 1978
17. $x = 3at/(1+t^3),\ y = 3at^2/(1+t^3)$.
18. $x = a(\cos t + \log \tan \frac{1}{2} t),\ y = a \sin t$. [Banars, 1979]
19. If $\sin y = x \sin(a+y)$, prove that
$$\frac{dy}{dx} = \frac{\sin^2(a+y)}{\sin a}$$ [Roorkee, Arch., 1978]
20. If $x^y = e^{x-y}$, prove that
$$\frac{dy}{dx} = \frac{\log x}{(1+\log x)^2}.$$ [Bhopal, 1978]
21. If $\sqrt{(y+x)} + \sqrt{(y-x)} = c$, show that
$$\frac{dy}{dx} = \frac{y}{x} - \sqrt{\left(\frac{y^2}{x^2} - 1\right)}$$ [Roorkee, 1973]
22. $\sqrt{(1-x^2)} + \sqrt{(1-y^2)} = a(x-y)$, prove that
$$\frac{dy}{dx} = \frac{\sqrt{(1-y^2)}}{\sqrt{(1-x^2)}}.$$ [Roorkee, 1979]

10.15 DIFFERENTIAL COEFFICIENT AS RATE OF CHANGE

Suppose a car starts from O and moves in a straight line along OP, accelerating all the time, Let its position at time t and $t+h$ seconds be P and O; and let the distance OP, OQ be s and s+k meters respectively. Then the distance PQ (=k) is covered in time h. Therefore k/h is the rate (per second at which the distance

Fig. 10.4

from P to Q covered. This is the average velocity (v_{av}) of the car during h seconds. Since the car is accelerating v_{av} will be greater than the car's instantaneous, v_{av} will be greater than the car's instantaneous velocity v at P,

If we take a smaller interval of time h, the corresponding k/h will determine an average velocity nearer v. So we define

$$v = \lim_{h \to 0} (k/h).$$

Let $s = f(t)$, so that $s + k = f(t + h)$, then

$$v = \lim_{h \to 0} \frac{f(t+h) - f(t)}{h} = \frac{d}{dt} f(t) = \frac{ds}{dt}.$$

We notice that $f(t + h) f(t)$ gives the increment in distance and h increment in time. Their ratio gives the rate r which distance increases with time. Hence ds/dt is equal to rate of increase of distance with time.

Similarly, $df(x)/dx$ gives the rate of increase of $f(x)$ with x. The increments in x and $f(x)$ are often denoted by δx and $\delta f(x)$. In terms of increments

$$\frac{dy}{dx} = \lim_{\delta x \to 0} \frac{\delta y}{\delta x}$$

For the function $f(x) = 1/x$, $f(2) = 0.5$ and $f(2.5) = 0.4$. So the increment, $f(2.5) - f(2) = -0.1$. Thus negative increment means a decreases in the function. In such a case the differential coefficient will also be negative.

It follows that $f(x)$ increase with x when $df(x)/dx$ is positive. If $df(x)/dx$ is negative $f(x)$ decrease when x increase.

Ex. A conical vessel of semi-vertical angle α, has its axis vertical and vertex downwards, and is being filled with a liquid at a uniform rate v. Find and expression for the rate at which the surface of the liquid is rising when the vessel is filled up to a height x. *[Roorkee, Arch., 1970]*

Let V be the volume of the liquid in the vessel when the height is x, then

$$V = \frac{1}{3} \pi y^2 x = \frac{1}{3} \pi (x \tan a)^2 x$$

$$= \frac{1}{3} \pi x^3 \tan^2 \alpha.$$

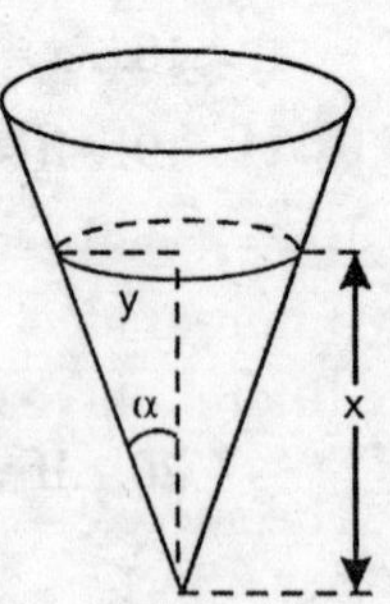

Fig. 10.5

Therefore

$$\frac{dV}{dt} = \pi x^2 \frac{dx}{dt} \tan^2 \alpha.$$

Now dV/dt is the rate of increase of volume, which is given to be v. Hence, the rate at which the surface of the liquid is rising

$$= \frac{dx}{dt} = \frac{1}{\pi x^2} \cot^2 \alpha . \frac{dV}{dt} = \frac{v \cot^2 \alpha}{\pi x^2}.$$

Examples:

1. A point moves in a straight line so that $x = \sqrt{t}$. Show that the acceleration is negative and proportional to the cube of the velocity.
2. A point moves along the x-axis such that its position at time t is given by $x = 3t + \cos 3t$. Prove that its velocity and acceleration both vanish together, and find how far it moves before coming to rest. [Roorkee, 1957]
3. An inverted cone has has a depth of 10 cm. and a base of radius 5 cm, Water is poured into it at rate of $1\frac{1}{2}$ c.c. per minute. Find the rate at which the level of the water in the cone is rising when the depth of water is 4 cm. *[Gorakhpur, 1978]*
4. The side cross-section of a trough 2 meters log is an equilateral triangle. If water flows in it at the rate of 1,000 cu. Cm. per sec. find the rate at which the depth is increasing when the water is 40 cm. deep. *[Roorkee, Arch., 1971]*

5. A conical vessel with its axis vertical and vertex upwards is being filled at a uniform rate with liquid through a hole at the vertex. Prove that when the vessel is eigth-ninths full the surface of the liquid is rising four times as fast as when the vessel is one-ninth full.
6. A rod AB, 100 cm. long moves with its ends A and B on two perpendicular lines OX and OY respectively. If A is 80 cm. from O and is moving away at the rate of 20 cm. per second, find the rate at which the end B is moving.
7. A ladder 5 cm. long standing on a horizontal floor, leans against a vertical wall. If the top slides downwards at the rate of 10 cm./sec., find the rate at which the angle between the floor and ladder is decreasing when the lower end of the ladder is 2 m. from the wall. [*Roorkee*, 1979]
8. Two straight railway tracks intersect at an angle of 60°. On one the train .is 8 km. from the junction and moving towards it at the rate of 40 kph. On the other a train is 12 km. from the junction and moving away at the rate at 20 kph. Find the rate at which the trains are approaching or separating.
9. A revolving search light on the top of a tower in the sea, b miles away from a straight shore, is making n revolutions per minute. Show that the velocity of the light along the shore is $(\pi nb/30)\ \text{cosec}^2\ \phi$ miles per minute when the ray of light makes an angle ϕ with the shore.
10. A point source of light is hung a feet directly above a straight horizontal path on which a boy b feet in height is walking. How fast is the boy's shadow lengthening when he is walking away from the light at the rate of c feet per minute? [*Roorkee*, 1957]
11. Prove that as x increases

    $(a \sin x + + b \cos x)/(c \sin x + d \cos x)$

    where a, b, c, d are constants, either increases for all values of x.

    or decreases for all values of x.
12. Find the range of values of x for which the function $x^3 - 6x^2 - 36x + 7$ increases with x.

10.16 SUCCESSIVE DIFFERENTIATION

If $y = f(x)$ gives a function of x, its differential coefficient dy/dx will give another function, which in general can be differentiated. The differential coefficient of dy/dx is called the second differential coefficient of y. Similarly, the differential coefficient of the second differential coefficient is called the third differential coefficient of y, and so on.

The successive differential coefficients of y are denoted by $\frac{dy}{dx'}\frac{d^2y}{dx^2}, \frac{d^3y}{dx^3}, \ldots$.

Alternative notations are $Dy, D^2y, D^3y, \ldots, y_1, y_2, y_3 \ldots$, or $y', y'', y''', \ldots$.

The nth differential coefficient of these notations are d^ny/dx^n, D^ny, y_n, $y(n)$ respectively.

The nth differential coefficient at x=a is denoted by $(y_n)x{=}a$, $(y_n)\ a$ or $f^{(n)}(a)$.

Use of partial fractions or trigonometry is sometimes helpful is successive differentiation.

***Example* 10.9.** Find the third differential coefficient of $x^2 \sin 3x + 5x^2$

Let $\quad y = x^2 \sin 3x + 5x^2,$

then
$$y_1 = x^2 \cos 3x.\ 3 + 2x \sin 3x + 5.\ 2x,$$
$$y_2 = 3\{x^2\ (-\sin 3x)\ 3 + 2x \cos 3x\} + 2\{x \cos 3x.\ 3 + 1.\ \sin 3x) + 10,$$
$$= -9x^2 \sin 3x + 12x \cos 3x + 2 \sin 3x + 10,$$
$$y_3 = -9\{x^2 \cos 3x.\ 3 + 2x \sin 3x\} + 12\{x(-\sin 3x)\ 3 + \cos 3x\} + 2 \cos 3x.\ 3 = 0$$
$$= -27x^2 \cos 3x - 54x \sin 3x + 18 \cos 3x$$
$$= (18 - 27x^2) \cos 3x - 54x \sin 3x.$$

Example **10.10. Find the fourth differential coefficient of**

(*i*) $\cos^4 x$,

(*ii*) $x/(x-1)(x-2)$.

(*i*) Let $y = \cos^4 x = \frac{1}{4}(2\cos^2 x)^2 = \frac{1}{4}(1+\cos 2x)^2$

$$= \tfrac{1}{4}(1 + 2\cos 2x + \cos^2 2x)$$

$$= \tfrac{1}{8}(2 + 4\cos 2x + 1 + \cos 4x).$$

Therefore
$$y_1 = \tfrac{1}{8}\{4(-\sin 2x)2 - \sin 4x.\ 4\}$$

$$= -\tfrac{1}{2}\{2\sin 2x + \sin 4x\},$$

$$y_2 = -\tfrac{1}{2}\{4\cos 2x + 4\cos 4x\},$$

$$y_3 = -2\{-\sin 2x.\ 2 - \sin 4x.\ 4\},$$

$$y_4 = 2\{4\cos 2x = 16\cos 4x\}$$

$$= 8(\cos 2x + 4\cos 4x).$$

(*ii*) **Breaking into partial factions.**

$$y = \frac{x}{(x-1)(x-2)} = \frac{2}{x-2} - \frac{1}{x-1}$$

Therefore
$$y_1 = 2(-1)(x-2)^{-2} - (-1)(x-1)^{-2},$$

$$y_2 = 2(-1)(-2)(x-2)^{-3} - (-1)(-2)(x-1)^{-3},$$

Similarly
$$y_3 = (-1)(-2)(-3)[2(x-2)^{-4} - (x-1)^{-4}],$$

and
$$y_4 = (-1)(-2)(-3)(-4)[2(x-2)^{-5} - (x-1)^{-5}], = 4!\left\{\frac{1}{(x-2)^5} - \frac{1}{(x-1)^5}\right\}.$$

NOTE: In simple cases the partial fractions can be obtained by inspection. In other cases, we may apply the methods given in Chapter 5.

10.17 STANDARD RESULTS

The student may shorten his work be remembering some of the following results.

(1) If
$$y = (ax+b)^m, \text{ then } y_1 = m\,.a(ax+b)^{m-1},$$

$$y_2 = m(m-1),\ a^2(ax+b)^{m-2}, \text{ etc. Hence}$$

$$D^n(ax+b)^m = m(m-1)(m-2)\ldots(m-n+1)\,a^n(ax=b)^{m-n}.$$

In particular

$$D^n(ax+b)^{-1} = (-1)^n n! a^n (ax+b)^{-n-1}.$$

If m is a positive integer, then (m = 1)the and all successive differential coefficients of $(ax+b)$ are zero.

(2) if
$$y = e^{ax} \text{ then } y_1 = ae^{ax},\ y_2 = a^2e^{ax}, \text{ etc.}$$

Therefore
$$D^n e^{ax} = a^n e^{ax}.$$

(3) If $y = \log(ax+b)$, then $y_1 = a(ax+b)^{-1}$. For further differentiation, we can use (1). Hence

$$D^n \log(ax+b) = \frac{(-1)^{n-1}(n-1)!a^n}{(ax+n)^n}$$

(4) If $y = \sin(ax+b)$,

then $y_1 = a\cos(ax+b) = a\sin(ax+b+\frac{1}{2}\pi)$,

$y_2 = a_2\cos(ax+b+\frac{1}{2}\pi) = a^2\sin(ax=b+\frac{1}{2}\pi+\frac{1}{2}\pi)$

$y_3 = a^3\sin(ax=b+\frac{3}{2}\pi)$; etc.

Hence $D^n\sin(ax+b) = a^n\sin(ax+\mathrm{b}+\frac{1}{2}n\pi)$.

(5) Similarly

$$D^n\cos(ax=b) = a^n\cos(ax+b+\tfrac{1}{2}n\pi).$$

10.18 LEIBNITZ THEOREM

If u and v be any two functions of x, then

$$D^n(uv) = (D^nu).v+{}^nC1D^{n-1}u.Dv+{}^nC_2D^{n-2}.D^2v+...+{}^nC_rD^{n-r}u.D^rv+...+u.D^nv.$$

We shall prove this theorem by mathematical induction. Assume that the above result is true for a particular value of n. Then differentiating with respect to x, we have

$D^{n+1}(uv) = \{(D^{n+1}u).v+D^nu.Dv\}+{}^nC_1\{D^nu.Dv+D^{n-1}u.D^2v\}+...+{}^nCr\{D^{n-r+1}u.D^rv+D^{n-r}u.D^{r+1}v\}+...$

$+\{Du.D^nv+u.D^{n+1}v\}$.

Rearranging, we have

$$D^{n+1}(uv) = (D^{n+1}u).v+(1=nC_1)(D^nu.Dv)+...+({}^nDr+{}^nC_{r+1})(D^{n-r}u.D^{r+1}v)+...+u.D^{n+1}v$$

But $nCr+{}^nC_{r+1} = {}^{n+1}C^{r+1}$

Hence $D^{n+1}(uv) = (D^{n+1}u).v+{}^{n+1}C_1D^nu.Dv+...+{}^{n+1}C_{r+1}D^{n-r}u.D^{r+1}v+...+u.D^{n+1}v$.

Therefore, if the theorem is true for any value of n, it is true for the next value of n.

But it is evident that it holes for $n = 1$, namely

$$D(uv) = (Du).v+u.Dv.$$

Hence the theorem must be true when $n = 2$, and so when $n = 3$; and so on. Thus it must be true for every positive integral values of n.

NOTE. ${}^nC_1$, nC_2, ... are the binomial coefficients. Thus ${}^nC_r = n!/r!(n-r)!$

***Example* 10.11.** Find the third differential coefficient of $x^2\sin 3x+5x^2$.

Using Leibnitx theorem, we have $D^3(x^2\sin 3x+5x^2)$.

$$= (D^3x^2)\sin 3x+3(D^2x^2)(D\sin 3x)+3(Dx^2)(D^2\sin 3x)+x^2(D^3\sin 3x)+D^3(5x^2)$$

$$= 0+3(2)(3\cos 3x)+3(2x)(-9\sin 3x)+x^2(-27\cos 3x)+0$$

$$= (18-27x^2)\cos 3x-54x\sin 3x.$$

***Example* 10.12.** If $y = e^{a\sin^{-1}x}$, prove that $(1-x^2)y_{n+2}-(2n+1)xy_{n+1}-(n^2+a^2)y_n = 0$, and hence find the value of y_n when $x = 0$. *[Roorkee, 1996]*

Since $y = e^a\sin^{-1}x$,

therefore $y_1 = ea\sin^{-1}x\,\frac{a}{\sqrt{(1-x^2)}} = \frac{ya}{\sqrt{(1-x^2}}$ ...(1)

*Named after the German mathematician Gottfried Wilhelm Leibnitz (1646 & 1716) who was an independent inventor of the calculus, and to whom our notation for differentiation is due.

Hence $$(1-x^2)y_1^{\,2} = a^2y^2.$$

By differentiation $(1-x^2)$. $$2y_1y_2 - 2xy_1^{\,2} = 2a^2yy_1,$$

or $$(1-x^2)y_2 - xy_1 - a^2y = 0. \qquad ...(2)$$

Differentiate this n times by Leibnitz theorem. Now

$$D^n\{1-x^2)y_2\} = (D^ny_2)(1-x^2) + n(D^{n-1})\{D1-x^2)\} + \tfrac{1}{2}n(n-1)\,(D^{n-2}y_2)\{D^2(1-x^2)\} + ...$$
$$= (1-x^2)\,y_{n+2} + n(-2x)y_{n+1} + \tfrac{1}{2}n(n-1)(-2)y_n,$$

the other terms being zero. Similarly

$$Dn\,\{xy_1\} = (D^ny_1)x + n(D^{n-1}y_1)(Dx)+0$$
$$= xy_{n+1} + ny_n,$$

Hence (2) gives

$$(1-x^2)y_{n+2} - 2nxy_{n+1} - n(n-1)yn - xy_{n+1} - ny_n - a^2y_n = 0,$$

or $$(1-x^2)y_{n+2} - (2n+1)xy_{n+1} - (n^2+a^2)y_n = 0.$$

Putting $x = 0$ in this, we obtain

$$(y_{n+2})_0 = (n^2+a^2)(y_n)_0.$$

A repeated application of this gives

$$(y_n)_0 = \{(n-2)^2 + a^2\}(y_{n-2})_0$$
$$= \{(n-2)^2 + a^2\}\{(n-4)^2 + a^2\}\,(y_{n-4})_0 = \text{etc.}$$

The last factor will be $(2^2 + a^2)\,(y_2)_0$ or $(1^2 = a^2)\,(y_1)_0$ according as n is even or odd.

We see from (1) and (2) that $(y_1)_0 = a$ and $(y_2)_0 = a_2$. Hence

$$(yn)_0 = \{n-2)^2 + a^2\}\,\{(n-4)^2 + a^2\}\,...\,(2^2 + a^2)a^2,\ n \text{ even}$$
$$= \{(n-2)^2 + a^2\}\{(n-4)^2 + a^2\}\,...\,(1 + a^2)\,a,\ n \text{ odd}.$$

Examples:

1. Find the second differential coefficient of
   (*i*) x^4e^{5x},
   (*ii*) $\tan e^{5x}$.
2. if $y = e^{ax}\sin bx$, prove that $y_2 - 2ay_1 + (a^2 + b^2)\,y = 0$.
3. If $y = x^2\cos x$, prove that $x^2\dfrac{d^2y}{dx^2} - 4x\dfrac{dy}{dx} + (x^2 = 6)y = 0$. [*Roorkee, Arch.*, 1978]
4. If $y = \sin(\sin x)$, prove that $\dfrac{d^2y}{dx^2} + \dfrac{dy}{dx}\tan x + y\cos^2 x = 0$. [*Delhi, Arch.*, 1970]

Find the nth differential coefficient of

5. $\log\{(ax+b)(cx+d)\}$
6. $\sin 2x \sin 3x$.
7. $(a^2 - x^2)^{-1}$
8. $e^{ax}\sin bx$.

9. $x^2 \sin 2x$. [*Roorkee, Arch.*, 1971]

10. $ex\,(2x+3)^3$ [*Roorkee, Arch.*, 1970]

11. $x^2 \log x$.

12. $ex \log x$

13. Prove that $\frac{d^n}{dx^n}(x^2e^x)\,\{x^2+2nx+n(n-1)\}e^x$. [*Roorkee, Arch.*, 1973]

14. If $y = a\cos(\log x) + b\sin(\log x)$, show that $x^2y_{n+2} + (2n=1)xy_{n+1} + (n^2+1)y_{n=0}$. [*Roorkee*, 1978]

15. If $y = \frac{\sin^{-1}x}{\sqrt{(1-x^2)}}$ prove that $(1-x^2)\frac{dy}{dx} = xy+1$.

    By applying Leibnitz theorem, show that $(1-x^2)y_{n+1} - (2n=1)xy_n - n^2y_{n-1} = 0$.
    Hence find the value of y_{n+1} when $x=0$. [*Roorkee*, 1973]

16. If $y = \sin^{-1} x$, find $(y_n)_0$. [*Gorakhpur*, 1996]

17. If $y = (\sin^{-1}x)^2$, show that $(1-x^2)y_{n+2} - (2n+1)xy_{n+1} - n^2y_{n=0}$, and find $(y_n)_0$. [*Allahabad*, 1970]

18. If $y = \sin(a\sin^{-1}x)$, prove that

    $(1-x^2)y^2 - xy_1 + a^2y = 0$, and $(1-x^2)y_{n+2} - (2n+1)xy_{n+1} - (n^2-a^2)y_n = 0$. [*Roorkee*, 1997]

19. If $\cos^{-1}(y/b) = \text{loge}(x/n)n$, show that $x^2y_{n+2} + (2n+1)xy_{n+1} + 2n^2y_n = 0$. [*AMIE*, 1978]

ORDINARY AND LINEAR DIFFERENTIAL EQUATIONS

This chapter is devoted to solution of exact differential equations reducible to exact equation, linear differential equations (of second and higher order), rules for finding complementary function, particular integral and complete solutions of linear differential equations, method of variation of parameters to find particular integral, Cauchy's homogeneous linear equations, Legendres's linear equation and simultaneous linear equation with constant coefficients.

10.19 EXACT DIFFERENTIAL EQUATIONS

A differential equation is an equation involving differentials or differential co-efficients. Thus

(*i*) $e^x\,dx + e^y\,dy = 0$

(*ii*) $\frac{dy}{dx} = x^2 - 1$

(*iii*) $\left(\frac{d^2y}{dx^2}\right) + 2\left(\frac{dy}{dx}\right)^2 + y = 0$

(*iv*) $y = x\frac{dy}{dx} + \frac{x}{dy/dx}$

(*v*) $(x = y^2 - 3y) = (x^2 + 3x + y)\,dy$ are all examples of differential equation.

A differential equation obtained from its primitive directly by differentiation, without any operation of multiplication, elimination or reduction etc, is said to be an exact differential equation.

Thus a differential equation of the form

$$M(x, y)dx + N(x, y)dy = 0$$

is an exact differential equation if it can be obtained directly by differentiation the equation $u(x, y) = c$, which is its primitive.

i.e., if
$$du = Mdx = Ndy$$

Theorem: The necessary and sufficient condition for the differential equation

$$Mdx + Ndy = 0 \text{ to be exact is}$$

$$\frac{\partial M}{\partial y} = \frac{\partial N}{\partial x}$$

Proof: Condition is Necessary : If the equation $Mdx + Ndy = 0$ is exact then $\frac{\partial N}{\partial y} = \frac{\partial N}{\partial x}$

The equation $Mdx + Ndy = 0$ will be exact, if

$$du = Mdx + Ndy \quad ...(a)$$

where u is some function of x and y,

but
$$du = \frac{\partial u}{\partial x}dx + \frac{\partial u}{\partial y}dy \text{ (u, being a function of x and y)} \quad ...(b)$$

$$M = \frac{\partial u}{\partial x} \text{ and } N = \frac{\partial u}{\partial y}$$

$\therefore$
$$\frac{\partial M}{\partial y} = \frac{\partial^2 u}{\partial y \partial x} \text{ and } \frac{\partial N}{\partial x} = \frac{\partial^2 u}{\partial x \partial}$$

But
$$\frac{\partial^2 u}{\partial y \partial x} = \frac{\partial^2 u}{\partial x \partial y}$$

$\therefore$
$$\frac{\partial M}{\partial y} = \frac{\partial N}{\partial x} \text{ which is the necessary condition for exactness}$$

Condition is Sufficient: If $\frac{\partial M}{\partial y} = \frac{\partial N}{\partial x}$ then $Mdx + Ndy = 0$ is exact.

Let
$$u = \int Mdx, \text{ where y is supposed constant while performing integration.}$$

$\therefore$
$$\frac{\partial u}{\partial x} = M \text{ and } \frac{\partial^2 u}{\partial y \partial x} = \frac{\partial M}{\partial y}$$

But
$$\frac{\partial^2 y}{\partial y \partial x} = \frac{\partial^2 u}{\partial x \partial y} \text{ and } \frac{\partial M}{\partial y} = \frac{\partial N}{\partial x} \text{ (given)}$$

$\therefore$
$$\frac{\partial N}{\partial x} = \frac{\partial^2 u}{\partial x \partial y} = \frac{\partial}{\partial x}\left(\frac{\partial u}{\partial y}\right)$$

Integrating both sides w.r.t x treating y as constant, we have

$$N = \frac{\partial u}{\partial y} + f(y)$$

$\therefore$
$$Mdx + Ndy = \frac{\partial u}{\partial x} + \left\{\frac{\partial u}{\partial y} + f(y)\right\} dy = \left(\frac{\partial u}{\partial x}dx + \frac{\partial u}{\partial y}dy\right) + f(y)\, dy$$

$$= du + f(y)\, dy = d\left[u + \int f(y)\, dy\right]$$

which shows that $Mdx + Ndy$ is an exact differential and hence $Mdx + Ndy = 0$ is an exact differential equation.

Method of solution

The equation
$$Mdx + Ndy = 0 \text{ becomes}$$

$$d\left[u + \int f(y)\,dy\right] = 0$$

Integrating $\quad u + \int f(y)\,dy = c$

But $\quad u = \int\limits_{y\text{–constant}} Mdx$

and $\quad f(y) = \text{terms of } N \text{ not containing } x$

$\therefore$ The solution of $\quad Mdx + Ndy = 0$ is $\int\limits_{y \text{ constant}} Mdx + \int (\textit{terms of N not conraining x})\, dy = c$

Provided $\dfrac{\partial M}{\partial y} = \dfrac{\partial N}{\partial x}$

***Example* 10.13.** Solve $(x^2 - ay)\,dx = (ax - y^2)\,dy = 0.$

Solution :

$$(x^2 - ay)dx - y^2)\,dy = 0$$

$$(x^2 - ay)dx + (y^2 - ax)dy = 0$$

Here $\quad M = x^2\; N = y^2 - ax$

$$\therefore \quad \frac{\partial M}{\partial y} = -a \text{ and } \frac{\partial N}{\partial x} = -a\; \frac{\partial M}{\partial y} = \frac{\partial M}{\partial x}$$

Thus the given equation is exact and its solution is

$$\int\limits_{y \text{ constant}} Mdx + \int (\textit{terms of N not containing x})\; dy = c$$

i.e.,
$$\int\limits_{y \text{ constant}} (x^2 + \int y^2\, dy = c$$

or
$$\frac{x^3}{3} - ayx\; \frac{y^3}{3} = c$$

***Example* 10.14.** Solve $(x^2 + y^2 - a^2)\,x\,dx + (x^2 - y^2 - b^2)\,y\,dy = 0$

Solution : Here

$$M = (x^2 + y^2 - a^2)x \text{ and } N = (x^2 - y^2 - b^2)y$$

$$\therefore \quad \frac{\partial M}{\partial y} = 2xy \text{ and } \frac{\partial N}{\partial x} = 2xy$$

$$\frac{\partial M}{\partial y} = \frac{\partial N}{\partial x}$$

Thus the equation is exact and its solution is

$$\int\limits_{y \text{ constant}} Mdx + \int (\textit{terms of N not containing x})\; dy = c$$

i.e.,
$$\int\limits_{y\text{–constant}} (x^3 + y^2\,x - a^2\,x)\,dx + \int (-y^3 - b^2 y\; y)$$

or
$$\frac{x^4}{4} + \frac{y^2x^2}{2} - \frac{a^2x^2}{2} - \frac{y^4}{4} - \frac{b^2y^2}{2} = c$$

or $$\frac{x^4 + 2y^2x^2 - 2a^2x^2 - y^4 - 2b^2y^2}{4} = c$$

or $$x^4 + 2y^2x^2 - 2a^2x^2 - y^4 - 2b^2y^2 = 4c$$

or $$x^4 + 2x^2y^2 - 2a^2x^2 - y^4 - 2b^2x^2 = c.$$

***Example* 10.15.** Solve $ye^{xy}\,dx + (xe^{xy} + 2y)dy = 0,$

Solution : Here $M = ye^{xy}$ and $N = xe^{xy} + 2y$

$$\therefore \quad \frac{\partial M}{\partial y} = e^{xy} + xy\,e^{xy} \text{ and } \frac{\partial N}{\partial x} = xy\,e^{xy} + e^{xy}$$

$$\frac{\partial M}{\partial y} = \frac{\partial N}{\partial x}$$

Thus the equation is exact and its solution is

$$\int_{y\ constant} (ye^{xy})\,dx + \int (2y)\,dy = c$$

or $$\frac{ye^{xy}}{y} + \frac{2y^2}{2} = c \text{ or } e^{xy} + y^2 = c$$

***Example* 10.16.** Solve $(x^2 - 4xy - 2y^2)\,dx + (y^2 - 4xy - 2x^2)dy = 0$

Solution : Here $$M = x^2 - 4xy - 2y^2$$

and $$N = y^2 - 4xy - 2x^2$$

$$\therefore \quad \frac{\partial M}{\partial y} = -4x - 4y \text{ and } \frac{\partial N}{\partial x} = -4y - 4x$$

$$\frac{\partial M}{\partial y} = \frac{\partial N}{\partial x}$$

Thus the equation is exact and its solution is

$$\int_{y\ constant} M dx + \int (\textit{terms of N not containing x})\,dy = c$$

i.e., $$\int_{y\ constant} (x^2 - 4xy - 2y^2)\,dx + \int (y^2)\,dy = c$$

or $$\frac{x^3}{3} - \frac{4x^2}{2}y - 2y^2x + \frac{y^3}{3} = c$$

or $$\frac{2x^3 - 12x^2y - 12y^2x + 2y^3}{6} = c$$

or $$2x^3 + 2y^3 - 12x^2y - 12y^2x = 6c$$

or $$x^3 + y^3 - 6x^2y - 6y^2x = 3c$$

or $$x^3 + y^3 - 6x^2y - 6y^2x = c'$$

***Example* 10.17.** Solve $(2x^3 - xy^2 - 2y + 3)\,dx - (x^2y + 2x)dy = 0.$

Solution : Here $M = 2x^3 - xy^2 - 2y + 3$ and $N = -x^2y - 2x$

$$\therefore \quad \frac{\partial M}{\partial y} = -2xy - 2 \text{ and } \frac{\partial N}{\partial x} = -2xy - 2$$

$$\frac{\partial M}{\partial y} = \frac{\partial N}{\partial x}$$

Thus the equation if exact and its solution is $\int\limits_{y\ constant} M dx + \int (\textit{terms of N containing x})\, dy = c$

i.e., $\int\limits_{y\ constant} (2x^3 - xy^2 - 2y + 3)\, dx + \int (0)\, dy = c$

or $$\frac{2x^3}{4} - y^2\frac{x^2}{2} - 2yx = 3x = c$$

or $$\frac{x^4}{2} - \frac{x^2y^2}{2} - 2yx + 3x = c$$

or $$x^4 - x^2y^2 - 4yx + 6x = 2c$$

or $$x^4 - y^2y^2 - 4xy + 6x = c'$$

***Example* 10.18.** Solve $(5x^4 + 3x^2y^2 - 2xy^3)\, dx + (2x^3y - 3x^2y^2 - 5y^4)\, dy = 0$.

Solution : Here $M = 5x^4 + 3x^2y^2 - 2xy^3$

and $N = 2x^3y - 3x^2y^2 - 5y^4$

$$\therefore \quad \frac{\partial M}{\partial y} = 6x^2y - 6xy^2 \text{ and } \frac{\partial N}{\partial x} = 6x^2y - 6y^2x$$

$$\frac{\partial M}{\partial y} = \frac{\partial N}{\partial x}$$

Thus the equation is exact and its solution is $\int\limits_{y\ constant} M dx + \int (\textit{terms of N not containing x})\, dy = c$

i.e., $\int\limits_{y\ constant} (5x^4 + 3x^2y^2 - 2xy^3)\, dx + \int (-5y^4)\, dy = c$

or $$\frac{5x^5}{5} + \frac{3x^3y^2}{3} - \frac{2x^2y^3}{2} - \frac{5y^5}{5} = c$$

or $$x^5 = x^3y^3 - x^2y^3 - y^5 = c.$$

***Example* 10.19.** Solve $(x^4 - 2xy^2 + y^4)dx - (2x^2y - 4xy^3 + \sin y)dy = 0$.

Solution : Here $M = x^4 - 2xy^2 + y^4$

and $N = -2x^2y + 4xy^3 - \sin y$

$$\therefore \quad \frac{\partial M}{\partial y} = -4xy + 4y^3 \text{ and } \frac{\partial N}{\partial x} = -4xy + 4y^3 \; \frac{\partial M}{\partial y} = \frac{\partial N}{\partial x}$$

Thus the equation is exact and its solution is $\int\limits_{y\ constant} M dx + \int (\textit{terms of N not containing x})\, dy = c$

i.e., $\int\limits_{y\ constant} (x^4 - 2xy^2 + y^4)\, dx + \int (-\sin y)\, dy = c$

or $$\frac{x^5}{5}-\frac{2x^2y^2}{2}+y^4x = \cos y = c$$

or $$\frac{x^5}{5}-x^2y^2 \times y^4x+\cos y = c$$

***Example* 10.20.** Solve $(3x^2 + 6xy^2)\,dx + (6x^2y + 4y^3)dy = 0$

Solution : Here $M = 3x^2 + 6xy^2$ and $N = 6x^2y + 4y^3$

$$\therefore \quad \frac{\partial M}{\partial y} = 12xy \text{ and } \frac{\partial N}{\partial x} = 12xy$$

$$\frac{\partial M}{\partial y} = \frac{\partial N}{\partial x}$$

Thus the equation is exact and its solution is $\int\limits_{y\ constant} Mdx + \int (\textit{terms of N not conataining x})\,dy = c$

i.e., $$\int\limits_{y\ constant} (3x^2 + 6xy^2)\,dx + \int (4y^3)\,dy = 0$$

or $$\frac{3x^3}{3}+\frac{6x^2y^2}{2}+\frac{4y^4}{4} = c$$

or $$x^3 + 3x^2y^2 + y^4 = c$$

***Example* 10.21.** Solve $\frac{2x}{y^3}\,dx + \frac{y^2-3x^2}{y^4}\,dy = 0.$

Solution : Here $$M = \frac{2x}{y^3} \text{ and } N = \frac{y^2-3x^2}{y^4}$$

$$\therefore \quad \frac{\partial M}{\partial y} = -6xy^{-4} \text{ and } \frac{\partial N}{\partial x} = -6xy^{-4}$$

$$\frac{\partial M}{\partial y} = \frac{\partial N}{\partial x}$$

Thus the equation is exact and its solution is $\int\limits_{y\ constant} Mdx + \int (\textit{terms of N not containing x})\,dy = c$

i.e., $$\int\limits_{y\ constant} \frac{2x}{y^3}\,dx + \int \left(\frac{1}{y^2}\right) dy = c$$

or $$\frac{2x^2}{2y^3}+\frac{y^{-1}}{(-1)} = c \text{ or } \frac{x^2}{y^3}-\frac{1}{y} = c$$

or $$x^2 - y^2 = cy^3$$

***Example* 10.22.** Solve $\left[y\left(1+\frac{1}{x}\right)+\cos y\right]dx + (x + \log x - x\sin y)\,dy = 0.$

Solution : Here $M = y\left(1+\frac{1}{x}\right) + \cos y$ and $N = x + \log x - x \sin y$

$$\therefore \quad \frac{\partial M}{\partial y} = \left(1+\frac{1}{x}\right) - \sin y \text{ and } \frac{\partial N}{\partial x} = \left(1+\frac{1}{x}\right) - \sin y \;\; \frac{\partial M}{\partial y} = \frac{\partial N}{\partial x}$$

This the equation is exact and its solution is $\int_{y \text{ consant}} M dx + \int (\textit{terms of N not containing x})\, dy = c$

$$\text{i.e.,} \quad \int_{y \text{ constant}} \left[y\left(1+\frac{1}{x}\right) + \cos y\right] dx + \int (0)\, dy = c$$

$$\text{or} \quad y(x + \log x) + x \cos y = c$$

***Example* 10.23.** Solve [cos x tan y + cos $(x + y)$ dx + [sin x $\sec^2 y$ + cos $(x + y)$]dy = 0.

Solution : Here $M = \cos x \tan + \cos(x + y)$

$$\text{and} \quad N = \sin x \sec^2 y + \cos (x + y)$$

$$\therefore \quad \frac{\partial M}{\partial y} = \cos x \sec^2 y - \sin (x + y)$$

$$\frac{\partial N}{\partial x} = \cos x \sec^2 y - \sin (x + y)$$

$$\frac{\partial M}{\partial y} = \frac{\partial N}{\partial x}$$

Thus the equation is exact and its solution is $\int_{y \text{ constant}} M dx + \int (\textit{terms of N not containing x})\, dy = c$

$$\text{i.e.,} \quad \int_{y \text{ constant}} [\cos x \tan y + \cos(x + y)]\, dx = c$$

or $\sin x \tan y + \sin (x + y) = c$

***Example* 10.24.** Solve $(\sec x \tan x \tan y - e^x)dx + \sec x \sec^2 y\, dy = 0$.

Solution : Here $M = \sec x \tan x \tan y - e^x$

$$\text{and} \quad N = \sec x \sec^2 y$$

$$\therefore \quad \frac{\partial M}{\partial y} = \sec x \tan x \sec^2 y \text{ and } \frac{\partial N}{\partial x} = \sec x \tan x \sec^2 y$$

$$\frac{\partial M}{\partial y} = \frac{\partial N}{\partial x}$$

Thus the equation is exact and its solution is

$$\text{i.e.,} \quad \int_{y \text{ constant}} (\sec x \tan x \tan y - e^x)\, dx + \int (0)\, dy = c$$

i.e. $\tan y \sec x - ex = c$

EXERCISE 10.1

Solve the following differential equations:

1. $(y^2 e^{xy2} + 4x^3)dx + (2xy\, e^{xy2} - 3y^2)dy = 0.$ **Ans.** $e^{xy2} + x^4 - y^3 = c$

2. $(2xy \cos x^2 - 2xy + 1)dx + (\sin x^2 - x^2)dy = 0.$ **Ans.** $y \sin x^2 - x^2y + x = c$

3. $(\sin x \cos y + e^{x2})dx + (\cos x \sin y + \tan y)dy = 0.$ **Ans.** $-\cos x \cos y + \frac{1}{2} e2x + \log \sec y = c$

4. $\dfrac{dy}{dx} + \dfrac{y \cos x + \sin y + y}{\sin x + x \cos y + x} = 0.$ **Ans.** $y \sin x = (\sin y + y)x = 0$

10.20 EQUATIONS REDUCIBLE TO EXACT DIFFERENTIAL EQUATION

Sometimes a differential equation which is not exact, can be made so, on multiplication by a suitable factor called an integrating factor. The rules for finding integrating factors of the equations

$$Mdx + Ndy = 0 \text{ are as follows:}$$

I.F. (Integrating factor) by inspection

In a number of problems, a little analysis help to find the integrating factor. The following differentials are useful in selecting a suitable integrating factor.

(*i*) $ydx + xdy = d(xy)$

(*ii*) $\dfrac{xdy - ydx}{x^2} = d\left(\dfrac{y}{x}\right)$

(*iii*) $\dfrac{ydx - xdy}{y^2} = d\dfrac{(x)}{y}$

(*iv*) $\dfrac{xdy - ydx}{x^2 + y^2} = d\left(\tan^{-1}\dfrac{y}{x}\right)$

(*v*) $\dfrac{xdy - ydx}{xy} = d\left[\log\left(\dfrac{y}{x}\right)\right]$

(*vi*) $\dfrac{ydx + xdy}{xy} = d\,[\log (xy)]$

(*vii*) $\dfrac{xdy + ydx}{x^2 + y^2} = d\left[\dfrac{1}{2}\log\dfrac{x+y}{x-y}\right]$

I.F. for a homogeneous equation

If $Mdx + Ndy = 0$ is a homogeneous equation in x and y, then $\dfrac{1}{Mx + ny}$ is an I.F. provided $Mx + Ny \neq 0$

Note: If $Mx + Ny$ consists of only one term, use the above method of I.F. otherwise, proceed by putting $y = vx$.

I.F for an equation of the form

$$f_1(xy)y\, dx + f_2(xy)x\, dy = 0$$

If $Mdx + Ndy = 0$ is of the form $f_1(xy)y\, dx + f_2(xy)x\, dy = 0$

Then $\dfrac{1}{Mx - Ny}$ is an I.F., provided $Mx - Ny \neq 0$

I.F. for the equation Mdx + Ndy = 0:

(a) If $\dfrac{\dfrac{\partial M}{\partial Y}-\dfrac{\partial N}{\partial x}}{N}$ is a function of x only, say $f(x)$, then $e^{\int g(y)dy}$ is an I.F. (for N to be a smaller term)

(b) If $\dfrac{\dfrac{\partial N}{\partial x}-\dfrac{\partial M}{\partial y}}{M}$ is a function of y only say $g(y)$, then $e^{\int g(y)dy}$ is an I.F.

(for M to be a smaller term)

I.F. for the equation of the form

$x^a y^b$ (my $dx + nx\, dy) + x^c y^d (py\, dx + qx\, dy) = 0$

where a, b, c, d, m, n, p, q are all constant is $x^h y^k$, where h, k are so chosen that after multiplication by $x^h y^k$ the equation becomes exact.

ILLUSTRATIVE EXAMPLES

***Example* 10.25.** Solve $y\,dx - x\,dy + 3x^2y^2 e^{x3}\,dx = 0$.

Solution : Since $3x^2\, ex^3 = d(e^{x3})$, the term $3x^2y^2e^{x3}$ should not involve y^2.

This suggest that $\frac{1}{y^2}$ may be an I.F.

Multiplying throughout by $\frac{1}{y^2}$, we have $\frac{ydx - xdy}{y^2} + 3x^2\, e^{x3}\, dx = 0$

$d\left(\frac{x}{y}\right) + d(e^{x3}) = 0$, which is exact,

Integrating, we get $\frac{x}{y} + e^{x3} = c$

***Example* 10.26.** Solve $xdy - ydx = x\sqrt{x^2 - y^2}\; dx$.

Solution : The given equation is $xdy - ydx = x^2 \sqrt{1-\left(\frac{y}{x}\right)^2}\; dx$

or
$$\frac{\frac{xdy - ydx}{x^2}}{\sqrt{\left(1-\frac{y}{x}\right)^2}} = dx \text{ or } d\left(\sin^{-1}\frac{y}{x}\right) = dx, \text{ which is exact}$$

Integrating, we get

$$\sin^{-1}\frac{y}{x} = x + c \text{ or } y = x\sin(x = c)$$

***Example* 10.27.** Solve $y(2xy + ex)dx = ex\, dy$.

Solution : It is easy to note that the terms $ye^x\, dx$ and dy should be put together

$$\therefore \qquad (y\, e^x\, dx - e^x\, dy) + 2xy^2\, dx = 0$$

now we observe that the term $2xy^2\, dx$ should not involve y^2. Thus suggests that $\frac{1}{y^2}$ may be I.F.

Multiplying throughout by $\frac{1}{y^2}$, it follows $\frac{ye^x dx - e^x dy}{y^2} + 2x\, dx = 0$ or $d\left(\frac{e^x}{y}\right) + 2x\, dx = 0$

Integrating, we get $\frac{e^x}{y} + x^2 = c$ which is the required equation.

Example **10.28.** Solve $x\,dy - y\,dx + a(x^2 + y^2)dx = 0$.

Solution :

$$\frac{xdy - ydx}{x^2 + y^2} + adx = 0$$

$$d\left(\tan^{-1}\frac{y}{x}\right) + adx = 0$$

Integrating, we get $\tan^{-1}\frac{y}{x} + ax = c$.

Example **10.29.** Solve $y\,dx - x\,dy + \log x\,dx = 0$

Solution : $-[x\,dy - y\,dx] + \log x\,dx = 0$.

Multiplying throughout by $\frac{1}{x^2}$, we get $-\left[\frac{xdy - ydx}{x^2}\right] + \frac{\log x}{x^2}\,dx = 0$ or $-d\left[\frac{y}{x}\right] + \frac{\log x}{x^2}\,dx = 0$

Integrating, we get $\frac{-y}{x} + \log x\,\frac{x^{-1}}{(-1)} - \frac{1}{x} = c$ or $\frac{-y}{x} - \frac{1}{x}\log x - \frac{1}{x} = c$

$$-y - \log x - 1 = cx \text{ or } y + \log x + 1 + cx = 0$$

Example **10.30.** Solve $\frac{dy}{dx} = \frac{x^3 + y^3}{xy^2}$

Solution : This equation $(x^3 + y^3)dx - xy^2\,dy = 0$ is homogeneous in x and y

$$\therefore \quad \text{I.F.} = \frac{1}{Mx + Ny}$$

$$= \frac{1}{(x^3 + y^3)x - (xy^2)y} = \frac{1}{x^4 + y^3x - xy^3} = \frac{1}{x^4}$$

Multiplying throughout by $\frac{1}{x^4}$' the equation becomes

$$\left(\frac{1}{x} + \frac{y^3}{x^4}\right)dx - \left(\frac{xy^2}{x^4}\right)dy = 0$$

Here

$$M = \frac{1}{x} + \frac{y^3}{x^4} \text{ and } N = \frac{-y^2}{x^3}$$

$$\therefore \quad \frac{\partial M}{\partial y} = \frac{3y^2}{x^4} \text{ and } \frac{\partial N}{\partial x} = \frac{3y^2}{x^4}$$

$$\frac{\partial M}{\partial y} = \frac{\partial N}{\partial x}$$

Thus the equation is exact and its solution is $\int\limits_{y\text{ constant}} Mdx + \int (\text{terms of } N \text{ not containing } x)\,dy = c$

i.e., $$\int_{y\ consant} \left(\frac{1}{x}+\frac{y^3}{x^4}\right) dx + \int (0)\, dy = c$$

or $$\log x - \frac{1}{3}\frac{y^3}{x^3} = c \text{ or } 3\log x - \left(\frac{y}{x}\right)^3 = 3c$$

or $$3\log x - \left(\frac{y}{x}\right)^3 = c'$$

Example **10.31.** Solve $(x^2y - 2xy^2)dx - (x^3 - 3x^2y)dy = 0$.

Solution : This equation is homogeneous in x and y

$\therefore$ $$\text{I.F.} = \frac{1}{Mx+Ny} = \frac{1}{(x^2y-2xy^2)x+(3x^2y-x^3)y}$$

$$= \frac{1}{x^3-2x^2y^2+3x^2y^2-x^3y} = \frac{1}{x^2y^2}$$

Multiplying throughout, $\frac{1}{x^2y^2}$, the equation becomes

$$\left(\frac{1}{y}-\frac{2}{x}\right)dx - \left(\frac{x}{y^2}-\frac{3}{y}\right)dy = 0$$

Here $$M = \frac{1}{y}-\frac{2}{x} \text{ and } N = \frac{3}{y}-\frac{x}{y^2}$$

$\therefore$ $$\frac{\partial M}{\partial y} = -y^2 \text{ and } \frac{\partial N}{\partial x} = \frac{-1}{y^2} = -y^{-2}$$

$$\frac{\partial M}{\partial y} = \frac{\partial N}{\partial x}$$

Thus the equation is exact and its solution is $\int_{y\ constant} M dx + \int (\text{terms of } N \text{ not containing } x)\, dy = c$

i.e., $$\int_{y\ constant} \left(\frac{1}{y}-\frac{2}{x}\right)dx + \int \frac{3}{y}\, dy = c$$

or $$\frac{x}{y} - 2\log x + 3\log y = c$$

Example **10.32.** Solve $(1 + xy)y\, dx + (1 - xy)\, x\, dy = 0$.

Solution : The given equation is of the form $f_1(xy)y\, dx + f_2(xy)x\, dy = 0$

Here $$M = (1+xy)\, y \text{ and } N = (1-xy)x$$

$$\text{I.F.} = \frac{1}{Mx-Ny} = \frac{1}{(1+xy)yx-(1-xy)xy} = \frac{1}{2x^2y^2}$$

Multiplying throughout by $\frac{1}{2x^2y^2}$, it becomes

$$\left(\frac{1}{2x^2y}+\frac{1}{2x}\right)dx + \left(\frac{1}{2xy^2}-\frac{1}{2y}\right)dy = 0$$

Here $M = \frac{1}{2x^2y} + \frac{1}{2x}$ and $N = \frac{1}{2xy^2} - \frac{1}{2y}$

$\therefore$ $\frac{\partial M}{\partial y} = -\frac{1}{2x^2y^2}$ and $\frac{\partial N}{\partial x} = \frac{-1}{2y^2x^2}$

$$\frac{\partial M}{\partial y} = \frac{\partial N}{\partial x}$$

Thus the equation is exact and its solution is

$$\int_{y \text{ constant}} M dx + \int (\text{terms of } N \text{ not containing } x)\, dy = c$$

i.e., $\int_{y \text{ constant}} \left(\frac{1}{2x^2y} + \frac{1}{2x}\right) dx + \int \left(\frac{-1}{2y}\right) dy = c$

$$= \frac{x^{-1}}{(-1)(2y)} + \frac{1}{2}\log x - \frac{1}{2}\log y \Rightarrow c$$

$$= \frac{-1}{2xy} + \frac{1}{2}\log x - \frac{1}{2}\log y \Rightarrow c = \frac{-1}{xy} + \log x - \log y = 2c$$

$$= \frac{-1}{xy} + \log\left(\frac{x}{y}\right) = 2c \Rightarrow \log\left(\frac{x}{y}\right) - \frac{1}{xy} = c'$$

***Example* 10.33.** Solve $y(xy + 2x^2y^2)dx + x(xy - x^2y^2)dy = 0$.

Solution : The given equation is of the form

$$f_1(xy)y\,dx + f_2(xy)x\,dy = 0$$

Here $M = xy^2 + 2x^2y^3$ and $N = x^2y - x^3y^2$

Now $Mx - Ny = x^2y^2 + 2x^3y^3 - x^2y^2 + x^3y^3 = 3x^3y^3 \neq 0$

$\therefore$ $\text{I.F.} = \frac{1}{Mx - Ny} = \frac{1}{3x^3y^3}$

Multiplying throughout by $\frac{1}{3x^3y^3}$, the given equations becomes

$$\left(\frac{1}{3x^2y} + \frac{2}{3x}\right) dx + \left(\frac{1}{3xy^2} + \frac{1}{3y}\right) dy = 0$$

Here $M = \frac{1}{3x^2y} + \frac{2}{3x}$ and $N = \frac{1}{3}\left(\frac{1}{xy^2} - \frac{1}{3y}\right)$

$\therefore$ $\frac{\partial M}{\partial y} = \frac{-1}{3x^2y^2}$ and $\frac{\partial N}{\partial x} = \frac{-1}{3y^2x^2}$

$$\frac{\partial M}{\partial y} = \frac{\partial N}{\partial x}$$

Thus the equation is exact and its solution is $\int_{y \text{ constant}} M dx + \int (\text{terms of } N \text{ not containing } x)$

i.e., $\int\limits_{y\text{ constant}} \left(\frac{1}{3x^2y}+\frac{2}{3x}\right) dx + \int -\frac{1}{3y}\, dy = c$

$\frac{-x^{-1}}{3y}+\frac{2}{3}\log x-\frac{1}{3}\log y = c$ or $-\frac{1}{3xy}+\frac{2}{3}\log x-\frac{1}{3}\log y = c$

$\frac{-1}{xy}+2\log x-\log y = 3c$ or $\frac{-1}{xy}+2\log x-\log y = c'$

***Example* 10.34.** Solve $(x^2y^2 + 1)\, x\, dy + (x^2y^2 - 1)y\, dx = 0$

Solution : $(x^2y^2 - 1)y\, dx = (x^2y^2 + 1)x\, dy = 0$

The given equation is of the form

$$f_1(xy)\, ydx + f_2(xy)x\, dy = 0$$

Here I.F. = $$\frac{1}{Mx - Ny} = \frac{1}{(x^2y^3 - y)x - (x^3y^2 + x)y}$$

$$= \frac{1}{x^3y^3 - xy - x^3y^3 + (-xy)} = \frac{-1}{2xy}$$

Multiplying throughout by $\frac{-1}{2xy}$, we get

$$-\frac{1}{2}\left[\left(xy^2\frac{1}{x}\right)dx + \left(x2y+\frac{1}{y}\right)dy\right] = 0$$

$$\left(xy^2-\frac{1}{x}\right) dx + \left(x^2y+\frac{1}{y}\right) dy = 0$$

Here $$M = xy^2 - \frac{1}{x} \text{ and } N = x^2y + \frac{1}{y}$$

$\therefore$ $$\frac{\partial M}{\partial y} = 2xy \text{ and } \frac{\partial N}{\partial x} = 2xy$$

$$\frac{\partial M}{\partial y} = \frac{\partial N}{\partial x}$$

Thus the equation is exact and its solution is $\int\limits_{y\text{ consant}} Mdx + \int (\text{terms of } N \text{ not containing } x)\, dy = c$

i.e., $\frac{x^2y^2}{2} - \log x + \log y = c$ or $\log\left(\frac{y}{x}\right) + \frac{1}{2}x^2y^2 = c$

***Example* 10.35.** Solve $(x^2y^2 + xy + 1)y\, dx + (x^2y^2 - xy \div 1)x\, dy = 0.$

Solution : The given equation is of the form

$$f_1(xy)y\, dx + f_2(xy)x\, dy = 0$$

Here $$M = x^2y^3 = xy^2 + y \text{ and } N = x^3y^2 - x^2y + x$$

i.e., $$\frac{1}{Mx - Ny} = \frac{1}{2x^2y^2}$$

Multiplying throughout by $\frac{1}{2x^2y^2}$, the given equation becomes

$$\left(y+\frac{1}{x}+\frac{1}{x^2y}\right)dx+\left(x-\frac{1}{y}+\frac{1}{xy^2}\right)dy = 0$$

Here $$M = y+\frac{1}{x}+\frac{1}{x^2y} \text{ and } N = x-\frac{1}{y}+\frac{1}{x^2y^2}$$

$\therefore$ $$\frac{\partial M}{\partial y} = 1-\frac{1}{x^2y^2} \text{ and } \frac{\partial N}{\partial x} = 1-\frac{1}{x^2y^2}$$

$$\frac{\partial M}{\partial y} = \frac{\partial N}{\partial x}$$

Thus the equation is exact and its solution is

$$\int_{y \text{ consant}} Mdx + \int (\textit{terms of N not containing x})\, dy = c$$

i.e $$\int_{y \text{ constant}} \left(y+\frac{1}{x}+\frac{1}{x^2y}\right)dx + \int\left(-\frac{1}{y}\right)dy = c$$

$$yx + \log x + \frac{x^{-1}}{y(-1)} - \log y = c \text{ or } xy + \log\left(\frac{x}{y}\right) - \frac{1}{xy} = c$$

***Example* 10.36.** Solve $(xy^2 - e^{x^{\frac{1}{3}}})\, dx - x^2y\, dy = 0$

Solution : Here $M = xy^2 - e^{x^{\frac{1}{3}}}$ and $N = -x^2y$

$$\frac{\frac{\partial M}{\partial y}-\frac{\partial N}{\partial x}}{N} = \frac{2xy-(-2xy)}{-x^2y} = -\frac{4}{x}$$

which is a function of x only

$\therefore$ $$\text{I.F.} = e^{\int \frac{-4}{x}dx} = e^{-4\log x} = \frac{1}{x^4}$$

Multiplying throughout by $\frac{1}{x^4}$, we have

$$\left(\frac{y^2}{x^3}-\frac{1}{x^4}e^{x^{\frac{1}{3}}}\right)dx - \frac{y}{x^2}dy = 0$$

which is exact.

The solution is $$\int_{y \text{ constant}} \left(\frac{y^2}{x^3}-\frac{1}{x^4}e^{x^{\frac{1}{3}}}\right)dx = c$$

or $$\frac{-y^2}{2x^2}+\frac{1}{3}\int\frac{-3}{x4}e^{x^{\frac{1}{3}}}dx = c$$

or $$\frac{-y^2}{2x^2}+\frac{1}{3}\int e^t dt = c' \text{ where } t = \frac{1}{x^3}$$

or $$\frac{y^2}{2x^2}+\frac{1}{3}e^t = c \text{ or } \frac{3y^2}{x^2}+2e^{x^{\frac{1}{3}}} = c$$

where $c' = 6c$

***Example* 10.37.** Solve $(xy^3 + y)dx + 2(x^2y^2 + x + y^4)dy = 0$.

Solution : Here $M = xy^3 + y$ and $N = 2[x^2y^2 + x + y^4]$

$$\therefore \quad \frac{1}{M}\left(\frac{\partial N}{\partial x} - \frac{\partial M}{\partial y}\right) = \frac{1}{y(xy^2+1)}[4xy^2 = 2 - 3xy^2 - 1]$$

$$= \frac{1}{y},$$

which is a function of y alone.

$$\therefore \quad \text{I.F.} = \int_e \frac{1}{4} dy = e^{\log y} = y$$

Multiplying throughout by y, it becomes

$$(xy^4 + y^2)\, dx + (2x^2y^3 + 2y^5)\, dy = 0$$

Here $M = xy^4 + y^2$ and $N = 2x^2y^3 + 2xy + 2y^5$

$$\therefore \quad \frac{\partial M}{\partial y} = 4xy^3 + 2y \text{ and } \frac{\partial N}{\partial x} = 4xy^3 + 2y$$

$$\frac{\partial M}{\partial y} = \frac{\partial N}{\partial x}$$

Thus the equation is exact and its solution is $\int_{y\text{ constant}} Mdx + \int (\text{terms of } N \text{ not conaining } x)\, dy = c$

i.e., $\int_{y\text{ constant}} (xy^4 + y^2)\, dx + \int 2y^5\, dy = c$

or $$\frac{x^2y^4}{2} + y^2x + \frac{2y^6}{6} = c \text{ or } \frac{x^2y^4}{2} + xy^2 + \frac{y^6}{3} = c$$

***Example* 10.38.** Solve $(y^4 + 2y)\, dx + (xy^3 + 2y^4 - 4x)dy = 0$.

Solution : Here $M = y^4 + 2y$ and $N = xy^3 + 2y^4 - 4x$

$$\frac{\dfrac{\partial N}{\partial x} - \dfrac{\partial M}{\partial y}}{M} = \frac{y^3 - 4 - 4y^3 - 2}{y^4 + 2y}$$

$$= \frac{-3y^3 - 6}{y^4 + 2y} = \frac{-3[y^3 + 2]}{y[y^3 + 2]} = \frac{-3}{y}$$

which is a function of y along

$$\therefore \quad \text{I.F.} = e^{\int \frac{3}{4} dy} = e^{-3\log y} = \frac{1}{y^3}$$

Multiplying throughout by $\frac{1}{y^3}$, the equation becomes

$$\left(y + \frac{2}{y^2}\right) dx + \left(x + 2y - \frac{4x}{y^3}\right) dy = 0$$

Here $$M = y + \frac{2}{y^2} \text{ and } N = x + 2y - \frac{4x}{y^3}$$

$$\frac{\partial M}{\partial y} = 1 - \frac{4}{y^3} \text{ and } \frac{\partial N}{\partial x} = 1 - \frac{4}{y^3}$$

$$\frac{\partial M}{\partial y} = \frac{\partial N}{\partial x}$$

Thus the equation of exact and it solution is $\int\limits_{y\ \text{constant}} M dx + \int (\text{terms of } N \text{ not containing } x)\ dy = c$

i.e., $$\int\limits_{y\ constant} \left(y + \frac{2}{y^2}\right) dx + \int 2y\, dy = c$$

or $$yx + \frac{2x}{y^2} + \frac{2y^2}{2} = c \text{ or } yx + \frac{2x}{y^2} + y^2 = c$$

or $$\left(y + \frac{2}{y^2}\right) x + y^2 = c$$

***Example* 10.39.** Solve $(2x^2y^2 + y)\, dx + (3x - x^3y)dy = 0$.

Solution : The equation can be written as

$$(2x^2y^2dx - x^3y\, dy) + (y\, dx + 3x\, dy) = 0$$

or $x^2y[2y\, dx - x\, dy] + x^0y^0\, [y\, dx + 3x\, dy] = 0$

The given equation is of the form

$x^ay^b\, (my\, dx + nx\, dy) + x^cy^d(py\, dx + qx\, dy) = 0$

Therefore, it has an I.F. of the form x^hy^k

Multiplying the given equation by x^hy^k, we have

$(2x^{h+2}\, y^{k+2} + x^hy^{k+1})dx + (3x^{h+1}\, yk - x^{h+3}\, y^{k+1})dy = 0$

For the equation to be exact, we must have

$$\frac{\partial M}{\partial y} = \frac{\partial N}{\partial x}$$

i.e., $2(k+2)x^{h+2}\, y^{k+1} + (k+1)x^hy^k = 3(h+1)x^hy^k - (h+3)x^{h+2}\, y^{k+1}$

which holds when

$$2(k+2) = -(h+3) \text{ and } (k+1)\ 3(h+1)$$

i.e., when $$h + 2k + 7 = 0 \text{ and } 3h - k + 2 = 0$$

Solving there equation, we have

$$h = \frac{11}{7} \text{ and } k = \frac{-19}{7}$$

$$\therefore \quad \text{I.F.} = x^{\frac{-11}{7}}\, y^{\frac{-19}{7}}$$

Multiplying the given equation by $x^{\frac{-11}{7}}\, y^{\frac{-19}{7}}$, we have

$$\left(2x^{\frac{3}{7}}y^{\frac{-5}{7}} + x^{\frac{-11}{7}}\, y^{\frac{-12}{7}}\right) + \left(3x^{\frac{-4}{7}}\, y^{\frac{-19}{7}} - x^{\frac{10}{7}}\, y^{\frac{-12}{7}}\right) dy = 0$$

which is exact. The solution is

$$\int_{y \text{ constant}} (2x^{3/7}y^{-5/7} + x^{-11/7}y^{-12/7})\, dx = c$$

or $$\frac{7}{5}x^{\frac{10}{7}}y^{\frac{-5}{7}} - \frac{7}{4}x^{\frac{-4}{7}}y^{\frac{-12}{7}} = c \quad \text{or} \quad \frac{4x^{10/7}y^{-5/7} - 5x^{-4/7}y^{-12/7}}{20} = \frac{c}{7}$$

or $$4x^{10/7}y^{-5/7} - 5x^{-4/7}y^{-12/7} = \frac{20c}{7}$$

or $$4x^{10/7}y^{-5/7} - 5x^{-4/7}y^{-12/7} = c'$$

Example **10.40.** Solve $y(xy + 2x^2y^2)dx + x(xy - x^2y^2)dy = 0$.

Solution : The equation can be written as

$$xy(y\, dx + x\, dy) + x^2y^2(2y\, dx - x\, dy) = 0$$

The given equation is of the form $x^ay^b\,(my\, dx + nx\, dy) + x^cy^d(py\, dx + qx\, dy) = 0$

Comparing with it, we get

$$a = 1, b = 1, m = 1, n = 1, c = 2, d = 2, p = 2 \text{ and } q = -1$$

$\therefore$ $$\text{I.F.} = x^hy^k$$

Where the value of h and k can be determined from the relations

$$\frac{a+h+1}{m} = \frac{b+k+1}{n} \text{ and } \frac{c+h+1}{p} = \frac{d+k+1}{q}$$

$\therefore$ $$\frac{1+h+1}{1} = \frac{1+k+1}{1}$$

or $$2 + h = 2 + k \text{ or } h = k \quad ...(1)$$

and $$\frac{2+h+1}{2} = \frac{2+k+1}{-1}$$

or $$-3 - h = 6 + 2k \text{ or } 6 + 2k = 3 + h = 0$$

or $$9 + 2k + h = 0 \quad ...(2)$$

Solving (1) and (2), we get $h = k = -3$

$\therefore$ $$\text{I.F.} = x^{-3}y^{-3} = \frac{1}{x^3y^3}$$

Multiplying throughout by $\frac{1}{x^3y^3}$, it becomes

$$\left(\frac{1}{x^2y} + \frac{2}{x}\right)dx + \left(\frac{1}{xy^2} + \frac{1}{y}\right)dy = 0$$

Here $$M = \frac{1}{x^2y} + \frac{2}{x} \text{ and } N = \frac{1}{xy^2} - \frac{1}{y}$$

$\therefore$ $$\frac{\partial M}{\partial y} = \frac{-1}{x^2y^2} \text{ and } \frac{\partial N}{\partial x} = \frac{-1}{x^2y^2}$$

$$\frac{\partial M}{\partial x} = \frac{\partial N}{\partial x}$$

Thus the equation is exact and its solution is $\int_{y \text{ constant}} M dx + \int (\text{terms of } N \text{ not containing } x)\, dy = c$

i.e., $$\int_{y\ \text{constant}} \left(\frac{1}{x^2y}+\frac{2}{x}\right) dx + \int\left(-\frac{2}{y}\right) dy = c$$

or $$\frac{(x^{-1})}{y(-1)} + 2 \log x + \log y = c \text{ or } \frac{-1}{xy} + 2 \log x - \log y = c$$

or $$2 \log x - \log y - \frac{1}{xy} = c$$

EXERCISE 10.2

Solve the following differential equations:

1. $x\,dy - y\,dx = xy^2\,dx.$ **Ans.** $\frac{x^2}{2} + \frac{x}{y} = c$
2. $(xy\,e^{x/y} + y^2)dx - x^2\,e^{x/y}\,dy = 0$ **Ans.** $e^{x/y} + \log x = c$
3. $(3xy^2 - y^3)dx - (2x^2y - xy^2)dy = 0.$ **Ans.** $3 \log x - 2 \log y + \frac{y}{x} = c$
4. $(x^2 = y^2 + 2x)dx + 2y\,dy = 0.$ **Ans.** $e^x(x^2 + y^2) = c$
5. $(x^2 + y^2 + 1)dx - 2xy\,dy = 0.$ **Ans.** $x - \frac{y^2}{2} - \frac{1}{x} = c$
6. $\left(y + \frac{y^3}{3} + \frac{x^2}{2}\right) dx + \frac{1}{4}\,(x + xy^2)dy = 0.$ **Ans.** $3x^4y + x^4y^3 + x^6 = c$
7. $(3xy - 2xy^2)dx + (x^2 - 2axy)dy = 0.$ **Ans.** $x^2(ay^2 - xy) = c$
8. $(2x^2y - 3y^4)dx + (3x^3 + 2xy^3)dy = 0.$ **Ans.** $5x^{\frac{-36}{13}}\,y^{\frac{24}{13}} - 12x^{\frac{-10}{13}}\,y^{\frac{15}{13}} = c$
9. $y(2xy + 1)dx = x(1 + 2xy - x^3y^3)dy = 0.$ **Ans.** $\frac{1}{x^2y^2} + \frac{1}{3x^3y^3} + \log y = c$

10.21 LINEAR DIFFERENTIAL EQUATION

A linear differential equation is that is which the dependent and its derivatives occurs only in the first degree and are not multiplied together. Thus, the general linear differential equation of nth order is of the form where P_1, P_2,..., P_{n-1}, P_n and X are function of x only.

$$\frac{d^n y}{dx^n} + P_1 \frac{d^{n-1}}{dx^{n-1}} + P_2 \frac{d^{n-2} y}{dx^{n-2}} + \ldots + P_{n-1} \frac{dy}{dx} + P_n y = X$$

A linear differential equation with constant coefficients is of the form where a_1, a_2,..., a_{n-1} an are constants and X is either a constant or a function of x only.

$$\frac{d^n y}{dx^n} + a_1 \frac{d^{n-1} y}{dx^{n-1}} + a_2 \frac{d^{n-2} y}{dx^{n-2}} + \ldots + a_{n-1} \frac{dy}{dx} + a_n y = X$$

10.22 THE OPERATOR

The part $\frac{d}{dx}$ of the symbol $\frac{dy}{dx}$ may be regarded as an operator such that when it operates on y, the result is the derivatives of y.

Similarly $\frac{d^2}{dx^2}, \frac{d^3}{dx^3}, ..., \frac{d^n}{dx^n}$, may be regarded as operators

For brevity, we write

$$\frac{d}{dx} = D, \frac{d^2}{dx^2} = D^2, ..., \frac{d^n}{dx^n} = D^n$$

Thus the symbol D is a differential operator or simply an operator written in symbolic form, the equation.

$$\frac{d^n y}{dx^n} + a_1 \frac{d^{n-1} y}{dx^{n-1}} + a_2 \frac{d^{n-2} y}{dx^{n-2}} + ... + a_{n-1} \frac{dy}{dx} + a_n y = X$$

becomes

$$(D^n + a_1 D^{n-1} + a_2 D^{n-2} + ... + a_{n-1} D + a_n)y = X$$

or $$f(D)y = X$$

where $$f(D) = D^n + a_1 D^{n-1} = a_2 D^{n-2} + ... + a_{n-1} D + a_n$$

i.e. $f(D)$ is a polynomial in D.

The operator D can be treated as an algebraic quantity.

Thus $$D(u + v) = Du + Dv \text{ or } D(\lambda u) = \lambda Du$$

$$D^p D^q u = D^{p+q} u \text{ or } D^p D^q u = D^q D^p u$$

The polynomial $f(D)$ can be factorized by ordinary rules of algebra and the factors may be written in any order.

10.23 THEOREMS

Theorem 1. If $y = y_1, y = y_2, ..., y = y_n$ are n linearly independent solutions of the differential equation

$$(D^n + a_1 D^{n-1} + a_2 D^{n-2} + ... + a_n) y = 0 \quad ...(1)$$

then $u = c_1 y_1 + c_2 y_2 + ... + c_n y_n$ is also its solution, where $c_1, c_2, ..., c_n$ are arbitrary constants

Proof: Since $y = y_1, y = y_2, ..., y = y_n$ are solution of equation (1)

$$\therefore \quad D^n y_1 + a_1 D^{n-1} y_1 + a_2 D^{n-2} y_1 + ... + a^n y_1 = 0$$

$$D^n y_2 + a_1 D^{n-1} y_2 + a_2 D^{n-2} y_2 + ... + a^n y_2 = 0$$

... ...

... ...

$$D^n y_n + a_1 D^{n-1} y_n + a_2 D^{n-2} y_n + ... + a_n y_n = 0$$

Now
$$\begin{aligned} D_n u + a_1 D^{n-1} u + a_2 D^{n-2} u + ... + a_n u \\ = Dn(c_1 y_1 + c_2 y_2 + ... + c_n y_n) + a_1 D^{n-1} (c_1 y_1 + c_2 y_2 + ... + c_n y_n) \\ + a_2 D^{n-2} (c_1 y_1 + c_2 y_2 + ... + c_n y_n) + ... \\ + a_n D^n (c_1 y_1 + c_2 y_2 + ... c_n y_n) \\ = c_1 (D^n y_1 + a_1 D^{n-1} y_1 + a_2 D^{n-2} + ... + a_n y_1) + c_2 (D^n y_2 + a_2 D^{n-1} y_2 \\ + a_2 D^{n-2} y_2 + ... + a_n y_2) + ... + c_n (D^n y_n + a_2 D^{n-1} y_n \\ + a_2 D^{n-2} y_n + ... + a_n y_n) \\ = c_1(0) + c_2(0) + ... + c_n(0) = 0 \end{aligned}$$

which show that $u = c_1 y_1 + ... + c_n y_n$ is also the solution of equation (1).

Since this contains n arbitrary constant, it is the general or complete solution of equation (1).

Theorem 2. If $y = u$ is the complete solution of the equation $f(D)y = 0$ and $y = v$ is a particular solution (containing no arbitrary constants) of the equation $f(D)y = X$, then the complete solution of the equation.

$$f(D)y = X \text{ is } y = u + v$$

Proof. Since $y = u$ is a complete solution of the equation $f(D)y = 0$

$$\therefore \quad f(D)u = 0 \qquad ...(2)$$

Also $y = v$ is a particular solution of the equation $f(D)y = x$

$$\therefore \quad f(D)v = X \qquad ...(3)$$

Adding (2) and (4)

$$f(D)\,(u + v) = X$$

Thus $y = u + v$ satisfied the equation (3), hence it is the complete solution because it contains n arbitrary constants. The part $y = u$ is called the complementary function (C.F.) and the part $y = v$ is called the particular integral of the equation (3).

$\therefore$ The complete solution of equation (3) is $y =$ C.F. + P.I.

Thus in order to solve the equation (3), we first find the C.F. *i.e.*, C.S. of equation (1) and then the P.I. *i.e.* a particular solution of equation (3).

10.24 AUXILIARY EQUATION (A.E.)

The equation obtained by equation to zero the symbolic coefficient of y is called the auxiliary equation, briefly written as A.E.

10.25 RULES FOR FINDING THE COMPLEMENTARY FUNCTION

To solve the equation

$$\frac{d^n y}{dx_n} k_1 \frac{d^{n-1} y}{dx^{n-1}} k_2 \frac{d^{n-2} y}{dx^{n-2}} + ... + k_n y = 0 \qquad ...(1)$$

where k's are constants

The equation (1) in symbolic form is

$$(D^n + k_1 D^{n-1} + k_2 D^{n-2} + ... + k_n)y = 0$$

Its symbolic coefficient equated to zero *i.e.* ...(2)

$$D^n + k_1 D^{n-1} + k_2 D^{n-2} + ... + k_n = 0$$

is called the auxiliary equation (A.E.). Let $m_1, m_2, ..., mn$ be its roots

Case 1. If all the roots be real and different, then (2) is equivalent to

$$(D - m_1)(D - m_2) \,...\, (D - m_n)y = 0 \qquad ...(3)$$

Now (3) will be satisfied by the solution of

$$(D - m_n)y = 0 \text{ i.e. by } \frac{dy}{dx} - m_n y = 0$$

This is a Leibnitz's linear equation

$$\text{I.F.} = e^{-m_n x}$$

$\therefore$ Its solutionis $ye^{-m_n x} = c_n$

i.e.
$$y = c_n e^{-m_n x}$$

Similarly, since the factors in (3) can be taken in any order, it will be satisfied by the solution of

$$(D - m_1)y = 0, (D - m_2) = 0 \text{ etc.}$$

i.e. by $\quad y - c_1 e^{m_1 x}, y = c_2 e^{m_2 x}$ etc.

Thus the complete solution of (1) is

$$y = c_1 e^{m_1 x} = c_2 e^{m_3 x} + ... + c_n e^{m_n x} \quad ...(4)$$

Case 2. If the two roots are equal say $m_1 = m_2$ hen equation (4) becomes

$$y = (c_1 + c_2)e^{m_1 x} + c_3 e^{m_3 x} + ... + c_n e^{m_n x}$$

$$y = c\, e^{m_1 x} + c_3 e^{m_3 x} + ... c_n e^{m_n x}$$

$[\because c_1 + c_2 =$ one arbitrary constant $c]$

It has only $n - 1$ arbitrary constant and is therefore, not the complete solution of (1). In this case we proceed as follows:

The part of the complete solution corresponding to the repeated root is the complete solution of $(D - m_1)(D - m_1)\, y = 0$

Putting $(D - m_1)y = z$, it becomes $(D - m_1)z = 0$

or $$\frac{dz}{dx} - m_1 z = 0$$

This is a Leibnitz's linear in z

$$\text{I.F.} = e^{-m_1 x}$$

$\therefore$ Its solution is $\quad z\, e^{-m_1 x} = c_1$ or $z = c_1 e^{+m_1 x}$

Thus $$(D - m_1)y = z = c_1 e^{m_1 x} \text{ or } \frac{dy}{dx} - m_1 y = c_1{}^{m_1 x} \quad ...(5)$$

Its I.F. being $e^{-m_1 x}$, the solution of (5) is

$$y\, e^{-m_1 x} = \int c_1 e^{m_1 x} e^{-m_1 x}\, dx + c_2 = c_1 x + c_2$$

or $$y = (c_1 x + c_2)e^{m_1 x}$$

Thus the complete solution of (1) is

$$y = (c_1 x + c_2)e^{m_1 x} + c_3 e^{m_3 x} + ... c_n e^{m_n x}$$

If, however, the A.E. has the three equal roots (*i.e.* $m_1 = m_2 = m_3$) then the complete solution is

$$y = (c_1 x^2 + c_2 x + c_3)\, e^{m_1 x} + c_4 e^{m_4 x} + ... + c_n e^{m_n x}$$

Case 4. If one pair of roots of imaginary *i.e.*,

$$m_1 = \alpha + i\beta$$

$m_2 = \alpha - i\beta$, then the complete solution is

$$y = c_1 e^{(\alpha + i\beta)x} + c_2 e^{(\alpha - i\beta)x} + c_3\, e^{m_3 x} + ... c_n e^{m_n x}$$

$$= e^{\alpha x}\, (c_1 e^{i\beta x} + c_2 e^{-i\beta x}) + c_3\, e^{m_3 x} + ...\, c_n e^{m_n x}$$

$$= e^{\alpha x}\, [c_1(\cos \beta x + i \sin \beta x) + c_2\, (\cos \beta x - i \sin \beta x)] + c_3\, e^{m_3 x} + ... c_n e^{m_n x}$$

$[\because$ by Euler's theorem, $e^{i\theta} = \cos\theta + i\sin\theta]$

$$= e^{\alpha x}[c_1 \cos \beta x + c_2 \cos \beta x + c_1 i \sin \beta x - c_2 i \sin \beta x)] + c_3 e^{m_3 x} + \ldots c_n e^{m_n x}$$

$$= e^{\alpha x}[(\cos \beta x)(c_1 + c_2) + \sin \beta x (ic_1 - ic_2)] + c_3 e^{m_3 x} + \ldots c_n e^{m_n x}$$

$$= e^{\alpha x}[\cos \beta x\, c_1 + c_2 \sin \beta x] + c_3 e^{m_3 x} + \ldots c_n e^{m_n x}$$

where $c_1 + c_2 = c_1, i(c_1 - c_2) = c_2$

$$= e^{\alpha x}[c_1 \cos \beta x + c_2 \sin \beta x] + c_3 e^{m_3 x} + \ldots c_n e^{m_n x}$$

Case 4. If two pairs of imaginary roots be equal *i.e.*,

$$m_1 = m_2 = \alpha + i\beta$$

$m_3 = m_4 = \alpha - i\beta$, then by case 2, the complete solution is

$$y = e^{\alpha x}[c_1 x + c_2) \cos \beta x + (c_3 x + c_4) \sin \beta x] + \ldots + c_n e^{m_n x}$$

ILLUSTRATIVE EXAMPLES

***Example* 10.41.** Solve $\dfrac{d^2y}{dx^2} + \dfrac{dy}{dx} - 2y = 0.$

Solution : Given equation in symbolic form is $(D^2 + D - 2)y = 0.$

It's A.E. is $D^2 + D - 2 = 0$

i.e. $(D + 2)(D - 1) = 0$ or $D = -2$ and $D = 1$

Hence the C.S. is $y = c_1 e^{-2x} + c_2 e^{x}$

***Example* 10.42.** Solve $\dfrac{d^4x}{dt^4} + 4x = 0.$

Solution : Given equations in symbolic form is $(D^4 + 4x)\, x = 0.$

$\therefore$ A.E. is $D^4 + 4 = 0$ or $D^4 + 4D^2 + 4 - 4D^2 = 0$

or $(D^2 + 2)^2 - (2D)^2 = 0$ or $(D^2 + 2D + 2)(D^2 + 2 - 2D) = 0$

$\therefore$ either $D^2 + 2D + 2 = 0$ or $D^2 - 2D + 2 = 0$

Whence $D = \dfrac{-2 \pm \sqrt{-4}}{2}$ and $\dfrac{2 \pm \sqrt{-4}}{2}$

i.e., $D = -1 \pm i$ and $1 \pm i$

Hence the required solution is

$$x = e^{-t}(c_1 \cos t + c_2 \sin t) + e^{t}(c_3 \cos t + c_4 \sin t)$$

***Example* 10.43.** Solve $\dfrac{d^2x}{dt^2} + 6\dfrac{dx}{dt} + 9x = 0.$

Solution : Given equation is symbolic form is

$$(D^2 + 6D + 9)x = 0$$

$\therefore$ A.E. is $D^2 = 6D = 9 = 0$ *i.e.*, $(D + 3)2 = 0$

Whence $D = -3, -3$

Hence the C.S. is $x = (c_1 + c_2 t)\, e^{-3t}$

Example **10.44.** Solve $(D^3 + D^2 + 4D + 4)\, y = 0$.

Solution : Here the A.E. is $D^3 + D^2 + 4D + 4 = 0$

i.e., $(D^2 + 4)(D + 1) = 0 \therefore D = -1, \pm 2i$

Hence the C.S. is $y = c_1 e^{-x} + e^{ox}(c_2 \cos 2x + c_3 \sin 2x)$

i.e. $y = c_1 e^{-x} + c_2 \cos 2x + c_3 \sin 2x$.

Example **10.45.** Solve $\frac{d^3y}{dx^3} - 7\frac{dy}{dx} - 6y = 0$.

Solution : Given equation is symbolic form is $(D^3 - 7D - 6)y = 0$

It's A.E. is $D^3 - 7D - 6 = 0$ or $(D + 1)(D + 2)(D - 3) = 0$

Whence $D = -1, -2, 3$

Hence the C.S. is $y = c_1 e^{-x} + c_2 e^{-2x} + c_3 e^{3x}$

Example **10.46.** Solve $(D^3 - 4D^2 + 4D)y = 0$.

Solution : The A.E. is $D^3 - 4D^2 + 4D = 0$

or $D(D^2 - 4D + 4) = 0$ or $D(D - 2)^2 = 0$

Whence $D = 0, 2, 2$

Hence the C.S. is $y = c_1 e^{ox} + (c_2 x + c_3)e^{2x}$ or $y = c_1 = (c_2 x + c_3)e^{2x}$.

Example **10.47.** Solve $\frac{d^4y}{dx^4} + 13\frac{d^2y}{dx^2} + 36y = 0$.

Solution : Given equation in symbolic form is

or $(D^4 = 13D^2 + 36)y = 0$

It's A.E. is $D^4 = 13D^2 + 36 = 0$

or $(D^2 = 4)(D^2 + 9) = 0$ or $D = \pm 2i, \pm 3i$

Hence the C.S. is $y = e^{0x}(c_1 \cos 2x + c_2 \sin 2x) + e^{0x}(c_3 \cos 3x + c^4 \sin 3x)$

or $y = c_1 \cos 2x + c_2 \sin 2x = c_3 \sin 2x + c_3 \cos 3x + c^4 \sin 3x$.

Example **10.48.** Solve $\frac{d^2x}{dt^2} + 3a\frac{dx}{dt} - 4a^2x = 0$.

Solution : Given equation in symbolic form is

$$(D^2 + 3aD - 4a^2)x = 0$$

It's A.E. is $D^2 + 3aD - 4a^2 = 0$.

i.e., $(D - a)(D + 4a) = 0$

Whence $D = a, D = -4a$

$\therefore$ The complete, solution is

$$x = c_1 e^{-4at} + c_2 e^{at}$$

Example **10.49.** Solve $y'' - 2y' + 10y = 0, y(0) = 4, y'(0) = 1$.

Solution : Given equation is symbolic form is

$$(D^2 - 2D + 10)y = 0$$

Its A.E. is $D^2 - 2D + 10 = 0$

i.e., $D = \frac{2 \pm 6i}{2} = 1 \pm 3i$

Hence the complete solution is

$$y = e^x[c_1 \cos 3x + c_2 \sin 3x]$$

Now $y(0) = 4\ y'(0) = 1$

$\therefore$ $c_1 = 4$

and $\frac{dy}{dx} = c_1[-e^x \sin 3x. 3 + \cos 3x.ex] + c_2 [\cos 3x. 3. e^x + \sin 3x. e^x]$

or $1 = c_1[\cos 3(0) + c_2 (3) \text{ or } 1 = 4 = 3c_2$

or $1 - 4 = 3c_2 \text{ or } -3 = 3C_2$

or $c_2 = -1 \therefore y = e^x [4 \cos 3x - \sin 3x]$

Example **10.50.** Solve $\frac{d^2y}{dx^2} + \frac{dy}{dx} = 0$.

Solution : Given equation in symbolic form is $(D^2 + D)y = 0$

Its A.E. is $D^2 + D = 0$ *i.e.* $D(D + 1) = 0$

Whence $D = 0, D = -1$

Hence the C.S. is $y = c_1e^{-x} + c_2 e^0 = c_1e^{-x} + c_2$

Example **10.51.** Solve $\frac{d^3y}{dx^3} + y = 0$

Solution : Given in symbolic form is $(D^3 + 1)y = 0$.

Its A.E. is $D^3 + 1 = 0$ *i.e.,* $(D + 1)(D^2 + 1 - D) = 0$

Whence $D = -1 \text{ and } D = \frac{1 \pm \sqrt{3}i}{2}$

Hence the complete solution is

$$y = c_1e^{-x} + e^{x/2}\left[c_2 \cos\frac{\sqrt{3}}{2}x + c_3 \sin\frac{\sqrt{3}}{2}x\right]$$

Example **10.52.** Solve $\frac{d^3y}{dx^3} - 3\frac{d^2y}{dx^2} + 3\frac{dy}{dx} - y = 0$.

Solution : $(D^3 - 3D^2 + 3D - 1)y = 0$

Its A.E. is $D^3 - 3D^2 + 3D - 1 = 0$

or $(D^3 - 1) - 3D(D - 1) = 0 \text{ or } (D - 1)(D^2 + 1 + D) - 3D(D - 1) = 0$

or $(D-1)(D^2+1+D-3D)=0$ or $(D-1)(D^2-2D+1)=0$

or $(D-1)(D^2-D-D+1)=0$ or $(D-1)[D(D-1)-(D-1)=0$

or $(D-1)(D-1)(D-1)=0$ or $D=1, 1, 1,$

$\therefore$ Its C.S. is $y = (c_1x^2+c_2x+c_3)\,e^x.$

***Example* 10.53.** Solve $\frac{d^4y}{dx^4}+8\frac{d^2y}{dx^2}+16y=0.$

Solution : Given equation is symbolic form is $(D^4+8D^2+16)y=0.$

Its A.E. is $D^4+8D^2+16 = 0$

or $D^4+4D^2+4D^2 = 16=0$

or $D^2[D^2+4]+4[D^2+4] = 0$ or $D=\pm 2i$

$\therefore$ Its C.S. is $y = (c_1x+c_2)\cos 2x+(c_3x+c_4)\sin 2x$

***Example* 10.54.** Solve $4y''+4y''+y'=0$

Solution : Given equation is symbolic form is

$$(4D^3+4D^2+D)y = 0$$

Its A.E. is $4D^3+4D^2+D=0$ or $D(4D^2+4D+1)=0$

or $D(2D+1)(2D+1) = 0$

or $D = 0, D=-\frac{1}{2},-\frac{1}{2}$

$\therefore$ Its C.S. is $y = c_1e^{0x}+(c_2+c_3x)\,e^{-x/2}=c_1+(c_2+c_3x)\,e^{-x/2}$

EXERCISE 10.3

1. Solve $\frac{d^2y}{dx^2}-3\frac{dy}{dx}-4y=0$ **Ans.** $y=c_1\,e^{4x}+C_2e^{-x}$

2. Solve $\frac{d^2y}{dx^2}+(a+b)\frac{dy}{dx}+aby=0$ **Ans.** $y=c_1\,e^{-ax}+c_2e^{-bx}$

3. Solve $\frac{d^2y}{dx^2}-\frac{4dy}{dx}+y=0$ **Ans.** $y=c_1\,e^{(2+\sqrt{3})}\,x+c_2e\,(2-\sqrt{3})\,x$

4. Solve $\frac{d3y}{dx3}+\frac{6d^2y}{dx^2}11\frac{dy}{dx}+6y=0$ **Ans.** $y=c_1\,e^{-x}+c_2e^{-2x}+c_3e^{-3x}$

5. Solve $\frac{d^4y}{dx^4}-5\frac{d^2y}{dx^2}+4y=0$ **Ans.** $y=c_1e^x+c_2e^{-x}+c_3e^{2x}+c_4e^{-2x}$

6. Solve $\frac{d^2y}{dt^2}-3\frac{dx}{dt}+2x=0$, given that when $t=0$, x and $\frac{dx}{dt}=0$. **Ans.** $x=0$

7. Solve $\frac{d^2y}{dt^2} + 4\frac{dx}{dt} + 29y = 0$, given that when $x = 0$, $y = 0$ and $\frac{dy}{dx} = 15$. **Ans.** $y = 3e^{-2x} \sin 5x$

8. If $\frac{d^4y}{dt^4} = m^4x$, show that $x = C_1 \cos mt + C_2 \sin mt + C_3 \cosh mt + C_4 \sinh mt$.

10.26 THE INVERSE OPERATOR $\frac{1}{f(D)}$

Definition: $\frac{1}{f(D)}$ X is that function of x, (free from arbitrary constants), which when operated upon by $f(D)$ gives X.

Thus $$f(D)\left\{\frac{1}{f(D)}X\right\} = X$$

$\therefore$ $f(D)$ and $\frac{1}{f(D)}$ are inverse operators.

Theorem 1. $\frac{1}{f(D)}$ X is the particular integral of $f(D)\,y = X$.

Proof. Putting $y = \frac{1}{f(D)}$ X in the given equation $f(D)y = X$, we get

$$f(D)\left\{\frac{1}{f(D)}X\right\} = X$$

or $$X = X \text{ which is true}$$

$\therefore$ $$y = \frac{1}{f(D)}X \text{ is a solution of the given equation.}$$

Theorem 2. $$\frac{1}{D}X = \int X\,dx$$

Proof. Let $$\frac{1}{D}X = y$$

Operating both sides by D, we have

$$D\left(\frac{1}{D}X\right) = Dy \text{ or } X = \frac{dy}{dx}$$

Integrating both sides w.r.t x

$$y = \int X\,dx$$

no arbitrary constant begin added since $y = \frac{1}{D}$ X constant no arbitrary constant

$\therefore$ $$\frac{1}{D}X = \int X\,dx$$

Theorem 3. $$\frac{1}{D-a}X = e^{ax}\int Xe^{-ax}\,dx$$

Proof. Let $$\frac{1}{D-a}X = y$$

Operating on both sides by (D – a), we get

$(D-a)$ $\left[\frac{1}{(D-a)}X\right] = (D-a)y$ or $X = \left(\frac{dy}{dx}\right) - ay$

i.e. $$\frac{dy}{dx} - ay = X$$

which is linear equation

$$\text{I.F.} = \int_e -adx = e^{-ax}$$

$\therefore$ Its solution is $ye^{-ax} \int Xe^{-ax}\, dx$, no constant begin added

or $$y = e^{ax} \int Xe^{-ax}\, dx$$

Hence $$\frac{1}{D-a} X = e^{ax} \int e^{-ax} X\, dx$$

10.27 RULES FOR FINDING THE PARTICULAR INTEGRAL

Consider the equation $\frac{d^n y}{dx^n} + k_1 \frac{d^{n-1} y}{dx^{n-1}} + k_2 \frac{d^{n-2} y}{dx^{n-2}} + ... + k_n y = X$ operating on both sides by $\frac{1}{f(D)}$, we get

Which in symbolic form is $(D^n + k_1 D^{n-2} + k_2 D^{n-2} + ... + k_n)y$

$\therefore$ $$\text{P.I.} = \frac{1}{D^n + k_1 D^{n-1} + k_2 D^{n-2} + ... + k_n} X$$

Case I. When $$X = e^{ax}$$

Since $$D e^{ax} = ae^{ax} ; D^2 e^{ax} = a^2 e^{ax} .... D^n e^{ax} = a^n e^{ax}$$

$\therefore$ $$(Dn + k_1 D^{n-1} + ... + k^n)\, e^{ax} = (a^n + k_1 a^{n-1} + ... + k_n)\, e^{ax}$$

i.e., $$f(D)e^{ax} = f(a)e^{ax}$$

$$\frac{1}{f(D)} f(D)e^{ax} = \frac{1}{f(D)} f(a)e^{ax}$$

or $$e^{ax} = f(a) \frac{1}{f(D)} e^{ax}$$

$\therefore$ dividing by $f(a)$ ( ) ( ) provided $f(a) \neq 0$ ...(1)

Case of Failure

If $f(a) = 0$, the above rule fails and we proceed further. Since a is a root of A.E.

$$f(D) = D^n + k_1 D^{n-1} + ... + k_n = 0$$

$\therefore$ $D - a$ is a facto of $f(D)$

Suppose $f(D) = (D-a)\, \phi(D)$, where $\phi(a) \neq 0$. Then

( ) $$= \frac{1}{D-a} \frac{1}{\phi(D)} e^{ax} = \frac{1}{D-a} \cdot \frac{1}{\phi(a)} e^{ax} \qquad \left[\because \frac{1}{f(D)} e^{ax} = \frac{1}{f(a)} e^{ax}\right]$$

$$= \frac{1}{\phi(a)} \cdot \frac{1}{D-a} e^{ax} = \frac{1}{\phi(a)} e^{ax} \int e^{ax} . e^{-ax} dx \qquad \left[\because \frac{1}{D-a} X = e^{ax} \int Xe^{-ax} dx\right]$$

$$= \frac{1}{\phi(a)} e^{ax} \int dx = x \frac{1}{\phi(a)} e^{ax}$$

i.e. $$\frac{1}{f(D)} e^{ax} = x \frac{1}{f'(a)} e^{ax} \quad ...(2)$$

$$\left[\begin{array}{l} \because f(D) = (D-a)\phi'(D) + 1.\phi(D) \\ \because f'(a) = 0 \times \phi'(a) + \phi(a) \end{array} \right]$$

If $f'(a) = 0$, then applying (2) again, we get operating on both sides by $\frac{1}{f(D^2)}$, we get

$$\frac{1}{f(D)} e^{ax} = x^2 \frac{1}{f''(a)} e^{ax}, \text{ provided } f''(a) \neq 0 \text{ and so on.}$$

Case 2. When $X = \sin(ax+b)$ or $\cos(ax+b)$

Since
$$D \sin(ax+b) = a\cos(ax+b)$$
$$D^2 \sin(ax+b) = -a^2 \sin(ax+b)$$
$$D^3 \sin(ax+b) = -a^3 \cos(ax+b)$$
$$D^4 \sin(ax+b) = a^4 \sin(ax+b)$$

or
$$(D^2)^2 \sin(ax+b) = (-a^2)^2 \sin(ax+b)$$

In general
$$(D^2)^r \sin(ax+b) = (-a^2)^r \sin(ax+b)$$

$\therefore$
$$f(D)^2 \sin(ax+b) = f(-a^2) \sin(ax+b)$$

or
$$\sin(ax+b) = f(-a^2) \frac{1}{f(D^2)} \sin(ax+b)$$

$\therefore$ dividing by $f(-a^2)$

$$\frac{1}{f(D^2)} \sin(ax+b) = \frac{1}{f(-a^2)} \sin(ax+b), \text{ provided } f(-a^2) \neq 0$$

Similarly
$$\frac{1}{f(D^2)} \cos(ax+b) = \frac{1}{f(-a^2)} \cos(ax+b), \text{ provided } f(-a^2) \neq 0.$$

Case of Failure

If $f(-a^2) = 0$, the above rule fails and we proceed further

Since $\cos(ax+b) + i\sin(ax+b) = e^{i(ax+b)}$

$\therefore$
$$\frac{1}{f(D^2)} [\cos(ax+b) + i\sin(ax+b)] = \frac{1}{f(D^2)} e^{i(ax+b)}$$

If we replace D by ia, then $f(D^2) = f(-a^2) = 0$, so that it is a case of failure

$$= x. \frac{1}{f'(D^2)} e^{i(ax+b)} = x. \frac{1}{f'(D^2)} [\cos(ax+b) + i\sin(ax+b)]$$

Equating real parts $\frac{1}{f(D^2)} \cos(ax+b) = x. \frac{1}{f'(D^2)} \cos(ax+b)$, provided $f'(-a^2) \neq 0$.

Equating imaginary parts $\frac{1}{f(D^2)} \sin(ax+b) = x^2. \frac{1}{f'(D^2)} \sin(ax+b)$ provided $f'(-a^2) \neq 0$. If $f'(-a^2) \neq 0$ then

$\frac{1}{f(D^2)} \sin(ax+b) = x^2 . \frac{1}{f''(D^2)} \sin(ax+b)$, provided $f''(a^2) \neq 0$.

$\frac{1}{f(D^2)}\cos(ax+b) = x^2 \cdot \frac{1}{f'(D^2)}\cos(ax=b)$, provided $f'(-a^2) \neq 0$ and so on.

Case 3. When $X = X^m$, m being a positive integer

Here $$\text{P.I.} = \frac{1}{f(D)} = [f(D)]^{-1} x^m$$

Expand $[f(D)]^{-1}$ in ascending powers of D as far as the term in D^m and operates on x^m term by term. Since the $(m+1)$. Thus the equation is exact and its solution is and higher derivatives of x^m are zero, we need not consider terms beyond D^m

Case 4. When $X = e^{ax} V$, V being a function of x.

If u is a function of x, then

$$D(e^{ax}u) = e^{ax}Du + ae^{ax}u = e^{ax}(D+a)u$$

$$D^2(e^{ax}u) = d[e^{ax}(D+a)u] = eax(D^2 - aD)u + ae^{ax}(D+a)u$$

$$= e^{ax}(D^2 + 2aD + a^2)u = e^{ax}(D+a)2u$$

Similarly $$D^3(e^{ax}u) = e^{ax}(D+a)^3 u$$

In general $$D^n(e^{ax}u) = e^{ax}(D+a)^n u$$

$$\therefore \quad f(D)(e^{ax}u) = e^{ax} f(D+a)u$$

Operating on both sided by $\frac{1}{f(D)}$' we get

$$\frac{1}{f(D)}[f(D)(e^{ax}u)] = \frac{1}{f(D)}[e^{ax} f(D)+a)u]$$

or $$e^{ax}u = \frac{1}{f(D)}[e^{ax} f(D+a)u] \quad ...(1)$$

Now let $$f(D+a)u = V \text{ i.e. } u = \frac{1}{f(D+a)}V$$

$\therefore$ From (1), we have $$e^{ax}\frac{1}{f(D+a)}V = \frac{1}{f(D)}(e^{ax}V)$$

or $$\frac{1}{f(D)}(e^{ax}V) = e^{ax}\frac{1}{f(D+a)}V$$

Thus e^{ax} whish in on the right of $\frac{1}{f(D)}$ may be taken out to the left provided D is replaced by $D + a$.

Case 5. When X is any other function of x. Resolve $f(D)$ into linear factors

Let $$f(D) = (D-m_1)(D-m_2)\ldots(D-m_n)X$$

Then $$\text{P.I.} = \frac{1}{f(D)}X = \frac{1}{(D-m_1)(D-m_2)\ldots(D-m_n)}X$$

$$= \left(\frac{A_1}{D-m_1} + \frac{A_2}{D-m_2} + \ldots + \frac{A_n}{D-m_n}\right) \times [\text{partial fraction}]$$

$$= A_1\frac{1}{D-m_1}X + A_2\frac{1}{D-m_2}X + \ldots + A_n\frac{1}{D-m_n}X$$

$$= A_1 e^{m_1 x}\int Xe^{-m_1 x}dx + A_2 e^{m_2 x}\int Xe^{-m_2 x}dx + \ldots + A_n e^{m_n x}\int Xe^{-m_n x}dx$$

$$\left[\because \frac{1}{D-m}X = e^{mx}\int Xe^{-mx}dx\right]$$

10.28 WORKING PROCEDURE TO SOLVE THE EQUATION

$\frac{d^n y}{dx^n} + k_1 \frac{d^{n-1} y}{dx^{n-1}} + ... + k_{n-1} \frac{dy}{dx} + k_n y = X$ of which the symbolic from is

$$(D^n + k_1 D^{n-1} + ... + k_{n-1} D + k_n) y = X$$

Step 1. To find the complementary function

1. Write the auxiliary function
   i.e., $D^n + k_1 D^{n-1} + ... + k_{n-1} D + k_n = 0$ and solve it for D.
2. Write the C.F. as follows:

- m_1, m_2, m_3 ... (real and different roots)
  C.F. $= c_1 e^{m_1 x} c_2 e^{m_2 x} c_3 e^{m_3 x} + ...$
- m_1, m_2, m_3 ... (two real and equal roots)
  C.F. $= (c_1 + c_2 x) e^{m1x} + c_3 e^{m_3 x} + ...$
- m_1, m_1, m_1, m_4 ... (three real and equal roots)
  C.F. $= (c_1 + c_2 x + c_3 x^2) e^{m1x} + c_4 e^{m_4 x} + ...$
- $\alpha + i\beta, \alpha - i\beta, m_3$ ... (a pair of imaginary roots)
  C.F. $= e^{ax} (c_1 \cos px + c_2 \sin \beta x) + c_3 e^{m_3 x} + ...$
- $\alpha \pm i\beta, \alpha - i\beta, m_3$ ... (2 pair of imaginary roots)
  C.F. $= e^{ax} [c_1 + c_2 x) \cos \beta x + (c_3 + c_4 x) \sin \beta x] + c_5 e^{m_3 x} +$

Step 1. To find the particular integral

From symbolic from P.I. $= \frac{X}{D^n + k_1 D^{n-1} + ... + k_{n-1} D + k_n}$

$$= \frac{1}{f(D)} \text{ or } \frac{1}{\phi(D^2)} X$$

• When $X = e^{ax}$

$$\text{P.I.} = \frac{1}{f(D)} e^{ax}, \text{ put } D = a [f(a) \neq 0]$$

$$= x \frac{1}{f(D)} e^{ax}, \text{ Put } D = a\ [f(a) = 0, f(0) \neq 0]$$

$$= x \frac{1}{f'(D)} e^{ax}, \text{ put } D = a\ [f(a) = 0, f'(0) \neq 0]$$

and so on

where $f(D)$ = differential coefficient of $f(D)$ w.r.t D

$f''(D)$ = differential coefficient of $f'(D)$ w.r.t D, etc.

• When $X = \sin (ax + b)$ or $\cos (ax + b)$

$$\text{P.I.} = \frac{1}{\phi(D^2)} \sin (ax + b) \text{ [or } \cos (ax + b)], \text{ put } D^2 = -a^2\ [\phi(-a^2) \neq 0]$$

$$= x\frac{1}{\phi'(D^2)}\sin(ax+b) \text{ [or } \cos(ax+b)\text{, put } D^2 = -a^2$$

$$[\phi(-a^2) = 0,\ \phi'(-a^2) \neq 0]]$$

$$= x^2\frac{1}{\phi''(D^2)}\sin(ax+b) \text{ [or } \cos(ax+b)\text{, put } D^2 = -a^2$$

$$[\phi'(-a) = 0,\ \phi''(-a^2) \neq 0] \text{ and so on}$$

where $\phi, (D^2)$ = differential coefficient $\phi(D)$ w.r.t, D

$\phi''(D^2)$ = differential coefficient $\phi'(D)$ w.r.t D, etc

• When $X = X^m$, m being a positive integer

$$\text{P.I.} = \frac{1}{f(D)}x^m = [f(D)]^{-1}x^m$$

To evaluate it, expand $[f(D)]^{-1}$ in ascending power s of D by binomial theorem as far as D^m and operate on x^m term by term

• When $X = e^{ex}V$, where V is function of x

$$\text{P.I.} = \frac{1}{f(D)} = e^{ax}V = e^{ax}\frac{1}{f(D+a)}V$$

And the evaluate $\frac{1}{f(D+a)}V$

• When X is any function of x

$$\text{P.I.} = \frac{1}{f(D)}X$$

Resolve $\frac{1}{f(D)}$ into partial fractions and operate each partial fraction on X remembering that

$$\frac{1}{D-a}X = e^{ax}\int Xe^{-ax}\,dx$$

Step 3. To find the complete solute

Then the complete solution is y = C.F. + P.I.

ILLUSTRATIVE EXAMPLES

Example **10.55.** Find the P.I. of $(4D^2 + 4D - 3y)y = e^{ax}$.

Solution :

$$\text{P.I.} = \frac{1}{4D^2+4D-3}e^{2x} = \frac{1}{4(2)^2+4(2)-3}e^{2x} = \frac{1}{21}e^{x}$$

Example **10.56.** Find the P.I. of$(D^2 + 3D + 2)y = 5$.

Solution :

$$\text{P.I.} = \frac{5e^{ax}}{D^2+3D+2} = \frac{5e^{ax}}{(0)^2+3(0)+2} = \frac{5}{2}$$

Example **10.57.** Find the P.I. of $(D^3 - 3D^2 + 4)y = e^{2x}$.

Solution :

$$\text{P.I.} = \frac{1}{D^3-3D^2+4}e^{2x}.$$

When the D is replaced by 2, the denominator vanishes. It is the case of failure.

$$\therefore \quad \text{P.I.} = x.\frac{1}{3D2-6D}e^{2x} = x\frac{1}{3(2)^2-6(2)}e^{2x} = x.\frac{1}{12-12}e^{2x}$$

It is again a case of failure

$$\therefore \quad \text{P.I.} = x^2\, x^2\frac{1}{6D\text{–}6}e^{2x} = \frac{x^2}{6}e^{2x}$$

***Example* 10.58.** Find the P.I. of $(D^3+1)y = \sin(2x+3)$.

Solution : P.I. $= \dfrac{1}{D^3+1}\sin(2x+3) = \dfrac{1}{D(-2^2)+1}\sin(2x+3)$

By putting $\quad D^2 = -2^2$

$$= \frac{1}{1\text{–}4D}\sin(2x+3)$$

Multiplying and dividing by $(1+4D)$

$$= \frac{1+4D}{(1-4D)(1+4D)}\sin(2x+3) = \frac{1+4D}{1-16D^2}\sin(2x+3)$$

$$= \frac{1+4D}{1-16D(-2^2)}\sin(2x+3) \text{ by putting } D^2 = -2^2$$

$$= \frac{1+4D}{65}[\sin 2x = 3)] = \frac{1}{65}\sin(2x+3) + \frac{4}{65}D\sin(2x+3)$$

$$= \frac{1}{65}\sin(2x+3) + \frac{4}{65}\cos(2x+3)^2$$

$$\because \quad D = \frac{d}{dx}$$

$$= \frac{1}{65}[\sin(2x+3) + 8\cos(2x=3)]$$

***Example* 10.59.** Find the P.I. of $\dfrac{d^3y}{dx^3}+\dfrac{dy}{dx} = \sin 2x$.

Solution : Given equation in symbolic form is

$$(D^3+4D)y = \sin 2x$$

$$\therefore \quad \text{P.I.} = \frac{1}{D^3+4D}\sin 2x = \frac{1}{\text{–}4D+4D}\sin 2x \text{ putting } D^2 = -2^2$$

by putting D3 = – 22, the denominator vanishes. It is a case of failure

$$\therefore \quad \text{P.I.} = x\frac{1}{3(-4)+4}\sin 2x = x\frac{1}{-12+4}\sin 2x = -\frac{x}{8}\sin 2x$$

***Example* 10.60.** Find the P.I. of $(D^2+5D+4)y = x^2+7x+9$.

Solution : $\quad$ P.I. $= \dfrac{1}{D^3+5D+4}(x^3+7x+9) = \dfrac{1}{4\left(1+\dfrac{5D}{4}+\dfrac{D^2}{4}\right)}(x^2+7x=9)$

$$= \frac{1}{4}\left[1+\left(\frac{5D}{4}+\frac{D^2}{4}\right)\right]^{-1} (x^2+7x=9)$$

$$= \frac{1}{4}\left[1-\left(\frac{5D}{4}+\frac{D^2}{4}\right)+\left(\frac{5D}{4}+\frac{D^2}{4}\right)^2-....\right](x^2+7x+9)$$

$$= \frac{1}{4}\left[1-\frac{5D}{4}-\frac{D^2}{4}+\frac{25D^2}{16}...\right](x^2+7x+9) \qquad \text{[retaining upto } D^2]$$

$$= \frac{1}{4}\left[x^2+7x+9-\frac{5}{4}D(x^2+7x+9)+\frac{21}{16}D^2(x^2+7x+9)\right]$$

$$= \frac{1}{4}\left[x^2+7x+9-\frac{5}{4}(2x+7)+\frac{21}{16}(2)\right]$$

$$= \frac{1}{4}\left(x^2+\frac{9}{2}x+\frac{23}{8}\right)$$

***Example* 10.61.** Find the P.I. of $(D^2-2D+4)y = e^x \cos x$.

Solution :

$$\text{P.I.} = \frac{1}{D^2-2D+4}e^x\cos x = e^x\frac{1}{(D+1)^2-2(D+1)+4}\cos x$$

Replacing D by D + 1

$$= e^x\frac{1}{D^2+1+2D-2D-2+4}\cos x = e^x\frac{1}{D^2+3}\cos x$$

$$= e^x\frac{1}{-1+3}\cos x = \frac{1}{2}ex\cos x \text{ By putting } D^2 = -12$$

***Example* 10.62.** Solve $(D^3-6D^2+11D-6)y = e^{-2x}+e^{-3x}$

Solution : Its A.E. is $D^3-6D^2+11D-6 = 0$

or $(D-1)(D-2)(D-3) = 0$

Whence $D = 1, 2, 3$

$\therefore$ $\text{C.F.} = c_1e^x+c_2e^{2x}+c^3e^3x$

$$\text{P.I.} = \frac{1}{D^3-6D^2+11D-6}(e^{-2x}+e^{-3x})$$

$$= \frac{1}{D^3-6D^2+11D-6}e^{-2x}+\frac{1}{D3-6D^2+11D-6}e^{-3x}$$

$$= \frac{1}{D^3-6D^2+11D-6}e^{-2x}+\frac{1}{D^3-6D^2+11D-6}e^{-3x}$$

$$= \frac{1}{60}e-2x - \frac{1}{120}e^{-3x} = \frac{1}{120}(2e^{-2x}+e^{-3x})$$

Hence the C.S. is $y = \text{C.F.} + \text{P.I.}$

$$= c_1e^x+c_2e^{2x} = c_3e^{3x} \sim \frac{1}{120}(2e^{-2x}+e^{-3x})$$

***Example* 10.63.** Solve $(D+2)(D-1)^2y = e^{-2x}+2\sin hx$.

Solution: Its A.E. is $(D+2)(D-1)^2 = 0$

So that $D = -2, 1, 1,$

$$\therefore \quad \text{C.F.} = c_1 e^{-2x} + (c_2 + c_3 x)e^x$$

$$\text{P.I.} = \frac{1}{(D+2)(D-1)^2}(e^{-2x} + 2 \sin hx)$$

$$= \frac{1}{(D+2)(D-1)}(e^{-2x} + e^x - e^{-x}) \left[\because \sin hx = \frac{e^x - e^{-x}}{2}\right]$$

Now

$$\frac{1}{(D+2)(D-1)^2} e^{-2x} = \frac{1}{(D+2)}\left[\frac{1}{(D-1)^2} e^{-2x}\right] = \frac{1}{(D+2)}\left[\frac{1}{(-2-1)^2}{}^{-2x}\right]$$

$$= \frac{1}{9} \cdot \frac{1}{D+2} e^{-2x} = \frac{1}{9} x . \frac{1}{1} e^{-2x} = \frac{x}{9} e^{-2x}$$

$$\frac{1}{(D+2)(D-1)^2} e^x = \frac{1}{(D-1)^2}\left[\frac{1}{D+2} e^x\right] = \frac{1}{(D-1)^2}\left[\frac{1}{3} e^x\right] = \frac{1}{3} \frac{1}{(D-1)^2} e^x$$

$$= \frac{1}{3} x \frac{1}{2(D-1)} e^x = \frac{x^2}{3} . \frac{1}{2} e^x = \frac{1}{6} x^2 e^x$$

$$\frac{1}{(D+2)(D-1)^2} e^{-x} = \frac{1}{4} e^{-x}$$

$$\therefore \quad \text{P.I.} = \frac{x}{9} e^{-2x} + \frac{x^2}{6} e^x - \frac{1}{4} e^{-x}$$

Hence the C.S. is $y = c_1 e^{-2x} + (c_2 + c_3 x)\, e^x + \frac{x}{9} e^{-2x} + \frac{x^2}{6} e^x - \frac{1}{4} e^{-x}$

***Example* 10.64.** Solve $\frac{d^2 y}{dx^2} + 4\frac{dy}{dx} + 5y = -2 \cosh x$. Also find y when $y = 0$, $\frac{dy}{dx} = 1$ at $x = 0$.

Solution : Its A.E. is $D^2 + 4D + 5 = 0$ or $D = -2 \pm i$

$\therefore$ C.F. is $e^{-2x}(c_1 \cos x + c_2 \sin x)$

$$\text{P.I.} = \frac{1}{D^2 + 4D + 5} - (e^x + e^{-x})$$

Solving separately, we get

$$\frac{(-e^x)}{D^2 + 4D + 5} = \frac{-e^x}{1 + 4 + 5} = \frac{-e^x}{10}$$

and

$$\frac{(-e^{-x})}{D^2 + 4D + 5} = \frac{-e^{-x}}{1 - 4 + 5} = \frac{-e^{-x}}{2}$$

$$\therefore \quad \text{C.S. is } y = e^{-2x}[c_1 \operatorname{cox} x + c_2 \sin x] - \frac{e^x}{10} - \frac{e^{-x}}{2}$$

$$y = 0 \text{ when } x = 0$$

$$\therefore \quad 0 = c_1 - \frac{1}{10} - \frac{1}{2}$$

or $$c_1 = \frac{1}{10}+\frac{1}{2} = \frac{6}{10} = \frac{3}{5}$$

$$\frac{dy}{dx} = 1 \text{ when } x = 0$$

$\therefore$ $$\frac{dy}{dx} = c_1[e^{-2x}(-\sin x) - e^{-2x}(2)\cos x] + c_2[e^{-2x}\cos x - 2e^{-2x}\sin x] - \frac{e^x}{10}+\frac{e^{-x}}{2}$$

$$1 = c_1(0) - c_1(2) + c_2 - \frac{1}{10}+\frac{1}{2}$$

or $$1 = -\frac{3}{5}(2) + c_2 - \frac{1}{10}+\frac{1}{2} = -\frac{6}{5}+c_2-\frac{1}{10}+\frac{1}{2}$$

or $$c_2 = 1+\frac{6}{5}+\frac{1}{10}-\frac{1}{2} = \frac{10+12+1-5}{10} = \frac{18}{10} = \frac{9}{5}$$

$\therefore$ $$y = e^{-2x}\left[\frac{3}{5}\cos x + \frac{9}{5}\sin x\right]\frac{e^x}{10}-\frac{e^{-x}}{2}$$

or $$y = \left[\frac{3}{5}e^{-2x}(\cos x + 3\sin x) - \frac{e^x}{10}-\frac{e^{-x}}{2}\right]$$

***Example* 10.65.** Solve $\frac{d^2x}{dt^2} + n^2x = k\cos(nt+\alpha)$.

Solution : Its A.E. is $D^2 + n^2 = 0$ or $D = \pm$ in

$\therefore$ C.S. is $C_1 \cos nt + C_2 \sin$ nt

$$\text{P.I.} = \frac{1}{D^2+n^2}k\cos(nt+\alpha) = \frac{1}{2D}kx\cos(nt+\alpha)$$

$$= \frac{kx}{2}\int \cos(nt+\alpha) = \frac{kx}{2n}\sin(nt+\alpha)$$

$\therefore$ C.S. is $$x = c_1\cos nt + c_2 \sin nt = \frac{kx}{2n}\sin(nt+\alpha)$$

***Example* 10.66.** Solve $\frac{d^2x}{dt^2} + \frac{2dx}{dt} + 3x = \sin t$.

Solution : Its A.E. is $D^2 + 2D + 3 = 0$ or $D = -1 \pm \sqrt{2}\,i$

$\therefore$ C.F. is $e^{-t}[C_1 \cos\sqrt{2}\,t + C_2 \sin\sqrt{2}\,t]$

$$\text{P.I.} = \frac{1}{D^2+2D+3}\sin t = \frac{1}{(-12)+2D+3}\sin t$$

$$= \frac{1}{-1+2D+3}\sin t = \frac{1}{2+2D}\sin t = \frac{1}{2(D+1)}\sin t$$

We know that $$\frac{1}{D-a}X = e^{ax}\int Xe^{-ax}\,dx$$

$\therefore$ $$\frac{1}{2(D+1)}\sin t = I = \frac{1}{2}e^{-t}\int \sin te^t\,dt = \frac{e^{-t}}{2}\left[\sin te^t - \int \cos te^t\,dt\right]$$

$$I = \frac{e^{-t}}{2}\left[\sin te^t - \cos te^t - \int \sin te^t\, dt\right]$$

$$= \frac{e^{-t}}{2}[\sin te^t - \cos te^t] - \frac{e^{-t}}{2}\int \sin te^t\, dt$$

$$= \frac{e^{-t}}{2}.\, e^t[\sin t - \cos t] - I$$

$$I + I = \frac{1}{2}(\sin t - \cos t)$$

$$2I = \frac{1}{2}(\sin t - \cos t)$$

$$I = \frac{1}{4}(\sin t - \cos t)$$

$\therefore$ $$\text{P.I.} = \frac{1}{4}(\sin t - \cos t)$$

Complete solution is $x = e^{-t}[c_1 \cos\sqrt{2}\, t + c_2 \sin\sqrt{2}\, t] + \frac{1}{4}(\sin t - \cos t)$

***Example* 10.67.** Solve $\frac{d^2y}{dx^2} + \frac{3dy}{dx} + 2y = 4\cos^2 x.$

Solution : Its A.E. is $D^2 + 3D + 2 = 0$ or $(D+2)(D+1) = 0$

or $$D = -2, D = -1$$

$\therefore$ C.F. is $c_1e^{-x} + c_2e^{-2x}$

$$\text{P.I.} = \frac{1}{D^2+3D+2} 4\cos^2 x = \frac{1}{D2+3D+2}4\left(\frac{\cos 2x + 1}{2}\right)$$

$$= \frac{1}{D^2+3D+2}2(\cos 2x + 1) = \frac{1}{D^2+3D+2}2(\cos 2x + e^{0x})$$

Solving separately, we get

$$\frac{1}{D^2+3D+2}(2e^{0x}) = \frac{2e^{0x}}{2} = 1$$

and $$\frac{1}{D^2+3D+2}2\cos 2x = \frac{1}{3D-2}2\cos 2x = \frac{2}{3}\frac{1}{\left(D-\frac{2}{3}\right)}\cos 2x$$

Let $$I = \frac{2}{3}e^{2x/3}\int \cos 2xe^{-2x/3}dx$$

$$I = \frac{2}{3}e^{2x/3}\left[-\frac{3}{2}\cos 2xe^{-2x/3} - \int(-\sin 2x).2.\frac{e^{-2x/3}}{-2}3\, dx\right]$$

$$I = \frac{2}{3}e^{2x/3}\left[-\frac{3}{2}^{-2x/3}\cos 2x + \frac{9}{2}\sin 2xe^{-2x/3} - 9\int \cos 2xe^{-2x/3}dx\right]$$

$$I = \frac{2}{3}e^{2x/3}\left[-\frac{3}{2}e^{-2x/3}\cos 2x + \frac{9}{2}e^{-2x/3}\sin 2x\right] - 9\,\frac{2}{3}e^{2x/3}\int \cos 2xe^{-2x/3}dx$$

$$I + 9I = \frac{2}{3}e^{2x/3}\left[-\frac{3}{2}e^{-2x/3}\cos 2x + \frac{9}{2}e^{-2x/3}\sin 2x\right]$$

$$10\,I = (-\cos 2x + 3\sin 2x)$$

$$I = \frac{1}{10}(-\cos 2x + 3\sin 2x)$$

Hence the C.S. is $y = c_1 e^{-x} + c_2 e^{-2x} + 1 + \frac{1}{10}(3\sin 2x - \cos 2x)$

***Example* 10.68.** Solve $\frac{d^3y}{dx^3} + 2\frac{d^2y}{dx^2} + \frac{dy}{dx} = e^{2x} + \sin 2x.$

Solution : Its A.E. is $D^3 + 2D^2 + D = 0$ or $D(D^2 + 2D + 1) = 0$

or $$D(D+1)(D+1) = 0 \text{ or } D = 0, D = -1, D = -1$$

$\therefore$ C.F. is $c_1 + (c_2 + c_3 x)e^{-x}$

$$\text{P.I.} = \frac{1}{D^3 + 2D^2 + D}(e^{2x} + \sin 2x)$$

$$= \frac{e^{2x}}{18} + \frac{3}{50}\cos 2x - \frac{2}{25}\sin 2x$$

$\therefore$ C.S. is $$y = c_1 + (c_2 + c_3 x)e^{-x} + \frac{e^{2x}}{18} + \frac{3}{10}\cos 2x - \frac{2}{25}\sin 2x$$

***Example* 10.69.** Solve $\frac{d^2y}{dx^2} + 4y = x^2 + \sin 2x.$

Solution : Its A.E. is $D_2 = 4 = 0$ or $D^2 = -4$

or $$D = \pm 2i$$

$\therefore$ C.F. is $(c_1 \cos 2x + c_2 \sin 2x)$

$$\text{P.I.} = \frac{1}{D^2 + 4}(x^2 + \sin 2x)$$

Solving separately, we get

$$\frac{1}{D^2+4}x^2 = \frac{1}{4}\left[1 + \frac{D^2}{4}\right]^{-1} x^2 = \frac{x^2}{4} - \frac{1}{8}$$

and $$\frac{1}{D^2+4}\sin 2x = \frac{1}{2D}x\sin 2x = \frac{1}{2}\int x \sin 2x\,dx$$

$$= \frac{1}{2}\left(\frac{-x}{2}\cos 2x\right) + \frac{1}{8}(\sin 2x) = \frac{-x}{4}\cos 2x + \frac{1}{8}(\sin 2x)$$

Hence C.S. is $$y = c_1 \cos 2x + c_2 \sin 2x + \frac{x^2}{4} - \frac{1}{8} - \frac{x}{4}\cos 2x + \frac{1}{8}(\sin 2x)$$

***Example* 10.70.** Solve $\frac{d^2y}{dx^2} - 4y = x^2.$

Solution : Its A.E. is $D^2 - 4 = 0$ or $D = \pm 2$

$\therefore$ C.F. is $c_1 e^{2x} + c_2 e^{-2x}$

$$\text{P.I.} = \frac{1}{D^2-4}x^2 = (D^2-4)^{-1}x^2 = \frac{1}{4}\left(\frac{D^2}{4}-1\right)^{-1} x^2$$

$$= -\frac{1}{4}\left[1-\frac{D2}{4}\right]^{-1} x^2 = \frac{-1}{4}\left[1+\frac{D^2}{4}\right]x^2 = \frac{-x^2}{4}-\frac{1}{8}$$

Hence the C.S. is $$y = c_1 e^{2x} + c_2 e^{-2x} - \frac{x^2}{4} - \frac{1}{8}$$

Example **10.71.** Solve $(D^3 - D)y = 2x + 1 + 4\cos x + 2e^x$.

Solution : Its A.E. is $D^3 - D = 0$ or $D(D^2 - 1) = 0$

or $$D = 0, D = \pm 1$$

$\therefore$ C.F. is $c_1 + c_2 e^x + c_3 e^{-x}$

$$\text{P.I.} = \frac{1}{D^3-D}(2x + 1 = 4\cos x + 2e^x)$$

Solving separately, we get

$$\frac{1}{D^2-D}2x = \frac{2}{D(D^2-1)}x = -\frac{2}{D}(1+D^2)x = -x^2$$

$$\frac{1}{D^3-D} = \frac{1}{D^3-D}e^{0x} = \frac{x}{3D-1}e^{0x} = -x$$

$$\frac{1}{D^3-D}4\cos x = -2\int \cos x\,dx = -2\sin x$$

$$\frac{1}{D^3-D}2e^x = 2x\frac{1}{3D^2-1}e^x = \frac{2x}{2}e^x = xe^x$$

Hence the complete solution is $y = c_1 + c_2 e^x + c_3 e^{-x} - x^2 - x - 2\sin x + xe^x$.

Example **10.72.** Solve $(D^4 - 1)y = e^x \cos x$.

Solution : Its A.E. is $D^4 - 1 = 0$ or $(D^2)2 - (1^2)^2 = 0$

or $(D^2 - 1)(D^2 + 1) = 0$ or $(D - 1)(D + 1)(D^2 + 1) = 0$

or $D = 1, |D = -1, D = \pm i$

$\therefore$ C.F. is $c_1 e^x + c_1 e^{-x} + c_3 \cos x + c_4 \sin x$

$$\text{P.I.} = \frac{1}{D^4-1}e^x\cos x = e^x\frac{1}{(D+1)^4-1}\cos x$$

$$= e^x\frac{1}{[(D+1)2]2-(1^2)^2}\cos x$$

$$= e^x\frac{1}{(D2+2D+1-1(D^2+2D+1+1)^2}\cos x$$

$$= e^x\frac{1}{(D^2+2D)(D^2+2D+2)}\cos x$$

$$= e^x\frac{1}{(2D-1)(2D+1)}\cos x$$

$$= e^x \frac{1}{4D^2 - 1} \cos x = \frac{-e^x}{5} \cos x$$

Hence C.S. is $\quad y = c_1 e^x = c_2 e^{-x} + c_3 x + c_4 \sin x - \frac{-e^x}{5} \cos x$

Example **10.73.** Solve $\frac{d^4y}{dx^4} - y = \cos x \cos h\, x.$

Solution : Given equation is symbolic form is

$$(D^4 - 1)y = \cos x \cos hx$$

Its A.E. is $\quad D^4 - 1 = 0 \text{ or } D = 1, -1, \pm i$

$\therefore$ C.S. is $c_1 e^x + c_2 e^{-x} + c_3 \cos \quad x = c_4 \sin x$

$$\text{P.I.} = \frac{1}{D^4 - 1} \cos x \cos hx = \frac{1}{D^4 - 1} \cos x \left(\frac{e^x + e^{-x}}{2}\right)$$

$$= \frac{1}{2}\left[\frac{1}{D^4 - 1} e^x \cos x + \frac{1}{D^4 - 1} e^{-x} \cos x\right]$$

$$= \frac{1}{2}\left[e^x \frac{1}{(D+1)^4 - 1} \cos x + e^{-x} \frac{1}{(D-1)^4 - 1} \cos x\right]$$

$$= \frac{1}{2}\left[e^x \frac{1}{(D4 + 4D^3 + 6D^2 + 4D} \cos x + e^{-x} \frac{1}{D^4 - 4D^3 + 6D^2 - 4D} \cos x\right]$$

$$= \frac{1}{2}\left[e^x \frac{1}{(-1^2)^2 + 4D(-1^2) + 6(-1^2) + 4D} \cos x\, e^x \frac{1}{(-1^2)^2 - 4D(-1^2) + 6(-1^2) - 4D} \cos x\right]$$

$$= \frac{1}{2}\left[\frac{e^x}{-5} \cos x + e^{-x}\left(\frac{1}{5}\right) \cos x\right]$$

$$= -\frac{1}{5}\left[\frac{e^x \cos x}{2} + \frac{e^{-x} \cos x}{2}\right] = -\frac{1}{5} \cos x \left[\frac{e^x + e^{-x}}{2}\right]$$

$$= -\frac{\cos x}{5} \cos hx$$

Hence the C.S. is $y = c_1 e^x + c_2 e^{-x} + c_3 \cos x + c_4 \sin x - \frac{1}{5} \cos x \cos hx$

Example **10.74.** Solve $\frac{d^2y}{dx^2} + 2y = x^2 e^3x + ex \cos 2x.$

Solution : Its A.E. is $D^2 + 2 = 0$ or $D = \pm \sqrt{2}\, i$

$\therefore$ C.F. is $(c_1 \cos \sqrt{2}\, x + c_2 \sin \sqrt{2}\, x$

$$\text{P.I.} = \frac{1}{D^2 + 2} [x^2 e^{3x} + e^x \cos 2x]$$

Solving separately, we get

$$\frac{1}{D^2 + 2} x^2 e^{3x} = e^{3x} \frac{1}{(D+3)^2 + 2} x^2 = e^{3x} \frac{1}{D^2 + 6D + 11} x^2$$

$$= \frac{e^{3x}}{11}\left[1+\left\{\frac{D^2}{11}+\frac{6D}{11}\right\}\right]^{-1} x^2$$

$$= \frac{e^{3x}}{11}\left[1-\left(\frac{D^2}{11}+\frac{6D}{11}\right)+\frac{(D^4+36D^2+12D^3)}{(11)^2}+\ldots\right]x^2$$

$$= \frac{e^{3x}}{11}\left[x^2-\frac{2}{11}-\frac{12x}{11}+\frac{72}{121}\right]=\frac{e^{3x}}{11}\left[x^2-\frac{12x}{11}+\frac{50}{121}\right]$$

and $$\frac{1}{D^2+2}\,e^x\cos 2x = e^x\frac{1}{(D+1)^2+2}\cos 2x + e^x\frac{1}{2D-1}\cos 2x$$

$$= e^x\frac{(2D+1)}{(2D-1)(2D+1)}\cos 2x = e^x\frac{(2D+1)}{4D2-1}\cos 2x$$

$$= e^x\frac{(2D+1)}{[4(-2^2)-1]}\cos 2x = e^x\frac{(2D+1)}{-16-1}\cos 2x$$

$$= -\frac{e^x}{17}[-2\sin 2x.2+\cos 2x] = \frac{e^x}{17}[4\sin 2x-\cos 2x$$

Hence the C.S. is

$$y = c_1\cos\sqrt{2}\,x+c_2\sin\sqrt{2}\,x+\frac{e^x}{17}[4\sin 2x-\cos 2x]+\frac{ex}{11}\left[x^2-\frac{12x}{11}+\frac{50}{121}\right]$$

***Example* 10.75.** Solve $(D^3+2D^2+D)y = x^2e^{2x}+\sin x$.

Solution : Its A.E. is $D^3+2D^2+D=0$ or $D(D^2+2D+1)=0$

or $$D = 0, D=-1, D=-1$$

$\therefore$ C.F. is $c_1+(c_2+c_3x)e^{-x}$

$$\text{P.I.} = \frac{1}{D^3+2D^2+D}(x^2e^{2x}+\sin x)$$

Solving separately, we get

$$\frac{1}{D^3+2D^2+D}x^2e^{2x} = e^{2x}\frac{1}{(D+2)^3+2(D+2)^2+(D+2)}x^2$$

$$= e^{2x}\frac{1}{D^3+8D^2+21D+18}x^2 = \left[1+\left\{\frac{D^3}{18}+\frac{8D^2}{18}+\frac{21D}{18}\right\}\right]^{-1}x^2$$

$$= \frac{e^{2x}}{18}\left[1-\left\{\frac{D^3}{18}+\frac{8D^2}{18}\frac{21D}{18}\right\}\right]+\frac{(D^2)[D^4+64D^2+(21)^2+2.8D.21+2D^2(8D+21)}{(18)^2}\Bigg\}+\ldots\Bigg]x^2$$

$$= \frac{e^{2x}}{18}\left[1-\frac{8(2)}{18}-\frac{21}{18}(2x)+\frac{(2)(21)^2}{(18)^2}\right]=\frac{e^{2x}}{18}\left[1-\frac{8}{9}-\frac{7}{3}x+\frac{49}{18}\right]$$

$$= \frac{e^{2x}}{18}\left[1-\frac{7}{3}x+\frac{(49-16)}{18}\right]=\frac{e^{2x}}{18}\left[1-\frac{7x}{3}+\frac{11}{6}\right]$$

and

$$\frac{1}{D^3+2D^2+D}\sin^2 x = \frac{1}{2(D^3+2D^2+D)}(1-\cos 2x)$$

$$= \frac{1}{2(D^3+2D^2+D)}(e^{0x}-\cos 2x)$$

$$= \frac{1}{2}\left[x - \frac{1}{-2^2 D + 2 - (2)^2 + D}\cos 2x\right]$$

$$= \frac{x}{2} - \frac{1}{2(-4D - 8 + D)}\cos 2x = \frac{x}{2} - \frac{1}{2(-3D - 8)}\cos 2x$$

$$= \frac{x}{2} + \frac{1}{2(3D + 8)}\cos 2x = \frac{x}{2} + \frac{1(3D - 8)}{2(3D + 8)(3D - 8)}\cos 2x$$

$$= \frac{x}{2} + \frac{1}{2}\frac{(3D - 8)}{(9D^2 - 64)}\cos 2x = \frac{x}{2} + \frac{1}{2(-100)}[-3 \sin 2x.\, 2 - 8 \cos 2x]$$

$$= \frac{x}{2} - \frac{1}{200}[-6 \sin 2x - 8 \cos 2x]$$

$$= \frac{x}{2} + \frac{1}{100}[3 \sin 2x + 4 \cos 2x]$$

Hence the C.S. is

$$y = c_1 + (c_2 + c_3 x)\, e^{-x} + \frac{1}{100}[3 \sin 2x + 4 \cos 2x] + \frac{e^{2x}}{18}\left[x^2 + \frac{7x}{3} + \frac{11}{6}\right]$$

***Example* 10.76.** Solve $\frac{d^2y}{dx^2} + 4y = x \sin x$.

Solution : Its A.E. is $D^2 + 4 = 0$ or $D_2 = -2^2$

$\therefore$ C.F. is $c_1 \cos 2x = c_2 \sin 2x$

$$\text{P.I.} = \frac{1}{D^2 + 4}\, x\, (\text{I.P. of } e_{1x})$$

$$= \text{I.P. of}\left[e^{ix}\frac{1}{\{(D+i)^2 + 4\}}x\right] = \text{I.P. of}\left[e^{ix}\frac{1}{D^2 + i^2 + 2Di + 4}x\right]$$

$$= \text{I.P. of } eix\left[\frac{1}{D^2 + 2Di + 3}\right] = \text{I.P. of}\left[\frac{e^{ix}}{3}\left\{1 + \left(\frac{D^2}{3} + \frac{2Di}{3}\right)\right\}^{-1} x\right]$$

$$= \text{I.P. of}\left[\frac{e^{ix}}{3}\left(1 - \frac{D^2}{3} + \frac{2Di}{3}\right)x\right] = \frac{1}{3}\,\text{I.P. of }\left[(\cos x + i \sin x)\left(x - \frac{2i}{3}\right)\right]$$

$$= \frac{1}{3}\,\text{I.P. of}\left[x \cos x + xi \sin x - \frac{2i}{3}\cos x - \frac{2i^2}{3}\sin x\right]$$

$$= \frac{1}{3}\,\text{I.P. of}\left[\left(x \cos x + \frac{2}{3}\sin x\right) + i\left(x \sin x - \frac{2}{3}\cos x\right)\right]$$

$$= \frac{1}{3}\left(x \sin x - \frac{2}{3}\sin x\right)$$

Hence the C.S. is

$$y = c_1 \cos 2x + c_2 \sin 2x + \frac{1}{9}(3\, x \sin x - 2 \cos x)$$

***Example* 10.77.** Solve $\frac{d^2y}{dx^2} + y = \text{cosec } x$.

Solution : Given equation in symbolic form is $(D^2 + 1)y = \text{cosec } x$

$\therefore$ C.F. is $c^1 \cos x + c^2 \sin x$

$$\text{P.I.} = \frac{1}{D^2+1}\operatorname{cosec} x = \frac{1}{(D+i)(D-i)}\operatorname{cosec} x$$

$$= \frac{1}{2i}\left(\frac{1}{D-i} - \frac{1}{D+i}\right)\operatorname{cosec} x$$

$$= \frac{1}{2i}\left(\frac{1}{D-i}\operatorname{cosec} x - \frac{1}{D+i}\operatorname{cosec} x\right)$$

Now $\frac{1}{D-i}\operatorname{cosec} x = e^{ix}\int \operatorname{cosec} x\, e^{-ix}\,dx = e^{ix}\int \operatorname{cosec} x(\cos x - i\sin x)\,dx$

$$= e^{ix}\int(\cot x - i)\,dx = e^{ix}(\log\sin x - ix)$$

changing i to $-i$, we have $\frac{1}{D+i}\operatorname{cosec} x = e^{-ix}(\log\sin x + ix)$

$\therefore$ $\text{P.I.} = \frac{1}{2i}[e^{ix}(\log\sin x - ix) - e^{-ix}(\log\sin x + ix)]$

$$= \log\sin x\left(\frac{e^{ix} - e^{ix}}{2i}\right) - x\left(\frac{e^{ix} + e^{ix}}{2}\right)$$

$$= \log\sin x.\sin x - x\cos x$$

Hence the C.S. is $y = c_1\cos x + c_2\sin x + \sin x\log\sin x - x\cos x$.

***Example* 10.78.** Solve $\frac{d^2y}{dx^2} + a^2y = \tan ax$.

Solution : Given equation in symbolic form is $(D^2 + a^2)y = \tan ax$.

Its A.E. is $D^2 + a^2 = 0$ or $D = \pm ia$

$\therefore$ $\text{C.F.} = c_1\cos ax + c_2\sin ax$

$$\text{P.I.} = \frac{1}{D^2+a^2}\tan ax = \frac{1}{(D+ia)(D-ia)}\tan ax$$

$$= \frac{1}{2ia}\left[\frac{1}{D-ia} - \frac{1}{D+ia}\right]\tan ax$$

$$= \frac{1}{2ia}\left[\frac{1}{D-ia}\tan ax - \frac{1}{D+ia}\tan ax\right]$$

Now $\frac{1}{D-ia}\tan ax = e^{iax}\int \tan ax.e^{-iax}\,dx = e^{iax}\int \tan ax(\cos ax - i\sin ax)\,dx$

$$= e^{iax}\int\left(\sin ax - i\frac{\sin^2 ax}{\cos ax}\right)dx = e^{iax}\int\left[\sin ax - i\frac{(1-\cos^2 ax)}{\cos ax}\right]dx$$

$$= e^{iax}\int[\sin ax - i(\sec ax - \cos ax)]\,dx$$

$$= e^{iax}\left[-\frac{\cos ax}{a} - \frac{i}{a}\log(\sec ax + \tan ax) + i\frac{\sin ax}{a}\right]$$

$$= \frac{-1}{a}e^{iax}[(\cos ax - i\sin ax) + i\log(\sec ax + \tan dx)]$$

$$= \frac{-1}{a}e^{iax}[e^{iax} + i\log(\sec ax + \tan ax)]$$

$$= \frac{-1}{a}\,[1 = ie^{iax} \log(\sec ax + \tan ax)]$$

Changing i to $-i$, we have

$$\frac{1}{D+ia}\tan ax = \frac{-1}{a}\,[1 - ie^{-iax} \log(\sec ax + \tan ax)]$$

$$\therefore \quad \text{P.I.} = \frac{1}{2ia}\left[\frac{-1}{a}\{1 + ie^{iax}\, Log(\sec ax + \tan ax)\} + \frac{1}{a}\{1 - ie^{-iax} \log(\sec ax + \tan ax)\}\right]$$

$$= \frac{-1}{a^2}\log(\sec ax + \tan ax)\left(\frac{e^{iax} + e^{-iax}}{2}\right)$$

$$= \frac{-1}{a^2}\log(\sec ax + \tan ax)\cos ax$$

Hence the complete solution is

$$y = c_1 \cos ax + c_2 \sin ax - \frac{1}{a^2}\log(\sec ax + \tan ax)\cos ax$$

EXERCISE 10.4

1. $(D-2)^2 y = 8(e^{2x} + \sin 2x + x^2)$ **Ans.** $y = (c_1 + c_2 x)\,e^{2x} + 4x^2 e^{2x} + \cos 2x + 2x^2 + 4x + 3$

2. $\frac{d^2y}{dx^2} - 4y = x \sin hx$ **Ans.** $y = c_1 e^{2x} = c_2 e^{-2x} - \frac{x}{3}\sin hx - \frac{2}{9}\cos hx$

3. $(D^2 - 4D + 3)y = \sin 3x \cos 2x.$ **Ans.** $y = c_1 e^x + c_2 e^{3x} + \frac{1}{884}(10\cos 5x - 11\sin 5x) + \frac{1}{20}(\sin x + 2\cos x)$

4. $\frac{d^2y}{dx^2} + 2\frac{dy}{dx} + y = e^{2x} - \cos^2 x.$ **Ans.** $y = (c_1 + e_2 x)\,e^{-x} - \frac{1}{2} + \frac{1}{50}(3\cos 2x - 4\sin 2x)$

5. $\frac{d^2y}{dx^2} + 5\frac{dy}{dx} + 6y = e^{-2x}\sin 2x.$ **Ans.** $y = c_1 e^{-2x} + c_2 e^{-3x} - \frac{e^{-2x}}{10}(\cos 2x = 2\sin 2x)$

6. $(D^2 + 4)y = e^x \sin^2 x.$ **Ans.** $y = c_1 \cos 2x + c_2 \sin 2x + \frac{e^x}{2}\left[\frac{1}{5} - \frac{1}{17}(4\sin 2x + \cos 2x\right]$

7. $(D^2 - 1)y = x \sin + (1 + x^2)e^x.$ **Ans.** $y = c_1 e^x + c_2 e^{-x} - \frac{1}{2}(x \sin x = \cos x) + \left(\frac{xe^x}{12}\right)(2x^2 - 3x + 9)$

8. $\frac{d^2y}{dx^2} + q1\frac{dy}{dx} = 2y = e^x$ **Ans.** $y = c_1 e^{-x} + c_2 e^{-2x} + c_3 e^{-3x} + e^{-2x} e^{ex}$

9. $(D^2 - 4D + 4)y = 8x^2 e^{2x}\sin 2x$ **Ans.** $y = (c_1 + c_2 x)e^{2x} - e^{2x}\,[4x \cos 2x = (2x^2 - 3)\sin 2x]$

10.29 METHOD OF VARIATION OF PARAMETERS TO FIND PARTICULAR INTEGRAL

Consider the linear equation of second order with constant coefficients

$$\frac{d^2y}{dx^2} + a_1\frac{dy}{dx} + a_2y = X \quad ...(1)$$

Let its C.F. be $y = c_1y_1 + c_2y_2$ so that y_1 and y_2 satisfy the equation

$$\frac{d^2y}{dx^2} + a_1\frac{dy}{dx} + a_2y = 0 \quad ...(2)$$

Now, let us assume that the P.I. of (1) is $y = uy_1 + vy_2$ ... (3)

where us and v are unknown function of x.

Differentiating (3) w.r.t x, we have

$$y' = uy'_1 + vy'_2 + u'y_1 = v'y_2 = uy'_1 + vy'_2 \quad ...(4)$$

Assuming that u, v satisfy the equation $u'y_1 + v'y_2 = 0$ ...(5)

Differentiating (4) w.r.t x we have

$$y'' = uy''_1 + u'y'_1 + vy''_2 + v'y'_2$$

Substituting the values of y, y' and y'' in (1), we get

$$(uy_1'' + u'y'_1 + vy''_2 + v'y'_2) + a_1(uy_1' + vy_2') + a_2(uy_1 + vy_2) = X$$

or $$u(y_1'' + a_1y_1' + a_2y_1) + v(y_2'' + a_1y_2' + a_2y_2) + u'y'_1 + v'y'_2 = X$$

or $$u'y'_1 + v'y'_2 = X \quad ...(6)$$

Since y_1 and y_2 satisfy

$$u' = \begin{vmatrix} 0 & y_2 \\ X & y'_2 \end{vmatrix} \div \begin{vmatrix} y_1 & y_2 \\ y'_1 & y_2' \end{vmatrix} = \frac{-y_2X}{W}$$

and $$v' = \begin{vmatrix} y_1 & 0 \\ y_1 & X \end{vmatrix} \div \begin{vmatrix} y_1 & y_2 \\ y'_1 & y_2' \end{vmatrix} = \frac{y_1X}{W}$$

where $$W = \begin{vmatrix} y_1 & y_2 \\ y'_1 & y_2' \end{vmatrix}$$ is called the wronskian of y_1, y_2

Integrating $$u = -\int \frac{y^2X}{W}dx \text{ and } v = \int \frac{y_1X}{W}dx$$

Substituting in (3), the P.I. is known as the solution is obtained by varying the arbitrary constants, c_2, c_2 of the C.F., the method is known as variation of parameter, method of variation of parameters is to be used if instructed to do so.

ILLUSTRATIVE EXAMPLES

Example **10.79.** Apply the method of variation of parameter to solve $\frac{d^2y}{dx^2} + 4y = 4\sec^2 2x$.

Solution : Given equation in symbolic form is

$$(D^2 + 4)y = 4\sec^2 2x$$

Its A.E. is $$D^2 + 4 = 0$$

So that $$D = \pm 2i$$

$\therefore$ C.F. is $$y = c_1\cos 2x + c_2\sin 2x$$

Here $y_1 = \cos 2x, y_2 = \sin 2x$ and $X = 4 \sec^2 2x$

$$\therefore \quad W = \begin{vmatrix} y_1 & y_2 \\ y_1' & y_2' \end{vmatrix} = \begin{vmatrix} \cos 2x & \sin 2x \\ -2\sin 2x & 2\cos 2x \end{vmatrix} = 2$$

$$\text{P.I.} = -y_1 \int \frac{y_2 X}{W} dx + y_2 \int \frac{y_1 X}{W} dx$$

$$= -\cos 2x \int \frac{\sin 2x. 4\sec^2 2x}{2} dx + \sin 2x \int \frac{\cos 2x. 4\sec^2 2x}{2} dx$$

$$= -2\cos 2x \int \sec 2x \tan 2x \, dx + 2\sin 2x \int sex \, 2x$$

$$= -2\cos 2x. \frac{\sec 2x}{2} + 2\sin 2x \frac{1}{2} \log(\sec 2x = \tan 2x)$$

$$= -1 + \sin 2x \log(\sec 2x = \tan 2x)$$

Hence the C.S. is $y = c_1 \cos 2x + c_2 \sin 2x - 1 + \sin 2x \log(\sec 2x + \tan 2x)$

***Example* 10.80.** Solve by the method of variation of parameters $\frac{d^2y}{dx^2} + y = \text{cosec } x$.

Solution : (*i*) To find C.F. Its A.E. is $D^2 + 1 = 0$ or $D^2 = -1$ or $D = \pm i$

$\therefore$ C.F. is $c_1 \cos x + c_2 \sin x$

(*ii*) To find P.I. $y_1 = \cos x; y_2 = \sin x$

$$W = \begin{vmatrix} \cos x & \sin x \\ -\sin x & \cos x \end{vmatrix} = \cos^2 x + \sin^2 x = 1$$

Thus

$$\text{P.I.} = -\cos x \int \frac{\sin x \,\text{cosec}\, x}{1} dx + \sin x \frac{\int \cos x. \text{cosec}\, x}{1} dx$$

$$= -x \cos x + \sin x \log(\sin x)$$

(*iii*) C.S. is $y = c_1 \cos x + c_2 \sin x - x \cos x + \sin x \log(\sin x)$

***Example* 10.81.** Solve by the method of variation of parameters $\frac{d^2y}{dx^2} + a^2 y = \sec ax$.

Solution : (*i*) To find C.F. Its A.E. is $D^2 + a^2 = 0$ or $D = \pm ai$

$\therefore$ C.F. is $c_1 \cos ax + c_2 \sin ax$

(*ii*) **To find P.I** $y_1 = \cos ax; y_2 = \sin ax$

$$W = \begin{vmatrix} \cos ax & \sin ax \\ -a \sin ax & a \cos ax \end{vmatrix} = a\cos^2 ax + a \sin^2 ax = 0$$

$$\textbf{P.I.} = -\cos ax \int \frac{\sin ax. \sec ax}{a} dx + \sin ax \int \frac{\cos ax. \sec ax}{a} dx$$

$$= -\frac{\cos ax}{a^2}(-\log \cos ax) + \sin ax. \frac{x}{a}$$

(*iii*) C.S. is $y = c_1 \cos ax + c_2 \sin ax + \frac{1}{a^2} \text{cox } ax \log(\cos ax) + \left(\frac{1}{a}\right) x \sin ax$

Example **10.82.** Solve $\frac{d^2y}{dx^2} + y = \tan x.$

Solution : (*i*) To find C.F. Its A.E. is $D^2 + 1 = 0$ or $D = \pm i$

$\therefore$ C.F. is $c_1 \cos x + c_2 \sin x$

(*ii*) To find P.I. $y_1 = \cos x \,;\, y_2 + \sin x$

$$W = \begin{bmatrix} \cos x & \sin x \\ \sin x & \cos x \end{bmatrix} = \cos^2 x + \sin^2 x = 1$$

Thus
$$\text{P.I.} = -\cos x \int \sin x . \tan x \, dx + \sin x \int \cos x \tan x \, dx$$

$$= -\cos x \int \frac{\sin^2 x}{\cos x} dx = \sin x \int \sin x \, dx$$

$$= -\cos x \int \frac{1 - \cos^2 x}{\cos x} dx - \sin x \cos x$$

$$= -\cos x \int (\sec x - \cos x) dx - \sin x \cos x$$

$$= -\cos x \, [\log (\sec x + \tan x) - \sin x) - \sin x] - \sin x \cos x$$

$$= -\cos x \log (\sec x + \tan x) + \cos x \sin x - \sin x \cos x$$

$$= -\cos x \log (\sec x + \tan x)$$

(*iii*) C.S. is
$$y = c_1 \cos x + c_2 \sin x - \cos x \log (\sec x + \tan x)$$

Example **10.83.** Solve by the method of variation of parameters $\frac{d^2y}{dx^2} + y = x \sin x.$

Solution : (*i*) **To find C.F.** Its A.E. is $D^2 + 1 = 0$ or $D = \pm i$ in C.F. is $c_1 \cos x + c_2 \sin x$

(*ii*) **To find P.I.** $y_1 = \cos x \,;\, y_2 = \sin x \,;\, W = 1$

Thus
$$\text{P.I.} = -\cos x \int x \sin x . \sin x \, dx + \sin x \int x \sin x \cos x \, dx$$

$$= -\cos x \int \frac{x}{2} (1 - \cos 2x) dx + \frac{\sin x}{2} \int x \sin 2x \, ds$$

$$= -\frac{\cos x}{2}\left[\frac{x^2}{2} - \int x \cos 2x \, ds\right] + \frac{\sin x}{2}\left[\frac{x}{2}(-\cos 2x) - \int -\frac{\cos 2x}{2} dx\right]$$

$$= -\frac{\cos x}{2}\left[\frac{x^2}{2} - \left\{\frac{x \sin 2x}{2} \int \frac{\sin 2x}{2} dx\right\}\right] + \frac{\sin x}{2}\left[\frac{-x}{2}\cos 2x + \int -\frac{\cos 2x}{2} dx\right]$$

$$= -\frac{\cos x}{2}\left[\frac{x^2}{2} - \frac{x}{2}\sin 2x - \frac{\cos 2x}{4}\right] + \frac{\sin x}{2}\left[\frac{-x}{2}\cos 2x + \frac{1}{4}\sin 2x\right]$$

$$= \frac{-x^2}{4} \cos x + \frac{x}{4} \cos x \sin 2x + \frac{1}{8} \cos x \cos 2x - \frac{x}{4} \sin x \cos 2x + \frac{1}{8} \sin x \sin 2x$$

(*iii*) C.S. is $y = c_1 \cos x + c_2 \sin x - \frac{x^2}{4} \cos x + \frac{x}{4} \sin x + \frac{1}{8} \cos x$

Example **10.84.** Solve $y'' - 6y' + 9y = e^{3x} / x^2.$

Solution : (*i*) To find C.F. $D^2 - 6D + 9 = 0$

or $(D-3)(D-3) = 0$ or $D = 3, D = 3$

$\therefore$ C.F. $= (c_1 + c_2 x)e^{3x} = c_1 e^{3x} + c_2 x\, e^{3x}$

(*ii*) To find P.I. $y_1 = e^{3x}$; $y_2 = xe^{3x}$

$$W = \begin{bmatrix} e^{3x} & x^{e3x} \\ 3x^{3x} & e^{3x} + 3xe^{3x} \end{bmatrix}$$

$$= e^{6x} + 3xe^{6x} - 3xe^{6x} = 6^{3x}$$

Thus P.I. $= -e^{3x} \int \frac{xe^{3x}.e^{3x}}{x^2 e^{6x}} dx = xe^{3x} \int \frac{e^{3x}.e^{3x}}{x^2 e^{6x}} dx$

$$= -e^{3x} \log(x) + x\, e^{3x}\left(\frac{-1}{x}\right)$$

(*iii*) C.S. is $y = c_1 e^{3x} + c_2 x\, e^{3x} - e^{3x} \log x - \frac{1}{x}(x\, e^{3x})$

or $y = (c_1 + c_{2x})e^{3x} - e^{3x}(\log x + 1)$

***Example* 10.85.** Solve $y'' - 2y' + 2y = e^x \tan x$.

Solution : (*i*) To find C.F. $D^2 - 2D + 2 = 0$

Roots are $\frac{+2 \pm 2i}{2}$ or $1 \pm i$

$\therefore$ C.F. $= e^x(c_1 \cos x + c_2 \sin x)$

(*ii*) To find P.I. $y_1 = \cos x\, e^x$; $y_2 = ex \sin x$

$$y_1' = -\sin x\, e^x + e^x \cos x\, ; y_2'\, x\, e^x\, e^x \sin x$$

$$W = \begin{bmatrix} \cos xe^x & e^x \sin x \\ e^x \cos x - \sin xe^x & \cos xe^x + e^x \sin x \end{bmatrix}$$

$$= e^{2x} \cos^2 x + e^{2x} \sin x \cos x - e^{2x} \sin x \cos x + e^{2x} \sin^2 x$$

$$= e^{2x}[\cos^2 x + \sin^2 x] = e^{2x}$$

Thus

P.I. $= -\cos x\, e^x \int \frac{e^x \sin xe^x \tan x}{e^{2x}} dx + e^x \sin x \int \frac{e^x \cos xe^x \tan}{e^{2x}} dx$

$$= -\cos x\, e^x \int \frac{\sin^2 x}{\cos x} dx + e^x \sin x \int \sin x\, dx$$

$$= -\cos x\, e^x \int \frac{1-\cos^2 x}{\cos x} dx - e^x \sin x \cos x$$

$$= -\cos x\, e^x \int (\sec x - \cos x)\, dx - e^x \sin x \cos x$$

$$= -\cos x\, e^x\, [\log(\sec x + \tan x) - e^x \sin x \cos x$$

$$= -e^x \cos x \log(\sec x + \tan x)$$

(*iii*) Hence the C.S. is $y = e^x(c_1 \cos x + c_2 \sin x) - e^x \cos x \log(\sec x + \tan x)$

Example **10.86.** Solve $\frac{d^2y}{dx^2} + 4y = \tan 2x$.

Solution : (*i*) To find C.F. Its A.E. is $D^2 + 4 = 0$ or $D = \pm 2i$

$\therefore$ C.F. is $c_1 \cos 2x + c_2 \sin 2x$

(*ii*) **To find P.I.** $y_1 = \cos 2x \,;\, y_2 = \sin 2x$

and $X = \tan 2x$

$$W = \begin{vmatrix} \cos 2x & \sin 2x \\ -2\sin 2x & 2\cos 2x \end{vmatrix} = 2$$

Thus

$$\text{P.I.} = -\cos 2x \int \frac{\sin 2x \tan 2x}{2}\, dx + \sin 2x \int \frac{\cos 2x\, r \tan 2x}{2}\, dx$$

$$= -\frac{\cos x}{2} \int \frac{\sin^2 2x}{\cos 2x}\, dx + \frac{\sin 2x}{2} \int \sin 2x\, dx$$

$$= -\frac{\cos x}{2} \int \frac{1-\cos^2 2x}{\cos 2x} + \frac{\sin 2x}{2}\left(-\frac{\cos 2x}{2}\right)$$

$$= -\frac{\cos 2x}{2} \int (\sec 2x - \cos 2x)\, dx - \frac{1}{4} \sin 2x \cos 2x$$

$$= -\frac{1}{4} \cos 2x \log(\sec 2x + \tan 2x)$$

(*iii*) The C.S. is

$$y = c_1 \cos 2x + c_2 \sin 2x - \frac{1}{4} \cos 2x \log(\sec 2x = \tan 2x)$$

10.30 CAUCHY'S HOMOGENEOUS LINEAR EQUATION

An equation of the form

$$x^n \frac{d^n y}{dx^n} + k_1 x^{n-1} \frac{d^{n-1}}{dx^{n-1}} + \ldots k_{n-1} x \frac{dy}{dx} + k_n y = X \qquad \ldots(1)$$

where X is a function of x, is called Cauchy' hormogencous linear equation.

Such equations can be reduces to linear differential equation with constant coefficients, by putting

$$x = e^t \text{ or } t = \log x$$

Then if

$$D = \frac{d}{dt}$$

$$\left(\frac{dy}{dx}\right) = \frac{dy}{dt} \cdot \frac{dt}{dx} = \frac{dy}{dt} \cdot \frac{1}{x}$$

i.e

$$x\frac{dy}{dx} = Dy$$

$$\frac{d^2y}{dx^2} = \frac{d}{dx}\left(\frac{1}{x}\frac{dy}{dx}\right) = \frac{-1}{x}\frac{dy}{dt} + \frac{1}{x}\frac{d}{dt}\left(\frac{dy}{dt}\right)\frac{dt}{dx}$$

$$= \frac{-1}{x}\frac{dy}{dt} + \frac{1}{x}\frac{d^2y}{dt^2}\frac{dt}{dx} = \frac{1}{x^2}\left(\frac{d^2y}{dt^2} - \frac{dy}{dt}\right)$$

i.e. $$x^2 \frac{d^2y}{dx^2} = D(D-1)y$$

Similarly $$x^3 \frac{d^3y}{dx^3} = D(D-1)(D-2)y \text{ and so on.}$$

After making these substitutions in (1), there results a linear equation with constant co-efficients which can be solved as before.

10.31 LEGENDER'S LINEAR EQUATION

An equation of the form

$$(ax+b)n \frac{d^ny}{dx^n} + k_1 (ax+b)^{n-1} \frac{d^{n-1}y}{dx^{n-1}} + ... + k_n y = X \quad ...(2)$$

where k's are constants and X is a function of x, is called Legendre's linear equation.

Such equations can be reduced to linear equation with constant coefficients by the substitution $ax + b = e^t$.

i.e. $$t = \log(ax+b)$$

Then, if $$D = \frac{d}{dt}$$

$$\frac{dy}{dx} = \frac{dy}{dx} \cdot \frac{dt}{dx} = \frac{a}{ax+b}\frac{dy}{dt}$$

i.e. $$(ax+b)\frac{dy}{dx} = a\, Dy$$

$$\frac{d^2y}{dx^2} = \frac{d}{dx}\left(\frac{a}{ax+b}\frac{dy}{dt}\right) = \frac{-a^2}{(ax+b)^2}\frac{dy}{dt} + \frac{a}{ax+b}\frac{d}{dt}\left(\frac{dy}{dt}\right)\frac{dt}{dx}$$

$$= \frac{a^2}{(ax+b)^2}\left(\frac{d^2y}{dt^2} - \frac{dy}{dt}\right)$$

i.e. $$(ax+b)^2 \frac{d^2y}{dx^2} = a^2 D(D-1)y$$

Similarly $(ax+b)^3 \frac{d^3y}{dx^3} = a^3 D(D-1)(D-2)y$ and so on.

After making these replacement in (2), there results a linear equation with constant coefficients.

ILLUSTRATIVE EXAMPLES

***Example* 10.87.** Solve $x^2\frac{d^2y}{dx^2} - x\frac{dy}{dx} + y = \log x$.

Solution : This is Cauchy's homogencous linear equation

Put $$x = e^t \text{ i.e. } t = \log x$$

So that $$x\frac{dy}{dx} = Dy \text{ or } x^2\frac{d^2y}{dx^2} = D(D-1)y \text{ where } D = \frac{d}{dt}$$

Then the given equation becomes

$$[D(D-1) - D + 1]\, y = t \text{ or } (D-1)^2 y = t \quad ...(1)$$

which is linear equation with constant coefficients

Its A.E. is $$(D-1)^2 = 0 \text{ whence } D = 1, 1$$

$\therefore$ $$\text{C.F.} = (c_1 + c_2 t)e^t$$

$$\text{P.I.} = \text{T} + \text{T}\frac{1}{(D-1)^2}t = (1-D)^{-2}t = (1 + 2D + 3D^2 + ...)t$$

$$= t + 2$$

Hence the solution of (1) is

$$y = (c_1 + c_2 t)\, e^t + t + 2$$

Putting $t = \log x$ and $e^t = x$, we get

$y = (c_1 + c_2 \log x)x + \log x + 2$ as the required solution of (1)

Example **10.88.** Solve $x^2\dfrac{d^2y}{dx^2} - 4x\dfrac{dy}{dx} + 6y = x^2$.

Solution : This is Cauchy's homogeneous linear equation

Put $$x = e^t \text{ or } x^2 = e^{2t}$$

So that $$x\frac{dy}{dx} = Dy \text{ or } x^2\frac{d^2y}{dx^2} = D(D-1) \text{ where } D = \frac{d}{dt}$$

The given equation becomes

$$[D(D-1) - 4D + 6]\, y = e^{2t} \text{ or } (D^2 - 5D + 6)y = e^{2t}$$

To find C.F. –

Its A.E. is $$D^2 - 5D + 6 = 0 \text{ or } D = 2, 3$$

$\therefore$ $$\text{C.F.} = c_1 e^{2t} + c_2 e^{3t}$$

To find P.I.

$$\text{P.I.} = \frac{1}{D^2 - 5D + 6}e^{2t} = \text{t}\frac{1}{2D\ 5}e^{2t} = \frac{te^{2t}}{(-1)} = -te^{2t}$$

Thus the C.S. is $$y = c_1 e^{2t} + c_2 e^{3t} - te^{2t}$$

or $$y = c_1 x^2 + c_2 x^3 - x^2 \log x$$

Example **10.89.** Solve $x^2\dfrac{d^2y}{dx^2} - 2x\dfrac{dy}{dx} - 4y = x^4$.

Solution : Put $$x = e^t\,;\, c^4 = e^{4t}$$

So that $$x\frac{dy}{dx} = Dy$$

$$x^2\frac{d^2y}{dx^2} = D(D-1)y \text{ where } D = \frac{d}{dt}$$

The given equation becomes

$$[D(D-1)-2D-4]y = e^{4t}$$

To find C.F.–

Its A.E. is $(D^2-3D-4) = 0$ or $(D-4)(D+1)=0$

$$D = 4, D = -1$$

$\therefore$ $\quad$ C.F. $= c_1e^{4t}+c_2e^{-t}$

To find P.I.– $\quad y = \dfrac{1}{D^2-3D-4}e^{4t} = t\dfrac{1}{2D-3}e^{4t} = \dfrac{te^{4t}}{5}$

$\therefore$ $\quad$ C.S. is $y = c_1e^{4t}+c2e^{-t}+\dfrac{te^{4t}}{5}$

or $\quad y = c_1x^4+c_3x^{-1}+\dfrac{1}{5}x^4\log x$

or $\quad y = c_1x^4+\dfrac{c2}{x}+\dfrac{x^4}{5}\log x$

Example **10.90.** Solve $x^2\dfrac{d^2y}{dx^2}-3x\dfrac{dy}{dx}+4y=(1-x)^2$.

Solution : Let $x = e^t$

$$(D(D-1)-3D+4) = 0 \text{ or } D^2-4D+4=0$$

or $\quad D^2-2D-2D+4 = 0$ or $D(D-2)-(D-2)=0$

or $\quad D = 2, D=2$

$\therefore$ $\quad$ C.F. $= (c_1+c_2{}^t)e^{2t}$

$$\text{P.I.} = \frac{1}{D^2-4D+4}(1+e^t)^2$$

$$= \frac{1}{D^2-4D+4}(1+e^{2t}+2e^t)$$

$$= \frac{1}{D^2-4D+4}e^{0t}+\frac{1}{D^2-4d+4}e^{2t}+2\frac{1}{D^2-4D+4}e^t$$

$$= \frac{1}{4}+\frac{t^2e^{2t}}{2}2e^t$$

$\therefore$ C.S. is $\quad y = (c_1+c_2t)e^{2t}+\dfrac{1}{4}+\dfrac{e^{2t}t^2}{2}+2e^t$

or $\quad y = (c_1+c_2\log x)\,x^2+\dfrac{1}{4}+2x+\dfrac{x^2}{2}(\log x)^2$

$$= c_1+c_2\log x)x^2+\frac{1}{4}+2x+\frac{x^2}{2}(\log x)^2$$

Example **10.91.** Solve $x\dfrac{d^2y}{dx^2}-\dfrac{2y}{x}=x+\dfrac{1}{x^2}$.

Solution : $$x^2\frac{d^2y}{dx^2} - 2y = x^2 + \frac{1}{x} \text{ (put } x = e^x)$$

To find C.F. – Its A.E. is $$D^2 - D - 2 = 0 \text{ or } D^2 - 2D + D - 2 = 0$$

or $$(D-2)(D+1) = 0 \text{ or } D = 2, D = -1$$

$\therefore$ $$\text{C.F.} = c_1e^{2t} + c_2e^{-t}$$

To find P.I.

$$\text{P.I.} = \frac{1}{D^2 - D - 2}(e^{2t} + e^{-t})$$

$$= \frac{1}{D^2 - D - 2}e^{2t} + \frac{1}{D^2 - D - 2}e^{-t} = \frac{te^{2t}}{3} - \frac{te^{-t}}{3}$$

$$= \frac{1}{3}[te^{2t} - te^{-t}]$$

$\therefore$ $$\text{C.S. is } y = c_1e^{2t} + c_2e^{-t} + \frac{1}{3}[te^{2t} - te^{-t}]$$

or $$y = c_1x^2 + c_2\frac{1}{x} + \frac{1}{3}\left[\log x . x^2 - \log x . \frac{1}{x}\right]$$

or $$y = c_1x^2 + c_2x^{-1} + \frac{1}{3}\left(x^2 - \frac{1}{x}\right)\log x$$

***Example* 10.92.** The radial displacement u in a rotating disc at a distance r from the axis is given by $r^2\frac{d^2u}{dr^2} + r\frac{du}{dr} - u + kr^3 = 0$, where k is a constant. Solve the equation under the conditions $u = 0$ when $r = 0$, $u = 0$ when $r = 0$, $u = 0$ when $r = a$.

Solution : Let $r = e^t$

To find $$\text{C.F. } D(D-1) + D - 1 = 0 \text{ or } D^2 - D + D - 1 = 0$$

or $$D^2 - 1 = 0 \text{ or } D = \pm 1$$

$\therefore$ $$\text{C.F.} = c_1e^t + c_2e^{-t}$$

To find $$\text{P.I.} = \frac{1}{D^2 - 1}(-kr^3) = -k\frac{1}{D^2 - 1}e^{3t} = \frac{-k}{8}e^{3t}$$

$\therefore$ $$\text{C.S. is } u = c_1e^t + c_2e^{-t} - \frac{k}{8}e^{3t} \text{ or } u = c_1r + c_2r^{-1} - \frac{k}{8}r^3$$

$$u = 0, \text{ when } r = 0$$

$\therefore$ $$0\,c_1(0) = c_2\left(\frac{1}{0}\right)$$

$$u = 0, \text{ when } r = a$$

$$0 = c_1a + \frac{c_2}{a} - \frac{ka^3}{8} \text{ or } c_1a + \frac{c_2}{a} = \frac{ka^3}{8}$$

Solving these, we get $$c_2 = 0,\ c_1 = \frac{ka^3}{8a} = \frac{ka^3}{8}$$

$\therefore$ $$u = \frac{ka^2}{8}(r) - \frac{kr^3}{8} = \frac{kr}{8}[a^2 - r^2]$$

Example **10.93.** Solve $\frac{d^2y}{dx^2} + \frac{1}{x}\frac{dy}{dx} = \frac{12 \log x}{x^2}$

Solution : $$x^2\frac{d^2y}{dx^2} + x\frac{dy}{dx} = 12 \log x$$

Put $$x = e^t$$

(*i*) **To find C.F.** $$D(D-1) + D = 0$$

or $$D^2 - D + D = 0 \text{ or } D^2 = 0$$

$\therefore$ $$\text{C.F.} = (c_1 + c_2 t)\, e^{0t} = (c_1 + c_2)$$

(*ii*) **To find P.I.**

$$\text{P.I.} = \frac{1}{D^2} 12t = 12\frac{1}{D^2} t = \frac{12t^3}{6} = 2t^3$$

$\therefore$ $$\text{C.S. is } y = (c_1 + c_2 t) + 2t^3$$

or $$y = (c_1 + c_2 \log x) + 2(\log x)^3$$

Example **10.94.** Solve $x^2\frac{d^2y}{dx^2} - 2x\frac{dy}{dx} - 4y = x^2 + 2\log x.$

Solution : (*i*) To find C.F. $D(D-1) - 2D - 4 = 0 \quad (x = e^x)$

or $$D^2 - 2D - D - 4 = 0 \text{ or } D^2 - 3D - 4 = 0$$

or $$D = -1, D = 4$$

$\therefore$ $$\text{C.F.} = C_1 e^{-t} + C_2 e^{4t}$$

(*iii*) To find $$\text{P.I.} = \frac{1}{D^2 - 3D - 4}(e^{2t} + 2t)$$

$$= \frac{-e^{2t}}{6} - \frac{1}{2}\left[1 + \left(\frac{-D^2}{4} + \frac{3D}{4}\right)\right]^{-1} t$$

$$= \frac{-e^{2t}}{6} - \frac{1}{2}\left(1 + \frac{D^2}{4} - \frac{3D}{4}\right)t$$

$$= \frac{-e^{2t}}{6} - \frac{1}{2}\left(t - \frac{3}{4}\right) = \frac{-e^{2t}}{6} - \frac{t}{2} + \frac{3}{8}$$

$\therefore$ $$\text{C.S. is } y = c_1 e^{-t} + c_2 e^{4t} + \frac{3}{8} - \frac{t}{2} - \frac{e^{2t}}{6}$$

or $$y = \frac{c^1}{x} + c_2 x^4 - \frac{\log x}{2} + \frac{3}{8} - \frac{x^2}{6}$$

Example **10.95.** Solve $x^3 \dfrac{d^3y}{dx^3} + 2x^2 \dfrac{d^2y}{dx^2} + 2y = 10\left(x + \dfrac{1}{x}\right)$.

Solution : Given equation is Cauchy's homogeneous linear equation

Put $$x = e^z \text{ i.e. } z = \log x$$

So that $$x\frac{dy}{dx} = Dy,\ x^2\frac{d^2y}{dx^2} = D(D-1)y$$

$$x^3\frac{d^3y}{dx^3} = D(D-1)(D-2)y, \text{ where } D = \frac{d}{dz}$$

Substituting these values in the given equation, it reduced to

$$[D(D-1)(D-2) + 2D(D-1) + 2]y = 10(e^x + e^{-z})$$

or $$(D^3 - D^2 + 2)y = 10(ex + e^{-z})$$

which is a linear equation with constant coefficients.

(*i*) To find C.F. Its A.E. is $D^3 - D^2 + 2 = 0$

or $$(D+1)(D^2 - 2D + 2) = 0$$

$\therefore$ $$D = -1, \frac{2 \pm \sqrt{4-8}}{2} \text{ or } D = -1, 1 \pm i$$

$\therefore$ $$\text{C.F.} = c_1e^{-z} + e^z(c_2 \cos z + c_3 \sin z)$$

$$= \frac{c_1}{x} + x[c_2 \cos(\log x) + c_3 \sin(\log x)]$$

(*ii*) To find P.I.

$$\text{P.I.} = 10\frac{1}{D^3 - D^2 + 2}(e^z + e^{-z})$$

$$= 10\left(\frac{1}{D^3 - D^2 + 2}e^z + \frac{1}{D^3 - D^2 + 2}e^{-z}\right)$$

$$= 10\left(\frac{1}{2}e^z\right) + \frac{10}{5}ze^{-z} = 5e^z + 2ze^{-z}$$

$$= 5x + 2(\log x)\frac{1}{x}$$

Hence the C.S. is $y = \dfrac{c_1}{x} + x[c_2 \cos(\log x) + c_3 \sin(\log x)] + 5x + \dfrac{2}{x}\log x$.

Example **10.96.** Solve $x^2\dfrac{d^2y}{dx^2} - x\dfrac{dy}{dx} - 3y = x^2 \log x$.

Solution : Given equation is a Cauchy's homogeneous linear equation

Put $$x = e^z \text{ i.e. } z = \log x$$

So that $$x\frac{dy}{dx} = Dy,\ x^2\frac{d^2y}{dx^2} = D(D-1)y \text{ where } D = \frac{d}{dz}$$

Substituting these values in the given equation, it reduces to

$$[D(D-1)-D-3]\,y = ze^{2z}$$

or $$[D^2-D-D-3]\,y = ze^{2e} \text{ or } (D^2-2D-3)y = ze^{2z}$$

which is a linear equation with constant co-efficients

(*i*) **To find C.F.** Its A.E. is $D^2-2D-3 = 0$

or $(D-3)(D+1) = 0$ or $D = 3,\ D = -1$

$\therefore$ $$\text{C.F.} = c_1e^{3z} + c_2e^{-z} = c_1x^3 + \frac{c^2}{x}$$

(*ii*) **To find P.I.**

$$\text{P.I.} = \frac{1}{D^2-2D-3}(e^{2z}.z) = e^{2z}\frac{1}{(D+2)^2-2(D+2)-3}z$$

$$= e^{2z}\frac{1}{D^2+2D-3}z = e^{2z}\frac{1}{-3\left(1-\frac{2D}{3}-\frac{D^2}{3}\right)}z$$

$$= \frac{-e^{2z}}{3}\left[1-\left(\frac{2D}{3}+\frac{D^2}{3}\right)\right]^{-1} z = -\frac{e^{2z}}{3}\left[1+\left(\frac{2D}{3}+\frac{D^2}{3}\right)+....\right]z$$

$$= -\frac{1}{3}e^{2z}\left(z+\frac{2}{3}\right) = -\frac{x^2}{3}\left(\log x\frac{2}{3}\right)$$

Hence the C.S. is $$y = c_1x^3 + \frac{c^2}{x} - \frac{x^2}{3}\left(\log x\frac{2}{3}\right)$$

Example **10.97.** Solve $x^2\dfrac{d^2y}{dx^2}.x\dfrac{dy}{dx}+y = \log x \sin(\log x)$

Solution : Given equation is Cauchy's homogeneous linear equation

Put $x = e^z$ *i.e.* $z = \log x$

So that $$x\frac{dy}{dx} = Dy,\ x^2\frac{d^2y}{dx^2} = D(D-1)y \text{ where } D = \frac{d}{dz}$$

Substituting these values in the given equation, it reduces to

$$[D(D-1)+D+1]\,y = z\sin z$$

or $$(D^2-D+D+1) = z\sin z \text{ or } (D^2+1)y = z\sin z$$

(*i*) **To find C.F.** – Its A.E. is $D^2+1 = 0$ or $D = \pm i$

$\therefore$ $$\text{C.F.} = c_1\cos z + c_2\sin z$$

or $$\text{C.F.} = c_1\cos(\log x) + c_2\sin(\log x)$$

(*ii*) **To find P.I.**

$$\text{P.I.} = \frac{1}{D^2+1}z\sin z = \text{Imaginary part of } \frac{1}{D^2+1}ze^{iz}$$

$$= \text{I.P. of } e^{iz}\frac{1}{(D+i)^2+1}z = \text{I.P. of } e^{iz}\frac{1}{D^2+2iD}z$$

$$= \text{I.P. of } e^{iz} \frac{1}{2iD\left(1+\frac{D}{2i}\right)} z = \text{I.P. of } e^{iz} \frac{1}{2iD\left(1-\frac{iD}{2}\right)} z$$

$$= \text{I.P. of } \frac{1}{2i} e^{iz} \frac{1}{D}\left(1-\frac{iD}{2}\right)^{-1} z = \text{I.P. of } \frac{1}{2i} e^{iz} \frac{1}{D}\left(1+\frac{iD}{2}+...\right) z$$

$$= \text{I.P. of } \frac{1}{2i} e^{iz} \frac{1}{D}\left(z+\frac{1}{2}\right) = \text{I.P. of } \frac{1}{2i} e^{iz} \int\left(z+\frac{i}{2}\right) dz$$

$$= \text{I.P. of } \frac{1}{2i} e^{iz}\left(\frac{z^2}{2}+\frac{1}{2}z\right) = \text{I.P. of } e^{iz}\left(\frac{-i}{4}z^2+\frac{z}{4}\right)$$

$$= \text{I.P. of } (\cos z + i \sin z)\left(\frac{-i}{4}z^2+\frac{z}{4}\right)$$

$$= \frac{-z^2}{4} \cos z + \frac{z}{4} \sin z$$

$$= -\frac{1}{4} (\log x)^2 \cos(\log x) + \frac{1}{4} \log x \sin(\log x)$$

Hence the C.S. is

$$y = c_1 \cos (\log x = c_2 \sin (\log x) - 1/4(\log x)^2 \cos (\log x) + 1/4 \log x \sin (\log x)$$

***Example* 10.98.** Solve $x^3 \frac{d^3y}{dx^3} .3x^2 \frac{d^2y}{dx^2} .x \frac{dy}{dx} + 8y = 65 \cos (\log x)$.

Solution : Given equation is a Cauchy's homogeneous linear equation

Put $$x = e^t \text{ i.e. } t = \log x$$

So that $$x\frac{dy}{dx} = Dy, \qquad x^2 \frac{d^2y}{dx^2} = D(D-1)$$

$$x^3 \frac{d^3y}{dx^3} = D(D-1)(D-2)y, \text{ where } D = \frac{d}{dt}$$

Substituting these values in the given equation, it reduced to

$[D(D-1)(D-2) + 3(D-1)D + D + 8]y = 65 \cos t$

(*i*) **To find C.F.** Its A.E. is

$$D(D-1)(D-2) + 3D(D-1) + D + 8 = 0$$

or $$D^3 + 8 = 0 \text{ or } D^3 + 2^3 = 0$$

or $$(D+2)(D^2+4-2D) = 0 \text{ or } D = -2, D = 1 \pm \sqrt{3}\, i$$

$\therefore$ $$\text{C.F.} = c_1 e^{-2t} + e^1(c_2 \cos \sqrt{3}\, t + c_3 \sin \sqrt{3}\, t)$$

(*ii*) **To find P.I.**

$$\text{P.I.} = \frac{1}{D^3+8}(65 \cos t) = 65 \frac{1}{8-D} \cos t = -\frac{1}{D-8} \cos t$$

$$= -65\, e^{8t} \int \cos t e^{-8t}\, dt$$

Let $$I = \int e^{-8t} \cos t\, dt = e^{-8t} \sin t - \int e^{-8t}(-8) \sin t\, dt$$

$$= e^{-8t} \sin t + 8 \int e^{-8t} \sin t\, dt$$

$$= e^{-8t} \sin t + 8\left[-e^{-8t} \cos t - \int e^{-8t}(-8)(-\cos t)\, dt\right]$$

$$= e^{-8t} \sin t - 8e^{-8t} \cos t - 64\, I$$

$$65I = e^{-8t}\,[\sin t - 8 \cos t]$$

$$I = \frac{e^{-8t}}{65}\,[\sin t - 8 \cos t]$$

$\therefore$ $$\text{P.I.} = -65\, e^{8t}\, \frac{e^{-8t}}{65}\,[\sin t - 8 \cos t] = 8 \cos t - \sin t$$

Hence the C.S. is $$y = c_1 e^{-2t} + e^t (c_2 \cos \sqrt{3}\, t + \sin \sqrt{3}\, t) + 8 \cos t - \sin t$$

or $$y = c_1 x^{-2} + x(c_2 \cos \sqrt{3} \log x + c_3 \sin \sqrt{3} \log x) + 8 \cos \log x - \sin \log x$$

or $$y = (c_1/x^2) + x[c_2 \cos(\sqrt{3} \log x) + c_3 \sin(\sqrt{3} \log x)] + 8 \cos(\log x) - \sin(\log x)$$

***Example* 10.99.** Solve $x^2 \dfrac{d^2y}{dx^2} + 2x \dfrac{dy}{dx} - 12y = x^3 \log x.$

Solution : This is Cauchy's homogeneous linear equation

Put $$x = e^t \quad i.e. \quad t = \log x$$

So that $$x\frac{dy}{dx} = Dy, \quad x^2 \frac{d^2y}{dx^2} = D(D-1)y \text{ where } D = \frac{d}{dt}$$

Then the given equation becomes

$$[D(D-1) + 2D - 12]y = e^{3t}\, t$$

(*i*) **To find C.F.** Its A.E. is $D^2 - D + 2D - 12 = 0$

or $D^2 + D - 12 = 0$ or $D = -4,\ D = 3$

$\therefore$ $\text{C.F.} = c_1 e^{-4t} + c_2 e^{3t}$

(*ii*) **To find P.I.**

$$\text{P.I.} = \frac{1}{D^2 + D - 12}\, te^{3t} = e^{3t} \frac{1}{(D+3)^2 + (D+3) - 12}\, t$$

$$= e^{3t} \frac{1}{D^2 + 7D}\, t = e^{3t} \frac{1}{7D\left(\frac{D}{7} + 1\right)}\, t$$

$$= \frac{e^{3t}}{7D}\left(1 + \frac{D}{7}\right)^{-1} t = \frac{e^{3t}}{7D}\left(1 - \frac{D}{7}\right) t = \frac{e^{3t}}{7D}\left(t - \frac{1}{7}\right) = \frac{e^{3t}}{7}\left(\frac{t^2}{2} - \frac{t}{7}\right)$$

$\therefore$ $$\text{C.S. is } y = c_1 e^{-4t} + c_2 e^{3t} + \frac{e^{3t}}{7}\left(\frac{t^2}{2} - \frac{t}{7}\right)$$

or $$y = c_1x^{-4} + c_2x^3 + \frac{x^3}{7}\left[\frac{(\log x)^2}{2} - \frac{(\log x)}{7}\right]$$

$$= \frac{c_1}{x^4} + c_2x^3 + \frac{x^3}{98}\log x\,[7\log x - 2]$$

***Example* 10.100.** Solve $(3x+2)\frac{d^2y}{dx^2} + 3(3x+2)\frac{dy}{dx} - 36y = 3x^2 + 4x = 1.$

Solution : Given equation is a Legenre's linear equation

Put $$3x + 2 = e^x \text{ i.e. } z = \log(3x = 2)$$

So that $$(3x = 2)\frac{dy}{dx} = 3Dy \quad (3x+2)^2\frac{d^2y}{dx^2} = 3^2D(D-1)y, \text{ where } D = \frac{d}{dz}$$

Substituting these values in the given equation, it reduces to

$$3^2D(D-1)y + 3.3Dy - 36y = 3\left(\frac{ex-2}{3}\right)^2 + 4\left(\frac{e^x-2}{3}\right) + 1$$

or $$(9D^2 - 9D - 36)y = \frac{1}{3}e^{2x} - \frac{1}{3}$$

or $$9(D^2-4)y = \frac{1}{3}(e^{2z}-1) \text{ or } (D^2-4)y = \frac{1}{27}(e^{2z}-1)$$

which is linear equation with constant coefficients.

(*i*) **To find C.F.** Its A.E. is $D^2 - 4 = 0$ or $D = \pm 2$

$\therefore$ $$\text{C.F.} = c_1e^{2x} + c_2e^{-2z}$$

(*ii*) **To find P.I.**

$$\text{P.I.} = \frac{1}{27(D^2-4)}(e^{2z}-1)$$

$$= \frac{1}{27}\left[\frac{1}{(D^2-4}e^{2z} - \frac{1}{D2-4}e^{0z}\right] = \frac{1}{27}\left[\frac{ze^{2z}}{4} + \frac{1}{4}\right]$$

$$= \frac{1}{108}(ze^{2z}+1)$$

$\therefore$ $$\text{C.S. is } y = c_1e^{2z} + c_2e^{-2z} + \frac{1}{108}(ze^{2z}+1)$$

or $$y = c_1(3x+2)^2 + c_2(3x+2)^{-2} + \frac{1}{108}[(3x+2)^2\log(3x+2) + 1]$$

***Example* 10.101.** Solve $(2x+3)\frac{d^2y}{dx^2} - (2x+3)\frac{dy}{dx} - 12y = 6x$

Solution : Given equation is a Legendre's linear equation

Put $$(2x = 3) = e^t \text{ i.e. } \log(2x+3) = t$$

So that $$(2x+3)\frac{dy}{dx} = 2Dy$$

$$(2x+3)^2\frac{d^2y}{dx^2} = 2^2 D(D-1)y \text{ where } D = \frac{d}{dt}$$

Substituting these in the given equation, it reduces to

$$[4D(D-1) - 2D - 12] \quad y = 6\left(\frac{e^t - 3}{2}\right)$$

(*i*) **To find C.F.** Its A.E. is $4D(D-1) - 2D - 12 = 0$

or $$4D^2 - 4D - 2D - 12 = 0 \text{ or } 4D^2 - 6D - 12 = 0$$

or $$2D^2 - 3D - 6 = 0$$

roots are $\dfrac{3 \pm \sqrt{57}}{4}$

$\therefore$ $$\text{C.F.} = c_1e^{at} + c_2e^{bt} \text{ where } a = \frac{3+\sqrt{57}}{4} \text{ and } b = \frac{3-\sqrt{57}}{4}$$

(*ii*) **To find P.I.**

$$\text{P.I.} = \frac{1}{4D^2 - 6D - 12}3(e^t - 3)$$

$$= 3\left[\frac{1}{4D^2 - 6D - 12}e^t - 3\frac{1}{4D^2 - 6D - 12}e^{01}\right]$$

$$= 3\left[\frac{-1}{14}e^t - 3\left(\frac{1}{-12}\right)\right] = \frac{-3}{14}e^t + \frac{3}{4}$$

$\therefore$ $$\text{C.S. is } y = c_1(2x+3)^a + c_2(2x+3)b - \frac{3}{14}(2x+3) + \frac{3}{4}$$

where $$a = \frac{3+\sqrt{57}}{4} \text{ and } b = \frac{3-\sqrt{57}}{4}$$

***Example* 10.102.** Solve $(1+x)^2\dfrac{d^2y}{dx^2} + (1+x)\dfrac{dy}{dx} + y = 4\log(1+x)$.

Solution : This is a Legendre's linear equation

$\therefore$ Put $$(1+x) = e^t \text{ } i.e. \text{ } t = \log(1+x)$$

So that $$(1+x)\frac{dy}{dx} = Dy \text{ and } (1+x)^2\frac{d^2y}{dx^2} = D(D-1)y \text{ where } D = \frac{d}{dt}$$

Substituting these values in the given equation, it reduces to

$$[D(D-1) + D + 1]\, y = 4\cos t$$

(*i*) **To find C.F.** Its A.E. is $(D^2 + 1) = 0$ or $D = \pm i$

$\therefore$ $$\text{C.F.} = c_1\cos t + c_2\sin t$$

(*ii*) **To find P.I.**

$$\text{P.I.} = \frac{1}{D^2+1}(4\cos t) = 4t\,\frac{1}{2D}\cos t = 2t\sin t$$

$$\therefore \quad \text{C.S. is } y = c_1\cos t + c_2\sin t + 2t\sin t$$

or
$$y = c_1\cos[\log(1+x)] + c_2\sin[\log(1+x)] + 2\log(1+x)\sin[\log(1+x)]$$

EXERCISE 10.5

1. $(3x+2)^2\dfrac{d^2y}{dx^2} + 5(3x+2)\dfrac{dy}{dx} - 3y = x^2+x+1$ **Ans.** $y = c_1(3x+2)^{1/2} + c_2(3x+2)^{-1} + \dfrac{1}{27}\left[\dfrac{(3x+2)^2}{15} - \dfrac{(2x+2)}{4} - 1\right]$

2. $(1+2x)^2\dfrac{d^2y}{dx^2} - 6(1+2x)\dfrac{dy}{dx} + 16y = 8(1+2x)^2$ **Ans.** $y = (1+2x)2[c_1 + c_2\log(1+2x) + \{\log(1+2x\}^2]$

3. $(2x-1)^2\dfrac{d^2y}{dx^2} + (2x-1)\dfrac{dy}{dx} - 2y = 0$ **Ans.** $y = (2x-1)\,[c_1 + c_2(2x-1)\sqrt{3/2} + c_3(2x-1) - \sqrt{3/2}\,]$

4. $(5+2x)\dfrac{d^2y}{dx^2} - 6(5=2x)\dfrac{dy}{dx} + 8y = 0$ **Ans.** $y = (5+2x)\,[c_1(5+2x)\sqrt{2}\,]$

5. $(1+x)^2\dfrac{d^2y}{dx^2} + (1+x)\dfrac{dy}{dx} + y = 4\cos[\log(1+x)]$

Ans. $y = c_1\cos[\log(1+x)] + c_2\sin[\log(1+x) + 2\log(1+x)\sin\log(1+x)]$

10.32 SIMULTANEOUS LINEAR EQUATION WITH CONSTANT COEFFICIENTS

Differential equations in which there is one independent variable and two or more than two dependent variables are called simultaneous linear equation. Here we shall deal with systems of linear equations with constant coefficients only. Such a system of equations is solved of equations is solved by eliminating all but one of the dependent variables and then solving the resulting equations as before. Each of the dependent variables is obtained in the similar manner.

ILLUSTRATIVE EXAMPLES

***Example* 10.103.** Solve $\dfrac{dx}{dt} = 5x + y,\ \dfrac{dy}{dt} = y - 4x.$

Solution : Putting $\dfrac{d}{dt} = D$, the given equations become

$$(D-5)x - y = 0 \qquad \text{...(1)}$$
$$(D-1)y + 4x = 0 \qquad \text{...(2)}$$

Eliminating x from equation (1), we get

$$(D-5)x = y \text{ or } x = \frac{y}{D-5}$$

Substituting the value of x in equation (2), we get

$$(D-1)y + 4\left(\frac{y}{D-5}\right) = 0 \text{ or } [(D-1)(D-5)+4]\,y = 0$$

Its A.E. is $$D^2 - 6D + 9 = 0 \text{ or } D^2 - 3D - 3D + 9 = 0$$

or $$(D-3)(D-3) = 0 \text{ or } D = 3, 3$$

$\therefore$ $$\text{C.F.} = (c_1 = c_2 t)e^{3t}$$

$$\text{P.I.} = 0$$

$\therefore$ $$y = (c_1 + c_2 t)e^{3t}$$

$$\frac{dy}{dx} = 3e^{3t}(c_1 + c_2 t) + e^{3t} c_2$$

Substituting these values in (2), we get

$$3e^{3t}(c_1 + c_2 t) + e^{3t} c_2 - (c_1 + c_2 t)e^{3t} + 4x = 0$$

$$c_1 3e^{3t} + c_2 t\, 3e^{3t} + c_2 e^{3t} - c_1 e^{3t} - c_2 t e^{3t} + 4x = 0$$

$$c_1[3e^{3t} - e^{3t}] + c_2[3te^{3t} + e^{3t} - te^{3t}] = -4x$$

$$2e^{3t} c_1 + c_2[2t\, e^{3t} + e^{3t}] = -4x$$

$$2e^{3t} c_1 + c_2 e^{3t}(2t+1) = -4x$$

or $$x = -\frac{1}{4}[2c_1 e^{3t} + e^{3t} c_2(1+2t)]$$

or $$x = -\frac{1}{4} e^{3t}[2c_1 + c_2(1+2t)]$$

***Example* 10.104.** Solve $\frac{dx}{dt} + y = \sin t$; $\frac{dy}{dt} + x = \cos t$; given that $x = 2$ and $y = 0$ when $t = 0$.

Solution : Putting $\frac{d}{dt} = D$, the given equation becomes

$$Dx + y = \sin t \quad \text{...(1)}$$

$$Dy + x = \cos t \quad \text{...(2)}$$

Solving for x form equation (2), we get

$$x = \cos t - Dy$$

Substituting the value of x equation (1), we get

$$y = \sin t - D(\cos t - Dy) \text{ or } y = \sin t - D\cos t + D^2 y$$

or $$y = \sin t + \sin t + D^2 y \text{ or } y - D^2 y = 2\sin t$$

or $$y(1 - D^2) = 2\sin t$$

Its auxiliary equations is $$1 - D^2 = 0$$

or $$1 = D^2 \text{ or } D = \pm 1$$

$\therefore$ C.F. is $c_1 e^t + c_2 e^t$

$$\text{P.I.} = 2\frac{1}{1-D^2}\sin t = \frac{2}{2}\sin t = \sin t$$

$\therefore$
$$y = c_1 e^t = c_2 e^{-t} + \sin t$$
$$\frac{dy}{dx} = c_1 e^t - c_2 e^{-t} + \cos t$$

by using equation (2), we get

$$c_1 e^t - c_2 e^{-t} + \cos t + x = \cos t$$
$$x = c_2 e^{-t} - c_1 e^t$$

now, we are given that $x = 2$ when $t = 0$ and $y = 0$ when $t = 0$.

$\therefore$ $$x = c_2 - c_1 \text{ and } y = c_1 + c_2$$

or $$2 = c_2 - c_1 \text{ and } 0 = c_1 + c_2$$

Solving these, we get

$$c_2 = 1 ; c_1 = -1$$

$\therefore$ $$x = e^{-t} + e^t$$

and $$y = -e^t + e^{-t} + \sin t$$

or $$y = e^{-t} - e^t + \sin t$$

Example **10.105.** Solve $\frac{dx}{dt} + 4x + 3y = t;\ \frac{dy}{dt} + 2x + 5y = e^t.$

Solution : Writing D for $\frac{d}{dt}$, the given equation become

$$(D + 4)x + 3y = t \qquad ...(1)$$

and $$2x + (D = 5)y = e^t \qquad ...(2)$$

Solving for y form equation (2).

$$y = \frac{e^t - 2x}{D+5}$$

Substitute the value of y in the equation (1)

$$(D + 4)x + 3\left(\frac{e^t - 2x}{D+5}\right) = t$$

or $$(D + 4)(D + 5)x + 3e^t - 6 = 1 + 5t$$

or $$(D^2 + 5D + 4D + 20 - 6)x = 1 + 5t - 3e^t$$

or $$(D^2 + 9D + 14)x = 1 + 5t - 3e^t$$

Its A.E. is $$D^2 + 9D + 14 = 0 ; (D + 2)(D + 7) = 0$$

$\therefore$ $$D = -2, D = -7$$

$\therefore$ C.F. is $c_1 e^{-2t} + c_2 e^{-7t}$

$$\text{P.I.} = \frac{1}{D^2+9D+14}(1+5t-3e^t)$$

$$= \frac{1}{D^2+9D+14}e^{0t} + 5\frac{1}{\left(1+\frac{9D}{14}+\frac{D^2}{14}\right)}t - 3\frac{1}{D^2+9D+14}e^t$$

$$= \frac{1}{14}+\frac{5}{14}\left[1+\left(\frac{9D}{14}+\frac{D^2}{14}\right)^{-1}\right]t - \frac{1}{8}e^t$$

$$= \frac{1}{14}+\frac{5}{14}\left[1-\left(\frac{9D}{14}+\frac{D^2}{14}\right)+\ldots\right]t - \frac{1}{8}e^t$$

$$= \frac{1}{14}+\frac{5}{14}\left(t-\frac{9}{14}\right) - \frac{1}{8}e^t = \frac{1}{14}+\frac{5}{14}t - \frac{45}{196} - \frac{1}{8}e^t$$

$$= \frac{5}{14}t - \frac{31}{196} - \frac{1}{8}e^t$$

$$\therefore \quad x = c_1e^{-2t}+c_2e^{-7t}+\frac{5}{14}t-\frac{31}{196}-\frac{1}{8}e^t$$

Now
$$\frac{dx}{dt} = -2c_1e^{-2t}-7c_2e^{-t}+\frac{5}{14}-\frac{1}{8}e^t$$

Substituting the values of x and $\frac{dx}{dt}$ in (1), we get

$$3y = t-\left(-2c_1e^{-2t}-7c_2e^{-7t}+\frac{5}{14}+\frac{1}{8}e^t\right)-4\left(c_1e^{-2t}+c_2e^{-7t}+\frac{5t}{14}+\frac{31}{196}+\frac{1}{8}e^t\right)$$

or
$$3y = -2c_1e^{-2t}+3c_2e^{-7t}-\frac{3}{7}t+\frac{27}{98}+\frac{5}{8}e^t$$

or
$$y = \frac{1}{3}\left[-2c_1e^{-2t}+3c_2e^{-7t}-\frac{3}{7}t+\frac{27}{98}+\frac{5}{8}e^t\right]$$

or
$$y = -\frac{2}{3}c_1e^{-2t}+c_2e^{-7t}-\frac{1}{7}t+\frac{9}{98}+\frac{5}{24}e^t$$

***Example* 10.106.** Solve $\frac{d^2x}{dt^2}+4x+5y=t^2;\ \frac{d^2y}{dx^2}+5x+4y=t+1.$

Solution : Writing D for $\frac{d}{dt}$, the given equation becomes

$$D^2x+4x = 5y = t^2 \quad \ldots(1)$$

and
$$D^2y+5x+4y = t+1 \quad \ldots(2)$$

Solving for y from equation (2), we get

$$(D^2+4)y = -5x+t+1 \text{ or } y = \frac{-5x+t+1}{D^2+4}$$

Putting the value of y in equation (1), we get

$$D^2x+4x+5\left(\frac{-5x+t+1}{D^2+4}\right) = t^2$$

or $(D^4 = 8D^2 - 9)x = 4t^2 - 5t - 3$

Its A.E. is $D^4 = 8D^2 - 9 = 0$ or $D^4 + 9D^2 - D^2 - 9 = 0$

or $D^2(D^2 + 9) - (D^2 + 9) = 0$ or $(D^2 + 9)(D^2 - 1) = 0$

or $D = \pm 3i$ and $D = \pm 1$

$\therefore$ C.F. $= c_1e^t = c_2e^{-t} + c_3 \cos 3t + c_4 \sin 3t$

$$\text{P.I.} = \frac{1}{D^4 + 8D^2 - 9}(4t^2 - 5t - 3) = \frac{1}{-9\left(1 - \frac{8D^2}{9} - \frac{D^2}{9}\right)}(4t^2 - 5t - 3)$$

$$= -\frac{1}{9}\left[1 - \left(\frac{8D^2}{9} + \frac{D^4}{9}\right)\right]^{-1}(4t^2 - 5t - 3)$$

$$= -\frac{1}{9}\left[1 + \left(\frac{8D^2}{9} + \frac{D^4}{9}\right) + \ldots\right](4t^2 - 5t - 3)$$

$$= -\frac{1}{9}\left[4t^2 - 5t - 3 + \frac{8}{9}(8)\right] = -\frac{1}{9}\left[4t^2 - 5t + \frac{37}{9}\right]$$

$$x = c_1e^t + c_2e^{-t} + c_3 \cos 3t + c_4 \sin 3t + c_4 \sin 3t - \frac{4}{9}t^2 + \frac{5}{9}t - \frac{37}{81}$$

now $$\frac{dx}{dt} = c_1e^t - c_2e^{-t} - 3c_3 \sin 3t + 3c_4 \cos 3t - \frac{8}{9}t + \frac{5}{9}$$

and $$\frac{d^2x}{dt^2} = c_1e^t + c_2e^{-t} - 9c_3 \cos 3t - 9c_4 \sin 3t - \frac{8}{9}$$

Substituting the values of x and $\frac{d^2x}{dt^2}$ in (1), we have

$$5y = t^2 - 4x - \frac{d^2x}{dt^2}$$

$$5y = t^2 - 4c_1e^t - 4c_2e^{-t} - 4c_3 \cos 3t - 4c_4 \sin 3t + \frac{16}{9}t^2 - \frac{20}{t}t + \frac{148}{81} - c1e^t - c_2e^{-t} + 9c_3 \cos 3t + 9c_4 \sin 3t + \frac{8}{9}$$

$\therefore$ $$y = \frac{1}{5}\left[-5c_1e^t - 5c_2e^{-t} + 5c_3 \cos 3t + 5c_4 \sin 3t + \frac{25}{9}t^2 - \frac{20}{9}t + \frac{220}{81}\right]$$

Hence $$x = c_1e^t + c_2e^{-t} + c_3 \cos 3t + c_4 \sin 3t - \frac{1}{9}\left(4t^2 - 5t + \frac{37}{9}\right)$$

and $$y = -c_1e^t - c_2e^{-t} + c_3 \cos 3t + c_4 \sin 3t + \frac{1}{9}\left(5t^2 - 4t + \frac{44}{9}\right)$$

***Example* 10.107.** Solve the simultaneous equation: $t\frac{dx}{dt} + y = 0$, $t\frac{dy}{dt} + x = 0$, given $x(1) = 1$, $y(-) = 0$

Solution : The given equation are $t\frac{dx}{dt} + y = 0$ ...(1)

$$t\frac{dy}{dt} + x = 0 \quad \ldots(2)$$

Differentiating (1) w.r.t t, we have

$$t\frac{d^2x}{dt^2}+\frac{dx}{dt}+\frac{dy}{dt} = 0$$

$$t^2\frac{d^2x}{dt^2}+t\frac{dx}{dt}-x = 0 \text{ [using (2)]} \quad ...(3)$$

which is Cauchy's homogeneous linear equation

Putting $t = e^u$ *i.e.* $u = \log t$, so that $t\frac{d}{dt} = \frac{d}{du} = D$, equation (3) becomes

$$[D(D-1)+D-1]x = 0 \text{ or } (D^2-1)x = 0$$

Its A.E. is $\quad D^2 - 1 = 0$ or $D = \pm 1$

$$\therefore \quad x = c_1e^u + c_2e^{-u} \text{ or } x = c_1t+\frac{c_2}{t} \quad ...(4)$$

From equation (1), we have

$$y = -t\frac{dx}{dt} = -t\frac{d}{dt}\left(c_1t+\frac{c_2}{t}\right)$$

$$= -t\left(c_1-\frac{c_2}{t^2}\right) = -c_1t = \frac{c_2}{t} \quad ...(5)$$

Since $\quad x(1) = 1$

$\therefore$ From equation (4), we have

$$1 = c_1+c_2$$

Also $\quad y(-1) = 0$

$\therefore$ From equation (5), we have

$$0 = c_1+c_2$$

Solving these, we get

$$c_1 = c_2 = \frac{1}{2}$$

Hence $\quad x = \frac{1}{2}\left(t+\frac{1}{t}\right)$ and $y = \frac{1}{2}\left(-t+\frac{1}{t}\right)$

***Example* 10.108.** Solve the following simultaneous equations: $\frac{dx}{dt} = 2y$, $\frac{dy}{dt} = 2z$, $\frac{dz}{dt} = 2x$

Solution : The given equation are

$$\frac{dx}{dt} = 2y \quad ...(1)$$

$$\frac{dy}{dt} = 2x \quad ...(2)$$

$$\frac{dz}{dt} = 2x \quad ...(3)$$

Differentiating (*i*) w.r.t *t*, we get

$$\frac{d^2x}{dt^2} = 2\frac{dy}{dt}$$

We know that $$\frac{dy}{dt} = 2z$$

$\therefore$ $$\frac{d^2x}{dt^2} = 4z$$

Again differentiating w.r.t. t, we get

$$\frac{d^3x}{dt^3} = 4\frac{dz}{dt}$$

Putting the value of $\frac{dz}{dt}$, we get

$$\frac{d^3x}{dt^3} = 4(2x) \text{ or } \frac{d^3x}{dt^3} - 8x = 0$$

or $$(D^3-8)x = 0 \text{ where } D = d/dt$$

Its A.E. is $$D^3 - 8 = 0 \text{ or } (D-2)(D^2+2D+4) = 0$$

or $$D = 2,\ D = \frac{-2 \pm 2i\sqrt{3}}{2} \text{ or } D = 2,\ D = -1 \pm i\sqrt{3}$$

$\therefore$ $$\text{C.F. } x = c_1e^{2t} + c_2e^{-t}\cos(\sqrt{3}\,t - c_3)$$

From equation (1), we get

$$y = \frac{1}{2}\frac{dx}{dt}$$

$\therefore$ $$y = \frac{1}{2}[2c_1e^{2t} - c_2e^{-t}\cos(\sqrt{3}\,t - c_3) - c_2\sqrt{3}\,e^{-t}\sin(\sqrt{3}\,t - c_3)]$$

or $$y = c_1e^{2t} + c_2e^{-t}\left[\cos\frac{2\pi}{3}\cos(\sqrt{3} - c_3) - \sin\frac{2\pi}{3}\sin(\sqrt{3}\,t - c_3)\right]$$

or $$y = c_1e^{2t}\ c^2e^{-t}\ c_2e^{-t}\cos\left(\sqrt{3}\,t - c_3 + \frac{2\pi}{3}\right)$$

From equation (2), we get

$$x = \frac{1}{2}\frac{dy}{dt}$$

$\therefore$ $$z = \frac{1}{2}\left[2c_1e^{2t} - c_2e^{-t}\cos\left(\sqrt{3}\,t - c_3 + \frac{2\pi}{3}\right) - c_2\sqrt{3}\,e^{-t}\sin\left(\sqrt{3}\,t - c_3 + \frac{2\pi}{3}\right)\right]$$

$$= c_1e^{2t} + c_2e^{-t}\left[\cos\frac{2\pi}{3}\cos\left(\sqrt{3}\,t - c_3\frac{2\pi}{3}\right) - \sin\frac{2\pi}{3}\sin\left(\sqrt{3}\,t - c_3 + \frac{2\pi}{3}\right)\right]$$

$$= c_1e^{2t} + c^2e^{-t}\cos\left(\sqrt{3}\,t - c_3 + \frac{4\pi}{3}\right)$$

Note : $c_1\cos\beta x + c_2\sin\beta x$ can be replaced by $c_1\cos(\beta x - c_2)$

or $c_1\cos(\beta x + c_2)$ or $c_1\sin(\beta x - c_2)$ or $c_1\sin(\beta x + c^2)$

Example **10.109.** Solve $\frac{dx}{dt} + 2x + 3y = 0; 3x + \frac{dy}{dt} = 2y = 2e^{2t}$.

Solution : Putting $\frac{d}{dt} = D$, we get

$$Dx + 2x + 3y = 0 \quad ...(1)$$

$$3x = Dy + 2y = 2e^{2t} \quad ...(2)$$

Obtaining y from equation (2), we get

$$y = \frac{2e^{2t} - 3x}{D = 2}$$

Substituting the value of in equation (1), we get

$$(D + 2)x + 3[(2e^{2t} - 3x)/(D + 2)] = 0$$

or $$(D_2 + 4 + 4D)x + 6e^{2t} - 9x = 0$$

or $$(D^2 + 4D - 5)x = -6e^{2t}$$

Its A.E. is $$D^2 + 4D - 5 = 0 \text{ or } D = -5, 1$$

$\therefore$ $$\text{C.F.} = c_1e^{-5t} + c_2e^t$$

$$\text{P.I.} = \frac{1}{D^2 + 4D - 5}(-6e^{2t}) = \frac{-6}{4 + 8 - 5}$$

$\therefore$ $$x = c^1e^{-5t} = c_2e^t - \frac{6}{7}e^{2t}$$

$$\frac{dx}{dt} = -5c_1e^{-5t} + c_2e^t - \frac{12}{7}e^{2t}$$

Substituting these values, we get

$$-5c1e^{-5t} + c_2e^t - \frac{12}{7}e^{2t} + 2c_1e^{-5t} + 2c_2e^t - \frac{12}{7}e^{2t} = -3y$$

or $$-3c_1e^{-5t} = 3c_2e^t - \frac{24}{7}e^{2t} = -3y$$

or $$c_1e^{-5t} - c_2e^t + \frac{8}{7}e^{2t} = y$$

Example **10.110.** Solve $\frac{dx}{dt} - 7x + y = 0; \frac{dy}{dt} - 2x - 5y = 0$.

Solution : Putting $\frac{d}{dt} = D$, we get

$$Dx - 7x + y = 0 \quad ...(1)$$

$$Dy - 2x - 5y = 0 \quad ...(2)$$

Obtaining from equation (2), we get

$$(D - 5)y = 2x \text{ or } y = \frac{2x}{D - 5}$$

Putting the value of y in equation (1), we get

$$(D-7)x + \frac{2x}{D-5} = 0$$

or $$[(D-7)(D-5)+2]x = 0$$

or $$[D^2 - 5D - 7D = 35 + 2] = 0 \text{ or } [D^2 - 12D + 37]x = 0$$

Its A.E. is $$D^2 - 12D + 37 = 0 \text{ or } D = \frac{12 \pm 2i}{2} = 6 \pm i$$

$\therefore$ $$\text{C.F.} = e^{6t}[c_1 \cos t = c_2 \sin t]$$

$$\text{P.I.} = 0$$

$$x = \text{C.F.} + \text{P.I.}$$

$\therefore$ $$x = e^{6t}[c_1 \cos t + c_2 \sin t]$$

$$\left(\frac{dx}{dt}\right) = e^{6t}[-c_1 \sin t + c_2 \cos t] + [c_1 \cos t + c_2 \sin]\, 6e^{6t}$$

Substituting these values, we get

$e^{6t}[-c_1 \sin t + c_2 \cos t] + 6e^{6t}[c_1 \cos t + c_2 \sin t] - 7e^{6t}(c_1 \cos t + c_2 \sin t) = -y$

or $y = -e^{6t}(-c_1 \sin t + c_2 \cos t) - 6e^{6t}(c_1 \cos t + c_2 \sin t) + 7e^{6t}(c_1 \cos t + c_2 \sin t)$

or $y = c_1 e^{6t} \sin t - e^{6t} c_2 \cos t - 6e^{6t} c_1 \cos t - 6e^{6t} c_2 \sin t + 7e^{6t} c_1 \cos t + 7e^{6t} c_2 \sin t$

or $y = c_1 e^{6t} \sin t + e^{6t} \cos t\, c_1 - e^{6t} c_2 \cos t + e^{6t} c_2 \sin t$

or $y = e^{6t} \sin t(c_1 + c_2) + \cos t e^{6t}(c_1 - c_2)$

or $y = e^{6t}[\sin t(c_1 + c_2) + \cos t(c_1 - c_2)$

***Example* 10.111.** Solve $\frac{dx}{dt} + 2x - 3y = 5t$; $\frac{dy}{dt} - 3x + 2y = 2e^{2t}$.

Solution : Putting $\frac{d}{dt} = D$, the given equation become

$$Dx + 2x - 3y = 5t \quad ...(1)$$

$$Dy - 3x + 2y = 2e^{2t} \quad ...(2)$$

Obtaining x form equation (2), we get

$$(D+2)y - 2e^{2t} = 3x \text{ or } \frac{(D+2)y - 2e^{2t}}{3} = x$$

Substituting the value of x in equation (1), we get

$$(D+2)\left[\frac{(D+2)y - 2e^{2t}}{3}\right] - 3y = 5t$$

$$(D+2)^2 y - 2(D+2)e^{2t} - 9y = 15t$$

$$[(D+2)^2 - 9]y = 15t = 2(D+2)e^{2t}$$

$$(D^2 + 4D - 5) = 15t = 2.2.e^{2t} + 4e^{2t}$$

$$(D^2 + 4D - 5)y = 15t + 8e^{2t}$$

Its A.E. is $D^2 + 4D - 5 = 0$

or $D^2 + 5D - D - 5 = 0$ or $(D+5) - (D+5) = 0$

or $(D+5)(D-1) = 0$ or $D = 1, D = -5$

$\therefore$ C.F. $= c_1e^t + c_2e^{-5t}$

$$\text{P.I.} = \frac{1}{(D^2+4D-5)}(15t = 8e^{2t})$$

$$= \frac{15}{-5\left(\frac{D^2}{5} - \frac{4D}{5} + 1\right)}t + \frac{8}{D^2+4D-5}e^{2t}$$

$$= -3\left[1 - \left(\frac{D^2}{5} + \frac{4D}{5}\right)\right]^{-1} t + \frac{8}{7}e^{2t}$$

$$= -3\left[1 + \frac{D^2}{5} + \frac{4D}{5}\right] t + \frac{8}{7}e^{2t} = -3\left[t + \frac{4}{5}\right] + \frac{8}{7}e^{2t}$$

$\therefore$ C.S. is $y = c_1e^t + c_2e^{-5t} - 3\left(t + \frac{4}{5}\right) = \frac{8}{7}e^{2t}$

or $y = c_1e^t + c_2e^{-5t} + -3t - \frac{12}{5} + \frac{8}{7}e^{2t}$

and $\frac{dy}{dt} = c_1e^t - 5c_2e^{-5t} - 3 + \frac{16}{7}e^{2t}$

Substituting these values in equation (2), we get

$$c_1e^t - 5c_2c^{-5t} - 3 + \frac{16}{7}e^{2t} - 3x + 2\left[c_1e^t + c_2e^{-5t} - 3t - \frac{12}{5} + \frac{8}{7}e^{2t}\right] = 2e^{2t}$$

or $3x = c_1e^t - 5c_2e^{-5t} - 3 = \frac{16}{7}e^{2t} + e2c_1e^t + 2c_2e^{-5t} - 6t - \frac{24}{5} + \frac{16}{7}e^{2t} - 2e^{2t}$

or $x = c_1e^t - c_2e^{-5t} - 2t - \frac{13}{5} + \frac{6}{7}e^{2t}$

***Example* 10.112.** Solve $(D-1)x + Dy = 2t + 1$; $(2D+1)x + 2Dy = t$.

Solution : The given equation are

$(D-1)x + Dy = 2t = 1$ ...(1)

$(2D+1)x + 2Dy = t$ ...(2)

Multiplying equation (1) by 2 and subtracting equation (2) from equation (1), we get

$2(D-1)x + 2Dy = 4t + 2$

$(2D+1)x + 2Dy = t$

– – –

$x[2(D-1) - (2D+1)] = 3t + 2$

$x[2D - 2 - 2D - 1] = 3t + 2$

$x(-3) = 3t + 2$ or $x = -t - \frac{2}{3}$

Substituting the value of x in equation (1), we get

$$-(D-1)\left(t+\frac{2}{3}\right)+Dy = 2t+1$$

or
$$Dy-Dt-D\left(\frac{2}{3}\right)+t+\frac{2}{3} = 2t+1$$

or
$$Dy-1-0+t+\frac{2}{3} = 2t+1 \text{ or } Dy-1+t+\frac{2}{3}=2t+1$$

or
$$Dy = t+\frac{4}{3} \quad \text{or} \quad dy=\left(t+\frac{4}{3}\right)dt$$

Integrating on both sides, we get

$$y = \frac{t^2}{2}+\frac{4}{3}t+c$$

EXERCISE 10.6

Solve the following simultaneous equations:

1. $\frac{dx}{dt}+5x-2y=t;\ \frac{dy}{dt}+2x+y=0$, being given $x=y=0$ when $t=0$.

Ans. $x=\frac{-1}{27}(1+6t)e^{-3t}+\frac{1}{27}(2+3t)e^{-3t}+\frac{2}{27}(2-3t)$

2. $(D+1)x+(2D+1)y=e^t;\ (D-1)x+(D+1)y=1.$

Ans. $x=\frac{1}{2}[e^t-1-ac_1e^{at}+bc_2eb^t],\ y=c_1ea^t+c_2b^t+\frac{1}{2}$, where $a=\frac{1}{2}(3+\sqrt{17}),\ b=\frac{1}{2}(3-\sqrt{17})$

3. $\frac{d^2x}{dt^2}+y=\sin t;\ \frac{d^2y}{dx^2}+x=\cos t.$

Ans. $x=c_1e^t+c_2e^{-t}+c_2\cos t=c_4\sin t+\frac{t}{4}(\sin t-\cos t)$

$y=-c_1e^t-c_2e^{-t}+c_3\cos t+c_4\sin t+\frac{1}{4}(2+t)(\sin t-\cos t)$

4. The small oscillations of a certain system with two degree of freedom are given by the equation $D^2x+3x-2y=0;\ D^2x=D^2y-3x+5y=0$ where $D=\frac{d}{dt}$. If $x=0,\ y=0,\ Dx=3,\ Dy=2$, when $t=0$, find x and y when $t=\frac{1}{2}$.

Ans. $x=14015;\ y=1.096$

5. A mechanical system with two degree of freedom satisfies the equation $2\frac{d^2x}{dt^2}+3\frac{dy}{dt}=4;\ 2\frac{d^2x}{dt^2}-3\frac{dx}{dt}=0.$

Obtain expressions for x and y in terms of t, given $x,\ y,\ dx/dt$ all vanish at $t=0$.

Ans. $x=-\frac{8}{9}\left(1-\cos\frac{3t}{2}\right);\ y=\frac{4}{3}t-\frac{8}{9}\sin\frac{3t}{2}$

Partial Differential Equations

The contents of this chapter are : partial derivatives, formation of partial differential equation, linear differential equations, non-linear differential equation of first oder, Charpits methods, homogenous linear equation with constant coefficient, method of separation of variables and its application to wave equation and one dimensional heat equation.

11.1 PARTIAL DERIVATIVE

The readers are already familiar with the notion of partial differential equation. In this chapter we shall investigate the solution of special types of partial differential equation of the first and higher order.

If $z = f(x, y)$ be a function of x and y which are independent variables and z is a dependent variable, we shall employ the following notations

$$\frac{\partial z}{\partial x} = p, \frac{\partial z}{\partial y} = q, \frac{\partial^2 z}{\partial x^2} = r, \frac{\partial^2 z}{\partial x \partial y} = s \text{ and } \frac{\partial^2 z}{\partial y^2} = t$$

11.2 FORMATION OF PARTIAL DIFFERENTIAL EQUATIONS

A partial differential equation is formed by two methods

(1) By eliminating arbitrary constants

(2) By eliminating arbitrary functions

The method is best illustrated by the following examples.

***Example* 11.1.** Derive a partial differential equation by eliminating the constants from the equation.

$$(x-a)^2 + (y-b)^2 + z^2 = c^2$$

Solution : Here arbitrary constants are a and b

Differentiating w.r.t. 'x',. we get

$$2(x-a) + 2z\frac{\partial z}{\partial x} = 0 \quad x - a + zp = 0$$

$$x - a = -zp \qquad ...(i)$$

Differentiating w.r.t. 'y' we get

$$2(y-b) + 2z\frac{\partial z}{\partial y} = 0$$

$$(y-b) + zq = 0$$

$$y - b = zq \qquad ...(ii)$$

Putting the values of $(x - a)$ and $(y - b)$ in the given equation we have

$$p^2z^2 + q^2z^2 + z^2 = c^2$$

$$z^2[p^2 + q^2 + 1] = c^2$$

which is the desired partial differential equation.

Examples **11.2.** Form the partial differential equation by eliminating arbitrary constants from

$$\frac{x^2}{a^2} + \frac{y^2}{b^2} + \frac{z^2}{c^2} = 1 \quad ...(1)$$

Solution : Differentiating (1) partially w.r.t. 'x' and 'y' we have

$$\frac{2x}{a^2} + \frac{2z}{c^2}\frac{\partial z}{\partial x} = 0 \quad ...(2)$$

$$\frac{2y}{b^2} + \frac{2z}{c^2}\frac{\partial z}{\partial y} = 0 \quad ...(3)$$

Differentiating (2) w.r.t. 'x' and (3) w.r.t. 'y' we get

$$\frac{2}{a^2} + \frac{2}{c^2}\left[\left(z\frac{\partial^2 z}{\partial x^2}\right) + \left(\frac{\partial z}{\partial x}\right)^2\right] = 0$$

$$\frac{2}{b^2} + \frac{2}{c^2}\left[\left(z\frac{\partial^2 z}{\partial y^2}\right) + \left(\frac{\partial z}{\partial y}\right)^2\right] = 0$$

or $$c^2 + a^2\left\{\frac{\partial z}{\partial x}\right\}^2 = a^2 z\frac{\partial^2 z}{\partial x^2} = 0 \quad ...(4)$$

and $$c^2 + b^2\left\{\frac{\partial z}{\partial y}\right\}^2 + b^2 z\frac{\partial^2 z}{\partial y^2} + 0 \quad ...(5)$$

Now putting the values of c^2 from (2) in (4)

$$\left(-a^2\frac{z}{x}\frac{\partial z}{\partial x}\right) + a^2\left(\frac{\partial z}{\partial x}\right)^2 + a^2 z\frac{\partial^2 z}{\partial x^2} = 0$$

$$\frac{-z}{x}\frac{\partial z}{\partial x} + \left(\frac{\partial z}{\partial x}\right)^2 + z\frac{\partial^2 z}{\partial x^2} = 0$$

$$-z\frac{\partial z}{\partial x} + x\left(\frac{\partial x}{\partial x}\right)^2 + xz\frac{\partial^2 z}{\partial x^2} = 0 \quad ...(6)$$

Similarly from (3) and (5) we have $-z\frac{\partial z}{\partial y} + y\left(\frac{\partial z}{\partial y}\right)^2 + yz\frac{\partial^2 z}{\partial y^2} = 0$ ...(7)

(6) and (7) are partial differential equation.

Example **11.3.** By eliminating arbitrary function form partial differential equation from

$$xyz = f(x + y + z) \quad ...(1)$$

Solution : Differentiating partially (1) w.r.t 'x' and 'y'

$$y\left[z + \frac{x\partial z}{\partial x}\right] = f'(x + y + z)\left[1 + \frac{\partial z}{\partial x}\right] \quad ...(2)$$

$$yz + xyp = (1 + p)\, f'(x + y + z)$$

Similarly $$x\left[z + y\frac{\partial z}{\partial y}\right] = f'(x + y + z)\left[1 + \frac{\partial z}{\partial y}\right] \quad ...(3)$$

$$xz + xyq = f(x+y+z)(1+q)$$

from (2) and (3) we get

$$\frac{yz + xyp}{xz + xyq} = \frac{\log x}{(1+\log x)^2}.$$

$$(yz + xyp)(1+q) = (xz + xyq)(1+q)$$

$$yz + xyp + yzq = xz + xyq + xzp$$

$$x(y-z)p + y(z-x)q = z(x-y)$$

Example **11.4.** Form the partial differential equation from $z = f\frac{dy}{dx} \times \frac{dx}{dy}$

Solution :

$$z = f\left(\frac{xy}{z}\right) \quad ...(1)$$

Differentiating partially (1) w.r.t. 'x' and 'y'

$$\frac{\partial z}{\partial x} = f'\left(\frac{xy}{z}\right)\left[y\left(\frac{1}{z} - \frac{x}{z^2}\frac{\partial z}{\partial x}\right)\right] \quad ...(2)$$

or

$$p = \frac{\partial z}{\partial x} = f'\left(\frac{xy}{z}\right)\left[\frac{y}{z} - \frac{xy}{z^2}p\right]$$

$$p = f'\left(\frac{xy}{z}\right)\left[\frac{y}{z} - \frac{xy}{z^2}p\right] \quad ...(3)$$

Similarly

$$q = f'\left(\frac{xy}{z}\right)\left[\frac{x}{z} - \frac{xy}{z^2}q\right] \quad ...(4)$$

Dividing (3), by (4) we get

$$\frac{p}{q} = \frac{yz - xyp}{zx - yxq}$$

or

$$pzx = qzy$$

$$px = qy$$

Example **11.5.** Form the partial differential equation by eliminating arbitrary function of $z = y^2 + 2f\left(\frac{1}{x} + \log y\right)$.

Solution :

$$z = y^2 + 2f\left(\frac{1}{x} + \log y\right) \quad ...(1)$$

Differentiating w.r.t 'x' and 'y'

$$\frac{\partial z}{\partial x} = 2f'\left(\frac{1}{x} + \log y\right)\left(-\frac{1}{x^2}\right) \quad ...(2)$$

$$\frac{\partial z}{\partial y} = 2y + 2f'\left(\frac{1}{x} + \log y\right)\left(\frac{1}{y}\right) \quad ...(3)$$

From (2)

$$px^2 = 2f'\left(\frac{1}{x} + \log y\right) \quad ...(4)$$

From (3)

$$yq - 2y^2 = 2f'\left(\frac{1}{x} + \log y\right) \quad ...(5)$$

Dividing (4) by (5) we get

$$\frac{px^2}{2y^2 - qp} = 1$$

$$\Rightarrow \quad px^2 = 2y^2 - qy$$

$$px^2 + qy = 2y^2$$

Example **11.6.** Form the partial differential equation by eliminating arbitrary function from

$$z = y\,f(x) + xg(y)$$

Solution : $$z = y\,f(x) + xg(y) \quad ...(1)$$

Differentiating (1) partially w.r.t. z and y.

$$p = \frac{\partial z}{\partial x} = y\,f(x) + g(y) \quad ...(2)$$

$$q = \frac{\partial z}{\partial y} = f(x) + xg'(y) \quad ...(3)$$

Differentiating equation (2) partially w.r.t. 'y'.

$$s = \frac{\partial^2 z}{\partial x \partial y} = f'(x) + g'(y) \quad ...(4)$$

From (2) and (3) $$p = yf'(x) + g(y) \quad ...(5)$$

$$q = f(x) + f(x) + xg'(y) \quad ...(6)$$

Multiplying (5) by x and (6) by y and adding

$$px + qy = xy\,[f'(x) + g'(y)] + xg(y) + yf(x)$$

$$= xy\,\frac{\partial^2 z}{\partial x \partial y} + z$$

$$px + gy - z = xys \quad \left[S = \frac{\partial^2 z}{\partial x \partial y}\right]$$

Example **11.7.** **Find the differential equation of all the sphere whose centres lie on z-axis.**

Solution : The equation of the gives spheres are

$$(x-0)^2 + (y-0)^2 + (z-\alpha)^2 = r^2$$

where α is the arbitrary constant 'r' being fixed]

$$x^2 + y^2 + (z-\alpha)^2 = r^2$$

Differentiating partially w.r.t. 'x' we get

$$2x + 2\,(z-\alpha)\,\frac{\partial z}{\partial x} = 0$$

$$x + (z-\alpha)\,p = 0$$

Differentiating partially w.r.t. 'y' we get

$$y + (z-\alpha)\,q = 0$$

From (2) and (3), we get

$$\frac{x}{y} = \frac{p}{q} \text{ or } qx = py$$

***Example* 11.8.** Find differential equation of all planes which are at a constant distance 'α' from the origin.

Solution : The equation of the plane in normal form is

$$lx + my + nz = p \quad ...(1)$$

where, l, m, n are the direction consines of the normal from the origin to the plane

Also $$l^2 + m^2 + n^2 = 1 \quad ...(2)$$

Differentiating (1) partially w.r.t. 'x' and 'y'

$$l + n\frac{\partial z}{\partial x} = 0 \qquad l + np = 0 \qquad l = -np \quad ...(3)$$

$$m + n\frac{\partial z}{\partial y} = 0 \qquad m + nq = 0 \qquad m = -np \quad ...(4)$$

Substituting (3) and (4) in (2) we get

$$n^2p^2 + n^2 q^2 + n^2 = 1 \text{ or } n^2 = \frac{1}{1+p^2+q^2}$$

$$n = \frac{1}{\sqrt{1+p^2+q^2}}$$

$$l = \frac{-p}{\sqrt{1+p^2+q^2}}$$

$$m = \frac{-q}{\sqrt{1+q^2+q^2}}$$

Substituting the value of l, m, n in (1) we get

$$\frac{-px}{\sqrt{1+p^2+q^2}} + \frac{-py}{\sqrt{1+p^2+q^2}} + \frac{z}{\sqrt{1+p^2+q^2}} = \alpha$$

$$z = px + py + \alpha\sqrt{1+p^2+q^2}$$

***Example* 11.9.** Form the partial differential equation from $F(x + y + z, x^2 + y^2 + z^2) = 0$

Solution : We have $F(x + y + z, x^2 + y^2 + z^2) = 0$

Let $$u = x + y + z$$

& $$v = x^2 + y^2 + z^2$$

$$F(u, v) = 0$$

Differentiating partially w.r.t. 'x' and 'y'

$$\frac{\partial F}{\partial u}\left[\frac{\partial u}{\partial x} + \frac{\partial u}{\partial z}\cdot\frac{\partial z}{\partial x}\right] + \frac{\partial F}{\partial v}\left[\frac{\partial v}{\partial x} + \frac{\partial v}{\partial z}\cdot\frac{\partial z}{\partial y}\right] = 0$$

$$\frac{\partial F}{\partial u}(1 + 1.p) + \frac{\partial F}{\partial v}(2x + 2zp) = 0 \quad ...(1)$$

$$(1+p)\frac{\partial F}{\partial u} + 2(x + pz)\frac{\partial F}{\partial v} = 0$$

Similarly, $$(1+q)\frac{\partial F}{\partial u} + 2(y + qz)\frac{1}{\pi x^2} = 0 \quad ...(2)$$

From (1) and (2) we get

or
$$y + qz + py + pqz = x + pz + px + pqz$$
$$y + pz + py = x + pz + px$$
$$(y - z)\, p + (z - x)\, q = x - y$$

EXERCISE 11.1

1. Form partial differential equation from the following equations
   (a) $z = e^{-p^2 t} \cos px$
   (b) $z = ax + a^2y^2 + b$
   (c) $z = a(x + y) + b\,(x - y) + abt + c$
2. Find the differential equation of all spheres passing through the origin and having centres in the x-axis.

Find the partial differential equation by eliminating arbitrary functions from

3. $z = f(x) + e^x g(x)$
4. $v = \frac{1}{r}[f(r - at) + f(r + at)]$
5. $y = f(x + at) + x\phi\,(x + at)$
6. $z = x^n f\left(\frac{y}{x}\right)$
7. $z = f(x^2 - y^2)$
8. $z = e^{ax + by} f(ax - by)$
9. $z = f(x^2 - y) + g\,(x^2 + y)$
10. $z = xy\, f(x^2 + y^2 + z^2)$
11. $f(x^2 + y^2 + z^2, z^2 - 2xy) = 0$
12. $z = f\left(\frac{y}{x}\right) + \phi\,(xy)$
13. $F(x + y + z, x^2 + y^2 - z^2) = 0$
14. $u = f(x^2 + y) + F\,(x^2 - y)$
15. $f(x^2 + y^2, z - xy) = 0$

Answers

1. (a) $\frac{\partial^2 z}{\partial x^2} = \frac{\partial z}{\partial t}$ (b) $q = 2y\,p^2$ (c) $\left(\frac{\partial z}{\partial x}\right)^2 - \left(\frac{\partial z}{\partial y}\right) = 4\frac{\partial y}{\partial t}$
2. $x^2 - y^2 - z^2 + 2px = 0$
3. $\frac{\partial^2 z}{\partial y^2} = \frac{\partial z}{\partial y}$
4. $\frac{\partial^2 v}{\partial t^2} = \frac{a^2}{r^2}\left(r^2 \frac{\partial v}{\partial r}\right)$

5. $\left(\frac{\partial}{\partial t} - u\frac{\partial}{\partial x}\right)^2 y = 0$
6. $px + qy = nz$
7. $py + aq = 0$
8. $bp + aq = 2ab^2$
9. $x\frac{\partial^2 z}{\partial x^2} - 4x^2\frac{\partial^2 z}{\partial y^2} - p = 0$
10. $x(y^2 + z^2)\,p - y\,(x^2 + z^2)\,q = z\,(y^2 - x^2)$
11. $z(p - q) + (x - y) = 0$
12.
13. $(x + y)\,p - (x + z)\,q = x - y$
14. $x\frac{\partial^2 z}{\partial x^2} - 4x^3\frac{\partial^2 z}{\partial y^2} = \frac{\partial z}{\partial x}$
15. $py - qx = y^2 - x^2$

11.3 EQUATIONS SOLVABLE BY DIRECT INTEGRATION

Those equations which contain only one partial derivative can be solved by direct integration. In place of the usual constants of integration, we choose arbitrary functions of the variable held fixed during integration.

Example **11.10.** Solve by direct integration : $\frac{\partial^2 z}{\partial y^2} = \tan^2 y$.

Solution :

$$\frac{\partial^2 z}{\partial y^2} = \tan^2 y$$

$$= \sec^2 y - 1$$

Integrating twice w.r.t, 'y' treating x as constant.

$$\frac{\partial z}{\partial y} = \tan y - y + \phi\,(x)$$

$$z = \log \cos y - \frac{y^2}{2} + y\,\phi(x) + \psi(x),$$ is the required result,

where $f(x)$ and $y(x)$ are arbitrary functions of x

Example **11.11.** Solve $\frac{\partial^2 z}{\partial y^2} = z$, given $y = 0$, $z = e^x$ and $\frac{\partial z}{\partial y} = e^{-x}$

Solution : If z was the function of y alone, then the solution have been, $z = A \sin y + B \cos y$, where A and B are constant. Since, z is a function of x and y, A and B can be arbitrary function of x. Hence the solution of given equation is :

$$z = f(x) \sin y + \phi(x).\cos y \qquad ...(1)$$

$$\frac{\partial z}{\partial y} = f(x) \cos y - \phi(x) \sin y$$

When $y = 0,\ z = e^x, \quad \therefore \quad e^x = \phi(x)$

Also, when $y = 0,\ \frac{\partial z}{\partial y} = e^{-x}$

$$\Rightarrow \qquad f(x) = e^{-x}$$

Substituting for $f(x)$ and $\phi(x)$ in (1). we get $z = e^{-x}\cos y - e^{x}\sin y$ which is the required solution.

Examples **11.12.** Solve $\log\left[\frac{\partial^2 z}{\partial x \partial y}\right] = x + y$

Solution :
$$\log\left[\frac{\partial^2 z}{\partial x \partial y}\right] = x + y$$

or
$$\frac{\partial^2 z}{\partial x \partial y} = e^{x+y} = e^x . e^y$$

Integrating w.r.t.y, treating x as constant,

$$\frac{\partial z}{\partial x} = e^x . e^y + \phi(x)$$

Integrating w.r.t. 'n' treating y as constant,

$$z = e^x . e^y + \int \phi(x)\, dx + \psi(y)$$

or
$$z = e^{x+y} + F(x) + \psi(y) \quad \text{where} \quad F(x) = \int \phi(x)\, dx$$

EXERCISE 11.2

Solve the following equation :

1. $\frac{\partial^2 z}{\partial x^2} + z = 0$ given that when $x = 0$, $z = e^y$, $\frac{\partial z}{\partial x} = 1$

2. $\frac{\partial^2 u}{\partial x \partial t} = e^{-t}\cos x$

3. $\frac{\partial^2 z}{\partial x^2} = a^2 z$ given that when $x = 0$, $\frac{\partial z}{\partial x} = a \sin y$, $\frac{\partial z}{\partial y} = 0$.

4. $\frac{\partial^2 z}{\partial y \partial x} = \frac{x}{y} + a$

5. $\frac{\partial^2 z}{\partial y \partial x} = e^{-y}\cos x$ given $z = 0$ when $y = 0$; $\left(\frac{\partial z}{\partial y}\right)_{x=0} = 0$

Answers

1. $z = \sin x + e^y \cos x$
2. $u = -e^{-t}\sin x + f(x) + g(t)$
3. $z = \sin x + e^y \cos x$
4. $z = \frac{x^2}{2}\log n + axy + \phi(x) + f(y)$
5. $z = e^{x+y} + \int \phi(x) dx + \psi(y)$

11.4 LINEAR PARTIAL DIFFERENTIAL EQUATION OF FIRST ORDER

A differential equation that involves first order partial derivatiues *i.e.,* p and q only is called a partial differential equation of the first order.

If p and q both occur in the first degree and are not multiplied together, then it is called a linear partial differential equation of first order.

11.5 LAGRANGE'S LINEAR EQUATION

The partial differential equation of the type.

$Pp + Qp = R$, where P, Q and R are function of x,y,z *and* $p = \frac{\partial z}{\partial x}, q = \frac{\partial z}{\partial y}$ is called Lagrange's Linear Equation.

11.6 SOLUTION OF LAGRANGE'S LINEAR EQUATION

For this we have following theorem.

Theorem 11.1 : If P and Q are independent of z and R is linear in z then $Pp + Qp = R$ is a linear equation. The general solution of Lagrange's linear P.D.E is given by the equation $F(u, v) = 0$.

Solution : Lagrange's Linear Equation $Pp + Qq = R$ ...(1)

is obtained by eliminating an arbitrary function ϕ from $\phi(u, v) = 0$ ...(2)

where u, v are some definite function of x, y, z.

$$\frac{\partial \phi}{\partial u}\left(\frac{\partial u}{\partial x} + \frac{\partial u}{\partial z}p\right) + \frac{\partial \phi}{\partial v}\left(\frac{\partial v}{\partial x} + \frac{\partial v}{\partial z}p\right) = 0$$

and
$$\frac{\partial \phi}{\partial u}\left(\frac{\partial u}{\partial y} + \frac{\partial u}{\partial z}q\right) + \frac{\partial \phi}{\partial v}\left(\frac{\partial v}{\partial y} + \frac{\partial v}{\partial z}q\right) = 0$$

Eliminating $\frac{\partial \phi}{\partial u}$ and $\frac{\partial \phi}{\partial v}$

$$\left(\frac{\partial u}{\partial x} + \frac{\partial u}{\partial z}p\right)\left(\frac{\partial v}{\partial y} + \frac{\partial v}{\partial z}q\right) - \left(\frac{\partial v}{\partial x} + \frac{\partial v}{\partial z}p\right)\left(\frac{\partial u}{\partial y} + \frac{\partial u}{\partial z}q\right) = 0$$

or
$$\left(\frac{\partial u}{\partial y}\cdot\frac{\partial v}{\partial z} - \frac{\partial u}{\partial z}\cdot\frac{\partial v}{\partial y}\right)p + \left(\frac{\partial u}{\partial z}\cdot\frac{\partial v}{\partial x} - \frac{\partial u}{\partial x}\cdot\frac{\partial v}{\partial z}\right)q. = \frac{\partial u}{\partial x}\cdot\frac{\partial v}{\partial y} - \frac{\partial u}{\partial y}\cdot\frac{\partial v}{\partial x}$$

Which is same as (1) *i.e.,* $Pp + Qq + R$, where

$$P = \frac{\partial u}{\partial y}\cdot\frac{\partial v}{\partial z} - \frac{\partial u}{\partial z}\cdot\frac{\partial v}{\partial y}$$

$$Q = \frac{\partial u}{\partial z}\cdot\frac{\partial v}{\partial x} - \frac{\partial u}{\partial x}\cdot\frac{\partial v}{\partial z}$$

$$R = \frac{\partial u}{\partial x}\cdot\frac{\partial v}{\partial y} - \frac{\partial u}{\partial y}\cdot\frac{\partial v}{\partial x}$$

Now,
$$\frac{\partial u}{\partial x}dx + \frac{\partial u}{\partial y}dy + \frac{\partial u}{\partial z}dz = du = 0 \quad \text{...(3)}$$

$$\frac{\partial v}{\partial x}dx + \frac{\partial v}{\partial y}dy + \frac{\partial v}{\partial z}dz = dv = 0 \quad \text{...(4)}$$

Solving (3) and (4), we get

$$\frac{dx}{\frac{\partial u}{\partial y}\times\frac{\partial v}{\partial z}-\frac{\partial u}{\partial z}\times\frac{\partial v}{\partial y}} = \frac{dy}{\frac{\partial u}{\partial z}\times\frac{\partial v}{\partial x}-\frac{\partial u}{\partial x}\times\frac{\partial v}{\partial x}} = \frac{dz}{\frac{\partial u}{\partial x}\times\frac{\partial v}{\partial y}-\frac{\partial u}{\partial y}\times\frac{\partial v}{\partial x}}$$

Or $$\frac{dx}{P} = \frac{dy}{Q} = \frac{dz}{R} \quad ...(5)$$

Solution of these equation are

$$u = a \text{ and } v = b$$

Equation (5) *i.e.*,

$$\frac{dx}{P} = \frac{dy}{Q} = \frac{dz}{R} \text{ can be reduced to}$$

$$Pp + Qq = R \text{ as follows :}$$

Since $$z = f(x, y)$$

$\therefore$ $$dz = pdx + qdy \quad ...(6)$$

Also we have $$\frac{dx}{P} = \frac{dy}{Q} = \frac{dz}{R} = t\ (say)$$

Substituting for dx, dy and dz in (6)

$$Rt = tpP + qQt$$

$\therefore$ $$R = \text{Pp} + \text{Qq}$$

$\therefore$ $u = a$, $v = b$ or $f(u, v) = 0$ is the solution of $pP = qQ = R$

11.7 WORKING RULE TO SOLVE LAGRANGE'S LINER EQUATION ($Pp + q = R$)

(1) Form the auxiliary equation, *i.e.*,

$$\frac{dx}{P} = \frac{dy}{Q} = \frac{dz}{R}$$

(2) Find two independent integrals of auxiliary equation say $u = a$, $v = b$

(3) Then $\phi(u, v)$ or $u = \phi(v)$ will be the complete solution.

***Example* 11.13.** Solve : $(z - y)\, p + (x - z)\, q = y - x$

Solution : The given Lagrange's linear equation is

$$(z - y)\, p + (x - z)\, q = y - x$$

The auxiliary equation are :

$$\frac{dx}{z-y} = \frac{dy}{x-z} = \frac{dz}{y-x} \quad ...(1)$$

Using multiplies 1, 1, 1, each fraction is equal to $\dfrac{dx+dy+dz}{0}$

$\Rightarrow$ $$dx + dy + dz = 0$$

Integrating, we get $$x + y + z = a \quad ...(2)$$

Again using multipliers x, y, z we get each fraction of (1) equal to $\dfrac{xdx + ydy + zdz}{0}$

This gives $$xdx + ydy + zdz = 0$$

Integrating we get

$$\frac{x^2}{2} + \frac{y^2}{2} + \frac{z^2}{2} = C$$

or $$x^2 + y^2 + z^2 = b \text{ where } b = 2c$$

The complete solution of given equation is

$$x^2 + y^2 + z^2 = f(x + y + z)$$

Example **11.14.** Solve : $(x^2 - yz)p + (y^2 - zx)q = z^2 - xy$

Solution : The auxiliary equation are :

$$\frac{dx}{x^2 - yz} = \frac{dy}{y^2 - zx} = \frac{dz}{z^2 - xy}$$

$$\Rightarrow \quad \frac{dx - dy}{x^2 - yz - y^2 + zx} = \frac{dy - dz}{y^2 - zx - z^2 + xy} = \frac{dz - dx}{z^2 - xy - x^2 + xy}$$

$$\Rightarrow \quad \frac{dx - dy}{(x - y)(x + y + z)} = \frac{dy - dz}{(y - z)(x + y + z)} = \frac{dz - dx}{(z - x)(x + y + z)}$$

$$\Rightarrow \quad \frac{dx - dy}{x - y} = \frac{dy - dz}{y - z} = \frac{dz - dx}{z - x} \qquad ...(1)$$

Integrating first two fraction of (1) we get

$$\log (x - y) = \log (y - z) + \log a$$

or $$\frac{x - y}{y - z} = a$$

Similarly integrating last two fraction of (1), we get

$$\log (y - z) = \log (z - x) + \log b$$

or $$\frac{y - z}{z - x} = b$$

Thus, the complete solution is :

$$f(a, b) = 0$$

i.e., $$f\left(\frac{x - y}{y - z}, \frac{y - z}{z - x}\right) = 0$$

Example **11.15.** Solve : $p - q = \log (x + y)$

Solution : The equation is $p - q = \log (x + y)$

The auxiliary equation are :

$$\frac{dx}{1} = \frac{dy}{-1} = \frac{dz}{\log (x + y)}$$

Considering $$\frac{dx}{1} = \frac{dy}{-1}$$

$\Rightarrow \quad dx + dy = 0$

Integrating $\quad x + y = a$

Also, $$\frac{dx}{1} = \frac{dz}{\log(x+y)} \quad ...(1)$$

or $$\frac{dx}{1} = \frac{dz}{\log a} \quad \text{(By (1) : } x + y = a)$$

Integrating $\quad x \log a = z + b$

or $\quad x \log(x+y) = b \quad ...(2)$

Hence, the complete solution is

$$x \log(x+y) - z = f(x+y)$$

Example **11.16.** Solve the following differential equation :

$$p \cos(x+y) + q \sin(x+y) = z$$

Solution : The auxiliary equation are

$$\frac{dx}{\cos(x+y)} = \frac{dy}{\sin(x+y)} = \frac{dz}{z}$$

Let $\quad x + y = t > dx + dy = dt \quad ...(1)$

$$\Rightarrow \quad \frac{dx}{\cos t} = \frac{dy}{\sin t} = \frac{dz}{z} \text{ or } \frac{d(x+y)}{\cos t + \sin t} = \frac{dz}{z}$$

$$\frac{dt}{\cos t + \sin t} = \frac{dz}{z}$$

or $$\frac{\frac{1}{\sqrt{2}} dt}{\frac{1}{\sqrt{2}} \cos t + \frac{1}{\sqrt{2}} \sin t} = \frac{dz}{z}$$

$$\frac{dt}{\sin\left(\frac{\pi}{4} + t\right)} = \sqrt{2} \frac{dz}{z}$$

or $$\text{cosec}\left(\frac{\pi}{4} + t\right) = \sqrt{2} \frac{dz}{z}$$

Integrating, we get

$$\log \tan\left(\frac{\pi}{8} + \frac{t}{2}\right) = \sqrt{2} \log z + \log a$$

$$\log \tan\left(\frac{\pi}{8} + \frac{t}{2}\right) = \log(z^{\sqrt{2}} a)$$

$$\tan\left(\frac{\pi}{8} + \frac{t}{2}\right) = z^{\sqrt{2}} a$$

or $$z^{-\sqrt{2}} \tan\left(\frac{\pi}{8} + \frac{t}{2}\right) = a \quad ...(2)$$

Also we have,

$$dy = dt - dx \quad \text{(From 1.)}$$

Substituting this in auxiliary equation and first two fractions, we get

$$\frac{dx}{\cos t} = \frac{dt - dx}{\sin t}$$

$$\sin t\, dx = \cos t\, dt - \cos t\, dx$$

or
$$\cos t\, dt = (\sin t + \cos t)\, dx$$

$$dx = \frac{\cos t}{\sin t + \cos t} dt$$

or
$$2dx = \frac{2\cos t}{\sin t + \cos t} dt = \frac{(\cos t + \sin t) + (\cos t - \sin t)}{\sin t + \cos t} dt$$

$$2dx = dt + \frac{\cos t - \sin t}{\sin t + \cos t} dt$$

Integrating we get

$$2x = t + \log(\sin t + \cos t) + \log C$$

$$2x = (x + y) + \log(\sin t + \cos t)\, C$$

$$x - y = \log\{[\sin(x + y) + \cos(x + y)]C\}$$

or
$$e^{x-y} = [\sin(x + y) + \cos(x + y)]\, C$$

or
$$\frac{1}{C} = e^{y-x}[\sin(x + y) + \cos(x + y)]$$

or
$$b = e^{y-x}[\sin(x + y) + \cos(x + y)] \quad \left(\text{where } b = \frac{1}{C}\right) \quad ...(3)$$

Combining (2) and (3), we get

$$f\left[e^{y-x}\{\sin(x + y) + \cos(x + y)\},\ z - \sqrt{2}\tan\left(\frac{\pi}{8} + \frac{x + y}{2}\right)\right]$$ which is the required solution.

Examples **11.17.** Solve the differential equation

$$x^2\frac{\partial z}{\partial x} + y^2\frac{\partial z}{\partial y} = (x + y)z$$

Solution : The given equation can be re-written as $x^2p + y^2q = (x + y)z$ which is lagrange's equation.
Auxiliary equation are :

$$\frac{dx}{x^2} = \frac{dy}{y^2} = \frac{dz}{x + y} \quad ...(1)$$

Considering first two fraction and integrating

$$\frac{-1}{x} = \frac{-1}{y} + a$$

or
$$\frac{1}{x} - \frac{1}{y} = a. \quad ...(2)$$

Also, (1) can be written as

$$\frac{\frac{dx}{x}}{x} = \frac{\frac{dy}{y}}{y} = \frac{\frac{dz}{z}}{x + y}$$

$$= \frac{\frac{dx}{x} + \frac{dy}{y} - \frac{dz}{z}}{(x + y) - (x + y)}$$

$$\Rightarrow \qquad \frac{dx}{x} + \frac{dy}{y} - \frac{dz}{z} = 0$$

On integration, we get

$$\log x + \log y - \log z = \log b$$

or $$b = \frac{xy}{z} \qquad ...(3)$$

From (2) and (3), we get

$$f\left(\frac{1}{x} - \frac{1}{y}, \frac{xy}{z}\right) = 0$$

***Examples* 11.18.** Solve : $px(z - 2y^2) = (z - qy)(z - y^2 - 2x^3)$

Solution : $px(z - 2y^2) = (z - qy)(z - y^2 - 2x^3)$

or $$px(z - 2y^2) + qy(z - y^2 - 2x^3) = z(z - y^2 - 2x^3)$$

The auxiliary equation are :

$$\frac{dx}{x(z - 2y^2)} = \frac{dy}{y(z - y^2 - 2x^2)} = \frac{dz}{z(z - y^2 - 2x^2)} \qquad ...(1)$$

Taking last two fraction of (1) we get

$$\frac{dy}{y} = \frac{dz}{z}$$

On integration, we get

$$\log y = \log z + \log a \Rightarrow y = az \qquad ...(2)$$

From (1), we have

$$\frac{dx}{x(z - 2y^2)} = \frac{dz}{z(z - y^2 - 2x^3)}$$

$$\frac{dx}{x(z - 2a^2z^2)} = \frac{dz}{z(z - y^2 - 2x^3)} \qquad \text{[From (2), } y = az\text{]}$$

$$\frac{dx}{x(1 - 2a^2z)} = \frac{dz}{z - a^2z^2 - 2x^3}$$

$$(xdz - zdx) - a^2(2xz\,dz - z^2\,dx) + 2x^3dx = 0$$

$$\Rightarrow \qquad \frac{x\,dz - zdx}{x^2} - \frac{a^2(axzdz - z^2dx)}{x^2} + 2x\,dx = 0$$

or $$d\left(\frac{z}{x}\right) - a^2.d\left(\frac{z^2}{x}\right) = 0$$

Integrating, we get

$$\frac{z}{x} - a^2\frac{z^2}{x} + x^2 = b \qquad ...(3)$$

Combining (2) and (3), we get

$$f\left(\frac{y}{z}, \frac{z}{x} - \frac{a^2z^2}{x} + x^2\right) = 0$$

EXERCISE 11.3

1. $xp - yq = x^2 - y^2$
2. $(y - z)p + (x - y)\, q = z - x$
3. $p \tan x + q \tan y = \tan z$
4. $yz \dfrac{\partial z}{\partial x} + zx \dfrac{\partial z}{\partial y} = xy$
5. $(x + y)\left(\dfrac{\partial z}{\partial x} + \dfrac{\partial z}{\partial y}\right) = z = 1$
6. $(z^2 - 2yz - y^2)\, p + (xy + zx)\, q = xy - zx$
7. $\left(\dfrac{1}{z} - \dfrac{1}{y}\right)p + \left(\dfrac{1}{x} - \dfrac{1}{z}\right)q = \dfrac{1}{y} - \dfrac{1}{x}$
8. $x(y^2 - z^2)p + y(z^2 - x^2)q = z\,(x^2 - y^2)$
9. $xp + yp = 3z$
10. $yp - zp = 0$
11. $x\dfrac{\partial u}{\partial x} + y\dfrac{\partial u}{\partial y} + \dfrac{\partial u}{\partial z} = xyz$

Answers

1. $xy = f(x^2 + y^2 - 2z)$
2. $f\left(x + y + z, \dfrac{x^2}{2} + yz\right) = 0$
3. $f\left(\dfrac{\sin x}{\sin y}, \dfrac{\sin y}{\sin z}\right) = 0$
4. $f(x^2 - y^2, x^2 - z^2) = 0$
5. $f\left(x - y, \dfrac{x + y}{(z + 1)^2}\right) = 0$
6. $f(x^2 + y^2 + z^2, y^2 - 2yz - z^2) = 0$
7. $f(x + y + z, xyz) = 0$
8. $f(x^2 + y^2 + z^2, xyz) = 0$
9. $\dfrac{dy}{dx} = f\left(\dfrac{y}{z^{1/3}}\right)$
10. $f(x^2 + y^2, z) = 0$
11. $f\left(\dfrac{x}{y}, \dfrac{y}{z}, xyz - 3u\right) = 0$

11.8 NON-LINEAR PARTIAL DIFFERENTIAL EQUATION OF 1st ORDER

Non-linear P.D.E of first order contains p and q of degree (power) order then one and/or product terms of p and q. Its complete solution is given by $f(x, y, z, a, b) = 0$ where a and b are any two arbitrary constant. Some special types of non-linear first order P.D.E are presented.

Form 1: $f(a, q) = 0$

i.e., equation contains only p and q or in other words x, y, z are absent.

Assume that $p = a$ then $f(a, q) = 0$

Solving $q = \phi(a)$

Consider
$$dz = \frac{\partial z}{\partial x}dx + \frac{\partial z}{\partial y}dy$$
$$dz = a\,dx + \phi(a)\,dy$$

Integrating $z = ax + \phi(a)\,y + c$ were, a, b, c are arbitrary constant.

Thus, the complete solution is :

$z = ax + by + c$. where a, b satisfy the equation $f(a, b) = 0$

i.e., $b = \phi(a)$.

***Example* 11.19.** Solve $x^2p^2 + y^2q^2 = z^2$

Solution : The Given equation can be written as :

$$\left(\frac{x}{z}\frac{\partial z}{\partial x}\right)^2 + \left(\frac{y}{z}\frac{\partial z}{\partial y}\right)^2 = 1 \quad \text{...(1)}$$

Putting
$$\frac{dx}{x} + \frac{1}{2} = dX, \frac{dy}{y} = dY, \frac{dz}{z} = dZ$$

so that
$$X = \log x, Y = \log y, Z = \log z$$
$$\frac{\partial z}{\partial X} = \frac{x}{z}\cdot\frac{\partial z}{\partial x} \text{ and } \frac{\partial z}{\partial y} = \frac{y}{z}\frac{\partial z}{\partial y}$$

Therefore (1) can be written as
$$\left(\frac{\partial z}{\partial X}\right)^2 + \left(\frac{\partial z}{\partial Y}\right)^2 = 1$$

or
$$P^2 = Q^2 = 1 \text{ were } P = \frac{\partial z}{\partial X}, Q\frac{\partial z}{\partial Y}$$

It is of the from $f(P, Q) = 0$

It complete solution is
$$z = aX + bY + c \quad \text{...(2)}$$

where
$$a^2 + b^2 = 1 \text{ or } b = \sqrt{1 - a^2}$$

∴ Form (2) the complete solution is
$$\log z = a\log x + \sqrt{1 - a^2}\,\log y + c$$

***Example* 11.20.** $(x + y)(p + q)^2 + (x - y)(p - q)^2 = 1$

Solution : To reduce this equation of $f(P, Q) = 0$ form, put $x + y = X^2$ and $x - y = Y^2$. Then
$$X = \sqrt{x + y} \text{ and } Y = \sqrt{x - y}$$

So that $$p = \frac{\partial z}{\partial x} = \frac{\partial z}{\partial X}\cdot\frac{\partial X}{\partial x} + \frac{\partial z}{\partial Y}\cdot\frac{\partial Y}{\partial x}$$

$$= \frac{1}{2x}\cdot\frac{\partial z}{\partial x} + \frac{1}{2y}\cdot\frac{\partial z}{\partial y}$$

Since $$q = \frac{\partial z}{\partial y} = \frac{\partial z}{\partial X}\cdot\frac{\partial X}{\partial y} + \frac{\partial z}{\partial Y}\cdot\frac{\partial y}{\partial y} = \frac{1}{2X}\cdot\frac{\partial z}{\partial x} - \frac{1}{2y}\cdot\frac{\partial z}{\partial y}$$

Adding $$p + q = \frac{\partial z}{\partial X}\cdot\frac{1}{X}$$

Also, $$p - q = \frac{\partial z}{\partial Y}\cdot\frac{1}{Y}$$

Thus, given equation reduce to :

$$X^2\left(\frac{\partial z}{\partial X}\cdot\frac{1}{X}\right)^2 + Y^2\left(\frac{\partial z}{\partial Y}\cdot\frac{1}{Y}\right)^2 = 1$$

or $$P^2 + Q^2 = 1$$

where $$P = \frac{\partial z}{\partial X},\ Q = \frac{\partial z}{\partial Y}$$

The complete solution is :

$$z = aX + bY + c$$

where $$a^2 + b^2 = 1 \text{ or } b = \sqrt{1-a^2}$$

$$z = a\sqrt{x+y} + \sqrt{1-a^2}\,\sqrt{x-y} + c$$

EXERCISE 11.4

Find the complete solution of :

1. $pq = k$
2. $pq = p + q$
3. $p = e^q$
4. $(y - x)(qy - px) = (p - q)^2$
5. $x^2p^2 + y^2q^2 = z$

Hint : $\left(\frac{x}{\sqrt{z}}\frac{\partial z}{\partial x}\right)^2 + \left(\frac{y}{\sqrt{2}}\frac{\partial z}{\partial y}\right)^2 = 1$, Put $\frac{dx}{x} = dX,\ \frac{dx}{y} = dY,\ \frac{dx}{\sqrt{z}} = dZ$

6. $pq = x^m y^n z^1$

Answers

1. $z = ax + k\frac{y}{a} + c$
2. $z = ax + \frac{a}{a-1}y + c$
3. $z = ax + y + tna + c$

4. $z = a(x + y) + \sqrt{a}\, xy + c$

5. $2\sqrt{z} = a \log x + \sqrt{1 + a^2} \log y + c$

6. $z^p = P\left(\frac{ax^q}{q} + \frac{1}{a}\frac{y^r}{r} + c\right)$ where $P = 1 - l/2.\ q = m + 1,\ r = n + 1.$

From II : $f(z, p, q) = 0$ *i.e.,* equation does not involve the independent variables x and y.

Assume $q = ap$. substituting q in the given equation $f(z, p, ap) = 0$ and solving for p, we get $p = \phi(z)$

Now
$$dz = pdx + qdy = pdx + apdy$$
$$dz = p(dx + ady) = \phi(z)\,(dx + ady)$$

Integrating $x + ay = \int \frac{dz}{\phi(z)} + b$ were, b are two arbitrary constants.

Example **11.21.** $zpq = p + q$

Solution : Assume $q = ap$

Substituting in given equation

$$zp(ap) = p + ap$$

Solving for p. we get
$$p = \frac{1+a}{az}$$

Substituting in
$$dz = pdx + qdy$$
$$= p\,(dx + a\,dy)$$
$$dz = \frac{1+a}{az}\,(dx + a\,dy)$$

or
$$az\,dz = (1 + a)\,(dx + ady)$$

Integrating
$$\frac{az^2}{2} = (1 + a)\,(x + ay) + b$$

Example **11.22.** $p^2z^2 + q^2 = p^2q$

Solution : Let $q = ap$. The the given equation reduces to

$$p^2z^2 + a^2p^2 = p^2.ap$$

Solving
$$p = (z^2 + a^2)/a$$

Then
$$dz = pdx + qdy = p(dx + ady)$$
$$dz = \left(\frac{z^2 + a^2}{a}\right)(dx + ady)$$

or
$$\frac{adz}{(z^2 + a^2)} = dx + ady$$

Integrating
$$\tan^{-1}\left(\frac{z}{a}\right) = x + ay + b.$$ Thus the complete solution is :

$$z = a \tan\,(x + ay + b)$$

Example **11.23.** $px^2 = z(z - qy)$

Solution : To reduce this equation to $f(z, p, q) = 0$ form

from out $X = \log x$, $Y = \log y$ so that

$$dX = \frac{dx}{x}, dY = \frac{dy}{y}$$

Rewriting the given equation :

$$\left(x\frac{\partial z}{\partial x}\right)^2 = z\left(z - y, \frac{\partial z}{\partial y}\right)$$

or $$\left(\frac{\partial z}{\partial X}\right)^2 = z\left(z - \frac{\partial z}{\partial Y}\right) \qquad \left\{\frac{dx}{x} = dX \text{ and } \frac{dy}{y} = dy\right\}$$

Let $$p^2 = \frac{\partial z}{\partial X}, Q = \frac{\partial z}{\partial Y}.. \text{ Thus given equation reduce to}$$

$$p^2 = z(z - Q), \text{ which is of the form}$$

$$f(z, P, Q) = 0$$

Let $$Q = qp.$$

Substituting in (1),we get

$$p^2 - z^2 - zQ = z^2 - zap$$

or $$p^2 + azp - z^2 = 0$$

Solving $$P = \frac{-az \pm \sqrt{a^2z^2 + 4z^2}}{2} = z.k \qquad \text{where } \left(k = \frac{-a \pm \sqrt{a^2 + 4}}{2}\right)$$

Now $$dz = PdX + QdY$$

$$= P[dX + adY]$$

$$dz = k.z.(dX + adY) \text{ since } P = kz$$

Replacing X and Y by $\log x$ and $\log y$.

$$\frac{1}{k}\frac{dz}{z} = d(\log x) + ad(\log y)$$

Integrating $\log z^{1/k} = \log x + \log y^a + \log b$

$$\therefore \quad z^{1/k} = xy^a b \text{ where } a, b \text{ are constant and } k = (-a \pm \sqrt{a^2 + 4})/2$$

EXERCISE 11.5

1. $p^2 = qz$
2. $p + q = z$
3. $z^2(p^2 + q^2 + 1) = a^2$
4. $z^2(p^2x^2 + q^2) = 1$
5. $z = p^2 + q^2$
6. $z^2(p^2 + q^2 + 1) = 1$
7. $q^2 = z^2p^2(1 - p^2)$

Answers

1. $z = be^{ax} + a^2 y$
2. $(1 + a) \log z = x + ay + b$
3. $(1 + b^2)(a^2 - z^2) = (x + by + c)^2$
4. $\sqrt{1 + a^2}\, z^2 = +2(\log x + ay) + c$
5. $4z(1 + a^2) = (x + ay + b)^2$
6. $(1 + a^2)(1 - z^2) = (x + ay + b)^2$
7. $(x + ay + b)^2 = z^2 - a^2$

Form III : $f(x, p) = g(y, q)$

i.e., x, p and y, q are separable.

Assume $f(x, p) = g(y, q) = a$ = constant.

Solving each equation for p and q, we get

$$p = f_1(x, a) \text{ and } q = g_1(y, a)$$

Now,

$$dz = pdx + qdy$$

$$= f_1(x, a)\, dx + g_1(y, a)\, dy$$

Integrating : $z = \int f_1(x, a)\, dx + \int g_1(y, a)\, dy + b.$

Example **11.24.** Solve $p^2 - q^2 = x - y$

Solution : The given equation is $p^2 - x = q^2 - y$, which is of the from $f_1(x, p) = f_z(y, q)$.

Let

$$p^2 - x = q^2 - y = a$$

Then $p^2 = x + a$ and $q^2 = y + a$ *i.e.*,

$$p = \sqrt{x + a} \text{ and } q = \sqrt{y + a}$$

Substituting these values in $dz = pdx + qdy$.

We get

$$dz = \sqrt{x + a}\, dx + \sqrt{y + a}\, dy.$$

Integrating $z\ \frac{2}{3}(x + a)^{3/2} + \frac{2}{3}(x + a)^{3/2} + b.$

Which is the required complete solution.

Example **11.25.** $yp + xq + pq = 0$

Solution : Rewriting the given equation :

$$(x + p)\, q = -yp$$

or

$$\frac{x + p}{p} = -\frac{y^6}{3} = a$$

Then solving for p and q

$$p = \frac{x}{a - 1} \text{ and } q = \frac{-y}{a}$$

Now $$dz = pdx + qdy = \frac{x}{a-1}\,dx + \left(\frac{-y}{a}\right) dy$$

Integrating,

$$z = \frac{1}{(a-1)}\frac{x^2}{2} - \frac{y^2}{2a} + b$$

***Example* 11.26.** $p^2q^2 + x^2y^2 = x^2q^2(x^2 + y^2)$

Solution : To transform this to the standard form, put $x^2 = X$ and $y^2 = Y$ so that

$$2x\,dx = dX,\ 2y\,dy = dY$$

Then $$p = \frac{\partial z}{\partial x} = \frac{\partial z}{\partial X}\cdot\frac{\partial X}{\partial x} = 2x\frac{\partial z}{\partial X} = 2xP \qquad \left[\text{where } P = \frac{\partial z}{\partial X}\right]$$

$$q = \frac{\partial z}{\partial y} = \frac{\partial z}{\partial Y}\cdot\frac{\partial Y}{\partial y} = 2y\frac{\partial z}{\partial Y} = 2yQ \qquad \left[\text{where } Q = \frac{\partial z}{\partial Y}\right]$$

Substituting these values in given equation, we get

$$4x^2 p^2 .\, 4y^2 Q^2 + x^2 y^2 = x^2 .\, 4y^2 Q^2 (x^2 + y^2)$$

$$4XP^2 .\, 4YQ^2 + XY = 4XYQ^2 (X + Y)$$

or $$16\,P^2Q^2 + 1 = 4Q^2\,(X + Y)$$

Now rewrite this as

$$16P^2Q^2 - 4XQ^2 = 4Q^2 - 1$$

$$(4P^2 - X) = \frac{4YQ^2 - 1}{4Q^2} = a^2 \quad \text{say}$$

This is of form : $$f(P, X) = g(Q, Y)$$

Solving for P and Q, we get

$$P = \frac{1}{2}(X + a^2)^{1/2}$$

$$Q = \frac{1}{2(y - {}^{a})y_2}$$

Now $$dz = PdX + QdY$$

$$= \frac{1}{2}(X + a^2)^{1/2}\,dX + \frac{1}{2}(Y - a^2)^{-1/2}\,dY$$

Integrating $$z = \frac{1}{3}(X + a^2)^{3/2} + (Y - a)^{1/2} + b$$

$$z = \frac{1}{3}(x^2 + a^2)^{3/2} + (y^2 - a^2)^{1/2} + b$$

where a, b are arbitrary constant.

EXERCISE 11.6

1. $p^2 + q^2 = x + y$.
2. $z^2\,(p^2 + q^2) = x^2 + y^2$

3. $\sqrt{p} + \sqrt{q} = x + y$

4. $p + q = \sin x + \sin y$.

5. $q(p - \cos x) = \cos y$.

6. $p^2 + q^2 = z^2(x + y)$.

7. $z(xp - yq) = y^2 - x^2$. $\left[\text{Hint} : x\left(z\frac{\partial z}{\partial x}\right) - y\left(z\frac{\partial z}{\partial y}\right) = y^2 - x^2 \right.$ Put $zdz = dZ$ $\Big]$

8. $q = xyp^2$.

9. $pe^y = qe^x$.

10. $x(1 + y)p = y(1 + x)q$.

Answers

1. $z = \frac{2}{3}(x + a)^{3/2} + \frac{2}{3}(y - a)^{3/2} + b$

2. $z^2 = {}^x\sqrt{x^2 + a} + y\,{}^y\sqrt{y^2 - a} + a \log \frac{x + \sqrt{x^2 + a}}{y + \sqrt{y^2 - a}} + c$

3. $z = \frac{1}{3}(x + a)^3 + \frac{1}{3}(y - a)^3 + b$

4. $z = ax - \cos x - \cos y - ay + b$

5. $z = ax + \sin x + \frac{1}{a} \sin y + b$

6. $\log z = \frac{2}{3}(x + a)^{3/2} + \frac{2}{3}(y - a)^{3/2} + b$

7. $z^2 = 2a \log xy - x^2 - y^2 + b$

8. $z = 2\sqrt{ax} + \frac{1}{2} ay^2 + b$

9. $z = ae^x + ae^y + b$

10. $z = a[\log(xy) + x + y] + b)$

Clairaut Equation

From IV : Equations of the form $z = px + qy + f(p, q)$

The complete solution of this equation is

$$z = ax + by + f(a, b)$$

which is obtained by replacing p by and q by b in the given Claricaut equation.

Example 11.27. Give the complete solution for $(p - q)(z - px - qy) = 1$

Solution : Rewriting this equation, we get

$$z = px + qy + \frac{1}{p - q}$$

The complete solution is

$$z = ax + by + \frac{1}{a-b}$$

Example **11.28.** Solve : $4xyz = pq + 2px^2y + 2qxy^2$

Solution : Let $x^2 = X$ and $y^2 = Y$

so that
$$p = \frac{\partial z}{\partial x} = \frac{\partial z}{\partial X} \cdot \frac{\partial X}{\partial x} = 2x\frac{\partial z}{\partial X}$$

$$q = \frac{\partial z}{\partial y} = \frac{\partial z}{\partial Y} \cdot \frac{\partial Y}{\partial y} = 2y\frac{\partial z}{\partial Y}$$

$\therefore$ The given equation becomes :

$$4xyz = 4xy\frac{\partial z}{\partial X} \cdot \frac{\partial z}{\partial Y} + 4x^3y\frac{\partial z}{\partial X} + 4xy^3\frac{\partial z}{\partial Y}$$

$$z = x^2\frac{\partial z}{\partial X} + y^2\frac{\partial z}{\partial Y} + \frac{\partial z}{\partial X} \cdot \frac{\partial z}{\partial Y}$$

$$= X\frac{\partial z}{\partial X} + Y\frac{\partial z}{\partial Y} + \frac{\partial z}{\partial X} \cdot \frac{\partial z}{\partial Y}$$

$$z = PX + QY + PQ \quad \text{where} \quad P = \frac{\partial z}{\partial X} \text{ and } Q = \frac{\partial z}{\partial Y}$$

It is of the form

$$z = PX + QY + f(P, Q)$$

It complete solution is :

$$z = aX + bY + ab$$

$$z = ax^2 + by^2 + ab$$

EXERCISE 11.7

1. $pqz = p^2(xq + p^2) + q^2(yp + q^2)$
2. $(p + q)(z - xp - yq) = 1$
3. $z = px + qy - 2\ 2\sqrt{ab}$
4. $z = px + qy + \sin(p + q)$
5. $2q(z - px - qy) = 1 + q^2$
6. $(pq - p - q)(z - px - qy) = pq$

Answers

1. $z = ax + by + p\left(\frac{a^2}{b} + \frac{b^3}{a}\right)$
2. $z = ax + by + \left(\frac{1}{a+b}\right)$
3. $z = ax + by - {}^2\sqrt{ab}$

4. $z = ax + by + \sin(a+b)$

5. $z = ax + by + \dfrac{b^2+1}{2b}$

6. $z = ax + by + \dfrac{ab}{ab - a - b}$

11.9 CHARPIT'S METHOD

This is general method for solving partial differential equation with two independent variables.

Let the general partial differential equation be

$$f(x, y, z, p, q) = 0 \qquad \text{...(1)}$$

Since x depends on x, y we have

$$dz = \frac{\partial z}{\partial x}dx + \frac{\partial z}{\partial y}dy$$

or

$$= pdx + qdy \qquad \text{...(2)}$$

Now, in order to solve (1), we find another relation between x, y, z, p, q say

$$F(x, y, z, p, q) = 0$$

such that when the values of p,q desired from it and equation (1) are substituted in (2) it becomes integrable and this the solution of equation (1).

Now, to find relation (3), we differentiate partially (1) and (3) w.r.t x giving

$$\frac{\partial f}{\partial x} + \frac{\partial f}{\partial z}\cdot\frac{\partial z}{\partial x} + \frac{\partial f}{\partial p}\cdot\frac{\partial p}{\partial x} + \frac{\partial f}{\partial q}\cdot\frac{\partial q}{\partial x} = 0$$

$$\frac{\partial F}{\partial x} + \frac{\partial F}{\partial z}\cdot\frac{\partial z}{\partial x} + \frac{\partial F}{\partial p}\cdot\frac{\partial p}{\partial x} + \frac{\partial F}{\partial q}\cdot\frac{\partial q}{\partial x} = 0, \text{ on differentiating w.r.t. '}x\text{'}$$

or

$$\frac{\partial f}{\partial x} + p\frac{\partial f}{\partial z} + \frac{\partial f}{\partial p}\cdot\frac{\partial p}{\partial x} + \frac{\partial f}{\partial q}\cdot\frac{\partial q}{\partial x} = 0 \qquad \text{...(4)}$$

$$\frac{\partial F}{\partial x} + p\frac{\partial F}{\partial z} + \frac{\partial F}{\partial p}\cdot\frac{\partial p}{\partial x} + \frac{\partial F}{\partial q}\cdot\frac{\partial q}{\partial x} = 0 \qquad \text{...(5)}$$

Similarly, on differentiating (1) and (3) w.r.t 'y' we have

$$\frac{\partial f}{\partial y} + q\frac{\partial f}{\partial z} + \frac{\partial f}{\partial p}\cdot\frac{\partial p}{\partial y} + \frac{\partial f}{\partial q}\cdot\frac{\partial q}{\partial y} = 0 \qquad \text{...(6)}$$

$$\frac{\partial F}{\partial y} + q\frac{\partial F}{\partial z} + \frac{\partial F}{\partial p}\cdot\frac{\partial p}{\partial y} + \frac{\partial F}{\partial q}\cdot\frac{\partial q}{\partial y} = 0 \qquad \text{...(7)}$$

Eliminating $\dfrac{\partial p}{\partial x}$ from (4) and (5), we have

$$\left(\frac{\partial f}{\partial x}\cdot\frac{\partial F}{\partial p} - \frac{\partial f}{\partial p}\frac{\partial F}{\partial p}\right) + p\left(\frac{\partial f}{\partial z}\cdot\frac{\partial F}{\partial p} - \frac{\partial f}{\partial p}\cdot\frac{\partial F}{\partial z}\right) + \frac{\partial q}{\partial x}\left(\frac{\partial f}{\partial q}\cdot\frac{\partial F}{\partial p} - \frac{\partial f}{\partial p}\cdot\frac{\partial F}{\partial q}\right) = 0 \qquad \text{...(8)}$$

Eliminating $\dfrac{\partial q}{\partial y}$ from (6) and (7) we have

$$\left(\frac{\partial f}{\partial y}\cdot\frac{\partial F}{\partial q} - \frac{\partial f}{\partial q}\frac{\partial F}{\partial y}\right) + q\left(\frac{\partial f}{\partial z}\cdot\frac{\partial F}{\partial q} - \frac{\partial f}{\partial q}\cdot\frac{\partial F}{\partial z}\right) + \frac{\partial p}{\partial y}\left(\frac{\partial f}{\partial p}\cdot\frac{\partial F}{\partial q} - \frac{\partial f}{\partial q}\cdot\frac{\partial F}{\partial p}\right) = 0 \qquad \text{...(9)}$$

Since $\frac{\partial q}{\partial x}=\frac{\partial^2 z}{\partial x \partial y}=\frac{\partial^2 z}{\partial y\,\partial x}=\frac{\partial p}{\partial y}$, the last terms in (8) and (9) differ in sign only. Adding (8) and (9), we get,

$$\left(\frac{\partial f}{\partial x}+p\frac{\partial f}{\partial z}\right)\frac{\partial F}{\partial p}+\left(\frac{\partial f}{\partial y}+q\frac{\partial f}{\partial z}\right)\frac{\partial F}{\partial q}+\left(\frac{-p\partial f}{\partial p}-\frac{q\partial f}{\partial q}\right).\frac{\partial F}{\partial z}+\left(-\frac{\partial f}{\partial p}\right).\frac{\partial F}{\partial x}+\left(\frac{-\partial f}{\partial q}\right).\frac{\partial F}{\partial y}=0 \quad ...(10)$$

This is a Lagrange's Linear Equation in x, y, z, p, q as independent variable and F as dependent variable. Its subsidiary equation are :

$$\frac{dx}{-\partial f/\partial p}=\frac{dy}{-\partial f/\partial q}=\frac{dz}{\left(p\frac{\partial f}{\partial p}-q\frac{\partial f}{\partial q}\right)}=\frac{dp}{\left(\frac{\partial f}{\partial x}+p\frac{\partial f}{\partial z}\right)}=\frac{dp}{\frac{\partial f}{\partial y}+q\frac{\partial f}{\partial z}}=\frac{\partial F}{0} \quad ...(11)$$

Any of the integrals of (11) will satisfy (10), such an inegral which involves p or q or both may also be taken as (3) and thus this relation and equation (1) gives values of p and q. The values of p and q are substituted in (2) and on integration we get the solution.

In practice, we choose the simplest of integrals of (11).

Example **11.29.** Solve : $px + qy = pq$

Solution : $f(x, y, z, p, p) = 0$ is $px + qy - pq = 0$ ...(1)

$$\frac{\partial f}{\partial x}=p,\ \frac{\partial f}{\partial y}=q,\ \frac{\partial f}{\partial z}=0,\ \frac{\partial f}{\partial p}=x-q,\ \frac{\partial f}{\partial q}=y-p$$

Charpits equation are :

$$\frac{dx}{-\partial f/\partial p}=\frac{\partial y}{-\partial f/\partial q}=\frac{dz}{-p\frac{\partial f}{\partial p}-q\frac{\partial f}{\partial q}}=\frac{dp}{\frac{\partial f}{\partial x}+p\frac{\partial f}{\partial z}}=\frac{dF}{0}$$

$$\frac{dx}{-(x-q)}=\frac{dy}{-(y-p)}=\frac{dz}{-p(x-q)-q(y-p)}=\frac{dp}{P}=\frac{dp}{q}=\frac{dF}{0}$$

We choose the simplest integrals involving p and q, *i.e.*

$$\frac{dp}{p}=\frac{dp}{q}\Rightarrow \log p = \log q+\log a \Rightarrow p = aq$$

Putting for p in the given equation (1), we get

$$q(ax+y) = aq^2$$

$$\therefore \quad q = \frac{y+ax}{a}$$

$$p = aq = (y+ax)$$

Now $$dz = pdx + qdy$$

Putting for p and q in (2), we get

$$dz = (y+ax)\,dx+\frac{y+ax}{a}dy$$

$$adz = a(y+ax)\,dx+(y+ax)\,dy$$

$$adz = (y+ax)\,(adx+dy)$$

Integrating, we get

$$az = \frac{(y+ax)^2}{2} + b, \text{ which is the required solution.}$$

Example **11.30.** Solve by Charpit's method : $z = p^2x + q^2y$

Solution : Here $f(x, y, z, p, q) = p^2x + q^2y - z$

$$\frac{\partial f}{\partial x} = p^2, \frac{\partial f}{\partial p} = 2\,px, \frac{\partial f}{\partial y} = q^2, \frac{\partial f}{\partial q} = 2qy, \frac{\partial f}{\partial z} = -1$$

Charpits equation are :

$$\frac{dx}{\partial f/\partial p} = \frac{\partial y}{\partial f/\partial q} = \frac{dz}{p\frac{\partial f}{\partial p} + q\frac{\partial f}{\partial q}} = \frac{-dp}{\frac{\partial f}{\partial x} + p\frac{\partial f}{\partial z}} = \frac{-dp}{\frac{\partial f}{\partial y} + p\frac{\partial f}{\partial z}}$$

$$\frac{dx}{2\,px} = \frac{dy}{2qy} = \frac{dz}{2p^2x + 2q^2y} = \frac{-dp}{p^2 + p(-1)} = \frac{-dp}{q^2 + q(-1)}$$

$$\frac{dx}{2\,px} = \frac{dy}{2\,qy} = \frac{dz}{2(p^2x + q^2y)} = \frac{-dp}{p^2 - p} = \frac{-dp}{q^2 - q}$$

From (1) we get

$$\frac{\frac{dx}{2x} - \frac{dy}{2y}}{p-q} = \frac{\frac{-dp}{p} + \frac{dp}{q}}{p-q}$$

or

$$\frac{dx}{2x} - \frac{dy}{2y} = \frac{-dp}{p} + \frac{dp}{q}$$

Integrating in get $\frac{1}{2}\log\frac{x}{y} = \log\frac{q}{p} + \log a$

or

$$\sqrt{\frac{x}{y}} = \frac{q}{P}a$$

$$q = \frac{P}{a}\sqrt{\frac{x}{y}}$$

Substituting in equation

$$z = q^2x + q^2y$$

$$z = p^2x + \frac{p^2}{a^2}\cdot\frac{x}{y}\,y = p^2\left[1 + \frac{1}{a^2}\right].x$$

$$p^2 = \frac{za^2}{x(1+a^2)} \Rightarrow p = \frac{\sqrt{z}\,a}{\sqrt{x}\sqrt{a^2+1}}$$

$$q = \frac{p}{a}\sqrt{\frac{x}{y}} = \frac{\sqrt{z}\,a}{a\sqrt{x}\sqrt{a^2+1}} \times \sqrt{\frac{x}{y}}$$

$$= \frac{\sqrt{z}}{\sqrt{y}}\,\frac{1}{\sqrt{a^2+1}}$$

Substituting in $$dz = pdx + qdy$$

$$dz = \frac{\sqrt{z}}{\sqrt{x}}\frac{a}{\sqrt{a^2+1}}dx + \frac{\sqrt{2}}{\sqrt{y}}\frac{\sqrt{z}}{\sqrt{a^2+1}}dy$$

$$\sqrt{a^2+1}\,\frac{dz}{\sqrt{z}} = a\frac{dx}{\sqrt{x}} + \frac{1}{\sqrt{y}}dy$$

Integrating, we get

$$\sqrt{a^2+1}\,\frac{\sqrt{z}}{1/2} = \frac{a\sqrt{x}}{1/2} + \frac{\sqrt{y}}{1/2} + \frac{c}{1/2}$$

$$\sqrt{(a^2+1)z} = {}^{a}\sqrt{x} + \sqrt{y} + c$$

***Examples* 11.31.** Solve $z^2 = pq\,xy$

Solution : Here $f(x, y, z, p, q) = pqxy - z^2$

$$\frac{\partial f}{\partial x} = pqy, \frac{\partial f}{\partial y} = qpx, \frac{\partial f}{\partial z} = -2z, \frac{\partial f}{\partial p} = q\,xy, \frac{\partial f}{\partial q} = P\,xy$$

Subsidiary Equations are :

$$\frac{dx}{-qxy} = \frac{dy}{-pxy} = \frac{dz}{-p(qxy) - q(xyp)} = \frac{dp}{pqy + p(-2z)} = \frac{dp}{pqx + q(-2z)}$$

$$\frac{dx}{q\,xy} = \frac{dy}{p\,xy} = \frac{dz}{2\,pq\,xy} = \frac{-dp}{p\,qy - 2\,zp} = \frac{-dp}{p\,qx - 2\,zp}$$

$$\frac{\frac{dx}{x}}{qy} = \frac{\frac{dy}{y}}{px} = \frac{dz}{2\,pq\,xy} = \frac{\frac{dp}{p}}{qy - 2z} = \frac{\frac{-dq}{q}}{px - 2z}$$

From first two and last two, we get

$$\frac{\frac{dx}{x} - \frac{dy}{y}}{qy - px} = \frac{\frac{-dp}{p} + \frac{dp}{q}}{qy - px}$$

i.e., $$\frac{dx}{x} - \frac{dy}{y} = \frac{-dp}{p} + \frac{dp}{q}$$

Integrating $\log\frac{x}{y} = \log\frac{q}{p} + \log a$

$$\frac{x}{y} = \frac{q}{p}a;\ p = \frac{qy\,a}{x}$$

Substituting in given equation :

$$z^2 = \left(\frac{qy\,a}{x}\right)q\,xy$$

$$z^2 = q^2.y^2.\,a \text{ or } q^2 = \frac{z^2}{y^2 a}$$

$$q = \frac{z}{y\sqrt{a}}$$

$$\therefore \qquad p = \frac{\sqrt{a}\,z}{x}$$

Substituting p and q in $pdx + pdy = dz$, we get

$$dz = \frac{\sqrt{a}\,.\,z}{x}dx + \frac{z}{y\sqrt{a}}dy$$

$$\frac{dz}{z} = \sqrt{a}\frac{1}{x}dx + \frac{1}{\sqrt{a}}\frac{dy}{y}$$

Integrating, $\qquad \log z = \sqrt{a}\log x + \frac{1}{\sqrt{a}}\log y + \log c$

or $\qquad z = x^{\sqrt{a}}\,y^{1\sqrt{a}}\,c$ which is required solution.

***Example* 11.32.** Solve Charpits Method :

$$2z + p^2 + qy + 2y^2 = 0$$

Solution : $f(x, y, z, p, q) = 2z + p^2 + qy + 2y^2$

$$\frac{\partial f}{\partial x} = 0, \frac{\partial f}{\partial y} = q + 4y, \frac{\partial f}{\partial z} = 2, \frac{\partial f}{\partial p} = 2p, \frac{\partial f}{\partial q} = y$$

Subsidiary equation are :

$$\frac{dx}{-2p} = \frac{dy}{-y} = \frac{dz}{-2p^2 + qy} = \frac{dp}{2p} = \frac{dp}{4y + 3p}$$

Considering : $\qquad \frac{dx}{-2p} = \frac{dp}{2p}$

$\Rightarrow \qquad -dx = dp$

On integration, we get

$$p = -x + a$$

Substituting is given equation :

$$q = \frac{1}{y}[-2z - 2y^2 - (a - x)^2]$$

$$dz = pdx + qdy$$

$$dz = (a - x)\,dx - \frac{1}{y}\,[2z + 2y^2 + (a - x)^2]dy$$

Multiplying both sides by $2y^2$, we get

$$2y^2dz = 2y^2\,(a - x)\,dx - 2y\,[2z + 2y^2 + (a - x)^2]dy$$

$$2y^2\,dz + 4yz\,dy = 2y^2\,(a - x)\,dx - 2y\,(a - x)2\,dy - 4y^3\,dy$$

$$d(2\,zy^2) = -d[y^2\,(a - x)^2] - 4y^3\,\text{dy}$$

Integrating, we get

$$2zy^2 = -\,y^2\,(a - x)^2 - y^2 + b$$

or $\qquad y^2\,[2z + (x - a)2 + y^2] = b$ is the required solution.

EXERCISE 11.8

1. $(p^2 + q^2) = z$
2. $pxy + qp + qy = yz$
3. $2zx - px^2 - 2pxy + qp = 0$
4. $q + xp = p^2$
5. $p = (z + qy)^2$
6. $1 + p^2 = qz$
7. $p\,(1 + q^2) + (b - z)\,q = 0$

Answers

1. $4z\,(1 + a^2) = (x + ay + b)^2$
2. $\log\,(z - ax) = y - a\log\,(a + y) + b$
3. $z - ay = b\,(x^2 - a)$
4. $z = axe^{-y} - \frac{1}{2}a^2 e^{-2y} + b$
5. $yz = ax + \sqrt{ay} + b$
6. $\left[\frac{z^2}{z} \pm \left(\frac{z^2}{2}\sqrt{a^2 - 4a^2} - 2a^2 \log\,(z + \sqrt{z^2 - 4a^2})\right)\right]$
7. $2\sqrt{C(z - b) - 1} = x + Cy + a$

11.10 APPLICATION OF SECOND ORDER PARTIAL DIFFERENTIAL EQUATIONS

1. Equation of Vibrating string or wave equation $\frac{\partial^2 y}{\partial t^2} = c^2\,\frac{\partial^2 y}{\partial x^2}$

2. One dimensional heat flow equation $\frac{\partial u}{\partial t} = c^2\,\frac{\partial^2 u}{\partial x^2}$.

11.11 METHOD OF SEPARATION OF VARIABLES

In this method we assume the dependent variable to be a product of two functions each of which involves only one of the independent variables. So two ordinary differential equation are formed. The following examples will make the method clear.

Example **11.33.** Using the method of separation of variables, solve $\frac{\partial u}{\partial x} = 2\frac{\partial u}{\partial t} + u$, where$(x, 0) = 6e^{-3x}$

Solution :

$$\frac{\partial u}{\partial x} = 2\frac{\partial u}{\partial t} + u \qquad ...(1)$$

Let

$$u = X(x)\,T(t) \qquad ...(2)$$

where X is the function of x only and T is function of t only. Putting the value of u in (1), we get

$$\frac{\partial (XT)}{\partial x} = 2\frac{\partial}{\partial T}\,(X.T) + X.\,T$$

$$T\frac{dX}{dx} = 2\,X\,\frac{dT}{dt} + X.T$$

$$TX' = 2XT' + X.T$$

Separating the variables, we get

$$TX' = (2T' + T)X$$

$$\frac{X'}{X} = \frac{2T' + T}{T}$$

$$= 2\frac{T'}{T} + 1 = x \text{ (say)}$$

(*a*) $\frac{X'}{X} = k$ on integrating $\log X = kr + \log c_1$

$$\therefore \quad X = c_1 e^{kx}$$

(*b*) $$\frac{2T'}{T} + 1 = \text{k or } \frac{T'}{T} = \frac{1}{2}(k-1)$$

Integrating

$$\log T = \frac{1}{2}(k-1)t + \log c_2$$

$$T = c_2 \, e^{1/2(k-1)t}$$

$$\therefore \quad u = XT = c_1 e^{kx} \, c_2 e^{1/2(k-1)t}$$

$$= Kx = \frac{1}{2}(K-1)t$$

$$u(x, t) = c_1 c_2 e$$

But $$u(x, 0) = 6e^{-3x}$$

$$\therefore \quad c_1 c_2 = 6, \text{ k} = -3 \qquad \text{...(3)}$$

Putting the values of $c_1 c_2$ and k in (3), we get

$$u(x, t) = 6e^{-3x + \frac{1}{2}(-3-t)t}$$

$$= 6\, e^{-3x-2t}$$

***Example* 11.34.** Use the method of separation of variables to solve equation $py^3 + qx^2 = 0$

Solution : $py^3 + qx^2 = 0$

$$\frac{\partial z}{\partial x} y^3 + \frac{\partial z}{\partial y} x^2 = 0$$

Assume $$Z = X(x)\, Y(y)$$

$$\frac{\partial z}{\partial x} = X'Y$$

$$\frac{\partial z}{\partial y} = XY'$$

Putting in equation (1), we get

$$X'Y.\, y^3 + XY'\, x^2 = 0$$

$$X'Y\, y^3 = -XY'\, x^2$$

Separating the variables

$$\frac{X'}{Xx^2} = -\frac{Y'}{Yy^3} = K \text{ (say)}$$

(a) $$\frac{X'}{X} = Kx^2$$

Integrating w.r.t. x

$$\log X = K\frac{x^3}{3} + \log c_1$$

$$X = c_1 e^{\frac{k}{3}x^3}$$

(b) $$-\frac{Y'}{Yy^3} = k$$

$$\frac{Y'}{Y} = -ky^3$$

Integrating w.r.t. y

$$\log Y = -k\frac{y^4}{4} + \log c_2$$

$$Y = c_2 e{-k}\frac{y^4}{4}$$

$$Z = X.\,Y = c_1\,c_2\,e^{+\frac{k}{3}x^3 - \frac{k}{4}y^4}$$

$$= c_1\,c_2\,e^{\frac{k}{12}[4x^2 - 3y^4]} = c_1 c_2\, e^{a}[4x^3 - 3y^4]$$

$$(where = \frac{k}{12})$$

***Examples* 11.35.** Using method of separation of variable solve the equation $\frac{\partial u}{\partial x} + \frac{\partial u}{\partial y} = 3u$ given $u = 3e^{-y} - e^{-5y}$ when $x = 0$.

Solution : The equation is

$$\frac{\partial u}{\partial x} + \frac{\partial u}{\partial y} = 3u \quad ...(1)$$

Let $$u = X(x)\,Y(y)$$

is where X is a function of x alone Y is a function of y alone

$$\frac{\partial u}{\partial x} = X'Y$$

and $$\frac{\partial u}{\partial y} = XY''$$

Substituting is equation (1), we get

$$X'Y + XY' = 3XY$$

or $$X'Y = X(3\,Y - Y')$$

Separating the variables

$$\frac{X'}{X} = \frac{3Y - Y'}{Y} = 3 - \frac{Y'}{Y} = k \text{ (say)}$$

(*a*) $\frac{X'}{X} = k$

Integrating $\log X = kx + \log c_1$

$$X = c_1 e^{kx}$$

(*b*) $3 - \frac{Y'}{Y} = k$

$$\frac{Y'}{Y} = 3 - k$$

Integrating $\log Y = (3 - k)\, y + \log c_2$

$$Y = c_2\, e^{(3-k)y}$$

$$u = XY = c_1\, c_2\, e^{Kx + (3-K)Y}$$

or $u(x, y) = c_1\, c_2\, e^{kx} + (3 - k)y$ ...(2)

Given $u = 3e^{-y} - e^{-5y}$ when $x = 0$ ...(3)

From (2) putting $u(0, y) = c_1\, c_2\, e^{(3-k)}y$

Also $u(0, y) = 3e^{-y} - e^{-5y}$ (given)...(4)

Comparing (3) and (4), we get

$c_1\, c_2 = 3$ $\quad 3 - k = -1 \Rightarrow \quad c_1 c_2 = 3. \quad k = 4.$

$c_1\, c_2 = -1$ $\quad 3 - k = -5 \Rightarrow \quad c_1 c_2 = -1, \quad k = 8$

Hence $u(x, y) = 3\, e^{4x} + (3 - 4)\, y - e^{8x} + (3 - 8)\, y$

$$= 3e^{4x - y} - e^{8x - 5y}$$

which is the required solution

EXERCISE 11.9

Solve the following equation by the method of separation of variables

1. $\frac{\partial^2 z}{\partial x^2} - 2\frac{\partial z}{\partial x} + \frac{\partial z}{\partial y} = 0$

2. $4\frac{\partial u}{\partial t} + \frac{\partial u}{\partial x} = 3u,\ u = 3e^{-x} - e^{-5x},\ t = 0$

3. $\frac{\partial u}{\partial t} = 4\frac{\partial^2 u}{\partial x^2}\ u\,(x, 0) = 4x - \frac{1}{2}x^2$

4. $\frac{\partial^2 y}{\partial x^2} = \frac{\partial u}{\partial y} + 2u,\ u = 0,\ \frac{\partial u}{\partial x} = 1 + e^{-3y},\ x = \text{for all } y.$

5. $2\frac{\partial u}{\partial x} + 3\frac{\partial u}{\partial y} + 5u = 0,\ u(0, y) = 2e^{-y}$

6. $\frac{\partial^2 u}{\partial x^2} + 3\frac{\partial u}{\partial t}$ if $u(x, 0) = x^2 (25 - x^2)$

7. $\frac{\partial^2 u}{\partial x \, \partial y} = \sin x \sin y$, given $u_y = -2 \sin y$ at $x = 0$ and $u = 0$ when y is an odd multiple of $\frac{\pi}{2}$

Answers

1. $z = \left[c1e^{[1+\sqrt{1+k}]}x + c_2 e^{[1-\sqrt{1+k}]x}\right]c_3 e^{-ay}$
2. $u = 3e^{t-x} - e^{2t-5x}$
3. $u = \left(4x - \frac{x^2}{2}\right)e^{-p^2}t$
4. $u = \frac{1}{\sqrt{2}} \sinh \sqrt{2}x + e^{-3y} \sin x$
5. $u = 2e^{-x-y}$
6. $u = x^2 (25 - x^2)\, e^c - p^3 t$
7. u = $\cos y\,(1 + \cos x)$

11.12 WAVE EQUATION-VIBRATION OF STRETCHED STRING

Let us consider an elastic strip of length '*l*' lightly stretched between two point A and B. Let us study the transverse vibration of the string when no external forces acts forces acts on it. Take end A as the origin and string in the equilibrium position as x-axis and a line through A and perpendicular to AB as y-axis. Now displace the string slightly in perpendicular directions and then released, the string will then vibrate in x-y plane. When the string vibrate in xy plane, the displacement y of any point of the string is a function of x and time t. Let $P(x, y)$, $Q(x + dx, y + dy)$ be two neighbouring points on the string. Let ψ and $\psi + \delta\psi$ be the angles made by the tangent at P and Q respectively within x-axis. Let m be the mass per unit length of string which is homogenous. Consider the motion of the infinitesimal clement PQ of the string. The vertical component of the force which the element is subjected to is

$$T \sin (\psi + \delta\psi) - T \sin \psi = T[\{y + \delta\psi\}] = T\,\delta\psi \text{ (approx)} \quad ...(1)$$

where T is the tension caused by stretched the streching between the points A and B. Tension T is very large compared with of the string and hence the gravitational force may be neglected.

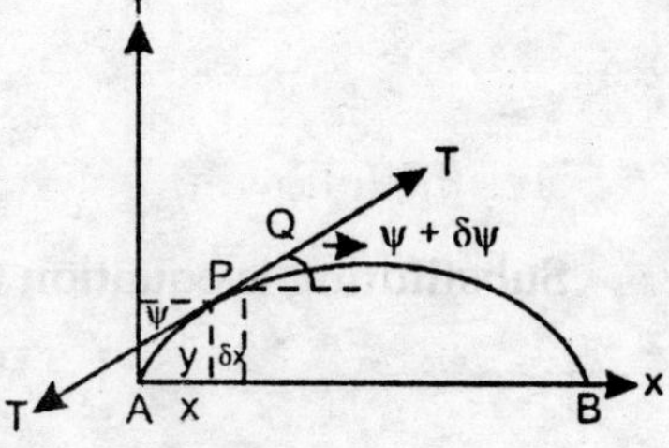

Fig. 11.1

Assumptions :

1. The effect of friction to be negligible
2. The string to be prefectly flexible. (It can transmit only tension but not bending or sharing)
3. The slope of deflection to be small at all point and at all times.

Now the acceleration of the elements in the direction of y-axis is $\frac{\partial^2 y}{\partial t^2}$. If the length PQ is δs then the mass of PQ is $m.\delta s$.

Hence by the second law of motion, equation of motion becomes

$$m\delta s \frac{\partial^2 y}{\partial t^2} = T\,\delta\psi \quad \text{[from equation (1)]}$$

$$\frac{\partial^2 y}{\partial t^2} = \frac{T}{m}\frac{\delta\psi}{\delta s}$$

which is called one dimensional wave equation.

As $Q \to P$, $\delta s \to 0$, therefore, taking limits as $\delta s \to 0$, the above equation becomes

$$\frac{\partial^2 y}{\partial t^2} = \frac{T}{m}\frac{\delta \psi}{\delta s}, \text{ where}$$

$$\frac{d\psi}{ds} = \text{curvature at } P \text{ of the deflection of curve}$$

$$\frac{\partial^2 y}{\partial t^2} = \frac{T}{m}\frac{\frac{\partial^2 y}{\partial x^2}}{\left[1+\left(\frac{\partial y}{\partial x}\right)^2\right]^{3/2}}$$

$$= \frac{T}{m}\cdot\frac{\partial^2 y}{\partial x^2} \text{ (approx) since } \left(\frac{\partial y}{\partial x}\right)^2 \text{ is negligible}$$

Putting $\frac{T}{m} = a^2$, we get

$$\frac{\partial^2 y}{\partial t^2} = a^2\frac{\partial^2 y}{\partial x^2}$$

which is called one dimensional wave equation.

11.13 SOLUTION OF WAVE EQUATION

The wave equation is $\frac{\partial^2 y}{\partial t^2} = a^2\frac{\partial^2 y}{\partial x^2}$ ...(1)

Assuming $y = X(x)\,T(t)$

where X is a function of x alone T is a function of t alone

Now $\frac{\partial^2 y}{\partial t^2} = XT'$

$$\frac{\partial^2 y}{\partial x^2} = X''T$$

Substituting is equation (1), we get

$$XT'' = a^2X''T$$

$$\frac{X''}{X} = \frac{1}{a^2}\frac{T''}{T} = K \text{ (say)} \quad ...(2)$$

(L.H.S. being a function of x along and R.H.S. being a function of t alone the equation (2) holds it each side is constant.)

$$X'' - kX = 0 \quad ...(3)$$

$$T'' - ka^2T = 0 \quad ...(4)$$

which are ordinary differential equations, solving (3) and (4), we get

(*i*) When k is positive and $= p^2$,

$$X = c_1 e^{px} + c_2 e^{-px}$$

$$T = c_3 e^{cpt} + c_4 e^{-cpt}$$

(*ii*) When k is negative and $-p^2$, $X = c_5 \cos px + c_6 \sin px$

$$T = c_1 \cos cpt + c_8 \sin cpt$$

(*iii*) When k = 0 $X = c_9 x + c_{10}$

$$T = c_{11} t + c_{12}$$

Thus the various possible solution are

$$y = (c_1 e^{vx} + c_2 e^{-px})(c_3 e^{cpt} + c_4 e^{-cpt}) \quad ...(5)$$

$$y = (c_5 \cos px + c_6 \sin px)(c_7 \cos cpt + c_8 \sin cpt) \quad ...(6)$$

$$y = (c_9 x + c_{10})(c_{11} t + c_{12}) \quad ...(7)$$

of these solutions, we have to choose that solution which is consistent with the physical nature of the problem. As the wave is periodic. so the only possible solution is (6).

i.e., $$y = (c_5 \cos px + c_6 \sin px)(c_7 \cos cpt + c_8 \sin cpt)$$

$$y = (c_1 \cos px + c_2 \sin px)(c_3 \cos cpt + c_4 \sin cpt)$$

Example **11.36.** A slightly stretched flexible string has its ends fixed at $x = 0$ and $x = 'l'$. At time $t = 0$, the string is given a shape defined by $f(x) = \mu x(l - x)$ where μ is a constant and then released. Find the displacement of any point x of the string at any time $t > 0$.

Solution : The wave equation is $\frac{\partial^2 y}{\partial t^2} = a^2 \frac{\partial^2 y}{\partial x^2}$ ...(1)

Solution of this equation is

$$y(x, t) = y = (c_1 \cos px + c_2 \sin px)(c_3 \cos cpt + c_4 \sin cpt) \quad ...(2)$$

Boundary condition are

$$y(0, t) = 0, y(l, t) = 0$$

$$f(x) = \mu x(l - x), \left(\frac{dy}{dt}\right)_{t=0} = 0$$

Applying condition $y(0, t) = 0$ to equation (2), we get

$$0 = c_1(c_3 \cos apt + c_4 \sin apt)$$

$$\Rightarrow \quad c_1 = 0$$

The reduced solution is

$$y(x, t) = c_2 \sin px\, (c_3 \cos apt + c_4 \sin apt) \quad ...(3)$$

Applying condition $y(l, t) = o$ to equation (3), we get

$$0 = c_2 \sin pl\, (c_3 \cos apt + c_4 \sin \text{apt})$$

$$\Rightarrow \quad \sin pl = 0, c_2 \neq 0$$

$$\therefore \quad pl = n\pi \text{ and } p = \frac{n\pi}{l}$$

The reduced solution becomes

$$y(x, t) = c_2 \sin \frac{n\pi}{l} x\left(c3 \cos \frac{an\pi}{l} + c4 \sin \frac{an\pi}{l}\right) \quad ...(4)$$

Apply the condition $\left(\frac{\partial y}{\partial t}\right)_{t=0} = 0$

For equation (4), we get

$$\frac{\partial y}{\partial t} = c_2 \sin \frac{n\pi}{l} x \left(-c_3 \frac{n\pi a}{l} \sin \frac{n\pi a}{l} t + c_4 \frac{n\pi a}{l} \cos \frac{n\pi a}{l} t\right)$$

Put $t = 0$

$$0 = c_2 \sin \frac{n\pi x}{l} \left(c_4 \frac{n\pi a}{l}\right)$$

$$\Rightarrow \qquad c_4 = 0$$

The reduced solution is

$$y(x, t) = c_2 c_3 \sin \frac{n\pi x}{l} \cos \frac{n\pi a}{l} t$$

$$= b_n \sin \frac{n\pi x}{l} \cos \frac{n\pi a}{l} t \qquad (b_n = c_2 c_3)$$

For different values of n we have different solution.

$$y(x, t) = \sum b_n \sin \frac{n\pi x}{l} \cos \frac{n\pi a}{l} t \qquad \ldots(5)$$

Apply the condition $f(x) = \mu x (l - x)$ or $y(x, 0) = \mu_n (l - x)$

The equation (5), we get

$$y(x, 0) = \sum b_n \sin \frac{n\pi x}{l}$$

which is Fourier range sin series

$$\mu_n (l - x) = \Sigma b_n \sin \frac{n\pi x}{l}$$

$$\therefore \qquad b_n = \frac{2}{l} \int_0^l \mu x (l - x) \sin \frac{n\pi x}{l} dx$$

$$= \frac{2\mu}{l}\left[(lx - x^2)\left\{\frac{-\cos\left(\frac{n\pi x}{l}\right)}{\frac{n\pi}{l}}\right\} + (l - 2x)\left\{\frac{\sin \frac{n\pi x}{l}}{\left(\frac{n\pi}{l}\right)^2}\right\} + (-2)\left\{\frac{\cos \frac{n\pi x}{l}}{\left(\frac{n\pi}{l}\right)^3}\right\}\right]$$

$$= \frac{2\mu}{l}\left[\frac{-2l^3}{n^3\pi^3} \cos np + \frac{2l^3}{n^3\pi^3}\right]$$

$$= \frac{4\mu}{n^3\pi^3 l} l^3 [1 - (-1)^n] = \frac{4\mu}{n^3\pi^3 l} l^3 [1 - (-1)^n]$$

$$= \begin{pmatrix} 0 \text{ when } n \text{ is even} \\ \frac{8\mu l^2}{n^3\pi^3} \text{ when } n \text{ is odd} \end{pmatrix}$$

$$= \frac{8\mu l^2}{(2m - 1)^3 \pi^3}$$

where $\qquad n = 2m - 1$ (n being odd)

The required solution is

$$y = \frac{8\mu l^2}{\pi^3} \Sigma \frac{1}{(2m-1)^3} \sin\left\{\frac{(2m-1)\pi x}{l}\right\} \cos\left\{\frac{(2m-1)\pi a t}{l}\right\}$$

Example **11.37.** Find the solution of the wave equation $\frac{\partial^2 y}{\partial t^2} = c^2 \frac{\partial^2 u}{\partial x^2}$ corresponding to the triangular initial deflection.

$$f(x) = \frac{2k}{l} x \text{ when } 0 < x < \frac{1}{2}$$

$$= \frac{2k}{l}(l - x),\ l/2 < x < l \text{ and initial velocity zero.}$$

Solution : The wave equation is $\frac{\partial^2 y}{\partial t^2} = c^2 \frac{\partial^2 u}{\partial x^2}$ ...(1)

Since wave motion is periodic, the solution of equation (1) will be of the form

$$y(x, t) = (c_1 \cos px + c_2 \sin px)(c_3 \cos cpt + c_4 \sin cpt) \quad ...(2)$$

Now the boundary conditions are

$$y(0, t) = 0 \quad ...(3)$$

$$y(l, t) = 0 \quad ...(4)$$

The initial conditions are

$$\left(\frac{\partial y}{\partial t}\right)_{t=0} = 0 \quad ...(5)$$

$$f(x) = \frac{2k}{l} x, 0 < x < \frac{1}{2} \quad ...(6)$$

$$= \frac{2k}{l}(l - x), \frac{l}{2} < x > l$$

As in the previous example, the solution of (*i*) satisfying boundary conditions (3), (4), (5) is

$$y(x, t) = b_n \sin \frac{n\pi x}{l} \cos \frac{cn\pi}{l} t \qquad (b_n = c_2 c_3)$$

Adding all these solution for different values of *n*, we get a general solution.

$$y(x, t) = \sum b_n \sin \frac{n\pi r}{l} \cos \frac{n\pi c}{l} t$$

Now apply the condition (6), we get

$$f(x) = \sum b_n \sin \frac{n\pi t}{l}$$

$$= \frac{2}{l}\left[\int_0^{l/2} \frac{2kx}{l} \sin \frac{n\pi x}{l} dx + \int_{l/2}^{l} \frac{2k}{l}(l - x) \frac{n\pi x}{l} dx\right]$$

$$= \frac{4k}{l^2}\left[\int_0^{l/2} x \sin \frac{n\pi x}{l} dx + \int_{l/2}^{l} (1 - x) \sin \frac{n\pi x}{l} dx\right]$$

$$= \frac{4k}{l}\left[\left\{-lx \frac{\cos np(x/l)}{n\pi} + \frac{l^2}{n^2\pi^2} \sin \frac{n\pi x}{l}\right\}_0^{l/2} + \left\{(l - x)\frac{1}{n\pi} \cos \frac{n\pi r}{l} - \frac{l^2}{n^2\pi^2} \sin \frac{n\pi x}{l}\right\}_{l/2}^{t}\right]$$

$$= \frac{8k}{n^2\pi^2} \sin \frac{n\pi}{2}$$

Hence the situation of (*i*) is

$$y(x, t) = \sum \frac{8k}{n^2\pi^2} \left(\cos \frac{n\pi ct}{l} \sin \frac{n\pi x}{l}\right)$$

***Example* 11.38.** A string is stretched and fastened to tow point '*t*' apart. Motion is started by displacing the string in the form. $y = a \sin\left(\frac{n\pi}{l}\right)$ from which it is released at time $t = 0$. Show that the displacement of any point at a distance x from one end at time 't' given by $y(x, t) = a \sin\left(\frac{\pi x}{l}\right) \cos\left(\frac{\pi cl}{l}\right)$.

Solution. The vibration of the string is given by

$$\frac{\partial^2 y}{\partial t^2} = c^2 \frac{\partial^2 y}{\partial x^2} \quad ...(1)$$

The boundary condition are

$$y(0, t) = 0 \quad ...(2)$$

As the end points of the string are fixed

$$y(l, t) = 0 \quad ...(3)$$

Also initially the transverse velocity of any point is zero

$$\left(\frac{\partial y}{\partial t}\right)_{t=0} = 0 \quad ...(4)$$

and

$$y(x, 0) = a \sin \frac{\pi x}{l} \quad ...(5)$$

Now we apply these initial and boundary conditions to solve the problem. Since the vibration of string is periodic therefore the solution of equation (1) is of the from.

$$y(x, t) = (c_1 \cos px + c_2 \sin px)(c_1 \cos cpt + c_4 \sin cpt)$$

Apply $y(0, t) = 0$, we get

$$0 = c_1(c_1 \cos cpt + c_1 \sin cpt) \Rightarrow c_1 = 0$$

The reduced solution now is

$$y(x, t) = c_2 \sin px\,(c_3 \cos cpt + c_4 \sin cpt)$$

Apply $(l, t) = 0$, we get

$$0 = c_2 \sin pl\,(c_3 \cos cpt + c_4 \sin cpt)$$

$$\Rightarrow \quad \sin pl = 0$$

x being integral

$$pl = n\pi \; p = \frac{n\pi}{l}$$

The reduced solution is

$$y(x, t) = c_2 \sin \frac{n\pi x}{l}\left(c_3 \cos \frac{cn\pi t}{l} + c_4 \sin \frac{cn\pi t}{l}\right)$$

Apply $\left(\frac{\partial y}{\partial t}\right)_{t=0} = 0$ we get

$$\frac{\partial y}{\partial t} = c_2 \sin \frac{n\pi x}{l}\left[-\frac{cn\pi}{l} c_3 \sin \frac{cn\pi}{l} + \frac{cn\pi}{l} c_4 \frac{\cos n\pi t}{l}\right]$$

$$\left(\frac{\partial y}{\partial t}\right)_{t=0} = 0 = c_2 \sin \frac{n\pi x}{l}\left[\frac{cn\pi}{l} c_4\right]$$

$$o = c_2 c_4 \sin \frac{n\pi x}{l} \cdot \frac{cn\pi}{l} = c_4 = 0$$

The reduced solution is

$$y(x, t) = c_2 \sin \frac{n\pi x}{l} c_3 \cos \frac{cnx}{l} t$$

$$= c_2 c_3 \sin \frac{n\pi x}{l} c_3 \cos \frac{cn\pi}{l} l$$

$$= b_n \sin \frac{n\pi x}{l} \cos \frac{cn\pi}{l} t \qquad \text{(where } b_n = c_2 c_3\text{)}$$

For different values of n, we have different solution.
Adding all these solution we get a general solution.

$$y(x, t) = \sum b_n \sin \frac{n\pi x}{l} \cos \frac{cn\pi}{l} t$$

Applying $y = a \sin \frac{\pi x}{l}$ to the above equation, we get

$$a \sin \frac{\pi x}{l} = \sum b_n \sin \frac{m\pi x}{l}$$

which is satisfied if we take $b_n = a$, $n = 1$
Hence the required solution is

$$y(x, t) = \sum a \sin \frac{\pi x}{l} \cos \frac{c\pi}{l} t$$

***Example* 11.39.** The points of trisection of a string are pulled aside through the same distance on opposite side of the position of equilibrium and the string is released from rest. Derive an expression for the displacement of the string at subsequent time and show that the mind point of the string remains at rest.

Solution. Let B, C be the points of trisection of the string OA = l. Initially the string is held in the form OB'C'A where BB' = CC' = a (say). The displacement $y(x, t)$ of any point of string is given by:

$$\frac{\partial^2 y}{\partial t^2} = c^2 \frac{\partial^2 y}{\partial x^2} \qquad ...(1)$$

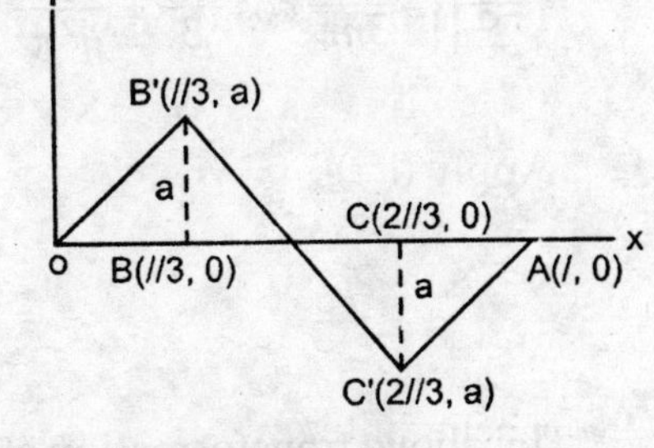

Fig. 11.2

and, the boundary condition are

$$y(0, t) = 0 \qquad ...(2)$$

$$y(l, t) = 0 \qquad ...(3)$$

$$\left(\frac{\partial y}{\partial t}\right)_{t=0} = 0 \qquad ...(4)$$

The fourth condition is obtained as below

$$\text{equation of B}' \text{ is } y = \frac{3a}{l} x$$

$$\text{equation of B}'\text{C is } y = \frac{3a}{l}(l - 2x)$$

$$\text{equation of C}'\text{A is } y = \frac{3a}{l}(x - l)$$

$\therefore$ The fourth condition is $y(x, 0) = \frac{3a}{l} x \qquad 0 \leq x \leq \frac{1}{3}$

$$= \frac{3a}{l}(l - 2x) \qquad \frac{1}{3} < x < \frac{2l}{3}$$

$$= \frac{3a}{l}(x-l) \qquad \frac{2l}{3} < x < l \qquad \text{...(5)}$$

As before on applying (2), (3), (4) condition to the solution

$$y(x, t) = (c_1 \cos px + c_2 \sin px)(c_3 \cos cpt + c_4 \sin cpt)$$

Of equation (1), we get the reduced solution

$$y(x, t) = b_n \sin \frac{n\pi x}{l} \cos \frac{n\pi ct}{l}$$

Adding all such solution for different values of n we get general solution.

$$y(x, t) = \sum_{n=l}^{\infty} b_n \sin \frac{n\pi x}{l} \cos \frac{n\pi cl}{l} \qquad \text{...(6)}$$

Apply fourth condition to this equation, we get, putting $t = 0$

$$y(x, 0) = \sum_{n=l}^{\infty} b_n \frac{n\pi x}{l}$$

$$\therefore \qquad b_n = \frac{2}{l}\left[\int_0^l y(x,0) \sin \frac{\pi x}{l}\, dx\right]$$

$$= \frac{2}{l}\left[\int_0^{l/3} \frac{3ax}{l} \sin \frac{\pi x}{l} dx + \int_{l/3}^{2l/3} \frac{3a}{l}(1-2x) \sin \frac{n\pi x}{l} dx + \int_{2l/3}^{l} \frac{3a}{l}(x-l) \sin \frac{n\pi x}{l} dx\right]$$

$$= \frac{6a}{l^2}\left[x\left\{-\frac{\cos \frac{n\pi x}{l}}{\frac{n\pi}{l}}\right\} + l\left\{\frac{\sin \frac{n\pi x}{l}}{\left(\frac{n\pi}{l}\right)^2}\right\}\right]_0^{l/3} + \left[(1-2x)\left\{-\frac{\cos \frac{n\pi x}{l}}{\frac{n\pi}{l}}\right\} - (-2)\left\{\frac{\sin \frac{n\pi x}{l}}{\left(\frac{n\pi}{l}\right)^2}\right\}\right]_{l/3}^{2l/3} + \left[(x-1)\left\{\frac{\cos \frac{n\pi x}{l}}{\frac{n\pi}{l}}\right\} - 1\left\{\frac{\sin \frac{n\pi x}{l}}{\left(\frac{n\pi}{l}\right)^2}\right\}\right]_{2l/3}^{l}$$

$$= \frac{6a}{l^2}\left[\left\{-\frac{l^2}{3n\pi} \cos \frac{n\pi}{3} + \frac{l^2}{n^2\pi^2} \sin \frac{n\pi}{3}\right\} + \left\{\frac{l^2}{3n\pi} \cos \frac{2\pi x}{3} - \frac{2l^2}{n^2\pi^2} \sin \frac{2n\pi}{3} + \frac{n\pi}{3}\right\} - \left\{\frac{l^2}{3n\pi} \cos \frac{n\pi}{3} + \frac{l^2}{n^2\pi^2} \sin \frac{2nx}{3}\right\}\right]$$

$$= \frac{6a}{l^2} \cdot \frac{3l^2}{n^2\pi^2}\left(\sin \frac{n\pi}{3} - \sin \frac{2n\pi}{3}\right)$$

$$= \frac{18a}{n^2\pi^2}\left[\sin \frac{n\pi}{3} \{1 + (-1)^n\}\right]$$

The above transformation stem from the fact that

$$\sin \frac{2n\pi}{3} = \sin\left(n\pi - \frac{n\pi}{3}\right) = (-1)^n \sin \frac{\pi}{3}$$

Thus

$$b_n = \begin{cases} 0 \text{ when } n \text{ is odd} \\ \frac{36a}{n^2\pi^2} \sin \frac{n\pi}{3}, \text{ if } n \text{ is even} \end{cases}$$

Hence equation (6) becomes

$$y(x, t) = \sum_{m=1}^{\infty}\left(\frac{36a}{4m^2\pi^2} \sin \frac{2mx}{3} \sin \frac{2m\pi x}{m} \cos \frac{2m\pi cl}{l}\right)$$

$$= \frac{9a}{\pi^2} \sum_{m=1}^{\infty} \frac{1}{m^2} \sin \frac{2m\pi}{l} \cos \frac{2m\pi cl}{l} \qquad \text{when } n = 2\text{m}$$

***Example* 11.40.** If a string of length 'l' is initially at rest in equilibrium position and each of the points is given the velocity $\left(\frac{\partial y}{\partial t}\right)_{t=0} = b_0 \sin^3 \frac{\pi x}{l}$. Find the displacement $y(x, t)$.

Solution. The vibration of string is given by the equation

$$\frac{\partial y}{\partial t} = c^2 \frac{\partial^2 y}{\partial x^2} \quad \text{...(1)}$$

Since the vibration of the string in periodic. Its solution is of the from

$$y(x, t) = (c_1 \cos px + c_2 \sin px)(c_1 \cos \text{cpt} + c_4 \sin cpt) \quad \text{...(2)}$$

The boundary condition are

$$y(0, t) = 0 \quad \text{...(3)}$$

$$y(l, t) = 0 \quad \text{...(4)}$$

Since the string is at rest initially, we have

$$y(x, 0) = 0 \quad \text{...(5)}$$

$$\left(\frac{\partial y}{\partial t}\right)_{t=0} = b_0 \sin^3 \frac{\pi x}{t} \quad \text{...(6)}$$

On apply (3), (4) conditions to equation (2), we get the reduced solution (as before)

$$y(x, t) = c_2 \sin \frac{n\pi x}{l}\left(c_1 \cos \frac{cn\pi}{l} t + c_4 \sin \frac{cn\pi}{l} t\right)$$

Now apply $y(x, 0) = 0$

$$0 = c_2 \sin \frac{n\pi x}{l} c_3$$

$$= c_2 c_3 \sin \frac{m\pi x}{l} \Rightarrow c_3 = 0$$

The reduced solution is

$$y(x, t) = c_2 c_4 \sin \frac{n\pi x}{l} \sin \frac{cn\pi}{l} l$$

$$= b_n \sin \frac{n\pi x}{l} \sin \frac{cn\pi}{l} t \qquad b_n = c_2 c_4$$

Adding all such solution of (1) we get the general solution

$$y(x, t) = \sum_{n=1}^{\infty} b_n \sin \frac{n\pi x}{l} \sin \frac{cn\pi}{l} t \quad \text{...(7)}$$

Now apply the last condition $\left(\frac{\partial y}{\partial t'}\right)_{t=0} = b_0 \sin^3 \frac{\pi x}{l}$

$$\left(\frac{\partial y}{\partial t}\right) = \sum_{n=1}^{\infty} b_n \frac{cn\pi}{l} t \cos \frac{cn\pi}{l} t \sin \frac{n\pi x}{l}$$

$$\left(\frac{\partial y}{\partial t}\right)_{t=0} = b_n \sin^3 \frac{\pi x}{l} = \sum_{n=1}^{\infty} b_n \frac{cn\pi}{l} \sin \frac{n\pi x}{l}$$

$$b_0 \sin^3 \frac{\pi x}{l} = \sum_{n=1}^{\infty} b_n \frac{cn\pi}{l} \sin \frac{n\pi x}{l}$$

or $$\frac{b_0}{4}\left[3\sin\frac{\pi x}{l}-\sin\left(\frac{\pi x}{l}\right)\right] = \sum b_n\frac{cn\pi}{l}\sin\frac{n\pi x}{l}$$

$$= \frac{c\pi}{l}b_1\sin\frac{\pi x}{l}+\frac{2c\pi}{l}b_2\sin\frac{2\pi r}{l}+\frac{3c\pi}{l}b_3\sin\frac{3\pi}{l}+\dots$$

Equating the coefficient from both sides, we get

$$\frac{3b_0}{4} = \frac{c\pi}{l}b_1;$$

$$0 = \frac{2c\pi}{l}b_2;$$

$$-\frac{b_0}{4} = \frac{3c\pi}{l}b_3 \dots$$

$$b_1 = \frac{3lb_0}{4};$$

$$b_3 = -\frac{lb_0}{12c\pi};$$

$$b_2 = b_4 = b_5,\dots = 0$$

Hence substituting in (7), we get

$$y(x,t) = \frac{lb_0}{12c\pi}\left\{9\sin\frac{\pi r}{l}\sin\frac{c\pi t}{l}-\sin\frac{3\pi x}{l}\sin\frac{3c\pi t}{l}\right\}$$

EXERCISE 11.10

1. Solve the wave equation $\frac{\partial^2 u}{\partial t^2}=c^2\frac{\partial^2 u}{\partial x^2}$ given $u = 0$, when $x = 0$ and $x = p$, $\frac{\partial u}{\partial t}=0$, when $t = 0$, $u(x, 0) = x$, $0 < x < \pi$.
2. A lightly stretched string with fixed end points $x = 0$ and $x = l$ is initially in a position given by $y = y_0 \sin^3\left(\frac{\pi x}{l}\right)$. If it is released from rest from this position, find the displacement $y(x, t)$.
3. Find the solution of wave equation $\frac{\partial^2 u}{\partial t^2}=c^2\frac{\partial^2 u}{\partial x^2}$ such that $y = p_o \cos pt$ (P_0 is constant) when $x = t$ and $y = 0$, when $x = 0$.
4. A tightly stretched string of length 'l' has its end fastened at $x = 0$, $x = l$. The mid point of the string is then taken to a height h and then reduced from rest in that position. Find the lateral displacement of *a* point of the string at time t from the instant of release.
5. A tightly string with fixed end $x = 0$ and $x = l$ is initially at rest in its equilibrium position. If it is set vibrating by giving to each its points a velocity $\lambda x(l - x)$, find the displacement of the string at any distance x from one end at any time t.
6. A tight string of 20 cms length, fastened at both ends is displacement from its position of equilibrium by imparting to each of its points a velocity given by $v = \begin{cases} x, & 0 \le x \le 10 \\ 20 - x, & 10 \le x \le 20 \end{cases}$ x being the distance from one end. Determine the displacement at any subsequent time.

7. Show that the wave equation $\frac{\partial^2 u}{\partial t^2} = c^2 \frac{\partial^2 u}{\partial x^2}$ under the condition $u(0, t) = 0$, $u(l, t) = 0$ for all t; $u(x, 0) = f(x)$, $\left(\frac{\partial u}{\partial t}\right)_{t=0} = g(x)$ has the solution of the from $u(x, t) = \sum_{n=1}^{\infty} (B_n \cos \lambda_n t + C_n \sin \lambda_n t) \sin \frac{n\pi x}{l}$

where $$B_n = \frac{2}{l}\int_0^1 f(x) \sin \frac{n\pi x}{l} dx \qquad \lambda_m = \frac{n\pi x}{l}$$

$$C_n = \frac{2}{cn\pi}\int_0^1 g(x) \sin \frac{n\pi x}{l} dx$$

Answer

1. $u = \frac{2\Sigma(-1)^{n+1}}{n} \sin nx \cos nct$

2. $y = \frac{y_0}{4}\left\{3 \sin \frac{\pi x}{l} \cos \frac{\pi ct}{l} - \frac{3\pi x}{l} \cos \frac{3\pi ct}{l}\right\}$

3. $y = -\frac{P_0}{\sin (P/c)l} \cos pt \sin \frac{P}{c} x$

4. $y = \frac{8h}{\pi^2}\left[\sin \frac{\pi x}{l} \cos \frac{\pi ct}{l} - \frac{1}{3^2} \sin \frac{3\pi x}{l} \cos \frac{3\pi ct}{l} + ...\right]$

5. $y = \frac{8\lambda l^3}{c\pi^4}\left\{\sum_{n=1}^{\infty} \frac{1}{(2n-1)^4} \sin \frac{(2n-1)}{l} \sin \frac{(2n-1)\pi ct}{l}\right\}$

6. $y = \frac{1600}{a\pi^3}\left\{\sin \frac{\pi x}{20} \sin \frac{\pi at}{20} - \frac{1}{3^3} \sin^3 \frac{3\pi at}{20} + ...\right\}$

7. $y(x, t) = \frac{ah}{\pi^2} \sum_{n=1}^{\infty} \frac{1}{n^2} \sin \frac{n\pi}{3} \sin \frac{n\pi x}{l} \cos \frac{n\pi ct}{l}$

11.14 ONE DIMENSIONAL HEAT FLOW

Let the heat be flowing along bar of uniform cross-section in the direction perpendicular to the cross section. Take one end of the bar as origin and the direction of heat flow as x-axis.

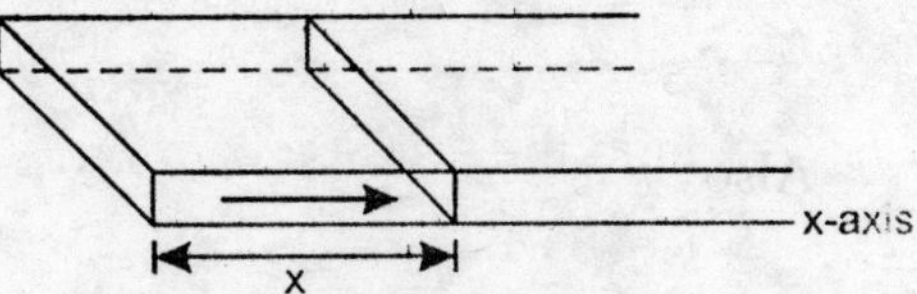

Fig. 11.3

Let the temperature of the bar at any time t at a point x distance from the origin be $u(x, t)$. Then the equation of the dimension heat flow is $\frac{\partial u}{\partial t} = c^2 \frac{\partial^2 u}{\partial x^2}$.

11.15 SOLUTION OF HEAT EQUATION

$$\frac{\partial u}{\partial t} = c^2 \frac{\partial^2 u}{\partial x^2} \qquad ...(1)$$

Let $u = XT$ be the solution of (1)

X is function of x along

T is function of 't' along

$$\frac{\partial u}{\partial t} = XT'$$

and $$\frac{\partial^2 u}{\partial x^2} = X''T$$

Put these values in (1), we get

$$XT' = c^2X''T$$

$$\frac{X''}{X} = \frac{1}{c^2}\frac{T'}{T} = k \text{ (constant)}$$

Since L.H.S. is a function of x alone and R.H.S. is a function of 't' alone so the equation holds only if each is constant say k.

Now we have three cases

Case (1). When k is positive say p^2, then

$$\frac{X''}{X} = p^2 \Rightarrow X'' - p^2 X = 0$$

Auxiliary equation is $D^2 - p^2 = 0$, $D = \pm p$ where $D = \frac{d}{dx}$

$$\therefore \quad X = c_1e^{px} + c_2e^{-px}$$

Also $$\frac{1}{c^2}\frac{T'}{T} = p^2 \quad \Rightarrow \quad \frac{T'}{T} = p^2c^2$$

$$\log T = p^2c^2t + \log c_3$$

$$\therefore \quad T = c_3\, e^{p^2c^2t}$$

In this case the solution of (1) is

$$u(x, t) = (c_1\, e^{px} + c_2e^{-px})\, c_3\, ep^2c^2t$$

Case II. When k is negative say $-p^2$, then

$$\frac{X''}{X} = -p^2 \qquad X'' + p^2x = 0$$

Auxiliary equation is $D^2 + p^2 = 0$; $D = \pm ip$

$$\therefore \quad X = (c_4 \cos px + c_5 \sin px)$$

Also $$\frac{T'}{c^2T} = p^2; \qquad \frac{T'}{T} = -p^2c2$$

$$\therefore \quad \text{lot } T = -p^2c^2t + \log c_6$$

$$\therefore \quad T = c_6\, e^{-}p^2c^2t$$

$\therefore$ This solution is $u(x, t) = (c_4 \cos px + c_5 \sin px)\, c_6\, e^{-p^2c_2t}$ ...(II)

Case III. When k is zero, $k = 0$, then

$$\frac{X''}{X} = 0; \text{ or } X''\ 0$$

$$\Rightarrow \quad X = c_7x + c_8$$

$$\frac{T'}{c^2T} = 0; \text{ or } T'\ 0$$

$$\Rightarrow \quad T = c9$$

$\therefore$ The solution is $u(x, t) = (c_7x + c_8)c_9$ ...(III)

$\therefore$ Out of the solution I, II, III of equation (I)

$u(x, t) = (c_4 \cos px + c_5 \sin px)\, c_6\, c{-}p^2c^2t$ is most suitable as it is consistent, with the physical nature of the problem. (Since as t increases temperature u decreases)

***Example* 11.41.** Solve the following boundary value problem $\frac{6u}{\partial t} = \alpha^2 \frac{\partial^2 u}{\partial x^2}$ $0 < x < 1$, $\frac{\partial u(0, t)}{\partial x} = 0$, $u(x, 0) = x$

Solution : The given heat equation is

$$\frac{\partial u(1, t)}{\partial x} = 0$$

$$\frac{\partial u}{\partial t} = \alpha^2 \frac{\partial^2 u}{\partial x^2} \quad 0 < x < 1 \qquad \text{...(1)}$$

Its general solution will be

$$u(x, t) = (c_1 \cos px + c_2 \sin px)\, e^{-p^2c^2t} \qquad \text{...(2)}$$

The boundary conditions are

(*i*) $\frac{\partial u(0, t)}{\partial x} = 0$

(*ii*) $\frac{\partial u(l, t)}{\partial x} = 0$

(*iii*) $u(x, 0) = x$

Apply condition (*i*) to (2), we get

$$\frac{\partial u(x, t)}{\partial x} = [-c_1 p \sin px + c_2 p \cos px]\, e^{-p^2c^2t}$$

$$\frac{\partial u(0, t)}{\partial x} = c_2\, pc^{-p^2c^2t} = 0 \Rightarrow c_2 = 0$$

The reduced solution is

$$u(x, t) = c_1 \cos px\, e^{-p^2c^2t} \qquad \text{...(3)}$$

Apply condition (*ii*) we get

$$\frac{\partial u(x, t)}{\partial x} = c_1 \cos px\, e^{-p^2c^2t}$$

$$\frac{\partial u(l, t)}{\partial x} = c_1\, p \sin 1p\, e^{-p^2c^2t} = 0 \Rightarrow \sin pl = 0$$

$$pl = n\pi \Rightarrow p = \frac{n\pi}{l}$$

The reduce solution is

$$u(x, t) = c_1 \cos \frac{n\pi x}{1}\, e^{-p^2c^2t} \qquad \text{...(4)}$$

$$= c_1 \cos \frac{n\pi r}{l}\, e^{-c^2n^2\frac{p^2}{l^2}t}$$

Adding all such solution of equation (1) for different values of n, we get the general solution

$$u(x, t) = \sum_{n=0}^{\infty} A_n \cos \frac{n\pi x}{l} e^{-\frac{c^2\pi^2n^2}{l^2}t}, \; A_n = c_1$$

$$= A_0 + \sum_{n=0}^{\infty} A_n \frac{n\pi x}{l} e^{-\frac{c^2\pi^2 n^2}{l^2}.t}$$

Now apply the condition (*iii*), we get

$$x = A_0 + \sum_{n=0}^{\infty} A_n \cos \frac{n\pi x}{l}$$

which is a cosine series in the interval (0, *l*)

Here
$$A_0 = \frac{a_0}{2}$$

$$a_0 = \frac{2}{l}\int_0^t x\, dx = 1$$

$$\therefore \quad A_0 = \frac{l}{2}$$

$$A_n = \frac{2}{l}\int_0^t x \cos \frac{n\pi x}{t} dx = \frac{n\pi r}{l} dx = \frac{2}{l}\left[\frac{1}{n\pi} x \sin \frac{l^2}{n^2\pi^2} \cos \frac{n\pi x}{l}\right]_0^l$$

$$= \frac{2}{l}\left[\left(\frac{l^2}{n\pi}\sin n\pi + \frac{l^2}{n^2\pi^2}\cos n\pi\right) - \left(0 + \frac{l^2}{n^2\pi^2}\right)\right]$$

$$= \frac{2}{l}\left[\frac{l^2}{n\pi}\sin n\pi + \frac{l^2}{n^2\pi^2}\right] - \left(0 + \frac{l^2}{n^2\pi^2}\right)\bigg]$$

$$= \frac{2}{l}[\cos n\pi - 1]\frac{l^2}{n^2\pi^2} = \frac{2l}{n^2\pi^2}[(-1)^n - 1]$$

$$= 0 \text{ when } n \text{ is even}$$

$$u(x, t) = \frac{1}{2} - \frac{4l}{\pi^2} \text{ when } n \text{ is odd}$$

$$u(x, t) = \frac{1}{2} - \frac{4l}{\pi^2}\sum_{n=13,4}^{\infty}\left(\frac{1}{n^2}\cos\frac{n\pi x}{1}e - \frac{c^2 n^2\pi^2}{l^2}t\right)$$

$$= \frac{l}{2} - \frac{4l}{\pi^2}\left\{\frac{1}{(2m-1)}\cos\frac{(2m-1)\pi x}{l}e^{-\frac{c^2(2m-1)^2\pi^2}{l^2}t}\right\}$$

***Example* 11.42.** A uniform rod of conducting material of length 100 cms has its ends kept at zero temperature and the initial temperature is

$$u(x, 0) = x \qquad 0 \le x = \le 50$$

$$= 100 - x \qquad 50 \le x \le 100$$

find the temperature $u(x, t)$ at any time.

Solution. One dimensional heat flow equation is

$$\frac{\partial u}{\partial t} = c^2 \frac{\partial^2 u}{\partial x^2} \qquad ...(1)$$

Its general solution is

$$u(x, t) = (c_1 \cos px + c_2 \sin px)\, e^{-c^2p^2t} \qquad ...(2)$$

which is in consistent with the physical nature of the problem

The boundary condition are

(i) $u(0, t) = 0$

(ii) $u(100, t) = 0$

(iii) $u(x, 0) = x \qquad 0 < x < 50$

$= 100 - x \qquad 50 \le x \le 100$

Apply condition (i) to equation (2)

$$0 = c_1 e^{-c^2p^2t} \Rightarrow c_1 = 0$$

The reduced solution is

$$u(x, t) = c_2 \sin px\, e^{-c^2p^2t} \qquad ...(3)$$

Apply condition (ii) to eqn. (3)

$$u(100, t) = c_2 \sin 100\, p\, e^{-c^2p^2t} = 0 \Rightarrow \sin 100\, p = 0$$

$$100\, p = np \Rightarrow p = \frac{n\pi}{100},\ c_2 \neq 0$$

The reduced solution is

$$u(x, t) = c_2 \sin \frac{n\pi x}{100} e^{-\frac{c^2n^2\pi^2}{(100)^2}} \qquad ...(4)$$

The general solution of (1) is obtained by adding such solution of (1). Hence the general solution is

$$u(x, t) = \Sigma b_n \sin \frac{n\pi x}{100} e^{\frac{-c^2\pi^2x^2}{(100)^2}} \qquad ...(5)$$

Now apply condition (iii), we get

$$u(x, 0) = \Sigma b_n \sin \frac{n\pi x}{100}$$

which is half range sine series in the interval (0, l)

$$\therefore \quad b_n = \frac{2}{100}\int_0^{100} u(x, 0) \sin \frac{n\pi x}{100}\, dx$$

$$= \frac{2}{100}\left[\int_0^{100} x \sin \frac{n\pi x}{100}\, dx + \int_{50}^{100} (100 - x) \sin \frac{n\pi x}{100} dx\right]$$

$$= \frac{1}{50}\left[\left\{x\left(-\frac{\cos n\pi x/100}{np/100}\right) - \frac{(\sin(npx/100))}{(np/100)^2}\right\}_0^{50} + \left\{(100 - x)\frac{\left(-\cos\frac{n\pi x}{100}\right)}{\frac{n\pi}{100}} - \frac{\left(\sin\frac{n\pi x}{100}\right)}{\left(\frac{n\pi}{100}\right)^2}\right\}_{50}^{100}\right]$$

$$= \frac{1}{50}\left[\left\{-\frac{100}{n\pi} x \cos \frac{n\pi x}{100} + \left(\frac{100}{n\pi}\right)^2 \sin \frac{n\pi x}{100}\right\}_0^{50} + \left\{-(100 - x)\frac{100}{n\pi} \cos \frac{n\pi x}{100} - \left(\frac{100}{n\pi}\right)^2 \sin \frac{n\pi x}{100}\right\}_{50}^{100}\right]$$

$$= \frac{1}{50}\left[\left\{\left(\frac{100}{n\pi}\right)^2 \sin \frac{n\pi}{2}\right\} + \left\{0 + 50\frac{100}{n\pi} \cos \frac{n\pi}{2} + \left(\frac{100}{n\pi}\right)^2 \sin \frac{n\pi}{2}\right\}\right]$$

$$= \frac{1}{50}\left[\left(\frac{100}{n\pi}\right)^2 \sin \frac{n\pi}{2} + \left(\frac{100}{n\pi}\right)^2 \sin \frac{n\pi}{2}\right] = \frac{2}{50}\left(\frac{100}{n\pi}\right)^2 \sin \frac{n\pi}{2}$$

The equation solution is

$$u(x, t) = \Sigma \frac{2}{50} \frac{(100)^2}{n^2\pi^2} \sin \frac{n\pi}{2} \sin \frac{n\pi x}{100} e^{\frac{-c^2n^2\pi^2}{(100)^2}t}$$

$$= \frac{(100)^2}{n^2\pi^2}\Sigma\frac{1}{n^2}\sin\frac{n\pi}{2}\sin e^{\frac{c^2n^2\pi^2}{(100)^2}t}$$

If n is even $\sin\frac{n\pi}{2} = 0$

If n is odd $\sin\frac{n\pi}{2} = (-1)^m$ where $n = 2m + 1$

$$u(x, t) = \frac{400}{\pi^2}\Sigma\frac{(-1)^m}{(2m+1)^2}\sin\frac{(2m+1)\pi x}{100}e^{-\frac{c^2(2m+1)^2\pi^2}{(100)^2}t}$$

Example **11.43.** An insulated rod of length 'l' has its ends. A and B maintained at 0°C and 100°C respect unit steady same condition prevail. If B is suddenly reduced to 0°C and maintained at 0°C, find the temperature at a distance x from A at time 't'.

Solution. Let the equation of the heat flow be

$$\frac{\partial u}{\partial t} = c^2\frac{\partial^2 u}{\partial x^2} \quad ...(1)$$

Its general solution is

$$u(x, t) = (c_1 \cos px + c_2 \sin px)c_3\, e^{-p^2c^2t} \quad ...(2)$$

which is consistent with the physical nature of the problem.

Prior to the change of temperature at B when $t = 0$ or for steady state conditions

$$\frac{\partial^2 u}{\partial t} = 0$$

∴ The equation (1) reduces to

$$\frac{\partial^2 u}{\partial x^2} = 0 \Rightarrow u = ax + b$$

Since $u = 0$ for $x = 0$; $u = 100$ for $x = l$, we have

$$b = 0 \text{ and } a = \frac{100}{l}$$

$$u = \frac{100x}{l} \quad ...(3)$$

or

$$u(x, 0) = \frac{100x}{l} \quad ...(4)\ \text{(Initial condition)}$$

Now the boundary conditions are

$$u(0, t) = 0 \quad ...(5)$$

$$u(l, t) = 0 \quad ...(6)$$

Apply the condition (5) to equation (2), we have

$$0 = c_1\, e^{-p^2c^2t} \Rightarrow c_1 = 0$$

∴ The reduced solution is

$$u(x, t) = c_2 \sin px\, e^{-p^2c^2t}$$

Apply condition (6), we get

$$0 = c_2 \sin pl\, px\, e^{-p^2c^2t} \Rightarrow \sin pl = 0$$

$$pl = n\pi \Rightarrow p = \frac{n\pi}{l}$$

The solution reduces to

$$u(x, t) = c_2 \sin \frac{n\pi x}{l} e^{-\frac{n^2\pi^2c^2}{l^2}t}$$

Adding all such solution of equation (1) we get the general solution

$$u(x, 0) = \Sigma b_n \sin \frac{n\pi x}{l} e^{-\frac{n^2\pi^2c^2}{l^2}} \qquad [c_2 = b_n]$$

Applying the condition (4)

$$u(x, 0) = u\,(x, 0)\; \Sigma b_n \sin \frac{n\pi x}{l}$$

$$\frac{100x}{l} = \Sigma b_n \sin \frac{n\pi x}{l}$$

which is a half range since series in $(0, l)$

$$\therefore \quad b_n = \frac{2}{l}\int_0^l \frac{100x}{l} \sin \frac{n\pi x}{l} dx = \frac{200}{l^2}\int_0^l \sin \frac{n\pi x}{l} dx$$

$$\therefore \quad = \frac{200}{l^2}\left[x\left(-\frac{\cos \frac{n\pi x}{l}}{\frac{n\pi}{l}}\right) - 1\left(\frac{-\sin \frac{n\pi x}{l}}{\left(\frac{n\pi}{l}\right)^2}\right)\right]_0^l$$

$$= \frac{200}{n\pi}(-1)^{n+1}$$

$$\therefore \quad u(x, t) = \frac{200}{\pi} S \frac{(-1)^{n+1}}{n} \sin \frac{n\pi x}{l} e^{-\frac{c^2n^2\pi^2}{l^2}t}$$

Example **11.44.** An insulated rod of length 10 cms has its end A and B maintained at 50°C and 100°C respectively until steady conditions prevail. The temperature at A is raised to 90°C and that of B is lowered to 60°C. Find the temperature distribution in the rod at time t.

Solution. Prior to change of temperature, when $t = 0$, the steady state conditions prevail and the temperature at every point of the rod was independent of time *i.e.*

$$\frac{\partial u}{\partial t} = 0; \frac{\partial^2 u}{\partial x^2} = 0 \; u = ax + b \qquad ...(1)$$

Initially when $x = 0$ $u = 50$ $\therefore a = 5$

$x = 10$ $u = 100$ $\therefore b = 50$

$$\therefore \quad u(x, 0) = 5x + 50 \qquad ...(2)$$

which is the initial temperature distribution

Similarly the final temperature distribution (when steady state again prevails) in given by

$x = 0;$ $u = 90;$ $a = -3$

$x = 10;$ $u = 60;$ $b = 90$

$$u(x, 0) = 90 - 3x \qquad ...(3)$$

To get $u(x, t)$ at any time t, we break up the temperature function as follows

$$u(x, t) = u_s(x) + u_1(x, t)$$

where $u_x(x)$ is the steady temperature distribution the rod (when the steady state again prevails) and $u_t(x, t)$ is the transient temperature distribution which decreases as t increases. Both $u_s(x)$ and $u_s(x, t)$ satisfy heat equation

$$\frac{\partial u}{\partial t} = c^2 \frac{\partial^2 u}{\partial x^2}$$

Hence $$u(x, t) = 90 - 3x + \Sigma(a_n \cos px + b_n \sin px)\, e^{-c^2 p^2 t} \quad ...(4)$$

Since $$u = 90 \text{ when } x = 0$$

$\therefore$ $$90 = 90 + \Sigma a_n\, e^{-c^2 p^2 t}$$

or $$0 = \Sigma a_n\, e^{-c^2 p^2 t} \Rightarrow a_n = 0$$

The solution (2) reduces to

$$u(x, t) = 90 - 3x + \text{Ebn} \sin px\, e^{-c^2 p^2 t}$$

Also $u = 60$ when $x = 10$

$$60 = 60 + \Sigma b_n \sin 10\, p\, e^{-c^2 p^2 t}$$

$$0 = \Sigma b_n \sin 10 \text{ pe } e^{-c^2 p^2 t} \Rightarrow \sin 10\, p = 0\ b_n \neq 0$$

$$10p = n\pi \Rightarrow \text{p} = \frac{n\pi}{10}$$

The reduced solution is

$$u(x, t) = 90 - 3x + \Sigma b_n \sin \frac{n\pi x}{10} e^{\frac{-c^2 n^2 \pi^2}{100} t}$$

Using initial condition *i.e.*,

$$u = 5\underline{x} + 50 \text{ when } t = 0, \text{ we get}$$

$$5x + 50 = 90 - 3x + \Sigma b_n \sin \frac{n\pi x}{10}$$

$$8x - 40 = \Sigma b_n \sin \frac{n\pi x}{10}$$

which is a half range since series in (0, 10)

$$b_n = \frac{2}{10} \int_0^{10} (8x - 40) \sin \frac{n\pi x}{10} dx$$

$$= \frac{1}{5} \left[(8x - 40) \frac{-\cos \frac{n\pi x}{10}}{\frac{n\pi}{10}} - 8 \frac{\left(-\sin \frac{n\pi x}{10}\right)}{\left(\frac{n\pi}{10}\right)^2} \right]_0^{10}$$

$$= \frac{1}{5}\left[-40 \times \frac{10}{n\pi} \cos n\pi - \left(40 \times \frac{10}{n\pi}\right) \right]$$

$$= -\frac{80}{n\pi}[(-1)^n + 1] = \frac{80}{n\pi}[1 + (-1)^n]$$

$$= \begin{cases} 0, \text{ when } n \text{ is} \\ -\frac{160}{n\pi} \text{ when } n \text{ is even} \end{cases}$$

$$u(x, t) = 90 - 3x - \sum_{n = 2, 4\, 6, \ldots}^{\infty} \frac{160}{n\pi} \sin \frac{n\pi x}{10} e^{-\frac{c^2 n^2 \pi^2}{100} t}$$

$$= 90 - 3x - \frac{160}{\pi}\sum_{n=1}^{\infty}\frac{1}{2m}\sin\frac{2m\pi x}{10}\, e^{-\frac{c^2 m^2\pi^2}{25}}$$

$$= 90 - 3x - \frac{80}{\pi}\sum_{n=1}^{\infty}\frac{1}{m}\sin\frac{2m\pi x}{10}\, e^{-\frac{c^2 m^2\pi^2}{25}}$$

***Example* 11.44.** A rod of length 'l' with insulated sides in initially at a temperature $u_0(x)$. Its ends are suddenly cooked to 0°C and are kept at that temperature. Prove that the temperature function $u(x, t)$ is given by

$$u(x, t) = \sum_{n=1}^{\infty} b_n \sin\frac{n\pi}{l} e^{-\frac{c^2 n^2 \pi^2}{l^2}t}$$

Solution. As proved, the general solution of heat equation

$$\frac{\partial u}{\partial t} = c^2\frac{\partial^2 u}{\partial x^2} \quad ...(1)$$

is

$$u(x, t) = (c_1 \cos px + c_2 \sin px)\, e^{-c^2 p^2 t} \quad ...(2)$$

Since the ends ($x = 0$, $x = l$) are cooled to 0°C and kept at that temperature through out, we have the boundary conditions

$$u(0, t) = 0,\ u(l, t) = 0 \text{ for all } t \quad ...(3)$$

Also initial condition is

$$u(x, 0) = u_0(x) \quad ...(4)$$

From (2) and (3), we get

$$c_1 e^{-c^2 p^2 t} = 0 \Rightarrow c_1 = 0$$

$$\sin pl = 0 \quad pl = n\pi, \quad p = \frac{n\pi}{l}$$

The reduced solution is

$$u(x, t) = c_2 \sin\frac{n\pi x}{l} e^{-\frac{c^2 n^2 \pi^2}{l^2}t}$$

$$= b_n \sin\frac{n\pi x}{l} e^{-\frac{c^2 n^2 \pi^2}{l^2}t} \quad [c_2 = b_n]$$

Adding all such solution of (1), we get the general solution

$$u(x, t) = \Sigma b_n \sin\frac{n\pi x}{l} e^{-\frac{c^2 n^2 \pi^2}{l^2}t}$$

Apply condition (4) *i.e.*, $u(x, 0) = u_0(x)$, we get

$$u_0(x) = \Sigma b_n \sin\frac{n\pi x}{l}$$

which is a half range since series in $(0, l)$

$$b_n = \frac{2}{l}\int_0^t u_0(x)\sin\frac{n\pi x}{l}dx$$

Hence

$$u(x, t) = \Sigma b_n \sin\frac{n\pi x}{l} e^{-\frac{c^2 n^2 \pi^2}{l^2}t}$$

Where

$$b_n = \frac{2}{l}\int_0^l u_0(x)\sin\frac{n\pi x}{l}dx$$

EXERCISE 11.11

1. Solve $\frac{\partial^2 u}{\partial x^2} = h^2 \frac{\partial u}{\partial t}$ where $u(0, t) = u(l, t) = u(l, t) = 0; = \sin\frac{n\pi}{l}$ by method of variable separable.

2. Solve $\frac{\partial u}{\partial t} = 3\frac{\partial^2 u}{\partial x^2}$ $t < 0, 0 < x < 2$ $u(0, t) = u(2, t) = 0$ and $u(x, 0) = x$.

3. Find the solution of $\frac{\partial \theta}{\partial t} = K\frac{\partial^2 \theta}{\partial x^2}$ such that

   (i) θ is not infinite when $t \to \infty$ 0.

   (ii) $\frac{\partial \theta}{\partial x} = 0$ when $x = 0$ for all t, $\theta = 0$, when $x = l$ for all t.

   (iii) $\theta = \theta_c$ when $t = 0$ for all values of x.

4. Solve the equation $\frac{\partial u}{\partial t} = \frac{\partial^2 u}{\partial x^2}$ with boundary condition $u(x, 0) = 2 \sin n\pi x$, $u(0, t) = u(l, t) =$ where $0 < x < 1$, $t > 0$.

5. Find the temperature $u(x, t)$ in a bar of length p which is perfectly insulated every where including the ends $x = 0$, $x = \pi$. This lends to the conditions $\frac{\partial u}{\partial x}(0, t) = 0$, $\frac{\partial u}{\partial x}(\pi, t)$. Further the initial conditions are given below

$$u(x, 0) = \begin{Bmatrix} x & 0 < x < \pi/2 \\ \pi x & \pi/2 < x < \pi \end{Bmatrix}$$

6. The temperature of one end of a bar, 50 cm long with insulated sides, is kept at 0°C and that the other end is kept at 100°C until steady state conditions prevail. The two ends are then suddenly, so that temperature gradient is zero, at each end there after. Find the temperature distribution.

7. Solve the equation $\frac{\partial u}{\partial t} = k\frac{\partial^2 u}{\partial x^2}$ where

   (i) u is finite when $t \to +\infty$.

   (ii) $\frac{\partial u}{\partial x} = 0$ for $x = 0$, $x = 1$

   (iii) $u = lx - x^2$ for $t = 0$, $0 < x <$.

8. An insulated rod of length l has its ends A and B maintained at 0°C and 100°C respectively until steady state condition prevail. If the temperature of A is raised to 20°C and that of B is lowered to 80°C, find the temperature at any time t.

Answer

1. $u(x, t) = \sin\frac{\pi x}{l} e^{-\frac{\pi^2 t}{h^2 l^2}}$

2. $u(x, t) = \frac{4}{\pi}\sum_{n=l}^{\infty}\left\{\frac{(-1)^{n+1}}{n}\sin\frac{n\pi x}{2}e^{-3p^2 t}\right\}\left(p^2 = \frac{n^2\pi^2}{4}\right)$

3. $\theta = \frac{4\theta}{\pi}\left[e^{-\left(\frac{\pi}{2l}\right)^2 kt}\cos\frac{\pi x}{2l} - \frac{1}{3}e^{-\left(\frac{3\pi}{2l}\right)^2}\cos\frac{3\pi}{2l} + ...\right]$

4. $u(x, t) = 3\sum_{1}^{\infty} e^{-n^2\pi^2 t}\sin(n\pi x)$

6. $u(x, t) = 50 - \frac{400}{\pi^2}\sum_{1}^{\infty} \frac{2}{(2n-1)^2} \cos\frac{(2n-1)\pi x}{50} e^{\frac{c^2 p^2 (2n-1)^2}{2500} t}$

7. $u(x, t) = \frac{l^2}{6} - \frac{l^2}{\pi^2}\sum_{n=1}^{\infty} \frac{1}{n^2} \cos\frac{2n\pi x}{l} e^{\frac{-4kn^2 x^2}{l^2} t}$

8. $u(x, t) = \frac{60x}{l} + 20 - \frac{40}{\pi}\sum_{n=1}^{\infty} \frac{1}{n} \sin\left(\frac{2n\pi x}{l}\right) e^{\frac{-4c^2 n^2 \pi^2 t}{l^2}}$

OBJECTIVE QUESTIONS

1. Form the partial differential equation of the family of surfaces $z = e^{mx} f(x + y)$
2. Find the partial differential equation obtained by eliminating arbitrary functions for $z = f(x + ky) + g(x - ky)$
3. Find the partial differential equation from $z = f(xy/z)$
4. How a partial differential equation of formed ?
5. Write down the auxiliary equation for charpit's method to solve $f(x, y, z, p, q) = 0$
6. Obtained the complete solution of : $q = xyp^2$
7. Solve. $q = e^{-p\alpha}$
8. How can a partial differential equation of the from $f(z, p, q) = 0$ be solved ?
9. Define non-linear partial differential equation.
10. Give the solution of a Lagrange's partial differential equation of the from $p_p + Qq = R$

Answers

1. $p - q = mz$

2. $\frac{\partial^2 z}{\partial y^2} = k^2 \frac{\partial^2 z}{\partial x^2}$

3. $px = qy$

5. $\frac{dx}{-\partial f/\partial p} = \frac{dy}{-\partial f/\partial q} = \frac{dz}{-\frac{q\partial f}{\partial p} - \frac{q\partial f}{\partial q}} = \frac{dp}{-\frac{\partial f}{\partial x} + \frac{q\partial f}{\partial z}} = \frac{dq}{-\frac{\partial f}{\partial y} + \frac{q\partial f}{\partial z}} = \frac{df}{0}$

6. $(2\tau - ay^2 - 2c)2 = 16\, ax$

7. $z = ax + e^{-a\alpha} y + c$

Function of a Complex Variable

This is chapter os devoted to the study of limit continuity and differentiability of function of a complex variable, analytic function C-R equations, Mile-Thomson method, complex integration, Cauchy's integral theorem and formula, singular points, residues, Cauchy's residue theorem, Taylor's and Laurent's series and applications of residues.

12.1 COMPLEX NUMBER

A complex number $z = x + iy$ is defined as an ordered pair (x, y) where x and y are real numbers and $i = \sqrt{-1}$. The real numbers x and y are called the real and imaginary parts of z.

The complex number $z = x + iy$ can be represented in the polar form as follows :

Let $P(x, y)$ represent the complex number $z = x + iy$. Let vector $\overrightarrow{OP}$ make an angle θ with the positive direction of OX and $OP = r$ and $< MOP = 1$, then $x = OM = OP \cos\theta = r\cos\theta$ where $OP = r$, is known as radius vector

and
$$y = MP = OP \sin\theta = r\sin\theta$$

$$x = r\cos\theta \text{ and } y = r\sin\theta$$

$$x^2 + y^2 = r^2 \text{ and } \tan\theta = \frac{y}{x}$$

$\therefore$ $r = \sqrt{x^2+y^2}$ is called modulus of z, denoted by $|z|$

and $\theta = \tan^{-1} y/x$ is called the amplitude or argument of z, denoted by arg z.

The value of θ satisfying $-\pi < \theta < \pi$ is called the principle value of the amplitude. Hence $z = x + iy$

$$= r(\cos\theta + i\sin\theta) = re^{i\theta}$$

(r, θ) form the complex number z and sis called the polar form of the complex number.

12.2 LIMIT OF A FUNCTION

Let $f(z)$ be a single valued function of z defined in a neighbourhood of $z = z_0$, then $f(z)$ is said to have the limit l as z approaches z_0 (along any path straight or curved) if given arbitrary real number $\in > 0$, however small $\ni$ a real number $\delta > 0$ such that $|f(z) - l| < \in$ whenever $0 < |z - z_0| < \delta$.

This means that values of $f(z)$ are as close to l as desired, for all values of z and sufficiently close to z_0 (except for $z = z_0$). We write as :

$$\lim_{z \to z_0} f(z) = l$$

12.3 CONTINUITY OF A FUNCTION

The function $f(z)$ is said to be continuous at z_0 if $\lim_{z \to z_0} f(z) = f(z_0)$, provided that the limit is independent of the direction of approach of z to z_0.

For $f(z)$ to be continuous, it must satisfy following conditions :

1. $f(z_0)$ must be defined.
2. $\lim_{z \to z_0} f(z)$ exists.
3. $\lim_{z \to z_0} f(z) = f(z_0)$.

If any of the above conditions is not satisfied then $f(z)$ is said to be discontinuous at $z = z_0$.

12.4 DIFFERENTIABILITY OF A FUNCTION OF A COMPLEX VARIABLE

Let $w = f(z)$ be a single-valued function of the complex variable $z = x + iy$. Then the derivative of w is defined as

$$\frac{dw}{dz} = f'(z) = \lim_{\delta z \to 0} \frac{f(z + \delta z) - f(z)}{\delta z}$$

Provided the limits exists, independent of the manner in which $\delta z \to 0$.

Example **12.1.** Show that $\lim_{z \to 0} \frac{[\text{Re}(z) - \text{Im}(z)]^2}{|z|^2}$ does not exist.

Solution : Let $z \to 0$ along the path $y = mx$, then

$$\lim_{z \to 0} f(z) = \lim_{z \to 0} \frac{[\text{Re}(z) - \text{Im}(z)]^2}{|z|^2} = \lim_{z \to 0} \frac{(x-y)^2}{x^2 + y^2}$$

$$= \lim_{z \to 0} \frac{(x - mx)^2}{x^2 + m^2 x^2}$$

$$= \lim_{z \to 0} \frac{(1-m)^2 x^2}{(1+m^2) x^2}$$

$$= \frac{(1-m)^2}{1+m^2}$$

Which depends on m. For different values of m, we have different paths and different limits. Hence limit does not exist.

Example **12.2.** Show that function $f(z)$ defined by

$$f(z) = \begin{cases} \dfrac{\text{Re}(z^2)}{|z|^2} & z \neq 0 \\ 0, & z = 0 \end{cases} \text{ is not continous}$$

Solution : Given $f(0) = 0$

Let $z = 0$ along the path $y = mx$, then

$$\lim_{z \to 0} f(z) = \lim_{z \to 0} \frac{\text{Re}(z^2)}{|z|^2} = \lim_{z \to 0} \frac{x^2 - y^2}{x^2 + y^2}$$

$$= \lim_{z \to 0} \frac{x^2 - m^2 x^2}{x^2 + m^2 x^2} = \lim_{z \to 0} \frac{(1-m^2) x^2}{(1+m^2) x^2} = \frac{1-m^2}{1+m^2}$$

which depends of m. For different values of m, we have different limits.

Hence limit does not exist and function is not continuous at $z = 0$.

EXERCISE 12.1

Show that the following limits do not exist :

1. $\lim\limits_{z\to 0} \dfrac{\text{Im}(z)^2}{|z|^2}$

2. $\lim\limits_{z\to 0} \dfrac{\text{Re}\, z}{|z|}$

Find the limits of :

3. $\lim\limits_{z\to -i} (2x^3 - 5y^3 i)$

Ans. $5i$

4. $\lim\limits_{z\to 1+i} \dfrac{z^3-1}{z^2-1}$

Ans. $\dfrac{7}{5} + \dfrac{4}{5}i$

5. Show that $f(z) = \sin z$ is continuous for all z.
6. Show that the function :

$$f(z) = \begin{cases} \dfrac{\text{Im}(z)}{|z|} & z \neq 0 \\ 0, & z = 0 \end{cases} \text{ is not continuous at } z = 0.$$

7. Examine the continuity of following functions :

(i) $f(z) = \begin{cases} \dfrac{z^2+1}{z+i}, & z \neq -i \\ 0, & z = -i \end{cases}$ at $z = -i$

(ii) $f(z) = \lim\limits_{z\to 1+i} \begin{cases} \dfrac{z^2 - iz^2 + z - i}{z-i} & z \neq i \\ 0, & z = i \end{cases}$ at $z = i$

Ans. (i) Not continuous (ii) Continuous.

12.5 ANALYTIC FUNCTION

If a single valued function $f(z)$ possesses a unique derivative at every point of a region R, then $f(z)$ is called an analytic function or a regular function or holomorphic function of z in R.

A point where the function ceases to be analytic is called a singular point.

12.6 NECESSARY AND SUFFICIENT CONDITIONS FOR $f(z)$ TO BE ANALYTIC

The necessary and sufficient conditions for the function

$$w = f(z) = u(x, y) + iv(x, y)$$

to be analytic in a region R, are

(i) $\dfrac{\partial u}{\partial x}, \dfrac{\partial u}{\partial y}, \dfrac{\partial v}{\partial x}, \dfrac{\partial v}{\partial y}$ are continuous functions of x and y in the region R.

(*ii*) $\frac{\partial u}{\partial x} = \frac{\partial v}{\partial y}, \frac{\partial u}{\partial y} = -\frac{\partial v}{\partial x}$

The conditions in (*ii*) are known as **Cauchy-Riemann equations** or briefly *C-R* equations.

Proof. (*a*) Necessary Condition. Let $w = f(z) = u(x, y) + iv(x, y)$ be analytic in a region *R*, then $\frac{dw}{dz} = f'(z)$ exists uniquely at every point of that region.

Let δx and δy be the increment in *x* and *y* respectively. Let δu, δv and δz be the corresponding increments in *u*, *v* and *z* respectively. Then,

$$f'(z) = \lim_{\delta z \to 0} \frac{f(z+\delta z) - f(z)}{\delta z} = \lim_{\delta z \to 0} \frac{(u+\delta u) + i(v+\delta v) - (u+iv)}{\delta z}$$

$$= \lim_{\delta z \to 0} \left(\frac{\delta u}{\delta z} + i\frac{\delta v}{\delta z}\right) \quad ...(1)$$

Since the function $w = f(z)$ is analytic in the region *R*, the limit (1) must exist independent of the manner in which $\delta z \to 0$, *i.e.*, along whichever path δx and $\delta y \to 0$.

First, let $\delta z \to 0$ along a line parallel to *x*-axis so that $\delta y = 0$ and $\delta z = \delta x$.

[Since $z = x + iy$, $z + \delta z = (x + \delta x) + i(y + \delta y)$ and $\delta z = \delta x + i\delta y$]

$\therefore$ From (1), $f'(z) = \lim_{\delta x \to 0} \left(\frac{\delta u}{\delta x} + i\frac{\delta v}{\delta x}\right) = \frac{\partial u}{\partial x} + i\frac{\partial v}{\partial x}$ ...(2)

Now, let $\delta z \to 0$ along a line parallel to *y*-axis so that $\delta x = 0$ and $\delta z = i\delta y$.

$\therefore$ From (1), $f'(z) = \lim_{\delta y \to 0} \left(\frac{\delta u}{i\delta y} + i\frac{\delta v}{i\delta y}\right) = \frac{1}{i}\frac{\partial u}{\partial y} + \frac{\partial v}{\partial y}$

$$= \frac{\partial v}{\partial y} - i\frac{\partial u}{\partial y} \quad ...(3)\left[\because \frac{1}{i} = -i\right]$$

From (2) and (3), we have $\frac{\partial u}{\partial x} + i\frac{\partial v}{\partial x} = \frac{\partial v}{\partial y} - i\frac{\partial u}{\partial y}$

equating the real and imaginary parts, $\frac{\partial u}{\partial x} = \frac{\partial v}{\partial y}$ and $\frac{\partial u}{\partial y} = -\frac{\partial v}{\partial x}$

Hence the necessary condition for $f(z)$ to be analytic is that the C-R equations must be satisfied.

(*b*) **Sufficient Conditions :** Let $f(z) = u + iv$ be a single-valued function possessing partial derivatives $\frac{\partial u}{\partial x}, \frac{\partial u}{\partial y}, \frac{\partial v}{\partial x}, \frac{\partial v}{\partial y}$ at each point of a region *R* and satisfying C-R equations.

i.e., $$\frac{\partial u}{\partial x} = \frac{\partial v}{\partial y} \text{ and } \frac{\partial u}{\partial y} = -\frac{\partial v}{\partial x}$$

We shall show the $f(z)$ is analytic, *i.e.*, $f'(z)$ exists at every point of the region *R*.

By Taylor's theorem for functions of two variables, we have on omitting second and higher degree terms of δx and δy.

$$f(z + \delta z) = u(x + \delta x, y + \delta y) + iv(x + \delta x, y + \delta y)$$

$$= \left[u(x, y) + \left(\frac{\partial u}{\partial x}\delta x + \frac{\partial u}{\partial y}\delta y\right)\right] + i\left[v(x, y) + \left(\frac{\partial v}{\partial x}\delta x + \frac{\partial v}{\partial y}\delta y\right)\right]$$

$$= [u(x, y) + iv(x, y)] + \left(\frac{\partial u}{\partial x} + i\frac{\partial v}{\partial x}\right)\delta x + \left(\frac{\partial u}{\partial y} + i\frac{\partial v}{\partial y}\right)\delta y$$

$$= f(z) + \left(\frac{\partial u}{\partial x} + i\frac{\partial v}{\partial x}\right)\delta x + \left(\frac{\partial u}{\partial y} + i\frac{\partial v}{\partial y}\right)\delta y$$

or
$$f(z+\delta z)-f(z) = \left(\frac{\partial u}{\partial x}+i\frac{\partial v}{\partial x}\right)\delta x+\left(\frac{\partial u}{\partial y}+i\frac{\partial v}{\partial y}\right)\delta y$$

$$= \left(\frac{\partial u}{\partial x}+i\frac{\partial v}{\partial x}\right)\delta x+\left(-\frac{\partial v}{\partial x}+i\frac{\partial u}{\partial x}\right)\delta y \quad \text{(Using } C\text{-}R \text{ equations)}$$

$$= \left(\frac{\partial u}{\partial x}+i\frac{\partial v}{\partial x}\right)\delta x+\left(\frac{\partial u}{\partial x}+i\frac{\partial v}{\partial x}\right)i\delta y \quad [\because 1=i^2]$$

$$= \left(\frac{\partial u}{\partial x}+i\frac{\partial v}{\partial y}\right)(\delta x+i\delta y)=\left(\frac{\partial u}{\partial x}+i\frac{\partial v}{\partial x}\right)\delta z \quad [\delta x+i\delta y=\delta z]$$

$$\Rightarrow \quad \frac{f(z+\delta z)-f(z)}{\delta z} = \frac{\partial u}{\partial x}+i\frac{\partial v}{\partial x}$$

$$\therefore \quad f'(z)=\lim_{\delta z\to 0}\frac{f(z+\delta z)-f(z)}{\delta z} = \frac{\partial u}{\partial x}+i\frac{\partial v}{\partial x}$$

Thus $f'(z)$ exits, because $\frac{\partial u}{\partial x}, \frac{\partial v}{\partial x}$ exist.

Hence $f(z)$ is analytic.

Note 1. The real and imaginary parts of an analytic function are called **conjugate functions.** Thus, if $f(z) = u(x, y) + iv(x, y)$ is an analytic function, then $u(x, y)$ and $v(x, y)$ are conjugate functions. The relation between two conjugate functions is given by *C-R* equations.

Note 2. When a function $f(z)$ is known to be analytic, it can be differentiated in the ordinary way as if z is a real variable.

Thus,
$$f(z) = z^2 \Rightarrow f'(z)=2z$$
$$f(z) = \sin z \Rightarrow f'(z)=\cos z \text{ etc.}$$

12.7 CAUCHY-RIEMANN EQUATIONS IN POLAR CO-ORDINATES

Let (r, θ) be the polar co-ordinates of the point whose cartesian coordinates (x, y), then

$$x = r\cos\theta,\ y=r\sin\theta,$$
$$z = x+iy=r(\cos\theta+i\sin\theta)=re^{i\theta}$$
$$\therefore \quad u+iv = f(z)=f(re^{i\theta}) \quad \text{...(1)}$$

Differentiating (1) partially w.r.t. r, we have

$$\frac{\partial u}{\partial r}+i\frac{\partial v}{\partial r} = f'(re^{i\theta}) \quad \text{...(2)}$$

Differentiating (1) partially w.r.t. θ, we have

$$\frac{\partial u}{\partial \theta}+i\frac{\partial v}{\partial \theta} = f'(re^{i\theta})\cdot ire^{i\theta}=ir\left(\frac{\partial u}{\partial r}+i\frac{\partial v}{\partial r}\right) \quad \text{[Using (2)]}$$

$$= -r\frac{\partial v}{\partial r}+ir\frac{\partial u}{\partial r}$$

Equating real and imaginary parts, we get

$$\frac{\partial u}{\partial \theta} = -r\frac{\partial v}{\partial r} \text{ and } \frac{\partial v}{\partial \theta}=r\frac{\partial u}{\partial r}$$

or
$$\frac{\partial u}{\partial r} = \frac{1}{r}\frac{\partial v}{\partial \theta} \text{ and } \frac{\partial v}{\partial r}=-\frac{1}{r}\frac{\partial u}{\partial \theta}$$
which is the polar form of *C-R* equations.

12.8 HARMONIC FUNCTIONS

Any solution of the Laplace's equation $\frac{\partial^2 \phi}{\partial x^2} + \frac{\partial^2 \phi}{\partial y^2} = 0$ is called a harmonic function.

Let $f(z) = u + iv$ be analytic in some region of the z-plane, then u and v satisfy C-R equations.

$$\therefore \quad \frac{\partial u}{\partial x} = \frac{\partial v}{\partial y} \quad \text{...(1)}$$

and
$$\frac{\partial u}{\partial y} = -\frac{\partial v}{\partial x} \quad \text{...(2)}$$

Differentiating (1) partially w.r.t. x and (2) w.r.t. y, we get

$$\frac{\partial^2 u}{\partial x^2} = \frac{\partial^2 v}{\partial x \partial y} \quad \text{...(3)}$$

and
$$\frac{\partial^2 u}{\partial y^2} = -\frac{\partial^2 v}{\partial y \partial x} \quad \text{...(4)}$$

Assuming $\frac{\partial^2 v}{\partial x \partial y} = \frac{\partial^2 v}{\partial y \partial x}$ and adding (3) and (4), we get

$$\frac{\partial^2 u}{\partial x^2} + \frac{\partial^2 u}{\partial y^2} = 0 \quad \text{...(5)}$$

Now, differentiating (1) partially w.r.t. y and (2) w.r.t. x, we get

$$\frac{\partial^2 u}{\partial y \partial x} = \frac{\partial^2 v}{\partial y^2} \quad \text{...(6)}$$

and
$$\frac{\partial^2 u}{\partial x \partial y} = -\frac{\partial^2 v}{\partial x^2} \quad \text{...(7)}$$

Assuming $\frac{\partial^2 u}{\partial y \partial x} = \frac{\partial^2 u}{\partial x \partial y}$ and subtracting (7) from (6), we get

$$\frac{\partial^2 v}{\partial x^2} + \frac{\partial^2 v}{\partial y^2} = 0 \quad \text{...(8)}$$

Equations (5) and (8) show that the real and imaginary parts u and v of an analytic function satisfy the Laplace's equations.

Hence, u and v are known as harmonic functions.

12.9 ORTHOGONAL SYSTEM

Every analytic function $f(z) = u + iv$ defines two families of curves $u(x, y) = c_1$ and $v(x, y) = c_2$, which form an orthogonal system.

Consider the two families of curves

$$u(x, y) = c_1 \quad \text{...(1)}$$

and
$$v(x, y) = c_2 \quad \text{...(2)}$$

Differentiating (1) w.r.t. x, we get

$$\frac{\partial u}{\partial x} + \frac{\partial u}{\partial y} \cdot \frac{\partial y}{\partial x} = 0 \text{ or } \frac{dy}{dx} = -\frac{\frac{\partial u}{\partial x}}{\frac{\partial u}{\partial y}} = m_1 \text{ (say)}$$

Fig. 12.1.

Similarly, from (2) we get
$$\frac{dy}{dx} = -\frac{\frac{\partial v}{\partial x}}{\frac{\partial v}{\partial y}} = m_2 \text{ (say)}$$

$$\therefore \quad m_1 m_2 = \frac{\frac{\partial u}{\partial x}\cdot\frac{\partial v}{\partial x}}{\frac{\partial u}{\partial y}\cdot\frac{\partial v}{\partial y}}$$

Since $f(z)$ is analytic, u and v satisfy *C-R* equations

i.e.,
$$\frac{\partial u}{\partial x} = \frac{\partial v}{\partial y} \text{ and } \frac{\partial u}{\partial y} = -\frac{\partial v}{\partial x}$$

$\therefore$ From (3),
$$m_1 m_2 = \frac{\frac{\partial v}{\partial y}\cdot\frac{\partial v}{\partial x}}{-\frac{\partial v}{\partial x}\cdot\frac{\partial v}{\partial y}} = -1$$

Thus, the product of the slopes of the curves (1) and (2) is – 1. Hence, the curves intersect at right angles, *i.e.,* they form an orthogonal system.

***Example* 12.3.** Prove that function cos *hz* is analytic.

Solution : Here
$$f(z) = u + iv = \cos h\,(x + iy)$$

$$\Rightarrow \quad u + iv = \cos hx \cos y + i \sin hx \sin y$$

$$\therefore \quad u = \cos hx \cos y,\ v = \sin hx \sin y$$

$$\frac{\partial u}{\partial x} = \sin hx \cos y,\ \frac{\partial v}{\partial x} = \cos hx \sin y$$

$$\frac{\partial u}{\partial y} = -\cos hx \sin y,\ \frac{\partial v}{\partial y} = \sin hx \cos y$$

$$\therefore \quad \frac{\partial u}{\partial x} = \frac{\partial v}{\partial y} \text{ and } \frac{\partial u}{\partial y} = -\frac{\partial v}{\partial x}$$

Thus *C-R* equations are satisfied. Since sinh x, cosh x, sin y and cos y are continuous function, $\frac{\partial u}{\partial x}, \frac{\partial u}{\partial y}, \frac{\partial v}{\partial x}, \frac{\partial v}{\partial y}$ are also continuous functions. Thus $f(z)$ = cosh z satisfies *C-R* theorem and hence is analytic.

***Example* 12.4.** Prove that the function $f(z)$ defined by :

$$f(z) = \frac{x^3(x+i) - y^3(1-i)}{x^2+y^2},\ z \neq 0 \text{ and } f(0) = 0$$ is continuous and *C-R* equations are satisfied at the origin, yet $f(0)$ does not exist.

Solution : Here,
$$f(z) = \frac{x^3(1+i) - y^3(1-i)}{x^2+y^2}$$

$$f(z) = u + iv = \frac{(x^3 - y^3) + i(x^3 + y^3)}{x^2+y^2},\ (z \neq 0)$$

$$\Rightarrow \quad u = \frac{x^3 - y^3}{x^2+y^2},\ v = \frac{x^3 + y^3}{x^2+y^2}$$

$\therefore$ u and v are rational function of x and y with non-zero denominators. Thus u, v are continuous functions when $z \neq 0$. Hence $f(z)$ is continuous. To check continuity at $z = 0$. Changing u, v to polar co-ordinate by putting $x = r\cos\theta$, $y = r\sin\theta$, we get

$$u = r(\cos^3\theta - \sin^3\theta),\ v = r(\cos^3\theta + \sin^3\theta)$$

When $z \to 0$, $r \to 0$

$$\therefore \quad \lim_{z\to 0} u = \lim_{r\to 0} r(\cos^3\theta - \sin^3\theta) = 0$$

Similarly,

$$\lim_{z\to 0} v = 0.$$

$$\lim_{z\to 0} f(z) = 0 = f(0).$$

$\Rightarrow f(z)$ is continuous at $z = 0$

Hence, $f(z)$ is continuous for all values of z. At origin, we have :

$$\frac{\partial u}{\partial x} = \lim_{x\to 0} \frac{u(x,0) - u(0,0)}{x} = \lim_{y\to 0} \frac{x-0}{x} = 1$$

$$\frac{\partial u}{\partial y} = \lim_{y\to 0} \frac{u(0,y) - u(0,0)}{y} = \lim_{y\to 0} \frac{-y-0}{y} = -1$$

$$\frac{\partial v}{\partial x} = \lim_{x\to 0} \frac{v(x,0) - v(0,0)}{x} = \lim_{x\to 0} \frac{x-0}{x} = 1$$

$$\frac{\partial v}{\partial y} = \lim_{y\to 0} \frac{v(0,y) - v(0,0)}{y} = \lim_{y\to 0} \frac{y-0}{y} = 1$$

$$u_x = v_y \text{ and } u_y = -v_x$$

Hence *C-R* equation are satisfied at origin.

$$f'(0) = \lim_{z\to 0} \frac{f(z) - f(0)}{z}$$

$$= \lim_{z\to 0} \frac{(x^3 - y^3) + i(x^3 + y^3)}{(x^2 + y^2)(x + iy)}$$

Let $z \to 0$ along the line $y = mx$. then

$$f'(0) = \lim_{z\to 0} \frac{(x^3 - im^3x^3) + i(x^3 + m^3x^3)}{x^3(1+m^2)(1+im)}$$

$$= \mathop{lt}_{x\to 0} x^3 \frac{[(1 - im^3) + i(1 + m^3)]}{x^3(1+m^2)(1+im)}$$

$$= \frac{1+i}{(1+m^2)(1+im)}$$

Thus limit has different values for different values of m. Hence $f(0)$ does not exist.

***Example* 12.5.** If $f(z) = \dfrac{x^3y(y - ix)}{x^6 + y^2}$ $(z \neq 0)$, $f(0)$ then prove that $\dfrac{f(z) - f(0)}{z} \to 0$ as $z \to 0$ along any radius vector but not as $z \to 0$ along $y = mx^3$.

Solution : Let us consider that $z \to 0$ along $y = mx$, then

$$f'(z) = \lim_{z \to 0} \frac{f(z) - f(0)}{z}$$

$$= \lim_{z \to 0} \frac{x^3 y(y - ix)}{(x^6 + y^2)(x + iy)}$$

$$= \lim_{x \to 0} \frac{x^3 (mx)[mx - ix]}{(x^6 + m^2 x^2)(x + imx)}$$

$$= \lim_{x \to 0} \frac{x^5 m(m - i)}{x^3 (x^4 + m^2)(1 + im)} = \lim_{x \to 0} \frac{x^2 m(m - i)}{(x^4 + m^2)(1 + im)} = 0$$

Now, let $z \to 0$ along $y = mx^3$, then

$$\lim_{z \to 0} \frac{f(z) - f(0)}{z} = \mathop{lt}_{x \to 0} \frac{x^3 y(y - ix)}{(x^6 + y^2)(x + iy)}$$

$$= \mathop{lt}_{x \to 0} \frac{mx^6 (mx^3 - ix)}{(x^6 + m^2 x^6)(x + imx^3)}$$

$$= \mathop{lt}_{x \to 0} \frac{mx^7 (mx^2 - i)}{x^7 (1 + m^2)(1 + imx^2)} = \frac{-mi}{1 + m^2}.$$

Thus limit has different values for different values of m. Therefore $\frac{f(z) - f(0)}{z} \neq 0$ as $z \to 0$ along $y = mx^3$ and $f'(0)$ does not exist.

***Example* 12.6.** Determine the analytic function whose real part is $e^x (x \cos y - y \sin y)$.

Solution : $f(z) = u + iv$ be an analytic function, where

$$u = e^x (x \cos y - y \sin y)$$

differentiating u partially w.r.t. x and y, we get

$$\frac{\partial u}{\partial x} = e^x [x \cos y - y \sin y] + e^x [\cos y]$$

$$= e^x [x \cos y - y \sin y + \cos y] \qquad ...(1)$$

$$\frac{\partial u}{\partial y} = e^x [-x \sin y - (y \cos y + \sin y)]$$

$$= -e^x [x \sin y + y \cos y + \sin y] \qquad ...(2)$$

Since the function $f(z)$ is analytic, v and u should satisfy C-R equations :

$$\frac{\partial u}{\partial x} = \frac{\partial v}{\partial y} \text{ and } \frac{\partial u}{\partial y} = -\frac{\partial v}{\partial x}$$

Now $$\frac{\partial v}{\partial y} = \frac{\partial u}{\partial x} = e^x [x \cos y - y \sin y + \cos y]$$

Integrating w.r.t. 'y' treating x as constant, we have

$$v = e^x [x \sin y - \{-y \cos y + \sin y\} + \sin y] + \phi(x)$$

$$= e^x [x \sin y + y \cos y - \sin y + \sin y] + \phi(x)$$

$$= e^x [x \sin y + y \cos y] \qquad ...(3)$$

where $\phi(x)$ is an arbitrary function of x.

$$\therefore \quad \frac{\partial v}{\partial x} = e^x [x \sin y + y \cos y] + e^x \sin y + \phi(x)$$

$$= e^x [x \sin y + y \cos y + \sin y] + \phi'(x)$$

But $$\frac{\partial v}{\partial x} = \frac{\partial u}{\partial y} = e^x [x \sin y + y \cos y + \sin y] \quad \text{(From 2)}$$

$$\Rightarrow \quad \phi'(x) = 0$$

i.e., $\phi(x) = c$, where c is an arbitrary const.

Substituting in (3) we get

$$v = e^x [x \sin y + y \cos y] + c$$

$$f(z) = u + iv$$

$$= e^x [x \cos y - y \sin y] + i\, e^x [x \sin y + y \cos y] + i\, c$$

$$= e^x [(x + iy) \cos y + i\,(x + iy) \sin y] + ic$$

$$= e^x (x + iy) [\cos y + i \sin y] + ic$$

$$= e^x [x + iy) e^{iy} + ic$$

$$= e^{x+iy} (x + iy) + ic$$

$$= e^z \,.\, z + ic \quad [z = x + iy]$$

12.10 MILNE-THOMSON'S METHOD

This method determines the analytic function $f(z)$ when either u or v is given.

Since $$z = x + iy, \; \bar{z} = x - iy$$

From here, $$x = \frac{z + \bar{z}}{2}, \; y = \frac{z - \bar{z}}{2i}$$

Now, $$f(z) = u(x, y) + iv\,(x, y)$$

$$= u\left(\frac{z + \bar{z}}{2}, \frac{z - \bar{z}}{2i}\right) + iv\left(\frac{z + \bar{z}}{2}, \frac{z - \bar{z}}{2i}\right)$$

Considering this as an identity in two independent variables z, $\bar{z}$ and putting $\bar{z} = z$, we get

$$f(z) = u(z, 0) + iv(z, 0)$$

which is same when x is replaced by z and y by 0.

The last example 12.6 cm can also be solved by Milne-Thomson method as follows :

Now $$f(z) = u + iv$$

$$\Rightarrow \quad f'(z) = \frac{\partial u}{\partial x} + i\frac{\partial v}{\partial x} = \frac{\partial u}{\partial x} - i\frac{\partial u}{\partial y}$$

$$= e^x [x \cos y - y \sin y + \cos y] - i\, e^x [-x \sin y - y \cos y - \sin y]$$

Replace x by z and y by 0.

$$f'(z) = e^x [z + 1] - i\, e^x [-0]$$

$$f'(z) = e^x [z + 1]$$

Integrating w.r.t. 'z'

$f(z) = e^x (z + 1) - e^x + iC$, considering the constant as imaginary since u is given and does not contain any constant.

$$f(z) = e^x z + iC$$

***Example* 12.7.** Find analytic function given,

$$f(z) = u(r, \theta) + iv(r, \theta) \text{ such that}$$

$$v(r, \theta) = r^2 \cos 2\theta - r\cos\theta + 2$$

Solution : Using C-R equations in polar coordinates which are :

$$\frac{\partial u}{\partial r} = \frac{1}{r}\frac{\partial v}{\partial \theta}, \frac{\partial v}{\partial r} = \frac{-1}{r}\frac{\partial u}{\partial \theta}$$

$$r\frac{\partial u}{\partial r} = \frac{\partial v}{\partial \theta} = -2r^2 \sin 2\theta + r\sin\theta \quad \text{...(1)}$$

$$-\frac{1}{r}\frac{\partial u}{\partial \theta} = \frac{\partial v}{\partial r} = 2r\cos 2\theta - \cos\theta \quad \text{...(2)}$$

From (1)

$$\frac{\partial u}{\partial r} = -2r\sin 2\theta + \sin\theta$$

Integrating w.r.t. 'r'

$$u = -r^2 \sin 2\theta + r\sin\theta + \phi(\theta)$$

where $\phi(\theta)$ is an arbitrary function :

$$\frac{\partial u}{\partial \theta} = -2r^2 \cos 2\theta + r\cos\theta + \phi'(\theta) \quad \text{...(3)}$$

From (2) and (3), we get

$$-2r^2 \cos 2\theta + r\cos\theta = \frac{\partial u}{\partial \theta} = -2r^2\, 2\theta + r\cos\theta + \phi'(\theta)$$

$$\phi'(\theta) = 0 \text{ or } \phi(\theta) = C$$

Thus

$$u = -r^2 \sin 2\theta + r\sin\theta + C$$

Hence

$$f(z) = u + iv = r^2(-\sin 2\theta + i\cos 2\theta) + r(\sin\theta - i\cos\theta) + C + 2i$$

$$= r^2(i^2 \sin 2\theta + i\cos 2\theta) - r(i\cos\theta - \sin\theta) + C + 2i$$

$$= i\,r^2 e^{i2\theta} - ire^{i\theta} + C + 2i$$

$$f(z) = i(r^2 e^{2i\theta} - re^{i\theta}) + C + 2i$$

***Example* 12.8.** Determine the analytic function.

$$w = u + iv \text{ if } v = \log(x^2 + y^2) + x - 2y$$

Solution :

$$v = \log(x^2 + y^2) + x - 2y$$

$$\frac{\partial v}{\partial x} = \frac{2x}{x^2 + y^2} + 1$$

$$\frac{\partial v}{\partial y} = \frac{2y}{x^2 + y^2} - 2$$

Since

$$w = u + iv$$

$$\frac{dw}{dz} = \frac{\partial u}{\partial x} + i\frac{\partial v}{\partial x} = \frac{\partial v}{\partial y} + i\frac{\partial v}{\partial x}$$

$$= \left(\frac{2y}{x^2 + y^2} - 2\right) + i\left(\frac{2x}{x^2 + y^2} + 1\right)$$

Replacing x by z and y by 0, we get

$$\frac{dw}{dz} = -2 + i\left(\frac{2}{z}+1\right) = (i-2) + \frac{2i}{z}$$

Integrating w.r.t. 'z' we have $w = (i-2)z + 2\,i\log z + C$.

***Example* 12.9.** If $f(z) = u + iv$ is an analytic function, find $f(z)$ given $u - v = e^x(\cos y - \sin y)$.

Solution : Given

$$f(z) = u + iv \qquad ...(1)$$

$$\text{if } f(z) = iu - v \qquad ...(2)$$

Adding (1) and (2), we get

$$(1+i)\,f(z) = (u-v) + i(u+v)$$

or

$$F(z) = U + iV$$

where

$$F(z) = (1+i)\,f(z),\ U = u - v,\ V = u + v.$$

$$\therefore \quad U = u - v = e^x(\cos y - \sin y). \qquad \text{(Given)}$$

Now, we proceed to find F(z) whose imaginary part is given.

$$\frac{\partial U}{\partial x} = e^x(\cos y - \sin y)$$

$$\frac{\partial U}{\partial y} = e^x(-\sin y - \cos y)$$

Since $f(z)$ is analytic, so is F(z).

$$F'(z) = \frac{\partial U}{\partial x} + i\frac{\partial V}{\partial x}$$

$$= \frac{\partial U}{\partial x} - i\frac{\partial U}{\partial y} \qquad \text{[By } C\text{-}R \text{ equations]}$$

$$= e^x(\cos y - \sin y) - i\,e^x(-\sin y - \cos y)$$

$$= e^x(\cos y - \sin y) + i\,e^x(\sin y + \cos y)$$

By Milne Thomson's method, replacing x by z and y by 0, we get

$$F'(z) = e^z(1-0) + i\,e^z(1)$$

$$F'(z) = e^z + ie^z$$

$$= (1+i)\,e^z$$

Integrating

$$F(z) = (1+i)\,e^z + C$$

$$(1+i)\,f(z) = (1+i)\,e^x + C$$

$$f(z) = e^z + a \qquad \text{where } \left(a = \frac{C}{1+i}\right)$$

***Example* 12.10.** If $f(z)$ is a regular function of z, prove that $\left(\frac{\partial^2}{\partial x^2} + \frac{\partial^2}{\partial y^2}\right)|f(z)|^2 = 4\,|f'(z)|^2$

Solution : Let $f(z) = u + iv$ so that $|f(z)| = \sqrt{u^2+v^2}$

$$|f(z)|^2 = u^2 + v^2 = \psi(x, y)$$

$$\frac{\partial \psi}{\partial x} = 2u\frac{\partial u}{\partial x} + 2v\frac{\partial v}{\partial x}$$

$$\frac{\partial^2 \psi}{\partial x^2} = 2\left[u\frac{\partial^2 u}{\partial x^2} + \left(\frac{\partial u}{\partial x}\right)^2 + v\frac{\partial^2 v}{\partial x^2} + \left(\frac{\partial v}{\partial x}\right)^2\right]$$

Similarly,

$$\frac{\partial^2 \psi}{\partial y^2} = 2\left[u\frac{\partial^2 u}{\partial y^2} + \left(\frac{\partial u}{\partial y}\right)^2 + v\frac{\partial^2 v}{\partial y^2} + \left(\frac{\partial v}{\partial y}\right)^2\right]$$

Adding, we get

$$\frac{\partial^2 \psi}{\partial x^2} + \frac{\partial^2 \psi}{\partial y^2} = 2\left[u\left(\frac{\partial^2 u}{\partial x^2} + \frac{\partial^2 u}{\partial y^2}\right) + \left(\frac{\partial u}{\partial x}\right)^2 + \left(\frac{\partial u}{\partial y}\right)^2 + v\left(\frac{\partial^2 v}{\partial x^2} + \frac{\partial^2 v}{\partial y^2}\right) + \left(\frac{\partial v}{\partial x}\right)^2 + \left(\frac{\partial v}{\partial y}\right)^2\right] \quad ...(1)$$

Since $f(z) = u + iv$ is a regular function of z, u and v satisfy C-R equations and Laplace's equation.

$$\frac{\partial u}{\partial x} = \frac{\partial v}{\partial y}, \frac{\partial u}{\partial y} = \frac{-\partial v}{\partial x} \text{ and } \frac{\partial^2 u}{\partial x^2} + \frac{\partial^2 u}{\partial y^2} = 0 = \frac{\partial^2 v}{\partial x^2} + \frac{\partial^2 v}{\partial y^2}$$

∴ From (1), we get

$$\frac{\partial^2 \psi}{\partial x^2} + \frac{\partial^2 \psi}{\partial y^2} = 2\left[0 + \left(\frac{\partial u}{\partial x}\right)^2 + \left(\frac{\partial v}{\partial x}\right)^2 + 0 + \left(\frac{\partial v}{\partial x}\right)^2 + \left(\frac{\partial u}{\partial x}\right)^2\right]$$

$$= 4\left[\left(\frac{\partial u}{\partial x}\right)^2 + \left(\frac{\partial v}{\partial x}\right)^2\right] \quad ...(2)$$

$$f(z) = u + iv$$

∴

$$f'(z) = \frac{\partial u}{\partial x} + i\frac{\partial v}{\partial x} \text{ and } |f'(z)|^2 = \left(\frac{\partial u}{\partial x}\right)^2 + \left(\frac{\partial v}{\partial x}\right)^2$$

substituting in (2) we get

$$\left(\frac{\partial^2 \psi}{\partial x^2} + \frac{\partial^2 \psi}{\partial y^2}\right) = 4\,|f'(z)|^2$$

or

$$\left(\frac{\partial^2}{\partial x^2} + \frac{\partial^2}{\partial y^2}\right)\psi = 4\,|f'(z)|^2$$

i.e.;

$$\left(\frac{\partial^2}{\partial x^2} + \frac{\partial^2}{\partial y^2}\right)|f(z)|^2 = 4\,|f'(z)|^2$$

***Example* 12.11.** If $f(z)$ is a holomorphic function of z, show that $\left\{\frac{\partial}{\partial x}|f(z)|\right\}^2 + \left\{\frac{\partial}{\partial y}|f(z)|\right\}^2 = |f'(z)|^2$.

Solution : Let $f(z) = u + iv$

Then

$$|f(z)| = (u^2 + v^2)^{\frac{1}{2}} = \psi(x, y)$$

$$\frac{\partial \psi}{\partial x} = \frac{1}{2}(u^2 + v^2)^{\frac{-1}{2}}\left[2u\frac{\partial u}{\partial x} + 2v\frac{\partial v}{\partial x}\right]$$

$$= \frac{1}{\sqrt{u^2 + v^2}}\left(u\frac{\partial u}{\partial x} + v\frac{\partial v}{\partial x}\right)$$

Similarly,

$$\frac{\partial \psi}{\partial y} = \frac{1}{\sqrt{u^2 + v^2}}\left(u\frac{\partial u}{\partial y} + v\frac{\partial v}{\partial y}\right)$$

$$= \frac{1}{\sqrt{u^2 + v^2}}\left(-u\frac{\partial v}{\partial x} + v\frac{\partial u}{\partial x}\right)$$

∴

$$\left(\frac{\partial \psi}{\partial x}\right)^2 + \left(\frac{\partial \psi}{\partial y}\right)^2 = \frac{1}{u^2 + v^2}\left[\left(u\frac{\partial u}{\partial x} + v\frac{\partial v}{\partial x}\right)^2 + \left(-u\frac{\partial v}{\partial x} + v\frac{\partial u}{\partial x}\right)^2\right]$$

$$= \frac{1}{u^2+v^2}\left[(u^2+v^2)\left(\frac{\partial u}{\partial x}\right)^2 + (u^2+v^2)\left(\frac{\partial v}{\partial x}\right)^2\right]$$

$$= \left(\frac{\partial u}{\partial x}\right)^2 + \left(\frac{\partial v}{\partial x}\right)^2$$

$$= |f'(z)|^2 \qquad \left[f'(z) = \frac{\partial u}{\partial x} + i\frac{\partial v}{\partial x}\right]$$

Hence proved.

Example 12.12. Prove that $u = x^2 - y^2 - 2xy - 2x + 3y$ is harmonic. Also find $f(z) = u + iv$ which is analytic.

Solution : Given $u = x^2 - y^2 - 2xy - 2x + 3y$

$$\therefore \quad \frac{\partial u}{\partial x} = 2x - 2y - 2$$

$$\frac{\partial^2 u}{\partial x^2} = 2 \qquad \text{...(1)}$$

$$\frac{\partial u}{\partial y} = -2y - 2x + 3$$

$$\frac{\partial^2 u}{\partial y^2} = -2$$

Adding (1) and (2)

$$\frac{\partial^2 u}{\partial x^2} + \frac{\partial^2 u}{\partial y^2} = 0 \text{ which proves that } u \text{ is harmonic.}$$

Now, let $f(z) = u + iv$

Then
$$f'(z) = \frac{\partial u}{\partial x} + i\frac{\partial v}{\partial x} \qquad \text{[C-R equations]}$$

$$= \frac{\partial u}{\partial x} - i\frac{\partial u}{\partial y}$$

$$= (2x - 2y - 2) - i(-2y - 2x + 3)$$

Replacing x by z and y by 0

$$f'(z) = (2z - 2) - i(-2z + 3)$$

Integrating w.r.t. 'z'

$$f(z) = (z^2 - 2z) - i(-z^2 + 3z) + C$$

$$= (1 + i)z^2 - (2 + 3i)z + C$$

Example 12.13. Using the polar form, of Cauchy-Riemann equations deduce that $\frac{\partial^2 u}{\partial r^2} + \frac{1}{r}\frac{\partial u}{\partial r} + \frac{1}{r^2}\frac{\partial^2 u}{\partial \theta^2} = 0$.

Solution : *C-R* equations in polar form are :

$$\frac{\partial u}{\partial r} = \frac{1}{r}\frac{\partial v}{\partial \theta} \qquad \text{...(1)}$$

and
$$\frac{\partial v}{\partial r} = \frac{-1}{r}\frac{\partial u}{\partial \theta} \qquad \text{...(2)}$$

Differentiating (1) partially w.r.t. 'r', we get

$$\frac{\partial^2 u}{\partial r^2} = \frac{-1}{r^2}\frac{\partial v}{\partial \theta} + \frac{1}{r}\frac{\partial^2 v}{\partial r \partial \theta} \qquad \text{...(3)}$$

From (1) we get

$$\frac{1}{r}\frac{\partial u}{\partial r} = \frac{1}{r^2}\frac{\partial v}{\partial \theta} \quad ...(4)$$

From (2),
$$\frac{\partial u}{\partial \theta} = -r\frac{\partial v}{\partial r}$$

Differentiating partially w.r.t. 'θ'.

$$\frac{\partial^2 u}{\partial \theta^2} = -r\frac{\partial^2 v}{\partial \theta \partial r} \Rightarrow \frac{1}{r^2}\frac{\partial^2 u}{\partial \theta^2} = \frac{-1}{r}\frac{\partial^2 v}{\partial \theta \partial r} \quad ...(5)$$

Adding (3), (4) and (5) we get

$$\frac{\partial^2 u}{\partial r^2} + \frac{1}{r}\frac{\partial u}{\partial r} + \frac{1}{r^2}\frac{\partial^2 u}{\partial \theta^2} = -\frac{1}{r^2}\frac{\partial v}{\partial \theta} + \frac{1}{r^2}\frac{\partial v}{\partial \theta} \quad \left[\frac{\partial^2 v}{\partial r \partial \theta} = \frac{\partial^2 v}{\partial \theta \partial r}\right]$$

$$= 0$$

EXERCISE 12.2

1. (*i*) Show that $f(z) = xy + iy$ is everyone continuous but not analytic.
   (*ii*) Show that $f(z) = z + 2\bar{z}$ is not analytic anywhere in the complex plane.
2. Examine the nature of function :
$$f(z) = \begin{bmatrix} \dfrac{x^2 y^5 (x+iy)}{x^4 + y^{10}}, & z \neq 0 \\ 0 & z = 0 \end{bmatrix}$$ in a region including the origin.
3. Show that $|z|^2$ is continuous everywhere but not analytic at any point.
4. Show that function $f(x) = z\,|z|$ is not analytic anywhere.
5. Prove that function $f(z) = \sinh z$ is analytic and find its derivative.
6. Determine the analytic function given :
   (*i*) $u = e^{2x}(x \cos 2y - y \sin 2y)$
   (*ii*) $u = \cos x$ and $\cosh y$
   (*iii*) $u = e^x [(x^2 - y^2)\cos y - 2xy \sin y]$
   (*iv*) $v = e^{-x}(x \cos y + y \sin y)$
   (*v*) $u + v = \dfrac{x}{x^2 + y^2}$, when $f(1) = 1$

Ans.

(*i*) $ze^{2z} + iC$ (*ii*) $\cot z + C$
(*iii*) $z^2e^2 + ic,$ (*iv*) $ize^{-z} + C$
(*v*) $1 - \frac{1}{1-i}\left(\frac{1}{z} - 1\right)$

7. If $f(z)$ is an analytic function of z, prove that $\left(\dfrac{\partial^2}{\partial x^2} + \dfrac{\partial^2}{\partial y^2}\right)|Rf(z)|^2 = 2\,|f'(z)|^2$

12.11 COMPLEX INTEGRATION

Let $f(z)$ be a continuous function of the complex variable $z = x + iy$ defined at all points of a curve C having end points A and B. Divide the curve C into n parts at the points.

$$A = S_0(z_0), S_1(z_1), \ldots\ldots\ldots S_i(z_i), \ldots\ldots\ldots S_n(z_n) = B$$

Let $\delta z_i = z_i - z_{i-1}$ and ξ_i be any point on the arc $S_{i-1} S_i$. Then the limit of the sum

$\sum_{i=1}^{n} f(\xi_i)\delta z_i$ as $n \to \infty$ and each $\delta z_i \to 0$, if it exists, is called the line integral of $f(z)$ along the curve C. We denote it by.

$$\int_C f(z)\,dz \text{ or } \int_{AB} f(z)\,dz$$

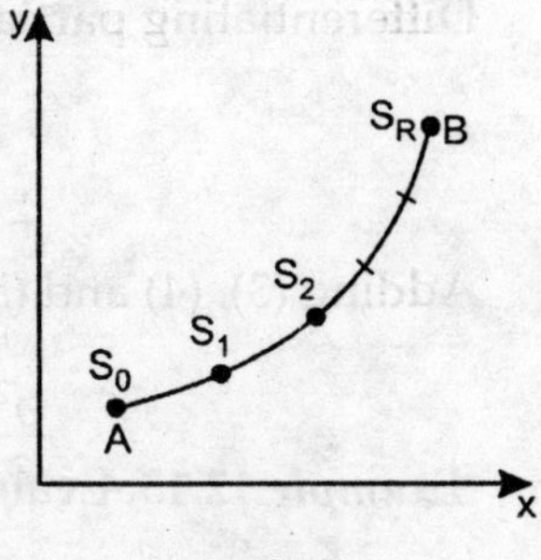

Fig. 12.2.

If $\qquad f(z) = u(x, y) + iv(x, y)$ and

Since $\qquad dz = dx + idy$

$$\therefore \quad \int_C f(z)\,dz = \int_C (u + iv)(dx + idy)$$

$$= \int_C (udx - vdy) + i\int_C (vdx + udy)$$

which shows that evaluation of line integral of a complex function can be reduced to the evaluation of two integrals of real functions.

Remarks :

1. If the function $f(z)$, is analytic then the line integral of $f(z)$ is independent of the path joining the points.
2. When the same path of integration is used in each integral, then

$$\int_a^b f(z)\,dx = -\int_b^a f(z)\,dz$$

3. If C is point in the arc joining a and b then

$$\int_a^b f(z)\,dz = \int_a^c f(z)\,dz + \int_c^b f(z)\,dz$$

***Example* 12.14.** Evaluate : $\int_0^{1+i} (x^2 - iy)\,dz$ along the paths.

(*a*) $y = x$ (*ii*) $y = x^2$

Solution : Along the line $y = x$

$$dy = dx \text{ and so } dz = dx + idy. \text{ Also } x \text{ varies from 0 to 1.}$$

$$= (1 + i)\,dx$$

$$\int_0^{1+i} (x^2 - iy)\,dz = \int_0^{1+i} (x^2 - ix)(1 + i)\,dx$$

$$= (1 + i)\left[\frac{x^3}{3} - \frac{ix^2}{2}\right]_0^1$$

$$= (1 + i)\left[\frac{1}{3} - \frac{i}{2}\right]$$

$$= \frac{(1 + i)(2 - 3i)}{6} = \frac{5}{6} - \frac{1}{6}i$$

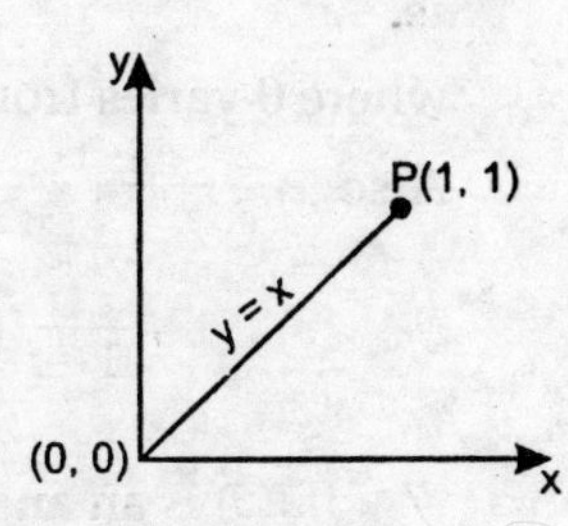

Fig. 12.3.

(*b*) Along the parabola $y = x^2$, $dy = 2x\,dx$

$\therefore$ $dz = (1 + 2ix)dx$ and x varies from 0 to 1 [as limits are $z = 0$ to $z = 1 + i$]

$$\int_0^{1+i} (x^2 - iy)\,dz = \int_0^1 (x^2 - ix^2)(1 + 2i)\,dx$$

$$= (1 - i)\int_0^1 x^2(1 + 2ix)\,dx$$

$$= (1 - i)\left[\frac{x^3}{3} + i\frac{x^4}{2}\right]_0^1$$

$$= (1 - i)\left[\frac{1}{3} + \frac{1}{2}i\right] = \frac{5}{6} + \frac{1}{6}i$$

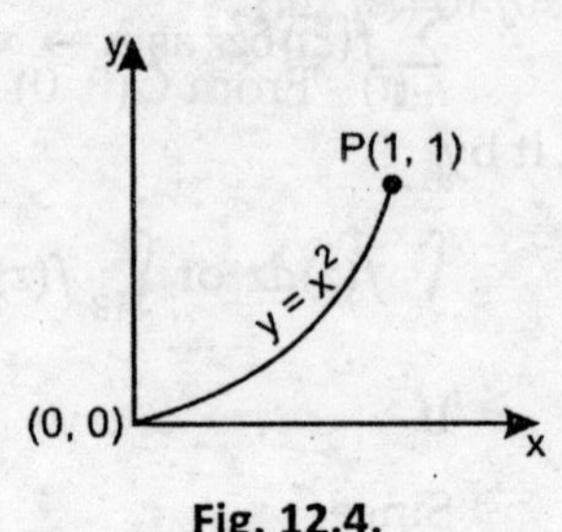

Fig. 12.4.

***Example* 12.15.** Evaluate $\int_0^{1+2i} \bar{z}\,dz$ along the curve $z = t^2 + it$.

Solution : Equations of path of integration are $x = t^2$ and $y = t$,

$\therefore$ $dx = 2t\,dt$, $dy = dt$ and z varies from $z = 0$ to $z = 4 + 2i$.

$\Rightarrow$ x varying from 0 to 4 implies t varies from 0 to ± 2 and y varying from 0 to 2 implies t varies from 0 to 2. Considering the common limit we get t varies from 0 to 2.

$$\therefore \int_0^{4+2i} (x - iy)(dx + i\,dy) = \int_0^2 (t^2 - it)(2t\,dt + dt)$$

$$= \int_0^2 (t^2 - it)(2t + 1)\,dt$$

$$= \int_0^2 (2t^3 + (1 - 2i)t^2 - it)\,dt$$

$$= \left|\frac{t^4}{2} + (1 - 2i)\frac{t^3}{3} - \frac{it^2}{2}\right|_0^2$$

$$= 8 + (1 - 2i)\frac{8}{3} - 2i$$

$$= \frac{32}{3} - \frac{22}{3}i.$$

***Example* 12.16.** Prove that $\oint \frac{dz}{(z - a)} = 2\pi i$ where C is the circle $|z - a| = r$.

Solution : The equation of the circle C is

$$|z - a| = r \text{ or } z - a = re^{i\theta}$$

where θ varies from 0 to 2π as z describes C once in anticlockwise direction.

Also $$dz = ire^{i\theta}\,d\theta$$

$$\oint_C \frac{dz}{(z - a)} = \int_0^{2\pi} \frac{ire^{i\theta}}{re^{i\theta}}\,d\theta = [i\theta]_0^{2\pi}$$

$$= 2\pi i$$

***Example* 12.17.** Evaluate $\int_C (z^2 + 3z + 2)\,dz$ where C is the are of the cycloid $x = a(\theta + \sin\theta)$, $y = a(1 - \cos\theta)$ between points (0, 0) and $(\pi a, 2a)$.

Solution : Since $f(z)$ is analytic between (0, 0) and $(\pi a, 2a)$ therefore integration is independent of the path. So considering integration first along x-axis *i.e.,* (0, 0) to $(\pi a, 0)$ and then along a line || to y-axis, from $(\pi a, 0)$ to $(\pi a, 2a)$.

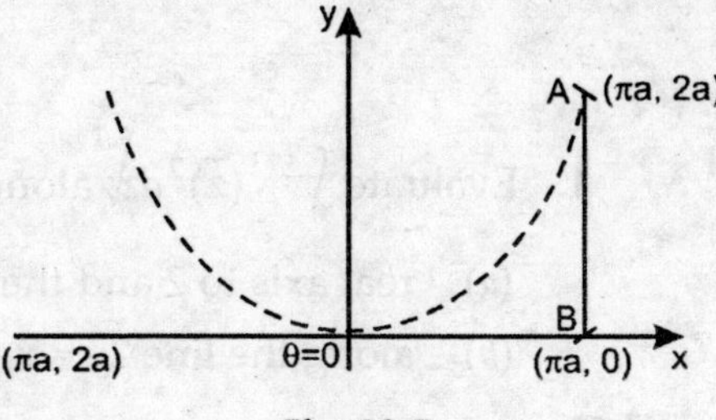

Fig. 12.5.

(*i*) From $O(0, 0)$ to $B(\pi a, 0)$: $y = 0$ and x varies from 0 to πa.

$$f(z) = z^2 + 3z + 2$$
$$= x^2 + 3x + 2$$

$$\int_{OB} f(z)\,dz = \int_0^{\pi a} (x^2 + 3x + 2)\,dx = \left[\frac{x^3}{3} + \frac{3x^2}{2} + 2x\right]_0^{\pi a}$$

$$= \frac{\pi^3 a^3}{3} + \frac{3\pi^2 a^2}{2} + 2\pi a = \frac{\pi a}{6}(2\pi^2 a^2 + 9\pi a + 12)$$

(*ii*) Along BA, $x = \pi a$ and y varies from 0 to $2a$.

$$z = x + iy = \pi a + iy$$
$$f(z) = z^2 + 3z + 2$$
$$= (\pi a + iy)^2 + 3(\pi a + iy) + 2$$

$$\int_{BA} f(z)\,dz = \int_0^{2a} [(\pi a + iy)^2 + 3(\pi a + iy) + 2]\,i\,dy$$

$$= \left[\frac{(\pi a + iy)^3}{3i} + \frac{3(\pi a + iy)^2}{2i} + 2y\right]_0^{2a} i$$

$$= \frac{1}{3}(\pi a + i2a)^3 + \frac{3}{2}(\pi a + i2a)^2 + 2i.2a$$

$$= \frac{a^3}{3}(\pi + 2i)^3 + \frac{3a^2}{2}(\pi + 2i)^2 + 4ia$$

$$\int_C f(z)\,dz = \int_{OB} f(z)\,dz + \int_{BA} f(z)\,dz$$

$$\int_C f(z)\,dz = \frac{\pi a}{6}(2\pi^2 a^2 + 9\pi a + 12) + \frac{a^3}{3}(\pi + 2i)^3 + \frac{3a^2}{2}(\pi + 2i)^2 + 4ia$$

Example **12.18.** Prove that $\oint_C \frac{1}{z}dz = -\pi i$ or πi according as C is the semicircular arc $|z| = 1$ from -1 to above or below the real axis.

Solution : The parametric equation of the circle $|z| = 1$ is

$$z = e^{i\theta} \Rightarrow dz = ie^{i\theta}, \text{ where } 0 \le \theta \le 2\pi.$$

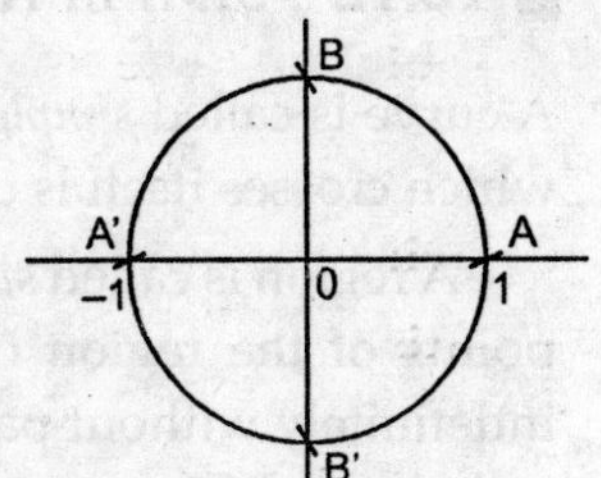

Fig. 12.6.

As z moves from -1 to 1 along the upper semicircular arc, θ varies from π to 0.

Therefore, we have

$$\int_{A'BA} \frac{dz}{z} = \int_\pi^0 \frac{ie^{i\theta}}{e^{i\theta}}d\theta = i\int_\pi^0 d\theta$$
$$= -\pi i$$

Also if z moves from -1 to 1 along the lower semicircle, θ varies from π to 2π, $\therefore$ we have

$$\int_{A'B'A} f(z)\,dz = \int_\pi^{2\pi} \frac{ie^{i\theta}}{e^{i\theta}}d\theta = i\int_\pi^{2\pi} d\theta = \pi i. \text{ Hence proved.}$$

EXERCISE 12.3

1. Evaluate $\int_0^{2+i} (\bar{z})^2 dz$, along

 (a) real axis to 2 and then vertically to 2 + i.

 (b) along the line 2y = x.

Ans.

 (a) $2 + \frac{11}{3}i$, (b) $\frac{10}{3} - \frac{5}{3}i$

2. Evaluate $\int_{1-i}^{2+i} (2x + iy + 1)\, dz$, along

 (a) The straight line joining (1 – i) to (2 + i)

 (b) The curve $x = 1 + t,\ y = 2t^2 - 1$.

Ans.

 (a) $4 + 8i$, (b) $4 + \frac{25}{3}i$

3. Evaluate $\oint_C (z - z^2)\, dz$, where C is the lower half of above circle. $|z| = 1$. What is the value of this integral if C is lower half of the above circle.

Ans. $\frac{2}{3}, \frac{-2}{3}$

4. Prove that :

$$\oint_C (z - a)^n dz = 0 \qquad [n \text{ is an integer} \neq -1, \text{ where C is the circle } |z - a| = r]$$

5. Show that for every path between the limits $\int_{-2}^{-2+i} (2 + z)^2 dz = \frac{-i}{3}$.

6. Evaluate $\oint_C |z|^2 dz$ around the square with vertices at (0, 0), (1, 0), (1, 1) and (0, 1).

12.12 SIMPLY AND MULTIPLY CONNECTED REGIONS

A curve is called *simple closed curve* if it does not cross itself (Fig. 12.7). A curve which crosses itself is called a *multiple curve* (Fig. 12.8).

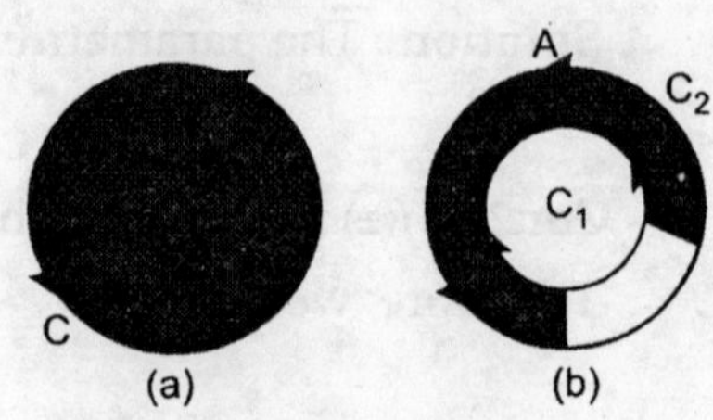

Fig. 12.7.

A region is called *simply connected* if every closed curve in the region encloses points of the region only, *i.e.*, every closed curve lying it can be contracted indefinitely without passing out of it. A region which is not simply connected is called a *multiply connected* region. In plain terms, a simply connected region in one which has no holes. Fig. 12.8 shows a multiply connected region *R* enclosed between two separate curves C_1 and C_2. (There can be more than two separate curves).

12.13 CAUCHY'S INTEGRAL THEOREM

Statement : If $f(z)$ is an analytic function $f'(z)$ is continuous at each point within and on a simple closed curve C, then

$$\oint_C f(z)\,dz = 0$$

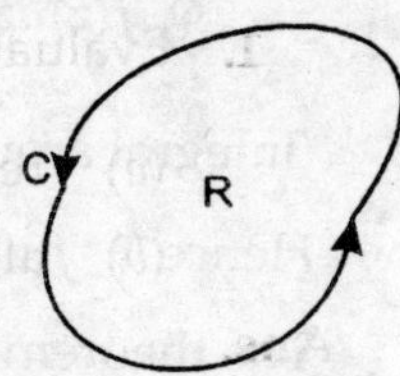

Fig. 12.8.

Proof : Let R be the region bounded by the curve C.

Let $$f(z) = u(x, y) + iv(x, y), \text{ then}$$

$$\oint_C f(z)\,dz = \oint_C (u + iv)(dx + idy)$$

$$= \oint_C (udx - vdy) + i\oint_C (vdx + udy) \qquad ...(1)$$

Since $f'(z)$ is continuous, the partial derivatives $\frac{\partial u}{\partial x}, \frac{\partial u}{\partial y}, \frac{\partial v}{\partial x}, \frac{\partial v}{\partial y}$ are also continuous in R. Hence by Green's Theorem, we have

$$\oint_C f(z)\,dz = \iint_R \left(-\frac{\partial v}{\partial x} - \frac{\partial u}{\partial y}\right)dx\,dy + i\iint_R \left(\frac{\partial u}{\partial x} - \frac{\partial v}{\partial y}\right)dx\,dy \qquad ...(2)$$

Now $f(z)$ being analytic at each point of the region R, by Cauchy-Riemann equations, we have

$$\frac{\partial u}{\partial x} = \frac{\partial v}{\partial y} \text{ and } \frac{\partial u}{\partial y} = -\frac{\partial v}{\partial x}$$

Thus, the two double integrals in (2) vanish.

Hence $$\int_C f(z)\,dz = 0.$$

Cor. 1. If $f(z)$ is analysis analytiç in a region R and P and Q are two points in R, then $\int_P^Q f(z)\,dz$ is independent of the path joining P and Q lying entirely in R.

Let PAQ and PBQ be any two paths joining P and Q.

By Cauchy's theorem,

$$\int_{PAQBP} f(z)\,dz = 0$$

$$\Rightarrow \quad \int_{PAQ} f(z)\,dz + \int_{QBP} f(z)\,dz = 0$$

$$\Rightarrow \quad \int_{PAQ} f(z)\,dz + \int_{PBQ} f(z)\,dz = 0$$

Hence $$\int_{PAQ} f(z)\,dz = \int_{PBQ} f(z)\,dz$$

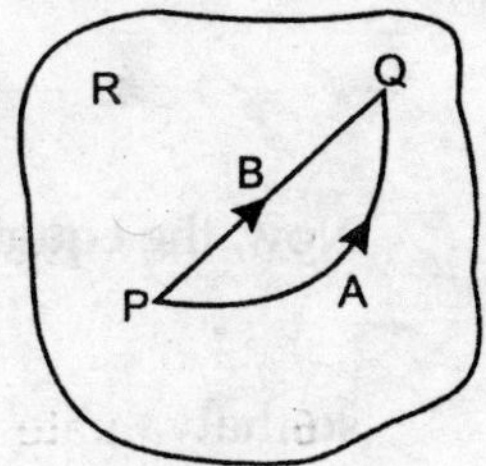

Fig. 12.9.

Cor 2. If $f(z)$ is the region bounded by two simple closed curves C and C_1 then

$$\oint_C f(z)\,dz = \oint_{C_1} f(z)\,dz$$

Let AB be a cross-cut joining the curves C and C_1, then the doubly connected region becomes simply connected.

By Cauchy's theorem $\int_{APQABDEBA} f(z)\,dz = 0$

$$\Rightarrow \quad \int_{APQA} f(z)\,dz + \int_{AB} f(z)\,dz + \int_{BDEB} f(z)\,dz + \int_{BA} f(z)\,dz = 0$$

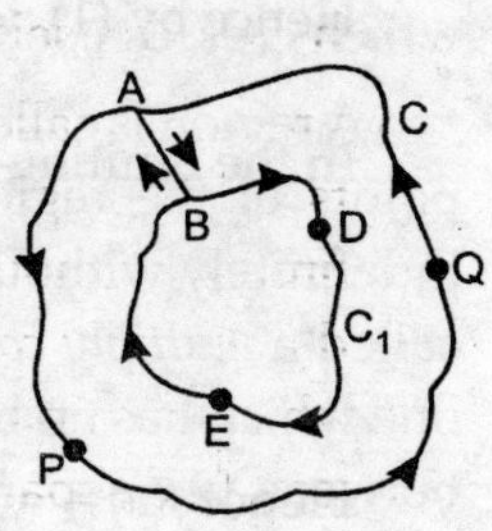

Fig. 12.10.

$$\Rightarrow \int_C f(z)\,dz + \int_{AB} f(z)\,dz - \int_{BEDB} f(z)\,dz - \int_{AB} f(z)\,dz = 0$$

$$\Rightarrow \oint_C f(z)\,dz - \oint_{C_1} f(z)\,dz = 0$$

(integral around a closed curve are taken positive when the curve is traversed in counter-clockwise direction)

Hence $\oint_C f(z)\,dz = \oint_{C_1} f(z)\,dz$

The theorem can be extended.

If a closed curve C contains non-intersecting closed curves $C_1, C_2, \ldots, C_n$, then by introducing cross-cuts, it can be shown that

$$\oint_C f(z)\,dz = \oint_{C_1} f(z)\,dz + \oint_{C_2} f(z)\,dz + \ldots + \oint_{C_3} f(z)\,dz$$

12.14 CAUCHY'S INTEGRAL FORMULA

Statement : If $f(z)$ is analytic within and on a closed curve C and a is any point within C, then

$$f(a) = \frac{1}{2\pi i}\oint_C \frac{f(z)}{z-a}\,dz$$

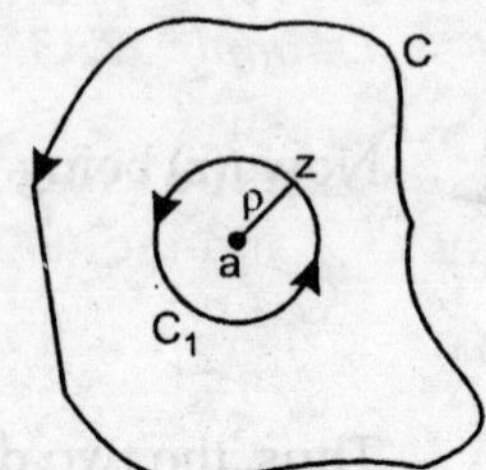

Fig. 12.11.

Proof. Consider the function $\frac{f(z)}{z-a}$, which is analytic at every point within C except at $z = a$.

Draw a circle C_1 with a as centre and radius ρ such that C_1 lies entirely inside C.

Thus $\frac{f(z)}{z-a}$ is analytic in the region between C and C_1.

∴ By Cauchy's theorem, we have

$$\oint_C \frac{f(z)}{z-a}\,dz = \oint_{C_i} \frac{f(z)}{z-a}\,dz \qquad \ldots(1)$$

Now, the equation of circle C_1 is

$$|z-a| = \rho \text{ or } z - a = re^{i\theta}$$

so that

$$dz = i\rho e^{i\theta}\,d\theta$$

$$\oint_{C_1} \frac{f(z)}{z-a}\,dz = \int_0^{2\pi} \frac{f(a+\rho e^{i\theta})}{\rho e^{i\theta}}\,.\,i\rho e^{i\theta} = i\int_0^{2\pi} f(a+\rho e^{i\theta})\,d\theta$$

Hence by (1), we have

$$\oint_C \frac{f(z)}{z-a}\,dz = i\int_0^{2\pi} f(a+\rho e^{i\theta})\,d\theta \qquad \ldots(2)$$

In the limiting form, as the circle C_1 shrinks to the point a, *i.e.*, $\rho \to 0$, then from (2),

$$\oint_C \frac{f(z)}{z-a}\,dz = \int_0^{2\pi} f(a)\,d\theta = f(a)\int_0^{2\pi} d\theta = 2\pi\,f(a)$$

Hence $f(a) = \frac{1}{2\pi i}\oint_C \frac{f(z)}{z-a}\,dz$ which is the required Cauchy's integral formula.

Cor. By Cauchy's integral formula, we have

$$f(a) = \frac{1}{2\pi i}\oint_C \frac{f(z)}{z-a} \qquad \ldots(1)$$

where a is any point within C and may be treated as a parameter.

Differentiating both sides of (1) w.r.t. *a*,

$$f'(a) = \frac{1}{2\pi i}\oint_C \frac{\partial}{\partial a}\left[\frac{f(z)}{z-a}\right]dz = \frac{1}{2\pi i}\oint_C \frac{f(z)}{(z-a)^2}dz$$

Similarly,
$$f''(a) = \frac{2!}{2\pi i}\oint_C \frac{f(z)}{(z-a)^3}dz$$

$$f'''(a) = \frac{3!}{2\pi i}\oint_C \frac{f(z)}{(z-a)^4}dz$$

and in general
$$f^n(a) = \frac{n!}{2\pi i}\oint_C \frac{f(z)}{(z-a)^{n+1}}dz$$

Thus, if a function of a complex variable has a first derivative in a simply connected region, all its higher derivatives exist in that region. This property is not exhibited by the functions of real variables.

***Example* 12.19.** $\oint_C \frac{3z^2+z}{z^2-1}dz$ where C is the circle $|z-1| = 1$

Solution : The integrand has singularities, where $z^2 - 1 = 0$, *i.e.*, at $z = 1$ and $z = -1$. The circle $|z-1| = 1$ has centre at $z = 1$ and radius 1 includes the point $z = 1$.

$$f(z) = 3z + z \text{ is an analytic function.}$$

Also,
$$\frac{1}{z^2-1} = \frac{1}{(z-1)(z+1)} = \frac{1}{2}\left(\frac{1}{(z-1)} - \frac{1}{z+1}\right)$$

$$\therefore \quad \oint_C \frac{3z^2+z}{(z^2-1)}dz = \frac{1}{2}\oint_C \frac{3z^2+z}{(z-1)}dz - \frac{1}{2}\oint_C \frac{3z^2+z}{(z+1)}dz$$

By Cauchy's integral formula

$$\oint_C \frac{3z^2+z}{(z-1)}dz = 2\pi i\, f(1), \text{ where } f(z) = 3z^2 + z$$

$$= 8\pi i.$$

By Cauchy's integral theorem,

$$\oint_C \frac{3z^2+z}{(z+1)}dz = 0 \qquad [\text{As } z = -1 \text{ lies outside the circle } |z-1| = 1]$$

$$\therefore \quad \oint_C \frac{3z^2+z}{(z-1)}dz = 4\pi i.$$

***Example* 12.20.** Evaluate : $\oint_C \frac{\cos \pi z^2}{(z-1)(z-2)}dz$, where C is the circle $|z| = 3$

Solution : The integrand has singularities at $(z-1)(z-2) = 0$ *i.e.* at $z = 1$ and $z = 2$

Both these points lie within the circle $|z| = 3$.

$$f(z) = \cos \pi z^2 \text{ is an analytic function.}$$

Also
$$\frac{1}{(z-1)(z-2)} = \frac{-1}{(z-1)} + \frac{1}{(z-2)}$$

$$\therefore \quad \oint \frac{\cos \pi z^2}{(z-1)(z-2)}dz = -\oint_C \frac{\cos \pi z^2}{(z-1)}dz + \oint_C \frac{\cos \pi z^2}{(z-2)}dz$$

$$= -2\pi i\, f(1) + 2\pi i\, f(2), \text{ where } f(z) = \cos \pi z^2$$
$$= -2\pi i \cos \pi + 2\pi i \cos 4\pi$$
$$= 2\pi i + 2\pi i$$
$$= 4\pi i.$$

***Example* 12.21.** Evaluate : $\oint_C \frac{\sin^6 z}{(z-\pi/6)^3} dz$ where C is $|z| = 1$.

Solution : The integrand has singularity at $z = \pi/6$ which lies within circle $|z| = 1$.

Now,
$$f^n(a) = \frac{n!}{2\pi i} \oint \frac{f(z)}{(z-a)^{n+1}} dz \quad ...(1)$$

Comparing (1) with given integrand, we get

$$a = \pi/6,\ n + 1 = 3 \text{ i.e. } n = 2,\ f(z) = \sin^6 z$$
$$f'(z) = 6 \sin^5 z \cos z$$

Also
$$f''(z) = 30 \sin^4 z \cos^2 z - 6 \sin^5 z \sin z$$
$$f''\left(\frac{\pi}{6}\right) = 30\left(\frac{1}{2}\right)^4\left(\frac{\sqrt{3}}{2}\right)^2 - 6\left(\frac{1}{2}\right)^6 = \frac{21}{16}$$

From (1) we have

$$f''\left(\frac{\pi}{6}\right) = \frac{2!}{2\pi i} \oint_C \frac{\sin^6 z}{\left(z-\frac{\pi}{6}\right)^3} dz$$

$$\oint_C \frac{\sin^6 z}{\left(z-\frac{\pi}{6}\right)^3} dz = \frac{21}{16}\pi i.$$

***Example* 12.22.** If $f(\xi) = \oint_C \frac{4z^2+z+5}{z-\xi} dz$, where C is ellipse $\frac{x^2}{4} + \frac{y^2}{9} = 1$, find $f(1), f(i), f'(-1), f''(-i)$.

Solution : The given curve C is an ellipse $\frac{x^2}{4} + \frac{y^2}{9} = 1$.

The points $z = 1, z = i, z = -1$ and $z = -i$ lie inside the closed curve C.

Let
$$\Phi(z) = 4z^2 + z + 5$$

By Cauchy's Integral Formula,

$$\Phi(\xi) = \frac{1}{2\pi i} \oint_C \frac{\Phi(z)}{z-\xi} dz$$

$$2\pi i\, \Phi(\xi) = \oint_C \frac{4z^2+z+5}{z-\xi} dz = f(\xi)$$

$$\Rightarrow \quad f(\xi) = 2\pi i\,(4\xi^2 + \xi + 5) = 2\pi i\,(1 + i) = 2\pi\,(i - l)$$
$$f(1) = 2\pi i\,(4 \times 1 + 1 + 5) = 20\,\pi i$$
$$f(i) = 2\pi i\,(-4 + i + 5)$$
$$f'(\xi) = 2\pi i\,(8\xi + 1)$$
$$f'(-1) = 2\pi i\,(-8 + 1) = -14\pi i$$
$$f''(\xi) = 2\pi i\,(8) = 16\pi i$$

EXERCISE 12.4

1. Evaluate using Cauchy integral formula :

$$\oint_C \frac{\sin^2 z}{\left(z-\frac{\pi}{6}\right)^3}\,dz, \text{ where C is } |z| = 1$$

Ans. πi

2. $\dfrac{\cos \pi z}{z^2-1}dz$ around a rectangle with vertices

(i) $2 \pm i, -2 \pm i$ (ii) $-1, 2-i, 2+i, i$

Ans.

(i) 0, (ii) $-\pi i$

3. Evaluate : $\oint_C \dfrac{z^3+z+1}{z^2-3z+2}dz$, where C is ellipse $4x^2 + 9y^2 = 1$

Ans. 0

4. Use Cauchys integral formula to evaluate :

(i) $\oint_C \dfrac{e^{2z}}{(z+1)^4}dz$, where C is the circle $|z| = 2$

(ii) $\oint_C \dfrac{e^z}{(z-1)(z-2)^2}dz$, where C is circle $|z| = 3$

(iii) $\oint_C \dfrac{e^{-z}}{z^2}dz$, where C is the circle $|z| = 2$

(iv) $\oint_C \dfrac{z^3-2z+1}{(z-i)^2}dz$ where C is $|z| = 3$

(v) $\oint_C \dfrac{z}{z^3-3z+2}dz$ where C is $|z-2| = \dfrac{1}{2}$

(vi) $\int \dfrac{\tan z}{z}dz$ where C is $|z| = 2$

Ans.

(i) $\dfrac{8\pi i}{3e^2}$, (ii) $2\pi ie$, (iii) $-2\pi i$, (iv) $10\pi i$, (v) 0

5. If $f(\xi) = \oint_C \dfrac{3z^2+7z+1}{z-\xi}dz$ where C is the circle $x^2 + y^2 = 4$, find the values of $f(3)$, $f'(1-i)$ and $f''(1-i)$

Ans. $0, 2\pi(6+13i), 12\pi i$

12.15 SINGULAR POINTS, RESIDUES

A point at which a function $f(z)$ ceases to be analytic is called a singular point or singularity of $f(z)$.

For example $z = a$ is a singular point of $f(z) = \dfrac{1}{z-a}$. A singular point $z = a$ of a function $f(z)$ is called an isolated singular point if there exists a circle with centre $z = a$ which contains no other singular point of $f(z)$.

For example z = – 1, 1 are two isolated singular points of the function $f(z) = \frac{z}{z^2-1}$.

The function $f(z) = \frac{1}{\sin(\pi z)}$ has an infinite number of isolated singular points at z = ± 1, ± 2, ... when z = *a* is an isolated singular point of *f*(z) in a Laurent's series about z = *a*.

Thus
$$f(z) = \sum_{n=0}^{\infty} a_n (z-a)^n + \sum_{n=1}^{\infty} a_{-n}(z-a)^{-1}$$
$$= \sum_{n=0}^{\infty} a_n (z-a)^n + [a_{-1}(z-a)^{-1} + a_{-2}(z-a)^{-2} + ... + a_{-m}(z-a)^{-m} + ...]$$

If the series with negative powers of terms say *m*, so that $a_{-m} \neq 0$; a_{-m-2} = ... = 0. Then z = *a* is called a pole of order *m*. A pole of order one is called a simple pole.

If the series with negative powers does not terminate *i.e.*, has an infinite number of terms, then z = *a* is called an essential singular point of *f*(z).

The coefficient of $(z-a)^{-1}$ in the expansion of *f*(z) around an isolated singularity is called the residue of *f*(z) at that point. Thus, in Laurent series expansion of *f*(z) around z = *a* *i.e.*,

$f(z) = a_0 + a_1(z-a) + a_2(z-a)^2 + ... + a_{-1}(z-a)^{-1} + a_{-2}(z-a) + ...$ the residue of *f*(z) at z = *a* is a_{-1}

$$\therefore \quad \text{Res}\,[f(z), a] = \frac{1}{2\pi i}\int_C f(z)\,dz$$

i.e.,
$$\int_C f(z)\,dz = 2\pi i\,\text{Res}\,[f(z), a]$$

12.16 RESIDUE THEOREM

If *f*(z) is analytic inside and on a curve C except at a finite number of singular points within C, then

$$\int_C f(z)\,dz = 2\pi i \times (\text{sum of the residues at singular points within C}).$$

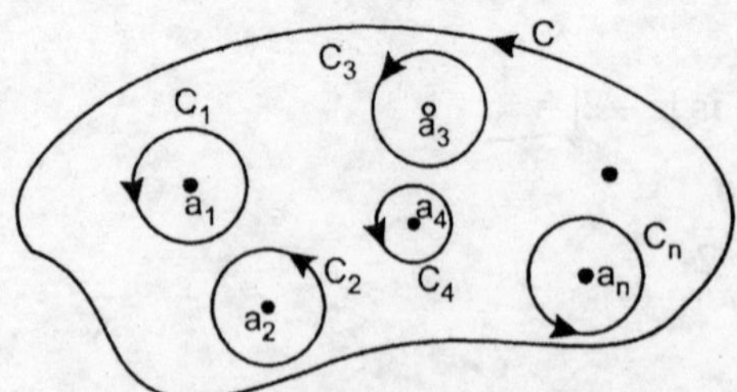

Fig. 12.12.

Let us surround each of the singular points a_1, a_2 .... a_n by small circles such that it encloses no other singular point.

Then these circles together with C, forms a multiply connected region in which *f*(z) is analytic.

∴ Applying Cauchy's theorem, we have

$$\oint_C f(z)\,dz = \oint_{C_1} f(z)\,dz + \oint_{C_2} f(z)\,dz + ... + \oint_{C_n} f(z)\,dz$$

$$\oint_C f(z)\,dz = 2\pi i\,[\text{Res}\,f(a_1) + \text{Res}\,f(a_2) + ... + \text{Res}\,f(a_n)]$$

12.17 CALCULATION OF RESIDUES

If $f(z)$ has pole of order n at $z = a$, then

$$\text{Res}\,[f(z), a] = \frac{1}{(n-1)!} \lim_{z \to a} \left\{ \frac{d^{n-1}}{dz^{n-1}} \{(z-a)^n f(z)\} \right\}.$$

***Example* 12.23:** Determine the poles of the function : $f(z) = \dfrac{z^2}{(z-1)^2(z+2)}$ and the residue at each pole. Hence evaluate $\oint_C f(z)\,dz$ where C is the circle $|z| = 2.5$.

Solution : $f(z) = \dfrac{z^2}{(z-1)^2(z+2)}$

$f(z)$ has a simple pole at $z = -2$ and a pole of order two at $z = 1$. Both these residues lie within C *i.e.*, $|z| = 2.5$

$$\text{Res}\, f(1) = \frac{1}{(2-1)!} \lim_{z \to 1} \left\{ \frac{d^{2-1}}{dz^{2-1}} (z-1)^2 f(z) \right\}$$

$$= \frac{d}{dz} \lim_{z \to 1} \left\{ (z-1)^2 \cdot \frac{z^2}{(z-1)^2(z+2)} \right\}$$

$$= \lim_{z \to 1} \left[\frac{(z+2)2z - z^2}{(z+2)^2} \right]$$

$$= \frac{6-1}{9} = \frac{5}{9}$$

$$\text{Res}\, f(-2) = \lim_{z \to -2} \left[(z+2) \frac{z^2}{(z+1)(z-1)^2} \right] = \frac{4}{9}$$

$$\oint_C f(z)\,dz = \oint_C \frac{z^2}{(z-1)^2(z+2)} dz \text{ where C is } |z| = 2.5$$

$$\therefore \quad \int_c \frac{z^2}{(z-1)^2(z+2)} dz = 2\pi i \times (\text{sum of the residues})$$

$$= 2\pi i \left(\frac{5}{9} + \frac{4}{9} \right) = 2\pi i$$

***Example* 12.24:** $\displaystyle\int_c \frac{z-3}{z^2+2z+5} dz$, where C is the circle

(*i*) $|z| = 1$, (*ii*) $|z + 1 - i| = 2$, (*iii*) $|z + 1 + i| = 2$.

Solution :

(*i*) $|z| = 1$

$$f(z) = \frac{z-3}{z^2+2z+5}$$

Poles of $f(z)$ are :

$$z^2 + 2z + 5 = 0$$

$$\Rightarrow \quad z = \frac{-2 \pm \sqrt{4-20}}{2} = \frac{-2 \pm 4i}{2} = -1 \pm 2i$$

$-1 + 2i$ and $-1 - 2i$ they both lie outside the circle $|z| = 1$

$\therefore$ By Cauchy's theorem, we have $\oint_C \frac{z-3}{z^2+2z+5} dz = 0$

(ii) $|z+1-i| = 2$

or $|z-(-1+i)| = 2$

$-1+2i$ lies inside the given circle and $-1-2i$ lies outside

Residue at
$$(-1+2i) = \lim_{z \to -1+2i} \left[(z+1-2i)\frac{z-3}{(z+1-2i)(z+1+2i)}\right]$$
$$= \frac{-1+2i-3}{-1+2i+1+2i} = \frac{2i-4}{4i}$$
$$= \frac{-i}{4}(2i-4)$$
$$= \frac{-1}{4}(-2-4i)$$
$$= \frac{1+2i}{2}$$

Fig. 12.13.

Hence by Residue-theorem
$$\int_c f(z)dz = 2\pi i \text{ [Sum of Residues]}$$
$$= 2\pi i\left(i+\frac{1}{2}\right)$$
$$= \pi(-2+i)$$
$$= \pi(i-2)$$

(iii) $|z+1+i| = 2$

$-1-2i$ lies within the origin.

$$\therefore \text{Res at } (-1-2i) = \lim_{z \to -1-2i}\left[(z+1+2i)\frac{z-3}{(z+1+2i)(z+1-2i)}\right]$$
$$= \frac{-1-2i-3}{-1-2i+1-2i}$$
$$= \frac{-(4+2i)}{-4i}$$
$$= -i\left(\frac{i}{2}+1\right)$$
$$= \left(\frac{1}{2}-i\right)$$

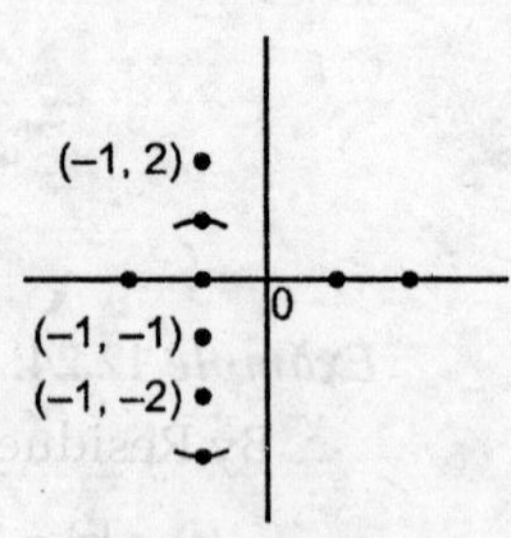

Fig. 12.14.

Hence by Residue theorem
$$\oint_c f(z)dz = 2\pi i \times \left(\frac{1}{2}-i\right)$$
$$= \pi(i+2)$$

***Example* 12.25.** Evaluate

(*i*) $\int_C \tan z\, dz$ where the circle $|z| = 2$

(*ii*) $\oint_C \frac{e^z}{z^2+1} dz$ where c is the circle $|z| = 2$

(*iii*) $\oint_C \frac{z \sec z}{(1-z)^2} dz$ where the circle $|z| = 2$

Solution :

(*i*) $f(z) = \tan z = \frac{\sin z}{\cos z}$

Poles of $f(z) \Rightarrow \cos z = 0$

$$\Rightarrow \quad z = (2n+1)\frac{\pi}{2}, \text{ where } n = 0, \pm 1, \pm 2$$

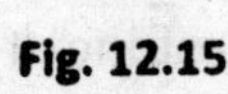

Fig. 12.15.

of these poles
$$z = \frac{\pi}{2} = \frac{22}{14} = 1.5714 \text{ and}$$

$$z = \frac{-\pi}{2} \text{ lie within C, } |z| = 2$$

$$\text{Residue at } \frac{\pi}{2} = \lim_{z \to \pi/2} \left(z - \frac{\pi}{2}\right) \frac{\sin z}{\cos z} \qquad \left[\frac{0}{0}\text{Form}\right]$$

$$= \lim_{z \to \pi/2} \left(\frac{\sin z + \left(z - \frac{\pi}{2}\right)\cos z}{-\sin z} \right) \qquad \text{[By L' Hospital's Rule]}$$

$$= \frac{1}{-1} = -1$$

$$\text{Residue at } \left(\frac{-\pi}{2}\right) = \lim_{z \to -\pi/2} \left(z + \frac{\pi}{2}\right) \frac{\sin z}{\cos z} \qquad \left[\frac{0}{0}\text{Form}\right]$$

$$= \lim_{z \to -\pi/2} \frac{\sin z + \left(z + \frac{\pi}{2}\right)\cos z}{-\sin z} = 1 \qquad \text{[By L' Hospital's Rule]}$$

$\therefore$ By Residue theorem, $\oint_C \tan z\, dz = (2\pi i)(-1+1) = 0$

(*ii*) $\oint_C \frac{e^z}{z^2+1} dz$ where c is $|z| = 2$ [Simple poles at $z = i$ and $z = i$, both singularities lie within C]

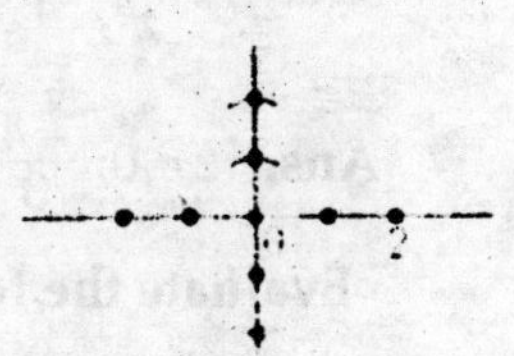
Fig. 12.16.

$$\text{Residue at } (z = i) = \lim_{z \to i} \frac{(z-i)e^z}{(z-i)(z+i)}$$

$$= \frac{e^i}{2i} = \frac{-ie^i}{2}$$

$$\text{Residue at } (z = -i) = \lim_{z \to -i} \frac{(z+i)e^z}{(z+i)(z-i)}$$

$$= \frac{e^{-i}}{-2i}$$

$$= \frac{i}{2}e^{-i}$$

$\therefore$ By Residue Theorem $\oint_C \frac{e^z dz}{z^2+1} = 2\pi i\left(\frac{-ie^i}{2} + \frac{i}{2}e^{-i}\right)$

$$= -\pi(-e^i + e^{-i})$$

$$= \pi(e^i - e^{-i})$$

(*iii*) $\oint_C \frac{z \sec z}{(1-z)^2} dz$ where C is $|z| = 2$.

Double pole at $z = 1$

Residue at $(z = 1) = \lim_{z \to 1} \frac{d}{dz} \frac{(z-1)^2 z \sec z}{(z-1)^2} = \lim_{z \to 1} \frac{d}{dz}(z \sec z)$

$$= \lim_{z \to 1} (\sec z + z \sec z \tan z)$$

$$= \sec 1 + \sec 1 \tan 1$$

By Residue Theorem, $\oint_C \frac{z \sec z}{(1-z)^2} dz = 2\pi i \sec 1 (1 + \tan 1)$

EXERCISE 12.5

Determine the poles and residue at each pole :

1. $\frac{z+1}{z^2(z-2)}$

Ans. $z = 0, 2; \frac{-3}{4}, \frac{3}{4},$

2. $\frac{z^2}{(z-1)(z-2)^2}$

Ans. $z = 1, 2; 1, 0,$

3. $\frac{1-e^{2z}}{z^4}$

Ans. $z = 0; \frac{-4}{3}$

Evaluate the following integrals (Using Residue Theorem) :

4. $\oint_C \frac{e^z}{\cos \pi z}$ where C is unit circle $|z| = 1$.

Ans. $-4i \sin h \frac{1}{2}$

5. $\oint_C \frac{1-2z}{z(z-1)(z-2)} dz$, where C is the circle $|z| = 1.5$.

Ans. $3\pi i$

6. $\oint_C \frac{12z-7}{(z-1)^2(2z+3)} dz$ where C is the circle $|z| = 2$.

Ans. 0

7. $\oint_C \frac{z \sec z}{(1-z)^2} dz$ where C is the circle $|z| = 2$.

Ans. $2\pi i \sec 1 \, (1 + \tan 1)$

8. Find the residue of $f(z) = \frac{z^3}{(z-1)^4(z-2)(z-3)}$ at its pole and hence evaluate $\int_C f(z)\, dz$, where C is the circle $|z| = \frac{5}{2}$.

Ans. $\frac{-27}{8}\pi i$

9. $\int_C \frac{\sin \pi z^2 + \cos \pi z^2}{(z-1)^2(z-2)} dz$, where C is the circle $|z| = 3$.

Ans. $4\pi(\pi + i)\, i$

10. $\oint_C \frac{1}{z^2 \sin z} dz$, where C is a triangle with vertices (0, 1), (2, – 2), (7, 1).

Ans. $\frac{2i(-1)^n}{n^2\pi}$

12.18 SERIES OF COMPLEX TERMS

A series of the form $(a_1 + ib_1) + (a_2 + ib_2) + \ldots\ldots + (a_n + ib_n) +$ where $a_1, b_1, a_2, b_2, \ldots\ldots$ are real number is called a series of complex terms and is expressed in the form $\sum_{n=1}^{\infty} a_n + i \sum_{n=1}^{\infty} b_n$ ...(1)

The series (1) is convergent if Σa_n and Σb_n both are convergent.

If Series Σa_n and Σb_n converge to a and b then series is said to converge to $a + ib$.

The necessary condition for convergence of series (1) is $\underset{n \to \infty}{lt} (a_n + ib_n) = 0$.

The series is said to be absolutely convergent if the series.

$|a_1 + ib_1| + |a_2 + ib_2| + |a_3 + ib_3| + \ldots\ldots\ldots\ldots + |a_n + ib_n| + \ldots\ldots\ldots\ldots$ is convergent.

Since $|a_n| \le |a_n + ib_n|$ and $|b_n| \le |a_n + ib_n|$, we conclude that an absolutely convergent series is convergent. But the converse is not true.

If the series of functions $u_1(z) + u_2(z) + \ldots\ldots\ldots\ldots + u_n(z) + \ldots\ldots\ldots\ldots = \sum_{n=1}^{\infty} u_n(z)$ ...(2)

convergence to $S(z)$ and $S_n(z)$ is the nth partial sum, then the series (2) is said to be uniformly convergent in a region R, if given any positive number ε, there exists a positive number N, depending on ε but not on z, such that for every z in R,

$|S(z) - S_n(z)| < \varepsilon$ for all $n >$ N.

A series of the form $a_0 + a_1(z-a) + a_2(z-a)^2 + \ldots\ldots + a_n(z-a)^n + \ldots\ldots\ldots$

$$= \sum_{n=1}^{\infty} a_n (z-a)^n \quad \ldots(3)$$

is called a power series in $(z - a)$.

If the series (3) is convergent for all values of z satisfying $|z - a| < R$, *i.e.*, for all values of z lying inside the circle with centre at $z = a$ and radius R, then this circle is called the circle of convergence and R is called the Radius of convergence.

12.19 TAYLOR'S SERIES

If $f(z)$ is analytic inside a circle C with centre at a, then for all z inside C,

$$f(z) = f(a) + (z-a)f'(a) + \frac{(z-a)^2}{2!}f''(a) + \dots\dots + \frac{(z-n)^2}{n!}f^n(a) + \dots\dots$$

Proof : Let z be any point inside the circle C. Draw a circle C_1 with centre at a and radius smaller than that of C such that z is an interior point of C_1. Let w be any point on C_1, then

$$|z-a| < |z-w| \text{ i.e., } \left|\frac{z-a}{z-w}\right| < 1$$

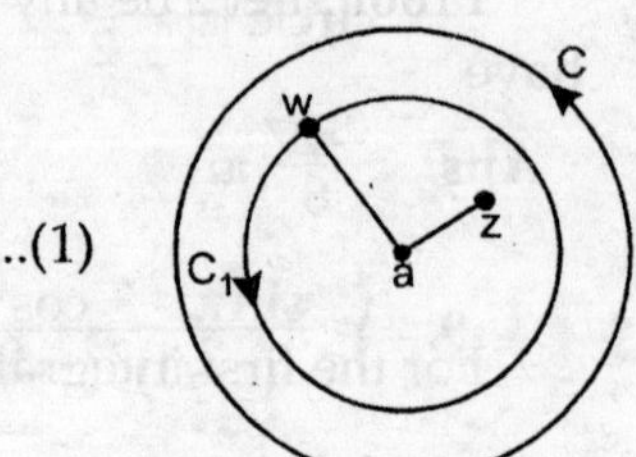

Fig. 12.17.

Now, $$\frac{1}{w-z} = \frac{1}{(w-a)-(z-a)} = \frac{1}{w-a}\left[1 - \frac{z-a}{w-a}\right]^{-1} \quad \dots(1)$$

expanding R.H.S. by binomial theorem as $\left|\frac{z-a}{w-a}\right| < 1$, we get

$$\frac{1}{w-z} = \frac{1}{(w-a)}\left[1 + \frac{z-a}{w-a} + \left(\frac{z-a}{w-a}\right)^2 + \dots + \left(\frac{z-a}{w-a}\right)^2 + \dots\right]$$

This series converges uniformly since $\left|\frac{z-a}{w-a}\right| < 1$. Multiplying both sides of (1) by $\frac{1}{2\pi} f(w)$ and integrating term by term w.r.t. w, over C_1 we get

$$\frac{1}{2\pi i}\oint_{C_1}\frac{f(w)}{w-z}dz = \frac{1}{2\pi i}\oint_{C_1}\frac{f(w)}{w-a}dw + \frac{z-a}{2\pi i}\oint_{C_1}\frac{f(w)}{(w-a)^2}dw + \frac{(z-a)^2}{2\pi i}\oint_{C_1}\frac{f(w)}{(w-a)^3}dw + \dots + \frac{(z-a)^n}{2\pi}\oint_{C_1}\frac{f(w)}{(w-a)^{n+1}}dw + \dots \quad \dots(2)$$

Since $f(w)$ is on and inside C_1, by Cauchy's formula,

$$\frac{1}{2\pi i}\oint_{C_1}\frac{f(w)}{w-z}dw = f(z)$$

Also $$f^n(a) = \frac{n!}{2\pi i}\oint\frac{f(w)}{(w-a)^{n+1}}dw \text{ i.e., } \frac{1}{2\pi i}\oint\frac{f(w)}{(w-a)^{n+1}}dw = \frac{f^n(a)}{n!}$$

$\therefore$ From (2), we have

$$f(z) = f(a) + (z-a)f'(a) + \frac{(z-a)^2}{2!}f'(a) + \dots + \frac{(z-a)^n}{n!}f'(a) + \dots \quad \dots(3)$$

which is the required Taylor's series for $f(z)$ about $z = a$.

Cor 1. Putting $z = a + h$ in (3), we get

$$f(a+h) = f(a) + hf'(a) + \frac{h^2}{2!}f''(a) + \dots + \frac{h^n}{n!}f^n(a) + \dots$$

Cor 2. If $a = 0$, the series (3) becomes

$$f(z) = f(0) + zf'(0) + \frac{z^2}{2!}f''(0) + \dots + \frac{z^n}{n!}f^n(0) + \dots$$

This series is called **Maclaurin's series.**

12.20 LAURENT'S SERIES

If $f(z)$ is analytic inside and on the boundary of the ring shaped region R bounded by two concentric circles C_1 and C_2 of radii r_1 and r_2 $(r_1 > r_2)$ respectively having centre at a, then for all z in R.

$$f(z) = a_0 + a_1(z-a) + a_2(z-a)^2 + \ldots + a_1(z-a)^{-1} + a_{-2}(z-a)^{-2} + \ldots$$

where $$a_n = \frac{1}{2\pi i}\oint_{c_1} \frac{f(w)}{(w-a)^{n+1}}dw;\ n = 0, 1, 2, \ldots\ldots$$

and $$a_{-n} = \frac{1}{2\pi i}\oint_{C_1} \frac{f(w)}{(w-a)^{n+1}}dw;\ n = 1, 2, 3, \ldots\ldots$$

Proof : Let z be any point in the region R, then by Cauchy's integral formula for double connected region, we have

$$f(z) = \frac{1}{2\pi i}\oint_{C_1} \frac{f(w)}{(w-z)}dw - \frac{1}{2\pi i}\oint_{C_2} \frac{f(w)}{w-2}dw \qquad \ldots(1)$$

For the first integral in (1), w lines on C_2

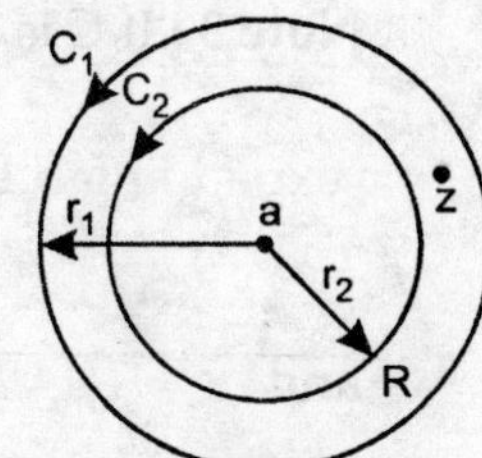

Fig. 12.18.

$\therefore\ |z-a| < |w-a|$ *i.e.,* $\left|\frac{z-a}{w-a}\right| < 1$

Now $$\frac{1}{w-a} = \frac{1}{(w-a)-(z-a)} = \frac{1}{w-a}\left(1-\frac{z-a}{w-a}\right)^{-1}$$

$$= \frac{1}{w-a}\left[1+\frac{z-a}{w-a}+\left(\frac{z-a}{w-a}\right)^2+\ldots\ldots\right]$$

Multiplying both sides by $\frac{1}{2\pi i}f(w)$ and integrating term by term w.r.t. w, along the circle C_1, we get

$$\frac{1}{2\pi i}\oint_{C_1}\frac{f(w)}{w-z}dw = \frac{1}{2\pi i}\oint_{C_1}\frac{f(w)}{w-a}dw + \frac{z-a}{2\pi i}\oint_{C_1}\frac{f(w)}{(w-a)^2}dw + \frac{(z-a)^2}{2\pi i}\oint_{C_2}\frac{f(w)}{(w-a)^3}dw + \ldots$$

$$= a_0 + a_1(z-a) + a_2(z-a)^2 + \ldots\ldots$$

$$= \sum_{n=0}^{\infty} a_n(z-a)^n \quad \ldots(2)\ \left[\because a_n = \frac{1}{2\pi i}\oint_{c_1}\frac{f(w)}{(w-a)^{n+1}}dw,\ n = 0, 1, 2\ldots\ldots\right]$$

For the second integral in (1), w lies on C_2

$\therefore\ |w-a| < |z-a|$ *i.e.,* $\left|\frac{w-a}{z-a}\right| < 1$

Now, $$\frac{1}{w-z} = \frac{1}{(w-a)-(z-a)} = \frac{1}{z-a}\left(1-\frac{w-a}{z-a}\right)^{-1}$$

$$= \frac{1}{z-a}\left[1+\frac{w-a}{z-a}+\left(\frac{w-a}{z-a}\right)^2+\ldots\ldots\right]$$

Multiplying both sides by $\frac{1}{2\pi i}f(w)$ and integrating term by term w.r.t. w, along the circle C_2, we get

$$-\frac{1}{2\pi i}\oint_{C_2}\frac{f(w)}{w-z}dw = \frac{1}{z-a}\cdot\frac{1}{2\pi i}\oint_{C_2}f(w)\,dw + \frac{1}{(z-a)^2}\cdot\frac{1}{2\pi i}\oint_{C_2}(w-a)f(w)\,dw$$

$$+\frac{1}{(z-a)^3}\cdot\frac{1}{2\pi i}\oint_{C_2}f(w-a)^2 f(w)\,dw + \ldots\ldots$$

$$= a_{-1}(z-a)^{-1} + a_{-2}(z-a)^{-2} + a_{-3}(z-a)^{-3} + \ldots\ldots$$

$$= \sum_{n=1}^{\infty} a_{-n}(z-a)^{-1} \quad \ldots(3) \quad \left[\because a_{-n} = \frac{1}{2\pi i}\oint_{C_2} \frac{f(w)}{(w-a)^{n+1}}dw,\ n = 1, 2, 3 \ldots\ldots\right]$$

Substituting from (2) and (3) in (1), we get

$$f(z) = \sum_{n=0}^{\infty} a_n(z-a)^n + \sum_{n=1}^{\infty} a_{-n}(z-a)^{-n}$$

Note 1 : In the statement of Laurent's series, $a_n = \frac{1}{2\pi i}\oint_{C_1} \frac{f(w)}{(w-a)^{n+1}}dw \neq \frac{f^n(a)}{n!}$ because $f(z)$ is not given to be analytic inside C_1.

Note 2 : In case $f(z)$ is analytic inside C_1, then $a_{-n} = 0$ and $a_n = \frac{1}{2\pi i}\oint_{C_1} \frac{f(w)}{(w-a)^{n+1}}dw = \frac{f^n(a)}{n!}$ and Laurent's series reduces to Taylor's series.

Note 3 : If C is any simple closed curve which lies in the ring-shaped

$$\oint_{C_1} \frac{f(w)}{(w-a)^{n+1}}dw = \oint_{C} \frac{f(w)}{(w-a)^{n+1}}dw$$

and

$$\oint_{C_2} \frac{f(w)}{(w-a)^{-n+1}}dw = \oint_{C} \frac{f(w)}{(w-a)^{-n+1}}dw$$

Fig. 12.19.

∴ Laurent's series can be written as

$$f(z) = \sum_{n=-\infty}^{\infty} a_n(z-a)^n, \text{ where } a_n = \frac{1}{2\pi i}\oint_{C} \frac{f(w)}{(w-a)^{n+1}}dw$$

Note 4 : The process of finding the coefficient a_n by complex integration is complicated. In practice, we expand the function $f(z)$ by binomial theorem or by some other method to obtain Taylor's or Laurent's series.

Example **12.26.** Expand $f(z) = \cos z$ in Taylors series about $z = \frac{\pi}{4}$

Solution : $f(z) = \cos z$ $\quad f\left(\frac{\pi}{4}\right) = \frac{1}{\sqrt{2}}$

$$f'(z) = -\sin z \qquad f'\left(\frac{\pi}{4}\right) = -\frac{1}{\sqrt{2}}$$

$$f''(z) = -\cos z \qquad f''\left(\frac{\pi}{4}\right) = -\frac{1}{\sqrt{2}}$$

$$f'''(z) = \sin z \qquad f'''\left(\frac{\pi}{4}\right) = \frac{1}{\sqrt{2}}$$

and so on.

$$\text{Hence, } \cos z = f(z) = f\left(\frac{\pi}{4}\right) + \left(z-\frac{\pi}{4}\right)f'\left(\frac{\pi}{4}\right) + \frac{\left(z-\frac{\pi}{4}\right)}{2!}f'\left(\frac{\pi}{4}\right) + \ldots\ldots$$

$$= \frac{1}{\sqrt{2}} + \left(z-\frac{\pi}{4}\right)\left(-\frac{1}{\sqrt{2}}\right) + \frac{\left(z-\frac{\pi}{4}\right)^2}{2!}\left(\frac{-1}{\sqrt{2}}\right) + \ldots\ldots$$

$$= \frac{1}{\sqrt{2}}\left[1 - \left(z-\frac{\pi}{4}\right) - \frac{1}{2!}\left(z-\frac{\pi}{4}\right)^2 + \frac{1}{3!}\left(z-\frac{\pi}{4}\right)^2 + \ldots\ldots\right]$$

Example **12.27.** Expand $\dfrac{1}{z^2-3z+2}$ in the region :

(*a*) $|z|<1$ (*b*) $1<|z|<2$ (*c*) $|z|<2$ (*d*) $0<|z-1|<1$

Solution : Here $f(z)=\dfrac{1}{z^2-3z+2}=\dfrac{1}{(z-1)(z-2)}=\dfrac{1}{z-2}-\dfrac{1}{(z-1)}$ ...(1)

(*i*) If $|z|<1$, then $\left|\dfrac{z}{2}\right|<1$

As we will using binomial theorem for expansion.

$\therefore$ arranging (1) suitable for binomial expansion *i.e.*, $\dfrac{1}{1-x}$ such that $|x|<1$

Now (1) can be written as

$$f(z)=\frac{1}{-2\left(1-\frac{z}{2}\right)}+\frac{1}{1-z}$$

$$=\frac{-1}{2}\left(1-\frac{z}{2}\right)^{-1}+(1-z)^{-1}$$

$$=\frac{-1}{2}\left[1+\frac{z}{2}+\frac{z^2}{4}+\frac{z^3}{8}+\ldots\ldots\right]+(1+z+z^2+z^3+\ldots\ldots)$$

[By Binomial Theorem]

$$=\frac{1}{2}+\frac{3}{4}z+\frac{7}{8}z^2+\frac{15}{16}z^3+\ldots\ldots \text{ which is required Taylor's series}$$

(*b*) $1<|z|<2$, then $\left|\dfrac{1}{z}\right|<1$ and $\left|\dfrac{z}{2}\right|<1$

$\therefore$

$$f(z)=\frac{1}{-2\left(1-\frac{z}{2}\right)}-\frac{1}{z\left(1-\frac{1}{z}\right)}=\frac{-1}{2}\left(1-\frac{z}{2}\right)^{-1}-\frac{1}{z}\left(1-\frac{1}{z}\right)^{-1}$$

$$=\frac{-1}{2}\left[1+\frac{z}{2}+\frac{z^2}{4}+\frac{z^3}{8}+\ldots\right]-\frac{1}{z}\left[1+\frac{1}{z}+\frac{1}{z^2}+\frac{1}{z^3}+\ldots\right]$$

$$=-\frac{1}{2}\left[1+\frac{z}{2}+\frac{z^2}{4}+\ldots\right]-[z^{-1}+z^{-2}+z^{-3}+\ldots\ldots]$$

which is Laurent's series.

(*c*) If $|z|>2$, then $\left|\dfrac{2}{z}\right|<1$ and $\left|\dfrac{1}{z}\right|<1$

$\therefore$

$$f(z)=\frac{1}{z\left(1-\frac{2}{z}\right)}-\frac{1}{z\left(1-\frac{1}{z}\right)}=\frac{1}{z}\left(1-\frac{1}{z}\right)^{-1}-\frac{1}{z}\left(1-\frac{1}{z}\right)^{-1}$$

$$=\frac{1}{z}\left(1+\frac{2}{z}+\frac{4}{z^2}+\frac{8}{z^3}+\ldots\ldots\right)-\frac{1}{z}\left[1+\frac{1}{z}+\frac{1}{z^2}+\frac{1}{z^3}+\ldots\ldots\right]$$

[By Binomial Theorem]

$$=z^{-2}+3z^{-3}+7z^{-4}+\ldots\ldots\ldots \text{ which is Laurent's series.}$$

(*d*) For $0<|z-1|<1$, we have

$$f(z)=\frac{1}{(z-1)-1}-\frac{1}{(z-1)}=\frac{-1}{1-(z-1)}-\frac{1}{(z-1)}$$

$$= -(z-1)^{-1} - [1-(z-1)]^{-1}$$

$$= -(z-1)^{-1} - [1+(z-1)+(z-1)^2+(z-1)^3+......]$$

Example **12.28.** Show that when $|z+1|<|$, $z^{-2} = 1 + \sum_{n=1}^{\infty}(n+1)(z+1)^n$

Solution : $f(z) = z^{-2} = \frac{1}{z^2} = \frac{1}{[(z+1)-1]^2} = \frac{1}{[1-(z+1)]^2}$

$$= [1-(z+1)]^{-2}$$

$$= 1 + 2(z+1) + 3(z+1)^2 + 4(z+1)^3 + ...$$

[By binomial Theorem, since $|z+1| < 1$]

$$= 1 + \sum_{n=1}^{\infty}(n+1)(z+1)^n$$

Example **12.29.** Expand in Laurent's series :

(*i*) $\frac{z}{(z-1)(z-3)}$ for $|z-1| < 2$

Solution : $\left|\frac{(z-1)}{2}\right| < 1$

Let $(z-1) = u$. Then $|z-1| < 2$ implies $|u| < 2$

And so $\frac{z}{(z-1)(z-3)} = \frac{u+1}{u(u-2)} = \left[\frac{-1}{2u} + \frac{3}{2(u-2)}\right]$

Hence, for $|u| < 2$, we have

$$f(z) = \frac{-1}{2u} - \frac{3}{4\left[1-\frac{u}{2}\right]}$$

$$= -\frac{1}{2u} - \frac{3}{4}\left[1-\frac{u}{2}\right]^{-1}$$

$$= \frac{-1}{2u} - \frac{3}{4}\left[1+\frac{u}{2}+\frac{u^2}{2^2}+\frac{u^3}{2^3}+......\right]$$

$$= \frac{-1}{2(z-1)} - \frac{3}{4}\left[1+\frac{z-1}{2}+\frac{(z-1)^2}{2^2}+\frac{(z-1)^3}{2^3}+......\right]$$

$$= \frac{-1}{2(z-1)} - \frac{3}{4}\sum_{n=0}^{\infty}\left(\frac{(z-1)}{2}\right)^n$$

(*ii*) $\frac{(z-1)}{z^2}$ for $|z-1| > 1$

Let $(z-1) = u \Rightarrow$ Given problem as $\frac{u}{(u+1)^2}$ for $|u| > 1$ or $\frac{1}{|u|} < 1$

$$= \frac{u}{u^2\left[1+\frac{1}{u}\right]^2} = \frac{1}{u\left(1+\frac{1}{u}\right)^2}$$

$$= \frac{1}{u}\left[1+\frac{1}{u}\right]^{-2}$$

$$= \frac{1}{u}\left[1 - \frac{2}{u} + \frac{4}{2u^2} - \frac{8}{3!u^3} + \ldots\ldots\right]$$

$$= \left[\frac{1}{u} - \frac{2}{u^2} + \frac{2}{u^3} - \frac{4}{3u^4} + \ldots\ldots\right]$$

$$= \left[\frac{1}{(z-1)} - \frac{2}{(z-1)^2} - \frac{2}{(z-1)^3} - \frac{4}{3(z-1)^4} + \ldots\ldots\right]$$

(*iii*) Expand $\frac{e^{2x}}{(z-1)^3}$ about the singularity $z = 1$ in Laurent's series.

To expand $f(z) = \frac{e^{2x}}{(z-1)^3}$ about $z = 1$ *i.e.*, in powers of $(z - 1)$, we put $z - 1 = t$, or, $z = t + 1$.

$$\therefore \quad f(z) = \frac{e^{2(t+1)}}{t^3} = \frac{e^2}{t^3}e^{2t}$$

$$= \frac{e^2}{t^3}\left[1 + 2t + \frac{(2t)^2}{2!} + \frac{(2t)^3}{3!} + \frac{(2t)^4}{4!} + \ldots\ldots\right]$$

$$= e^2\left[\frac{1}{t^3} + \frac{2}{t^2} + \frac{2}{t} + \frac{4}{3} + \frac{2}{3}t + \ldots\ldots\right]$$

$$= e^2\left[\frac{1}{(z-1)^3} + \frac{2}{(z-1)^2} + \frac{2}{(z-1)} + \frac{4}{3} + \frac{2}{3}(z-1) + \ldots\ldots\right]$$

$$= e^2\left[(z-1)^{-3} + 2(z-1)^{-2} + 2(z-1)^{-1} + \frac{4}{3} + \frac{2}{3}(z-1) + \ldots\ldots\right]$$

which is required Laurent's series.

***Example* 12.30.** Find Laurent's series expansion of function :

(*i*) $\frac{z^2-1}{z^2+5z+6}$, about $z = 0$ in region $2 < |z| < 3$.

Solution. $\frac{z^2-1}{(z+3)(z+2)}$

$$= 1 - \frac{5z+7}{(z^2+5z+6)}$$

$$= 1 + \frac{3}{z+2} - \frac{8}{z+3}$$

For $2 < |z| < 3$, (1) Reduces to

$$= 1 + \frac{3}{z\left(1+\frac{2}{z}\right)} - \frac{8}{3\left(1+\frac{z}{3}\right)}$$

$$= 1 + \frac{3}{z}\left(1+\frac{2}{z}\right)^{-1} - \frac{8}{3}\left(1+\frac{z}{3}\right)^{-1}$$

$$= 1 + \frac{3}{z}\left(1 - \frac{2}{z} + \frac{2^2}{z^2} - \frac{2^3}{z^3} + \ldots..\right)\frac{-8}{3}\left(1 - \frac{z}{3} + \frac{z^2}{3^2} - \frac{z^3}{3^3} + \ldots.\right)$$

$$= 1 + \frac{3}{z}\sum_{n=0}^{\infty}(-1)^n \cdot \frac{2^n}{z^n} - \frac{8}{3}\sum_{n=0}^{\infty}(-1)^n \cdot \frac{z^n}{3^n}$$

(ii) $\dfrac{z^2 - 6z - 1}{(z-1)(z-3)(z+2)}$ in region $3 < |z+2| < 5$.

Let $z + 2 = u$. So $3 < |u| < 5$.

$$\frac{(u-2)^2 - 6(u-2) - 1}{(u-3)(u-5)u.} = \frac{u^2 - 10u + 15}{u(u-3)(u-5)} = \frac{A}{u} + \frac{B}{(u-3)} + \frac{C}{(u-5)}$$

Resolving into partial fractions.

$u^2 - 10u + 15 = A(u-3)(u-5) + Bu(u-5) + Cu(u-3)$

for $u = 0, A = -1$

$u = 3, -6B = -6$ or $B = +1$

$u = 5, 10C = -10$ or $C = -1$

$$\frac{u^2 - 10u + 15}{u(4-3)(4-5)} = \frac{+1}{u} + \frac{1}{u-3} - \frac{1}{u-5} \Rightarrow \frac{1}{u} + \frac{1}{u-3} - \frac{1}{u-5}$$

For $3 < |u| < 5$

$$= \frac{+1}{u} + \frac{1}{u\left[1 - \frac{3}{u}\right]} + \frac{1}{-5\left[1 - \frac{u}{5}\right]}$$

$$= \frac{+1}{u} + \frac{1}{u}\left[1 - \frac{3}{u}\right]^{-1} - \frac{1}{5}\left[1 - \frac{u}{5}\right]^{-1}$$

$$= \frac{+1}{u} + \frac{1}{u}\left[1 + \frac{3}{u} + \frac{3^2}{u^2} + \frac{3^2}{u^3} + \ldots\right] - \frac{1}{5}\left[1 + \frac{u}{5} + \frac{u^2}{25} + \frac{u^3}{5^3} + \ldots\right]$$

$$= \frac{+1}{u} + \frac{1}{u}\sum_{n=0}^{\infty}\frac{3^n}{u^n} - 1\frac{1}{5}\sum_{n=0}^{\infty}\frac{u^n}{5^n}$$

$$= \frac{+1}{z+2} + \frac{1}{z+2}\sum_{n=0}^{\infty}\frac{3^n}{(z+2)^n} - \frac{1}{5}\sum_{n=0}^{\infty}\frac{(z+2)^n}{5^n}$$

***Example* 12.31:** Find the Laurent's expansion of

(i) $\dfrac{z}{(z^2-1)(z^2+4)}$ for $|z| < 1$

Solution : $\dfrac{z}{(z^2-1)(z^2+4)} = \dfrac{1}{3}\left[\dfrac{-z}{z^2-1} + \dfrac{z}{z^2-4}\right]$

$$= \frac{1}{3}\left[\frac{+z}{[1-z^2]} - \frac{z}{4\left[1 - \frac{z^2}{4}\right]}\right]$$

$$= \left\{\frac{1}{3}z[1-z^2]^{-1} - \frac{z}{12}\left[1 - \frac{z^2}{4}\right]^{-1}\right\}$$

$$= \frac{1}{3}z[1 + (z^2) + (z^2)^2 + \ldots..] - \frac{z}{12}\left[1 + \frac{z^2}{4} + \left(\frac{z^2}{4}\right)^2 + \left(\frac{z^2}{4}\right)^3 + \ldots.\right]$$

$$= \left(\frac{z}{3} - \frac{z}{12}\right) + \left(\frac{1}{3} - \frac{1}{48}\right) z^3 + \left(\frac{1}{3} - \frac{1}{192}\right) . z^2 + ...$$

$$= \frac{z}{4} - \frac{5}{16} z^3 + \frac{21}{64} z^5 + ...$$

(ii) $1 < |z| < 2$

$$= \frac{1}{3}\left[\frac{-z}{z^2 - 1} + \frac{z}{z^2 - 4}\right]$$

$$= \frac{1}{3}\left\{\frac{-z}{z^2\left[1 - \frac{1}{z^2}\right]} + \frac{z}{-4\left[1 - \left(\frac{z}{2}\right)^2\right]}\right\}$$

$$= \frac{1}{3}\left[\frac{-1}{z}\left(1 + \frac{1}{z^2} + \frac{1}{z^4} + \frac{1}{z^6} + ...\right)\right] - \frac{z}{4}\left[1 + \frac{z^2}{4} + \frac{z^3}{16} + ...\right]$$

***Example* 12.32:** Expand $f(z) = \dfrac{1}{z^2(z-i)}$ as Laurents series about i and hence find the residue there at.

Solution : Let $z - i = t$, $z = t + i$

$$f(z) = \frac{1}{(t+i)^2 t} = \frac{1}{i^2 t}\left(1 + \frac{t}{i}\right)^{-2}$$

$$= \frac{-1}{t}\left[1 - \frac{2t}{i} + \frac{(-2)(-3)}{2!}\left(\frac{t}{i}\right)^2 + ...\right]$$

$$= \frac{-1}{t}\left[1 - \frac{2t}{i} - 3t^2 + ...\right]$$

$$= \left[\frac{-1}{t} + \frac{2}{i} + 3t + ...\right]$$

$$= \left[-\frac{1}{t} - 2i + 3t + ...\right]$$

EXERCISE 12.6

Expand the following function as a Taylor's series :

1. $\dfrac{\sin z}{z - \pi}$ about $z = \pi$.

Ans. $-1 + \dfrac{(z-\pi)^2}{3!} - \dfrac{(z-\pi)^4}{5!} + ...$

2. $\log(1 + z)$ about $z = 0$.

Ans. $z - \dfrac{z^2}{2} + \dfrac{z^3}{3} - \dfrac{z^4}{4} + ...$

3. $\sin z$ about $z = \dfrac{\pi}{4}$.

Ans. $\dfrac{1}{\sqrt{2}}\left[1 + \left(z - \frac{\pi}{4}\right) - \frac{1}{2i}\left(z - \frac{\pi}{4}\right)^2\right] - \frac{1}{3i}\left(z - \frac{\pi}{4}\right)^3 + ...$

Expand the following function as Laurent's series :

4. $\dfrac{1}{(z^2+1)(z^2+2)}$

(*i*) $0 < |z| < 1$ (*ii*) $1 < |z| < \sqrt{2}$ (*iii*) $|z| > \sqrt{2}$.

Ans.

(*i*) $\frac{1}{2}\left[1 - \sum_{n=1}^{\infty} (-1)^n \left(\frac{1}{2^n} - 2\right) z^{2n}\right]$, (*ii*) $\frac{-1}{2} - \sum_{n=1}^{\infty} (-1)^n \left[\frac{1}{z^{2n}} + \frac{z^{2n}}{2^{n+1}}\right]$

(*iii*) $\sum_{n=1}^{\infty} (-1)^{n+1}, \frac{2^n - 1}{(z^2)^{n+1}}$

5. $\dfrac{1}{z(z-1)(z-2)}$ for $|z| > 2$.

Ans. $z^{-3} + 3z^{-4} + 7z^{-5} + \ldots\ldots + (2^{n+1} - 1)z^{-(n+1)} + \ldots\ldots$

6. Expand $f(z) = \dfrac{(z-2)(z+2)}{(z+1)(z+4)}$ in the region :

(*i*) $1 < |z| < 4$ (*ii*) $|z| > 4$

Ans.

(*i*) $\sum_{n=1}^{\infty} \frac{(-1)^{n+1}}{4^n} z^n + \sum_{n=1}^{\infty} (-1)^n z^{-n}$ (*ii*) $1 + \sum_{n=1}^{\infty} (-1)^n (1 + 4^n) z^{-n}$

12.21 APPLICATION OF RESIDUES TO EVALUATE REAL INTEGRALS

The residue theorem provides a simple method for evaluating many important definite integral of real variables. Some of these are :

(1) Integrals of the type $\int_0^{2\pi} F(\cos\theta, \sin\theta)\, d\theta$ where $F(\cos\theta, \sin\theta)$ is a rational function of $\cos\theta$ and $\sin\theta$.

such integrals can be reduced to complex line integrals by the substitution, $z = e^{i\theta}$, so that

$$dz = i.e^{i\theta} d\theta,\ d\theta = \frac{dz}{iz}$$

Also

$$\cos\theta = \frac{e^{i\theta} + e^{-i\theta}}{2} = \frac{1}{2}\left(z + \frac{1}{z}\right)$$

$$\sin\theta = \frac{e^{i\theta} - e^{i\varphi}}{2i} = \frac{1}{2i}\left(z - \frac{1}{z}\right)$$

As θ varies from 0 to 2π, z moves once round the unit circle in the anti-clockwise direction.

$$\therefore \int_0^{2\pi} F(\cos\theta, \sin\theta)\, d\theta = \oint_C F\left(\frac{z + z^{-1}}{2}, \frac{z - z^{-1}}{2i}\right) . \frac{dz}{iz}$$

C is the unit circle $|z| = 1$.

The integral on the right can be evaluated by using the residue theorem.

Example 12.33: $\int_0^{2\pi} \frac{d\theta}{2+\cos\theta}$

Solution : Put $z = e^{i\theta}$ so that $\cos\theta = \frac{1}{2}\left(z+\frac{1}{z}\right)$ and $d\theta = \frac{dz}{iz}$

Thus $$\int_0^{2\pi} \frac{d\theta}{2+\cos\theta} = \oint_C \frac{1}{2+\frac{1}{2}\left(z+\frac{1}{z}\right)}\frac{dz}{iz}$$

$$= \frac{2}{i}\oint_C \frac{1}{z^2+4z+1}dz, \text{ where C is the circle } |z| = 1$$

The poles of the integral are the roots of $z^2 + 4z + 1 = 0$ which are $z = 2 \pm \sqrt{3}$. Of the two poles. only $z = -2 + \sqrt{3}$ lies inside the circle C. Residue at $z = -2 + \sqrt{3}$ is

$$= \lim_{z\to\alpha} (z-\alpha)\frac{1}{(z^2+4z+1)} \text{ where } \alpha = -2+\sqrt{3}$$

$$= \lim_{z\to\alpha} \frac{1}{(2z+4)} = \frac{1}{(2\alpha+4)} = \frac{1}{2\sqrt{3}}$$

$\therefore$ By Residue theorem,

$$\oint_C \frac{1}{z^2+4z+1} = 2\pi i\left(\frac{1}{2\sqrt{3}}\right) = \frac{\pi i}{\sqrt{3}}$$

Hence, $$\int_0^{2\pi} \frac{d\theta}{2+\cos\theta} = \frac{2}{i}\oint \frac{dz}{z^2+4z+1} = \frac{2}{i}\left(\frac{\pi i}{\sqrt{3}}\right) = \frac{2\pi}{\sqrt{3}}$$

Example **12.34.** Using residue to prove that $\int_0^{2\pi} \frac{\cos 2\theta\, d\theta}{1-2p\cos\theta+p^2} = \frac{\pi p^2}{1-p^2}$ $(0 < p < 1)$

Solution : $$I = \int_0^{\pi} \frac{\cos 2\theta d\theta}{1-2p\cos+p^2} = \frac{1}{2}\int_0^{2\pi} \frac{\cos 2\theta d\theta}{1-p(e^{i\theta}+e^{-i\theta})+p^2}$$

$$= \frac{1}{2} \text{ real part of } \int_0^{2\pi} \frac{e^{2i\theta}}{(1-pe^{i\theta})(1-pe^{-i\theta})}d\theta$$

$$= \frac{1}{2} \text{ real part of } \oint_C \frac{z^2}{(1-pz)\left(1-\frac{p}{z}\right)}\frac{dz}{iz} \qquad \left[\begin{array}{l} z = e^{i\theta} \\ d\theta = \frac{dz}{iz}\end{array}\right]$$

$$= \frac{1}{2} \text{ real part of } \oint_c \frac{-iz^2}{(1-pz)(z-p)}dz$$

$$= \frac{1}{2} \text{ real part of } \oint_C f(z)\,dz, \text{ where } f(z) = \frac{-iz^2}{(1-pz)(z-p)}$$

Poles of $f(z)$ are : $z = \frac{1}{p}$ and $z = p$, out of which $z = p$ lies within the unit circle C as $p < 1$.

The residue of $f(z)$ at $z = p$ is

$$= \lim_{z\to p} (z-p)\left[\frac{-iz^2}{(1-pz)(z-p)}\right] = \frac{-ip^2}{1-p^2}$$

Hence, by C-R theorem,

$$\oint_C f(z)\,dz = 2\pi i \text{ (sum of residues within C)}$$

$$= 2\pi i\left[\frac{-ip^2}{1-p^2}\right]$$

$$= \frac{2\pi p^2}{(1-p^2)} \text{ which is purely real.}$$

Hence, $$I = \frac{1}{2}\left(\frac{2\pi p^2}{1-p^2}\right) = \frac{\pi p^2}{1-p^2}$$

Example **12.35.** Evaluate $\int_0^{2\pi} \frac{d\theta}{1-2p\sin\theta+p^2}, 0<p<1.$

Solution : Put $\sin\theta = \frac{e^{i\theta}-e^{-i\theta}}{2i}$, $\therefore \sin\theta = \frac{z-\frac{1}{z}}{2i}$

$$z = e^{i\theta},\ dz = iz\,d\theta \Rightarrow d\theta = \frac{dz}{iz}$$

$$\therefore \quad \int_0^{2\pi} \frac{d\theta}{1-2p\sin\theta+p^2} = \oint_C \frac{-dz}{pz^2-i(p^2+1)z-p}$$

$$= \oint_C \frac{-dz}{(z-ip)(pz-i)}$$

Hence poles are $$z = ip \text{ and } z = \frac{i}{p}$$

Since $p<1$, $z = ip$ lie within the unit circle C. So residue at $z = pi$ is

$$= \lim_{z\to ip}(z-ip)\frac{1}{(z-ip)(pz-i)}$$

$$= \frac{1}{p^2 i - i} = \frac{1}{i(p^2-1)}$$

$\therefore$ By Residue Theorem,

$$-\oint_C \frac{dz}{pz^2-(p^2+1)zi-p} = -2\pi i\left[\frac{1}{i(p^2-1)}\right] = \frac{2\pi}{1-p^2}$$

12.22 INTEGRALS OF TYPE

Integrals of type $\int_{-\infty}^{\infty} \frac{f(x)}{f(x)}dx$, where $f(x)$ and $f(x)$ are polynomials in x such that $\frac{xf(x)}{F(x)} \to 0$ as $x \to \infty$ and $F(x)$ has no zeros on the real axis.

Consider the integral $\oint_c \frac{f(z)}{F(z)}dz$ over the closed contour C consisting of the real axis from $-R$ to R and the semicircle C_r of radius R in the upper half plane.

We take R large enough so that all the poles of $\frac{f(z)}{F(z)}$ in the upper half plane in lie within C.

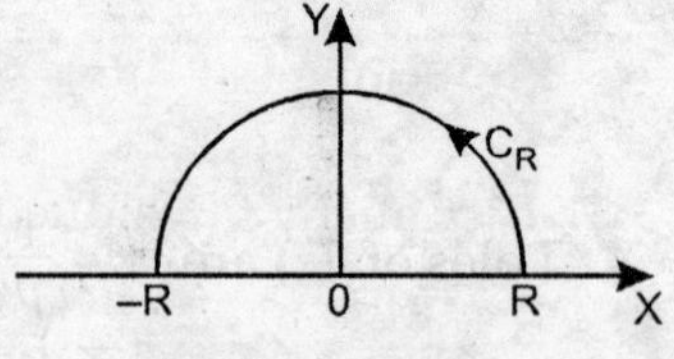

Fig. 12.20

By Residue theorem, we have

$$\oint_{C_1} \frac{f(z)}{F(z)} dz + \int_{-R}^{R} \frac{f(x)}{F(x)} dx = 2\pi i \quad \text{[Sum of residue of } \frac{f(z)}{F(z)} \text{ in upper half plane]}$$

[$\because$ on real axis, $z = x$]

Finally making $R \to \infty$, we have value of $\int_{-\infty}^{\infty} f(x)\,dx$ provided $\int_{C_1} \frac{f(z)}{f(z)} dz \to 0$ when $R \to \infty$.

$$\therefore \int_{-\infty}^{\infty} \frac{f(x)}{F(x)} dx = 2\pi i \text{ (sum of residues of } \frac{f(z)}{F(z)} \text{ in the upper half planes)}$$

***Example* 12.36:** $\int_{-\infty}^{\infty} \frac{x^2\,dx}{(x^2+1)(x^2+4)}$.

Solution : Consider $\phi(z) = \int_{-\infty}^{\infty} \frac{z^2\,dz}{(z^2+1)(z^2+4)} = \oint_c f(z)\,dz$

Where C is the contour consisting of semicircle CR of radius R together with the part of real axis from $-R$ to R.

The poles $z = \pm i$ and $z = \pm 2i$. Of these $z = i$, $2i$ lie within C.

$\therefore$ Res of $f(z)$ at $z = i$ is : $\lim_{z \to i} (z - i), f(z)$.

$$= \lim_{z \to i} (z - i) \frac{z^2}{(z^2+1)(z^2+4)}$$

$$= \lim_{z \to i} \frac{z^2}{(z+i)(z^2+4)} = \frac{i^2}{2i(i^2+4)}$$

$$= \frac{-1}{2i(-1+4)} = \frac{-1}{6i}$$

Residue of $f(z)$ at $z = 2i$ is $\lim_{z \to 2i} (z - 2i) f(z)$

$$= \lim_{z \to 2i} \frac{z^2}{(z^2+1)(z+2i)} = \frac{-2}{2i(-3)} = \frac{1}{3i}$$

$$\therefore \int_{-\infty}^{\infty} \frac{x^2\,dx}{(x^2+1^2)(x^2+4)} = 2\pi i \text{ (Sum of Residue of } f(z) \text{ in upper half of } z\text{-plane)}$$

$$= 2\pi i \left[\frac{-1}{6i} + \frac{1}{3i}\right]$$

$$= 2\pi i \left[\frac{3}{18i}\right] = \frac{\pi}{3}.$$

***Example* 12.37.** Evaluate $\int_0^{\infty} \frac{x^2}{(x^6+1)} dx$, by applying calculus of residues.

Solution : Let $f(z) = \frac{z^2}{z^6+1}$

By Cauchy's Residue theorem,

$$\oint_C f(z)\,dz = \int_{-R}^{R} \frac{x^2}{x^6+1} dx + \oint_C \frac{z^2}{z^6+1} dz$$

Since $\lim_{R\to\infty} zf(z) = 0, \therefore \int_c \frac{z^2}{z^6+1}dz$

$$\lim_{R\to\infty}\int_{-R}^{R}\frac{x^2dx}{(x^6+1)} = 2\pi i \text{ (Sum of residues)}$$

i.e., $$\int_{-\infty}^{\infty}\frac{x^2dx}{(x^6+1)}dx = 2\pi i \text{ (Sum of Residues)}$$

Solving $z^6+1 = 0$, we get $z = e^{i(2n+1)\pi/6}$

where $n = 0, 1, 2, 3, 4, 5.$

Thus $z = e^{\frac{i\pi}{6}}, e^{\frac{i\pi}{2}}, e^{\frac{i5\pi}{6}}, e^{\frac{i3\pi}{2}}, e^{\frac{i11\pi}{6}}$, are simple poles of $f(z)$.

Out of these, first three lie inside C.

Residue at $z = e^{\frac{i\pi}{6}}$ is $\lim_{z\to e^{i\pi/6}}\frac{(z-e^{i\pi/6})z^2}{z^6+1}$ $\left[\frac{0}{0}\text{ Form}\right]$

$$= \lim_{z\to e^{i\pi/6}}\frac{2(z-e^{i\pi/6})+z}{6z^4}$$ [By L' Hospital's Rule]

$$= \frac{1}{6}e^{\frac{-3\pi}{6}} = \frac{1}{6}e^{\frac{-i\pi}{2}} = \frac{1}{6}(-i)$$

Similarly residues at $e^{\frac{i\pi}{2}}, e^{\frac{i5\pi}{6}}$ are $\frac{-1}{6}e^{\frac{i3\pi}{2}}, \frac{-1}{6}e^{\frac{i5\pi}{2}}$

Hence sum of Residue at poles is

$$= \frac{-1}{6}\left[e^{\frac{i\pi}{2}}+e^{\frac{i3\pi}{2}}+e^{\frac{i5\pi}{2}}\right]$$

$$= \frac{-1}{6}[i-i+i] = \frac{-1}{6}i$$

Hence $$\int_{-\infty}^{\infty}\frac{x^2}{x+x^6}dx = 2\pi i\left[\frac{-1}{6}i\right] = \frac{\pi}{3}$$

Hence $$\int_0^{\infty}\frac{x^2}{1+x^6}dx = \frac{\pi}{6}$$

EXERCISE 12.7

1. $\int_0^{2\pi}\frac{\sin^2\theta}{a+b\cos\theta}d\theta, (0<b<a).$

Ans. $\frac{2\pi}{b^2}(a-\sqrt{a^2-b^2})$

2. $\int_0^{2\pi}\frac{d\theta}{5-3\cos\theta}$

Ans. $\frac{\pi}{2}$

3. $\int_0^{2\pi} \frac{d\theta}{a + b\cos\theta}$ $(a > b > 0)$

Ans. $\frac{2\pi}{\sqrt{a^2 - b^2}}$

4. $\int_0^{2\pi} \frac{d\theta}{a + b\sin\theta}$, $a > |b|$.

Ans. $\frac{2\pi}{\sqrt{a^2 - b^2}}$

5. $\int_0^{2\pi} \frac{\cos 2\theta}{1 - 2a\cos\theta + a^2} d\theta$, $0 > a < 1$.

Ans. $\frac{2\pi a^2}{1 - a^2}$

6. $\int_{-\infty}^{\infty} \frac{x^2}{(x^2 + a^2)(x^2 + b^2)} d\theta$, $(a > 0, b > 0)$

Ans. $\frac{\pi}{a + b}$

7. $\int_0^{\infty} \frac{dx}{x^4 + 1}$

Ans. $\frac{\pi\sqrt{2}}{4}$

8. $\int_0^{\infty} \frac{dx}{(x^6 + 1)}$

Ans. $\frac{\pi}{3}$

9. $\int_0^{\infty} \frac{x^2 dx}{(x^2 + 9)(x^2 + 4)^2}$

Ans. $\frac{\pi}{200}$

10. $\int_0^{\infty} \frac{dx}{(x^2 + 1)^2}$

Ans. $\frac{\pi}{4}$

OBJECTIVE QUESTIONS

1. What is the value of $\oint_C \frac{z^2 - z + 1}{(z - 1)} dz$ where $C : |z| = \frac{1}{2}$

Ans. 0

2. $f(z) = \frac{\sin\sqrt{z}}{\sqrt{z}}$ has which singularity at $z = 0$

Ans. Removable singularity

3. Find the value of $\oint_c \frac{1}{1+z^2}$ where $C: x^2+y^2=9$

Ans. 0

4. Find the values of the following :

(i) $\oint_C \frac{dz}{z^2+9} dz$, C is $x^2+4y^2=16$ (ii) $\oint_C \frac{e^{2z}}{(z-1)^3} dz$ where C is $|z|=1.5$.

(iii) $\int_C \frac{\sin 6}{(z-2)^{10}} dz$: C is $|z|=1$ (iv) $\underset{z \to e^{\pi i/3}}{Lt} \frac{(z-e^{\pi i/3})z}{z^3+1}$

(v) $\oint_C \frac{e^z}{(z+1)^2} dz$: C is circle $|z-1|=3$

Ans.

(i) 0 (ii) $4\pi ie^2$ (iii) 0 (iv) $\frac{1}{6}-\frac{i\sqrt{3}}{6}$

(v) $\frac{2\pi i}{e}$

5. Is function $f(z)=z^2\bar{z}$ analytic at $z=0$.

Ans. No, it does not satisfy C - R equations.

6. What is the period of $f(z)=e^{iz}$

Ans. 2π

7. Name the singularity in following functions and give the order, if $f(z)$ has poles.

(i) $f(z)=\frac{z-\sin z}{z^3}$ (ii) $f(z)=\frac{1}{(1-z)^2}\sin\left(\frac{1}{1-z}\right)$ at $z=1$

Ans.

(i) Pole of order 3 (ii) Essential singularity

8. Find the residues of

(i) $f(z)=\frac{1+z}{(a-z)^2}$ (iii) $f(z)=\frac{1+e^z}{\sin z+z\cos z}$ at $z=0$

Ans. (i) 1 (ii) 1 (iii) $\frac{5}{9}$ and $\frac{4}{9}$

9. Evaluate :

(i) $\underset{z \to 2+i}{Lt} \frac{z^2-2iz}{z^2+4}$ (ii) $\underset{z \to \pi/2}{Lt} \left(\frac{\sin z}{z}\right)$

Ans. (i) $\frac{7-4i}{13}$ (ii) $\frac{2}{\pi}$

10. If $f(z)=\frac{z}{1+z}$, find $f'(z)$ and determine where $f(z)$ is non-analytic.

Ans. $\frac{1}{(1+z)^2}$, $z=-1$

Laplace Transforms and its Applications

This chapter covers the topics related to Laplace transform, transforms of elementary functions, properties of Laplace transforms, transform of derivatives and integrals, multiplication by t^n, division by t, inverse transforms, convolution theorem, application to differential equations, simultaneous linear equations will constant coefficients, unit step function, unit impulse function and periodic functions.

13.1 DEFINITION OF LAPLACE TRANSFORM

Laplace transform in solving different equations with boundary conditions without finding the general solution and the values of arbitrary constants.

***Definition*:**

Let $f(t)$ be a function of 't' defined for all values of $t \geq 0$. Then the Laplace transform of $f(t)$ denoted by $L[f(t)]$ and defined as

$$F(s) = L[f(t)] = \int_0^{\infty} e^{-st} f(t)\, dt$$

provided the integral exists. Here s is a parameter which may be real or complex L is called Laplace transform operator.

The Laplace transform changes or transforms, the function $f(t)$ into a different function $f(s)$. Note that $f(t)$ is a function of t and $F(s)$ is a function of s.

13.2 SOME IMPORTANT RESULTS

1. $$L(1) = \int_0^{\infty} e^{-st} dt = \left[-\frac{e^{-st}}{s}\right]_0^{\infty} = \frac{1}{s}, \text{ if } s > 0$$

2. $$L(t^n) = \int_0^{\infty} e^{-st} t^n dt$$

Put

$$st = x$$

$$t = \frac{x}{s}, dt = \frac{dx}{s}$$

$$= \int_0^\infty e^{-x}\left(\frac{x}{s}\right)^n \frac{ds}{s} = \frac{1}{s^{n+1}}\int_0^\infty e^{-x}x^n\,dx = \frac{1}{s^{(n+1)}}\,\Gamma(n+1) = \frac{n!}{s^{n+1}}$$

where n is a positive integer.

3. $$L(e^{at}) = \int_0^\infty e^{-st}e^{at}\,dt = \int_0^\infty e^{-(s-a)t}\,dt = \left[\frac{e^{-(s-a)}}{-(s-a)}\right]_0^\infty = \frac{1}{s-a},\ \text{if } s > a$$

4. $$L[\cosh(at)] = L\left[\frac{(e^{at}+e^{-at})}{2}\right] = \frac{1}{2}\,[L(e^{at}) + L(e^{-at})]$$

$$= \frac{1}{2}\left[\frac{1}{s-a} + \frac{1}{s+a}\right] = \frac{s}{s^2-a^2},\ s > |a|$$

5. $$L[\sin h(at)] = L\left[\frac{(e^{at}-e^{-at})}{2}\right] = \frac{1}{2}\,[L(e^{at}) - L(e^{-at})] = \frac{a}{s^2-a^2},\ s > |a|$$

6. $$L[\sin(at)] = L\left[\frac{(e^{iat}-e^{-iat})}{2i}\right] = \frac{1}{2i}\,[L(e^{iat}) - L(e^{-iat})]$$

$$= \frac{1}{2i}\left[\frac{1}{s-ia} + \frac{1}{s+ia}\right] = \frac{a}{s^2+a^2},\ s > 0$$

7. $$L[\cos(at)] = L\,\frac{(e^{iat}+e^{-iat})}{2} = \frac{1}{2}\,[L(e^{iat}) + L(e^{-iat})]$$

$$= \frac{1}{2}\left[\frac{1}{s-ia} + \frac{1}{s+ia}\right] = \frac{s}{s^2+a^2},\ s > 0$$

13.3 PROPERTIES OF LAPLACE TRANSFORMS

1. **Linearity Property :** If a, b are constants and $f_1(t)$, $f_2(t)$ are functions of t, then

$$L[a\,f_1(t) + b\,f_2(t)] = a\,L[f_1(t)] + b\,L[f_2(t)]$$

Proof :

$$L[af_1(t) + bf_2(t)] = \int_0^\infty e^{-st}[af_1(t) + bf_2(t)]\,dt \qquad \text{(By def.)}$$

$$= a\int_0^\infty e^{-st}f_1(t)\,dt + b\int_0^\infty e^{-st}f_2(t)\,dt$$

$$= a\,L[f(t)] + b\,L[f_2(t)].$$

2. **First Shifting Property :** If $L[f(t)] = F(s)$, then

$$L[e^{at}\,f(t)] = F(s-a)$$

Proof : By definition,

$$L[e^{at}\,f(t)] = \int_0^\infty e^{-st}e^{at}f(t)\,dt$$

$$= \int_0^\infty e^{-(s-a)t} f(t)\, dt$$

$$= \int_0^\infty e^{-pt} f(t)\, dt \text{ where } p = s - a$$

$$= F(p) = F(s-a)$$

With the help of this property, we have :

1. $L[e^{at} . t^n] = \dfrac{n!}{(s-a)^{n+1}}$ where n is a positive integer $\left[\because L(t^n) = \dfrac{n!}{S^{n+1}}\right]$

2. $L[e^{at} \cosh bt] = \dfrac{s-a}{(s-a)^2 - b^2}$

3. $L[e^{at} \sin h\, bt] = \dfrac{b}{(s-a)^2 - b^2}$

4. $L[e^{at} \cos bt] = \dfrac{s-a}{(s-a)^2 + b^2}$

5. $L(e^{at} \sin bt) = \dfrac{b}{(s-a)^2 + b^2}$

Example **13.1.** Find the Laplace transform of

(*i*) $e^{at} - t^{-1/2} + 1$

(*ii*) $\sin^3 t$

(*iii*) $\sin 2t \cos 3t$

(*iv*) $3 \cosh 2t - 5 \sin h\, 3t$

Solution : (*i*) $L[e^{at} - t^{-1/2} + 1] = L(e^{at}) - L(t^{-1/2}) + L(1)$ (By Linearity)

$$= \frac{1}{s-a} - \frac{\Gamma\left(-\frac{1}{2}+1\right)}{s^{\frac{-1}{2}+1}} + \frac{1}{s} = \frac{1}{s-a} - \frac{\Gamma\left(\frac{1}{2}\right)}{\sqrt{s}} + \frac{1}{s} = \frac{1}{s-a} - \frac{\sqrt{\pi}}{\sqrt{s}} + \frac{1}{s}$$

(*ii*) $$L(\sin^3 t) = \frac{1}{4} L(3 \sin t - \sin 3t)$$

$$= \frac{1}{4} [3\, L(\sin t) - L(\sin 3t)]$$

$$= \frac{1}{4}\left[\frac{3}{s^2+1} - \frac{3}{s^2+9}\right]$$

(*ii*) $$L(\sin 2t \cos 3t) = \frac{1}{2} L(2 \sin 2t \cos 3t)$$

$$= \frac{1}{2} L(\sin 5t + \sin(-t))] = \frac{1}{2} L(\sin 5t - \sin t)$$

$$= \frac{1}{2} [L(\sin 5t) - L(\sin 5t)] = \frac{1}{2}\left[\frac{5}{s^2+25} - \frac{1}{s^2+1}\right]$$

(*iv*) $L[3\cosh 2t - 5\sinh 3t] = 3L(\cosh 2t) - 5L(\sinh 3t)$

$$= 3\frac{s}{s^2-4} - 5\frac{3}{s^2-9} = \frac{3s}{s^2-4} - \frac{15}{s^2-9}$$

Example 13.2. Find the Laplace transform defined as

$$f(t) = \begin{cases} \cos\left(t-\frac{2\pi}{3}\right), & t > \frac{2\pi}{3} \\ 0, & t < \frac{2\pi}{3} \end{cases}$$

Solution :

$$L(t) = \int_0^\infty f(t)e^{-st}dt = \int_0^{2\pi/3} f(t)e^{-st}dt + \int_{2\pi/3}^\infty f(t)e^{-st}dt$$

$$= \int_0^{2\pi/3} 0\, e^{-st}dt + \int_{2\pi/3}^\infty \cos\left(t-\frac{2\pi}{3}\right)e^{-st}dt$$

$$= \int_{2\pi/3}^\infty \cos\left(t-\frac{2\pi}{3}\right)e^{-st}dt \quad \text{Put } t-\frac{2\pi}{3} = x \qquad \because dt = dx$$

$$= \int_0^\infty \cos x e^{-s\left[x+\frac{2\pi}{3}\right]}dx$$

$$= \int_0^\infty \cos x e^{-st}.e^{\frac{-2x}{3}}dx = e^{\frac{-2x}{3}s}\int_0^\infty e^{-st}\cos x\, dx$$

$$\left[\int e^{ax}\cos bx\, dx = \frac{e^{ax}}{a^2+b^2}(a\cos bx + b\sin bx)\right]$$

$$= e^{\frac{-2\pi s}{3}}\left(\frac{s}{1+s^2}\right)$$

Example 13.3. Find the Laplace transform of $t^2 \cos at$

Solution :

$$L(t^2\cos at) = L\left[t^2\frac{(e^{iat}+e^{-iat})}{2}\right]$$

$$= \frac{1}{2}[L(t^2e^{iat}) + L(t^2e^{-iat})]$$

$$= \frac{1}{2}\left[\frac{2!}{(s-ia)^3} + \frac{2!}{(s+ia)^3}\right] = \left[\frac{(s+ia)^3+(s-ia)^3}{(s^2+a^2)^3}\right] = \frac{2s(s^2-3a^2)}{(s^2+a^2)^3}$$

Example 13.4. Find the Laplace transform of $e^{-3t}(2\cos 5t - 3\sin 5t)$

Solution : $L[e^{-3t}(2\cos 5t - 3\sin 5t)]$

$$= 2L(e^{-3t}\cos 5t) - 3L(e^{-3t}\sin 5t)$$

$$= 2\frac{s+3}{(s+3)^2+5^2} - 3\frac{5}{(s+3)^2+5^2}$$

$$= \frac{2s + 6 - 15}{(s+3)^2 + 5^2} = \frac{2s - 9}{(s+3)^2 + 25}$$

$$= \frac{2s - 9}{s^2 + 9 + 6s + 25} = \frac{2s - 9}{s^2 + 6s + 34}$$

Example **13.5.** Show that

(*i*) $$L\,(t \sin at) = \frac{2as}{(s^2 + a^2)^2}$$

(*ii*) $$L\,(t \cos at) = \frac{(s^2 - a^2)}{(s^2 + a^2)^2}$$

Solution : Consider $$L(t) = \frac{1}{s^2}$$

$$L\,(t\, e^{iat}) = \frac{1}{(s - ia)^2} = \frac{s^2 - a^2 + 2ias}{(s^2 + a^2)^2} = \frac{s^2 - a^2}{(s^2 - a^2)^2} + \frac{i2as}{(s^2 - a^2)^2}$$

Equating real and imaginary parts

$$L\,(t \sin at) = \frac{2as}{(s^2 + a^2)^2}$$

$$L\,(t \cos at) = \frac{s^2 - a^2}{(s^2 + a^2)^2}$$

Example **13.6.** Find Laplace transform of $f(t)$ defined as

$$f(t) = \begin{cases} \frac{t}{T}, & \text{when } 0 < t < T \\ 1, & \text{when } t > T \end{cases}$$

Solution : $$L\{f(t)\} = \int_0^\infty e^{-st} f(t)\,dt = \int_0^T e^{-st} \frac{t}{T} dt + \int_T^\infty e^{-st} . 1 dt$$

$$= \frac{1}{T}\left[\left(t . \frac{e^{-st}}{-s}\right)_0^T - \int_0^T 1 . \frac{e^{-st}}{-s} dt\right] + \left[\frac{e^{-st}}{-s}\right]_T^\infty$$

$$= \frac{1}{T}\left[-\frac{T}{s} e^{-sT} + \frac{1}{s}\left\{\frac{e^{-st}}{-s}\right\}_0^T\right] + \frac{e^{-st}}{s} = \frac{1 - e^{-st}}{Ts^2}$$

Example **13.7.** Find Laplace transforms of

(*i*) $3t$

(*ii*) $e^{3 \log t}$

(*iii*) $a^t + t^3$

Solution : (*i*) $$L(3^t) = L(e^{\log 3})^t\; L(e^{t \log 3}) = \frac{1}{s - \log 3}$$

(*ii*) $$L(e^{3 \log t}) = L(e^{\log t^3}) = L(t^3) = \frac{3!}{s^4}$$

(*iii*)
$$L(a^t + t^3) = L(a^t) + L(t^3)$$
$$= L(e^{\log a}) + \frac{3!}{s^4} = L(e^{t\log a}) + \frac{3!}{s^4}$$
$$= \frac{1}{s - \log a} + \frac{3!}{s^4}$$

***Example* 13.8.** Find Laplace transforms of

(*i*) $a^t - 2\,(e^{1/2\,at}) \cos \frac{\sqrt{3}}{2}\, at$

(*ii*) $e^{-2t} \sin^2 t \cos^2 t$

(*iii*) $e^{-4t} \,.\, t \sin 3t$

Solution : $L\left[a^t - 2\left(e^{\frac{1}{2}at}\right)\cos\frac{\sqrt{3}}{2}at\right]$

$$= L(a^t) - 2L\left(e^{\frac{1}{2}at}\cos\frac{\sqrt{3}}{2}at\right) = L(e^{\log}\text{ at}) - 2\,L\left(e^{\frac{1}{2}at}\cos\frac{\sqrt{3}}{2}at\right)$$
$$= L\,(e^{t\log a}) - 2L\left(e^{\frac{1}{2}at}\cos\frac{\sqrt{3}}{2}at\right)$$
$$= \frac{1}{s - \log a} - 2\frac{s - \frac{1}{2}a}{\left(s - \frac{1}{2}a\right)^2 + \left(\frac{\sqrt{3}}{2}\right)^2 a^2}$$
$$= \frac{1}{s - \log a} - \frac{2s - a}{\left(s - \frac{1}{2}a\right)^2 + \frac{3}{4}a^2}$$
$$= \frac{1}{s - \log a} - \frac{2s - a}{s^2 + \frac{1}{4}a^2 - sa + \frac{3}{4}a^2}$$
$$= \frac{1}{s - \log a} - \frac{2s - a}{s^2 - sa + a^2}$$

(*ii*)
$$L(e^{-2t}\sin^2 t\cos^2 t) = \frac{1}{4}\,L(e^{-2t}\sin^2 2t) = \frac{1}{4}\,L\left[e^{-2t}\,\frac{(1-\cos 4t)}{2}\right]$$
$$= \frac{1}{8}\,L[e^{-2t} - e^{-2t}\cos 4t] = \frac{1}{8}\left[\frac{1}{s+2} - \frac{s+2}{(s+2)^2 + 4^2}\right]$$
$$= \frac{1}{8}\left[\frac{(s+2)^2 + 16 - (s+2)^2}{(s+2)(s^2+4s+20)}\right] = \frac{2}{(s+2)(s^2+4s+20)}$$

(*iii*) Since
$$L(t) = \frac{1}{s^2}$$

$\therefore$ By first shifting theorem
$$L[te^{3it}] = \frac{1}{(s-3i)^2} = \frac{(s+3i)^2}{[(s-3i)(s+3i)]^2}$$

or $$L[t\,(\cos 3t + i\sin 3t)] = \frac{(s^2-9)+6is}{(s^2+9)^2}$$

Equating Imaginary Parts, we get

$$L[t\sin 3t] = \frac{6s}{(s^2+9)^2}$$

Again Applying First Shifting Theorem, we have

$$L[e^{-4t}\,t\sin 3t] = \frac{6(s+4)}{\{(s+4)^2+9\}^2} = \frac{6s+24}{(s^2+8s+25)^2}$$

EXERCISE 13.1

Find the Laplace transforms of the following :

1. $t^3\,e^{-2t}$

Ans. $\dfrac{6}{(s^2+4)}$

2. $\sinh^3 t$

Ans. $\dfrac{6}{(s^2-1)(s^2-9)}$

3. $\cos at \sinh at$

Ans. $\dfrac{1}{2}\left[\dfrac{s-a}{(s-a)^2+a^2} - \dfrac{s+a}{(s+a)^2+a^2}\right]$

4. $\sin 2t \cos 3t$

Ans. $\dfrac{2(s^2-5)}{(s^2+1)(s^2+25)}$

5. $e^{-3t}\cos^3 t$

Ans. $\dfrac{(s+3)(s^2+6s+16)}{(s^2+6s+10)(s^2+6s+18)}$

6. $t^n\,e^{-at}$

Ans. $\dfrac{n!}{(s+a)^{n+1}}$

7. $4 + 2\sqrt{t} - \dfrac{1}{\sqrt{t}}$

Ans. $\dfrac{4\sqrt{s} - (1-s)\sqrt{\pi}}{s^{3/2}}$

8. $2e^t \sin 4t \cos 2t$

Ans. $\dfrac{8(s^2-2s+13)}{(s^2-2s+37)(s^2-2s+5)}$

9. $f(t) = |t-1| + |t+1|\ t \geq 0$

Ans. $\dfrac{2}{s}\left(1+\dfrac{e^{-s}}{s}\right)$

10. $f(t) = \begin{cases} \cos(t-\alpha) & 1 \geq \alpha \\ 0 & t < \alpha \end{cases}$

Ans. $\dfrac{2}{s^3} - \dfrac{e^{-2x}}{s^3}(2+3s+3s^2) - \dfrac{3^{-3s}}{s^2}(1-5s)$

11. $f(t) = \begin{cases} \dfrac{t}{k} & 0 < t < k \\ 1 & t > k \end{cases}$

Ans. $\dfrac{1-e^{-sk}}{ks^2}$

12. $f(t) = \begin{cases} t, & 0 < t < 1 \\ 2-t, & 1 \leq t \leq 2 \\ 0, & t \geq 2 \end{cases}$

Ans. $\dfrac{(1-e^{-s})^2}{s^2}$

13.4 CONDITIONS FOR EXISTENCE OF LAPLACE TRANSFORMS

If $f(t)$ is a real valued function of t, $t > 0$ such that

(*a*) If $f(t)$ is continuous in every closed finite interval $[0, a]$ $a > 0$

(*b*) If $f(t)$ is of exponential order. So *i.e.* $e^{-at} \cdot f(t) \to 0$ as $t \to \infty$ for $s > s_0$ then integral $\int_0^\infty e^{-st} f(t)\,dt$ exists for all $s > s_0$, $s_0 > 0$, *i.e.,* $L[f(t)]$ exists for all $s > s_0$

13.5 LAPLACE TRANSFORM OF t^n f(t)

(*i*) $$L[t\,f(t)] = -\frac{d}{ds}F(s)$$

(*ii*) $$L[t^n f(t)] = (-1)^n \frac{d^n}{ds^n} F(s)$$

Proof :

(*i*) $$F(s) = L[f(t)] = \int_0^\infty e^{-st} f(t)\,dt$$

$$\frac{d}{ds}[F(s)] = \frac{d}{ds}\int_0^\infty e^{-st} f(t)\,dt = \int_0^\infty \frac{\partial}{\partial s}[e^{-st} f(t)]\,dt$$

[Using Leibnitz's rule for differentiation under the integral sign]

$$= -\int_0^\infty [t\, e^{-st} f(t)]\, dt = -\int_0^\infty e^{-st}\{t\, f(t)\}\, dt = -L\{t\, f(t)\}$$

$$L[t\, f(t)] = (-1)\frac{d}{ds}[F(s)]$$

(*ii*) To prove $L[t^n f(t)] = (-1)^n \dfrac{d^n}{ds^n} F(s)$ we use mathematical induction.

From (*i*) $L[t\, f(t)] = \dfrac{-d}{ds} F(s)$ *i.e.* the result is true for $n = 1$

Now assume the theorem to be true for $n = m$, so that

$$L[t^m f(t)] = (-1)^m \frac{d^m}{ds^m}[F(s)]$$

or
$$\int_0^\infty e^{-st} t^m f(t)\, dt = (-1)^m \frac{d^m}{ds^m}[F(s)]$$

Differentiating w.r.t. s both sides, we have

$$\frac{d}{ds}\int_0^\infty e^{-st} t^m f(t) = (-1)^m \frac{d^{m+1}}{ds^{m+1}}[F(s)]$$

$$\int_0^\infty \frac{\partial}{\partial s}[e^{-st} t^m f(t) dt]dt = (-1)^m \frac{d^{m+1}}{ds^{m+1}}[F(s)]$$

$$\int_0^\infty -te^{-st} f(t)\, dt = (-1)^m \frac{d^{m+1}}{ds^{m+1}}[F(s)]$$

$$-\int_0^\infty e^{-st}[t^{m+1} f(t)]\, dt = (-1)^m \frac{d^{m+1}}{ds^{m+1}}[F(s)]$$

or
$$L[t^{m+1} f(t)] = (-1)^{m+1} \frac{d^{m+1}}{ds^{m+1}}[F(s)]$$

which shows that the theorem is true for $n = m + 1$

Hence by mathematical induction, the theorem is true for all positive integers n.

13.6 LAPLACE TRANSFORM OF $\left[\frac{f(t)}{t}\right]$

If $L[f(t)] = F(s)$, then $L\left[\dfrac{f(t)}{t}\right] = \displaystyle\int_s^\infty F(s)\, ds$, provided the integral exists.

Proof :
$$\int_s^\infty F(s)\, ds = \int_s^\infty [L\, f(t)]\, ds = \int_0^\infty \left[\int_0^\infty e^{-st} f(t)\, dt\right] ds$$

Since s and t are independent, the order of integration can be changed

$$\int_s^{\infty} F(s)\,ds = \int_0^{\infty}\left[\int_s^{\infty} e^{-st} f(t)\,ds\right] dt = \int_0^{\infty}\left[f(t)\int_s^{\infty} e^{-st}\,ds\right] dt$$

$$= \int_0^{\infty} f(t)\left[\frac{e^{-st}}{-t}\right]_s^{\infty} dt = \int_0^{\infty} \frac{f(t)}{t} e^{-st}\,dt$$

$$= \int_0^{\infty} e^{-st}\left[\frac{f(t)}{t}\right] dt = L\left(\frac{f(t)}{t}\right)$$

$$L\left(\frac{f(t)}{t}\right) = \int_s^{\infty} F(s)\,ds$$

13.7 SECOND SHIFTING THEOREM

If $\quad L[f(t)] = F(s)$ and $g(t) = \begin{cases} f(t-a), & t > a \\ 0, & t < a \end{cases}$

Then $\quad L[g(t)] = e^{-as} F(s)$

Proof :

$$L[g(t)] = \int_0^{\infty} e^{-st} g(t) = \int_0^{\infty} e^{-st} g(t)\,dt + \int_0^{\infty} e^{-st} g(t)\,dt$$

$$= \int_0^{\infty} e^{-st} f(t-a)\,dt \quad \text{Put } t - a = v;\ dt = dv$$

$$= \int_0^{\infty} e^{-s(a+v)} f(v)\,dv = e^{-as}\int_0^{\infty} e^{-sv} f(v)\,dv$$

$$= e^{-ax}\int_0^{\infty} e^{-st} f(t)\,dt = e^{-as} F(s)$$

$$L[g(t)] = e^{-as} F(s)$$

***Example* 13.9.** Find the Laplace transform of $t \sin^2 3t$

Solution :

$$L[t \sin^2 3t] = L\left[t\frac{(1-\cos 6t)}{2}\right]$$

$$= \frac{1}{2}[L(t) - L(t\cos 6t)]$$

$$= \frac{1}{2}\left[\frac{1}{s^2} - L(t\cos 6t)\right] = \frac{1}{2s^2} - \frac{1}{2}\left\{-\frac{d}{ds}\frac{s}{s^2+36}\right\}$$

$$= \frac{1}{2s^2} + \frac{1}{2}\frac{(s^2+36).1 - s(2s)}{(s^2+36)^2} = \frac{1}{2s^2} + \frac{1}{2}\frac{36-s^2}{(s^2+36)^2}$$

Example **13.10.** Find the Laplace transform of *t* sin at

Solution : Now,

$$L(\sinh at) = \frac{a}{s^2 - a^2}$$

$$L(t \sinh at) = -\frac{d}{ds}\left(\frac{a}{s^2 - a^2}\right) = \frac{2as}{(s^2 - a^2)^2}$$

Example 6.11 Find Laplace transform of $\frac{1}{t}$ [cos 2*t* – cos 3*t*]

Solution : $L\left[\frac{1}{t}(\cos 2t - \cos 3t)\right]$

Now, $$L[\cos 2t - \cos 3t] = \left[\frac{s}{s^2+4} - \frac{s}{s^2+9}\right]$$

$$L\left[\frac{1}{t}(\cos 2t - \cos 3t)\right] = \frac{1}{2}\left[\int_s^\infty \frac{2s}{s^2+4}ds - \int_s^\infty \frac{2s}{s^2+9}ds\right]$$

$$= \frac{1}{2}[\log(s^2+4) - \log(s^2+9)]_s^\infty = \frac{1}{2}\left[\log\frac{s^2+4}{s^2+9}\right]_s^\infty$$

$$= \frac{1}{2}\left[\log\frac{1+4/s^2}{1+9/s^2}\right]_s^\infty = \frac{1}{2}\left[\log 1 - \log\frac{s^2+4}{s^2+9}\right]$$

$$= -\frac{1}{2}\log\frac{s^2+4}{s^2+9}$$

Example **13.12.** Find Laplace transform of $(1 + te^{-t})^2$

Solution : We have $$(1 + t\,e^{-t})^2 = (1 + t^2 e^{-2t} + 2t\,e^{-t})$$

$$L(1 + te^{-t})^2 = L(1 + t^2 e^{-2t} + 2t\,e^{-1})$$

$$= L(1) + L(t^2 e^{-2t}) + 2L(t\,e^{-t})$$

$$= \frac{1}{s} + (-1)^2 \frac{d^2}{ds}\left(\frac{1}{s+2}\right) + 2(-1)\frac{d}{ds}\left(\frac{1}{s+1}\right)$$

$$= \frac{1}{s} + \frac{(-1)(-2)}{(s+2)^3} - 2\frac{(-1)}{(s+1)^2} = \frac{1}{s} + \frac{2}{(s+2)^3} + \frac{2}{(s+1)^2}$$

Example **13.13.** Find Laplace transform of $\frac{1-\cos t}{t}$

Solution : We have $$L(1 - \cos t) = \frac{1}{s} - \frac{s}{s^2+1}$$

$$L\frac{(1-\cos t)}{t} = \int_s^\infty \frac{1}{s}ds - \int_s^\infty \frac{s}{s^2+1}ds = \left[\log s - \frac{1}{2}\log(s^2+1)\right]_s^\infty$$

$$= \left[\log s - \log\sqrt{s^2+1}\,\right]_s^\infty = \left[\log\frac{s}{\sqrt{s^2+1}}\right]_s^\infty$$

$$= \left[\log\frac{1}{\sqrt{1+\frac{1}{s^2}}}\right]_s^\infty = -\log\frac{s}{\sqrt{s^2+1}}$$

Example **13.14.** Find Laplace transform of $\dfrac{1-\cos t}{t^2}$

Solution :

$$L\left[\frac{(1-\cos t)}{t^2}\right] = \int_s^\infty \left[-\log s + \frac{1}{2}\log(s^2+1)\right]ds$$

$$= -\int_s^\infty \log s\, ds + \frac{1}{2}\int_s^\infty \log(s^2+1)\,ds$$

$$= [s(1-\log s)]_s^\infty + \frac{1}{2}[s\log(s^2+1)]_s^\infty - \int_s^\infty \frac{s\,.\,2s}{1+s^2}ds$$

$$= [s - s\log)]_s^\infty + \frac{1}{2}[s\log(s^2+1)]_s^\infty - \int_s^\infty \frac{(s^2+1)-1}{s^2+1}ds$$

$$= [s - s\log s)]_s^\infty + \frac{1}{2}[s\log(s^2+1)]_s^\infty - \int_s^\infty \left(1+\frac{1}{1+s^2}\right)ds$$

$$= [s - s\log s)]_s^\infty + \frac{1}{2}[s\log(s^2+1)]_s^\infty - [s - \tan^{-1}s]_s^\infty$$

$$= \left[s - s\log s + \frac{1}{2}s\log(s^2+1) - s + \tan^{-1}s\right]_s^\infty$$

$$= \left[-s\log s + s\log\sqrt{s^2+1} + \tan^{-1}s\right]_s^\infty$$

$$= \left[s\log\frac{\sqrt{s^2+1}}{s} + \tan^{-1}s\right]_s^\infty$$

$$= -s\log\frac{\sqrt{s^2+1}}{s} + \tan^{-1}\infty - \tan^{-1}s$$

$$= -s\log\frac{\sqrt{s^2+1}}{s} + \frac{\pi}{2} - \tan^{-1}s$$

$$= \cot^{-1}s - s\log\sqrt{\frac{s^2+1}{s^2}} = \cot^{-1}s - \frac{1}{2}s\log(1+s^2)$$

Example **13.15.** Given $L\left(2\sqrt{\frac{t}{\pi}}\right)=\frac{1}{s^{3/2}}$ show that $\frac{1}{s^{1/2}}=L\left(\frac{1}{\sqrt{\pi t}}\right)$

Solution :

$$L\left(\frac{1}{\sqrt{\pi t}}\right)=\frac{1}{\sqrt{\pi}}L\left(\frac{1}{\sqrt{t}}\right)=\frac{1}{\sqrt{\pi}}\left[\frac{1}{t}\sqrt{t}\right]=\frac{1}{\sqrt{\pi}}\int_s^\infty\frac{\sqrt{\pi}}{2s^{3/2}}ds$$

$$=\frac{1}{2}\int_s^\infty\frac{1}{s^{3/2}}ds=\frac{1}{s^{1/2}}\qquad\left[L(\sqrt{t})=\frac{\Gamma\frac{3}{2}}{s^{3/2}}=\frac{\frac{1}{2}\Gamma\frac{1}{2}}{s^{3/2}}=\frac{\sqrt{\pi}}{2s^{3/2}}\right]$$

$$\frac{1}{s^{1/2}}=L\left(\frac{1}{\sqrt{\pi t}}\right)$$

Example **13.16.** Find Laplace transforms of $\frac{\sin at}{t}$

Solution : Now

$$L(\sin at)=\frac{a}{s^2+a^2}$$

$$L\left(\frac{\sin at}{t}\right)=\int_s^\infty\frac{a}{s^2+a^2}ds=a\times\frac{1}{a}\left[\tan^{-1}\frac{s}{a}\right]_s^\infty=\frac{\pi}{2}-\tan^{-1}\frac{s}{a}=\cot^{-1}\frac{s}{a}$$

Example **13.17.** Evaluate $L\left[e^{-4t}\frac{\sin 3t}{t}\right]$

Solution : $L(\sin 3t)=\frac{3}{s^2+3^2}$

$$L\left(\frac{\sin 3t}{t}\right)=\int_s^\infty\frac{3}{s^2+9}ds=\frac{3}{3}\left[\tan^{-1}\frac{s}{3}\right]_s^\infty$$

$$=\frac{\pi}{2}-\tan^{-1}\frac{s}{3}=\cot^{-1}\frac{s}{3}$$

$$L\left(e^{-4t}\frac{\sin 3t}{t}\right)=\cot^{-1}\frac{s+4}{3}=\tan^{-1}\frac{3}{s+4}$$

EXERCISE 13.2

Find the Laplace transform of the following :

1. $\frac{e^{-at}-e^{-bt}}{t}$

Ans. $\log\frac{s+b}{s+a}$

2. $e^{-2t}\dfrac{\sin^2 t}{t}$

Ans. $\frac{1}{2}\log 2$

3. $\dfrac{\cos at - \cos bt}{t}$

Ans. $\frac{1}{2}\log\dfrac{s^2+b^2}{s^2+a^2}$

4. $\frac{1}{t}\sinh t$

Ans. $\frac{1}{2}\log\dfrac{s-1}{s+1}$

5. $\int_0^\infty \dfrac{\sin t}{t}dt$

Ans. $\dfrac{\pi}{2}$

6. $\dfrac{\sin 2t}{t}$

Ans. $\cos^{-1}\dfrac{s}{2}$

13.8 LAPLACE TRANSFORM OF DERIVATIVES

Theorem : If $f'(t)$ be continuous and $L[f(t)] = F(s)$, then $L[f'(t)] = sF(s) - f(0)$, provided $\underset{t\to\infty}{Lt} = 0$

Proof :

$$L[f'(t)] = \int_0^\infty e^{-st}.\, f'(t)\,dt$$

Integrating by parts, we get

$$= [e^{-st}f(t)]_0^\infty - \int_0^\infty - s\,e^{-st}, f(t)dt$$

$$= \underset{t\to\infty}{Lt}\ e^{-st}.\, f(x) - f(0) + s\int_0^\infty e^{-st}, f(t)$$

$$= 0 - f(0) + sL\,[f(t)] = sF(s) - f(0) \qquad \text{...(i)}$$

Note : The above can be generalised for order n as follows :
If $f(t)$ and its first $(n-1)$ derivatives be continuous then

$$L[f^n(t)] = S^n F(s) - s^{n-1}F(0) - s^{n-2}F'(0) - s^{n-3}F''(0) \ldots F^{n-1}(0)$$

$$L[f^n(t)] = \int_0^\infty e^{-st} f^n(t)\,dt$$

$$= [e^{-st} f^{n-1}(t) - (-s)\,e^{-st} f^{n-2}(t) + (-s)^2 e^{-st} f^{n-3}(t) - \ldots\ldots + (-1)^{n-1}(-s)^{n-1} e^{-st} f(t)]_0^\infty + (-1)^n(-s)^n \int_0^\infty e^{-st} f(t)\,dt$$

$$= -f^{n-1}(0) - S f^{n-2}(0) - S^2 f^{n-3}(0) \ldots\ldots - S^{n-1} f(0) + S^n L[f(t)]$$

$$= S^n F(s) - S^{n-1} f(0) - S^{n-2} f'(0) - S^{n-3} f''(0) \ldots\ldots - f^{n-1}(0) \qquad \ldots(ii)$$

From (*ii*) we get,

$$L[f'(t)] = s^2F(s) - s\,f(0) - f(0) \qquad [n = 2]$$

$$L[f''(t)] = s^3F(s) - s^2f(0) - sf'(0) - f''(0) \text{ and so on.}$$

The results will be used in the solution of different equations.

13.9 LAPLACE TRANSFORM OF INTEGRALS

If $\quad L[f(t)] = F(s)$, then $L\left[\int_0^t f(t)\,dt\right] = \dfrac{F(s)}{s}$

Proof : Let $\quad g(t) = \int_0^t f(t)\,dt$, then $g'(t) = f(t)$ and $g(0) = 0$

$$L[g'(t)] = sL[g(t)] - g(0) = sL[g(t)]$$

$$\frac{1}{s} L[g'(t)] = L[g(t)] = L\left[\int_0^t f(t)\,dt\right]$$

$$\frac{1}{s} L[f(t)] = L\left[\int_0^t f(t)\,dt\right]$$

$$\frac{F(s)}{s} = L\left[\int_0^t f(t)\,dt\right]$$

***Example* 6.18.** Find $L[f(t)]$, given $f(0) = 0$, $f'(0) = 2$, $f''(0) = -1$ and $f'''(t) = (f)'(t)$

Solution :

$$L[f'''(t)] = s^3 F(s) - s^2 f(0) - s\,f'(0) - f'(0)$$

$$= s^3 F(s) - 2s + 1$$

$$L[f'(t)] = [s^3 F(s) - 2s + 1]$$

$$f'''(t) = f'(t)$$

$$\therefore \quad L[f'''(t)] = L[f'(t)]$$

Also
$$L[f'(t)] = s\,F(s) - f(0) = s\,F(s)$$
$$s\,F(s) = s^3\,F(s) - 2s + 1$$
$$(s^3 - s)\,F(s) = 2s - 1$$
$$F(s) = \frac{2s-1}{s(s^2-1)}$$
$$L[f(t)] = \frac{2s-1}{s(s^2-1)}$$

***Example* 13.19.** If $E(t) = \int_t^{\infty} \frac{e^{-x}}{x}$ then show that $L[E(t)] = \frac{\log(s+1)}{s}$

Solution : Let
$$E(t) = F(t) = \int_t^{\infty} \frac{e^{-x}}{x}\,dx$$

Then
$$\underset{t \to \infty}{Lt}\ F(t) = 0,$$
$$F(t) = \int_t^{\infty} \frac{e^{-x}}{x}\,dx$$

so that
$$F'(t) = -\left[\frac{e^{-t}}{t}\right]$$
$$t\,F'(t) = -e^{-t}$$
$$L[t\,F'(t)] = -L[e^{-1}] = -\frac{1}{s+1}$$
$$-\frac{d}{ds}\,[LF'(t)] = -\frac{1}{s+1}$$
$$\frac{d}{ds}\,[s\,[LF(t)] - f(0)] = +\frac{1}{s+1}$$
$$\frac{d}{ds}\,[s\,f(s)] = +\frac{1}{s+1} \qquad \{LF(t) = f(s)\}$$
$$\frac{d}{ds}\,[s\,f(s)] = +\frac{1}{s+1} \qquad \{F(0) = \text{constant}\}$$

Integrating
$$s\,f(s) = \log\,(s+1) + c$$

When
$$s = 0,\ c = 0$$
$$s\,f(s) = \log\,(1+s)$$
$$f(s) = \frac{\log(1+s)}{s}$$
$$L[F(t)] = \frac{\log(1+s)}{s}$$

$$L[E(t)] = \frac{\log(1+s)}{s}$$

Example **13.20.** Find $L\left(\int_0^t \frac{\sin t}{t} dt\right)$

Solution : We have $$f(t) = \frac{\sin t}{t}$$

$$L[f(t)] = L\left[\frac{\sin t}{t}\right]$$

$$\therefore \quad L[f(t)] = F(s)$$

Now $$L[\sin t] = \frac{1}{1+s^2}$$

$$L\left[\frac{\sin t}{t}\right] = \int_s^\infty \frac{1}{1+s^2} ds = (\tan^{-1} s)_s^\infty = \frac{\pi}{2} - \tan^{-1} s = \cot^{-1} s = F(s)$$

$$\therefore \quad L\left[\int_0^t \frac{\sin t}{t} dt\right] = \frac{\cot^{-1} s}{s} \qquad \left[L\left(\int_0^t f(t)\, dt\right) = \frac{F(s)}{s}\right]$$

EXERCISE 13.3

1. If $c(t) = \int_1^\infty \frac{\cos x}{x} dx$, then $L\{C(t)\} = \frac{\log(s^2+1)}{2s}$

2. If $L\,(t \sin at) = \frac{2as}{(s^2+a^2)^2}$, show that

   (*i*) $L\,[at \cos at + \sin at] = \frac{2as^2}{(s^2+a^2)^2}$

   (*ii*) $L\,[2 \cos at - at \sin at] = \frac{2s^3}{(s^2+a^2)^2}$

3. If $3f'(t) - 5f(t) + 7f(t) = \sin 2t$ and $f(0) = 4$, $f'(0) = 6$ find $L[f(t)]$.

4. If $f'(t) + 3'(t) + 2f(t) = 0$, $f(0) = 1$ and $f(0) = -2$ find $L\, f(t)$.

5. Find the *L.T.* : $\int_0^t \frac{(1-e^{-2x})}{x} dx$

Ans. $\frac{1}{5}\log\left(1+\frac{2}{3}\right)$

6. Find the *L.T.* : $\int_0^t e^t \frac{\sin t}{t} dt$

Ans. $\frac{1}{5}\cot^{-1}(s-1)$

Example 13.21 Evaluate $L\left(\int_0^t e^t \frac{\sin t}{t} dt\right)$

Solution :

$$L(\sin t) = \frac{1}{s^2+1}$$

$$L\left(\frac{\sin t}{t}\right) = \int_s^\infty \frac{ds}{s^2+1} = [\tan^{-1} s]_s^\infty = \frac{\pi}{2} - \tan^{-1} s = \cot^{-1} s$$

$$L\left(e^t \frac{\sin t}{t}\right) = \cot^{-1}(s-1) = F(s)$$

$$L\left(\int_0^t e^t \frac{\sin t}{t}\right) = \frac{F(s)}{s} = \frac{\cot^{-1}(s-1)}{s}$$

Example **13.22.** Evaluate :

(*i*) $\int_0^\infty t^3 e^{-t} \sin t\, dt$

(*ii*) $\int_0^\infty \frac{e^{-2t}\sin^2 t}{t} dt$

Solution : (*i*)

$$\int_0^\infty t^3 e^{-t} \sin t\, dt = \int_0^\infty e^{-st}(t^3 \sin t)\, dt$$

$$= L(t^3 \sin t)$$

$$= (-1)^3 \frac{d^3}{ds^3}\left(\frac{1}{s^2+1}\right) = \frac{d^2}{ds^2}\left(\frac{2s}{(s^2+1)^2}\right)$$

$$= \frac{d}{ds}\left[\frac{(s^2+1)^2 - 2s\,(s(s^2+1)\times 2s)}{(s^2+1)^4}\right]$$

$$= \frac{d}{ds}\left[\frac{2(s^2+1) - 8s^2}{(s^2+1)^3}\right] = \frac{d}{ds}\left[\frac{2(1-3s^2)}{(s^2+1)^3}\right]$$

$$= \frac{24s(s^2-1)}{(s^2+1)^4} = 0 \quad \text{as} \quad s \to 1$$

(*ii*)

$$\int_0^\infty \frac{e^{-2t}\sin^2 t}{t} dt = \int_0^\infty e^{-2t}\left(\frac{\sin^2 t}{t}\right) dt$$

$$= L\left(\frac{\sin^2 L}{t}\right) = L\left[\frac{1-2\cos 2t}{2t}\right]$$

$$L[(1-\cos 2t)] = \frac{1}{s} - \frac{s}{s^2+4}$$

$$L\left[\frac{(1-\cos 2t)}{2t}\right] = \frac{1}{2}\int_s^\infty \left(\frac{1}{s} - \frac{s}{s^2+4}\right) ds$$

$$= \frac{1}{2}\left[\log s - \frac{1}{2}\log(s^2+4)\right]_s^\infty = \frac{1}{2}\left[\log s - \log\sqrt{s^2+4}\right]_s^\infty$$

$$= \frac{1}{2}\left[\log\frac{s}{\sqrt{s^2+4}}\right]_s^\infty = \frac{1}{2}\left[\log\frac{1}{\sqrt{1+\frac{4}{s^2}}}\right]$$

$$= \frac{1}{2}\left[\log 1 - \log\frac{s}{\sqrt{s^2+4}}\right] = \frac{1}{2}\left[\log\frac{\sqrt{s^2+4}}{s}\right]$$

$$= \frac{1}{4}\left[\log\frac{s^2+4}{s^2}\right] = \frac{1}{4}\log 2 \quad s \to 2$$

13.10 INVERSE LAPLACE TRANSFORM

Here we Introduce the idea of Inverse Laplace Transform

If Laplace Transform of a function $f(t)$ is $f(s)$, then inver transform of $F(s)$ is $f(t)$.

If Symbolically

If $$L[f(t)] = F(s)$$

Then $$L^{-1}[F(s)] = F(t)$$

Now we give below the results which follow directly from the results already proved for Laplace transform of various functions. The inverse Laplace transform is very useful from the application point of view.

13.11 IMPORTANT FORMULAE OF INVERSE LAPLACE TRANSFORM

1. $$L^{-1}\left(\frac{1}{s}\right) = 1$$

2. $$L^{-1}\left(\frac{1}{s-a}\right) = e^{at}$$

3. $$L^{-1}\left(\frac{1}{s^n}\right) = \frac{t^{n-1}}{(n-1)!} \quad n = 1, 2, 3, \ldots$$

4. $$L^{-1}\left(\frac{s}{s^2-a^2}\right) = \cosh at$$

5. $$L^{-1}\left(\frac{1}{s^2+a^2}\right) = \frac{1}{a}\sinh at$$

6. $$L^{-1}\left(\frac{1}{s^2+a^2}\right) = \frac{1}{a}\sin at$$

7. $$L^{-1}\left(\frac{1}{s^2+a^2}\right) = \cos at$$

8. $$L^{-1}[F(s-a)] = e^{at} f(t)$$

9. $$L^{-1}\left(\frac{1}{(s-a)^2+b^2}\right) = \frac{1}{b} e^{at} \sin bt$$

10. $$L^{-1} \frac{s-a}{(s-a)^2-b^2} = e^{at} \cosh bt$$

11. $$L^{-1}\left(\frac{1}{(s-a)^2-b^2}\right) = \frac{1}{b} e^{at} \sinh bt$$

12. $$L^{-1}\left(\frac{s-a}{(s-a)^2-b^2}\right) = e^{at} \cosh bt$$

13. $$L^{-1}\left(\frac{1}{(s^2+a^2)^2}\right) = \frac{1}{2a^2}[\sin at - at \cos at]$$

14. $$L^{-1}\left(\frac{s}{(s^2+a^2)^2}\right) = \frac{1}{2a} t \sin at$$

15. $$L^{-1}\left(\frac{s^2-a^2}{(s^2+a^2)^2}\right) = t \cos at$$

16. $$L^{-1}\left(\frac{s^2}{(s^2+a^2)^2}\right) = \frac{1}{2a^3}[\sin at + at \cos at]$$

***Example* 13.23.** **Find the inverse Laplace transform of the following :**

(*i*) $\frac{1}{s-3}$

(*ii*) $\frac{1}{s^2-4}$

(*iii*) $\frac{s}{s^2-16}$

(*iv*) $\frac{s}{s^2+9}$

(*v*) $\frac{s}{s^2+16}$

(*vi*) $\frac{1}{(s-1)^2+1}$

(*vii*) $\frac{s-2}{(s-2)^2+9}$

(*viii*) $\frac{1}{(s-3)^2+4}$

(*ix*) $\frac{s+3}{(s-3)^2+25}$

Solution : (*i*) $L^{-1}\left(\frac{1}{s-3}\right) = e^{3t}$

(*ii*) $$L^{-1}\left(\frac{1}{s^2-4}\right) = L^{-1}\left(\frac{1}{2}\frac{2}{s^2-2^2}\right) = \frac{1}{2}\sinh 2t$$

(*iii*) $$L^{-1}\left(\frac{s}{s^2-16}\right) = \cosh 4t$$

(*iv*) $$L^{-1}\left(\frac{1}{s^2+9}\right) = \frac{1}{3}\sin 3t$$

(*v*) $$L^{-1}\left(\frac{s}{s^2+16}\right) = \cos 4t$$

(*vi*) $$L^{-1}\left(\frac{1}{(s-1)^2+1}\right) = et \sin t$$

(*vii*) $$L^{-1}\left(\frac{s-2}{(s-2)^2+9}\right) = e^{2t}\cos 3t$$

(*viii*) $$L^{-1}\left(\frac{1}{(s-3)^2+4}\right) = e^{3t}\,\frac{1}{2}\sin 2t$$

(*ix*) $$L^{-1}\left(\frac{s+3}{(s+3)^2+25}\right) = e^{-3t}\cos 5t$$

***Example* 13.24.** Evaluate $L^{-1}\left(\frac{s}{s^4+s^2+1}\right)$

Solution :

$$L^{-1}\left(\frac{s}{s^4+s^2+1}\right) = L^{-1}\frac{s}{(s^2+1)^2-s^2} = L^{-1}\left[\frac{s}{(s^2+1+s)(s^2+1-s)}\right]$$

$$= L^{-1}\frac{1}{2}\left[\frac{1}{s^2-s+1} - \frac{1}{s^2+s+1}\right]$$

$$= \frac{1}{2}\left[L^{-1}\left(\frac{1}{s^2-s+1}\right) - L^{-1}\left(\frac{1}{s^2+s+1}\right)\right]$$

$$= \frac{1}{2}\left[L^{-1}\left(\frac{1}{\left(s-\frac{1}{2}\right)+\left(\frac{\sqrt{3}}{2}\right)^2}\right) - L^{-1}\left(\frac{1}{\left(s+\frac{1}{2}\right)^2+\left(\frac{\sqrt{3}}{2}\right)^2}\right)\right]$$

$$= \frac{1}{2}\left[e^{\frac{1}{2}}\,L^{-1}\left(\frac{1}{s^2+\left(\frac{\sqrt{3}}{2}\right)^2}\right) - e^{\frac{1}{2}t}\,L^{-1}\left(\frac{1}{s^2+\left(\frac{\sqrt{3}}{2}\right)^2}\right)\right]$$

$$= \frac{1}{2}\left[\left(e^{\frac{1}{2}t} - e^{\frac{1}{2}t}\right)\left(L^{-1}\left(\frac{1}{s^2+\left(\frac{\sqrt{3}}{2}\right)^2}\right)\right)\right]$$

$$= \left[\sinh\frac{1}{2}t\left(\frac{1}{\frac{\sqrt{3}}{2}}\sin\frac{\sqrt{3}}{2}t\right)\right] = \frac{2}{\sqrt{3}}\sinh\frac{t}{2}\sin\frac{\sqrt{3}}{2}t$$

Example **13.25.** Evaluate $L^{-1}\left(\dfrac{3s-2}{s^2-4s+20}\right)$

Solution :
$$L^{-1}\left(\frac{3s-2}{(s-2)^2+4^2}\right) = 3L^{-1}\left[\frac{s-\frac{2}{3}}{(s-2)^2+4^2}\right] = 3L^{-1}\left[\frac{(s-2)+\frac{4}{3}}{(s-2)^2+4^2}\right]$$
$$= 3\,e^{2t}\left\{L^{-1}\left(\frac{s}{s^2+4^2}+\frac{4/3}{s^2+4^2}\right)\right\}$$
$$= 3\,e^{2t}\left[\cos 4t+\frac{4}{3}\times\frac{1}{4}\sin 4t\right]$$
$$= 3\,e^{2t}\cos 4t+e^{2t}\sin 4t$$

Example **13.26.** Evaluate $L^{-1}\left(\dfrac{1}{s^2-6s+9}\right)$

Solution : $L^{-1}\left(\dfrac{1}{s^2-6s+9}\right)=L^{-1}\left(\dfrac{1}{(s-3)^2}\right)=e^{3t}\,L^{-1}\left(\dfrac{1}{s^2}\right)=e^{3t}\,.\,\mathrm{t}$

Example **13.26.** Evaluate

(*i*) $L^{-1}\left(\dfrac{1}{s^2+10s+50}\right)$

(*ii*) $L^{-1}\left(\dfrac{1}{s^2+8s+16}\right)$

Solution : (*i*)
$$L^{-1}\left(\frac{1}{(s+5)^2+5^2}\right) = e^{-5t}\,L^{-1}\left(\frac{1}{s^2+5^2}\right)=e^{-5t}\,\frac{1}{5}\sin 5t$$

(*ii*)
$$L^{-1}\left(\frac{1}{s^2+8s+16}\right) = L^{-1}\left(\frac{1}{(s+4)^2}\right)=e^{-4t}\,L^{-1}\left(\frac{1}{s^2}\right)=e^{-4t}\,.\,t$$

Example **13.27.** Evaluate $L^{-1}\left(\dfrac{3s+9}{s^2+2s+10}\right)$

Solution :
$$L^{-1}\left(\frac{3s+9}{(s+1)^2+3^2}\right) = 3L^{-1}\left(\frac{s+3}{(s+1)^2+3^2}\right)$$
$$= 3\left[L^{-1}\left\{\frac{(s+1)}{(s+1)^2+3^2}\right\}+L^{-1}\left\{\frac{2}{(s+1)^2+3^2}\right\}\right]$$
$$= 3\left[e^{-t}\,L^{-1}\left\{\frac{s}{s^2+3^2}\right\}+2e^{-t}\,L^{-1}\left\{\frac{1}{s^2+3^2}\right\}\right]$$
$$= 3\,e^{-t}\left[\cos 3t+\frac{2}{3}\sin 3t\right]$$
$$= e^{-t}\,[3\cos 3t+2\sin 3t]$$

Example **13.28.** Evaluate $L^{-1}\left(\dfrac{s^2+s+2}{s^{3/2}}\right)$

Solution :
$$L^{-1}\,\frac{s^2+s+2}{s^{3/2}} = L^{-1}\left(\frac{1}{s^{-\frac{1}{2}}}\right)+L^{-1}\left(\frac{1}{s^{\frac{1}{2}}}\right)+2L^{-1}\left(\frac{1}{s^{3/2}}\right)$$

We know that
$$L(t^n) = \frac{\Gamma\left(\frac{1}{2}\right)}{s^{n+1}}$$

$$\frac{t^n}{\Gamma(n+1)} = L^{-1}\frac{1}{s^{n+1}}$$

Now,
$$L^{-1}\left(\frac{1}{s^{-1/2}}\right) = \frac{t^{\frac{-1}{2}-1}}{\Gamma\left(\frac{-1}{2}\right)} = \frac{1}{\Gamma\left(\frac{-1}{2}\right)}\cdot\frac{1}{t^{3/2}}$$

$$L^{-1}\left(\frac{1}{s^{+1/2}}\right) = \frac{t^{\frac{1}{2}-1}}{\Gamma\left(\frac{1}{2}\right)} = \frac{1}{\sqrt{\pi t}} \qquad \left[\Gamma\left(\frac{1}{2}\right) = \sqrt{\pi}\right]$$

$$L^{-1}\left(\frac{1}{s^{3/2}}\right) = \frac{t^{\frac{1}{2}}}{\Gamma\left(\frac{3}{2}\right)} = \frac{t^{1/2}}{\frac{1}{2}\sqrt{\pi}}$$

$$L^{-1}\,\frac{s^2+s+2}{s^{3/2}} = \frac{1}{\Gamma\left(\frac{-1}{2}\right)t^{3/2}} + \frac{1}{\sqrt{\pi t}} + 4\frac{t^{1/2}}{\sqrt{\pi}}.$$

***Example* 13.29.** Find the inverse transform of $\left(\frac{5s+3}{(s-1)(s^2+2s+5)}\right)$

Solution :
$$\frac{5s+3}{(s-1)(s^2+2s+3)} = \frac{A}{s-1} + \frac{Bs+C}{s^2+2s+5}$$

$$5s+3 = A(s^2+2s+5) + (s-1)(Bs+C)$$

Put
$$s = 1$$

$$8 = A[1+2+5] = 8A \quad A = 1$$

Equating constant terms
$$3 = 5A - C \quad 3 = 5 - C \quad C = 2$$

Equating coefficients of s^2
$$0 = A + B \quad B = -1$$

$$\frac{5s+3}{(s-1)(s^2+2s+5)} = \frac{1}{s-1} + \frac{2-s}{s^2+2s+5} = \frac{1}{s-1} - \frac{s+1-3}{(s+1)^2+4}$$

$$= \frac{1}{s-1} + \frac{s+1}{(s+1)^2+2^2} + \frac{3}{(s+1)^2+2^2}$$

$$L^{-1}\left[\frac{5s+3}{(s-1)(s^2+2s+5)}\right] = L^{-1}\left(\frac{1}{s-1}\right) - L^{-1}\left(\frac{s+1}{(s+1)^2+2^2}\right) + 3L^{-1}\left(\frac{1}{(s+1)^2+2^2}\right)$$

$$= e^t - e^{-t}L^{-1}\left(\frac{1}{s^2+2^2}\right) + 3e^{-1}L^{-1}\left(\frac{1}{s^2+2^2}\right)$$

$$= e^t - e^{-t}\cos 2t + \frac{3}{2}e^{-t}\sin 2t$$

EXERCISE 13.4

Find inverse Laplace transforms of the following :

1. $\dfrac{1}{2s-5}$

Ans. $\dfrac{1}{3}\,e^{5t/2}$

2. $\dfrac{3s+1}{(s-1)^2(s^2+1)}$

Ans. $2e^t - 2\cos t + \sin ts$

3. $\dfrac{4s+15}{16s^2-25}$

Ans. $\dfrac{1}{4}\left(\cosh\dfrac{5}{4}t - 3\sinh\dfrac{5}{4}t\right)$

4. $\dfrac{s}{(s^2-1)^2}$

Ans. $\dfrac{1}{2}\,t\cosh t$

5. $\dfrac{s^2+2}{s(s^2+4)}$

Ans. $\sinh 2t - 2t$

6. $\dfrac{1}{s(s^2+4)}$

Ans. $\dfrac{1}{4}\,(1-\cos 2t)$

7. $\dfrac{s+1}{(s^2+2s+2)^2}$

Ans. $\dfrac{1}{2}\,e^{-t}\sin t$

8. $\dfrac{e^{-ax}}{s^2}\quad a>0$

Ans. $t-a$ when $t>a$ and 0 otherwise

Evaluate the following :

9. $L^{-1}\left[\dfrac{4p+5}{(p-1)^2(p+2)}\right]$

Ans. $\dfrac{1}{3}\,[(9t+1)(e^t - e^{-2t})$

10. $L^{-1}\left[\dfrac{s^2}{(s^2+a^2)(s^2+b^2)}\right]$

Ans. $\dfrac{a\sin at - b\sin bet}{a^2-b^2}$ if $a^2 - b^2,\ ab \neq 0$

11. $L^{-1}\left[\dfrac{2s-1}{s^3-s}\right]$

Ans. $1 - \dfrac{3}{2}e^{-t} + \dfrac{1}{2}e^{t}$

Prove the following :

12. $L^{-1}\left[\dfrac{3s-2}{s^2-4s+20}\right] = 3e^{-2t}\cos 4t + e^{2t}\sin 4t$

13. $L^{-1}\left[\dfrac{1}{s^2+8s+16}\right] = t\,e^{-4t}$

14. $L^{-1}\left[\dfrac{1}{s^2(s^2+1)(s^2+4)}\right] = \dfrac{1}{24}(6t - 8\sin t + \sin 2t)$

15. $L^{-1}\left[\dfrac{s}{(s^2-2s+2)(s^2+2s+2)}\right] = \dfrac{1}{2}\sin t \sinh t$

16. $L^{-1}\left[\dfrac{s+3}{(s^2+6s+13)^2}\right] = \dfrac{1}{4}te^{-3t}\sin 2t$

17. $L^{-1}\left[\dfrac{2s^2-4}{(s+1)(s-2)(s-3)}\right] = \dfrac{-1}{6}e^{-t} - \dfrac{4}{3}e^{2t} + \dfrac{7}{2}e^{3t}$

18. $L^{-1}\left[\dfrac{9(s^2-2a^2)}{s^4+4a^4}\right] = \cos at \sinh at$

19. $L^{-1}\left[\dfrac{s^3}{s^4-a^4}\right] = \dfrac{1}{2}(\cosh at + \cos at)$

20. $L^{-1}\left[\dfrac{1}{s^3-a^3}\right] = \dfrac{1}{3a^2}\left[e^{at} - e^{\frac{1}{2}at}\left\{\cos\dfrac{\sqrt{3}}{2}at + \sqrt{3}\sin\dfrac{\sqrt{3}}{2}at\right\}\right]$

13.12 METHODS OF FINDING INVERSE LAPLACE TRANSFORMS

(A) *Multiplication by Powers of s :*

If $L^{-1}[F(s)] = f(t)$ then $L^{-1}[s\,F(s)] = \dfrac{d}{dt}f(t)$

provided $f(0) = 0$

Also $L^{-1}[s^n F(s)] = \dfrac{d^n}{dt^n}[f(t)]$ provided $f(0) = f'(0) = f^{n-1}(0) = 0$

These results follow from the result already obtained in case of lap lace transform.

(B) *Division by Powers of s* :

If $$L^{-1}[F(s)] = f(t), \text{ then}$$

$$L^{-1}\left[\frac{F(s)}{s}\right] = \int_0^t f(t)\,dt$$

$$L^{-1}\left[\frac{F(s)}{s^2}\right] = \int_0^t\left[\int_0^t f(t)\,dt\right] \text{ and so on}$$

13.13 CONVOLUTION THEOREM

If $$L^{-1}[F(s)] = f(t) \text{ and } L^{-1}[G(s)] = g(t), \text{ then}$$

$$L^{-1}[F(s)\,.\,G(s)] = \int_0^t f(u)g(t-u)\,du = f*g$$

[$f*g$ is called the convolution of $f(t)$ and $g(t)$]

Proof : Let $$\phi(t) = \int_0^t f(u)g(t-u)\,du$$

$$L\{\phi(t)\} = \int_0^\infty e^{-st}\left\{\int_0^t f(u)g(t-u)\,du\right\}dt$$

$$= \int_0^\infty\int_0^t e^{-st} f(u)g(t-u)\,du\,dt$$

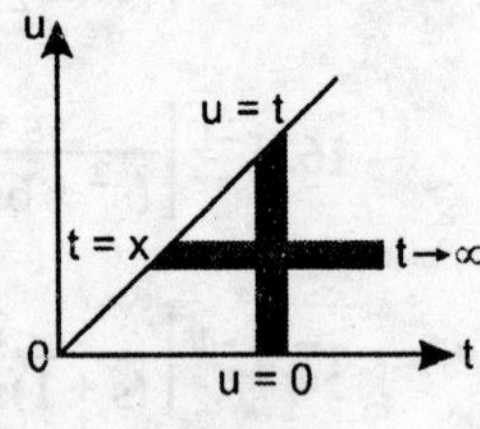

Fig. 13.1

On changing the order of integration

$$L\{\phi(t)\} = \int_0^\infty\int_u^\infty e^{-st} f(u)g(t-u)\,dt\,du$$

$$= \int_0^\infty e^{-su} f(u)\left[\int_u^\infty e^{-s(t-u)} g(t-u)\,dt\right]du$$

$$= \int_0^\infty e^{-su} f(u)\left[\int_0^\infty e^{-su} g(v)\,dv\right]du \text{ on putting } t-u=v$$

$$= \int_0^\infty e^{-su} f(u)G(s)\,du = G(s)\int_0^\infty e^{-su} f(u)\,du$$

$$= G(s)\,F(s)$$

$\therefore$ $$L^{-1}\{F(s)\,G(s)\} = \phi(t) = \int_0^t f(u)g(u-t)\,du$$

***Example* 13.30.** Find the inverse Laplace transform of

(*i*) $\dfrac{1}{s(s^2+9^2)}$

(*ii*) $\dfrac{1}{s(s-2)^3}$

(*iii*) $\dfrac{1}{s^2(s^2+25)}$

Solution : (*i*) Since $\quad L^{-1}\left(\dfrac{1}{s^2+3^2}\right) = \dfrac{1}{3}\sin 3t$

$$\therefore \quad L^{-1}\left[\frac{1}{s(s^2+3^2)}\right] = \int_0^t \frac{1}{3}\sin 3t\, dt = \frac{1}{9}[-\cos 3t]_0^t = \frac{1}{9}\,[1-\cos 3t].$$

(*ii*) Since $\quad L^{-1}\left[\dfrac{1}{(s-2)^3}\right] = e^{2t}\,\dfrac{t^2}{2!}$

$$L^{-1}\left[\frac{1}{s(s-2)^3}\right] = \int_0^t e^{2t}\,\frac{t^2}{2!}dt$$

$$= \frac{1}{2}\left[\left(t^2\,\frac{e^{2t}}{2}\right)_0^t - \int_0^t \frac{e^{2t}}{2}2t\,dt\right]$$

$$= \frac{1}{2}\left[\frac{t^2}{2}e^{2t} - \left\{\left(t\,.\,\frac{e^{2t}}{2}\right)_0^t - \int_0^t \frac{e^{2t}}{2}dt\right\}\right]$$

$$= \frac{1}{2}\left[\frac{t^2}{2}e^{2t} - t\frac{e^{2t}}{2} + \left(\frac{e^{2t}}{4}\right)_0^t\right]$$

$$= \frac{1}{2}\left[\frac{t^2}{2}e^{2t} - t\frac{e^{2t}}{2} + \left(\frac{e^{2t}}{4} - \frac{1}{4}\right)\right]$$

$$= \frac{1}{8}\,[2t^3\,e^{2t} - 2t\,e^{2t} + e^{2t} - 1]$$

$$= \frac{1}{8}\,[(2t^2 - 2t + 1)e^{2t}] - \frac{1}{8}$$

(*iii*) Since $\quad L^{-1}\left[\dfrac{1}{s^2+25}\right] = \dfrac{1}{5}\sin 5t$

$$L^{-1}\left[\frac{1}{s(s^2+25)}\right] = \frac{1}{5}\int_0^t \sin 5t\, dt = \frac{1}{25}\,(1-\cos 5t)$$

$$L^{-1}\left[\frac{1}{s^2(s^2+25)}\right] = \frac{1}{25}\int_0^t (1-\cos 5t)\,dt$$

$$= \frac{1}{25}\left[t - \frac{\sin 5t}{5}\right]_0^t$$

***Example* 13.31.** Find inverse transform of

(*i*) $\log \frac{(s-1)}{(s+1)}$

(*ii*) $\log\left(1-\frac{4}{s^2}\right)$

(*iii*) $\frac{s^2-a^2}{(s^2+a^2)^2}$

(*iv*) $\log\left(\frac{s^2}{s^2+1}\right)$

Solution : (*i*) $L^{-1}\left[\log\frac{(s-1)}{(s+1)}\right]$

Here $$F(s) = \log\frac{(s-1)}{(s+1)} = \log(s-1) - \log(s+1)$$

$$\frac{d}{ds}[F(s)] = \frac{1}{s-1} - \frac{1}{s+1} = \frac{2}{s^2-1}$$

$$L^{-1}\left[\frac{d}{ds}[F(s)]\right] = 2\sinh t \quad \text{...(1)}$$

But $$L^{-1}\left[\frac{d}{ds}[F(s)]\right] = -t f(t) \quad \text{...(2)}$$

From (1) and (2)

$$t f(t) = -2\sinh t$$

$$f(t) = -\frac{2\sinh t}{t} = \frac{e^{-t}-e^{t}}{t}$$

(*ii*) $$L^{-1}\left\{\log\left(1-\frac{4}{s^2}\right)\right\} = L^{-1}\left\{\log\left(1-\frac{2}{s}\right)\left(1+\frac{2}{s}\right)\right\}$$

$$= L^{-1}\left\{\log\left(1-\frac{2}{s}\right) + \log\left(1+\frac{2}{s}\right)\right\}$$

Here, $$F(s) = \log\left(1-\frac{2}{s}\right) + \log\left(1+\frac{2}{s}\right)$$

$$\frac{d}{ds}[F(s)] = \frac{1}{1-\frac{2}{s}}\left(\frac{2}{s^2}\right) + \frac{1}{1+\frac{2}{s}}\left(-\frac{2}{s^2}\right)$$

$$= \frac{2}{s^2}\left[\frac{s}{s-2} - \frac{s}{s+2}\right] = \frac{2s}{s^2}\left[\frac{4}{s^2-2^2}\right]$$

$$= \frac{8}{s(s^2-2^2)}$$

$$L^{-1}\left[\frac{d}{ds}[F(s)]\right] = 8\,L^{-1}\left[\frac{1}{s(s^2-2^2)}\right]$$

Since $$L^{-1}\left[\frac{1}{s^2-2^2}\right] = \frac{1}{2}\sinh 2t$$

$\therefore$ $$8\,L^{-1}\left(\frac{1}{s(s^2-2^2)}\right) = 8\int_0^t\left(\frac{1}{2}\sinh 2t\right)dt = 4\left[\frac{\cosh 2t}{2}\right]_0^t = 2[\cosh 2t]_0^t$$

$$= 2\,[\cosh 2t - 1]$$

From (1) $$-t\,f(t) = 2\,[\cosh 2t - 1]$$

$$f(t) = 2\,\frac{[1-\cosh 2t]}{t}$$

(*iii*) $L^{-1}\left[\frac{s^2-a^2}{(s^2+a^2)^2}\right]$

Here $$F(s) = \frac{s^2-a^2}{(s^2+a^2)} = \frac{(s^2+a^2)-2a^2}{(s^2+a^2)^2} = \frac{1}{s^2+a^2} - \frac{2a^2}{(s^2+a^2)^2}$$

$$L^{-1}\,[F(s)] = L^{-1}\left[\frac{1}{s^2+a^2}\right] - 2a^2\,L^{-1}\left[\frac{1}{(s^2+a^2)^2}\right]$$

$$= \frac{1}{a}\sin at - 2a^2\,L^{-1}\,\frac{1}{(s^2+a^2)^2}$$

But by convolution theorem

$$L^{-1}\left[\frac{1}{(s^2+a^2)^2}\right] = L^{-1}\left\{\frac{1}{(s^2+a^2)}\cdot\frac{1}{(s^2+a^2)}\right\}$$

$$= \int_0^t \frac{1}{a}\sin au\,\frac{1}{a}\sin a(t-u)\,du = \frac{1}{2a^2}\int_0^1 2\sin au\sin(t-u)\,du$$

$$= \frac{1}{2a^2}\int_0^t \{\cos a(2u-t) - \cos at\}\,du$$

$$= \frac{1}{2a^2}\left[\frac{\sin a(2u-t)}{2a} - u\cos at\right]_0^t$$

$$= \frac{1}{2a^2}\left[\frac{2\sin at}{2a} - t\cos at\right] = \frac{1}{2a^3}\,[\sin at - at\cos at]$$

$\therefore$ $$L^{-1}\left[\frac{s^2-a^2}{(s^2+a^2)^2}\right] = \frac{1}{a}\sin at - \frac{2a^2}{2a^3}\,[\sin at - at\cos at]$$

$$= \frac{1}{a}\sin at - \frac{1}{a}\sin at + t\cos at = t\cos at$$

(*iv*) $$L^{-1}\left[\log\frac{s^2}{s^2+1}\right] = L^{-1}\{2\log s - \log(s^2+1)\}$$

Here $$F(s) = 2\log s - \log(s^2+1)$$

$$\frac{d}{ds}[F(s)] = \frac{2}{s} - \frac{1}{s^2+1}\times 2s = \frac{2}{s} - \frac{2s}{s^2+1} = \frac{2s^2+2-2s^2}{s(s^2+1)}$$

$$= \frac{2}{s(s^2+1)}$$

$\therefore$ $$L^{-1}\left[\frac{d}{ds}[F(s)]\right] = 2\int_0^t \sin t\,dt = 2[(-\cos t)]_0^t = 2[1-\cos t]$$

$$-t f(t) = 2(1-\cos t)$$

$$f(t) = \frac{2(\cos t - 1)}{t}$$

***Example* 13.32.** Find the inverse Laplace transform

(*i*) $\dfrac{s}{(s^2+a^2)^2}$

(*ii*) $\dfrac{2s+1}{s^2(s+1)^2}$

(*iii*) $\log\dfrac{s^2+a^2}{s^2+b^2}$

(*iv*) $\dfrac{s+3}{(s^2+6s+13)^2}$

Solution : (*i*) L–1 $\left[\dfrac{s}{(s^2+a^2)^2}\right]$

Let $$F(s) = \frac{s}{(s^2+a^2)}$$

$\Rightarrow$ $$L^{-1}[F(s)] = \cos at = f(t)$$

$$G(s) = \frac{1}{s^2+a^2}$$

$\Rightarrow$ $$L^{-1}G(s) = \frac{1}{a}\sin at = g(t)$$

By convolution theorem

$$L^{-1}\{F(s)\,G(s)\} = \int_0^t f(u)g(t-u)\,du = \int_0^t \cos au\,\frac{1}{a}\sin a(t-u)\,du$$

$$= \frac{1}{2a}\int_0^t 2\sin(at-au)\cos au\,du$$

$$= \frac{1}{2a}\int_0^t \sin at + \sin(t-2u)a\,du$$

$$= \frac{1}{2a}\left[u\sin at + \cos\frac{(t-2u)a}{-2a}\right]_0^t$$

$$= \frac{1}{2a}\left[t\sin at + \frac{\cos(at)}{-2a} - \frac{\cos at}{-2a}\right]$$

$$= \frac{1}{2a}\,t\sin at$$

$$L^{-1}\left[\frac{s}{(s^2+a^2)^2}\right] = \frac{1}{2a}\,t\sin at$$

Alternate :

$$L^{-1}\left[\frac{s}{(s^2+a^2)^2}\right] = f(t)$$

$$L[f(t)] = \frac{s}{(s^2+a^2)}$$

$$L\left[\frac{f(t)}{t}\right] = \int_s^\infty \frac{s}{(s^2+a^2)}ds = \frac{1}{2}\int_s^\infty 2s(s^2+a^2)^{-2}ds$$

$$= \frac{1}{2}\left[\frac{-1}{s^2+a^2}\right]_s^\infty = +\frac{1}{2}\left[\frac{1}{s^2+a^2}\right]$$

$$\frac{f(t)}{t} = \frac{1}{2}L^{-1}\left[\frac{1}{s^2+a^2}\right] = \frac{1}{2a}\sin at$$

$$f(t) = \frac{1}{2a}\,t\sin at.$$

(*ii*)

$$L^{-1}\left[\frac{2s+1}{s^2(s+1)^2}\right] = L^{-1}\left[\frac{s+1+s}{s^2(s+1)^2}\right] = L^{-1}\left[\frac{1}{s^2(s+1)}\right] + L^{-1}\left[\frac{1}{s(s+1)^2}\right]$$

Consider

$$L^{-1}\left(\frac{2s+1}{(s+1)^2}\right) = L^{-1}\left[\frac{2(s+1)-1}{(s+1)^2}\right] = L^{-1}\left(\frac{2}{s+1}\right) - L^{-1}\left(\frac{1}{(s+1)^2}\right) = 2e^{-t} - te^{-t}$$

$$L^{-1}\left(\frac{2s+1}{(s+1)^2}\right) = \int_0^t (2e^{-t} - te^{-t})\,dt = \left(\frac{2e^{-t}}{-t}\right)_0^t - \left[\frac{te^{-t}}{-1} - \frac{e^{-t}}{(-1)^2}\right]_0^t$$

$$= -2\{e^{-t}-1\} - \{-te^{-t} - e^{-t} + 1\} = -2e^{-t} + 2 + te^{-t} + e^{-t} - 1$$

$$= t\,e^{-t} - e^{-t} + 1$$

$$L^{-1}\left[\frac{(2s+1)}{s^2(s+1)^2}\right] = \int_0^t (te^{-t} - e^{-t} + 1)\,dt = \left[t\frac{e^{-t}}{t} - \frac{e^{-t}}{(t)^2} - \frac{e^{-t}}{t} + t\right]_0^t$$

$$= [t\,e^{-t} - e^{-t} + e^{-t} + t]_0^t$$

$$= [t + t\,e^{-t}]_0^t = t\{1 + e^{-t}\}.$$

(*iii*)
$$L^{-1}\left[\log\frac{s^2+a^2}{s^2+b^2}\right] = f(t)$$

$$F(s) = \log(s^2+a^2) - \log(s^2+b^2)$$

$$\frac{d}{ds}\{F(s)\} = \frac{2s}{s^2+a^2} - \frac{2s}{s^2+b^2} = 2\left[\frac{s}{s^2+a^2} - \frac{s}{s^2+b^2}\right]$$

$$L^{-1}\left[\frac{d}{ds}\{F(s)\}\right] = 2L^{-1}\left[\frac{s}{s^2+a^2}\right] - 2L^{-1}\left[\frac{s}{s^2+b^2}\right]$$

$$= 2\cos at - 2\cos bt = 2[\cos at - \cos bt]$$

$$-t f(t) = 2[\cos at - \cos bt]$$

$$f(t) = 2\left[\frac{\cos bt - \cos at}{t}\right].$$

(*iv*)
$$L^{-1}\left[\frac{s+3}{s^2+6s+13}\right] = L^{-1}\left[\frac{s+3}{[(s+3)^2+2^2]^2}\right]$$

Now
$$L^{-1}\left[\frac{s}{(s^2+2^2)^2}\right] = L^{-1}\left(\frac{s}{s^2+2^2}\cdot\frac{1}{s^2+2^2}\right)$$

Let
$$F(s) = \frac{s}{s^2+2^2}, \text{ so } f(t) = \cos 2t$$

$$G(s) = \frac{1}{s^2+2^2}, \text{ so } g(t) = \frac{1}{2}\sin 2t$$

By convolution theorem

$$L^{-1}[F(s)\,.\,G(s)] = \int_0^t (\cos 2u)\left[\frac{1}{2}\sin 2(t-u)\,du\right]$$

$$= \frac{1}{2}\int_0^t \sin 2(t-u)\cos 2u\,du$$

$$= \frac{1}{4}\int_0^t 2\sin 2(t-u)\cos 2u\,du$$

$$= \frac{1}{4}\int_0^t [\sin 2t + \sin 2(t-2u)]\,du$$

$$= \frac{1}{4}\left[u\sin 2t + \frac{\cos 2(t-2u)}{+4}\right]_0^t$$

$$= \frac{1}{4}\left[t\sin 2t + \frac{1}{4}\cos 2t - \frac{\cos 2t}{4}\right] = \frac{1}{4}t\sin 2t$$

$$L^{-1}\left[\frac{s}{(s^2+2^2)^2}\right] = \frac{1}{4}t\sin 2t$$

$$L^{-1}\left[\frac{s+3}{(s+3)^2+2^2}\right] = \frac{1}{4}e^{-3t}\,t\sin 2t$$

EXERCISE 13.5

1. $\log\left(\frac{1+s}{s}\right)$

Ans. $\frac{1}{t}(1-e^{-t})$

2. $\frac{1}{s^3(s^2+1)}$

Ans. $\frac{1}{2}t^2+\cos t-1$

3. $\log\left(1+\frac{1}{s^2}\right)$

Ans. $\frac{2(1-\cos t)}{t}$

4. $\log\left(\frac{s(s+1)}{s^2+4}\right)$

Ans. $\frac{2\cos 2t-1-e^{-t}}{t}$

5. $\frac{1}{(s+2)(s+b)}$

Ans. $\frac{e^{-at}-e^{-at}}{a-b}$

6. $\cot^{-1}\left(\frac{s}{a}\right)$

Ans. $\frac{\sin at}{t}$

7. $\tan^{-1}\left(\frac{2}{s}\right)$

Ans. $\frac{\sin 2t}{t}$

8. $\frac{s^2}{(s^2+a^2)(s^2+b^2)}$

Ans. $\frac{a\sin at-b\sin bt}{a^2-b^2}$

9. $\frac{4s+5}{(s+2)(s-1)^2}$

Ans. $\frac{1}{3}e^t+3te^t-\frac{1}{3}e^{-2t}$

10. $\cot^{-1}(s+1)$

Ans. $e^{-t}\dfrac{\sin t}{t}$

13.14 UNIT STEP FUNCTION OR (Heaviside's Unit Function)

A number of discontinuous function can be expressed in terms of unit step function which is defined as follows :

$$u(t-0) = \begin{cases} 0 & t < a \\ 1 & t \geq a \end{cases} \text{ where } a > 0$$

The graph of this function is given below (Fig. 13.2)

If a_1 and a_2 are two real (positive) numbers such that $a_1 < a_2$, then

$$u(t-a_1) - u(t-a_2) = \begin{cases} 1 & a_1 \leq t \leq a_2 \\ 0 & \text{otherwise} \end{cases}$$

The graph of the function is also given as below (Fig. 13.3)

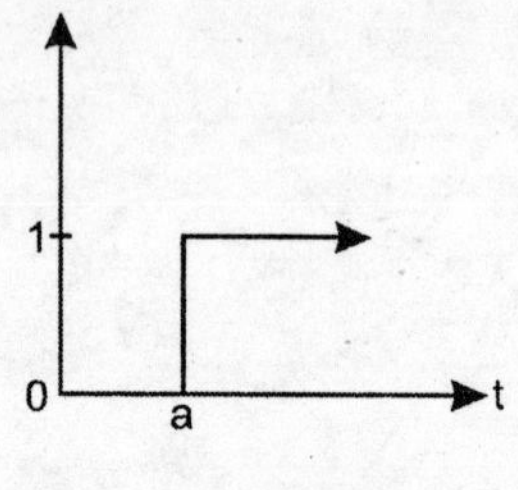

Fig. 13.2.

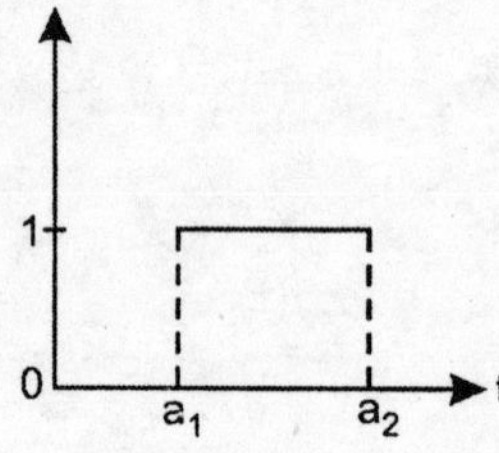

Fig. 13.3

13.15 LAPLACE TRANSFORM OF UNIT STEP FUNCTION

Unit step function is

$$u(t-a) = \begin{cases} 0 & t < a \\ 1 & t \geq a \end{cases}$$

Now
$$L[u(t-a)] = \int_0^\infty e^{-st}u(t-a)\,dt = \int_0^a e^{-st}0\,dt + \int_a^\infty e^{-st}.1\,dt$$

$$= 0 + \int_0^\infty e^{-st}\,dt = \left[\frac{e^{-st}}{-s}\right]_0^\infty = \frac{1}{s}e^{-as}$$

In particular
$$L[u(t)] = \frac{1}{s}$$

***Example* 13.33.** Find Laplace transform of the function

$$f(t) = \begin{cases} 0, & t < 2 \\ 6, & t \geq 2 \end{cases}$$

Solution : We have

$$f(t) = 6\begin{cases} 0, & t < 2 \\ 1, & t \geq 2 \end{cases}$$

$$L\,f(t) = 6\,L\,u(t-2) = \frac{6}{s}\,e^{-2s}$$

***Example* 13.34.** Find Laplace transform of

$$f(t) = \begin{cases} 7, & t < 1 \\ 5, & t \geq 1 \end{cases}$$

Solution :

$$f(t) = \begin{cases} 7, & t < 1 \\ 5, & t \geq 1 \end{cases}$$

$$= 7 - 2\begin{cases} 0, & t < 1 \\ 1, & t \geq 1 \end{cases}$$

$$L\,f(t) = L(7) - 2u(t-1)$$

$$= \frac{7}{s} - 2\,\frac{e^{-s}}{s} = \frac{7 - 2e^{-t}}{s}$$

***Example* 13.35.** Find the Laplace transform of

$$f(t) = \begin{cases} \pi, & t < 2 \\ \pi + 2, & t \geq 2 \end{cases}$$

Solution :

$$f(t) = \begin{cases} \pi, & t < 2 \\ \pi + 2, & t \geq 2 \end{cases}$$

$$= \pi + 2\begin{cases} 0, & t < 2 \\ 1, & t \geq 2 \end{cases}$$

$$L\,f(t) = L(\pi) + L[u(t-2)]$$

$$= \frac{\pi}{s} + \frac{2}{s}\,e^{-2s} = \frac{\pi + 2e^{-2s}}{s}$$

13.16 SECOND SHIFTING PROPERTY FOR INVERSE LAPLACE

If $L^{-1}\,[F(s)] = f(t)$, then $L^{-1}\,\{e^{-as}\,F(s)\} = f(t-a)\,u(t-a)$

Proof : We have,

$$L[f(t-a)\,u(t-a)] = \int_0^\infty e^{-st} f(t-a)\,u(t-a)\,dt$$

$$= \int_0^a e^{-st} f(t-a)\,0\,dt + \int_a^\infty e^{-st} f(t-a)\,.\,1dt$$

$$= \int_0^\infty e^{-s(u+a)} f(u)\,du \quad \text{Put } t - a = u;\ dt = du$$

$$= e^{-sa} \int_0^\infty e^{-su} f(u)\,du = e^{-sa}\,F(s)$$

$$f(t-a)\,u(t-a) = L^{-1}[e^{-as}F(s)]$$

Cor 1. $$L^{-1}[e^{-as}f(s)] = f(t-a)\,u(t-a)$$

Cor 2. If $$a = 0,\quad f(s) = L\{f(t)\,u(t)\}$$

Example 13.36. Find the Laplace transform of $(t-1)^2\,u(t-1)$

Solution : Comparing $(t-1)^2\,u(t-1)$ with $f(t-a)\,u(t-a)$, we have

$$a = 1 \text{ and } f(t) = t^2$$

$$\therefore \quad F(s) = L[f(t)] = \frac{2}{s^3}$$

$\therefore$ By second shifting property

$$L[(t-1)^2\,u(t-1)] = e^{-s}F(s) = e^{-s}\cdot\frac{2}{s^3} = \frac{2e^{-s}}{s^3}$$

Example 13.37. Find Laplace transform of $\sin t\,u(t-\pi)$

Solution : Comparing $\sin t\,u(t-\pi)$ with $f(t-a)\,u(t-a)$ we have

$$a = \pi$$

Also $$\sin t = -\sin(t-\pi)$$

$$\therefore \quad f(t) = -\sin t$$

$$\therefore \quad F(s) = L[f(t)] = +L[-\sin t] = -\frac{1}{1+s^2}$$

By second shifting property

$$L[\sin t\,u(t-\pi) = e^{-as}F(s) = -e^{-\pi s}\frac{1}{1+s^2}$$

Example 13.38. Find the inverse transform of

(*i*) $\dfrac{e^{-2s}}{s-3}$

(*ii*) $\dfrac{se^{-as}}{s^2-v^2}$

(*iii*) $\dfrac{se^{-s/2}+\pi e^{-s}}{s^2+\pi^2}$

Solution : (*i*) Here $$F(s) = \frac{1}{s-3},\text{ then } f(t) = e^{3t}$$

Now $$L^{-1}\left\{\frac{e^{-2s}}{s-3}\right\} = L^{-1}\{e^{-2s}F(s)\} = f(t-2)\,u(t-2)$$

$$= e^{3(t-2)}\,u(t-2)$$

(*ii*) $$F(s) = \frac{s}{s^2-v^2}$$

$$\therefore \qquad f(t) = \cosh vt$$

Now $$L^{-1}\left[\frac{se^{-as}}{s^2 - v^2}\right] = L^{-1}\{e^{-as} F(s)\}$$

$$= f(t-a)\,u(t-a) = \cosh v(t-a)\,u(t-a)$$

(*iii*) $$L\left[\frac{se^{-s/2} + \pi e^{-s}}{s^2+\pi^2}\right] = L^{-1}\left[\frac{se^{-s/2}}{s^2+\pi^2}\right] + L^{-1}\left[\frac{\pi e^{-s}}{s^2+\pi^2}\right]$$

Now, $$L^{-1}\left[\frac{s}{s^2+\pi^2}\right] = \cos \pi t$$

$$L^{-1}\left[\frac{\pi}{s^2+\pi^2}\right] = \sin \pi t$$

$$L^{-1}\left\{\frac{se^{-s/2}}{s^2+\pi^2}\right\} = f\left(t-\frac{1}{2}\right) u\left(t-\frac{1}{2}\right) = \cos \pi \left(t-\frac{1}{2}\right) u\left(t-\frac{1}{2}\right)$$

$$= \sin \pi t\, u\left(t-\frac{1}{2}\right)$$

$$L^{-1}\left\{\frac{\pi e^{-s}}{s^2+\pi^2}\right\} = f(t-1)\,u(t-1)$$

$$= \sin \pi(t-1)\,u(t-1) = -\sin \pi t\, u(t-1)$$

$$L^{-1}\left\{\frac{se^{-s/2} + \pi e^{-s}}{s^2+\pi^2}\right\} = \sin \pi t\, u\left(t-\frac{1}{2}\right) - \sin \pi t\, u(t-1)$$

$$= \sin \pi t \left\{u\left(t-\frac{1}{2}\right) - u(t-1)\right\}$$

***Example* 13.39.** Find the Laplce transform of

$$f(t) = \begin{cases} t-1, & 1 < t < 2 \\ 3-t, & 2 < t < 3 \end{cases}$$

Solution : $$f(t) = \begin{cases} t-1, & 1 < t < 2 \\ 3-t, & 2 < t < 3 \end{cases}$$

$$= (t-1)\{u(t-1) - u(t-2)\} + (3-t)\{u(t-2) - u(t-3)\}$$

$$= (t-1)\,u(t-1) - 2(t-2)\,u(t-2) + (t-3)\,u(t-3)$$

$$L[f(t)] = L[(t-1)\,u(t-1)] - L[2(t-2)\,u(t-2)] + L[(t-3)\,u(t-3)]$$

$$= \frac{e^{-s}}{s^2} - 2\frac{e^{-2s}}{s^2} + \frac{e^{-3s}}{s^2}$$

***Example* 13.40.** Find Laplace transform of

$$f(t) = \begin{cases} \sin t, & t < \pi \\ t, & t \geq \pi \end{cases}$$

Solution : $$f(t) = \begin{cases} \sin t + 0, & t < \pi \\ t - \sin t + \sin t, & t \geq \pi \end{cases}$$

$$f(t) = \sin t + \begin{cases} 0, & t < \pi \\ t - \sin t, & t \geq \pi \end{cases}$$

$$= \sin t + (t - \sin t)\, u(t - \pi)$$

$$f(t) = \sin t + t\, u(t - \pi) - \sin t\, u(t - \pi)$$

$$f(t) = \sin t + (t - \pi + \pi)\, u(t - \pi) + \sin (t - \pi)\, u(t - \pi)$$

$$L[f(t)] = L[\sin t] + L[(t - \pi)\, u(t - \pi)] + \pi\, L[u(t - \pi) + L[\sin (t - \pi)\, u(t - \pi)]$$

$$= \frac{1}{s^2 + 1} + \frac{e^{-\pi s}}{s^2} + \pi \frac{e^{-\pi s}}{s} + \frac{e^{-\pi s}}{s^2 + 1}$$

***Example* 13.41.** Find Laplace transform of $f(t) = \begin{cases} \cos\left(t - \frac{2\pi}{3}\right), & t \geq \frac{2\pi}{3} \\ 0, & t < \frac{2\pi}{3} \end{cases}$

Solution :

$$f(t) = \begin{cases} \cos\left(t - \frac{2\pi}{3}\right), & t \geq \frac{2\pi}{3} \\ 0, & t < \frac{2\pi}{3} \end{cases}$$

$$= \cos\left(t - \frac{2\pi}{3}\right) \begin{cases} 0, & t < 2\pi/3 \\ 1, & t \geq 2\pi/3 \end{cases}$$

$$= \cos\left(t - \frac{2\pi}{3}\right) u\left(t - \frac{2\pi}{3}\right)$$

$$L[f(t)] = L\left[\cos\left(t - \frac{2\pi}{3}\right) u\left(t - \frac{2\pi}{3}\right)\right]$$

$$= \frac{s}{1 + s^2}\, e^{(-2\pi/3)s}$$

***Example* 13.42.** Find the inverse Laplace transform of $\frac{se^{-as}}{s^2 - \omega^2}$ $a > 0$

Solution : Let

$$F(s) = \frac{s}{s^2 - w^2}, \text{ then } f(t) = \cosh wt$$

$$L^{-1}\left[\frac{se^{-as}}{s^2 - w^2}\right] = L^{-1}\,[e^{-as}\, F(s)] = f(t - a)\, u(t - a)$$

$$= \cosh w(t - a)\, u(t - a)$$

13.17 UNIT IMPULSE FUNCTION (OR DIRAC-DELTA FUNCTION)

In mechanics we come across problems such as bending of heavy beams where a very larger force acts for a very small time. To deal with such problems we require the help of unit impulse function of Dirac-Delta function which is defined as :

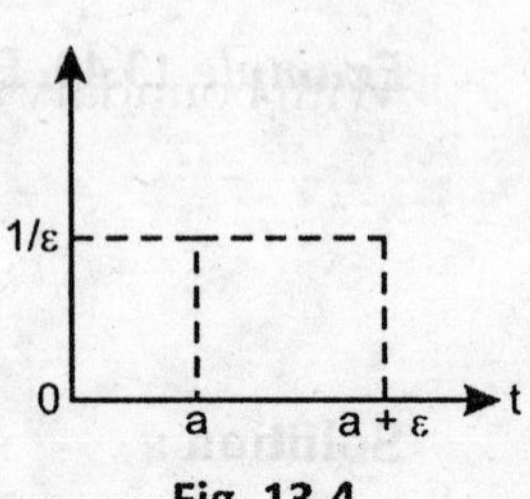

Fig. 13.4

$$\delta(t - a) = \lim_{\epsilon \to 0} \delta_{\epsilon}\,(t - a) = \lim_{\epsilon \to 0} \begin{cases} \frac{1}{\epsilon}, & a \leq t \leq a + \epsilon \\ 0, & \text{otherwise} \end{cases}$$

Graphically, the function is represented as below :

$$\text{Area of the graph} = \int_0^\infty \delta_\in(t-a)\,dt = \int_0^{a+\in} \frac{1}{\in}dt = \frac{1}{\in}[t]_0^{a+\in} = 1$$

As shown in the Fig. 13.4 the value of the function is height of the strip and becomes infinite as $\in \to 0$ and the area of the strip is unity.

Thus unit impulse function is defined as

$$\delta(t-a) = \begin{cases} \infty & t = a \\ 0 & t \neq a \end{cases}$$

where
$$\int_0^\infty \delta(t-a)\,dt = 1$$

13.18 LAPLACE TRANSFORM OF UNIT IMPULSE FUNCTION

$$L[\delta(t-a)] = \int_0^\infty e^{-st}\delta(t-a)\,dt = \lim_{\in\to 0}\int_a^{a+\in} e^{-st}\frac{1}{\in}dt$$

$$= \lim_{\in\to 0}\frac{1}{\in}\left[\frac{e^{-st}}{-s}\right]_a^{a+\in} = \lim_{\in\to 0} e^{-as}\left[\frac{e^{-s\in}-1}{-s\in}\right]$$

$$= e^{-as}\,.\,1 = e^{-as}$$

Laplace transform unit impulse function $= e^{-as}$

Cor 1.
$$L[\delta(t)] = e^{-s\times 0}$$

$$L[f(t)] = (1 + e^{-sa} + e^{-2sa} + \ldots)\int_0^a e^{-st}f(t)\,dt$$

$$= \frac{1}{1-e^{-as}}\int_0^a e^{-st}f(t)\,dt$$

$$L\,f(t) = \frac{1}{1-e^{-as}}\int_0^a e^{-st}f(t)\,dt$$

***Example* 13.43.** A beam is simply supported at its ends at $x = 0$ and $x = l$. A concentrated load P_0 acts vertically downward at the point $x = \frac{l}{3}$. Find deflection at any point of the beam and also at $x = \frac{l}{3}$.

Solution : The different equation of the beam is

$$\frac{d^4y}{dx^4} = \frac{P_0}{EI}\delta\left(x - \frac{l}{3}\right)$$

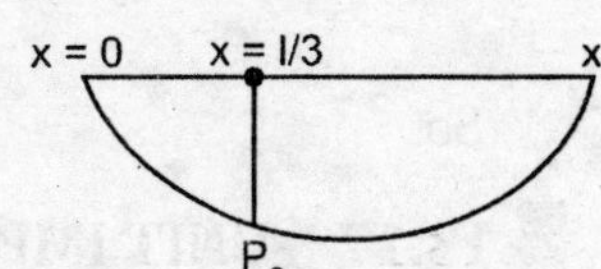

Fig. 13.5

With boundary conditions

$$y(0) = y''(0) = 0 \text{ and}$$
$$y(l) = y''(l) = 0$$

Taking Laplace transform of equation we get

$$L[D^4(y)] = \frac{P_0}{EI}L\left[\delta\left(x-\frac{l}{3}\right)\right]$$

$$= s^4\overline{y} - s^3y(0) - s^2\,y'(0) - sy''(0) - y'''(0) = \frac{P_0}{EI}e^{\frac{ls}{3}}$$

$$s^4\overline{y} = as^2 + b + \frac{P_0}{EI}e^{\frac{ls}{3}}$$

Taking $y'(0) = a$

$$y'''(0) = b$$

$$\overline{y} = \frac{a}{s^2} + \frac{b}{s^4} + \frac{P_0}{EI}\frac{e^{-ls/3}}{s^4}$$

Taking Laplace transform (inverse)

$$y = L^{-1}\overline{y} = ax + \frac{bx^3}{3!} + \frac{P_0}{EI}\frac{\left[x-\frac{l}{3}\right]^3}{3!}\,a\left[x-\frac{l}{3}\right]$$

$$\Rightarrow \qquad y = \begin{cases} ax + b\dfrac{x^3}{6} & 0 < x < \dfrac{l}{3} \\ ax + b\dfrac{x^3}{6} + \dfrac{P_0}{6EI}\left(x-\dfrac{l}{3}\right)^3 & \dfrac{l}{3} < x < l \end{cases}$$

$$y' = a_0 + \frac{bx^2}{2} + \frac{P_0}{2EI}\left(x-\frac{1}{3}\right)^2$$

$$y'' = bx + \frac{P_0}{EI}\left(x-\frac{l}{3}\right)$$

Using the condition $y(l) = 0$

$$y''(l) = 0$$

We have,

$$al + \frac{bl^3}{6} + \frac{4P_0l^3}{8IEI} = 0$$

and

$$bl + \frac{2P_0l}{3EI} = 0$$

$$\Rightarrow \qquad b = \frac{2P_0}{3EI}$$

So

$$al - \frac{P_0l^3}{9EI} + \frac{4P_0l^3}{81EI} = 0$$

$$\Rightarrow \qquad a = \frac{5P_0l^2}{81EI}$$

Deflection at any point of the beam is given by

$$y = \begin{cases} \dfrac{5P_0l^2}{81EI}x - \dfrac{P_0}{9EI}x^3 & 0 < x < l/3 \\ \dfrac{5P_0l^2}{81EI}x - \dfrac{P_0}{9EI}x^3 + \dfrac{P_0}{6EI}\left(x-\dfrac{l}{4}\right)^3 & \dfrac{l}{3} < x < l \end{cases}$$

Deflection at = $\frac{l}{3}$ is

$$y = \frac{5P_0}{81EI}\left\{t^2\left(\frac{l}{3}\right)\right\} - \frac{P_0}{9EI}\left\{\frac{t^3}{27}\right\} = \frac{4P_0 l^3}{243EI}$$

13.19 DEFINITION OF PERIODIC FUNCTION

Definition : The periodic Function of(t) of period 'a' is defined as

$$f(t+a) = f(t) \qquad a > 0$$

In general

$$f(t) = f(t+a) = f(t+2a) = f(t+3a) = \ldots \qquad a > 0$$

13.20 LAPLACE TRANSFORM OF A PERIODIC FUNCTION f(t)

The Laplace transform of a periodic function $f(t)$ is

$$L[f(t)] = \int_0^a \frac{e^{-st} f(t)}{1 - e^{-sa}} dt$$

Proof : We have

$$L[f(t)] = \int_0^\infty e^{-st} f(t)\, dt = \int_0^a e^{-st} f(t)\, dt + \int_a^{2a} e^{-st} f(t)\, dt + \int_{2a}^{3a} e^{-st} f(t)\, dt + \ldots \qquad \ldots(1)$$

Put $t = u + na$ $\qquad dt = du$

We have

$$\int_{na}^{(n+1)a} e^{-st} f(t)\, dt = \int_0^a e^{-s(u+na)} f(u+na)\, du = e^{-nsa} \int_0^a e^{-su} f(u+an)\, du$$

$$= e^{-n\,sa} \int_0^a e^{-su} f(u)\, du \qquad [f(t) \text{ being periodic of period } a]$$

$$= e^{-n\,sa} \int_0^a e^{-st} f(t)\, dt$$

$$\therefore \quad \int_{na}^{(n+1)a} e^{-st} f(t)\, dt = e^{-n\,sa} \int_0^a e^{-st} f(t)\, dt$$

Hence equation (1) becomes

$$L\, f(t) = \int_0^a e^{-st} f(t)\, dt + e^{-sa} \int_0^a e^{-sa} f(t)\, dt + e^{-2sa} \int_0^a e^{-st} f(t)\, dt + \ldots$$

$$= (1 + e^{-as} + e^{-2as} + \ldots) \int_0^a e^{-st} f(t)\, dt$$

$$= \frac{1}{1 - e^{-sa}} \int_0^a e^{-st} f(t)\, dt$$

***Example* 13.44.** Find Laplace transform of the waveform $f(t) = \left(\frac{2t}{3}\right)$, $0 \le t \le 3$

Solution : We have

$$L[f(t)] = \frac{1}{1-e^{-as}} \int_0^a e^{-st} f(t)\, dt$$

Hence the period is 3,

$$f(t) = \left(\frac{2t}{3}\right)$$

$$\therefore \quad L\left(\frac{2t}{3}\right) = \frac{1}{1-e^{-3s}} \int_0^3 e^{-st}\left(\frac{2t}{3}\right) dt = \frac{1}{1-e^{-3t}}\left[\frac{2}{3}\int_0^3 e^{-3t} dt\right]$$

$$= \frac{1}{1-e^{-3s}}\left[t \cdot \frac{e^{-s}}{-s} - (1)\frac{e^{-st}}{(-s)^2}\right]_0^3$$

$$= \frac{2}{3}\frac{1}{1-e^{-3s}}\left[\frac{3e^{-3s}}{-s} - \frac{e^{-3s}}{s^2} + \frac{1}{s^2}\right]$$

$$= \frac{2}{3}\frac{1}{1-e^{-3s}}\left[\frac{3e^{-3s}}{-s} + \frac{1-e^{-3s}}{s^2}\right] = \frac{2e^{-3s}}{-s(1-e^{-3s})} + \frac{2}{3s^2}$$

***Example* 13.45.** Find the Laplace transform of the function $f(t)$ of period 2π which on the interval $(0, 2\pi)$ is given by $f(t) = \begin{cases} \sin t, & 0 \le t \le \pi \\ 0, & \pi \le t \le 2\pi \end{cases}$

Solution : We have

$$L[f(t)] = \frac{1}{1-e^{-as}} \int_0^{2\pi} e^{-st} f(t)\, dt$$

$$= \frac{1}{1-e^{-2\pi s}}\left[\int_0^{\pi} e^{-st} f(t)\, dt + \int_{\pi}^{2\pi} e^{-st} f(t)\, dt\right]$$

$$= \frac{1}{1-e^{-2\pi s}}\left[\int_0^{\pi} e^{-st} \sin t\, dt\right]$$

$$= \frac{1}{1-e^{-2\pi s}}\left[\frac{e^{-st}}{s^2+1}\{-s \sin t - \cos t\}\right]_0^{\pi}$$

$$= \frac{1}{1-e^{-\pi 2s}}\left[\frac{e^{-\pi s}}{s^2+1} - \frac{1}{s^2+1}(-1)\right] = \frac{1}{1-e^{-2\pi x}}\left[\frac{e^{-s\pi}+1}{s^2+1}\right]$$

$$= \left[\frac{1+e^{-\pi x}}{1-e^{-2\pi x}}\right]\frac{1}{s^2+1} = \frac{1}{(1-e^{-\pi s})}\frac{1}{s^2+1}$$

***Example* 13.46.** Find the Laplace transform of the square wave form of period $2c$ given by

$$f(t) = \begin{cases} 1, & 0 < t < c \\ -1, & c < t < 2c \end{cases}$$

Solution : Here period is $2c$

$$L[f(t)] = \frac{1}{1-e^{-2cs}}\int_0^{2c} e^{-st} f(t)\,dt$$

$$= \frac{1}{1-e^{-2cs}}\left[\int_0^{c} e^{-st}\,.\,1dt + \int_c^{2c} e^{-st}\,.\,(-1)\,dt\right]$$

$$= \frac{1}{1-e^{-2cs}}\left[\int_0^{c} e^{-st}\,.\,dt - \int_c^{2c} e^{-st}\,.\,dt\right]$$

$$= \frac{1}{1-e^{-2cs}}\left[\left\{\frac{e^{-st}}{-s}\right\}_0^c - \left\{\frac{e^{-st}}{-s}\right\}_c^{2c}\right]$$

$$= \frac{1}{1-e^{-2cs}}\left[\left(\frac{e^{-sc}}{-s} - \frac{1}{-s}\right) - \left(\frac{e^{-2sc}}{-s} - \frac{e^{-sc}}{-sc}\right)\right]$$

$$= \frac{1}{1-e^{-2cs}}\left[-\frac{e^{-cs}}{s} + \frac{1}{s} + \frac{e^{-2cs}}{s} - \frac{e^{-cs}}{s}\right]$$

$$= \frac{1}{1-e^{-2cs}}\left[\frac{e^{-2cs} - 2e^{-cs} + 1}{s}\right] = \frac{1}{1-e^{-2cs}}\,\frac{[1-e^{-cs}]^2}{s}$$

$$= \frac{1}{1-e^{-2cs}}\,\frac{[1-e^{-cs}]^2}{s} = \frac{1}{s}\left(\frac{1-e^{-cs}}{1-e^{-cs}}\right)$$

***Example* 13.47.** Find the Laplace transform of the function $f(t)$ given by

$$f(t) = \begin{cases} \sin wt, & 0 < t < \pi/w \\ 0, & \pi/w < t < 2\pi/w \end{cases}$$

given that $f(t)$ is periodic with period $\frac{2\pi}{\omega}$.

Solution :

$$L[f(t)] = \frac{1}{1-e^{-as}}\int_0^{a} e^{-st} f(t)\,dt$$

Hence period is $\frac{2\pi}{w}$

$$L[f(t)] = \frac{1}{1-e^{-\frac{2\pi s}{\omega}}}\int_0^{2\pi/\omega} e^{-st} f(t)\,dt$$

$$= \frac{1}{1-e^{-\frac{2\pi s}{\omega}}}\left[\int_0^{\pi/\omega} e^{-st}\sin\omega t\,dt + \int_{\pi/\omega}^{2\pi/\omega} e^{-st}(0)\,dt\right]$$

$$= \frac{1}{1-e^{-\frac{2\pi s}{\omega}}}\left[\int_0^{\pi/\omega} e^{-st}\sin\omega\,t\,dt\right]$$

$$= \frac{1}{1-e^{-\frac{2\pi s}{\omega}}}\left[\frac{e^{-st}}{s^2+w^2}\{-s\sin\omega t - w\cos\omega t\}\right]_0^{\pi/\omega}$$

$$= \frac{1}{1-e^{-\frac{2\pi s}{\omega}}}\left[\frac{we^{-\frac{\pi s}{w}}+w}{s^2+w^2}\right] = \frac{w[1+e^{-\frac{\pi s}{w}}]}{(s^2+w^2)\left(1-e^{-\frac{2\pi s}{w}}\right)}$$

$$= \frac{w}{(s^2+w^2)(1-e^{-\frac{\pi s}{w}})}$$

***Example* 13.48.** Draw the graph of the following periodic function and find its Laplace transform given by

$$f(t) = \begin{cases} t, & \text{for } 0 < t \le a \\ 2a - t, & \text{for } a < t < 2a \end{cases}$$

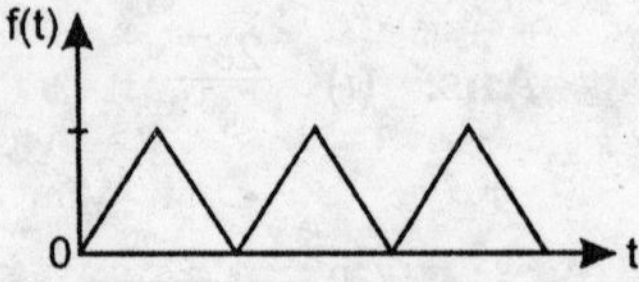

Fig. 13.6.

Solution : The given function is period $2a$

$$L[f(t)] = \frac{1}{1-e^{-2as}}\int_0^{2a} f(t)e^{-st}dt$$

$$= \frac{1}{1-e^{-2as}}\left[\int_0^a e^{-st}f(t)dt + \int_a^{2a} e^{-st}f(t)dt\right]$$

$$= \frac{1}{1-e^{-2as}}\left[\int_0^a t.e^{-st}dt + \int_a^{2a}(2a-t)e^{-st}dt\right]$$

$$= \frac{1}{1-e^{-2as}}\left[\left\{t.\frac{e^{-st}}{s} - (1)\frac{e^{-st}}{(-s)^2}\right\}_0^a + \left\{(2a-t)\frac{e^{-st}}{-s} - (-1)\frac{e^{-st}}{(-s)^2}\right\}_a^{2a}\right]$$

$$= \frac{1}{1-e^{-2as}}\left[\left\{t.\frac{e^{-st}}{-s} - \frac{e^{-st}}{s^2}\right\}_0^a + \left\{(2a-t)\frac{e^{-st}}{-s} + \frac{e^{-st}}{s^2}\right\}_a^{2a}\right]$$

$$= \frac{1}{1-e^{-2as}}\left[-a\frac{e^{-as}}{s} - \frac{e^{-as}}{s^2} + \frac{1}{s^2} + \frac{e^{-2as}}{s^2} + a\frac{e^{-as}}{s} - \frac{e^{-as}}{s^2}\right]$$

$$= \frac{1}{1-e^{-2as}}\left[-a\frac{e^{-as}}{s} - \frac{e^{-as}}{s^2} + \frac{1}{s^2} + \frac{e^{-2as}}{s^2} + a\frac{e^{-as}}{s} - \frac{e^{-as}}{s^2}\right]$$

$$= \frac{1}{s^2(1-e^{-2as})}[e^{-2as} - 2e^{-as} + 1]$$

$$= \frac{1}{s^2[1-e^{-2as}]}[1-e^{-as}]^2$$

$$= \frac{[1-e^{-as}]}{s^2[1+e^{-as}]} = \frac{1}{s^2}\frac{e^{\frac{as}{2}} - e^{\frac{as}{2}}}{e^{\frac{as}{2}} + e^{\frac{as}{2}}} = \frac{1}{s^2}\tanh\frac{as}{2}$$

EXERCISE 13.6

1. Find the Laplace transform of

(i) $(t-1)^2\, u(t-1)$ (ii) $\sin t\, u(t-\pi)$

(iii) $e^{-3t}\, u(t-2)$ (iv) $\sin 2t\, u(t-\pi)$

(v) $e^{t-1}\, u(t-1)$.

Ans. (i) $\dfrac{2e^{-s}}{s^3}$

(ii) $-\dfrac{e^{-\pi s}}{s^2+1}$

(iii) $\dfrac{e^{-2(s+3)}}{s+3}$

(iv) $-2\,\dfrac{e^{-\pi s}}{s^2+4}$

(v) $\dfrac{e^{-s}}{s-1}$

2. Find the inverse transform of

(i) $\dfrac{e^{-\pi s}}{s^2+1}$

(ii) $\dfrac{e^{-\pi s}}{s^2}$

(iii) $\dfrac{se^{-2s}}{s^2-1}$

(iv) $\dfrac{e^{-cs}}{s^2(s+a)}\quad c>0$

(v) $\dfrac{3(1+e^{-\pi s})}{s^2+9}$

(vi) $\dfrac{e^{-s}}{(s+1)^3}$

Ans. (i) $-\sin t\,.\,u(t-\pi)$

(ii) $(t-\pi)\,.\,u(t-\pi)$

(iii) $\cosh(t-2)\,.\,u(t-2)$

(iv) $\dfrac{1}{a^2}\,[a(t-c)-1+e^{-a(t-c)}]\,.\,u(t-c)$

(v) $\sin 3t\,[1-u(t-\pi)]$

(vi) $\dfrac{1}{2}\,e^{-(t-1)}\,.\,(t-1)^2\, u(t-1)$

3. Express the following as unit step functions and then find their Laplace transform

(i) $f(t) = \begin{cases} 3 & 1 \le t < 2 \\ 4 & 2 \le t < 3 \\ 5 & t \ge 3 \end{cases}$

(ii) $f(t) = \begin{cases} 2t, & 0 < t < \pi \\ 1, & t > \pi \end{cases}$

(iii) $f(t) = \begin{cases} t^2 & 0 < t < 2 \\ 0 & t > 2 \end{cases}$

Ans. (i) $3u(t-1) + u(t-2) + u(t-3)\ \dfrac{e^{-s}}{s}(3 + e^{-s} + e^{-2s})$,

(ii) $(1-2t)\,u(t-\pi) + 2t\,u(t)\ .\ \dfrac{1}{s^2}[2 + e^{-\pi s}(s - 2\pi s - 2)]$,

(iii) $t^2[u(t) - u(t-2)]\ .\ \dfrac{2(1-e^{-2s})}{s^3} + \dfrac{4(1+s)e^{-2s}}{s^2}$

4. Find the Laplace transform of the periodic function $f(t) = e^t$ for $0 < t < 2\pi$.

Ans. $\dfrac{e^{2(1-s)\pi} - 1}{(1-s)(1-e^{-2\pi s})}$

5. Find the Laplace transform of $f(t) = t^2$, $0 < t < 2$, $f(t+2) = f(t)$.

Ans. $\dfrac{2 - e^{-2s} - 4se^{-2s} - 4s^2e^{-2s}}{s^2(1-e^{-2s})}$

6. Find the Laplace transform of $f(t) = \begin{cases} \cos wt & 0 < t < \pi/\omega \\ 0 & \pi/\omega < t < 2\pi/\omega \end{cases}$

Ans. $\left[\dfrac{s}{(s^2 + w^2)(1 - e^{-\frac{\pi s}{w}})}\right]$

7. A beam of length 'l' has its ends $x = 0$ and $x = l$ hinged. If concentrated load W acts at the point $x = \dfrac{l}{3}$, show that the deflection is given by

$$Ely = \frac{W}{81}x(5l^2 - 9x^2) + \frac{W}{6}\left(x - \frac{l}{3}\right)^2 . u\left(x - \frac{l}{3}\right).$$

8. The deflection of a beam of length l_1 champed horizontally at both end and loaded at $x = \dfrac{l}{4}$ by a weight w is given by $EI = \dfrac{d^4y}{dx^4} = W\delta\left(x - \dfrac{l}{4}\right)$.

13.21 APPLICATION OF LAPLACE TRANSFORMS TO SOLVE ORDINARY DIFFERENTIAL EQUATIONS WITH CONSTANT AND SIMULTANEOUS DIFFERENTIAL EQUATIONS

By the use of Laplace transforms ordinary differential equations with constant coefficients and simultaneous differential equations can be solved.

Method :

1. Take Laplace transforms of both sides of the given differential equation, using initial conditions. This gives an algebraic equation.

2. Solve the algebraic equation thus obtained in (1) to get $\overline{y}$ in terms of s.
3. Take inverse Laplace transform of both sides to set y as a function of t so obtained.

***Remember :**

$$L\{f^n(t)\} = s^n\,\overline{y} - s^{n-1}\,y(0) - s^{n-2}\,y'(0) - s^{n-3}\,y''(0) \ldots y^n-1\,(0)$$

$$\overline{y} = L[f(t)] = L[y(t)]$$

***Example* 13.49.** Using Laplace transform, solve $y''(t) + y(t) = t$ given that $y'(0) = 1$ $y(\pi) = 0$.

Solution : Here $L[y(t)] = \overline{y}$. The given equation is :

$$y'' + y = t$$

Taking Laplace tranform on both sides of the given equation

$$s^2\overline{y} - sy(0) + \overline{y} = \frac{1}{s^2}$$

$$(s^2+1)\,\overline{y} = sa + 1 + \frac{1}{s^2}$$

Putting $y'(0) = 1$; $y(\pi) = 0$; $y(0) = a$ (say)

$$\overline{y} = \frac{s}{s^2+1}a + \frac{1}{s^2+1} + \frac{1}{s^2(s^2+1)}$$

Taking inverse transform

$$y = a\cos t + \sin t + L^{-1}\left[\frac{1}{s^2(s^2+1)}\right]$$

$$L^{-1}\left[\frac{1}{s^2+1}\right] = \sin t$$

$$L^{-1}\left[\frac{1}{s^2(s^2+1)}\right] = \int_0^t \sin u\,du = 1 - \cos t$$

$$L^{-1}\left[\frac{1}{s^2(s^2+1)}\right] = \int_0^t (1-\cos u)\,du = t - \sin t$$

$$y = a\cos t + \sin t + t - \sin t$$

$$= a\cos t + t$$

When $t = \pi$; $y = 0$ we get $a = \pi$

$$y = \pi\cos t + t$$

***Example* 13.50.** Solve $x'' + 2x' + 5x = e^{-t}\sin t$ where $x(0) = 0$; $x'(0) = 1$

Solution : The given equation is

$$x'' + 2x' + 5x = e^{-t}\sin t$$

Taking the Laplace transform of both sides

$$[s^2\overline{x} - sx(0) - x'(0)] + 2[s\overline{x} - x(0)] + 5\overline{x} = \frac{1}{(s+1)^2+1}$$

Using the given conditions,

$$s^2\bar{x} - 1 + 2s\bar{x} + 5\bar{x} = \frac{1}{(s+1)^2+1}$$

$$(s^2+2s+5)\,\bar{x} = \frac{1}{(s+1)^2+1} + 1$$

$$\bar{x} = \frac{1}{(s^2+2s+5)\{(s+1)^2\}} + \frac{1}{(s^2+2s+5)}$$

$$= \frac{1}{(s^2+2s+2)(s^2+2s+5)} + \frac{1}{(s^2+2s+5)}$$

$$= \frac{1}{3}\left[\frac{1}{s^2+2s+2} - \frac{1}{s^2+2s+5}\right] + \frac{1}{s^2+2s+5}$$

$$= \frac{1}{3}\left[\frac{1}{s^2+2s+2} + \frac{2}{s^2+2s+5}\right]$$

Taking inverse Laplace transform

$$x = \frac{1}{3}\left[e^{-t}\sin t + 2.\frac{1}{2}e^{-t}\sin 2t\right] = \frac{1}{3}\,e^{-t}\,[\sin t + \sin 2t]$$

***Example* 13.51.** Solve the equation $y'' + 4y' + 3y = e^{-t}$; $y(0) = y'(0) = 1$ using Laplace transform.

Solution : Taking Laplace transform of the given equation using $L(y) = \bar{y}$

$$[s^2\bar{y} - sy(0) - y'(0)] + 4[s\bar{y} - y(0)] + 3y = L(e^{-t})$$

$$(s^2+4s+3)\,\bar{y} - s - 5 = \frac{1}{s+1}$$

$$(s^2+4s+3)\,\bar{y} = \frac{1}{s+1} + s + 5 = \frac{s^2+6s+6}{s+1}$$

$$(s+1)(s+3)\,\bar{y} = \frac{s^2+6s+6}{s+1}$$

$$\bar{y} = \frac{s^2+6s+6}{(s+1)^2(s+3)}$$

$$\bar{y} = \frac{7}{4(s+1)} + \frac{1}{2(s+1)^2} - \frac{3}{4(s+3)}$$

Taking inverse Laplace transform

$$y = \frac{7}{4}\,e^{-t} + \frac{1}{2}\,te^{-t} - \frac{3}{4}\,e^{-3t}$$

***Example* 13.52.** Solve $\frac{d^2x}{dt^2} + 9x = \cos 2t$, given $x(0) = 1$, $x(\pi/2) = -1$

Solution : Taking Laplace transform of the given equation, we get

$$L\left[\frac{d^2x}{dt^2}\right]+9L(x) = L[\cos 2t] \text{ where } L(x)=\bar{x}$$

$$[x^2\bar{x}-sx(0)-x'(0)]+9\bar{x} = \frac{s}{s^2+4}$$

$$(s^2+9)\,\bar{x}-s-c = \frac{s}{s^2+4} \qquad x'(0)=c \text{ (suppose)}$$

$$(s^2+9)\,\bar{x} = \frac{s}{s^2+4}+s+c$$

$$\bar{x} = \frac{s}{(s^2+4)(s^2+9)}+\frac{s}{s^2+9}+\frac{c}{s^2+9}$$

$$= \frac{s}{5(s^2+4)}-\frac{s}{5(s^2+9)}+\frac{s}{s^2+9}+\frac{c}{s^2+9}$$

Taking inverse Laplace transform, we get

$$x = \frac{1}{5}L^{-1}\left[\frac{s}{s^2+4}\right]+\frac{4}{5}L^{-1}\left[\frac{s}{s^2+9}\right]+cL^{-1}\left[\frac{1}{s^2+9}\right]$$

$$= \frac{1}{5}\cos 2t+\frac{4}{5}\cos 3t+\frac{c}{3}\sin 3t$$

Using the condition $x\left(\frac{\pi}{2}\right)=-1$

$$-1 = -\frac{1}{5}-\frac{c}{3} \Rightarrow c=\frac{12}{5}$$

$$x = \frac{1}{5}\cos 2t+\frac{4}{5}\cos 3t+\frac{4}{5}\sin 3t$$

Example **13.53.** Solve $(D^3-3D^2+3D-1)\,y=t^2e^{2t}$ given $y(0)=1;\ y'(0)=0;\ y''(0)=-2$

Solution : Taking Laplace transform of the given equation

$$L(y''')-3L(y'')+3L(y')-L(y) = L(t^2e^{2t})$$

$$[s^3\bar{y}-s^2y(0)-s\,y'(0)-y''(0)]-3[s^2\bar{y}-sy(0)-y'(0)]+3[s\,\bar{y}-y(0)]-\bar{y} = \frac{2!}{(s-2)^3}$$

$$\Rightarrow \quad (s^3-3s^2+3s-1)\,\bar{y}-s^2+2+3s-3 = \frac{2}{(s-2)^3}$$

$$(s-1)^3\,\bar{y} = \frac{2}{(s-2)^3}+s^2-3s+1$$

$$\bar{y} = \frac{2}{(s-1)^3(s-2)^3}+\frac{(s-1)^2}{(s-1)^3}-\frac{s}{(s-1)^3}$$

Taking partial fraction of $\frac{2}{(s-1)^3(s-2)^3}$, we get

$$= \frac{12}{s-2}-\frac{6}{(s-2)^2}+\frac{2}{(s-2)^3}-\frac{12}{s-1}-\frac{6}{(s-1)^2}-\frac{2}{(s-1)^3}$$

$$\bar{y} = \frac{12}{s-2} - \frac{6}{(s-2)^2} + \frac{2}{(s-2)^3} - \frac{12}{s-1} - \frac{6}{(s-1)^2} - \frac{2}{(s-1)^3}$$

$$y = L^{-1}(\bar{y}) = 12\,e^{2t} - 6t\,e^{2t} + 2\,\frac{t^2}{2!}\,e^{2t} - 11\,e^t - 7t\,e^t - \frac{3t^2}{2!}\,e^t$$

$$= e^{2t}\,[12 - 6t + t^2] - e^t\left[11 + 7t + \frac{3t^2}{2}\right]$$

***Example* 13.54.** Using Laplace transforms, find the solution of the initial value problem $y'' + 9y = 9u(t-3)$. $y(0) = y'(0) = 0$ where $u(t-3)$ is the unit step function.

Solution : Taking Laplace transforms of the given equation

$$y'' + 9y = 9\,u(t-3), \text{ we get}$$

$$L[y''] + 9\,L[y] = 9\,L[u(t-3)]$$

$$s^2\bar{y} - s\,y(0) - y'(0) + 9\,\bar{y} = 9\,\frac{e^{-3s}}{s}$$

$$s^2\bar{y} + 9\bar{y} = 9\,\frac{e^{-3s}}{s}$$

$$(s^2 + y^2)\,\bar{y} = 9\,\frac{e^{-3s}}{s}$$

$$\therefore \quad \bar{y} = \frac{9e^{-3s}}{s(s^2+9)}$$

$$\therefore \quad y = L^{-1}\left[\frac{9e^{-3s}}{s(s^2+9)}\right]$$

Now
$$L^{-1}\left[\frac{3}{s^2+9}\right] = \sin 3t$$

$$L^{-1}\,\frac{3}{s(s^2+9)} = \int_0^t \sin 3t\,dt = -\,[\cos 3t]_0^t = 1 - \cos 3t$$

$$\therefore \quad y = 3L^{-1}\left[\frac{3e^{-3s}}{s(s^2+9)}\right] = 3\,[1 - \cos 3(t-3)]\,u(t-3)$$

***Example* 13.55.** A resistance R in series with inductance L is connected with *e.m.f.* $E(t)$. The current is given by $L\,\frac{di}{di} + Ri = E(t)$. If the switch is connected at $t = 0$ and disconnected at $t = a$, find the current i in terms of t.

Solution : Conditions under which current i flows are $i = 0$, $t = 0$

$$E(t) = \begin{cases} E & 0 < t < a \\ 0 & t > a \end{cases}$$

Given equation is

$$L\,\frac{di}{di} + Ri = E(t) \qquad \text{...(1)}$$

Taking Laplace transforms of (1)

$$L[s\,\bar{i} - i(0)] + R\,\bar{i} = \int_0^\infty e^{-st} E(t)\,dt$$

$$L s\,\bar{i} + R\,\bar{i} = \int_0^\infty e^{-st} E(t)\,dt \qquad [i(0) = 0]$$

$$(Ls + R)\,\bar{i} = \int_0^a e^{-st} E\,dt + \int_\alpha^\infty e^{-st}(0)\,dt$$

$$= E\left[\frac{e^{-st}}{-s}\right]_0^a = \frac{E}{s}[1 - e^{-as}] = \frac{E}{s} - \frac{e^{-as}}{s}$$

$$\bar{i} = \frac{E}{s(Ls+R)} - \frac{E\,e^{-as}}{s[Ls+R]}$$

$$\therefore \quad \bar{i} = L^{-1}\left[\frac{E}{s(Ls+R)}\right] - L^{-1}\left[\frac{Ee^{-as}}{s[Ls+R]}\right]$$

Now we have

$$L^{-1}\left[\frac{E}{s(Ls+R)}\right] = \frac{E}{L} L^{-1}\left[\frac{1}{s\left(s+\frac{R}{L}\right)}\right] = \frac{E}{L}\cdot\frac{L}{R} L^{-1}\left[\frac{1}{s} - \frac{1}{s+\frac{R}{L}}\right]$$

$$= \frac{E}{R}[1 - e^{\frac{R}{L}t}]$$

Also $$L^{-1}\frac{Ee^{-as}}{s(Ls+R)} = \frac{E}{R}[1 - e^{\frac{R}{L}(t-a)}]\,u(t-a)$$

Hence $$i = \frac{E}{R}[1 - e^{-(R/L)t}] - \frac{E}{R}[1 - e^{(-R/L)(t-a)}]\,u(t-a)$$

$$\frac{E}{R}[1 - e^{-(R/L)t}] \quad \text{for } 0 < t < a \; [u(t-a) = 0]$$

$$\frac{E}{R}[1 - e^{-(R/L)t}] - \frac{E}{R}[1 - e^{(-R/L)(t-a)}] \quad \text{for } t > 0$$

$$\frac{E}{R}[e^{-\frac{R}{L}(t-a)} - e^{-\frac{R}{L}t}] = \frac{E}{R}e^{-\frac{R}{L}t}[e^{\frac{Ra}{L}} - 1]$$

***Example* 13.56.** Solve by Laplace transform, the differential equation $(D^2 + p^2)\,x = a\sin(pt + \alpha)$; $x = Dx = 0$ at $t = 0$.

Solution : Taking Laplace transforms of the given equation

$$[s^2\bar{x} - sx(0) - x'(0)] + p^2\bar{x} = a\,L\sin(pt + \alpha)$$

$$= a\,L[\sin pt\cos\alpha + \cos pt\sin\alpha]$$

$$= a\left[\frac{p}{s^2+p^2}\cos\alpha + \frac{s}{s^2+p^2}\sin\alpha\right]$$

$$= a\left[\frac{[p\cos\alpha + s\sin\alpha}{s^2+p^2}\right]$$

But $\quad x = 0 = x'(0) = 0$

$$s^2\bar{x} + p^2\bar{x} = a\left[\frac{p\cos\alpha + s\sin\alpha}{s^2+p^2}\right]$$

$$(s^2+p^2)\,\bar{x} = a\left[\frac{p\cos\alpha + s\sin\alpha}{s^2+p^2}\right]$$

$$\bar{x} = \frac{ap\cos\alpha + as\sin\alpha}{(s^2+p^2)^2}$$

$$= a\cos\alpha\,\frac{p}{(s^2+p^2)^2} + a\sin\alpha\,\frac{s}{(s^2+p^2)^2}$$

But $\quad L^{-1}\left[\frac{s}{(s^2+p^2)^2}\right] = \frac{-1}{2}\frac{d}{ds}\left[\frac{1}{(s^2+p^2)}\right]$, where $\frac{1}{s^2+p^2} = F(s) \Rightarrow f(t) = \frac{1}{p}\sin pt$

$\therefore \quad L^{-1}\left[\frac{s}{(s^2+p^2)^2}\right] = \frac{t}{2p}\sin pt \;\ldots\; [t\,f(t)] = L^{-1}\left[\frac{-d}{ds}F(s)\right]$

Again,

$$L^{-1}\left[\frac{s}{(s^2+p^2)^2 . s}\right] = \int_0^t \frac{u}{2p}\sin pu\,du$$

$$L^{-1}\left[\frac{s}{(s^2+p^2)^2}\right] = \frac{1}{2p}\left[-\frac{u}{p}\cos pu + \frac{1}{p^2}\sin pu\right]$$

$$= \frac{1}{2p}\left[-\frac{1}{p}\cos pt + \frac{1}{p^2}\sin pt\right]$$

Taking inverse Laplace transform, we get

$$x = \frac{pa\cos\alpha}{2p}\left[-\frac{t}{p}\cos pt + \frac{1}{p^2}\sin pt\right] + a\sin\alpha\left[\frac{t}{2p}\sin pt\right]$$

$$= \frac{a}{2p^2}\,[pt\sin a\sin pt - pt\cos pt\cos\alpha + \sin pt\cos\alpha]$$

$$= \frac{a}{2p^2}\,[\sin pt\cos\alpha - pt\cos(pt+\alpha)]$$

Example **13.57.** Solve the initial value problem $2y'' + 5y' + 2y = e^{-2t}$; $y(0) = 1$; $y'(0) = 1$ using Laplace transform.

Solution : $\quad 2y'' + 5y' + 2y = e^{-2t};\ y(0) = 1;\ y'(0) = 1$

Taking the Laplace transform of both sides, we get

$$2[s^2\bar{y} - sy(0) - y'(0)] + 5[\bar{y} - y(0)] + 2\bar{y} = \frac{1}{s+2}$$

Substituting $y(0) = 1$; $y'(0) = 1$, we have

$$2[s^2\bar{y} - s - 1] + 5[s\bar{y} - 1] + 2\bar{y} = \frac{1}{s+2}$$

$$[2s^2 + 5s + 2]\,\overline{y} - 2s - 2 - 5 = \frac{1}{s+2}$$

$$\overline{y} = \frac{1}{(s+2)(2s^2+5s+2)} + \frac{2s+7}{2s^2+5s+2}$$

$$= \frac{2s^2+11s+15}{(2s+1)(s+2)^2} = \frac{4/9}{2s-1} - \frac{11/9}{s+2} - \frac{1/3}{(s+2)^2}$$

$$= \frac{4}{9}\,\frac{1}{2}\,\frac{1}{s+1/2} - \frac{11}{9}\,\frac{1}{s+2} - \frac{1}{3(s+2)^2}$$

Taking inverse Laplace transform, both sides, we get

$$y = \frac{20}{9}e^{-\frac{1}{2}t} - \frac{11}{9}e^{-2t} - \frac{1}{3}te^{-2t}$$

Example **13.58.** Solve $\frac{d^4y}{dx^4} - k^4y = 0$, when k is constant given $y(0) = 1$ and $\frac{dy}{dt} = \frac{d^4y}{dt^2} = \frac{d^3y}{dt^3} = 0$ when $t = 0$.

Solution : Taking Laplace transform of both sides, we get

$$s^4\overline{y} - s^3y(0) - s^2y'(0) - s\,y''(0) - y'''(0) - k^4\overline{y} = 0$$

$$\overline{y} = \frac{s^3}{s^4 - k^4} \quad \text{Since } y(0) = 1\ y'(0) = y''(0) = y'''(0) = 0$$

$$= \frac{s^3}{(s^2+k^2)(s-k)(s+k)} = \frac{1}{4}\left(\frac{1}{s-k}\right) + \frac{1}{4}\left(\frac{1}{s+4}\right) + \frac{1}{2}\left(\frac{s}{s^2+k^2}\right)$$

$$y(t) = \frac{1}{4}e^{kt} + \frac{1}{4}e^{-kt} + \frac{1}{2}\cos kt$$

Example **13.59.** Solve $x'' - 3x' + 2x = 1 - e^{2t}$, $x(0) = 1$, $x'(0) = 0$ by Laplace transform.

Solution :
$$x'' - 3x' + 2x = 1 - e^{2t} \quad ...(1)$$

$$x(0) = 1;\ x'(0) = 0$$

Taking Laplace transform on both sides of equation (1)

$$[s^2\overline{x} - sx(0) - x'(0)] - 3[s\overline{x} - x(0)] + 2\overline{x} = \frac{1}{s} - \frac{1}{s-2}$$

Putting $x(0) = 1$; $x'(0) = 0$, we get

$$s^2\overline{x} - s - 3s\overline{x} + 3 + 2\overline{x} = \frac{1}{s} - \frac{1}{s-2}$$

$$(s^2 - 3s + 2)\,\overline{x} = \frac{1}{s} - \frac{1}{s-2} + s - 3$$

$$\overline{x} = \frac{1}{s(s-1)(x-2)} - \frac{1}{(s-1)(s-2)^2} + \frac{s-3}{(s-1)(s-2)}$$

$$= \frac{1}{s(s-1)(x-2)} - \frac{1}{(s-1)(s-2)^2} + \frac{s-2-1}{(s-1)(s-2)}$$

$$= \frac{1}{s(s-2)(x-2)} - \frac{1}{(s-1)(s-2)^2} + \frac{1}{s-1} - \frac{1}{(s-1)(s-2)}$$

$$= \frac{1}{2s} - \frac{1}{s-1} + \frac{1}{2(s-2)} - \frac{1}{s-1}$$

$$+ \frac{1}{s-2} - \frac{1}{(s-2)^2} + \frac{1}{s-1} - \frac{1}{s-2} + \frac{1}{s-1}$$

$$\bar{x} = \frac{1}{2s} + \frac{1}{2(s-2)} - \frac{1}{(s-2)^2}$$

Taking inverse Laplace transform

$$x = \frac{1}{2} + \frac{1}{2}e^{2t} - te^{2t} = \frac{1}{2} + \left(\frac{1}{2} - t\right)e^{2t}$$

13.22 SIMULTANEOUS DIFFERENTIAL EQUATIONS

The Laplace transform can be used to solve two or more simultaneous ordinary differential equations. We consider the following simultaneous differential equation :

***Example* 13.60.** Solve $\frac{dx}{dt} + 3x + y = 0$ ...(1)

$$\frac{dy}{dt} - x + y = 0 \quad ...(2)$$

subject to the condition $x(0) = y(0) = 1$.

Solution : Taking the Laplace transform of (1) and (2), we get

$$s\bar{x} - x(0) + 3\bar{x} + \bar{y} = 0 \quad \text{where} \quad \bar{x} = L(x) \quad ...(3)$$

and

$$s\bar{y} - y(0) - \bar{x} + \bar{y} = 0 \quad \text{where} \quad \bar{y} = L(y) \quad ...(4)$$

which becomes on simplification

$$(s+3)\,\bar{x} + \bar{y} = 1 \quad \text{or} \quad (s+3)\,\bar{x} + \bar{y} = 1 \quad ...(5)$$

$$(s+1)\,\bar{y} - \bar{x} = 1 \qquad -\bar{x} + (s+1)\,\bar{y} = 1 \quad ...(6)$$

Solving (5) and (6), we have

$$\bar{x} = \frac{s}{(s+2)^2} = \frac{1}{s+2} - \frac{2}{(s+2)^2}$$

Taking inverse Laplace transform

$$x(t) = e^{-2t} - 2t\,e^{-2t} = (1-2t)\,e^{-2t}$$

From (1)

$$y(t) = -\left[\frac{dx}{dt} + 3x\right]$$

$$= -[2e^{-2t} - 2e^{-2t} + 4te^{-2t}] - 3[e^{-2t} - 2t\,e^{-2t}]$$

$$= e^{-2t}(1+2t).$$

***Example* 13.61.** Solve $\frac{dx}{dt} - y = e^t$ and $\frac{dy}{dt} + x = \sin t$ given $x(0) = 1$; $y(0) = 0$.

Solution :

$$x' - y = e^t \quad ...(1)$$

$$y' + x = \sin t \quad \text{...(2)}$$

Taking Laplace transform of (1) and (2), we get

$$x\bar{x} - x(0) - \bar{y} = \frac{1}{s-1} \quad \text{...(3)}$$

$$s\bar{y} - y(0) + \bar{x} = \frac{1}{s^2+1} \quad \text{...(4)}$$

Putting $x(0) = 1; y(0) = 0$, we get

$$s\bar{x} - 1 - \bar{y} = \frac{1}{s-1} \quad \text{...(5)}$$

$$s\bar{y} + \bar{x} = \frac{1}{s^2+1} \quad \text{...(6)}$$

On solving (5) and (6), we get

$$\bar{x} = \frac{s^4 + s^2 + s - 1}{(s-1)(s^2+1)^2} = \frac{1}{2}\frac{1}{s-1} + \frac{1}{2}\frac{s+1}{s^2+1} + \frac{1}{(s^2+1)^2}$$

$$\bar{y} = \frac{-s^2 + s^2 - 2s}{(s-1)(s^2+1)^2} = -\frac{1}{2}\frac{1}{s-1} + \frac{1}{2}\frac{s-1}{s^2+1} + \frac{s}{(s^2+1)^2}$$

On inversion, we have

$$x = \frac{1}{2}e^t + \frac{1}{2}\cos t + \frac{1}{2}\sin t + \frac{1}{2}(\sin t - t\cos t)$$

$$= \frac{1}{2}[e^t + \cos t + 2\sin t - t\cos t]$$

$$y = -\frac{1}{2}e^t - \frac{1}{2}\cos t - \frac{1}{2}\sin t + \frac{1}{2}t\sin t$$

$$= \frac{1}{2}[-e^t - \sin t + \cos t + t\sin t]$$

***Example* 13.62.** Solve $\frac{d^2x}{dt^2} + \frac{dy}{dt} + 3x = 15e^{-t}$ ...(1)

$$\frac{d^2y}{dt^2} - 4\frac{dy}{dt} + 3y = 15\sin 2t \quad \text{...(2)}$$

given $x(0) = 35; x'(0) = -48; y(0) = 27; y'(0) = -55.$

Solution : Taking Laplace transforms, we have

$$s^2\bar{x} - sx(0) - x'(0) + s\bar{y} - y(0) + 3\bar{x} = 15\frac{1}{s+1} \quad \text{...(3)}$$

$$s^2\bar{y} - sy(0) - y'(0) - 4s\bar{x} + 4x(0) + 3\bar{y} = 15 \times \frac{2}{s^2+4} \quad \text{...(4)}$$

From (3) and (4), we get

$$s^2\overline{x} - 35s + 48 + s\overline{y} - 27 + 3\overline{x} = \frac{15}{5+1}$$

$$(s^2+3)\,\overline{x} + 5\overline{y} = \frac{15}{s+1} + 35s - 21 \qquad ...(5)$$

$$s^2\overline{y} - 27s + 55 - 4[\overline{x} - 35] + 3\overline{y} = \frac{30}{s^2+4}$$

$$(s^2+3)\,\overline{y} - 4\overline{x}s = \frac{30}{s^2+4} + 27s - 55 - 140$$

$$= \frac{30}{s^2+4} + 27s - 195 \qquad ...(6)$$

Multiplying (5) by s^2+3 and (6) by s and subtract, we get

$$\overline{x}\,\{(s^2+3)^2 + 4s^2\} = \frac{15(s^2+3)}{s+1} + 35s\,(s^2+3) - 21(s^2+3) - \frac{30s}{s^2+4} - 27s^2 + 195s - 63$$

$$= \frac{15s^2+45}{s+1} - \frac{30s}{s^2+4} + 35s^3 - 48s^2 + 300s - 63$$

$$\overline{x}\,[s^2 + 9 + 10s^2] = \frac{15s^2+45}{s+1} - \frac{30s}{s^2+4} + 35s^2 - 48s^2 + 300s - 63$$

$$\overline{x}\,[s^2+9][s^2+1] = \frac{15s^2+45}{s+1} - \frac{30s}{s^2+4} + 35s^3 - 48s^2 + 300s - 63$$

$$\overline{x}\,[s^2+9][s^2+1] = (35s^3 - 48s^2 + 300s - 63) + \frac{15(s^2+3)}{s+1} - \frac{30s}{s^2+4}$$

$$\overline{x} = \frac{35s^3 - 48s^2 + 300s - 63}{(s^2+9)(s^2+1)} + \frac{15(s^2+3)}{(s+1)(s^2+1)(s^2+9)} - \frac{30s}{(s^2+1)(s^2+4)(s^2+9)}$$

Breaking into partial fraction

$$\overline{x} = \frac{30s}{s^2+1} - \frac{45}{s^2+9} + \frac{3}{s+1} + \frac{2s}{s^2+4} \qquad ...(7)$$

Multiply (5) by $4s$ and (6) by (s^2+3) gives and add

$$\overline{y} = \frac{7s^3 - 55s^2 - 3s - 585}{(s^2+1)(s^2+9)} + \frac{60s}{(s+1)(s^2+1)(s^2+9)}$$

$$= \frac{7s^3 - 55s^2 - 3s - 585}{(s^2+1)(s^2+9)} + \frac{60s}{(s+1)(s^2+1)(s^2+9)} + \frac{30(s^2+2)}{(s^2+1)(s^2+4)(s^2+9)}$$

Breaking into partial fractions

$$\overline{y} = \frac{30s}{s^2+1} - \frac{45}{s^2+9} + \frac{3}{s+1} + \frac{2s}{s^2+4} \qquad ...(8)$$

Multiply (5) by $4s$ and (6) by (s^2+3) gives and add

$$\bar{y} = \frac{7s^3 - 55s^2 - 3s - 585}{(s^2+1)(s^2+9)} + \frac{60s}{(s+1)(s^2+1)(s^2+9)} + \frac{30(s^2+2)}{(s^2+1)(s^2+4)(s^2+9)}$$

Breaking into partial fractions

$$\bar{y} = \frac{30s}{s^2+9} - \frac{60}{s^2+1} - \frac{3}{s+1} + \frac{2}{s^2+4} \quad \text{... (9)}$$

Taking inverse Laplace transform of (7) and (8), we get

$$x = 30 \cos t - 15 \sin 3t + 3e^{-t} + 2 \cos 2t$$

$$y = 30 \cos 3t - 60 \sin t - 3e^{-t} + \sin 2t$$

EXERCISE 13.7

Solve the following differential equations :

1. $\frac{d^2y}{dx^2} + y = 0$ where $y = 1, \frac{dy}{dx} = 1$ at $x = 0$
2. $\frac{d^2y}{dx^2} + \frac{dy}{dx} - 2y = 1 - 2x$, gives $y = 0, \frac{dy}{dx} = 4$ at $x = 0$
3. $\frac{d^3y}{dx^3} + 2\frac{d^2y}{dx^2} - \frac{dy}{dx} - 2y = 0$, where $y = 1, \frac{dy}{dx} = 2, \frac{d^2y}{dx^2} = 2$ at $x = 0$
4. $y'' + 2'y - y = t$ when $y(0) = 0, y'(0) = 1$
5. $\frac{d^2y}{dx^2} + y = 3 \cos 2x$ where $y = \frac{dy}{dx} = 0$ at $x = 0$
6. $y'' + 4y' + 3y = t, t > 0$ where $y(0) = 0$ and $y'(0) = 1$
7. A condenser of capacity C is charged potential E and discharged at $t = 0$ through an inductance 'L' and resistance 'R'. The change q at time t is governed by the differential equation $L\frac{d^2q}{dt^2} + R\frac{dq}{dt} + \frac{q}{C} = E$, using Laplace transform, show that charge q is given by $q = \frac{CE}{\eta} e^{-\mu t} [\mu \sin \eta + \eta \cos \eta t]$ where $\mu = \frac{R}{2L}$ and $\eta^2 = \frac{1}{LC} - \frac{R^2}{4L^2}$
8. Use Laplace transforms, solve $(D + 1) y_1 + (D - 1)y_2 = e^{-t}; (D + 2)y_1 + (D + 1)y_2 = e^t$
9. Solve $\frac{dx}{dt} = y = \sin t, \frac{dy}{dt} + x = \cos t$ given that $x = 2$ and $y = 0$ when $t = 0$
10. Solve $3\frac{dx}{dt} + 3\frac{dy}{dt} + 5x = 25 \cos t; 2\frac{dx}{dt} - 3\frac{dy}{dt} = 5 \sin t$ with $x(0) = 2; y(0) = 3$
11. Using Laplace transform solve the equation $(D - 2) x - (D + 1) y = 6e^{3t}, (2D - 3) x + (D - 3)y = 6e^{3t}$ given $x = 3, y = 0$ when $t = 0$
12. Using Laplace transform solve $(D^2 - 1)x + 5Dy = t; - 2Dx + (D^2 - 4)y = - 2$ given $x = 0, Dx = 0, y = 0, Dy = 0$

when $t = 0$

13. Solve $(D-2)x + 3y = 0$; $2x + (D-1)y = 0$; $x(0) = 8$, $y(0) = 3$

14. Using Laplace transform solve $(D^2 - 2)x - 3y = e^{2t}$; $(D^2 + 2)y = x = 0$ where $x = 1$, $y = 1$, $Dx = Dy = 0$ when $t = 0$

15. Solve $Dx + Dy = t$; $D^2x - y = e^{-t}$ when $x(0) = 3$, $x'(0) = -2$, $y(0) = 0$

Answers

1. $y = \sin x + \cos x$
2. $y = e^x - e^{2x} + x$
3. $y = \frac{5}{3}e^x - e^{-x} + \frac{1}{3}e^{-2x}$
4. $y = \frac{t}{3} + \frac{4}{3\sqrt{3}} \sin h\sqrt{3}\, t$
5. $[y = \cos x - \cos 2x]$
6. $y = \frac{4}{9} + \frac{t}{6} + e^{-t} - \frac{5}{9}e^{-3t}$

8. $y_1 = e^{3t}$; $y_2 = \sinh t + e^{-t} \sin h\, 2t$
9. $x = e^t + e^{-t}$, $y = e^{-t} - e^t + \sin t$
10. $x = 2\cos t + 3\sin t$; $y = 3\cos t + 2\sin 2t$
11. $y = \sinh t + \cosh t - e^{-3t} - t\, e^t$; $x = e^t + 2t\, e^t + 2e^{3t}$
12. $x = -t + 5\sin t - 2\sin 2t$; $y = 1 - 2\cos t + \cos 2t$
13. $x = 5e^{-t} + 3e^{4t}$; $y = 5e^{-t} + 2e^{4t}$
14. $y = \frac{19}{10}\cos t - \frac{1}{5}\sin t - \frac{1}{15}e^t - \frac{1}{4}e^t - \frac{7}{12}e^{-t}$

    $x = \frac{3}{4}e^t + \frac{7}{4}e^{-t} + \frac{2}{5}e^{2t} + \frac{1}{10}(2\sin t - 19\cos t)$
15. $y = 1 - \frac{1}{2}e^{-1} + \frac{3}{2}\sin t - \frac{1}{2}\cos, t$; $x = 2 + \frac{1}{2}[t^2 + \cos t - 3\sin t + e^{-t}]$

Important Formulae

1. $L(1) = \frac{1}{s}$
2. $L(t^n) = \frac{n!}{s^{n+1}}$ where $n = 0, 1, 2, 3, \ldots,$
3. $L(e^{at}) = \frac{1}{s-a}$
4. $L(\cosh at) = \frac{s}{s^2 - a^2}$

5. $L(\sinh at) = \frac{a}{s^2 - a^2}$

6. $L(\sin at) = \frac{a}{s^2 + a^2}$

7. $L(\cos at) = \frac{s}{s^2 + a^2}$

Results of Theorem

1. $L(t^n f(t)) = (-1)^n \frac{d^n}{ds^n}[F(s)]$

2. $L\left[\int_t^\infty \left(\frac{e^{-x}}{x}\right)dx\right] = \frac{\log(s+1)}{s}$

3. Division by *t*

If $L[f(t)] = F(s)$, then $L\left[\frac{1}{t}f(t)\right] = \int_s^\infty F(s)\,dxs$

4. (*a*) The unit step function $u[t-a]$ is defined as $u(t-a)$ is defined as $u(t-a) = \begin{cases} 0 \text{ when } & 1 < 0 \\ 1 & t \geq a \end{cases}$

(*b*) L.T. of unit step function $L[u(t-a)] = \frac{e^{-as}}{s}$

(*c*) Second shifting theorem.

If $L[f(t)] = F(s)$, then $L[f(t-a).\,u(t-a)] = e^{-as} f(s)$.

5. Convolution theorem. If $L[f(t)] = F(s)$ and $L[g(t)] = G(s)$.

Then $$L\left[\int_0^t f(x)g(t-x)\,dx\right] = F(s).G(s)$$

6. Inverse Laplace transform:

1. $$L^{-1}\left(\frac{1}{s}\right) = 1$$

2. $$L^{-1}\left[\frac{1}{s^n}\right] = \frac{t^{n-1}}{n-1!}$$

3. $$L^{-1}\left[\frac{1}{s-a}\right] = e^{at}$$

4. $$L^{-1}\left[\frac{s}{s^2 - a^2}\right] = \cosh at$$

5. $$L^{-1}\left[\frac{1}{s^2 - a^2}\right] = \frac{1}{u}\sinh at$$

6. $L^{-1}\left[\frac{1}{s^2+a^2}\right] = \frac{1}{a}\sin at$

7. $L^{-1}\left[\frac{s}{s^2+a^2}\right] = \cos at$

8. $L^{-1}[F(s-a)] = e^{at} f(t)$

9. $L^{-1}\left[\frac{1}{(s^2+a^2)^2}\right] = \frac{1}{2a^2}(\sin at - at\cos at)$

10. $L^{-1}\left[\frac{s}{(s^2+a^2)^2}\right] = \frac{1}{2a}\, t\sin bt$

11. $L^{-1}\left[\frac{s^2-a^2}{(s^2+a^2)^2}\right] = t\cos at$

12. $L^{-1}[(1)] = s(t)$

13. $L^{-1}\left[\frac{1}{(s-a)^2+b^2}\right] = \frac{1}{b} e^{at}\sin bt$

14. $L^{-1}\left[\frac{1}{(s-a^2)-b^2}\right] = \frac{1}{b} e^{at}\sin bt$

7. Multiplication by *s*

$$L^{-1}[sF(s)] = \frac{1}{dt} f(t)$$

8. Division by *s*

$$L^{-1}\left[\frac{F(s)}{s}\right] = \int_0^t f(t)\,dt$$

9. 1st shifting property

If $L^{-1} F(s) = f(t)$ then $L^{-1} F(s+a) = e^{-at} L^{-1} [F(s)]$.

10. Second shifting property

$$L^{-1}[e^{-as} F(s)] = f(t-a)\,u(t-a)$$

11. $L^{-1}[F(s)] = t\,L^{-1}\left[\int_s^{\infty} F(s)\,ds\right]$

12. Inverse Convolution Theorem :

$$L^{-1}[F(s)\,G(s)] = \int_0^t f(x)\,g(t-x)\,dx$$

13. $L\{f^n(t)\} = s^n f(s) - s^{n-1} f(0) - s^{n-2} f(0) \ldots\ldots -f^{n}-1(0)$

OBJECTIVE QUESTIONS

1. Give the Laplace transform of :

(i) $\sin^2 3t$ (ii) $t^2 e^t$

(iii) $e^{-t} \sin ht$ (iv) $e^{-3t} \cos 3t$

(v) $\sqrt{t}$

Ans. (i) $\frac{18}{s(s^2+36)}$

(ii) $\frac{2}{(s-2)^3}$

(iii) $\frac{1}{s(s+2)}$

(iv) $\frac{s+3}{s^2+6s+18}$

(v) $\frac{\sqrt{\pi}}{2s^{3/2}}$

2. Give the inverse Laplace transform of :

(i) 1 (ii) $\frac{1}{s^2+a^2}$

(iii) $\frac{e^{-s}}{s^3}$ (iv) $\frac{se-\pi}{s^2+9}$

Ans. (i) $\delta(t)$ (ii) $\cos at$

(iii) $u(t-1)\,.\,\frac{(t-1)^2}{2}$ (iv) $-\cos 3t\,.\,u(t-\pi)$

3. If $L[f(t)] = F(s)$, then what is $L[u(t-a)f(t)]$?

Ans. $e^{-as}\,.\,L\,[f(t+a)]$

4. If $L[f(t)] = \frac{s+1}{s^2+3}$, find $L[f(2, t)]$.

Ans. $\frac{s+2}{s^2+12}$

5. Find Laplace of :

(i) $L\left[\int_0^t \cos u\, du\right]$ (ii) $L\left[\frac{\sin t}{t}\right]$

Ans. (i) $\frac{1}{s^2+1}$ (ii) $\cot^{-1} s$.

6. Given $L^{-1}\left[\frac{1}{s^2+a^2}\right] = \frac{\sin at}{a}$, then find $L^{-1}\left[\frac{s}{s^2+a^2}\right]$

Ans. $\cos at$

7. Find the Laplace transform of :

(i) $(\sin t + \cos t)^2$ (ii) $\sin h + 4t + \sin 4t$

Ans. (i) $\frac{(s^2+2s+4)}{s(s^2+4)}$ (ii) $\frac{8s^2}{s^4-256}$

8. Prove that $\int_0^\infty \frac{\sin t\, dt}{t} = \frac{\pi}{2}$

9. Find $L[t \sin t]$.

Ans. $\frac{2s}{(1+s^2)^2}$

10. Using convolution theorem find the inverse Laplace transform of : $\frac{a}{s(s-a)}$

Ans. $e^{at}-1$

Fourier Transform and its Application

This chapter is devoted to the study of Fourier Integrals, Fourier sine and cosine transforms, finite Fourier sine and cosine transform, Parseval's identity, relation between Fourier and Laplace transform Fourier transform of the derivatives of functions

14.1 FOURIER TRANSFORM

Previously we have studied Fourier series in which we confined ourselves to periodic function. To a periodic function we assigned a linear combination a linear combination of sines and cosines waves of various frequencies and amplitudes. Laplace transform was used in the solution of ordinary differential equations. Out aim in this chapter is to develop a concept called Fourier transform and their properties which all useful in the study of partial differential equation. The theory of integral transform affords mathematical devices with solution of many boundary value problems *i.e.* conduction of heat, transverse vibrations of a string, free and forced vibrations of a membrane etc. can be obtained.

14.2 FOURIER INTEGRAL TRANSFORM OF A FUNCTION

Integral transform of a function $f(x)$ is defined as

$$I[f(x)] = \overline{F}(s) = \int_a^b f(x)k(s,x)\,dx$$

where $k(s, x)$ is a function of s and x and is called the kernel of the transforms, s is he parameter which may be real of imaginary.

Simple example of Kernel are as follows:

1. When $k(s, x) = e^{-ax}$ it leads to Laplace transform of $f(x)$ *i.e.* $\overline{F}(s) = \int_0^\infty f(x)e^{-ax}\,dx$
2. When $k(s, x) = e^{isx}$ it leads to Fourier transform of $f(x)$ *i.e.* $F(s) = \int_0^\infty f(x)e^{ixs}\,dx$
3. When $k(s, x) = x^{s-1}$, it gives Mellin transform of $f(x)$ *i.e.* $M(s) = \int_0^\infty f(x)\,x^{s-1}\,dx$

Theorem 14.1 Fourier Integral Theorem : Consider a function $f(x)$ which satisfies the Drichlet's conditions in every interval (–C, C) so that, we have

$$f(x) = \frac{a0}{2} + \sum_{n=1}^{\infty} \left(a_n \cos\frac{m\pi x}{c} + b_n \sin\frac{m\pi x}{c} \right) \quad ...(1)$$

where $$a_0 = \frac{1}{c}\int_{-c}^{c} f(t)\,dt,\; a_n = \frac{1}{C}\int_{-C}^{C} f(t)\frac{n\pi t}{C}dt$$

and $$b_n = \frac{1}{c}\int_{-c}^{c} f(t)\sin\frac{n\pi x}{c}dt$$

Substituting the values of a_0, a_n and b_n in (1), we have

$$f(x) = \frac{1}{2c}\int_{-c}^{c} f(t)\,dt + \frac{1}{c}\sum_{n=1}^{\infty}\int_{-c}^{c} f(t)\cos\frac{n\pi(t-x)}{c}dt \quad ...(2)$$

If we assume that $\int_{-\infty}^{\infty} | f(x) |\, dx$ converges, the first term on the right side of (2) approaches 0 as $c \to \infty$, since

$$\left|\frac{1}{2c}\int_{-c}^{c} f(t)\,dt\right| \le \frac{1}{2c}\int_{-\infty}^{\infty} | f(t)\,dt$$

The second term on the right side of (2) tends to

$$\underset{cc\to\infty}{Lt}\frac{1}{c}\sum_{n=1}^{\infty}\int_{-\infty}^{\infty} f(t)\cos\frac{n\pi(t-x)}{c}dt$$

Let $\frac{\pi}{c} \triangleq \delta\lambda$ $$= \underset{\delta\lambda\to 0}{Lt}\frac{1}{\pi}\sum_{n=1}^{\infty}\delta\lambda\int_{-\infty}^{\infty} f(t)\cos n\,\delta\lambda\,(t-x)\,dt$$

This is of the form $\underset{\delta\lambda\to 0}{Lt}\sum_{n=1}^{\infty} F(n\delta\lambda)$; *i.e.* $\int_{0}^{\infty} F(\lambda)\,d\lambda$

Thus as $c \to \infty$, (2) become $f(x) = \frac{1}{\pi}\int_{0}^{\infty}\int_{-\infty}^{\infty} f(t) \cos\lambda\,(t-x)\,dt\,d\lambda$ which is known as the Fourier integral of $f(x)$.

14.3 FOURIER INTEGRAL FORMULA

1. Fourier integral of $f(x)$ is given by:

$$f(x) = \frac{1}{\pi}\int_{0}^{\infty}\int_{-\infty}^{\infty} f(t)\cos\lambda\,(t-x)\,dt\,d\lambda$$

2. Fourier since integral:

$$f(x) = \frac{2}{\pi}\int_{\pi}^{\infty}\sin\lambda x\int_{-\infty}^{\infty} f(t)\sin\lambda t\,dt d\lambda$$

3. Fourier cosine integral:

$$f(x) = \frac{2}{\pi}\int_0^\infty \cos\lambda x \int_0^\infty f(t)\cos\lambda t \, dt \, d\lambda$$

4. Complex form of Fourier integral:

$$f(x) = \frac{1}{2\pi}\int_{-\infty}^{\infty}\int_{-\infty}^{\infty} f(t)e^{i\lambda(t-x)} \, dt \, d\lambda$$

***Example* 14.1.** Express $f(x) = \begin{cases} 1 \text{ for } |x| < 1, \\ 0 \text{ for } |x| > 1, \end{cases}$ as Fourier Integral. Hence evaluate $\int_0^\infty \frac{\sin\lambda\cos\lambda x}{\lambda} d\lambda$

Solution : Fourier Integral for $f(x)$ is:

$$= \frac{1}{\pi}\int_0^\infty \int_{-\infty}^{\infty} f(t)\cos\lambda(t-x) \, dt \, d\lambda$$

$$= \frac{1}{\pi}\int_0^\infty \left[\int_{-\infty}^{-1} 0.\cos\lambda(t-x)\,dt + \int_{-1}^{1} 1.\cos\lambda(t-x)\,dt + \int_1^\infty 0.\cos\lambda(t-x)\,dt\right] d\lambda$$

$$= \frac{1}{\pi}\int_0^\infty \int_{-1}^{1} \cos\lambda(t-x) \, dt \, d\lambda$$

$$= \frac{1}{\pi}\int_0^\infty \left[\frac{\sin\lambda(t-x)}{\lambda}\right]_{-1}^{1} d\lambda = \frac{1}{\pi}\int_0^\infty \frac{\sin\lambda(1-x) - \sin\lambda(-1-x)}{\lambda} d\lambda$$

$$= \frac{1}{\pi}\int_0^\infty \frac{\sin\lambda(1+x) + \sin\lambda(1-x)}{\lambda} d\lambda$$

$$= \frac{2}{\pi}\int_0^\infty \frac{\sin\lambda\cos\lambda x}{\lambda} d\lambda$$

$$\therefore \quad \int_0^\infty \frac{\sin\lambda\cos\lambda x}{\lambda} d\lambda = \frac{1}{2}\pi f(x) = \begin{cases} \frac{1}{2}\pi \text{ for } |x| < 1, \\ 0 \text{ for } |x| > 1, \end{cases}$$

At $|x| = 1$, *i.e.* $x = \pm 1$, $f(x)$ is discontinuous and the integral had the value

$$= \frac{1}{2}[f(1+0) + f(1-0)]$$

$$= \frac{1}{2}\left[0 + \frac{\pi}{2}\right]$$

$$= \frac{\pi}{4}$$

Putting $x = 0$, we get $\int_0^\infty \frac{\sin\lambda}{\lambda} d\lambda = \frac{\pi}{2}$ or $\int_0^\infty \frac{\sin x}{x} dx = \frac{\pi}{2}$

Example **14.2.** Find the Fourier cosine integral of e^{-ax}. Hence show that

$$\int_0^\infty \frac{\cos \lambda x}{1+\lambda^2}\, d\lambda = \frac{\pi}{2} e^{-x},\ x \ge 0$$

Solution :

$$f(x) = \frac{2}{\pi}\int_0^\infty \cos \lambda x \int_0^\infty f(t) \cos \lambda t \, dt\, d\lambda$$

$$= \frac{2}{\pi}\int_0^\infty \cos \lambda x \int_0^\infty e^{-at} \cos \lambda t \, dt\, d\lambda$$

$$= \frac{2}{\pi}\int_0^\infty \cos \lambda x \left[\frac{e^{-at}}{a^2+\lambda^2}(-a\cos \lambda t + \lambda \sin \lambda t)\right]_0^\infty d\lambda$$

$$= \frac{2}{\pi}\int_0^\infty \cos \lambda x - \frac{1}{a^2+\lambda^2}(-a)\, d\lambda$$

$$= \frac{2}{\pi}\int_0^\infty \cos \lambda x - \frac{a}{(a^2+\lambda^2)}\, d\lambda$$

$$= \frac{2a}{\pi}\int_0^\infty \frac{\cos \lambda x}{a^2+\lambda^2}\, d\lambda \quad ...(1)$$

is the Fourier cosine integral of e^{-ax}.

From (1)

$$\int_0^\infty \frac{\cos \lambda x}{a^2+\lambda^2}\, d\lambda = \frac{\pi}{2a} e^{-ax}$$

Put

$$a = 1$$

$$\int_0^\infty \frac{\cos \lambda x}{1^2+\lambda^2}\, d\lambda = \frac{\pi}{2} e^{-x},\ x \ge 0$$

Example **14.3.** Express $f(x) \begin{cases} 1 \\ 0 \end{cases}$ for $0 \le x < \pi$ as Fourier sine integral and hence evaluate $\int_0^\infty \frac{1-\cos(\pi\lambda)}{\lambda} \sin(x\lambda)\, d\lambda$

Solution : Fourier sine Integral is given by:

$$f(x) = \frac{2}{\pi}\int_0^\infty \sin \lambda x \int_0^\infty f(t) \sin \lambda t \, dt\, d\lambda$$

$$= \frac{2}{\pi}\int_0^\infty \sin \lambda x \left[\int_0^\pi f(t) \sin \lambda t \, dt + \int_\pi^\infty f(t) \sin \lambda t \, dt\right] d\lambda$$

$$= \frac{2}{\pi}\int_0^\infty \lambda x \int_0^\pi \sin \lambda t \, dt\, d\lambda$$

$$= \frac{2}{\pi}\int_0^\infty \sin \lambda x \left|\frac{-\cos \lambda t}{\lambda}\right|_0^\pi d\lambda$$

$$= \frac{2}{\pi}\int_0^\infty \frac{1-\cos\pi\lambda}{\lambda}\sin\lambda x\,d\lambda \qquad ...(1)$$

From (1)

$$\int_0^\infty \frac{1-\cos\pi\lambda}{\lambda}\sin\lambda x\,d\lambda = \frac{\pi}{2}f(x)$$

$$= \begin{Bmatrix}\pi/2, 0 \le x < \pi \\ 0, x > \pi\end{Bmatrix}$$

Since $x = p$ is the point of discontinuity of $f(x)$, the value of above integral at $x = \pi$ is given by

$$= \left[\frac{f(\pi-0)+f(\pi+0)}{2}\right]$$

$$= \frac{\pi}{4}$$

***Example* 14.4.** Using the Fourier Integral transform show that

(*i*) $\displaystyle\int_0^\infty \frac{w\sin xw}{1+w^2}dw = \frac{\pi}{2}e^{-x}, x > 0$

(*ii*) $\displaystyle\int_0^\infty \frac{\sin w\cos wx}{2}dw = \frac{\pi}{2}, 0 \le x \le 1$

Solution : (*i*) Let $f(x) = e^{-x}$

Using Fourier sine Integral, we have

$$f(x) = \frac{2}{\pi}\int_0^\infty \sin\lambda x\int_0^\infty e{-}t\sin\lambda t\,dt\,d\lambda$$

$$\frac{\pi}{2}f(x) = \int_0^\infty \sin\lambda x\int_0^\infty e^{-t}\sin\lambda t\,dt\,d\lambda$$

$$= \int_0^\infty \sin\lambda x\left\{\frac{e^{-t}}{1+\lambda^2}(-\sin\lambda t - \lambda\cos\lambda t\right\}_0^\infty d\lambda$$

$$= \int_0^\infty \lambda x\left(\frac{\lambda}{1+\lambda^2}\right)d\lambda$$

$$= \int_0^\infty \frac{\lambda\sin\lambda x}{1+\lambda^2}d\lambda$$

Replace λ by w,

$$\int_0^\infty \frac{w\sin wx}{1+w^2}dw = \frac{\pi}{2}e^{-x}$$

(*ii*) Let $f(x) = e^{0x} = 1 \Rightarrow f(t) = 1,\ 0 \le t < 1$

Using Fourier Cosine Integral, we have

$$f(x) = \frac{2}{\pi}\int_0^\infty \cos\lambda x \int_0^\infty f(t)\cos\lambda t\, dt\, d\lambda$$

$$= \frac{2}{\pi}\int_0^\infty \cos\lambda x \int_0^1 1.\cos\lambda t\, dt\, d\lambda$$

$$= \frac{2}{\pi}\int_0^\infty \cos\lambda x \left|\frac{\sin\lambda t}{\lambda}\right|_0^1 d\lambda$$

$$f(x) = \frac{2}{\pi}\int_0^\infty \frac{\cos\lambda x \sin\lambda}{\lambda} d\lambda$$

or
$$\int_0^\infty \frac{\cos\lambda x\sin\lambda}{\lambda} d\lambda = \frac{\pi}{2} f(x)$$

$$= \frac{\pi}{2}e^{0x} = \frac{\pi}{2}$$

14.4 INFINITE FOURIER TRANSFORM AND INVERSE FORMULA

1. Fourier sine transform of $f(x)$ is defined as:

$$F_s[f(x)] = F_s(s) = \int_0^\infty f(x)\sin sx\, dx,\ 0 < x < \infty$$

Inverse sine transform is given by:

$$f(x) = \frac{2}{\pi}\int_0^\infty F_s(s)\sin sx\, dx$$

2. Fourier cosine transform of $f(x)$ is defined as

$$F_c[f(x)] = F_c(s) = \int_0^\infty f(x)\cos sx\, dx,\ 0 < x < \infty$$

Inverse cosine transform is given by

$$f(x) = \frac{2}{\pi}\int_0^\infty F_c(s)\cos sx\, dx$$

3. Fourier transform of $f(x)$ is defined as:

$$F[f(x)] = F(s) = \int_{-\infty}^\infty F(s)e^{-is\,x.}\, dx$$

Inverse Fourier Transform is given by :

$$f(x) = \frac{1}{2\pi}\int_{2\pi}^{\infty} F(s)\, e^{-isx}\, ds$$

***Example* 14.5.** Find the fourier transform of $f(x)$, where

$$f(x) = \begin{cases} 1, & |\, x \,| < a \\ 0, & |\, x \,| > a \end{cases}$$

Hence Evaluate:

(*i*) $\int_{-\infty}^{\infty} \frac{\sin ax \cos sx}{s}\, ds$

(*ii*) $\int_{0}^{\infty} \frac{\sin x}{x}\, dx = \frac{\pi}{2}$

Fourier transform of $f(x)$ is given by

$$F\{f(x)\} = \int_{-\infty}^{\infty} f(x)\, e^{isx}\, dx$$

$$= \int_{-\infty}^{-a} f(x)\, e^{isx}\, dx + \int_{-a}^{a} f(x)\, e^{isx}\, dx + \int_{a}^{\infty} e^{isx}\, f(x)\, dx$$

$$= \int_{-\infty}^{-a} 0.e^{isx}\, dx + \int_{-a}^{a} I.e^{isx}\, dx + \int_{a}^{\infty} 0.e^{isx}\, dx$$

$$= \left[\frac{e^{isx}}{is}\right]_{-a}^{a} = \frac{e^{ias} - e^{-ias}}{is}$$

$$= \frac{2}{5}\left(\frac{e^{ias} - e^{ias}}{2i}\right) = \frac{2 \sin as}{s},\ |s \neq 0|$$

For $s = 0$, we find $F\{f(x)\} = 2a$

By inversion formula for Fourier Transform, we have

$$f(x) = \frac{1}{2\pi}\int_{-\infty}^{\infty} F\{f(x)\}\, e^{-isx}\, dx$$

$$= \frac{1}{2\pi}\int_{-\infty}^{\infty} \frac{2 \sin as}{s}\, e^{-isx}\, ds$$

$$= \frac{1}{\pi}\int_{-\infty}^{\infty} \frac{\sin as\, (\cos sx - i \sin sx)}{x}\, ds$$

$$= \frac{1}{\pi}\int_{-\infty}^{\infty} \frac{\sin ax \cos sx}{s}\, dx - \frac{i}{\pi}\int_{-\infty}^{\infty} \frac{\sin as \sin sx}{s}\, ds$$

$$= \frac{1}{\pi}\int_{-\infty}^{\infty}\frac{\sin ax\cos sx}{s}ds - 0 \quad \text{[second integral is an odd function of s]}$$

$$\int_{-\infty}^{\infty}\frac{\sin as\cos sx}{s}ds = \pi f(x) = \begin{cases}\pi, & |x| < a\\ 0, & |x| > a\end{cases} \qquad ...(1)$$

Since the intergrade in (1) is an even function of s, we have

$$2\int_{0}^{\infty}\frac{\sin as\cos sx}{s}ds = \begin{cases}\pi, & |x| < a\\ 0, & |x| > a\end{cases}$$

$$\Rightarrow \int_{0}^{\infty}\frac{\sin as\cos sx}{s}ds = \begin{cases}\pi/2 & |x| < a\\ 0, & |x| < a\end{cases}$$

Put $x = 0$,

$$\int_{0}^{\infty}\frac{\sin as}{s}ds = \begin{cases}\pi/2, a > 0\\ 0, a < 0\end{cases}$$

Put $a = 1$,

$$\int_{0}^{\infty}\frac{\sin s}{s}ds = \frac{\pi}{2}$$

$$\Rightarrow \int_{0}^{\infty}\frac{\sin x}{x}dx = \frac{\pi}{2}$$

***Example* 14.6.** Find the Fourier transform of

$$f(x) = \begin{Bmatrix}1 - x^2, if\ |x| < 1\\ 0\ , if\ |x| > 1\end{Bmatrix}$$

and use it to evaluate $\int_{0}^{\infty}\left(\frac{x\cos x - \sin x}{x^3}\right)\cos\frac{x}{2}dx$

Solution : Fourier Transform of $f(x)$ is given by

$$F\{f(x)\} = \int_{-\infty}^{\infty} f(x)\, e^{isx}\, dx$$

$$= \int_{-\infty}^{-1} f(x)e^{isx}\, dx + \int_{-1}^{1} f(x)e^{isx}\, dx + \int_{1}^{\infty} f(x)e^{isx}\, dx$$

$$= \int_{-\infty}^{-1} 0.dx + \int_{-1}^{1}(1 - x^2)e^{isx}\, dx + \int_{1}^{\infty} 0.dx$$

$$= \left[(1 - x^2)\frac{e^{isx}}{is} - (-2x)\frac{e^{isx}}{(is)^2} + (-2)\frac{e^{isx}}{(is)^3}\right]_{-1}^{1}$$

$$= \left[\frac{-2}{s^2}(e^{is}+e^{-is})+\frac{2}{is^3}(e^{is}-e^{-is})\right]$$

$$= \left[\frac{-4}{s^2}\left(\frac{e^{is}+e^{-is}}{2}\right)+\frac{-4}{s^3}\right](s\cos s-\sin s)$$

$$= 4\left[-\frac{\cos s}{x^2}+\frac{\sin s}{s^2}\right]=\frac{-4}{s^3}(s\cos s-\sin s)$$

By inversion formula for Fourier Transform, we have

$$f(x) = \frac{1}{2\pi}\int_{-\infty}^{\infty} F\{f(x)\}e^{-isx}\,ds$$

$$= \frac{-2}{\pi}\int_{-\infty}^{\infty}\left(\frac{s\cos s-\sin s}{s^3}\right)(\cos sx - i\sin sx)dx$$

$$= \frac{-2}{\pi}\int_{-\infty}^{\infty}\left(\frac{s\cos s-\sin s}{s^3}\right)\cos sx + \frac{2i}{\pi}\int_{-\infty}^{\infty}\left(\frac{s\cos s-\sin s}{s^3}\right)\sin sx\,ds$$

$$= \frac{-4}{\pi}\int_{0}^{\infty}\left(\frac{s\cos s-\sin s}{s^3}\right)\cos sx\,dx = \frac{-\pi}{4}\,i(x)$$

{Since it is an even function and second integral is odd function]

$$\Rightarrow \int_0^{\infty}\left(\frac{s\cos s-\sin s}{s^3}\right)\cos sx\,dx = \frac{-\pi}{4}f(x)$$

$$= \begin{cases}\frac{-\pi}{4}(1-x^2), & \text{if } |x|<1\\ 0, & \text{if } |x|>1\end{cases}$$

Put $x = 1/2$, we have $\int_0^{\infty}\left(\frac{s\cos s-\sin s}{s^3}\right)\cos\frac{s}{2}\,ds = \frac{-\pi}{4}\left(1-\frac{1}{4}\right)$

Hence $\int_0^{\infty}\left(\frac{x\cos x-\sin x}{x^3}\right)\cos\frac{x}{2}\,dx = -\frac{3\pi}{16}$

Example **14.7.** Find the Fourier sine transform of $e^{-|x|}$. Hence evaluate $\int_0^{\infty}\frac{x\sin mx}{1+x^2}dx$

Solution : $f(x) = e^{-|x|} = e^{-x}$, as x is positive in $(0, \infty)$

$$F_s\{f(x)\} = \int_0^{\infty} f(x).\sin sx\,dx = \int_0^{\infty} e^{-x}\sin sx\,dx$$

$$= \left[\frac{e^{-x}}{1+s^2}(-\sin sx - s\cos sx)\right]_0^{\infty} = \frac{s}{1+s^2}$$

Using inversion formula for Fourier sine transform, we get

$$f(x) = \frac{2}{\pi}\int_0^{\infty} F_s\{f(x)\}\sin sx\,ds$$

or $$e^{-x} = \frac{2}{\pi}\int_0^{\infty} \frac{s}{1+s^2} \sin sx\, ds$$

Replacing x by m, we have

$$e^{-m} = \frac{2}{\pi}\int_0^{\infty} \frac{s}{1+s^2} \sin ms\, ds$$

$$= \frac{2}{\pi}\int_0^{\infty} \frac{x \sin mx}{1+x^2} dx$$

Hence $$\frac{2}{\pi}\int_0^{\infty} \frac{x \sin mx}{1+x^2}\, dx = \frac{\pi}{2} e^{-m}$$

***Example* 14.8.** Find the Fourier sine transform of $\frac{e^{-ax}}{x}$

Solution : Fourier sine transform of $f(x)$ is given by

$$F_s\}\{f(x)\} = \int_0^{\infty} f(x) \sin sx\, dx$$

$$= \int^{\infty} \frac{e^{-ax}}{x} \sin sx\, dx = I \quad ...(1)$$

Differentiating w.r.t. s, we have

$$\frac{dI}{ds} = \int_0^{\infty} \frac{\partial}{\partial s}\left(\frac{e^{-ax}}{x} \sin sx\right) dx \quad \text{[By Leibnit'z Rule]}$$

$$= \int_0^{\infty} \frac{e^{-ax}}{x} x \cos x\, dx = \int_0^{\infty} e^{-ax} \cos sx\, dx$$

$$= \left[\frac{e^{-ax}}{a^2+s^2}(-a \cos sx + s \sin sx)\right]_0^{\infty}$$

$$= 0 - \frac{1}{a^2+s^2}(-a) = \frac{a}{s^2+a^2}$$

Integrating w.r.t. s, we get

$$I = \tan^{-1} \frac{s}{a} + c \quad ...(2)$$

From (1) when $s = 0$, I = 0

substituting in (2) $0 = 0 + c \Rightarrow c = 0$

$$\therefore \quad I = \tan^{-1} \frac{s}{a} \text{ or } F_s\left[\frac{e^{-ax}}{x}\right] = \tan^{-1} \frac{s}{a}$$

Example **14.9.** Find the Fourier cosine transform of $f(x) = \frac{1}{1+x^2}$. Hence derive Fourier sine transform of $\phi(x) = \frac{1}{x(1+x^2)}$.

Solution : Fourier cosine transform of$(x) = \frac{1}{1+x^2}$ is

$$F_c\{f(x)\} = \int_0^\infty \frac{\cos sx}{1+x^2}\,dx = I \qquad ...(1)$$

Differentiating w.r.t. s, we have

$$\frac{dI}{ds} = \int_0^\infty \frac{-x \sin sx}{1+x^2}\,dx = \int_0^\infty \frac{-x^2 \sin sx}{x(1+x^2)}\,dx$$

$$= \int_0^\infty \frac{[1-(1+x^2)\sin sx}{x(1+x^2)}\,dx = \int_0^\infty \frac{\sin sx}{x(1+x^2)}\,dx - \int_0^\infty \frac{\sin sx}{x}\,dx$$

$$= \int_0^\infty \frac{\sin sx}{x(1+x^2)}\,dx - \frac{\pi}{2} \qquad ...(2)$$

Differentiating again w.r.t. s, we have

$$\frac{d^2I}{ds^2} = \int_0^\infty \frac{x \cos sx}{x(1+x^2)}\,dx = \int_0^\infty \frac{\cos sx}{1+x^2}\,dx = \mathrm{I} \qquad ...(3)$$

$\Rightarrow$ $\qquad (D^2 - I)\,I = 0$, where $D = \frac{d}{ds}$

Auxiliary Equation is $\qquad D^2 - 1 = 0 \Rightarrow D \pm 1$

$\therefore \qquad \mathrm{I} = c_1 e^s + c_2 e^{-s}$

$$\frac{dI}{ds} = c_1 e^s - c_2 e^{-s} \qquad ...(4)$$

From (1) when $\qquad s = 0,\ I = \int_0^\infty \frac{dx}{1+x^2} = [\tan^{-1} x]^\infty{}_0 = \frac{\pi}{2}$

From (3), $\qquad I = c_1 + c_2$

$$\therefore \qquad c_1 + c_2 = \frac{\pi}{2} \qquad ...(5)$$

From (2) when $\qquad s = 0,\ \frac{dI}{ds} = \frac{-\pi}{2}$

From (4) $\qquad \frac{dI}{ds} = c_1 - c_2$

$$\therefore \qquad c_1 - c_2 = \frac{-\pi}{2} \qquad ...(6)$$

Solving (5) and (6), $c_1 = 0,\ c_2 = \frac{\pi}{2}$

$$\therefore \qquad I = \frac{\pi}{2}e^{-s} \Rightarrow F_c\{f(x)\} = \frac{\pi}{2}e^{-s}$$

Putting the values of c_1 and c_2 in (4)

$$\frac{dI}{ds} = \frac{-\pi}{2}e^{-s}$$

$\therefore$ From (2)

$$\frac{-x}{2}e^{-s} = \int_0^\infty \frac{\sin sx}{x(1+x^2)}\,dx - \frac{\pi}{2}$$

$$\int_0^\infty \frac{\sin sx}{x(1+x^2)}\,dx = \frac{\pi}{2}(1-e^{-s})$$

$$\Rightarrow \qquad F_s\{\phi(x)\} = \frac{\pi}{2}(1-e^{-s})$$

Example **14.10.** Find the Fourier cosine transform e^{-x}

Solution :

$$F_c\{f(x)\} = \int_0^\infty f(x).\cos sx\,dx$$

$$= \int_0^\infty e^{-x^2}\cos sx\,dx = I$$

Differentiating (1) w.r.t. 's', we get

$$\frac{dI}{ds} = \int_0^\infty -xe^{-x2}\sin sx\,dx$$

$$= \frac{1}{2}\int_0^\infty (-2xe^{-x2})\sin sx\,dx$$

Integrating by parts considering sin sx as first function, we get

$$= \frac{1}{2}\left[\sin sx\, e^{-x^2}\Big|_0^\infty - s\int_0^\infty \cos sx\, e^{-x^2}\,dx\right]$$

$$= \frac{-s}{2}\int_0^\infty e^{-x2}\cos sx\,dx = \frac{-s}{2}I$$

$$\frac{dI}{I} = \frac{-s}{2}\,ds$$

Integrating we have $\log I = \frac{-s^2}{4} + \log A$

or $$I = Ae^{-s2}/4$$

From (1), when $s = 0$,

$$I = \int_0^\infty e^{-x^2}\, dx = \int_0^\infty \frac{e^{-t}}{2\sqrt{t}}\, dt \text{ where } t = x^2$$

$$= \frac{1}{2}\int_0^\infty t^{-1/2} e^{-t}\, dt = \frac{1}{2}\Gamma\left(\frac{1}{2}\right) = \frac{\sqrt{\pi}}{2}$$

$\therefore$ From (2), $\quad \dfrac{\sqrt{\pi}}{2} = A$

Hence $\quad I = F_c\{e - x^2\} = \dfrac{\sqrt{\pi}}{2}\, e^{-s^2}/4$

Note : In above example we derived $\displaystyle\int_0^\infty e^{-x^2}\, dx = \frac{\sqrt{\pi}}{2}$

From here $\displaystyle\int_{-\infty}^\infty e^{-x^2} dx = \sqrt{\pi}$. These values of integral can be used directly.

Example 14.11. Find Fourier sine and cosine transform of x^{m-1}, $m > 0$.

Solution : Let $f(x) = x^{m-1}$

Fourier sine transform is given by:

$$F_s\{f(x)\} = F_s(s) = \int_0^\infty x^{m-1}.\sin sx\, dx$$

and $\quad F_c\{f(x)\}\ F_c(s) = \displaystyle\int_0^\infty x^{m-1}.\cos sx\, dx$

Consider $\quad \displaystyle\int_0^\infty e^{-is\,x} x^{m-1} dx = \frac{\Gamma(m)}{(is)^m}$

$$\Rightarrow \quad \int_0^\infty (\cos sx - i \sin sx)\, x^{m-1}\, dx = \left(\frac{1}{i}\right)^m \frac{\Gamma(m)}{s^m} = (-i)^m\, \frac{\Gamma(m)}{s^m}$$

$$= \left(\cos\frac{\pi}{2} - i \sin\frac{\pi}{2}\right)^m \frac{\Gamma(m)}{s^m}$$

$$= \left(\cos\frac{m\pi}{2} - i \sin\frac{m\pi}{2}\right)\frac{\Gamma(m)}{s^m}$$

Equating real and imaginary parts on both sides, we get

$$\int_0^\infty x^{m-1} \cos sx\, dx = \frac{\Gamma(m)}{s^m}\cos\frac{m\pi}{2}$$

and $\quad \displaystyle\int_0^\infty x^{m-1}.\sin sx\, dx = \frac{\Gamma(m)}{s^m}\sin\frac{m\pi}{2}$

$\therefore$ From (1)

$$F_s(x^{m-1}) = \frac{\Gamma(m)}{s^m}\sin\frac{m\pi}{2}$$

and From (2), $$F_c(x^{m-1}) = \frac{\Gamma(m)}{s^m}\cos\frac{m\pi}{2}$$

Example **14.12.** Find Fourier sine and cosine transform of $f(x) = x^n e^{-ax}$

Solution : Fourier sine transform is

$$F_s\{f(x)\} = \int_0^\infty f(x)\sin sx\,dx$$

$$= \int_0^\infty x^n e^{-ax}\sin sx\,dx$$

and Fourier cosine transform is $F_c\{f(x)\} = \int_0^\infty f(x)\cos sx\,dx$

$$= \int_0^\infty x^n e^{-ax}\cos sx\,dx$$

Consider $\int_0^\infty x^n e^{-ax}.e^{isx}\,dx = \int_0^\infty e^{-(a-is)x}x^{(n+1)-1}\,dx$

$$= \frac{\Gamma(n+1)}{(a-is)^{n+1}} \qquad \text{...(1)}$$

Put $a = r\cos\theta$, $s = r\sin\theta \Rightarrow r = (a^2 = s^2)^{1/2}$

(1) reduces to

$$= \frac{\Gamma(n+1)}{(r\cos\theta - ir\sin\theta)^{n+1}} = \frac{\Gamma(n+1)}{r^{n+1}\,e^{-i(n+1)\theta}}$$

$$\int_0^\infty x^n e^{-as}\,e^{isx}\,dx = \frac{\Gamma(n+1)}{(a^2+s^2)^{(n+1)/2}}e^{i(n+1)\theta}$$

Equating real and imaginary parts, we get

$$\int_0^\infty x^n e^{-ax}\cos sx\,dx = \frac{\Gamma(n+1)}{(a^2+s^2)^{(n+1)/2}}\cos(n+1)\theta$$

$$\int_0^\infty x^n e^{-ax}\sin sx\,dx = \frac{\Gamma(n+1)}{(a^2+s^2)^{(n+1)/2}}\sin(n+1)\theta$$

Example **14.13.** Prove that Fourier transform of $e^{-x^2}/2$ is $\sqrt{2\pi}\,e^{-s^2}/2$

Solution : $$F\{f(x)\} = \int_{-\infty}^{\infty} e^{-x^2}/2\,e^{isx}\,dx$$

$$= \int_{-\infty}^{\infty} e^{\frac{x^2}{2}+isx}\,dx$$

$$= \int_{-\infty}^{\infty} e^{\frac{1}{2}} (x^2 - 2\, isx + i^2 s^2) + \frac{1}{2} i^2 s^2 \, dx$$

$$= e^{-s^2/2} \int_{-\infty}^{\infty} e^{-\left(\frac{x-is}{\sqrt{2}}\right)^2} dx$$

Let $\frac{x-is}{\sqrt{2}} = t$

$\Rightarrow \quad dx = \sqrt{2}\, dt$

$$F\{f(x)\} = e^{-s^2/2} \int_{0}^{\infty} \sqrt{2}\, e^{t^2} .dt$$

$$= \sqrt{2}\, e^{-s^2/2} \int_{-\infty}^{\infty} e^{-t^2} .dt$$

$$= \sqrt{2}\, e^{-s^2/2}.\sqrt{\pi} \text{ or } \sqrt{2\pi}\, e^{-s^2/2}$$

Remember:

(i) $\int_{0}^{\infty} e^{-x^2} dx = \frac{\sqrt{\pi}}{2}$

(ii) $\int_{-\infty}^{\infty} e^{-x^2} dx = \sqrt{\pi}$

(iii) $\int_{-\infty}^{\infty} e^{-x} .x^{n-1}\, dx = \Gamma(n)$

(iv) $\Gamma(n+1) = n!$, is n is an integer

(v) $\int_{0}^{\infty} \frac{\sin x}{x} dx = \frac{\pi}{2}$.

EXERCISE 14.1

1. Find the Fourier integral representation of function $f(x) = \begin{bmatrix} 0, x < 0 \\ e^{-x}, x > 0 \end{bmatrix}$
2. Using Fourier integral, show that

   (i) $\int_{0}^{\infty} \frac{\sin \pi\lambda \sin \lambda\pi}{1-\lambda^2} d\lambda = \begin{bmatrix} \frac{1}{2}\pi \sin x, 0 \le x \le \pi \\ 0, \quad x > \pi \end{bmatrix}$

   (ii) $\int_{0}^{\infty} \frac{m^3 \sin mx}{m^4 + 4} dm = \frac{\pi}{2} e^{-x}, x \ge 0$

3. Find the Fourier transform of $f(x) = \begin{bmatrix} 1, & |x| < 1 \\ 0, & |x| > 1 \end{bmatrix}$. Hence evaluate $\int_0^\infty \frac{\sin x}{x} dx$

4. Find the Fourier sine transform of $\frac{1}{x(x^2 + a^2)}$

5. Solve the integral equation:

$$\int_0^\infty f(x) \cos x \, dx = \begin{Bmatrix} 1-\lambda, & 0 \le \lambda \le 1 \\ 0, & \lambda > 1 \end{Bmatrix}$$

Hence deduce that

$$\int_0^\infty \frac{\sin^2 x}{x^2} dx = \frac{\pi}{2}$$

6. Find the sine transform of $f(x) = \begin{cases} x, & 0 < x < 1 \\ 2-x, & 1 < x < 2 \\ 0, & x > 2 \end{cases}$

7. Find the Fourier transform of $f(x) = \begin{cases} x^2, & |x| < a \\ 0, & |x| > a \end{cases}$

8. Find Fourier cosine transform of $f(x) = e^{-x} + e^{-2x}, x > 0$

Answers:

1. $f(x) = \frac{1}{\pi} \int_0^\infty \frac{\lambda \sin \lambda x + \cos \lambda x}{1 + \lambda^2} d\lambda$

3. $\frac{2 \sin s}{s}$

4. $\frac{\pi}{2a^2} (1 - a^{-as})$

5. $\frac{2(1 - \cos x)}{\pi r^2}$

6. $\frac{2(1 - \cos s) \sin s}{s}$

7. $\frac{2}{s^3} [(a^2 s^2 - 2) \sin as + 2as \cos as$

8. $\frac{6 + 3s^2}{4 + 5s^2 + s^4}$

14.5 FINITE TRANSFORM AND INVERSION

(1) Finite Fourier sine transform of $f(x)$ in $0, x < l$ is defined as

$$F_s(n) = \int_0^l f(x) . \sin \frac{n p x}{l} dx \text{ where } n \text{ is an integer.}$$

$f(x)$ is called the inverse Fourier sine transform of $F_s(n)$ and is given by

$$f(x) = \frac{2}{l}\sum_{n=1}^{\infty} F_s(n)\sin\frac{m\pi x}{l}$$

(2) Finite Fourier cosine transform of $f(x)$ in $0 < x < 1$ is given by:

$$F_c(n) = \int_0^1 f(x).\cos\frac{n\pi x}{l}\, dx \text{ where } n \text{ is an integer.}$$

$f(x)$ is called the inverse finite Fourier cosine transform of $F_c(n)$ ad is given by

$$f(x) = \frac{1}{l}F_c(0) + \frac{2}{l}\sum_{n=1}^{\infty} F_c(n)\cos\frac{n\pi x}{1}$$

***Example* 14.14.** Find the finite sine transform of $f(x) = \sin ax$ where 'a' is an integer

Solution :

$$f(x) = \sin ax$$

$$F_s(n) = \int_0^c f(x).\sin\frac{n\pi x}{c}\, dx \qquad [\text{Here } c = \pi]$$

$$= \int_0^\pi \sin ax \sin nx\, dx$$

$$= \frac{1}{2}\int_0^\pi 2(\sin nx.\sin ax)\, dx \qquad ...(1)$$

$$= \frac{1}{2}\int_0^\pi [\cos (a-n)x - \cos (a+n)x]\, dx$$

$$= \frac{1}{2}\left[\frac{\sin(a-n)x}{(a-n)} - \frac{\sin(a+n)x}{(a+n)}\right]_0^\pi$$

$$= \frac{1}{2}(0),\ a \neq n \qquad [\sin m\pi = 0 \text{ where } m \text{ is an integer}]$$

$$= 0,\ a \neq 0$$

When $a = n$, then from (1)

$$[F_s(n) = \frac{1}{2}\int_0^\pi (1 - \cos 2nx)dx$$

Hence,

$$F_s[n] = \begin{bmatrix} 0, & \text{if } a\,\pi\, n \\ \pi/2 & \text{if } a \neq n \end{bmatrix}$$

***Example* 14.15.** Find the finite cosine transform of $\left(1-\frac{x}{\pi}\right)^2$ in $0 < x < \pi$

Solution : Let $f(x) = \int_0^\pi f(x)\cos\left(\frac{m\pi x}{\pi}\right) dx = \int_0^\pi \left(1-\frac{x}{\pi}\right)^2 \cos nx\, dx$

$$= \left[\frac{\sin nx}{n}\left(1-\frac{x}{\pi}\right)^2\right]_0^\pi - \left[-2\left(1-\frac{x}{\pi}\right)\left(\frac{-\cos nx}{n^2}\right)\right]_0^\pi + \left[2\left(\frac{-\sin nx}{n}\right)\right]_0^\pi$$

$$= \frac{2}{s^2\pi}\left[1-\frac{1}{\pi}\left(\frac{\sin nx}{n}\right)\right]_0^\pi = \frac{2}{n^2\pi}[1-0]$$

$$F_c(n) = \frac{2}{\pi n^2} \text{ if } s > 0$$

If $n = 0$, then

$$F_c(0) = \int_0^\pi \left(1-\frac{x}{\pi}\right)^2 \cos ox\, dx$$

$$= \left[\frac{1}{3}\left(1-\frac{x}{\pi}\right)^3\left(\frac{1}{-1/\pi}\right)\right]_0^\pi$$

Hence $$F_c(n) = \begin{bmatrix} \frac{\pi}{3}, & \text{if } n = 0 \\ \frac{2}{\pi n^2} & \text{if } n = 1,2,3,\dots \end{bmatrix}$$

***Example* 14.16.** Find the finite Fourier sine and cosine transform of $f(x) = x^2$, $0 < x < 4$

Solution : (*i*) $$F_s(n) = \int_0^l f(x)\sin\frac{n\pi x}{c}dx$$

$$= \int_0^4 x^2 \sin\frac{n\pi x}{4}dx$$

$$= \left[\left|x^2\left(\frac{-\cos n\pi x}{4}\right)\right|_0^4 - \left|2x\left(\frac{16}{n^2\pi^2}\right)\left(\frac{-\sin n\pi x}{4}\right)\right|_0^4 + \left|\frac{2\times 64}{n^3\pi^3}\left(\cos\frac{n\pi x}{4}\right)\right|_0^4\right]$$

$$= \frac{-4^3}{n\pi}\cos n\pi + \frac{128}{n^3\pi^3}(\cos n\pi - 1)$$

$$F_s(n) = \frac{-64}{n\pi}\cos n\pi + \frac{128}{n^3\pi^3}(\cos n\pi - 1)$$

(*ii*) Finite Fourier cosine transform, is given by

$$F_c(n) = \int_0^4 f(x)\cos\frac{n\pi x}{4}dx$$

$$= \int_0^4 x^2 \sin\frac{n\pi x}{4}dx$$

$$= \left[\left|\frac{4x^2}{n\pi}\left(\frac{\sin n\pi x}{4}\right)\right|_0^4 - \left|2x\left(\frac{-\cos n\pi x}{4}\right).\frac{16}{n^2\pi^2}\right|_0^4 + \left|2\left(\frac{-\sin n\pi x}{4}\right).\frac{64}{n^3\pi^3}\right|_0^4\right]$$

$$= 0 - \frac{8}{n\pi}\left[\frac{-4x}{n\pi}\cos\frac{n\pi x}{4} + \frac{16}{n^2\pi^2}\sin\frac{n\pi x}{4}\right]_0^4$$

$$= \frac{128}{n^2\pi^2}\cos n\pi$$

EXERCISE 14.2

1. Find finite cosine transform of $f(x) = x,\ 0 < x < \pi$

2. Find finite Fourier sine transform and cosine transform of:

   (i) $f(x) = \begin{Bmatrix} kx, & 0 \le x \le \pi/2 \\ k(1-x), & \pi/2 \le x \le x \end{Bmatrix}$

   (ii) $f(x) = x(1 - x),\ 0 \le x \le 1.$

3 Find $f(x)$ if $F_c(n) = \begin{cases} \dfrac{\sin(n\pi/2}{2n}, & n = 1, 2, 3, \ldots \\ \pi/4, & n = 0 \end{cases}$ where $0 < x < 2\pi$

4. Find finite cosine transform of (x) if $f(x) = \begin{cases} 1, & 0 < x < \pi/2 \\ -1, & \pi/2 < x < \pi \end{cases}$

5. Find $F_s(n)$ if $f(x) = \cos x,\ 0 \le x \le \pi$

Answers

1. $\frac{-16}{n\pi}(-1)^n$

2. (i)$F_s(n) = \frac{-16}{n\pi}(-1)^n \frac{2k}{n^2}\sin\frac{n\pi}{2}, Fc(n) = \begin{cases} \frac{k}{n^2}\left[2\cos\frac{n\pi}{2} - (-1)^n - 1,\right] & n \neq 0 \\ \frac{k\pi^2}{4} & n = 0 \end{cases}$

   (ii) $F_s(n) = \frac{2l^3}{n^3\pi^3}$ [1–(–1)n. $F_c(n) = \begin{cases} \frac{-l^3}{n^2\pi^2}[(-1)n + 1], & n \neq 0 \\ \frac{l^3}{6}, & n = 0 \end{cases}$

3. $F_c(n) = 4\sum_{n=1}^{\infty}\frac{(-1)^{n-1}}{n^3}\sin\frac{n\pi x}{8}$

4. $\frac{1}{4} + \frac{1}{2\pi}\sum_{n=1}^{\infty}\frac{\sin(nx/2)}{n}\cos nx$

5. $\frac{n}{(n^2-1)}[1 + (-1)^n]$

14.6 PROPERTIES OF FOURIER TRANSFORM

1. Linearity Property : If $F(s)$ and $G(s)$ are Fourier transforms of $f(x)$ and $g(x)$ respectively, then $F[af(x) + bg(x)] = aF(s) + bG(s)$, where a and b are constants.

Proof : By definition of Fourier transform, we have

$$F(s) = F\{f(x)\} = \int_{-\infty}^{\infty} f(x)e^{isx}dx$$

and

$$G(s) = F[g(x)\} = \int_{-\infty}^{\infty} g(x)e^{isx}\,dx$$

$\therefore$

$$F[af(x)+bg(x)] = \int_{-\infty}^{\infty}[af(x)+bg(x)]e^{isx}dx$$

$$= \int_{-\infty}^{\infty} af(x)e^{isx}\,dx + \int_{-\infty}^{\infty} bg(x)e^{isx}\,dx$$

$$= aF(s) + bG(s).$$

Cor 1. If $F_s(s)$ and $G_s(s)$ are the Fourier sine transforms of $f(x)$ and $g(x)$, then

$F_s[af(x)+bg(x)] = aF_s(s) + bG_s(s)$ where a, b are constants.

Cor 2. If $F_c(s)$ and $G_c(s)$ are the Fourier cosine transforms of $f(x)$ and $g(x)$ respectively, then

$Fc[af(x)+bg(x)] = aF_c(s) + bG_s(s)$ where a, b are constants.

2. Change of Scale Property : It $F(s)$ is the complex Fourier Transforms of $f(x)$, then

$$F[f(ax)] = \frac{1}{a}F\left(\frac{s}{a}\right),\ a \neq 0$$

Proof : By definition of complex Fourier Transforms, we have

$$F(s) = F[f(x)] = \int_{-\infty}^{\infty} f(x).e^{isx}\,dx$$

$$F[f(x) = \int_{-\infty}^{\infty} f(ax).e^{isx}\,dx$$

Putting $ax = t$ *i.e.* $x = \frac{t}{a}$, we have $dx = \frac{1}{a}dt$

When $x \to -\infty$, $t \to \infty$ and when $x \to \infty$, $t \to \infty$

$\therefore$ From (2), we have

$$F[f(ax)] = \int_{-\infty}^{\infty} f(t)e^{ist/a}.\frac{dt}{a}$$

$$= \frac{1}{a}\int_{-\infty}^{\infty} f(t)e^{i(s/a)t}dt = \frac{1}{a}F\left(\frac{s}{a}\right)$$

Cor. (*i*) If $F_s(s)$ is the Fourier sine transform of $f(x)$, then

$$F_s[f(ax)] = \frac{1}{a}F_s\left(\frac{s}{a}\right)$$

Cor. (*ii*) If $F_c(s)$ is the Fourier cosine transform of $f(x)$, then

$$F_c[f(ax)] = \frac{1}{a} F_c\left(\frac{s}{a}\right)$$

3. Shifting Property : If $F(s)$ is the complex Fourier transform of $f(x)$, then

$$F[f(x-a)] = e^{isa} F(s)$$

Proof : By definition of complex Fourier Transform, we have

$$F(s) = F[f(x)] = \int_{-\infty}^{\infty} f(x).e^{isx}\, dx$$

$$\therefore \quad F\{f(x-a) = \int_{-\infty}^{\infty} f(x-a)e^{isx}\, dx \qquad ...(1)$$

Putting $x - a = t$ *i.e.* $x = a + t$, we have $ds = dt$

When $x \to -\infty$, $t \to -\infty$ and when $x \to \infty$, $t \to \infty$

$\therefore$ From (1), we have

$$F[(x-a)] = \int_{-\infty}^{\infty} f(t)e^{is(a+t)}dt = \int_{-\infty}^{\infty} f(t).e^{ist}dt$$

$$= e^{isa} \int_{-\infty}^{\infty} f(t).e^{ist}dt$$

$$= e^{isa} F(s)$$

3(*a*). Shifting on Time Axis : If $F(s)$ is the complex Fourier Transforms of $f(t)$ and t_0 is any real number, then

$$F[f(t-t_0)] = e^{ist_0} F(s)$$

Proof : By definition of complex Fourier Transform,

$$F(s) = F\{f(t)\} = \int_{-\infty}^{\infty} f(t).e^{ist}\, dt$$

Putting $t \sim t_0 = T$ *i.e.* $t = t_0 + T$, we have $dt = dT$.

When $t \to \sim\infty$, $T \to \sim\infty$ and when $t \to \infty$, $T \to \infty$

$\therefore$ From (1), we have

$$F[f(t-t_0) = \int_{-\infty}^{\infty} f(T)e^{is(t_0+T)}dt$$

$$= \int_{-\infty}^{\infty} f(T)\, e^{ist_0}\, e^{isT}\, dT = e^{ist0} \int_{-\infty}^{\infty} f(T)\, e^{isT}\, dT$$

$$= e^{ist_o} F(s)$$

Remark : Inverse Fourier Transform of $e^{ia\,t0}\, F(s)$ is $f(t - t_0)$

3(*b*) Shifting on Frequency Axis : If $F(s)$ is the complex Fourier transform of $f(t)$, then s_0 is any real number, then

$$F[e^{is_0t} f(t)] = F(s + s_0)$$

By definition of Fourier transform, we have

$$F(s) = F[f(t)] = \int_{-\infty}^{\infty} f(t)e^{ist}\, dt$$

$$F[e^{is_0t} f(t)] = \int_{-\infty}^{\infty} e^{is_0t} f(t)e^{ist}\, dt$$

$$= \int_{-\infty}^{\infty} e^{i(s+s_0)t}\, dt = F(s + s_o)$$

Theorem 14.2 Modulation Theorem : If $F(s)$ is the complex Fourier Transform of $f(x)$, then

$$F[f(x) \cos ax] = \frac{1}{2}[F(s + a) + F(s - a)]$$

Proof : By definition of complex Fourier Transform, we have

$$F(s) = F[f(x)] = \int_{-\infty}^{\infty} f(x)e^{isx}\, dx$$

$$\therefore \quad F[f(x) \cos ax] = \int_{-\infty}^{\infty} f(x) \cos ax . e^{isx}\, dx$$

$$= \int_{-\infty}^{\infty} f(x)\left(\frac{e^{iax} + e^{-iax}}{2}\right)e^{isx}\, dx$$

$$= \frac{1}{2}\left[\int_{-\infty}^{\infty} f(x)e^{i(s+a)x}\, dx + \int_{-\infty}^{\infty} f(x).e^{i(s-a)x}\, dx\right]$$

$$= \frac{1}{2}[F(s + a) + F(s - a)]$$

Cor: If $F_s(s)$ and $F_c(s)$ are Fourier sine and cosine transform $f(x)$ respectively, then

(*i*) $F_s[f(x) \cos ax] = \frac{1}{2}[F_s(s + a) + F_s(s - a)]$

(*ii*) $F_s[f(x) \sin ax] = \frac{1}{2}[F_c(s - a) + F_c(s + a)]$

(*iii*) $F_c[f(x) \cos ax] = \frac{1}{2}[F_c(s + a) + F_c(s - a)]$

(*iv*) $F_c[f(x) \sin ax] = \frac{1}{2}[F_s(s + a) + F_s(s - a)]$

Example **14.17.** Find the fourier transform of e^{-x^2}. Hence find the Fourier transform of

(*i*) $e^{-ax^2}, (a > 0)$

(*ii*) $e^{-x^2/2}$

(*iii*) $e^{-4}(x-3)^2$

(*iv*) $e^{-x^2}\cos 2x$

Solution : Fourier Transform of $f(x) = e^{-x^2}$ is given by

$$F\{f(x)\} = \int_{-\infty}^{\infty} f(x)\,.\,e^{isx}\,dx = \int_{-\infty}^{\infty} e^{-x2}.e^{isx}\,dx$$

$$= \int_{-\infty}^{\infty} e^{-(x^2-isx)}\,dx = \int_{-\infty}^{\infty} e^{-\left\{\left(x-\frac{is}{2}\right)^2+\frac{s^2}{4}\right\}}dx$$

$$= e^{-s^2/4}\int_{-\infty}^{\infty} e^{-\left(x-\frac{is}{2}\right)^2}dx$$

$$= e^{-s^2/4}\int_{-\infty}^{\infty} e^{-z^2}\,dz \text{ where } z = x - \frac{is}{2}$$

$$= 2e^{-s^2/4}\int_{-\infty}^{\infty} e^{-z^2}\,dz$$

$$= 2e^{-s^2/4}\frac{\sqrt{\pi}}{2}$$

$$= \sqrt{\pi}\,e^{-s^2/4} = F(s)$$

(*i*) $e^{-ax^2} = e^{-(\sqrt{ax})^2} = f(\sqrt{ax})$

By change of scale property, we have

$$F\{f(\sqrt{ax})\} = \frac{1}{\sqrt{a}}F\left(\frac{s}{\sqrt{a}}\right)$$

$$\Rightarrow \qquad F(e^{-ax^2}) = \frac{1}{\sqrt{a}}\sqrt{\pi}\,e^{\frac{1}{4}\left(\frac{s}{\sqrt{a}}\right)^2} = \sqrt{\frac{\pi}{a}}\,e^{-s^2/4a}$$

(*ii*) Putting $a = \frac{1}{2}$ in deduction (*i*),

$$F(e^{-x^2/2}) = \sqrt{2}\,\pi\,e^{-s^2/2}$$

(*iii*) $$e^{-4x^2} = e^{-(2x)^2} = f(2x)$$

By change of scale property,

$$F[f(2x)] = \frac{1}{2}F\left(\frac{s}{2}\right)$$

$$F[e^{-4x^2}] = \frac{\sqrt{\pi}}{2}e^{\frac{1}{4}\left(\frac{s}{2}\right)^2} = \frac{\sqrt{\pi}}{2}e^{-s^2/16}$$

By shifting property $F\{f(x-a)\} = e^{isa}F(s)$

$$\therefore \quad F[e^{-4(x-3)2}] = e^{3is}\frac{\sqrt{\pi}}{2}e^{-s^2}/4$$

$$= \frac{\sqrt{\pi}}{2}e^{\left(3is\frac{s^2}{a}\right)}$$

(*iv*) By Modulation Theorem :

$$F[f(x)\cos ax] = \frac{1}{2}[F(s+a)+F(s-a)]$$

$$\therefore \quad F[e^{-x^2}\cos 2x] = \frac{1}{2}\left[\sqrt{\pi}\,e^{\frac{1}{4}(s+2)^2}+\sqrt{\pi}\,e^{\frac{1}{4}(s-2)^2}\right]$$

$$= \frac{\sqrt{\pi}}{2}\left[e^{\frac{1}{4}(s+2)^2}+e^{\frac{1}{4}(s-2)^2}\right]$$

14.7 DEFINITION OF CONVOLUTIONS

The convolution of two functions $f(x)$ and $g(x)$ over the interval$(-\infty, \infty)$ is defined as

$$f(x)^* g(x) = \int_{-\infty}^{\infty} f(u)\,g(x-u)\,du$$

Theorem 14.3 Convolution Theorem for Fourier Transforms

The Fourier transform of the convolution of $f(x)$ and $g(x)$ is equal to the product of the Fourier transforms of $f(x)$ and $g(x)$, *i.e,*

$$F[f(x) * g(x)] = F[f(x)].\,F[g(x)]$$

$$= F(s).\,G(s)$$

Proof : By definition of Fourier transform

$$F(s) = F[f(x)] = \int_{-\infty}^{\infty} f(x).\,e^{isx}\,dx$$

$$G(s) = F[g(x)] = \int_{-\infty}^{\infty} g(x).e^{isx}\,dx$$

and by definition of convolution,

$$f(x) * g(x) = \int_{-\infty}^{\infty} f(u)\,g(x-u)\,du$$

$$\therefore \quad F[f(x) * g(x)] = \int_{-\infty}^{\infty} [f(x) * g(x)e^{isx}\,dx$$

$$= \int_{-\infty}^{\infty}\left[\int_{-\infty}^{\infty} f(u)g(x-u)\,du\right]e^{isx}\,dx$$

$$= \int_{-\infty}^{\infty} f(u)\left[\int_{-\infty}^{\infty} g(x-u)e^{isx}\,dx\right]du$$

[By changing the order of integration]

Putting $x - u = t$, *i.e.*, $x = u + t$, we have $dx = dt$

When $x \to -\infty$, $t \to -\infty$ and when $x \to \infty$, $t \to \infty$

$$\therefore \quad F[f(x) * g(x)] = \int_{-\infty}^{\infty} f(u)\left[\int_{-\infty}^{\infty} g(t)e^{is(u+t)}\,dt\right]$$

$$= \int_{-\infty}^{\infty} f(u)\left[\int_{-\infty}^{\infty} g(t).e^{isu}.e^{ist}\,dt\right]$$

$$= \left[\int_{-\infty}^{\infty} f(u)e^{isu}\,du\right]\left[\int_{-\infty}^{\infty} g(t)e^{ist}\,dt\right]$$

$$= \left[\int_{-\infty}^{\infty} f(x)e^{isx}\,dx\right]\left[\int_{-\infty}^{\infty} g(x)e^{isx}\,dx\right]$$

$$= F[f(x)]\,F[g(x)] = F(s).\,G(s)$$

Remark: The following properties can be proved:

(*i*) $f(x) * g(x) = g(x) * f(x)$

(*ii*) $f(x) * [g(x) * h(x)] = [f(x) * g(x)] * h(x)$

(*iii*) $f(x) * [g(x) + h(x)] = f(x) * g(x) + f(x) * h(x)$

***Example* 14.18.** Verify convolution theorem for $f(x) = g(x) = e^{-x^2}$

Solution : $F[f(x)] = \int_{-\infty}^{\infty} e^{-x^2} e^{isx}\,dx$

$$= \int_{-\infty}^{\infty} e^{-(x^2-isx)}\,dx = \int_{-\infty}^{\infty} e^{-\left[\left(x-\frac{is}{2}\right)^2+\frac{s^2}{4}\right]}\,dx$$

$$= \int_{-\infty}^{\infty} e^{-\left(x-\frac{is}{2}\right)^2} e^{-s^2/4}\,dx$$

$$= e^{-s^2}/4 \int_{-\infty}^{\infty} e^{-y^2}\,dy \text{ where } y = x - \frac{is}{2} \Rightarrow dy = dx$$

$$= e^{-s^2}/4.\ \sqrt{\pi}$$

Similarly $\quad F[g(x) = \sqrt{\pi}\ e^{-s^2}/4$

$$F[f(x).\,g(x)] = \pi e^{-s^2/2} \quad ...(1)$$

Now, $$F[f(x) * g(x)] = F\left[\int_{-\infty}^{\infty} f(u).g(x-u)\,du\right]$$

$$= F\left[\int_{-\infty}^{\infty} e^{-u^2}.e^{-(x-u)^2}\,du\right]$$

$$= \left[\int_{-\infty}^{\infty} e^{-u^2}.e^{-(x-u)^2}\,du\right]dx$$

Changing the order of integration

$$= \int_{-\infty}^{\infty} e^{-u^2}\left[\int_{-\infty}^{\infty} e^{-(x-u)^2} e^{isx}\,dx\right]du$$

$$= \int_{-\infty}^{\infty} e^{-u^2}\left[\int_{-\infty}^{\infty} e^{-t^2} e^{is(t+u)}\,dt\right]du \qquad [x-u=t \Rightarrow x=u+t\ dx=dt]$$

$$= \int_{-\infty}^{\infty} e^{-u^2} e^{isu}\,du \int_{-\infty}^{\infty} e^{-r^2} e^{ist}\,dt$$

$$= \left[e^{\frac{s^2}{4}} \int_{-\infty}^{\infty} e^{\left(u\frac{is}{2}\right)^2}du\right]\left[e^{\frac{s^2}{4}} \int_{-\infty}^{\infty} e^{-\left(t\frac{is}{2}\right)^2}dt\right]$$

$$= (e^{-s^2/2}\sqrt{\pi})(e^{-s^2/2}\sqrt{\pi})$$

$$= \pi e^{-s^2/2} \text{ which is same as (1)}$$

Hence convolution theorem is verified.

14.8 RELATION BETWEEN FOURIER AND LAPLACE TRANSFORMS

$$\text{If } f(t) = \begin{cases} e^{-xt} & t>0 \\ 0, & t<0 \end{cases}, \text{ then } F[f(t)] = L[g(t)]$$

Proof : By definition of Fourier transform,

$$F[f(t)] = \int_{-\infty}^{\infty} f(t)\,e^{ist}\,dt$$

$$= \int_{-\infty}^{0} 0.e^{ist}\,dt + \int_{0}^{\infty} e^{-st}.g(t)e^{ist}\,dt$$

$$= \int_{0}^{\infty} e^{-(x-is)t}\,g(t)\,dt$$

$$= \int_{0}^{\infty} e^{-Pt}\,g(t)\,dt \text{ where } P = x - is$$

$$= L[g(t)]$$

Theorem 14.4 Parseval's identity for Fourier Transforms

If the Fourier transforms of $f(x)$ and $g(x)$ are $F(s)$ and $G(s)$ respectively, then

(i) $\frac{1}{2\pi}\int_{-\infty}^{\infty} F(s)\overline{G}(s)\,ds = \int_{-\infty}^{\infty} f(x)\overline{g}(x)\,dx$

(ii) $\frac{1}{2\pi}\int_{-\infty}^{\infty} |F(s)|^2 ds = \int_{-\infty}^{\infty} |f(x)|^2\, dx$ where bar stands for the complex conjugate

Proof : (*i*) By definition of Fourier Transform,

$$F(s) = F[f(x)] = \int_{-\infty}^{\infty} f(x).e^{isx}\,dx$$

and

$$G(s) = F[g(x)] = \int_{-\infty}^{\infty} g(x)e^{isx}\,dx$$

By inversion formula for Fourier Transform

$$f(x) = \frac{1}{2\pi}\int_{-\infty}^{\infty} F(s)e^{-isx}\,dx$$

By inversion formula for Fourier transform

$$f(x) = \frac{1}{2\pi}\int_{-\infty}^{\infty} F(s)e^{-isx}\,dx$$

$$g(x) = \frac{1}{2\pi}\int_{-\infty}^{\infty} G(s).e^{-isx}\,ds$$

$$\therefore \quad \int_{-\infty}^{\infty} f(x)\overline{g}(x)\,dx = \int_{-\infty}^{\infty} f(x)\left[\frac{1}{2\pi}\int_{-\infty}^{\infty}\overline{G}(s).e^{isx}\,ds\right]dx$$

$$= \frac{1}{2\pi}\int_{-\infty}^{\infty}\overline{G}(s)\left[\int_{-\infty}^{\infty} f(x)e^{isx}\,dx\right]ds \quad \text{[Changing the order of integration]}$$

$$= \frac{1}{2\pi}\int_{-\infty}^{\infty}\overline{G}(s).F(s)\,ds$$

$$= \frac{1}{2\pi}\int_{-\infty}^{\infty}(s).\overline{G}(s)\,ds.$$

(*ii*) By definition of Fourier transform

$$F(s) = F[f(x)] = \int_{-\infty}^{\infty} f(x)e^{isx}\,dx$$

By inversion formula for Fourier transform

$$f(x) = \frac{1}{2\pi}\int_{-\infty}^{\infty} F(s)e^{-isx}\,dx$$

$$\int_{-\infty}^{\infty} |f(x)|^2 dx = \int_{-\infty}^{\infty} f(x)\,\overline{f}(x)\,dx$$

$$= \int_{-\infty}^{\infty} f(x)\left[\frac{1}{2\pi}\int_{-\infty}^{\infty} \overline{F}(s).e^{isx}\,ds\right]dx$$

$$= \frac{1}{2\pi}\int_{-\infty}^{\infty} \overline{F}(s)\left[\int_{-\infty}^{\infty} f(x)e^{isx}\,dx\right]ds$$ [Changing order of integration]

$$= \frac{1}{2\pi}\int_{-\infty}^{\infty} \overline{F}(s).[F(s)]\,ds$$

$$= \frac{1}{2\pi}\int_{-\infty}^{\infty} |F(s)|^2 ds$$

Remark : Following Parseval's identify,

(*i*) $\frac{2}{\pi}\int_0^{\infty} F_s(s).G_s(s)\,ds = \int_0^{\infty} f(x)g(x)\,dx$

(*ii*) $\frac{2}{\pi}\int_0^{\infty} [F_s(s)]^2\,ds = \int_0^{\infty} [f(x)]^2\,dx$

(*iii*) $\frac{2}{\pi}\int_0^{\infty} F_c(s)\,Gc(s)\,ds = \int_0^{\infty} f(x)g(x)\,dx$

(*iv*) $\frac{2}{\pi}\int_{\pi}^{\infty} [F_c(s)]^2 ds = \int_0^{\infty} [f(x)]^2 dx$

***Example* 14.19.** Using Parseval's identities, prove that

(*i*) $$\int_0^{\infty} \frac{dt}{(a^2+t^2)(b^2+t^2)} = \frac{\pi}{2ab(a+b)}$$

$$F_c(s) = \frac{a}{a^2+s^2},\ G_c(s) = \frac{b}{b^2+s^2}$$

Let $$f(x) = e^{-ax},\ g(x) = e^{-bx}$$

$$F_c(s) = \frac{a}{a^2+s^2},\ G_c(s) = \frac{b}{(b^2+s^2)}$$

Using Parseval's identity for Fourier cosine transforms,

$$\frac{2}{\pi}\int_0^\infty F_c(s)\,G_c(s)\,ds = \int_0^\infty f(x)\,g(x)\,dx$$

We have

$$\frac{2}{\pi}\int_0^\infty \frac{ab}{(a^2+s^2)(b^2+s^2)}\,ds = \int_0^\infty e^{-(a+b)x}\,dx$$

or

$$\frac{2ab}{\pi}\int_0^\infty \frac{ds}{(a^2+s^2)(b^2+s^2)} = \left|\frac{e^{-(a+b)x}}{-(a+b)}\right|_0^\infty = \frac{1}{a+b}$$

Thus

$$\int_0^\infty \frac{dt}{(a^2+t^2)(b^2+t^2)} = \frac{\pi}{2ab\,(a+b)}$$

(*ii*) Prove that

$$\int_0^\infty \frac{\sin at}{t(a^2+t^2)}\,dt = \frac{\pi\,(1-e^{-a^2})}{a^2}$$

Let

$$f(x) = e^{-ax} \text{ and } g(x) = \begin{bmatrix} 1, & 0<x<a \\ 0, & x>a \end{bmatrix}$$

$$F_c(s) = \frac{a}{s^2+a^2},\; G_c(s) = \frac{\sin as}{s}$$

Using

$$\frac{2}{\pi}\int_0^\infty F_c(s)\,G_c(s)\,ds = \int_0^\infty f(x)\,g(x)\,dx$$

$$\frac{2}{\pi}\int_0^\infty \frac{a\sin as}{(a^2+s^2).s}\,ds = \int_0^a e^{-ax}.\,dx = \left(\frac{1-e^{-a^2}}{a}\right)$$

Thus,

$$\int_0^\infty \frac{\sin at}{t\,(at^2+t^2)}\,dt = \frac{\pi}{2a^2}\,(1-e^{-a^2})$$

Note: If the given integral to be evaluated involves cosine term then we apply Parsevals identity for Fourier sine transform and if sine term is involved we apply Parseval's identity for cosine. This will be more clear in next example.

***Example* 14.20.** Using Parseval's identity, prove that,

(*i*) $\int_0^\infty \left(\frac{1-\cos x}{2}\right)^2 dx = \frac{\pi}{2}$

(*ii*) $\int_0^\infty \frac{\sin^4 t}{t^2}\,dt = \frac{\pi}{4}$

Solution : (*i*) Let $f(x) = \begin{cases} 1, & 0<x<1 \\ 0, & x>1 \end{cases}$. Since given interal involves cosine functions, we consider Fourier sine transform

$$F_s(s) = \frac{1-\cos s}{s}$$

$$F_s(s) = \int_0^\infty f(x).\sin sx\,dx = \int_0^\infty 1.\sin sx\,dx$$

$$= \left|\frac{-\cos sx}{s}\right|_0^\infty = \left|\frac{-\cos sx}{s}\right|_0^1$$

$$= \frac{-\cos s}{s} + \frac{1}{s} = \left(\frac{1-\cos s}{s}\right)$$

$$\frac{2}{\pi}\int_0^\infty |F_s(s)|^2\,ds = \int_0^\infty |f(x)|^2\,dx$$

$$\frac{2}{\pi}\int_0^\infty \left(\frac{1-\cos s}{s}\right)^2 ds = \int_0^\infty |f(x)|^2\,dx$$

$$= \int_0^\infty 1^2\,dx = |x|_0^1 = 1$$

$$\int_0^\infty \left(\frac{1-\cos s}{2}\right)^2 = \frac{\pi}{2}.$$

(*ii*) We know $2\sin^2\frac{x}{2} = (1-\cos x)$

Substituting in (*i*), we get

$$\int_0^\infty \left(\frac{2\sin^2 s/2}{2}\right)^2 = \int_0^\infty \frac{4\sin^4\frac{s}{2}}{s^2}\,ds = \frac{\pi}{2}$$

Let $\frac{s}{2} = t,\ ds = 2\,dt$

$$4\int_0^\infty \frac{\sin^4 t}{4t^2}2dt = \frac{\pi}{2} \Rightarrow \int_0^\infty \frac{\sin^4 t}{t^4}\,dt = \frac{\pi}{4}$$

14.9 FOURIER TRANSFORMS OF THE DERIVATIVES

The Fourier transform of the function $u(x, t)$ is given by

$$F[u(x,t) = \int_{-\infty}^{\infty} u\,e^{isx}\,dx$$

Fourier transform of $\frac{\partial u}{\partial x}$:

Suppose $u \to 0$ as $x \to \pm\infty$, then Fourier transform of $\frac{\partial u}{\partial x}$ is given by.

$$F\left[\frac{\partial u}{\partial x}\right] = \int_{-\infty}^{\infty} \frac{\partial u}{\partial x}.e^{isx}dx$$

$$= \int_{-\infty}^{\infty} e^{isx} \cdot \frac{\partial u}{\partial x} dx$$

Integrating by parts,

$$= [e^{isx} u]_{-\infty}^{\infty} - \int_{-\infty}^{\infty} is\, e^{isx} . u\, dx$$

$$= 0 - is \int_{-\infty}^{\infty} u.e^{isx} dx \qquad [\because u \to 0 \text{ as } x \to \pm \infty]$$

$$= -is\, F(u)$$

Hence $\qquad F\left[\frac{\partial u}{\partial x}\right] = -\text{is } F(u)$

(*ii*) Fourier Transform of $\frac{\partial^2 u}{\partial x^2}$

Suppose u and $\frac{\partial u}{\partial x} \to 0$ as $x \to \pm \infty$, then the Fourier transform of $\frac{\partial^2 u}{\partial x^2}$ is given by

$$F\left[\frac{\partial^2 u}{\partial x^2}\right] = \int_{-\infty}^{\infty} \frac{\partial^2 u}{\partial x^2} e^{isx} dx = \int_{-\infty}^{\infty} e^{isx} \cdot \frac{\partial^2 u}{\partial x^2} dx$$

$$= \left[e^{isx} \frac{\partial u}{\partial x}\right]_{-\infty}^{\infty} - is \int_{-\infty}^{\infty} e^{isx} \frac{\partial u}{\partial x} dx$$

$$= 0\,[-is\, e^{isx} .u\,]_{-\infty}^{\infty} + (is)^2 \int_{-\infty}^{\infty} e^{isx} .u\, dx$$

$$= 0 - s^2 \int_{-\infty}^{\infty} e^{isx} .u\, dx$$

$$= -s^2 F(u)$$

Similarly we find Fourier transform of $\frac{\partial^n u}{\partial x^n}$

Suppose $u, \frac{\partial u}{\partial x}, \frac{\partial^2 u}{\partial x^2}, \ldots . \frac{\partial^{n-1} u}{\partial x^{n-1}} \to 0$ as $x \to \pm\infty$, then the Fourier transform of $\frac{\partial^n u}{\partial x^n}$ is given by,

$$F\left[\frac{\partial^n u}{\partial x^n}\right] = \int_{-\infty}^{\infty} \frac{\partial^n u}{\partial x^n} e^{isx} .dx$$

$$= \int_{\infty}^{\infty} e^{isx} \frac{\partial^n u}{\partial x^n} dx$$

$$= \left[e^{isx} \frac{\partial^{n-1} u}{\partial x^{n-1}} - is\, e^{isx} \frac{\partial^{n-2}}{\partial x^{n-2}} + (is)^2 e^{is} \frac{\partial^{n-3} u}{\partial x^{n-3}} + \ldots . (-is)^{n-1} u\right]_{-\infty}^{x}$$

$$+(-is)^n \int_{-\infty}^{\infty} e^{isx}.u\,dx$$

$$= 0+(-is)^n \int_{-\infty}^{\infty} u.e^{isx}\,dx = (-is)^n.F(u)$$

Hence, $$F\left[\frac{\partial^n u}{\partial x^n}\right] = (-is)^n F(u)$$

Fourier sine and cosine transforms of $\frac{\partial^2 u}{\partial x^2}$:

The Fourier sine and cosine transform of function $u(x, t)$ are given by,

$$F_s[u(x,t)] = \int_0^{\infty} u \sin sx\,dx$$

$$F_c[u(x,t)] = \int_0^{\infty} u \cos sx\,dx$$

Fourier sine transform of $\frac{\partial^2 u}{\partial x^2}$:

Suppose u and $\frac{\partial u}{\partial x} \to 0$ as $x \to \infty$, then the Fourier sine transform of $\frac{\partial^2 u}{\partial x^2}$ is given by

$$F_s\left[\frac{\partial^2 u}{\partial x^2}\right] = \int_0^{\infty} \frac{\partial^2 u}{\partial x^2} \sin sx\,dx = \int_0^{\infty} \sin sx \frac{\partial^2 u}{\partial x^2} dx$$

$$= \left[\sin sx \frac{\partial u}{\partial x}\right]_0^{\infty} - \int_0^{\infty} s \cos sx \frac{\partial u}{\partial x} dx$$

$$= 0 - s\int_0^{\infty} (\cos sx)\frac{\partial u}{\partial x} dx \qquad \left[\because \frac{\partial u}{\partial x} \to 0 \text{ as } x \to \infty\right]$$

$$= -s[(\cos sx.u)]_0^{\infty} - \int_0^{\infty} (-s \sin sx).u\,dx$$

$$= -s\left[0-(u)_{x=0} + s\int_0^{\infty} u \sin sx\,dx\right] \qquad [\because u \to 0 \text{ as } x \to \infty]$$

$$= s(u)_{x=0} - s^2 F_s(u).$$

(*ii*) Fourier cosine transform $\frac{\partial^2 u}{\partial x^2}$

Suppose u and $\frac{\partial u}{\partial x} \to 0$ as $x \to \infty$, then the Fourier transform of $\frac{\partial^2 u}{\partial x^2}$ is given by

$$F^c\left[\frac{\partial^2 u}{\partial x^2}\right] = \int_0^{\infty} \frac{\partial^2 u}{\partial x^2} \cos sx\,dx$$

$$= \left[\cos sx \frac{\partial u}{\partial x}\right]_0^\infty - \int -s \sin sx \frac{\partial u}{\partial x} dx$$

$$= \left[0 - \left(\frac{\partial u}{\partial x}\right)_{x=0}\right] + s \int_0^\infty \sin sx \frac{\partial u}{\partial x} dx$$

$$= -\left(\frac{\partial u}{\partial x}\right)_{x=0} + s \int_0^\partial \sin sx. \frac{\partial u}{\partial x} dx$$

$$= -\left(\frac{\partial u}{\partial x}\right)x = 0 + s\left[\{\sin sx.u\}_0^\infty - \int_0^\infty s \cos sx\, u\, dx\right]$$

$$= -\left(\frac{\partial u}{\partial x}\right)_{x=0} - s^2 \int_0^\infty u \cos sx\, dx$$

$$= -\left(\frac{\partial u}{\partial x}\right)_{x=0} - s^2 F_c(u)$$

Hence $$F_c\left[\frac{\partial^2 u}{\partial x^2}\right] = -\left(\frac{\partial u}{\partial x}\right)_{x=0} - s^2 F_c(u)$$

EXERCISE 14.3

1. Using Parseval's identities, prove that

(i) $\int_0^\infty \frac{d\lambda}{(4+\lambda^2)(9+\lambda^2)} = \frac{\pi}{60}$

(ii) $\int_0^\infty \frac{\lambda^2}{(4+\lambda^2)(9+\lambda^2)} = \frac{\pi}{10}$

(iii) $\int_0^\infty \frac{dx}{(x^2+1)^2} = \frac{\pi}{4}$

(iv) $\int_0^\infty \frac{x^2}{(x^2+1)^2} dx = \frac{\pi}{4}$

2. If $f(x) = \begin{cases} 1, & |x| < a \\ 0, & |x| > a \end{cases}$ and $F(s) = \frac{2 \sin as}{s}$ $(s \neq 0)$, then prove that $\int_0^\infty \frac{\sin^2 ax}{x^2} dx = \frac{\pi a}{2}$

3. Using Parseval's identity, prove that

(i) $\int_0^\infty \frac{\sin^2 x}{x^2} = \frac{\pi}{2}$

(ii) $\int_0^\infty \frac{1 - \cos 2x}{x^2} dx = \pi$

14.10 UNIT STEP FUNCTION

Unit step function if defined as

$$u(t) = \begin{Bmatrix} 0, & t < 0 \\ 1, & t \geq 0 \end{Bmatrix}$$

Also
$$u(t-a) = \begin{Bmatrix} 0, & t < 0 \\ 1, & t \geq 0 \end{Bmatrix}$$

For $a < b$, unit step function if given by:

$$u(t-a) - u(t-b) = \begin{Bmatrix} 0, & t < a \\ 1, & a < t < b \\ 0, & t \geq b \end{Bmatrix}$$

14.11 UNIT IMPULSE OR DIRAC-DELTA FUNCTIONS

Direc-Delta function is defined as

$$\delta_\varepsilon(t-a) = \begin{Bmatrix} 0, & t < a \\ \frac{1}{\varepsilon}, & a \leq t \leq a+\varepsilon \\ 0, & t \geq a+\varepsilon \end{Bmatrix}$$

When a = 0,

$$\delta_\varepsilon(t) = \begin{Bmatrix} 0, & t < 0 \\ \frac{1}{\varepsilon}, & 0 \leq t \leq \varepsilon \\ 0, & t \geq \varepsilon \end{Bmatrix}$$

14.12 FOURIER TRANSFORM OF DIRAC-DELTA FUNCTION

Fourier Transform of Dirac-Delta function is given by:

$$F[\delta(t-a)] = \int_{-\infty}^{\infty} \delta(t-a) e^{ist}\, dt$$

$$= \lim_{\varepsilon \to 0} \int_a^{a+\varepsilon} \frac{1}{\varepsilon} e^{ist}\, dt$$

$$= \lim_{\varepsilon \to 0} \frac{1}{\varepsilon} \int_a^{a+\varepsilon} e^{ist}\, dt$$

$$= \lim_{\varepsilon \to 0} \frac{1}{\varepsilon} \left[\frac{e^{ist}}{is} \right]_a^{a+\varepsilon}$$

$$= \lim_{\varepsilon \to 0} \frac{e^{is(a+\varepsilon)} - e^{isa}}{is\varepsilon}$$

$$= \lim_{\varepsilon \to 0} e^{isa} \left(\frac{e^{is\varepsilon} - 1}{is\varepsilon} \right)$$

$$= e^{isa} \times 1 \qquad \left[\lim_{x \to 0} \frac{e^{x-1}}{x} = 1 \right]$$

$$= e^{isa}.$$

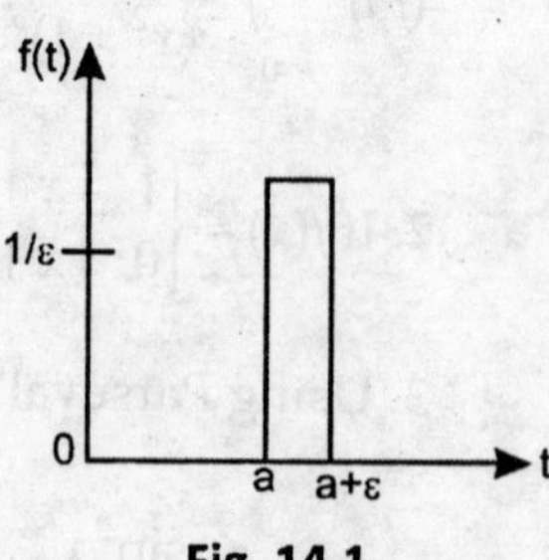

Fig. 14.1

OBJECTIVE QUESTIONS

1. If $f(t) = \begin{cases} 1, & |t| \le 1 \\ 0, & |t| > 1 \end{cases}$ find $F\{f(t)\}$.

2. If $F\{f(t)\} = F(s)$ then what is $F[f(at)]$ equals to

3. Find $F_s(s)$ if $f(x) = e^{-|x|}$

4. Give the formula for $F^{-1}\{1\}$

5. What is the value of $\int_0^\infty \frac{\sin^4 x}{x^2}dx$

6. Find $F_s(n)$ for $f(t) = \pi - t,\ 0 < t < \pi$.

7. Find $F_c(n)$ of $f(t) = t,\ 0 < t < \pi$ for $n = 0$

8. Find $F_s(n)$ for $f(t) = \frac{1}{t},\ t > 0$

9. Find $F_c(s)$ for $(t) = \frac{1}{k} f(t/k)\ ; t > 0,\ k > 0$

10. Find the fourier transform of $f(t) = \begin{cases} 1, & -a \le t \le a \\ 0, & \text{otherwise} \end{cases}$

Answers

1.	Sin s	2.	$F\{f(at)\} = \frac{1}{2} F(s/a),\ a \neq 0$
3.	$\frac{s}{1+s^2}$	4.	$\delta(t)$
5.	$\frac{\pi}{2}$	6.	$\frac{\pi}{n}$
7.	$\frac{\pi^2}{2}$	8.	$\frac{\pi}{2}$
9.	$F_c(sk)$	10.	$2\frac{\sin a\, s}{s}$

Empirical Laws and Curve Fitting

15.1 INTRODUCTION

In engineering and science, it is often required to express a given data obtained from observations in the form of law relating to the two variables involved. Such a law is called the *empirical law.*

In order to find the mathematical relation for the given data, we first plot the given data on sheet of graph paper and then sketch a line that virtually conform to the given data. The curve, thus obtained is called the approximating curve and the algebraic equation of the curve is called Empirical equation. We can draw different types of such approximating curves to the given data and thus equations of different types can be obtained to express the given data.

The curve which is most suitable for predicting the values of an unknown is called the curve of the best fit. The process to find the equation of the curve of 'best fit' is known as *curve fitting.*

If it is desired to obtained an equation representing the given data, we can use any of the four methods, i.e. graphical method, method of group average, method of moments and method of least-square. The graphical method and method of average fail to give the values of the unknown constants uniquely and accurately while the other methods can do. The method of least-square is probably the best to fit a unique curve to a given data.

15.2 GRAPHICAL METHOD

When the curve representing the given data is linear relation $y = mx + c$, we plot the given points on the graph paper using suitable scale. Then draw a line of 'best fit' such that the points are evenly distributed about this line, some on one side and some on the other side. Taking any two points (x_1, y_1) and (x_2, y_2) on the line, calculate m, the slope of the line, i.e. $m = \frac{y_2 - y_1}{x_2 - x_1}$ and c, the intercept on y-axis can be found easily by substituting any point on the line. By substituting the values of m and c, we get the straight line required.

When the points representing the observed values do not approximate to a straight line, a smooth curve is drawn through them and from the shape of the curve, we try to infer the law of the curve and then reduce it to the form $y = mx + c$.

15.3 LAWS REDUCIBLE TO LINEAR LAWS

If the laws of relationship between the variables x and y are not linear laws, reduce them to linear laws by suitable substitutions. Given below are some of the laws of common use to reduce them to linear laws.

1. $y = mx^n + c$, the law which is not linear.

   In order to reduce it to linear law put

$$xn = X \text{ and } y = Y, \text{ getting}$$
$$Y = mX + c, \text{ which is a linear law.}$$

2. $y = ax^n$, the law which is not linear.

Taking logarithms of both sides, it becomes

$$\log_{10}y = \log_{10}a + n\log_{10}x$$

Putting $\log_{10}x = X, \log_{10}a = c$ and $\log_{10}y = Y$

$$Y = c + nx \text{ which is a linear law.}$$

3. $y = ax^n + b \log x$, the law which is not linear,

writes as $\dfrac{y}{\log x} = a\dfrac{x^n}{\log x} + b$

Taking $\dfrac{x^n}{\log x} = X$ and $\dfrac{y}{\log x} = Y$

$$Y = aX + b \text{ which is a linear law.}$$

4. $y = ae^{bx}$, law which is not linear

Taking logarithms,

$$\log_{10}y = (b\log_{10}e)\,x + \log_{10}a$$
$$x = X, \log_{10}y = Y$$
$$Y = mX + c \text{ which is a linear}$$

5. $xy = ax = by$ which is not linear.

We have $y = a + by/x$

Putting $y = Y\,y/x = X$

$$Y = bX + a \text{ which is a linear law.}$$

***Example* 15.1.** Fit a curve of the form $y = ax^2 + b$ to the following data.

x	12	16	20	22	24	26	30
y	6.0	7.5	9.0	9.8	10.7	11.5	14.0

Solution : The given relation is table for X and Y is

$$y = ax^2 = b$$

Putting $x^2 = X, y = Y$

$$Y = aX + b$$

x	**$X = x^2$**	**$y = Y$**
12	144	6.0
16	256	7.5
20	400	9.0
22	484	9.8
24	576	10.7

26	676	11.5
30	900	14.0

Selecting two points (144, 6.0) (900, 14.0)

Slope $$a = \frac{14-6}{900-144} = \frac{8}{756} = 0.0106$$

Hence equation becomes

$$Y = (0.0106)\,X + b$$

Since (144, 6.0) lies on this line

$$6 = (0.0106)\,(144) + b$$

$$b = 4.48 \text{ (approx).}$$

Hence the required from is

$$y = (0.0106)x^2 + 4.48$$

Example **15.2.** The values of x and y obtained in an experiment are as follows

x	2.30	3.10	4.00	4.92	5.91	7.20
y	33.0	39.1	50.3	67.2	85.6	125.0

The probable law is $y = ae^{bx}$. Test graphically the accuracy of this law and if the law holds good, find the best value of the constants.

Solution : Given law is $y = ae^{bx}$.

Taking the logarithm to base 10, we get

$$\log_{10}y = \log_{10}a + (b\log_{10}e)x$$

Putting $$x = X,\ \log_{10}y = Y, \text{ it becomes}$$

$$Y = (b\log_{10}e)\,X + \log_{10}a$$

$$Y = A + BX$$

where $$B = b\log_{10}e,\ A = \log_{10}a$$

Tables of values of X and Y is as follows.

$X = x$	2.30	3.10	4.00	4.92	5.91	7.20
$Y = \log_{10}y$	1.52	1.59	1.71	1.83	1.93	2.1
Points	P_1	P_2	P_3	P_4	P_5	P_6

Scale: 1 small division along x-axis = 0.1

10 small divisions along y-axis = 0.1

Plot these points and draw the line of best fit. As all these point lies along a straight line, the given law is nearly accurate (Fig. 15.1)

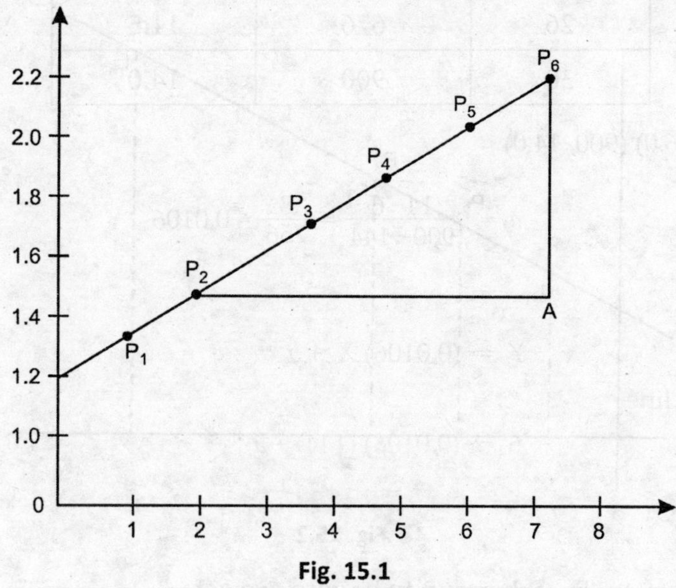

Fig. 15.1

Now slope of this line is

$$B = \frac{P_6A}{P_2A} = \frac{2.1-1.59}{7.2-3.1} = 0.1244$$

Since $P_2(3.1, 1.59)$ lies on (2)

$$1.59 = A + (0.1244)(3.1)$$

$\Rightarrow$ $$A = 1.2044$$

But $$b = B / \log_{10}e = 0.1244 / 0.4343 = 0.2864$$

and $$\log_{10}a = A, \therefore a = \text{anti} \log_{10}A = 16.01$$

Hence the curve of 'best fit' is $y = 16.01\, e^{0.2864x}$.

15.4 METHOD OF GROUP AVERAGES

Let $(x_1, y_1), (x_2, y_2),..., (x_n, y_n)$ be given n sets of observations of related data and let the line be

$$y = ax + b \quad ...(15.1)$$

when $x = x_1$, the observed value of $y = y_1 = M_1P_1$ and from Eq. (15.1)

$$y = ax_1 + b = M_1L_1$$

Whish is known as the expected value or calculated value of y ar M_1.

Then $$e_1 = \text{observed value at } M_1 - \text{expected value at } M_1$$

$$= y_1 - (ax_1 + b) = L_1P_1$$

$$... = ...$$

$$e_n = y_n - (ax^n + b) = L_nP_n$$

Some of the errors may be + ve and some are – ve.

The method of group average is based on the assumption that the sum of the residuals at all points is zero. Since there are two unknowns a, b, we require two equations to find them. So we divide the given data into two groups. Let the first group contains k observations $(x_1, y_1), (x_2, y_2),..., (x_n, y_n)$ and the second group contains the remaining $(n - k)$ observations $(x_{k+1}, y_{k+1}),..., (x_k, y_k)$ (Fig. 15.2).

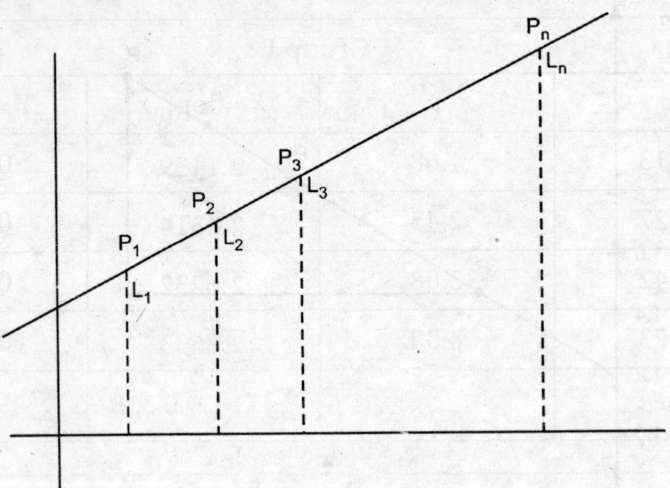

Fig. 15.2

Assuming the sum of the errors in each group to be zero, we get

$$\{y_1 - (ax_1 + b)\} = \{y_2 - (ax_2 + b)\} + ... + \{y_k - (ax_k + b) = 0$$

$$\{y_{k+1} - (ax_{k+1} + b)\} = \{y_{k+2} - (ax_{k+2} + b)\} + ... + \{yn - (ax_n + b)\} = 0$$

On Simplification, we get

$$\frac{y_1 + y_2 + ... + y_k}{k} = b + a\frac{x_1 + x_2 + ... + x_k}{k} \quad ...(15.2)$$

$$\frac{y_{k+1} + y_{k+2} + ... + y_{k+n}}{n-k} = b + a\frac{x_{k+1} + x_{k+2} + ... + x_n}{n-k} \quad ...(15.3)$$

In Eq. (15.2), $\frac{1}{k}(x_1 + x_2 + ... + x_k)$ and $\frac{1}{k}(y_1 + y_2 + ... + y_k)$ are the average values of x' s and y's of the 1st group. Hence the Eqs (15.2) and (15.3) are obtained from Eq. (15.1) by replacing x and y by their respective average of the two groups. Solving Eqs (15.2) and (15.3), we get a and b.

Note: The main drawback of this method is that a different grouping of observations will give different values of a and b, so in practice, we divide the data is such a way that each group contains an even number of observations.

***Example* 15.3:** Using the method of group averages, fir he law of the form $y = ax^b$ to the following data

x	0.013	0.027	0.042	0.073	0.108	0.151	0.233	0.341
y	1.68	2.45	3.08	4.09	4.97	5.95	7.39	9.00

Solution: Given $$y = ax^b \quad ...(i)$$

Taking logarithm on both sides

$$\log_{10}y = \log_{10}a + b\log_{10}x$$

$$y = A + bX \quad ...(ii)$$

where $$Y = \log_{10}y, A = \log_{10}a, X = \log_{10}x$$

We form the following tables:

Group I

x	y	$X = \log_{10}x$	$Y = \log_{10}y$
0.013	1.68	$\bar{2}.1139$	0.2253
0.027	2.45	$\bar{2}.4314$	0.3892
0.042	3.08	$\bar{2}.6232$	0.4886
0.073	4.09	$\bar{2}.8633$	0.6117
		$\sum X = \bar{6}.0.318$	$\sum Y = 1.7148$

Group II

x	y	$X = \log_{10}x$	$Y = \log_{10}y$
0.108	4.97	$\bar{1}.0.334$	0.6964
0.151	5.95	$\bar{1}.1790$	0.7745
0.233	7.32	$\bar{1}.3674$	0.8686
0.341	9.00	$\bar{1}.5328$	0.9542
		$\sum X = \bar{3}.1126$	$\sum Y = 3.2937$

Substituting the average of x and y of two groups in the given relation in Eq. (*ii*)

$$\frac{1.7148}{4} = A + b\,\frac{\bar{6}.0318}{4}$$

$$\frac{3.2937}{4} = A = b\,\frac{\bar{3}.1126}{4}$$

$$4A + (\bar{6}.0318)b = 1.7148$$

$$4A + (\bar{3}.1126)b = 3.2937$$

Solving, we get $A = 1.1897$, $b = 0.51$

This implies a = antilog 1.1897 = 15.49

Hence the equation is $y = (15.49)c(0.51)$

***Example* 15.4.** The latent heat of vapourisation of steam r is given in the following able at different temperature t.

t	40	50	60	70	80	90	100	110
r	1069.1	1063.6	1058.2	1052.7	1049.3	1041.8	1036.3	1030.8

For this range of temperature, a relation of the form $r = a + bt$ is known to fit the data.

Find the value of a and b by he method of group averages.

Solution: Let us divide the data into two groups each containing four readings, then we have

t	r	t	r
40	1069.1	80	0149.3
50	1063.3	90	1041.8
60	1058.2	100	1036.3
70	1052.7	11	1030.8
$\sum t = 220$	$\sum r = 4243.6$	$\sum t = 380$	$\sum r = 4158.2$

Substituting the average of t's and r's of the groups in the give relation, $r = a + bt$

$$\frac{4243.6}{4} = a + b\,\frac{220}{4} \Rightarrow 1060.9 = a + 55b \qquad ...(i)$$

$$\frac{4158.2}{4} = a + b\,\frac{380}{4} \Rightarrow 1039.55 = a + 95b \qquad ...(ii)$$

Solving Eqs (*i*) and (*ii*), we get

$$a = 1090.26$$

$$b = -0.534$$

15.5 LAWS CONTAINING THREE CONSTANTS

So far, we have applied that above methods to fit the data to laws involving two constants only. Now, we shall discuss the laws involving three constants, i.e. $y = a + bx = cx^2$, $y = a + bx^c$ and $a + be^{cx}$. To fit such laws to a set of observations, we proceed as follows to reduce these laws to linear laws.

1. Equating $\qquad y = a + bx + cx^2 \qquad ...(15.4)$

Let (x_1, y_1) be a points on the curve in Eq. (15.4)

$$y_1 = a + bx_1 + cx_1^2 \qquad ...(15.5)$$

From Eqs (15.4) and (15.5), we get

$$y - y_1 = b\,(x - x_1) + c\,(x^2 - x_1^2)$$

$$\frac{y - y_1}{x - x_1} = b + c\,[(x^2 - x_1^2)/\,(x - x_1)] = b + c\,(x + x_1)$$

Putting $\qquad \dfrac{y - y_1}{x - x_1} = Y,\ x + x_1 = X$, we get

$Y = b + cX$. Now b and c can be found by the graphical method.

2. Equating $\qquad y = a + bx^c \qquad ...(15.6)$

It can be written as $y - a = bx^c$

To find a, let (x_1, y_1), (x_2, y_2), (x_3, y_3) by three points on the curve in Eq. (15.4), such that x_1, x_2, x_3, are in geometrical progression

$$\text{i.e.} \quad x_1 x_3 = x_2^2$$

$$\text{Then} \quad (y_1 - a) = bx_1^c$$

$$(y_2 - a) = bx_2^c$$

$$\text{and} \quad (y_3 - a) = bx_3^c$$

$$(y_1 - a)(y_3 - a) = b^2 (x_1 x_3)^c = b^2 x_2^2 c = (y_2 - a)^2$$

$$\text{or} \quad a(y_1 + y_3 - 2y_2) = y_1 y_3 - y_2^2 \Rightarrow a = \frac{y_1 y_3 - y_2^2}{y_1 + y_3 - 2y_2}$$

which gives a, Now Eq. (15.4) reduce to law having two constant b and c only.

Taking logarithms, Eq. (15.4) becomes which can be written as

$$\log_{10} (y - a) = \log_{10} b + c \log_{10} x$$

$$\text{or} \quad Y = B + cX \qquad \text{...(15.7)}$$

$$\text{where} \quad X = \log_{10} x,\ Y = \log_{10}(y - a),\ B = \log_{10} b$$

Hence, we can find b and c as before from Eq. (15.7).

3. Equating

$$y = a + be^{cx}$$

$$y - a = be^{cx}$$

To find a, let (x_1, y_1), (x_2, y_2), (x_3, y_3) be three particular points on the curve in Eq. (15.4), such that x_1, x_2, x_3 are in arithmetical progression,

$$x_1 + x_3 = 2x_2$$

$$\text{Thus} \quad y_1 - a = be^{cx1}$$

$$y_2 - a = be^{cx2}$$

$$y_2 - a = be^{cx2}$$

$$y_3 - a = be^{cx3}$$

$$\text{and} \quad (y_1 - a)(y_3 - a) = b^2 e_c(x_1 \ x_3)$$

$$= (be^{cx2})^2 = (y_2 - a)^2$$

$$a(y_1 + y_3 - 2y_2) = y_1 y_3 - y_2^2$$

i.e. which gives a, now reduce Eq. (15.4) to a low containing two constants a and b taking logarithm in Eq. (15.4) becomes

$$\log_{10}(y - a) = \log_{10} b + cx \log_{10} e$$

$$Y = B + Cx \qquad \text{...(15.8)}$$

$$\text{where} \quad Y = \log_{10} (y - a),\ B = \log_{10} b,\ C = c\log_{10} e$$

Hence, we can find b and c as before from Eq. (15.7).

***Example* 15.5.** Using the method of average, fit a parabola $y = a + bx = cx^2$ to the following data

x	20	40	60	80	100	120
y	5.5	9.1	14.9	22.8	33.3	46.0

Solution : Let the point (20, 5.5) be on the parabola

$$y = a + bx + cx^2 \quad \text{...(i)}$$

$$5.5 = a + 20b + 400c \quad \text{...(ii)}$$

From Eqs (*i*) and (*ii*)

$$\frac{y - 5.5}{x - 20} = b + c(x + 20)$$

$$Y = b + cX$$

$$Y = \frac{y - 5.5}{x - 20}, \; X = x + 20$$

Choosing first three observations as one group and next three observations as another group, we make the following table for the values of X and Y.

For Group I					
x	y	$x-20$	$y-5.5$	$X = x+20$	$Y = y - 5.5/x - 20$
20	5.5	0	0	–	–
40	9.1	20	3.6	60	0.18
60	14.9	40	9.4	80	0.235
				$\sum X = 140$	$\sum Y = 0.415$

$$\overline{X}_1 = \frac{140}{2} = 70, \; \overline{Y}_1 = \frac{0.415}{2} = 0.2075$$

For Group II					
x	y	$x-20$	$y-5.5$	X	Y
80	22.9	60	17.3	100	0.288
100	33.3	80	27.8	120	0.348
120	46	100	40.5	140	0.405
				$\sum X = 360$	$\sum Y = 1.041$

$$\overline{X}_2 = \frac{360}{3} = 120, \; \overline{Y}_2 = \frac{1.041}{3} = 0.347$$

Substituting the average in Eq. (15.7)

$$\overline{Y}_1 = b + c\,\overline{X}_1$$

$$0.2075 = b + c\,(70)$$

$$\overline{Y}_2 = b + c\,\overline{X}_2$$

$$0.347 = b + c\,(120)$$

Solving these $\quad b = 0.0122, \; c = 0.00279$

Substituting in Eq. (15.7) whish is the required curve.

$$\frac{y-5.5}{x-20} = 0.0122 + 0.00279\,(x+20)$$

or

$$y = 0.00279x^2 + 0.0122x + 4.14$$

Example **15.6** In a magnetic arc at constant arc length, voltage v consumed by the arc is observed for values of he current i.

i	0.5	1	2	4	8	12
v	160	120	94	75	62	56

If v and i are connected by he relation of the form $v = ai^b + c$, find a, b and c and hence the curve by the method of group averages.

Solution: Take $i_1 = 1$, $i2 = 2$, $i_3 = 4$ which are in GP

Then $v_1 = 120$, $v_2 = 94$, $v_3 = 75$

$$c = \frac{v_1 v_3 - v_2^2}{v_1 + v_3 - 2v_2} = \frac{(120)(75)-(94)^2}{120+75-2\times 94} = 23.4$$

$$v = ai^b + c \Rightarrow v - 23.4 = ai^b$$

Taking logarithm on both sides which is linear in X and Y.

$$\log_{10}(v = 23.4) = \log_{10}a + b\log_{10}i$$

Putting

$$\log_{10}(v - 23.4) = y;\ \log_{10}a = A;\ \log_{10} i = X$$

$$Y = A + bk \qquad \text{...}(i)$$

To fit Eq (*i*), we use the method of group averages. So we form the table for X and Y.

Group I				
i	v	$v - 23.4$	$Y = \log_{10}(v - 23.4)$	$X = \log_{10} i$
0.5	160	136.6	2.1354	
1	120	96.6	1.9850	0.0000
2	94	70.6	1.8488	0.3010
			$\sum Y = 5.9662$	$\sum X = 0.3010$

Group II				
i	v	$v - 23.4$	$Y = \log_{10}(v - 23.4)$	$X = \log_{10} i$
4	75	51.6	1.7126	0.6021
8	62	38.6	1.5866	0.9031
12	56	32.6	1.5132	0.0792
			$\sum Y = 4.8124$	$\sum X = 2.5844$

Substituting the average of X's and Y's in Eq. (i), we get

$$\frac{5.9692}{3} = A + b\,(0) \Rightarrow 3A = 5.992 \qquad ...(ii)$$

and

$$\frac{4.8124}{3} = A + b \times \frac{2.5844}{3} \qquad ...(iii)$$

$$\Rightarrow \quad 3A + 2.5844b = 4.8124$$

Solving Eqs (ii) and (iii), we get

$$A = 1.9897,\ b = -0.4477$$

i.e.

$$a = 10^A = 10^{1.9897} = 97.66$$

Therefore the required equations is

$$v = (97.66)i^{-(0.4477)} + 23.4$$

***Example* 15.7** The variables s and t are connected by the relation $s = a + be^{ni}$ and their corresponding values are given in the following data.

t	1	2	6	8	11
s	12.71	12.46	11.65	11.34	10.99

Fit the curve by the method of group averages.

Solution : The given curve can be written as

$$s - a = b\,e^{nt} \qquad ...(i)$$

$$Y + B + Nt \qquad ...(ii)$$

where $Y = \log_{10}(s - a)$, $B = \log_{10}b$ and $N = n\log_{10}e$

Take $t_1 = 1$, $t_2 = 6$, $t_3 = 11$ such that $x_1 + x_3 = 2x_2$.

The corresponding values of $s_1 = 12.71$, $s_2 = 11.65$ and $s_3 = 10.99$

$$a = \frac{s_1 s_3 - s_2^2}{s_1 + s_3 - 2s_2} = \frac{(12.71)(10.99) - (11.65)^2}{12.71 + 10.99 - 2(11.65)}$$

$$= \frac{3.9604}{0.4} = 9.091$$

Group I		
t	s	$Y = \log10\ (s - 9.901)$
1	12.71	0.4486
2	12.46	0.4081
6	11.65	0.2448
$\sum t = 9$		$\sum Y = 1.0995$

Group I		
t	s	Y
8	11.34	0.1581
11	10.99	0.0370
$\sum t = 19$		$\sum Y = 0.1951$

Average of the first group are and the average of the second group are

$$\overline{t_1} = \frac{\sum t}{3} = \frac{9}{3} = 3 \ \overline{Y_1} = \frac{\sum Y}{3} = \frac{1.0995}{3} = 0.3665$$

$$\overline{t_2} = \frac{\sum t}{2} = \frac{19}{2} = 9.5 \ \overline{Y_1} = \frac{\sum Y}{2} = \frac{0.1951}{2} = 0.0976$$

Substituting the average of t's and Y's of the group in Eq. (*ii*), we get

$$0.3665 = B + 3N \text{ and } 0.0976 = B + 9.5N$$

On solving, we get $B = 0.4907$, $N = -0.0414$

$$b = \text{antilog}_{10} B = 3.097 \text{ and } n = \frac{N}{\log_{10} e}$$

Hence the required equations is

$$s = 9.901 + 3.097e^{-0.0953t}$$

PROBLEMS

1. Fit a curve of the form $y = ae^{bx}$ to the following data:

x	1	2	3	4	5	6
y	14	27	40	55	68	300

[Ans. $y = 7.973e^{0.5419}x$]

2. The resistance R of carbon filament lamp was measured at various values of voltage V and the following observations were made

V	62	70	78	84	92
R	73	70.7	69.2	67.8	66.3

Assuming a law of the form, find by graphical method of best value of a and b.

[Ans. a 1120, $b = 55.1$]

3. R is resistance to motion of a train at speed V. Find a low of the type $R = aV^2 + b$ connecting R and using the following data

R (kg/ton)	8	10	15	21	30
V (km/h)	10	20	30	40	50

[Ans. $a = 0.0085$, $b = 7.35$]

4. The following table gives corresponding value of x and y. Obtain an equation of the form $y = a + bx$ using the method of grouping

x	0	5	10	15	20	24
y	12	15	17	22	24	30

5. Fit a curve of the form $y = abx$ using the method of group average for the following data.

x	2	4	6	8	10	12
y	7.32	8.24	9.20	10.19	11.01	12.05

[Ans. $y = (6.7468)(1.0505)^x$]

6. Convert the equation $y = b/x\,(x - a)$ to a linear form and hence determine a and b which will fit the following data using the method of group average.

x	8	10	15	20	30	40
y	13	14	15.4	16.3	17.2	17.8

[Ans. $a = 0.2039$, $b = 0.051$]

15.6 PRINCIPLE OF LEAST SQUARES

Curve fitting by (*i*) graphical method, (*ii*) method of group average has already been discussed. But these method have the drawback of being unable to give the unique curve of fit. The value of the constant in the first method depends upon the choice of points on the line and in the second method on grouping the observations. Hence, both the method fails to give the unique curve of fit. The principle of least squares, however, provides a unique set of values to the constant and hence suggests a curve of best fit to the given data.

Let (x_i, y_i) $i = 1, 2,..., n$ be n sets of observations and let $y = f(x)$ be the proposed relation between x and y when $x = x_i$, the observed value of y_i is $y_i = p_iM_i$ and expected value, i.e. $L_iM_i = f(x_i)$.

It is obvious that some value of e_i's may be + ve and – ve. We square each of them and form the sum of the squares for giving equal weightage to each residual, *i.e.*

$$E = \sum_{i=1}^{n} e_i^2 = \sum_{i=1}^{n} [y_i - f(x_i)]^2$$

Now if $E = 0$, then $y_i = f(x_i)$, all points be on the curve. If $E \neq 0$, then the minimum of E results the best fitting curve to the date. Then the curve of best fit is that for which e's are as small as possible, i.e. the sum of the square of the errors is minimum (Fig. 15.3). This is known as the principle of least squares. Using the principle, we shall fir the following curves:

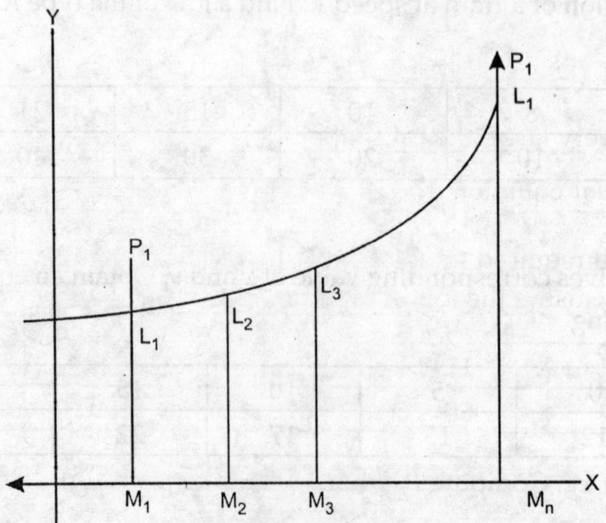

Fig. 15.3

(*i*) A straight line, $y = ax + bx$

(*ii*) A parabola $y = ax^2 + bx + c$

(*iii*) The exponential curve $y = ax^{bx}$ and

(*iv*) The curve $y = ax^b$.

15.7 FITTING A STRAIGHT LINE

Let (x_i, y_i) $i = 1, 2, \ldots, n$ be set of observations of the related data and $y = ax + b$ be the straight line to be fitted. Now, we have to find a and b when $x = x^i$, the residual ax $x = x_i$ is

$$e_i = y_i - f(x_i) = y_i - (ax_i - b); \; i = 1, 2, \ldots, n$$

$$E = \sum_{i=1}^{n} e_i^2 = \sum_{i=1}^{n} [y_i - (ax_i + b)]^2$$

By the principle of least squares, E is minimum

$$\frac{\partial E}{\partial a} = 0, \; \frac{\partial E}{\partial b} = 0$$

$$\sum_{i=1}^{n} 2\,(y_i - ax_i - b)(-x_i) = \sum_{i=1}^{n} (y_i - ax_i - bx_i) = 0$$

and
$$\sum_{i=1}^{n} 2\,(y_i - ax_i - b)(-1) = \sum_{i=1}^{n} (y_i - ax_i - b) = 0$$

$$\therefore \quad \sum_{i=1}^{n} y_i x_i = a \sum_{i=1}^{n} (x_i^2 + b \sum_{i=1}^{n} x_i \qquad \text{...(15.9)}$$

and
$$\sum_{i=1}^{n} y_i = a \sum_{i=1}^{n} x_i + nb \qquad \text{...(15.9a)}$$

Since x_i, y_i are known, on solving Eqs (15.9) and (15.9a), we can find a and b.

Note: 1. Equations (15.9) and (15.9a) are called the normal equation and by dropping the suffixes, they can be written as

$$\sum y = a \sum x + nb$$

$$\sum xy = a \sum x^2 + b \sum x$$

2. If the values x and y are very large, the calculation can be made easy by using the transformation $X = \frac{x-p}{h}$, $Y = \frac{y-q}{k}$ which reduce the linear equation $y = ax + b$ to form $Y = AX + B$.

***Example* 15.8** If P is the pull required to fit a load W by means of a pulley block, find a linear law of the form $P = mW + c$ connecting P and W, using the following data

P	12	15	21	25
W	50	70	100	120

when P and W are taken in kg. Compute P, when $W = 150$ kg.

Solution: The law is $\qquad P = mW + c$

The corresponding normal equation are

$$\sum P = 4c + m\sum W \ (i)$$

Here $$n = 4 \sum WP = c\sum W + m\sum W^2 \ (ii)$$

The values of $\sum W$ and relative value can be calculated by means of following table.

W	P	W^2	WP
50	12	2500	600
70	15	4900	1050
100	21	10000	2100
120	25	14400	3100

Substituting in Eqs (*i*) and (*ii*), we have

$$73 = 4c + 340m \qquad ...(iii)$$

and $$6750 = 340c + 31800 \qquad ...(iv)$$

From Eqs (*iii*) and (*iv*), we get

$$m = 0.1879, c = 2.2785$$

Hence the line of best fit is

$$P = 2.2785 + 0.1879W$$

When $W = 150$ kg, $$P = 2.2785 + 0.1879 \times 150 = 30.4635 \text{ kg.}$$

Example **15.9** Fit a straight line to the following, using method of least square.

x	1	2	3	4	5
y	14	27	40	55	68

Solution: Method I: Let the straight line of best fit be

$$y = ax + b \qquad ...(i)$$

The normal equations are

$$\sum xy = a\sum x^2 + b\sum x \qquad ...(ii)$$

$$\sum y = a\sum x + 5b \qquad ...(iii)$$

The value of $\sum x, \sum y, \sum x^2, \sum xy$ are calculated as below:

x	y	x^2	xy
1	14	1	14
2	27	4	54
3	40	9	120
4	55	16	220
5	68	25	340
$\sum x = 15$	$\sum y = 204$	$\sum x^2 = 55$	$\sum xy = 748$

Normal Eqs (*ii*) and (*iii*) becomes

$$748 = 55a = 15b$$

$$204 = 15a + 5b$$

Solving, we get $a = 13.6, b = 0$

Putting these values in Eq. (*i*), we get the line of best fit as $y = 13.6x$

Method II: Let $X = x - 3$, $Y = y - 40$ for making the calculation simpler.

Let the straight line is new variables be

$$Y = AX + B$$

Now the normal equations are

$$\sum XY = A\sum X^2 + B\sum X$$

$$\sum Y = A\sum X + 5B$$

The value of $2\sum X, \sum Y, \sum X^2, \sum XY$ are calculated as below:

x	y	$X = x - 3$	$Y = y - 40$	X^2	XY
1	14	–2	–26	4	52
2	27	–1	–13	1	13
3	40	0	0	0	0
4	55	1	15	1	15
5	68	2	28	4	56
		$\sum X = 0$	$\sum Y = 4$	$\sum X^2 = 10$	$\sum XY = 136$

Substituting these values, we get

$$136 = 10A$$

$$4 = 5B$$

or $A = 13.6, B = 0.8$

The equation is $Y = 13.6X + 0.8$

$$y - 40 = 13.6\,(x - 3) + 0.8$$

$$y = 13.6x \text{ which is the same as above.}$$

PROBLEMS

1. Fit a straight line to the following data by the method of least squares.

x	–4	1	2	3
y	4	6	10	8

[**Ans.** $y = 0.7x + 6.65$]

2. Fit a straight line to the following data by the method of least square.

x	5	10	15	20	25
y	16	19	23	26	30

[Ans. $y = 0.7x + 123$]

3. The table below give the temperature T (in °C) and length l (in mm) of heated rod. If $l = 10T + a_1$, find the best value of a_0, a_1.

x	20	30	40	50	60	70
y	800.3	800.4	800.6	800.7	800.9	801.0

[Ans. $0.0146T + 800$]

4. Fit a straight line to the following data.

x	71	68	73	69	67	65	66	67
y	69	72	70	70	68	67	68	64

Also find y at $x = 70$ [Ans. $y = 0.424x + 37.908$, $y = 67.6030$]

15.8 FITTING A PARABOLA OR A SECOND DEGREE CURVE

Let (x_i, y_i), $[i = 1, 2, 3,...,n]$ be the given n sets of observation of the related variables x, y and let $y = ax^2 + bx + c$ be the equation which fits the observations best. We have to find a, b, c,.

For $x = x_i$, the expected value of y is $ax_i^2 + bx_i + c$ and observed value of y is y_i.

The residual $e = y_i - (ax_i^2 + bx_i + c)$

Let E denote the sum of the squares of the residuals

$$E = \sum_{i=1}^{n} [y_i - (ax_i^2 + bx_i + c)]^2 \quad ...(15.10)$$

By the principle of least squares, E should be minimum for which the conditions are

$$\frac{\partial E}{\partial a} = 0, \frac{\partial E}{\partial b} = 0, \frac{\partial E}{\partial c} = 0$$

Therefore, differentiate on both sides of Eq. (15.10) partially w.r.t. a, b and c, we get

$$\sum_{i=1}^{n} 2[y_i - (ax_i^2 + bx_i + c)](-x_i^2) = 0$$

$$\sum_{i=1}^{n} 2[y_i - (ax_i^2 + bx_i + c)](-x_i) = 0$$

$$\sum_{i=1}^{n} 2[y_i - (ax_i^2 + bx_i + c)]\,(-1) = 0$$

i.e.
$$\sum_{i=1}^{n} (x_i^2 y_i - ax_i^4 - bx_i^3 - cx_i^2) = 0$$

$$\sum_{i=1}^{n}(x_i y_i - a x_i^3 - b x_i^2 - c x_i) = 0$$

$$\sum_{i=1}^{n}(y_i - a x_i^2 - b x_i - c) = 0$$

Simplifying and dropping the suffixes, we get

$$\Sigma y = a\Sigma x^2 + b\Sigma x + \Sigma c$$

$$\Sigma xy = a\Sigma x^3 + b\Sigma x^2 + c\Sigma x$$

$$\Sigma x^2 y = a\Sigma x^4 + b\Sigma x^3 + c\Sigma x^2$$

Solving the above normal equations, we get the values of a, b and c. Substituting these values of a, b and c in $y = ax^2 + bx + c$, we get the least fitting parabola.

***Example* 15.10.** Fit a second degree parabola to the data

x	1929	1930	1931	1932	1933	1934	1935
y	352	356	357	358	360	361	361

Solution : Let be the best $y = ax^2 + bx + c$ ...(*i*)

Since the values of x and y are large, we choose the origins for x and y at 1932 and 357 respectively. In other words, we transform x and y.

$$X = x - 1932,\ Y = y - 357, \text{ then Eq. } (i) \text{ can be reduced to}$$

$$Y = aX^2 + bX + c$$

The normal equations are

$$\Sigma Y = a\Sigma X^2 + b\Sigma X + nc$$

$$\Sigma XY = a\Sigma X^3 + b\Sigma X^2 + c\Sigma X$$

$$\Sigma X^2 Y = a\Sigma X^4 + b\Sigma X3 + c\Sigma X^2$$

We form the table:

x	y	$X = x - 1932$	$Y = y - 357$	X^2	X^3	X^4	XY	X^2Y
1929	352	− 3	− 5	9	− 27	81	15	− 45
1930	356	− 2	− 1	4	− 8	16	2	− 4
1931	357	− 1	0	1	− 1	1	0	0
1932	358	0	1	0	0	0	0	0
1933	360	1	3	1	1	1	3	3
1934	361	2	4	4	8	16	8	16
1935	361	3	4	9	27	81	12	36
		$\Sigma X = 0$	$\Sigma Y = 5$	$\Sigma X^2 = 28$	$\Sigma X^3 = 0$	$\Sigma X^4 = 196$	$\Sigma XY = 40$	$\Sigma X^2 Y = 6$

Hence the normal equations became

$$28a + 7c = 6$$

$$28b = 40$$

$$196a + 28c = 6$$

On solving, we get $b = 1.4286$, $a = -0.21429$, $c = 1.7143$.

Equation (*ii*) becomes

$$Y = -0.21429X^2 + 1.4286X + 1.7143$$

$$y - 357 = -0.21429(x - 1932)^2 + 1.4286(x - 1932) + 1.7143$$

$$y = -0.21429x^2 + 829.445x - 802265.33 \text{ is the required equation.}$$

***Example* 15.11.** Find a second degree parabola $y = a + bx + x^2$ to the following data :

x	1	3	5	7	9
y	2	7	10	11	9

Solution : Method I : Normal equations are

$$\Sigma y = 5a + b\Sigma x + c\Sigma x^2$$

$$\Sigma xy = a\Sigma x + b\Sigma x^3 + c\Sigma x^3$$

$$\Sigma x^2 y = a\Sigma x^2 + b\Sigma x^3 + c\Sigma x^4$$

We form the table:

x	y	x^2	x^3	x^4	xy	x^2y
1	2	1	1	1	2	2
3	7	9	27	81	21	63
5	10	25	125	625	50	250
7	11	49	343	2401	77	539
9	9	81	729	6561	81	729
$\Sigma x = 25$	$\Sigma y = 39$	$\Sigma x^2 = 165$	$\Sigma x3 = 1225$	$\Sigma x^4 = 9669$	$\Sigma xy = 231$	$\Sigma x^2 y = 1583$

Putting the values in equations (normal), we obtain a, b, c

$$a = -1.556,\ b = 3.757,\ c = -0.2857$$

Hence the fitted equation is

$$y = -1.556 + 3.757x - 0.2857x^2$$

Method II: The above can also be solved by using the change of origin and scale. Thus, we have to take

$$X = \frac{x-5}{2} \text{ and the fitted line will be}$$

$$y = a_1 + b_1 X + c_1 X^2$$

$$= a_1 + b_1\left(\frac{x-5}{2}\right) + c_1\left(\frac{x-5}{2}\right)^2$$

$$y = A + BX + CX^2$$

Proceed as above and find A, B and C.

***Example* 15.12.** Fit a second degree parabola to the following data using method of least squares.

x	**0**	**1**	**2**	**3**	**4**
y	1	1.8	1.3	2.5	6.3

Solution : Let $y = ax^2 + bx + c$ be the parabola

The normal equations are

$$\Sigma y = a\Sigma x^2 + b\Sigma x + 5c$$

$$\Sigma xy = a\Sigma x^3 + b\Sigma x^2 + c\Sigma x$$

$$\Sigma x^2 y = a\Sigma x^4 + b\Sigma x^3 + c\Sigma x^2$$

The values Σx, Σx^2, ... etc. are calculated by means of the following table:

x	y	x^2	x^3	x^4	xy	x^2y
0	1	0	0	0	0	0
1	1.8	1	1	1	1.8	1.8
2	1.3	4	8	16	2.6	5.2
3	2.5	9	27	81	7.5	22.5
4	6.3	16	64	256	25.2	100.8
$\Sigma x = 10$	$\Sigma y = 12.9$	$\Sigma x^2 = 30$	$\Sigma x3 = 100$	$\Sigma x^4 = 354$	$\Sigma xy = 37.1$	$\Sigma x^2 y = 130.3$

Substituting the obtained values in the normal equations, we get

$$12.9 = 30a + 10b + 5c$$

$$37.1 = 100a + 30b + 10c$$

$$130.3 = 354a + 100b + 30c$$

On solving, we get $a = 0.55, b = -1.07, c = 1.42.$

The required parabola is

$$y = 0.55x^2 - 1.07x + 1.42$$

***Example* 15.13.** Fit a second degree parabola to the following data.

x	**1.0**	**1.5**	**2.0**	**2.5**	**3.0**	**3.5**	**4.0**
y	1.1	1.3	1.6	2.0	2.7	3.4	4.1

Solution : We shift the origin to (2.5, 0). This results in change of the variable x to X by the relation

$$X = \frac{x - 2.5}{0.5} = \frac{x - \frac{5}{2}}{\frac{1}{2}} = 2x - 5$$

Let the parabola of fit be $y = a + bX + cX^2$.

The values of ΣX and other corresponding values are calculated as below.

x	X	y	Xy	X^2	X^2y	X^3	X^4
1.0	− 3	1.1	− 3.3	9	9.9	− 27	81
1.5	− 2	1.3	− 2.6	4	5.2	− 8	16
2.0	− 1	1.6	− 1.6	1	1.6	− 1	1
2.5	0	2.0	0.0	0	0.0	0	0

3.0	1	2.7	2.7	1	2.7	1	1
3.5	2	3.4	6.8	4	13.6	8	16
4.0	3	4.1	12.3	9	36.9	27	81
$\Sigma X = 0$	$\Sigma y = 16.2$	$\Sigma Xy = 14.3$	$\Sigma X^2 = 28$	$\Sigma X^2 y = 69.9$	$\Sigma X^3 = 0$	$\Sigma X^4 = 196$	

The normal equations are

$$7a + 28c = 16.2$$

$$28b = 14.3$$

$$28a + 196c = 69.9$$

Solving these equations, we get

$$a = 2.07, b = 0.511, c = 0.061$$

$$y = 2.07 + 0.511X + 0.061X^2$$

Replacing X by $2x - 5$, we get

$$y = 2.07 + 0.511\,(2x - 5) + 0.061\,(2x - 5)2$$

or

$$y = 1.04 - 0.198x + 0.244x^2$$

This is the required parabola of 'best fit'.

15.9 FITTING OF OTHER CURVES

$$y = ax^b$$

Taking logarithm $\log_{10} = \log_{10} a + b \log_{10} x$

i.e. $Y = A + bX$, where $Y = \log_{10} y$, $A = \log_{10} a$, $X = \log_{10} x$

The normal equations are therefore

$$\Sigma Y = \Sigma A + b\Sigma X$$

$$\Sigma XY = A\Sigma X + b\Sigma X^2$$

from which A and b can be determined. Then a can be calculated from $A = \log_{10} a$.

1. $$y = ae^{bx}$$

Taking logarithm $\log_{10} y = \log_{10} a + bx \log_{10} e$

or $$Y = A + Bx$$

where $$Y = \log_{10} y, A = \log_{10} a, B = b \log_{10} e$$

Hence the normal equations are

$$\Sigma Y = nA + B\Sigma X$$

$$\Sigma xY = A\Sigma x + B\Sigma x^2$$

from which A and B can be found and consequently a, b can be calculated.

2. $$xy^a = b$$

Taking logarithm, we get

$$\log_{10} x + a \log_{10} y = \log_{10} b$$

or $$\log_{10} y = 1/a \log_{10} b - 1/a \log_{10} x$$

This is of the form $$Y = A + BX$$

where $$y = \log_{10} y, X = \log_{10} x, A = 1/a \log_{10} b, B = -1/a$$

Here the normal equations are

$$\Sigma Y = nA + B\Sigma X$$

$$\Sigma XY = A\Sigma X + B\Sigma X^2$$

from which A and B can be determined and hence a, b are calculated.

***Example* 15.14.** Apply the method of least squares to determine the constants a and b such that $y = ae^{bx}$ fits the following data.

x	0.0	0.5	1.0	1.5	2.0	2.5
y	0.10	0.45	2.15	9.15	40.35	180.75

Solution : The curve to be fitted is $y = ae^{bx}$

or $Y = A + Bx$ when $Y = \log_{10} y$

$$A = \log_{10} a,\ B = b \log_{10} e$$

The normal equations are

$$\Sigma Y = 6A + B\Sigma x$$

$$\Sigma xY = A\Sigma x + B\Sigma x^2$$

x	y	$Y = \log_{10} y$	x^2	xY
0	0.10	– 1	0	0
0.5	0.45	– 0.3468	0.25	– 0.1734
1.0	2.15	0.3324	1.0	0.3324
1.5	9.15	0.9614	2.25	1.4421
2.0	40.35	1.6058	4.00	3.2116
2.5	180.75	2.2571	6.25	5.6428
$\Sigma x = 7.5$		$\Sigma Y = 3.8099$	$\Sigma x^2 = 13.75$	$\Sigma xY = 10.4555$

Substituting the values in normal equations, we get

$$3.8099 = 6A + 7.5B \text{ and } 10.4555 = 7.5A + 13.755B$$

On solving and the curve is $A = -0.9916,\ B = 13013$

$$a = \text{antilog}_{10} A = 0.1019\ b = B/\log_{10} e = 2.9963$$

$$y = 0.1019 \exp(2.9963x)$$

***Example* 15.15.** Fit $y = ab^x$ by the method of least square to the data given below.

x	0	1	2	3	4	5	6	7
y	10	21	35	59	92	200	400	610

Solution : The curve to be fitted is $y = ab^x$ or $Y = A + Bx$

where $A = \log_{10} a,\ B = \log_{10} b$ and $Y = \log_{10} y$

The normal equations are

$$\Sigma Y = 8A + B\Sigma x$$

$$\Sigma xY = A\Sigma x + B\Sigma x^2$$

x	y	$Y = \log_{10} y$	x^2	xY
0	10	1.0000	0	0
1	21	1.3222	1	1.3222
2	35	1.5441	4	3.0882
3	59	1.7708	9	5.3124
4	92	1.9638	16	7.8552
5	200	2.3010	26	11.5050
6	400	2.6021	36	15.6126
7	610	2.7853	49	19.4971
$\Sigma x = 28$		$\Sigma Y = 15.2893$	$\Sigma x^2 = 140$	$\Sigma xY = 64.1927$

Substituting the values in normal equations, we get

$$15.2893 = 8A + 28B,\ 64.1927 = 28A + 140B$$

On solving

$$A = 1.02115,\ B = 0.2542892$$

$$a = \text{antilog}_{10} A = 10.499 \text{ and } b = \text{antilog } B = 1.7959$$

Hence the required curve is $y = 10.499\ (1.7959)^x$.

***Example* 15.16.** The pressure and volume of a gas related by equation $pV^r = kr$, k being constant. Fit the equation to the following set of observations.

p **(kg/cm²)**	0.5	1.0	1.5	2.0	2.5	3.0
V **(litres)**	1.62	1.00	0.75	0.62	0.52	0.46

Solution : The law is $pV^r = k$

Taking logarithm on both sides, we get

$$\log_{10} p + r \log_{10} V = \log_{10} k$$

$$\log_{10} V = 1/r \log_{10} k - 1/r \log_{10} p \text{ or } Y = A + BX$$

where

$$X = \log_{10} p,\ Y = \log_{10} V,\ A = 1/r \log_{10} k,\ B = -1/r$$

The normal equations are

$$\Sigma Y = 6A + B\Sigma X \quad ...(i)$$

$$\Sigma XY = A\Sigma X + B\Sigma X^2 \quad ...(ii)$$

Now ΣX are calculated as follows.

p	V	$X = \log_{10} p$	$Y = \log_{10} V$	XY	X^2
0.5	1.62	– 0.3010	0.2095	– 0.0640	0.0906
1.0	1.00	0.0000	0.000	– 0.0000	0.0000
1.5	0.75	0.1761	– 0.1249	– 0.0220	0.0310
2.0	0.062	0.3010	– 0.2076	– 0.0625	0.0906
2.5	0.52	0.3979	– 0.2840	0.1130	0.1583
3.0	0.46	0.4771	– 0.3372	– 0.1699	0.2276
		$\Sigma X = 1.0511$	$\Sigma Y = -0.7442$	$\Sigma XY = -0.4214$	$\Sigma X^2 = 0.5981$

Equation (*i*) and (*ii*) becomes

$$6A + 1.0511B = -0.7442$$

$$1.0511A + 0.5981B = -0.4214$$

On solving, we get $A = 0.0132,\ B = -0.7836,\ r = -\frac{1}{B} = 1.276$

$$k = \text{antilog}\ (Ar) = \text{antilog}\ (0.0168) = 1.039$$

Hence the equation of the best fit is

$$pV^{1.276} = 1.039$$

***Example* 15.17.** Determine the constants a and b by the method of least squares such that $y = ae^{bx}$ fits the following data.

x	2	4	6	8	10
y	4.077	11.084	30.128	81.897	222.62

Solution : The given relation is $y = ae^{bx}$.

Taking logarithm on both sides, we get

$$\log_{10} y = \log_{10} a + bx$$

Setting $\log_{10} y = Y,\ x = X$

$$\log_{10} a = A \text{ and } b = B$$

We get $Y = A + BX$ which is strong line.

We have the following table.

$X = x$	$Y = \log_{10} y$	X^2	XY
2	1.405	4	2.810
4	2.405	16	9.620
6	3.405	36	20.430
8	4.405	64	35.240
10	5.405	100	54.050
$\Sigma X = 30$	$\Sigma Y = 17.025$	$\Sigma X^2 = 220$	$\Sigma XY = 122.150$

Using normal equations, we get

$$5A + 30B = 17.025$$

$$30A + 220B = 122.150$$

On solving, we get $A = 0.405,\ B = 0.5$

Hence $a = e^A = e^{0.405} = 1.4999$

$$b = B = 0.5$$

***Example* 15.18.** Fit the curve of the form $y = ax^b$ to the data:

x	1	2	3	4	5	6
y	1200	900	600	200	110	50

Solution : Given $y = ax^b$

Taking logarithm on both sides, we get

$$\log_{10} y = \log_{10} a + \log_{10} x$$

Putting $Y = \log_{10} y,\ X = \log_{10} x,\ A = \log_{10} a$, we get

$Y = A + bX$ which is linear in X and Y

The normal equations are

$$\Sigma Y = 6A + b\Sigma X \quad ...(i)$$

$$\Sigma XY = A\Sigma X + b\Sigma X^2 \quad ...(ii)$$

Now ΣX are calculated as follows:

x	y	$X = \log_{10} x$	$Y = \log_{10} y$	X^2	Y^2
1	1200	0.0000	3.0792	0.0000	0.0000
2	900	0.3010	2.9542	0.0906	0.8892
3	600	0.4771	0.7782	0.2276	1.3255
4	200	0.6021	2.3010	0.3625	1.3854
5	110	0.6990	2.0414	0.4886	1.4269
6	50	0.7781	1.6990	0.6064	1.3220
		$\Sigma X = 2.8573$	$\Sigma Y = 14.8530$	$\Sigma X^2 = 1.7747$	$\Sigma XY = 6.3490$

The normal equation become

$$14.8530 = 6A + 2.8573b$$

$$6.3490 = 2.8573A + 1.7747b$$

On solving, we get $A = 3.3086,\ b = -1.749$

$$a = \text{antilog}\ (3.3086) = 2035$$

The required equation is

$$y = (2035)x^{-1.7494}.$$

PROBLEMS

1. Fit a parabola to the data:

x	0	0.1	0.2	0.3	0.4	0.5	0.6	0.7	0.8	0.9
y	3.1950	3.2299	3.2532	3.2611	3.2516	3.2282	3.1807	3.1266	3.0594	2.9759

[Ans. $y = -0.7653x^2 + 0.4425x + 3.1951$]

2. Fit a parabola to the data:

x	2	4	6	8	10
y	3.07	12.85	31.47	57.38	57.38

[Ans. $y = 0.99x^2 - 83x + 0.61$]

3. Fit a parabola to the data:

x	1	2	3	4	5
y	5	12	26	60	97

[Ans. $y = 5.7x^2 - 11x + 10.3$]

4. Fit a parabola to the data:

x	1	2	3	4	5
y	2	3	5	8	10

[Ans. $y = 0.403 + 0.388x - 2.79x^2$]

5. Fit a parabola to the data:

Year x	**1911**	**1912**	**1913**	**1914**	**1915**
Commodity y	10	12	8	10	14

[Ans. $y = 9.4 + 0.6\,(x - 1913) + 0.7\,(x - 1913)^2$]

6. Fit a parabola to the data:

Speed v (ft/min)	**350**	**400**	**500**	**600**
Life t (min.)	61	26	7	2.6

Find the best values of a and b in the law $v = ae^{bt}$ by the method of least squares.

[Ans. $V = (553.4)e^{-(0.0083)t}$]

7. Fit a law of the type $y = ae^{bx}$ to the data:

x	0	1	2	3
y	1.05	2.10	3.85	8.30

[Ans. $y = 4.64\, e^{0.46x}$]

8. Fit a curve of the form $y = ax^b$ to the given data:

x	61	21	7	2.6
y	350	400	500	600

[Ans. $y = (682.3)x^{-(0.1588)}$]

9. Fit a curve of the form $y = ax^b$ to the following data:

x	2	3	4	5	6
y	144	172.8	207.4	248.8	298.5

[Ans. $100\,(1.2)^x$]

15.10 CURVE FITTING BY A SUM OF EXPONENTS

A frequently encountered problem in engineering and physics is that of fitting a sum of exponentials of the form to a set of data points, say $(x_1, y_1), (x_2, y_2), ..., (x_n, y_n)$. Assuming that n is known and $a_1, a_2, ..., a_n, \lambda_1, \lambda_2, ..., \lambda_n$ are unknowns and which are to be determined.

$$y = f(x) = a_1 e^{\lambda_1 x} + a_2 e^{\lambda_2 x} + ... + a_n e^{\lambda nx} \quad ...(15.11)$$

It is clear that $f(x)$ satisfies a differential equation of the type where $A'_i s$ are all unknown.

$$\frac{d^n y}{dx^n} + A_1 \frac{d^{n-1} y}{dx^{n-1}} + A_2 \frac{d^{n-2} y}{dx^{n-2}} + ... + A_n y = 0 \quad ...(15.12)$$

To solve the Eq. (15.12), Frogberg suggested a method, in which we numerically evaluate the derivative $\frac{d^n y}{dx^n}$, $\frac{d^{n-1} y}{dx^{n-1}}$ at the n data points and substituting them in Eq. (15.12), we therefore, obtain a system of n linear equations for the n unknowns $A_1, A_2, ..., A_n$ which can then be solved.

Also, it can be verified that $\lambda_1, \lambda_2, ..., \lambda_n$ are the roots of the algebraic equation which when solved enables us to determine $a_1, a_2, ..., a_n$ from Eq. (15.11) by the method of least squares.

$$\lambda^n + A_1 \lambda^{n-1} + A_2 \lambda^{n-2} + ... + A_n = 0 \quad ...(15.13)$$

Disadvantage: Due to the numerical evaluations of higher derivatives whose accuracy increases as we evaluate higher order derivatives and leading to unreliable results.

In 1974, Moore described a computational technique which leads to more reliable results. Now demonstrate the method for the case $n = 2$.

Let the function to be fitted to a given data of the form which satisfies a differential equation where A_1, A_2 are to be determined.

$$Y = a_1 e^{\lambda_1 x} + a_2 e^{\lambda_2 x} + ... \quad ...(15.14)$$

$$\frac{d^2 y}{dx^2} = A_1 \frac{dy}{dx} + A_2 y \quad ...(15.15)$$

Assuming that x_0 is the initial value of x, we integrating Eq. (15.15) from x_0 to x, we get

$$y'(x) - y'(x_0) = A_1 y(x) - A_1 y(x_0) + A_2 \int_{x_0}^{x} y(x)\,dx \quad ...(15.16)$$

where $$y' = \frac{dy}{dx}$$

Again integrating Eq. (15.16) from x_0 to x, we get

$$y(x) - y(x_0) - (x - x_0)\, y'(x_0) = A_1 \int_{x_0}^{x} y(x)\,dx - A_1 (x - x_0)\, y(x_0) + A_2 \int_{x_0}^{x}\int_{x_0}^{x} y(x)\,dx... \quad ...(15.17)$$

Since we know that

$$\int_{x_0}^{x}\int_{x_0}^{x} f(x)\,dx...dx = \frac{1}{(n-1)!}\int_{x_0}^{x} (x-t)^{n-1} f(t)\,dt \quad ...(15.18)$$

Using Eq. (15.18), Eq. (15.17) becomes

$$y(x) - y(x_0) - (x - x_0)\, y'(x_0) = A_1 \int_{x_0}^{x} y(x)\,dx - A_1 (x - x_0)\, y(x_0) + A_2 \int_{x_0}^{0} (x - t)\, y(t)\,dt \quad ...(15.19)$$

In order to use the Eq. (15.19) to set up a linear system for A_1 and A_2, we eliminate $y'(x_0)$ in the following way. Choose two data points x_1 and x_2 such that

$$x_0 - x_1 = x_2 - x_0, \text{ we obtain from Eq. (15.59)}$$

$$y(x_1) - y(x_0) - (x_1 - x_0)\, y'(x_0) = A_1 \int_{x_0}^{x_1} y(x)\,dx - A_1 (x - x_0) y(x_0) + A_2 \int_{x_0}^{x_1} (x_1 - t) y(t)\,dt \quad ...(15.20)$$

and $$y(x_2) - y(x_2) - (x_2 - x_0)\, y'(x_0) = A_1 \int_{x_0}^{x_1} y(x)\,dx - A_1 (x_2 - x_0) y(x_0) + A_2 \int_{x_0}^{x_1} (x_2 - t) y(t)\,dt \quad ...(15.21)$$

Adding Eqs (15.20) and (15.21) and satisfying, we get

$$y(x_1) + y(x_2) - 2y(x_0) = A_1 \left[\int_{x_0}^{x_1} y(x)\,dx + \int_{x_0}^{x_2} y(x)\,dx \right] + A_2 \left[\int_{x_0}^{x_1} (x_1 - t) y(t)\,dt + \int_{x_0}^{x_2} (x_2 - t) y(t)\,dt \right]$$

...(15.22)

In Eq. (15.22), we get a linear system of equations for A_1 and A_2 and then we obtain λ_1 and λ_2 from the characteristic equation

$$\lambda^2 = A_1 \lambda + A_2$$

Finally, a_1 and a_2 can be obtained by the averaging technique.

Example 15.19: Fit a function of the form $y = a_1 e^{\lambda_1 x} + a_2 e^{\lambda_2 x}$ to the data given by

x	1.0	1.1	1.2	1.3	1.4	1.5	1.6	1.6	1.8
y	1.54	1.67	1.81	1.97	2.15	2.35	2.58	2.83	2.83

Solution : Choosing $x_1 = 1.0$, $x_0 = 1.2$, $x_2 = 1.4$, then Eq. (15.22) becomes

$$0.07 = A_1 \left[\int_{1.2}^{1.0} y(x)\,dx + \int_{1.2}^{1.4} y(x)\,dx \right] + A_2 \left[\int_{1.2}^{1.0} (1.0 - t) y(t)\,dt + \int_{1.2}^{1.4} (1.4 - t)(t)\,dt \right]$$

Evaluating the integrals by Simpson's rule and simplifying, the above equation becomes

$$1.81 A_1 + 2.18 A_2 = 2.10 \quad ...(i)$$

Again choosing $x_1 = 1.4$, $x_0 = 1.6$ and $x_2 = 1.8$.

then Eq. (*i*) after simplifying, we get

$$2.88 A_1 + 3.10 A_2 = 3 \quad ...(ii)$$

Solving Eqs (*i*) and (*ii*), we get

$$A_1 = 0.03204, A_2 = 0.9364$$

Equation (15.22*a*) becomes

$$\lambda^2 - 0.03204\lambda - 0.9364 = 0 \quad ...(iii)$$

Solving Eq. (*iii*), we get

$$\lambda_1 = 0.99, \lambda_2 = -0.96$$

Using least squares method, we get

$$a_1 = 0.499, a_2 = 0.491$$

Hence $$y = 0.499\, e^{0.99x} + 0.491\, e^{-0.96x}$$

In order to use the Eq. (15.9) to set up a linear system for A_1 and A2, we eliminate $y'(x_0)$ in the following way.

Solution to Numerical, Algebraic and Transcendental Equation

16.1 INTRODUCTION

The following equation

$$f(x) = a_0x^n + a_1x^{n-1} + \dots + a_{n-1}\,x + a_n$$

where a_is are constants ($a_0 \neq 0$) and n is a positive integer called a polynomial in x of degree n, and the equation $f(x) = 0$ is called an algebraic equation of degree n.

If $f(x)$ contains some other functions like exponential, trigonometric, logarithmic, etc. then $f(x) = 0$ is called a transcendental equation, e.g.

$$x^3 - 3x + 6 = 0,\ x^5 - 7x^4 + 3x^2 + 36x + 7 = 0$$

are algebraic equations of third and fifth degree, whereas $x^2 - 3\cos x + 1 = 0$, $x^2e^x - 2 = 0$, $x \log_{10} x = 1.2$, etc. are transcendental equations.

In both the cases, if the coefficients are of pure numbers, they are called numerical equations. For the algebraic equations of degree two, three or four, methods are available to solve them. But the need often arises to solve higher degree or transcendental equation for which no direct method exists. Such equations can be solved by approximate methods.

Intermediate Value Property

If $f(x)$ is a continuous function in $[a, b]$ and $f(a)$, $f(b)$ have opposite signs, then the equation $f(x) = 0$ must have at least one root between $x = a$ and $x = b$.

Note: Every equation of odd degree has at least one real root.

Descartes' Rule of Signs

The equation $f(x) = 0$ cannot have more positive roots than change of signs in $f(x) = 0$, and more –ve roots than the change of signs in $f(-x)$, e.g.

$$f(x) = 2x^5 - 3x^4 + 5x^3 - 8x^2 + 1 = 0 \qquad (\text{signs of } f(x) = + - + - +)$$

As $f(x)$ has four changes of signs from (+ve to –ve or –ve to +ve), above equation cannot have more than four positive roots.

Also

$$f(x) = 2(-x)^5 - 3(-x)^4 + 5(-x)^3 - 8(-x)^2 + 1$$
$$= -2x^5 - 3x^4 - 5x^3 - 8x^2 + 1$$

This shows that $f(x)$ has one change of sign. It cannot have more than one negative root.

Existence of Imaginary Roots

If an equation of nth degree has (at the most) p positive roots and q negative roots, then it follows that the equation has at least $n - (p + q)$ imaginary roots.

E.g.

$$f(x) = x^7 - 3x^4 + 2x^3 - 1 = 0 \text{ has at least four imaginary roots.}$$

$$f(x) = x^7 - 3x^4 + 2x^3 - 1 \text{ has three possible changes or at the most has three positive roots and no negative root}$$

$$f(-x) = -x^7 - 3x^4 - 2x^3 - 1$$

$$\text{Degree of equation} = 7$$

$\therefore$ at least 4 imaginary roots

16.2 BISECTION METHOD (BOLZANO METHOD OR INTERVAL HALVING METHOD)

It is based on repeated application of intermediate value property for root to exist in the given interval $[a, b]$.

Let $f(x)$ be a continuous function in $[a, b]$

and $\quad f(a) < 0$ and $f(b) > 0$

or $\quad f(a)\,f(b) < 0$

Then first approximation to the root is

$$x_1 = \frac{a+b}{2}$$

If $f(x_1) = 0$, then x_1 is a root of $f(x)$, otherwise root lies between a and x_1 or x_1 and b as $f(x_1) > 0$ or $f(x_1) < 0$ (Fig. 16.1).

Then we proceed as before and continue till the desired accuracy is achieved.

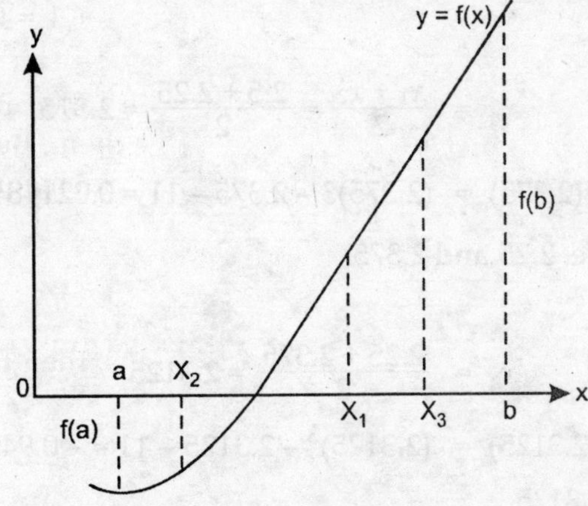

Fig. 16.1

Advantages

1. The main advantage of this method is that if $f(x)$ is continuous in $[a, b]$ and this interval actually contains a root, then this method is guaranteed to work.
2. The number of iterations required to achieve a specific accuracy is almost known in advance. Since the interval $[a, b]$ is halved each time, the last value of x differ from the true root by less than 1/2 the last interval or we can say error after the nth iteration $< \left|\frac{b-a}{2^n}\right|$.

   In other words, after 'n' bisection, the length of the subinterval which contain x^n is $\frac{b-a}{2^n}$. If the error is to be made less than a small quantity $\in$ say

$$\frac{b-a}{2^n} < \in, \text{ i.e. } 2^n > \frac{b-a}{\varepsilon}$$

The number of iterations should be greater than $\dfrac{\log \frac{b-a}{\in}}{\log 2}$.

Disadvantages

1. Its rate of convergence is slow.
2. When there are multiple root, interval halving (bisection method) may not be applicable.

Example **16.1.** Find a root of the equation $x^3 - x - 11 = 0$ correct to four decimal points using bisection method.

Solution : Let

$$f(x) = x^3 - x - 11$$

$$f(2) = -5 < 0$$

$$f(3) = 13 > 0$$

∴ root lies between 2 and 3, therefore, first approximation to the root is

$$x_1 = \frac{(2+3)}{2} = 2.5$$

Now $\quad f(2.5) = (2.5)^3 - 2.5 - 11 = 2.215 \text{ (+ ve)} > 0$

∴ root lies between 2 and 2.5.

2nd approximation

$$x_2 = \frac{2+2.5}{2} = 2.25$$

Now $\quad f(2.25) = (2.25)3 - 2.25 - 11 = -1.859375 < 0$

∴ root lies between 2.5 and 2.25.

3rd approximation

$$x_3 = \frac{x_1 + x_2}{2} = \frac{2.5 + 2.25}{2} = 2.375$$

Now $\quad f(2.375) = (2.375)3 - 2.375 - 11 = 0.0214843 > 0$

∴ root lies between x_2 and x_3, i.e. 2.25 and 2.375.

4th approximation

$$x_4 = \frac{2.25 + 2.375}{2} = 2.3125$$

$$f(2.3125) = (2.3125)^3 - 2.3125 - 11 = -0.9460449 < 0$$

∴ root lies between 2.375 and 2.3125.

5th approximation

$$x_5 = \frac{x_3 + x_4}{2} = \frac{2.375 + 2.3125}{2} = 2.34375$$

$$= -0.469137 < 0$$

∴ root lies between x_3 and x_5.

6th approximation

$$x_6 = \frac{x_3 + x_5}{2} = \frac{2.375 + 2.34375}{2} = 2.359375$$

Now $\quad f(2.359375) = (2.359375)3 - 2.359375 - 11 = -0.225592 < 0$

∴ root lies between x_3 and x_6.

7th approximation

$$x_7 = \frac{x_3 + x_6}{2} = \frac{2.375 + 2.359375}{2} = 2.3671875$$

$$f(2.3671875) = (2.3671875)^3 - 2.3671875 - 11 = -0.1024708 < 0$$

$\therefore$ root lies between x_3 and x_7.

8th approximation

$$x_8 = \frac{2.375 + 2.3671875}{2} = 2.3710938$$

$$f(2.3710938) = (2.3710938)^3 - 2.3710938 - 11 = -0.040601 < 0$$

$\therefore$ root lies between 2.375 and 2.3710938.

9th approximation

$$x_9 = \frac{2.375 + 2.3710938}{2} = 2.3730469$$

$$f(2.3730469) = (2.3730469)^3 - 2.3730469 - 11 = -9.585468 \times 10^{-3} < 0$$

$\therefore$ root lies between x_3 and x_9.

10th approximation

$$x_{10} = \frac{2.375 + 2.3730469}{2} = 2.3740235$$

$$f(2.3740235) = (2.3740235)^3 - 2.3790235 - 11 = 5.942661 \times 10^{-3} > 0$$

$\therefore$ root lies between x_9 and x_{10}.

11th approximation

$$x_{11} = \frac{2.3730469 + 2.3740235}{2} = 2.3735352$$

$$f(2.3735352) = (2.3735352)^3 - 2.3735352 - 11 = -1.823101 \times 10^{-3} < 0$$

$\therefore$ root lies between x_{10} and x_{11}.

12th approximation

$$x_{12} = \frac{x_{10} + x_{11}}{2} = \frac{2.3740235 + 2.3735352}{2} = 2.3737794$$

$$f(2.3737794) = (2.3737794)3 - 2.3737794 - 11 = 2.059952 \times 10^{-3} > 0$$

$\therefore$ root lies between x_{11} and x_{12}.

Following the same procedure, we have

$$x_{13} = 2.3736573$$

$$x_{14} = 2.3735963$$

$$x_{15} = 2.3736268 \quad \therefore \quad x_{14} = x_{15}$$

Hence, the root correct to four decimal places is 2.3736.

Example **16.2:** Using bisection method, find the negative root of $x^3 - x + 11 = 0$.

Solution : Let

$$f(x) = x^3 - x + 11$$

$$f(-x) = -x^3 + x + 11$$

The negative root of $f(x) = 0$ is positive root of $f(-x) = 0$.

Therefore, we will find the positive root of $f(-x) = 0$

or

$$\phi(x) = x^3 - x + 11 = 0$$

In last example, we got $x = 2.3736$, hence, the negative root is -2.3736.

Example **16.3.** Find a positive root of the equation $xe^x = 1$, which lies between 0 and 1.

Solution :

$$f(x) = xe^x - 1$$
$$f(0) = -1 < 0$$
$$f(1) = 1.718 > 1$$

$\therefore$ root lies between 0 and 1.

n	*a*	*b*	*x*	*f(x)*
1	0	1	0.5	– ve
2	0.5	1	0.75	+ ve
3	0.5	0.75	0.625	+ ve
4	0.5	0.625	0.5625	– ve
5	0.5625	0.625	0.59375	+ ve
6	0.5625	0.59375	0.5781	+ ve
7	0.5625	0.5781	0.5703	+ ve
8	0.5625	0.5703	0.5664	– ve
9	0.5664	0.5703	0.5684	+ ve
10	0.5664	0.5684	0.5674	+ ve
11	0.5664	0.5674	0.5669	
			≈ 0.567	– ve
12	0.5669	0.5674	0.5671	
			≈ 0.567	+ ve

The required root is 0.567, which is correct to three decimal places.

Example **16.4.** Find one root of $e^x - 3x = 0$ correct to two decimal places.

Solution :

$$f(x) = e^x - 3x$$
$$f(0) = 1 = \text{positive}$$
$$f(1) = e - 3 = \text{negative}$$
$$f(1.5) = e^{1.5} - 4.5 = \text{negative}$$
$$f(2) = e^2 - 6 = \text{positive}$$
$$f(1.6) = e^{1.6} - 4.8 = \text{positive}$$

$\therefore$ root lies between 1.5 and 1.6, therefore, first approximation to the root is:

$$x_1 = \frac{1.5 + 1.6}{2} = 1.55$$

$$f(x_1) = 0.06 + \text{ve}$$

$\therefore$ root lies between 1.5 and 1.55.

2nd approximation

$$x_2 = \frac{1.5 + 1.55}{2} = 1.525$$

$$f(x_2) = 0.02 = +\text{ ve}$$

$\therefore$ root lies between 1.5 and 1.525.

3rd approximation

$$x_3 = \frac{1.5 + 1.525}{2} = 1.5125$$

$$f(1.5125) = 0.00056 = +\text{ ve}$$

$\therefore$ root lies between 1.5 and 1.5125.

4th approximation

$$x_4 = \frac{1.5 + 1.5125}{2} = 1.5062$$

$$f(1.5062) = -0.00904 = -\text{ ve}$$

$\therefore$ root lies between 1.5062 and 1.55.

$$x_5 = \frac{1.5062 + 1.5125}{2} = 1.50935$$

The required root is 1.50 which is correct to two decimal places.

Example 16.5. Find the root of $x^3 - 4x - 9 = 0$ using bisection method in four stages.

Solution :

$$f(x) = x^3 - 4x + 9 = 0$$

$$f(1) = \text{negative}$$

$$f(2) = \text{negative}$$

$$f(3) = \text{positive}$$

$\therefore$ root lies between 2 and 3, therefore, first approximation to the root is:

$$x_1 = \frac{2+3}{2} = 2.5$$

$$f(2.5) = (2.53)^2 - 4(2.5) - 9 = -3.375 \text{ (–ve)}$$

$\therefore$ root lies between 2.5 and 3.

2nd approximation

$$x_2 = \frac{3 + 2.5}{2} = 2.75$$

$$f(2.75) = (2.75)^2 - 4(2.75) - 9 = 0.7969 > 0$$

$\therefore$ root lies between 2.75 and 2.5.

3rd approximation

$$x_3 = \frac{2.75 + 2.5}{2} = 2.625$$

$$f(2.625) = (2.625)^3 - 4(2.625) - 9 = -1.4121 \text{ (–ve)}$$

$\therefore$ root lies between 2.625 and 2.75.

4th approximation

$$x_4 = \frac{2.625 + 2.75}{2} = 2.6875$$

Hence fourth approximation = 2.6875

Example 16.6. Find the root of $\tan x + x = 0$ up to two decimal places which lies between 2 and 2.1.

Solution :

$$f(x) = \tan x + x$$

$$f(2) = -0.18 \text{ (–ve)}$$

$$f(2.1) = 0.39 \text{ (+ve)}$$

$\therefore$ root lies between 2.0 and 2.1, therefore, first approximation to the root is:

$$x_1 = \frac{2 + 2.1}{2} = 2.05$$

$$f(2.05) = 0.12 \text{ (+ve)}$$

$\therefore$ root lies between 2 and 2.05.

2nd approximation

$$x_2 = \frac{2 + 2.05}{2} = 2.025$$

$$f(2.025) = -0.023$$

$\therefore$ root lies between 2.025 and 2.05.

3rd approximation

$$x_3 = \frac{2.025 + 2.05}{2} = 2.0375$$

$$f(2.0375) = 0.054 = +\text{ve}$$

$\therefore$ root lies between 2.025 and 2.0375.

4th approximation

$$x_4 = \frac{2.025 + 2.0375}{2} = 2.03125$$

The required root is 2.03 which is correct to two decimal places.

16.3 METHOD OF FALSE POSITION (REGULA-FALSI METHOD)

This method (also known as Regula-Falsi method) is used for finding the real root of an equation $f(x) = 0$ and is some what similar to the bisection method.

Consider the equation $f(x) = 0$. Let a and b $(a < b)$ be two values of s such that $f(a)$ and $f(b)$ are of opposite signs. Then the graph of $y = f(x)$ crosses the x-axis at some point between a and b (Fig. 16.2).

Therefore, the equation of chord joining the two points $A\,[a, f(a)]$ and $B\,[b, f(b)]$ is

$$y - f(a) = \frac{f(b) - f(a)}{b - a}(x - a) \qquad \text{...(16.1)}$$

Let in the interval (a, b) the graph of the function be considered as a straight line. So the intersection of line given by Eq. (16.1) with x-axis will give an approximate value of the root. Putting $y = 0$ in Eq. (16.1), we get

$$f(a) = \frac{f(b) - f(a)}{b - a}(x - a)$$

$$x = \frac{af(b) - bf(a)}{f(b) - f(a)}$$

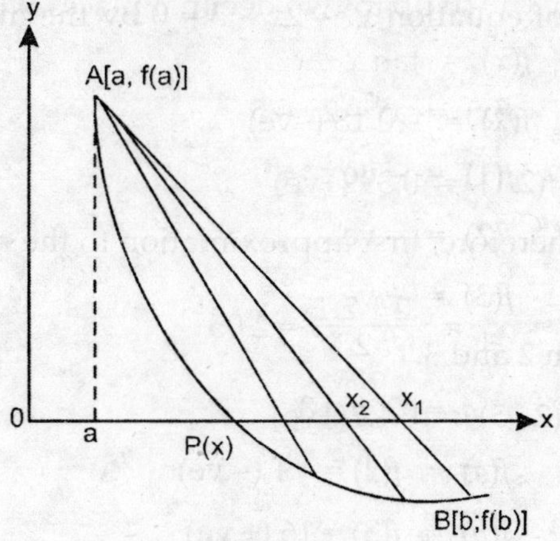

Fig. 16.2

Hence the first approximation to the root is given by

$$x_1 = \frac{af(b) - bf(a)}{f(b) - f(a)} \qquad \text{...(16.2)}$$

Now if $f(x_1)$ and $f(a)$ are of opposite sign, then the root lies in between a and x_1. So, we replace b by x_1 in Eq. (16.2) and get the next approximation x_2. But if $f(x_1)$ and $f(a)$ are of same sign, then $f(x_1)$ and $f(b)$ will be opposite sign, therefore, the root lies between x_1 and b. Hence we replace a by x_1 in Eq. (16.2) and get the next approximate x_2. The process is to be repeated till the root is found to be of desired accuracy.

***Example* 16.7.** Find root of the equation $x \log_{10} x - 1.2 = 0$ correct to three decimal places.

Solution :

$$f(x) = x \log_{10} x - 1.2$$
$$f(1) = -1.2 = -\text{ ve}$$
$$f(2) = -0.5979 = -\text{ ve}$$
$$f(3) = 0.23136 = +\text{ ve}$$

$\therefore$ root lies between 2 and 3.

Taking

$$a = 2, b = 3$$
$$f(a) = f(2) = -0.59795$$
$$f(b) = f(3) = 0.23136$$

By method of false position

1st iteration

$$x_1 = \frac{af(b) - bf(a)}{f(b) - f(a)} = \frac{2(0.23136) - 3(-0.59794)}{0.23136 + 0.59794} = 2.72102$$

$$f(2.72102) = -0.0170863 \text{ is the root lies between 2.72102 and 3.}$$

Now

$$a = 2.72102, b = 3$$
$$f(2.72102) = -0.01709$$
$$f(3) = 0.23136$$

2nd iteration

$$x_2 = \frac{af(b) - bf(a)}{f(b) - f(a)} = \frac{2.72101 \times 0.23136 - 3 \times -0.01709}{0.23136 + 0.01709} = 2.74021$$

$$f(0.74021) = -0.0003802$$

The root lies between 2.74021 and 3.

Proceeding in the same way, we get 3rd and 4th iterations

$$x_3 = 2.74024, x_4 = 2.74063$$

$\therefore$ root is 2.740 correct to three decimal places.

Example **16.8.** Find the real root of equation $x^3 - 2x - 5 = 0$ by the method of false position correct to three decimal places.

Solution :

$$f(x) = x^3 - 2x - 5$$
$$f(1) = -\text{ve}$$
$$f(2) = -\text{ve}$$
$$f(3) = +\text{ve}$$

$\therefore$ the root of $f(x) = 0$ lies between 2 and 3.

Taking $a = 2, b = 3$

$$f(a) = f(2) = -1\ (-\text{ve})$$
$$f(b) = f(3) = 16\ (+\text{ve})$$

1st approximation
$$x_1 = \frac{af(b) - bf(a)}{f(b) - f(a)} = \frac{2 \times 16 - 3(-1)}{16 + 1} = 2.0588$$

$$f(2.0588) = (2.055)^3 - 2(2.0588) - 5 = -0.391\ (-\text{ve})$$

$\therefore$ the root lies between 2.0588 and 3.

Now $a = 2.0588, f(a) = -0.391, b = 3, f(b) = 16$

2nd approximation
$$x_2 = \frac{2.0588(16) - 3(-0.391)}{16 - (-0.391)} = 2.08125$$

$$f(2.08125) = -0.147\ (-\text{ve})$$

$\therefore$ the root lies between 2.08125 and 3.

Now $a = 2.08125, f(a) = -0.147, b = 3, f(b) = 16$

3rd approximation
$$x_3 = \frac{2.08125(16) - 3(-0.147)}{16 + 0.147} = 2.0896$$

$$f(2.0896) = -0.0551\ (-\text{ve})$$

$\therefore$ the root lies between 2.0896 and 3.

4th approximation
$$x_4 = \frac{2.0896(16) - 3(-0.0551)}{16 + 0.0551} = 2.0927$$

$$f(2.0927) = -0.0206\ (-\text{ve})$$

$\therefore$ the root lies between 2.0927 and 3.

5th approximation
$$x_5 = \frac{2.0927(16) - 3(-0.0206)}{16 + 0.0206} = 2.0939$$

$$f(2.0939) = -0.00726\ (-\text{ve})$$

$\therefore$ the root lies between 2.0939 and 3

6th approximation
$$x_6 = \frac{2.0939(16) - 3(-0.00726)}{16 + 0.00726} = 2.0943$$

$$f(x_6) = -0.0028$$

$\therefore$ the root lies between 2.0943 and 3.

7th approximation
$$x_7 = \frac{2.0943(16) - 3(-0.0028)}{16 + 0.0028} = 2.0944$$

Therefore 2.094 is correct to three decimal places.

***Example* 16.9.** Find real root of $xe^x = \cos x$ correct up to four decimal places.

Solution :

$$f(x) = \cos x - xe^x = 0$$

$$f(0) = 1$$

$$f(1) = -2.17798$$

$\therefore$ the root lies between 0 and 1.

$$a = 0, b = 1$$

$$f(0) = f(1), f(b) = -2.17798$$

1st approximation $x_1 = \dfrac{af(b) - bf(a)}{f(b) - f(a)} = \dfrac{0(-2.17798) - 1}{-2.17798 - 1} = 0.31467$

$$f(0.31467) = 0.51987 \text{ (+ ve)}$$

$\therefore$ the root lies between 0.31467 and 1.

$$f(a) = f(0.31467) = 0.51987$$

$$f(b) = f(1) = -2.17798$$

2nd approximation $x_2 = \dfrac{0.31467(-2.17798) - 1(0.51987)}{-2.17798 - 0.51987} = 0.44673$

$$f(x_2) = 0.20356 \text{ (+ ve)}$$

$f(0.44673)$ and $f(1)$ are of opposite sign.

$$f(0.44673) = 0.20356$$

$$f(1) = -2.17798$$

3rd approximation $x_3 = \dfrac{0.44673(-2.17798) - 1 \times (0.20356)}{-2.17798 - 0.20356} = 0.49402$

$$f(x_3) = 0.070788$$

$\therefore$ root lies between 0.49402 and 1.

Proceeding in the same way, we have

$$x_4 = 0.50995$$

$$x_5 = 0.51520$$

$$x_6 = 0.51692$$

$$x_7 = 0.51748$$

$$x_8 = 0.51767$$

$$x_9 = 0.51775$$

Hence, root is 0.5177 correct to four decimal places.

***Example* 16.10.** Find the real root of $xe^x - 3 = 0$ correct to three decimal places.

Solution :

$$f(x) = xe^x - 3, f(0) = -3 \text{ (– ve)}$$

$$f(1) = e - 3 = -0.28172 \text{ (– ve)}$$

$$f(1.5) = 1.5e^{1.5} - 3 = 3.72253 \text{ (+ ve)}$$

$\therefore$ the root lies between 1 and 1.5.

$$a = 1, b = 1.5$$

$$f(a) = f(1) = -0.28172$$

$$f(b) = f(1.5) = 3.72253$$

1st approximation $$x_1 = \frac{af(b) - bf(a)}{f(b) - f(a)} = \frac{1(3.72253) - 1.5(-0.28172)}{3.72253 - (-0.28172)} = 1.035$$

$$f(x_1) = -0.0864 \ (-\text{ve})$$

$\therefore$ the root lies between 1.5 and 1.035.

$$a = 1.035, b = 1.5$$

$$f(a) = -0.028648$$

$$f(b) = 3.72253$$

2nd approximation $$x_2 = \frac{af(b) - bf(a)}{f(b) - f(a)} = \frac{(1.035)(3.72253) - 1.5(-0.028648)}{3.72253 - (-0.0286485)} = 1.045$$

$$f(x_2) = -0.0286485 \ (-\text{ve})$$

$\therefore$ the root lies between 1.045 and 1.5.

$$f(1.045) = -0.0286485$$

$$f(1.5) = 3.72253$$

3rd approximation $$x_3 = \frac{(1.045)(3.72253) - 1.5(-0.0286485)}{3.72253 + 0.0111652} = 1.048$$

$$f(1.0480) = -0.0111652 \ (-\text{ve})$$

$\therefore$ the root lies between 1.0480 and 1.5.

$$f(1.0480) = -0.0111652$$

$$f(1.5) = 3.72253$$

4th approximation $$x_4 = \frac{1.0480(3.72253) - 1.0480(-0.0111652)}{3.72253 + 0.0111652} = 1.048$$

$$f(1.049) = -5.320155 \times 10^{-3} \ (-\text{ve})$$

$\therefore$ the root lies between 1.049 and 1.5.

Similarly $$x_5 = 1.0496$$

$$x_6 = 1.0498$$

Hence, the root is 1.05.

Example **16.11.** Find the real root of $x^3 - 18 = 0$.

Solution : Given $$x^3 - 18 = 0 = f(x)$$

$$f(2) = -10 \ (-\text{ve})$$

$$f(3) = 9 \ (+\text{ve})$$

$\therefore$ the root lies between 2 and 3.

$$a = 2, b = 3$$

1st approximation $$x_1 = \frac{af(b) - bf(a)}{f(b) - f(a)} = \frac{2 \times 9 - 3 \times (-10)}{9 + 10} = 2.5263$$

$$f(2.5263) = -1.87666$$

$\therefore$ the root lies between 2.5263 and 3.

2nd approximation $x_2 = \dfrac{2.5263(9) - 3(-1.87666)}{9 + 1.87666} = 2.6080$

$$f(2.6080) = -2.61260$$

$\therefore$ the root lies between 2.6080 and 3.

3rd approximation $x_3 = \dfrac{2.6080(9) - 3(-2.61260)}{9 + 2.61260} = 2.6961$

$$f(2.6961) = 1.5998$$

$\therefore$ the root lies between 2.6080 and 2.6961.

$$f(a) = -2.61260, f(b) = 1.5998$$

4th approximation $x_4 = \dfrac{2.6080(1.5998) - 2.6961(-2.61260)}{1.5998 + 2.61260} = 2.6626$

$$f(6.6626) = 8.7784$$

$\therefore$ the root lies between 2.6080 and 2.6626.

5th approximation $x_5 = \dfrac{2.6080(8.7784) - 2.6626(2.61260)}{8.7784 + 2.6126} = 2.6148$

$$f(x_5) = -12.1580$$

$\therefore$ the root lies between 2.648 and 6.6626.

$$f(2.6148) = -12.1580, f_2(2.6626) = 8.7784$$

6th approximation $x_6 = \dfrac{2.6148(8.7784) + 2.6626(12.1580)}{20.9364} = 2.642$

Hence, the root is 2.6.

16.4 NEWTON-RAPHSON METHOD

Let x_0 be an approximate root of the equation $f(x) = 0$. If $x_1 = x_0 + h$ be the exact root, and h being a small quantity, then $f(x_1) = 0$.

Expanding it by Taylor's theorem, we get

$$f(x_1) = f(x_0 + h) = 0$$

$$f(x_0) + hf'(x_0) + \frac{h^2}{21} f''(x_0) + \ldots = 0$$

Since h is small, we can neglect h^2 and higher power of h,

Hence, $f(x_0) + hf'(x_0) = 0$

or $h = -\dfrac{f(x_0)}{f'(x_0)}, f'(x_0) \neq 0$

Hence $x_1 = x_0 + h = x_0 - \dfrac{f(x_0)}{f'(x_0)}$

$$x_2 = x_1 - \frac{f(x_1)}{f'(x_1)}$$

$$\ldots = \ldots$$

$$x_{n+1} = x_n - \frac{f(x_n)}{f'(x_n)} \quad (n = 0, 1, 2, ...)$$

Notes :

1. This method is useful in case of large values of $f'(x)$, i.e. when the graph of $f(x)$ which crosses x-axis is almost vertical, then h will be small and hence, the root can be calculated in less than or in short time.
2. Newton's formula provides quick root if the initial approximation x_0 is chosen sufficiently close to the root.

16.5 CONVERGENCE OF NEWTON-RAPHSON METHOD

Newton-Raphson formula

$$x_{n+1} = x_n - \frac{f(x_n)}{f'(x_n)}$$

or

$$x_{n+1} = \phi(x_n) = x_n - \frac{f(x_n)}{f'(x_n)}$$

In general

$$\phi(x) = x_n - \frac{f(x)}{f'(x)}$$

which gives

$$\phi'(x) = 1 - \frac{[f'(x)]^2 - f(x)f''(x)}{[f'(x)]^2} = 1 - 1 + \left[\frac{f(x)f''(x)}{[f'(x)]^2}\right] = \frac{f(x)f''(x)}{[f(x)]^2}$$

Since iteration method converges if $|\phi'(x)| < 1$.

Newton's method will converge if

$$\left|\frac{f(x)f''(x)}{[f(x)f''(x)]^2}\right| < 1$$

$$|f(x)\, f''(x)| < [f'(x)]^2$$

in the interval considered.

The interval containing the root 'a' of $f(x) = 0$ should be selected in which the above conditions is satisfied.

Notes :

1. The process will fail if $f'(x) = 0$ is in neighbourhood of the root. In such case, Regula-Falsi method should be used.
2. If the initial approximation to the root is not given, choose the two values of x, say a and b such that $f(a)$ and $f(b)$ are of opposite sign. If $|f(a)| < |f(b)|$, then take 'a' as the initial approximation.
3. Newton-Raphson method is also referred to as method of tangents.

***Example* 16.12.** Find the negative root of $x^3 - 21x + 3500 = 0$ correct to two decimal places.

Solution : Let

$$f(x) = x^3 - 21x + 3500$$

$$f(-15) = (-15)^3 + 21 \times 15 + 3500$$

$$= -3375 + 315 + 3500 = 440 > 0$$

$$f(-16) = -260 < 0$$

since $f(-16) < f(-15)$ (numerically)

$\therefore$ root lies near -16.

$$x_0 = -15.8$$

$$x_{n+1} = x_n - \frac{f(x_n)}{f'(x_n)} = x_n - \frac{x_n^3 - 21x_n + 35}{3x_n^2 - 21} = \frac{2x_n^3 - 3500}{3x_n^2 - 21}$$

$$x_1 = \frac{2(-15.8)^3 - 3500}{3(-15.8)^2 - 21} = -15.645$$

$$x_2 = \frac{2(-15.64)^3 - 3500}{3(-15.64)^2 - 21} = -15.643$$

$$x_3 = \frac{2(-15.643)^3 - 3500}{3(-15.643)^2 - 21} = -15.643$$

Hence, the root is – 15.64 correct to two decimal places.

***Example* 16.13.** Using Newton-Raphson, solve $x \sin x + \cos x = 0$ which is near to $x = \pi$.

Solution : Let

$$f(x) = x \sin x + x \cos x$$

$$f'(x) = x \cos x + \sin x - \sin x = x \cos x$$

$$x_{n+1} = x_n - \frac{f(x_n)}{f'(x_n)} = \frac{x_n^2 \cos x_n - x_n \sin x_n - \cos x_n}{x_n \cos x_n}$$

$$x_1 = \frac{\pi^2 \cos \pi - \pi \sin \pi - \cos \pi}{\pi \cos \pi} = \frac{\pi^2 + 1}{-\pi} = 2.8232$$

$$x_2 = \frac{2.8232 \cos(2.8232) - 2.8232 \sin(2.8232) - \cos(2.8232)}{2.8232 \cos(2.8232)} = 2.79819$$

$$x_3 = \frac{2.79819 \cos(2.79819) - 2.79819 \sin(2.79819) - \cos(2.79819)}{2.79819 \cos(2.79819)}$$

$$= 2.79819$$

Hence, the required root is 2.79819.

***Example* 16.14.** Find a real root of equation $x = e^{-x}$ by Newton-Raphson method.

Solution :

$$f(x) = xe^x - 1 = 0$$

$$f'(x) = e^x + xe^x = (1 + x)e^x, f(0) = -1, f(1) = e^{-1} = +\text{ve}$$

$\therefore$ the root lies between 0 and 1.

Let $x_0 = 1$

Thus $x_1 = 1 - \frac{e-1}{2e} = \frac{1}{2}\left(1 + \frac{1}{e}\right) = 0.6839397$

$$f(x_1) = 0.3553424$$

$$f'(x_1) = 3.337012$$

so that $x_2 = 0.6839397 - \frac{0.3553424}{3.337012} = 0.577545$

$$x_3 = 0.5672297$$

$$x_4 = 0.56714323$$

The required root is 0.567.

***Example* 16.15.** Solve by Newton's method, the root of equation $e^x = 4x$ which is approximately correct to three decimal places.

Solution :

$$f(x) = e^x - 4x$$

$$f(2) = e^2 - 8 = -0.610944 \; (-\text{ve})$$

$$f(3) = e^3 - 12 = 8.085537 \; (+\text{ve})$$

$f(2)\,f(3) < 0$

The root of $f(x) = 0$ lies between 2 and 3; $|f(2)| < |f(3)|$

$\therefore$ the root lies nearer to 2.

Let us take

$$x_0 = 2.1$$

$$f(x) = e^x - 4x$$

$$f'(x) = e^x - 4$$

$$f(x_0) = f(2.1) = e^{2.1} - 4(2.1) = 8.16617 - 8.4 = -0.23383$$

$$f'(x_0) = f'(2.1) = e^{2.1} - 4 = 4.16617$$

1st approximation $\quad x_1 = x_1 - \dfrac{f(x_0)}{f'(x_0)} = 2.1 - \dfrac{(-0.23383)}{4.16617} = 2.1 + 0.0561258 = 2.1561$

2nd approximation $\quad x_2 = x_1 - \dfrac{f(x_1)}{f'(x_0)} = 2.1561 - \dfrac{f(2.1561)}{f'(2.1561)} = 2.1561 - \dfrac{0.0129861}{4.6373861} = 2.1533$

3rd approximation $\quad x_3 = x_2 - \dfrac{f(x_2)}{f'(x_2)} = 0.1533 - \dfrac{(0.0013484)}{4.6106516} = 2.1532$

The value of root correct to three decimal places = 2.1532.

Example **16.16.** Solve $x^4 - 5x^3 + 20x^2 - 40x + 60 = 0$ by Newton-Raphson method given that all the roots are complex.

Solution :

$$f(x) = x^4 - 5x^3 + 20x^2 - 40x + 60$$

$$f'(x) = 4x^3 - 15x^2 + 40x - 40$$

Using Newton-Raphson formula, we have

$$x_{n+1} = x_n - \frac{f(x_n)}{f'(x_n)} = x_n - \frac{x_n^4 - 5x_n^3 + 20x_n^2 - 40x_n + 60}{4x_n^3 - 15x_n^2 + 40x_n - 40}$$

$$= \frac{3x_n^4 - 10x_n^3 + 20x_n^2 - 60}{4x_n^3 - 15x_n^2 + 40x_n - 40}$$

Putting $n = 0$ and taking $x_0 = 2(1 + i)$ by trial, we get

$$x_1 = \frac{3(2+2i)^4 - 10(2+2i)^3 + 20(2+2i)^2 - 60}{4(2+2i)^3 - 15(2+2i)^2 + 40(2+2i) - 40} = 1.92(1+i)$$

$$x_2 = \frac{3(1.92+1.92i)^4 - 10(1.92+1.92i)^3 + 20(1.92+1.92i)^2 - 60}{4(1.92+1.92i)^3 - 15(1.92+1.92i)^2 + 40(1.92+1.92i) - 40}$$

$$= 1.915 + 1.908i \qquad \text{...}(i)$$

Since imaginary roots occur in conjugate pairs so the root of i are $1.915 \pm 1.908i$ up to three decimal places. Assuming that the other pair of roots of i is $\alpha + i\beta$, we have

$$\text{Sum of the roots} = (\alpha + i\beta) + (\alpha - i\beta) \text{ or } (1.915 + 1.908i) + (1.915 - 1.908i) = 5$$

$$2\alpha + 3.83 = 5 \text{ or } \alpha = 0.585$$

Also product of roots $= (\alpha^2 + \beta^2)$ or $\{(1.915)^2 + (1.908)^2\} = 60$

which gives $\beta = 2.805$

Hence the other roots are $0.585 \pm 2.805i$.

***Example* 16.17.** Using Newton-Raphson method, establish iteration $x_{n+1} = \frac{1}{2}\left[x_n + \frac{N}{x_n}\right]$, where N is a positive number.

Solution : Let

$$\sqrt{N} = x$$

$$x^2 = N$$

$$x^2 - N = 0$$

Let

$$f(x) = x^2 - N$$

$$f'(x) = 2x$$

Now, Newton-Raphson formula

$$x_{n+1} = x_n - \frac{f(x_n)}{f'(x_n)} = x_n - \frac{x_n^2 - N}{2x_n} = \frac{1}{2}\left[x_n + \frac{N}{x_n}\right], N = 0, 1, 2, 3, \ldots$$

***Example* 16.18.** Using Newton-Raphson method, establish the formula where N is a positive number.

$$x_{n+1} = \frac{1}{3}\left[2x_n + \frac{N}{x_n^2}\right]$$

Solution : Let

$$x = N^{1/3}$$

$$x^3 = N$$

$$x^3 - N = 0$$

$$f(x) = x^3 - N$$

$$= 3x^2$$

Newton-Raphson formula is

$$x_{n+1} = x_n - \frac{f(x)}{f'(x)} = x_n - \frac{x_n^3 - N}{3x_n^2} = \frac{3x_n^2 - x_n^3 + N}{3x_n^2} = \frac{2x_n^3 + N}{3x_n^2}$$

$$= \frac{1}{3}\left[2x_n + \frac{N}{x_n^2}\right] N = 0, 1, 2, \ldots$$

***Example* 16.19.** Find cube root of 12.

Solution : Let

$$x = 12^{1/3}$$

$$x^3 - 12 = 0$$

$$f(x) = x^3 - 12 = 0 \qquad (\therefore x = 12^{1/3})$$

$$f(1) = -\text{ve}$$

$$f(2) = -\text{ve} = 8 - 12 = -4$$

$$f(3) = +\text{ve} = 21 - 12 = 9$$

$\therefore$ the root lies between 2 and 3 $|f(2)| < |f(3)|$

$\therefore$ the root lies near to 2.

Let us take $x_0 = 2.1$

By Newton-Raphson method

$$x_{n+1} = \frac{1}{3}\left[2x_n + \frac{N}{x_n^2}\right]$$

$$x_1 = \frac{1}{3}\left[2x_0 + \frac{N}{x_0^2}\right] = \frac{1}{3}\left[2 \times 2.1 + \frac{12}{(2.1)^2}\right] = 2.307$$

$$x_2 = 2.2901.$$

$$x_3 = 2.29$$

Hence, the root is 2.29 correct to two decimal places.

Example **16.20.** Find the smallest root of equation $e^{-x} = \sin x$ correct to four decimal places.

Solution : $f(x) = e^{-x} - \sin x,\ f'(x) = e^{-x} - \cos x$

$$x_{n+1} = x_n + \frac{e^{-x_n} - \sin x_n}{e^{-x_n} + \cos x_n}$$

$$f(0) = 1 - \sin 0 = 1$$

$$f(1) = e^{-1} - \sin 1 = -0.47359$$

$|f(0)| > |f(1)|$, root is near to 1

Let us take $x_0 = 0.6$

$$x_1 = 0.6 + \frac{e^{-0.6} - \sin 0.6}{e^{-0.6} + \cos 0.6} = 0.6 - \frac{0.01583}{1.3745} = 0.6 - 0.1151 = 0.588548$$

$$x_2 = 0.58848 + \frac{e^{-0.58848} - \sin(0.58848)}{e^{-0.58848} + \cos(0.58848)} = 0.588559287$$

Hence, desired root is 0.5885.

Example **16.21.** Find by Newton-Raphson method, the real root of equation $3x = \cos x + 1$ correct to four decimal places.

Solution : Let $f(x) = 3x - \cos x - 1 = 0$

$$f(0) = -1 - 1 = -2\ (-\text{ ve})$$

$$f(1) = 3 - 0.5403 - 1 = 1.4597 = +\text{ ve}$$

Root lies between 0 and 1, $|f(0)| > |f(1)|$.

Let us take $x_0 = 0.6$

Newton-Raphson method

$$x_{n+1} = x_n - \frac{f(x_n)}{f'(x_n)} = x_n - \frac{3x_n - \cos x_n - 1}{3 + \sin x_n}$$

$$= \frac{x_n \sin x_n + \cos x_n + 1}{3 + \sin x_n}$$

$n = 0, 1, 2, 3, \ldots$, we get successive approximation

1st approximation $x_1 = \dfrac{x_0 \sin x_0 + \cos x_0 + 1}{3 + \sin x_0} = \dfrac{(0.6)\sin(0.6) + \cos(0.6) + 1}{3 + \sin(0.6)} = 0.6071$

2nd approximation $x_2 = \dfrac{0.6071 \times 0.57049 + 0.831311}{3 + 0.57049} = 0.6071$

Hence, the approximate correct root to four decimal places is 0.6071.

***Example* 16.22.** Find the real root of $xe^x = 2$ correct to two decimal places.

Solution : Given
$$f(x) = xe^x - 2$$
$$f(0) = -2$$
$$f(1) = e - 2 = 0.71828 = +\text{ve}$$

Root lies between 0 and 1

$|f(0)| > |f(0)|$ $\therefore$ root lies nearest to 1

Let us take $x_0 = 1$

$$x_{n+1} = x_n - \frac{f(x_n)}{f'(x_n)}$$
$$x_1 = x_0 - \frac{f(x_0)}{f'(x_0)} = \frac{e+2}{2e} = 0.867879$$
$$x_2 = 0.867879 - \frac{0.06716}{4.44902} = 0.85278$$
$$x_3 = 0.853$$

Hence, the root is 0.853 correct to two decimal places.

***Example* 16.23.** Find a root of equation $x \sin x + \cos x = 0$.

Solution :
$$f(x) = x \sin x + \cos x$$
$$f'(x) = x \cos x$$
$$f(0) = 1\ (+\text{ve})$$
$$f\left(\frac{\pi}{2}\right) = \frac{\pi}{2}\ (+\text{ve})$$
$$f(\pi) = -1\ (-\text{ve})$$

$\therefore$ root lies between $= \frac{\pi}{2}, \pi \left|f\left(\frac{\pi}{2}\right)\right| > |f(\pi)|$

Root lies between near to π.

Taking $x_0 = \pi$, we have

$$x_{n+1} = x_n - \frac{x_n \sin x_n + \cos x_n}{x_n \cos x_n} = 3.1416 - \frac{3.1416 \sin(3.1416) + \cos(3.1416)}{3.1416 \cos(3.1416)} = 2.8233$$
$$x_2 = 2.7986$$
$$x_3 = 2.7984$$

Hence, the root is 2.798 correct to three decimal places.

***Example* 16.24.** Using Newton-Raphson method, find the root of equation $x \log_{10} x = 1.2$.

Solution : Let
$$f(x) = x \log_{10} x - 1.2$$
$$f'(x) = \log_{10} x + x(\log_{10} e/x) \left[\frac{d}{dx}\log_{10} x = \frac{1}{x}\log_a e\right]$$
$$= \log_{10} x + 0.4343$$
$$f(2.5) = -0.2051499 < 0$$
$$f(3) = 0.2313636 > 0$$

$\therefore$ root lies between 2.5 and 3.

Root lies nearer to 3.

Let us take $x_0 = 27$

By Raphson's method

$$x_{n+1} = x_n - \frac{x_n \log_{10} x_n - 1.2}{\log_{10} x_n + 0.4343} = \frac{0.4343 x_n + 1.2}{\log_{10} x_n + 0.4343}$$

$$x_1 = \frac{0.4343(2.7) + 1.2}{\log_{10} 2.7 + 0.4343} = 2.7407986$$

$$x_2 = 2.7406461$$

$$x_3 = 2.7406461$$

Hence, the root is 2.7406461.

***Example* 16.25.** Solve $\sin x = 1 - x^3$, find the – ve root using Newton-Raphson method.

Solution : Let

$$f(x) = \sin x - 1 - x^3, f'(x) = \cos x - 3x^2$$

$$f(1) = \sin 1 - 1 - 1 = -1.1585$$

$$f(1) = -9.09070$$

$$f(-1) = \sin(-1) - 1 + 1 = -0.8414709 < 0$$

$$f(-2) = 6.0907026 > 0$$

$\therefore$ root lies between – 1 and – 2, $|f(-1)| < |f(-2)|$

$$x_{n+1} = \frac{x_n \cos x_n - \sin x_n - 2x_n^3 + 1}{\cos x_n - 3x_n^2}$$

Root lies near to – 1, take $x_0 = -1$

$$x_1 = \frac{x_0 \cos x_0 - \sin x_0 - 2x_0^3 + 1}{\cos x_0 - 3x_0^2} = -1.2763653$$

$$x_2 = -1.2497465$$

$$x_3 = -1.2490526$$

$$x_4 = -1.2490522$$

$$x_5 = -1.2490521$$

$$x_6 = -1.2490522$$

Hence, the root is – 1.249052 correct to six decimal places.

16.6 GENERALIZED NEWTON'S METHOD FOR MULTIPLE ROOTS

If a be a root of $f(x) = 0$ which is repeated m times, then

$$x_{n+1} = x_n - m\frac{f(x_n)}{f'(x_n)}$$

which is called generalized Newton's formula.

Notes :

1. If a is a root of $f(x) = 0$ with multiplicity m, then it is also a root of $f'(x) = 0$ with multiplicity $m - 1$ of $f''(x) = 0$ with multiplicity $m - 2$ and so on. Hence, if initial approximation x_0 is sufficiently close to root, then the expression will have the same value.

$$x_0 - m\frac{f(x_0)}{f'(x_0)}, x_0 - (m-1)\frac{f'(x_0)}{f''(x_0)}, x_0 - (m-2)\frac{f''(x_0)}{f'''(x_0)}$$

2. Generalized Newton's formula has a second order of convergence for determining multiple root.

***Example* 16.26.** Find the double root of the equation $x^3 - x^2 - x + 1 = 0$.

Solution : Let $\quad x^3 - x^2 - x + 1 = f(x)$

$$f(x) = 0 \text{ and } f'(x) = 0 \text{ will have the same root}$$

$$f(x) = x^3 - x^2 - x + 1$$

$$f'(x) = 3x^2 - 2x - 1 \quad f(x) = 0 \text{ and } f'(x) = 0 \text{ will have the same root}$$

$$f''(x) = 6x - 2$$

Starting with $x_0 = 0.9$, we have

1st approximation $\quad x_1 = x_0 - m\dfrac{f(x_0)}{f'(x_0)} = 0.9 - \dfrac{2 \times 0.19}{-0.37} = 1.003$

and $\quad x_1 = x_0 - (m-1)\dfrac{f'(x_0)}{f''(x_0)} = 0.9 - (2-1)\dfrac{(-0.37)}{3.4} = 1.009$

The closeness of these values implies that there is a double root near $x = 1$.

Choosing $x_1 = 1.01$ for the next approximation, we get

$$x_2 = x_1 - m\frac{f(x_1)}{f'(x_1)} = 1.01 - \frac{2 \times 0.0002}{0.0403} = 1.0001$$

$$= x_1 - (m-1)\frac{f'(x_1)}{f''(x_1)} = 1.01 - \frac{0.0403}{4.06} = 1.0001$$

This show that there is a double root at $x = 1.000$ which is quite near the actual root $x = 1$.

***Example* 16.27.** Find the multiple root of $x^4 - 6x^3 + 12x^2 - 10x + 3 = 0$ with multiplicity 3, by modified Newton-Raphson method with an initial guess of $x_0 = 0$.

Solution :

$$f(x) = x^4 - 6x^3 + 12x^2 - 10x + 3$$

$$f'(x) = 4x^3 - 18x^2 + 24x - 10$$

Given the initial value $x_0 = 0$ and $m = 3$

$$x_1 = x_0 - m\frac{f(x_0)}{f'(x_0)} = 0 - 3\frac{3}{(-10)} = 0.9$$

$$x_2 = 0.9 - 3\frac{f(0.9)}{f'(0.9)} = 0.9$$

$$= 0.9 - 3\left[\frac{(0.9)^4 - 6(0.9)^3 + 12(0.9)^2 - 10(0.9) + 3}{4(0.9)^3 - 18(0.9)^2 + 24(0.9) - 10}\right] = 0.998$$

$$x_3 = 0.998 - 3\left[\frac{f(0.998)}{f'(0.998)}\right] = 1.096$$

$$x_4 = 1.096 - 3\left[\frac{f(1.096)}{f'(1.096)}\right] = 1.0965$$

Hence, $x = 1.0965$ is the required root.

***Example* 16.28.** Find the double root of $x^3 - 5.4x^2 + 9.24x - 5.096 = 0$ given that it is nearer to 1.5.

Solution : Let $\quad f(x) = x^3 - 5.4x^2 + 9.24x - 5.096$

$$f'(x) = 3x^2 - 10.8x + 9.24$$

$$x_{n+1} = x_n - m\frac{f(x_n)}{f'(x_n)} = x_n - (2)\frac{f(x_n)}{f'(x_n)}$$

$$= x_n (2)\frac{[x_n^3 - 5.4x_n^2 + 9.24x_n - 5.096]}{[3x_n^2 - 10.8x_n + 9.24]} = \frac{x_n^3 - 9.24x_n + 10.192}{3x_m^2 - 10.8x + 9.24}$$

Given the initial value $x_0 = 1.5$

$$x_1 = \frac{(1.5)^3 - (9.24 \times 1.5) + 10.192}{3 \times (1.5)^2 - (10.8 \times 1.5) + 9.24} = 1.3952$$

$$x_2 = \frac{(1.3952)^3 - (9.24 \times 1.3952) + 10.192}{3 \times (1.3952)^2 - (10.8 \times 1.3952) + 9.24} = 1.3966$$

$$x_3 = \frac{(1.3966)^3 - (9.24 \times 1.3966) + 10.192}{3 \times (1.3966)^2 - (10.8 \times 1.3966) + 9.24} = 1.4024$$

$$x_4 = \frac{(1.4024)^3 - (9.24 \times 1.4024) + 10.192}{3 \times (1.4024)^2 - (10.8 \times 1.4024) + 9.24} = 1.4211$$

Hence, $x = 1.4$ correct to one decimal place.

16.7 SECANT METHOD

Although the Newton-Raphson method is often very efficient, there are situations where it performs poorly, and it is not convenient for polynomials and many other functions, there are certain functions whose derivative may be extremely difficult or inconvenient to evaluate. For these cases, the derivative can be approximated by a finite divided difference as shown in Fig. 16.3.

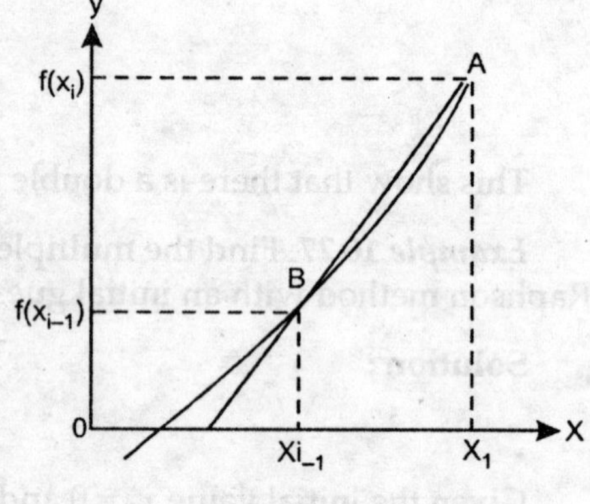

Fig. 16.3

$$f'(x) = \frac{f(x_n) - f(x_{n-1})}{x_n - x_{n-1}}$$

The approximation can be substituted in the Newton-Raphson method to yield the following iterative equation which is called the formula for the secant method.

$$x_{n+1} = x_n - \frac{x_n - x_{n-1}}{f(x_n) - f(x_{n-1})} f(x_n)\ n > 1 \geq 1$$

This method is an improvement over the method of false position as to condition $f(x_0) f(x_1) < 0$ is not required. At each iteration that the interval must obtain the root.

Note: If somehow $f(x_n) = f(x_{n-1})$, then this method fails and hence doesn't converge. This is the drawback of secant method over the method of position, which always converges. But if secant method converges, its rate of convergence is faster than Regula-Falsi method.

***Example* 16.29.** Solve $e^{-x} - x = 0$ by secant method.

Solution :

$$f(x) = e^{-x} - x$$

$$f(0) = e^0 - 0 = 1\ (+\text{ve})$$

$$f(1) = e^{-1} - 1 = -0.6321\ (-\text{ve})$$

$\therefore$ A root lies between 0 and 1.

$$x_0 = 0,\ x_1 = 1$$

$$x_2 = x_1 - \frac{x_1 - x_0}{f(x_1) - f(x_0)} f(x_1) = 1 - \frac{(1-0)(-0.6321)}{(1) - (0.6321)} = 0.61270$$

$$f(x_2) = (0.61270) = -0.07081 \ (-\text{ve})$$

$$x_3 = x_2 - \frac{x_2 - x_1}{f(x_2) - f(x_1)} f(x_2)$$

$$= 0.61270 - \frac{(0.61270 - 1)}{(-0.07081) + 0.63212} (-0.07081) = 0.56384$$

$$f(x_3) = f(0.56384) = 0.00518 \ (+\text{ve})$$

$$x_4 = 0.56384 - \frac{(0.00518)(0.56384 - 0.61270)}{0.00518 + 0.07081} = 0.56717$$

Hence, $x = 0.56$ is correct to two decimal places.

Example 16.30. Find a root of $xe^x = \cos x$ correct to four decimal places using secant method.

Solution : Let $f(x) = \cos x - xe^x = 0$

Let $x_0 = 0, x_1 = 1$

$$f(x_0) = 1, f(x_1) = \cos 1 - e = -2.17798$$

By Secant method

$$x_2 = x_1 - \frac{x_1 - x_0}{f(x_1) - f(x_0)} f(x_1)$$

$$f(x_1) = 1 - \frac{1 - 0}{-2.17798 - 1} (-2.17798) = 0.31467$$

$$f(x_2) = 0.51987$$

$$x_3 = x_2 - \frac{(x_2 - x_1)}{f(x_2) - f(x_1)} f(x_2)$$

$$= 0.31467 - \frac{(0.31467 - 1)}{0.51987 + 2.17798} (0.51987) = 0.44673$$

$$f(x_3) = 0.20354$$

$$x_4 = x_3 - \frac{x_3 - x_2}{f(x_3) - f(x_2)} f(x_3) = 0.53171$$

$$x_5 = 0.51690$$

$$x_6 = 0.51775, x_7 = 0.51776 \text{ etc.}$$

Hence 0.5177 is the required root.

16.8 FIXED-POINT ITERATION METHOD

In this method, we express $f(x) = 0$ in the form $x = \phi(x)$.

Let $x = x_0$ be an initial approximation of the desired root. Then the first approximation to the root is given by

$$x_1 = \phi(x_0)$$

$$x_2 = \phi(x_1)$$

$$\ldots = \ldots$$

$$\ldots = \ldots$$

$$x_n = \phi(x_{n-1})$$

If the sequences of approximation $x_1, x_2, ..., x_n$ converges to α, it is taken as root of the equation $f(x) = 0$. For convergence purpose, the initial approximation x_0 is to be done carefully. The choice of x_0 is determined according to sufficient condition.

16.8.1 *Sufficient Conditions for Convergence*

(*i*) If α be a root of $f(x) = 0$ which is equivalent to $x = \phi(x)$

(*ii*) I be any interval containing the point $x = \alpha$.

(*iii*) $|\phi'(x)| < 1$ $(x \in I)$ then the sequence approximation $x_0, x_1, ..., x_n$ will converge to α provided the initial approximation x_0 is chosen in I.

Proof: Since α is a root of $x = f(x)$, we have $\alpha = \phi(\alpha)$.

If x_{n-1} and x_n be two successive approximation to α, we have

$$x_n = \phi(x_{n-1}) \quad ...(16.3)$$

$$x_n - \alpha = \phi(x_{n-1}) - \phi(\alpha)$$

By mean value theorem

$$\frac{\phi(x_{n-1}) - \phi(\alpha)}{x_{n-1} - \alpha} = \phi'(\xi), \text{ where } x_{n-1} < \xi < \alpha$$

Hence Eq. (16.3) becomes $x_n - \alpha = (x_{n-1} - \alpha)\,\phi'(\xi)$

If $|\phi'(x_i)| \le K \le 1$ for all i, then

$$|x_n - \alpha| \le K\,|x_{n-1} - \alpha| \quad ...(16.4)$$

Similarly

$$|x_{n-1} - \alpha| \le K\,|x_{n-2} - \alpha|$$

$$|x_n - \alpha| \le K^2\,|x_{n-2} - \alpha|$$

Proceeding in this way, we have

$$|x_n - \alpha| \le K^n\,|x_0 - \alpha|$$

As $n \to \infty$, the RHS tends to zero, therefore, the sequence of approximation converges to the root α.

Notes:

1. The smaller the value of $\phi'(x)$, the more rapid will be the convergence.
2. This method of iteration is particularly useful for finding the real roots of an equation in the form of an infinite series.

Example **16.31.** Solve $e^{-x} = 10x$ using iteration method to correct to four decimal points.

Solution : Let $\qquad f(x) = e^{-x} - 10x = 0$

Then $\qquad f(0) = 1, f(1) = -9.6321$

$\therefore$ the root lies between 0 and 1.

Rewriting $\qquad x = \frac{1}{10}e^{-x} = \phi(x)$

$$\phi'(x) = \frac{-1}{10}e^{-x}$$

$$|\phi'(x)| = \frac{e^{-x}}{10} = \frac{1}{e^x.10} < 1 \;\forall\; x \in (0, 1)$$

Hence, this method can be applied.

Let $\qquad x_0 = 0,$

then
$$x_1 = \phi(x_0) = 1/10 = 0.1$$
$$x_2 = \frac{e^{-0.1}}{10} = 0.09048$$
$$x_3 = 0.091349$$
$$x_4 = 0.091274$$

Hence required root is 0.0913.

***Example* 16.32.** Find the root of the equation $f(x) = x^3 + x^2 - 1 = 0$ by iteration method.

Solution : $f(0) = -1, f(1) = 1$

$\therefore$ root lies between 0 and 1.

Rewriting the given equation

$$x^2(x+1) = 1$$
$$x = \frac{-1}{2\sqrt{x+1}} = \phi(x)$$
$$\phi'(x) = \frac{-1}{2(x+1)^{3/2}}$$

as $|\phi'(x)| < 1$, for all $x \in (0, 1)$.

Hence, method of iteration is applicable here.

Taking initial approximation as $x_0 = 1$

$$x_1 = \phi(x_0) = \frac{1}{\sqrt{1+x_0}} = \frac{1}{\sqrt{2}} = 0.70711, \phi(x_1) = -0.14644$$
$$x_2 = \phi(x_1) = \frac{1}{\sqrt{1+0.70711}} = 0.76537, f(x_2) = 0.3414$$
$$x_3 = 0.75263, \quad f(x_3) = 7.2213 \times 10^{-3}$$
$$x_4 = 0.75536 \quad f(x_4) = 1.5565 \times 10^{-3}$$
$$x_5 = 0.75477 \quad f(x_5) = -3.44323 \times 10^{-4}$$
$$x_6 = 0.7549 \quad f(x_6) = 7.38295 \times 10^{-5}$$

Hence, required root is 0.7549.

Note: The given equation can be written in many ways $x^2 = 1 - x^3$

or
$$x = \sqrt{(1-x^3)} = \phi(x)$$
$$\phi'(x) = \frac{3x^2}{2(1-x^3)^{3/2}} > 1 \text{ in } (0, 1)$$

The condition $|\phi'(x)| < 1$ is not satisfied.

***Example* 16.33.** Solve $x^3 + x^2 - 100 = 0$ using iteration method.

Solution : Let
$$x^2(x+1) = 100$$
$$f(4) = -20$$
$$f(5) = 125 + 25 - 100 = 50$$

$\therefore$ root lies between 4 and 5.

$$x = \frac{10}{\sqrt{x+1}} = \phi(x)$$

$$\phi'(x) = \frac{-10}{2(x+1)^{3/2}}$$

$$|\phi'(x)| < 1, \text{ for all } x \in (4, 5)$$

Let $x_0 = 4$

$$\frac{10}{\sqrt{x_0+1}} = x_1 = \frac{10}{\sqrt{5}} = 4.472$$

$$x_2 = \frac{10}{\sqrt{5.472}} = 4.2749$$

$$x_3 = \frac{10}{\sqrt{4.2749+1}} = 4.354$$

$$x_4 = \frac{10}{\sqrt{4.354+1}} = 4.3217$$

$$x_5 = \frac{10}{\sqrt{4.3217+1}} = 4.3348$$

$$x_6 = \frac{10}{\sqrt{4.3348+1}} = 4.3295$$

$$x_7 = \frac{10}{\sqrt{4.3295+1}} = 4.33168$$

$$x_8 = 4.33079$$

$$x_9 = 4.331159$$

$$x_{10} = 4.33100$$

Hence, required root is 4.331 correct to three decimal places.

***Example* 16.34.** Find the root of equation $x = \frac{1}{2} + \sin x = f(x)$ by iteration method.

Solution :

$$f(x) = \frac{1}{2} + \sin x - x = 0$$

$$f(0) = \frac{1}{2}$$

$$f(1) = \frac{1}{2} + \sin 1 - 1 = 0.3414 > 0$$

$$f(2) = -0.09070 < 0$$

$\therefore$ root lies between 1 and 2.

Rewriting the above equation

$$\phi(x) = \frac{1}{2} + \sin x$$

$$\phi'(x) = \cos x$$

$$|f'(x)| < \text{ for all } x \in (1, 2)$$

Let

$$x_0 = 1$$

$$x_1 = \frac{1}{2} + \sin 1 = 1.3414$$

$$x_2 = \frac{1}{2} + 0.4738 = 1.4738, \; x_3 = 1.4952$$

$$x_4 = 1.49715, \; x_5 = 1.4972$$

Hence, the required root is 1.497.

***Example* 16.35.** Find $\sqrt{30}$ by iteration method.

Solution : Let

$$x = \sqrt{30}$$

$$f(x) = x^2 - 30$$

$$f(5) = -5$$

$$f(6) = -6$$

$\therefore$ root lies between 5 and 6.

$$x^2 - 30 = 0$$

$$x^2 + x - x - 30 = 0$$

$$x(x+1) = x + 30$$

$$x = \frac{x+30}{x+1} = 1 + \frac{29}{x+1} = \phi(x)$$

$$\phi'(x) = \frac{-29}{(x+1)^2}$$

$$|\phi'(x)| = \frac{29}{(x+1)^2} < 1; \; 5 < x < 6$$

Hence, iteration method is applicable taking $x_0 = 5$.

$$x_1 = \phi(x_0) = 1 + \frac{29}{6} = 5.83$$

$$x_2 = \phi(x_1) = 1 + \frac{29}{6.83} = 5.2439$$

$$x_3 = \phi(x_2) = 1 + \frac{29}{6.245} = 5.64453$$

$x_4 = 5.36449$ $x_5 = 5.55653$ $x_6 = 5.42307$

$x_7 = 5.514974$ $x_8 = 5.451284073$ $x_9 = 5.495229$

$x_{10} = 5.464815$ $x_{11} = 5.48582$ $x_{12} = 5.47129$

$x_{13} = 5.48133$ $x_{14} = 5.47439$ $x_{15} = 5.47918$

$x_{16} = 5.47587$ $x_{17} = 5.47816$ $x_{18} = 5.47658$

$x_{19} = 5.47767$ $x_{20} = 5.47691$ $x_{21} = 5.47744$

$x_{22} = 5.477077$

Hence, the root is 5.477.

Example **16.36:** Find a real root of the equation $x - \frac{x^3}{3} + \frac{x^5}{5} - \frac{x^7}{42} + \frac{x^9}{216} - \frac{x^{11}}{1320} + ... = 0.433$ by iteration method.

Solution :

$$x = 0.433 + \frac{x^3}{3} - \frac{x^5}{5} + \frac{x^7}{42} - \frac{x^9}{216} + \frac{x^{11}}{1320}$$

Let

$$x_0 = 0.433$$

$$x_1 = 0.433 + \frac{(0.433)^3}{3} - \frac{(0.433)^3}{3} + \frac{(0.433)^7}{42} - \frac{(0.433)^9}{216} + \frac{(0.433)^{11}}{1320} = 0.45708$$

$$x_2 = 0.433 + \frac{(0.45708)^3}{3} - \frac{(0.45708)^5}{5} + \frac{(0.45708)^7}{42} = 0.460939$$

$$x_3 = 0.433 + \frac{0.46094}{3} - \frac{(0.46094)^5}{5} + \frac{(0.46094)^7}{42} = 0.46158$$

$$x_4 = 0.461695$$

Hence, the root is 0.461 correct to three decimal places.

Example 16.37. Solve by iteration method $x^2 + 1 = x^3$.

Solution :

$$f(x) = x^3 - x^2 - 1$$

Root lies between 1 and 2.

$$f(1.4) = 1.784$$

$$f(2) = 3$$

Numerically

$$f(1.4) < f(2)$$

Root lies near to 1.4

$$x = (x^2 + 1)^{1/3} = \phi(x)$$

$$|\phi'(x)| = |\phi'(1.4)| = 0.388 < 1 \text{ and } |\phi'(x)| = |\phi'(2)| = 0.455 < 1.$$

We shall consider the interval (1.4, 2)

Let

$$x_0 = 1.4$$

$$x_1 = \phi(x_0) = [1 + (1.4)^2]^{1/3} = 1.435$$

$$x_2 = [1 + (1.435)^2]^{1/3} = 1.4252$$

$$x_3 = [1 + (1.422)^2]^{1/3} = 1.459$$

$$x_4 = [1 + (1.459)^2]^{1/3} = 1.462$$

$$x_5 = [1 + (1.462)^2]^{1/3} = 1.464$$

$$x_6 = [1 + (1.464)^2]^{1/3} = 1.464$$

Hence, required root is 1.464.

Example 16.38. Find a real root of $2x - \log_{10} x = 7$ using iteration method.

Solution :

$$f(x) = 2x - \log_{10} x - 7 = 0$$

$$f(3) = 6 - \log_{10} 3 - 7 = -1.4471$$

$$f(4) = 8 - 0.602 - 7 = 0.398$$

The root lies between 3 and 4.

$$x = \frac{1}{2}(\log_{10} x + 7) = \phi(x)$$

$$\phi'(x) = \frac{1}{2}\left(\frac{1}{x}\log_{10} e\right)$$

$$|\phi'(x)| < 1 \text{ when } 3 < x < 4$$

Iteration can be applied x_3 and x_4 are correct to four decimal places.

$$x_0 = 3.3 \text{ (root near to 4)}$$

$$x_1 = \frac{1}{2}(\log_{10} 3.6 + 7) = 3.77815$$

$$x_2 = \frac{1}{2}(\log_{10} 3.77815 + 7) = 3.78863$$

$$x_3 = \frac{1}{2}(\log_{10} 3.78863) = 3.78924$$

$$x_4 = \frac{1}{2}(\log_{10} 3.78924 + 7) = 3.78927$$

Hence, the root is 3.7892 correct to four decimal places.

Example **16.39.** Find the root of $xe^x = 1$ by iteration method if the root lies between 0 and 1.

Solution :

$$x = e^{-x} = \phi(x)$$

$$\phi'(x) = -e^{-x}$$

$$|f'(x)| = |e^{-x}| < 1 \ \forall\ 0 < x < 1$$

Let us take $x_0 = 1$

$x_1 = 0.3678794$ $x_2 = 0.692206$ $x_3 = 0.5004735$

$x_4 = 0.6062435$ $x_5 = 0.5453957$ $x_6 = 0.5796123$

$x_7 = 0.5601154$ $x_8 = 0.5711431$ $x_9 = 0.5684793$

$x_{10} = 0.5684287$ $x_{11} = 0.5664147$ $x_{12} = 0.5675566$

$x_{13} = 0.5669087$ $x_{14} = 0.5672762$ $x_{15} = 0.5670679$

$x_{16} = 0.5671860$ $x_{17} = 0.567119$ $x_{18} = 0.567157$

$x_{19} = 0.5671354$ $x_{20} = 0.5671477$

Hence the root is 0.5671.

16.9 MULLER'S METHOD

Method I: In this method, $f(x)$ is approximated by a second degree curve in the vicinity of a root. The roots of the quadratic are then assumed to be the approximations to the root of the equation $f(x) = 0$.

Let x_{i-2}, x_{i-1}, x_i be three distinct approximations to a root of $f(x) = 0$ and let y_{i-2}, y_{i-1}, y_i be the corresponding value of $y = f(x)$. Assuming that

$$P(x) = A(x - x_i)^2 + B(x - x_i) + y_i \quad \text{...(16.5)}$$

is the parabola passing through the points (x_{i-2}, y_{i-2}), (x_{i-1}, y_{i-1}) and (x_i, y_i), we have

$$y_{i-1} = A(x_{i-1} - x_i)^2 + B(x_{i-1} - x_i) + y_i \quad \text{...(16.5a)}$$

$$y_{i-2} = A(x_{i-2} - x_i)^2 + B(x_{i-2} - x_i) + y_i \quad \text{...(16.5b)}$$

from Eqs (16.5a) and (16.5b), we get

$$y_{i-1} - y_i = A(x_{i-1} - x_i)^2 + B(x_{i-1} - x_i)$$

$$y_{i-2} - y_i = A(x_{i-2} - x_i)^2 + B(x_{i-2} - x_i)$$

Solve the above equations, we get

$$A = \frac{(x_{i-2} - x_i)(y_{i-1} - y_i) - (x_{i-1} - x_i)(y_{i-2} - y_i)}{(x_{i-2} - x_{i-1})(x_{i-1} - x_i)(x_{i-2} - x_i)}$$

$$B = \frac{(x_{i-2} - x_i)^2(y_{i-1} - y_i) - (x_{i-1} - x_i)^2(y_{i-2} - y_i)}{(x_{i-2} - x_{i-1})(x_{i-1} - x_i)(x_{i-2} - x_i)}$$

With the values of A and B given above, the quadratic equation

$P(x) = A(x - x_i)^2 + B(x - x_i) + y_i = 0$ give the next approximation x_{i+1}

$$x_{i+1} - x_i = \frac{(-B \pm \sqrt{B^2 - 4Ay_i})}{2} \qquad \text{...(16.6)}$$

A direct solution from Eq. (16.6) gives inaccurate result, so we can rewrite Eq. (16.6) as

$$A + \frac{B}{x - x_i} + \frac{y_i}{(x - x_i)^2} = 0$$

$$\frac{1}{x - x_i} = \frac{-B \pm \sqrt{B^2 - 4Ay_i}}{2y_i}$$

$$x - x_i = \frac{-2y_i}{(B \mp \sqrt{B^2 - 4Ay_i})}$$

$$x_{i+1} = x_i - \frac{-2y_i}{(B \mp \sqrt{B^2 - 4Ay_i})} \qquad \text{...(16.7)}$$

The sign in the denominator should be chosen so that it will be the largest in magnitude (i.e. if $B > 0$, choose plus, if $B < 0$, choose minus, if $B = 0$, choose either).

Note: Muller's method like Newton-Raphson's method will find a complex. Of course, the computation must use complex arithmetic root if given complex starting values. This method converges at the same rate as Newton-Raphson's method.

Method II: This method is developed by writing a quadratic equation that fits through three points in the vicinity of root, in the form $av^2 + bv + c$.

Let $\qquad h_1 = x_1 - x_0$

and $\qquad h_2 = x_0 - x_2$

We find all the coefficient by evaluating $P_2(v)$ at three points

$$v = 0,\ a(0)^2 + b(0) + c = f_0$$

$$= h_1 \Rightarrow ah_1^2 + bh_1 + c = f_1$$

$$= -h_2 \Rightarrow ah_2^2 - bh_2 + c = f_2$$

From the first equation $c = f_0$. Let $h_2/h_1 = v$, we can solve the other equations for a and b.

$$a = \frac{vf_1 - f_0(1 + v) + f_2}{vh_1^2(1 + v)},\ b = \frac{f_1 - f_0 - ah_1^2}{h_1}$$

After computing a, b and c, we solve for the root of $av^2 + bv + c = 0$ by quadratic formula, choosing the root nearest to middle point x_0, putting this value

$$\text{root} = x_0 - \frac{2c}{b \mp \sqrt{b^2 - 4ac}}$$

With the sign in the denominator taken to give the largest absolute value of the denominator (i.e. $b > 0$, choose plus, if $b < 0$, choose minus, if $b = 0$, choose either).

Example **16.40.** Find a root of $\cos x = xe^x$ which lies between 0 and 1.

Solution : Let $y = \cos x - xe^x$

Let $x_{i-2} = -1, x_{i-1} = 0, x_i = 1$

$$y_{i-2} = 0.9082, y_{i-1} = 1, y_i = -2.178$$

$$A = \frac{(x_{i-2} - x_i)(y_{i-1} - y_i) - (x_{i-1} - x_i)(y_{i-2} - y_i)}{(x_{i-2} - x_{i-1})(x_{i-1} - x_i)(x_{i-2} - x_i)}$$

$$B = \frac{(x_{i-2} - x_i)^2(y_{i-1} - y_i) - (x_{i-1} - x_i)^2(y_{i-2} - y_i)}{(x_{i-2} - x_{i-1})(x_{i-1} - x_i)(x_{i-2} - x_i)}$$

Substituting the value of $x_{i-2}, x_{i-1}, x_i, y_{i-2}, y_{i-1}, y_i$ in above formula, we have

$$A = \frac{(-1-1)(1+2.178) - (0-1)(0.9082 + 2.178)}{(0+1)(0-1)(-1-1)} = -1.6349$$

$$B = \frac{(-1-1)^2(1+2.178) - (0-1)^2(0.9082 + 2.178)}{(-1-0)(0-1)(-1-1)}$$

The quadratic equation is given by

$$-1.6349(x-1)^2 - 4.8129(x-1) - 2.178 = 0$$

The next approximation to the desired root is

$$x_{i+1} = x_i - \frac{2y_i}{B \pm \sqrt{B^2 - 4Ay_i}}$$

$$= 1 - \frac{2(-2.178)}{-4.8129 - \sqrt{(-4.8129)^2 - 4(1.6349)(-2.178)}} = 0.4415$$

For the next iteration, we have

$$x_{i-2} = 0, y_{i-2} = 1$$

$$x_{i-1} = 0.4415, y_{i-1} = 0.21756$$

$$x_i = 1, y_i = -2.178$$

$$A = \frac{(0-1)(0.21756 + 2.178) - (0.4415 - 1)(1 + 2.178)}{(0.4415 - 0)(0.4415 - 1)(0 - 1)}$$

$$= \frac{-2.3956 + 1.775}{(0.4415)(-1)(-0.5585)} = \frac{-0.6206}{0.2465} = -2.517$$

$$B = \frac{(1)^2(0.21756 + 2.178) - (-0.5585)^2(1 + 2.178)}{(-0.4415)(-0.5585)(-1)}$$

$$= \frac{2.3956 - (0.31192)(3.178)}{-0.2466} = \frac{1.4043}{-0.2466} = -5.695$$

Next approximation

$$x_{i+1} = x_i - \frac{2y_i}{(B \pm \sqrt{B^2 - 4Ay_i})} = 1 - \frac{2(-2.178)}{-5.695 - \sqrt{10 - 5.0}}$$

$$= 1 - \frac{(-4.356)}{-5.695 - 3.24} = 1 - \frac{(-4.356)}{(-8.925)}$$

$$= 1 - 0.4875 = 0.5124$$

After a third iteration, we get 0.5177 as the value for the root.

PROBLEMS

1. Find the positive root of $x \cos x = 0$ by bisection method correct to three decimal places. **[Ans.** $x = 0.738$**]**
2. Find the positive root of the following equations by bisection method.
   (*a*) $x^3 - x - 1 = 0$ **[Ans.** 1.32475**]**
   (*b*) $3x - \cos x - 1 = 0$ **[Ans.** 66664**]**
   (*c*) $x^4 - x^2 - 2x^2 - 6x = 4$ **[Ans.** 2.7065**]**
   (*d*) $x \log_{10} x = 1.2$ which lies between 2 and 3 **[Ans.** 2.74**]**
   (*e*) $3x + \sin x = e^x$ **[Ans.** 0.34375**]**
3. Find a real root of the equation x3 – 2x – 5 = 0 by the Regula-Falsi method to correct to three decimal places. **[Ans.** 2.0944**]**
4. Find the real root of $xe^x = 2$ by Regula-Falsi method correct to three decimal places. **[Ans.** 0.85261**]**
5. Find the real root of the following equation correct to four decimal places using Regula-Falsi method.
   (*a*) $x^3 - 4x - 1 = 0$ **[Ans.** 0.2541**]**
   (*b*) $x = \cos x$ **[Ans.** 0.7391**]**
   (*c*) $e^x \sin x = 1$ **[Ans.** 0.5885**]**
   (*d*) $2x - \log_{10} x = 7$ **[Ans.** 3.7892**]**
   (*e*) $x^3 + 1 = 9x$ **[Ans.** 2.943**]**
   (*f*) $2x - 3 \sin x = 5$ **[Ans.** 1.5236**]**
   (*g*) $x - e^{-x} = 0$ **[Ans.** 0.567**]**
6. Using Newton-Raphson method, find a root correct to three decimal places of the following equations.
   (*a*) $x^x = 1000$ **[Ans.** 3.592**]**
   (*b*) $x(1 - \log_e x) = 0.5$ **[Ans.** 0.187**]**
   (*c*) $2x \tan x = 1$ **[Ans.** 0.653**]**
   (*d*) $\sin x = 1 - x$ **[Ans.** 0.511**]**
   (*e*) $x^3 - 3x^2 + 7x - 8 = 0$ **[Ans.** 1.675**]**
   (*f*) $2x - 3 \sin x = 5$ **[Ans.** 1.5236**]**

7. Use Muller's method to find a root of the following equations.

(a) $x^3 - 3x - 5 = 0$ **[Ans. 2.26]**

(b) $x^3 - x^2 - x - 1$ **[Ans. 1.839]**

8. Use Secant method to find a root of the following equations

(a) $2x^3 - 3x - 6 = 0$ **[Ans. 1.78376]**

(b) $x^3 - 6x + 4 = 0$ **[Ans. 0.73205]**

(c) $x = \cos x$ **[Ans. 0.73908]**

(d) $3x = \cos x + 1$ **[Ans. 0.60710]**

Numerical Differentiation and Integration

17.1 INTRODUCTION

Consider a set of values (x_i, y_i) of a function. The process of computing the derivative or derivatives of that function at some value of x from the given set of values is called *numerical differentiation*. This may be done by first approximating the function by a suitable interpolation formula and then differentiating it as many times as desired.

If the value of x are equispaced and the derivative is required near the beginning of the table, we employ *Newton forward interpolation formula*. If it is required near the end of the table, we used *Newton backward interpolation formula*.

For values near the middle of the table, dy/dx is calculated by means of Stirling's or Bessel's formula.

If the values of x are not equispaced, we use Newton's divided differences or Lagrange's interpolation formula to get the derivative value.

Note: While using these formula, it must be observed that the table of values defines the function at these points only and does not completely defines the function and the function may not be differentiable at all.

As such the process of numerical differentiation should be used only if the tabulated values are such that differences of some order are constant otherwise, errors are bound to creep in and go on increasing as derivatives of higher order are found. This is due to the fact that the difference between $f(x)$ and the approximating polynomial $\phi(x)$ may be small at the data point $f'(x) - \phi'(x)$ may be large.

17.2 FORMULA FOR DERIVATIVES

Consider the function $y = f(x)$ which is tabulated for the values

$$x_i = x_0 + ih, i = 0, 1, 2, ..., n$$

17.2.1 *Derivatives Using Newton's Forward Difference Formula*

Newton's forward interpolation formula

$$y = y_0 + p\Delta y_0 + \frac{p(p-1)}{2!}\Delta^2 y_0 + \frac{p(p-1)(p-2)}{3!}\Delta^3 y_0 + ...$$

Method I: Differentiating both sides w.r.t. p, we have

$$\frac{dy}{dp} = \Delta y_0 + \frac{2p-1}{2!}\Delta^2 y_0 + \frac{3p^2-6p+2}{3!}\Delta^3 y_0 + ...$$

Since $$p = \frac{x-x_0}{h}, \text{ therefore } \frac{dp}{dx} = \frac{1}{h}$$

Now

$$\frac{dy}{dx} = \frac{dy}{dp} \cdot \frac{dp}{dx}$$

$$= \frac{1}{h}\left[\Delta y_0 + \frac{2p-1}{2!}\Delta^2 y_0 + \frac{3p^2-6p+2}{3!}\Delta^3 y_0 + \frac{4p^3-18p^2+22p-6}{4!}\Delta^4 y_0 + \ldots\right] \quad \ldots(17.1)$$

At $x = x_0$, $p = 0$, hence putting $p = 0$, we have

$$\left(\frac{dy}{dx}\right)_{x_0} = \frac{1}{h}\left[\Delta y_0 - \frac{1}{2}\Delta^2 y_0 + \frac{1}{3}\Delta^3 y_0 - \frac{1}{4}\Delta^4 y_0 + \frac{1}{5}\Delta^5 y_0 - \frac{1}{6}\Delta^6 y_0 + \ldots\right] \quad \ldots(17.2)$$

Again differentiating Eq. (17.1) w.r.t. x, we get

$$\frac{d^2 y}{dx^2} = \frac{d}{dp}\left(\frac{dy}{dx}\right)\frac{dp}{dx}$$

$$= \frac{1}{h^2}\left[\frac{2}{2!}\Delta^2 y_0 + \frac{6p-6}{3!}\Delta^3 y_0 + \frac{12p^2-36p+22}{4!}\Delta^4 y_0 + \ldots\right]$$

Putting $p = 0$, we obtain

$$\left(\frac{d^2 y}{dx^2}\right) = \frac{1}{h^2}\left[\Delta^2 y_0 - \Delta^3 y_0 + \frac{11}{12}\Delta^4 y_0 - \frac{5}{6}\Delta^5 y_0 + \frac{137}{180}\Delta^6 y_0 + \ldots\right] \quad \ldots(17.3)$$

Similarly $\left(\frac{d^3 y}{dx^3}\right) = \frac{1}{h^3}\left[\Delta^2 y_0 - \frac{3}{2}\Delta^4 y_0 + \ldots\right]$ and so on.

Method II: We know that

$$1 + \Delta = E = e^{hD}$$

$$hD = \log(1+\Delta) = \Delta - \frac{1}{2}\Delta^2 + \frac{1}{3}\Delta^3 - \frac{1}{4}\Delta^4 + \ldots$$

$$D = \frac{1}{h}\left[\Delta - \frac{1}{2}\Delta^2 + \frac{1}{3}\Delta^3 - \frac{1}{4}\Delta^4 + \ldots\right]$$

$$D^2 = \frac{1}{h^2}\left[\Delta - \frac{1}{2}\Delta^2 + \frac{1}{3}\Delta^3 - \frac{1}{4}\Delta^4 + \ldots\right]^2$$

$$= \frac{1}{h^2}\left[\Delta^2 - \Delta^3 + \frac{11}{12}\Delta^4 - \frac{5}{6}\Delta^5 + \ldots\right]$$

$$D^3 = \frac{1}{h^3}\left[\Delta^3 - \frac{3}{2}\Delta^4 + \frac{7}{4}\Delta^5 + \ldots\right]$$

Now applying the above identities to y_0, we get

$$Dy_0 = \left(\frac{dy}{dx}\right)_{x_0} = \frac{1}{h}\left[\Delta y_0 - \frac{1}{2}\Delta^2 y_0 + \frac{1}{3}\Delta^2 y_0 - \frac{1}{4}\Delta^4 y_0 + \frac{1}{5}\Delta^5 y_0 - \frac{1}{6}\Delta^6 y_0 + \ldots\right]$$

$$D^2_{y_0} = \left(\frac{d^2 y}{dx^2}\right)_{x_0} = \frac{1}{h^2}\left[\Delta^2 y_0 - \Delta^3 y_0 + \frac{11}{12}\Delta^4 y_0 - \frac{5}{6}\Delta^5 y_0 + \frac{137}{180}\Delta^6 y_0 + \ldots\right]$$

$$D^3_{y_0} = \left(\frac{d^3 y}{dx^3}\right)_{x_0} = \frac{1}{h^2}\left[\Delta^3 y_0 - \frac{3}{2}\Delta^4 y_0 + \ldots\right]$$

17.2.2 *Derivatives Using Backward Difference Formula*

Newton's backward interpolation formula

$$y = y_n + p\nabla y_n + \frac{p(p+1)}{2!}\nabla^2 y_n + \frac{p(p+1)(p+2)}{3!}\nabla^3 y_n + \ldots$$

Method I: Differentiating both sides w.r.t. p, we get

$$\frac{dy}{dp} = \nabla y_n + \frac{2p+1}{2!}\nabla^2 y_n + \frac{3p^2+6p+2}{3!}\nabla^3 y_n + \ldots$$

Since $p = \frac{x-x_n}{h}$, therefore $\frac{dp}{dx} = \frac{1}{h}$

$$\frac{dy}{dx} = \frac{dy}{dp}\cdot\frac{dp}{dx}$$

$$= \frac{1}{h}\left[\nabla y_n + \frac{2p+1}{2!}\nabla^2 y_n + \frac{3p^2+6p+2}{3!}\nabla^3 y_n + \ldots\right] \quad \ldots(17.4)$$

Put $x = x_n$, $p = 0$, here putting $p = 0$ in Eq. (17.4), we get

$$\left(\frac{dy}{dx}\right)_{x_n} = \frac{1}{h}\left[\nabla y_n + \frac{1}{2}\nabla^2 y_n + \frac{1}{3}\nabla^3 y_n + \frac{1}{4}\nabla^4 y_n + \ldots\right] \quad \ldots(17.5)$$

Again differentiating Eq. (17.5) w.r.t. x, we have

$$\frac{d^2y}{dx^2} = \frac{d}{dp}\left(\frac{dy}{dx}\right)\frac{dp}{dx}$$

$$= \frac{1}{h}\left[\nabla^2 y_n + \frac{6p+6}{3!}\nabla^3 y_n + \frac{6p^2+18p+11}{12}\nabla^2 y_n + \ldots\right]$$

Putting $p = 0$, we obtain and so on.

$$\left(\frac{d^2y}{dx^2}\right)_{x_n} = \frac{1}{h^2}\left[\nabla^2 y_n + \nabla^3 y_n + \frac{11}{12}\nabla^4 y_n + \frac{5}{6}\nabla^5 y_n + \frac{137}{180}\nabla^6 y_n + \ldots\right]$$

Method II: We know that $1 - \nabla = E^{-1} = e^{-hD}$

$$-hD = \log(1-\nabla) = -\left[\nabla + \frac{1}{2}\nabla^2 + \frac{1}{3}\nabla^3 + \frac{1}{4}\nabla^4 + \ldots\right]$$

$$D = \frac{1}{h}\left[\nabla + \frac{1}{2}\nabla^2 + \frac{1}{3}\nabla^3 + \frac{1}{4}\nabla^4 + \ldots\right]$$

$$D^2 = \frac{1}{h^2}\left[\nabla + \frac{1}{2}\nabla^2 + \frac{1}{3}\nabla^3 + \ldots\right]^2 = \frac{1}{h^2}\left[\nabla^2 + \nabla^3 + \frac{11}{12}\nabla^4 + \ldots\right]$$

Similarly $D^3 = \frac{1}{h^3}\left[\nabla^3 + \frac{3}{2}\nabla^4 + \frac{7}{4}\nabla^5 + \ldots\right]$

Applying these identities to y_n, we get

$$Dy_n = \left(\frac{dy}{dx}\right)_{x_n} = \frac{1}{h}\left[\nabla y_n + \frac{1}{2}\nabla^2 y_n + \frac{1}{3}\nabla^3 y_n + \frac{1}{4}\nabla^4 y_n + \ldots\right]$$

$$D^2 y_n = \left(\frac{d^2 y}{dx^2}\right)_{x_n} = \frac{1}{h^2}\left[\nabla^2 y_n + \nabla^3 y_n + \frac{11}{12}\nabla^4 y_n + ...\right]$$

$$\left(\frac{d^3 y}{dx^3}\right)_{x_n} = \frac{1}{h^3}\left[\nabla^3 y_n + \frac{3}{2}\nabla^4 y_n + ...\right]$$

17.2.3 *Derivatives Using Central Difference Formula*

$$y_p = y_0 + \frac{p}{1!}\left[\frac{\Delta y_0 + \Delta y_{-1}}{2}\right] + \frac{p^2}{2!}\Delta^2 y_{-1} + \frac{p(p^2-1)}{3!}\left[\frac{(\Delta^3 y_{-1} + \Delta^3 y_{-2})}{2}\right] + \frac{p^2(p^2-1)}{4!}\Delta^4 y_{-2} + ... \quad (17.6)$$

Differentiating Eq. (17.6), both sides w.r.t. p, we get

$$\frac{dy}{dp} = \left[\frac{\Delta y_0 + \Delta y_{-1}}{2}\right] + \frac{2p}{2!}\Delta^2 y_{-1} + \frac{3p^2-1}{3!}\left[\frac{(\Delta^3 y_{-1} + \Delta^3 y_{-2})}{2}\right] + \frac{4p^3-2p}{4!}\Delta^4 y_{-2} + ...$$

Since

$$p = \frac{x - x_0}{h}, \frac{dp}{dx} = \frac{1}{h}, \frac{dy}{dx} = \frac{dy}{dp}\cdot\frac{dp}{dx}$$

$$\frac{dy}{dx} = \frac{1}{h}\left[\left(\frac{y_0 + \Delta y_{-1}}{2}\right) + p\Delta^2 y_{-1} + \frac{3p^2-1}{6}\left(\frac{\Delta^3 y_{-1} + \Delta^3 y_{-2}}{2}\right) + \frac{2p^3-p}{12}\Delta^4 y_{-2} + ...\right] \quad ...(17.7)$$

At $x = x_0$, $p = 0$, hence putting $p = 0$ in Eq. (17.7), we get

$$\left(\frac{dy}{dx}\right)_{x_0} = \frac{1}{h}\left[\frac{\Delta y_0 + \Delta y_{-1}}{2} - \frac{1}{6}\left(\frac{\Delta^3 y_{-1} + \Delta^3 y_{-2}}{2}\right) + \frac{1}{30}\left(\frac{\Delta^5 y_{-2} + \Delta^5 y_{-3}}{2}\right)\right]$$

Similarly

$$\left(\frac{d^2 y}{dx^2}\right) = \frac{1}{h^2}\left[\Delta^2 y_{-1} - \frac{1}{12}\Delta^4 y_{-2} + \frac{1}{90}\Delta^6 y_{-3} + ...\right] \text{ and so on.}$$

17.3 MAXIMA AND MINIMA OF A TABULATED FUNCTION

Newton's forward interpolation formula:

$$y = y_0 + p\Delta y_0 + \frac{p(p-1)}{2!}\Delta^2 y_0 + \frac{p(p-1)(p-2)}{3!}\Delta^3 y_0 + ...$$

Differentiating w.r.t. p, we get

$$\frac{dy}{dp} = \Delta y_0 + \frac{2p-1}{2}\Delta^2 y_0 + \frac{3p^2-6p+2}{6}\Delta^3 y_0 + ... \quad (17.8)$$

For maxima or minima $\frac{dy}{dp} = 0$. Hence, equating the right hand side of Eq. (17.8) to zero, retaining only up to third differences, we obtain

$$\Delta y_0 + \frac{2p-1}{2}\Delta^2 y_0 + \frac{3p^2-6p+2}{6}\Delta^3 y_0 = 0 \quad (17.9)$$

$$\text{or } \left(\frac{1}{2}\Delta^3 y_0\right)p^2 + (\Delta^2 y_0 - \Delta^3 y_0)p + \left(\Delta y_0 - \frac{1}{2}\Delta^2 y_0 + \frac{1}{3}\Delta^3 y_0\right) = 0$$

Substituting the value of Δy_0, $\Delta^2 y_0$, $\Delta^3 y_0$ from the difference table, we solve this quadratic for p. Then the corresponding value of x are given by $x = x_0 + ph$ at which y is maximum or minimum.

***Example* 17.1. Find the first, second and third derivatives of $f(x)$ at $x = 1.5$ if**

x	1.5	2.0	2.5	3.0	3.5	4.0
$f(x)$	3.375	7.000	13.625	24.000	38.875	59.000

Solution : We have to find the derivative at point $x = 1.5$ which is at the beginning of the given data. Therefore, we use here the derivatives of Newton's forward interpolation formula.

x	$y = f(x)$	Δy	$\Delta^2 y$	$\Delta^3 y$	$\Delta^4 y$
1.5	3.375				
		3.625			
2.0	7.000		3.000		
		6.625		0.750	
2.5	13.625		3.750		0
		10.375		0.750	
3.0	24.000		4.500		0
		14.875		0.750	
3.5	38.875		5.250		
		20.125			
4.0	59.000				

Here $\quad x_0 = 1.5,\ y_0 = 3.375,\ \Delta y_0 = 3.625,\ \Delta^2 y_0 = 3,\ \Delta^3 y_0 = 0.75,\ h = 0.5$

$$\left(\frac{dy}{dx}\right)_{x=x_0} = f'(x) = \frac{1}{h}\left[\Delta y_0 - \frac{1}{2}\Delta^2 y_0 + \frac{1}{3}\Delta^3 y_0 - \frac{1}{4}\Delta^4 y_0 + \ldots\right]$$

$$f'(1.5) = \frac{1}{0.5}\left[3.625 - \frac{1}{2}(3) + \frac{1}{3}(0.75)\right] = 4.75$$

$$\left(\frac{d^2y}{dx^2}\right)_{x=x_0} = f''(x_0) = \frac{1}{h^2}\left[\Delta^2 y_0 - \Delta^3 y_0 + \frac{11}{12}\Delta^4 y_0 + \ldots\right]$$

$$f''(1.5) = \frac{1}{(0.5)^2}[3 - 0.75] = 9$$

$$\left(\frac{d^3y}{dx^3}\right)_{x=y_0} = f'''(x) = \frac{1}{h^3}\left[\Delta^3 y_0 - \frac{3}{2}\Delta^4 y_0 + \ldots\right]$$

$$f''(1.5) = \frac{1}{(0.5)^3}[0.75] = 6$$

***Example* 17.2.** The population of a certain town is shown in the following table.

Year	1951	1961	1971	1981	1991
Population	19.96	36.65	58.81	77.21	94.61

Find the rate of growth of population in the year 1981.

Solution : We have to find the derivative at point $x = 1981$ which is near the end of the given data. Therefore, we use here the derivatives of Newton's backward interpolation formula. The table of differences is as follows:

x	$y = f(x)$	Δy	$\Delta^2 y$	$\Delta^3 y$	$\Delta^4 y$
1951	19.96				
		16.69			
1961	36.65		5.47		
		22.16		−9.23	
1971	58.81		−3.76		11.99
		18.40		2.76	
1981	77.21		−1		
		17.40			
1991	94.61				

Here $h = 10, x_n = 1991, \nabla y_n = 17.4, \nabla^2 y_n = -1, \nabla^3 y_n = 2.76, \nabla^4 y_n = 11.99$

$$\left(\frac{dy}{dx}\right)_{x=x_0} = \frac{1}{h}\left[\nabla y_n + \frac{2p+1}{3}\nabla^2 y_n + \frac{3p^2+6p+2}{6}\nabla^3 y_n + ...\right]$$

$$p = \frac{1981-1991}{10} = -1$$

But $p = -1, h = 10$

$$y'(1981) = \frac{1}{10}\left[17.4 + \frac{2(-1)+1}{2}(-1) + \frac{3(-1)^2+6(-1)+2}{6}(2.76) + \frac{2(-1)^3+9(-1)^2+11(-1)+3}{12}(11.99)\right]$$

$$= 1.6440833$$

We have to find out the rate of growth of population in the year 1981, i.e. $\left(\frac{dy}{dx}\right)_{x=x_n}$ now $x_n + ph = 1981$

Example 17.3. Obtain the value of $f'(90)$ using Stirling's formula to the following data.

x	60	75	90	105	120
$f(x)$	28.2	38.2	43.2	40.9	37.7

Also find the maximum value of function from the data.

Solution : Since $x = 90$ is in the middle of the table, we use central difference formula. We use Stirling's formula.

x	$y = f(x)$	Δy	$\Delta^2 y$	$\Delta^3 y$	$\Delta^4 y$
60	28.2				
		10			
75	38.2		−5		
		5		−2.3	
90	43.2		−7.3		8.7
		−2.3		6.4	
105	40.9		−0.9		

		−3.2			
120	37.7				

Here $x_0 = 90$, $y_0 = 43.2$, $\nabla y_0 = -2.3$, $\nabla_{y-1} = 5$, $\nabla^3 y_{-1} = -2.3$, $\nabla^3 y_{-2} = 6.4$ and $h = 15$.

$$\left(\frac{dy}{dx}\right)_{x=y_0} = \frac{1}{h}\left[\left\{\frac{\Delta y_0 + \Delta_{y-1}}{2}\right\} - \frac{1}{6}\left\{\frac{\Delta^3 y_{-1} + \Delta^3 y_{-2}}{2}\right\} + \ldots\right]$$

$$f(90) = \frac{1}{15}\left[\left(\frac{-2.3+5}{2}\right) - \frac{1}{6}\left(\frac{-2.3+6.4}{2}\right)\right] = 0.0672222$$

To find the maximum value of tabular function by Stirling's formula

$$y = y(x_0 + ph) = y_0 + \frac{p}{2}(\Delta y_0 + \Delta y_{-1}) + \frac{p^2}{2!}\Delta^2 y_{-1}$$

$$+ \frac{p(p^2-1)}{3!}\left[\frac{(\Delta^3 y_{-1} + \Delta^3 y_{-2})}{2}\right] + \frac{p^2(p^2-1)}{4!}\Delta^4 y_{-2} + \ldots$$

Substituting the values from table, we get after simplification

$$y = 43.2 + 1.35p - 3.65p^2 + 0.3417\,(p^3 - p)$$

$$= 0.3417p^3 - 3.65p^2 + 1.0083p + 43.2$$

If y is maximum, $\frac{dy}{dp} = 0$,

i.e. $1.0251p^2 - 7.3p + 1.0083 = 0$

$$p = \frac{7.3 + \sqrt{(7.3)^2 - 4(1.0251)(1.0083)}}{2.(1.0251)}$$

$$= 6.9803 \text{ or } 0.1409$$

$p = 6.9803$ is out of range

$\therefore \; p = 0.1409$ because p lies between $-1/4$ and $1/4$.

Hence $x = x_0 + ph = 90 + 15(0.1409) = 92.1135$

and maximum value of $y = 0.3417\,(0.1409)^3 - 3.65\,(0.1409)^2 + 1.0083\,(0.1409) + 43.2$

$= 43.27$

***Example* 17.4.** Using Bessel's formula, find $f(7.5)$ from the following data:

x	7.47	7.48	7.49	7.50	7.51	7.52	7.53
$f(x)$	0.193	1.195	0.198	0.201	0.203	0.206	0.208

Solution : Taking $x_0 = 7.50$, $h = 0.01$, we have

p	x	y_p	Δy	$\Delta^2 y$	$\Delta^3 y$	$\Delta^4 y$	$\Delta^5 y$	$\Delta^6 y$
−3	7.47	0.193						
			0.002					
−2	7.48	0.195		0.001				
			0.003		−0.001			

– 1	7.49	0.198		0.000		0.000		
			0.003		– 0.001		0.003	
0	7.50	0.201		– 0.001		0.003		– 0.010
			0.002		0.002		– 0.007	
1	7.51	0.203		0.001		– 0.004		
			0.003		– 0.002			
2	7.52	0.206		– 0.001				
			0.002					
3	7.53	0.208						

Bessel's formula is

$$y_p = y_0 + p\Delta y_0 + \frac{p(p-1)}{2!}\left[\frac{\Delta^2 y_{-1} + \Delta^2 y_0}{2}\right] + \frac{\left(p-\frac{1}{2}\right)p(p-1)}{3!}\Delta^3 y_{-1}$$

$$+ \frac{(p+1)p(p-1)(p-2)}{4!}\left[\frac{\Delta^4 y_{-2} + \Delta^4 y_{-1}}{2}\right] + \frac{\left(p-\frac{1}{2}\right)(p+1)p(p-1)(p-2)}{5!}\Delta^5 y_{-2}$$

$$+ \frac{(p+2)(p+1)p(p-1)(p-2)(p-3)}{6!}\left[\frac{\Delta^6 y_{-3} + \Delta^6 y_{-2}}{2}\right] \quad ...(i)$$

Since

$$p = \frac{x - x_0}{h}, \frac{dp}{dx} = \frac{1}{h}$$

$$\frac{dy}{dx} = \frac{dy}{dx} \cdot \frac{dp}{dx} = \frac{1}{h}\frac{dy}{dp}$$

$$\left.\frac{dy}{dx}\right]_{x=x_0} = \frac{1}{h} \cdot \left[\frac{dy}{dp}\right]_{p=0}$$

Differentiate Eq. (*i*) w.r.t. p and put $p = 0$

$$\frac{dy}{dx} = \frac{1}{h}\left[\begin{array}{l}\Delta y_0 - \frac{1}{4}(\Delta^2 y_{-1} + \Delta^2 y_0) + \frac{1}{12}\Delta^3 y_{-1} + \frac{1}{24} \\ (\Delta^4 y_{-2} + \Delta^4 y_{-1}) - \frac{1}{120}\Delta^5 y_{-2} - \frac{1}{240}(\Delta^6 y_{-3} + \Delta^6 y_{-2})\end{array}\right]$$

$$\left(\frac{dy}{dx}\right)_{7.5} = \frac{1}{0.01}\left[\begin{array}{l}0.002 - \frac{1}{4}(-0.001 + 0.001) + \frac{1}{12}(0.002) + \frac{1}{24} \\ (-0.004 + 0.003) - \frac{1}{120}(-0.007) - \frac{1}{240}(-0.010 + 0)\end{array}\right]$$

$$= 0.223$$

Example **17.5.** A slider in a machine moves along a straight rod. Its distance x cm along the rod is given below for various values of the time t seconds. Find the velocity of the slider and its acceleration when $t = 0.3$ seconds.

t	0	0.1	0.2	0.3	0.4	0.5	0.6
x	30.13	31.62	32.87	33.64	33.95	33.81	33.24

Solution : Taking $x_0 = 7.50$, $h = 0.01$, we have

t	x	Δ	Δ^2	Δ^3	Δ^4	Δ^5	Δ^6
0	30.13						
		1.49					
0.1	31.62		– 0.24				
		1.25		– 0.24			
0.2	32.87		– 0.48		0.26		
		0.77		0.02		– 0.27	
0.3	33.64		– 0.46		– 0.01		0.29
		0.31		0.01		0.02	
0.4	33.95		– 0.45		0.01		
		– 0.14		0.02			
0.5	33.81		– 0.43				
		– 0.57					
0.6	33.24						

As the derivatives are required near the middle of the table, we use Stirling's formula

$$\left(\frac{dx}{dt}\right)_{t_0} = \frac{1}{h}\left[\left(\frac{\Delta x_0 + \Delta x_{-1}}{2}\right) - \frac{1}{6}\left(\frac{\Delta^3 x_{-1} + \Delta^3 x_{-2}}{2}\right) + \frac{1}{30}\left(\frac{\Delta^5 x_{-2} + \Delta^5 x_{-3}}{2}\right) + \ldots\right]$$

$$\left(\frac{d^2x}{dt^2}\right)_{t_0} = \frac{1}{h^2}\left[\Delta^2 x_{-1} - \frac{1}{12}\Delta^4 x_{-2} + \frac{1}{90}\Delta^6 x_{-3} + \ldots\right]$$

Here $h = 0.1$, $t_0 = 0.3$, $\Delta x_0 = 0.31$, $\Delta x_{-1} = 0.77$, $\Delta^2 x_{-1} = -0.46$ etc.

Put the values

$$\left(\frac{dx}{dt}\right)_{0.3} = \frac{1}{0.1}\left[\left(\frac{0.31 + 0.77}{2}\right) - \frac{1}{6}\left(\frac{0.01 + 0.02}{2}\right) + \frac{1}{30}\left(\frac{0.02 - 0.27}{2}\right) + \ldots\right]$$

$$\left(\frac{d^2x}{dt^2}\right)_{0.3} = \frac{1}{(0.1)^2}\left[-0.46 - \frac{1}{12}(-0.01) + \frac{1}{90}(0.29) + \ldots\right]$$

$$= -45.6$$

velocity is 5.33 cm/sec and acceleration is – 45.6 cm/sec^2.

Example **17.6:** Find $f(10)$ from the following data:

x	3	5	11	27	34
$f(x)$	– 13	23	899	17315	35606

Solution : As the value of x are not equispaced, we shall use Newton's divided difference formula.

The divided difference table:

x	$f(x)$	1st divided difference	2nd divided difference	3rd divided difference	4th divided difference
3	− 13				
		18			
5	23		16		
		146		0.998	
11	899		29.96		0.0002
		1025		1.003	
27	17315		69.04		
		2613			
34	35606				

Newton's divided difference formula:

$$\begin{aligned} f(x) &= f(x_0) + (x - x_0)\, f(x_0, x_1) + (x - x_0)(x - x_1)\, f(x_0, x_1, x_2) \\ &\quad + (x - x_0)(x - x_1)(x - x_2)\, f(x_0, x_1, x_2, x_3) \\ &\quad + (x - x_0)(x - x_1)(x - x_2)(x - x_3)\, f(x_0, x_1, x_2, x_3, x_4) \end{aligned}$$

Differentiating w.r.t. x, we get

$$\begin{aligned} f'(x) &= f(x_0, x_1) + (2x - x_0 - x_1)\, f(x_0, x_1, x_2) + [3x^2 - 2x\,(x_0 + x_1 + x_2) \\ &\quad + x_0x_1 + x_1x_2 + x_2x_0]\, f(x_0, x_1, x_2, x_3) + [4x^2 - 3x^2\,(x_0 + x_1 + x_2 + x_3) \\ &\quad + 2x(x_0x_1 + x_1x_2 + x_2x_3 + x_2x_0 + x_1x_3 + x_0x_2) \\ &\quad - (x_0x_1x_2 + x_2x_2x_3 + x_2x_3x_0 + x_0x_1x_3)]\, f(x_0, x_1, x_2, x_3, x_4) \end{aligned}$$

Put $x_0 = 3$, $x_1 = 5$, $x_2 = 11$, $x_3 = 27$ and $x = 10$

$$\begin{aligned} f'(10) &= 18 + 12 \times 16 + 23 \times 0.998 - 426 \times 0.002 \\ &= 232.869 \end{aligned}$$

***Example* 17.7.** **Given the following data, find the maximum value of y:**

x	− 1	1	2	3
$f(x)$	− 21	15	12	3

Solution : Since the arguments are not equispaced, we will form the divided difference table.

x	y	Δy	$\Delta^2 y$	$\Delta^3 y$
− 1	− 21			
		18		
1	15		− 7	
		− 3		1
2	12		− 3	
		− 9		
3	3			

Using Newton's divided difference formula

$$y = y_0 + (x - x_0)\,\Delta y_0 + (x - x_0)(x - x_1)\,\Delta^2 y_0 + (x - x_0)(x - x_1)(x - x_2)\,\Delta^3 y_0$$

$$= -21 + (x + 1)(18) + (x + 1)(x - 1)(-7) + (x + 1)(x - 1)(x - 2)(1)$$

$$= x^3 - 9x^2 + 17x + 6$$

for maximum value, $\frac{dy}{dx} = 0 \Rightarrow 3x^2 - 18x + 17 = 0$

$$x = \frac{+18 \pm \sqrt{(-18)^2 - 4(3)(17)}}{6}$$

$$= 4.8257 \text{ or } 1.1743$$

But $x = 4.8257$ is out of range

$\therefore$ $= 1.1743$ is the value giving the maximum of y

Maximum of y (at $x = 1.1743$) $= (1.743)^3 - 9\,(1.1743)^2 + 17\,(1.1743) + 6$

$$= 15.171612$$

PROBLEMS

1. From the following table of values, estimate $y'(1.05)$ and $y''(1.25)$.

x	1.00	1.05	1.10	1.15	1.20	1.25	1.30
y	1.0000	1.0247	1.0488	1.0724	1.0954	1.1180	1.1402

[Ans. $y'(1.05) = 0.4876$, $y''(1.25) = -0.1583$]

2. Find $f'(8)$ given $f(6) = 1.556$, $f(7) = 1.908$, $f(12) = 2.158$.

[Ans. $f'(8) = 0.109$]

3. Find the values of $f'(0.5)$ using Stirling's formula from the following data.

x	0.35	0.40	0.45	0.50	0.55	0.60	0.65
$f(x)$	1.521	1.506	1.488	1.467	1.444	1.418	1.389

[Ans. – 0.44]

4. Find the value of cos (1.74) from the following table.

x	1.7	1.74	1.78	1.82	1.86
sin x	0.9916	0.9857	0.9857	0.9691	0.9584

[Ans. – 0.17125]

5. Find $y'(1.2)$ using Bessel's formula from the following data.

x	1.0	1.1	1.2	1.3	1.4
y	43.1	47.7	52.1	56.4	60.8

[Ans. 43.1666]

6. The following data gives corresponding value of pressure and specific volume of a superheated steam:

v	1.0	1.2	1.4	1.6	1.8	2.0
p	0	0.128	0.544	1.296	2.432	4.00

(*i*) Find the rate of change of pressure with respect to volume when $V = 2$

(*ii*) Find the rate of change of volume with respect to pressure

Hint: $\left(\frac{dv}{dp} = \frac{1}{\frac{dp}{dv}}\right)$

[Ans. (*i*) – 52.4, (*ii*) – 0.01908**]**

7. If $y = f(x)$ y_n denotes $f(x_0 + nh)$. Prove that if powers of h above h^6 be neglected.

$$\left(\frac{dy}{dx}\right)_{x=x_0} = \frac{3}{4h}\left[(y_1 - y_{-1}) - \frac{1}{5}(y_2 - y_{-2}) + \frac{1}{45}(y_3 - y_{-3})\right]$$

8. From the following table, find the maximum value of y:

x	0	2	4	6
y	3	3	11	27

[Ans. $x = 1, y = 2$**]**

17.4 CUBIC SPLINE METHOD

The cubic spline derived in interpolation with cubic spline can conveniently be used to compute the first and second derivatives of a function. For a natural cubic spline, the recurrence formula are given in Eqs. (17.11) and (17.12).

$$\frac{h_i}{6}M_{i-1} + \frac{1}{3}(h_i + h_{i+1})M_i + \frac{h_{i+1}}{6}M_{i+1} = \frac{y_{i+1} - y_i}{h_{i+1}} - \frac{y_i - y_{i-1}}{h_i} \quad (i = 1, 2, 3, ..., N-1) \qquad ...(17.10)$$

For equal intervals, we have $h_i = h_{i+1} = h$ and equation simplifies to may be used to compute the spline second derivatives depending upon the choice of subdivisions gives the spline in the interval of interest from which the first derivatives can be computed.

$$M_{i-1} + 4M_i + M_{i+1} = \frac{6}{h^2}(y_{i+1} - 2y_i + y_{i-1})(i = 1, 2, 3, ..., N-1) \qquad ...(17.11)$$

Thus,

$$S_{i+1}(x) = \frac{1}{h_{i+1}}\left[\begin{array}{l}\frac{(x_{i+1} - x)^3}{6}M_i + \frac{(x - x_i)^3}{6}M_{i+1} \\ + \left(y_i - \frac{h_{i+1}^2}{6}M_i\right)(x_{i+1} - x) + \left(y_{i+1} - \frac{h_{i+1}^2}{6}M_{i+1}\right)(x - x_i)\end{array}\right] \qquad ...(17.12)$$

For the first derivatives at the tabular points, it would of course, be easier to use the formula given below directly if on the other hand, end conditions involving the first derivatives are given

$$S'_i(x_i) = \frac{h}{2}M_i - \frac{1}{h_i}\left(y_{i-1} - \frac{h_i^2}{6}M_{i-1}\right) + \frac{1}{h_i}\left(y_i - \frac{h_i^2}{6}M_i\right)$$

$$= \frac{1}{h_i}(y_i - y_{i-1}) + \frac{h_i}{6}M_{i-1} + \frac{h_i}{3}M_i \quad (i = 1, 2, 3, ..., N-1)$$

$$S'_{i+1}(x_i) = \frac{1}{h_{i+1}}(y_{i+1} - y_i) - \frac{h_{i+1}}{3} M_i - \frac{h_{i+1}}{6} M_{i+1} \; (i = 1, 2, ..., N-1)$$

Thus the recurrence formula is for equally spaced knots, may be used to compute the remaining first derivatives.

$$\frac{1}{h} m_{i-1} + 2\left(\frac{1}{h_i} + \frac{1}{h_{i+1}}\right) m_i + \left(\frac{1}{h_{i+1}}\right) m_{i+1} = \frac{3}{h_{i+1}^2}(y_{i+1} - y_i) + \frac{3}{h_i^2}(y_i - y_{i-1})(i = 1, 2, ..., N-1)$$

i.e.
$$m_{i-1} + 4m_i + m_{i+1} = \frac{3}{h}(y_{i+1} - y_{i-1})(i = 1, 2, ..., N-1)$$

Example 17.8. We consider the function $f(x) = \sin x$ in $[0, \pi]$.

Solution : $M_0 = M_N = 0$

Let $N = 2$, i.e. $h = \pi/2$

The

x	0	$\pi/2$	π
y	0	1	0

$$M_0 = M_2 = 0$$

Using formula of Eq. (7.11), we obtain

$$M_0 + 4M_1 + M_2 = \frac{6}{h^2}(y_0 - 2y_1 + y_2)$$

$$M_1 = \frac{-12}{\pi^2}$$

Equation (7.12) now gives the spline in each interval thus, for $0 \le x \le \pi/2$, we obtain which gives

$$S(x) = \frac{2}{\pi}\left[0 + \frac{(x-0)^3}{6} \times \left(\frac{-12}{\pi^2}\right) + \frac{3x}{2}\right] = \frac{2}{\pi}\left[-\frac{2x^3}{\pi^2} + \frac{3x}{2}\right]$$

$$S'(x) = \frac{2}{\pi}\left[\frac{-2}{\pi^2}(3x^2) + \frac{3}{2}\right]$$

$$S'\left(\frac{\pi}{4}\right) = \frac{2}{\pi}\left[\frac{-6}{\pi^2}\left(\frac{\pi^2}{16}\right) + \frac{3}{2}\right] = \frac{9}{4\pi} = 0.71619725$$

Exact value of $S'\left(\frac{\pi}{4}\right) = \cos\frac{\pi}{4} = \frac{i}{\sqrt{2}} = 0.70710681$

The percentage error in the computed value of $S'\left(\frac{\pi}{4}\right)$ is 1.28%.

From (*i*)
$$S'(x) = \frac{-24}{\pi^3} x$$

$$S''\left(\frac{\pi}{4}\right) = \frac{-24}{\pi^3} \cdot \frac{\pi}{4} = \frac{-6}{\pi^2} = -0.60792710$$

Since the exact value is $\frac{-1}{\sqrt{2}}$ the percentage error in this result is 14.03%.

We now consider value of $y = \sin x$ in interval of 10° from $x = 0$ to π. To obtain the spline second derivatives, we have used a computer and the results are given in the table below (up to $x = 90°$).

x in degrees	*y''(x)* Exact	Cubic spline
10	– 0.173648178	– 0.174089426
20	– 0.342020143	– 0.342889233
30	– 0.500000000	– 0.501270524
40	– 0.642787610	– 0.644420964
50	– 0.766044443	– 0.767990999
60	– 0.866025404	– 0.868226016
70	– 0.939692621	– 0.942080425
80	– 0.984807753	– 0.987310197
90	– 1.0000000000	– 1.002541048

It is seen that there is a greatest accuracy in the values of spline second derivatives.

17.5 NUMERICAL INTEGRATION

The process of computing $\int_a^b y\,dx$, where $y = f(x)$ is given by a set of tabulated values $[x_i, y_i]$, $i = 1, 2, ..., n$; $a = x_0$ and $b = x_n$ is called *numerical integration.*

This process when applied to a function of a single variable $y = f(x)$ is called quadrature.

17.6 QUADRATURE FORMULA

Let us consider an interval (a, b) divided into n equal parts of width h (Fig. 17.1), so that

$$x_0 = a,\ x_0 + h = x_1$$

$$x_0 + 2h = x_2, ...; x_0 + nh = x_n = b$$

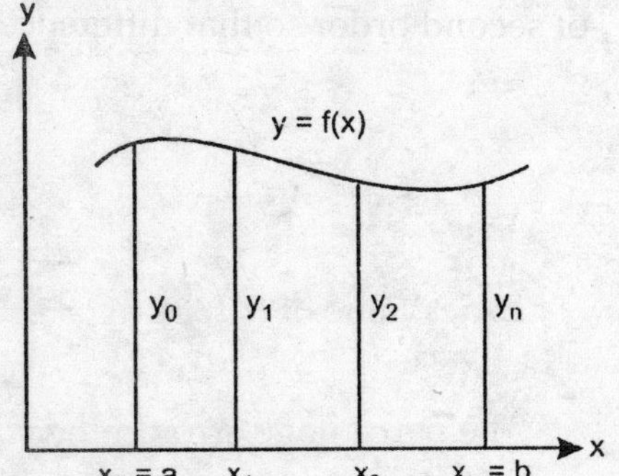

Fig. 17.1

Now $$I = \int_{x_0}^{x_n+nh} f(x)\,dx$$

Put $$x = x_0 + ph, \text{ so that } dx = h\,dp$$

or $$I = h\int_0^n f(x_0 + ph)\,dp$$

Replacing $f(x_0 + ph)$ by Newton's forward interpolation formula, we get

$$I = h\int_0^n \left(y_0 + p\Delta y_0 + \frac{p(p-1)}{2!}\Delta^2 y_0 + ...\right)dp$$

Integrating each term by term, we have after substituting the limit

$$\int_{x_0}^{x_0+nh} f(x)\,dx = h\left[\begin{array}{l} ny_0 + \dfrac{n^2}{2}\Delta y_0 + \dfrac{1}{2}\left(\dfrac{n^3}{3} - \dfrac{n^2}{2}\right)\Delta^2 y_0 + \dfrac{1}{6}\left(\dfrac{n^4}{4} - n^3 + n^2\right) \\ \Delta^3 y_0 + \dfrac{1}{24}\left(\dfrac{n^5}{5} - \dfrac{3n^4}{2} + \dfrac{11n^3}{3} - 3n^2\right)\Delta^4 y_0 + ... \end{array}\right] \quad ...(17.13)$$

This is known as Newton-Cotes' quadrature formula, from the general formula, we deduce the following important quadrature rules by taking $n = 1, 2, 3, ...$

17.6.1 *Trapezoidal Rule*

Put $n = 1$ in Eq. (17.13) and neglecting second and higher order differences, we get

$$\int_{x_0}^{x_0+h} f(x)\,dx = h\left[y_0 + \frac{1}{2}\Delta y_0\right]$$

$$= h\left[y_0 + \frac{1}{2}(y_1 - y_0)\right] = \frac{h}{2}[y_0 + y_1]$$

$$\int_{x_0+h}^{x_0+2h} f(x)\,dx = \frac{h}{2}[y_1 + y_2] \qquad \text{...(17.14)}$$

$$\int_{x_0+2h}^{x_0+3h} f(x)\,dx = \frac{h}{2}[y_2 + y_3] \qquad \text{...(17.15)}$$

$$\int_{x_0+(n-1)h}^{x_0+nh} f(x)\,dx = \frac{h}{2}[y_{n-1} + y_n] \qquad \text{....(17.16)}$$

Adding

$$\int_{x_0}^{x_0+nh} f(x)\,dx = \frac{h}{2}[y_0 + y_n + 2(y_1 + y_2 + \ldots + y_{n-1})] \qquad \text{...(17.17)}$$

Note: The area of each strip (trapezium) is found separately. Then the area under the curve and the ordinates x_0 and $x_0 + nh$ is approximately equal to sum of the area of n trapeziums.

17.6.2 *Simpson's 1/3 Rule*

Put $n = 2$ in Eq. (7.13) above and taking the curve through (x_0, y_0), (x_1, y_1) and (x_2, y_2) as a parabola, i.e. a polynomial of second order so that differences of order higher than second order will vanish, we get

$$\int_{x_0}^{x_0+2h} f(x)\,dx = h\left[2y_0 + 2\Delta y_0 + \frac{1}{2}\left(\frac{8}{3} - 2\right)\Delta^2 y_0\right]$$

$$= h\left[2y_0 + 2(y_1 - y_0) + \frac{1}{3}(y_2 - 2y_1 + y_0)\right] = \frac{h}{3}[y_0 + 4y_1 + y_2]$$

$$\int_{x_0+2h}^{x_0+4h} f(x)\,dx = \frac{h}{3}(y_2 + 4y_3 + y_3)$$

$$\ldots = \ldots$$

$$\int_{x_0+(n-2)h}^{x_0+nh} f(x)\,dx = \frac{h}{3}(y_{n-2} + 4y_{n-1} + y_n), \ n \text{ being even}$$

Adding

$$\int_{x_0}^{x_0+nh} f(x)\,dx = \frac{h}{3}[(y_0 + y_n) + 4(y_1 + y_3 + \ldots + y_{n-1}) + 2(y_2 + y_4 + \ldots + y_{n-2})]$$

This is known as Simpson's 1/3 rule or Simpson's rule and is most commonly used.

Note: While applying Simpson's 1/3rd rule, the given interval must be divided into even number of equal subintervals, since we find the area of two strips at a time.

17.6.3 *Simpson's 3/8 Rule*

Putting $n = 3$ and in Eq. (7.13) neglecting all differences above the third order, we have

$$\int_{x_0}^{x_0+3h} f(x)\,dx = h\left[3y_0 + \frac{9}{2}\Delta y_0 + \frac{1}{2}\left(\frac{27}{3} - \frac{9}{2}\right)\Delta^2 y_0 + \ldots\right]$$

$$= h\left[3y_0 + \frac{9}{2}(y_1 - y_0) + \frac{9}{4}(y_2 - 2y_1 + y_0) + \frac{3}{8}(y_3 - 3y_2 + 3y_1 - y_0)\right]$$

$$= \frac{3h}{8}[y_0 + 3y_1 + 3y_2 + y_3]$$

$$\int_{x_0+3h}^{x_0+6h} f(x)\,dx = \frac{3h}{8}(y_3 + 3y_4 + 3y_5 + y_6)$$

$$\ldots = \ldots$$

$$\int_{x_n+(n-3)h}^{x_0+nh} f(x)\,dx = \frac{3h}{8}(y_{n-3} + 3y_{n-2} + 3y_{n-1} + y_n) \quad \text{(if } n \text{ is a multiple of 3)}$$

$$\int_{x_0}^{x_0+nh} f(x)\,dx = \frac{3h}{8}[y_0 + y_n + 3(y_1 + y_2 + y_4 + y_5 + \ldots + y_{n-1}) + 2(y_3 + y_6 + y_9 + \ldots + y_{n-3})$$

This is known as Simpson's 3/8 rule.

Note: Application of Simpson's rule: If various ordinates in a table represents equispaced cross-sectional area then Simpson's rule gives the volume of the solid. If ordinates denote the velocities at equal intervals of time, then Simpson's rule gives the distance travelled.

17.6.4 *Boole's Rule*

Put $n = 4$ in Eq. (7.13) and neglecting all differences above fourth order, we have

$$\int_{x_0}^{x_0+4h} f(x)\,dx = 4h\left[y_0 + 2\Delta y_0 + \frac{5}{3}\Delta^2 y_0 + \frac{2}{3}\Delta^3 y_0 + \frac{7}{90}\Delta^4 y_0\right]$$

$$= \frac{2h}{5}[7y_0 + 32y_1 + 12y_2 + 32y_3 + 7y_4]$$

$$\int_{x_0+4h}^{x_0+8h} f(x)\,dx = \frac{2h}{45}[7y_4 + 32y_5 + 12y_6 + 32y_7 + 7y_8]$$

Adding all these integrals from x_0 to $x_0 + nh$, where n is multiple of 4, we get

$$\int_{x_0}^{x_0+nh} f(x)\,dx = \frac{2h}{45}[7y_0 + 32y_1 + 12y_2 + 32y_3 + 14y_4 + 32y_5 + 12y_6 + 32y_7 + 14y_8 + \ldots]$$

This is known as Boole's rule.

Note: The number of subintervals should be taken as a multiple of 4.

17.6.5 Weddle's Rule

Put $n = 6$ in Eq. (17.13) and neglecting all differences above sixth order, we obtain

$$\int_{x_0}^{x_0+6h} f(x)\,dx = h\left[6y_0 + 18\Delta y_0 + 27\Delta^2 y_0 + 24\Delta^3 y_0 + \frac{123}{10}\Delta^4 y_0 + \frac{33}{10}\Delta^5 y_0 + \frac{41}{140}\Delta^6 y_0\right]$$

If we replace the coefficient of $\Delta^6 y_0$ by $\frac{42}{140}$ (the error made will be negligible) and the difference in term of y's, we get

$$\int_{x_0+3h}^{x_0+6h} f(x)\,dx = \frac{3h}{10}[y_0 + 5y_1 + y_2 + 6y_3 + y_4 + 5y_5 + y_6]$$

Similarly
$$\int_{x_0+6h}^{x_0+12h} f(x)\,dx = \frac{3h}{10}[y_6 + 5y_7 + y_8 + 6y_9 + y_{10} + 5y_{11} + y_{12}]$$

Adding all these integrals and rearranging them in an order

$$\int_{x_0}^{x_0+nh} f(x)\,dx = \frac{3h}{10}[(y_0 + y_n) + (5y_1 + y_2 + 6y_3 + y_4 + 5y_5) + (2y_6 + 5y_7 + y_8 + 6y_9$$
$$+ y_{10} + 5y_{11}) + (2y_{n-6} + 5y_{n-5} + y_{n-4} + 6y_{n-3} + y_{n-2} + 5y_{n-1})]$$

This formula is used when number of subintervals is a multiple of 6.

Note: Weddle's rule is generally more accurate than any others. Of the two Simpson rules, the 1/3 rule is better. Simpson rule is useful to civil engineers for calculating the amount of earth that must be moved to fill a depression or make a dam.

In the above formula, the coefficients may be remembered in groups of six.

First group: Coefficients: 1, 5 1, 6 1, 5

All interior group: Coefficients: 2, 5 1, 6 1, 5

Last group: Coefficients: 2, 5 1, 6 1, 5.

Note: If there are only 7 ordinates, the coefficients are 1, 5, 1, 6, 1, 5, 1.

1. In trapezoidal rule $f(x)$ is a linear function of x, *i.e.* of the form $f(x) = ax + b$. It is the simplest rule but least accurate.
2. In Simpson's 1/3 rule, $f(x)$ is polynomial of second degree, *i.e.* $f(x) = ax^2 + bx + c$. To apply this rule, the number of intervals n must be even, in other words, the number of ordinates must be odd.
3. In Simpson's 3/8 rule, $f(x)$ is a polynomial of third degree, *i.e.* $f(x) = ax^3 + bx^2 + cx + d$. To apply this rule, the number of intervals n must be a multiple of 3.
4. In Weddle's rule, $f(x)$ is a polynomial of degree 6.

To apply this rule, the number of intervals n must be a multiple of 6, but it requires at least seven consecutive values of the function.

17.7 NEWTON-COTES' FORMULA

As the Lagrange's formula is based on the same assumption as an approximate integration formula, viz. $f(x)$ is a polynomial, it can be used in finding an approximate formula.

Let y_0, y_n be values of the function $y = f(x)$ corresponding to the argument $x_0, x_1, ..., x_n$. Let the values of the arguments be at equal intervals, say h,

i.e. $$x_i = x_0 + ih$$

$$x_1 = x_0 + h, x_2 = x_0 + 2h, \text{... i.e. } x_n = x_0 + nh$$

Now by Lagrange's formula

$$f(x) \approx P(x) = \sum_{K=1}^{n} L_K(x) y_k$$

where $$L_k(x) = \frac{(x-x_0)(x-x_1)(x-x_2)...(x-x_{k-1})(x-x_{k+1})(x-x_n)}{(x_k-x_0)(x_k-x_1)(x_k-x_2)...(x_k-x_{k-1})(x_k-x_{k+1})(x_k-x_n)}$$

Thus $$I = \int_{x_0}^{x_0+nh} f(x)\,dx = \int_{x_0}^{x_0+nh} P(x)\,dx$$

$$= \int_{x_0}^{x_0+nh} \sum_{k=0}^{n} L_k(x) y_k\,dx = \int_{x_0}^{x_0+nh} [L_0(x) y_0 + ... L_n(x) y_n]\,dx$$

$$= y_0 \int_{x_0}^{x_0+nh} L_0(x)\,dx + ... + y_n \int_{x_0}^{x_0+nh} L_n(x)\,dx = \sum_{K=0}^{n} y_k \int_{x_0}^{x_0+nh} L_k(x)\,dx$$

Substituting $x = x_0 + uh \Rightarrow dx = hdu$, we get

$$I = h \sum_{K=0}^{n} y_k \int_0^n L_k du = nh \sum y_k \frac{1}{n} \int_0^n L_k du \quad ...(17.18)$$

where $$L_k = L_k(x_0 + u_h)$$

$$= \frac{(x_0+uh-x_0)(x_0+uh-x_1)...(x_0+uh-x_{k-1})(x_0+uh-x_{k+1})...(x_0+uh-x_n)}{(x_k-x_0)(x_k-x_1)...(x_k-x_{k-1})(x_k-x_{k+1})...(x_k-x_n)}$$

Putting $x_k = x_0 + kh$, we get

$$L_k = \frac{u(u-1)...(u-k+1)(u-k-1)...(u-n)}{k(k-1)...(+1)(-1)...(k-n)} \quad ...(17.19)$$

Now putting $$\frac{1}{n}\int_0^n L_k du = C_k^n, \text{ we get}$$

$$I = nh \sum_{k=0}^{n} y_k C_k^n \quad ...(17.20)$$

This is the required formula, it is known as Cotes' formula.

The numbers C_k^n, $0 \le k \le n$ are called Cotes' numbers. We can evaluate I from Eq. (17.20) in case we know the values of Cotes's numbers.

Properties of Cotes' numbers:

1. $C_k^n = C_{n-k}^n$

2. $\sum_{k=0}^{n} C_k^n = 1$

Proof: We have shown that

$$C_k^n = \frac{1}{n}\int_0^n L_k du$$

$$= \frac{1}{n}\int_0^n \frac{u(u-1)\dots[u-(k-1)][u-(k+1)]\dots(u-n)}{k(k-1)\dots(+1).(-1)(k-n)}du$$

$$= \frac{1}{n}\int_0^n \frac{u(u-1)\dots[u-(k-1)][u-(k+1)]\dots(u-n)}{k!(-1)^{n-k}(n-k)!}du$$

Again $$C_k^n = \frac{1}{n}\int_0^n \frac{u(u-1)\dots[u-(n-k-1)][u-(n-k+1)]\dots(u-n)}{(n-k)!k!(-1)^k}du$$

Putting $u-n=-t$ or $u=n-t$, $du=-dt$, we get

$$C_k^n = \frac{1}{n}\int_0^n \frac{(n-t)(n-t-1)\dots(-t+k-1)(-t+k-1)(-t)}{k!(n-k)!(-1)^k}(-dt)$$

$$= \frac{1}{n}\frac{(-1)^n(t-1)\dots(t-k+1)(t-k-1)(t-n)}{k!(n-k)!(-1)^k}$$

$$= \frac{1}{n}\int_0^n \frac{(-1)^{n-k}t(t-1)\dots[t-(k-1)][t-(k+1)\dots(t-n)}{k!(n-k)!}(dt)$$

$$= \frac{1}{n}\int_0^n \frac{t(t-1)\dots[t-(k-1)][(t-(k+1)\dots(t-n)}{k!(n-k)!(-1)^{n-k}}(dt)$$

[Multiplying numerator and denominator by $(-1)^{n-k}$]

$$= C_{n-k}^n$$

$$\int_a^b f(x)dx = \int_a^b f(t)dt, \text{ by the properties of definite integrals}$$

Some deduction from Cotes' formula: On putting $n = 1$ in Eq. (7.20), we get

$$I = \int_{x_0}^{x_1} f(x)dx\ 1\,.\,h\sum_{k=0}^{n} C_k^1 y_k = h\left[C_0^1 y_0 + C_1^1 y_1\right]$$

Again by $$C_k^n = \frac{1}{n}\int_0^n L_k du, \text{ we have}$$

$$C_0^1 = \frac{1}{1}\int_0^1 L_0 du = \int_0^1 \frac{(u-1)}{0-1}du$$

$$= (-1)\int_0^1 (u-1)du = (-1)\left(\frac{u^2}{2}-u\right)_0^1 = \frac{1}{2}$$

$$C_1^1 = \frac{1}{1}\int_0^1 L_1 du = \int_0^1 \frac{(u-0)}{(1-0)}du = \int_0^1 u\,du = \frac{1}{2}$$

$$I = \frac{h}{2}[y_0 + y_1]. \qquad \text{This is trapezoidal rule}$$

Again putting $n = 2$ in Eq. (7.20), we get

$$I = \int_{x_0}^{x_0+nh} f(x)\,dx = 2h \sum_{k=0}^{2} C_k^2 y_k = 2h[y_0 C_0^2 + y_1 C_1^2 + y_2 C_2^2]$$

Now
$$C_0^2 = \frac{1}{2}\int_0^2 L_0\,du = \frac{1}{2}\int_0^2 \frac{(u-1)(u-2)}{(0-1)(0-2)}du = \frac{1}{6}$$

$$C_1^2 = \frac{1}{2}\int_0^2 L_1\,du = \frac{1}{2}\int_0^2 \frac{u(u-2)}{(1-0)(1-2)}du = \frac{2}{3}$$

$$C_2^2 = \frac{1}{6} \text{ by using } (C_k^n = C_{n-k}^n)$$

Hence
$$I = 2h\left(\frac{1}{6}y_0 + \frac{2}{3}y_1 + \frac{1}{6}y_2\right) = \frac{h}{3}(y_0 + 4y_1 + y_2)$$

This is Simpson's 1/3 rule.

Similarly putting $n = 3$ in Eq. (7.20), we can find the Simpson's 3/8 rule.

Putting $n = 4$ in Eq. (7.20), we can find the Boole's rule and putting $n = 6$ in Eq. (7.20), we can find the Weddle' rule.

Below, we give the Cotes' numbers for some values of n.

	C_0^n	C_1^n	C_2^n	C_3^n	C_4^n	C_5^n	C_6^n
$n = 1$	1/2	1/2					
$n = 2$	1/6	4/6	1/6				
$n = 3$	1/8	3/8	3/8	1/8			
$n = 4$	7/90	32/90	12/90	32/90	7/90		
$n = 5$	19/208	75/288	50/288	50/88	75/288	19/288	
$n = 6$	41/840	216/840	27/840	278/840	27/840	216/840	41/800

Example **17.9:** Evaluate $\int_0^4 e^x\,dx$ by Simpson's rule using $e^1 = 2.72$, $e^2 = 7.39$, $e^3 = 20.09$, $e^4 = 54.60$.

Solution : Here $y = f(x) = e^x$. Divide the interval (0.4) into four equal intervals (0, 1), (1, 2), (2, 3), (3, 4) which are even and $h = 1$, we apply 1.3 Simpson's rule.

x	$f = f(x)$	*Odd ordinates*	*Even ordinates*	*Extreme ordinates*
$x_0 = 0$	$y_0 = e^0 = 1$			1
$x_1 = 1$	$y_1 = e^1 = 2.72$	2.72		
$x_2 = 2$	$y_2 = e^2 = 7.39$		7.39	
$x_3 = 3$	$y_3 = e^3 = 20.09$	20.09		
$x_4 = 4$	$y_4 = e^4 = 54.60$			54.60
	Total	22.81	7.397	55.60

By 1/3 Simpson's rule

$$\int_0^4 e^x dx = \frac{h}{3}\text{ [sum of extreme ordinates + 4 (odd ordinate) + 2 (even ordinate)]}$$

$$= \frac{1}{3}[55.60 + 4(22.81) + 2(7.397)] = 53.872$$

Example **17.10.** Calculate $\int_0^{\frac{\pi}{2}} e^{\sin x} dx$ correct to four decimal places, using Simpson's 3/8th rule.

Solution : Let us divide the interval into three equal subintervals by the ordinates $x_0 = 0$, $x_1 = \pi/6$, $x_2 = 2\pi/6$, $x_3 = 3\pi/6$.

$$y = f(x)$$

$$x_0 = 0 \quad y_0 = e^0 = 1 \qquad \text{Hence } h = \pi/6$$

$$x_1 = \pi/6 \quad y_1 = e^{\sin \pi/6} = e^{1/2} = 1.64872$$

$$x_2 = \pi/3 \quad y_2 = e^{\sin \pi/3} = e^{\sqrt{3}/2} = 2.3632$$

$$x_3 = \pi/2 \quad y_3 = e^{\sin \pi/2} = e = 2.71828$$

By simpson's 3/8th rule.

$$\int_0^{\frac{\pi}{2}} e^{\sin x} dx = \frac{3h}{8}[(y_0 + y_3) + 3(y_1 + y_2)]$$

$$= \frac{3 \times \frac{\pi}{2}}{8}[(1.271828) + 3(1.64872 + 2.3622)] = 0.091121$$

Example **17.11.** Evaluate $\int_0^1 e^{-x^2} dx$ by Simpson's rule.

***Solution* :** Divide the interval (0, 1) into ten equal subintervals.

x	x^2	$y = e^{-x^2}$	*Odd ordinates*	*Even ordinates*	*Extreme ordinates*
$x_0 = 0.0$	0.0	$y_0 = e^0 = 1$			1.00000
$x_1 = 0.1$	0.01	$y_1 = e^{-0.01} = 0.99005$	0.99005		
$x_2 = 0.2$	0.04	$y_2 = e^{-0.04} = 0.96080$		0.96080	
$x_3 = 0.3$	0.09	$y_3 = e^{-0.09} = 0.91391$	0.91391		
$x_4 = 0.4$	0.016	$y_4 = e^{-0.16} = 0.85214$		0.85214	
$x_5 = 0.5$	0.25	$y_5 = e^{-0.25} = 0.77880$	0.77880		
$x_6 = 0.6$	0.36	$y_6 = e^{-0.36} = 0.69768$		0.69768	
$x_7 = 0.7$	0.49	$y_7 = e^{-0.49} = 0.61263$	0.61263		
$x_8 = 0.8$	0.64	$y_8 = e^{-0.64} = 0.5273$		0.52730	
$x_9 = 0.9$	0.81	$y_9 = e^{-0.81} = 0.4449$	0.44492		
$x_{10} = 1.0$	1.0	$y_{10} = e^{-1} = 0.36788$			0.36788
		Total	3.74031	3.03792	1.36788

By Simpson's 1/3rd rule, here $h = 0.1$

$$\int_0^1 e^{-x^2} dx = \frac{h}{3}[1.36788 + 4(3.74031) + 2(3.0379)]$$

$$= \frac{1}{3}[1.36788 + 14.96124 + 6.07584] = 0.746832$$

***Example* 17.12.** The velocity v (km/min) of a moped which starts from rest at fixed intervals of time t (minute) as follows.

t	2	4	6	8	10	12	14	16	18	20
v	10	18	25	29	32	20	11	5	2	0

Find the distance travelled in 20 minutes.

Solution : We have
$$\frac{ds}{dt} = v$$

$$[s]_0^{20} = \int_0^{20} v\,dt$$

Let us divide the interval (0, 20) into 10 equal subintervals.

t	v	*Extreme ordinate*	*Odd ordinates*	*Even ordinates*
$t_0 = 0$	$v_0 = 0$	0		
$t_1 = 2$	$v_1 = 10$		10	
$t_2 = 4$	$v_2 = 18$			18
$t_3 = 6$	$v_3 = 25$		25	
$t_4 = 8$	$v_4 = 29$			29
$t_5 = 10$	$v_5 = 32$		32	
$t_6 = 12$	$v_6 = 20$			20
$t_7 = 14$	$v_7 = 11$		11	
$t_8 = 16$	$v_8 = 5$			5
$t_9 = 18$	$v_9 = 2$		2	
$t_{10} = 20$	$v_{10} = 0$	0		
	Total	0	80	72

By simpson's 1/3rd rule, here $h = 2$

$$[s]_0^{20} = \int_0^{20} v\,dt = \frac{h}{3}[0 + 4 \times 80 + 2 \times 72] = \frac{2}{3}[320 + 144] = \frac{2}{3}[464] = \frac{298}{3}$$

$$\text{Distance} = 309.3 \text{ km}$$

***Example* 17.13.** A curve is drawn to pass through the following points:

x	1	1.5	2	2.5	3	3.5	4
y	2	2.4	2.7	2.8	3	2.6	2.1

Estimate the area bounded by the curve, x-axis and the lines $x = 1$, $x = 4$. Also find the volume of solid generating by revolving this area using Weddle's rules.

Solution : The area bounded by the curve, x-axis and the lines $x = 1$, $x = 4$ given by which show that the integral found by Weddle's rule is the nearer to the actual value than other.

$$A = \int_1^4 y\, dx = \frac{3h}{10}[y_0 + 5y_1 + y_2 + 6y_3 + y_4 + 5y_5 + 2y_6]$$

$$= 3 \times \frac{0.5}{10}[2 + 5(2.4) + 2.7 + 6(2.8) + 3 + 5(2.6) + 2(2.1)]$$

$$= 8.055 \text{ square units}$$

Volume of the solid generated

$$V = \pi \int_1^4 y^2 dx = \pi \frac{3h}{10}[y_0^2 + 5y_1^2 + y_2^2 + 6y_3^2 + y_4^2 + 5y_5^2 + 2y_6^2]$$

$$= \pi. \frac{3(0.5)}{10}[4 + 5(2.4)^2 + (2.7)^2 + 6(2.8)^2 + 9 + 5(2.6)^2 + 2(2.1)^2]$$

Example 17.14. Evaluate $\int_0^{10} \frac{dx}{1+x^2}$ by using

1. Trapezoidal rule
2. Simpson's 1/3rd rule
3. Simpson's 3/8th rule
4. Weddle's rule

Compare the result with the actual values.

Solution : Taking $h = 1$, divide the whole range of integration [0, 10] into ten equal parts by 10 points $x_0 = 0$, $x_1 = 1$, $x_2 = 2$, $x_3 = 3$, $x_4 = 4$, $x_5 = 5$, $x_6 = 6$, $x_7 = 7$, $x_8 = 8$, $x_9 = 9$, $x_{10} = 10$.

x	$y = \frac{1}{1+x^2}$
0	$y_0 = 1$
1	$y_1 = 0.5$
2	$y_2 = 0.2$
3	$y_3 = 0.1$
4	$y_4 = 0.0588235$
5	$y_5 = 0.0384615$
6	$y_6 = 0.027027$
7	$y_7 = 0.02$
8	$y_8 = 0.0153846$
9	$y_9 = 0.0121951$
10	$y_{10} = 0.0099009901$

1. Trapezoidal rule

$$\int_0^{10} \frac{1}{1+x^2} dx = \frac{h}{2} [(y_0 + y_{10}) + 2(y_1 + y_2 + y_3 + y_4 + y_5 + y_6 + y_7 + y_8 + y_9)]$$

$$= \frac{1}{2} [(1 + 0.009900990) + 2(0.5 + 0.2 + 0.1 + 0.0588235 + 0.0384615$$

$$+ 0.027027 + 0.02 + 0.0153846 + 0.0121951)]$$

$$= 1.4768422$$

2. Simpson's 1/3rd rule

$$\int_0^{10} \frac{1}{1+x^2} dx = \frac{h}{2} [(y_0 + y_{10}) + 4(y_1 + y_3 + y_5 + y_7 + y_9) + 2(y_2 + y_4 + y_6 + y_8)]$$

$$= \frac{1}{3} [(1 + 0.009900990) + 4(0.5 + 0.1 + 0.0384615 + 0.02 + 0.012951)$$

$$+ 2(0.2 + 0.0588235 + 0.027027\ 0.0153846)]$$

$$= 1.4316659$$

3. Simpson's 3/8th rule

$$\int_0^{10} \frac{1}{1+x^2} dx = \frac{3h}{8} [(y_0 + y_{10}) + 3(y_1 + y_2 + y_4 + y_5 + y_7 + y_8) + 2(y_9)]$$

$$= \frac{3}{8} [(1 + 0.009900990) + 3(0.5 + 0.2 + 0.0588235 + 0.0384615$$

$$+ 0.02 + 0.0153846) + 2(0.1 + 0.027027 + 0.0121951)]$$

$$= 1.4198828$$

4. Weddle's rule

$$\int_0^{10} \frac{1}{1+x^2} dx = \frac{3h}{8} [(y_0 + y_{10}) + (5y_1 + y_2 + 6y_3 + y_4 + 5y_5 + 2y_6 + 5y_7 + y_8 + 6y_9)]$$

$$= \frac{3}{10} [(1 + 0.009900990) + 5(0.5) + 0.2 + 6(0.1) + 0.0588235$$

$$+ 5(0.0384615) + 2(0.027027) + 5(0.02) + 0.0153846 + 6(0.0121951)] \backslash$$

$$= 1.4410924$$

Now $\int_0^{10} \frac{1}{1+x^2} dx = [\tan^{-1} x]_0^{10} = \tan^{-1} 10 = 1.4711237$

Which show that the integral found by Weddle's rule is the nearer to the actual value than other.

Example 17.15. Evaluate $\int_0^{10} e^x dx$ by Weddle's rule given that

$$e^0 = 1, e^1 = 2.72, e^2 = 7.39, e^3 = 20.09, e^4 = 54.60, e^5 = 148.41$$

$$e^6 = 403.43, e^7 = 1096.63, e^8 = 2980.96, e^9 = 8103.08, e^{10} = 22026.47$$

Solution : Here $h = 1$

By Weddle's rule

$$\int_0^{10} e^x dx = \frac{3h}{10}[(y_0 + y_{10}) + 5y_1 + y_2 + 6y_3 + y_4 + 5y_5 + 2y_6 + 5y_7 + y_8 + 6y_9]$$

$$= \frac{3}{10}[(1 + 22026.47) + 5(2.72) + 7.39 + 6(20.09) + 54.60 + 5(148.41)$$

$$+ 2(403.43) + 5(1096.63) + 2980.96 + 6(8103.08)]$$

$$= 24256.53$$

Example **17.16.** Evaluate $\int_{0.2}^{1.4}(\sin x - \log_e x + e^x)dx$ by

1. Trapezoidal rule
2. Simpson's 1/3rd rule
3. Simpson's 3/8th rule
4. Weddle's rule

Solution : Let $y = f(x) = \sin x - \log_e x + e^x$. Divide the interval (0.2, 1.4) into six equal parts by the ordinates at point.

$$x_0 = 0.2, x_1 = 0.4, x_2 = 0.6, x_3 = 0.8,$$

$$x_4 = 1.00, x_5 = 1.2, x_6 = 1.4$$

so that

x	$\sin x$	$-\log_e x$	e^x	y	*Trapezoidal rule*	*Simpson's 1/3rd rule*	*Simpson's 1/3rd rule*	*Weddle's 3/8th rule*
$x_0 = 0.2$	0.1987	– 1.6095	1.2214	$y_0 = 3.0296$	$y_0 = 3.0296$	$y_0 = 3.0296$	$y_0 = 3.0296$	$y_0 = 3.0296$
$x_1 = 0.4$	0.3894	– 0.9163	1.4918	$y_1 = 2.79752$	$y_1 = 5.5950$	$4y_1 = 11.1900$	$3y_1 = 8.3925$	$5y_1 = 13.9875$
$x_2 = 0.6$	0.5646	– 0.5108	1.8221	$y_2 = 2.8975$	$2y_2 = 5.7950$	$2y_2 = 5.7889$	$3y_2 = 8.6925$	$y_2 = 2.8975$
$x_6 = 0.8$	0.7174	– 0.2232	2.2255	$y_3 = 3.1661$	$2y_3 = 6.3322$	$4y_3 = 12.6644$	$2y_3 = 6.3322$	$6y_3 = 18.9966$
$x_4 = 1.0$	0.8415	– 0.000	2.7183	$y_4 = 3.5598$	$2y_4 = 7.1196$	$2y_4 = 7.1196$	$2y_4 = 10.6794$	$y_4 = 3.5598$
$x_5 = 1.2$	0.9320	– 0.1824	3.3201	$y_5 = 4.0698$	$2y_5 = 16.2792$	$4y_5 = 16.2792$	$3y_5 = 12.2074$	$5y_5 = 10.3490$
$x_6 = 1.4$	0.9855	0.3365	4.0552	$y_6 = 4.7042$	$y_6 = 4.7042$	$y_6 = 4.7042$	$y_6 = 4.7042$	$y_6 = 4.7042$
					40.7152	60.7820	54.0398	67.5242

1. Trapezoidal rule

$$\int_{0.2}^{1.4} f(x)dx = \frac{h}{2}[40.7152] = \frac{0.2}{2}[40.7152] = 4.0715$$

2. Simpson's 1/3rd rule

$$\int_{0.2}^{1.4} f(x)dx = \frac{h}{3}[64.7820] = \frac{0.2}{3}[64.7820] = 4.0521$$

3. Simpson's 3/8th rule

$$\int_{0.2}^{1.4} f(x)dx = \frac{3h}{8}[54.0398] = \frac{3}{8}(0.2)[54.0398] = 4.0529$$

4. Weddle's rule

$$\int_{0.2}^{1.4} f(x)dx = \frac{3h}{10}[67.5242] = \frac{3}{10} \times 0.2\,[67.5242] = 4.0515$$

Example **17.17.** A rocket is launched from the ground. Its acceleration is registered during the first 80 seconds and is given in the table below. Using Simpson's 1/3rd rule, find the velocity of the rocket at $t = 80$.

$t(s)$	0	10	20	30	40	50	60	70	80
f(cm/sec^2)	30	31.63	33.64	35.47	37.75	40.33	43.50	46.69	50.67

Solution : Now

$$\frac{dv}{dt} = f, h = 10$$

$$= \frac{h}{3}[(f_0 + f_8) + 4(f_1 + f_3 + f_5 + f_7) + 2(f_2 + f_4 + f_6)]$$

$$= \frac{10}{3}[(30 + 50.67) + 4(31.63 + 35.47 + 40.33 + 46.69)$$

$$+ 2(33.64 + 37.75 + 43.50)]$$

$$= \frac{10}{3}[80.67 + 616.76 + 228.68] = \frac{926.11}{3}$$

$$= 3087.3 \text{ cm/sec}^2 = 30.87 \text{ m/sec}^2$$

Example **17.18.** A reservoir discharge water through sluices at a depth h feet below the water surface area A for various values of h as given below.

$h(f)$	10	11	12	13	14
A(in sq ft)	950	1070	1200	1350	1530

If t denotes time in minutes, the rate of fall of the surface is given by $\frac{dh}{dt} = \frac{-48\sqrt{h}}{A}$. Estimate the time taken for the water level to fall from 14 to 10 feet above his sluices.

Solution :

$$\frac{A}{\sqrt{h}} = \frac{950}{\sqrt{10}}, \frac{1070}{\sqrt{11}}, \frac{1200}{\sqrt{12}}, \frac{1350}{\sqrt{13}}, \frac{1530}{\sqrt{14}}$$

$$= 300.41, 322.61, 346.41, 374.47, 408.90$$

Now

$$t = \int_{14}^{10} \frac{A}{-48\sqrt{h}} dh$$

By Simpson's 1/3rd rule, we have

$$h = -1$$

$$t = \int_{14}^{10} \frac{A}{-48\sqrt{h}} dh$$

$$= -\frac{1}{48} \cdot \frac{-1}{3}[(300.41 + 408.90) + 4(322.61 + 374.41) + 2(346.41)]$$

$$= \frac{1}{144}[709.31 + 2788.08 + 692.82] = \frac{4190.21}{144} = 29.09$$

$$= 29 \text{ minutes (approx)}$$

Time taken for water to fall from 14 to 10 feet above the sluices = 29 minutes (approx.)

Example **17.19.** Integrate numerically $\int_0^{\frac{\pi}{2}} \sqrt{\cos\theta}\, d\theta$

Solution : $x = 0$, $y = \sqrt{\cos\theta}$

x	y
$x_0 = 0$	$y_0 = 1$
$x_1 = \frac{\pi}{6}$	$y_1 = 0.930$
$x_2 = \frac{2\pi}{6}$	$y_2 = 0.707$
$x_3 = \frac{3\pi}{6}$	$y_3 = 0$

$$\int_0^{\frac{\pi}{2}} \sqrt{\cos\theta}\, d\theta = \frac{3}{8}(h)\,[(y_0 + y_3) + 3(y_1 + y_2)]$$

$$= \frac{3}{8}\left(\frac{\pi}{6}\right)[1 + 3(1.637)] = \frac{\pi}{16}[5.911] = 1.1600$$

Example **17.20.**

$x = 1$	1.5	2	2.5	3	3.5	4
$y = 2$	2.4	2.7	2.8	3.0	2.6	2.1

Estimate the area bounded by the curve, x-axis to the lines $x = 1$, $x = 4$

Solution : Area $= \int_1^4 y dx = \frac{3}{8} h\,[(y_0 + y_6) + 3(y_1 + y_2 + y_4 + y_5) + 2 \times y_3]$

$$= \frac{3}{8}(0.5)\,[(2 + 2.1) + 3(2.4 + 2.7 + 3 + 2.67) + 2(2.8)]$$

$$= \frac{3}{8}(0.5)\,[4.1 + 3(10.7) + 5.6] = \frac{3}{8}(0.5)\,[41.8] = 7.8375$$

Example **17.21.** A curve is given by

x	0	1	2	3	4	5	6
y	0	2	2.5	2.3	2	1.7	1.5

The x-coordinate of CG of the area bounded by the curve, the end coordinates and x-axis is given by $A\bar{x} = \int_0^6 xydx$, where A is the area. Find $\bar{x}$ by using Simpson's rule.

Solution : Now the area $\qquad A = \int_0^6 y\, dx$

By Simpson's 1/3rd rule

$$= \frac{1}{3}[(y_0 + y_6) + 4(y_1 + y_3 + y_5) + 2(y_2 + y_4)]$$

$$= \frac{1}{3}[(0.0+1.5)+4(2+2.3+1.7)+2(2.5+2)]$$

$$= \frac{1}{3}[1.5+24+9] = \frac{34.5}{3} = 11.5$$

Now

x	0	1	2	3	4	5	6
xy	0	2	6	6.9	8	8.5	9

By Simpson's 1/3rd rule

$$\int_0^6 xydy = \frac{1}{3}[(0+9)+4(17.4)+2(13)] = \frac{1}{3}[140.6] = 34.866$$

$$A\overline{x} = \int_0^6 xydx$$

$$(11.5)\overline{x} = 34.886$$

$$\overline{x} = \frac{34.886}{11.5} = 3.0318$$

***Example* 17.22.** A body is in the form of a solid of revolution. The diameter D in cm of its section at distance x cm from one end are given below. Estimate volume of the solid.

x	0	2.5	5.0	7.4	10.0	12.5	15.0
D	5	5.5	6.0	6.75	6.25	15.5	4.0
y	25	30.25	36	45.56	39.06	36.25	16.0

Solution : Volume of revolution $= \pi \int_0^{15} r^2 dx$

$$\pi\int_0^{15}\left(\frac{D}{2}\right)^2 dx = \frac{\pi}{5}\int_0^{15} D^2 dx \quad h = 2.5$$

By Simpson's 1/3rd rule

$$\text{Volume} = \frac{\pi}{4}\left(\frac{2.5}{3}\right)[(25+16)+4(30.25+45.56+36.25)+2(36+39.06)]$$

$$= \frac{\pi}{12}(2.5)[41+424.24+150.12] = \frac{\pi}{12}\times 2.5\,[615.36] = 402.5$$

PROBLEMS

1. Find the integral $\int_{1.0}^{1.8} \frac{e^x + e^{-x}}{2} dx$ using Simpson's 1/3 rule by taking $h = 0.2$

[Ans. 1.7670]

2. Find the integral $\int_0^6 [f(x)]^2 dx$ using Simpson's 1/3 rule, given that

x	0	1	2	3	4	5	6
$f(x)$	1	0	1	4	9	16	25

[Ans. 626]

3. Compute the integral using Simpson's 3/8 rule $\int_0^{\frac{\pi}{2}} \frac{dx}{\sin^2 x + \frac{1}{4}\cos^2 x}$.

[Ans. 3.1416]

4. Compute the integral $I = \sqrt{\frac{2}{\pi}} \int_0^1 e^{-\frac{x^2}{2}} dx$ using Simpson's 1/3 rule taking $h = 0.125$.

[Ans. 0.6827]

5. A river is 80 feet wide. Depth d in feet at a distance of x feet from one bank is given by the following table.

x	0	10	20	30	40	50	60	70	80
$y = d$	0	4	7	9	12	15	14	8	3

Find the approximate area of the cross-section.

[Ans. 710 sq feet]

6. When a train is moving at 30 minutes an hour, stream is shut off and brakes are applied. The speed of the train in miles per hour after t second is given below.

t	0	5	10	15	20	24	30	35	40
y	30	24	19.5	16	13.6	11.7	10.0	8.5	7.0

Determine how far the train has moved in 40 seconds.

[Ans. 296.7 yards]

7. A curve passes through the points (1, 0.2) (2, 0.7) (3, 1) (4, 1.3) (5, 1.5) (6, 1.7) (7, 1.9) (8, 2.1) (9, 2.3) using Weddle's rule, estimate the volume generated by revolving the area between the curve, x-axis and the ordinates $x = 1$ and $x = 9$ about the x-axis.

[Ans. 59.68 cu units]

8. The following tables gives the velocity v of a particle at time t.

t (seconds)	0	2	4	6	8	10	12
v (m/sec)	4	6	16	34	60	94	136

Find the distance moved by the particle in 12 seconds and also the acceleration at $t = 2$ sec.

[Ans. 552 m, 4 m/sec^2]

9. Evaluate by (*i*) Trapezoidal rule (*ii*) Simpson's 1/3 rule (*iii*) Simpson's 3/8 rule, (*iv*) Weddle's rule $\int_4^{5.2} \log x \, dx$. Given that

x	4.0	4.2	4.4	4.6	4.8	5.0	5.2
Log x	1.3863	1.4351	1.4816	1.5261	1.5686	1.6094	1.6484

[Ans. 1.8276551, 1.8278472, 1.8278470, 1.8278474]

17.8 ERRORS IN QUADRATURE FORMULA

$P(x)$ is a polynomial representing $y = f(x)$ in the interval $[a, b]$.

17.8.1 *Errors in Trapezoidal Rule*

Divide the interval $[a, b]$ into n subintervals $(x_0, x_1), (x_1, x_2), \ldots, (x_{n-1}, x_n)$, where $a = x_0$ and $b = x_n$.

Error in the interval $(x_0, x_1) = \int_{x_0}^{x_1} f(x)\,dx - \int_{x_0}^{x_1} P(x)\,dx$

Expanding $y = f(x)$ by Taylor's series around $x = x_0$, we have

$$y = f(x) = f(0) + (x - x_0) f'(0) + \frac{(x-x_0)^2}{2!} f''(0) + \frac{(x-x_0)^3}{3!} f'''(0) + \ldots$$

$$= y_0 + (x - x_0)\, y'(0) + \frac{(x-x_0)^2}{2!} y''(0) + \frac{(x-x_0)^3}{3!} y''(0) + \ldots$$

$$\int_{x_0}^{x_1} f(x)\,dx = \int_{x_0}^{x_0+h} \left[y_0 + (x - x_0) y'_0 + \frac{(x-x_0)^2}{2!} y''_0 + \frac{(x-x_0)^3}{3!} y'''_0 + \ldots \right] dx$$

$$= \left[y_0 + \frac{(x-x_0)^2}{2!} y'_0 + \frac{(x-x_0)^3}{3!} y''_0 + \frac{(x-x_0)^4}{4!} y'''_0 + \ldots \right]_{x_0}^{x_0+h}$$

$$= y_0 h + \frac{h^2}{2!} y'_0 + \frac{h^3}{3!} y''_0 + \frac{h^4}{4!} y'''_0 + \ldots$$

Also $\int_{x_0}^{x_1} P(x)\,dx = \frac{h}{2}[y_0 + y_1]$ by trapezoidal rule

$$= \frac{h}{2}[f(x_0) + (x_0 + h)] = \frac{h}{2}\left[y_0 + y_0 + hy'_0 + \frac{h^2}{2!} y''_0 + \frac{h^3}{3!} y'''_0 + \ldots \right]$$

$$= \frac{h}{2}\left[2y_0 + hy'_0 + \frac{h^2}{2!} y''_0 + \frac{h^3}{3!} y'''_0 + \ldots \right]$$

$$= hy_0 + \frac{h^2}{2!} y'_0 + \frac{h^3}{2.2!} y''_0 + \frac{h^4}{3!} y''_0 + \ldots$$

$$\text{Error in } (x_0, x_1) = \int_{x_0}^{x_1} f(x)\,dx - \int_{x_0}^{x_1} P(x)\,dx$$

$$= h^3\left(\frac{1}{3!} - \frac{1}{2.2!}\right) y''_0 + \ldots = -\frac{1}{12} h^2 y''_0 + \ldots$$

Principal part of the error in $[x_0, x_1] = -\frac{1}{12} h^3 y''_0$

Principal part of the error in $[x_1, x_2] = -\frac{1}{12} h^3 y''_1$

Principal part of the error in $[x_2, x_3] = -\frac{1}{12} h^3 y''_2$

Principal part of the error in $[x_{n-1}, x_n] = -\frac{h^3}{12} [y''_{n-1}]$

Let $y''(x)$ be the largest of $y''_0 + y''_1 + ... + y''_{n-1}$

Total error $\quad Z < -\frac{1}{12} n y''(x) h^3 = -\frac{1}{12}(b-a) h^2 y''(x) \quad [nh = b - a]$

Hence the error in the trapezoidal rule is of the order h^2.

17.8.2 *Errors in Simpson's 1/3 Rule*

Error in the interval (x_0, x_2) is $\int_{x_0}^{x_0+2h} f(x)\,dx - \int_{x_0}^{x_0+2h} P(x)\,dx$.

Now by Taylor's series

$$y = f(x) = y_0 + (x - x_0)y'_0 + \frac{(x-x_0)^2}{2!} y''_0 + \frac{(x-x_0)^3}{3!} y'''_0 + ...$$

$$\int_{x_0}^{x_2} f(x)\,dx = \int_{x_0}^{x_0+2h} \left[y_0 + (x-x_0)y'_0 + \frac{(x-x_0)^2}{2!} y''_0 + \frac{(x-x_0)^3}{3!} y'''_0 + \frac{(x-x_0)^4}{4!} y''''_0 + ...\right] dx$$

$$= 2hy_0 + \frac{4h^2}{2!} y'_0 + \frac{8h^3}{3!} y''_0 + \frac{16h^4}{4!} y''_0 + \frac{32h^5}{5!} y''_0 + ...$$

Also

$$\int_{x_0}^{x_0+2h} P(x)\,dx = \frac{h}{3}[y_0 + 4y_1 + y_2] \text{ by Simpson's 1/3rd rule}$$

$$= \frac{h}{3}[y_0 + 4f(x_0 + h) + f(x_0 + 2h)]$$

$$+ \left(y_0 + 2hy'_0 + \frac{4h^2}{2!} y''_0 + \frac{8h^3}{3!} y'''_0 + \frac{16h^4}{4!} y''''_0\right)$$

$$= \frac{h}{3}\left[6y_0 + y'_0 h(4+2) + y''_0 \frac{h^2}{2!}(4+4) + y'''_0 + \frac{h^3}{3!}(4+8) + y''''_0 \frac{h^4}{4!}(4+16) + ...\right]$$

$$= \frac{h}{3}\left[6y_0 + 6y'_0 h + 4y''_0 h^2 + 2y'''_0 h^3 + \frac{20}{24} y''''_0 h^4 + ...\right]$$

$$= 2y_0 h + 2y'_0 h^2 + \frac{4}{3} y''_0 h^3 + \frac{2}{3} y'''_0 h^4 + \frac{5}{18} y''_0 h^5$$

$$E_1 = \int_{x_0}^{x_0+2h} f(x)\,dx - \int_{x_0}^{x_0+2h} P(x)\,dx = \left(\frac{32}{5!} - \frac{5}{18}\right) y''_0 h^5$$

$$= \left(\frac{4}{15} - \frac{5}{18}\right) h^5 y''_0 + \ldots = -\frac{h^5}{90} y''_0 + \ldots$$

Principal part of the error in $(x_0, x_2) = -\frac{h^5}{90} y''_0$

Principle part of the error in $(x_2, x_4) = -\frac{h^5}{90} y''_0$

Principle part of error in $(x_{n-2}, x_n) = -\frac{h^5}{90} y''''_{n-2}$

$$\text{Total error} = -\frac{h^5}{90}[y''_0 + y''_2{}'' \ldots y''''_{n-2}] \cong \frac{h^5}{90} ny''''(x)$$

17.8.3 *Errors in Simpson's 3/8 Rule*

$$\int_{x_0}^{x_0+3h} f(x)\,dx - \int_{x_0}^{x_0+3h} P(x)\,dx = \left[xy_0 + \frac{(x-x_0)^2}{2!} \times y'_0 + \frac{(x-x_0)^3}{3!} y'''_0 + \frac{(x-x_0)^4}{4!} y''_0\right]_{x_0}^{x_0+3h}$$

$$-\frac{3h}{8}(y_0 + 3y_1 + 3y_2 + y_3)$$

$$= 3hy_0 + \frac{(3h)^2}{2!} y'_0 + \frac{(3h)^3}{3!} y''_0 + \frac{(3h)^4}{4!} y'''_0 + \ldots$$

$$-\frac{3h}{8}[y_0 + 3f(x_0 + h) + 3f(x_0 + 2h) + f(x_0 + 3h)]$$

$$= 3hy_0 + \frac{(3h)^2}{2!} y'_0 + \frac{(3h)^3}{4!} y''_0 + \frac{(3h)^4}{4!} y'''_0 + \ldots$$

$$-\frac{3h}{8}\left[y_0 + 3\left(y_0 + hy'_0 + \frac{h^2}{2!} y''_0 + \frac{h^3}{3!} y'''_0 + \ldots\right)\right.$$

$$+3\left(y_0 + 2hy'_0 + \frac{(2h)^2}{2!} y''_0\right)$$

$$-\frac{3h}{8}\left[y_0 + 3\left(y_0 + hy'_0 + \frac{h^2}{2!} y''_0 + \frac{h^3}{3!} y'''_0 + \ldots\right)\right.$$

$$+3\left(y_0 + 2hy'_0 + \frac{(2h)^2}{2!} y''_0 + \frac{(2h)^3}{3!} y'''_0 + \ldots\right)$$

$$\left. + y_0 + 3hy'_0 + \frac{(3h)^2}{2!} y''_0 + \frac{(3h)^3}{3!} y'''_0 + \ldots\right]$$

On simplifying, we get

$$\int_{x_0}^{x_0+3h} f(x)\,dx - \int_{x_0}^{x_0+3h} P(x)\,dx = -\frac{3h^5}{80} y''_0 + \ldots$$

$$\int_{x_0+3h}^{x_0+6h} f(x)\,dx - \int_{x_0+3h}^{x_0+3h} P(x)\,dx = -\frac{3h^5}{80} y''_3{}'' + \ldots$$

Adding, we get total error $Z < -\frac{3h^5}{80} y''''(x)$

Similarly, Boole's and Weddle's rule can be established.

17.8.4 *Error in Boole's Rule*

In this case, the principal part of the error in the interval

$$[x_0, x_4] = \frac{-8h^7}{945} y''''$$

17.8.5 *Error in Weddle's Rule*

In this case, principal part of the error in the interval

$$[x_0, x_6] = \frac{-h^7}{140} y''''$$

17.9 ROMBERG· METHOD OF INTEGRATION

This method provides a simple modification to the quadrature formula for finding their better approximation. Let us improve upon the value of integral

$$I = \int_a^b f(x)\,dx$$

By trapezoidal rule: If I_1, I_2 be the value of I with subintervals of width h_1 and h_2 and E_1 and E_2 their corresponding errors respectively. Then

$$E_1 = \frac{-(b-a)}{12} h_1^2 y''(\overline{X})$$

$$E_2 = \frac{-(b-a)}{12} h_2^2 y''(\overline{X})$$

where $y''(\overline{X})$ is also the largest value as $y''(X)$, we can assume that $y''(X)$ and $y''(\overline{X})$ are nearly equal.

$$\therefore \quad \frac{E_1}{E_2} = \frac{h_1^2}{h_2^2} \text{ or } \frac{E_1}{E_2 - E_1} = \frac{h_1^2}{h_2^2 - h_1^2} \qquad \ldots(17.21)$$

$$\frac{E_2}{E_1} - 1 = \frac{h_2^2}{h_1^2} - 1$$

As $$I = I_1 + E_1 = I_2 + E_2 \text{ and } E_2 - E_1 = I_1 - I_2 \qquad \ldots(17.22)$$

From Eqs (17.21) and (17.22), we have which is a better approximation of I.

$$E1 = \frac{h_1^2}{h_2^2 - h_1^2}(I_1 - I_2)$$

Hence $$I = I_1 + E_1 = I_1 + \frac{h_1^2}{h_2^2 - h_1^2}(I_1 - I_2)$$

$$= \frac{I_1 h_2^2 - I_2 h_1^2}{h_2^2 - h_1^2} \qquad ...(17.23)$$

To find I systematically, we take

$$h_1 = h \text{ and } h_2 = \frac{1}{2} h.$$

So that Eq. (17.23) gives

$$I = \frac{I_1\left(\frac{h}{2}\right)^2 - I_2 h_2^2}{\left(\frac{h}{2}\right)^2 - h^2} = \frac{4I_2 - I_1}{3}$$

i.e.

$$I\left(h, \frac{h}{2}\right) = \frac{1}{3}\left[4I\left(\frac{h}{2}\right) - I(h)\right] \qquad ...(17.24)$$

We got the result by applying trapezoidal rule.

Now we use the trapezoidal rule several times successively halving h and apply Eq. (17.24) to each pair of values as per the following scheme:

$I(h)$			
	$I(h, h/2)$		
$I(h/2)$		$I(h, h/2, h/4)$	
	$I(h/2, h/4)$		$I(h, h/2, h/4, h/8)$
$I(h/4)$		$I(h/2, h/4, h/8)$	
	$I(h/4, h/8)$		
$I(h/8)$			

The computation is continued till successive values are close to each other. This method is called Richardson's deferred approach to the limit and its systematic refinement is called Romberg's method.

***Example* 17.23.** Use Romberg's method to compute $\int_0^1 \frac{dx}{1+x}$. Hence find $\log_e 2$ correct to two decimal places.

Solution : We take $h = 0.5, 0.25, 0.125$ successively and evaluate the given integral using trapezoidal rule.

(*i*) When $h = 0.5$, the value of $y = (1 + x)^{-1}$ are

x	0	0.5	1
$1/1 + x = y$	1	0.666	0.5

Trapezoidal rule

$$I(0.5) = \int_0^1 \frac{dx}{1+x} = \frac{0.5}{2}\,[1 + 5 + 2(0.666)] = 0.708$$

(*ii*) When $h = 0.25$, the value of $y = (1 + x)^{-1}$ are

x	0	0.25	0.5	0.75	1
y	1	0.8	0.666	0.571	0.5

$$I(0.25) = \frac{0.25}{2}\,[(1 + 0.5) + 2(0.8 + 0.666 + 0.571)] = 0.69675$$

(*iii*) When $h = 0.125$, the value of $y = (1 + x)^{-1}$ are

x	0	0.125	0.25	0.375	0.5	0.625	0.75	0.875	1
y	1	0.88	0.8	0.7272	0.66	0.615	0.571	0.53	0.5

$$I(0.125) = \frac{0.125}{2}[(1 + 0.5) + 2(0.88 + 0.8 + 0.7272 + 0.66 + 0.615 + 0.571 + 0.53)]$$

$$= 0.69165$$

Thus, we have

$$I(h) = 0.708,\ I(h/2) = 0.69675,\ I(h/4) = 0.69165$$

$$I\left(h, \frac{h}{2}\right) = \frac{1}{3}\left[4I\left(\frac{h}{2}\right) - I(h)\right] = \frac{1}{3}[4 \times 0.69675 - 0.708] = 0.693$$

$$I\left(\frac{h}{2}, \frac{h}{4}\right) = \frac{1}{3}\left[4I\left(\frac{h}{4}\right) - I\left(\frac{h}{2}\right)\right] = \frac{1}{3}[4 \times 0.69165 - 0.69675] = 0.68995$$

$$I\left(h, \frac{h}{2}, \frac{h}{4}\right) = \frac{1}{3}\left[4I\left(\frac{h}{2}, \frac{h}{4}\right) - I\left(h, \frac{h}{2}\right)\right] = \frac{1}{3}[4 \times 0.68995 - 0.693] = 0.68893$$

The value of the integral = 0.6893.

***Example* 17.24.** Use Romberg's method to find $\int_0^1 \frac{dx}{1 + x^2}$ correct to four decimal places.

Solution : We take $h = 0.5, 0.25, 0.125$ successively and find the given integral using trapezoidal rule.

1. When $h = 0.5$, the value of $y = (1 + x^2)^{-1}$ are

x	0	0.5	1.0
y	1	0.8	0.5

$$I = \frac{0.5}{2}[1 + 0.5 + 2 \times 0.8] = 0.775$$

2. When $h = 0.25$, the value of $y = (1 + x^2)^{-1}$ are

x	0	0.25	0.5	0.75	1.0
y	1	0.9412	0.8	0.64	0.5

$$I = \frac{0.25}{2}[1 + 0.5 + 2(0.9412 + 0.8 + 0.64)] = 0.7828$$

3. When $h = 0.125$, we find that $I = 0.7848$

Thus, we have $I(h) = 0.7750$, $I(h/2) = 0.7828$, $I(h/4) = 0.7848$

$$I\left(h, \frac{h}{2}\right) = \frac{1}{3}\left[4I\left(\frac{h}{2}\right) - I(h)\right] = \frac{1}{3}[3.1312 - 0.775] = 0.7854$$

$$I\left(\frac{h}{2}, \frac{h}{4}\right) = \frac{1}{3}\left[4I\left(\frac{h}{4}\right) - I\left(\frac{h}{2}\right)\right] = \frac{1}{3}[3.1392 - 0.7828] = 0.7855$$

$$I\left(h, \frac{h}{2}, \frac{h}{4}\right) = \frac{1}{3}\left[4I\left(\frac{h}{2}, \frac{h}{4}\right) - I\left(h, \frac{h}{2}\right)\right] = \frac{1}{3}[3.142 - 0.7534] = 0.7855$$

The value of integral = 0.7855

PROBLEMS

Use Romberg's method to evaluate the following, up to four decimal places.

1. $\int_0^1 e^x dx$ **[Ans.** 1.7183**]**

2. $\int_0^1 \cos x \, dx$ **[Ans.** 0.8415**]**

17.10 EULER-MACLAURIN FORMULA

Taking $\Delta F(x) = f(x)$, we define the inverse operator Δ^{-1} as where $x_0, x_1, ..., x_n$ are the $(n + 1)$ equispaced value of x with difference h.

$$F(x) = \Delta^{-1} f(x) \quad ...(17.25)$$

Now $$F(x_1) - F(x_0) = \Delta F(x_0) = f(x_0)$$

Similarly $$F(x_2) - F(x_1) = \Delta F(x_1) = f(x_1)$$

$$... = ...$$

$$... = ...$$

$$... = ...$$

$$F(x_n) - F(x_{n-1}) = \Delta F(x_{n-1}) = f(x_{n-1})$$

Adding all these, we get

$$F(x_n) - F(x_0) = \sum_{i=0}^{n-1} f(x_i) \quad ...(17.26)$$

From Eq. (17.25), we have

$$F(x) = \Delta^{-1} f(x) = (E-1)^{-1} f(x) = (e^{hD} - 1)^{-1} f(x) \quad [E = e^{hD}]$$

$$= \left[\left(1 + hD + \frac{h^2D^2}{2!} + \frac{h^3D^3}{3!} + ...\right) - 1\right]^{-1} f(x)$$

$$= (hD)^{-1}\left[1 + \frac{hD}{2!} + \frac{h^2D^2}{3!} + ...\right]^{-1} f(x)$$

$$= \frac{1}{h} D^{-1}\left[1 - \frac{hD}{2} + \frac{h^2D^2}{12} - \frac{h^4D^4}{720} + ...\right] f(x)$$

$$= \frac{1}{h}\int f(x)\,dx - \frac{1}{2} f(x) + \frac{h}{12} f'(x) - \frac{h^3}{720} f'''(x) + ... \quad ...(17.27)$$

Putting $x - x_n$ and $x - x_0$ in Eq. (7.27) and then subtracting, we get

$$F(x_n) - F(x_0) = \frac{1}{h}\int_{x_0}^{x_n} f(x)\,dx - \frac{1}{2}[f(x_n) - f(x_0)]$$

$$+ \frac{h}{12} [f'(x_n) - f'(x_0)] - \frac{h^3}{720} [f'''(x_n) - f'''(x_0)] + \ldots \qquad \ldots(17.28)$$

From Eqs (17.25) and (17.28), we have

$$\sum_{i=0}^{n-1} f(x_i) = \frac{1}{h} \int_{x_0}^{x} f(x)\,dx - \frac{1}{2} [f(x_n) - f(x_0)]$$

$$+ \frac{h}{12} [f'(x_n) - f'(x_0)] - \frac{h^3}{720} [f'''(x_n) - f'''(x_0)] + \ldots$$

$$\frac{1}{h} \int_{x_0}^{x_n} f(x)\,dx = \sum_{i=0}^{n-1} f(x_1) + \frac{1}{2} [f(x_n) - f(x_0)] - \frac{h}{12} [f'(x_n) - f'(x_0)]$$

$$+ \frac{h^3}{720} [f'''(x_n) - f'''(x_0)] + \ldots$$

$$= \frac{1}{2} [f(x_0) + 2f(x_1) + 2f(x_2) + \ldots + 2f(x_{n-1}) + f(x_n)]$$

$$- \frac{h}{12} [f'(x_n) - f'(x_0)] + \frac{h^3}{720} [f'''(x_n) - f'''(x_0)] + \ldots$$

We can write as which is called Euler-Maclaurin formula.

$$\int_{x_0}^{x_0+nh} y\,dx = \frac{h}{2} [y_0 + 2y_1 + 2y_2 + \ldots + 2y_{n-1} + y_n]$$

$$- \frac{h^2}{12} (y'_n - y'_0) + \frac{h^4}{720} (y_n''' - y_0''') + \ldots$$

Note: This formula is often used to find the sum of a series of the form

$$y(x_0) + y(x_0 + h) + \ldots + y(x_0 + nh)$$

Example **17.25.** Apply Euler-Maclaurin formula to find $\frac{1}{51^2} + \frac{1}{53^2} + \frac{1}{55^2} + \ldots + \frac{1}{99^2}$.

Solution : Taking $y = \frac{1}{x^2}$, $x_0 = 51$, $h = 2$, $n = 24$, we have $y = \frac{-2}{x^3}$, $y''' = \frac{-24}{x^5}$.

Then Euler-Maclaurin formula gives

$$\int_{51}^{99} \frac{dx}{x^2} = \frac{2}{5} \left[\frac{1}{51^2} + \frac{2}{53^2} + \frac{2}{55^2} + \ldots + \frac{2}{97^2} + \frac{1}{99^2} \right]$$

$$- \frac{(2)^2}{12} \left[\frac{-2}{99^3} - \frac{(-2)}{51^3} \right] + \frac{(2)^4}{720} \left(\frac{-24}{99^5} - \frac{(-24)}{51^5} \right)$$

$$\frac{1}{51^2} + \frac{1}{53^2} + \frac{1}{55^2} + \ldots + \frac{1}{99^2} = \frac{1}{2} \int_{51}^{99} \frac{dx}{x^2} + \left(\frac{1}{51^2} + \frac{1}{99^2} \right) + \frac{1}{3} \left(\frac{1}{51^3} - \frac{1}{99^3} \right)$$

$$- \frac{8}{30} \left(\frac{1}{51^5} - \frac{1}{99^5} \right) + \ldots$$

$$= \frac{1}{2}\left|-\frac{1}{x}\right|_{51}^{99} + 0.000243 + 0.0000022$$

$$= 0.00499 \text{ approx.}$$

Example 17.26. Use Euler-Maclaurin's formula to prove that

$$\sum_{1}^{n} x^2 = \frac{n(n+1)(2n+1)}{6}$$

Solution : Euler-Maclaurin's formula is given by

$$\frac{1}{h}\int_{x_0}^{x_0+nh} f(x)\,dx = \sum_{i=0}^{n} f(x_i) - \frac{1}{2}\,[f(x_n) + f(x_0)]$$

$$- \frac{h}{12}\,[f'(x_0+nh) - f'(x_0)] + \frac{h^3}{720}\,[f'''(x_0+nh) - f'''(x_0)] + \dots$$

Now putting $f(x) = x^2$, $f'(x) = 2x$, $x_0 = 0$, $h = 1$, $x_0 + nh = n$, $x_i = x_0 + ih = i$, we easily obtain

$$\int_0^n x^2\,dx = \sum_{i=0}^{n} i^2 - \frac{1}{2}\,[n^2 + 0] - \frac{1}{12}\,[2n - 0]$$

[third and higher order derivatives of x^2 are zero]

$$\frac{n^3}{3} = \sum_{i=0}^{n} i^2 - \frac{n^2}{2} - \frac{n}{6}$$

$$\sum_{i=0}^{n} i^2 = \frac{n^3}{3} + \frac{n^2}{n} + \frac{n}{6} = \frac{n(2n^2+3n+1)}{6} = \frac{n(n+1)(2n+1)}{6}$$

$$\sum_{x=1}^{n} x^2 = \frac{n(n+1)(2n+1)}{6}$$

Example 17.27. Find the sum of the fourth powers of the first n natural numbers by means of the Euler-Maclaurin formula.

Solution : The general form of the Euler-Maclaurin formula is

$$\frac{1}{h}\int_{x_0}^{x_0+nh} f(x)\,dx = \sum_{0}^{n} f(x+nh) - \frac{1}{2}\,[f(x+nh) + f(x_0)]$$

$$- \frac{h}{12}\,[f'(x_0+nh) - f'(x_0)] + \frac{h^3}{720}\,[f'''(x_0+nh) - f'''(x_0)]$$

Putting $x_0 = 0$, $h = 1$, $f(x) = x_i$, we get

$$\int_0^n x^4\,dx = \sum_{0}^{n} x^4 - \frac{1}{12}\,[n^4 + 0^4] - \frac{1}{12}\,[5n^3 - 4(0)^3] + \frac{1}{720}\,[24n - 24(0)]$$

$$\frac{n^5}{5} = \sum_{0}^{n} x^4 - \frac{n^4}{2} - \frac{n^3}{3} + \frac{n}{40}$$

$$\sum x^4 = \frac{n^5}{5} + \frac{n^4}{2} + \frac{n^3}{3} - \frac{n}{30}$$

PROBLEMS

1. Evaluate $\int_0^{\frac{\pi}{2}} \sin x dx$ using the Euler-Maclaurin formula.

   (*i*) $h = \pi/8$ **[Ans.** 1.000003]

   (*ii*) $h = \pi/4$ **[Ans.** 0.837336]

2. Sum the series by using Euler-Maclaurin summation formula.

   (*i*) $\frac{1}{400} + \frac{1}{402} + \ldots + \frac{1}{498} + \frac{1}{505}$ **[Ans.** 0.11382114]

   (*ii*) $\frac{1}{100} + \frac{1}{101} + \frac{1}{102} + \frac{1}{103} + \frac{1}{104}$ **[Ans.** 0.0490291]

   (*iii*) $\frac{1}{201^2} + \frac{1}{203^2} + \ldots + \frac{1}{299^2}$ **[Ans.** 0.0008333290]

3. Find the sum of the fifth power of the first n natural numbers by means of Euler-Maclaurin formula.

   [Ans. $\frac{n^6}{6} + \frac{n^5}{2} + \frac{5n^4}{12} - \frac{n^2}{12}$]

4. If $f(x)$ and its derivative $f'(x)$ are continuously increasing in the range $0 < x < n$, show that $\int_0^n f(x)dx < \frac{1}{2}$ $[f(0) + f(n)] + \sum_{x=0}^{n-1} f(x)$.

   [Hint: In Euler-Maclaurin formula, put $a = 0$, $h = 1$]

5. Obtain the approximate quadrature formula $\int_0^1 \frac{dx}{1+x^2} = \frac{1}{4a}\left(1+\frac{1}{6a}\right) + \sum_{x=1}^{a} \frac{a}{a^2+x^2}$.

   [Hint: Use Euler-Maclaurin formula with $f(x) = \frac{a}{a^2+x^2}$ and range $(0, a)$ put $x = ay$ in the integral]

$$\int_0^a \frac{a}{a^2+x^2}dx = \sum_{x=0}^{a} \frac{a}{a^2+x^2} - \frac{1}{2}\left(\frac{1}{a}+\frac{1}{2a}\right) + \frac{1}{2a} \cdot \frac{1}{2a^2}$$

6. Evaluate $\int_0^1 \frac{dx}{1+x}$ correct to five decimal places by using Euler-Maclaurin formula. **[Ans.** 0.69315]

17.11 GAUSSIAN INTEGRATION

Earlier, some integration formula were derived to evaluate the integral $I = \int_a^b f(x)dx$ when values of the function were known as equispaced points of the intervals. Gauss derived a formula which uses the same number of functional values but with different spacing and yields better results.

Gauss's formula is expressed in the form

$$\int_{-1}^{+1} f(u)du = w_1F(u_1) + w_2F(u_2) + \ldots + w_nF(u_n)$$

where the w_i and u_i are called weights and abscissae respectively. An advantage of this formula is that the abscissae and weights are symmetrical with respect to the middle points of the interval. In Eq. (7.29) there exist $2n$ unknowns.

$$= \sum_{i=0}^{n} w_i F(u_i) \qquad ...(17.29)$$

Thus $2n$ relations between them are necessary which can be obtained such that the formula is exact for all polynomials of degree not exceeding $(2n - 1)$. Thus, we may consider

$$F(u) = C_0 + C_1 u + C_2 u^2 + C_3 u^3 + ... + C_{2n-1} u^{2n-1} \qquad ...(17.30)$$

Now from the left hand side of Eq. (17.29), we readily get

$$\int_{-1}^{+1} F(u)\, du = \int_{-1}^{1} [C_0 + C_1 u + C_2 u^2 + C_2 u^3 + ... + C_{2n-1} u^{2n-1}]\, du$$

$$= 2C_0 + \frac{2}{3} C_2 + \frac{2}{5} C_3 + ... \qquad ...(17.31)$$

By setting $u = u_i$ in Eq. (17.30), we obtain

$$F(ui) = C_0 + C_1 u_i + C_2 u_i^2 + C_3 u_i^3 + ... + C_{2n-1} u_i^{2n-1} \qquad ...(17.32)$$

Now substituting these values on the right hand side of Eq. (17.29), we get

$$\int_{-1}^{+1} F(u_i)\, du = w_i [C_0 + C_1 u_1 + C_2 u_1^2 + C_3 u_1^3 + ... + C_{2n-1} u_i^{2n-1}$$

$$+ w_2 [C_0 + C_1 u_1 + C_2 u_1^2 + C_3 u_2^3 + ... + C_{2n-1} u_i^{2n-1}] + w_3 [C_0 + C_1 u_3 + C_2 u_3^3$$

$$+ ... + C_{2n-1} u_3^{2n-1}] + ... + w_n [C_0 + C_1 u_n + C_2 y_n^3 + ... + C_{2n-1} u_n^{2n-1}]$$

$$\int_{-1}^{+1} F(u_i)\, du = C_0 [w_1 + w_2 + w_3 + ... + w_n] + C_1 [w_1 u_1 + w_2 u_2 + w_3 u_3 + ... + w_n u_n]$$

$$+ C_1 [w_1 u_1^2 + w_2 u_2^2 + w_3 u_3^2 + ... + w_n u_n^2]$$

$$+ C_{2n-1} [w_1 u_1^{2n-1} + w_2 u_2^{2n-1} + w_3 u_3^{2n-1} + ... + w_n u_n^{2n-1}] \qquad ...(17.32a)$$

But the Eqs (17.31) and (17.32a) are identical for all C_i, hence comparing coefficients of C_i, we obtain $2n$ equation in $2n$ unknown in $2n$ unknown w_i and u_i $(i = 1, 2, 3, ..., n)$, yielding

$$w_1 + w_2 + w_3 + ... + w_n = 2$$

$$w_1 u_1 + w_2 u_2 + w_3 u_3 + ... + w_n u_n = 0$$

$$w_1 u_1^2 + w_2 u_2^2 + w_3 u_3^2 = ... + w_n u_n^2 = \frac{2}{3} \qquad ...(17.33)$$

$$... = ...$$

$$w_1 u_1^{2n-1} + w_2 u_2^{2n-1} + w_3 u_3^{2n-1} + ... + w_n u_n^{\ 2n-1} = 0$$

Theoretically, we can obtain n unknown w_is and u'_is by solving the above of non-linear equations by the usual algebraic methods but practically it is very difficult to solve even for small values of n.

As an illustration, we consider the case $n = 2$, then the formula is

$$\int_{-1}^{1} F(u)\, du = w_1 F(u_1) + w_2 F(u_2)$$

Since this formula is exact when $F(u)$ is a polynomial of degree not exceeding 3, we put successively $F(u) = 1$, u, u^2 and u^3. Then Eq. (17.32) gives (four equations)

$$w_1 + w_2 = 2$$

$$w_1u_1^2 + w_2u_2^2 = \frac{2}{3} \qquad ...(17.34)$$

$$w_1u_1 + w_2u_2 = 0$$

$$w_1u_1^3 + w_2u_2^3 = 0$$

Solving the above equations, we have

$$w_1 + w_2 = 1$$

$$u_2 = -u_1 = \frac{1}{\sqrt{3}} \qquad ...(17.35)$$

This method when applied to the general system above, will be extremely complicated and difficult, and an alternate method must be chosen to solve the nonlinear system.

It can be shown that u_i are the zeros of the $(n+1)^{\text{th}}$ Legendre polynomials $P_{n+1}(u)$ which can be generated using the recurrence relation where $P_0(u) = 1$ and $P_1(u) = u$. The first Legendre polynomials are given by

$$(n+1)\, P_{n+1}(u) = (2n+1)\, u\, P_n(u) - nP_{n-1}(u)$$

$$P_0(u) = 1$$

$$P_1(u) = u$$

$$P_2(u) = \frac{1}{2}(3u^2 - 1) \qquad ...(17.36)$$

$$P_3(u) = \frac{1}{2}(5u^3 - 3u)$$

$$P_4(u) = \frac{1}{8}(35u^4 - 30u^2 + 3)$$

It can also be shown that the corresponding weights w_i are given by where the u_1 are the abscissae.

$$w_i = \int_{-1}^{1} j = 0\left(\frac{u - u_j}{u_i - u_j}\right) du \qquad ...(17.37)$$

As an example, when $n = 1$, we solve $P_2(u) = 0$, i.e. which gives the two abscissae

$$\frac{1}{2}(3u^2 - 1) = 0$$

$$u_0 = -\frac{1}{\sqrt{3}} = -\frac{\sqrt{3}}{3} \text{ and } u_1 = \frac{1}{\sqrt{3}} = \frac{\sqrt{3}}{3}$$

The corresponding weights are given by which give the four abscissae

$$w_0 = \int_{-1}^{1} \frac{u - u_1}{u_0 - u_1} du = \frac{1}{u_0 - u_1}\left[\frac{u^2}{2} - u_1 u\right]_{-1}^{1} = 1$$

$$w_1 = \int_{-1}^{1} \frac{u - u_0}{u_1 - u_0} du = \frac{1}{u_1 - u_0}\left[\frac{u^2}{2} - u_0 u\right]_{-1}^{1} = 1$$

Similarly for $n = 3$, we solve $P_4(u) = 0$

$$\frac{1}{8}(35u^2 - 30u^2 + 3) = 0$$

$$u_1 = \pm\left(\frac{15 \pm 2\sqrt{30}}{35}\right)^{\frac{1}{2}}$$

The weight w_i can then be found from Eq. (17.37). It should be noted, however, that the abscissae u_i and weight w_i are extensively tabulated for different values of n. We list below, in table, the abscissae and weight for values of n up to $n = 6$.

In general case, the limits of the integral in $\int_a^b f(x)\,dx$ have to be changed to those in Eq. (17.29) by means of the transformation

$$x = \frac{1}{2}u(b-a) + \frac{1}{2}(a+b)$$

Here, we tabulate the abscissae u_i and the weight w_i for the different value of n ($n = 2, 3, 4, 5, 6$)
Abscissae and weight for Gaussian integration is given below.

n		ui	wi		
2	0.57735	0.2692	1.0000000		
3		0.0			0.888888889
	0.77459	6692	0.55555	55556	
4	0.33998	10436	0.65214	51549	
	0.86113	63116	0.34785	48451	
5		0.0	0.56888	88889	
	0.53846	93101	0.47862	86705	
	0.90617	98459	0.23692	68851	
6	0.23861	91861	0.46791	39346	
	0.66120	93865	0.36076	15730	
	0.93246	95142	0.17132	44924	

***Example* 17.28.** Find $I = \int_0^1 x\,dx$ by Gauss's formula with $n = 4$ up to five decimal places.

Solution : First we change the limits of integration with the help of $x = \frac{1}{2}(u+1)$.

This gives $$I = \frac{1}{4}\int_{-1}^{1}(u+1)\,dx = \frac{1}{4}\sum_{i=1}^{n} w_i F(u_i)$$

where $$F(u_i) = u_i + 1$$

Now we choose $n = 4$ and use the abscissae and weights corresponding $n = 4$ in standard table and obtain where the abscissae and weights have been rounded to five decimal places.

$$I \approx \frac{1}{4}\,[(-0.86114 + 1)(0.34785) + (-0.33998 + 1)(0.65214)(0.3398 + 1)(0.65214) + (0.86114 + 1)(0.34785)]$$

$$= 0.49999$$

17.12 NUMERICAL DOUBLE INTEGRATION

17.12.1 *Trapezoidal Rule*

Suppose we have $I = \int_a^b \int_c^d f(x, y)\,dx\,dy$, the region of integration for which the area is bounded by the lines $x = c$, $x = d$, $y = a$, $y = b$ in the xy-plane.

Now we consider the double integral

$$I = \int_{y_i}^{y_{j+1}} \int_{x_i}^{y_{i+1}} f(x, y)\,dx\,dy \quad \text{...(17.38)}$$

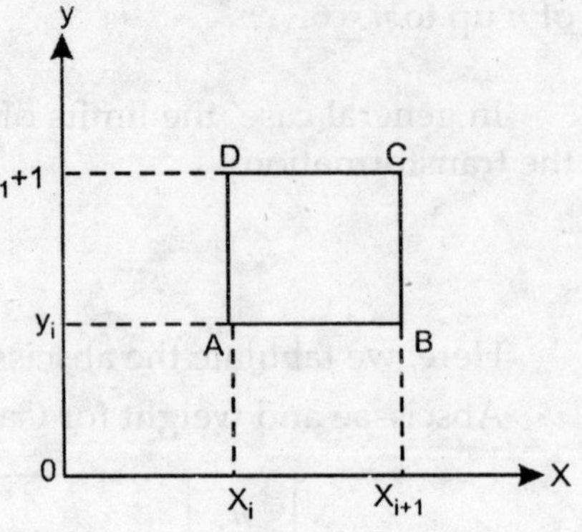

Fig. 17.2

where $x_{i+1} = x_i + h$ and $y_{k+1} = y_j + k$.

Also here $x_{i+1} - x_i = h$ and $y_{j+1} - y_j = k$ for all i, j.

Here I denotes the area of the region ABCD as shown in Fig. 17.2.

Now let us apply trapezoidal rule repeated to get the value of I in Eq. (17.38), we get

$$I = \int_{y_i}^{y_{j+1}} \frac{h}{2}\,[f(x_i + y) + f(x_{i+1}, y)]\,dy$$

$$= \frac{h}{2}\left[\int_{y_i}^{y_{i+1}} f(x_i, y)\,dy + \int_{y_i}^{y_{j+1}} f(x_{i+1}, y)\,dy\right]$$

$$= \frac{h}{2}\left[\frac{k}{2}[f(x_i, y_{j+1}) + f(x_i, y_j)] + \frac{k}{2}[f(x_{i+1}, y_{j+1}) + f(x_{i+1}, y_j)]\right]$$

$$I = \frac{hk}{4}\,[f(x_i, y_{j+1}) + f(x_i, y_j)\,f(x_{i+1}, y_{j+1}) + f(x_{i+1}, y_j)]$$

$$= \frac{hk}{4}\,[(f_{i,\,j+1} + f_{i,\,j} + f_{i+1,\,j+1} + f_{i+1,\,j})] \quad \text{...(17.38a)}$$

where $f_{i,\,j} = \mathrm{f}(x_i, y_j)$.

Using Fig. 7.2, we have

$$I = \frac{hk}{4}\,[\text{sum the value of } f(x, y) \text{ at the four corner points}]$$

i.e.

$$= \frac{hk}{4}\,[f_A + f_B + f_C + f_D] \quad \text{...(17.39)}$$

Suppose we require $\int_c^d \int_a^b f(x, y)\,dxdy$, where a, b, c, d are constants. The double integral denotes the area of the region bounded by the line $x = a$, $x = b$, $y = x$ and $y = d$ as shown in Fig. 17.3.

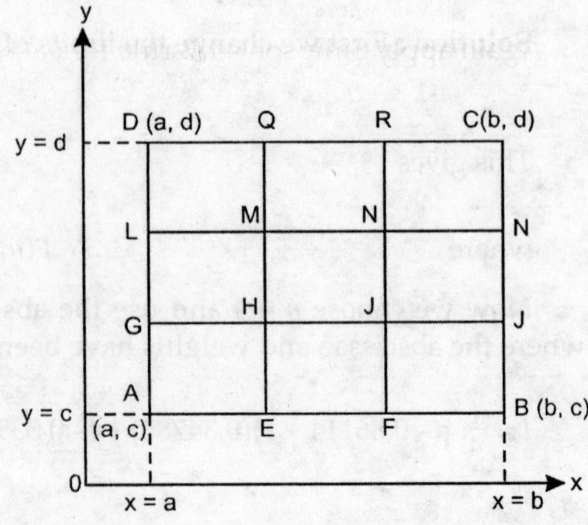

Fig. 17.3

First divide the region of integration (rectangle) $ABCD$ into meshes (or series subsquares) by dividing AB into three equal parts (in general into equal parts), each part length being h, and AD into three equal parts (in general n equal parts) each part length being k. Here, the integral over the whole rectangle is equal is to sum of the integrals over each mesh.

$$I = \int_c^d \int_a^b f(x, y)\,dx\,dy$$

$$= \iint_{AEHG} + \iint_{EFJH} + \iint_{FBKJ} + \iint_{GHML} + \iint_{HJNM} + \iint_{JKPN} + \iint_{LMQD} + \iint_{MNRQ} + \iint_{NPCR}$$

$$= \frac{hk}{4}[(f_A + f_E + f_H + f_G) + (f_E + f_F + f_J + f_H) + \ldots] \text{ [using Eq. (17.39)]}$$

$$= \frac{hk}{4}[(f_A + f_B + f_C + f_D) + 2(f_E + f_K + f_F + f_P + f_R + f_Q + f_L + f_G)$$

$$+ 4(f_H + f_J + f_M + f_N)] \text{ [on simplification]}$$

$$I = \frac{hk}{4}[(\text{sum of the values of } f \text{ at the four corners}) + 2\,(\text{sum of the value of } f \text{ at the remaining nodes on the boundary}) + 4\,(\text{sum of the value of f at the interior nodes})].$$

This equation is known as trapezoidal rule for double integration.

17.12.2 *Simpson's Rule for Double Integration*

Apply the Simpson's 1/3 rule, the number of intervals must be even, i.e. number of ordinates must be odd.

Now we consider the integral

$$I = \int_{y_i}^{y_{j+1}} \int_{x_i}^{x_{i+2}} f(x, y)\, dx\, dy \qquad \ldots(17.40)$$

where $y_{j+1} = y_j + k$, $x_{i+1} = x_i + h$.

The region of integration is shown in Fig. 17.4.

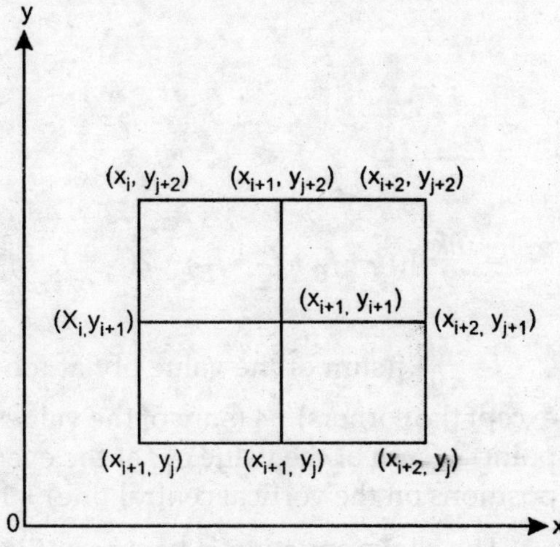

Fig. 17.4

Let us apply Simpson's 1/3 rule to the integral in Eq. (17.40), we get

$$I = \int_{y_i}^{y_{j+2}}\left[\int_{x_i}^{x_{i+2}} f(x, y)\, dx\right] dy$$

$$= \int_{y_j}^{y_{j+2}} \frac{h}{3}[f(x_i, y_j) + f(x_{i+2}, y_j) + 4f(x_{i+1}, y_j)]\, dy$$

Again apply Simpson's 1/3 rule for each integral in y direction

$$= \frac{h}{3}\left[\frac{k}{3} f(x_i, y_j) + f(x_i, y_{j+2}) + 4f(x_i, y_{j+1})\right]$$

$$+ \frac{k}{3} f(x_{i+2}, y_j) + f(x_i, y_{j+2}) + 4f(x_{i+2}, y_{j+1})]$$

$$+ \frac{4k}{3}[f(x_{i+1}, y_j) + f(x_{i+1}, y_{j+2}) + 4f(x_{i+1}, y_{j+1})]$$

$$= \frac{hk}{9}[(f_{i,j} + f_{i,j+2} + f_{i+2,j} + f_{i+2,j+2}) + 4(f_{i,j+1} + f_{i+1,j} + f_{i+1,j+2} + f_{i+2,j+1}) + 16 f_{i+1,j+1}]$$

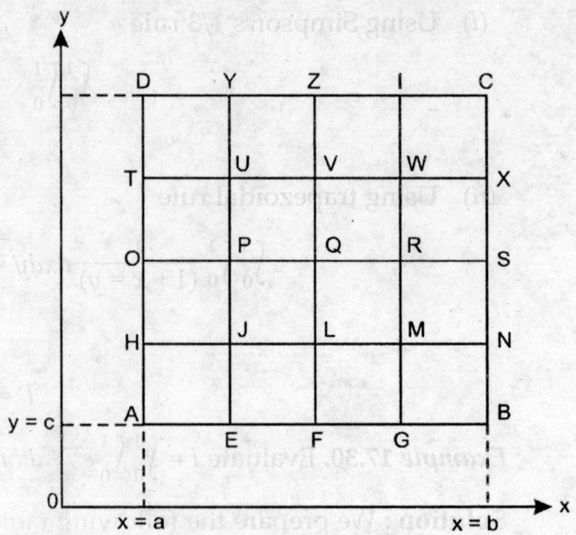

Fig.17.5

i.e. $I = \frac{hk}{9}$ (sum of the value of f at four corners) + 4 (sum of the values of f at the remaining nodes in the boundary) + 16 (value of f at the central point).

Suppose we require $I = \int_c^d \int_a^b f(x, y)\,dxdy$, where the limits are constant. Here the region of integral is bounded by $x = a$, $x = b$, $y = c$ and $y = d$ as shown in Fig. 17.5.

Firstly divide the interval (a, b) into four (in general number) equal parts and the interval (c, d) into four (in general even number) equal parts. Here length of each interval in (a, b) is h and in (c, d) is k.

Apply Simpson's 1/3 rule given by Eq. (17.41) to the four rectangles AFQO, FBSQ, SCZQ and ZDOQ. Hence we have

$$I = \frac{hk}{9}[(f_A + f_F + f_Q + f_O) + 4(f_E + f_L + f_P + f_H) + 16f_J]$$

$$+ [(f_F + f_B + f_S + f_Q) + 4(f_G + f_N + f_R + f_L) + 16f_M]$$

$$+ [(f_S + f_C + f_Z + f_Q) + 4(f_X + f_I + f_V + f_R) + 16f_W]$$

$$+ [(f_Z + f_D + f_O + f_Q) + 4(f_Y + f_T + f_P + f_V) + 16f_U]$$

$$= \frac{hk}{9}[(f_A + f_B + f_C + f_D) + 2(f_F + f_S + f_Z + f_O) + 4(f_E + f_G + f_N + f_X + f_T + f_H) + 4(f_Q + 8(f_P + f_R) + 16(f_J + f_M + f_W + f_V)]$$

$I = \frac{hk}{9}$ [(sum of the value of f at four corners) + 2 (sum of the value of f at the odd positions on the boundary except the corners) + 4 (sum of the values of f at the even positions on the boundary) + 4 (the values of f at the centre point) + (sum of the value of f at the even positions on the horizontal central line) + 8 (sum of the value of f at even positions on the vertical central line) + 16 (sum of values of f at the even position on even rows of the matrix)].

The above equation is known as Simpson's rule for double integrals.

***Example* 17.29.** Using (*i*) Simpson's 1/3 rule (*ii*) Trapezoidal rule, evaluate taking $h = k = 0.5$.

$$\int_0^1 \int_0^1 \frac{1}{(1 + x + y)} dxdy,$$

Solution : Here

$$f(x, y) = \frac{1}{1 + x + y}$$

We prepare the following table

y \ x	0	0.5	1
0	1	0.66	0.50
0.5	0.66	0.50	0.40
1	0.50	0.40	0.33

(*i*) Using Simpson's 1/3 rule

$$\int_0^1\int_0^1 = \frac{(0.5)(0.5)}{9}[(1+0.5+0.5+0.33)+4(0.66+0.4+0.4+0.66)+16(0.5)]$$

$$= (0.0278)[0.33+8.48+8] = 0.5229$$

(*ii*) Using trapezoidal rule

$$\int_0^1\int_0^1 \frac{1}{(1+x=y)}dxdy = \frac{(0.5)(0.5)}{9}[(1+0.5+0.5+0.33)$$

$$= +2(0.66+0.4+0.4+0.66)+4(0.5)]$$

$$= (0.0625)[2.33+4.24+2] = 0.5356$$

***Example* 17.30.** Evaluate $I = \int_0^1\int_0^1 e^{x+y}dx\,dy$ using the trapezoidal and Simpson's rules with $h = k = 0.5$.

Solution : We prepare the following table

y \ x	0	0.5	1
0	1	1.6487	2.7183
0.5	1.6487	2.7183	4.4817
1	2.7183	4.4817	7.3891

(*i*) Using the trapezoidal rule, we obtain

$$I = \frac{0.25}{4}[1.0+4(1.6487)+6(2.7183)+4(4.4817)+7.389] = 3.0762$$

(*ii*) Using Simpson's rule, we obtain

$$I = \frac{0.25}{4}[1.0+2.7183+7.3891+2.7183+4(1.6387+4.4817+4.4817+1.6487)+16(2.7183)]$$

$$= \frac{26.59042}{9} = 2.9545$$

The exact value of the double integral is 2.9525 and therefore, it can be verified that the result given by Simpson's rule is about sixty times more accurate than that given by trapezoidal rule.

***Example* 17.31.** Evaluate $\int_2^{2.4}\int_4^{4.4} xy\,dx\,dy$ using (*i*) trapezoidal rule (*ii*) Simpson's rule. Verify your result by actual integration.

Solution : To evaluate $\int_2^{2.4}\int_4^{4.4} xy\,dx\,dy$

Here $f(x, y) = xy$

(*ii*) To use trapezoidal rule, prepare the following table [i.e. value of $f(x, y)$]

y \ x	4	4.4
2	8	8.8
2.4	9.6	10.56

Using trapezoidal rule

$$\int_{2}^{2.4}\int_{4}^{4.4} xy\,dx\,dy = \frac{hk}{4}\text{ [sum of the value of } f \text{ at the four corner points]}$$

$$= \frac{(0.4)(0.4)}{4}[9.6 + 10.56 + 8.8 + 8] = (0.04)(36.96) = 1.4784$$

(*ii*) To apply Simpson's rule, the range of x and y in two equal parts

$$h = \frac{4.4-4}{2} = 0.2,\ k = \frac{2.4-2}{2} = 0.2$$

We prepare the following table for the value of $f(x, y)$

y \ x	4	4.2	4.4
2	8	8.4	8.8
2.2	8.8	9.24	9.68
2.4	9.6	10.08	10.56

Using Simpson's rule

$$\int_{2}^{2.4}\int_{4}^{4.4} xy\,dx\,dy = \frac{(0.2)(0.2)}{9}[(8 + 9.6 + 10.56 + 8.8) + 4(8.8 + 10.08 + 9.68 + 8.4) + 16(9.24)]$$

$$= (0.004)\,[36.96 + 147.84 + 147.84] = 1.33056$$

By actual integration

$$\int_{2}^{2.4}\int_{4}^{4.4} xy\,dx\,dy = \int_{2}^{2.4} y\,dy \int_{4}^{4.4} x\,dx$$

$$= \left[\frac{y^2}{2}\right]_2^{2.4}\left[\frac{x^2}{2}\right]_4^{4.4} = \frac{1}{4}[(2.4)^2 - 2^2]\,[(4.4)^2 - 4^2]$$

$$= \frac{1}{4}(3.36)\times(1.76) = 1.4784$$

The actual value and the value of trapezoidal rule are equal.

PROBLEMS

1. Using trapezoidal rule, evaluate $\int_{1}^{2}\int_{1}^{2}\frac{dy\,dx}{x+y}$ taking four subintervals. **[Ans.** 0.3407]

2. Evaluate $\int_{0}^{1}\int_{0}^{1} xe^{y}\,dx\,dy$ using trapezoidal rule ($h = k = 0.5$) **[Ans.** 0.876]

3. Apply trapezoidal rule to find $\int_{1}^{5}\int_{1}^{5}\frac{dx\,dy}{\sqrt{x^2+y^2}}$ taking two subintervals. **[Ans.** 4.134]

4. Evaluate $\int_{1}^{2.6}\int_{2.}^{3.2}\frac{dx\,dy}{x+y}$ using Simpson's rule. **[Ans.** 0.49]

5. Find the value of double integral $\int_1^{1.4}\int_1^{1.4} \frac{dx\,dy}{x+y}$ by trapezoidal rule. **[Ans.** 0.0349**]**

6. Evaluate $\int_0^{\frac{\pi}{2}}\int_0^{\frac{\pi}{2}} \sin(x+y)\,dxdy$ using Simpson's rule. **[Ans.** 2.0080**]**

Using both trapezoidal and Simpson's rules, evaluate the following integrals from 7 to 10. Check your answer by direct integration.

7. Find $\int_0^1\int_0^1 e^{2x+3y}\,dx\,dy$.

8. Find $\int_3^{3.8}\int_4^{4.8} \frac{1}{x^2+y^2}\,dx\,dy$ taking $h = k = 0.2$.

9. Find $\int_1^{1.8}\int_2^{2.8} (x^2+y^2)\,dx\,dy$ using $h = k = 0.2$.

10. Find $\int_0^1\int_0^1 e^{-(x+y)}\,dx\,dy$

11. Find the double integral $I = \int_0^1\int_0^2 \frac{2xy}{(1+x^2)(1+y^2)}\,dy\,dx$ by using trapezoidal rule with step length $h = k = 0.25$. **[Ans.** 0.5582**]**

12. Compute numerically $\iint_D \frac{dx\,dy}{x^2+y^2}$, where D is the square with corners at (1, 1), (2, 1), (2, 2), (1, 2).

[Ans. 0.2313**]**

Numerical Solution of Ordinary Differential Equations

18.1 INTRODUCTION

The analytical methods for solution with which we are familiar can be applied to a limited class of differential equations. If differential equations appearing in the physical problem do not belong to any of these familiar type, we resort to numerical methods. A variety of numerical methods have been evolved for solving differential equation of order one or more than use.

To describe various numerical methods of solving ordinary differential equations, we consider the first order differential equation

$$dy / dx = f(x, y); \text{ given } y(x_0) = y_0$$

In most of these methods, we replace the differential equation by a difference equation and solve it. These solutions yield solutions in one of the two forms:

(*i*) A series of y in terms of power of x, from which the value of y can be obtained by direct substitution.

(*ii*) A set of tabulated values of x and y.

The methods of Taylor's and Picard belongs to class (*i*) of solutions, whereas those of Euler, Runge-Kutta, Adams-Bashforth, etc. belong to class (*ii*). These methods are called step-by-step methods or marching methods because the values of y are computed by steps ahead for equal intervals 'h' of the independent variables. In methods of Euler and Runge-Kutta, the interval 'h' should be kept small, and therefore, these methods are used for computing y over a limited range of values of x, whereas Milne's Adams-Bashforth methods are applied for finding y over a wide range of values of x. Milne's Adams-Bashforth methods requires starting values which are usually obtained by Picard's, Taylor's series or Runge-Kutta methods.

(*a*) **Initial value problems:** Problems involving first order differential equations in which all the conditions are specified at the initial point only are called initial value problems such as $dy/dx = f(x, y)$, $y(x_0) = y_0$.

(*b*) **Boundary value problems:** Problems involving second and higher order differential equations in which all the conditions are specified at two or more points are called boundary value problems.

18.2 PICARD'S METHOD

Consider the first order differential equation with initial condition

$$\frac{dy}{dx} = f(x, y), \ y(x_0) = y_0 \qquad ...(18.1)$$

Integrating Eq. (18.1), we have

$$\int_{y_0}^{0} dy = \int_{x_0}^{x} f(x, y)\, dx$$

$$y = y_0 + \int_{x_0}^{x} f(x, y)\, dx \qquad ...(18.2)$$

As a first approximation y_1, we put $y = y_0$ in $f(x, y)$ and then integrate, we get to get a sequence giving better approximation and hence

$$y_1 = y_0 + \int_{x_0}^{x} f(x, y_0)\, dx$$

As a second approximation y_2, we put $y = y_1$ in $f(x, y)$ and then integrate, we get

$$y_2 = y_0 + \int_{x_0}^{0} f(x, y_1)\, dx \text{ and so on}$$

$$y_n = y_0 + \int_{x_0}^{x} f(x, y_{n-1})\, dx$$

This is known as Picard's method.

Example **18.1.** Employ Picard's method to find the solution of $\frac{dy}{dx} = x^2 + y^2$ for $x = 0.4$ given that $y = 0$ when $x =$ 0 up to four decimal places.

Solution : Here $x_0 = 0, y_0 = 0, f(x, y) = x^2 + y^2$

By Picard's method

$$y = y_0 + \int_{x_0}^{x} f(x, y)\, dx$$

$$= 0 + \int_{0}^{x} (x^2 + y^2)\, dx$$

For the first approximation y_1, put $y = 0$ in $f(x, y)$

$$y_1 = \int_0^x x^2\, dx$$

For the second approximation y_2, put $y = \frac{x^3}{3}$

$$y_2 = \int_0^x \left[x^2 + \left(\frac{x^3}{3}\right)^2\right] dx = \int_0^x \left(x^2 + \frac{x^6}{9}\right) dx = \frac{x^3}{3} + \frac{x^7}{63}$$

For the third approximation y_3, put $y = \frac{x^3}{3} + \frac{x^7}{63}$

$$y_3 = \int_0^x \left[x^2 + \left(\frac{x^3}{3} + \frac{x^7}{63}\right)^2\right] dx$$

$$= \int_0^x \left(x^2 + \frac{x^6}{9} + \frac{x^{14}}{(63)^2} + \frac{2x^{10}}{3 \times 63}\right) dx = \frac{x^3}{3} + \frac{x^7}{63} + \frac{x^{15}}{59535} + \frac{2x^{11}}{2079}$$

For next approximation, the integral becomes very difficult to evaluate, so we stop up to y_3.

Now when $x = 0.4$

$$y_1 = \frac{(0.4)^3}{3} = 0.021$$

$$y_2 = \frac{(0.4)^3}{3} + \frac{(0.4)^7}{63} = 0.02133333 + 0.00006260 = 0.02139593$$

$$y_3 = \frac{(0.4)^3}{3} + \frac{(0.4)^7}{63} + \frac{(0.4)^{15}}{59535} + \frac{2(0.4)^{11}}{2079}$$

$$= 0.21333333 + 0.00000107 + 0.0000839 + 0.000000403$$

$$= 0.021418341$$

Hence (y_2) (y_3), we have $y(0.4) = 0.0214$.

***Example* 18.2.** Use Picard's method to approximate y when $x = 0.1$ and 0.2 for $dy/dx = x + y^2$, where $y = 0$ when $x = 0$.

Solution : Method I: Here $y = 0$ when $x = 0$, $x_0 = 0$, $y_0 = 0$.

$$f(x, y) = x + y^2$$

By Picard's method

$$y = 0 + \int_0^x (x + y^2)\,dx \quad \text{...(i)}$$

For first approximation y_1, put $y = 0$ in RHS of Eq. (*i*)

$$y_1 = \int_0^x x\,dx = \frac{x^2}{2}$$

For second approximation y_2, put $y = x^2/2$ in RHS of Eq. (*i*)

$$y_2 = \int_0^x \left(x + \frac{x^4}{4}\right)dx = \frac{x^2}{2} + \frac{x^5}{20}$$

For third approximation y_3, put $y = x^2/2 + x^5/20$ in RHS of Eq. (*i*)

$$y_3 = \int_0^x \left[x + \left(\frac{x^2}{2} + \frac{x^5}{20}\right)^2\right]dx = \int_0^x x + \left(\frac{x^4}{4} + \frac{1}{400}x^{10} + \frac{1}{20}x^7\right)dx$$

$$= \frac{x^2}{2} + \frac{1}{20}x^5 + \frac{1}{4400}x^{11} + \frac{1}{1600}x^8$$

For $x = 0.1$

$$y_1 = \frac{(0.1)^2}{2} = \frac{0.01}{2} = 0.005$$

$$y_2 = \frac{0.01}{2} + \frac{0.00001}{20} = 0.005 + 0.000005 = 0.00500$$

$$y_3 = \frac{0.01}{2} + \frac{0.00001}{20} + \frac{0.00000001}{160} + \ldots = 0.00500$$

There is no difference between y_2 and y_3

$$\therefore \quad y(0.1) = 0.00500$$

For $x = 0.2$, we may take $x_0 = 0.1$, $y_0 = 0.005$, we first approximate y_1 and then y_2

$$y_1 = 0.005 + \int_{0.1}^{x}(x + 0.000025)\,dx = 0.005 + \left[\frac{x^2}{2} + 0.000025x\right]_{0.1}^{x}$$

$$= 0.005 + \frac{x^2}{2} + 0.000025x - \frac{(0.1)^2}{2} - 0.000025(0.1)$$

$$= 0.005 - 0.005 - 0.0000025 + \frac{1}{2}x^2 + \frac{25}{10^6}x$$

$$= \frac{0.04}{2} + \frac{25}{10^6}(0.2) - 0.0000025 = 0.02 + 0.0000025 = 0.0200$$

$$y_2 = 0.005 + \int_{0.1}^{x}(x + 0.0004)\,dx$$

$$= 0.005 + \left[\frac{x^2}{2} + 0.0004x\right]_{0.1}^{x} = \frac{x^2}{2} + \frac{4}{10^4}x - \frac{4}{10^5}$$

For $x = 0.2$ $\quad y_2 = 0.02 + 0.00008 - 0.00004 = 0.02004$

y_1 and y_2 are approximately the same up to four decimal places

$$\therefore \quad y(0.2) = 0.02004$$

Method II: Put $x = 0.2$, then

$$y_2 = \frac{x^2}{2} + \frac{x^5}{20}$$

$$y_3 = \frac{x^2}{2} + \frac{x^5}{20} + \frac{x^{11}}{4400} + \frac{x^8}{1600}, \text{ we get}$$

$$y_2 = 0.0200 \quad y_3 = 0.02004$$

Hence $y(0.2) = 0.0200$ correct to four decimal places.

***Example* 18.3.** Use Picard's method to solve $\frac{dy}{dx} = y$, $y(0) = 1$ and compare it with the exact solution.

Solution : Here $x_0 = 0$, $y_0 = 1$, $f(x, y) = y$

First approximation which is the same as Picard's solution.

$$y_1 = 1 + \int_0^x y_0\,dx = 1 + x$$

Second approximation

$$y_2 = 1 + \int_0^x (1 + x)\,dx = 1 + x + \frac{x^2}{2!}$$

Third approximation

$$y_3 = 1 + \int_0^x \left(1 + x + \frac{x^2}{2!}\right)dx = 1 + x + \frac{x^2}{2!} + \frac{x^3}{3!}$$

Fourth approximation

$$y_4 = 1+\int_0^x \left(1+x+\frac{x^2}{2!}+\frac{x^3}{3!}\right)dx = 1+x+\frac{x^2}{2!}+\frac{x^3}{3!}+\frac{x^4}{4!}$$

$$y_5 = 1+\int_0^x \left(1+x+\frac{x^2}{2!}+\frac{x^3}{3!}+\frac{x^4}{4!}\right)dx = 1+x+\frac{x^2}{2!}+\frac{x^3}{3!}+\frac{x^4}{4!}+\frac{x^5}{5!}$$

Similarly $$y_6 = 1+x+\frac{x^2}{2!}+\frac{x^3}{3!}+\frac{x^4}{4!}+\frac{x^5}{5!}+\frac{x^6}{6} \text{ and so on.}$$

Exact solution of $$\frac{dy}{dx} = y,\ x=0,\ y=1$$

$$dy = y\,dx$$

$$\frac{dy}{y} = dx$$

$$y = e^{x+h}$$

When $x = 0, y = 1$

$$1 = e^h$$

$$y = e^x = 1+x+\frac{x^2}{2!}+\frac{x^3}{3!}+\frac{x^4}{4!}$$

Example **18.4.** Using Picard's method, find the solution of the equation up to third approximation $\frac{dy}{dx} = 1 + xy$, $y(0) = 0$.

Solution : We have $$y = y_0 + \int_{x_0}^{0} f(x, y)\,dx$$

Here $$y_0 = 0,\ x_0 = 0$$

$$y = 0+\int_0^x f(x,y)\,dx = 0+\int_0^x (1+xy)\,dx \quad ...(i)$$

First approximation putting $y = 0$

$$y_1 = 0+\int_0^x (1+x\,.\,0)\,dx = x$$

Second approximation putting $y = x$ in Eq. (*i*)

$$y_2 = \int_0^x (1+x\,.\,x)\,dx = x+\frac{x^3}{3}$$

Third approximation putting $y = x + x^3/3$ in Eq. (*i*)

$$y_3 = 0+\int_0^x 1+x\left(x+\frac{x^3}{3}\right)dx = x+\frac{x^3}{3}+\frac{x^5}{15}$$

Example **18.5.** Solve the equation $\frac{dy}{dx} = \frac{x^2}{y^2+1}$ with initial condition $y = 0$ when $x = 0$. Use Picard's method to obtain y for $x = 0.25, 0.5, 1.0$ correct to three decimal places.

Solution : We have
$$y = y_0 + \int_{x_0}^{0} f(x, y)\,dx$$
$$= 0 + \int_0^x \frac{x^2}{y^2+1}dx \quad ...(i)$$

Putting $y = 0$ in Eq. (*i*) to get first approximation y_1

First approximation
$$y_1 = 0 + \int_0^x x^2\,dx = \frac{1}{3}x^3$$

Second approximation putting $y = \frac{1}{3}x^3$ in Eq. (*i*), we have
$$y_2 = 0 + \int_0^x \frac{x^2}{\frac{1}{9}x^6+1}dx = \tan^{-1}\left(\frac{1}{3}x^3\right) = \frac{1}{3}x^3 - \frac{1}{81}x^9 + ...$$

Neglecting $\frac{1}{81}x^9$ and higher powers as 0.25, 0.5, 1 are small.

Hence
$$y(0.25) = \frac{1}{3}(0.25)^3 = 0.005$$
$$y(0.5) = \frac{1}{3}(0.5)^3 = 0.042$$
$$y(1.0) = \frac{1}{3} - \frac{1}{81} = 0.321 \qquad \left(\text{taking into consideration } \frac{1}{81}x^9\right)$$

***Example* 18.6.** Use Picard's method to approximate the value of y when $x = 0.1$ given that $y = 1$ when $x = 0$ and $\frac{dy}{dx} = \frac{y-x}{y+x}$.

Solution :
$$\frac{dy}{dx} = f(x, y) = \frac{y-x}{y+x} \cdot (y_0 = 1, x_0 = 0) \quad ...(i)$$

First approximation
$$y_1 = y_0 + \int_{x_0}^{0} f(x, y_0)\,dx = y_0 + \int_{x_0}^{x} \frac{y-x}{y+x}dx$$
$$= 1 + \int_0^x \frac{1-x}{1+x}dx = 1 + \int_0^x \left(\frac{2}{1+x} - 1\right)dx$$
$$= 1 + [2\log(1+x) - x]_0^x = 1 + 2\log(1+x) - x$$
$$= 1 - x + 2\log(1+x)$$

Second approximation
$$y_2 = 1 + \int_0^x \frac{y_1 - x}{y_1 + x}dx$$
$$= 1 + \int_0^x \frac{1 - x + 2\log(1+x) - x}{1 - x + 2\log(1+x) + x}dx \quad ...(ii)$$
$$= 1 - 2\int_0^x \frac{x}{1 + 2\log(1+x)}dx + x$$

which is quite difficult to integrate putting $x = 0.1$ in Eq. (*ii*), we get

First approximation $$y_1 = 1 - 0.1 + 2\log(1 + 0.1) = 0.9828.$$

***Example* 18.7.** Use Picard's method to solve the equations

$$\frac{dx}{dt} = -y$$

$$\frac{dy}{dt} = x$$

Given that $x = 1$, $y = 0$ when $t = 0$.

Solution :

$$\frac{dx}{dt} = -y \qquad ...(i)$$

$$\frac{dy}{dt} = x \qquad ...(iii)$$

Integrating Eq. (*i*) w.r.t. t between $t = 0$ to t

$$[x]_0^x = -\int_0^t y\,dt$$

$$x - 1 = -\int_0^t y\,dt$$

$$x = 1 - \int_0^t y\,dt \qquad ...(iii)$$

Integrating Eq. (*ii*) w.r.t. t from $t = 0$ to t, we get

$$[y]_1^y = \int_0^t x\,dt$$

$$y - 0 = \int_0^t x\,dt$$

$$y = \int_0^t x\,dt \qquad ...(iv)$$

Replacing y by 0 in Eq. (*iii*) and x by 1 in Eq. (*iv*), we get

$$x = 1 - \int_0^t 0\,dt \text{ and } y = \int_0^t 1dt = t$$

$$= 1 - \int_0^t t\,dt = 1 - \frac{t^2}{2} \text{ and } y = \int_0^t \left(1 - \frac{t^2}{2}\right)dt = t - \frac{t^3}{6}$$

$$= 1 - \int_0^t \left(t - \frac{t^3}{6}\right)dt = 1 - \frac{t^2}{2} + \frac{t^4}{24}$$

$$y = \int_0^t \left(1 - \frac{t^2}{2} + \frac{t^4}{24}\right)dt = t - \frac{t^3}{6} + \frac{t^5}{120}$$

PROBLEMS

1. Using Picard's method, find $y(0.2)$ given that $y' = x - y$; $y(0) = 1$. **[Ans.** $y(0.2) = 0.837$]
2. Using Picard's method, obtain a solution up to fifth approximation to the equation $y' = y + x$, such that $y(0) = 1$. Check your answer by finding the exact particular solution. Also find $y(0.1)$ and $y(0.2)$.

   [Ans. $y(0.1) = 1.1103$, $y(0.2) = 1.2428$]
3. Using Picard's method, find $y(0.2)$ and $y(0.4)$ given that $y' = 1 + y^2$ and $y(0) = 0$.

   [Ans. $y(0.2) = 0.2027$, $y(0.4) = 0.4227$]
4. Using Picard's method to approximate the value of y when $x = 0.1$ given that $y(0) = 1$ and $y' = 3x + y^2$.

   [Ans. $y(0.1) = 1.127$]
5. Using Picard's method obtain the second approximation of the solution to the given equation $y'' - x^3y' - x^3y = 0$ so that $y(0) = 1$, $y'(0) = 0.5$. **[Ans.** $1 + \frac{1}{2}x + \frac{3}{40}x^5$]

18.3 TAYLOR'S SERIES METHOD

Consider the equation

$$\frac{dy}{dx} = f(x, y) \qquad ...(18.3)$$

$$y' = f(x, y)$$

$$\frac{d^2y}{dx^2} = \frac{df}{dx} = \frac{\partial f}{\partial x} + \frac{\partial f}{\partial y}\frac{dy}{dx} \qquad ...(18.4)$$

$$y'' = f_x + f_y\frac{dy}{dx} = f_x + f_y f$$

Differentiating this successively, we get y'', y''' so on.

Now the Taylor's series about $x = x_0$ is

$$y = y_0 + (x - x_0)\, y_0' + \frac{(x - x_0)^2}{2!}\, y_0'' + \frac{(x - x_0)^3}{3!}\, y_0''' + ... \qquad ...(18.5)$$

Putting $x = x_0$ and $y = y_0$ in y', y'', y''', we get the values of y_0', y_0'', y_0'''.

Substituting these values in Taylor's series, we get the value of y for every x for which Eq. (18.5) converges. On finding the value y_1 for $x = x_1$ from Eq. (18.5), y', y'', y''' can be obtained at $x = x_1$ by means of Eqs (18.3), (18.4) etc. Then y can be expanded about $x = x_1$. In this way, the solution can be extended beyond the range of convergence of series in Eq. (18.5).

Example **18.8.** From Taylor's series for $y(x)$, find $y(0.1)$ correct to four decimal places if $y(x)$ satisfies $\frac{dy}{dx} = x - y^2$, $y(0) = 1$.

Solution : The Taylor's series for $y(x)$ is given by

$$y = y_0 + (x - x_0)\, y_0' + \frac{(x - x_0)^2}{2!}\, y_0'' + \frac{(x - x_0)^3}{2!}\, y_0''' + ...$$

Now $x_0 = 0$, $y_0 = 1$

Then $$y = 1 + xy'_0 + \frac{x^2}{2} y''_0 + \frac{x^3}{6} y'''_0 + \frac{x^4}{24} y''''_0 + \ldots$$

Now $$\frac{dy}{dx} = x - y^2$$

$$y' = x - y^2 \qquad y'_0 = -1$$

$$y'' = 1 - 2yy' \qquad y''_0 = 1 - 2(-1) = 3$$

$$y''' = 0 - 2yy'' - 2y'^2 \qquad y'''_0 = -8$$

$$y'''' = -2yy''' - 2y'y'' \qquad y''''_0 = 34$$

$$y(0.1) = 1 - 0.1 + \frac{3}{2}(0.1)^2 - \frac{4}{3}(0.1)^3 - \frac{17}{12}(0.1)^4 \ldots$$

$$= 0.9138$$

Example **18.9.** Find by Taylor's series method, the values of y at $x = 0.1$, $x = 0.1$, $x = 0.2$ to five decimal places from $\frac{dy}{dx} = x^2 y - 1$, $y(0) = 1$.

Solution : Here $x = 0$, $y = 1$

$$y' = x^2 y - 1 \qquad y'_0 = -1$$

$$y'' = 2xy + x^2 \frac{dy}{dx} \qquad y''_0 = 0$$

$$y''' = 2y + 4xy' + x^2 y'' \qquad y'''_0 = 2$$

$$y'''' = 6y' + 6xy'' + x^2 y'' \qquad y''''_0 = -6 \text{ etc.}$$

Putting these values in Taylor's series, we have

$$y = 1 + x(-1) + \frac{x^2}{2!}(0) + \frac{x^3}{3!}(2) + \frac{x^4}{4!}(-6) + \ldots$$

$$= 1 - x + \frac{x^3}{3} - \frac{x^4}{4} + \ldots$$

$$y(0.1) = 1 - (0.1) + \frac{(0.1)^3}{3} - \frac{(0.1)^4}{4} + \ldots = 0.90033$$

$$y(0.2) = 1 - (0.2) + \frac{(0.2)^3}{3} - \frac{(0.2)^4}{4} + \ldots = 0.80227$$

Example **18.10.** Apply Taylor's series method to integrate

$$y' = 2t + 3y,\ y(0) = 1,\ t \in [0, 0.1],\ h = 0.1$$

Solution : $$\frac{dy}{dt} = 2t + 3y \qquad \left(\frac{dy}{dt}\right)_{t=0} = 3, \text{ when } t = 0, y = 1$$

$$\frac{d^2y}{dt^2} = 2 + 3\frac{dy}{dt} \qquad \left(\frac{d^2y}{dt^2}\right)_{t=0} = 2 + 3 \times 3 = 11$$

$$\frac{d^3y}{dt^3} = 3\frac{d^2y}{dt^2} \quad \left(\frac{d^3y}{dt^3}\right)_{t=0} = 3 \times 11 = 33$$

$$\frac{d^4y}{dt^4} = 3\frac{d^3y}{dt^3} \quad \left(\frac{d^4y}{dt^4}\right)_{t=0} = 3 \times 33 = 99$$

$$\frac{d^5y}{dt^5} = 3\frac{d^4y}{dt^4} \quad \left(\frac{d^5y}{dt^5}\right)_{t=0} = 3 \times 99 = 297$$

Taylor's series about $t = t_0 = 0$

$$y = y_0 + (t - t_0)\left(\frac{dy}{dt}\right)_{t=0} + \frac{(t-t_0)^2}{2!}\left(\frac{d^2y}{dt^2}\right)_{t=0} + \frac{(t-t_0)^3}{3!}\left(\frac{d^3y}{dt^3}\right)_{t=0} + \frac{(t-t_0)^4}{4!}\left(\frac{d^4y}{dt^4}\right)_{t=0}$$

$$= 1 + 3t + \frac{11}{2!}t^2 + \frac{33}{3!}t^3 + \frac{99}{4!}t^4 + \frac{297}{5!}t^5 + \ldots$$

$$y(0, 0.1) = 1 + 3(0.1) + \frac{11}{2}(0.2)^2 + \frac{33}{6}(0.1)^3 + \frac{99}{24}(0.1)^4 + \frac{297}{120}(0.1)^5 + \ldots$$

$$= 1 + 0.3 + 0.22 + 0.0055 + 0.0004125 + 0.00002475$$

$$= 1.525937$$

***Example* 18.11.** Find the solutions $y(0.1)$ and $y(0.2)$ of the initial value problem $y' = x(1 - 2y^2)$, $y(0) = 1$, using the first two nonzero terms of the Taylor's series and $h = 0.1$.

Solution :

$$y' = x(1 - 2y^2)$$

$$y'' = (1 - 2y^2) + x(-4yy')$$

$$y''' = -4yy' - 4yy' + x(-4y'^2 - 4yy'')$$

$$y'''' = 8y'^2 - 8yy'' + (-4y'^2 - 4yy'') + x(-8y'y'' - 4y'y'' - 4yy'')$$

Putting $x = 0,\ y = 1$

$$y' = 0$$

$$y_0'' = 1 - 2(1)^2 = -1$$

$$y_0''' = 0 + 0 = 0$$

$$y_0'''' = 0 - 8(1)(-1) + [0 - 4(-1)] + 0 = 12$$

Now the Taylor's series about $x = x_0$ is

$$y = y_0 + xy_0' + \frac{x^2}{2!}y_0'' + \frac{x^3}{3!}y_0''' + \frac{x^4}{4!}y_0'''' + \ldots$$

$$y(0.1) = 1 + 0 + \frac{(0.1)^2}{2!}(-1) + \frac{(0.1)^3}{3!} \times 0 + \frac{(0.1)^4}{4!} \times 12$$

$$= 1 + -0.005 + 0.00005 = 0.99505$$

for $x_1 = 0.1,\ y_1 = 0.99505$

Now Taylor's series about $x = x_1$

$$y = y_1 + (x - x_1)\, y_1' + \frac{(x - x_1)^2}{2!} y_1'' + \frac{(x - x_1)^3}{3!} y_1''' + \frac{(x - x_1)^4}{4!} y_1'''' + \ldots$$

$$y_2 = y_1 + h y_1' + y_1'' + \frac{h^4}{4!} y_1'''' + \ldots$$

Now

$$y_1' = 0.1\,[1 - 2(0.99505)^2] = -\,0.098024$$

$$y_1'' = (1 - 2\,(0.99505)^2 + 0.1\,[-4 \times 0.99505 \times -\,0.98024] = -\,0.94122$$

$$y(0.2) = 0.99505 + 0.1\,[-\,0.098024] + \frac{0.01}{2}\,[-\,0.94122]$$

$$= 0.99505 - 0.0098024 - 0.005 \times 0.04122 + \ldots$$

$$= 0.9805415$$

Example **18.12.** Employ Taylor's series method to obtain the approximate value of y at $x = 0.2$ for $y' = 2y + 3e^x$, $y(0) = 0$. Compute the approximation value with the exact value.

Solution : Here $x_0 = 0 \quad y_0 = 1$

$$y' = 2y + 3e^x \quad y_0' = 3$$

$$y'' = 2y' + 3e^x \quad y_0'' = 6 + 3 = 9$$

$$y''' = 2y'' + 3e^x \quad y_0''' = 18 + 3 = 21$$

$$y'''' = 2y''' + 3e^x \quad y_0'''' = 42 + 3 = 45$$

Now by Taylor's series expansion of $y = f(x)$ about $x = x_0 = 0$.

We have $y = y_0 + (x - 0)\, y_0' + \dfrac{(x - 0)^2}{2!} y_0'' + \dfrac{(x - 0)^3}{3!} y_0''' + \dfrac{(x - 0)^4}{4!} y_0'''' + \ldots$

$$= y_0 + x y_0' + \frac{x^2}{2!} y_0'' + \frac{x^3}{3!} y_0''' + \frac{x^4}{4!} y_0'''' + \ldots$$

$$= 3x + \frac{x^2}{2!} \times 9 + \frac{x^3}{3!} \times 21 + \frac{x^4}{4!} \times 45 + \ldots$$

$$= 3x + \frac{9x^2}{2} + \frac{7x^3}{2} + \frac{15x^4}{8} + \ldots$$

when $x = 0.2$

$$y(0.2) = 3(0.2) + \frac{9(0.2)^3}{2} + \frac{7(0.2)^3}{2} + \frac{15(0.2)^4}{8} + \ldots$$

$$= 0.6 + 0.18 + 0.028 + 0.0030 = 0.811$$

Exact value

$$\frac{dy}{dx} = 2y + 3e^x$$

$$\frac{dy}{dx} - 2y = 3e^x$$

$$\text{I.F.} = e^{-2\int dx} = e^{-2x}$$

$$e^{-2x}y = \int 3e^{-x}dx + C = -3e^{-x} + C$$

$$y = -3e^{x} + Ce^{+2x}$$

when $x = 0, y = 0$

$$0 = -3 + C, C = 3$$

$$y = -3e^{x} + 3e^{2x} = 3e^{x}(e^{x} - 1]$$

when $x = 0.2, y = 3e^{2}(e^{2} - 1)]$

Hence, the value of y at $x = 0.2$ is 0.811.

Example **18.13.** Use Taylor's series method to solve $\frac{dy}{dx} = x^2 - y, x = 0, y = 1$ at $x = 0.1, x = 0.2, x = 0.3, h = 0.1$.

Solution : Here

$$x_0 = 0 \qquad y_0 = 1$$

$$y' = x^2 - y \qquad y'_0 = x_0^2 - y_0 = -1$$

$$y'' = 2x - y' \qquad y''_0 = 2x_0 - y'_0 = 1$$

$$y''' = 2 - y'' \qquad y'''_0 = 2 - y''_0 = 1$$

$$y'''' = -y''' \qquad y'''' = -y'''_0 = -1$$

Now by Taylor's series expand $y = f(x)$ about $x = x_0$, we have

$$y = y_0 + (x - x_0)\,y'_0 + \frac{(x - x_0)^2}{2!}\,y''_0 + \frac{(x - x_0)^3}{3!}\,y'''_0 + \ldots$$

From $x_1 = 0.1$

$$y_1 = y_0 + (x_1 - x_0)\,y'_0 + \frac{(x_1 - x_0)^2}{2!}\,y''_0 + \frac{(x_1 - x_0)^3}{3!}\,y'''_0 + \ldots$$

$$= y_0 + hy'_0 + \frac{h^2}{2!}\,y''_0 + \frac{h^3}{3!}\,y'''_0 + \ldots$$

$$= 1 + 0.1\,(-1) + \frac{(0.1)^2}{2!}\,(1) + \frac{(0.1)^3}{3!}\,(1) + \ldots = 0.905125$$

$$x_1 = 0.1, y_1 = 0.905125$$

Now

$$y'_1 = x_1^2 - y_1 = (0.1)^2 - 0.905125 = -0.895125$$

$$y''_1 = 2x_1 - y'_1 = 2(0.1) + (0.895725) = 1.905125$$

$$y'''_1 = 2 - y''_1 = 2 - (1.095125) = 0.904875$$

$$y''''_1 - y'''_1 = -0.904875$$

Expanding by Taylor's series about $x = x_1$, we have

$$y = y_1 + (x - x_1)\,y'_1 + \frac{(x - x_1)^2}{2!}\,y''_1 + \ldots$$

For $x_2 = 0.2$ $\quad y_2 = y_1 + hy'_1 + \frac{h^2}{2!}\,y''_1 + \frac{h^3}{3!}\,y'''_1 + \ldots$

$$= (0.905125) + (0.1)(0.895125) + \frac{(0.1)^2}{2!}(1.095125) + \frac{(0.1)^3}{3!}(0.904875) + ...$$

$$= 0.8212354$$

$$x_2 = 0.2, \quad y_2 = 0.8212352$$

$$y'_2 = -0.7812352$$

$$y''_2 = 1.1812332$$

$$y'''_3 = 0.8187648$$

Expanding by Taylor's series about $x = x_2$

$$y = y_2 + (x - x_2)\, y'_2 + \frac{(x - x_2)^2}{2!}\, y''_2 + ...$$

For $x_3 = 0.3$

$$y_3 = y_2 + hy'_2 + \frac{h^2}{2!}\, y''_2 + \frac{h^3}{3!}\, y'''_2 + ...$$

$$= (0.8212352) + (0.1)(-0.7812352) + \frac{(0.1)^2}{2!}(1.1812332)$$

$$+ \frac{(0.1)^3}{3!}(0.8187645) + ...$$

$$= 0.7491509$$

PROBLEMS

1. Using Taylor's series method, solve $y' = xy + y^2$, $y(0) = 1$ at $x = 0.1, 0.2, 0.3$.

   [Ans. $y(0.1) = 1.1167$, $y(0.2) = 1.2767$, $y(0.3) = 1.5023$]

2. Use Taylor's series method to compute $y(0.2)$ to three decimal places given $\frac{dy}{dx} = 1 - 2xy$ that $y(0) = 0$.

   [Ans. 2.0206]

3. Solve $y' = y^2 + x$, $y(0) = 1$, using Taylor's series method and compute $y(0.1)$ and $y(0.2)$.

   [Ans. 1.1164, 1.2725]

4. Evaluate $y(0.1)$ correct to six decimal places by Taylor's series method if $y(x)$ satisfies $y' = xy + 1$, $y(0) = 1/$

   [Ans. 1.053425]

5. Using Taylor's series method, find $y(1.1)$, $y(1.2)$, given $\frac{dy}{dx} = xy^{1/3}$ and $y(1) = 1$. **[Ans.** 1.106808, 0.809333]

6. Solve $\frac{dy}{dx} = x_2 + y_2$, $y(0.1) = 1$. Determine the first three nonzero terms and hence obtain $y(1)$.

   [Ans. $\frac{1}{3}x^3 + \frac{1}{63}x^7 + \frac{2}{2079}x^{11} + ...$, $y(1) = 0.3502$]

18.4 EULER'S METHOD

Consider the equation

$$\frac{dy}{dx} = f(x, y) \quad ...(18.6)$$

given that $y(x_0) = y_0$. Its curve of solution through $P(x_0, y_0)$ is shown dotted in Fig. 18.1. Now we have to find the ordinate of any other point Q on this curve.

Let us divide ZM into n subintervals each of width h at $Z_1, Z_2, ...$ so that h is quite small.

In the interval ZZ, we approximate the curve by the tangent at P. If the ordinate through Z_1 meets the tangent in $P_1(x_0 + h, y_1)$, then

$$y_1 = Z_1P_1 = ZP + R_1P_1 = y_0 + PR_1, \tan\theta$$

$$= y_0 + h\left(\frac{dy}{dx}\right)_P = y_0 + hf(x_0, y_0) \quad ...(18.7)$$

Let P1Q1 be the curve of solution of Eq. (18.6) through P_1 and its tangent at P_1 meet the ordinate through Z_2 in P_2 $(x_0 + 2h, y_2)$. Then

$$y_2 = y_1 + hf(x_0 + h, y_1) \quad ...(18.7a)$$

Repeating this process n times, we finally reach on an approximation MP_n of MQ given by

$$y_n = y_{n-1} + hf(x_0 + (n-1)\,h, y_{n-1})$$

This is Euler's method of finding an approximate solution of Eq. (18.5).

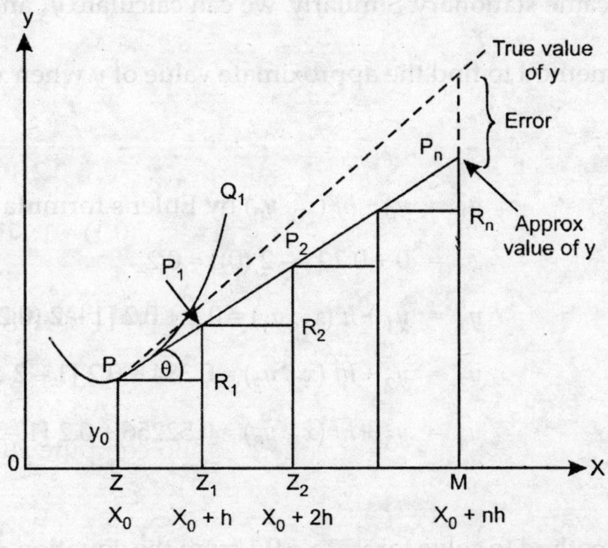

Fig. 18.1

Note: In Euler's method, we approximate the curve of the solution by the tangent in each interval, i.e. by a sequence of short lines. Unless h is small, the error is bound to be quite significant. This sequence of lines may also deviate considerably from the curve of solution. As such, the method is very slow and hence there is a modification in this method.

18.5 MODIFIED EULER'S METHOD

In the Euler's method, the curve of solution in the interval ZZ_1 is approximately by the tangent at P (Fig. 18.1), such that at P_1, we have

$$y_1 = y_0 + hf(x_0, y_0) \quad ...(18.8)$$

Then the slope of the curve of solution through P_1, *i.e.* is computed and the tangent at P_1, and P_1Q is drawn meeting the ordinate through Z_2 in P_2 $(x_0 + 2h, y_2)$, we find better approximation y_1 of $y(x_0 + h)$ by taking the curve as the mean of slopes of the tangents at P and P_1, etc.

$$\left(\frac{dy}{dx}\right)_{P_1} = f(x_0 + h, y_1)$$

$$y_1^{(1)} = y_0 + \frac{h}{2}[f(x_0, y_0) + f(x_0 + h, y_1)] \quad ...(18.9)$$

Similarly a better value of y_1 can be obtained

$$y_1^{(2)} = y_0 + \frac{h}{2}[f(x_0, y_0) + f(x_0 + h, y_1^{(1)})]$$

we repeat this step till two consecutive values of y agree.

This is then taken as the starting point for the next interval Z_1Z_2.

Once y_1 is obtained to desired degree of accuracy y corresponding to Z_2 found from Euler's

$$y_2 = y_1 + hf(x_0 + h, y_1)$$

and a better approximation $y_2^{(1)}$ can be obtained by

$$y_2^{(1)} = y_1 + \frac{h}{2}[f(x_0 + h, y_1) + f(x_0 + 2h, y_2)]$$

we repeat this step till y_2 became stationary. Similarly, we can calculate y_3 and better approximate.

***Example* 18.14.** Use Euler's method to find the approximate value of y when $x = 0.8$, if $\frac{dy}{dx} = 1 - 2xy$, given that $y = 0$ when $x = 0$ and $h = 0.2$.

Solution : $y_0 = 0$, $x_0 = 0$.

For $x_1 = 0.2$ $\quad y_1 = y_0 + hf(x_0, y_0)$ by Euler's formula

$$= 0 + 0.2[1 - 2(0)] = 0.2$$

For $x_2 = 0.4$ $\quad y_2 = y_1 + h(x_1, y_1) = 0.2 + 0.2[1 - 2(0.2)(0.2)] = 0.384$

For $x_3 = 0.6$ $\quad y_3 = y_2 + hf(x_2, y_2) = 0.384 + 0.2[1 - 2(0.4)(0.384)] = 0.52256$

For $x_4 = 0.8$ $\quad y_4 = y_3 + hf(x_3, y_3) = 0.52256 + 0.2[1 - 2(0.6)(0.52256)] = 0.5971456$

when $x = 0.8$, $y = 0.5971456$.

***Example* 18.15.** Use Euler's method to solve for y at $x = 0.1$ from the equation $\frac{dy}{dx} = x + y + xy$ and $y(0) = 1$, taking step size = 0.025.

Solution : By Euler's method $x_0 = 0$, $x_1 = 0.025$, $x_2 = 0.05$, $x_3 = 0.075$, $x_4 = 0.1$.

For $x_1 = 0.025$ $\quad y_1 = y_0 + hf(x_0, y_0)$

$$= 1 + 0.025(0 + 1 + 0)) = 1.025$$

For $x_2 = 0.05$ $\quad y_2 = 1.025 + 0.025[0.025 + 1.025 + 0.025 \times 1.025]$

$$= 1.025 + 0.025[1.05 + 0.025625] = 1.025 + 0.02689 = 1.05189$$

For $x_3 = 0.075$ $\quad y_3 = 1.05189 + 0.025[0.05 + 1.05189 + 0.05 \times 1.05189]$

$$= 1.05189 + 0.025[1.10189 + 0.05259]$$

$$= 1.05189 + 0.025 \times 1.15455$$

$$= 1.05189 + 0.0287125 = 1.0806$$

Similarly $x_4 = 0.1$

$$y_4 = 1.0806 + 0.025\,[0.075 + 1.0806 + 0.075 + 1.0806]$$

$$= 1.0806 + 0.025\,[1.1556 + 0.08105]$$

$$= 1.0806 + 0.025 \times 1.23665 = 1.0806 + 0.0309 = 1.1115$$

***Example* 18.16. Using modified Euler's method, find the approximate value of y at $x = 2$ in steps of 0.2 from $\frac{dy}{dx} = 2 + \sqrt{xy}$, $y(1) = 1$.**

Solution : $\frac{dy}{dx} = 2 + \sqrt{xy}$, $y(1) = 1$, $h = 0.2$, $y_0 = 1$, $x_0 = 1$, $x_1 = 1.2$, $x_2 = 1.4$, $x_3 = 1.6$, $x_4 = 1.8$, $x_5 = 2$.

By Euler's method

$$y_1 = y_0 + hf(x_0, y_0)$$

$$= 1 + 0.2[2 + \sqrt{1}\,] = 1 + 0.2(3) = 1.6$$

By modified Euler's method

For $x_1 = 1.2$

$$y_1^{(1)} = y_0 + \frac{h}{2}\,[f(x_0, y_0) + f(x_1, y_1)]$$

$$= 1 + 0.1[3 + 2 + \sqrt{(1.2)(1.6)}\,] = 1.63856$$

$$y_1^{(2)} = y_0 + \frac{h}{2}\,[f(x_0, y_0) + f(x_1, y_1^{(1)})]$$

$$= 1 + 0.1[3 + 2 + \sqrt{(1.2)(1.63856)}\,] = 1.6402$$

$$y_1^{(3)} = y_0 + \frac{h}{2}\,[f(x_0, y_0) + f(x_1, y_1^{(2)})]$$

$$= 1.6402$$

Hence, when $x_1 = 1.2$, $y_1 = 1.6402$.

By Euler's method

$$= 1.6402 + 0.2[2 + \sqrt{(1.2)(1.6402)}\,]$$

$$= 1.6402 + 0.2[2 + 1.4025] = 2.32078$$

Apply modified Euler's method

$$y_2^{(1)} = y_1 + \left[\frac{h}{2} f(x_1, y_1) + f(x_2, y_2)\right]$$

$$= 1.6402 + 0.1[3.4029 + 2\,\sqrt{(1.4)(2.32078)}\,] = 2.36074$$

$$y_2^{(2)} = y_1 + \frac{h}{2}\,[f(x_1, y_1) + f(x_2, y_2^{(1)})]$$

$$= 1.6402 + 0.1[3.4029 + 2\,\sqrt{(1.4)(2.36074)} = 2.36228$$

$$y_2^{(3)} = y_1 + \frac{h}{2}\,[f(x_1, y_1) + f(x_2, y_2^{(2)})]$$

$$= 1.6402 + 0.1[3.4029 + 2\sqrt{(1.4)(2.36228)} = 2.36234$$

$$y_2^{(4)} = 1.6402 + 0.1[3.4029 + 2\sqrt{(1.4)(2.36234)} = 2.36234$$

Hence $x_2 = 1.4, y_2 = 2.36234$

$$x_3 = 1.6 \text{ by Euler's method } y_3 = y_2 + hf(x_2, y_2).$$

$$= 2.36234 + 0.2\,[2 + \sqrt{(1.4)(2.36234)}\,] = 3.126058$$

Apply modified Euler's method

$$y_3^{(1)} = y_2 + \frac{h}{2}\,[f(x_2, y_2) + f(x_3, y_3)]$$

$$= 2.36234 + 0.1[3.81859 + 2 + \sqrt{(1.6)(3.126058)}\,] = 3.16784$$

$$y_3^{(2)} = y_2 + \frac{h}{2}\,[f(x_2, y_2) + f(x_3, y_1^{(3)})]$$

$$= 2.36234 + 0.1[3.81859 + 2 + \sqrt{(1.6)(3.16784)}\,] = 3.16933$$

$$y_3^{(3)} = 2.36234 + 0.1[3.81859 + 2 + \sqrt{(1.6)(3.16933)}\,] = 3.1693$$

when $x_3 = 1.6, y_3 = 3.1693$

$$x_4 = 1.8 \text{ by Euler's method } y_4 = y_3 + hf(x_3, y_3)$$

$$= 3.1693 + 0.2[2 + \sqrt{3.1693)(1.6)}\,]$$

$$= 3.1693 + 0.2\,(4.25186] = 4.0196$$

Apply modified Euler's method

$$y_4^{(1)} = 3.1693 + 0.1[4.25186 + 2 + \sqrt{(1.8)(4.0196)}\,] = 4.06347$$

$$y_4^{(2)} = 3.1693 + 0.1[4.25186 + 2 + \sqrt{(1.8)(4.06347)}\,] = 4.06493$$

$$y_4^{(3)} = 3.1693 + 0.1[4.25186 + 2 + \sqrt{(1.8)(4.0693)}\,] = 4.06493$$

Hence, $x_4 = 1.8, y_4 = 4.06493$

$$x_5 = 2.00 \text{ by Euler's method } y_5 = y_4 + hf(x_4, y_4)$$

$$= 4.06493 + 0.2[2 + \sqrt{(1.9)(4.06491)}\,] = 5.00589$$

By modified Euler's method

$$y_5^{(1)} = 4.0649 + 0.1[4.0.7049 + 2 + \sqrt{2 \times 5.00589}\,] = 0.051803$$

$$y_5^{(2)} = 4.0649 + 0.1[4.0.7049 + 2 + \sqrt{2 \times 5.051803}\,] = 5.05325$$

$$y_5^{(3)} = 5.05329$$

when $x_5 = 2, y_5 = 5.0532$,

hence $x = 2, y = 5$.

$$= 1.05189 + 0.025 \times 1.15455$$

$$= 1.05189 + 0.0287125 = 1.0806$$

Similarly $x_4 = 0.1$ $\quad y_4 = 1.0806 + 0.025\,[0.075 + 1.0806 + 0.075 + 1.0806]$

$$= 1.0806 + 0.025\,[1.1556 + 0.08105]$$

$$= 1.0806 + 0.025 \times 1.23665 = 1.0806 + 0.0309 = 1.1115$$

***Example* 18.16.** Using modified Euler's method, find the approximate value of y at $x = 2$ in steps of 0.2 from $\frac{dy}{dx} = 2 + \sqrt{xy}$, $y(1) = 1$.

Solution : $\frac{dy}{dx} = 2 + \sqrt{xy}$, $y(1) = 1$, $h = 0.2$, $y_0 = 1$, $x_0 = 1$, $x_1 = 1.2$, $x_2 = 1.4$, $x_3 = 1.6$, $x_4 = 1.8$, $x_5 = 2$.

By Euler's method

$$y_1 = y_0 + hf(x_0, y_0)$$

$$= 1 + 0.2[2 + \sqrt{1}\,] = 1 + 0.2(3) = 1.6$$

By modified Euler's method

For $x_1 = 1.2$ $\quad y_1^{(1)} = y_0 + \frac{h}{2}\,[f(x_0, y_0) + f(x_1, y_1)]$

$$= 1 + 0.1[3 + 2 + \sqrt{(1.2)(1.6)}\,] = 1.63856$$

$$y_1^{(2)} = y_0 + \frac{h}{2}\,[f(x_0, y_0) + f(x_1, y_1^{(1)})]$$

$$= 1 + 0.1[3 + 2 + \sqrt{(1.2)(1.63856)}\,] = 1.6402$$

$$y_1^{(3)} = y_0 + \frac{h}{2}\,[f(x_0, y_0) + f(x_1, y_1^{(2)})]$$

$$= 1.6402$$

Hence, when $x_1 = 1.2$, $y_1 = 1.6402$.

By Euler's method

$$= 1.6402 + 0.2[2 + \sqrt{(1.2)(1.6402)}\,]$$

$$= 1.6402 + 0.2[2 + 1.4025] = 2.32078$$

Apply modified Euler's method

$$y_2^{(1)} = y_1 + \left[\frac{h}{2}f(x_1, y_1) + f(x_2, y_2)\right]$$

$$= 1.6402 + 0.1[3.4029 + 2\,\sqrt{(1.4)(2.32078)}\,] = 2.36074$$

$$y_2^{(2)} = y_1 + \frac{h}{2}\,[f(x_1, y_1) + f(x_2, y_2^{(1)})]$$

$$= 1.6402 + 0.1[3.4029 + 2\,\sqrt{(1.4)(2.36074)} = 2.36228$$

$$y_2^{(3)} = y_1 + \frac{h}{2}\,[f(x_1, y_1) + f(x_2, y_2^{(2)})]$$

By Euler's method $\quad y_2 = y_1 + hf(x_1, y_1) = 1.2309 + 0.2[0.2 + \sqrt{1.2309}]$

$$= 1.2309 + 0.2[0.2 + 1.109] = 1.4590$$

By modified Euler's method

$$y_2^{(1)} = 1.2309 + \frac{0.2}{2}[0.2 + \sqrt{(1.2309)} + 0.4 + \sqrt{(1.4590)}]$$

$$= 1.2309 + 0.1[0.6 + 1.109 + 1.222] = 1.5240$$

$$y_2^{(2)} = 1.2309 + 0.1[(0.2 + \sqrt{1.2309}) + (0.4 + \sqrt{1.5240})]$$

$$= 1.2309 + 0.1[0.6 + 1.109 + 1.235] = 1.5253$$

$$y_2^{(3)} = 1.2309 + 0.1[(0.2 + \sqrt{1.2309}) + (0.4 + \sqrt{1.5253})] = 1.5253$$

Hence, we have $y_2 = 1.5253$, $x_2 = 0.4$.

Again by Euler's method

$$y_3 = y_2 + 0.2\,(x_2 + \sqrt{y_2})$$

$$= 1.5253 + 0.2[0.4 + \sqrt{1.5253}]$$

$$= 1.5253 + 0.2[0.4 + \sqrt{1.235}] = 1.78057$$

By modified Euler's method

$$y_3^{(1)} = y_2 + \frac{0.2}{2}[f(x_2, y_2) + f(x_3, y_3)]$$

$$= 1.5253 + 0.1[(0.4 + \sqrt{1.5253}) + (0.6 + \sqrt{1.78057})]$$

$$= 1.5253 + 0.1[1 + 1.235 + 1.361] = 1.8849$$

$$y_3^{(2)} = 1.5253 + 0.1[1 + \sqrt{1.5253}) + \sqrt{1.8849}]$$

$$= 1.5253 + 0.1[1 + 1.235 + 1.373] = 1.8851$$

$$y_3^{(3)} = 1.8851$$

Hence $x = 0.6$, when $y = 1.8851$.

***Example* 18.19.** Using Euler's method, obtain the solution of the equation $\frac{dy}{dx} = x + y$ with initial condition for the range $0 \le x \le 0.3$ in step of 0.1, $y_0 = 1$, $x_0 = 0$.

Solution : For $x_1 = 0.1$ by Euler's method,

$$y_1 = y_0 + hf(x_0, y_0)$$

$$= 1 + 0.1\,[1] = 1.1$$

Apply modified Euler's method

$$y_1^{(1)} = y_0 + \frac{1}{2}h\,[f(x_0, y_0) + f(x_1, y_1)]$$

$$= 1 + \frac{0.1}{2}[f(0.1) + f(0.1, 1.1)] = 1 + 0.05[(0 + 1) + f(0.1, 1.1)]$$

$$= 1 + 0.05(2.2) = 1 + 0.1[1.1] = 1 + 0.11 = 1.11$$

$$y_1^{(2)} = 1 + \frac{0.1}{2}[f(0.1) + f(0.1, 1.11)] = 1 + \frac{0.1}{2}[1 + 0.1 + 1.11]$$

$$= 1 + \frac{0.1}{2}[2.2] = 1 + 0.1[1.105] = 1 + 0.1105 = 1.1105$$

$$y_1^{(3)} = 1 + \frac{0.1}{2}[1 + 0.1 + 1.1105] = 1 + \frac{0.1}{2}[2.2105]$$

$$= 1 + 0.1[1.105] = 1 + 0.1105 = 1.1105$$

Hence $y_1 = 1.1105$ when $x_1 = 0.1$

For $x_2 = 0.2$ by Euler's method

$$y_2 = y_1 + hf(x_1, y_1)$$

$$= 1.1105 + 0.1[x_1 + y_1] = 1.1105 + 0.1[0.1 + 1.1105]$$

$$= 1.1105 + 0.1[1.2105]$$

$$= 1.1105 + 0.12105 = 1.23155 = 1.2316$$

First better approximation to y_2

$$y_2^{(1)} = y_1 + \frac{h}{2}[f(x_1, y_1) + f(x_2, y_2)$$

$$= 1.1105 + \frac{0.1}{2}[1.2105 + 1.4316]$$

$$= 1.1105 + \frac{0.1}{2}[2.6421] = 1.1105 + 0.1[1.32105]$$

$$= 1.1105 + 0.1321 = 1.2426$$

$$y_2^{(2)} = y_1 + \frac{h}{2}[f(x_1, y_1) + f(x_2, y_2^{(1)})$$

$$= 1.1105 + \frac{0.1}{2}[1.2105 + 1.4426]$$

$$= 1.1105 + \frac{0.1}{2}[2.6531] = 1.1105 + 0.1[1.3266]$$

$$= 1.1105 + 0.13266 = 1.24316$$

$$y_2^{(3)} = 1.1105 + \frac{0.1}{2}[1.2105 + 1.44316]$$

$$= 1.1105 + \frac{0.1}{2}[2.65366] = 1.1105 + 0.1[1.3268]$$

$$= 1.1105 + 0.13268 = 2.4318$$

Hence $y_2 = 2.43216$, when $x = 0.2$.

For $x = 0.3$,

$$y_3 = y_2 + hf(x_2, y_2)$$

$$= 2.4316 + (0.1)[0.2 + 2.4316] = 2.4316 + (0.1)[2.6316]$$

$$= 2.4316 + 0.26316 = 2.69476$$

$$y_3^{(1)} = y_2 + \frac{h}{2}[f(x_2, y_2) + f(x_3, y_3)]$$

$$= 2.4316 + \frac{0.1}{2}[2.6.316 + 0.3 + 2.6958]$$

$$= 2.4316 + \frac{0.1}{2}[5.5264] = 2.4316 + 0.1[2.7632]$$

$$= 2.4316 + 0.2763 = 2.6679$$

Hence $y_3 = 2.6679$, when $x = 0.3$.

***Example* 18.20.** Using Euler's method, find an approximate value of y when $x = 0.1$ given that $\frac{dy}{dx} = x + y$ and $y = 1$ when $x = 0$.

Solution : The problem can be solved as shown in the table. Taking $h = 0.1$, the various calculations are arranged as follows, here h is very small.

x	y	$x + y = \frac{dy}{dx}$	**old** y **+ 0.1** $\left(\frac{dy}{dx}\right)$ **= new** y
0.0	1.00	1.00	$1.00 + 0.1\,(1.00) = 1.10 = y_1$
0.1	1.10	1.20	$1.10 + 0.1\,(1.20) = 1.22 = y_2$
0.2	1.22	1.42	$1.22 + 0.1\,(1.42) = 1.36 = y_3$
0.3	1.36	1.66	$1.36 + 0.1\,(1.66) = 1.53 = y_4$
0.4	1.53	1.93	$1.53 + 0.1\,(1.93) = 1.72 = y_5$
0.5	1.72	2.22	$1.72 \div 0.1\,(1.22) = 1.94 = y_6$
0.6	1.94	1.54	$1.94 + 0.1\,(2.54) = 2.19 = y_7$
0.7	2.19	2.89	$2.19 + 0.1\,(2.89) = 2.48 = y_8$
0.8	2.48	3.26	$2.48 + 0.1\,(3.26) = 2.81 = y_9$
0.9	2.81	3.71	$2.81 + 0.1\,(3.71) = 1.18 = y_{10}$
1.0	3.18	—	—

Thus, the required approximate value of $y = 3.18$.

***Example* 18.21.** Solve the following by Euler's modified method $\frac{dy}{dx} = \log(x + y)$, $y(0) = 2$ at $x = 1.2$ and 1.4 with $h = 0.2$.

Solution : The problem can also be solved by Euler's modified method as shown in the following table.

x	$\log(x+y) = y'$	Mean slope	Old y + 0.2 (mean slope) = new y
0.0	log (0 + 2)	—	2 + 0.2 (0.301) = 2.0602
0.2	log (0.2 + 2.0602)	½ (0.301 + 0.3541)	2 + 0.2 (0.3276) = 2.0655
0.2	log (0.2 + 2.0655)	½ (0.301 + 0.3552)	2 + 0.2 (0.3281) = 2.0656
0.0	0.3552	—	2.066 + 0.2 (0.3552) = 2.1366
0.4	log (0.4 + 2.1366)	½ (0.3552 + 0.4042)	2.0656 + 0.2 (0.3797) = 2.1415
0.4	log (0.4 + 2.1415)	½ (0.3552 + 0.4051)	2.0656 + 0.2 (0.3801) = 2.1416
0.4	0.4051	—	2.1416 + 0.2 (0.4051) = 2.2226
0.6	log (0.6 + 2.2226)	½ (0.4051 + 0.4506)	2.1416 + 0.2 (0.4279) = 2.2272
0.6	log (0.6 + 2.2272)	½ (0.4051 + 0.4514)	2.1416 + 0.2 (0.4282) = 2.2272
0.6	0.4514	—	2.2272 + 0.2 (0.4514) = 2.3175
0.8	log (0.8 + 2.3175)	½ (0.4514 + 0.4938)	2.2272 + 0.2 (0.4726) = 2.3217
0.8	log (0.8 + 2.3217)	½ (0.4514 + 0.4943)	2.2272 + 0.2 (0.4727) = 2.3217
0.8	0.4943	—	2.3217 + 0.2 (0.4943) = 2.4206
1.0	log (1 + 2.4206)	½ (0.4943 + 0.5341)	2.3217 + 0.2 (0.5142) = 2.4245
1.0	log (1 + 2.4245)	½ (0.4943 + 0.5346)	2.3217 + 0.2 (0.5144) = 2.4545
1.0	0.5346	—	2.4245 + 0.2 (0.5346) = 2.5314
1.2	log (1.2 + 2.5314)	½ (0.5643 + 0.5719)	2.4245 + 0.2 (0.5532) = 2.4245
1.2	log (1.2 + 2.5351)	½ (0.5346 + 0.5723)	2.4245 + 0.2 (0.5534) = 2.5351
1.2	0.5723	—	2.5351 + 0.2 (0.5723) = 2.6496
1.4	log (1.4 + 2.6496)	½ (0.5723 + 0.6074)	2.5354 + 0.2 (0.5898) = 2.6531
1.4	log (1.4 + 2.6531)	½ (0.5723 + 0.6078)	2.5351 + 0.2 (0.5900) = 2.6531

Hence $y(1.2) = 2.5351$, $y(1.4) = 2.6531$ approximately.

Notes:

1. We obtain the exact value of y as 3.44, whereas by Euler's method $y = 3.18$ in example 8.20. If the interval is divided in small intervals, the accuracy would be increased considerable but at the cost of double labour. The Euler's method is though simple, not the best.
2. The modified Euler's method is an improvement over the Euler's method and the result is very near to the exact result.

18.6 RUNGE-KUTTA (RK) METHODS

This method as an advantage over Taylor's series method where we have to calculate higher order derivatives. In this method, we are not required to calculate higher order derivatives and give greater accuracy. These methods agree with Taylor series solution up to term in h^r where r differs from method and is called the order of that method.

1. **First Order RK Method**

By Euler's method, we have

$$y_1 = y_0 + hf(x_0, y_0) = y_0 + hy'_0 \qquad [\because y' = f(x, y)]$$

Expanding by Taylor's series

$$y_1 = y(x_0 + h) = y_0 + hy'_0 + \frac{h^2}{2!} y''_0 + \ldots$$

It shows that Euler's method agrees with Taylor's series solution up to terms in h. Hence Euler's method is RK method of first order.

2. **Second Order RK Method**

The modified Euler's method gives

$$y_1 = y_0 + [f(x_0, y_0) + f(x_0 + h, y_1^{(1)})] \qquad \ldots(18.10)$$

where $$y_1^{(1)} = y_0 + hf(x_0, y_0)$$

i.e. $$y_1 = y_0 + \frac{h}{2} f(x_0, y_0) + f(x_0 + h, y_0 + hf(x_0, y_0))] \qquad \ldots(18.11)$$

Now, by Taylor's series

$$y_1 = y(x_0 + h) = y_0 + hy'_0 + \frac{h^2}{2!} y''_0 + \ldots \quad \text{(by Taylor's series)} \qquad \ldots(18.12)$$

Now expanding $f[x_0 + h, y_0 + hf(x_0, y_0)]$ by Taylor's series for a function of two variables.

$$y_1 = y_0 + \left[\begin{array}{l}\frac{h}{2} f(x_0, y_0) + f\left\{(x_0, y_0) + g\left(\frac{\partial f}{\partial x}\right)_{(x_0, y_0)} + hf(x_0, y_0)\left(\frac{\partial f}{\partial x}\right)_{(x_0, y_0)}\right\} \\ + \text{term containing second and higher power of } h\end{array}\right]$$

$$= y_0 + \frac{1}{2}\left[\begin{array}{l}hf(x_0, y_0) + hf(x_0, y_0) + h^2\left\{\left(\frac{\partial f}{\partial x}\right)_{(x_0, y_0)} + \left(\frac{\partial f}{\partial y}\right)_{(x_0, y_0)} f(x_0, y_0)\right\} \\ + O(h^3)(\text{term containing } h^3 \text{ onwards})\end{array}\right]$$

$$= y_0 + \frac{1}{2}\left[2hf(x_0, y_0) + h^2\left\{\left(\frac{\partial f}{\partial x}\right)_{(x_0, y_0)} + f(x_0, y_0)\left(\frac{\partial f}{\partial y}\right)_{(x_0, y_0)}\right\}\right] + O(h^3)$$

$$= y_0 + hf_0 + \frac{h^2}{2} f_0 + O(h^3) \qquad \left[\because \frac{df}{dx} = \frac{\partial f}{\partial x} + f\frac{\partial f}{y}\right]$$

$$= y_0 + hy'_0 + \frac{h^2}{2} y''_0 + O(h^3) \qquad \ldots(18.13)$$

Comparing Eqs. (18.12) and (18.13), it follows that modified Euler's method is RK method of second order. These agree up to third term.

Working rule:

$$y_1 = y_0 + \frac{1}{2}(k_1 + k_2)$$

$$k_1 = hf(x_0, y_0)$$

$$k_2 = h[f(x_0 + h, y_0 + k_1)]$$

3. Third Order RK Method

The third order RK formula is

$$y_1 = y_0 + \frac{1}{6}(k_1 + 4k_2 + k_3)$$

where
$$k_1 = hf(x_0, y_0)$$

$$k_2 = hf(x_0 + h, y_0 + k_1)$$

$$k_3 = hf(x_0 + h, y_0 + k_2)$$

$$k_4 = hf\left(x_0 + \frac{h}{2}, y_0 + \frac{k_1}{2}\right)$$

$$k = \frac{1}{6}(k_1 + 4k_4 + k_3)$$

$$y_1 = y_0 + k$$

4. Fourth Order RK Method

This method is commonly used.

Working rule: Consider

$$\frac{dy}{dx} = f(x, y), y(x_0) = y_0$$

$$k_1 = hf(x_0, y_0)$$

$$k_2 = hf\left(x_0 + \frac{h}{2}, y_0 + \frac{k_1}{2}\right)$$

$$k_3 = hf\left(x_0 + \frac{h}{2}, y_0 + \frac{k_2}{2}\right)$$

$$k_4 = hf(x_0 + h, y_0 + k_3)$$

$$k = \frac{1}{6}(k_1 + 2k_2 + 2k_3 + k_4)$$

k is weighted mean of k_1, k_2, k_3, k_4

Then
$$y_1 = y_0 + k$$

and
$$x_1 = x_0 + h$$

The increment in y in second interval is computed in a similar manner by means of the following formula

$$k_1 = hf(x_1, y_1)$$

$$k_2 = hf\left(x_1 + \frac{h}{2}, y_1 + \frac{k_1}{2}\right)$$

$$k_3 = hf\left(x_1 + \frac{h}{2}, y_1 + \frac{k_1}{2}\right)$$

$$k_4 = hf(x_1 + h, y_1 + k_3)$$

$$k = \frac{1}{6}(k_1 + 2k_2 + 2k_3 + k_4)$$

$$y_2 = y_1 + k \text{ and } x_2 = x_1 + h$$

and so on for succeeding intervals.

Note: One of the advantage of these methods in that the operation is identical whether the differential equation is linear or nonlinear.

***Example* 18.22.** Using RK method of fourth order, find $y(0.2)$ for the equation $\frac{dy}{dx} = \frac{y-x}{y+x}$, $y(0) = 1$. Take $h = 0.2$.

Solution : Given
$$y' = f(x, y) = \frac{y-x}{y+x},\ x_0 = 0,\ y_0 = 1,\ h = 0.2.$$

$$k_1 = hf(x_0, y_0) = 0.2\left[\frac{1-0}{1+0}\right] = 0.2$$

$$k_2 = h\left[f\left(x_0 + \frac{h}{2}, y_0 + \frac{k_1}{2}\right)\right] = 0.2\left[\frac{\left(y_0 + \frac{k_1}{2}\right) - \left(x_0 + \frac{h}{2}\right)}{\left(y_0 + \frac{k_1}{2}\right) + \left(x_0 + \frac{h}{2}\right)}\right]$$

$$= 0.2\left[\frac{1 + \frac{0.2}{2} - 0 - \frac{0.2}{2}}{1 + \frac{0.2}{2} + 0 - \frac{0.2}{2}}\right] = 0.166$$

$$k_3 = hf\left(x_0 + \frac{h}{2}, y_0 + \frac{k_2}{2}\right) = 0.2\left[\frac{\left(y_0 + \frac{k_1}{2}\right) - \left(x_0 + \frac{h}{2}\right)}{\left(y_0 + \frac{k_1}{2}\right) + \left(x_0 + \frac{h}{2}\right)}\right]$$

$$= 0.2\left[\frac{\left(1 + \frac{0.1666}{2}\right) - \left(0 + \frac{0.2}{2}\right)}{\left(1 + \frac{0.1666}{2}\right) + \left(0 + \frac{0.2}{2}\right)}\right] = 0.16619$$

$$k_4 = h[f(x_0 + h, y_0 + k_3)] = 0.2\left[\frac{y_0 + k_3 - (x_0 + h)}{y_0 + k_3 + (x_0 + h)}\right]$$

$$= 0.2\left[\frac{1 + 0.16619 - 0.2}{1 + 0.16619 + 0.2}\right] = 0.2 \times \frac{0.96619}{1.36619} = 0.14144$$

$$k = \frac{1}{6}(0.2 + 0.1666 + 0.16619 + 0.14144] = 0.16783$$

$$y(0.2) = y_0 + k = 1 + 0.16783 = 1.16783$$

***Example* 18.23.** Using RK method of order four, compute $y(0.2)$ and $y(0.4)$ from $10\frac{dy}{dx} = x^2 + y^2$, $y(0) = 1$, taking $h = 0.1$.

Solution : Given
$$y' = f(x, y) = x_2 + y_2,\ x_0 = 0,\ y_0 = 1,\ h = 0.1$$

Here
$$k_1 = (0.1)\left(\frac{x_0^2 + y_0^2}{10}\right) = 0.1$$

$$k_2 = \frac{0.1}{10}\left(x_0 + \frac{h}{2}\right)^2 + \left(y_0 + \frac{k_1}{2}\right)^2 = 0.01012525$$

$$k_3 = \frac{0.1}{10}\left(x_0 + \frac{h}{2}\right)^2 + \left(y_0 + \frac{k_2}{2}\right)^2 = 0.0101265$$

$$k_4 = \frac{0.1}{10}\left[(x_0 + h)^2 + (y_0 + k_3)^2\right] = 0.010303$$

$$k = \frac{1}{6}\left[0.01 + 2(0.01012525) + 2(0.0101265) + 0.0101344\right] = 0.0101344$$

$$y_1 = y_0 + 0.101344 = 1.010344$$

For $y(0.2)$,

$$x_1 = x_0 + h = 0 + 0.1 = 0.1$$

$$y_1 = 1.0101344,\ h = 0.1$$

$$k_1 = \frac{0.1}{10}\left[(0.1)^2 + (1.010134)^2\right] = 0.010303$$

$$k_2 = hf\left(x_1 + \frac{h}{2}, y_1 + \frac{k_1}{2}\right)$$

$$= 0.1f\left(0.1 + \frac{0.1}{2}, 1.010134 + \frac{0.0101303}{2}\right) = 0.010533$$

$$k_3 = hf\left(x_1 + \frac{h}{3}, y_1 + \frac{k_2}{2}\right) = 0.1f\left(0.1 + \frac{0.1}{2}, 1.01034 + \frac{0.010533}{2}\right)$$

$$= \frac{0.1}{10}\left[\left(0.1 + \frac{0.1}{2}\right)^2 + \left(1.010134 + \frac{0.010533}{2}\right)^2\right] = 0.0105354$$

$$k_4 = \frac{0.1}{10}\left[(0.1 + 0.1)^2 + (1.010134 + 0.0105354)^2\right] = 0.010817$$

$$k = \frac{0.1}{10}\left[0.010303 + 2(0.010533 + 0.0105354) + 0.010817\right] = 0.01054$$

$$y_2 = y_1 + k = 1.01034 + 0.01054 = 1.0206066$$

For $y(0.3)$,

$$x_2 = x_1 + h = 0.1 + 0.1 = 0.2$$

$$y_2 = 1.0206066,\ h = 0.1$$

$$k_1 = hf(x_2, y_2) = 0.1\left(\frac{x_2^2 + y_2^2}{10}\right) = 0.010816$$

$$k_2 = hf\left(x_2 + \frac{h}{2}, y_2 + \frac{k_1}{2}\right) = \frac{0.1}{10}\left[\left(x_2 + \frac{h}{2}\right)^2 + \left(y_2 + \frac{k_1}{2}\right)^2\right] = 0.011152$$

$$k_3 = hf\left(x_2 + \frac{h}{2}, y_2 + \frac{k_2}{2}\right) = \frac{0.1}{10}\left[\left(x_2 + \frac{h}{2}\right)^2 + \left(y_2 + \frac{k_1}{2}\right)^2\right] = 0.0111553$$

$$k_4 = hf(x_2 + h, y_2 + k_3) = \frac{1.0}{10}\left[(x_2 + h)^2 + (y_2 + k_3)^2\right] = 0.011545$$

$$k = \frac{1}{6}(k_1 + 2k_2 + 2k_3 + k_4) = 0.0111626$$

$$y_2 = 1.0206066 + 0.0111626 = 1.03176$$

For $y(0.4)$, $\quad x_3 = x_2 + h = 0.2 + 0.1 = 0.3,\ y_3 = 1.03176,\ h = 0.1$

$$k_1 = hf(x_3, y_3) = 0.1\left(\frac{x_3^2 + y_3^2}{10}\right) = 0.0115453$$

$$k_2 = hf\left(x_3 + \frac{h}{2}, y_3 + \frac{k_1}{2}\right) = \frac{0.1}{10}\left[\left(x_3 + \frac{h}{2}\right)^2 + \left(y_3 + \frac{k_2}{2}\right)^2\right] = 0.011989$$

$$k_3 = hf\left(x_3 + \frac{h}{2}, y_3 + \frac{k_2}{2}\right) = \frac{0.1}{10}\left[\left(x_3 + \frac{h}{2}\right)^2 + \left(y_3 + \frac{k_2}{2}\right)^2\right] = 0.0119943$$

$$k_4 = hf(x_3 + h, y_3 + k) = \frac{0.1}{10}\left[(x_3 + h)^2 + (y_3 + k_3)^2\right] = 0.012494$$

$$k = \frac{1}{6}(k_1 + 2k_2 + 2k_3 + k_4) = 0.01200093$$

$$y_4 = y_3 + k = 1.03176 + 0.01200093 = 1.04376$$

***Example* 18.24.** Using RK method of fourth order find $y(0.2)$ given that $\frac{dy}{dx} = 3x + \frac{y}{2}$, $y(0) = 1$ taking $h = 0.1$.

Solution : For $y(0.1)$ $\quad k_1 = 0.1\,(3x_0 + y_0/2) = 0.1 \times 0.5 = 0.05.$

$$k_2 = hf\left(x_0 + \frac{h}{2}, y_0 + \frac{k_1}{2}\right)$$

$$= 0.1\left[\frac{3h}{2} + \frac{y_0 + \frac{k_1}{2}}{2}\right] = 0.1\left[\frac{3(0.1)}{2} + \frac{1 + \frac{0.05}{2}}{2}\right] = 0.06625$$

$$k_3 = hf\left(x_0 + \frac{h}{2}, y_0 + \frac{k_2}{2}\right) = 0.1\left[\frac{3h}{2} + \frac{y_0 + \frac{k_2}{2}}{2}\right] = 0.06665625$$

$$k_4 = hf(x_0 + h, y_0 + k_3) = 0.1\left[3h + \frac{y_0 + k_3}{2}\right] = 0.0833328$$

$$k = \frac{1}{6}\left[0.05 + 0.06625 + 0.06665625 + 0.833328\right] = 0.0665242$$

$$y_1 = 1 + 0.0665242 = 1.0665242$$

For $y(0.2)$ $\quad x_1 = 0 + 0.1 = 0.1,\ y_1 = 1.0665242,\ h = 0.1$

$$k_1 = 0.1\left(3x_1 + \frac{y_1}{2}\right) = 0.1\left[3(1) + \frac{1.0665242}{2}\right] = 0.08332621$$

$$k_2 = 0.1\left[3\left(x_1 + \frac{h}{2}\right) + \frac{y_1 + \frac{k_1}{2}}{2}\right] = 0.10040925$$

$$k_3 = 0.1\left[3\left(x_1 + \frac{0.1}{2}\right) + \frac{y_1 + \frac{k_1}{2}}{2}\right] = 0.100836441$$

$$k_4 = 0.119497636$$

$$k = \frac{1}{6}[k_1 + 2k_2 + 2k_3 + k_4] = 0.10841655$$

$$y_2 = y_1 + k = 1.0665242 + 0.10841655 = 1.1749$$

***Example* 18.25.** For the equation $\frac{dy}{dx} = x - y$, $y(0) = 1$, find the value of y when $x = 0.1, 0.2$ taking $h = 0.1$

Solution : Given $x_0 = 0$, $y_1 = 1$, $f(x_0, y_0) = -1$.

$$k_1 = hf(x_0, y_0) = -0.1$$

$$k_2 = 0.1\left(x_0 + \frac{h}{2}\right) - \left(y_0 + \frac{k_1}{2}\right) = 0.1\left[\frac{0.1}{2} - 1 + \frac{0.1}{2}\right] = -0.09$$

$$k_3 = 0.1\left(x_0 + \frac{h}{2}\right) - \left(y_0 + \frac{k_2}{2}\right) = 0.1\left[\frac{0.1}{2} - 1 + \frac{0.09}{2}\right] = -0.0905$$

$$k_4 = 0.1[0 + 0.1 - 1 + 0.0905] = -0.8085$$

$$k = \frac{1}{6}[k_1 + 2k_2 + 2k_3 + k_4] = -0.090325$$

$$y_1(0.1) = y_0 + k = 1 + (-0.090325) = 0.909675$$

When $x_1 = x_0 + h = 0 + 0.1$, $y_1 = 0.909675$, $h = 0.1$

$$k_1 = 0.1[0.1 - 0.90968] = -0.80968$$

$$k_2 = 0.1\left[0.1 + \frac{0.1}{2} - 0.90968 + \frac{0.80968}{2}\right] = -0.0719195$$

$$k_3 = 0.1\left[0.1 + \frac{0.1}{2} - 0.90968 + \frac{0.07192}{2}\right] = -0.07237$$

$$k_4 = 0.1[0.1 + 0.1 - 0.90968 + 0.07237] = -0.06373$$

$$k = -0.072213$$

$$y_2(0.2) = y_1 + k = 0.90968 - 0.072213 = 0.981893$$

18.7 PREDICTOR-CORRECTOR METHODS

Consider the differential equation

$$\frac{dy}{dx} = f(x, y);\ y(x_0) = y_0$$

By Euler's method, we have

$$y_i = y_{i-1} + hf[x_0 + (\overline{i-1})\,h,\ y_{i-1}],\ i = 1, 2, 3$$

or $$= y_{i-1} + hf(x_{i-1}, y_{i-1}) \quad ...(18.14)$$

By modified Euler's method

$$y_i = y_{i-1} + \frac{h}{2}[f(x_{i-1}, y_{i-1}) + f(x_i, y_i)] \quad ...(18.15)$$

Substitute the value of y_i from Eq. (18.14) in Eq. (18.15), we get better approximate value of y_i and again substitute y_i in Eq. (18.2), we get still better approximation of y_i. This process is repeated till two values of y_i are same. This technique of refining an initially crude y_i by means of a more accurate formula is known as **Predictor-Corrector method**.

Equation (18.14) is called **Predictor** and Eq. (18.15) is called **Corrector**.

In predictor-corrector methods, four prior values are needed for finding the value of y at x_i. Though slightly complex, these methods have the advantage of giving an estimate of error from successive approximation of y_i.

In this section, we will study two such methods, namely

(*i*) Milne's method

(*ii*) Adams-Bashforth method

18.8 MILNE'S METHOD

Given $$\frac{dy}{dx} = f(x, y) \quad ...(18.16)$$

and $$y = y_0, x = x_0$$

To find an approximate value of y for $x = x_0 + nh$ by Milne's method, we proceed as follows.

The value $y_0 = y(x_0)$ being given, we compute

$$y = y(x_0 + h), y_2 = y(x_0 + 2h), y_3 = y(x_0 + 3h)$$

by Picard's or Taylor's series method

Next we calculate $$f_0 = f(x_0, y_0), f_1 = f(x_0 + h, y_1)$$

$$f_2 = f(x_0 + 2h, y_2), f_3 = f(x_0 + 3h, y_3)$$

Then to find $y_4 = y(x_0 + 4h)$, we substitute Newton's forward interpolation formula

$$f(x, y) = f_0 + n\Delta f_0 + \frac{n(n-1)}{2}\Delta^2 f_0 + \frac{n(n-1)(n-2)}{6}\Delta^3 f_0 + ...$$

Integrating Eq. (18.16) with its limits (x_0, x_4)

$$\int_{x_0}^{x_4} \frac{dy}{dx}.dx = \int_{x_0}^{x_4} f(x, y)\,dx$$

$$y_4 - y_0 = \int_{x_0}^{x_4} f(x, y)\,dx$$

$$y_4 = y_0 + \int_{x_0}^{x_0+4h} f(x, y)\,dx$$

$$= y_0 + \int_{x_0}^{x_0+4h} \left(f_0 + n\Delta f_0 + \frac{n(n-1)}{2}\Delta^2 f_0 + ...\right)dx$$

Put $x = x_0 + nh$, $dx = hdn$

$$y_4 = y_0 + h\int_0^4 \left(f_0 + n\Delta f_0 + \frac{n(n-1)}{2}\Delta^2 f_0 + ...\right)dn$$

$$= y_0 + h\left(4f_0 + 8\Delta f_0 + \frac{20}{3}\Delta^2 f_0 + \frac{8}{3}\Delta^3 f_0 + \ldots\right)$$

Neglecting fourth and higher order differences and expressing in terms of function value, we get

$$y_4 = y_0 + \frac{4h}{3}(2f_1 - f_2 + 2f_3)$$

[we know that $\Delta f_0 = f_1 - f_0,\ \Delta^2 f = f_2 - 2f_1 + f_0,\ \Delta^3 f_0 = f_3 - 3f_2 + 3f_1 + f_0$]

$$= 4f_0 + 8\Delta f_0 + \frac{20}{3}\Delta^2 f_0 + \frac{8}{3}\Delta^3 f_0$$

$$= 4f_0 + 8(f_1 - f_0) + \frac{20}{3}(f_2 - 2f_1 + f_0) + \frac{8}{3}(f_3 - 3f_2 + 3f_1 + f_0)$$

$$= \frac{4}{3}(2f_1 - f_2 + 2f_3)$$

Hence $y_4 = y_0 + \frac{4h}{3}(2f_1 - f_2 + 2f_3)$ which is called **predictor.**

In general predictor

$$y_{n+1} = y_{n-3} + \frac{4h}{3}[2f_{n-2} - f_{n-1} + 2f_n]$$

Then an improved value of f_4 is computed and again the corrector method is applied **to find a better value of** y_4, we repeat this step until y_4 remains unchanged.

Hence, Milne's corrector formula is

$$y_{n+1} = y_{n-1} + \frac{h}{3}[f_{n-1} + 4f_n + f_{n+1}]$$

Then y_5 can be obtained from the predictor as

$$y_5 = y_1 + \frac{4h}{3}[2f_2 - f_3 + 2f_4]$$

Then f_5 is calculated and a better value of

$$y_5 = y_3 + \frac{h}{3}[f_3 + 4f_4 + f_5]$$

This step is to be repeated till two values of y_5 are same, then calculate y_6 and so on. **This is Milne's predictor-corrector method.**

Example 18.26. *Find* $y(2.0)$ if $y(x)$ is the solution of $\frac{dy}{dx} = \frac{1}{2}(x + y) = f(x, y)$, assuming $y(0) = 2$, $y(0.5) = 2.636$, $y(1.0) = 3.595$, $y(1.5) = 4.968$ using Milne's predictor-corrector method.

Solution : By Milne's predictor-corrector

$$y_{n+1} = y_{n-3} + \frac{4h}{3}[2f_{n-2} - f_{n-1} + 2f_n]$$

$$y_4 = y_0 + \frac{4h}{3}[2f_1 - f_2 + 2f_3]$$

x	y	$f(x, y) = (x + y)/2$
$x_0 = 0$	2	$1 = f_0$
$x_1 = 0.5$	2.636	$1.568 = f_1$
$x_2 = 1.0$	3.595	$2.2975 = f_2$
$x_3 = 1.5$	4.968	$3.234 = f_3$

$$y_4 = y_0 + \frac{4h}{3}[2f_1 - f_2 + 2f_3]$$

$$= 2 + \frac{4(0.5)}{3}[2(1.568) - 2.2975 + 2(3.234)] = 6.871$$

$$f_4 = \frac{(x+y)}{2} = \frac{2 + 6.871}{2} = 4.4355$$

By corrector $\quad y_{n+1} = y_{n-1} + \frac{h}{3}[f_{n-1} + 4f_n + f_{n+1}]$

$$y_4 = y_2 + \frac{h}{3}[f_2 + 4f_3 + f_4]$$

$$= 3.595 + \frac{0.5}{3}[2.2975 + 4(3.234) + 4.4355] = 6.873346667$$

or $\quad f_4 = 4.43658333$

Again by correct method $\quad y_4 = y_2 + \frac{h}{3}[f_2 + 4f_3 + f_4]$

$$= 3.595 + \frac{0.5}{3}[2.2975 + 4(3.234) + 4.43658]$$

$= 6.873346667$ **a value which is same as before.**

Hence $\quad y_4 = 6.8733$ **correct to four decimal places.**

Example 18.27. Using RK method of order four, find y for $x = 0.1, 0.2, 0.3$ given $\frac{dy}{dx} = x - y^2$, $y(0) = 1$, then continue the solution at $x = 0.4$ using Milne's method.

Solution : Given $\quad f(x, y) = x - y^2 = \frac{dy}{dx}, x_0 = 0, y_0 = 1, h = 0.1.$

By RK method of fourth order

To find y_1: $\quad k_1 = hf(x_0, y_0) = (0.1)(x_0 - y_0^2) = (0.1)(0 - 1^2) = -0.1$

$$k_2 = h\left[f\left(x_0 + \frac{h}{2}, y_0 + \frac{k_1}{2}\right)\right] = 0.1\left[\frac{h}{2} - \left(y_0 + \frac{k_1}{2}\right)^2\right]$$

$$= 0.1\left[\frac{0.1}{2} - \left(1 - \frac{0.1}{2}\right)^2\right] = -0.08525$$

$$k_3 = hf\left(x_0 + \frac{h}{2}, y_0 + \frac{k_2}{2}\right) = (0.1)\left[\frac{h}{2} - \left(y_0 + \frac{k_2}{2}\right)^2\right]$$

$$= (0.1)\left[\frac{0.1}{2}\left(1 - \frac{0.08525}{2}\right)^2\right] = -0.0866567$$

$$k_4 = hf(x_0 + h, y_0 + k_3) = (0.1)[1.0 - (1 - 0.0866567)^2] = -0.0734196$$

$$k = \frac{1}{6}(k_1 + 2k_2 + 2k_3 + k_4) = -0.0862054$$

$$y_1 = y_0 + k = 1 - 0.0862054 = 0.9137956 = 0.91375$$

To find y_2

$$y_2 = y(0.2)$$

$$x_1 = x_0 + h = 0 + 0.1 = 0.1$$

$$k_1 = (0.1)[x_1 - y_1^2] = (0.1)[0.1 - (0.9137946)^2] = -0.073502057$$

$$k_2 = (0.1)\left[x_1 + \frac{h}{2} - \left(y_1 + \frac{k_1}{2}\right)^2\right]$$

$$= (0.1)\left[\left(0.1 + \frac{0.1}{2}\right) - \left(0.913795 - \frac{0.073502}{2}\right)^2\right]$$

$$= 0.0619206$$

$$= y_0 + \frac{h}{24}[19(0.15342) + 9(0.208170368) - 5(0.10101) + 0.050125]$$

$$k_3 = (0.1)\left[x_1 + \frac{h}{2} - \left(y_1 + \frac{k_2}{2}\right)^2\right]$$

$$= (0.1)\left[\left(0.1 + \frac{0.1}{2}\right) - \left(0.913795 - \frac{0.0619206}{2}\right)^2\right]$$

$$= -0.0629398$$

$$k_4 = (0.1)[x_1 + h - (y_1 + k_3)^2]$$

$$= (0.1)[0.1 + 0.1 - (0.0913795 - 0.0629398)^2] = -0.0522395$$

$$k = \frac{1}{6}(k_1 + 2k_2 + 2k_3 + k_4) = 0.0626025$$

$$y_2 = 0.913795 - 0.0626025 = 0.8511925$$

To find y_3:

$$y_3 = y(0.3) = ?,\ x_2 = x_1 + h = 0.1 + 0.1 = 0.2$$

$$k_1 = hf(x_2, y_2) = (0.1)[0.2 - (0.8511925)^2] = -0.052453$$

$$k_2 = h\left[x_2 + \frac{h}{2} - \left(y_2 + \frac{k}{2}\right)^2\right]$$

$$= (0.1)\left[0.2 + \frac{0.1}{2} - \left(0.8511925 - \frac{0.052453}{2}\right)^2\right] = -0.043057$$

$$k_3 = (0.1)\left[0.25 - \left(0.8511925 - \frac{0.043057}{2}\right)\right] = 0.0438342$$

$$k_4 = (0.1)[(0.2 + 0.1) - (0.8511925 - 0.0438342)^2] = -0.0351827$$

$$k = 0.0435697$$

$$y_3 = y_2 - 0.0435697 = 0.8511925 - 0.0435697 = 0.8076228$$

x	y	$f = x^2 - y^2$
$x_0 = 0$	1	$-1 = f_0$
$x_1 = 0.1$	0.913795	$-0.73502 = f_1$
$x_2 = 0.2$	0.8511925	$-0.52453 = f_2$
$x_3 = 0.3$	0.8076228	$-0.35225 = f_3$

To find y_4

$$y_4 = y_0 + \frac{4h}{3}[2f_1 - f_2 + 2f_3]$$

$$= 1 + \frac{4(0.1)}{3}[2(-0.73502) + 0.52453 + 2(-0.35225)] = 0.779999$$

$$f_4 = 0.4 - (0.779)^2 = -0.208398$$

By corrector

$$y_4 = y_2 + \frac{h}{3}[f_2 + 4f_3 + f_4]$$

$$= 0.8511925 + \frac{0.1}{3}[-0.52453 + 4(-0.35225) - 0.208398] = 0.779794$$

Hence $y(0.4) = 0.779$.

Example 18.28: Use Milne's method to find $y(0.3)$ from $y' = x^2 + y^2$, $y(0) = 1$, find initial value $y(0.1)$, $y(0.2)$ from Taylor's series method.

Solution : By Taylor's series

$$y = y_0 + (x - x_0)y'_0 + \frac{(x-x_0)^2}{2!}y''_0 + \frac{(x-x_0)^3}{3!}y'''_0 + \ldots$$

$$y' = x_2 + y_2,\ y'(0) = 1$$

$$y'' = 2x + 2yy',\ y''(0) = 0 + 2y_0y'_0 = 2 \times 1 \times 2 = 2$$

$$y''' = 2 + 2y'^2 + 2yy'',\ y'''(0) = 2 + 2 + 2 \times 1 \times 2 = 8$$

$$y = 1 + x(1) + \frac{x^2}{2!}(2) + \frac{x^3}{3!}(8) + \ldots = 1 + x + x^2 + \frac{4}{3}x^3 + \ldots$$

$$y(0.1) = 1 + (0.1) + (0.1)^2 + \frac{4(0.1)^3}{3} + \ldots = 1.11133$$

$$y(0.2) = 1 + (0.2) + (0.2)^2 + \frac{4(0.2)^3}{3} + \ldots = 1.25066$$

$$y(-0.1) = 1 + (-0.1) + (-0.1)^2 + \frac{4(-0.1)^3}{3} + \ldots = 0.908667$$

x	y	$f = x^2 + y^2$
-1	0.908667 (y_1)	$f_1 = 0.835675$

0	1 (y_0)	$f_0 = 1$
0.1	1.11133 (y_2)	$f_2 = 1.245054$
0.2	1.25066 (y_3)	$f_3 = 1.6041504$

Using predictor formula

$$y_4 = y_0 + \frac{4h}{3}(2f_1 - f_2 + 2f_3)$$

$$= 1 + \frac{4(0.1)}{3}[2(0.835675) - (1.245054) + 2(1.6041504)] = 1.484612907$$

$$\therefore \quad f_4 = x_4^2 + y_4^2 = (0.3)^2 + (1.484612907)^2 = 2.298579613$$

Using corrector formula

$$y_4 = y_2 + \frac{h}{3}(f_2 + 4f_3 + f_4)$$

$$= 1.11133 + \frac{0.1}{3}(1.245054 + 4(1.6041504) + 2.298579613) = 1.44333784$$

$$f_4 = x_4^2 + y_4^2 = (0.3)^2 + (1.44333784)^2 = 2.173224122$$

Again using corrector formula

$$y_4 = 1.11133 + \frac{0.1}{3}[1.245054 + 4(1.6041504) + 2.173224122] = 1.439159324$$

$$f_4 = 2.16117956$$

Again applying corrector formula

$$y_4 = 1.438757839$$

$$f_4 = 2.160024118$$

Hence $y(0.3) = 1.43$.

Example **18.29.** Using Milne's predictor-corrector method to obtain the solution of $\frac{dy}{dx} = x - y^2$ at $x = 0.8$ given

x	0	0.2	0.4	0.6
y	0.0000	0.0200	0.0795	0.1762

Solution : Given $f(x, y) = x - y^2$, $h = 0.2$, $x_0 = 0$, $y_0 = 0$, $f_0 = 0$, $f_1 = 0.1996$, $f_2 = 0.39368$, $f_3 = 0.56895$.

By using predictor formula

$$y_4 = y_0 + \frac{4h}{3}[2f_1 - f_2 + 2f_3]$$

$$= 0 + \frac{4(0.2)}{3}[2(0.1996) - 0.39368 + 2(0.56895)] = 0.304912$$

$$f_4 = x_4 - y_4^2 = 0.8 - (0.304912)^2 = 0.707029$$

By using corrector formula

$$y_4 = y_2 + \frac{h}{3}[f_2 + 4f_3 + f_4]$$

$$= 0.02 + \frac{0.2}{2}[0.39368 + 4(0.56895) + 0.707029]$$

$$f_4 = (0.8) - (0.3576509)^2 = 0.672086$$

$$y_4 = 0.02 + \frac{0.2}{3}[0.39368 + 4(0.56895) + 0.672086] = 0.3541566$$

$$f_4 = (0.8) - (0.3541566)^2 = 0.674573$$

$$y_4 = 0.02 + \frac{0.2}{3}[0.39368 + 4(0.56895) + 0.674573] = 0.3544053$$

Hence $y(0.8) = 0.354$.

***Example* 18.30.** Solve the initial value problem $\frac{dy}{dx} = 1 + xy^2$, $y(0) = 1$ for $x = 0.4$ using Milne's method, given

x	0.1	0.2	0.3
y	1.105	1.223	1.355

Solution : Using the predictor

$$y_4 = y_0 + \frac{4h}{3}[2f_1 - f_2 + 2f_3]$$

$$f_1 = 1 + x_1 \cdot y_1^2 = 1.1221$$

$$f_2 = 1 + (0.2)(1.233)^2 = 1.2991$$

$$f_3 = 1 + (0.3)(1.355)^2 = 1.5508$$

$$f_4 = ?$$

$$y = 1 + \frac{4(0.1)}{3}[2(1.1221) - 1.2991 + 2(1.5508)] = 1.56593$$

$$f_4 = 1 + (0.4)(1.56593)^2 = 1.98085$$

By applying corrector formula

$$y_4 = y_2 + \frac{h}{3}[f_2 + 4f_3 + f_4]$$

$$= 1.223 + \frac{0.1}{3}[1.2991 + 4(1.5508) + 198085] = 1.539105$$

$$f_4 = 1 + (0.4)(1.539105)^2 = 1.94754$$

Again using the corrector formula

$$y_4 = 1.224 + \frac{0.1}{3}[1.2991 + 4(1.5508) + 1.94754] = 1.5385$$

$$f_4 = 1 + (0.4)(1.5385)^2 = 1.946879$$

Again using the corrector formula

$$y_4 = 1.537973$$

$$f_5 = 1.946144$$

Again using the corrector formula

$$y_4 = 1.5379 \approx 1.5$$

Hence $y(0.4) = 1.5$.

18.9 ADAMS-BASHFORTH METHOD

Given $\qquad \dfrac{dy}{dx} = f(x, y)$ and $y_0 = y(x_0)$

We calculate $\qquad y_{-1} = y\,(x_0 - h), y_{-2} = y\,(x_0 - 2h), y_{-3} = f\,(x_0 - 3h)$

by Taylor's series method or Euler's method.

Next we calculate $\qquad f_{-1} = f(x_0 - h, y_{-1}), f_{-2} = f(x_0 - 2h, y_{-2}), f_{-3} = f(x_0 - 3h, y_{-3})$

Then to find y_1, use Newton's backward interpolation formula, *i.e.*

$$f(x, y) = f_0 + n\nabla f_0 + \frac{n(n+1)}{2!}\nabla^2 f_0 + \frac{n(n+1)(n+2)}{3!}\nabla^3 f_0 + \ldots$$

$$y_1 = y_0 + \int_{x_0}^{x_0+h} f(x, y)\,dx$$

$$= y_0 + \int_{x_0}^{x_0+h}\left(f_0 + n\nabla f_0 + \frac{n(n+1)}{2!}\nabla^2 f_0 + \ldots\right)dx$$

[Put $x = x_0 + nh$ so that $dx = hdn$]

$$= y_0 + h\int_0^1 (f_0 + n\nabla f_0 + \ldots)\,dn$$

$$= y_0 + h\left[f_0 + \frac{1}{2}\nabla f_0 + \frac{5}{12}\nabla^2 f_0 + \frac{3}{8}\nabla^3 f_0 + \ldots\right]$$

Neglecting fourth and higher order differences expressing $\nabla f_0, \nabla^2 f_0, \nabla^3 f_0$ in terms of function value, we get

$$y_1 = y_0 + \frac{h}{24}[55f_0 - 59f_{-1} + 37f_{-2} - 9f_{-3}] \qquad \ldots(18.17)$$

This is called Adam-Bashforth predictor formula.

Having found y_1, we find

$$f_1 = f(x_0 + h, y_1)$$

Then to find a better value of y_1, we derive a corrector formula by substituting Newton's backward formula as f_1, *i.e.*

$$f(x, y) = f_1 + n\nabla f_1 + \frac{n(n+1)}{2!}\nabla^2 f_1 + \ldots$$

$$y_1 = y_0 + \int_{x_0}^{x_1} f(x, y)\,dx$$

Put $\qquad x = x_1 + nh, dx = hdn$

$$= y_0 + h \int_{-1}^{0} \left(f_1 + n \nabla f_1 + \frac{n(n+1)}{2!} \nabla^2 f_1 + \ldots \right) dn$$

$$= y_0 + h \left[f_1 - \frac{1}{2} \nabla f_1 - \frac{1}{12} \nabla^2 f_1 - \frac{1}{24} \nabla^3 f_1 + \ldots \right]$$

Neglecting fourth and higher order differences and expressing ∇f_1, $\nabla^2 f_1$, $\nabla^2 f_1$ in terms of their function value, we have

$$y_1 = y_0 + \frac{h}{24} [9f_1 + 19f_0 - 5f_{-1} + f_{-2}] \quad \ldots(18.18)$$

which is called Adams-Moulton corrector formula.

Example **18.31:** Using Adams-Bashforth formula, determine $y(0.4)$ given $\frac{dy}{dx} = \frac{1}{2} xy$ and the following data

x	0	0.1	0.2	0.3
y	1	1.0025	1.0101	1.0228

Solution : Here $f(x, y) = xy$, $h = 0.1$

$x = 0$ $\quad y_{-3} = 1$ $\quad f_{-3} = \frac{1}{2} \times 0 \times 1 = 0$

$x = 0.1$ $\quad y_{-2} = 1.0025$ $\quad f_{-2} = \frac{1}{2}(0.1)(1.0025) = 0.050125$

$x = 0.2$ $\quad y_{-1} = 1.0101$ $\quad f_{-1} = \frac{1}{2}(0.2)(1.0101) = 0.10101$

$x = 0.3$ $\quad y_0 = 1.0228$ $\quad f_0 = \frac{1}{2}(0.3)(1.0228) = 0.15342$

Using the predictor

$$y_1 = y_0 + \frac{h}{24}(55f_0 - 59f_{-1} + 37f_{-2} - 9f_{-3})$$

$$= 1.0228 + \frac{0.1}{24}[55(0.15342) - 59(0.10101) + 37(0.050125) - 9(0)]$$

$$= 1.04085184$$

$$x = 0.4,\ y_1 = 1.04085184$$

$$f_1 = \frac{1}{2}(0.4)(0.04085184) = 0.208170368$$

Using the corrector

$$y_1 = y_0 + \frac{h}{24}[19(0.15342) + 9(0.208170368) - 5(0.10101) + 0.050125]$$

$$= 1.040856328$$

Hence $y(0.4) = 1.04085$.

Example **18.32.** Given $y' = x^2 - y$, $y(0) = 1$ and the starting values $y(0.1) = 0.90516$, $y(0.2) = 0.82127$, $y(0.3) = 0.74918$, find $y(0.4)$ using Adams-Bashforth method.

Solution : Here $f(x, y) = x^2 - y$, $h = 0.1$

$x = 0$ $\quad y_{-3} = 1.0$ $\quad f_{-3} = (0)^2 - 1 = -1$

$x = 0.1$ $\quad y_{-2} = 0.90516$ $\quad f_{-2} = (0.1)^2 - 0.90516 = -0.89516$

$x = 0.2$ $\quad y_{-1} = 0.82127$ $\quad f_{-1} = (0.2)^2 - 0.82127 = 0.78127$

$x = 0.3$ $\quad y_0 = 0.74918$ $\quad f_0 = \frac{1}{2}(0.3)^2\ 0.74918 = -0.65918$

Using the predictor

$$y_1 = y_0 + \frac{h}{24}[55f_0 - 59f_{-1} + 37f_{-2} - 9f_{-3}]$$

$$= 0.74918 + \frac{(0.1)}{24}[55(-0.65918) - 59(0.78127) + 37(-0.85916) - 9(-1)]$$

$$= 0.6758013$$

$$x = 0.4,\ y_1 = 0.6758013,\ f_1 = (0.4)^2 - 0.6758013 = -0.05158013$$

Using the corrector

$$y_1 = y_0 + \frac{h}{24}[55f_0 - 59f_{-1} + 37f_{-2} - 9f_{-3}]$$

$$= 0.74918 + \frac{(0.1)}{24}[9(-0.5158013) + 19(-0.65918) - 5(0.78127) + (-0.89516)]$$

$$= 0.690199$$

$$x = 0.4,\ y_1 = 0.690199,\ f_1 = (0.4)^2 - 0.690199 = -530199$$

Using the corrector

$$y_1 = 0.74918 + \frac{(0.1)}{24}[9(-0.530199) + 19(-0.65918) - 5(0.78127) + (-0.89516)]$$

$$= 0.689659$$

$$= 0.4,\ y_1 = 0.689659,\ f_1 = (0.4)^2 - 0.689659 = -0.529659$$

Using the corrector

$$y_1 = 0.74918 + \frac{(0.1)}{24}[9(-0.529659) + 19(-0.65918) - 5(0.78127) + (-0.89516)]$$

$$= 0.68968$$

Hence $y(0.4) = 0.6896$

***Example* 18.33.** Using Adams-Bashforth method, obtain the solution $\frac{dy}{dx} = x - y^2$ at $x = 0.8$ using the given values.

x	0	0.2	0.4	0.6
y	0	0.0200	0.07915	0.1762

Solution : Here $f(x, y) = x - y^2$, $h = 0.2$

$x = 0.0 \quad y_{-3} = 0.0 \quad f_{-3} = 0 - (0)^2 = 0$

$x = 0.2 \quad y_{-2} = 0.0200 \quad f_{-2} = 0.2 - (0.0200)^2 = 0.1996$

$x = 0.4 \quad y_{-1} = 0.07915 \quad f_{-1} = 0.4 - (0.0795)^2 = 0.3937$

$x = 0.6 \quad y_0 = 0.1762 \quad f_0 = 0.6 - (0.1762)^2 = 0.5689$

Using the predictor

$$y_1 = y_0 + \frac{h}{24}[55f_0 - 59f_{-1} + 37f_{-2} - 9f_{-3}]$$

$$= 0.1762 + \frac{(0.2)}{24}[55(0.5689) - 59(0.3937) + 37(0.1996) - 9(0)] = 0.30492$$

$$x = 0.8, y_1 = 0.30492, f_1 = 0.8 - (0.30492)^2 = 0.707023$$

Using the corrector

$$y_1 = y_0 + \frac{h}{24}[9f_1 + 19f_0 - 5f_{-1} + f_{-2}]$$

$$= 0.1762 + \frac{(0.2)}{24}[9(0.7070237) + 19(0.5689) - 5(0.3937) + (0.1996)]$$

$$= 0.3056177$$

$$x = 0.8, y_1 = 0.30456177, f_1 = 0.8 - (0.30456177)^2 = 0.701051$$

Again using the corrector

$$y_1 = 0.1762 + \frac{(0.2)}{24}[9(0.701051) + 19(0.5689) - 5(0.3937) + (0.1996)]$$

$$= 0.304114$$

$$f_1 = 0.8 - (0.30414)^2 = 0.70751$$

Again using the corrector

$$y_1 = 0.1762 + \frac{(0.2)}{24}[9(0.70751) + 19(0.5689) - 5(0.3937) + (0.1996)]$$

$$= 0.304598$$

Hence $y(0.8) = 0.304$.

PROBLEMS

Euler's Method

1. Evaluate the integral of $y' - y^2 = 0$ by Euler's method at $x = 0.1, 0.2, 0.3$ given $y(0) = 1$.

   [Ans. $y(0.1) = 1.1, y(0.2) = 1.221, y(0.3) = 1.37$**]**

2. Evaluate by Euler's method $y' = y - x$, $y(0) = 2$, at $x = 0.2, 0.4, 0.6$.

   [Ans. $y(0.2) = 2.4, y(0.4) = 2.84, y(0.6) = 3.38$**]**

3. Apply Euler's method to solve $dy/dx = x + y$, $y(0) = 0$, find out $y(0.2)$. **[Ans.** $y(0.2) = 1.18318081$**]**

4. Consider the initial value problem $dy/dx = y - x^2 - 1$, $y(0) = 0.5$. Find $y(0.2)$ by Euler's method.

   [Ans. $y(0.2) = 0.8$**]**

Modified Euler's Method

1. Solve $\frac{dy}{dx} = 1 - y$, $y(0) = 0$ by modified Euler's method at $x = 0.1, 0.2, 0.3$. Compare your result with exact solution. **[Ans.** $y(0.1) = 0.095, y(0.2) = 0.18098, y(0.3) = 0.258787$**]**

2. Use modified Euler's method to find y at $x = 0.1, 0.2$, given $\frac{dy}{dx} = x^2 + y^2$, $y(0) = 1$.

   [Ans. $y(0.1) = 1.1105, y(0.2) = 1.25026$**]**

3. Solve $\frac{dy}{dx} = \log(x + y)$, $y(0) = 2$ by modified Euler's method and find y at $x = 1.2, 1.4$.

[Ans. $y(1.2) = 2.5351$, $y(1.4) = 2.6531$]

4. Solve the modified Euler's method $y' = x + y$, $y(0) = 1$, $h = 0.1$ at $x = 0.1, 0.2, 0.3$.

[Ans. $y(0.1) = 1.1105$, $y(0.2) = 1.2432$, $y(0.3) = 1.4004$]

5. Given $\frac{dy}{dx} = \frac{y - x}{y + x}$ with $y(0)$, find y for $x = 0.1$ by (a) Euler's method (b) Modified Euler's method.

[Ans. (a) $y(0.1) = 1.0928$, (b) $y(0.1) = 1.0928$]

6. Determine the value of y correct to 4 decimal places using Euler's method and modified Euler's method when $x = 0.4$, given that $y' = y + x^2$, $y(0) = 1$ with $h = 0.2$ and compare this result with the exact result.

[Ans. Euler's = 1.4480, Modified Euler's = 1.5141, exact = 1.5155]

7. Given $\frac{dy}{dx} = \sqrt{xy + 1}$, $y(0) = 1$, $h = 0.025$. Determine the value of y correct to three decimal places, using Euler's modified method when $x = 0.075$. **[Ans.** $y = 1.076$]

Runge-Kutta Method

1. Given $\frac{dy}{dx} + y = 0$, $y(0) = 1$, find the value of y at $x = 0.1, 0.2$, using Runge-Kutta method of (*i*) second order (*ii*) third order (*iii*) fourth order. **[Ans.** (*i*) 0.905, 0.819 (*ii*) 0.91, 0.8234 (*iii*) 0.9048, 0.818731]

2. Tabulate by RK method, the numerical solution of $\frac{dy}{dx} = \frac{y^2 - 2x}{y^2 + x}$ and $y(0) = 1$, find y for $x = 0.1, 0.2, 0.3, 0.4, 0.5$. **[Ans.** $y(0.1) = 1.0911$, $y(0.2) = 1.1677$, $y(0.3) = 1.235$, $y(0.5) = 1.338$]

3. Use RK method of order four to find y when $x = 0.4$ in steps of 0.2 given $\frac{dy}{dx} = 1 + y^2$, $y(0) = 0$.

[Ans. $y(0.4) = 0.4$]

4. Solve $y' = 1 + \sqrt{xy}$, $y(0) = 1$ by RK method of second order choosing $h = 0.25$, find $y(1)$.

[Ans. $y(1) = 2.9118$]

5. For the initial value problem $y' = x + y^2$, $y(0) = 1$, $h = 0.2$. Compare the solution obtained for $y(0.4)$ by modified Euler's method, RK method of order four and exact solution.

[Ans. MEM = 1.51408, RKM = 1.51547, exact = 1.51547]

6. Solve $y' = \frac{1 + xy}{x + y}$, $y(1) = 1.2$ by RK method of order two choosing $h = 0.1$, find $y(1.2)$.

[Ans. $y(1.2) = 1.4028$]

7. Using RK method of order four, solve $y'' - 2y' + 2y = e^{2x} \sin x$ with $y(0) = 0$, $y'(0) = -0.6$ to find $y(0.2)$.

[Ans. $y(0.2) = -0.5207$]

Milne's Predictor-Corrector Method

1. Solve $\frac{dy}{dx} = 2e^x - y$ at $x = 0.4$ and 0.5 by Milne's predictor-corrector method given values at the four points $x = 0, 0.1, 0.2$ and 0.3 as $y_0 = 2.0$, $y_1 = 2.010$, $y_2 = 2.040$, $y_3 = 2.090$.

   [Ans. $y(0.4) = 2.1621$, $y(0.5) = 2.2447$**]**

2. Given $y'' + xy + y = 0$, $y(0) = 1$, obtain y for $x = 0, 0.1, 0.2$ and 0.3 by Taylor's series method and find the solution for $y(0.4)$ by Milne's predictor-corrector method. **[Ans.** $y(0.4) = -0.3692$**]**

3. Consider the problem $y' = y + x_2$, $y(0) = 1$, using Milne's predictor-corrector method, find $y(0.8)$, $h = 0.2$.

   [Ans. $y(0.8) = 2.4366$**]**

4. Using Milne's method or RK method to get $y(0.4)$, given $y' = xy + y^2$, $y(0) = 1$, use Taylor's series method or Euler's method or RK method to get $y(0.1)$, $y(0.2)$ and $y(0.3)$. **[Ans.** $y(0.4) = 2.1119$**]**

5. Using Milne's method, find $y(0.8)$ and $y(1.0)$ given $y' = \frac{1}{x+y}$, $y(0) = 2$, $y(0.2) = 2.0933$, $y(0.4) = 2.175$, $y(0.6) = 2.2493$. **[Ans.** $y(0.8) = 1.83299$, $y(1.0) = 1.83698$**]**

Adam-Bashforth Method

1. Given $y' = 1 + y^2$, $y(0) = 0$, find $y(0.2)$ by Taylor's method, $y(0.4)$ by modified Euler's method, $y(0.6)$ by RK method hence find $y(0.8)$ by both Adam's and Milne's method. **[Ans.** 0.0035, 0.0069, 0.0104, 0.0139**]**

2. Solve $2y' - x - y = 0$, given $y(0) = 2$, $y(0.5) = 2.6.636$, $y(1) = 3.595$, $y(1.5) = 4.968$ to get $y(2)$ by Adam's method.

   [Ans. $y(2) = 6.8731$**]**

3. Using Adam's method to find $y(0.4)$, given $\frac{dy}{dx} = \frac{xy}{2}$, $y(0) = 1$, $y(0.1) = 1.01$, $y(0.2) = 1.022$, $y(0.3) = 1.023$.

   [Ans. 1.04081, 1.04101**]**

4. Solve by ABM for $y(0.4)$, $y(0.5)$ given $y' = -2x - y$, $y(0) = -1$, $y(0.1) = -0.9145122$, $y(0.2) = -0.8561923$, $y(0.3) = -0.8224547$. **[Ans.** at 0.4, predicted value = – 0.810968, Correction value = – 0.810965**]**

5. Solve $y' = x^3 + y^2$, $h = 0.1$ find y at $y(0.8)$, $y(0.9)$ initial values are $y(0.2) = 0.0004$, $y(0.4) = 0.0064$, $y(0.6) = 0.0325$. **[Ans.** $y(0.8) = 0.1035$, $y(0.9) = 0.1669$**]**

Numerical Solution of Partial Differential Equations

19.1 INTRODUCTION

Partial differential equations occur very frequently in science, engineering and applied mathematics. Many partial differential equations cannot be solved by analytical methods in closed form solution. In most of the research work in fields like, applied elasticity, theory of plates and shells, hydrodynamics, quantum mechanics, etc. the research problem is reduced to partial differential equations since analytical solutions are not available, we go in for numerical solutions of the partial differential equations by various methods. Certain types of boundary value problems can be solved by replacing the differential equation by the corresponding difference equation and then solving the latter by a process of iteration. This method was devised and first used by LF Richardson and it was later improved by H Liebmann.

19.2 FINITE DIFFERENCE APPROXIMATION TO VARIOUS DERIVATIVES

If $y(x)$ and its derivatives are single valued continuous function of x. Then by Taylor's expansion, we have

$$y(x+h) = y(x) + hy'(x) + \frac{h^2}{2!} y''(x) + \frac{h^3}{3!} y'''(x) + \ldots \quad \ldots(19.1)$$

and

$$y(x-h) = y(x) - hy'(x) + \frac{h^2}{2!} y''(x) - \frac{h^3}{3!} y'''(x) + \ldots \quad \ldots(19.2)$$

Equation (19.1) gives

$$y'(x) = \frac{1}{h}[y(x+h) - y(x)] - \frac{h^2}{2!} y''(x) + \ldots$$

i.e.

$$= \frac{1}{h}[y(x+h) - y(x)] + O(h)$$

which is forward difference approximation of $y'(x)$ with an error of order h.

Similarly Eq. (19.2) gives

$$y'(x) = \frac{1}{h}[y(x) - y(x-h)] + O(h)$$

which is backward difference approximation

Subtracting Eq. (19.2) from Eq. (19.1), we have

$$y(x+h) - y(x-h) = 2h\,[y'(x)] + O(h^2)$$

or

$$y'(x) = y(x+h) - y(x-h)/2\text{h} + O(h^2)$$

which is central difference approximation of $y'(x)$ with an error of order h^2.

Adding Eqs (19.1) and (19.2)

$$y(x+h) + y(x-h) = 2\left[y(x) + \frac{h^2}{2!}y''(x)\right] + \frac{2h^4}{4!}\,y''''(x)$$

$$\frac{y(x+h) - 2y(x) + y(x-h)}{h^2} = y''(x) + O(h^2)$$

Central differences are better approximations than forward and backward approximations.

19.3 FINITE DIFFERENCE APPROXIMATION TO PARTIAL DERIVATIVES

Let xy plane be divided into a series of rectangles whose sides are parallel to x- and y-axes such that $\Delta x = h$ and $\Delta y = k$. The points of intersection of these families of lines are called mesh points, lattice points or grid points.

If (x_i, y_j) is any grid point, $x_i = x_0 + ih$, $y_j = y_0 + jk$, and we take one corner as origin, then

$$x_i = ih,\; y_j = jk,\; i, j = 0, 1, 2, \ldots$$

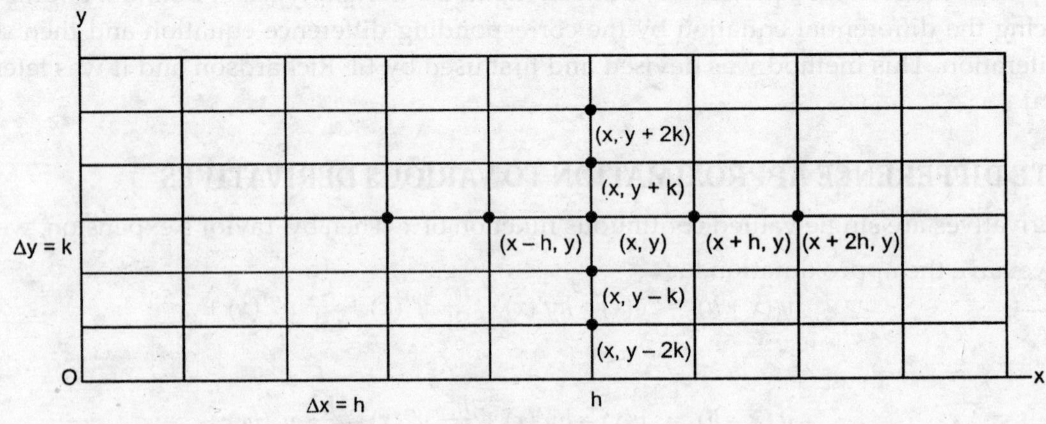

Fig. 19.1

Thus finite difference approximation for the partial derivative in x direction

$$\frac{\partial u}{\partial x} = \frac{u(x+h, y) - u(x, y)}{h} + O(h) \qquad \text{[forward difference]}$$

$$= \frac{u(x, y) - u(x-h, y)}{h} + O(h) \qquad \text{[backward difference]}$$

$$= \frac{u(x+h, y) - u(x-h, y)}{h} + O(h^2) \qquad \text{[central difference]}$$

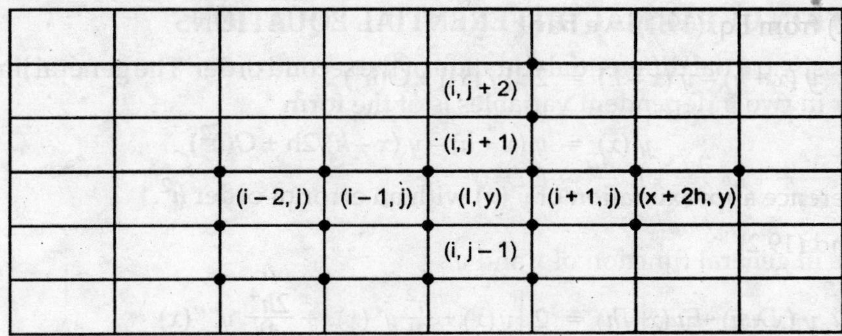

Fig. 19.2

$$\frac{\partial^2 u}{\partial x^2} = \frac{u(x-h, y) - 2u(x, y) + u(x+h, y)}{h^2} + O(h^2)$$

Writing $u(x, y) = u(ih, jk)$ as simply $u_{i, j}$

$$\therefore \quad u_x = \frac{u_{i+1, j} - u_{i,j}}{h} + O(h) \qquad \text{[forward difference]}$$

$$= \frac{u_{i, j} - u_{i-1, j}}{h} + O(h) \qquad \text{[backward difference]}$$

$$= \frac{u_{i+1, j} - u_{i-1, j}}{2h} + O(h^2) \qquad \text{[central difference]}$$

$$u_{xx} = \frac{u_{i-1, j} - u_{i, j} + u_{i+1, j}}{h^2} + O(h^2)$$

Similarly, we have the approximation for the derivative w.r.t. y

$$u_y = \frac{u_{i, j+1} - u_{i,j}}{k} + O(k) = \frac{u_{i, j} - u_{i, j-1}}{k} + O(k)$$

$$= \frac{u_{i, j+1} - u_{i, j-1}}{2k} + O(k^2)$$

and
$$u_{yy} = \frac{u_{i, j-1} - 2u_{i, j} + u_{i, j+1}}{k^2} + O(k^2)$$

The third and fourth derivatives are as

$$y_1''' = \frac{1}{2h^3}(y_{i+2} - 2y_{i+1} + 2y_{i-1} - y_{i-2})$$

$$y_1'''' = \frac{1}{h^4}(y_{i+2} - 4y_{i+1} + 6y_i - 4y_{i-1} + y_{i-2})$$

Note: The accuracy of this method depends on the size of subinterval h and also the order of approximation. As we reduce h, the accuracy improves but the number of equations to be solved also increases.

19.4 CLASSIFICATION OF PARTIAL DIFFERENTIAL EQUATIONS

In this section, we will classify partial differential equation of the second order. The general linear partial differential equation of second order in two independent variables is of the form

$$A\frac{\partial^2 u}{\partial x^2} + B\frac{\partial^2 u}{\partial x \partial y} + C\frac{\partial^2 u}{\partial y^2} + D\frac{\partial u}{\partial x} + E\frac{\partial u}{\partial y} + Fu = 0$$

where A, B, C, D, E, F are in general function of x and y.

$$\text{or} \quad Au_{xx} + Bu_{xy} + Cu_{yy} + Du_x + Eu_y + Fu = 0 \qquad \text{...(19.3)}$$

The above equation of second order (linear) Eq. (19.3) is said to be

(i) Elliptic at a point (x, y) in the plane if $B^2 - 4AC < 0$

(ii) Parabolic if $B^2 - 4AC = 0$

(iii) Hyperbolic if $B^2 - 4AC > 0$

Note: The same differential equation may be elliptic in one region, parabolic in another and hyperbolic in some other region. For example, $u_{xx} + u_{yy} = 0$ is elliptic if $x > 0$, hyperbolic if $x < 0$ and parabolic if $x = 0$.

Examples:

	Elliptic type	Parabolic type	Hyperbolic type
1.	$\frac{\partial^2 u}{\partial x^2} + \frac{\partial^2 u}{\partial y^2} = 0$	$\frac{\partial^2 u}{\partial x^2} = \frac{1}{\alpha^2}\frac{\partial u}{\partial t}$	$\frac{\partial^2 u}{\partial x^2} = \frac{1}{\alpha^2}\frac{\partial^2 u}{\partial t^2}$
	(Laplace equation in two dimensions)	(One dimensional) heat equation)	(One dimensional wave equation)
2.	$\frac{\partial^2 u}{\partial x^2} + \frac{\partial^2 u}{\partial y^2} = f(x, y)$		
	(Poisson's equation)		

19.5 ELLIPTIC EQUATION

An important equation of the ellipse is

$$\frac{\partial^2 u}{\partial x^2} + \frac{\partial^2 u}{\partial y^2} = 0, \text{ i.e. } u_{xx} + u_{yy} = 0 \qquad \text{...(19.4)}$$

This equation is called Laplace's equation.

Replacing the derivatives by the corresponding difference expressions in Eq. (19.4), we get

$$\frac{u_{i-1,j} - 2u_{i,j} + u_{i+1,j}}{h^2} + \frac{u_{i,j-1} - 2u_{i,j} + u_{i,j+1}}{k^2} = 0$$

Taking a square mesh and putting $h = k$, we get

$$u_{i,j} = \frac{1}{4}[u_{i+1,j} + u_{i-1,j} + u_{i,j+1} + u_{i,j-1}] \qquad \text{...(19.5)}$$

i.e. the value of u at any interior mesh point is the arithmetic mean of its values at the four neighbouring mesh points to the left, right, below and above. This is called standard five point formula (SFPF).

Instead of Eq. (19.5), we may also use the formula

$$u_{i,j} = \frac{1}{4}[u_{i-1,j-1} + u_{i+1,j-1} + u_{i-1,j+1} + u_{i+1,j+1}] \quad ...(19.6)$$

which shows that the value of $u_{i,j}$ is the arithmetic mean of its values at four neighbouring diagonal mesh points. This is called the diagonal five point formula (DFPF).

The SFPE and DFPF are represented in Figs 19.1 and 19.2.

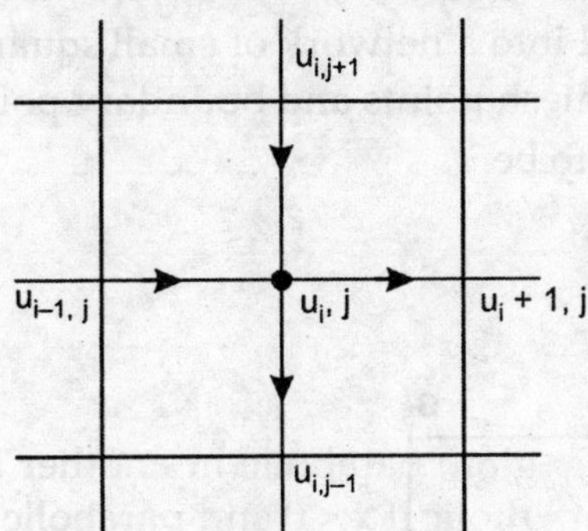

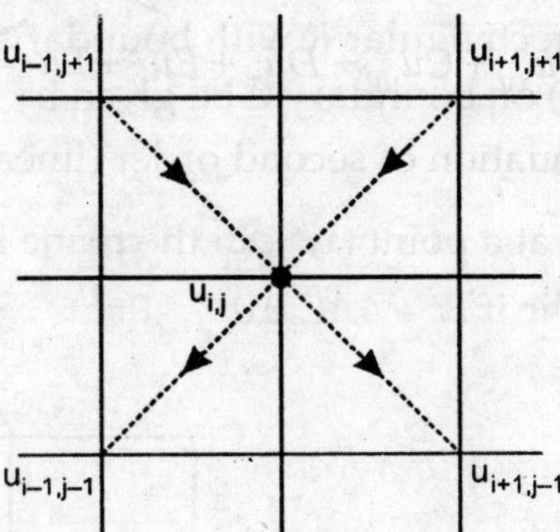

Fig. 19.3

Although Eq. (19.6) is less accurate than Eq. (19.5), yet it serves as a reasonably good approximation.

Note: The DFPF is valid since we know that Laplace equation remains invariant when coordinate axes are rotated through 45°, but the error in DFPF is four times the error in SFPF, therefore we prefer SFPF.

19.5.1 *Jacobi's Method*

Let (n) denote the nth iteration value of $u_{i,j}$, the iterative formula to solve Eq. (19.5) is

$$u_{i,j}^{(n+1)} = \frac{1}{4}\left[u_{i-1,j}^{(n)} + u_{i+1,j}^{(n)} + u_{i,j+1}^{(n)} + u_{i,j-1}^{(n)}\right] \quad ...(19.7)$$

It gives improved value of $u_{i,j}$ at interior mesh points is called five point Jacobi's formula.

19.5.2 *Gauss-Seidal Method*

The iterative formula is Gauss-Seidal method converges twice as fast as Jacobi's method.

$$u_{i,j}^{(n+1)} = \frac{1}{4}\left[u_{i-1,j}^{(n+1)} + u_{i+1,j}^{(n)} + u_{i,j+1}^{(n+1)} + u_{i,j-1}^{(n)}\right] \quad ...(19.8)$$

The method uses the latest iterative values available and scans the mesh points systematically from left to right along successive rows. Gauss-Seidal method converges twice as fast as Jacobi's method.

19.5.3 *Successive Over Relaxation Method (SOR Method)*

Equation (19.8) can be written as which shows that $\frac{1}{4}R_{i,j}$ is the change in the value $u_{i,j}$ for one Gauss-Seidal iteration. In the SOR method, if there is a large change then this is given to $u_{i,j}^{(n)}$, and the iteration formula is written as

$$u_{i,j}^{(n+1)} = u_{i,j}^{(n)} + \frac{1}{4}\left[u_{i-1,j}^{(n+1)} + u_{i+1,j}^{(n)} + u_{i,j-1}^{(n+1)} + 4u_{i,j}^{(n)}\right]$$

$$= u_{i,j}^{(n)} + \frac{1}{4}R_{i,j}$$

$$u_{i,j+1}^{(n+1)} = u_{i,j}^{(n)} + \frac{1}{4}w_{Ri,j}$$

$$= \frac{1}{4} w\left[u_{i-1,j}^{(n+1)} + u_{i+1,j}^{(n)} + u_{i,j-1}^{(n+1)} + u_{i,j+1}^{(n)}\right] + (1-w)\, u_{i,j}^{(n)} \quad ...(19.9)$$

The rate of convergence of Eq. (19.9) depends on the choice of w, which is called the accelerating factor and lie between 1 and 2.

19.6 SOLUTION OF LAPLACE EQUATION BY LIEBMANN'S ITERATION PROCESS

Consider the Laplace solution with the given boundary condition. For simplicity, we assume that function $u(x, y)$ is required over a rectangular R with boundary C. Let R be divided into a network of small squares of side h. Let the values of $u(x, y)$ on boundary C be given by C_i and the interior mesh points and boundary points be as shown in the figure below.

$$\frac{\partial^2 u}{\partial x^2} + \frac{\partial^2 u}{\partial y^2} = 0$$

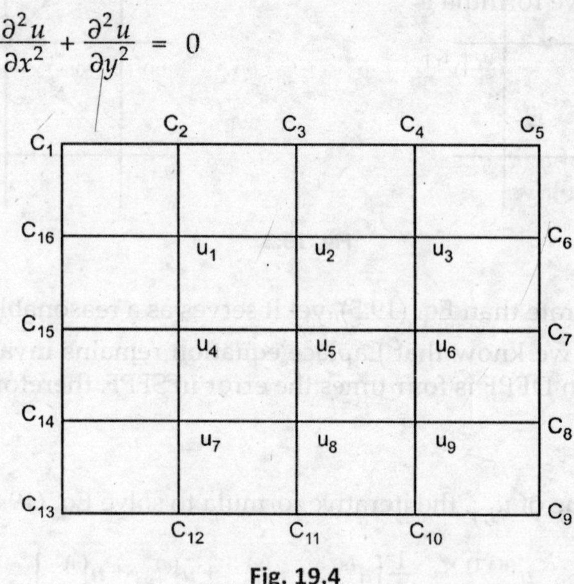

Fig. 19.4

The value of $u(x, y)$ satisfying the Laplacian equation can be replaced by either SFPF or DFPF. To start the iteration process, initially we find rough values at interior points and then improve them by process mostly using SFPF.

We first find u_4 at the centre of the square by taking average of four boundary values

$$u_5 = \frac{1}{4}[c_{15} + c_7 + c_3 + c_{11}]$$

Next, we find the initial values at the centre of the four large inner squares using DFPF

$$u_1 = \frac{1}{4}[c_1 + u_5 + c_3 + c_{15}]$$

$$u_3 = \frac{1}{4}[u_5 + c_5 + c_3 + c_7]$$

$$u_7 = \frac{1}{4}[u_5 + c_{15} + c_{13} + c_{11}]$$

$$u_9 = \frac{1}{4}[u_5 + c_9 + c_7 + c_{11}]$$

The values at the remaining interior points are obtained by SFPF

$$u_2 = \frac{1}{4}[u_1 + u_3 + c_3 + u_5]$$

$$u_4 = \frac{1}{4}[c_{15} + u_5 + u_1 + u_7]$$

$$u_6 = \frac{1}{4}[u_5 + c_7 + u_3 + u_9]$$

$$u_8 = \frac{1}{4}[u_7 + u_9 + u_5 + c_{11}]$$

Now we have got all the boundary values of u and rough values at every mesh grid points in the interior of region R, we proceed with an iteration process to improve their accuracy. We start with u_1 and iterate it using the latest available values of the four adjacent points. We iterate all the mesh points systematically from left to right along successive rows. The iterative formula is

$$u_{i,j}^{(n+1)} = \frac{1}{4}\left[u_{i-1,j}^{n+1} + u_{i+1,j}^{(n)} + u_{i,j-1}^{(n)} + u_{i,j+1}^{n+1}\right]$$

Example **19.1.** Find the value of $u(x, y)$ satisfying Laplace equation $\Delta^2 u = 0$ at the pivotal point of system mesh with boundary values as shown below:

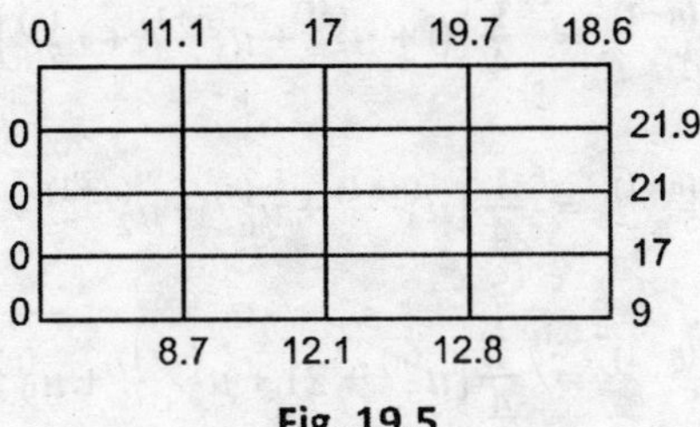

Fig. 19.5

Solution :

$$u_5^{(0)} = \frac{1}{4}[0 + 21 + 17 + 12.1] = \frac{49.1}{4} = 12.275 \text{ (SFPF)}$$

$$u_1^{(0)} = \frac{1}{4}[0 + 12.275 + 17 + 0] = \frac{29.275}{4} = 7.0687 \text{ (DFPF)}$$

$$u_3^{(0)} = \frac{1}{4}[18.6 + u_5 + 17 + 21] = \frac{1}{4}[18.6 + 12.275 + 38] = \frac{68.875}{4}$$

$$= 17.219 \text{ (DFPF)}$$

$$u_7^{(0)} = \frac{1}{4}[0 + 12.1 + u_5 + 0] = \frac{24.375}{4} = 6.0937 \text{ (DFPF)}$$

$$u_9^{(0)} = \frac{1}{4}[9 + u_5 + 33.1] = \frac{54.375}{4} = 13.5937 \text{ (DFPF)}$$

$$u_2^{(0)} = \frac{1}{4}[u_1 + u_3 + 17 + u_5] = \frac{53.5627}{4} = 13.3907 \text{ (SFPF)}$$

$$u_4^{(0)} = \frac{1}{4}[0 + u_5 + u_1 + u_7] = \frac{1}{4}(25.4374) = 6.3593 \text{ (SFPF)}$$

$$u_6^{(0)} = \frac{1}{4}[u_5 + 21 + 19.7 + u_9] = \frac{1}{4}(68.5687] = 17.1422 \text{ (SFPF)}$$

$$u_8^{(0)} = \frac{1}{4}[u_5 + 12.1 + u_7 + u_9] = \frac{45.0624}{4} = 11.2656$$

Now by Gauss-Seidal method

$$u_{i,j}^{(n+1)} = \frac{1}{4}\left[u_{i-1,j}^{(n+1)} + u_{i+1,j}^{(n)} + u_{i,j+1}^{(n+1)} + u_{i,j-1}^{(n)}\right]$$

$$u_1^{(n+1)} = \frac{1}{4}[0 + u_2^{(n)} + 11.1 + u_4^{(n)}]$$

$$u_2^{(n+1)} = \frac{1}{4}[u_1^{(n+1)} + u_3^{(n)} + u_5^{(n)} + 17]$$

$$u_3^{(n+1)} = \frac{1}{4}[u_2^{(n+1)} + 21.9 + 19.7 + u_6^{(n)}]$$

$$u_4^{(n+1)} = \frac{1}{4}[0 + u_5^{(n)} + u_1^{(n+1)} + u_7^{(n)}]$$

$$u_5^{(n+1)} = \frac{1}{4}[u_4^{(n+1)} + u_6^{(n)} + u_2^{(n+1)} + u_8^{(n)}]$$

$$u_6^{(n+1)} = \frac{1}{4}[u_5^{(n)} + 21 + u_3^{(n+1)} + u_9^{(n)}]$$

$$u_7^{(n+1)} = \frac{1}{4}[0 + u_8^{(n)} + u_4^{(n+1)} + 8.7]$$

$$u_8^{(n+1)} = \frac{1}{4}[0 + u_7^{(n+1)} + u_5^{(n+1)} + 12.1 + u_9^{(n)}]$$

$$u_9^{(n+1)} = \frac{1}{4}[u_8^{(n+1)} + 17 + u_6^{(n+1)} + 12.8]$$

Put $n = 0$

$$u_1^{(1)} = \frac{1}{4}[0 + 13.3907 + 6.3593 + 11.1] = \frac{30.8500}{4} = 7.7125$$

$$u_2^{(1)} = \frac{1}{4}[7.7125 + 17.2190 + 12.275 + 17] = \frac{54.2065}{4} = 13.5515$$

$$u_3^{(1)} = \frac{1}{4}[13.5516 + 41.6 + 17.1422] = \frac{72.2938}{4} = 18.0734$$

$$u_4^{(1)} = \frac{1}{4}[0 + 12.275 + 7.7125 + 6.0937] = \frac{26.5203}{4} = 6.5203$$

$$u_5^{(1)} = \frac{1}{4}[6.5203 + 17.1422 + 13.5516 + 11.2656] = \frac{48.4979}{4} = 12.1199$$

$$u_6^{(1)} = \frac{1}{4}[12.1199 + 21 + 18.0734 + 13.5937] = \frac{64.7870}{4} = 16.1967$$

$$u_7^{(1)} = \frac{1}{4}[8.7 + 11.2656 + 6.5203] = \frac{26.4859}{4} = 6.6215$$

$$u_8^{(1)} = \frac{1}{4}[6.6215 + 12.1199 + 12.1 + 13.5937] = \frac{44.4351}{4} = 11.1088$$

$$u_9^{(1)} = \frac{1}{4}[11.1088 + 29.8 + 16.1967] = \frac{57.1055}{4} = 14.2764$$

Put $n = 1$

$$u_1^{(2)} = \frac{1}{4}[0 + 13.5516 + 6.5203 + 11.1] = \frac{31.1719}{4} = 7.793$$

$$u_2^{(2)} = \frac{1}{4}[7.793 + 18.0734 + 12.1199 + 17] = \frac{54.9863}{4} = 13.7466$$

$$u_3^{(2)} = \frac{1}{4}[41.6 + 13.7466 + 16.1967] = \frac{71.5433}{4} = 17.8858$$

$$u_4^{(2)} = \frac{1}{4}[12.1199 + 7.7131 + 6.6215] = \frac{26.5344}{4} = 6.6336$$

$$u_5^{(2)} = \frac{1}{4}[6.6336 + 16.1967 + 13.7466 + 11.1088] = \frac{47.5857}{4} = 11.8964$$

$$u_6^{(2)} = \frac{1}{4} + [11.8964 + 21 + 17.8858 + 14.2764] = \frac{65.0586}{4} = 16.2646$$

$$u_7^{(2)} = \frac{1}{4}[8.7 + 11.1088 + 6.6336] = \frac{26.4424}{4} = 6.6106$$

$$u_8^{(2)} = \frac{1}{4}[6.6106 + 11.8964 + 12.1 + 14.2764] = \frac{44.8834}{4} = 11.2208$$

$$u_9^{(2)} = \frac{1}{4}[11.2208 + 29.8 + 16.2696] = \frac{57.2854}{4} = 14.3213$$

Put $n = 2$

$$u_1^{(3)} = \frac{1}{4}[0 + 13.7466 + 6.6336 + 11.1] = 7.8700$$

$$u_2^{(3)} = \frac{1}{4}[7.8700 + 17.8858 + 11.8964 + 17] = 13.6630$$

$$u_3^{(3)} = \frac{1}{4}[41.6 + 13.663 + 16.2646] = 71.8819$$

$$u_4^{(3)} = \frac{1}{4}[71.8964 + 7.8700 + 6.6106] = 6.5942$$

$$u_5^{(3)} = \frac{1}{4}[6.5942 + 16.2646 + 13.6630 + 11.2208] = 11.9356$$

$$u_6^{(3)} = \frac{1}{4}[11.9356 + 21 + 17.8819 + 14.3213] = 16.2597$$

$$u_7^{(3)} = \frac{1}{4}[8.7 + 112208 + 6.5942] = 6.6288$$

$$u_8^{(3)} = \frac{1}{4}[6.6288 + 11.9356 + 12.1 + 14.3213]$$

$$u_9^{(3)} = \frac{1}{4}[29.8 + 11.2464 + 16.2597] = 14.3265$$

Put $n = 3$

$$u_1^{(4)} = 7.8393\ u_1^{(5)} = 7.143$$

$$u_2^{(4)} = 13.6642\ u_2^{(5)} = 13.6663$$

$$u_3^{(4)} = 17.8810\ u_3^{(5)} = 17.8885$$

$$u_4^{(4)} = 6.6009\ u_4^{(5)} = 6.6052$$

$$u_5^{(4)} = 11.9428\ u_5^{(5)} = 11.9214$$

$$u_6^{(4)} = 16.2876\ u_6^{(5)} = 12.2876$$

$$u_7^{(4)} = 6.6368\ u_7^{(5)} = 6.6329$$

$$u_8^{(4)} = 11.2265\ u_8^{(5)} = 11.2457$$

$$u_9^{(4)} = 14.3285\ u_9^{(5)} = 14.3326$$

The value of $u_1^{(4)}, u_2^{(4)}, u_3^{(4)}, u_4^{(4)}, u_5^{(4)}, u_6^{(4)}, u_7^{(4)}, u_8^{(4)}, u_9^{(4)}$, and $u_1^{(5)}, u_2^{(5)}, u_3^{(5)}, u_4^{(5)}, u_5^{(5)}, u_6^{(5)}, u_7^{(5)}, u_8^{(5)}, u_9^{(5)}$, are approximately the same.

Hence $u_1 = 7.84$, $u_2 = 13.66$, $u_3 = 17.88$, $u_4 = 6.60$, $u_5 = 11.92$, $u_6 = 16.28$, $u_7 = 6.63$, $u_8 = 11.29$, $u_9 = 14.33$.

***Example* 19.2.** Solve the elliptic equation $u_{xx} + u_{yy} = 0$, for the following square mesh with boundary values as shown.

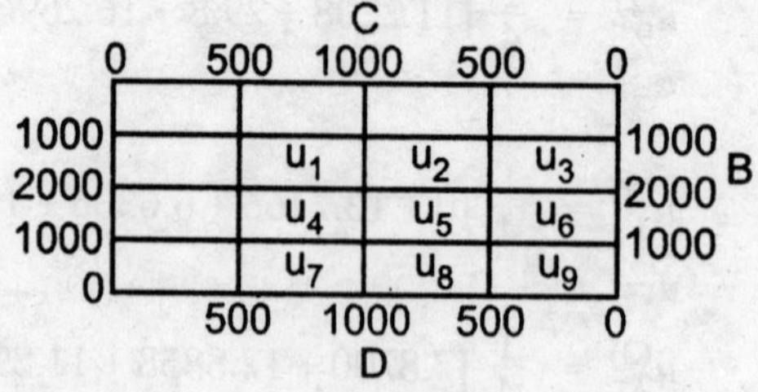

Fig. 19.6

Solution : Let $u_1, u_2, ..., u_9$ be the values of the interior mesh points from the given figure, we can see that the boundary values of u are symmetrical about AB.

$$\therefore \quad u_7 = u_1,\ u_8 = u_2,\ u_9 = u_3$$

Also the values of u being symmetrical about CD are

$$u_3 = u_1, u_6 = u_4, u_9 = u_7$$

$\therefore$ It is sufficient to find the values u_1, u_2, u_4 and u_5, the initial values of these are given below.

$$u_5^{(0)} = \frac{1}{4}[2000 + 2000 + 1000 + 1000] = 1500 \text{ (SFPF)}$$

$$u_1^{(0)} = \frac{1}{4}[0 + 1500 + 1000 + 2000] = 1125 \text{ (DFPF)}$$

$$u_2^{(0)} = \frac{1}{4}[1125 + 1125 + 1000 + 1500] = 1187.5 \text{ (SFPF)}$$

$$u_4^{(0)} = \frac{1}{4}[2000 + 1500 + 1125 + 1125] = 1437.5 \text{ (SFPF)}$$

Now we carry out the iteration using the following formula by SFPF

$$u_1^{(n+1)} = \frac{1}{4}[1000 + u_2^{(n)} + 500 + u_4^{(n)}]$$

$$u_2^{(n+1)} = \frac{1}{4}[u_1^{(n+1)} + u_3^{(n+1)} + 1000 + u_5^{(n)}]$$

$$u_4^{(n+1)} = \frac{1}{4}[2000 + u_5^{(5)} + u_1^{(n+1)} + u_1^{(n+1)}]$$

$$u_5^{(n+1)} = \frac{1}{4}[u_4^{(n+1)} + u_4^{(n+1)} + u_2^{(n+1)} + u_2^{(n+1)}]$$

First iteration ($n = 0$)

$$u_1^{(1)} = \frac{1}{4}[1000 + 1187.5 + 500 + 1437.5] = 1031.25$$

$$u_2^{(1)} = \frac{1}{4}[1031.25 + 1031.25 + 1000 + 1500] = 1140.625$$

$$u_4^{(1)} = \frac{1}{4}[2000 + 1500 + 1031.25 + 1031.25] = 1390.625$$

$$u_5^{(1)} = \frac{1}{4}[1390.25 + 1390.25 + 1140.625 + 1140.625] = 1265.625$$

Second iteration ($n = 1$)

$$u_1^{(2)} = \frac{1}{4}[1000 + 1140.625 + 500 + 1390.625] = 1007.8125$$

$$u_2^{(2)} = \frac{1}{4}[1007.8125 + 1007.8125 + 1000 + 1265.625] = 1070.3125$$

$$u_4^{(2)} = \frac{1}{4}[2000 + 1265.625 + 1007.8125 + 1007.8125] = 1320.3125$$

$$u_5^{(2)} = \frac{1}{4}[1320.3125 + 1320.3125 + 1070.3125 + 1070.3125] = 1195.3125$$

Third iteration ($n = 2$)

$$u_1^{(3)} = \frac{1}{4}[1000 + 1070.3125 + 500 + 1320.3125] = 972.65625$$

$$u_2^{(3)} = \frac{1}{4}[972.65625 + 972.65625 + 1000 + 1195.3125] = 1035.1563$$

$$u_4^{(3)} = \frac{1}{4}[2000 + 1195.3125 + 972.65625 + 972.65625] = 1285.1563$$

$$u_5^{(3)} = \frac{1}{4}[2(1285.1563) + 2(1035.1563)] = 1160.1563$$

Fourth iteration ($n = 3$)

$$u_1^{(4)} = \frac{1}{4}[1000 + 1035.1563 + 500 + 1285.1563] = 955.07815$$

$$u_2^{(4)} = \frac{1}{4}[955.07815 + 955.07815 + 1000 + 1160.1563] = 1017.5782$$

$$u_4^{(4)} = \frac{1}{4}[2000 + 1160.1563 + 955.07815 + 955.07815] = 1267.5782$$

$$u_5^{(4)} = \frac{1}{4}[2 \times 1267.5782 + 2 \times 1017.5782] = 1142.5782$$

Fifth iteration ($n = 4$)

$$u_1^{(5)} = \frac{1}{4}[1000 + 1017.5782 + 500 + 1267.5782] = 946.2891$$

$$u_2^{(5)} = \frac{1}{4}[946.2891 + 946.2891 + 1000 + 1142.5782] = 1008.7891$$

$$u_4^{(5)} = \frac{1}{4}[2000 + 1142.5782 + 2(946.2891)] = 1258.7891$$

$$u_5^{(5)} = \frac{1}{4}[2(1258.7891) + 2(1008.7891)] = 1133.7891$$

Sixth iteration ($n = 5$)

$$u_1^{(6)} = \frac{1}{4}[1000 + 1008.7891 + 500 + 1258.7891] = 941.89455$$

$$u_2^{(6)} = \frac{1}{4}[2(941.89455) + 1000 + 1133.7891] = 1004.3946$$

$$u_4^{(6)} = \frac{1}{4}[2000 + 1133.7891 + 2(941.89455)] = 1254.3496$$

$$u_5^{(6)} = \frac{1}{4}[1254.3946 + 1254.3946 + 2(1004.3946)] = 1129.3946$$

Seventh iteration ($n = 6$)

$$u_1^{(7)} = \frac{1}{4}[1000 + 1004.3946 + 500 + 1254.3946] = 939.6973$$

$$u_2^{(7)} = \frac{1}{4}[2(939.6973) + 1000 + 1129.3946] = 1002.1973$$

$$u_4^{(7)} = \frac{1}{4}[2000 + 1129.3946 + 2(939.6973)] = 1252.1973$$

$$u_5^{(7)} = \frac{1}{4}[2(1252.1973) + 2(1002.1973)] = 1127.1973$$

Eighth iteration ($n = 7$)

$$u_1^{(8)} = \frac{1}{4}[1000 + 1002.1972 + 500 + 1252.1973] = 938.59863$$

$$u_2^{(8)} = \frac{1}{4}[2(939.59863) + 1000 + 1127.1973] = 1001.0987$$

$$u_4^{(8)} = \frac{1}{4}[2000 + 1127.1973 + 2(938.59685)] = 1251.0987$$

$$u_5^{(8)} = \frac{1}{4}[2(1251.0987) + 2(1001.0987)] = 1126.0987$$

Ninth iteration ($n = 8$)

$$u_1^{(9)} = \frac{1}{4}[1000 + 1001.0987 + 500 + 12521.0987] = 938.04935$$

$$u_2^{(9)} = \frac{1}{4}[2(938.04935) + 1000 + 1126.09871] = 1000.549\ 4$$

$$u_4^{(9)} = \frac{1}{4}[2000 + 1126.0987 + 2(938.04935)] = 1250.5494$$

$$u_5^{(9)} = \frac{1}{4}[2(1250.5494) + 2(1000.5494)] = 1125.5494$$

From the eighth and ninth iterations, we see there is a negligible difference between the values

$$\therefore \quad u_1 = 939,\ u_2 = 1001,\ u_4 = 1251, \text{ and } u_5 = 1126$$

$$u_3 = 939,\ u_6 = 1251,\ u_7 = 939,\ u_8 = 1001,\ u_9 = 939$$

***Example* 19.3.** Given the values of $u(x, y)$ on the boundary of the square in the following figure. Evaluate the function $u(x, y)$ satisfying Laplace equation $\nabla^2 u = 0$ as the pivot point on this figure by (a) Jacobi method (b) Gauss-Seidal method.

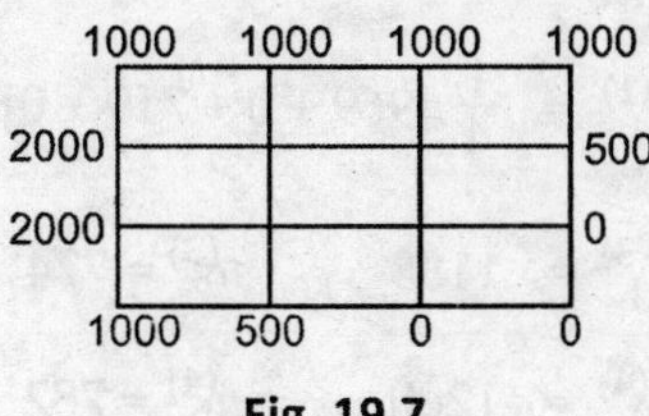

Fig. 19.7

Solution : Let $u_4 = 0$ then

$$u_1 = \frac{1}{4}[1000 + 0 + 1000 + 2000] = 1000 \text{ (DFPF)}$$

$$u_2 = \frac{1}{4}[1000 + 500 + 1000 + 0] = 625 \text{ (SFPF)}$$

$$u_3 = \frac{1}{4}[2000 + 0 + 1000 + 500] = 875 \text{ (SFPF)}$$

$$u_4 = \frac{1}{4}[874 + 0 + 625 + 0] = 375 \text{ (SFPF)}$$

(*a*) Using Jacobi formula

$$u_1^{(n+1)} = \frac{1}{4}[2000 + u_2^{(n)} + 1000 + u_3^{(n)}]$$

$$u_2^{(n+1)} = \frac{1}{4}[u_1^{(n)} + 5000 + 1000 + u_4^{(n)}]$$

$$u_3^{(n+1)} = \frac{1}{4}[2000 + u_4^{(n)} + u_1^{(n)} + 500]$$

$$u_4^{(n+1)} = \frac{1}{4}[u_3^{(n)} + 0 + u_2^{(n)} + 0]$$

First iteration ($n = 0$)

$$u_1^{(1)} = \frac{1}{4}[2000 + 625 + 1000 + 875] = 1125$$

$$u_2^{(1)} = \frac{1}{4}[1000 + 500 + 1000 + 375] = 719$$

$$u_3^{(1)} = \frac{1}{4}[2000 + 375 + 1000 + 500] = 969$$

$$u_4^{(1)} = \frac{1}{4}[875 + 0 + 625 + 0] = 375$$

Second iteration ($n = 1$)

$$u_1^{(1)} = \frac{1}{4}[2000 + 719 + 1000 + 969] = 1172$$

$$u_2^{(1)} = \frac{1}{4}[1125 + 500 + 1000 + 375] = 750$$

$$u_3^{(1)} = \frac{1}{4}[2000 + 375 + 1125 + 500] = 1000$$

$$u_4^{(1)} = \frac{1}{4}[969 + 0 + 719 + 0] = 422$$

Similarly $\quad u_1^{(3)} = 1188 \quad u_2^{(3)} = 774 \quad u_3^{(3)} = 1024 \quad u_4^{(3)} = 438$

$u_1^{(4)} = 1200 \quad u_2^{(4)} = 782 \quad u_3^{(4)} = 1032 \quad u_4^{(4)} = 450$

$u_1^{(5)} = 1204 \quad u_2^{(5)} = 788 \quad u_3^{(5)} = 1038 \quad u_4^{(5)} = 454$

$u_1^{(6)} = 1206.5 \quad u_2^{(6)} = 790 \quad u_3^{(6)} = 1040 \quad u_4^{(6)} = 456.6$

$$u_1^{(7)} = 1208 \quad u_2^{(7)} = 791 \quad u_3^{(7)} = 1041 \quad u_4^{(7)} = 458$$

$$u_1^{(8)} = 1208 \quad u_2^{(8)} = 791.5 \quad u_3^{(8)} = 1041.5 \quad u_4^{(8)} = 458$$

Hence $u_1 = 1208$, $u_2 = 791.5$, $u_3 = 1041.5$, $u_4 = 258$.

(*b*) Using Gauss-Seidal method

$$u_1^{(n+1)} = \frac{1}{4}[2000 + u_2^{(n)} + 1000 + u_3^{(n)}]$$

$$u_2^{(n+1)} = \frac{1}{4}[u_1^{(n+1)} + 500 + 1000 + u_4^{(n)}]$$

$$u_3^{(n+1)} = \frac{1}{4}[2000 + u_4^{(n)} + u_1^{(n+1)} + 500]$$

$$u_4^{(n+1)} = \frac{1}{4}[u_3^{(n+1)} + 0 + u_2^{(n+1)} + 0]$$

First iteration ($n = 0$)

$$u_1^{(1)} = \frac{1}{4}[2000 + 625 + 1000 + 875] = 1125$$

$$u_2^{(1)} = \frac{1}{4}[1125 + 500 + 1000 + 375] = 750$$

$$u_3^{(1)} = \frac{1}{4}[2000 + 375 + 1125 + 500] = 1000$$

$$u_4^{(1)} = \frac{1}{4}[1000 + 0 + 750 + 0] = 438$$

Second iteration ($n = 1$)

$$u_1^{(1)} = \frac{1}{4}[2000 + 750 + 1000 + 1000] = 1188$$

$$u_2^{(2)} = \frac{1}{4}[1188 + 500 + 1000 + 438] = 782$$

$$u_3^{(2)} = \frac{1}{4}[2000 + 438 + 1188 + 500] = 1032$$

$$u_4^{(2)} = \frac{1}{4}[1032 + 0 + 782 + 0] = 454$$

Similarly

$$u_1^{(3)} = 1204 \quad u_2^{(3)} = 789 \quad u_3^{(3)} = 1040 \quad u_4^{(3)} = 458$$

$$u_1^{(4)} = 1207 \quad u_2^{(4)} = 791 \quad u_3^{(4)} = 1041 \quad u_4^{(4)} = 458$$

$$u_1^{(5)} = 1208 \quad u_2^{(5)} = 791.5 \quad u_3^{(5)} = 1041.5 \quad u_4^{(5)} = 458.2$$

Hence $u_1 = 1208$, $u_2 = 791.5$, $u_3 = 1041.5$, $u_4 = 458.25$.

Example **19.4.** Solve the elliptic equation $u_{xx} + u = 0$ for the following square mesh with boundary values as shown in the figure. Iterate until the maximum difference between two successive values at any point is less than 0.001.

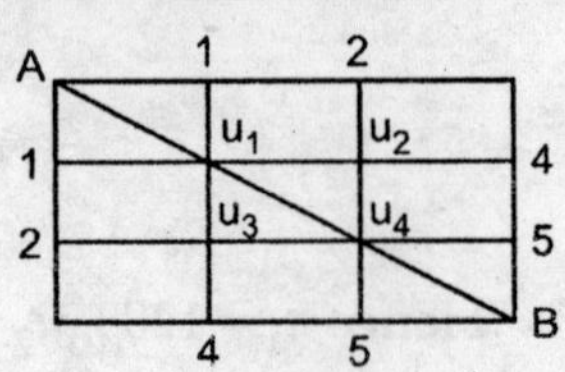

Fig. 19.8

Solution :

The figure is symmetrical about AB.

$$u_2 = u_3 = 0 \text{ (say)}$$

$$u_1^{(0)} = \frac{1}{4}[1 + u_2 + 1 + u_3] = 0.5 \text{ (SFPF)}$$

$$u_2^{(0)} = \frac{1}{4}[0.5 + 2 + 4 + 2.5] = 2.25 \text{ (SFPF)}$$

$$u_4^{(0)} = \frac{1}{4}[u_3 + 5 + u_2 + 5] = 2.5 \text{ (SFPF)}$$

Now using Gauss-Seidal formula

$$u_1^{(n+1)} = \frac{1}{4}[1 + u_2^n + 1 + u_3^n]$$

$$u_2^{(n+1)} = \frac{1}{4}[u_1^{(n+1)} + 4 + 2 + u_4^n]$$

$$u_4^{(n+1)} = \frac{1}{4}[u_3^{(n+1)} + 5 + u_2^{(n+1)} + 5]$$

First iteration ($n = 0$)

$$u_1^{(1)} = \frac{1}{4}[2 + 2 \times 2.25] = \frac{1}{2}[1 + 2.25] = 1.625$$

$$u_2^{(1)} = \frac{1}{4}[u_1^{(1)} + 4 + 2 + u_4^{(0)}] = \frac{1}{4}[1.625 + 6 + 2.5] = 2.53125$$

$$u_4^{(1)} = \frac{1}{4}[u_3^{(1)} + 5 + u_2^{(1)} + 5] = \frac{1}{4}[2.53125 + 3] = 3.765625$$

Second iteration ($n = 1$)

$$u_1^{(2)} = \frac{1}{4}[1 + u_2^1 + 1 + u_1^3] = \frac{1}{4}[2 + 2 \times 2.53125] = \frac{1}{2}[1 + 2.53125]$$

$$= 1.765625$$

$$u_2^{(2)} = \frac{1}{4}[6 + 1.765625 + 3.765625] = 2.8828125$$

$$u_4^{(2)} = \frac{1}{2}[2.8828125] = 3.9414063$$

Third iteration ($n = 2$)

$$u_1^{(3)} = \frac{1}{2}[1 + 2.8828125] = 1.9414063$$

$$u_2^{(3)} = \frac{1}{6}[6 + 1.9414063 + 3.9414063] = 2.9707031$$

$$u_4^{(3)} = \frac{1}{2}[2.970703 + 5] = 3.9853516$$

Fourth iteration ($n = 3$)

$$u_1^{(4)} = \frac{1}{2}[1 + 2.9707031] = 1.9853516$$

$$u_2^{(4)} = \frac{1}{6}[6 + 1.9853516 + 3.9853516] = 2.9926758$$

$$u_4^{(4)} = \frac{1}{2}[2.9926758 + 5] = 3.9963379$$

Fifth iteration ($n = 4$)

$$u_1^{(5)} = \frac{1}{2}[1 + 2.9926578] = 1.9963289$$

$$u_2^{(5)} = \frac{1}{4}[6 + 1.99263289 + 3.9963379] = 2.9981667$$

$$u_4^{(5)} = \frac{1}{2}[2.9918667 + 5] = 3.9990834$$

Sixth iteration ($n = 5$)

$$u_1^{(6)} = \frac{1}{2}[1 + 2.9981667] = 1.9990834$$

$$u_2^{(6)} = \frac{1}{4}[6 + 1.9990839 + 3.9990834] = 2.9995417$$

$$u_4^{(6)} = \frac{1}{2}[2.9995417 + 5] = 3.9997709$$

Hence $u_1 = 1.999$, $u_2 = 2.999$, $u_4 = 3.999$.

***Example* 19.5.** $u_{xx} + u_{yy} = 0$ in $0 \le x \le 4$, $0 \le y \le 4$, given that $u(0, y) = 0$; $u(4, y) = 8 + 2y$, $u(x, 0) = \frac{x^2}{2}$ and $u(x, 4) = x^2$, take $h = k = 1$, and obtain the result correct to one decimal.

Solution : Let us divide the region R, i.e. $0 \le x \le 4$, $0 \le y \le 4$ into 16 squares meshes. The numerical values of the boundary using the given analytical expressions are calculated and exhibited in the figure below.

Let $u_1, u_2, u_3, \ldots, u_9$ be the values of u at the interior mesh points. Now the initial values of u's are calculated either by SFPF or DFPF as given below.

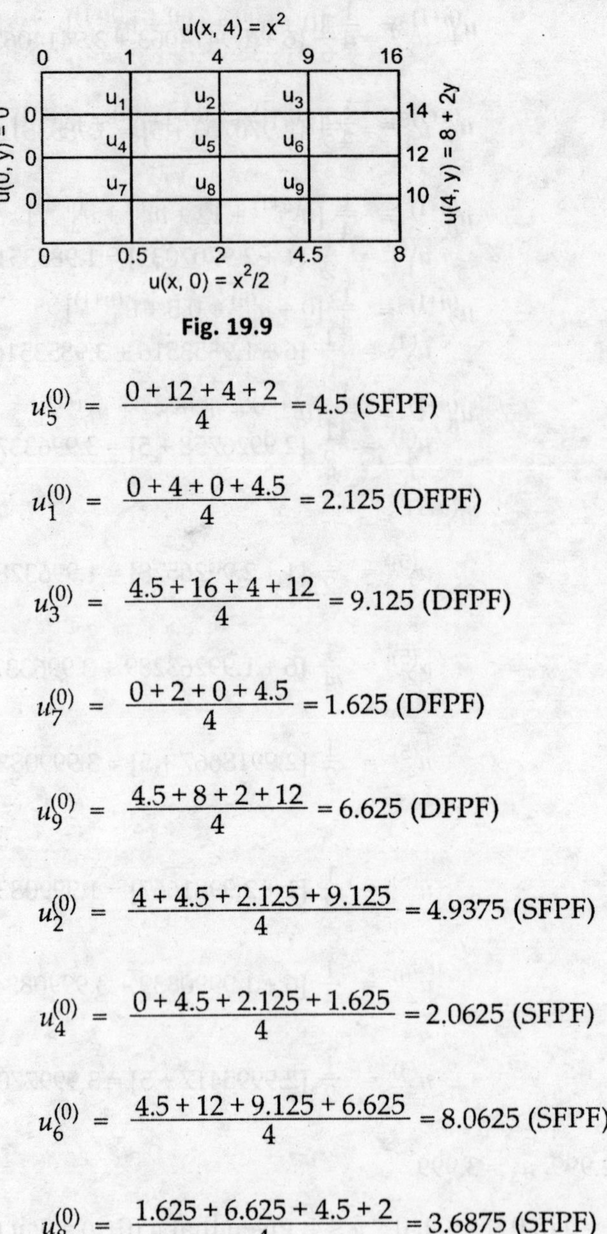

Fig. 19.9

$$u_5^{(0)} = \frac{0+12+4+2}{4} = 4.5 \text{ (SFPF)}$$

$$u_1^{(0)} = \frac{0+4+0+4.5}{4} = 2.125 \text{ (DFPF)}$$

$$u_3^{(0)} = \frac{4.5+16+4+12}{4} = 9.125 \text{ (DFPF)}$$

$$u_7^{(0)} = \frac{0+2+0+4.5}{4} = 1.625 \text{ (DFPF)}$$

$$u_9^{(0)} = \frac{4.5+8+2+12}{4} = 6.625 \text{ (DFPF)}$$

$$u_2^{(0)} = \frac{4+4.5+2.125+9.125}{4} = 4.9375 \text{ (SFPF)}$$

$$u_4^{(0)} = \frac{0+4.5+2.125+1.625}{4} = 2.0625 \text{ (SFPF)}$$

$$u_6^{(0)} = \frac{4.5+12+9.125+6.625}{4} = 8.0625 \text{ (SFPF)}$$

$$u_8^{(0)} = \frac{1.625+6.625+4.5+2}{4} = 3.6875 \text{ (SFPF)}$$

Now we use Gauss-Seidal formula

$$u_{i,j}^{(n+1)} = \frac{1}{4}\left[u_{i-1,j}^{(n+1)} + u_{i+1,j}^{(n)} + u_{i,j-1}^{(n)} + u_{i,j+1}^{(n+1)}\right]$$

To improve the above result, we proceed as follows :

$$u_1^{(n+1)} = \frac{1}{4}\left[0 + u_2^{(n)} + u_4^{(n)} + 1\right]$$

$$u_2^{(n+1)} = \frac{1}{4}\left[u_1^{(n+1)} + u_3^{(n)} + u_5^{(n)} + 4\right]$$

$$u_3^{(n+1)} = \frac{1}{4}\left[u_2^{(n+1)} + 14 + u_6^{(n)} + 9\right]$$

$$u_4^{(n+1)} = \frac{1}{4}[0 + u_5^{(n)} + u_7^{(n)} + u_1^{(n+1)}]$$

$$u_5^{(n+1)} = \frac{1}{4}[u_4^{(n+1)} + u_6^{(n)} + u_8^{(n)} + u_2^{(n+1)}]$$

$$u_6^{(n+1)} = \frac{1}{4}[u_5^{(n+1)} + 12 + u_9^{(n)} + u_3^{(n+1)}]$$

$$u_7^{(n+1)} = \frac{1}{4}[0 + u_8^{(n)} + 0.5 + u_4^{(n+1)}]$$

$$u_8^{(n+1)} = \frac{1}{4}[u_7^{(n+1)} + u_9^{(n)} + 2 + u_5^{(n+1)}]$$

$$u_9^{(n+1)} = \frac{1}{4}[u_8^{(n+1)} + 10 + 4.5 + u_6^{(n+1)}]$$

First iteration ($n = 0$)

$$u_1^{(1)} = \frac{1}{4}[0 + u_2^{(0)} + u_4^{(0)} + 1] = \frac{1}{4}[0 + 4.9375 + 2.0625 + 1] = 2$$

$$u_2^{(1)} = \frac{1}{4}[u_1^{(1)} + u_3^{(0)} + u_5^{(0)} + 4] = \frac{1}{4}[2 + 9.125 + 4.5 + 4] = 4.90625$$

$$u_3^{(1)} = \frac{1}{4}[u_2^{(1)} + 14 + u_6^{(0)} + 9] = \frac{1}{4}[4.90625 + 14 + 8.0625 + 9] = 8.9921875$$

$$u_4^{(1)} = \frac{1}{4}[0 + u_5^{(0)} + u_1^{(0)} + u_7^{(0)}] = \frac{1}{4}[0 + 4.5 + 1.625 + 2] = 2.03125$$

$$u_5^{(1)} = \frac{1}{4}[u_4^{(1)} + u_6^{(0)} + u_8^{(0)} + u_2^{(1)}]$$

$$= \frac{1}{4}[2.03125 + 8.0625 + 3.6875 + 4.90625] = 4.67185$$

$$u_6^{(1)} = \frac{1}{4}[u_5^{(1)} + 12 + u_9^{(0)} + u_3^{(1)}]$$

$$= \frac{1}{4}[4.671875 + 12 + 6.625 + 8.9921875] = 8.0772656$$

$$u_7^{(1)} = \frac{1}{4}[0 + u_8^{(0)} + 0.5 + u_4^{(1)}] = \frac{1}{4}[0 + 3.6875 + 0.5 + 2.03125] = 1.5546875$$

$$u_8^{(1)} = \frac{1}{4}[u_7^{(1)} + u_9^{(0)} + 2 + u_5^{(1)}] = \frac{1}{4}[1.5546875 + 6.625 + 2 + 4.671875]$$

$$= 3.7128906$$

$$u_9^{(1)} = \frac{1}{4}[u_8^{(1)} + 10 + 4.5 + u_6^{(1)}] = \frac{1}{4}[3.7128906 + 10 + 4.5 + 8.0722656]$$

$$= 6.5712891$$

Second iteration ($n = 1$)

$$u_1^{(2)} = \frac{1}{4}[0 + u_2^{(1)} + u_4^{(1)} + 1] = \frac{1}{4}[0 + 4.90625 + 2.03135 + 1] = 1.984375$$

$$u_2^{(2)} = \frac{1}{4}[u_1^{(2)} + u_2^{(1)} + u_5^{(1)} + 4] = \frac{1}{4}[1.984375 + 8.9921875 + 4.645875 + 4]$$

$$= 4.9121094$$

$$u_3^{(2)} = \frac{1}{4}[u_2^{(2)} + 14 + u_6^{(1)} + 9] = \frac{1}{4}[23 + 4.9121094 + 8.0722656] = 8.9960937$$

$$u_4^{(2)} = \frac{1}{4}[0 + u_5^{(1)} + u_7^{(1)} + u_1^{(2)}] = \frac{1}{4}[0 + 4.671875 + 1.5546875 + 1.984375]$$

$$= 2.0527344$$

$$u_5^{(2)} = \frac{1}{4}[u_4^{(2)} + u_6^{(1)} + u_8^{(1)} + u_2^{(2)}]$$

$$= \frac{1}{4}[2.0527344 + 8.0722656 + 3.7128906 + 4.9121094] = 4.6875$$

$$u_6^{(2)} = \frac{1}{4}[u_5^{(2)} + 12 + u_9^{(1)} + u_3^{(2)}] = \frac{1}{4}[4.6875 + 6.7512891 + 8.9960937]$$

$$= 8.063720$$

$$u_7^{(2)} = \frac{1}{4}[0 + u_8^{(1)} + 0.5 + u_4^{(2)}] = \frac{1}{4}[0.5 + 3.7128906 + 2.0527344] = 1.5664063$$

$$u_8^{(2)} = \frac{1}{4}[u_7^{(2)} + u_9^{(1)} + 2 + u_5^{(2)}] = \frac{1}{4}[1.5664063 + 6.571289 + 4.6875]$$

$$= 3.7062988$$

$$u_9^{(2)} = \frac{1}{4}[u_8^{(2)} + 10 + 4.5 + u_6^{(2)}] = \frac{1}{4}[3.7062988 + 14.5 + 8.063207]$$

$$= 2.5675049$$

Third iteration ($n = 2$)

$u_1^{(3)} = 1.991211$ $\quad u_2^{(3)} = 4.91181012$ $\quad u_3^{(3)} = 8.9956055$

$u_3^{(3)} = 2.0612793$ $\quad u_5^{(3)} = 4.6875$ $\quad u_6^{(3)} = 8.0626526$

$u_7^{(3)} = 1.5668945$ $\quad u_8^{(3)} = 3.7054749$ $\quad u_9^{(3)} = 6.56707319$

Hence $u_1 = 1.99$, $u_2 = 4.91$, $u_3 = 8.99$, $u_4 = 2.06$, $u_5 = 4.69$, $u_6 = 8.06$, $u_7 = 1.57$, $u_8 = 3.71$, $u_9 = 6.57$.

PROBLEMS

1. **Solve the equation $u_{xx} + u_{yy} = 0$ for the squares mesh with the boundary values as shown.**

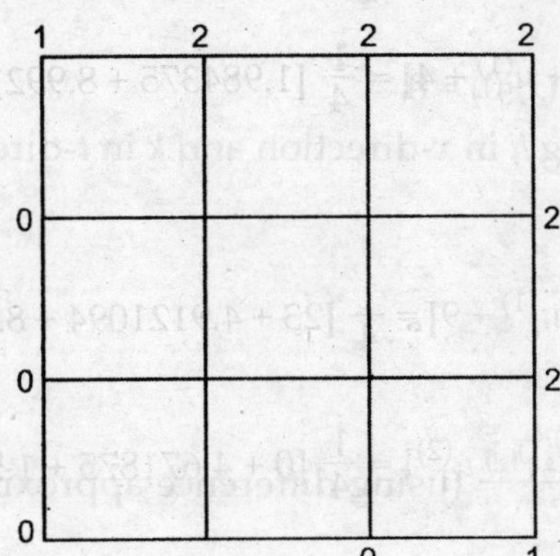

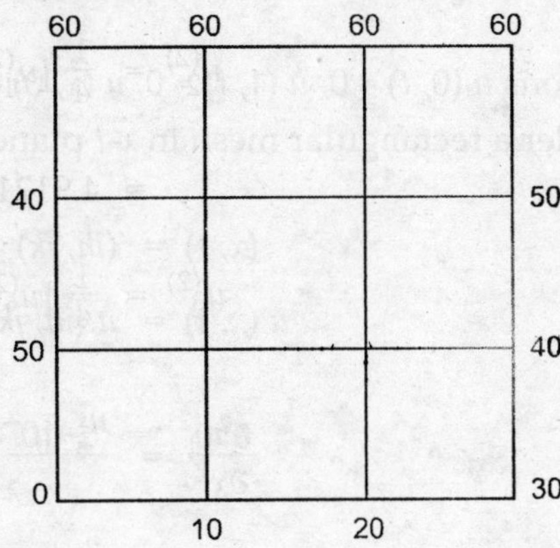

Fig. 19.10

[**Ans.** (*a*) $u_1 = 1.999$, $u_2 = 2.999$, $u_3 = 3.999$, $u_4 = 2.999$,
(*b*) $u_1 = 26.66$, $u_2 = 33.33$, $u_3 = 43.33$, $u_4 = 46.66$]

2. **Solve the equation $\nabla^2 u$ for the square mesh given below by (a) Jacobi method (b) Gauss-Seidal method.**

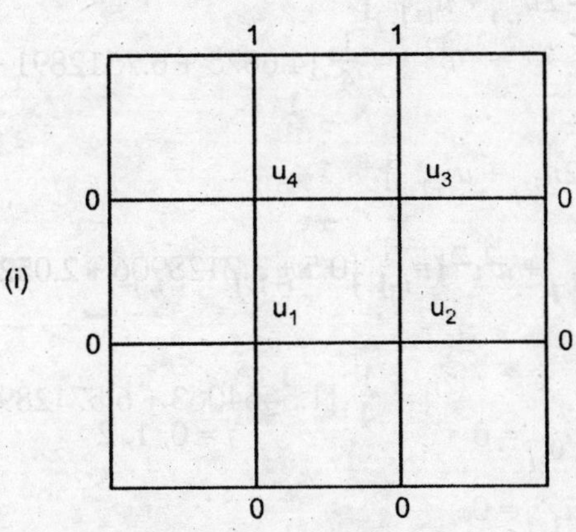

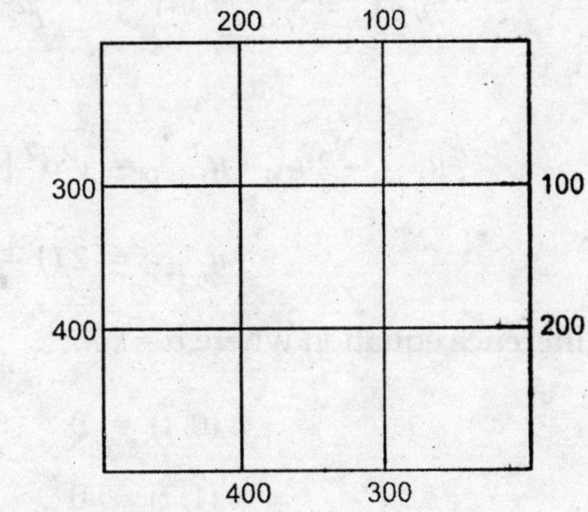

Fig. 19.11

[**Ans.** (*a*) $u_1 = 0.126$, $u_2 = 0.126$, $u_3 = 0.376$, $u_4 = 0.376$,
(*b*) $u_1 = u_4 = 250$, $u_2 = 175$, $u_3 = 3251$]

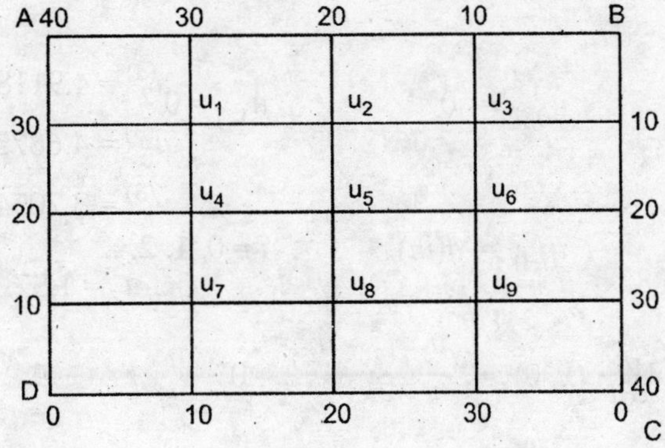

Fig. 19.12

[**Ans.** (*a*) $u_1 = u_9 = 25$, $u_3 = u_7 = 15$, $u_2 = 20$, $u_5 = 20$]

19.7 WAVE EQUATION

Consider
$$\frac{\partial^2 u}{\partial t^2} = c^2 \frac{\partial^2 u}{\partial x^2} \quad \text{...(19.10)}$$

Given the conditions $u(0, t) = 0$, $u(1, t) = 0$, $u(x, 0) = f(x)$ $u_t(x, 0) = 0$.

Solution : Consider a rectangular mesh in x-t plane spacing h in x-direction and k in t-direction. Denoting the mesh point

$$(x, t) = (ih, jk)$$

$$u(x, t) = u(ih, jk) = u_{i,j}$$

We have
$$\frac{\partial^2 u}{\partial x^2} = \frac{u_{i-1,j} - 2u_{i,j} + u_{i+1,j}}{h^2} \text{ (using difference approximation)}$$

$$\frac{\partial^2 u}{\partial t^2} = \frac{u_{i,j-1} - 2u_{i,j} + u_{i,j+1}}{k^2}$$

Substituting in Eq. (19.10), we have

$$u_{i,j-1} - 2u_{i,j} + u_{i,j+1} = c^2 \frac{k^2}{c^2} [u_{i-1,j} - 2u_{i,j} + u_{i+1,j}]$$

Put $\alpha = k/h$

$$u_{i,j-1} - 2u_{i,j} + u_{i,j+1} = c^2\alpha^2 [4_{i-1,j} - 2u_{i,j} + u_{i+1,j}]$$

$$u_{i,j+1} = 2(1 - c^2\alpha^2)\, u_{i,j} + a^2c^2 (u_{i-1,j} + u_{i+1,j}) - u_{i,j-1}$$

which is a difference equation where $\alpha = k/h$.

Now
$$u(0, t) = 0 \qquad u_{0,j} = 0 \qquad j = 0, 1, 2$$

$$u(1, t) = 0 \qquad u_{1,j} = 0$$

$$u(x, 0) = f(x) \qquad u_{i,0} = f(x) = f(ih) \qquad i = 0, 1, 2$$

$$u_t(x, 0) = 0 \qquad \frac{\partial}{\partial t} u(x, 0) = 0$$

The boundary conditions are

$$u_{0,j} = 0 \qquad u_{1,j} = 0, \qquad j = 0, 1, 2, \ldots$$

The initial conditions are

$$u_{i,0} = f(ih), \qquad i = 0, 1, 2, \ldots$$

Also
$$\frac{\partial u}{\partial t}(x, 0) = \frac{u_{i,j+1} - u_{i,j}}{k} = 0$$

or
$$u_{i,j+1} = u_{i,j}$$

$$u_{i,1} = u_{i,0} = f(ih)$$

Now the above equation gives the values of u on the first two rows $j = 0$, $j = 1$ which are the same.

Putting $j = 1$ in the difference equation

$$u_{i,\,j+1} = 2\,(1 - c^2\alpha^2)\,u_{i,\,j} + \alpha^2 c^2\,(u_{i+1,\,j} + u_{i-1,\,j}) - u_{i,\,j-1}$$

we have

$$u_{i,\,2} = 2\,(1 - \alpha^2 c^2)\,u_{i,\,1} + \alpha^2 c^2\,(u_{i+1,\,1} + u_{i-1,\,1}) - u_{i,\,0},\ i = 1, 2, 3, \ldots$$

We get the values in the third row. Proceeding in the same way

i/j	0	1	2	3	.....	1
0	0	0	0	0	.....	0
1	0	0	0	0	.....	0
2	0	0	0	0	.....	0
...	...	...	...	...	.....	...

Putting $j = 2$, we get

$$u_{i,\,2} = 2\,(1 - \alpha^2 c^2)\,u_{i,\,2} + a^2 c^2\,(u_{i+1,\,2} + u_{i-1,\,2}) - u_{i,\,1}$$

Putting $i = 1, 2, 3, \ldots$, we get the values of u in the fourth row. Hence proceeding in this way, we get all the rows filled.

This is the scheme for the solution of the wave equation.

***Example* 19.6.** Evaluate the pivotal values of the equation $u_{tt} = 16u_{xx}$ taking $\Delta x = 1$ up to $t = 1.25$, the boundary conditions are $u\,(0, t) = u\,(5, t) = 0$, $u_t\,(x, 0) = 0 = x^2(5 - x)$.

Solution : Here $c^2 = 16$, $h = 1$, $t = 1.25$, $jk = 1.25$.

The difference equation is

$$u_{i,\,j+1} = 2\,(1 - 16\alpha^2)\,u_{i,\,j} + 16\alpha^2\,(u_{i-1,\,j}) + u_{i+1,\,j} - u_{i,\,j-1}$$

We choose α in such a way that coefficient of $u_{i,\,j} = 0$

i.e.

$$1 - 16\alpha^2 = 0,\ \alpha^2 = 1/16$$

$$\alpha = \frac{1}{4} = \frac{k}{h} = \frac{k}{1} \quad \therefore \quad k = \frac{1}{4}$$

Now

$$jk = \frac{5}{4},\ k = \frac{1}{4}$$

$$j \times \frac{1}{4} = \frac{5}{4},\ j = 5$$

j goes from 0 to 5

j goes from 0 to 5

Now

$$\left.\begin{aligned} u(0, t) = 0,\ u_{0,\,j} = 0 \\ u(5, t) = 0,\ u_{5,\,j} = 0 \end{aligned}\right\} \text{ for all } j$$

The entries in the first and last columns are

i/j	0	1	2	3	4	5
0	0	4	12	18	16	0
1	0	4	12	18	16	0
2	0	8	10	10	2	0
3	0	6	6	– 6	– 6	0
4	0	– 2	– 10	– 10	– 8	0
5	0	– 16	– 18	– 12	– 4	0

Also $$u(x, 0) = x^2(5 - x)$$

$$u_{i,0} = i^2(5 - i)$$

$$u_{1,0} = 4$$

$$u_{2,0} = 12$$

$$u_3 = 18$$

$$u_{4,0} = 16$$

Also $$u_t(x, 0) = 0 \text{ or } \frac{u_{i,j+1} - u_{i,j}}{k}$$

$$u_{i,1} = u_{i,0}$$

The entries in first two rows are the same.

Again we have the difference equation putting $\alpha^2 = \frac{1}{16}$

$$u_{i,j+1} = u_{i-1,j} + u_{i+1,j} - u_{i,j-1}$$

Putting $j = 1$, we get

$$u_{i,2} = u_{i-1,j} + u_{i+1,j} - u_{i,0}$$

Taking $i = 1, 2, 3, 4, \ldots$ respectively, we obtain

$$i = 1\ u_{1,2} = u_{0,1} + u_{2,1} - u_{1,0} = 0 + 12 - 4 = 8$$

$$= 2\ u_{2,2} = u_{1,1} + u_{3,1} - u_{2,0} = 4 + 18 - 12 = 10$$

$$= 3\ u_{3,2} = u_{2,1} + u_{4,1} - u_{3,0} = 12 + 16 - 18 = 10$$

$$= 4\ u_{4,2} = u_{3,1} + u_{5,1} - u_{4,0} = 18 + 0 - 16 = 2$$

These are the entries in the third row.

Again putting j = 2, we get

$$u_{i,3} = u_{i-1,2} + u_{i+1;2} - u_{i,1}$$

Putting $i = 1, 2, 3, 4$ $$= 1\ u_{1,3} = u_{0,2} + u_{2,2} - u_{1,1} = 0 + 10 - 4 = 6$$

$$= \ 2 \ u_{2,3} = u_{1,2} + u_{3,2} - u_{2,1} = 8 + 10 - 12 = 6$$

$$= \ 3 \ u_{3,3} = u_{2,2} + u_{4,2} - u_{3,1} = 10 + 2 - 18 = -6$$

$$= \ 4 \ u_{4,3} = u_{3,2} + u_{5,2} - u_{4,1} = 10 + 0 - 16 = -6$$

Similarly putting $j = 3, 4, 5$; we get the remaining rows filled as in table.

***Example* 19.7.** Find the nodal values of the wave equation $\frac{\partial^2 u}{\partial t^2} = 16 \frac{\partial^2 u}{\partial x^2}$ given that $u(0, t) = u(5, t) = 0$, $u(x, 0) = x^2(5 - x)$, $u_t(x, 0) = 0$, taking $h = 1$ and up to one half of the period of vibration.

Solution : Here $h = 1$, $c^2 = 16$, $c = 4$.

Period of vibration is $\frac{2l}{c} = \frac{2 \times 5}{4} = \frac{5}{2}$

Half period of vibration $= \frac{5}{4}$

up to $t = \frac{5}{4} = 1.25$

Follow the same procedure as in Exampl 19.6.

***Example* 19.8.** Solve $u_{tt} = 4u_{xx}$ with the boundary conditions $u(0, t) = 0$, $u(4, t)$, $u_t(x, 0) = x(4 - x)$, taking $h = 1$ and $k = 0.5$.

Solution : Given equation is $u_{tt} = 4u_{xx}$.

Here $c^2 = 4$, $c = 2$.

Now the initial condition are $u(0, t) = 0$, i.e. $u_{0,j} = 0$

$$u(4, t) = u_{4,j} = 0 \text{ for all } j$$

Hence entries in the first and last column are zero.

Also $u(x, 0) = x(4 - x)$

$$u_{i,0} = i(4 - i)$$

$$u_{1,0} = 3$$

$$u_{2,0} = 4$$

$$u_{3,0} = 3$$

$$u_{4,0} = 0$$

Hence the entries in the first row are 0, 3, 4, 3, 0.

Also $u_t(x, 0) = 0 \Rightarrow u_{i,1} = u_{i,0}$

Entries in the first row and second row are the same.

j \ i	0	1	2	3	4
0	0	3	4	3	0
1	0	3	4	3	0
2	0	1	2	1	0

3	0	−1	−2	−1	0
4	0	−3	−4	−1	0

Also we have difference equation

$$u_{i,\, j+1} = u_{i+1,\, j} + u_{i-1,\, j} - u_{i,\, j-1}$$

Putting $j = 1$

$$u_{i,\, 2} = u_{i+1,\, 1} + u_{i-1,\, 1} - u_{i,\, 0}$$

Putting $i = 1, 2, 3$, we get

$u_{1,\, 2} = u_{2,\, 1} + u_{0,\, 1} - u_{1,\, 0} = 4 + 0 -$ [illegible] $= 1$

$$u_{2,\, 2} = u_{3,\, 1} + u_{1,\, 1} - u_{2,\, 0} = 3 + 3 - 4 = 2$$

$$u_{2,\, 2} = u_{4,\, 1} + u_{2,\, 1} - u_{3,\, 0} = 0 + 4 - 3 = 1$$

Putting $j = 2$ in difference equation, we get

$$u_{i,\, 3} = u_{i+1,\, 1} + u_{i-1,\, 2} - u_{i,\, 1}$$

Putting $i = 1, 2, 3$;

$$u_{1,\, 3} = u_{2,\, 2} + u_{0,\, 2} - u_{1,\, 1} = 2 + 0 - 3 = -1$$

$$u_{2,\, 3} = u_{3,\, 2} + u_{1,\, 2} - u_{2,\, 1} = 1 + 1 - 4 = -2$$

$$u_{3,\, 3} = u_{4,\, 2} + u_{2,\, 2} - u_{3,\, 1} = 0 + 2 - 3 = 1$$

Putting $j = 3$ in the difference equation, we get fourth row filled.

Example 19.9. Solve $u_{tt} = u_{xx}$ up to $t = 0.5$ with spacing of 0.1, given $u(0, t) = 0 = u(1, t)$, $u_t(x, 0) = 0$, $u(x, 0) = 10 + x(1 - x)$.

Solution : Here $u_{tt} = u_{xx}$, $c^2 = 1$, $t = 0.5$, $k = 0.1$, j goes from 0 to 5.

We have the difference equation

$$u_{i,\, j+1} = 2(1 - \alpha^2 c^2)\, u_{i,\, j} + \alpha^2 c^2 (u_{i,\, -1,\, j} + u_{i+1,\, j}) - u_{i,\, j-1}$$

We choose t in such a way that the coefficient of $u_{i,\, j} = 0$

$$\therefore \quad 1 - \alpha^2 c^2 = 0,\ 1 - \alpha^2 = 0,\ \alpha^2 = 1,\ \alpha = 1$$

But

$$\alpha = k/\text{h}$$

$$1 = 0.1/h,\ h = 0.1,\ i \text{ goes from 0 to 10}$$

	x	0	0.1	0.2	0.3	0.4	0.5	0.6	0.7	0.8	0.1	1
t	i/j	0	1	2	3	4	5	6	7	8	9	10
0	0	0	10.09	10.16	10.21	10.24	10.25	10.24	10.21	10.16	10.09	0
1	1	0	10.09	10.16	10.21	10.24	10.25	10.25	10.21	10.16	16.09	0
2	2	0	0.07	10.14	10.19	10.22	10.25	10.22	10.19	10.14	0.07	0
3	3	0	0.05	0.1	10.15	10.18	10.19	10.18	10.15	0.1	0.05	0
4	4	0	0.06	0.06	0.09	10.12	10.13	10.12	10.[illegible]	0.06	0.03	0
5	5	0	0.01	0.02	0.03	0.04	10.05	0.04	0.03	0.02	0.01	0

Initial conditions are entries

$$u(0, t) = 0 \quad u_{0,j} = 0$$

$$u(1, t) = 0 \quad u_{1,j} = 0 \text{ for all } j$$

∴ in the first and last columns are zero

$$[u(1, t) = u(10 \times 0.1, jk) = u(10h, jk) = u(10h, jk) = u_{10, j}]$$

Boundary conditions are

$$u(x, 0) = 10 + x(1 - x)$$

$$u_{i,0} = 10 + i/10\,(1 - i/10)$$

$$u_{1,0} = 10 + \frac{1}{10}\left(1 - \frac{1}{10}\right) = 10.09$$

$$u_{2,0} = 10 + \frac{2}{10}\left(1 - \frac{2}{10}\right) = 10.16$$

$$u_{3,0} = 10 + \frac{3}{10}\left(1 - \frac{3}{10}\right) = 10.21$$

$$u_{4,0} = 10 + \frac{4}{10}\left(1 - \frac{4}{10}\right) = 10.24$$

$$u_{5,0} = 10 + \frac{5}{10}\left(1 - \frac{5}{10}\right) = 10.25$$

Similarly $u_{6,0} = 10.24$

$$u_{7,0} = 10.21$$

$$u_{8,0} = 10.16$$

$$u_{9,0} = 10.09$$

Also $u_t(x, 0) = 0 \Rightarrow u_{i,0} = u_{i,1}$

The entries in the first two rows are the same.

Now, the difference equation is

$$u_{i,j+1} = u_{i-1,j} + u_{i+1,j} - u_{i,j-1}$$

Putting $j = 1$

$$u_{i,2} = u_{i-1,1} + u_{i+1,1} - u_{i,0}$$

Taking $i = 1, 2, 3, 4, 5, 6, 7, 8, 9$

$$u_{1,2} = u_{0,1} + u_{2,1} + u_{1,0} = 0 + 10.16 - 10.07 = 0.07$$

$$u_{2,2} = u_{1,1} + u_{3,1} + u_{2,0} = 10.09 + 10.21 - 10.16 = 10.14$$

$$u_{3,2} = u_{2,1} + u_{4,1} + u_{3,0} = 10.16 + 10.24 - 10.21 = 10.19$$

$$u_{4,2} = u_{3,1} + u_{5,1} + u_{4,0} = 10.21 + 10.25 - 10.24 = 10.22$$

$$u_{5,2} = u_{4,1} + u_{6,1} + u_{5,0} = 10.24 + 10.24 - 10.25 = 10.23$$

Similarly we get $u_{6,2} = 10.22$, $u_{7,2} = 10.19$, $u_{8,2} = 10.14$, $u_{9,2} = 0.07$.

Put $j = 2, 3, 4$, we get all the rows filled.

PROBLEMS

1. Solve the equation $\frac{\partial^2 u}{\partial t^2} = \frac{\partial^2 u}{\partial x^2}$ given the boundary conditions $u(x, 0) =$ for $0 \leq x \leq 1$,

$$u(0, t) = u(1, t) = 0 \text{ for } t \geq 0$$

Use the explicit scheme to calculate u for

$$x = 0(0.1), 1 \text{ and } t = 0(0.1)0.5.$$

Hint:

j \ i	0	0.1	0.2	0.3	0.4	0.5
0.1	0	0.37	0.07	0.096	0.113	0.119
0.2	0	0.31	0.059	0.082	0.096	0.101
0.3	0	0.23	0.043	0.059	0.07	0.073
0.4	0	0.12	0.023	0.031	0.037	0.039
0.5	0	0	0	0	0	0

2. Solve the hyperbolic partial differential equation (vibration of strings) for one half period of oscillation taking h = 1, $u_{tt} = 25u_{xx}$, $u(0, t) = u(5, t) = 0$

$$u_i(x, 0) = 0, \; u(x, 0) = \begin{cases} 2x \text{ for } 0 \leq x \leq 2.5 \\ 10 - 2x \text{ for } 2.5 \leq x \leq 5 \end{cases}$$

Hint:

j \ i	0	1	2	3	4	5
0	0	2	4	4	2	0
1	0	2	3	3	2	0
2	0	1	1	1	1	0
3	0	– 1	– 1	– 1	– 1	0
4	0	– 2	– 3	– 3	– 2	0
5	0	– 2	– 4	– 4	– 2	0

3. Approximate the solution to the equation $u_{tt} = 16u_{xx}$ taking $\Delta x = 1$ up to $t = 1.25$. The boundary conditions are $u(0, t) = u(5, t) = 0$, $u_t(x, 0) = 0$ and $u(x, 0) = x^2(5 - x)$.

Hint:

j \ i	0	1	2	3	4	5
0	0	4	12	18	16	0
1	0	4	12	18	16	0
2	0	8	10	10	2	0

3	0	6	6	− 6	− 6	0
4	0	− 2	− 10	− 10	− 8	0
5	0	− 16	− 18	− 12	− 4	0

19.8 SOLUTION OF ONE-DIMENSIONAL HEAT EQUATION

Consider
$$\frac{\partial u}{\partial t} = c^2 \frac{\partial^2 u}{\partial x^2} \qquad ...(19.11)$$

where $c^2 = k/s\rho$ is the diffusivity of the substance (cm^2/sec).

19.9.1 *Bender-Schmidt Method*

Consider a mesh (rectangular) in x-t plane with spacing h along x-direction and k along t-direction, denoting mesh point

$$(x, t) = (ih, jk)$$

$\therefore$
$$u(x, t) = u(ih, jk) = u_{i, j}$$

We have
$$\frac{\partial u}{\partial t} = \frac{u_{i, j+1} - u_{i, j}}{h}$$

$$\frac{\partial^2 u}{\partial x^2} = \frac{u_{i+1, j} - 2u_{i, j} + u_{i-1, j}}{h^2}$$

i − 1, j + 1   i, j + 1   i + 1, j + 1 (j + 1)th level

i − 1, j   i, j   i + 1, j   (j)th level

0   i, j −1

Fig. 19.13

Substituting in Eq. (9.11), we get

$$u_{i, j+1} = \frac{c^2 k}{h^2}[u_{i-1, j} - 2u_{i, j} + u_{i+1, j}] + u_{i, j}$$

$$= \alpha u_{i-1, j} + (1 - 2\alpha)\, u_{i, j} + \alpha u_{i+1, j} \qquad ...(19.12)$$

where
$$\alpha = \frac{c^2 k}{h^2}$$

This formula enables us to determine the value of u at the $(i, j + 1)$th mesh point in terms of known values at the point x_{i-1}, x_i, x_{i+1} at the instant t_j. It is a relation between the values at the two levels $j + 1, j$. Therefore, it is called a two-level function.

Equation (9.12) is called Schmidt explicit formula which is valid only for $0 < \alpha < \frac{1}{2}$.

If
$$\alpha = \frac{1}{2}$$

$$u_{i, j+1} = \frac{1}{2}\,[u_{i-1, j} + u_{i+1, j}] \qquad \text{...(19.13)}$$

This relation is called Bender-Schmidt recurrence relation.

19.9.2 *Crank-Nicholson Method*

According to this method $\frac{\partial^2 u}{\partial x^2}$ is replaced by the average of its finite difference approximation on the jth and $(j + 1)$th level of rows. Thus

$$\frac{\partial^2 u}{\partial x^2} = \frac{1}{2}\left[\frac{u_{i-1, j} - 2u_{i, j} + u_{i+1, j}}{h^2} + \frac{u_{i-1, j+1} - 2u_{i, j+1} + u_{i-1, j+1}}{h^2}\right]$$

$$\frac{u_{i, j+1} - u_{i, j}}{h} = c^2\,\frac{1}{2}\left[\frac{u_{i-1, j} - 2u_{i, j} + u_{i+1, j}}{h^2} + \frac{u_{i-1, j+1} - 2u_{i, j+1} + u_{i-1, j+1}}{h^2}\right]$$

or $$-\alpha u_{i-1, j+1} + (2 + 2\alpha)\, u_{i, j+1} - \alpha u_{i+1, j+1} = \alpha u_{i-1, j} + (2 - 2\alpha)\, u_{i, j} + \alpha u_{i+1, j} \qquad \text{...(19.14)}$$

where $$\alpha = \frac{c^2 k}{h^2}$$

i – 1, j + 1 i, j + 1 i + 1, j + 1 (j + 1)th level
i – 1, j i – 1, j (j)the level
0 i, j –1 i – 1, j

Fig. 19.14

Clearly the left side of Eq. (19.14) contains three unknown values of u at the $(j + 1)$th level while all the three values on the right are known at the jth level.

Thus Eq. (19.14) is a two-level implict and is known as Crank-Nicholson formula.

If there are n internal mesh point on each row then the relation in Eq. (19.14) gives n simultaneous equations for the n unknown values in terms of known boundary values. These equations can be solved to obtain the values at these mesh points.

19.9.3 *Du Fort and Frankel Method*

If we replace derivatives in Eq. (19.11) by the central difference approxmations

$$\frac{\partial u}{\partial t} = \frac{u_{i, j+1} - u_{i, j-1}}{2k}$$

and $$\frac{\partial^2 u}{\partial x^2} = \frac{u_{i-1, j} - 2u_{i, j} + u_{i+1, j}}{h^2}$$

we obtain $$u_{i, j+1} - u_{i, j-1} = \frac{2kc^2}{h^2}\,[u_{i-1, j} - 2u_{i, j} + u_{i+1, j}]$$

$$u_{i, j+1} = u_{i, j-1} + 2\alpha\,[u_{i-1, j} + 2u_{i, j} + u_{i+1, j}] \qquad \text{...(19.15)}$$

where $\alpha = \dfrac{c^2 k}{h^2}$. This difference equation is called the *Richardson scheme* which is a three-level method.

If we replace $u_{i,\,j}$ by means of the values $u_{i,\,j-1}$ and $u_{i,\,j+1}$, i.e.

$$u_{i,\,j} = \frac{1}{2}\,[u_{i,\,j-1} + u_{i,\,j+1})\text{ in Eq. (19.15)}$$

we get

$$u_{i,\,j+1} = u_{i,\,j-1} + 2\alpha\,[u_{i,\,j-1} - (u_{i,\,j-1} + u_{i,\,j+1}) + u_{i+1,\,j}]$$

On simplification, we have

$$u_{i,\,j+1} = \frac{1-2\alpha}{1+2\alpha}\,u_{i,\,j-1} + \frac{2\alpha}{1+2\alpha}\,(u_{i-1,\,j} + u_{i+1,\,j})$$

This difference scheme is called Du Fort-Frankel scheme which is a three level explicit method, its computation model is given below.

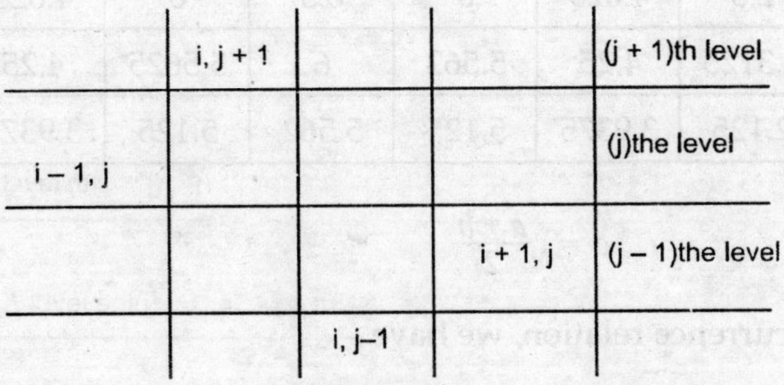

i, j + 1 (j + 1)th level

(j)the level

i – 1, j

i + 1, j (j – 1)the level

i, j–1

Fig. 19.15

Example **19.10.** Find the value of $u\,(x,\,t)$ satisfying the parabolic equation $\dfrac{\partial u}{\partial t} = 4\,\dfrac{\partial^2 u}{\partial x^2}$ and the boundary condition $u\,(0,\,t) = 0 = u\,(8,\,t)$ and $u\,(x,\,0) = 4x - \dfrac{1}{2}\,x^2$ at the points $x = i$, $i = 0, 1, 2, \ldots, 7$ and $t = \dfrac{1}{8}\,j$; $j = 0, 1, 2, \ldots, 5$.

Solution : The equation is $\dfrac{\partial u}{\partial t} = 4\,\dfrac{\partial^2 u}{\partial x^2}$ with boundary conditions $u\,(0,\,t) = 0$, $u\,(8,\,t) = 0$

$$u\,(x,\,0) = 4x - \frac{1}{2}\,x^2 \text{ at the points } x = i,\ i = 0, 1, 2, 3, \ldots, 7$$

and

$$t = \frac{1}{8}\,j;\ j = 0, 1, 2, 3, \ldots, 5$$

Here

$$c^2 = 4,\ t = kj = \frac{1}{8}\,j;\ k = \frac{1}{8}$$

$$x = ih = i,\ h = 1 \text{ also } \alpha = \frac{kc^2}{\alpha^2} = \frac{1}{2}$$

We have Bendra-Schmidt recurrence relation

$$u_{i,\,j+1} = \frac{1}{2}\,[u_{i-1,\,j} + u_{i+1,\,j}]$$

Now since $u\,(0,\,t) = 0$, $u_{0,\,j} = 0$, $u\,(8,\,t) = 0$, $u_{8,\,j} = 0$ for all values of j.

The entries in the first and last column are zero.

Also
$$u(x, 0) = 4x - \frac{1}{2}x^2$$
$$u_{i,0} = 4i - \frac{1}{2}i^2$$

Putting $i = 1, 2, 3, 4, 5, 6, 7$, $u_{1,0} = 3.5$, $u_{2,0} = 6$, $u_{3,0} = 7.5$, $u_{4,0} = 8$, $u_{5,0} = 7.5$, $u_{6,0} = 6$, $u_{7,0} = 3.5$.

j \ i	0	1	2	3	4	5	6	7	8
0	0	3.5	6	7.5	8	7.5	6	3.5	0
1	0	3	5.5	7	7.5	7	7.5	3	0
2	0	2.75	5	6.5	7	6.5	5	2.75	0
3	0	2.5	4.625	6	6.5	6	4.625	2.5	0
4	0	2.3125	4.25	5.562	6	5.5625	4.25	2.3125	0
5	0	2.125	3.9375	5.125	5.562	5.125	3.9375	2.125	0

This means
$$c = \frac{a+b}{2}$$

Now by Bendre-Schmidt recurrence relation, we have
$$u_{i,j+1} = \frac{1}{2}[u_{i-1,j} + u_{i+1,j}]$$

Put $j = 0$
$$u_{i,1} = \frac{1}{2}[u_{i-1,0} + u_{i+1,0}]$$

Putting $i = 1, 2, 3, 4, 5, 6, 7$
$$u_{1,1} = \frac{1}{2}[u_{0,0} + u_{2,0}] = \frac{1}{2}[0 + 6] = 3$$
$$u_{2,1} = \frac{1}{2}[u_{1,0} + u_{3,0}] = \frac{1}{2}[3.5 + 7.5] = 5.5$$
$$u_{3,1} = \frac{1}{2}[u_{2,0} + u_{4,0}] = \frac{1}{2}[6 + 8] = 7$$
$$u_{4,1} = \frac{1}{2}[u_{3,0} + u_{5,0}] = \frac{1}{2}[7.5 + 7.5] = 7.5$$
$$u_{5,1} = 7,\ u_{6,1} = 5.5,\ u_{7,1} = 3$$

There are entries in the second row.

Put $j = 1$, we have
$$u_{i,2} = \frac{1}{2}[u_{i-1,1} + u_{i+1,1}]$$

Putting $i = 1, 2, 3, 4, 5, 6, 7$, we get $u_{1,2} = 2.275$, $u_{2,2} = 5$, $u_{3,2} = 6.5$, $u_{4,2} = 7$, $u_{5,2} = 5$, $u_{6,2} = 5$, $u_{7,2} = 2.75$.

Put $j = 2$
$$u_{i,3} = \frac{1}{2}[u_{i-1,2} + u_{i+1,2}]$$

$$u_{1,3} = \frac{1}{2}[0+5] = 2.5$$

$$u_{2,3} = \frac{1}{2}[2.75+6.5] = 4.625$$

$$u_{3,3} = \frac{1}{2}[5+7] = 6$$

$$u_{4,3} = \frac{1}{2}[6.5+6.5] = 6.5$$

$$u_{5,3} = \frac{1}{2}[7+5] = 6$$

$$u_{6,3} = \frac{1}{2}[6.5+2.75] = 4.625$$

$$u_{7,3} = \frac{1}{2}[5+0] = 2.5$$

Similarly we can obtain entries in fourth and fifth rows.

***Example* 19.11.** Solve the equation $\frac{\partial u}{\partial t} = \frac{\partial^2 u}{\partial x^2}$ subject to the condition $u(x, 0) = \sin x\pi$, $0 \le x \le 1$: $u(0, t) = u(1, t) = 0$ using (1) Schmidt method (2) Crank-Nicholson method (3) Du fort-Frankel method. Carry out computation for two levels taking $h = 1/3$, $k = 1/36$.

Solution : Here $c^2 = 1$, $h = \frac{1}{3}$, $k = \frac{1}{36}$, $\alpha = c^2 \frac{k}{h^2} = \frac{1 \times \frac{1}{36}}{\frac{1}{9}} = \frac{1}{4}$.

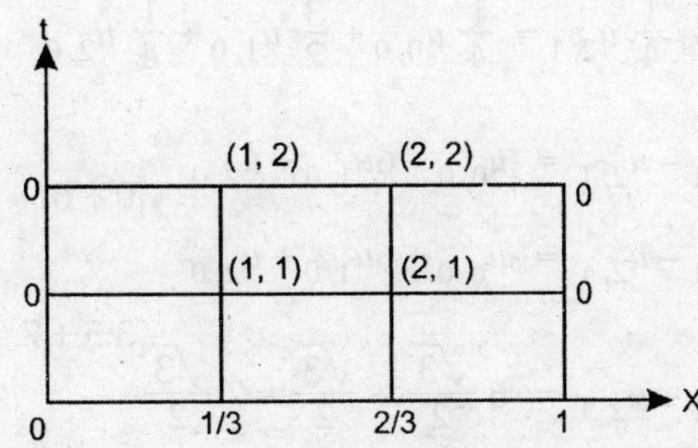

Fig. 19.16

$$u(x, 0) = \sin \pi x,\ u(ih, 0) = \sin \pi i/3$$

$$u_{1,0} = \sin \frac{\pi}{3} = \sqrt{\frac{3}{2}},\ u_{i,0} = \sin \pi \frac{i}{3}$$

$$u_{2,0} = \sin \frac{2\pi}{3} = \sqrt{\frac{3}{2}},\ u_{3,0} = 0$$

and all the other boundary values are zero.

1. Schmidt formula

$$u_{i,j+1} = \alpha u_{i-1} + (1-2\alpha)\, u_{i,j} + 2u_{i+1,j}$$

$$= \frac{1}{4}(u_{i-1,\,j} + 2u_{i,\,j} + u_{i+1,\,j}) \qquad \text{...}(i)$$

Put $j = 0$ $\quad u_{i,\,1} = \frac{1}{4}(u_{i-1,\,0} + 2u_{i,\,0} + u_{i+1,\,0})$

Putting $i = 1, 2$ $\quad u_{1,\,1} = \frac{1}{4}(u_{0,\,0} + 2u_{1,\,0} + u_{2,\,0}) = 0.65$

$$u_{2,\,1} = \frac{1}{4}(u_{1,\,0} + 2u_{2,\,0} + u_{3,\,0}) = 0.65$$

Put $j = 1$ in Eq. (*i*)

$$u_{i,\,2} = \frac{1}{4}(u_{i-1,\,1} + 2u_{i,\,0} + u_{i+1,\,1})$$

Putting $i = 1, 2$ $\quad u_{1,\,2} = \frac{1}{4}(u_{0,\,1} + 2u_{1,\,1} + u_{2,\,1}) = 0.49$

$$u_{2,\,2} = \frac{1}{4}(u_{1,\,1} + 2u_{2,\,1} + u_{3,\,1}) = 0.49$$

2. Crank-Nicholson formula

$$-\frac{1}{4}u_{i-1,\,j+1} + \frac{5}{2}u_{i,\,j+1} - \frac{1}{4}u_{i+1,\,j+1} = \frac{1}{4}u_{i-1,\,j} + \frac{3}{2}u_{i,\,j} + \frac{1}{4}u_{i+1,\,j}$$

Put $j = 0$ $\quad -\frac{1}{4}u_{i-1,\,1} + \frac{5}{2}u_{i,\,1} - \frac{1}{4}u_{i+1,\,1} = \frac{1}{4}u_{i-1,\,0} + \frac{3}{2}u_{i,\,0} + \frac{1}{4}u_{i+1,\,0}$

Putting $i = 1, 2 - \frac{1}{4}u_{0,\,1} + \frac{5}{2}u_{1,\,1} - \frac{1}{4}u_{2,\,1} = \frac{1}{4}u_{0,\,0} + \frac{3}{2}u_{1,\,0} + \frac{1}{4}u_{2,\,0}$

$$-u_{0,\,1} + 10u_{1,\,1} - u_{2,\,1} = u_{0,\,0} + 6u_{1,\,0} + u_{2,\,0}$$

$$10u_{1,\,1} - u_{2,\,1} = u_{0,\,0} + 6u_{1,\,0} + u_{2,\,0}$$

$$10u_{1,\,1} - u_{2,\,1} = 6\frac{\sqrt{3}}{2} + \frac{\sqrt{3}}{2} = 7\frac{\sqrt{3}}{2} \qquad \text{...}(ii)$$

$$-\frac{1}{4}u_{1,\,1} + \frac{5}{2}u_{2,\,1} - \frac{1}{4}u_{3,\,1} = \frac{1}{4}u_{1,\,0} + \frac{3}{2}u_{2,\,0} + \frac{1}{4}u_{1,\,0}$$

$$-u_{1,\,1} + 10u_{2,\,1} = 7\frac{\sqrt{3}}{2} \qquad (iii)$$

$$u_{1,\,1} = u_{2,\,1} = 0.67$$

Solving Eqs (*ii*) and (*iii*), we get

Put $j = 1$ $\quad -\frac{1}{4}u_{i-1,\,2} + \frac{5}{2}u_{i,\,2} - \frac{1}{4}u_{i+1,\,2} = \frac{1}{4}u_{i-1,\,2} + \frac{3}{2}u_{i,\,1} + \frac{1}{4}u_{i-1,\,1}$

Putting $i = 1, 2$ we get

$$-u_{0,2} + 10u_{1,2} - u_{2,2} = u_{0,1} + 6u_{1,1} + u_{2,1}$$

$$10u_{1,2} - u_{2,2} = 4.69 \qquad ...(iv)$$

$$-u_{1,2} + 10u_{2,2} - u_{3,2} = u_{1,1} + 6u_{2,1} + u_{3,1}$$

$$-u_{1,2} + 10u_{2,2} = 4.69$$

Solving Eqs (*iii*) and (*iv*), we get

$$u_{1,2} = u_{2,2} = 0.52$$

3. Du Fort-Frankel formula

$$u_{i,j+1} = \frac{1-2\alpha}{1+2\alpha} u_{i,j-1} + \frac{2\alpha}{1+2\alpha} (u_{i-1,j} + u_{i+1,j})$$

Now $\alpha = \frac{1}{4}$

$\therefore$ $u_{i,j+1} = \frac{1}{3} [u_{i,j-1} + u_{i-1,j} + u_{i+1,j}], j \neq 0$

Now put $j = 1$ $u_{i,j+1} = \frac{1}{3} [u_{i,0} + u_{i-1,1} + u_{i+1,1}]$

Putting $i = 1, 2$ $i = 1, u_{1,2} = \frac{1}{3} [u_{1,0} + u_{0,1} + u_{2,1}] = \frac{1}{3} \left[\frac{\sqrt{3}}{2} + 0 + u_{2,1}\right]$

$$= 2, u_{2,2} = \frac{1}{3} [u_{2,0} + u_{1,1} + u_{3,1}]$$

$$= \frac{1}{3} \left[\frac{\sqrt{3}}{2} + u_{1,1} + 0\right] = \frac{1}{3} \left[\frac{\sqrt{3}}{2} + 0 + u_{2,1}\right]$$

Now $u_{2,1}$ and $u_{1,2}$ will be obtained by Schmidt method as $u_{2,1} = u_{1,1} = 0.65$

$$u_{1,2} = \frac{1}{3} \left[\frac{\sqrt{3}}{2} + 0.65\right] = 0.5$$

$$u_{2,2} = \frac{1}{3} \left[\frac{\sqrt{3}}{2} + 0.65\right] = 0.5$$

***Example* 19.12.** Find the solution of the parabolic equation $u_{xx} = 2u_t$

where $u(0, t) = u(4, t) = 0$

$$u(x, 0) = x(4 - x), h = 1.$$

Find the values up to $t = 5$.

Solution : Here $c^2 = \frac{1}{2}, t = 5, jk = 5$.

Let us choose k such that $\alpha = 1/2$

$$\alpha = \frac{c^2 k}{h^2} = \frac{1}{2}k, \frac{1}{2} = \frac{1}{2}k, k = 1, h = 1$$

j goes from 0 to 5

i goes from 0 to 4

Also $u(0, t) = 0, u_{0, j} = 0$

$u(4, t) = 0, u_{4, j} = 0$ for all j

$u(x, 0) = x(4 - x)\ u_{i, 0} = i(4 - i)$

Since $u_{0, j} = 0, u_{4, j} = 0$ for all j

$\therefore$ all the entries in the first and last columns are zero as shown.

j \ i	0	1	2	3	4
0	0	3	4	3	0
1	0	2	3	2	0
2	0	1.5	2	1.5	0
3	0	1	1.5	1	0
4	0	0.75	1	0.75	0
5	0	0.5	0.75	0.5	0

Also
$$u_{i, 0} = i(4 - i)$$
$$u_{1, 2} = 3$$
$$u_{2, 2} = 4$$
$$u_{3, 2} = 3$$
$$u_{4, 2} = 0$$

Therefore, the entries in the first row are 3, 4, 3, 0.

Since $\alpha = \frac{1}{2}$, Bendre-Schmidt recurrence relation can be applied, i.e.

$$u_{i, j+1} = \frac{1}{2}[u_{i-1, j} + u_{i+1, j}]$$

Put $j = 0$
$$u_{i, 1} = \frac{1}{2}[u_{i-1, 0} + u_{i+1, 0}]$$

Putting $i = 1, 2, 3$
$$u_{1, 1} = \frac{1}{2}[u_{0, 0} + u_{2, 0}] = \frac{1}{2}[0 + 4]$$
$$u_{2, 1} = [u_{1, 0} + u_{3, 0}] = [3 + 3] = 3$$
$$u_{3, 1} = [u_{2, 0} + u_{4, 0}] = [4 + 0] = 2$$

Put $j = 1$
$$u_{i, 2} = \frac{1}{2}[u_{i-1, 1} + u_{i+1, 1}]$$

Putting $i = 1, 2, 3$

$$u_{1,2} = \frac{1}{2}[u_{0,1} + u_{2,1}] = \frac{1}{2}[0 + 3] = 1.5$$

$$u_{2,2} = \frac{1}{2}[u_{1,1} + u_{3,1}] = \frac{1}{2}[2 + 2] = 2$$

$$u_{3,2} = \frac{1}{2}[u_{2,1} + u_{4,1}] = [3 + 0] = 1.5$$

Similarly, we can find other values and put in the table as shown.

Put $j = 2$

$$u_{1,3} = \frac{1}{2}[0 + 2] = 1$$

$$u_{2,3} = \frac{1}{2}[1.5 + 1.5] = 1.5$$

$$u_{3,3} = \frac{1}{2}[2 + 0] = 1$$

Put $j = 3$

$$u_{1,4} = \frac{1}{2}[0 + 1.5] = 0.75$$

$$u_{2,4} = \frac{1}{2}[1 + 1] = 1$$

$$u_{3,4} = \frac{1}{2}[0 + 1.5] = 0.75 \text{ etc.}$$

***Example* 19.13.** Solve the equation $\frac{\partial^2 u}{\partial x^2} = \frac{\partial u}{\partial t}$ with the conditions $u(0, t) = u(1, t) = 0$, $u(x, 0) = x(1 - x)$. Assume $h = 0.1$. Tabulate u for $t = k, 2k, 3k$ choosing an appropriate value of k.

Solution :

x	0	1	2	3	4	5	6	7	8	9	10
j \ i	0	1	2	3	4	5	6	7	8	9	10
0	0	0.9	0.16	0.21	0.24	0.25	0.24	0.21	0.16	0.09	0
1	0	0.08	0.15	0.20	0.23	0.24	0.23	0.20	0.15	0.08	0
2	0	0.075	0.14	0.19	0.22	0.23	0.22	0.19	0.14	0.075	0
3	0	0.07	0.1325	0.18	0.21	0.22	0.21	0.18	0.1325	0.07	0

Here

$$c^2 = 1, h = 0.1$$

$$t = k, 2k, 3k$$

$$jk = k, 2k, 3k$$

$$j = 1, 2, 3$$

$$x = ih = 0.1 \times i$$

i.e.

$$x = 0, 0.1, 0.2, 0.3, 0.4, \ldots, 1$$

and j goes from 0 to 3.

Also $$u(0, t) = 0, \text{ i.e. } u_{0,j} = 0$$

$$u(1, t) = 0,\ u_{10,j} = 0 \text{ for all } j$$

$\therefore$ the entries in the first and last column are zero.

Since $u(x, 0) = x(1 - x)$, $u(ih, 0) = ih(i - ih)$

$$u_{i,0} = \frac{i}{10}\left(1 - \frac{i}{10}\right)$$

$$u_{1,0} = \frac{1}{10}\left(1 - \frac{1}{10}\right) = \frac{1}{10} \times \frac{9}{10} = 0.09$$

$$u_{2,0} = \frac{2}{10}\left(1 - \frac{2}{10}\right) = \frac{2}{10} \times \frac{8}{10} = 0.16$$

$$u_{3,0} = \frac{3}{10}\left(1 - \frac{3}{10}\right) = \frac{1}{10} \times \frac{7}{10} = 0.21$$

$$u_{4,0} = \frac{4}{10}\left(1 - \frac{4}{10}\right) = \frac{1}{10} \times \frac{6}{10} = 0.24$$

$$u_{5,0} = \frac{5}{10}\left(1 - \frac{5}{10}\right) = \frac{1}{10} \times \frac{5}{10} = 0.25$$

$$u_{6,0} = \frac{6}{10}\left(1 - \frac{6}{10}\right) = 0.24$$

$$u_{7,0} = 0.21,\ u_{8,0} = 0.16,\ u_{9,0} = 0.09$$

Let us choose $$\alpha = \frac{1}{2}$$

$$\frac{1}{2} = 100\,k,\ k = \frac{1}{200} = 0.005$$

We choose $$k = 0.005$$

Since $\alpha = \frac{1}{2}$, we apply Bendre-Schmidt recurrence relation,

i.e. $$u_{i,j+1} = \frac{1}{2}\left[u_{i-1,j} + u_{i+1,j}\right]$$

Put $j = 0$ $$u_{i,1} = \frac{1}{2}\left[u_{i-1,0} + u_{i+1,0}\right]$$

Put $i = 1, 2, 3, 4, 5$ $$u_{1,1} = \frac{1}{2}\left[u_{0,0} + u_{2,0}\right] = \frac{1}{2}\left[0 + 0.16\right] = 0.08$$

$$u_{2,1} = \frac{1}{2}\left[u_{1,0} + u_{3,0}\right] = \frac{1}{2}\left[0.09 + 0.21\right] = 0.15$$

$$u_{3,1} = \frac{1}{2}\left[u_{2,0} + u_{4,0}\right] = \frac{1}{2}\left[0.16 + 0.24\right] = 0.20$$

$$u_{4,1} = \frac{1}{2}[u_{3,0} + u_{5,0}] = \frac{1}{2}[0.21 + 0.25] = 0.23$$

$$u_{5,1} = \frac{1}{2}[u_{4,0} + u_{6,0}] = \frac{1}{2}[0.24 + 0.24] = 0.24$$

Going in the reverse order, we get

$$i = 6 \quad u_{6,1} = 0.23$$

$$i = 7 \quad u_{7,1} = 0.20$$

$$i = 8 \quad u_{8,1} = 0.15$$

$$i = 9 \quad u_{9,1} = 0.08$$

Putting $j = 1$ $\qquad u_{i,2} = \frac{1}{2}[u_{i-1,1} + u_{i+1,1}]$

Put $i = 1$ $\qquad u_{1,2} = \frac{1}{2}[u_{0,1} + u_{2,1}] = \frac{1}{2}[0 + 0.15] = 0.075$

$i = 2$ $\qquad u_{2,2} = \frac{1}{2}[u_{1,1} + u_{3,1}] = \frac{1}{2}[0.08 + 0.20] = 0.14$

$i = 3$ $\qquad u_{3,2} = \frac{1}{2}[u_{2,1} + u_{4,1}] = \frac{1}{2}[0.15 + 0.23] = 0.19$

$i = 4$ $\qquad u_{4,2} = \frac{1}{2}[u_{3,1} + u_{5,1}] = \frac{1}{2}[0.20 + 0.24] = 0.22$

$i = 5$ $\qquad u_{5,2} = \frac{1}{2}[u_{4,1} + u_{6,1}] = \frac{1}{2}[0.23 + 0.23] = 0.23$

Going in the reverse order, we get

$$i = 6 \quad u_{6,2} = 0.22$$

$$i = 7 \quad u_{7,2} = 0.19$$

$$i = 8 \quad u_{8,2} = 0.14$$

$$i = 9 \quad u_{9,2} = 0.075$$

Putting $j = 2$ and $i = 1, 2, 3, 4, 5, 6, 7, 8, 9, 10$, we get the last row filled.

PROBLEMS

1. Given $\frac{\partial^2 f}{\partial x^2} = \frac{\partial f}{\partial t}$, $f(0, t) = f(5, t) = 0$, $f(x, 0) = x^2(25 - x^2)$, find the values of f for $x = ih$ ($i = 0, 1, ..., 5$), $t = jk$ ($j = 0, 1, 2, ..., 6$) with $h = 1$ and $k = \frac{1}{2}$ using the explicit method.

Ans. The values of f ($i = 0, 1, 2, 3, 4, 5$ and $j = 1, 2, 3, 4, 5$, $h = 1$, $k = 1/2$) are

j \ i	0	1	2	3	4	5
0	0	24	84	144	144	0
1	0	42	78	57	57	0
2	0	39	67.5	39	39	0
3	0	30	49.5	33.75	33.75	0
4	0	26.6	43.5	24.75	24.75	0
5	0	19.9	32.25	21.75	21.75	0

2. Solve boundary value problem $u_t = u_{xx}$ under the conditions $u(0, t) = u(1, t) = 0$ and $u(x, 0) = \sin \pi x$ $0 \le x \le 1$ by taking $h = 0.2$ by Schmidt method $\alpha = \frac{1}{2}$.

Ans.

j \ i	0	1	2	3	4	5
0	0	0.59	0.95	0.95	0.59	0
1	0	0.475	0.77	0.77	0.475	0
2	0	0.38	0.62	0.62	0.38	0
3	0	0.31	0.50	0.50	0.31	0
4	0	0.25	0.41	0.41	0.25	0
5	0	0.20	0.33	0.33	0.20	0

3. Using Crank-Nicholson method, solve $u_{xx} = 16u_t$, $0 < x < 1$, $t > 0$ given $u(x, 0) = 0$, $u(1, t) = 50$ and $u(0, t) = 0$. Compute u for two steps in t-direction taking $h = \frac{1}{4}$.

Ans.

j \ i	0	0.25	0.50	0.75	1
0	0	0	0	0	0
1	0	41	42	43	50
2	0	43	44	45	100

4. Solve $u_t = u_{xx}$ subjected to the conditions $u(0, t) = u(1, t) = 0$ and $u(x, 0) = 2x$ for $0 \le x \le \frac{1}{2}$ $2(1 - x)$ for $\frac{1}{2} \le x \le 1$, take $h = \frac{1}{4}$ and k according to Bender-Schmidt equation.

Ans.

j \ i	0	1	2	3	4
0	0	5	1	5	0
1	0	5	5	5	0
2	0	25	5	25	0
3	0	25	25	25	0

ADVANCED ENGINEERING MATHEMATICS

REENA GARG, CHANDRIKA PRASAD

FEEDBACK FORM

1. Which topics from your examination point of view are inadequately or not covered in this book?

____

____

2. Is there any misprints/ mistakes/ factual inaccuracy in this book? If so, please specify.

____

____

3. What is your assessment of this book as regards the presentation of the subject-matter, expression, precision and price in relation to the other recommended books available on this subject?

____

____

4. Which competing books you regard as better than this? Please mention their author and publishers.

____

____

5. Any further suggestions/comments you would like to make for the improvement of the book?

____

____

PS: If needed, please attach a separate sheet.

Name : ____________________

College : ____________________

Address : ____________________

Email : ____________________

Mobile : ____________________

Please mail the feedback form at:

KHANNA BOOK PUBLISHING CO. (P) LTD.

4C/4344, Ansari Road, Daryaganj, New Delhi-110002

You can also email this feedback form at contact@khannabooks.com or Whatsapp at +91 – 99109 09320

BEST FEEDBACK WILL BE REWARDED SUITABLY